System-on-Chip Test Architectures

The Morgan Kaufmann Series in Systems on Silicon
Series Editor: Wayne Wolf, Georgia Institute of Technology

The Designer's Guide to VHDL, Second Edition
Peter J. Ashenden

The System Designer's Guide to VHDL-AMS
Peter J. Ashenden, Gregory D. Peterson, and Darrell A. Teegarden

Modeling Embedded Systems and SoCs
Axel Jantsch

ASIC and FPGA Verification: A Guide to Component Modeling
Richard Munden

Multiprocessor Systems-on-Chips
Edited by Ahmed Amine Jerraya and Wayne Wolf

Functional Verification
Bruce Wile, John Goss, and Wolfgang Roesner

Customizable and Configurable Embedded Processors
Edited by Paolo Ienne and Rainer Leupers

Networks-on-Chips: Technology and Tools
Edited by Giovanni De Micheli and Luca Benini

VLSI Test Principles & Architectures
Edited by Laung-Terng Wang, Cheng-Wen Wu, and Xiaoqing Wen

Designing SoCs with Configured Processors
Steve Leibson

ESL Design and Verification
Grant Martin, Andrew Piziali, and Brian Bailey

Aspect-Oriented Programming with the e Verification Language
David Robinson

System-on-Chip Test Architectures
Edited by Laung-Terng Wang, Charles Stroud, and Nur Touba

Coming Soon...

Reconfigurable Computing
Edited by Scott Hauck and Andre DeHon

Verification Techniques for System-Level Design
Masahiro Fujita, Indradeep Ghosh, Mukul Prasad

SYSTEM-ON-CHIP TEST ARCHITECTURES

NANOMETER DESIGN FOR TESTABILITY

Edited by

Laung-Terng Wang

Charles E. Stroud

Nur A. Touba

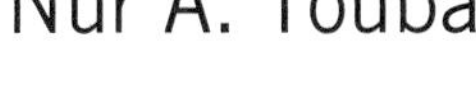

ELSEVIER

AMSTERDAM • BOSTON • HEIDELBERG • LONDON
NEW YORK • OXFORD • PARIS • SAN DIEGO
SAN FRANCISCO • SINGAPORE • SYDNEY • TOKYO

Morgan Kaufmann Publishers is an imprint of Elsevier

MORGAN KAUFMANN PUBLISHERS

Publishing Director	Joanne Tracy
Publisher	Denise E.M. Penrose
Senior Acquisitions Editor	Charles Glaser
Publishing Services Manager	George Morrison
Senior Production Editor	Dawnmarie Simpson
Associate Editor	Michele Cronin
Assistant Editor	Matthew Cater
Production Assistant	Lianne Hong
Cover Design	Joanne Blank
Cover Illustration	© Paul Vismara/Veer
Composition	Integra Software Services
Copyeditor	Karen Carriere
Proofreader	Phyllis Coyne et al. Proofreading Service
Indexer	Broccoli Information Management
Interior printer	The Maple-Vail Book Manufacturing Group
Cover printer	Phoenix Color Corporation

Morgan Kaufmann Publishers is an imprint of Elsevier.
30 Corporate Drive, Suite 400, Burlington, MA 01803, USA

This book is printed on acid-free paper.

Library of Congress Cataloging-in-Publication Data

System-on-chip test architectures: nanometer design for testability / edited by
Laung-Terng Wang, Charles Stroud, and Nur Touba.
p. cm.
Includes bibliographical references and index.
ISBN 978-0-12-373973-5 (hardcover : alk. paper)
1. Systems on a chip—Testing. 2. Integrated circuits—Very large scale integration—Testing.
3. Integrated circuits—Very large scale integration—Design.
I. Wang, Laung-Terng. II. Stroud, Charles E. III. Touba, Nur.
TK7895.E42S978 2007
621.39′5–dc22

2007023373

ISBN: 978-0-12-373973-5

For information on all Morgan Kaufmann publications, visit our Web site at www.mkp.com or www.books.elsevier.com

Printed in the United States.
07 08 09 10 5 4 3 2 1

Contents

2 Digital Test Architectures 41

Laung-Terng (L.-T.) Wang

3 Fault-Tolerant Design 123

Nur A. Touba

4 System/Network-on-Chip Test Architectures 171

Chunsheng Liu, Krishnendu Chakrabarty, and Wen-Ben Jone

5 SIP Test Architectures 225

Philippe Cauvet, Michel Renovell, and Serge Bernard

6 Delay Testing 263

Duncan M. (Hank) Walker and Michael S. Hsiao

8 Coping with Physical Failures, Soft Errors, and Reliability Issues 351

Laung-Terng (L.-T.) Wang, Mehrdad Nourani, and T. M. Mak

10 Design for Debug and Diagnosis 463

T. M. Mak and Srikanth Venkataraman

14 High-Speed I/O Interfaces 653

Mike Peng Li, T. M. Mak, and Kwang-Ting (Tim) Cheng

16 RF Testing 745

Soumendu Bhattacharya and Abhijit Chatterjee

PREFACE

Testing has become the number one challenge to nanometer *system-on-chip* (SOC) designs as a result of the unprecedented levels of design complexity and operating frequency made possible by advances in semiconductor manufacturing technologies. The vast demands and sophisticated applications in areas like consumer electronics have been the primary driving force for both the advances and the challenges in testing. The issues we face today range from *digital* to *memory* to *analog and mixed-signal* (AMS) testing and how these circuits interact with one another in the nanometer SOC design.

In digital testing, the requirement to operate circuits composed of tens to hundreds of millions of gates in the GHz range imposes severe challenges during manufacturing test. Because of the need for longer battery life or simply consuming less power, recent advances in low-power design methodologies require testing for these features without generating excessive heat. Multiple voltages and multiple frequencies have also stressed traditional test methodology. Challenges involve testing of these domains as well as synchronizers and level shifters. Test methodologies and solutions must be updated to reflect the constraints imposed by these new design methods and technologies. Furthermore, they must continue to provide quality and reliability measures that will improve manufacturing yield and defect level.

In memory testing, many novel *built-in self-test* (BIST) techniques have been developed in academia and practiced in industry to test large embedded memories, such as *static random-access memory* (SRAM) and *dynamic random-access memory* (DRAM) cores. Facing the trend that 80% of an SOC design could contain embedded memories in the nanometer design era, the need for advanced memory test techniques such as *built-in self-diagnosis* (BISD) and *built-in self-repair* (BISR) of memory defects is apparent and critical.

A common problem that arises in AMS testing within an SOC is that 10% of an SOC design that contains analog circuits could contribute to 90% of the total test cost during manufacturing test. Whereas the industry has found effective and efficient test solutions to significantly reduce the test cost for memory and digital logic circuits, the dominant analog test cost has erased the gain in other areas. Therefore, advanced architectures and new technologies to test the analog portion of the AMS, especially *radiofrequency* (RF), circuits must now also be developed to constrain the total test cost so that it stays within budget.

From the SOC testing point of view, test solutions must also address new fault models and failure mechanisms caused by manufacturing defects at the

65-*nanometer* (nm) process node and below. Based on the information extracted during layout, *defect-based tests* must supplement *structural tests* and *functional tests* in order to model and detect nontraditional manufacturing defects such as small delay defects and bridges. Advanced error and defect tolerance techniques are needed to cope with physical failures and tolerate *soft errors*. Test solutions must also be developed for silicon debug and diagnosis that can effectively and accurately help to localize fault sites. Better test solutions are needed for high-speed *input/output* (I/O) interfaces. Finally, specialized test solutions are required to target diverse *network-on-chip* (NOC), *system-in-package* (SIP), *field programmable gate array* (FPGA), and *microelectromechanical system* (MEMS) applications to ensure that testing does not become the bottleneck of SOC design and development.

As *complementary metal oxide semiconductor* (CMOS) technology scales toward 65 nm and below, more challenges and difficulties in circuit design and manufacturing will become prevalent as physical scaling is reaching the limit and quantum effects become more dominant. To sustain Moore's law at the nanotechnology level (under 10-nm feature size), new device structures and circuits will be needed to either replace or augment the conventional CMOS technology. Some of the possible device structures are *quantum-dot cellular automata* (QCA), *carbon nanotubes* (CNTs), and silicon nanowires. They offer hope of very high logic density and performance, yet low power consumption at the sub-10-nm scale. However, many of these nanodevices are believed to have high defect rates and thus will impose new and extremely severe test challenges when compared to testing conventional CMOS devices. Therefore, test solutions are required before we can begin to look into the possibility of utilizing these nanodevices.

When we embarked on a mission in 2005 to write a textbook to cover advanced *very-large-scale integration* (VLSI) testing and *design-for-testability* (DFT) architectures not presented in other then-available textbooks, it became clear that it would be impossible to cover all topics in a comprehensive, yet concise manner. Although there are a number of ways to address this problem, we decided to present these advances in testing in two books that collectively address most, if not all, of the current test issues and solutions that have been developed in academia and industry.

Our first book, *VLSI Test Principles and Architectures: Design for Testability*, was published in 2006 as a fundamental textbook for undergraduate and graduate students and as a reference book for researchers and practitioners. That text covered basic VLSI test principles and DFT architectures with details on topics currently used in industry, including logic and fault simulation, test generation, logic BIST, test compression, logic diagnosis, memory testing, boundary scan and core-based testing, as well as analog and mixed-signal testing. Although that text included a brief overview of some advanced topics in Chapter 12 (Test Technology Trends in the Nanometer Age), it left aside many of the details of these advanced approaches. These advanced topics included delay testing, coping with physical failures, soft errors and reliability issues, FPGA testing, MEMS testing, high-speed I/O (link) testing, and RF testing.

This book, *System-on-Chip Test Architectures*, is the second textbook in our series and focuses on the details of the aforementioned advanced topics. Given the fact that test technology must cover techniques from the chip to board to system level, including the nanotechnology scale, we continue to face the same dilemma as we did with our first book—that is, the inclusion of all topics. Thus, we have decided to focus this text on SOC applications. In addition, the text includes chapters devoted to other topics that are relevant to SOC. These include system/network-on-chip testing, system-in-package testing, low-power testing, design for debug and diagnosis, design for manufacturability and yield, and software-based self-testing. Topics beyond this scope as well as ideas that are still under development and have not yet reached industrial SOC applications are left as subjects of future books; however, we give an overview of some of these promising techniques in many of the chapters. Memory testing as well as the Institute of Electrical and Electronics Engineers (IEEE) boundary scan and core-based test standards used in SOC testing (including 1149.1, 1149.4, 1149.6, and 1500) were covered extensively in the first book and, as a result, are not repeated in this book; instead the reader is referred to the first book and its associated references for details on these topics. It should be noted, however, that new material related to these topics is included in a number of chapters in this book.

The advanced topics covered in this book can also be categorized into multiple sections, with each section consisting of multiple chapters. They are as follows:

1. DFT Architectures for

   Digital Logic Testing (Chapter 2)

   System/Network-on-Chip Testing (Chapter 4)

   System-in-Package Testing (Chapter 5)

   FPGA Testing (Chapter 12)

   High-Speed I/O Interfaces (Chapter 14)

   Analog and Mixed-Signal Testing (Chapter 15)

2. New Fault Models and Advanced Techniques for

   Delay Testing (Chapter 6)

   Low-Power Testing (Chapter 7)

   Coping with Physical Failures, Soft Errors, and Reliability Issues (Chapter 8)

   Software-Based Self-Testing (Chapter 11)

   RF Testing (Chapter 16)

3. Yield and Reliability Enhancement

   Fault-Tolerant Design (Chapter 3)

   Design for Manufacturability and Yield (Chapter 9)

   Design for Debug and Diagnosis (Chapter 10)

4. Nanotechnology Testing Aspects

   MEMS Testing (Chapter 13)

   Resonant Tunneling Diodes, Quantum-Dot Cellular Automata, Hybrid CMOS/Nanowires/Nanodevices, and Carbon Nanotubes (Chapter 17)

Each chapter of this book follows a specific format. The subject matter of the chapter is first introduced, with a historical perspective provided, if applicable. Related methods are explained in detail next. Then, industry practices, if applicable, are described before concluding remarks. Each chapter (except Chapter 17) contains a variety of exercises to allow this book to be used as a textbook for an advanced course in testing. Every chapter concludes with acknowledgment to contributors and reviewers and a list of references.

Chapter 1 introduces *system-on-chip* (SOC) *testing*. It begins with a discussion of the importance of testing as a requisite for achieving manufacturing quality and then identifies test challenges of the nanometer design era. This is followed by a brief overview of some of the IEEE boundary scan and core-based test standards that are widely used within industry (including 1149.1, 1149.4, 1149.6, and 1500). SOC examples practiced in industry are shown to illustrate the test challenges we face today.

Chapter 2 provides an overview of the most important test architectures for *digital logic testing*. Three basic *design-for-testability* (DFT) techniques widely used in industry are covered first: *scan design*, *logic built-in self-test* (BIST), and *test compression*. For each DFT technique, fundamental and advanced test architectures suitable for low-power and at-speed applications are discussed. The remainder of the chapter is devoted to *random-access scan,* a promising alternative to scan design for test power reduction.

Chapter 3 covers *fault-tolerant design* techniques that are applicable to both SOC designs and system applications. As the topic is quite broad, care is taken to describe widely used coding methods and fault tolerance schemes in an easy-to-grasp manner with extensive illustrations and examples. The chapter lists applications where the discussed techniques can be utilized.

Chapter 4 is devoted to both *system-on-chip* (SOC) and *network-on-chip* (NOC) *test architectures*. Various techniques for test access and test scheduling are thoroughly examined and presented. The chapter includes a discussion of the similarities and differences between the two as well as examples of each. Industrial designs are studied to show how these techniques are applicable to SOC and NOC testing.

Chapter 5 describes important test cost and product quality aspects of packing multiple dies in a *system-in-package* (SIP). After an introduction to the basic technologies, specific test challenges are presented. A number of bare-die test techniques to find *known-good-dies* are subsequently described. Functional system test and embedded component test techniques are then presented to test the SIP at the system level. The chapter ends with a brief discussion of future SIP design and test challenges related to nanometer technologies.

Chapter 6 addresses the *testing of delay faults*. The main focus of this chapter is on testing defect-based delay faults, often called *small delay defect testing*. Without loss of generality, however, conventional yet efficient delay fault simulation and test generation techniques for transition, gate-delay, and path-delay faults are first described. Advanced fault simulation and test pattern generation techniques associated with defect-based delay faults are then explained in detail.

Chapter 7 is devoted to *low-power testing*. After providing the motivations for reducing power during testing, power modeling and terminology used in the chapter are given. The main issues of excessive test power are then described. The remainder of the chapter is devoted to providing an overview of structural and algorithmic solutions that can be used to alleviate the issues raised by excessive power consumption during test application for digital nanometer designs.

Chapter 8 covers the full spectrum of *defect-based test* methods to cope with physical failures, soft errors, and reliability issues. First, new fault models are developed and solutions are presented to deal with noise-induced signal integrity issues. Defect-based tests are then discussed to further screen new defect-induced manufacturing faults. Finally, the rest of the chapter is devoted to illustrating adaptive designs and error-resilient architectures to tolerate soft errors and manufacturing faults.

Chapter 9 delves into the emerging hot topics of *design for manufacturability* (DFM) and *design for yield* (DFY). The chapter first describes in detail how lithography and variability during the manufacturing process can affect yield and induce defects. Then, innovative DFM and DFY techniques to improve yield and reduce defect level are explained in detail.

Chapter 10 is devoted to *silicon debug and diagnosis,* with heavy emphasis on design-for-debug architectures at the logic, circuit, and layout levels. This is complemented by an overview of common probing and diagnosis technologies for both wirebond and flip-chip packaging. The chapter also touches on system-level debug so as to link system issues back to silicon implementations. Finally, some of the future challenges unique to debug and diagnosis are also presented.

Chapter 11 provides a comprehensive discussion of *software-based self-testing*. The idea is to use on-chip programmable resources such as embedded processors to perform self-test and self-diagnosis. After explaining the basic concepts, various software-based self-test techniques are described to target processor cores, global interconnects, nonprogrammable cores, and *analog and mixed-signal* (AMS) circuits. Self-diagnosis techniques are also covered.

Chapter 12 addresses *testing field programmable gate arrays* (FPGAs) beginning with an overview of general FPGA architectures and operation. Following a discussion of the test challenges associated with FPGAs, various test approaches for FPGAs are described. The remainder of the chapter focuses on BIST and diagnosis of the programmable logic and routing resources in FPGAs. The chapter also presents new techniques for testing specialized cores such as configurable memories as well as new directions in FPGA testing using embedded processor-based on-chip reconfiguration.

Chapter 13 covers the *testing of microelectromechanical systems* (MEMS) *devices* that present new and interesting challenges as compared to the testing of microelectronics. This is partially because MEMS devices are designed to physically interact with the environment in which they operate. MEMS testing considerations, methods, and examples are presented, along with DFT and BIST techniques that have been proposed and implemented in commercially available MEMS devices.

Chapter 14 is devoted to *high-speed parallel/serial I/O link testing* at both component and system levels. This chapter starts with a discussion on signaling properties, such as jitter, noise, and *bit error rate* (BER), which impact the choice of high-speed I/O architectures. At the component level, instrumentation-based test methods for I/O characterization and DFT-assisted test methods for manufacturing test are first explained in detail. Novel DFT approaches for testing emerging circuits at signaling rates over 1 GHz, such as equalization and compensation, are also covered. At the system level, interconnect test methods using the IEEE 1149.1 and 1149.6 standards as well as the *interconnect BIST* (IBIST) method are then included.

Chapter 15 addresses *testing analog and mixed-signal* (AMS) *circuits* that are more frequently being incorporated in an SOC. The first book presented many of the basic issues and techniques for testing AMS circuits along with examples of testing discrete analog circuits. Although some of these basics are repeated in this book, this chapter focuses on mixed-signal BIST architectures that can be included in SOC implementations to test the analog cores and modules.

Chapter 16 extends AMS testing concepts to issues and techniques associated with *testing radiofrequency* (RF) *circuits*. This chapter outlines key test specifications for RF circuits and systems as well as covers industry practices for such devices. In addition, this chapter explains the operating principles of various test instrumentations widely used for AMS and RF testing and describes general *automatic test equipment* (ATE) architecture. From a production test perspective, concepts related to accuracy and repeatability are also discussed.

Chapter 17 is devoted to test technology trends for emerging nanotechnologies that are beyond the conventional CMOS. It introduces novel devices, circuits, architectures, and systems that have been proposed as alternatives to the CMOS at nanoscale dimensions, such as *resonant tunneling diodes* (RTDs), *quantum-dot cellular automata* (QCA), *silicon nanowires*, *single electron transistors*, and *carbon nanotubes*(CNTs). Defect characterization, fault modeling, test generation techniques, and the built-in self-test of systems built using such nanodevices, particularly for RTDs, QCA, and *crossbar arrays*, are discussed. Defect tolerance techniques for *carbon nanotube field effect transistors* (CNFETs) are also covered.

In the Classroom

This book is designed to be used as an advanced text for seniors and graduate students in computer engineering, computer science, and electrical engineering. It is also intended for use as a reference book for researchers and practitioners. The book is self-contained with most topics covered extensively from fundamental concepts to the current techniques used in research and industry. However, we assume that students have had basic courses in logic design, computer science, probability theory, and the fundamental testing and DFT techniques. Attempts are made to present algorithms, where possible, in an easily understood format.

To encourage self-learning, the instructor or reader is advised to check the Elsevier companion Web site (www.books.elsevier.com/companions) to access up-to-date software and lecture slides. Instructors will have additional privileges to access the Solutions directory for all exercises given in each chapter by visiting www.textbooks.elsevier.com and registering a username and password.

Laung-Terng (L.-T.) Wang
Charles E. Stroud
Nur A. Touba

ACKNOWLEDGMENTS

The editors would like to acknowledge many of their colleagues who helped create this book. Foremost are the 39 chapter/section contributors listed here. Without their strong commitments to contributing the chapters and sections of their specialty to the book in a timely manner, it would not have been possible to publish this book.

We also would like to thank the external contributors and reviewers for providing invaluable materials and feedback to improve the contents of this book. We wish to thank Alexandre de Morais Amory (UFRGS, Porto Alegre, Brazil), Dr. Jean-Marie Brunet (Mentor Graphics, Wilsonville, OR), Prof. Erika Cota (UFRGS, Brazil), C. Grecu (University of British Columbia, Canada), Dr. Vikram Iyengar (IBM, Burlington, VT), Dr. Ming Li (Siemens, Shanghai, China), Ke Li (University of Cincinnati), Dr. Lars Liebmann (IBM, Yorktown Heights, NY), Erik Jan Marinissen (NXP Semiconductors, Eindhoven, The Netherlands), Prof. David Pan (University of Texas, Austin, TX), Dr. Anuja Sehgal (AMD, Sunnyvale, CA), Jing Wang (Texas A&M University), Zheng Wang (Texas A&M University), Lei Wu (Texas A&M University), and Prof. Dan Zhao (University of Louisiana) for their contributions of materials, exercises, and figures to the book. We also wish to thank Prof. R. D. Shawn Blanton (Carnegie Mellon University), Prof. Erika Cota (UFRGS, Brazil), Prof. R. Dandapani (University of Colorado at Colorado Springs), Prof. Joan Figueras (University Politècnica de Catalunya, Spain), Prof. Dimitris Gizopoulos (University of Piraeus, Greece), Prof. Yinhe Han (Chinese Academy of Sciences, China), Claude E. Shannon Prof. John P. Hayes, (University of Michigan), Prof. Shi-Yu Huang (National Tsing Hua University, Taiwan), Prof. Sungho Kang (Yonsei University, Korea), Prof. Erik Larsson (Linköping University, Sweden), Prof. James C.-M. Li (National Taiwan University, Taiwan), Prof. Subhasish Mitra (Stanford University), Prof. Kartik Mohanram (Rice University), Prof. Saraju P. Mohanty (University of North Texas), Prof. Nicola Nicolici (McMaster University, Canada), Prof. Sule Ozev (Duke University), Prof. Partha Pande (Washington State University), Prof. Ian Papautsky (University of Cincinnati), Prof. Irith Pomeranz (Purdue University), Prof. Kewal K. Saluja (University of Wisconsin, Madison), Prof. Li-C. Wang (University of California, Santa Barbara), Prof. H.-S. Philip Wong (Stanford University), Prof. Tomokazu Yoneda (Nara Institute of Science and Technology, Japan), Prof. Xiaoyang Zeng (Fudan University, China), Dr. Florence Azais (LIRMM, Montpellier, France), Dr. Jayanta Bhadra (Freescale, Austin, TX), Dr. Yi Cai (Agere, Allentown, PA), Dr. Jonathan T.-Y. Chang (Intel, Santa Clara, CA),

Karthik Channakeshava (Virginia Tech), Dr. Li Chen (Intel, Hillsboro, OR), Dr. Bernard Courtois (CMP, Grenoble, France), Frans de Jong (NXP Semiconductors, Eindhoven, The Netherlands), Dr. Avijit Dutta (University of Texas, Austin), Herbert Eichinger (Infineon Technologies, Villach, Austria), François-Fabien Ferhani (Stanford University), Dr. Anne Gattiker (IBM, Austin, TX), Dhiraj Goswami (Mentor Graphics, Wilsonville, OR), Dr. Xinli Gu (Cisco, San Jose, CA), Dr. Dong Hoon Han (Texas Instruments, Dallas, TX), Dr. Mokhtar Hirech (Synopsys, Mountain View, CA), Tushar Jog (WiQuest, Allen, TX), Dr. Rohit Kapur (Synopsys, Mountain View, CA), Dr. Brion Keller (Cadence Design Systems, Endicott, NY), Dr. Haluk Konuk (Broadcom, Santa Clara, CA), Dr. Ajay Kumar (Texas Instruments, Dallas, TX), Dr. Christian Landrault (LIRMM, Montpellier, France), Dr. Ming Li (Siemens, Shanghai, China), Dr. Richard (Rick) Livengood (Intel, Santa Clara, CA), Dr. Shih-Lien Lu (Intel, Hillsboro, OR), Dr. Anne Meixner (Intel, Hillsboro, OR), Dr. Anurag Mittal (ARM, Sunnyvale, CA), Anandshankar S. Mudlapur (Intel, Folsom, CA), Dr. Benoit Nadeau-Dostie (LogicVision, Ottawa, Canada), Phil Nigh (IBM, Essex Junction, VT), Dr. Harry Oldham (ARM, Cambridge, United Kingdom), Peter O'Neill (Avago Technologies, Fort Collins, CO), Praveen K. Parvathala (Intel, Chandler, AZ), Jie Qin (Auburn University), Dr. Phil Reiner (Stanley Associates, Huntsville, AL), John Rogers (Harris, Melbourne, FL), Dr. Yasuo Sato (Hitachi, Tokyo, Japan), Rajarajan Senguttuvan (Georgia Institute of Technology), Masashi Shimanuchi (Credence, Milpitas, CA), Dr. Peilin Song (IBM, Yorktown Heights, NY), Michael Spica (Intel, Boise, ID), Dr. Franco Stellari (IBM, Yorktown Heights, NY), Derek Strembicke (AEgis Technologies Group, Huntsville, AL), Dr. Kun-Han Tsai (Mentor Graphics, Wilsonville, OR), Dr. Pramod Variyam (WiQuest, Allen, TX), Dr. Erik H. Volkerink (Verigy, Cupertino, CA), Dr. Seongmoon Wang (NEC Labs, Princeton, NJ), Dr. Yuejian Wu, (Nortel, Ottawa, Canada), Dr. Takahiro Yamaguchi (Advantest, Japan), Shrirang Yardi (Virginia Tech), Dr. Qing Zhao (Texas Instruments, Dallas, TX), and all chapter/section contributors for cross-reviewing the manuscript. Special thanks also go to many colleagues at SynTest Technologies (Sunnyvale, CA), including Dr. Ravi Apte, Boryau (Jack) Sheu, Dr. Zhigang Jiang, Zhigang Wang, Jianping Yan, Johnson Guo, Xiangfeng Li, Fangfang Li, Feng Liu, Yiqun Ding, Lizhen Yu, Ginger Qian, Jiayong Song, Jim Ma, Sammer Liu, Jongjoo Park, Jinwoo Cho, Paul Hsu, Karl Chang, Yi-Chih Sung, Tom Chao, Josef Jiang, Brian Wang, Renay Chang, and Teresa Chang who helped review the manuscript, solve exercises, develop lecture slides, and draw figures and tables.

The editors are indebted to many colleagues at Elsevier (Burlington, MA) who have been very helpful and patient with us during the preparation and production of this book, in particular, the senior acquisitions editor Charles B. Glaser; copyeditor Karen Carriere; senior production editor Dawnmarie Simpson; and associate editor Michele Cronin. Finally, we would like to acknowledge the generosity of SynTest Technologies (Sunnyvale, CA) for allowing Elsevier to put an exclusive version of the company's most recent VLSI testing and DFT software on the Elsevier companion Web site (www.books.elsevier.com/companions) for readers to use in conjunction with the book to become acquainted with DFT practices.

CONTRIBUTORS

Robert C. Aitken, R&D Fellow (Chapter 9)
ARM Ltd., Sunnyvale, California

R. Iris Bahar, Associate Professor (Chapter 17)
Division of Engineering, Brown University, Providence, Rhode Island

Serge Bernard, CNRS Researcher (Chapter 5)
LIRMM/CNRS, Montpellier, France

Soumendu Bhattacharya, Post-Doctoral Fellow (Chapter 16)
School of Electrical and Computer Engineering, Georgia Institute of Technology, Atlanta, Georgia

Philippe Cauvet, Senior Principal Engineer (Chapter 5)
NXP Semiconductors, Caen, France

Krishnendu Chakrabarty, Professor (Chapters 4 and 13)
Department of Electrical and Computer Engineering, Duke University, Durham, North Carolina

Abhijit Chatterjee, Professor, IEEE Fellow (Chapter 16)
School of Electrical and Computer Engineering, Georgia Institute of Technology, Atlanta, Georgia

Xinghao Chen, Associate Professor (Chapter 2)
Department of Electrical Engineering, The Grove School of Engineering
City College and Graduate Center of The City University of New York, New York

Kwang-Ting (Tim) Cheng, Chair and Professor, IEEE Fellow (Chapters 11 and 14)
Department of Electrical and Computer Engineering, University of California, Santa Barbara, California

F. Foster Dai, Professor (Chapter 15)
Department of Electrical and Computer Engineering, Auburn University, Auburn, Alabama

Robert Dean, Assistant Professor (Chapter 13)
Department of Electrical and Computer Engineering, Auburn University, Auburn, Alabama

William Eklow, Distinguished Manufacturing Engineer (Chapter 1)
Cisco Systems, Inc., San Jose, California; Chair, IEEE 1149.6 Standard Committee

John (Marty) Emmert, Associate Professor (Chapter 15)
Department of Electrical Engineering, Wright State University, Dayton, Ohio

Patrick Girard, CNRS Research Director (Chapter 7)
LIRMM/CNRS, Montpellier, France

Pallav Gupta, Assistant Professor (Chapter 17)
Department of Electrical and Computer Engineering, Villanova University, Villanova, Pennsylvania

Michael S. Hsiao, Professor and Dean's Faculty Fellow (Chapter 6)
Bradley Department of Electrical and Computer Engineering, Virginia Tech, Blacksburg, Virginia

Jiun-Lang Huang, Assistant Professor (Chapter 11)
Graduate Institute of Electronics Engineering, National Taiwan University, Taipei, Taiwan

Niraj K. Jha, Professor, IEEE Fellow and ACM Fellow (Chapter 17)
Department of Electrical Engineering, Princeton University, Princeton, New Jersey

Wen-Ben Jone, Associate Professor (Chapters 4 and 13)
Department of Electrical & Computer Engineering, University of Cincinnati, Cincinnati, Ohio

Kuen-Jong Lee, Professor (Chapters 1 and 2)
Department of Electrical Engineering, National Cheng Kung University, Tainan, Taiwan

Mike Peng Li, Chief Technology Officer (Chapter 14)
Wavecrest Corp., San Jose, California

Xiaowei Li, Professor (Chapter 2)
Institute of Computing Technology, Chinese Academy of Sciences, Beijing, China

Albert Lin, Ph.D. Student (Chapter 17)
Department of Electrical Engineering, Stanford University, Stanford, California

Chunsheng Liu, Assistant Professor (Chapter 4)
Department of Computer and Electronic Engineering, University of Nebraska-Lincoln, Omaha, Nebraska

T. M. Mak, Senior Researcher (Chapters 8, 10, and 14)
Intel Corp., Santa Clara, California

Yiorgos Makris, Associate Professor (Chapter 8)
Departments of Electrical Engineering and Computer Science, Yale University, New Haven, Connecticut

Mehrdad Nourani, Associate Professor (Chapter 8)
Department of Electrical Engineering, University of Texas at Dallas, Richardson, Texas

Nishant Patil, Ph.D. Student (Chapter 17)
Department of Electrical Engineering, Stanford University, Stanford, California

Ramesh Ramadoss, Assistant Professor (Chapter 13)
Department of Electrical and Computer Engineering, Auburn University, Auburn, Alabama

Michel Renovell, CNRS Research Director (Chapter 5)
LIRMM/CNRS, Montpellier, France

Chauchin Su, Professor (Chapter 1)
Department of Electrical and Control Engineering, National Chiao Tung University, Hsinchu, Taiwan

Mehdi Baradaran Tahoori, Assistant Professor (Chapter 17)
Department of Electrical and Computer Engineering, Northeastern University, Boston, Massachusetts

Mohammad H. Tehranipoor, Assistant Professor (Chapter 17)
Department of Electrical and Computer Engineering, University of Connecticut, Storrs, Connecticut

Srikanth Venkataraman, Principal Engineer (Chapter 10)
Intel Corp., Hillsboro, Oregon

Duncan M. (Hank) Walker, Professor (Chapter 6)
Department of Computer Science, Texas A&M University, College Station, Texas

Xiaoqing Wen, Professor (Chapters 2 and 7)
Graduate School of Computer Science and Systems Engineering, Kyushu Institute of Technology, Fukuoka, Japan

Cheng-Wen Wu, Tsing Hua Chair Professor, IEEE Fellow (Chapter 1)
Department of Electrical Engineering, National Tsing Hua University, Hsinchu, Taiwan

Shianling Wu, Vice President of Engineering (Chapter 2)
SynTest Technologies, Inc., Princeton Junction, New Jersey

Xingguo Xiong, Assistant Professor (Chapter 13)
Department of Electrical and Computer Engineering, University of Bridgeport, Bridgeport, Connecticut

ABOUT THE EDITORS

Laung-Terng (L.-T.) Wang, Ph.D., is chairman and chief executive officer (CEO) of SynTest Technologies (Sunnyvale, California). He received his BSEE and MSEE degrees from National Taiwan University in 1975 and 1977, respectively, and his MSEE and EE Ph.D. degrees under the Honors Cooperative Program (HCP) from Stanford University in 1982 and 1987, respectively. He worked at Intel (Santa Clara, California) and Daisy Systems (Mountain View, California) from 1980 to 1986 and was with the Department of Electrical Engineering of Stanford University as Research Associate and Lecturer from 1987 to 1991. Encouraged by his advisor, Professor Edward J. McCluskey, a member of the National Academy of Engineering, he founded SynTest Technologies in 1990. Under his leadership, the company has grown to more than 50 employees and 250 customers worldwide. The design for testability (DFT) technologies Dr. Wang has developed have been successfully implemented in thousands of application-specific integrated circuit (ASIC) designs worldwide. He has filed more than 25 U.S. and European patent applications in the areas of scan synthesis, test generation, at-speed scan testing, test compression, logic built-in self-test (BIST), and design for debug and diagnosis, of which 13 have been granted. Dr. Wang's work in at-speed scan testing, test compression, and logic BIST has proved crucial to ensuring the quality and testability of nanometer designs, and his inventions are gaining industry acceptance for use in designs manufactured at the 90-nanometer scale and below. He spearheaded efforts to raise endowed funds in memory of his NTU chair professor, Dr. Irving T. Ho, cofounder of the Hsinchu Science Park and vice chair of the National Science Council, Taiwan. Since 2003, he has helped establish a number of chair professorships, graduate fellowships, and undergraduate scholarships at Stanford University, National Taiwan University and National Tsing Hua University in Taiwan, as well as Xiamen University, Tsinghua University, and Shanghai Jiaotong University in China. Dr. Wang co-authored and co-edited an internationally used DFT textbook titled *VLSI Test Principles and Architectures: Design for Testability,* published in 2006. He received a Meritorious Service Award from the IEEE Computer Society in 2007 and is a member of Sigma Xi.

Charles E. Stroud, Ph.D., is a professor in the Department of Electrical and Computer Engineering at Auburn University in Alabama. He received his BSEE and MSEE degrees from the University of Kentucky in 1976 and 1977, respectively. He spent 15 years at AT&T Bell Laboratories where he was a distinguished member of technical staff designing VLSI devices and printed circuit boards for telecommunications and computer systems. Of the 21 production VLSI devices he designed, 16 incorporated built-in self-test (BIST), including the first BIST for random-access memories, the first completely self-testing chip using circular BIST, and the first BIST for mixed-signal systems. He received his Ph.D. in EE&CS from the University of Illinois at Chicago in 1991. He left Bell Labs in 1993 and has been in academia since that time where his accomplishments include the first BIST for field programmable gate arrays (FPGAs). He holds 16 U.S. patents for various BIST approaches for VLSI and FPGAs, has published more than 130 journal and conference papers with two Best Paper awards (1988 ACM/IEEE Design Automation Conference and 2001 IEEE Automatic Test Conference), and he authored *A Designer's Guide to Built-In Self-Test,* published in 2002. He has received seven teaching awards including two college-level and five department-level awards for undergraduate teaching. He has served on the editorial boards for IEEE Transactions on VLSI Systems, IEEE Design & Test of Computers, and the *Journal of Electronic Testing: Theory & Applications*. He has also served on the program committees for the IEEE International Test Conference, IEEE International On-Line Test Symposium, IEEE North Atlantic Test Workshop, IEEE International ASIC Conference, ACM International Symposium on FPGAs, and ACM/IEEE International Workshop on Hardware/Software Co-Design. He is a member of Tau Beta Pi and Eta Kappa Nu and is a fellow of the IEEE.

Nur A. Touba, Ph.D., is a professor in the Department of Electrical and Computer Engineering at the University of Texas at Austin. He received his BSEE degree from the University of Minnesota in 1990 and MSEE and Ph.D. degrees from Stanford University in 1991 and 1996, respectively. At Stanford, he worked with Professor Edward J. McCluskey in the Center for Reliable Computing where their joint research on logic built-in self-test (BIST) has won them a U.S. patent. He has been with the University of Texas at Austin since 1996 and has published more than 90 journal and conference papers. He has received a number of honors and awards, including a National Science Foundation (NSF) Early Faculty CAREER Award in 1997, a College of Engineering Foundation Faculty Award in 2001, the Best Paper Award at the VLSI Test Symposium in 2001, the Best Panel Award at the International Test Conference in 2005, and a General Motors Faculty Fellowship in 2006. His research interests are in design-for-testability and fault-tolerant design. He serves on the program committees for the IEEE International Test Conference, IEEE International Conference on Computer Design, IEEE/ACM Design Automation and Test in Europe Conference, IEEE Defect and Fault Tolerance Symposium, IEEE European Test Symposium, IEEE Asian Test Symposium, IEEE International On-Line Test Symposium, IEEE International Test Synthesis Workshop, IEEE International Workshop on Open Source Test Technology Tools, and IEEE Microprocessor Test and Verification Workshop. He is a senior member of the IEEE.

CHAPTER 1

INTRODUCTION

Laung-Terng (L.-T.) Wang
SynTest Technologies, Inc., Sunnyvale, California

Charles E. Stroud
Auburn University, Auburn, Alabama

Nur A. Touba
University of Texas, Austin, Texas

ABOUT THIS CHAPTER

Over the past three decades, we have seen the semiconductor manufacturing technology advance from 4 microns to 45 nanometers. This shrinkage of feature size has made a dramatic impact on design and test. Now we find ***system-on-chip*** (SOC) and ***system-in-package*** (SIP) designs that embed more than 100 million transistors running at operating frequencies in the *gigahertz* range. Within this decade, there will be designs containing more than a billion transistors. These designs can include all varieties of *digital*, *analog*, *mixed-signal*, *memory*, *optical*, ***microelectromechanical systems*** (MEMS), ***field programmable gate array*** (FPGA), and ***radiofrequency*** (RF) circuits. Testing designs of this complexity is a significant challenge, if not a serious problem. Data have shown it is beginning to require more than 20% of the development time to generate production test patterns of sufficient fault coverage to detect manufacturing defects.

Additionally, when the SOC design is operated in a system, soft errors induced by *alpha-particle radiation* can adversely force certain memory cells or storage elements to change their states. These soft errors can cause the system to malfunction. As ***complementary metal oxide semiconductor*** (CMOS) scaling continues, the combined manufacturing defects and soft errors start to threaten the practicality of these nanometer SOC designs.

In this chapter, we first describe the importance of SOC testing and review the design and test challenges reported in the ***International Technology Roadmap for Semiconductors*** (ITRS). Next, we outline the *Institute of Electrical and Electronics Engineers* (IEEE) standards used for testing SOC designs. These include the 1149.1 and 1149.6 boundary-scan standards, the 1500 core-based test and maintenance standard, and the 1149.4 analog boundary-scan standard. Some SOC design examples, including a ***network-on-chip*** (NOC) design, are then illustrated. Finally, we provide an overview of the book in terms of the chapters that discuss how to test various components and aspects of these highly complex nanometer SOC designs. The book concludes with an invited survey chapter on testing aspects of nanotechnology trends, which covers four of the most promising nanotechnologies: ***resonant tunneling diodes*** (RTDs), ***quantum-dot cellular automata*** (QCA), ***hybrid CMOS/nanowires/nanodevices***, and ***carbon nanotubes*** (CNTs).

1.1 IMPORTANCE OF SYSTEM-ON-CHIP TESTING

In 1965, Gordon Moore, Intel's cofounder, predicted that the number of transistors integrated per square inch on a die would double every year [Moore 1965]. In subsequent years, the pace slowed, but the number of transistors has continued to double approximately every 18 months for the past two decades. This has become the current definition of Moore's law. Most experts expect that Moore's law will hold for at least two more decades. Die size will continue to grow larger, but, at the same time, minimum feature size will continue to shrink. Although smaller transistor size can result in smaller circuit delay, a smaller feature size for interconnects does not reduce the signal propagation delay; thus, the signal propagation delay in interconnects has been the dominant factor in determining the delay of a circuit [Dally 1998]. To alleviate this problem, interconnects are made thicker to reduce the sheet resistance. Unfortunately, this induces **crosstalk noises** between adjacent interconnects because of capacitive and inductive coupling. This is referred to as a **signal integrity** problem, and it is extremely difficult to detect [Chen 2002]. As the clock frequency has been pushed up into the gigahertz range and supply voltage has also been scaled down along with device scaling, the power supply voltage drop caused by $L(di/dt)$ can no longer be ignored. This has caused a **power integrity** problem that again is extremely difficult to solve because finding test patterns with maximum current changes is quite difficult [Saxena 2003].

As the manufacturing technology continues to advance, precise control of the silicon process is becoming more challenging. For example, it is difficult to control the effective channel length of a transistor such that the circuit performance, including power and delay, exhibits much larger variability. This is a **process variation** problem, and it can make delay testing extremely complex [Wang 2004]. To reduce the leakage power dissipation, many low-power design techniques have been widely used. Unfortunately, low-power circuits might result in new fault models that increase the difficulty of fault detection; for example, a **drowsy cache** that can be supplied by low voltage (*e.g.*, 0.36 V) when it is idle has been proposed recently to reduce the leakage current [Kim 2004]. Though the leakage current can be reduced by several orders of magnitude, a new fault model called a **drowsy fault** can occur that causes a memory cell to fall asleep forever. Unfortunately, testing drowsy faults requires excessively long test application times, as it is necessary to drive the memory cells to sleep and then wake them up. As we move into the nanometer age and in order to keep up with Moore's law, many new nanotechnologies and circuit design techniques must be developed and adopted, all of which pose new test challenges

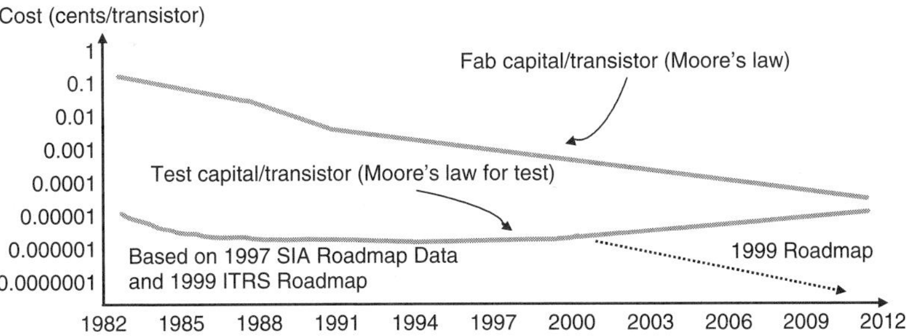

■ **FIGURE 1.1**

Fabrication capital *versus* test capital. (Courtesy of [Cheng 2006].)

that must be addressed concurrently. Otherwise, the cost of test would eventually surpass the cost of silicon manufacturing, as illustrated in Figure 1.1, according to roadmap data given in [SIA 1997] and [SIA 1999].

In 2004, the ***Semiconductor Industry Association*** (SIA) published an ***International Technology Roadmap for Semiconductors*** (ITRS), which includes an update to the test and test equipment trends for nanometer designs through the year 2010 and beyond [SIA 2004]. The ITRS is an assessment of the semiconductor technology requirements with the objective of ensuring advancements in the performance of integrated circuits. This assessment, also known as a **roadmap**, is a cooperative effort of the global industry manufacturers and suppliers, government organizations, consortia, and universities.

The ITRS identifies the technological challenges and needs facing the semiconductor industry through the end of the next decade. Difficult near-term and long-term test and test equipment challenges were reported in [SIA 2004] and are listed in Tables 12.1 and 12.2 of [Wang 2006]. The near-term challenges through 2010 for nanometer designs with feature size ≥45 nm include high-speed device interfaces, highly integrated designs, reliability screens, manufacturing test cost, as well as modeling and simulation. The long-term challenges beyond 2010 for nanometer designs with feature size <45 nm include the ***device under test*** (DUT) to ***automatic test equipment*** (ATE) interface, test methodologies, defect analysis, failure analysis, and disruptive device technologies. These difficult challenges encompass a full spectrum of test technology trends imperative for nanometer designs, including (1) developing new ***design for testability*** (DFT) and ***design for manufacturability*** (DFM) methods for digital circuits, analog circuits (including RF and audio circuits as well as high-speed serial interfaces), MEMS, and sensors; (2) developing the means to reduce manufacturing test costs as well as enhance device reliability and yield; and (3) developing techniques to facilitate defect analysis and failure analysis. The ITRS [SIA 2004] further summarizes the design test challenges, as shown in Table 12.3 of [Wang 2006]. These include (1) effective speed testing with increasing core frequencies and widespread proliferation of multi-GHz serial ***input/output*** (I/O) protocols; (2) capacity gap between design complexity and DFT, test generation, and fault grading tools; (3) quality and yield impact resulting from test process diagnostic limitations; (4) signal integrity testability and new fault models; (5) *system-on-chip* (SOC) and *system-in-package* (SIP) test including integrated self-test for heterogeneous SOCs and SIPs; (6) diagnosis, reliability screens, and yield improvement; and (7) fault-tolerance and online testing.

In [SIA 2005] and [SIA 2006], these difficult challenges were further refined and split into key drivers and difficult challenges. A future opportunities section was also added. The key drivers (not in any particular order) include (1) device trends such as increasing device integration (SOC, SIP, ***multichip packaging*** [MCP], and three-dimensional [3D] packaging) and integration of emerging and nondigital CMOS technologies (RF, analog, optical, and MEMS); (2) increasing test process complexity such as "distributed test" to maintain cost scaling; and (3) continued economic scaling of test such as managing (logic) test data volume. Difficult challenges (in order of priority) include (1) test for yield learning that is critically essential for fabrication process and device learning below optical device dimensions; (2) screening for reliability (*e.g.*, causing erratic, nondeterministic, and intermittent device behavior); (3) increasing systemic defects such as detecting symptoms and effects of line width variations, finite dopant distributions, and systemic process defects; and (4) potential yield losses caused by tester inaccuracies (*e.g.*, timing, voltage, current, temperature control), over-testing (*e.g.*, delay faults on nonfunctional paths), etc. Future opportunities (not in any particular order) include (1) test program automation (not ***automatic test pattern generation*** [ATPG]); (2) simulation and modeling of test interface hardware and instrumentation seamlessly integrated to the device design process; and (3) convergence of test and system reliability solutions between test (DFT), device, and system reliability (error detection, reporting, and correction).

A circuit defect may lead to a fault, a fault can cause a circuit error, and a circuit error can result in a system failure. Two major defect mechanisms can cause the SOC design to malfunction: manufacturing defects and soft errors. **Manufacturing defects** are physical (circuit) defects introduced during manufacturing that cause the design to fail to function properly in the device, on the ***printed circuit board*** (PCB), or in the system or field. These manufacturing defects can result in *static faults* (such as stuck-at faults) or *timing faults* (such as delay faults). There is general consensus with the **rule of ten**, which says that the cost of detecting a faulty device increases by an order of magnitude as we move through each stage of manufacturing, from device level, to board level, to system level, and finally to system operation in the field [Wang 2006]. **Soft errors**, also referred to as ***single event upsets*** (SEUs), are *transient faults* induced by environmental conditions, such as *alpha-particle radiation*, which cause a fault-free design to malfunction when deployed on-board, in-system, or in-field [May 1979] [Baumann 2005]. The probability of the occurrence of soft errors increases with decreasing feature size. For example, the probability of SEUs increased by a factor of more than 21 when moving from a feature size of 0.6 microns to 0.35 microns [Ohlsson 1998]. Transient faults are nonrepeatable temporary faults and thus cannot be detected during manufacturing. However, these defect mechanisms must be screened during manufacturing or tolerated in the design in order to enhance device reliability and yield, reduce defect level and test costs, and improve system reliability and system availability.

1.1.1 Yield and Reject Rate

Some percentage of the manufactured devices are expected to be faulty because of manufacturing defects. The **yield** of a manufacturing process is defined as the percentage of acceptable parts among all parts that are fabricated:

$$yield = \frac{number\ of\ acceptable\ parts}{total\ number\ of\ parts\ fabricated}$$

There are two types of yield loss: catastrophic and parametric. Catastrophic yield loss is due to random defects, and parametric yield loss is due to process variations. Automation of and improvements in an ***integrated circuit*** (IC) fabrication process line drastically reduce the particle density that creates random defects over time; consequently, parametric variations resulting from process fluctuations become the dominant reason for yield loss.

Methods to reduce process variations during fabrication are generally referred to as ***design for yield*** (DFY). The circuit implementation methods to avoid random defects are generally referred to as ***design for manufacturability*** (DFM). Broadly speaking, any DFM method helps to increase manufacturing yield and thus can be considered as a DFY method. Manufacturing yield relates to the failure rate λ. The **bathtub curve** shown in Figure 1.2 is a typical device (or system) failure chart indicating how early failures, wearout failures, and random failures contribute to the overall device (or system) failures.

The **infant mortality** period (with decreasing failure rate) occurs when a product is at its early production stage. Failures are mostly due to poor process or design which leads to poor product quality and thus the product should not be shipped during this period to avoid massive field returns. The **working life** period (with constant failure rate) represents the product "working life." Failures during this period tend to occur randomly. The **wearout** period (with increasing failure rate) indicates the "end of life" of the product. Failures during this period are due to age defects, such as metal fatigue, hot carriers, electromigration, dielectric breakdown, etc. For electronic products, this period is of less concern because they often would not enter this region because of technology advances and obsolescence.

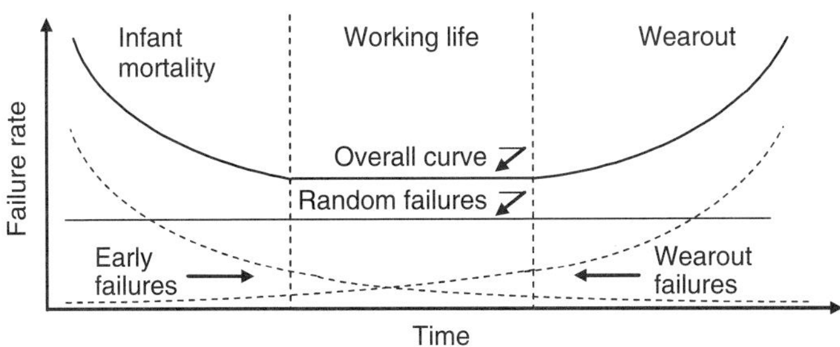

■ FIGURE 1.2

Bathtub curve.

When ICs are tested, the following two undesirable situations may occur:

1. A faulty device appears to be a good part passing the test.
2. A good device fails the test and appears as faulty.

These two outcomes are often due to a poorly designed test or the lack of ***design for testability*** (DFT). As a result of the first case, even if all products pass acceptance test, some faulty devices will still be found in the manufactured electronic system. When these faulty devices are returned to the IC manufacturer, they undergo ***failure mode analysis*** (FMA) for possible improvements to the IC development and manufacturing processes [Amerasekera 1987]. The ratio of field-rejected parts to all parts passing quality assurance testing is referred to as the **reject rate**, also called the **defect level**:

$$\textit{reject rate} = \frac{\textit{number of faulty parts passing final test}}{\textit{total number of parts passing final test}}$$

For a given device, the authors in [McCluskey 1988] showed that defect level *DL* is a function of process yield *Y* and fault coverage *FC*:

$$DL = 1 - Y^{(1-FC)}$$

The defect level provides an indication of the overall quality of the testing process [Williams 1981] [Bushnell 2000] [Jha 2003]. Generally speaking, a defect level of 500 ***parts per million*** (PPM) may be considered to be acceptable, whereas 100 PPM or lower represents high quality. The goal of **six sigma** manufacturing, also referred to as **zero defects**, is 3.4 PPM or less.

Example 1.1

Assume the process yield is 50% and the fault coverage for a device is 90% for the given test sets. By the preceding equation, we obtain $DL = 1 - 0.5^{(1-0.9)} = 0.067$. This means that 6.7% of shipped parts will be defective or the defect level of the product is 67,000 PPM. On the other hand, if a *DL* of 100 PPM is required for the same process yield of 50%, then the fault coverage required to achieve the PPM level is $FC = 1 - (\log(1 - DL)/\log(Y)) = 0.99986$. Because it could be extremely difficult, if not impossible, to generate tests that have 99.986% fault coverage, improvements over process yield might become mandatory in order to meet the stringent PPM goal.

1.1.2 Reliability and System Availability

Traditionally, **component reliability**, also called **device reliability**, is measured by acceptable defect level (defective PPM), failure rate per 1000 hours, noise characteristics, etc. Because process yield and fault coverage affect defect level, reliability screen methods, such as **stress testing** (through burn-in) and $\mathbf{I_{DDQ}}$ **testing** (by measuring the device leakage currents), are often used to accelerate the failure rate with pre-selected test sets. These methods, called **reliability screens**, are mainly developed to weed out weak devices before mass production so as to reduce **test escapes** that will cause field returns. Once weak devices are found, FMA is performed to analyze, debug, locate, and correct the failures so the process yield can be increased later.

When a manufactured electronic system is shipped to the field, it may also undergo testing as part of the installation process to ensure that the system is fault-free before placing the system into operation. During system operation, a number of events can result in a system failure; these events include *single-event upsets* (SEUs), electromigration, and material aging. Suppose the state of system operation is represented as S, where $S=0$ means the system operates normally and $S=1$ represents a system failure. Then S is a function of time t, as shown in Figure 1.3.

Suppose the system is in normal operation at $t = 0$, it fails at t_1, and the normal system operation is recovered at t_2 by some software modification, reset, or hardware replacement. Similar failure and repair events happen at t_3 and t_4. The duration of normal system operation (T_n), for intervals such as $t_1 - t_0$ and $t_3 - t_2$, is generally assumed to be a random number that is exponentially distributed. This is known as the **exponential failure law**. Hence, the probability that a system will operate normally until time t, referred to as **system reliability**, is given by:

$$P(T_n > t) = e^{-\lambda t}$$

where λ is the failure rate. Because a system is composed of a number of components, the overall failure rate for the system is the sum of the individual failure rates (λ_i) for each of the k components:

$$\lambda = \sum_{i=0}^{k} \lambda_i$$

The ***mean time between failures*** (MTBF) is given by:

$$MTBF = \int_0^{\infty} e^{-\lambda t} dt = \frac{1}{\lambda}$$

Similarly, the ***repair time*** (R) is also assumed to obey an exponential distribution and is given by:

$$P(R > t) = e^{-\mu t}$$

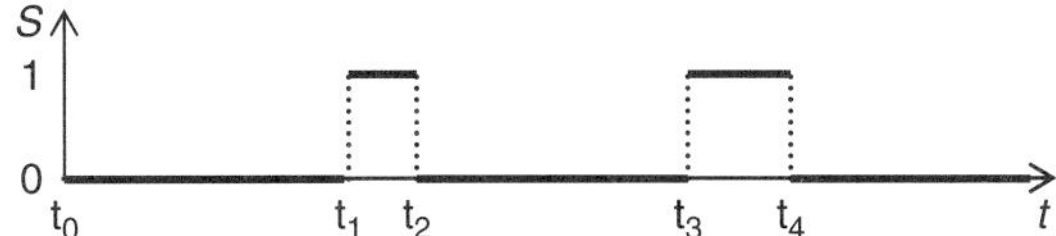

FIGURE 1.3

System operation and repair.

where μ is the repair rate. Hence, the ***mean time to repair*** (MTTR) is given by:

$$MTTR = \frac{1}{\mu}$$

The fraction of time that a system is operating normally (failure-free) is called the **system availability** and is given by:

$$system\ availability = \frac{MTBF}{MTBF + MTTR}$$

This formula is widely used in reliability engineering; for example, telephone systems are required to have system availability of 0.9999 (simply called **four nines**), whereas high-reliability systems may require seven nines or more.

In general, system reliability requires the use of spare components or units (made of components) in a fault-tolerant system to increase the system availability [Siewiorek 1998]. Device reliability, on the other hand, depends on reliability screens in ***very-large-scale-integration*** (VLSI) or SOC devices to improve process yield and hence the device defect level. Unfortunately, existing reliability screens are becoming either quite expensive (as in the case of burn-in) or ineffective (as in the case of using I_{DDQ} testing) for designs manufactured at 90 nm or below. Fundamentally new long-term solutions must be developed for reliability screens and may include significant on-die hardware for stressing or special reliability measurements, as indicated in [SIA 2004].

Today, little evidence shows that higher device reliability can prolong the working life of a device, but in theory, it could extend the availability of the system when embedded with such devices. Fault-tolerant architectures commonly found in high-reliability systems are now applied to robust SOC designs to tolerate soft errors and to make them **error-resilient**. These **error resilience** (or **defect tolerance**) schemes are now referred to as ***design for reliability*** (DFR).

Example 1.2

The number of failures in 10^9 hours is a unit (abbreviated FITS) that is often used in reliability calculations. For a system with 500 components where each component has a failure rate (λ_i) of 1000 FITS, the system failure rate (λ) becomes $(1000/10^9) \times 500 = 5 \times 10^{-4}$. Because MTBF = $1/\lambda$, the mean time between failures of the system is 2000 hours. Suppose that the availability of the system must be at least 99.999% each year. Then, the repair time allocated for system repair each year should be less than $t = T \times$ (1–system availability) $= 1 \times 365 \times 24 \times 60 \times 60 \times (1 - 0.99999) = 315$ seconds, which is about 5 minutes. This implies that only fault tolerance with ***built-in self-repair*** (BISR) capability can meet the system availability requirement.

1.2 BASICS OF SOC TESTING

The various DFM, DFY, and DFR methods that have been proposed in academia and industry are mainly used to improve the manufactured device quality and to extend the availability of the system once the manufactured devices are used in the field. When an SOC design fails as a chip, on a board, or in the system, the ability to find the root cause of the failure in a timely manner becomes critical. In this section, we briefly discuss several IEEE standards (including 1149.1, 1149.4, 1149.6, and 1500) and other techniques that ease silicon test and debug as well as system-level test and diagnosis. For detailed descriptions, the reader is referred to key references cited.

1.2.1 Boundary Scan (IEEE 1149.1 Standard)

The success of scan-based DFT techniques from the mid-1970s through the mid-1980s led to their adaptation for testing interconnect and solder joints on surface mount PCBs. This technique, known as **boundary scan**, eventually became the IEEE 1149.1 standard [IEEE 1149.1-2001] and paved the way for floating vias, microvias, and mounting components on both sides of PCBs to reduce the physical size of electronic systems. Boundary scan provides a generic test interface not only for interconnect testing between ICs but also for access to DFT features and capabilities within the core of an IC as illustrated in Figure 1.4 [IEEE 1149.1-2001] [Parker 2001]. The boundary-scan interface includes four mandatory ***input/output*** (I/O) pins for ***Test Clock*** (TCK), ***Test Mode Select*** (TMS), ***Test Data Input*** (TDI), and ***Test Data Output*** (TDO). A ***test access port*** (TAP) controller is included to access the boundary-scan chain and any other internal features designed into the device, such as access to internal scan chains, ***built-in self-test*** (BIST) circuits, or, in the case

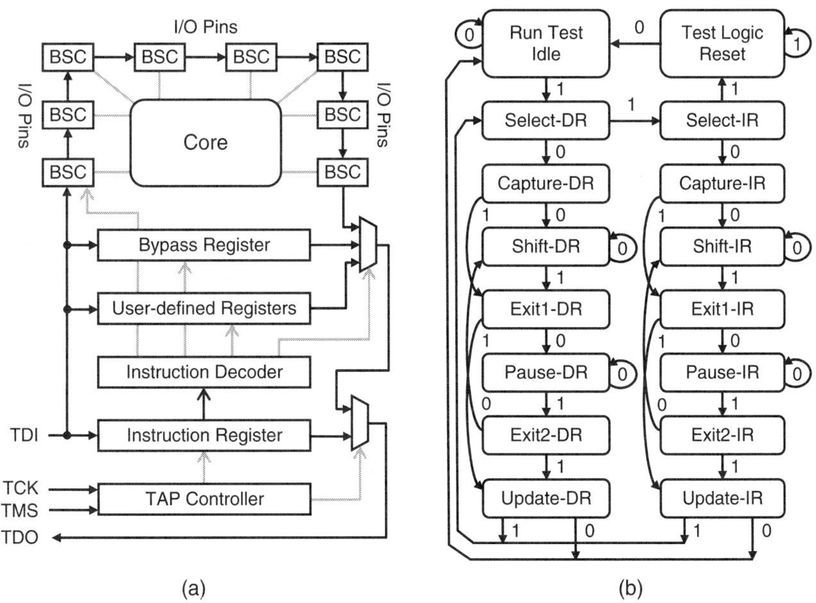

■ **FIGURE 1.4**

Boundary-scan interface: (a) boundary-scan implementation and (b) TAP controller state diagram.

of ***field programmable gate arrays*** (FPGAs), access to the configuration memory. The TAP controller is a 16-state ***finite state machine*** (FSM) with standardized state diagram illustrated in Figure 1.4b where all state transitions occur on the rising edge of TCK based on the value of TMS shown for each edge in the state diagram. Instructions for access to a given feature are shifted into the ***instruction register*** (IR) and subsequent data are written to or read from the ***data register*** (DR) specified by the instruction (note that the IR and DR portions of the state diagram are identical in terms of state transitions and TMS values). An optional ***Test Reset*** (TRST*) input can be incorporated to asynchronously force the TAP controller to the Test Logic Reset state for application of the appropriate values to prevent back driving of bidirectional pins on the PCB during power up. However, this input was frequently excluded because the Test Logic Reset state can easily be reached from any state by setting TMS = 1 and applying five TCK cycles.

The basic ***boundary-scan cell*** (BSC) used for testing interconnect on a PCB is illustrated in Figure 1.5a along with its functional operation. A more complex construction of the triple BSC bidirectional I/O buffer is illustrated in Figure 1.5b. The bidirectional buffer illustrates the need for the double-latched scan chain design of the basic BSC to prevent back driving of other bidirectional buffers on a PCB while shifting in test patterns and shifting out test responses. The Update latch holds all values (including tri-state control value) stable at the pads during the shifting process. Once the test pattern is shifted into the boundary-scan chain via the Capture flip-flops, the Update DR state of the TAP controller (see Figure 1.4b) transfers the test pattern to the Update latches for external applications to the PCB interconnect testing or access to DFT features and capabilities within the core of an IC.

The design of the basic BSC shown in Figure 1.5a also facilitates application of test patterns at the input buffers to the internal core of a device, as well as

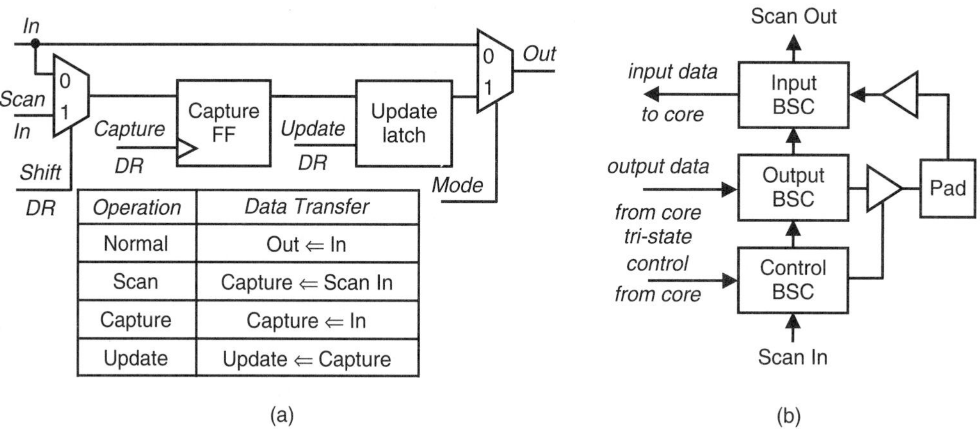

Operation	Data Transfer
Normal	Out ⇐ In
Scan	Capture ⇐ Scan In
Capture	Capture ⇐ In
Update	Update ⇐ Capture

■ **FIGURE 1.5**

Boundary-scan cells: (a) boundary-scan cell (BSC) and operation modes and (b) bidirectional buffer with BSCs.

capture of output responses from the internal logic at the output buffers. This internal test, referred to as **INTEST** in the IEEE 1149.1 standard, is an optional but valuable instruction, although it is not always implemented in an effort to reduce the area overhead and performance penalty associated with boundary scan. The external test of PCB interconnects, referred to as **EXTEST**, is a mandatory instruction to test the PCB interconnects for which boundary scan was intended. Another mandatory feature is the **BYPASS** instruction and uses the bypass register (shown in Figure 1.4a and consisting of a single flip-flop), which allows the entire boundary-scan chain to be bypassed to provide faster access to other devices on the board with the daisy-chained boundary-scan interface.

The boundary-scan interface is significant for several reasons. Obviously, it provides an approach to testing interconnects on digital PCBs and overcomes the test problems encountered with surface mount technology emerging in the late 1980s. Although boundary scan does not directly address test logic internal to the devices on a PCB, it does provide a standardized interface that can be used to access internal test mechanisms, such as scan chains and BIST circuits, designed specifically for the internal logic testing. Perhaps more important, it provides a proven solution to the test problems that would later be encountered with SOC and SIP implementations. Because the basic idea is to incorporate a large number of complex cores in a single chip (in the case of an SOC) or a single package (in the case of an SIP), testing interconnects between these modules can be performed in a manner similar to boundary scan. Hence, the IEEE 1149.1 standard can be, and is, extended and adapted to test the interconnections between the cores.

1.2.2 Boundary Scan Extension (IEEE 1149.6 Standard)

Although the IEEE 1149.1 boundary-scan standard has been the mainstay for board-level interconnect testing since the early 1990s, it is becoming ineffective to test modern-day high-speed serial networks that are **AC-coupled** (a "blocking" capacitor is placed in the net between the driver and receiver). Also, most high-speed serial networks use differential I/O pads.

The IEEE 1149.6 standard [IEEE 1149.6-2003] was introduced to address these problems with multigigabit I/Os. It is an extension of the IEEE 1149.1 standard to deal with this differential, AC-coupled interface as illustrated in Figure 1.6. When testing the two nets using the IEEE 1149.1 standard, the presence of the coupling

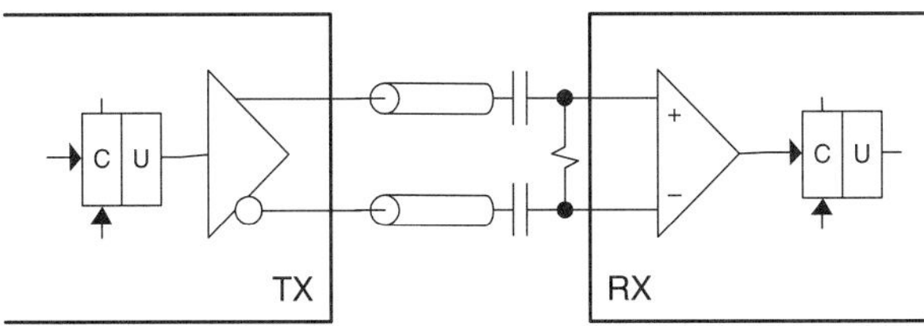

■ **FIGURE 1.6**

Differential, AC-coupled network.

capacitor in AC-coupled networks "blocks" DC signals. As a result, the DC level that is applied to the net during a boundary-scan EXTEST instruction decays over time to an undefined logic level. This places a minimum frequency requirement on TCK that the 1149.1 standard cannot support. The IEEE 1149.6 standard addresses this problem by capturing the edges of data transitions instead of capturing data levels; hence, the minimum TCK frequency requirement is removed.

IEEE 1149.6 addresses the problem of differential signaling by adding a single, boundary-scan cell, internal to the driver and boundary-scan cells on each input to the functional, differential receiver. The single boundary-scan cell on the driver side minimizes the loading and performance impact. Boundary-scan cells on each input of the receiver provide better coverage than a single boundary-scan cell implementation.

The IEEE 1149.6 circuit is composed of four parts: (1) the **analog test receiver**, (2) the **digital driver logic**, (3) the **digital receiver logic**, and (4) the **1149.6 *test access port*** (TAP). Each part is discussed in the subsequent paragraphs.

The analog test receiver is the most critical part of the 1149.6 implementation because it is the test receiver that is able to capture transition edges. The test receiver uses a "self-referenced" comparator, along with voltage and delay hysteresis to capture a valid edge and filter any unwanted noise. The test receiver uses a low-pass filter to create a delayed reference signal.

The digital driver logic is a simple extension to the IEEE 1149.1 driver. Unlike the 1149.1 driver, the 1149.6 driver is required to drive a pulse when it is executing the (1149.6) **EXTEST_PULSE** (or **EXTEST_TRAIN**) instruction. The EXTEST_PULSE instruction is used to drive the output signal to the opposite state, wait for the signal to fully decay, and then drive the signal to the correct value (this is the value that gets captured). By allowing the signal to fully decay, the maximum voltage swing is generated on the next driven edge, allowing for better capture by the analog test receiver. In rare cases a continuous waveform may be required for some high-speed logic. In this case, the EXTEST_TRAIN instruction is used instead of the EXTEST_PULSE to generate a continuous waveform based on TCK. The digital driver logic must also support the 1149.1 EXTEST instruction. It simply extends the 1149.1 logic by multiplexing the 1149.6 signal into the 1149.1 shift/update circuit, after the update flip-flop.

The digital receiver logic takes the output of the analog test receiver and sets a capture flip-flop to a corresponding logical zero or one. The digital test receiver logic also ensures that a valid transition has been captured on every test vector by initializing the state of the capture memory before the transition is driven onto the net. Without this initialization, it would be impossible to determine if two sequential transitions in the same direction (positive or negative) occurred, or if only one transition occurred (*i.e.*, if a positive transition occurs and is captured in the memory, and the subsequent test vector also generates a positive transition, there is no way to determine if the second transition occurred without clearing the contents of the capture memory before the second transition occurs).

Changes were made to the 1149.1 TAP to allow the 1149.6 driver logic to generate pulses. It was determined that the 1149.6 TAP would require an excursion through the Run-test/Idle state to allow for the generation of the pulse or pulses required by the EXTEST_PULSE and EXTEST_TRAIN instructions. Entry into the Run-test/Idle state during the execution of either EXTEST_PULSE or EXTEST_TRAIN would generate an "AC Test" signal. This would in turn cause the data which were driven onto the net during the Update-DR state to be inverted upon entry into the Run-test/Idle state (on the first falling edge of TCK) and to be inverted again on the exit from Run-test/Idle (on the first falling edge of TCK in the Select-DR state). As previously mentioned, the data signal is inverted, and then allowed to fully decay, in order to guarantee the maximum transition from the driver.

In summary, the IEEE 1149.6 standard is an extension of the IEEE 1149.1 standard; as a result, the 1149.6 standard must comply with all 1149.1 rules. The 1149.6 logic allows for testing of AC-coupled networks by capturing edges of pulses that are generated by 1149.6 drivers. A special, analog test receiver is used to capture these edges. The 1149.6 receiver logic is placed on both inputs of the differential receiver logic. Special hysteresis logic filters out noise and captures only valid transitions. These extensions allow for an equivalent level of testing (to 1149.1) for high-speed digital interconnects.

1.2.3 Boundary-Scan Accessible Embedded Instruments (IEEE P1687)

The use of BIST circuits and other embedded instruments is becoming more prevalent as devices become larger, faster, and more complex. These instruments can increase the device's test coverage, decrease test and debug time, and provide correlation between the system and ATE environments. Examples of some of these instruments are logic BIST, memory BIST (for both internal and external memory), built-in ***pseudo-random bit sequence*** (PRBS) or bit-error-rate testing for ***serializer/deserializer*** (SerDes), power management and clock control logic, and scan register dump capabilities. Currently, many of these instruments are not documented or are documented in an ad hoc manner. This makes access to the instruments difficult and time consuming at best, and oftentimes can make access impossible. The lack of a standard interface (through the IEEE 1149.1 TAP) to these instruments makes automation virtually impossible as well.

The proposed standard **P1687 (IJTAG)** will develop a standard methodology to access embedded test and debug features via the IEEE 1149.1 TAP [IEEE P1687-2007]. The proposed standard does not try to define the instruments themselves but instead tries to standardize the description of the embedded features and the protocols required to communicate with the embedded instrument. The proposed standard may also define requirements for the interface to the embedded features. This proposed standard is an extension to IEEE 1149.1 and uses the 1149.1 TAP to manage configuration, operation, and collection of data from the embedded instrument. More information can be found on the IEEE P1687 Web site (http://grouper.ieee.org/groups/1687).

1.2.4 Core-Based Testing (IEEE 1500 Standard)

There are a number of important differences between PCB and SOC (or SIP) implementations that must be considered when testing cores of an SOC [Wang 2006].

Cores can be deeply or hierarchically embedded in the SOC requiring a ***test access mechanism*** (TAM) for each core. The number and types of cores can be diverse and provided by different vendors with different types of tests and test requirements, and in the case of ***intellectual property*** (IP) cores, there may be little, if any, detailed information about the internal structure of the core. Although the clock rate of information transfer between internal cores is typically much high than that of the I/O pins of the SOC, additional ports to a core are much less costly compared to additional pins on a package. As a result, boundary scan alone does not provide a complete solution to the problem of testing the cores in an SOC. Therefore, the IEEE 1500 standard was introduced to address the problems associated with testing SOCs [Seghal 2004] [IEEE 1500-2005].

The most important feature of the IEEE 1500 standard is the provision of a "wrapper" on the boundary (I/O terminals) of each core to standardize the test interface of the core. An overall architecture of an SOC with N cores, each wrapped by an IEEE 1500 wrapper, is shown in Figure 1.7, and the structure of a 1500 wrapped core is given in Figure 1.8. The ***wrapper serial port*** (WSP) is a set of I/O terminals of the wrapper for serial operations, which consists of the ***wrapper serial input*** (WSI), the ***wrapper serial output*** (WSO), and several ***wrapper serial control*** (WSC) terminals. Each wrapper has a ***wrapper instruction register*** (WIR) to store the instruction to be executed in the corresponding core, which also controls operations in the wrapper including accessing the ***wrapper boundary register*** (WBR), the ***wrapper bypass register*** (WBY), or other user-defined function registers. The WBR consists of ***wrapper boundary cells*** (WBCs) that can be as simple as a single storage element (flip-flop for observation only), similar to the BSC shown in Figure 1.5a, or a complex cell with multiple storage elements on its shift path.

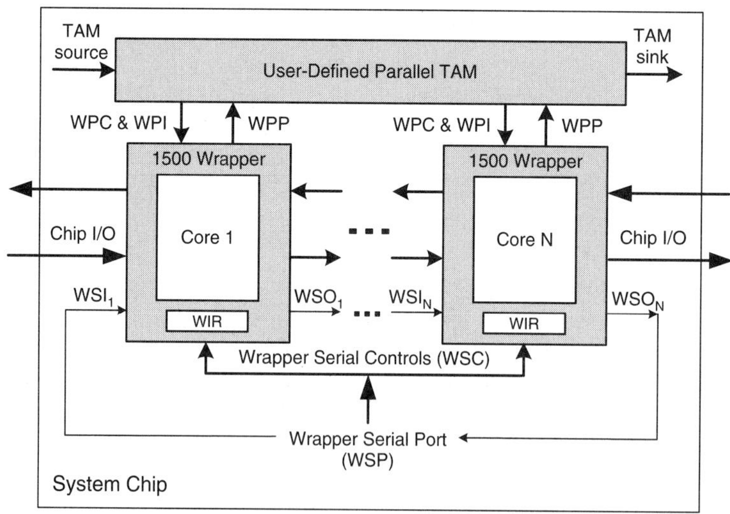

■ **FIGURE 1.7**

Overall architecture of a system per the IEEE 1500 standard.

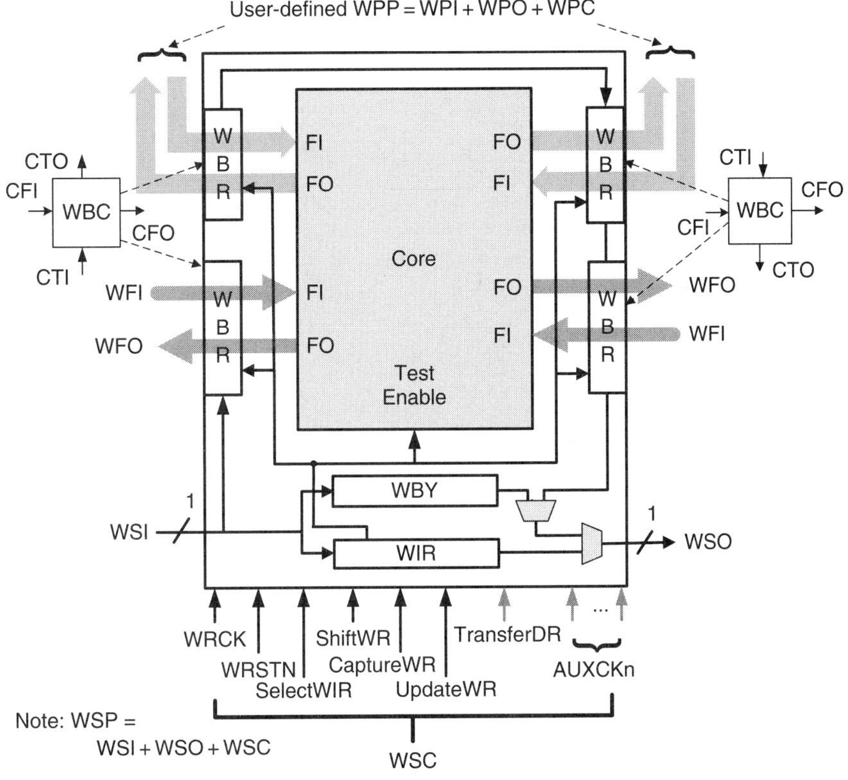

FIGURE 1.8

A core with the IEEE 1500 wrapper.

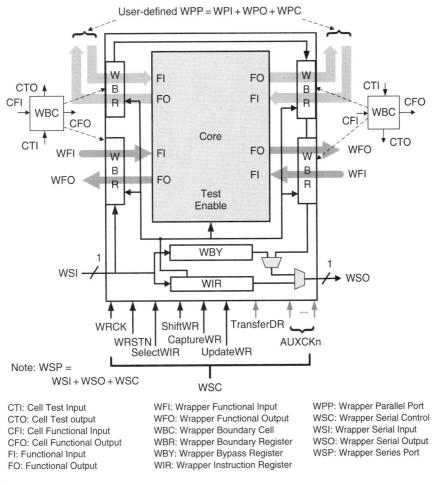

The WSP supports the serial test mode similar to that in the boundary-scan architecture, but without using a TAP controller. This implies that the serial control signals of 1500 can be directly applied to the cores and hence provide more test flexibility. For example, delay testing that requires a sequence of test patterns to be consecutively applied to a core can be supported by the 1500 standard. As shown in Table 1.1, in addition to the series I/O data signals WSI and WSO, the WSC consists of six mandatory terminals (WRCK, WRSTN, SelectWIR, CaptureWR, UpdateWR, and ShiftWR), one optional terminal (TransferDR), and a set of optional clock terminals (AUXCKn). The functions of CaptureWR, UpdateWR, and ShiftWR are similar to those of CaptureDR, UpdateDR, and ShiftDR of boundary scan, respectively (see Figure 1.5a). The SelectWIR is used to determine whether to select the WIR or not. This is required because no TAP controller is available for the 1500 standard. The TransferDR is used to transfer test data to correct positions on the shift path of a WBC when the shift path contains multiple storage elements. This enables multiple test data to be stored in consecutive positions of the shift path and hence allows delay testing to be carried out. Finally the AUXCKn terminals can be used to test cores with multiple clock domains.

In addition to the serial test mode, the 1500 standard also provides an optional parallel test mode with a user-defined, parallel ***test access mechanism*** (TAM). Each core can have its own ***wrapper parallel control*** (WPC), ***wrapper parallel input*** (WPI), and ***wrapper parallel output*** (WPO) signals. A user-defined parallel TAM, as summarized in the final three entries of Table 1.1 [IEEE 1500–2005], can transport test signals from the TAM-source (either inside or outside the chip) to the cores through WPC and WPI, and from the cores to the TAM-sink through WPO in a parallel manner; hence, it can greatly reduce the test time.

A variety of architectures can be implemented in the TAM for providing parallel access to control and test signals (both input and output) via the ***wrapper parallel port*** (WPP) [Wang 2006]. Some of these architectures are illustrated in Figure 1.9, including (1) multiplexed access where the cores time-share the test control and data ports, (2) daisy-chained access where the output of one core is connected to the input of the next core, and (3) direct access to each core.

Although it is not required or suggested in the 1500 standard, a chip with 1500-wrapped cores may use the same four mandatory pins as in the IEEE 1149.1 standard for chip interface so that the primary access to the IEEE 1500 architecture is via the boundary scan. An on-chip test controller with the capability of the TAP controller in the boundary-scan standard can be used to generate the WSC for each core. This on-chip test controller concept can also be used to deal with the testing of hierarchical cores in a complex system [Wang 2006].

TABLE 1.1 ■ IEEE 1500 Wrapper Interface Signals

Signal	Function
Wrapper serial input (WSI)	Serial data input to wrapper
Wrapper serial output (WSO)	Serial data output from wrapper
Wrapper clock (WRCK)	Clock for wrapper functions
Wrapper reset (WRSTN)	Resets wrapper to normal system mode
SelectWIR	Determines whether to select WIR
CaptureWR	Enables capture operation of selected register
ShiftWR	Enables shift operation of selected register
UpdateWR	Enables update operation of selected register
TransferDR	Optional transfer operation of selected register
AUXCKn	Up to n optional clocks for wrapper functions
Wrapper parallel in (WPI)	Optional parallel test data from TAM
Wrapper parallel control (WPC)	Optional parallel control from TAM
Wrapper parallel out (WPO)	Optional parallel test response to TAM

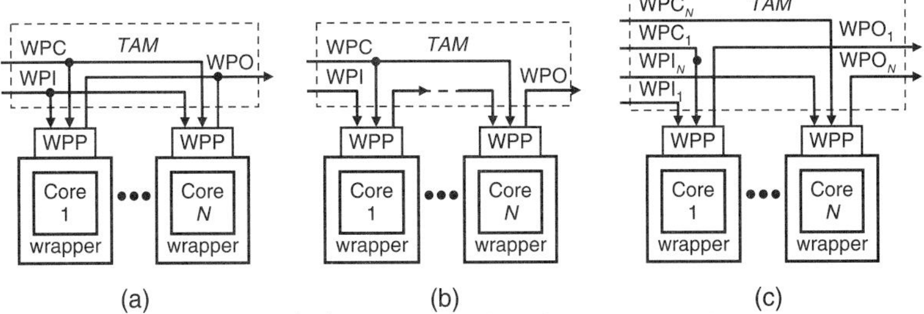

■ **FIGURE 1.9**

Example of user-defined parallel TAM architectures: (a) multiplexed, (b) daisy-chain, and (c) direct access.

An additional problem encountered with the increasing complexity of SOCs is the ability to verify or debug a design once it has been implemented in its final medium, for example silicon. The IEEE 1500 standard can be used for core-level testing with some capability for debug [Marinissen 2002] [Zorian 2005]. However, the process of post-silicon qualification or debug is time consuming and expensive, requiring as much as 35% to 70% of the total time-to-market interval. This is due in part to limited internal observability but primarily to the fact that internal clock frequencies are generally much higher than can be accessed through IEEE 1500 circuitry. One solution to this problem of ***design for debug and diagnosis*** (DFD) is the insertion of a reconfigurable infrastructure used initially for silicon validation or debug, with later application to system-level test and performance measurement. The infrastructure includes a wrapper fabric similar in some ways to the 1500 wrapper but reconfigurable in a manner similar to the programmable logic and routing resources of an FPGA. In addition, a signal monitoring network is incorporated along with functions analogous to an embedded logic analyzer for in-system, at-speed triggering and capture of signals internal to the SOC [Abramovici 2006].

1.2.5 Analog Boundary Scan (IEEE 1149.4 Standard)

There are also mixed-signal SOC implementations that include analog circuitry not addressed by either the IEEE 1149.1 or 1500 standard. Boundary scan for PCBs was extended to include mixed-signal systems with both digital and analog components in the IEEE 1149.4 standard [IEEE 1149.4-1999]. The purpose of the IEEE 1149.4 standard is "to define, document, and promote the use of a standard mixed-signal test bus that can be used at the device and assembly levels to improve the controllability and observability of mixed-signal designs and to support mixed-signal built-in test structures in order to reduce both test development time and test cost, and to improve test quality." Figure 1.10 shows a typical 1149.4 equipped IC and its PCB-level environment. At the chip level, an 1149.4 IC has two internal analog buses (AB1/AB2) connected to its analog pins. At the board level, AB1 and AB2 are connected to two board-level analog buses through two analog testability ports (AT1/AT2). Hence, one is able to access analog pins through the testability construct.

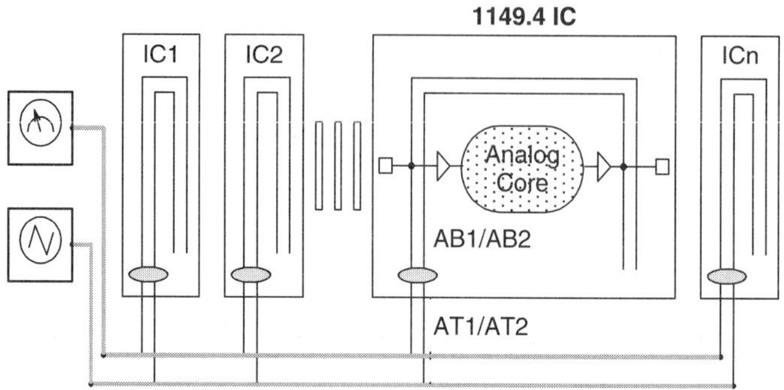

■ **FIGURE 1.10**

IEEE 1149.4 internal test configuration.

The IEEE 1149.4 standard defines test features to provide standardized approaches to (1) **interconnect testing** that tests the open/short of simple interconnects, (2) **parametric testing** that measures electrical parameters of extended interconnects with passive devices, and (3) **internal testing** that tests the functionality of on-chip analog cores.

An 1149.4-compliant design has a typical circuit architecture shown in Figure 1.11. In addition to boundary-scan circuitry for the digital domain, the standard includes an ***analog test access port*** (ATAP) which consists of a minimum of one analog test input and one analog test output connection. Connections between

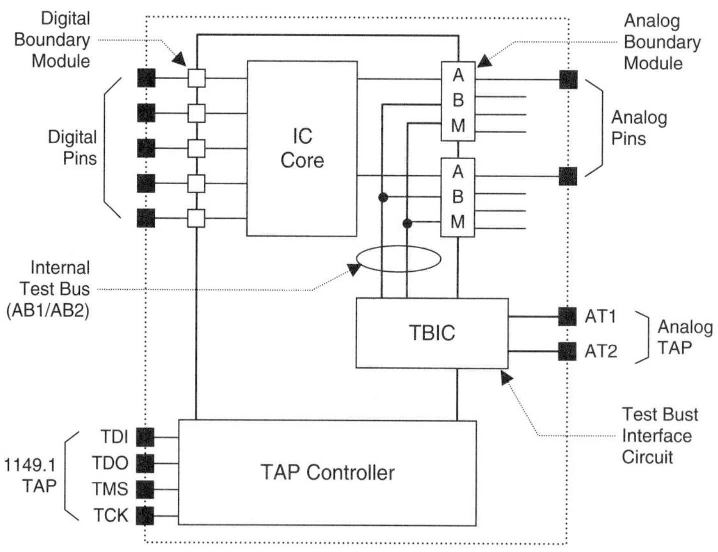

■ **FIGURE 1.11**

Typical IEEE 1149.4-compliant chip architecture.

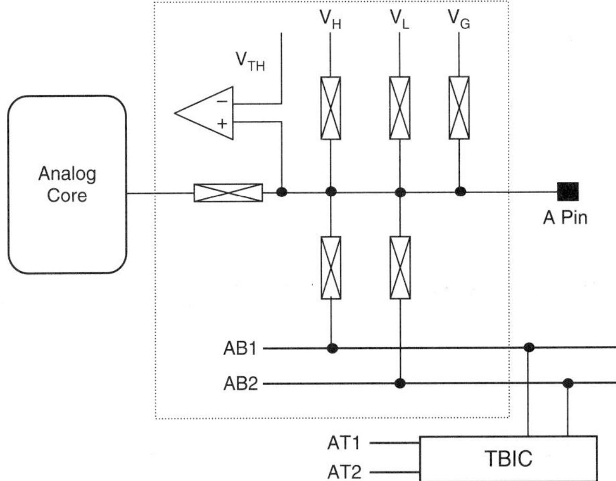

■ **FIGURE 1.12**

IEEE 1149.4 analog boundary module (ABM).

the two internal analog test buses (AB1 and AB2) and the ATAP are controlled by the ***test bus interface circuit*** (TBIC). The analog test buses provide connections to the ***analog boundary modules*** (ABMs), which are analogous to the BSCs in the digital domain. Each ABM (see Figure 1.12) includes switches that allow connection to AB1, AB2, a high voltage (V_H), a low voltage (V_L), a reference voltage (V_G), or an I/O pin of an analog core. This allows not only interconnect testing of analog signal nets on the PCB for opens and shorts, but perhaps more important, it allows for the test and measurement of passive components such as resistors and capacitors commonly used for filtering and coupling analog signal nets on a PCB [Wang 2006].

The IEEE 1149.4 standard is a superset of the IEEE 1149.1 standard, so each component responds to the mandatory instructions defined in 1149.1 [IEEE 1149.1-2001]. Herein, three types of instruction are defined: (1) mandatory instructions, (2) optional instructions, and (3) user-defined instructions.

Mandatory instructions include (1) **BYPASS** that isolates the chip from the entire DFT construct; (2) **SAMPLE/PRELOAD** that captures a digitized snapshot of the analog signal on the pin or loads a digital data pattern to specify the operation of the ABM; (3) **EXTEST** that disconnects analog pins from the core to perform open/short and parametric testing of the interconnects; and (4) **PROBE** that connects analog pins to internal buses for allowing access to the pins from edge connectors of the board.

Optional instructions recommended in the standard include (1) **INTEST** for internal core testing based on the **PROBE** instruction; (2) **RUNBIST** for the application of BIST circuitry; (3) **CLAMP** for properly conditioning the pin to V_H, V_L, or V_G; (4) **HIGHZ** for isolating the pin; and (5) device identification register such as **IDCODE** and the **USERCODE** as defined in the IEEE 1149.1 standard. In addition, users can define their own instructions, which will be treated as the extension of optional instructions.

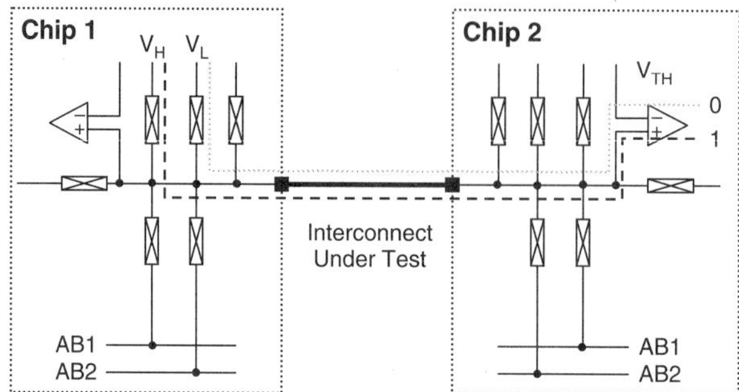

■ **FIGURE 1.13**

Example for interconnect testing using the IEEE 1149.4 bus.

Based on the preceding instructions, open/short interconnect testing, extended interconnect measurement, and internal analog core testing can be executed. Let us use interconnect testing as an example to show how 1149.4 works. Without external instrument, the ABM can perform interconnect testing in general and open/short in particular, as shown in Figure 1.13. The tested wire (shown in bold in the figure) connects analog Pin1 of Chip1 to Pin2 of Chip2. The three-step test procedure is as follows:

1. Switch V_H to Pin1 to detect a 1 at the comparator of Pin2.
2. Switch V_L to Pin1 to detect a 0 at the comparator of Pin2.
3. Switch V_H to Pin1 to detect a 1 at the comparator of Pin2.

Normally, V_H and V_L are set to V_{DD} and V_{SS}, and V_{TH} is set to $0.5V_{DD}$. The procedure detects static short faults and 1-to-0 and 0-to-1 transitions of open faults.

The most serious drawback of the 1149.4 standard is the parasitic effect associated with the long buses. A typical pin has a stray capacitance of 2~4 pF, a via has 0.5~1 pF, and a 1cm wire has 0.25~0.5 pF. Therefore, the bandwidth is severely limited by the stray capacitance of the bus and the stray resistance of the switches (in KΩ range). In that case, the standard recommends the replacement of passive switches with active **current buffers** and **voltage buffers**.

1.3 BASICS OF MEMORY TESTING

Advances in deep-submicron process technology have driven VLSI designs toward *system-on-chip* (SOC) applications that integrate *intellectual property* (IP) cores from various sources. Memory is one of the most universal cores in that almost all SOC devices contain some type of embedded memories. Nowadays embedded *static random-access memories* (SRAMs) are widely used, because by merging memory with logic, data bandwidth is increased and hardware cost can be reduced. For pad-limited, multimillion-gate designs, embedded *dynamic random-access memories* (DRAMs) (and **pseudo SRAMs** or **1T SRAMs** which contain only one transistor in each memory cell [Leung 2000]) are also becoming an attractive solution because of their compact memory density. However, with the rapid increase in capacity and density of these memory cores, the ability to detect, diagnose, and even repair all defective memories has quickly become a difficult and challenging problem, resulting in an increase in test cost and yield loss.

Fundamental topics in memory testing have been extensively covered in [van de Goor 1991]. In [Wang 2006], the authors first discuss the industry-wide use of memory fault models and (**March**) test algorithms (patterns), with special emphasis on memory fault simulation and test algorithm generation. A great deal of discussions is then centered on memory BIST and ***built-in self-repair*** (BISR), as well as **memory diagnosis** and **failure analysis**. The introduction of nanotechnologies and SOC devices brings forth new problems in semiconductor memory testing. Both the number of embedded memory cores and area occupied by memories are rapidly increasing in SOC devices. In addition, memories have been widely used as the technology driver—that is, they are often designed with a density that is at the extremes of the process technology. Therefore, the yield of on-chip memories usually determines chip yield. Go/no-go testing is no longer enough for embedded memories in the SOC era—memory diagnosis and failure analysis is quickly becoming a critical issue, as far as manufacturing yield and time-to-volume of SOC devices are concerned.

When memory cores (the number can be hundreds or even thousands) are embedded in an SOC and are surrounded by logic blocks, proper DFT methodology should be provided for core isolation and tester access, and a price has to be paid for the resulting hardware overhead, performance penalty, and noise and parasitic effects. Even if these are manageable, memory testers for full qualification and testing of the embedded memories will be much more expensive because of their increased speed and I/O data bandwidth, and if we also consider engineering change the overall investment in test equipment will be even higher. Fortunately, BIST has become a widely accepted practical solution to this dilemma. With BIST, the external tester requirement can be minimized, and tester time can be greatly reduced because parallel testing at the memory bank and chip levels is feasible. Another advantage of BIST is that it also is a good approach to protecting IP—that is, the IP (memory cores in this case) provider needs only deliver the BIST activation and response sequences for testing and diagnosis purposes without disclosure of the design details.

There are, however, some important issues that a pure BIST scheme does not solve, such as diagnosis and repair. High-density, high-operating clock rate, and deep-submicron technology are introducing more new failure modes and faults in memory circuits. Conventional memory testers designed for mass production tests provide only limited information for failure analysis that usually is insufficient for fast debugging. Extensive test data logging with real-time analysis, screening, compression, and diagnosis are still prohibited by high cost. Designers need a diagnosis-support mechanism within the BIST circuit and even a BISR scheme to increase product quality, reliability, and yield. The mechanism should be easy to deploy and use—tools for automatic generation and insertion of the BIST/BISR circuits as well as accompanying software and scripts are expected.

Memory repair has long been an important technique that is used to avoid yield loss. Memory repair requires **redundant elements** or **spare elements** such as spare rows, columns, or blocks of storage cells. The **redundancy** is added so that most faulty cells can be repaired or, more specifically, replaced by spare cells [Cenker 1979] [Smith 1981] [Benevit 1982] [Wang 2006]. *Redundancy*, however, adds to cost in another form. Analysis of redundancies to maximize yield (after repair) and minimize cost is a key process during manufacturing [Huang 2003]. **Redundancy analysis** using expensive memory testers is becoming inefficient (and therefore not cost-effective) as chip density continues to grow. Therefore, ***built-in redundancy analysis*** (BIRA) and BISR are now among the top items to be integrated with memory cores.

We illustrate the combined operation of BIST, BIRA, and BISR with the following example. Figure 1.14, taken from [Wang 2006], depicts the block diagram of a BISR scheme, including the BIST module, BIRA module, and **test wrapper** for the memory. The BIST circuit detects the faults in the main memory and spare memory and is programmable at the March element level [Huang 1999] [Wang 2006]. The BIRA circuit performs redundancy allocation. The test wrapper switches the memory between test/repair mode and normal mode. In test/repair mode, the memory is accessed by the BIST module, whereas in normal mode the wrapper selects the data outputs either from the main memory or the spare memory (replacing the faulty memory cells) depending on the control signals from the BIRA module. This BISR is a **soft repair** scheme; therefore, the BISR module will perform testing, analysis, and repair upon every power up. As Figure 1.14 indicates, the BIST circuit is activated by the ***power-on reset*** (POR) signal. When we turn on the power, the BIST module starts to test the spare memory.

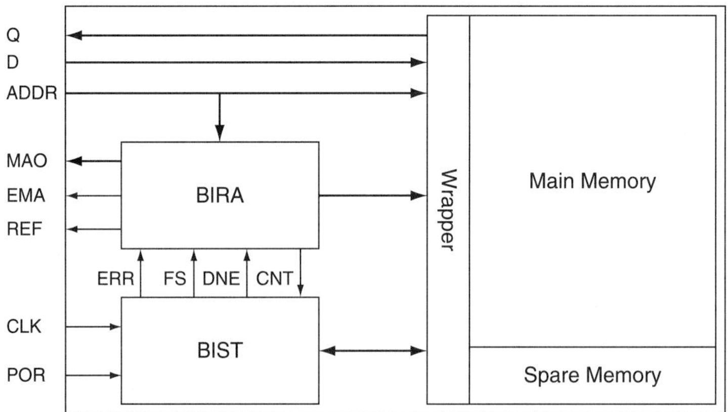

■ **FIGURE 1.14**

Block diagram of a BISR scheme [Wang 2006].

Once a fault is detected, the BIRA module is informed to mark the defective spare row or column as faulty through the *error* (ERR) and *fault syndrome* (FS) signals. After finishing the spare memory test, the BIST circuit tests the main memory. If a fault is detected (ERR outputs a pulse), the test process pauses and the BIST module exports FS to the BIRA module, which then performs the redundancy analysis procedure. When the procedure is completed and the memory testing is not yet finished, the BIRA module issues a *continue* (CNT) signal to resume the test process. During the redundancy analysis procedure, if a spare row is requested but there are no more spare rows, the BIRA module exports the faulty row address through the *export mask address* (EMA) and *mask address output* (MAO) signals. The memory will then be operated in a downgraded mode (*i.e.*, with smaller usable capacity) by software-based address remapping. If downgrade mode is not allowed, MAO is removed and EMA indicates whether the memory is repairable. When the main memory test and redundancy analysis are finished, the *repair end flag* (REF) signal goes high and the BIRA module switches to the normal mode. The BIRA module then serves as the address remapper, and the memory can be accessed using the original address bus (ADDR). When the memory is accessed, ADDR is compared with the fault addresses stored in the BIRA module. If ADDR is the same as any of the fault addresses, the BIRA module controls the wrapper to remap the access to spare memory.

Although BIST schemes, such as the one in the previous example, are promising, a number of challenges in memory testing must be considered. For example, BIST cannot replace external memory testers entirely if the BIST schemes used are only for functional testing. Even BIST with diagnosis support is insufficient because of the large amount of diagnosis data that must be transferred to an external tester, typically through a channel with limited bandwidth. Furthermore, memory devices normally require burn-in to reduce field failure rate. For logic devices I_{DDQ} is frequently used during burn-in to detect the failing devices, but I_{DDQ} for memories is difficult. What, then, should be done to achieve the same reliability requirement when we merge memory with logic? The combination of **built-in current sensors** and BIST is one possible approach, and the memory burn-in by BIST logic is another.

Yet another challenge is **timing qualification** or **AC testing**. With the shrinkage of feature size in process nodes, an increasing number of parametric failures caused by process variation are becoming critical as far as yield is concerned. In memories, these subtle defect types may cause additional circuit delays and even a system failure. The current practice of screening and diagnosing parametric defects and failures in the industry is AC testing to test the timing-critical parameters, but this is a time-consuming process as many parameters need to be tested. The lack of AC test capability is one of the reasons that BIST circuits have yet to replace traditional memory testers. For example, consider testing an asynchronous memory with synchronous BIST logic. The BIST timing resolution would not be able to compete with that of a typical external memory tester. It is hoped that this problem can be solved by proper delay fault models that may result from a systematic investigation of the relationship between delay faults and memory failures.

The success of BIST in SRAM does not guarantee its success in DRAM, **flash memory**, ***content addressable memory*** (CAM), etc. For embedded DRAM, as an example, the need for an external memory tester cannot be eliminated unless redundancy analysis and repair can be done on-chip, in addition to AC testing by BIST. The problems increase when merging logic with memory such as DRAM and flash memory [Wu 1998]. In addition to process technology issues, there are problems of guaranteeing the performance, quality, and reliability of the embedded memory cores in a cost-effective manner. Once again, new failure modes or fault models have to be tested because March algorithms, such as those used in SRAM BIST schemes, are considered insufficient for DRAM or flash memory.

Flash memory is by far the most popular **nonvolatile memory**, and it has been widely used in portable devices such as PDAs, cell phones, MP3 players, and digital cameras. The advent of deep-submicron IC manufacturing technology and SOC design methodology has greatly increased the use of embedded flash memory in those applications. For commodity flash memory, it is customary to test and repair the memories using a probe station during wafer-level test. However, for embedded flash memory, BIST/BISR has been considered the most cost-effective solution that addresses this issue, though more research is needed to make it practical. It should be also noted that for flash memory, there are special faults and redundancy architectures that are more complicated than SRAM [Yeh 2007].

Although flash memory is widely used today, it suffers from problems such as the use of high voltage for program and erase operations, as well as reliability issues. The industry has been trying to find a new nonvolatile memory to replace flash memory. Possible candidates include ***magnetic random-access memory*** (MRAM), ***ferroelectric random-access memory*** (FeRAM), and ***ovonic universal memory*** (OUM). Among them, MRAM has the advantages of high speed, high density, low power consumption, and almost unlimited read/write endurance [Su 2004, 2006]. The data storing/switching mechanism of MRAM is based on the resistance change of the ***magnetic tunnel junction*** (MTJ) device in each cell [Su 2004, 2006]. In [Su 2006], some MRAM testing issues are addressed including fault modeling, testing, and diagnosis. The ***write disturbance fault*** (WDF) model for MRAM is stressed, which is a fault that affects the data stored in the MRAM cell because of the excessive magnetic field during the write operation. The faulty behavior is mainly represented by the variation of the MTJ cell operating region.

In summary, although there have been many advances in memory testing, there are also many new challenges for the future. The basics of memory testing were covered extensively in [van de Goor 1991] and [Wang 2006], and some of the new memory test challenges have been summarized here. Memory testing is an important topic and will continue to be discussed throughout various chapters of this book in terms of how it relates to that particular topic area.

1.4 SOC DESIGN EXAMPLES

Numerous SOC designs are developed and manufactured each year. Today, designing a complex SOC circuit is no longer a mission-impossible task. Designers can

TABLE 1.2 ■ Popular IP Cores and Major IP Suppliers

IP Core	Major IP Suppliers
Central processing unit (CPU)	ARM, MIPS, Tensilica, ARC
Digital signal processor (DSP)	Texas Instruments
Static random-access memory (SRAM)	Virage Logic, ARM, MoSys
Dynamic random-access memory (DRAM)	Virage Logic, ARM
Structured ASIC (FPGA)	—
Digital-to-analog converter (DAC)	Analog Devices
Analog-to-digital converter (ADC)	Analog Devices
Universal serial bus (USB)	—
Phase-locked loop (PLL)	ARM, Virage Logic

often pick commercially available IP cores and integrate them together to fit the design requirements. These IP cores range from digital to memory to analog and mixed-signal cores. Table 1.2 lists a few popular IP cores and some of the major IP suppliers.

In this section, we illustrate a few representative SOC designs. More SOC designs and architectures can be found in [Dally 2004] and [De Micheli 2006]. The SOC designs selected here mainly cover the whole spectrum of IP cores, which can represent significant test challenges in the nanometer design era. Our objective is to convey to the reader that unless these SOC design-related test issues were solved in a timely manner, few billion-transistor SOC designs and safety-critical nanoscale devices would come to life to benefit the whole society.

1.4.1 BioMEMS Sensor

The wireless bioMEMS sensor developed at National Taiwan University is the first wearable prototype to detect ***C-reactive protein*** (CRP) based on nanomechanics [Chen 2006]. The CRP concentration in human serum is below 1μg/mL for a healthy person but may rise up to 100 or even 500 times in response to infection in the body. In [Blake 2003], the authors showed that high levels of CRP in the bloodstream raise the risk of a heart attack. One of the greatest benefits of this biosensor is that cardiovascular-event-related proteins, such as CRP, can be dissociated from anti-CRP by applying a low-frequency AC electric field (with 0.2 Hz 1 V electrical signal) to the sensor. This allows the design of a wireless label-free detection system for disease-related CRP using a **microcantilever** [Arntz 2003] with a safe "reusable" feature.

The CRP sensing mechanism is shown in Figure 1.15 [Chen 2006]. First, CMOS compatible silicon nitride is deposited on the silicon substrate. Next, a MEMS cantilever is fabricated by photolithography followed by micromachining technology. The optimum sensor size of a length of 200μm with each leg 40μm wide is determined by the balance among spring constant, bio-induced stress and stability in the flow field. On the top side of the cantilever, chromium (*Cr*), gold (*Au*), bio-linker, and anti-CRP are deposited. Chromium is adopted to improve the adhesion of gold to silicon nitride, whereas the gold layer is used to immobilize the bio-linker.

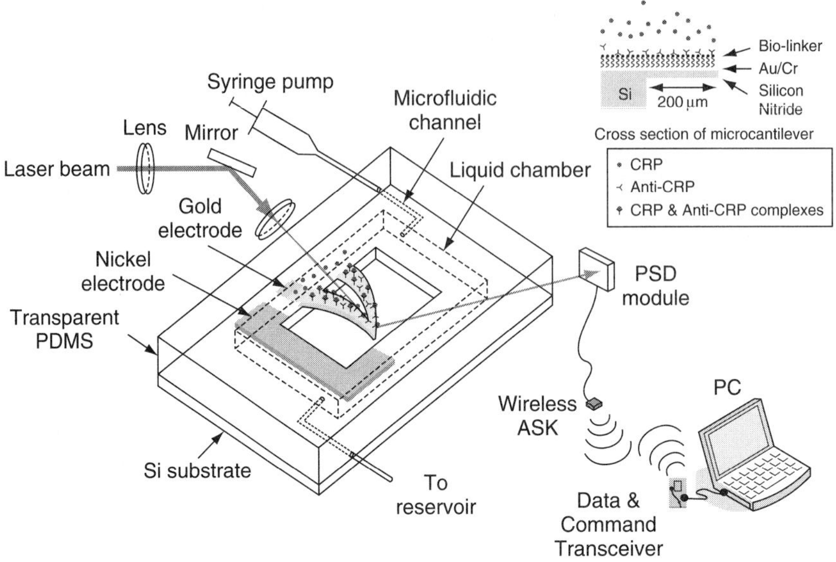

■ **FIGURE 1.15**

CRP sensing mechanism using microcantilever.

The anti-CRP is bonded to the bio-linker for probing CRP. Specific biomolecular interactions between CRP and anti-CRP alter the intermolecular nanomechanical interactions within the bio-linker layer. As a result, the cantilever is bent. The bending of the cantilever can be measured by optical beam deflection or piezoresistive techniques. The authors adopted the former approach because of its excellent minimum detection limit of 0.1 nm.

To demonstrate how the prototype works, the authors in [Chen 2006] further set up an experiment, as shown in Figure 1.15. After placing a *poly-dimethyl-siloxane* (PDMS)-based "cap" above the functionalized cantilever, two microfluidic channels and one liquid chamber for reaction are formed. The reagents are injected into the channels and chamber using a syringe pump. A laser beam through the optically transparent PDMS cap is focused onto the tip of the cantilever with the help of a ***charge-coupled device*** (CCD) camera, thereby ensuring the alignment among the laser beam, the cantilever, and the ***position-sensitive detector*** (PSD). The function blocks of the wireless CRP measurement system are shown in Figure 1.16 [Chen 2006]. Custom commands from a ***personal computer*** (PC) are received by a 0.45 V operated ASK receiver consisting of a common-source amplifier, gain stages, and a diode-RC demodulator. The 0.45 V operation is achieved by using low-threshold (0.2 V) transistors. After decoding the commands, the ***microcontroller unit*** (MCU) activates the 8-bit charge-redistributed ***analog-to-digital converter*** (ADC) and preamplifier to convert the analog bio-signals measured by the commercially available PSD into digital values. Finally, an ASK transmitter comprising a ring oscillator and a common-source *class-C* power amplifier transmits the digital values to the PC for analysis.

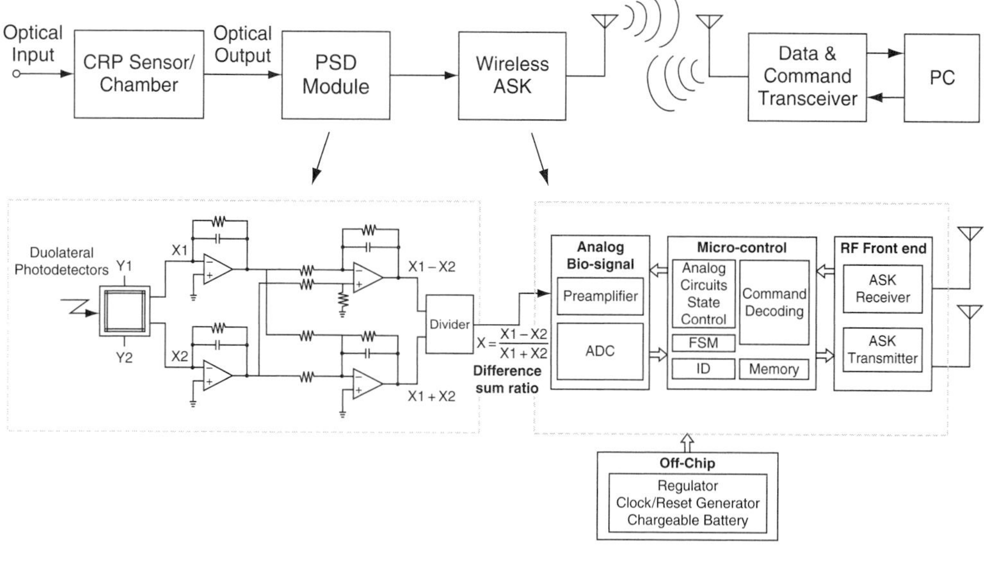

■ **FIGURE 1.16**

Function blocks of the wireless CRP measurement system.

Fabricated at a 180-nm process node, the wireless ASK chip is operated at 500 KHz and has a die size of 3.4 mm^2. Although this chip is small, the wireless CRP measurement system presents many test challenges, because it includes a microfluidic bioMEMS sensor, a PSD module, an RF transceiver (transmitter and receiver), an analog preamplifier, an ADC, and an MCU embedded with memory and analog circuits state control logic. Today, semiconductor lasers [Arntz 2003] and microlenses used in a ***compact disc*** (CD) player can be placed on top of the PDMS cap to replace the *He-Ne* laser, the reflected mirror, and the lens. A standard CMOS poly-resistor layer can be also used as the cantilever for the detection of the piezoresistive change caused by bending. Thus, the authors anticipate that the bioMEMS sensor, the semiconductor laser, and the microlenses can be also integrated with the PSD module and the wireless ASK circuit on the same die. This will make testing of this SOC device more challenging. *Self-diagnosis* and *self-calibration* might be needed to ensure the correct operation of the device all the time. When the device becomes *implantable*, it will require BISR.

1.4.2 Network-on-Chip Processor

The Cell processor [Pham 2005, 2006] co-designed by Sony, Toshiba, and IBM is one of the earliest ***network-on-chip*** (NOC) processors developed to address high-performance distributed computing. Built as the first-generation multiprocessor with a vision of bringing supercomputer power to everyday life, the Cell processor supports multiple operating systems including Linux and is designed for natural human interactions including photorealistic, predictable real-time response, and virtualized resource for concurrent activities. The processor architecture includes one *power processor element* (PPE), eight *synergistic processor elements* (SPEs), an *element interconnection bus* (EIB), a *memory interface controller* (MIC), a *bus interface controller* (BIC), a *pervasive unit* (PU) that supports extensive test, monitoring, and debug functions, a *power management unit* (PMU), and a *thermal management unit* (TMU). The chip was fabricated at a 90-nm process node and can operate at more than 4 GHz at a nominal voltage of 1.1 V or higher. The high-level chip diagram is shown in Figure 1.17.

The PPE is a 64-bit dual-threaded processor based on the Power Architecture [Rohrer 2004]. The processor contains a *power execution unit* (PXU), 32-KB instruction and data caches (L1), and a 512-KB cache (L2). Each SPE contains a *synergistic execution unit* (SXU), which is an independent processor, and a 256-KB *local store* (LS). The EIB connects the PPE, the eight SPEs, the MIC, and the BIC. The EIB is central to the Cell processor. This coherent bus transfers up to 96 bytes per processor cycle. The bus is organized as four 16-byte-wide (half-rate) rings, each of which supports up to three simultaneous data transfers. A separate address and command network manages bus requests and the coherence protocol. With a dual-threaded PPE and eight SPEs, this Cell processor is capable of handling 10 simultaneous threads and over 128 outstanding memory requests.

The MIC supports two Rambus XDR memory banks. For added reliability, the MIC supports *error correcting code* (ECC) and a periodic ECC scrub of memory. For debug and diagnosis, the MIC also allows the processor to be stopped and its system state scanned out. The BIC provides two Rambus RRAC flexible I/O (FlexIO) interfaces with varying protocols and bandwidth capabilities to address differing system requirements. The interfaces can be configured as either two I/O interfaces (IOIF 0/1) or as an I/O and a coherent SMP interface (IOIF and BIF). Both MIC and BIC are operated asynchronously to the processor; hence, each bus contains

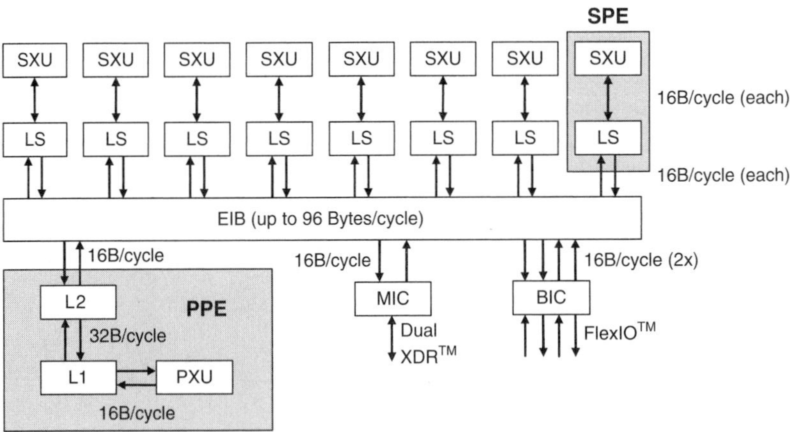

■ **FIGURE 1.17**

The EIB in the Cell processor.

speed-matching SRAM buffers and logic. The processor side operates at half the global processor clock rate, whereas the XDR side operates at one half the rate of the XDR interface and the FlexIO side operates at 1/2 or 1/3 the rate of the RRAC transceivers.

The PU (not shown in Figure 1.17) contains all of the global logic needed for basic functional operation of the chip, lab debug, and manufacturing test. To support basic functional operation of the chip, the PU contains a ***serial peripheral interface*** (SPI) to communicate with an external controller during normal operation, clock generation and distribution logic to enable the correct ***phase-locked loop*** (PLL) functions on the chip, and ***power-on-reset*** (POR) that systematically initializes all the units of the processor. The POR engine has a debug mode that allows its embedded 32 instructions to be single-stepped, skipped, or performed out of order. For lab debug and bring up, the PU contains chip-level global fault isolation registers to allow the operating system to quickly determine which unit generated an error condition and a *performance monitor* (PFM), which consists of a centralized unit connected to all functional units on the chip via a trace/debug bus to assist with performance analysis. An on-board programmable *trace logic analyzer* (TLA) captures and stores internal signals while the chip is running at full speed to assist with debug. An on-chip control processor, with an IEEE 1149.1 boundary-scan interface, is also available to help with debug. For manufacturing test, the PU supports 11 different test modes, including the ***array built-in self-test*** (ABIST), commonly referred to as ***memory BIST*** (MBIST), which tests all memories in parallel to reduce test time, and the ***logic BIST*** (LBIST) that includes a centralized controller in the PU and 15 LBIST satellites elsewhere in the design. *At-speed BIST* is provided on both ABIST and LBIST, and *at-speed scan* is implemented on internal scan chains. The PU also provides the logic used for programming and testing ***electronic fuses*** (eFuses), which are used for array repair and chip customization during the manufacturing test.

The PMU and TMU (not shown in Figure 1.17) work in tandem to manage chip power and avoid permanent damage to the chip because of overheating. For power reduction, the PMU provides a mechanism to allow software controls to reduce chip power when the full processing capabilities are not needed. The PMU also allows the operating system to throttle (by adjusting instruction issue rate), pause, or stop single or multiple elements to manage chip power. The processor embeds a *linear sensor* (linear diode) to monitor the global chip temperature for controlling external cooling mechanisms. Ten digital thermal sensors are distributed on the chip to monitor temperatures in critical local regions (*hot spots*). One ***digital thermal sensor*** (DTS) is located in each element, and one is adjacent to the linear sensor. The TMU continuously monitors each DTS and can be programmed to dynamically control the temperature of each element and interrupt the PPE when a temperature specified for each element is observed. Software controls the TMU by setting four temperature values and the amount of throttling for each sensor in the TMU. With increasing temperature, the first temperature specifies when the throttling of an element stops, the second when throttling starts, the third when the element is completely stopped, and the fourth when the chip's clocks are shut down.

In contrast to bus interconnect technology commonly found in SOC designs, NOC-platform-based SOC designs tend to contain two or more processors and a programmable crossbar switch (for dynamic routing) to improve on-chip communication efficiency [Dally 2004] [Jerraya 2004] [McNairy 2005] [Naffziger 2005, 2006] [De Micheli 2006]. The multiprocessor architecture and crossbar switch have represented additional test challenges and spurred eminent DFT needs for *at-speed BIST* [Wang 2006], *silicon debug and diagnosis* [Abramovici 2006] [Wang 2006], *online self-checking and fault tolerance* [Siewiorek 1998], *software-based self-testing* [Cheng 2006], *FPGA testing* [Stroud 2002], and *high-speed I/O interfaces testing* [Gizopoulos 2006].

1.5 ABOUT THIS BOOK

The subsequent chapters present promising techniques to address critical ITRS needs and challenges for testing nanometer SOC designs [SIA 2005, 2006]. These techniques include specific DFT architectures for testing digital, memory, as well as analog and mixed-signal circuits. As the test needs and challenges facing the semiconductor industry in the nanometer age are so broad and difficult in nature, it is becoming evident that further research must be conducted and better solutions have to be found on all subjects mentioned here.

1.5.1 DFT Architectures

SOC designs contain a variety of components, including digital, memory, as well as analog and mixed-signal circuits, all of which need to be tested. DFT is essential for reducing test costs and improving test quality.

Scan is widely used for digital logic testing. However, one of the challenges in the nanometer design era is dealing with the rapidly growing amount of scan data required for testing complex SOC designs. The bandwidth between an external ATE and the device under test is limited, creating a bottleneck on how fast the chip can be tested. This has led to the development of test compression and logic BIST architectures, which reduce the amount of data that needs to be transferred between the ATE and device under test. Another key issue for scan testing is applying at-speed tests, which are crucial for detecting delay faults. Chapter 2 reviews the basics of scan testing and presents test compression and logic BIST architectures as well as architectures for applying at-speed tests.

SOC designs contain numerous embedded cores, each of which has a set of tests that need to be applied. This presents challenges in terms of providing test access to the embedded cores and dealing with high test data volumes. Chapter 4 presents DFT architectures that facilitate low-cost modular testing of SOCs. Optimization techniques are described for wrapper designs and test access mechanisms to facilitate efficient test scheduling and reduce test data volume. ***Network-on-chips*** (NOCs), which provide a packet-based mechanism for transferring data between cores, are becoming increasingly attractive as nanometer design complexity increases. Chapter 4 discusses DFT architectures for testing NOCs including the interconnect, router, and network interface.

A *system-in-package* (SIP) is a combination of semiconductors, passives, and interconnect integrated into a single package. The use of SIPs is growing rapidly. A SIP contains multiple dies in the same package that presents additional test issues beyond those of an SOC. The assembly of the SIP needs to be tested, and the SIP may include MEMS and RF components. Chapter 5 provides background on SIPs and discusses the test challenges along with some solutions.

Field programmable gate arrays (FPGAs) are widely used and have grown in complexity to where they are able to implement very large systems. The reprogrammability of FPGAs makes them an attractive choice, not only for rapid prototyping of digital systems but also for low-to-moderate volume or fast time-to-market systems. Testing FPGAs requires loading numerous test configurations, which can be time consuming. Chapter 12 provides background on the architecture and operation of FPGAs and presents approaches for testing FPGAs as well as RAM cores embedded in FPGAs.

High-speed I/O interfaces involve using source-synchronous serial links that can have speeds exceeding 1 Gb/s. These interfaces are often used on digital systems and, for economic reasons, it is necessary to test them using a digital tester. Chapter 14 describes high-speed I/O architectures and techniques for testing them.

Analog ICs and the analog portions of mixed-signal circuits require different test techniques than those applied to digital components. Analog testing involves specification-based approaches as opposed to the fault-model-based and defect-based approaches used for digital testing. Chapter 15 reviews analog testing and then focuses on test architectures that can be used for on-chip testing and measurement of the analog portion(s) of SOC designs.

1.5.2 New Fault Models and Advanced Techniques

Nanometer technology is more susceptible to process variations and smaller defects, which often manifest themselves as timing-related problems. Conventional stuck-at fault testing becomes increasingly less effective, and new fault models need to be considered. Moreover, the lower voltage levels and smaller noise margins in nanometer technology make it increasingly susceptible to transient errors, thereby creating new challenges in terms of dealing with soft errors and reliability issues.

Delay faults are prevalent in nanometer technology and must be detected to achieve high quality. Delay faults can arise because of global process variations or local defects. Testing for delay faults typically requires two-pattern tests, which add an additional complication for test application in scan circuits. Chapter 6 describes delay test approaches including new defect-based delay fault models along with simulation and test generation techniques.

Power dissipation is typically much higher during testing than during normal operation because there is much greater switching activity. Excessive average power during testing can result in overheating and hot spots on the chip. Excessive peak power during testing can cause ground bounce and power supply droop, which may cause a good part to fail unnecessarily. Chapter 7 presents techniques for reducing power dissipation during testing, including low-power scan and BIST techniques.

New defect mechanisms are evolving in nanometer technology such as copper-related defects, optical defects, and design-related defects caused by threshold voltage variations and the use of variable power supply voltages in low power designs. Defects do not necessarily manifest themselves as a single isolated problem such as an open or a short. A circuit parameter out of specification can cause increased susceptibility to other problems (temperature effects, crosstalk, etc.). Moreover, the lower voltage levels and smaller noise margins in nanometer technology result in increased susceptibility to radiation-induced soft errors. Chapter 8 describes techniques for coping with the physical failures, soft errors, and reliability issues that arise in nanometer technology.

SOCs typically contain one or more processors. Software running on a processor can be used to perform a self-test of the SOC. This offers a number of advantages including reducing the amount of DFT circuitry required, providing functional at-speed tests, and avoiding the problem of excessive power consumption that arises in structural tests. Chapter 11 describes software-based self-testing techniques that can target the different components of an SOC, including the processor itself, global interconnect, nonprogrammable cores, and *analog and mixed-signal* (AMS) circuits.

As wireless devices are becoming increasingly prevalent, RF testing has been gaining importance. RF testing has different considerations than conventional analog and mixed-signal testing and requires the use of a different set of instruments to perform the measurements. Noise is a particularly important factor in RF measurements. Chapter 16 provides background on RF devices and RF testing methods. Advanced techniques for testing RF components along with future directions in RF testing are also described.

1.5.3 Yield and Reliability Enhancement

Nanometer technology is increasingly less robust because of less precise lithography, greater process variations, and greater sensitivity to noise. To cope with these challenges, features must be added to the design to enhance yield and reliability. How to do this in a systematic and effective manner is a topic of great interest and research.

One way to improve reliability is to have a fault-tolerant design that can continue operation in the presence of a fault. Fault tolerance requires the use of redundancy and hence adds area, performance, and power overhead to a design. In the past, fault-tolerant design has mainly been used only for mission-critical applications (medical, aviation, banking, etc.) where the cost of failure is very high. Most low-cost systems have not incorporated much fault tolerance. However, as failure rates continue to rise in nanometer SOC designs, which are increasingly susceptible to noise, it is becoming necessary to incorporate fault tolerance even in mainstream low-cost systems. Chapter 3 reviews the fundamentals of fault-tolerant design and describes many of the commonly used techniques.

Design for manufacturability (DFM) involves making layout design changes to improve any aspect of manufacturability, from mask making through lithography and chemical-mechanical processing. DFM is increasingly important in nanometer technology. ***Design for yield*** (DFY) involves techniques that are specifically targeted toward improving yield. Coming up with metrics for yield that could be optimized during design is a major challenge because the manufacturing process is not static; hence the relative importance of different factors impacting yield is constantly changing as process improvements are made. Chapter 9 discusses DFM and DFY, gives examples, and highlights the issues and challenges.

An important step for ramping up yield is to be able to debug and diagnose silicon parts that fail. By identifying the root cause of a failure, it is possible to correct any design bugs that may exist and improve the manufacturing process. Debug and diagnosis can be time consuming. Chapter 10 describes ***design for debug and diagnosis*** (DFD) techniques that involve adding features to a design to help speed up the process. These techniques permit engineers to more rapidly obtain information to aid in the process of finding the root cause of a failure.

1.5.4 Nanotechnology Testing Aspects

Although CMOS technology is not predicted to reach fundamental scaling limits for another decade, alternative emerging technologies are being researched in hopes of launching a new era in nanoelectronics. Future nanoelectronic devices are expected to have extremely high defect rates, making test and fault tolerance key factors for obtaining working devices. Another emerging technology is ***microelectromechanical systems*** (MEMS) devices, which are miniature electromechanical sensors and actuators fabricated using VLSI processing techniques. Testing MEMS is an interesting topic as MEMS physically interact with the environment, rather than only electrical interfaces like traditional ICs, as illustrated in the bioMEMS sensor example presented in this chapter.

MEMS has many application arenas, including accelerometers, pressure sensors, microoptics, inkjet nozzles, optical scanners, and fluid pumps, and has the potential to be used in many other applications. Chapter 13 describes MEMS devices and discusses the techniques that are used to test and characterize them. In addition, DFT and BIST techniques are presented that have been proposed as well as implemented on commercially available MEMS devices.

Several promising nanoelectronic technologies are emerging, including ***resonant tunneling diodes*** (RTDs), ***quantum-dot cellular automata*** (QCA), **silicon nanowiressingle electron transistors**, and ***carbon nanotubes*** (CNTs). These are all described in Chapter 17, along with the research work that has begun to develop test techniques for them.

1.6 EXERCISES

1.1 **(Defect Level)** Assuming the process yield (Y) is 90% and the device's fault coverage (FC) is 80%, 90%, or 99%, calculate their respective defect levels in terms of defective *parts per million* (PPM).

1.2 **(Defect Level)** Consider a system that contains five *printed circuit boards* (PCBs) each embedded with 40 ASIC devices. Assume that the cost to repair a device on a PCB is \$1000. If each ASIC device is manufactured with a process yield of 80% and tested with fault coverage of 90%, what is the manufacturing rework cost to repair the system?

1.3 **(Fault Coverage)** What's the fault coverage (*FC*) required to achieve 500 PPM for a given process yield (*Y*) of 90%?

1.4 **(Process Yield)** What's the process yield (*Y*) for a given device to arrive at 100 PPM when the device's fault coverage (*FC*) is 95%?

1.5 **(Mean Time between Failures)** Suppose a computer system has 10,000 components each with a failure rate of 0.01% per 1000 hours. What is the period of 99% system reliability?

1.6 **(Mean Time between Failures)** For a system containing *N* identical subsystems each with a constant failure rate λ, what is the overall system reliability? If the system contains three identical subsystems, what is the *mean time between failures* (MTBF) of the system?

1.7 **(System Availability)** Repeat Exercise 1.6. Suppose each of the three subsystems anticipates 20 failures per 1,000 hours and we need system availability of at least 99%. What is the repair time it would take to repair the system each year?

1.8 **(Boundary Scan)** How many mandatory and optional boundary-scan pins are required according to the IEEE 1149.1 and 1149.6 standards? Can they be shared with existing primary input or output pins? List three reasons for implementing boundary scan in an ASIC device.

1.9 **(Boundary-Scan Extension)** Explain why the IEEE 1149.1 standard cannot test high-speed, AC-coupled, differential networks and how the IEEE 1149.6 standard solves this problem. Use timing diagrams to show the problem with AC-coupled nets and how the IEEE 1149.6 standard solves the problem. Draw a timing diagram that shows how an 1149.6 Update/Capture cycle might look (including all relevant TAP actions), and show data driven and captured by the 1149.6 logic.

1.10 **(Boundary-Scan Accessible Embedded Instruments)** Give an example in which the proposed IEEE P1687 boundary-scan accessible embedded instruments standard (IJTAG) may aid in automating design and test reuse for an embedded test instrument.

1.11 **(Core-Based Testing)** How many mandatory and optional core-based test pins are required according to the IEEE 1500 standard? Can they be shared with existing primary input or output pins? List three reasons for implementing core-based wrappers in an ASIC device.

1.12 **(Analog Boundary Scan)** How many mandatory and optional core-based test pins are required according to the IEEE 1149.4 standard? Can they be shared with existing primary input or output pins? List three reasons for implementing analog boundary scan in an ASIC device.

1.13 **(Analog Boundary Scan)** Figure 1.18 shows a typical two-port network connecting (P1 P2) of chip A and (P3 P4) of chip B. Assume that P2 and P4 are grounded. The hybrid (H) parameters are defined by the equations given here:

$$h_{11} = \left.\frac{V_1}{I_1}\right|_{V_2=0} \qquad h_{21} = \left.\frac{I_2}{I_1}\right|_{V_2=0}$$

$$h_{12} = \left.\frac{V_1}{V_2}\right|_{I_1=0} \qquad h_{22} = \left.\frac{I_2}{V_2}\right|_{I_1=0}$$

a. Fill the following configuration table for the H parameter measurement.

b. Repeat (a) for the measurements of Y, Z, and G parameters.

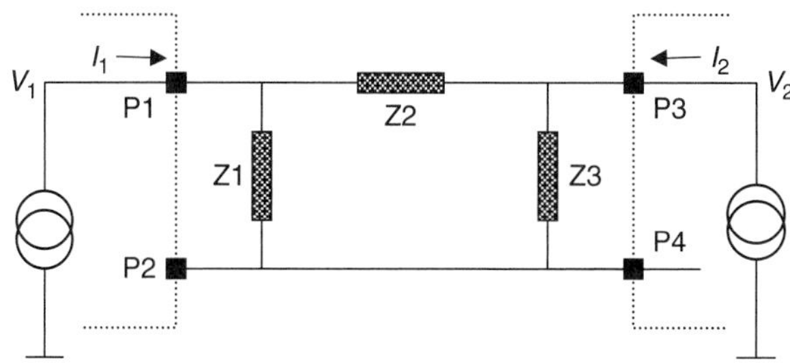

■ **FIGURE 1.18**

Testing two-port network.

Note: V_s: voltage source; V_m: voltage meter; I_s: current source; I_m: current meter.

H	P1	P3	
h_{11}	I_s, V_m	GND	$h_{11} = V_m/I_s$
h_{12}			$h_{12} =$
h_{21}			$h_{21} =$
h_{22}			$h_{22} =$

- Each entry in the table shall be filled with: V_s, I_s, V_m, I_m, GND, and Open.
- The example in h_{11} indicates that P1 is connected to the current source and the voltage meter; P3 is connected to ground (GND); h_{11} is obtained by the equation in the last column.

1.14 **(Memory Testing)** Define the following terms about memory testing: BIST, BIRA, and BISR. What are the advantages and disadvantages of soft repair over hard repair?

Acknowledgments

The authors wish to thank Professor Kuen-Jong Lee of National Cheng Kung University for contributing a portion of the IEEE 1149.1 and 1500 sections; William Eklow of Cisco Systems for contributing the IEEE 1149.6 and P1687 (IJTAG) sections; Professor Chauchin Su of National Chiao Tung University for contributing the IEEE 1149.4 section; Professor Cheng-Wen Wu of National Tsing Hua University for contributing the Basics of Memory Testing section; Professor Shey-Shi Lu of National Taiwan University for providing the bioMEMS sensor material; Professor Xinghao Chen of The City College and Graduate Center of The City University of New York for providing valuable comments; and Teresa Chang of SynTest Technologies for drawing a portion of the figures.

References

R1.0 Books

[Abramovici 1994] M. Abramovici, M. A. Breuer, and A. D. Friedman, *Digital Systems Testing and Testable Design*, Revised Printing, IEEE Press, Piscataway, NJ, 1994.

[Amerasekera 1987] E. A. Amerasekera and D. S. Campbell, *Failure Mechanisms in Semiconductor Devices*, John Wiley & Sons, London, United Kingdom, 1987.

[Bushnell 2000] M. L. Bushnell and V. D. Agrawal, *Essentials of Electronic Testing for Digital, Memory & Mixed-Signal VLSI Circuits*, Springer Science, New York, 2000.

[Cheng 2006] K.-T. Cheng, Embedded software-based self-testing for SoC design, in Chapter 28, *Embedded System Handbook*, R. Zurawski, editor, CRC Press, Boca Raton, FL, 2006.

[Dally 1998] W. J. Dally and J. W. Poulton, *Digital Systems Engineering*, Cambridge University Press, London, 1998.

[Dally 2004] W. J. Dally and B. Towles, *Principles and Practices of Interconnection Networks*, Morgan Kaufmann, San Francisco, 2004.

[De Micheli 2006] G. De Micheli and L. Benini, *Networks on Chips: Technology and Tools*, Morgan Kaufmann, San Francisco, 2006.

[Gizopoulos 2006] D. Gizopoulos, editor, *Advances in Electronic Testing: Challenges and Methodologies*, Springer Science, New York, 2006.

[Jerraya 2004] A. Jerraya and W. Wolf, *Multiprocessor Systems on Chips: Technology and Tools*, Morgan Kaufmann, San Francisco, 2004.

[Jha 2003] N. Jha and S. Gupta, *Testing of Digital Systems*, Cambridge University Press, London, 2003.

[Parker 2001] K. Parker, *The Boundary Scan Handbook*, Springer Science, New York, 2001.

[Siewiorek 1998] D. Siewiorek and R. S. Swarz, *Reliable Computer Systems: Design and Evaluation,* Third Edition, AK Peters, Boston, 1998.

[Stroud 2002] C. E. Stroud, *A Designer's Guide to Built-In Self-Test*, Springer Science, New York, 2002.

[van de Goor 1991] A. J. van de Goor, *Testing Semiconductor Memories: Theory and Practice*, John Wiley & Sons, Chichester, England, 1991.

[Wang 2006] L.-T. Wang, C.-W. Wu, and X. Wen, *VLSI Test Principles and Architectures: Design for Testability*, Morgan Kaufmann, San Francisco, 2006.

R1.1 Importance of System-on-Chip Testing

[Baumann 2005] R. Baumann, Soft errors in advanced computer systems, *IEEE Design & Test of Computers*, pp. 258–266, May/June 2005.

[Breuer 2004] M. Breuer, S. Gupta, and T. M. Mak, Defect and error tolerance in the presence of massive numbers of defects, *IEEE Design & Test of Computers*, pp. 216–227, May/June 2004.

[Chen 2002] W. Y. Chen, S. K. Gupta, and M. A. Breuer, Analytical models for crosstalk excitation and propagation in VLSI circuits, *IEEE Trans. on Computer-Aided Design*, 21(10), pp. 1117–1131, October 2002.

[Kim 2004] N. S. Kim, K. Flautner, D. Blaauw, and T. Mudge, Circuit and microarchitectural techniques for reducing cache leakage power, *IEEE Trans. on Very Large Scale Integration (VLSI) Systems*, 12(2), pp. 167–184, February 2004.

[May 1979] T. C. May and M. H. Woods, Alpha-particle-induced soft errors in dynamic memories, *IEEE Trans. on Electron Devices*, ED-26(1), pp. 2–9, January 1979.

[McCluskey 1988] E. J. McCluskey and F. Buelow, IC quality and test transparency, in *Proc. IEEE Int. Test Conf.*, pp. 295–301, September 1988.

[Moore 1965] G. Moore, Cramming more components onto integrated circuits, *Electronics*, pp. 114–117, April 19, 1965.

[Ohlsson 1998] M. Ohlsson, P. Dyreklev, K. Johansson and P. Alfke, Neutron single event upsets in SRAM-based FPGAs, in *Proc. IEEE Nuclear and Space Radiation Effects Conf.*, pp. 177–180, July 1998.

[Saxena 2003] J. Saxena, K. M. Butler, V. B. Jayaram, S. Kundu, N. V. Arvind, P. Sreeprakash, and M. Hachinger, A case study of IR-drop in structured at-speed testing, in *Proc. IEEE Int. Test Conf.*, pp. 1098–1104, September 2003.

[SIA 1997] SIA, *The National Technology Roadmap for Semiconductors*, Semiconductor Industry Association, San Jose, CA, 1997.

[SIA 1999] SIA, *The International Technology Roadmap for Semiconductors: 1999 Edition*, Semiconductor Industry Association, San Jose, CA (http://public.itrs.net), 1999.

[SIA 2004] SIA, *The International Technology Roadmap for Semiconductors: 2004 Update*, Semiconductor Industry Association, San Jose, CA (http://public.itrs.net), 2004.

[SIA 2005] SIA, *The International Technology Roadmap for Semiconductors: 2005 Edition*, Semiconductor Industry Association, San Jose, CA (http://public.itrs.net), 2005.

[SIA 2006] SIA, *The International Technology Roadmap for Semiconductors: 2006 Update*, Semiconductor Industry Association, San Jose, CA (http://public.itrs.net), 2006.

[Wang 2004] L.-C. Wang, J. J. Liou, and K.-T. Cheng, Critical path selection for delay fault testing based upon a statistical timing model, *IEEE Trans. on Computer-Aided Design*, 23(11), pp. 1550–1565, November 2004.

[Williams 1981] T. W. Williams and N. C. Brown, Defect level as a function of fault coverage, *IEEE Trans. on Computers*, 30(12), pp. 987–988, December 1981.

R1.2 Basics of SOC Testing

[Abramovici 2006] M. Abramovici, P. Bradley, K. Dwarakanath, P. Levin, G. Memmi, and D. Miller, A reconfigurable design-for-debug infrastructure for SoCs, in *Proc. ACM/IEEE Design Automation Conf.*, pp. 7–12, July 2006.

[IEEE 1149.4-1999] IEEE Std. 1149.4-1999, *IEEE Standard for a Mixed-Signal Test Bus*, IEEE Press, New York, 1999.

[IEEE 1149.1-2001] IEEE Std. 1149.1-2001, *IEEE Standard Test Access Port and Boundary Scan Architecture*, IEEE Press, New York, 2001.

[IEEE 1149.6-2003] IEEE Std. 1149.6-2003, *IEEE Standard for Boundary Scan Testing of Advanced Digital Networks*, IEEE Press, New York, 2003.

[IEEE 1500-2005] IEEE Std. 1500-2005, *IEEE Standard for Embedded Core Test*, IEEE Press, New York, 2005.

[IEEE P1687-2007] IEEE P1687-2007 Proposal, *IEEE Internal Boundary-Scan Proposal for Embedded Test and Debug*, IEEE Press, New York, 2007. (http://group.ieee.org/groups/1687).

[Marinissen 2002] E. Marinissen, R. Kapur, M. Lousberg, T. McLaurin, M. Ricchetti, and Y. Zorian, On IEEE P1500's standard for embedded core test, *J. Electronic Testing: Theory and Applications*, 18, pp. 365–383, 2002.

[Seghal 2004] A. Seghal, S. Goel, E. Marinissen, and K. Chakrabarty, IEEE P1500-compliant test wrapper design for hierarchical cores, in *Proc. IEEE Int. Test Conf.*, pp. 1203–1212, October 2004.

[Zorian 2005] Y. Zorian and A. Yessayan, IEEE 1500 utilization in SOC test and design, in *Proc. IEEE Int. Test Conf.*, pp. 1203–1212, October 2004.

R1.3 Basics of Memory Testing

[Benevit 1982] C. A. Benevit, J. M. Cassard, K. J. Dimmler, A. C. Dumbri, M. G. Mound, F. J. Procyk, W. Rosenzweig, and A. W. Yanof, A 256K dynamic random-access memory, *IEEE J. of Solid-State Circuits*, 17(5), pp. 857–862, May 1982.

[Cenker 1979] R. P. Cenker, D. G. Clemons, W. P. Huber, J. P. Petrizzi, F. J. Procyk, and G. M. Trout, A fault-tolerant 64K dynamic random-access memory, *IEEE Trans. on Electron. Devices*, 26(6), pp. 853–860, June 1979.

[Huang 1999] C.-T. Huang, J.-R. Huang, C.-F. Wu, C.-W. Wu, and T.-Y. Chang, A programmable BIST core for embedded DRAM, *IEEE Design & Test of Computers*, 16(1), pp. 59–70, January/March 1999.

[Huang 2003] C.-T. Huang, C.-F. Wu, J.-F. Li, and C.-W. Wu, Built-in redundancy analysis for memory yield improvement, *IEEE Trans. on Reliability*, 52(4), pp. 386–399, April 2003.

[Leung 2000] W. Leung, F.-C. Hsu, and M.-E. Jones, The ideal SoC memory: 1T SRAM, in *Proc. IEEE Int. ASIC/SOC Conf.*, pp. 32–36, September 2000.

[Smith 1981] R. T. Smith, J. D. Chipala, J. F. M. Bindels, R. G. Nelson, F. H. Fischer, and T. F. Mantz, Laser programmable redundancy and yield improvement in a 64K DRAM, *IEEE J. of Solid-State Circuits*, 16(5), pp. 506–514, May 1981.

[Su 2004] C.-L. Su, R.-F. Huang, C.-W. Wu, C.-C. Hung, M.-J. Kao, Y.-J. Chang, and W.-C. Wu, MRAM defect analysis and fault modeling, in *Proc. IEEE Int. Test Conf.*, pp. 124–133, October 2004.

[Su 2006] C.-L. Su, C.-W. Tsai, C.-W. Wu, C.-C. Hung, Y.-S. Chen, and M.-J. Kao, Testing MRAM for write disturbance fault, in *Proc. IEEE Int. Test Conf.*, Paper 3.4, October 2006.

[Wu 1998] C.-W. Wu, Testing embedded memories: Is BIST the ultimate solution?, in *Proc. IEEE Asian Test Symp.*, pp. 516–517, November 1998.

[Yeh 2007] J.-C. Yeh, K.-L. Cheng, Y.-F. Chou, and C.-W. Wu, Flash memory testing and built-in self-diagnosis with March-like test algorithms, *IEEE Trans. on Computer-Aided Design*, 26(6), pp. 1101–1113, June 2007.

R1.4 SOC Design Examples

[Abramovici 2006] M. Abramovici, P. Bradley, K. Dwarakanath, P. Levin, G. Memmi, and D. Miller, A reconfigurable design-for-debug infrastructure for SoCs, in *Proc. ACM/IEEE Design Automation Conf.*, pp. 7–12, July 2006.

[Arntz 2003] Y. Arntz, J. D. Seelig, H. P. Lang, J. Zhang, P. Hunziker, J. P. Ramseyer, E. Meyer, M. Hegner, and C. Gerber, Label-free protein assay based on a nanomechanical cantilever array, *Nanotechnology*, 14, pp. 86–90, 2003.

[Blake 2003] G. J. Blake, N. Rifai, J. E. Buring, and P. M. Ridker, Blood pressure, C-reactive protein, and risk of future cardiovascular events, *Circulation*, 108, pp. 2993–2999, December 2003.

[Chen 2006] C.-H. Chen, R.-Z. Hwang, L.-S. Huang, S. Lin, H.-C. Chen, Y.-C. Yang, Y.-T. Lin, S.-A. Yu, Y.-H. Wang, N.-K. Chou, and S.-S. Lu, A wireless Bio-MEMS sensor for C-reactive protein detection based on nanomechanics, *Digest of Papers, IEEE Int. Solid-State Circuits Conf.*, 1, pp. 562–563 & 673, February 2006.

[McNairy 2005] C. McNairy and R. Bhatia, Montecito: A dual-core, dual-thread itanium processor, *IEEE Micro*, 25(2), pp. 10–20, March/April 2005.

[Naffziger 2005] S. Naffziger, B. Stackhouse, and T. Grutkowski, The implementation of a 2-core multi-threaded itanium-family processor, *Digest of Papers, IEEE Int. Solid-State Circuits Conf.*, 1, pp. 182–183 & 592, February 2005.

[Naffziger 2006] S. Naffziger, B. Stackhouse, T. Grutkowski, D. Josephson, J. Desai, E. Alon, and M. Horowitz, The implementation of a 2-core multi-threaded itanium family processor, *IEEE J. of Solid-State Circuits Conf.*, 41(1), pp. 197–209, January 2006.

[Pham 2005] D. Pham, S. Asano, M. Bolliger, M. N. Day, H. P. Hofstee, C. Johns, J. Kahle, A. Kameyama, J. Keaty, Y. Masubuchi, M. Riley, D. Shippy, D. Stasiak, M. Suzuoki, M. Wang, J. Warnock, S. Weitzel, D. Wendel, T. Yamazaki, and K. Yazawa, The design and implementation of a first-generation CELL processor, *Digest of Papers, IEEE Int. Solid-State Circuits Conf.*, 1, pp. 184–185 & 592, February 2005.

[Pham 2006] D. C. Pham, T. Aipperspach, D. Boerstler, M. Bolliger, R. Chaudhry, D. Cox, P. Harvey, P. M. Harvey, H. P. Hofstee, C. Johns, J. Kahle, A. Kameyama, J. Keaty, Y. Masubuchi, M. Pham, J. Pille, S. Posluszny, M. Riley, D. L. Stasiak, M. Suzuoki, O. Takahashi, J. Warnock, S. Weitzel, D. Wendel, and K. Yazawa, Overview of the architecture, circuit design, and physical implementation of a first-generation Cell processor, *IEEE J. of Solid-State Circuits*, 41(1), pp. 179–196, January 2006.

[Rohrer 2004] N. Rohrer, M. Canada, E. Cohen, M. Ringler, M. Mayfield, P. Sandon, P. Kartschoke, J. Heaslip, J. Allen, P. McCormick, T. Pflüger, J. Zimmerman, C. Lichtenau, T. Werner, G. Salem, M. Ross, D. Appenzeller, and D. Thygesen, PowerPC 970 in 130nm and 90nm technologies, *Digest of Papers, IEEE Int. Solid-State Circuits Conf.*, 1, pp. 68–69, February 2004.

R1.5 About This Book

[SIA 2005] SIA, *The International Technology Roadmap for Semiconductors: 2005 Edition*, Semiconductor Industry Association, San Jose, CA (http://public.itrs.net), 2005.

[SIA 2006] SIA, *The International Technology Roadmap for Semiconductors: 2006 Update*, Semiconductor Industry Association, San Jose, CA (http://public.itrs.net), 2006.

CHAPTER 2

DIGITAL TEST ARCHITECTURES

Laung-Terng (L.-T.) Wang
SynTest Technologies, Inc., Sunnyvale, California

ABOUT THIS CHAPTER

Design for testability (DFT) has become an essential part for designing ***very-large-scale integration*** (VLSI) circuits. The most popular DFT techniques in use today for testing the digital logic portion of the VLSI circuits include **scan** and **scan-based logic *built-in self-test*** (BIST). Both techniques have proved to be quite effective in producing testable VLSI designs. Additionally, **test compression**, a supplemental DFT technique to scan, is growing in importance for further reducing test data volume and test application time during manufacturing test.

To provide readers with an in-depth understanding of the most recent DFT advances in scan, logic BIST, and test compression, this chapter covers a number of fundamental and advanced digital test architectures to facilitate the testing of modern digital circuits. These architectures are required to improve the product quality and reduce the defect level, test cost, and test power of a digital circuit while at the same time simplifying the test, debug, and diagnosis tasks.

In this chapter, we first describe fundamental scan architectures followed by a discussion on advanced low-power and at-speed scan architectures. Next we present a number of fundamental and advanced logic BIST architectures that allow the digital circuit to perform self-test on-chip, on-board, or in-system. We then discuss test compression architectures designed to reduce test data volume and test application time. This includes a description of advanced low-power and at-speed test compression architectures practiced in industry. Finally, we explore promising **random-access scan** architectures devised to further reduce test power dissipation and test application time while retaining the benefits of scan and logic BIST.

2.1 INTRODUCTION

With advances in semiconductor manufacturing technology, VLSI circuits can now contain tens to hundreds of millions of transistors running in the gigahertz range. The production and usage of these VLSI circuits has run into a variety of test challenges during wafer probe, wafer sort, pre-ship screening, incoming test of chips and boards, test of assembled boards, system test, periodic maintenance, repair test, etc. The semiconductor industry heavily relies on two techniques for testing digital circuits: *scan* and *logic built-in self-test* (BIST) [McCluskey 1986] [Abramovici 1994]. **Scan** converts a digital sequential circuit into a scan design and then uses ***automatic test pattern generation*** (ATPG) software [Bushnell 2000] [Jha 2003] [Wang 2006a] to detect faults that are caused by manufacturing defects (physical failures) and manifest themselves as errors, whereas logic BIST requires using a portion of the VLSI circuit to test itself on-chip, on-board, or in-system. To keep up with the design and test challenges [SIA 2003, 2006], more advanced ***design-for-testability*** (DFT) techniques have been developed to further address the test cost, delay fault, and test power issues [Gizopoulos 2006] [Wang 2006a]. The evolution of important DFT techniques for testing digital circuits is shown in Figure 2.1.

Scan design is implemented by first replacing all selected storage elements of the digital circuit with **scan cells** and then connecting them into one or more shift registers, called **scan chains,** to provide them with external access. With external access, one can now control and observe the internal states of the digital circuit by simply shifting test stimuli into and test responses out of the shift registers during scan testing. The DFT technique has proved to be quite effective in improving the product quality, testability, and diagnosability of scan designs [Crouch 1999] [Bushnell 2000] [Jha 2003] [Gizopoulos 2006] [Wang 2006a]. Although scan has offered many benefits during manufacturing test, it is becoming inefficient to test deep submicron or nanometer VLSI designs. The reasons mostly relate to the facts that (1) traditional test schemes using ATPG software to target single faults have become quite expensive and (2) sufficiently high fault coverage for these deep submicron or nanometer VLSI designs is hard to sustain from the chip level to the board and system levels.

To alleviate these test problems, the scan approach is typically combined with **logic BIST** that incorporates BIST features into the scan design at the design stage [Bushnell 2000] [Mourad 2000] [Stroud 2002] [Jha 2003]. With logic BIST, circuits that generate test patterns and analyze the output responses of the functional circuitry are embedded in the chip or elsewhere on the same board where the chip resides to test the digital logic circuit itself. Typically, pseudo-random patterns are applied to the *circuit under test* (CUT) while their test responses are compacted in a ***multiple-input signature register*** (MISR) [Bardell 1987] [Rajski 1998a]

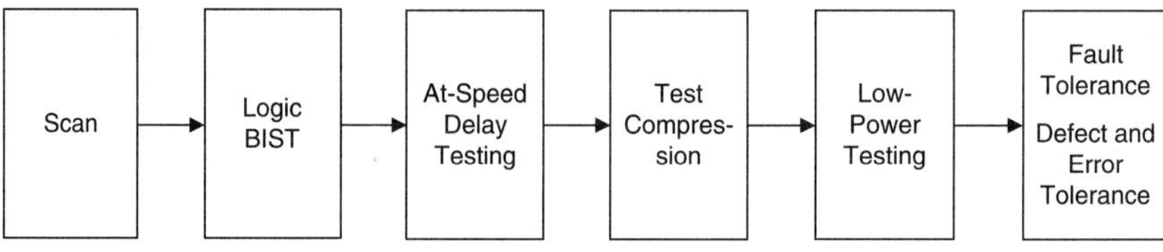

■ **FIGURE 2.1**

Evolution of DFT advances in digital circuit testing.

[Nadeau-Dostie 2000] [Stroud 2002] [Jha 2003] [Wang 2006a]. Logic BIST is crucial in many applications, in particular, for safety-critical and mission-critical applications. These applications commonly found in the aerospace/defense, automotive, banking, computer, health care, networking, and telecommunications industries require on-chip, on-board, or in-system self-test to improve the reliability of the entire system, as well as the ability to perform remote diagnosis.

Since the early 2000s, **test compression**, a supplemental DFT technique to scan, is gaining industry acceptance to further reduce test data volume and test application time [Touba 2006] [Wang 2006a]. Test compression involves compressing the amount of test data (both test stimulus and test response) that must be stored on ***automatic test equipment*** (ATE) for testing with a deterministic (ATPG-generated) test set. This is done by using **code-based schemes** or adding additional on-chip hardware before the scan chains to decompress the test stimulus coming from the ATE and after the scan chains to compress the test response going to the ATE. This differs from logic BIST in that the test stimuli that are applied to the CUT form a deterministic (ATPG-generated) test set rather than pseudo-random patterns.

Although scan design has been widely adopted for use during manufacturing test to ensure product quality, the continued increase in circuit complexity of a scan design has started reaching the limit of test power dissipation, which in turn threatens to damage the devices that are under test. As a result, ***random-access scan*** (RAS) design, as an alternative to scan design, is gaining momentum in addressing the test power dissipation issue [Ando 1980] [Baik 2005a] [Mudlapur 2005] [Hu 2006]. Unlike scan design that requires serially shifting data into and out of a scan cell through adjacent scan cells, random-access scan allows each scan cell to be randomly and uniquely addressable, similar to storage cells in a *random-access memory* (RAM).

In this chapter, we first cover three commonly used DFT techniques: scan, logic BIST, and test compression. For each DFT technique, we present a number of DFT architectures practiced in industry. Fundamental DFT architectures along with advanced DFT architectures suitable for low-power testing and at-speed testing, which are growing in importance for nanometer VLSI designs, are examined. All of these DFT architectures are applicable for testing, debugging, and diagnosing scan designs. Then, we describe some promising DFT architectures using random access scan to reduce test power dissipation and test application time. For more information on basic VLSI test principles and DFT architectures, refer to [Bushnell 2000], [Jha 2003], and [Wang 2006a]. Advances in fault tolerance, at-speed delay testing, low-power testing, and defect and error tolerance are further discussed in Chapters 3, 6, 7, and 8, respectively.

2.2 SCAN DESIGN

Scan design is currently the most widely used structured DFT approach. It is implemented by connecting selected storage elements of a design into one or more shift registers, called **scan chains**, to provide them with external access. Scan design accomplishes this task by replacing all selected storage elements with **scan cells**, each having one additional ***scan input*** (SI) port and one shared/additional ***scan output*** (SO) port. By connecting the SO port of one scan cell to the SI port of the next scan cell, one or more scan chains are created.

The scan-inserted design, called scan design, is now operated in three modes: **normal mode**, **shift mode**, and **capture mode**. Circuit operations with associated clock cycles conducted in these three modes are referred to as normal operation, shift operation, and capture operation, respectively.

In normal mode, all test signals are turned off, and the scan design operates in the original functional configuration. In both shift and capture modes, a **test mode** signal *TM* is often used to turn on all test-related fixes in compliance with scan design rules. A set of **scan design rules** that can be found in [Cheung 1996] and [Wang 2006a] are necessary to simplify the test, debug, and diagnosis tasks, improve fault coverage, and guarantee the safe operation of the device under test. These circuit modes and operations are distinguished using additional test signals or test clocks. Fundamental and advanced scan architectures are described in the following subsections.

2.2.1 Scan Architectures

In this subsection, we first describe a few fundamental scan architectures. These fundamental scan architectures include (1) *muxed-D scan design*, where storage elements are converted into muxed-D scan cells; (2) *clocked-scan design*, where storage elements are converted into clocked-scan cells; (3) *LSSD scan design*, where storage elements are converted into *level-sensitive scan design* (LSSD) *shift register latches* (SRLs); and (4) *enhanced-scan design*, where storage elements are converted into enhanced-scan cells each comprised of a D latch and a muxed-D scan cell.

2.2.1.1 Muxed-D Scan Design

Figure 2.2 shows a sequential circuit example with three D flip-flops. The corresponding muxed-D full-scan circuit is shown in Figure 2.3. An edge-triggered **muxed-D scan cell** design is shown in Figure 2.3a. This scan cell is composed of a D flip-flop and a multiplexer. The multiplexer uses a ***scan enable*** (*SE*) input to select between the ***data input*** (*DI*) and the ***scan input*** (*SI*.) The three D flip-flops, FF_1, FF_2, and FF_3, as shown in Figure 2.2, are replaced with three muxed-D scan cells, SFF_1, SFF_2, and SFF_3, shown in Figure 2.3b.

In Figure 2.3a, the data input *DI* of each scan cell is connected to the output of the combinational logic as in the original circuit. To form a scan chain, the scan inputs *SI* of SFF_2 and SFF_3 are connected to the outputs *Q* of the previous scan cells, SFF_1 and SFF_2, respectively. In addition, the scan input *SI* of the first scan cell SFF_1 is connected to the primary input *SI*, and the output *Q* of the last scan cell SFF_3 is connected to the primary output *SO*. Hence, in shift mode, *SE* is set to 1, and the scan cells operate as a single scan chain, which allows us to shift any combination of logic values into the scan cells. In capture mode, *SE* is set to 0, and

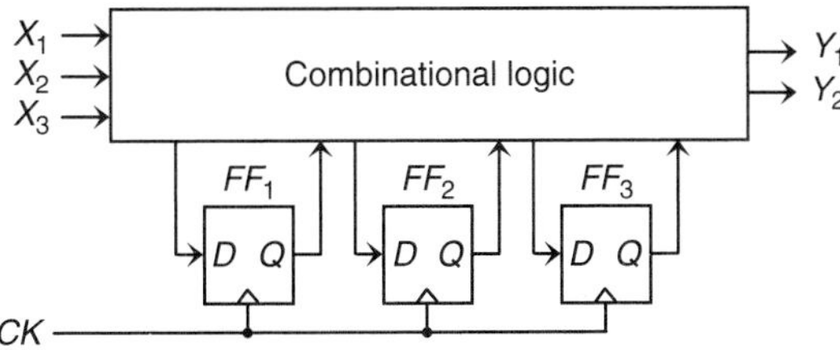

■ **FIGURE 2.2**

Sequential circuit example.

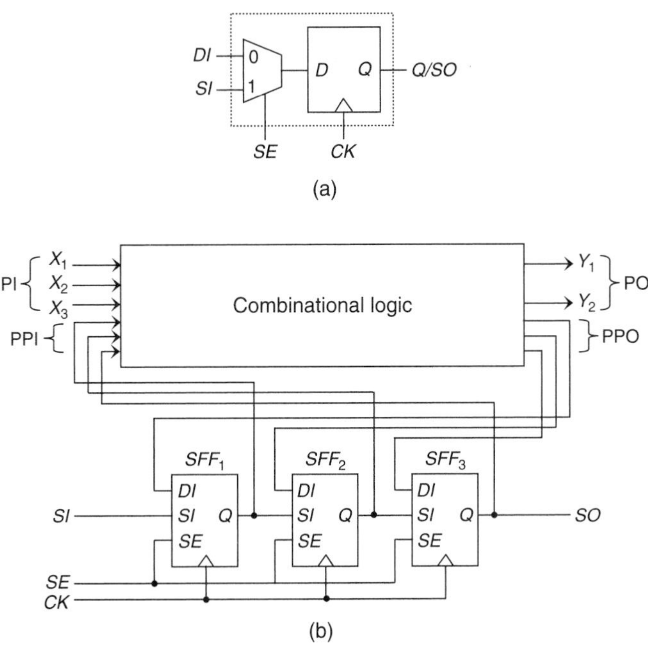

■ **FIGURE 2.3**

Muxed-D scan design: (a) muxed-D scan cell and (b) muxed-D scan design.

the scan cells are used to capture the test response from the combinational logic when a clock is applied.

In general, combinational logic in a full-scan circuit has two types of inputs: ***primary inputs*** (PIs) and ***pseudo primary inputs*** (PPIs). Primary inputs refer to the external inputs to the circuit, whereas pseudo primary inputs refer to the scan cell outputs. Both PIs and PPIs can be set to any required logic values. The only difference is that PIs are set directly in parallel from the external inputs, whereas PPIs are set serially through scan chain inputs. Similarly, the combinational logic in a full-scan circuit has two types of outputs: ***primary outputs*** (POs) and ***pseudo***

primary outputs (PPOs). Primary outputs refer to the external outputs of the circuit, and pseudo primary outputs refer to the scan cell inputs. Both POs and PPOs can be observed. The only difference is that POs are observed directly in parallel from the external outputs, and PPOs are observed serially through scan chain outputs.

2.2.1.2 Clocked-Scan Design

An edge-triggered ***clocked-scan cell*** can also be used to replace a D flip-flop in a scan design [McCluskey 1986]. Similar to a muxed-D scan cell, a clocked-scan cell also has a data input *DI* and a scan input *SI*; however, in the clocked-scan cell, input selection is conducted using two independent clocks, data clock *DCK* and shift clock *SCK*, as shown in Figure 2.4a.

Figure 2.4b shows a clocked-scan design of the sequential circuit given in Figure 2.2. This clocked-scan design is tested using shift and capture operations, similar to a muxed-D scan design. The main difference is how these two operations are distinguished. In a muxed-D scan design, a scan enable signal *SE* is used, as shown in Figure 2.3a. In the clocked scan shown in Figure 2.4, these two operations are distinguished by properly applying the two independent clocks *SCK* and *DCK* during shift mode and capture mode, respectively.

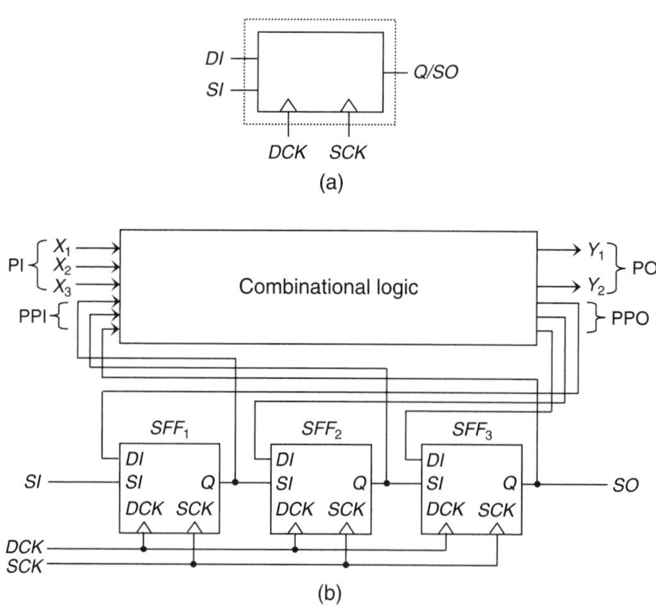

■ **FIGURE 2.4**

Clocked-scan design: (a) clocked-scan cell and (b) clocked-scan design.

2.2.1.3 LSSD Scan Design

Figure 2.5 shows a polarity-hold ***shift register latch*** (SRL) design described in [Eichelberger 1977] that can be used as an LSSD scan cell. This scan cell contains two latches: a master two-port D latch L_1 and a slave D latch L_2. Clocks C, A, and B are used to select between the data input D and the scan input I to drive $+L_1$ and $+L_2$.

LSSD scan designs can be implemented using either a **single-latch design** or a **double-latch design**. In single-latch design [Eichelberger 1977], the output port $+L_1$ of the master latch L_1 is used to drive the combinational logic of the design. In this case, the slave latch L_2 is used only for scan testing. Because LSSD designs use latches instead of flip-flops, at least two system clocks C_1 and C_2 are required to prevent combinational feedback loops from occurring. In this case, combinational logic driven by the master latches of the first system clock C_1 are used to drive the master latches of the second system clock C_2, and vice versa. For this to work, the system clocks C_1 and C_2 should be applied in a nonoverlapping fashion. Figure 2.6a shows an LSSD single-latch design using the polarity-hold SRL shown in Figure 2.5.

Figure 2.6b shows an example of LSSD **double-latch design** [DasGupta 1982]. In normal mode, the C_1 and C_2 clocks are used in a nonoverlapping manner, where the C_2 clock is the same as the B clock. The testing of an LSSD scan design is conducted using shift and capture operations, similar to a muxed-D scan design. The main difference is how these two operations are distinguished. In a muxed-D scan design, a scan enable signal SE is used, as shown in Figure 2.3a. In an LSSD scan design, these two operations are distinguished by properly applying nonoverlapping clock pulses to clocks C_1, C_2, A, and B. During the shift operation, clocks A and B are applied in a nonoverlapping manner, and the scan cells $SRL_1 \sim SRL_3$ form a single scan chain from SI to SO. During the capture operation, clocks C_1 and C_2 are applied in a nonoverlapping manner to load the test response from the combinational logic into the scan cells.

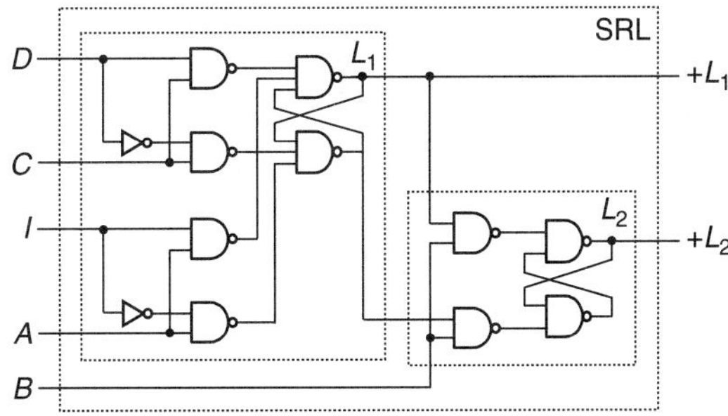

■ FIGURE 2.5

Polarity-hold shift register latch (SRL).

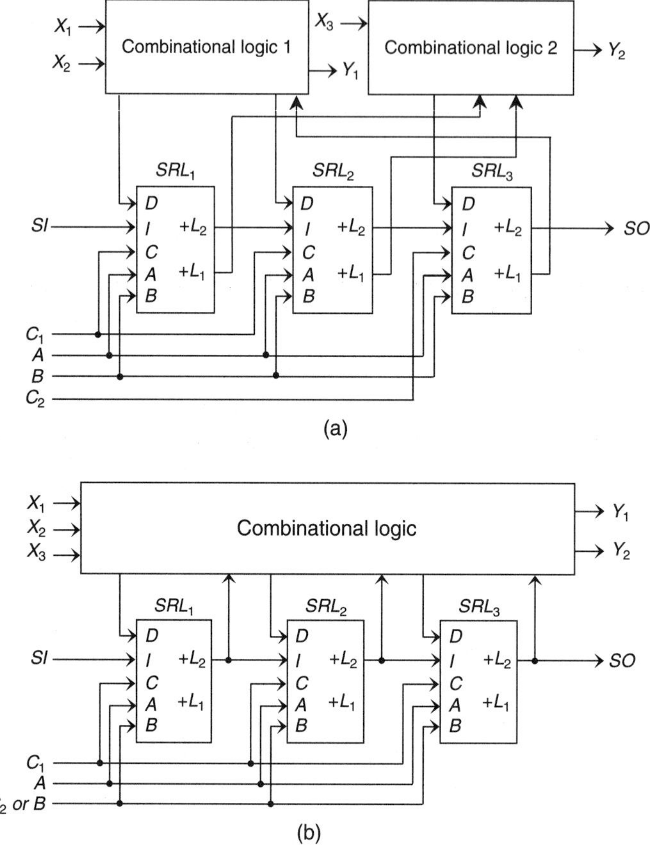

FIGURE 2.6

LSSD designs: (a) LSSD single-latch design and (b) LSSD double-latch design.

The operation of a polarity-hold SRL is race-free if clocks C and B as well as A and B are nonoverlapping. This characteristic is used to implement LSSD circuits that are guaranteed to have race-free operation in normal mode as well as in test mode.

2.2.1.4 Enhanced-Scan Design

Testing for a delay fault requires applying a pair of test vectors in an at-speed fashion. This is used to generate a logic value transition at a signal line or at the source of a path, and the circuit response to this transition is captured at the circuit's operating speed. Applying an arbitrary pair of vectors as opposed to a functionally dependent pair of vectors, generated through the combinational logic of the circuit

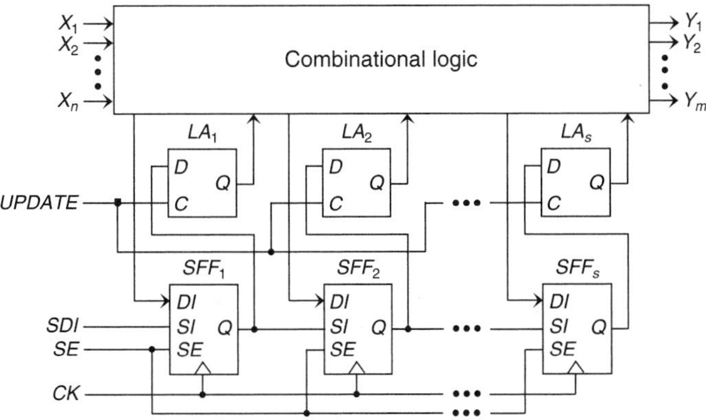

■ **FIGURE 2.7**

Enhanced-scan design.

under test, allows us to maximize the delay fault detection capability. This can be achieved using **enhanced scan** [Malaiya 1983] [Glover 1988] [Dervisoglu 1991]. The enhanced-scan or **hold-scan** test circuit was implemented in the 90-nm Intel Pentium 4 processor [Kuppuswamy 2004].

Enhanced scan increases the capacity of a typical scan cell by allowing it to store two bits of data that can be applied consecutively to the combinational logic driven by the scan cells. For a muxed-D scan cell or a clocked-scan cell, this is achieved through the addition of a D latch.

Figure 2.7 shows a general enhanced-scan architecture using muxed-D scan cells. In this figure, in order to apply a pair of test vectors $< V_1, V_2 >$ to the design, the first test vector V_1 is first shifted into the scan cells ($SFF_1 \sim SFF_s$) and then stored into the additional latches ($LA_1 \sim LA_s$) when the *UPDATE* signal is set to 1. Next, the second test vector V_2 is shifted into the scan cells while the *UPDATE* signal is set to 0 in order to preserve the V_1 value in the latches ($LA_1 \sim LA_s$). Once the second vector V_2 is shifted in, the *UPDATE* signal is applied, in order to change V_1 to V_2 while capturing the output response at-speed into the scan cells by applying *CK* after exactly one clock cycle.

The main advantage of enhanced scan is that it allows us to achieve high delay fault coverage by applying any arbitrary pair of test vectors, that otherwise would have been impossible. The disadvantages, however, are that each enhanced-scan cell needs an additional scan-hold D latch and that maintaining the timing relationship between *UPDATE* and *CK* for at-speed testing may be difficult. An additional disadvantage is that many **false paths,** instead of functional data paths, may be activated during test, causing an **over-test** problem. To reduce over-test, the conventional **launch-on-shift** (also called *skewed-load* [Savir 1993]) and **launch-on-capture** (also called *broad-side* in [Savir 1994] or double-capture in [Wang 2006a]) delay test techniques using normal scan chains can be used.

2.2.2 Low-Power Scan Architectures

Scan design can be classified as **serial scan design,** as test pattern application and test response acquisition are both conducted serially through scan chains. The major advantage of serial scan design is its low routing overhead, as scan data are shifted through adjacent scan cells. Its major disadvantage, however, is that individual scan cells cannot be controlled or observed without affecting the values of other scan cells within the same scan chain. High switching activities at scan cells during shift or capture can cause excessive test power dissipation, resulting in circuit damage, low reliability, or even test-induced yield loss.

Low-power scan architectures are scan designs targeting test power reduction. Test power is related to dynamic power. Dynamic power on a circuit node is measured as $0.5CV_{DD}{}^2f$, where C is the effective load capacitance, V_{DD} is the supply voltage, and f is the node's switching frequency [Girard 2002] [Jha 2003]. Thus, test power is proportional to $V_{DD}{}^2f$.

Many approaches can be used to reduce test power [Girard 2002]. Typically, these approaches can result in a reduction of 2X to 10X in test power (shift power, capture power, or both). A number of representative low-power scan architectures are described in this subsection. These scan architectures are all applicable to muxed-D, clocked, and LSSD scan designs. If achieving a 100X reduction in shift power is required, one may consider using random-access scan design given in Section 2.5 or the advanced techniques detailed in Chapter 7.

2.2.2.1 Reduced-Voltage Low-Power Scan Design

A simple approach to reducing test power is to reduce the supply voltage. By reducing the supply voltage by 2X, a reduction of 4X in test power can be immediately achieved. The problem with this approach is that the circuit may not be designed to function at the reduced supply voltage.

2.2.2.2 Reduced-Frequency Low-Power Scan Design

Another approach is to slow down the shift clock frequency [Chandra 2001]. By reducing the shift clock frequency by 10X, a reduction of 10X in test power can be immediately achieved. The drawback of this approach is that test application time is increased by 10X, as test application time is mainly dominated by shift clock frequency. This can result in a dramatic increase in test cost.

2.2.2.3 Multi-Phase or Multi-Duty Low-Power Scan Design

One common approach to reduce test power is to apply shift clocks in a multi-phase (nonoverlapping) or multi-duty (skewed) order [Bonhomme 2001] [Saxena 2001] [Yoshida 2003] [Rosinger 2004]. The **multi-phase clocking** technique splits the shift clock into a number of nonoverlapping clock phases each for driving a small scan segment of scan cells. Thus, test power is reduced, but test application time may be increased. To avoid test application time increase, the scan inputs and

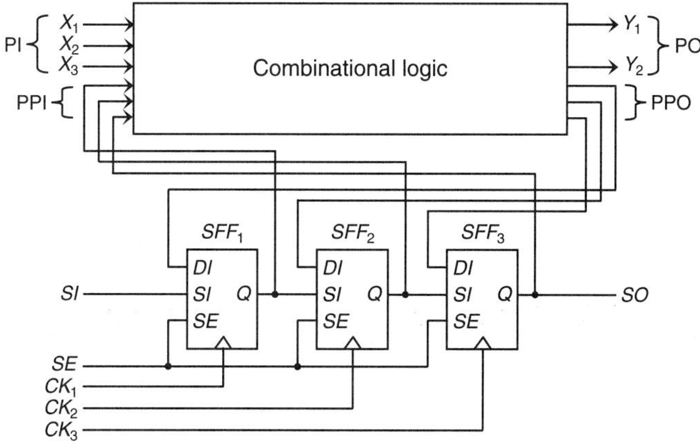

■ **FIGURE 2.8**

Multi-phase or multi-duty low-power scan design.

the scan outputs of all scan segments can be tied together and multiplexed with the original shift clock, respectively [Bonhomme 2001] [Saxena 2001] [Rosinger 2004]. The low-power scan design described in [Yoshida 2003] uses a **multi-duty clocking** technique to avoid test application time increase. This is done by adding delays to the shift clock so a skewed clock phase is applied to a small scan segment of scan cells. This technique also helps reduce peak power, but total energy consumption and heat dissipation may not change. A multi-phase or multi-duty low-power scan design reconfigured from Figure 2.3b is shown in Figure 2.8 where the clock *CK* shown in Figure 2.3b is split (or skewed) into three clock phases: CK_1, CK_2, and CK_3. Using this scheme, up to 3X reduction in test power can be achieved. The disadvantage of this approach is increased routing overhead and complexity during ***clock tree synthesis*** (CTS).

2.2.2.4 Bandwidth-Matching Low-Power Scan Design

It is also possible to reduce test power by splitting each scan chain into multiple scan chains and reducing the shift clock frequency. This is accomplished by using pairs of serial-in/parallel-out shift register and parallel-in/serial-out shift register for bandwidth matching [Whetsel 1998] [Khoche 2002]. Consider a design with 16 scan chains running at a shift clock frequency of 10 MHz. Each scan chain is split into 10 subscan chains with the SI and SO ports of each 10 subscan chains connected to a serial-in/parallel-out shift register and a parallel-in/serial-out shift register, respectively. In this case, the 16 pairs of shift registers run at 10 MHz, whereas all 160 subscan chains can now be shifted at 1 MHz. As a result, because test power is proportional to the shift clock frequency, a reduction of 10X in test power is achieved, without a corresponding increase in test time. Figure 2.9 shows the bandwidth-matching low-power scan design. The ***time-division demultiplexer***

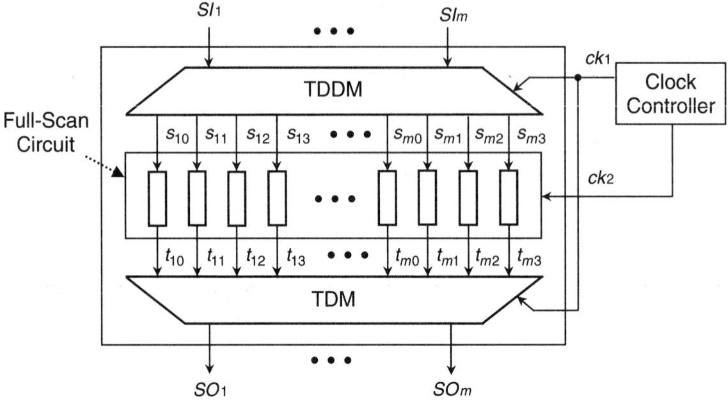

■ **FIGURE 2.9**

Bandwidth-matching low-power scan design.

(TDDM) is a serial-in/parallel-out shift register, whereas the ***time-division multiplexer*** (TDM) is a parallel-in/serial-out shift register. The main drawback of this approach is the induced area overhead.

2.2.2.5 Hybrid Low-Power Scan Design

Any of the above-mentioned low-power scan designs can typically achieve 2X to 10X reduction in test power (either shift power or capture power). When combined, further test power reduction is possible. In cases where a 100X reduction in shift power is required, one can consider using random-access scan designs as detailed in Section 2.5 or resort to a hybrid approach that combines two or more low-power test techniques. These advanced techniques are discussed in Chapter 7.

2.2.3 At-Speed Scan Architectures

Although scan design is commonly used in the industry for slow-speed stuck-at fault testing, its real value is in providing **at-speed testing** for high-speed and high-performance circuits. These circuits often contain multiple clock domains, each running at an operating frequency that is either synchronous or asynchronous to the other clock domains. Two clock domains are said to be **synchronous** if the active edges of both clocks controlling the two clock domains can be aligned precisely or triggered simultaneously. Two clock domains are said to be **asynchronous** if they are not synchronous.

There are two basic capture-clocking schemes for testing multiple clock domains at-speed: (1) *skewed-load* (also called *launch-on-shift*) and (2) *double-capture* (also called *launch-on-capture* or *broad-side*). Both schemes can test path-delay faults and transition faults within each clock domain (called **intra-clock-domain faults**) or across clock domains (called **inter-clock-domain faults**). Skewed-load uses the last

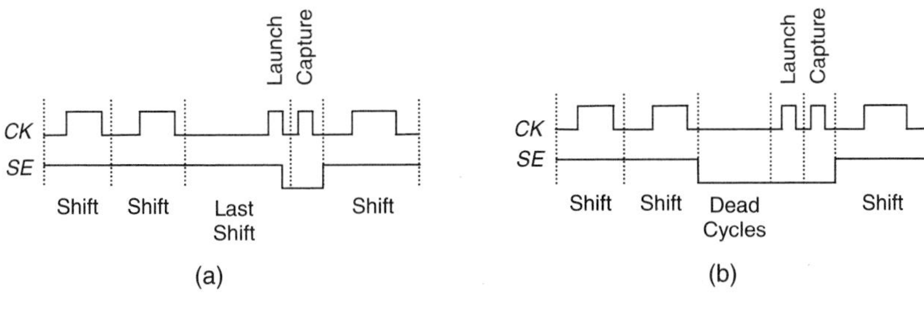

■ **FIGURE 2.10**

Basic at-speed test schemes: (a) skewed-load and (b) double-capture.

shift clock pulse followed immediately by a capture clock pulse to launch the transition and capture the output test response, respectively. Double-capture uses two consecutive capture clock pulses to launch the transition and capture the output test response, respectively. In both schemes, the second capture clock pulse must be running at the domain's operating speed or at-speed. The difference is that skewed-load requires the domain's scan enable signal *SE* to switch its value between the launch and capture clock pulses making *SE* act as a clock signal. Figure 2.10 shows sample waveforms using the basic skewed-load and double-capture at-speed test schemes.

Because scan designs typically include many clock domains, which do not interact with each other, clock grouping can be used to reduce test application time and test data volume during ATPG. **Clock grouping** is a process used to analyze all data paths in the scan design in order to determine all independent or noninteracting clocks that can be grouped and applied simultaneously.

An example of the clock grouping process is shown in Figure 2.11. This example shows the results of performing a circuit analysis operation on a scan design in order to identify all clock interactions, marked with an arrow, where a data transfer

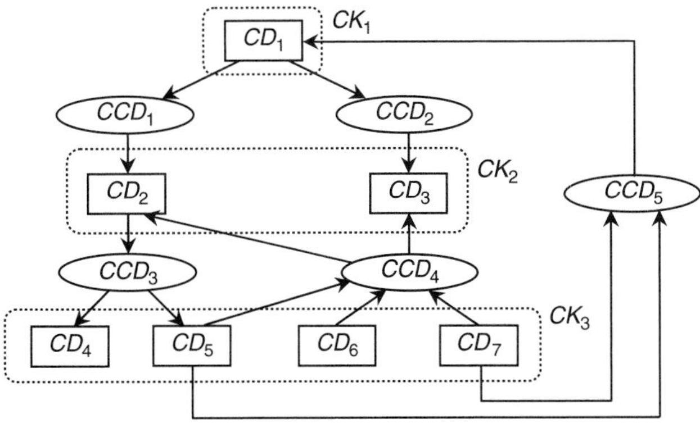

■ **FIGURE 2.11**

Clock grouping example.

from one clock domain to a different clock domain occurs. As Figure 2.11 illustrates, the circuit in this example has seven clock domains ($CD_1 \sim CD_7$) and five crossing-clock-domain data paths ($CCD_1 \sim CCD_5$). This example shows that CD_2 and CD_3 are independent from each other; hence, their related clocks can be applied simultaneously during test as CK_2. Similarly, clock domains CD_4 through CD_7 can also be applied simultaneously during test as CK_3. Therefore, in this example, three grouped clocks instead of seven individual clocks can be used to test the circuit during the capture operation.

To guarantee the success of the capture operation, additional care must be taken in terms of the way the grouped clocks are applied. This is mainly because the clock skew between different clock domains is typically large. A data path originating in one clock domain and terminating in another might result in a mismatch when both clocks are applied simultaneously, and the clock skew between the two clocks is larger than the data path delay from the originating clock domain to the terminating clock domain. To avoid the mismatch, the timing governing the relationship of such a data path shown in the following equation must be observed:

$$\text{clock skew} < \text{data path delay} + \text{clock-to-Q delay (originating clock)}$$

If this is not the case, a mismatch may occur during the capture operation. To prevent this from happening, grouped clocks can be applied sequentially (using the **staggered clocking** scheme [Wang 2005a, 2007]) such that any clock skew that exists between the clock domains can be tolerated during the test generation process. It is also possible to apply only one grouped clock during each capture operation using the **one-hot clocking** scheme. Most modern ATPG programs can also automatically mask off unknown values (*X's*) at the originating scan cells or receiving scan cells across clock domains. In this case, all grouped clocks can also be applied simultaneously using the **simultaneous clocking** scheme [Wang 2007]. During simultaneous clocking, if the launch clock pulses [Rajski 2003] [Wang 2006a] or the capture clock pulses [Nadeau-Dostie 1994] [Wang 2006a] can be aligned precisely, which applies only for synchronous clock domains, then depending on the ATPG capability, maybe there is no need to mask off unknown values across these synchronous clock domains. These clocking schemes are illustrated in Figure 2.12.

In general, one-hot clocking produces the highest fault coverage at the expense of generating many more test patterns than the other two schemes. Simultaneous clocking can generate the smallest number of test patterns but may result in high fault coverage loss because of unknown (X) masking. The staggered clocking scheme is a happy medium for its ability to generate a test pattern count close to simultaneous clocking and fault coverage close to one-hot clocking. For large designs, it is no longer uncommon for transition fault ATPG to take longer than 2 to 4 weeks to complete. To reduce test generation time while at the same time obtaining the highest fault coverage, modern ATPG programs tend to either (1) run simultaneous clocking followed by one-hot clocking or (2) use staggered clocking followed by one-hot clocking. As a result, modern **at-speed scan architectures** now start supporting a combination of at-speed clocking schemes for test circuits comprising multiple synchronous and asynchronous clock domains. Some programs can even generate test patterns by mixing skewed-load and double-capture schemes.

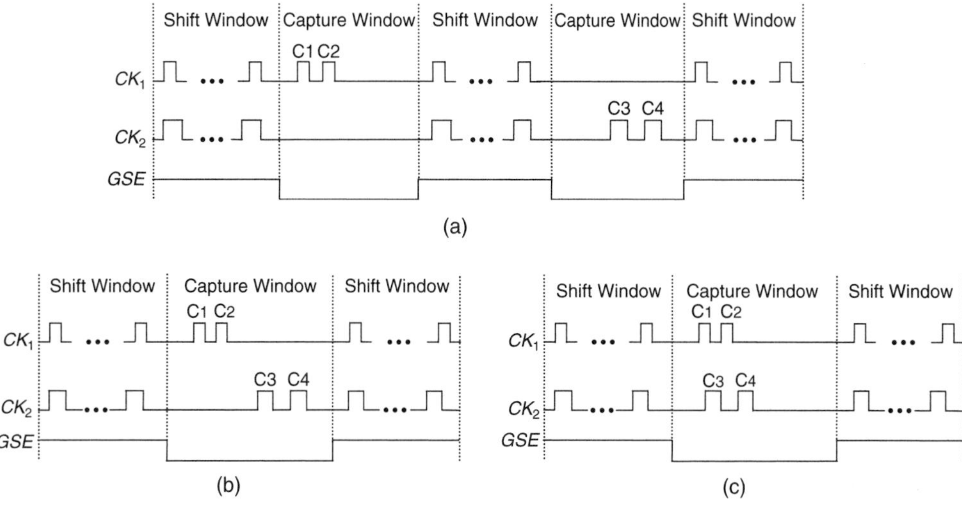

■ **FIGURE 2.12**

At-speed clocking schemes for testing two interacting clock domains: (a) one-hot clocking, (b) staggered clocking, and (c) simultaneous clocking.

In these modern at-speed scan architectures, the launch clock pulse and capture clock pulse can be either directly supplied from the tester or internally generated by the ***phase-locked loop*** (PLL) associated with each clock domain. Although it is easy to supply the clock pulses directly from the tester, the test cost associated with the use of an expensive tester and its limited high-frequency channels may hinder the approach from being practical. To use internal PLLs, additional on-chip clock controllers are required. When the skewed-load scheme is employed, it may be also necessary to perform *clock tree synthesis* (CTS) on the scan enable signal *SE* controlling each clock domain. Alternatively, the *SE* signal can be pipelined to avoid CTS. An example of a pipelined *SE* design to drive both positive-edge and negative-edge scan cells is shown in Figure 2.13 [Gizopoulos 2006]. Figure 2.14a

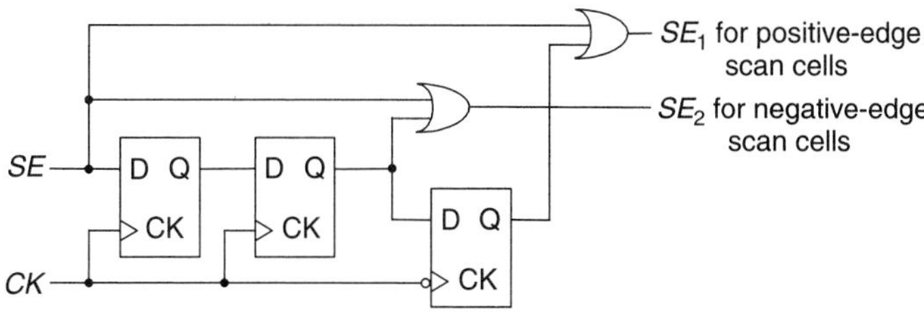

■ **FIGURE 2.13**

Pipelined scan enable design.

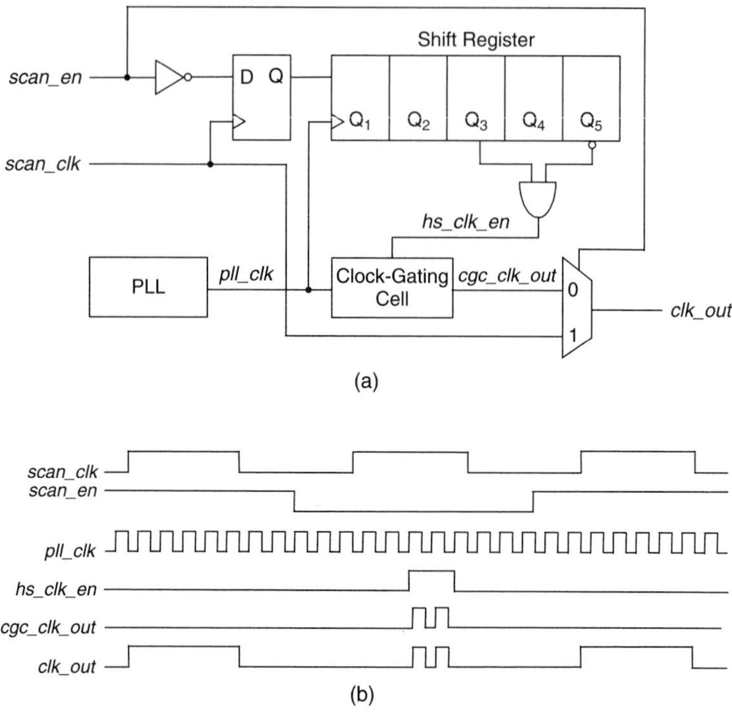

■ **FIGURE 2.14**

An on-chip clock controller for generating two capture clock pulses: (a) example on-chip clock controller and (b) waveform.

shows an on-chip clock controller for generating two capture clock cycles using the double-capture scheme [Beck 2005]. When *scan_en* is set to 1, *scan_clk* is directly connected to *clk_out*; when *scan_en* is set to 0, the output of the clock-gating cell is directly connected to *clk_out*. The implementation of the clock-gating cell makes sure that no glitches or spikes appear on *clk_out*. The clock-gating cell is enabled by the signal *hs_clk_en* that is generated from the five-bit register. The shift register is clocked by *pll_clk*. According to Figure 2.14b, a single *scan_clk* pulse is applied after *scan_en* is set to 0. This clock pulse generates a 1 that is latched by the D flip-flop and shifted through the shift register. After two *pll_clk* cycles, *hs_clk_en* is asserted for the next two *pll_clk* cycles. As the clock-gating cell is enabled during that period, exactly two PLL clock pulses are transmitted from the PLL to *clk_out*.

A test clock controller for detecting inter-clock-domain delay faults by using an internal PLL and the double-capture clocking scheme can be also found in [Furukawa 2006]. The authors in [Iyengar 2006] further presented an on-chip clock controller that can generate high-speed launch-on-capture as well as launch-on-shift clocking without the need to switch *SE* at-speed.

2.3 LOGIC BUILT-IN SELF-TEST

Figure 2.15 shows a typical logic *built-in self-test* (BIST) system. The ***test pattern generator*** (TPG) automatically generates test patterns for application to the inputs of the ***circuit under test*** (CUT). The ***output response analyzer*** (ORA) automatically compacts the output responses of the CUT into a *signature*. Specific BIST timing control signals, including scan enable signals and clocks, are generated by the **logic BIST controller** for coordinating the BIST operation among the TPG, CUT, and ORA. The logic BIST controller provides a pass/fail indication once the BIST operation is complete. It includes comparison logic to compare the *final signature* with an embedded *golden signature*, and it often encompasses **diagnostic logic** for fault diagnosis. As compaction is commonly used for output response analysis, it is required that all storage elements in the TPG, CUT, and ORA be initialized to known states before self-test and no unknown (X) values be allowed to propagate from the CUT to the ORA. In other words, the CUT must comply with more stringent **BIST-specific design rules** [Wang 2006a] in addition to those *scan design rules* required for scan design.

For BIST pattern generation, in-circuit TPGs are commonly constructed from ***linear feedback shift registers*** (**LFSRs**) [Golomb 1982] or **cellular automata** [Hortensius 1989] to generate test patterns or test sequences for exhaustive testing, pseudo-random testing, and pseudo-exhaustive testing [Bushnell 2000] [Wang 2006a]. **Exhaustive testing** always guarantees 100% single-stuck and multiple-stuck fault coverage. This technique requires all possible 2^n test patterns to be applied to an n-input combinational CUT, which can take too long for combinational circuits where n is huge; therefore, **pseudo-random testing** [Bardell 1987] is often used for generating a subset of the 2^n test patterns and uses fault simulation to calculate the exact fault coverage. The TPG is often referred to as a ***pseudo-random pattern generator*** (PRPG). In some cases, this might become quite time consuming, if not infeasible. To eliminate the need for fault simulation while at the same time maintaining 100% single-stuck fault coverage, we can

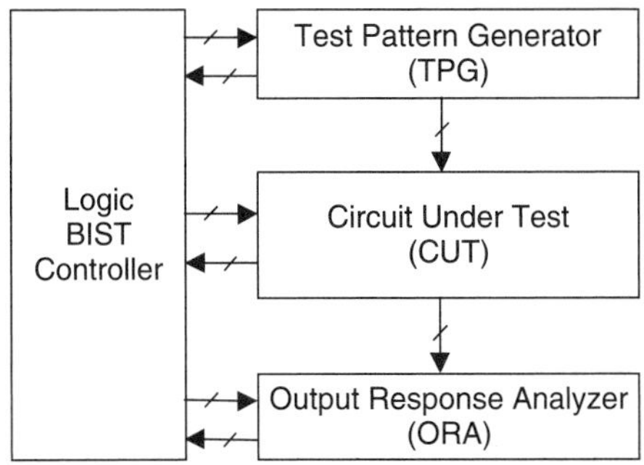

FIGURE 2.15

A typical logic BIST system.

use **pseudo-exhaustive testing** [McCluskey 1986] [Wang 2006a] to generate 2^w or $2^k - 1$ test patterns, where $w < k < n$, when each output of the n-input combinational CUT at most depends on w inputs. For testing delay faults, hazards must also be taken into consideration.

For output response compaction, the ORAs are commonly constructed from ***multiple-input signature registers*** (MISRs). The MISR is basically an LFSR that uses an extra XOR gate at the input of each LFSR stage for compacting the output responses of the CUT into the LFSR during each shift operation. Oftentimes, to further reduce the hardware overhead of the ORA, a *linear phase compactor* comprised of a network of XOR gates is connected to the MISR inputs.

2.3.1 Logic BIST Architectures

Several architectures for incorporating **offline BIST** techniques into a design have been proposed. These BIST architectures can be classified into two classes: (1) those using the **test-per-scan BIST** scheme and (2) those using the **test-per-clock BIST** scheme. The *test-per-scan BIST* scheme takes advantage of the already built-in scan chains of the scan design and applies a test pattern to the CUT after a shift operation is completed; hence, the hardware overhead is low. The *test-per-clock BIST* scheme, however, applies a test pattern to the CUT and captures its test response every system clock cycle; hence, the scheme can execute tests much faster than the test-per-scan BIST scheme but at an expense of more hardware overhead.

In this subsection, we only discuss two representative BIST architectures, one for each class. Although pseudo-random testing is commonly adopted in both BIST schemes, the exhaustive and pseudo-exhaustive test techniques are applicable for designs using the test-per-clock BIST scheme. For a more comprehensive survey of these BIST architectures, refer to [McCluskey 1985], [Bardell 1987], [Abramovici 1994], and [Wang 2006a].

2.3.1.1 Self-Testing Using MISR and Parallel SRSG (STUMPS)

A test-per-scan BIST design was presented in [Bardell 1982]. This design, shown in Figure 2.16, contains a PRPG (parallel ***shift register sequence generator*** [SRSG]) and a MISR. The scan chains are loaded in parallel from the PRPG. The system clocks are then triggered and the test responses are shifted to the MISR for compaction. New test patterns are shifted in at the same time while test responses are being shifted out. This BIST architecture using the test-per-scan BIST scheme is referred to as ***self-testing using MISR and parallel SRSG*** **(STUMPS)** [Bardell 1982].

Because of the ease of integration with traditional scan architecture, the **STUMPS** architecture is the only BIST architecture widely used in industry to date. To further reduce the lengths of the PRPG and MISR and improve the randomness of the PRPG, a STUMPS-based architecture that includes an optional linear phase shifter and an optional linear phase compactor is often used in industrial applications [Nadeau-Dostie 2000] [Cheon 2005]. The linear phase shifter and linear

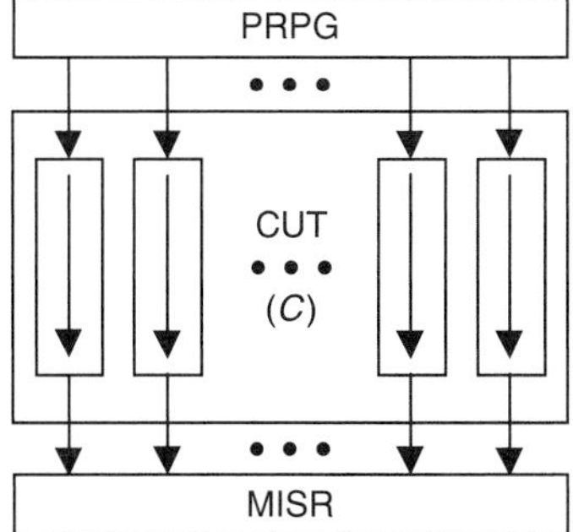

■ **FIGURE 2.16**

STUMPS.

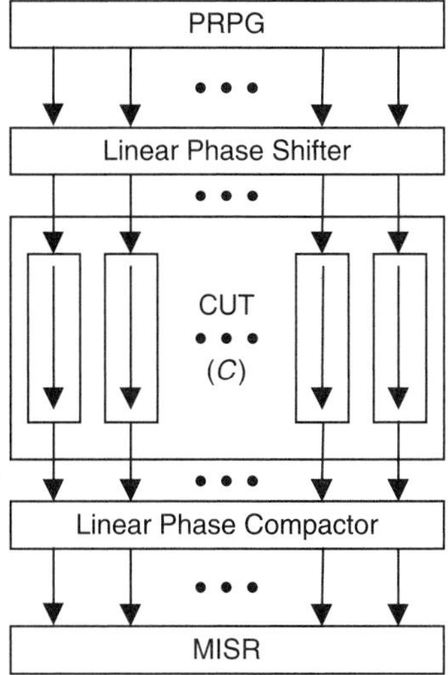

■ **FIGURE 2.17**

A STUMPS-based architecture.

phase compactor typically comprise a network of XOR gates. Figure 2.17 shows the STUMPS-based architecture.

2.3.1.2 Concurrent Built-In Logic Block Observer (CBILBO)

STUMPS is the widely adopted logic BIST architecture for scan-based designs. The acceptance of this STUMPS architecture is mostly because of the ease with which the BIST circuitry is integrated into a scan design. The efforts required to implement the BIST circuitry and the loss of the fault coverage for using pseudo-random patterns, however, have prevented the STUMPS-based logic BIST architecture from being widely used across all industries.

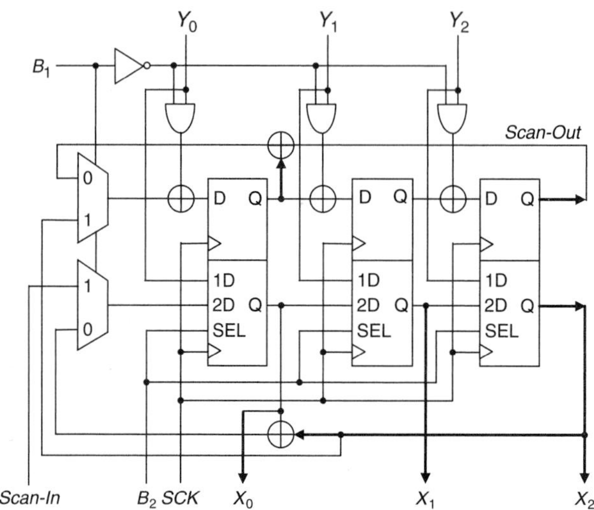

■ **FIGURE 2.18**

A three-stage concurrent BILBO (CBILBO).

One solution to solve the fault coverage loss problem is to use the ***concurrent bulit-in logic block observer*** (CBILBO) approach [Wang 1986]. The CBILBO is based on the test-per-clock BIST scheme and uses two registers to perform test generation and signature analysis simultaneously. A CBILBO design is shown in Figure 2.18, where only three modes of operation are considered: normal, scan, and test generation and signature analysis. When $B_1 = 0$ and $B_2 = 1$, the upper D flip-flops act as a MISR for signature analysis, whereas the lower two-port D flip-flops form a TPG for test generation. Because signature analysis is separated from test generation, an exhaustive or pseudo-exhaustive pattern generator

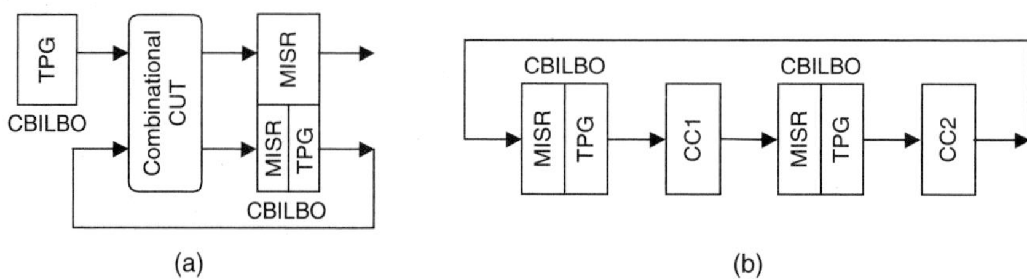

■ **FIGURE 2.19**

CBILBO architectures: (a) for testing a finite-state machine and (b) for testing a pipelined-oriented circuit.

(EPG/PEPG) can now be used for test generation; therefore, no fault simulation is required, and it is possible to achieve 100% single-stuck fault coverage using the CBILBO architectures for testing designs shown in Figure 2.19. However, the hardware cost associated with using the CBILBO approach is generally higher than for the STUMPS approach.

2.3.2 Coverage-Driven Logic BIST Architectures

In *pseudo-random testing*, the fault coverage is limited by the presence of ***random-pattern resistant*** (RP-resistant) faults. If the fault coverage is not sufficient, then four approaches can be used to enhance the fault coverage: (1) weighted pattern generation, (2) test point insertion, (3) mixed-mode BIST, and (4) hybrid BIST. The first three approaches are applicable for in-field coverage enhancement, whereas the fourth approach is applicable for manufacturing coverage enhancement.

Weighted pattern generation inserts a combinational circuit between the output of the PRPG and the CUT to increase the frequency of occurrence of one logic value while decreasing the other logic value. **Test point insertion** adds control points and observation points for providing additional controllability and observability to improve the detection probability of RP-resistant faults so they can be detected during pseudo-random testing. **Mixed-mode BIST** involves supplementing the pseudo-random patterns with some deterministic patterns that detect RP-resistant faults and are generated using on-chip hardware. When BIST is performed during manufacturing test where a tester is present, **hybrid BIST** involves combining BIST and external testing by supplementing the pseudo-random patterns with deterministic data from the tester to improve the fault coverage. This fourth option is not applicable when BIST is used in the field, as the tester is not present. Each of these approaches is described in more detail in the following subsections.

2.3.2.1 Weighted Pattern Generation

Typically, **weighted pseudo-random patterns** are used to increase the circuit's fault coverage. A **weighted pattern generation technique** employing an LFSR and a combinational circuit was first described in [Schnurmann 1975]. The combinational circuit inserted between the output of the LFSR and the CUT is to increase the frequency of occurrence of one logic value while decreasing the other logic value. This approach may increase the probability of detecting those faults that are difficult to detect using the typical LFSR pattern generation technique.

Implementation methods for realizing this scheme are further discussed in [Chin 1984]. The weighted pattern generation technique described in that paper modifies the maximum-length LFSR to produce an equally weighted distribution of 0's and 1's at the input of the CUT. It skews the LFSR probability distribution of 0.5 to either 0.25 or 0.75 to increase the chance of detecting those faults that are difficult to detect using just a 0.5 distribution. Better fault coverage was also found in [Wunderlich 1987], where probability distributions in a multiple of 0.125 (rather than 0.25) are used. For some circuits, several programmable probabilities or weight sets are required to further increase each circuit's fault

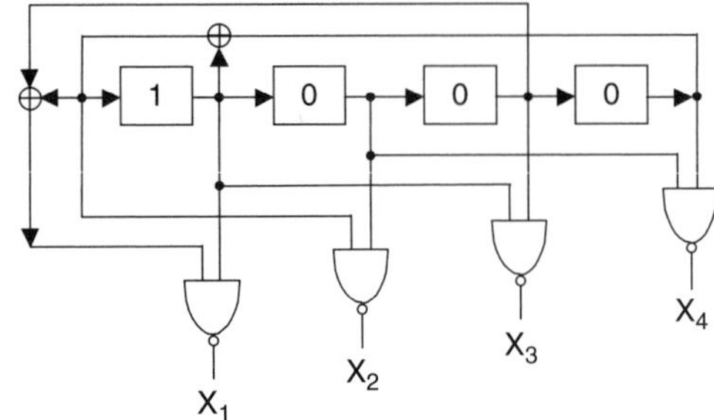

■ **FIGURE 2.20**

Example weighted LFSR as PRPG.

coverage [Waicukauski 1989] [Bershteyn 1993] [Kapur 1994] [Lai 2005]. Additional discussions on weighted pattern generation can be found in [Rajski 1998a] and [Bushnell 2000]. Figure 2.20 shows a four-stage weighted (maximum-length) LFSR with probability distribution 0.75 [Chin 1984].

2.3.2.2 Test Point Insertion

Although weighted pattern generation is simple in design, achieving adequate fault coverage for a BIST circuit remains a problem. Test points can then be used to increase the circuit's fault coverage to a desired level. Figure 2.21 shows two typical types of test points that can be inserted. A **control point** can be connected to a primary input, an existing scan cell output, or a dedicated scan cell output. An **observation point** can be connected to a primary output through an additional multiplexer, an existing scan cell input, or a dedicated scan cell input.

Figure 2.22b shows an example where one control point and one observation point are inserted to increase the detection probability of a 6-input AND-gate given in Figure 2.22a. By splitting the six-input AND gate into two fewer-input AND gates and placing a control point and an observation point between the two fewer-input AND gates, we can increase the probability of detecting faults in the original

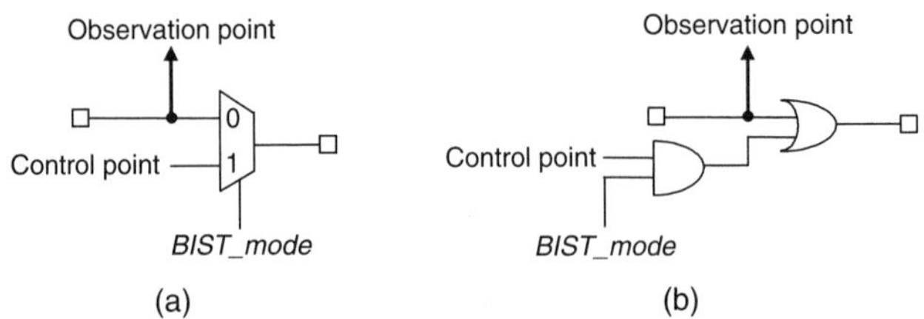

■ **FIGURE 2.21**

Typical test points inserted for improving a circuit's fault coverage: (a) test point with a multiplexer and (b) test point with AND-OR gates.

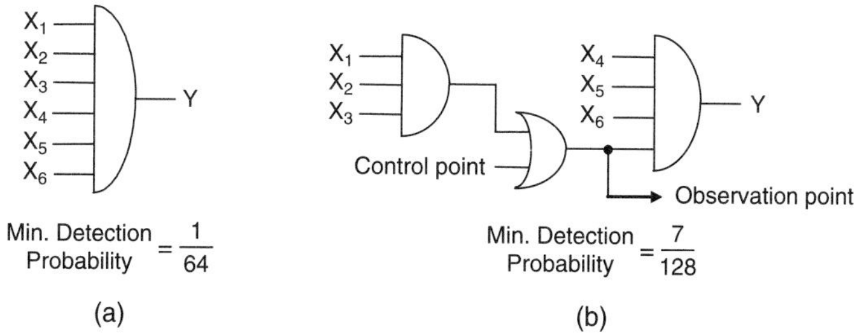

■ **FIGURE 2.22**

Example of inserting test points to improve detection probability: (a) an output RP-resistant stuck-at-0 fault and (b) example of inserted test points.

six-input AND gate, (*e.g.*, output Y stuck-at-0 and any input X_i stuck-at-1), thereby making the circuit more RP testable. After the test points are inserted, the most difficult fault to detect is the bottom input of the four-input AND gate stuck-at-1. In that case, one of inputs X_1, X_2, and X_3 must be 0, the control point must be 0, and all inputs X_4, X_5, and X_6 must be 1, resulting in a detection probability of 7/128 $(= 7/8 \times 1/2 \times 1/2 \times 1/2 \times 1/2)$.

Test Point Placement

Because test points add area and performance overhead, an important issue for test point insertion is where to place the test points in the circuit to maximize the coverage and minimize the number of test points required. Note that it is not sufficient to only use observation points, as some faults require control points in order to be detected. Optimal placement of test points in circuits with reconvergent fanout has been shown to be NP-complete [Krishnamurthy 1987]. Several approximation techniques for placement of test points have been developed using either fault simulation [Iyengar 1989] [Touba 1996] or testability measures to guide them [Seiss 1991] [Tamarapalli 1996] [Zhang 2000]. **Timing-driven test point insertion** techniques [Tsai 1998] have also been developed to avoid adding delay on a critical timing path. The number of test points that must be added can be reduced by using the **almost-full-scan BIST** technique proposed in [Tsai 2000] that excludes a small number of scan cells from the scan chains during BIST operation.

Control Point Activation

Once the test points have been inserted, the logic that drives the control points must be designed. When a control point is activated, it forces the logic value at a particular node in the circuit to a fixed value. During normal operation, all control points must be deactivated. During testing, there are different strategies as to when and how the control points are activated. One approach is **random activation,** where the control points are driven by the pseudo-random pattern generator. The drawback of this approach is that when a large number of control points are inserted, they can interfere with each other and may not improve the fault coverage as much as desired. An alternative to *random activation* is to use **deterministic activation**. The technique in [Tamarapalli 1996] divides the BIST into phases and deterministically activates some subset of the control points in each phase. The technique in [Touba 1996] uses *pattern decoding logic* to activate the control points only for certain patterns where they are needed to detect RP-resistant faults.

2.3.2.3 Mixed-Mode BIST

A major drawback of test point insertion is that it requires modifying the circuit under test. In some cases this is not possible or not desirable (*e.g.*, for hard cores, macros, hand-crafted designs, or legacy designs). An alternative way to improve fault coverage without modifying the CUT is to use **mixed-mode BIST**. Pseudo-random patterns are generated to detect the RP-testable faults, and then some additional deterministic patterns are generated to detect the RP-resistant faults. There are a number of ways for generating deterministic patterns on-chip. Three approaches are described next.

ROM Compression

The simplest approach for generating deterministic patterns on-chip is to store them in a *read-only-memory* (ROM). The problem with this approach is that the size of the required ROM is often prohibitive. Although several **ROM compression** techniques have been further proposed for reducing the size of the ROM, the industry seems to still shy away from using this approach [Agarwal 1981] [Aboulhamid 1983] [Dandapani 1984] [Edirisooriya 1992].

LFSR Reseeding

Instead of storing the test patterns themselves in a ROM, techniques have been developed for storing LFSR seeds that can be used to generate the test patterns [Könemann 1991]. The LFSR that is used for generating the pseudo-random patterns is also used for generating the deterministic patterns by reseeding it with computed seeds. The seeds can be computed with linear algebra as described in [Könemann 1991]. Because the seeds are smaller than the test patterns themselves, they require less ROM storage. One problem is that for an LFSR with a fixed characteristic (feedback) polynomial, it may not always be possible to find a seed that will efficiently generate the required deterministic test patterns. A solution to this problem was proposed in [Hellebrand 1995a] in which a ***multiple-polynomial LFSR*** (MP-LFSR), as illustrated in Figure 2.23, is used. An MP-LFSR is an LFSR with a reconfigurable feedback network. A *polynomial identifier* is stored with each seed to select the characteristic polynomial that will be used for that seed. Techniques for further reductions in storage can be achieved by using variable-length seeds [Rajski 1998b], a special ATPG algorithm [Hellebrand 1995b], folding counters [Liang 2001], and seed encoding [Al-Yamani 2005].

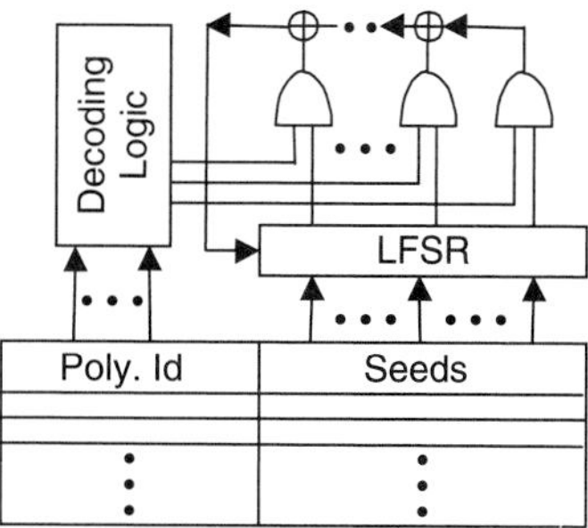

■ **FIGURE 2.23**

Reseeding with multiple-polynomial LFSR.

Embedding Deterministic Patterns

A third approach for mixed-mode BIST is to embed the deterministic patterns in the pseudo-random sequence. Many of the pseudo-random patterns generated during pseudo-random testing do not detect any new faults, so some of those "useless" patterns can be transformed into deterministic patterns that detect RP-resistant faults [Touba 1995]. This can be done by adding *mapping logic* between the scan chains and the CUT [Touba 1995] or in a less intrusive way by adding the *mapping logic* at the inputs to the scan chains to either perform bit-fixing [Touba 2001] or bit-flipping [Kiefer 1998]. Figure 2.24 shows a bit-flipping BIST scheme taken from [Kiefer 1998]. A bit-flipping function detects these "useless" patterns and maps them to deterministic patterns through the use of an XOR gate that is inserted between the LFSR and each scan chain.

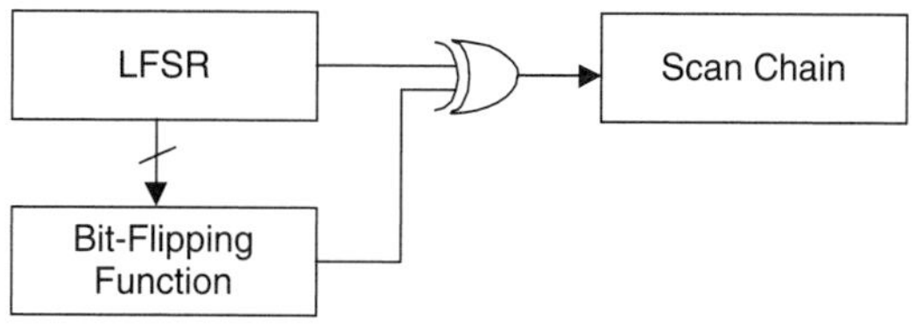

■ **FIGURE 2.24**

Bit-flipping BIST.

2.3.2.4 Hybrid BIST

For manufacturing fault coverage enhancement where a tester is present, deterministic data from the tester can be used to improve the fault coverage. The simplest approach is to perform top-up ATPG for the faults not detected by BIST to obtain a set of deterministic test patterns that "top-up" the fault coverage to the desired level and then store those patterns directly on the tester. In a system-on-chip, test scheduling can be done to overlap the BIST run time with the transfer time for loading the deterministic patterns from the tester [Sugihara 1998] [Jervan 2003]. More elaborate hybrid BIST schemes have been developed, which attempt to store the deterministic patterns on the tester in a compressed form and then make use of the existing BIST hardware to decompress them. Such techniques are described in [Das 2000], [Dorsch 2001], [Ichino 2001], [Krishna 2003a], [Wohl 2003a], [Jas 2004], and [Lei 2005]. More discussions on test compression can be found in the following section.

2.3.3 Low-Power Logic BIST Architectures

Test power consumption in logic BIST designs tends to become more serious than that in scan designs. One major reason is that unlike scan designs in which test power can be reduced by simply using software ATPG approaches [Girard 2002] [Wen 2006], test power in logic BIST designs can only be reduced using hardware.

However, there are still quite a few hardware approaches that can be used to reduce test power. The low-power scan architectures discussed in Section 2.2.2 are mostly applicable for BIST designs. Three approaches are further described next. For more information, refer to Chapter 7.

2.3.3.1 Low-Transition BIST Design

One simple approach is to design a **low-transition PRPG** that generates test patterns with low switching activity. [Wang 1999] belongs to this category. The *low-transition random test pattern generator* (LT-RTPG) described in [Wang 1999] and used as a PRPG that is shown in Figure 2.25 inserts an AND gate and a toggle (T) flip-flop at the scan input of the scan chain. The inputs of the AND gate are connected to a few outputs of the LFSR. If the output of the AND gate in the LT-RTPG is 0 for k cycles, then identical values are applied at the scan input for k clock cycles. Hence, the switching activity is reduced. This approach is less design-intrusive and entails no performance degradation. It also requires low hardware overhead. The drawback of this approach is low fault coverage or long test sequence when required to achieve adequate fault coverage.

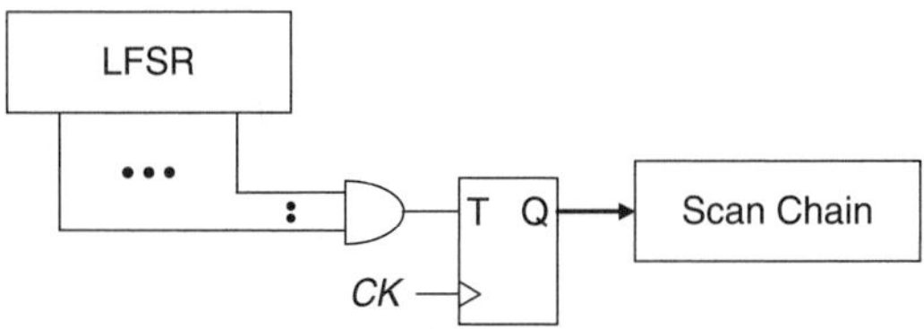

■ **FIGURE 2.25**

Low-transition random test pattern generator (LT-RTPG) as PRPG.

2.3.3.2 Test-Vector-Inhibiting BIST Design

Another approach is to inhibit the LFSR-generated pseudo-random patterns, which do not contribute to fault detection from being applied to the *circuit under test* (CUT). This **test-vector-inhibiting** technique can reduce test power while achieving the same fault coverage as the original LFSR. [Manich 2000] belongs to this category. A *test-vector-inhibiting RTPG* (TVI-RTPG) is shown in Figure 2.26 for use as a PRPG. When a pseudo-random pattern generated by the PRPG does not detect any faults, the pattern is not transmitted to the CUT. For this purpose, a decoding logic is connected to the output of the LFSR and outputs a 0 to inhibit the pseudo-random pattern from passing through the transmission gate network to the CUT. A transmission gate can be an XOR gate. Whereas this approach targets test-per-clock BIST, it is applicable for test-per-scan BIST designs. The drawback of this approach is high area overhead and impact on circuit performance.

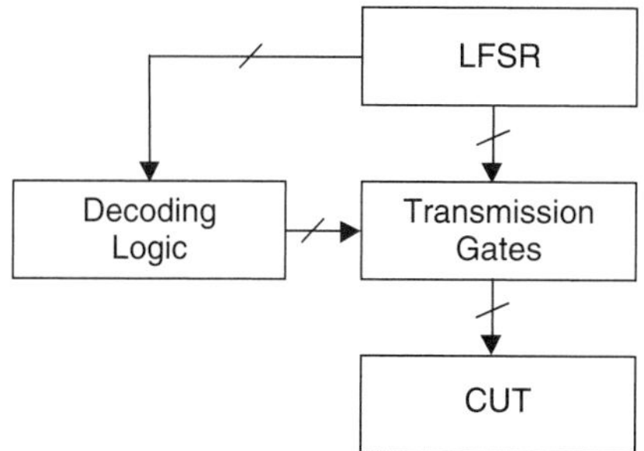

■ **FIGURE 2.26**

Test-vector-inhibiting RTPG (TVI-RTPG) as PRPG.

2.3.3.3 Modified LFSR Low-Power BIST Design

The third approach is to use a modified LFSR structure, composed of two separated or interleaved $n/2$-stage LFSRs, to drive the *circuit under test* (CUT). The two $n/2$-stage LFSRs would activate only one part of the CUT in a given time interval. [Girard 2001] belongs to this category. It was demonstrated in the paper that shorter test length to reach target fault coverage can be achieved with the proposed modified LFSR structure as shown in Figure 2.27. A test clock module is used to generate the two nonoverlapping clocks, CK_1 and CK_2, for driving LFSR-1 and LFSR-2, respectively. Because only one part of the CUT is activated at any given time, this BIST scheme provides high percentage of power (and energy) reduction and results in no performance degradation and test time increase. The drawback of this approach is the requirement of constructing special clock trees.

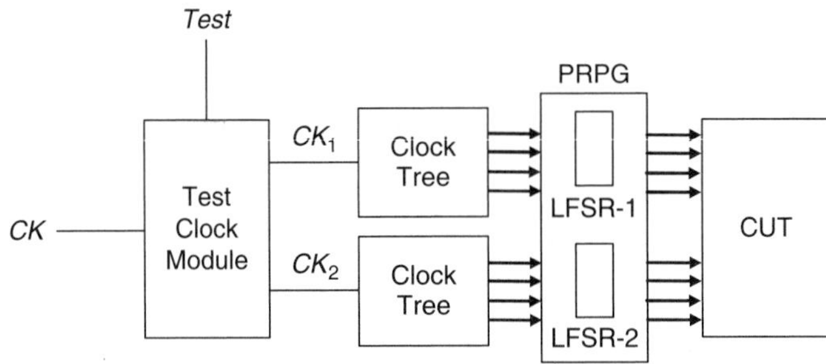

■ **FIGURE 2.27**

Two $n/2$-stage LFSRs as PRPG.

2.3.4 At-Speed Logic BIST Architectures

There are three basic capture-clocking schemes that can be used for testing multiple clock domains: (1) single-capture, (2) skewed-load, and (3) double-capture. We will illustrate with BIST timing control diagrams how to test synchronous and asynchronous clock domains using these schemes. In this section, we first discuss the three basic capture-clocking schemes and then briefly describe the logic BIST architectures practiced by the *electronic design automation* (EDA) vendors. Throughout this section, we will assume that a STUMPS-based architecture is used and that each clock domain contains one test clock and one scan enable signal. The faults we will consider include **structural faults**, such as stuck-at faults and bridging faults, as well as timing-related **delay faults**, such as path-delay faults and transition faults.

2.3.4.1 Single-Capture

Single-capture is a **slow-speed test** technique in which only one capture pulse is applied to each clock domain. It is the simplest for testing all intra-clock-domain and inter-clock-domain structural faults. Two approaches can be used: (1) one-hot single-capture and (2) staggered single-capture.

One-Hot Single-Capture

Using the **one-hot single-capture** approach, a capture pulse is applied to only one clock domain during each capture window, while all other test clocks are held inactive. A sample timing diagram is shown in Figure 2.28. In the figure, because only one capture pulse (C1 or C2) is applied during each capture window, this scheme can only test intra-clock-domain and inter-clock-domain structural faults. The main advantage of this approach is that the designer does not have to worry about clock skews between the two clock domains during self-test, as each clock domain is tested independently. The only requirement is that delays $d1$ and $d2$

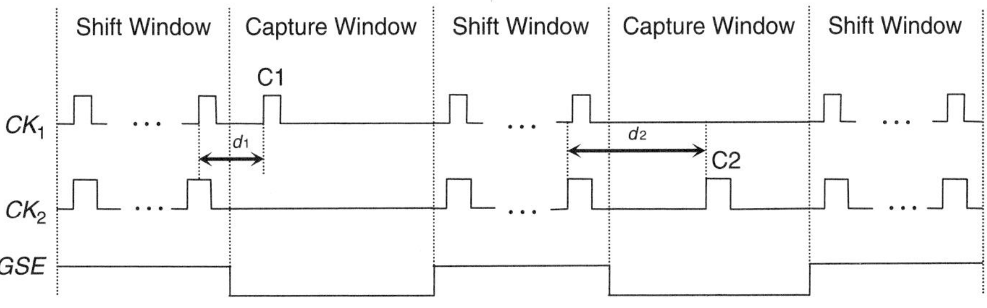

■ **FIGURE 2.28**

One-hot single-capture.

be properly adjusted; hence, this approach can be used for **slow-speed testing** of both synchronous and asynchronous clock domains. Another benefit of using this approach is that a single, slow-speed *global scan enable* (*GSE*) signal can be used for driving both clock domains, which makes it easy to integrate with scan. A major drawback is longer test time, as all clock domains have to be tested one at a time.

Staggered Single-Capture

The long test time problem using one-hot single-capture can be solved using the **staggered single-capture** approach [Wang 2006b]. A sample timing diagram is shown in Figure 2.29. In this approach, capture pulses C1 and C2 are applied in a sequential or staggered order during the capture window to test all intra-clock-domain and inter-clock-domain structural faults in the two clock domains. For clock domains that are synchronous, adjusting *d*2 will allow us to detect inter-clock-domain delay faults between the two clock domains at-speed. In addition, because *d*1 and *d*3 can be as long as desired, a single, slow-speed *GSE* signal can be used. This significantly simplifies the logic BIST physical implementation for designs with multiple clock domains. There may be some structural fault coverage loss between clock domains if the ordered sequence of capture clocks is fixed for all capture cycles.

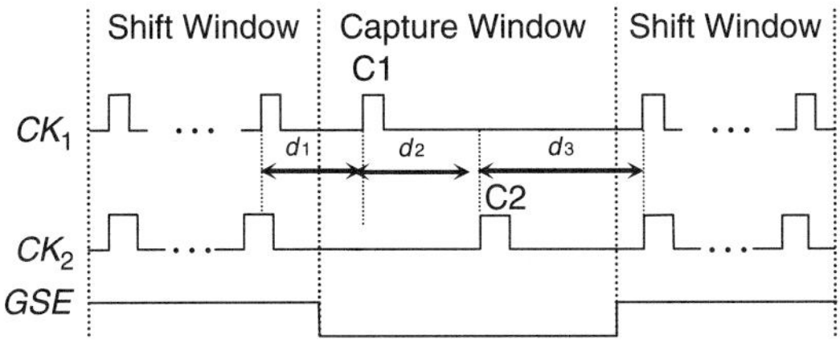

■ **FIGURE 2.29**

Staggered single-capture.

2.3.4.2 Skewed-Load

Skewed-load is an **at-speed delay test** technique in which a last shift pulse followed immediately by a capture pulse, running at the test clock's operating frequency, are used to launch the transition and capture the output response [Savir 1993]. It is also referred to as **launch-on-shift**. This technique addresses the intra-clock-domain delay fault detection problem, which cannot be tested using single-capture schemes. Skewed-load uses the value difference between the last shift pulse and the next-to-last-shift pulse to launch the transition and uses the capture pulse to capture the output response. For the last shift pulse to launch the transition, the scan enable signal associated with the clock domain must be able to switch operations from shift to capture in one clock cycle. Three approaches can be used: (1) one-hot skewed-load, (2) aligned skewed-load, and (3) staggered skewed-load.

One-Hot Skewed-Load

Similar to one-hot single-capture, the **one-hot skewed-load** approach tests all clock domains one by one [Bhawmik 1997]. A sample timing diagram is shown in Figure 2.30. The main differences are (1) it applies shift-followed-by-capture pulses (S1-followed-by-C1 or S2-followed-by-C2) to detect intra-clock-domain delay faults, and (2) each scan enable signal (SE_1 or SE_2) must switch operations from shift to capture within one clock cycle (*d*1 or *d*2). Thus, this approach can only be used for **at-speed testing** of intra-clock-domain delay faults in both synchronous and asynchronous clock domains. The disadvantages are (1) it cannot be used to detect inter-clock-domain delay faults, (2) it has a long test time, and (3) it is incompatible with scan, as a single, slow-speed *GSE* signal can no longer be used.

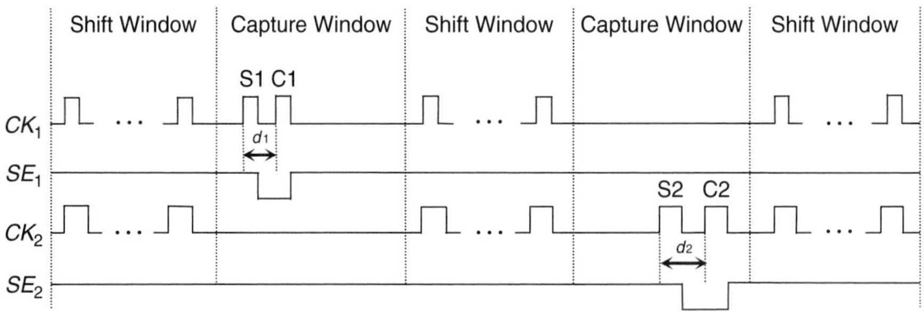

■ FIGURE 2.30

One-hot skewed-load.

Aligned Skewed-Load

The disadvantages of one-hot skewed-load can be resolved by using the aligned skewed-load scheme. One **aligned skewed-load** approach that aligns all capture edges together is illustrated in Figure 2.31 [Nadeau-Dostie 1994] [Nadeau-Dostie

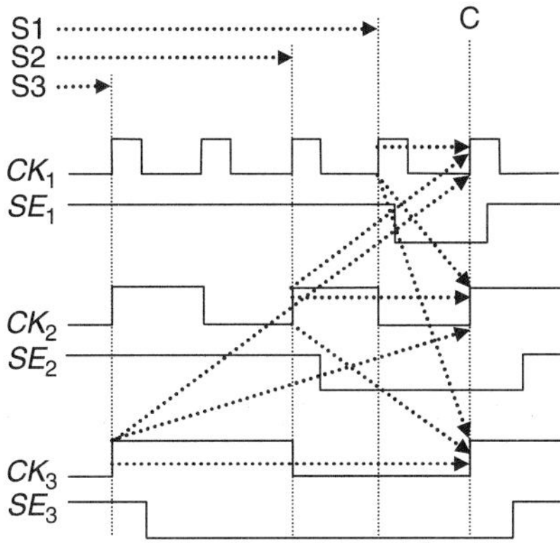

■ **FIGURE 2.31**

Capture aligned skewed-load.

2000]. The approach is referred to as **capture aligned skewed-load**. The major advantage of using this approach is that all intra-clock-domain and inter-clock-domain faults can be tested. The arrows shown in Figure 2.31 indicate the delay faults that can be tested. For example, the three arrows from S1 (CK_1) to C are used to test all intra-clock-domain delay faults in the clock domain controlled by CK_1, and all inter-clock-domain delay faults from CK_1 to CK_2 and CK_3. The remaining six arrows shown from S2 (CK_2) to C, and S3 (CK_3) to C are used to test all the remaining delay faults.

Because the active edges (rising edges) of the three capture pulses (see dash line C) must be aligned precisely, the circuit must contain one reference clock, and the frequency of all remaining test clocks must be derived from the reference clock. In the example given here, CK_1 is the reference clock operating at the highest frequency, and CK_2 and CK_3 are derived from CK_1 and designed to operate at 1/2 and 1/4 the frequency, respectively; therefore, this approach is only applicable for **at-speed testing** of intra-clock-domain and inter-clock-domain delay faults in synchronous clock domains.

A similar **aligned skewed-load** approach that aligns all last shift edges, rather than capture edges, is shown in Figure 2.32 [Hetherington 1999] [Rajski 2003]. This approach is referred to as **launch aligned skewed-load**. Similar to capture aligned skewed-load, it is also only applicable for **at-speed testing** of intra-clock-domain and inter-clock-domain delay faults in synchronous clock domains.

Consider the three clock domains, driven by CK_1, CK_2, and CK_3, again. The eight arrows among the dash line S and the three capture pulses (C1, C2, and C3) indicate the intra-clock-domain and inter-clock-domain delay faults that can be tested. Unlike in Figure 2.31, however, to test the inter-clock-domain delay faults from CK_1 to CK_3, a special shift pulse S1 (when SE_1 is set to 1) is required. As this method requires a much more complex timing-control diagram, a **clock suppression** circuit is used to enable or disable selected shift or capture pulses [Rajski 2003]. The dotted clock pulses shown in the figure indicate the suppressed shift pulses.

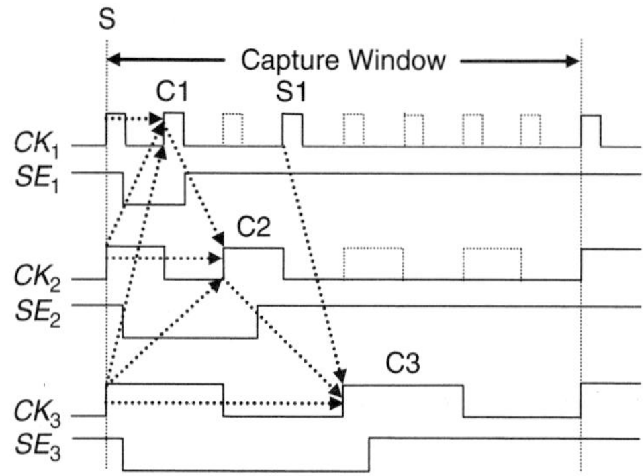

■ **FIGURE 2.32**

Launch aligned skewed-load.

Staggered Skewed-Load

Although the aligned skewed-load approaches can test all intra-clock-domain and inter-clock-domain faults in synchronous clock domains, their physical implementation is extremely difficult. There are two main reasons. First, to effectively align all active edges in either capture or last shift, the circuit must contain a reference clock. This reference clock must operate at the fastest clock frequency, and all other clock frequencies must be derived from the reference clock; such designs rarely exist. (2) For any two edges that cannot be aligned precisely because of clock skews, we must either resort to a one-hot skewed-load approach or add capture-disabling circuitry on the functional data paths of the two clock domains to prevent the cross-domain logic from interacting with each other during capture. This increases the circuit overhead, degrades the functional circuit performance, and reduces the ability to test inter-clock-domain faults.

The **staggered skewed-load** approach shown in Figure 2.33 relaxes these conditions [Wang 2005b]. For test clocks that cannot be precisely aligned, a delay $d3$ is inserted, to eliminate the clock skew interaction between the two clock domains.

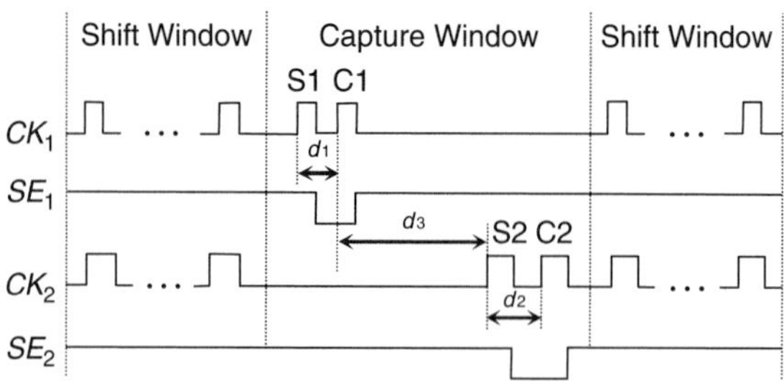

■ **FIGURE 2.33**

Staggered skewed-load.

The two last shift pulses (S1 and S2) are used to create transitions at the outputs of some scan cells, and the output responses to these transitions are captured by the following two capture pulses (C1 and C2), respectively. Both delays *d*1 and *d*2 are set to their respective clock domains' operating frequencies; hence, this scheme can be used to test all intra-clock-domain faults and inter-clock-domain structural faults in **asynchronous clock domains**. A problem still exists, as each clock domain requires an at-speed scan enable signal, which complicates physical implementation.

2.3.4.3 Double-Capture

The physical implementation difficulty using skewed-load can be resolved by using the double-capture scheme. **Double-capture** is another **at-speed test** technique in which two consecutive capture pulses are applied to launch the transition and capture the output response. It is also referred to as **broad-side** [Savir 1994] or **launch-on-capture**. The double-capture scheme can achieve **true at-speed test** quality for intra-clock-domain and inter-clock-domain faults in any synchronous or asynchronous design and it is easy for physical implementation. Here, true at-speed testing is meant to (1) allow detection of intra-clock-domain faults within each clock domain at its own operating frequency and detection of inter-clock-domain structural faults or delay faults, depending on whether the circuit under test is synchronous, asynchronous, or a mix of both, and (2) ease physical implementation for seamless integration with the conventional scan/ATPG technique.

One-Hot Double-Capture

Similar to one-hot skewed-load, the **one-hot double-capture** approach tests all clock domains one by one. A sample timing diagram is shown in Figure 2.34. The main differences are (1) two consecutive capture pulses are applied (C1-followed-by-C2 or C3-followed-by-C4) at their respective clock domains' frequencies (of period *d*1 or *d*2) to test intra-clock-domain delay faults, and (2) a single, slow-speed *GSE* signal is used to drive both clock domains. Hence, this scheme can

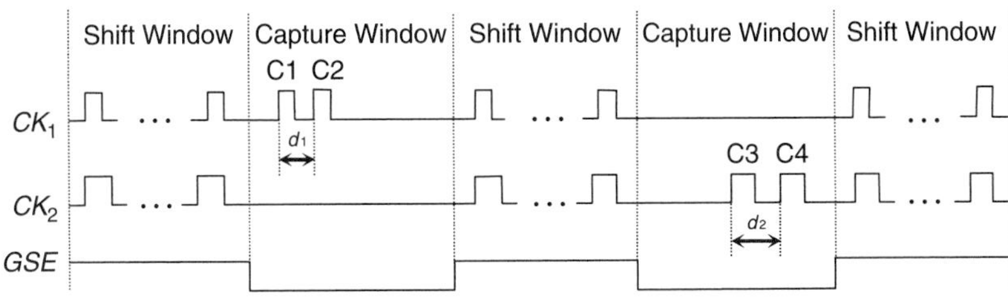

FIGURE 2.34

One-hot double-capture.

be used for **true at-speed testing** of intra-clock-domain delay faults in both synchronous and asynchronous clock domains. Two drawbacks remain: (1) it cannot be used to detect inter-clock-domain delay faults, and (2) it has a long test time.

Aligned Double-Capture

The drawbacks of the one-hot double-capture scheme can be resolved by using an **aligned double-capture** approach. Similar to the aligned skewed-load approach, the aligned double-capture scheme allows all intra-clock-domain faults and inter-clock-domain faults to be tested [Wang 2006b]. The main differences are (1) two consecutive capture pulses are applied, rather than shift-followed-by-capture pulses, and (2) a single, slow speed *GSE* signal is used. Figures 2.35 and 2.36 show two sample timing diagrams. This scheme can be used for **true at-speed testing** of **synchronous clock domains**. One major drawback is that precise alignment of the capture pulses is still required. This complicates physical implementation for designs with asynchronous clock domains.

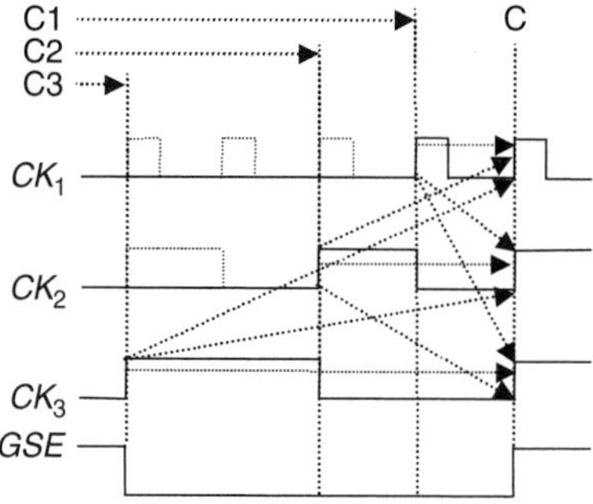

■ FIGURE 2.35

Capture aligned double-capture.

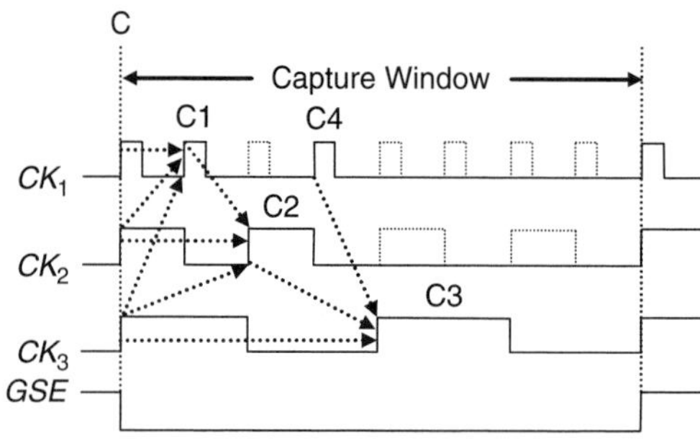

■ FIGURE 2.36

Launch aligned double-capture.

Staggered Double-Capture

The capture alignment problem in the aligned double-capture approach can finally be relaxed by using the **staggered double-capture** scheme [Wang 2005a, 2006b]. A sample timing diagram is shown in Figure 2.37. During the capture window, two capture pulses are generated for each clock domain. The first two capture pulses (C1 and C3) are used to create transitions at the outputs of some scan cells, and the output responses to the transitions are captured by the second two capture pulses (C2 and C4), respectively. Both delays *d*2 and *d*4 are set to their respective domains' operating frequencies. Because *d*1, *d*3, and *d*5 can be adjusted to any length, we can simply use a single, slow-speed *GSE* signal for driving all clock domains; hence, **true at-speed testing** is guaranteed using this approach for asynchronous clock domains. Because a single *GSE* signal is used, this scheme significantly eases physical implementation and allows us to integrate logic BIST with scan/ATPG easily to improve the circuit's manufacturing fault coverage.

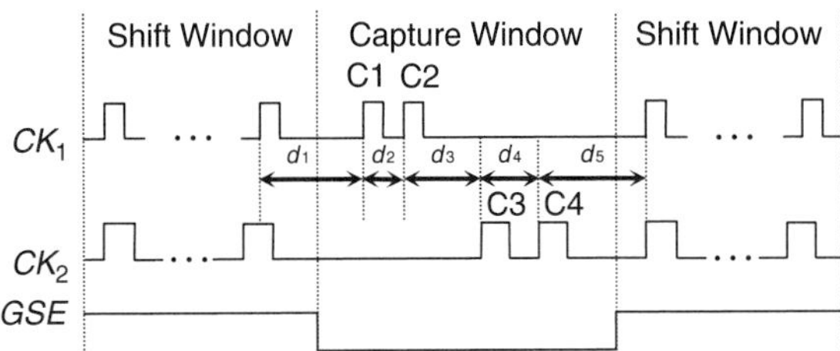

■ **FIGURE 2.37**

Staggered double-capture.

2.3.5 Industry Practices

Logic BIST has a history of more than 30 years since its invention in the 1970s. Although it is only a few years behind the invention of scan, logic BIST has yet to gain strong industry support. The worldwide market is estimated to be close to 10% of the scan market. The logic BIST products available in the marketplace include **Encounter Test** from Cadence Design Systems [Cadence 2007], **ETLogic** from LogicVision [LogicVision 2007], **LBIST Architect** from Mentor Graphics [Mentor 2007], and **TurboBIST-Logic** from SynTest Technologies [SynTest 2007]. The logic BIST product offered in Encounter Test by Cadence currently includes support for test structure extraction, verification, logic simulation for signatures, and fault simulation for coverage. Unlike all three other BIST vendors that provide their own logic BIST structures in their respective products, Cadence offers a service to insert custom logic BIST structures or to use any customer inserted logic BIST structures; the service includes working with the customer to have custom on-chip clocking for logic BIST. A similar case arises in ETLogic from LogicVision when using the double-capture clocking scheme.

All these commercially available logic BIST products support the STUMPS-based architectures. Cadence supports a weighted-random spreading network

TABLE 2.1 ■ Summary of Industry Practices for At-Speed Logic BIST

Industry Practices	Skewed-Load	Double-Capture
Encounter Test	Through service	Through service
ETLogic	√	Through service
LBIST Architect	√	√
TurboBIST-Logic		√

(XOR network) for STUMPS with multiple weight selects [Foote 1997]. For at-speed delay testing, ETLogic [LogicVision 2007] uses a **skewed-load-based at-speed BIST architecture**, TurboBIST-Logic [SynTest 2007] implements the **double-capture-based at-speed BIST architecture**, and LBIST Architect [Mentor 2007] adopts a **hybrid at-speed BIST architecture** that supports both skewed-load and double-capture. In addition, all products provide inter-clock-domain delay fault testing for synchronous clock domains. On-chip clock controllers for testing these inter-clock-domain faults at-speed can be found in [Rajski 2003], [Furukawa 2006], [Nadeau-Dostie 2006], and [Nadeau-Dostie 2007]. Table 2.1 summarizes the capture-clocking schemes for at-speed logic BIST used by the EDA vendors.

2.4 TEST COMPRESSION

Test compression can provide 10X to 100X reduction or even more in the amount of test data (both test stimulus and test response) that must be stored on the ***automatic test equipment*** (ATE) [Touba 2006] [Wang 2006a] for testing with a deterministic ATPG-generated test set. This greatly reduces ATE memory requirements; even more important, it reduces test time because fewer data have to be transferred across the limited bandwidth between the ATE and the chip. Moreover, test compression methodologies are easy to adopt in industry because they are compatible with the conventional design rules and test generation flows used for scan testing.

Test compression is achieved by adding some additional on-chip hardware before the scan chains to decompress the test stimulus coming from the tester and after the scan chains to compact the response going to the tester. This is illustrated in Figure 2.38. This extra on-chip hardware allows the test data to be stored on the tester in a compressed form. Test data are inherently highly compressible because typically only 1% to 5% of the bits on a test pattern that is generated by an ATPG program have specified (*care*) values. Lossless compression techniques can thus be used to significantly reduce the amount of test stimulus data that must be stored on the tester. The on-chip **decompressor** expands the compressed test stimulus back into the original test patterns (matching in all the care bits) as they are shifted into the scan chains. The on-chip **compactor** converts long output response sequences into short signatures. Because the compaction is lossy, some fault coverage can be lost because of unknown (*X*) values that might appear in the output sequence or

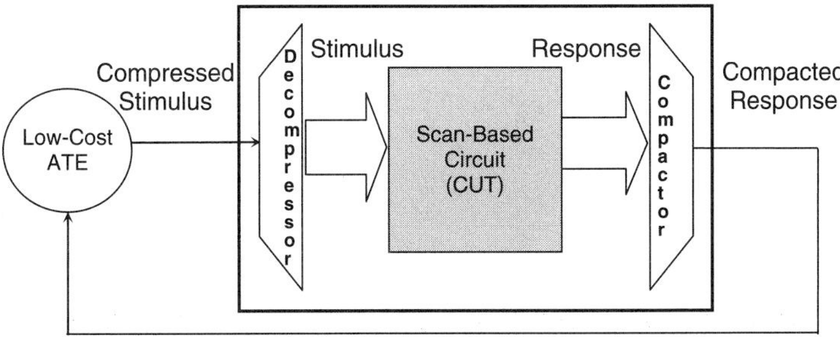

■ FIGURE 2.38

Architecture for test compression.

aliasing where a faulty output response signature is identical to the fault-free output response signature. With proper design of the *circuit under test* (CUT) and the compaction circuitry, however, the fault coverage loss can be kept negligibly small.

2.4.1 Circuits for Test Stimulus Compression

A ***test cube*** is defined as a deterministic test vector in which the bits that are not assigned values by the ATPG procedure are left as "don't cares" (*X*'s). Normally, ATPG procedures perform *random fill* in which all the *X*'s in the test cubes are filled randomly with 1's and 0's to create fully specified test vectors; however, for test stimulus compression, random fill is not performed during ATPG so the resulting test set consists of incompletely specified test cubes. The *X*'s make the test cubes much easier to compress than fully specified test vectors.

As mentioned earlier, test stimulus compression should be an information lossless procedure with respect to the specified (care) bits in order to preserve the fault coverage of the original test cubes. After decompression, the resulting test patterns shifted into the scan chains should match the original test cubes in all the specified (care) bits.

Many schemes for compressing test cubes have been surveyed in [Touba 2006] and [Wang 2006a]. Two schemes based on linear decompression and broadcast scan are described here in greater detail mainly because the industry has favored both approaches over code-based schemes from area overhead and compression ratio points of view. These industry practices can be found in [Wang 2006a].

2.4.1.1 Linear-Decompression-Based Schemes

A class of test stimulus compression schemes is based on using **linear decompressors** to expand the data coming from the tester to fill the scan chains. Any decompressor that consists of only XOR gates and flip-flops is a ***linear decompressor*** [Könemann 1991]. Linear decompressors have a very useful property: their *output space* (*i.e.*, the space of all possible test vectors that they can generate) is a linear subspace that is spanned by a Boolean matrix. In other words, for any linear decompressor that expands an m-bit compressed stimulus from the tester into an n-bit stimulus (test vector), there exists a Boolean matrix $\boldsymbol{A}_{nxm}$ such that the set of test vectors that can be generated by the linear decompressor is spanned by $\boldsymbol{A}$. A test vector Z can be compressed by a particular linear decompressor if and only if there exists a solution to a system of linear equations, $\boldsymbol{A}X = Z$, where $\boldsymbol{A}$ is the characteristic matrix of the linear decompressor and X is a set of *free variables* stored on the tester (every bit stored on the tester can be thought of as a "free variable" that can be assigned any value, 0 or 1).

The characteristic matrix for a linear decompressor can be obtained by symbolic simulation where each free variable coming from the tester is represented by a symbol. An example is shown in Figure 2.39, where a sequential linear decompressor containing an LFSR is used. The initial state of the LFSR is represented by free variables $X_1 - X_4$, and the free variables $X_5 - X_{10}$ are shifted in from two channels as the scan chains are loaded. After symbolic simulation, the final values in the scan chains are represented by the equations for $Z_1 - Z_{12}$. The corresponding system of linear equations for this linear decompressor is shown in Figure 2.40.

The symbolic simulation goes as follows. Assume that the initial seed $X_1 - X_4$ has been already loaded into the flip-flops. In the first clock cycle, the top flip-flop is loaded with the XOR of X_2 and X_5, the second flip-flop is loaded with X_3, the third flip-flop is loaded with the XOR of X_1 and X_4, and the bottom flip-flop is loaded with the XOR of X_1 and X_6. Thus, we obtain $Z_1 = X_2 \oplus X_5$, $Z_2 = X_3$, $Z_3 = X_1 \oplus X_4$, and $Z_4 = X_1 \oplus X_6$. In the second clock cycle, the top flip-flop is loaded with the XOR

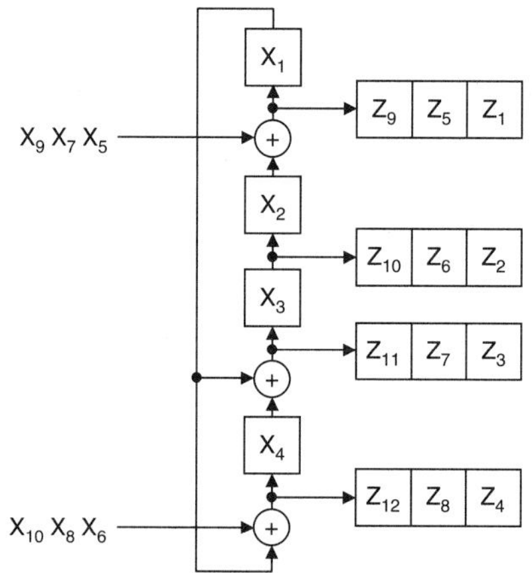

$Z_9 = X_1 \oplus X_4 \oplus X_9$	$Z_5 = X_3 \oplus X_7$	$Z_1 = X_2 \oplus X_5$
$Z_{10} = X_1 \oplus X_2 \oplus X_5 \oplus X_6$	$Z_6 = X_1 \oplus X_4$	$Z_2 = X_3$
$Z_{11} = X_2 \oplus X_3 \oplus X_5 \oplus X_7 \oplus X_8$	$Z_7 = X_1 \oplus X_2 \oplus X_5 \oplus X_6$	$Z_3 = X_1 \oplus X_4$
$Z_{12} = X_3 \oplus X_7 \oplus X_{10}$	$Z_8 = X_2 \oplus X_5 \oplus X_8$	$Z_4 = X_1 \oplus X_6$

■ **FIGURE 2.39**

Example of symbolic simulation for linear decompressor.

$$
\begin{pmatrix}
0&1&0&0&1&0&0&0&0&0\\
0&0&1&0&0&0&0&0&0&0\\
1&0&0&1&0&0&0&0&0&0\\
1&0&0&0&0&1&0&0&0&0\\
0&0&1&0&0&0&1&0&0&0\\
1&0&0&1&0&0&0&0&0&0\\
1&1&0&0&1&1&0&0&0&0\\
0&1&0&0&1&0&0&1&0&0\\
1&0&0&1&0&0&0&0&1&0\\
1&1&0&0&1&1&0&0&0&0\\
0&1&1&0&1&0&1&1&0&0\\
0&0&1&0&0&0&1&0&0&1
\end{pmatrix}
\begin{pmatrix}
X_1\\ X_2\\ X_3\\ X_4\\ X_5\\ X_6\\ X_7\\ X_8\\ X_9\\ X_{10}
\end{pmatrix}
=
\begin{pmatrix}
Z_1\\ Z_2\\ Z_3\\ Z_4\\ Z_5\\ Z_6\\ Z_7\\ Z_8\\ Z_9\\ Z_{10}\\ Z_{11}\\ Z_{12}
\end{pmatrix}
$$

■ **FIGURE 2.40**

System of linear equations for the decompressor in Figure 2.39.

of the contents of the second flip-flop (X_3) and X_7, the second flip-flop is loaded with the contents of the third flip-flop ($X_1 \oplus X_4$), the third flip-flop is loaded with the XOR of the contents of the first flip-flop ($X_2 \oplus X_5$) and the fourth flip-flop ($X_1 \oplus X_6$), and the bottom flip-flop is loaded with the XOR of the contents of the first flip-flop ($X_2 \oplus X_5$) and X_8. Thus, we obtain $Z_5 = X_3 \oplus X_7$, $Z_6 = X_1 \oplus X_4$, $Z_7 = X_1 \oplus X_2 \oplus X_5 \oplus X_6$, and $Z_8 = X_2 \oplus X_5 \oplus X_8$. In the third clock cycle, the top flip-flop is loaded with the XOR of the contents of the second flip-flop ($X_1 \oplus X_4$) and X_9, the second flip-flop is loaded with the contents of the third flip-flop ($X_1 \oplus X_2 \oplus X_5 \oplus X_6$); the third flip-flop is loaded with the XOR of the contents of the first flip-flop ($X_3 \oplus X_7$) and the fourth flip-flop ($X_2 \oplus X_5 \oplus X_8$), and the bottom flip-flop is loaded with the XOR of the contents of the first flip-flop ($X_3 \oplus X_7$) and X_{10}. Thus, we obtain $Z_9 = X_4 \oplus X_9$, $Z_{10} = X_1 \oplus X_6$, $Z_{11} = X_2 \oplus X_5 \oplus X_8$, and $Z_{12} = X_3 \oplus X_7 \oplus X_{10}$. At this point, the scan chains are fully loaded with a test cube, so the simulation is complete.

Combinational Linear Decompressors

The simplest linear decompressors use only combinational XOR networks. Each scan chain is fed by the XOR of some subset of the channels coming from the tester [Bayraktaroglu 2001, 2003] [Könemann 2003] [Wang 2004] [Mitra 2006]. The advantage compared with sequential linear decompressors is simpler hardware and control. The drawback is that, to encode a test cube, each **scan slice** must be encoded using only the free variables that are shifted from the tester in a single clock cycle (which is equal to the number of channels). The worst-case most highly specified scan slices tend to limit the amount of compression that can be achieved because the number of channels from the tester has to be sufficiently large to encode the most highly specified scan slices. Consequently, it is very difficult to obtain a high encoding efficiency (typically it will be less than 0.25); for the other less specified scan slices, a lot of the free variables are wasted because those scan slices could have been encoded with many fewer free variables.

One approach for improving the encoding efficiency of combinational linear decompressors, proposed in [Krishna 2003b], is to dynamically adjust the number of scan chains that are loaded in each clock cycle. So for a highly specified scan slice, four clock cycles could be used in which 25% of the scan chains are loaded in each cycle; for a lightly specified scan slice, only one clock cycle can be used in which 100% of the scan slices are loaded. This allows a better matching of the number of free variables with the number of specified bits to achieve a higher encoding efficiency. Note that it requires that the scan clock be divided into multiple domains.

Sequential Linear Decompressors

Sequential linear decompressors are based on linear finite-state machines such as LFSRs, cellular automata, or ring generators [Mrugalski 2004]. The advantage of a sequential linear decompressor is that it allows free variables from earlier clock cycles to be used when encoding a scan slice in the current clock cycle. This provides greater flexibility than combinational decompressors and helps avoid the problem of the worst-case most highly specified scan slices limiting the overall compression. The more flip-flops that are used in the sequential linear decompressor, the greater the flexibility that is provided. [Tobua 2006] classified the sequential linear decompressors into two classes:

1. **Static reseeding.** The earliest work in this area was based on static LFSR reseeding, a technique that computes a seed (an initial state) for each test cube [Touba 2006]. This seed, when loaded into an LFSR and run in autonomous mode, produces the test cube in the scan chains [Könemann 1991]. This technique achieves compression by storing only the seeds instead of the full test cubes.

   One drawback of using static reseeding for compressing test vectors on a tester is that the tester is idle while the LFSR is running in autonomous mode. One way around this is to use a shadow register for the LFSR to hold the data coming from the tester while the LFSR is running in autonomous mode [Volkerink 2003] [Wohl 2003b].

   Another drawback of static reseeding is that the LFSR must be at least as large as the number of specified bits in the test cube. One way around this is to only decompress a *scan window* (a limited number of scan slices) per seed [Krishna 2002] [Volkerink 2003] [Wohl 2005].

2. **Dynamic reseeding.** [Könemann 2001], [Krishna 2001], and [Rajski 2004] proposed dynamic reseeding approaches. Dynamic reseeding calls for the injection of free variables coming from the tester into the LFSR as it loads the scan chains [Touba 2006]. Figure 2.41 shows a generic example of a sequential linear decompressor that uses b channels from the tester to continuously inject free variables into the LFSR as it loads the scan chains through a combinational linear decompressor, which typically is a combinational XOR network. This network expands the LFSR outputs to fill n scan chains. The advantages of dynamic reseeding compared with static reseeding are that it allows continuous flow operation in which the tester is always shifting in data as fast as it can and is never idle, and it allows the use of a small LFSR.

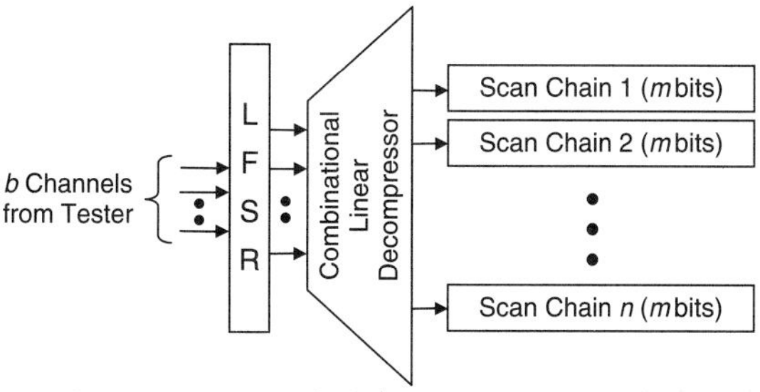

FIGURE 2.41

Typical sequential linear decompressor.

In [Rajski 2004], the authors described a methodology for scan vector compression based on a sequential linear decompressor. Instead of using an LFSR, this work uses a ring generator [Mrugalski 2004], which improves encoding flexibility and provides performance advantages. A fixed number of free variables are shifted in when decompressing each test cube. In this case, the control logic is simple because this methodology decompresses every test cube in exactly the same way. Constraining the ATPG generates test cubes that are encodable using the fixed number of free variables.

In [Könemann 2001], the authors described a methodology for scan vector compression in which the number of free variables used to encode each test cube varies. The method requires having an extra channel from the tester to gate the scan clock. For a heavily specified scan slice, this extra gating channel stops the scan shifting for one or more cycles, allowing the LFSR to accumulate a sufficient number of free variables from the tester to solve for the current scan slice before proceeding to the next one. This approach makes it easy to control the number of free variables that the decompressor uses to decompress each test cube. However, the additional gating channel uses some test data bandwidth.

2.4.1.2 Broadcast-Scan-Based Schemes

Another class of test stimulus compression schemes is based on broadcasting the same value to multiple scan chains. This was first proposed in [Lee 1998] and [Lee 1999]. Because of its simplicity and effectiveness, this method has been used as the basis of many test compression architectures, including some commercial ***design for testability*** (DFT) tools.

Broadcast Scan

To illustrate the basic concept of **broadcast scan**, first consider two independent circuits C_1 and C_2. Assume that these two circuits have their own test sets $T_1 = < t_{11}, t_{12}, \ldots, t_{1k} >$ and $T_2 = < t_{21}, t_{22}, \ldots, t_{2l} >$, respectively. In general, a test set may consist of random patterns and deterministic patterns. In the beginning

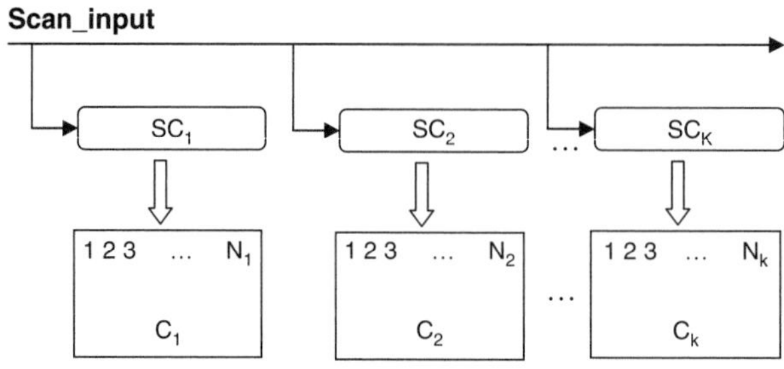

■ **FIGURE 2.42**

Broadcasting to scan chains driving independent circuits.

of the ATPG process, usually random patterns are initially used to detect the easy-to-detect faults. If the same random patterns are used when generating both T_1 and T_2 then we may have $t_{11} = t_{21}, t_{12} = t_{22}, \ldots$, up to some ith pattern. After most faults have been detected by the random patterns, deterministic patterns are generated for the remaining difficult-to-detect faults. Generally these patterns have many "don't care" bits. For example, when generating $t_{1(i+1)}$, many don't care bits may still exist when no more faults in C_1 can be detected. Using a test pattern with bits assigned so far for C_1, we can further assign specific values to the don't care bits in the pattern to detect faults in C_2. Thus, the final pattern would be effective in detecting faults in both C_1 and C_2.

The concept of pattern sharing can be extended to multiple circuits as illustrated in Figure 2.42. One major advantage of using *broadcast scan* for independent circuits is that all faults that are detectable in all original circuits will also be detectable with the broadcast structure. This is because if one test vector can detect a fault in a stand-alone circuit then it will still be possible to apply this vector to detect the fault in the broadcast structure. Thus, the broadcast scan method will not affect the fault coverage if all circuits are independent. Note that broadcast scan can also be applied to multiple scan chains of a single circuit if all subcircuits driven by the scan chains are independent.

Illinois Scan

If *broadcast scan* is used for multiple scan chains of a single circuit where the subcircuits driven by the scan chains are not independent, then the property of always being able to detect all faults is lost. The reason for this is that if two scan chains are sharing the same channel, then the ith scan cell in each of the two scan chains will always be loaded with identical values. If some fault requires two such scan cells to have opposite values in order to be detected, it will not be possible to detect this fault with broadcast scan.

To address the problem of some faults not being detected when using broadcast scan for multiple scan chains of a single circuit, the **Illinois scan architecture** was proposed in [Hamzaoglu 1999] and [Hsu 2001]. This scan architecture consists

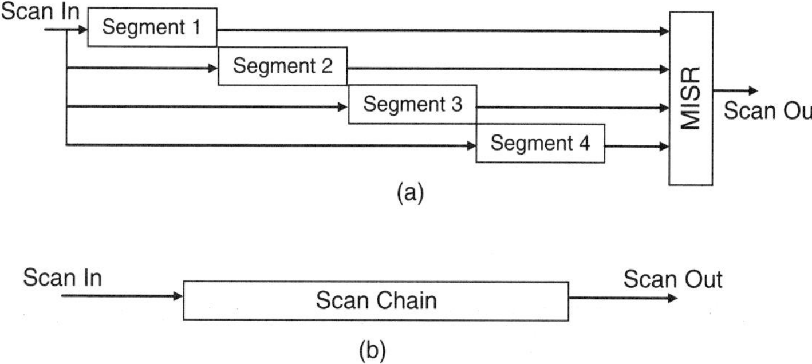

FIGURE 2.43

Two modes of Illinois scan architecture: (a) broadcast mode and (b) serial chain mode.

of two modes of operations, namely a *broadcast mode* and a *serial scan mode,* which are illustrated in Figure 2.43. The *broadcast mode* is first used to detect most faults in the circuit. During this mode, a scan chain is divided into multiple subchains called *segments* and the same vector can be shifted into all segments through a single shared scan-in input. The response data from all subchains are then compacted by a MISR or other space/time compactor. For the remaining faults that cannot be detected in broadcast mode, the *serial scan mode* is used where any possible test pattern can be applied. This ensures that complete fault coverage can be achieved. The extra logic required to implement the Illinois scan architecture consists of several multiplexers and some simple control logic to switch between the two modes. The area overhead of this logic is typically quite small compared to the overall chip area.

The main drawback of the Illinois scan architecture is that no test compression is achieved when it is run in *serial scan mode*. This can significantly degrade the overall **compression ratio** if many test patterns must be applied in serial scan mode. To reduce the number of patterns that need to be applied in serial scan mode, multiple-input broadcast scan or reconfigurable broadcast scan can be used. These techniques are described next.

Multiple-Input Broadcast Scan

Instead of using only one channel to drive all scan chains, a **multiple-input broadcast scan** could be used where there is more than one channel [Shah 2004]. Each channel can drive some subset of the scan chains. If two scan chains must be independently controlled to detect a fault, then they could be assigned to different channels. The more channels that are used and the shorter each scan chain is, the easier to detect more faults because fewer constraints are placed on the ATPG. Determining a configuration that requires the minimum number of channels to detect all detectable faults is thus highly desired with a multiple-input broadcast scan technique.

Reconfigurable Broadcast Scan

The *multiple-input broadcast scan* may require a large number of channels to achieve high fault coverage. To reduce the number of channels that are required, a **reconfigurable broadcast scan** method can be used. The idea is to provide the capability to reconfigure the set of scan chains that each channel drives. Two possible reconfiguration schemes have been proposed, namely **static reconfiguration** [Pandey 2002] [Samaranayake 2003] and **dynamic reconfiguration** [Li 2004] [Sitchinava 2004] [Wang 2004] [Han 2005c]. In *static reconfiguration,* the reconfiguration can only be done when a new pattern is to be applied. For this method, the target fault set can be divided into several subsets, and each subset can be tested by a single configuration. After testing one subset of faults, the configuration can be changed to test another subset of faults. In *dynamic reconfiguration,* the configuration can be changed while scanning in a pattern. This provides more reconfiguration flexibility and hence can in general lead to better results with fewer channels. This is especially important for hard cores when the test patterns provided by core vendor cannot be regenerated. The drawback of dynamic reconfiguration *versus* static reconfiguration is that more control information is needed for reconfiguring at the right time, whereas for static reconfiguration the control information is much less because the reconfiguration is done only a few times (only after all the test patterns using a particular configuration have been applied).

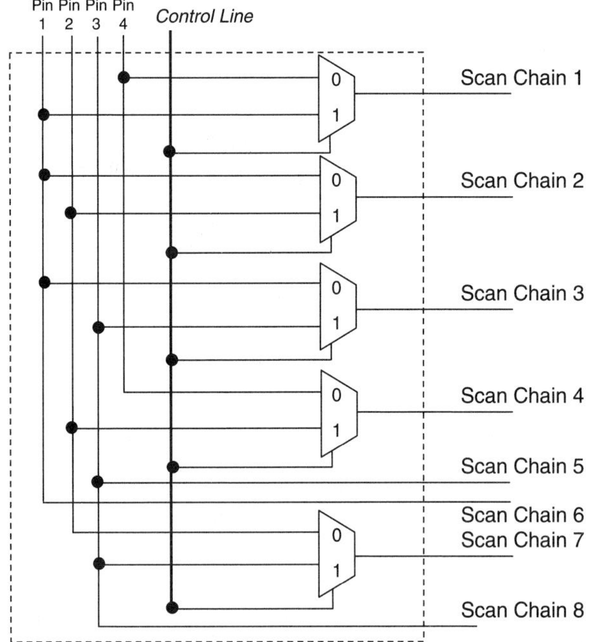

■ **FIGURE 2.44**

Example MUX network with control line(s) connected only to select pins of the multiplexers.

Figure 2.44 shows an example *multiplexer* (MUX) network, which can be used for dynamic configuration. When a value on the control line is selected, particular data at the four input pins are broadcasted to the eight scan chain inputs. For instance, when the control line is set to 0 (or 1), the scan chain 1 output will receive input data from pin 4 (or pin 1) directly.

Virtual Scan

Rather than using MUX networks for test stimulus compression, combinational logic networks can also be used as decompressors. The combinational logic network can consist of any combination of simple combinational gates, such as buffers, inverters, AND/OR gates, MUXs, and XOR gates. This scheme, referred to as **virtual scan**, is different from *reconfigurable broadcast scan* and *combinational linear decompression* where pure MUX and XOR networks are allowed, respectively. The combinational logic network can be specified as a set of constraints or just as an expanded circuit for ATPG. In either case, the test cubes that ATPG generates are the compressed stimuli for the decompressor itself. There is no need to solve linear equations, and *dynamic compaction* can be effectively utilized during the ATPG process.

The *virtual scan* scheme was proposed in [Wang 2002] and [Wang 2004]. In these papers, the decompressor was referred to as a **broadcaster**. The authors also proposed to add additional logic, when required, through *VirtualScan inputs* to reduce or remove the constraints imposed on the decompressor (broadcaster), thereby yielding little or no fault coverage loss caused by test stimulus compression.

In a broad sense, *virtual scan* is a generalized class of broadcast scan, Illinois scan, multiple-input broadcast scan, reconfigurable broadcast scan, and combinational linear decompression. The advantage of using virtual scan is that it allows the ATPG to directly search for a test cube that can be applied by the decompressor and allows very effective dynamic compaction. Thus, virtual scan may produce shorter test sets than any test stimulus compression scheme based on solving linear equations; however, because this scheme may impose XOR or MUX constraints directly on the original circuit, it may take more time than schemes based on solving linear equations to generate test cubes or compressed stimuli. Two examples of virtual scan decompression circuits are shown in Figure 2.45.

2.4.1.3 Comparison

In this section, we compare the **encoding flexibility** for different types of combinational decompression techniques: Illinois scan using a pure buffer network, reconfigurable broadcast scan using MUX networks, and linear combinational decompression using only one-level 2-input XOR gates or one-level, three-input XOR gates [Dutta 2006].

Consider the bits coming from the tester each clock cycle as a **tester slice**. Then the tester slice gets expanded every clock cycle to fill the scan slice, which equals the number of scan chains. The authors in [Dutta 2006] did some experiments to measure the encoding flexibility of different ways of expanding the tester slices into scan slices. Figure 2.46 shows the percentage of all possible scan slices with

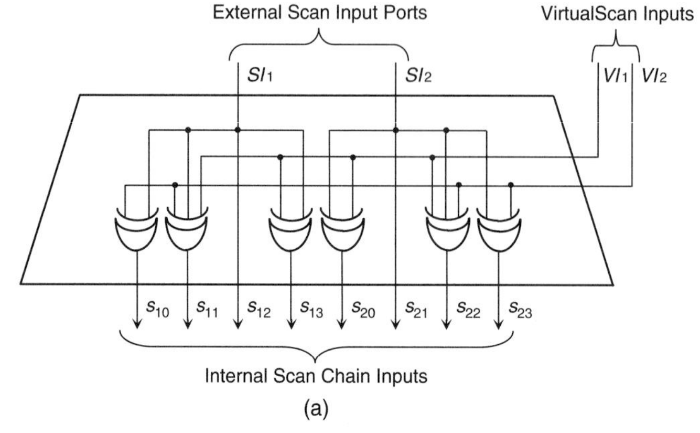

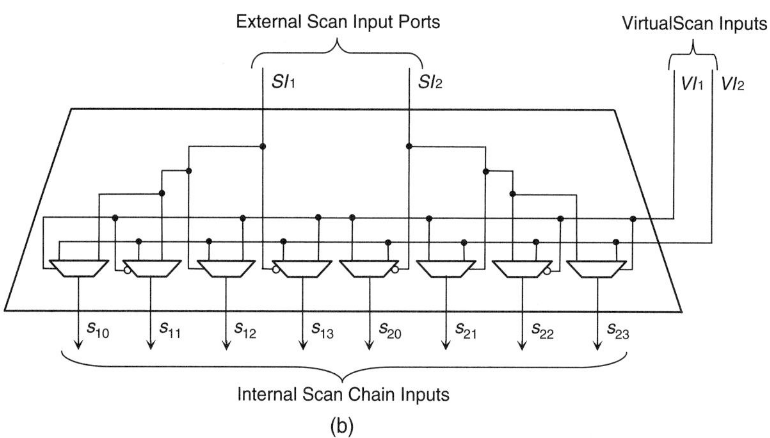

FIGURE 2.45

Example virtual scan decompression circuits: (a) broadcaster using an example XOR network with additional VirtualScan inputs to reduce coverage loss and (b) broadcaster using an example MUX network with additional VirtualScan inputs that can be also connected to data pins of the multiplexers.

different number of specified bits that can be encoded in each case for expanding a 16-bit tester slice to a 160-bit scan slice with an expansion ratio (or split ratio) of 10. The x-axis is the number of specified bits in the scan slice, and the y-axis is the percentage of all possible combinations of that number of specified bits that can be encoded. As the graph shows, all the decompression networks can always encode one specified bit. However, as the number of specified bits increases, the probability of being able to encode the scan slice drops. Because Illinois scan has the least encoding flexibility, it has the lowest probability of being able to encode a scan slice. The results for using MUXs are shown for two cases. One is where the control and data lines are separated (*i.e.*, one of the tester channels is dedicated to driving the select line and the other 15 tester channels are used to drive the data lines). The other is where combinations of all 16 tester channels are used to drive

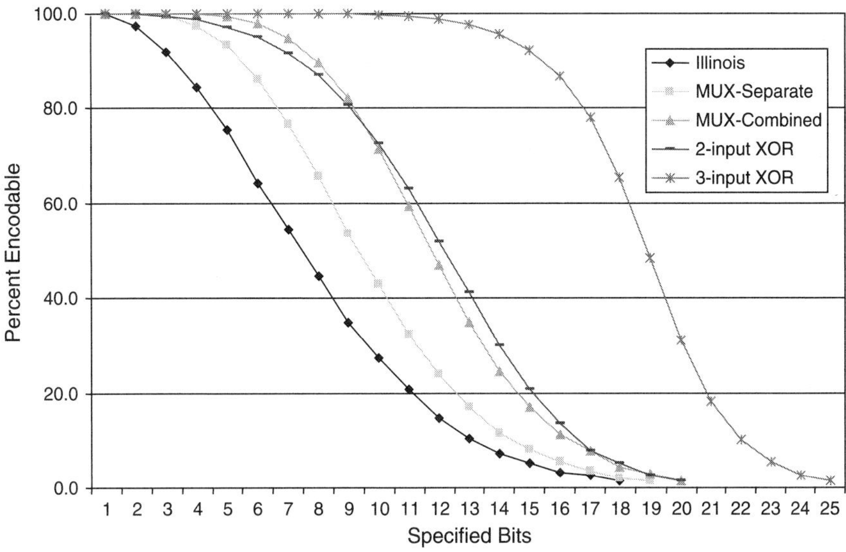

■ **FIGURE 2.46**

Encoding flexibility among combinational decompression schemes.

either the select or data lines of the MUXs. The results indicate that greater encoding flexibility can be obtained by not having a separate control line. Another interesting result is that using two-input XOR gates is not as good as using MUXs for low numbers of specified bits, but it becomes better than MUXs when the number of specified bits is equal to 10 or more. Using three-input XORs provides considerably better encoding flexibility, although it comes with the tradeoff of adding more complexity to the ATPG than the others.

The experiments indicate that using the combinational XOR network for test stimulus decompression provides the highest encoding flexibility and hence can provide better compression than using other broadcast-scan-based schemes. The more inputs used per XOR gate, the better the encoding flexibility. Better encoding flexibility allows a more aggressive expansion ratio and allows the ATPG to perform more dynamic compaction resulting in better compression.

2.4.2 Circuits for Test Response Compaction

Test response compaction is performed at the outputs of the scan chains. The purpose is to reduce the amount of test response that needs to be transferred back to the tester. Whereas test stimulus compression must be lossless, test response compaction can be lossy. A large number of different test response compaction schemes have been presented and described to various extents in the literature [Wang 2006a]. The effectiveness of each compaction scheme depends on its ability to avoid *aliasing* and tolerate *unknown test response bits* or *X*'s. These schemes can be grouped into three categories: (1) space compaction, (2) time compaction, and (3) mixed space and time compaction.

Typically, a **space compactor** using the space compaction scheme comprises XOR gates, whereas a **time compactor** using the time compaction scheme is a MISR. A **mixed space and time compactor** typically feeds a space compactor to a time compactor. The difference between space compaction and time compaction is that a space compactor compacts an m-bit-wide output pattern to a p-bit-wide output pattern (where $p < m$), whereas a time compactor compacts n output patterns to q output patterns (where $q < n$). This section presents some widely used compaction schemes in industry. Promising techniques for tolerating X's are also included.

2.4.2.1 Space Compaction

A space compactor is a combinational circuit for compacting m outputs of the circuit under test to n test outputs, where $n < m$. Space compaction can be regarded as the inverse procedure of *linear expansion* (which was described in Section 2.4.1.2). It can be expressed as a function of the input vector (*i.e.*, the data being scanned out) and the output vector (the data being monitored):

$$Y = \Phi(X)$$

where X is an m-bit input vector and Y is an n-bit output vector, $n < m$. Because each output sequence can contain unknown values (X's), the space compaction scheme in use shall have the capability to mask off or tolerate unknowns in order to prevent faults from going undetected.

X-Compact

X-compact [Mitra 2004] is an **X-tolerant response compaction** technique and has being used in several designs. The combinational compactor circuit designed using the X-compact technique is called an **X-compactor**. Figure 2.47 shows an

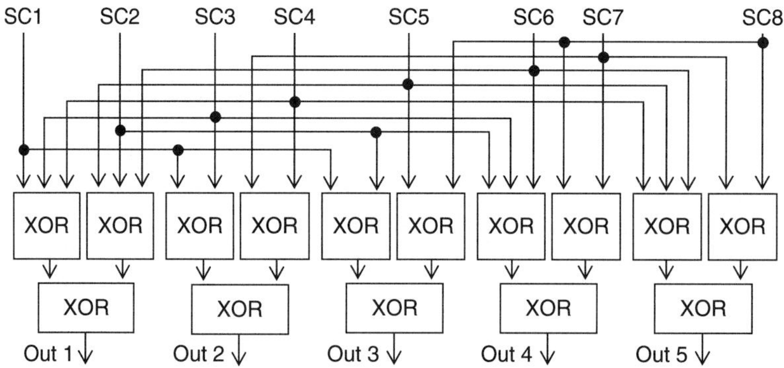

■ **FIGURE 2.47**

An X-compactor with eight inputs and five outputs.

example of the X-compactor with eight inputs and five outputs. It is composed of 4 three-input XOR gates and 11 two-input XOR gates.

The X-compactor can be represented as a binary matrix (matrix with only 0's and 1's) with n rows and k columns; this matrix is called the *X-compact matrix*. Each row of the X-compact matrix corresponds to a scan chain, and each column corresponds to an X-compactor output. The entry in row i and column j of the matrix is 1 if and only if the jth X-compactor output depends on the ith scan chain output; otherwise, the matrix entry is 0. The corresponding X-compact matrix M of the X-compactor shown in Figure 2.47 is as follows:

$$M = \begin{bmatrix} 1&1&1&0&0 \\ 1&0&1&1&0 \\ 1&1&0&1&0 \\ 1&1&0&0&1 \\ 1&0&1&0&1 \\ 1&0&0&1&1 \\ 0&1&0&1&1 \\ 0&0&1&1&1 \end{bmatrix}$$

For a conventional sequential compactor, such as a MISR, there are two sources of *aliasing*: error masking and error cancellation. **Error masking** occurs when one or more errors captured in the compactor during a single cycle propagate through the feedback path and cancel out with errors in the later cycles. **Error cancellation** occurs when an error bit captured in a shift register is shifted and eventually cancelled by another error bit. The error cancellation is a type of aliasing specific to multiple-input sequential compactor. Because the X-compactor is a combinational compactor, it only results in error masking. To handle aliasing, the following theorems provide a basis for systematically designing X-compactors:

Theorem 2.1

If only a single scan chain produces an error at any scan-out cycle, the X-compactor is guaranteed to produce errors at the X-compactor outputs at that scan-out cycle if and only if no row of the X-compact matrix contains all 0's.

Theorem 2.2

Errors from any one, two, or odd number of scan chains at the same scan-out cycle are guaranteed to produce errors at the X-compactor outputs at that scan-out cycle if every row of the X-compact matrix is nonzero, distinct, and contains an odd number of 1's.

If all rows of the X-compact matrix are distinct and contain an odd number of 1's, then a bitwise XOR of any two rows is nonzero. Also, the bitwise XOR of any odd number of rows is also nonzero. Hence, errors from any one or any two or any odd number of scan chains at the same scan-out cycle are guaranteed to produce errors at the compactor outputs at that scan-out cycle. Because all rows of the X-compact matrix of Figure 2.47 are distinct and odd, by Theorem 2.2, simultaneous errors from any two or odd scan chains at the same scan-out cycle are guaranteed to be detected.

The X-compact technique is nonintrusive and independent of the test patterns used to test the circuit. Insertion of X-compactor does not require any major change to the ATPG flow; however, the X-compactor cannot guarantee that errors other than those described in Theorem 2.1 and Theorem 2.2 are detectable.

X-Blocking

Instead of tolerating *X*'s on the response compactor, *X*'s can also be blocked before reaching the response compactor. During design, these potential X-generators (X-sources) can be identified using a scan design rule checker. When an X-generator is likely to reach the response compactor, it must be fixed [Naruse 2003] [Patel 2003]. The process is often referred to as **X-blocking** or **X-bounding**.

In X-blocking, the output of an X-source can be blocked anywhere along its propagation paths before *X*'s reach the compactor. In case the X-source has been blocked at a nearby location during test and will not reach the compactor, there is no need to fix further; however, care must be taken to ensure that no observation points are added between the X-source and the location at which it is blocked.

X-blocking can ensure that no *X*'s will be observed; however, it does not provide a means for observing faults, which can only propagate to an observable point through the now-blocked X-source. This can result in fault coverage loss. If the number of such faults for a given bounded X-generator justifies the cost, one or more observation points can be added before the X-source to provide an observable point to which those faults can propagate. These X-blocking or X-bounding methods have been discussed extensively in [Wang 2006a].

X-Masking

Although it may not result in fault coverage loss, the X-blocking technique does add area overhead and may impact delay because of the inserted logic. It is not surprising to find that, in complex designs, more than 25% of scan cycles could contain one or more *X*'s in the test response. It is difficult to eliminate these residual *X*'s by DFT; thus, an encoder with high X-tolerance is very attractive. Instead of blocking the *X*'s where they are generated, the *X*'s can also be masked off right before the response compactor [Wohl 2004] [Han 2005a] [Volkerink 2005] [Rajski 2005, 2006]. An example X-masking circuit is shown in Figure 2.48. The mask controller applies a logic value 1 at the appropriate time to mask off any scan output that contains an *X*.

Mask data are needed to indicate when the masking should take place. These mask data can be stored in compressed format and can be decompressed using on-chip hardware. Possible compression techniques are *weighted pseudo-random LFSR reseeding* or *run-length encoding* [Volkerink 2005].

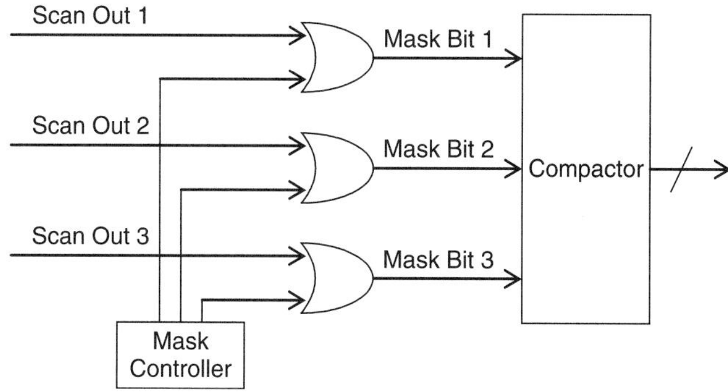

■ **FIGURE 2.48**

An example X-masking circuit.

X-Impact

Although X-compact, X-blocking, and X-masking each can achieve a significant reduction in fault coverage loss caused by *X*'s present at the inputs of a space compactor, the **X-impact** technique described in [Wang 2004] is helpful in that it simply uses ATPG to algorithmically handle the impact of residual *X*'s on the space compactor without adding any extra circuitry.

Example 2.1

An example of algorithmically handling X-impact is shown in Figure 2.49. Here, *SC*1 to *SC*4 are scan cells connected to a space compactor composed of XOR gates *G*7 and *G*8. Lines *a, b,..., h* are internal signals, and line *f* is assumed to be connected to an X-source (memory, non-scan storage element, etc.). Now consider the detection of the *stuck-at-0* (SA0) fault *f*1. Logic value 1 should be assigned to both lines *d* and *e* in order to activate *f*1. The fault effect will be captured by scan cell *SC*3. If

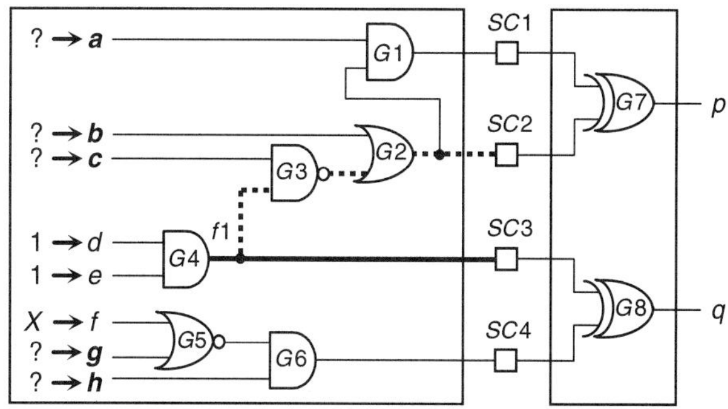

■ **FIGURE 2.49**

Handling of X-impact.

the X on f propagates to $SC4$, then the compactor output q will become X and $f1$ cannot be detected. To avoid this outcome, ATPG can try to assign either 1 to line g or 0 to line h in order to block the X from reaching $SC4$. If it is impossible to achieve this assignment, ATPG can then try to assign 1 to line c, 0 to line b, and 0 to line a in order to propagate the fault effect to $SC2$. As a result, fault $f1$ can be detected. Thus, X-impact is avoided by algorithmic assignment without adding any extra circuitry.

Example 2.2

It is also possible to use the X-impact approach to reduce aliasing. An example of algorithmically handling aliasing is shown in Figure 2.50. Here, $SC1$ to $SC4$ are scan cells connected to a compactor composed of XOR gates $G7$ and $G8$. Lines a, $b, \ldots, h$ are internal signals. Now consider the detection of the stuck-at-1 fault $f2$. Logic value 1 should be assigned to lines c, d, and e in order to activate $f2$, and logic value 0 should be assigned to line b in order to propagate the fault effect to $SC2$. If line a is set to 1, then the fault effect will also propagate to $SC1$. In this case, aliasing will cause the compactor output p to have a fault-free value, resulting in an undetected $f2$. To avoid this outcome, ATPG can try to assign 0 to line a in order to block the fault effect from reaching $SC1$. As a result, fault $f2$ can be detected. Thus, aliasing can be avoided by algorithmic assignment without any extra circuitry.

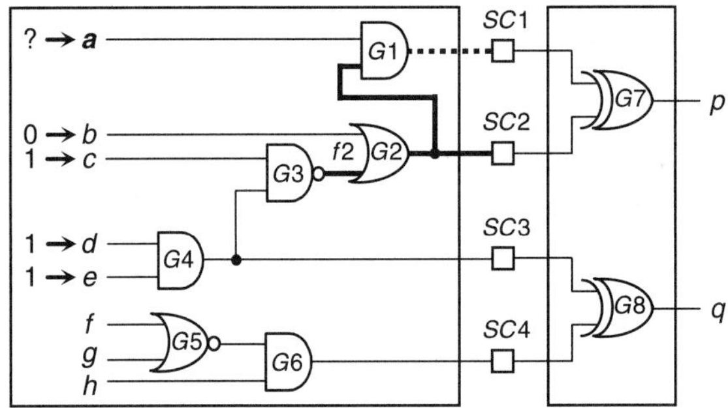

■ FIGURE 2.50

Handling of aliasing.

2.4.2.2 Time Compaction

A time compactor uses sequential logic (whereas a space compactor uses combinational logic) to compact test responses. Because sequential logic is used, one must make sure that no unknown (X) values from the circuit under test will reach the compactor. If that happens, X-bounding or X-masking must be employed.

The most widely adopted response compactor using time compaction is the ***multiple-input signature register*** (MISR). The MISR uses m extra XOR gates for compacting each m-bit-wide output sequence into the LFSR simultaneously. The final contents stored in the MISR after compaction is called the (final) *signature* of the MISR. For more information on signature analysis and the MISR design, refer to [Wang 2006a].

2.4.2.3 Mixed Time and Space Compaction

The previous two sections introduced different kinds of compactors for space compaction and time compaction independently. This section introduces mixed time and space compactors [Saluja 1983]. A mixed time and space compactor combines the advantages of a time compactor and a space compactor. Many mixed time and space compactors have been proposed in the literature, including OPMISR [Barnhart 2002], convolutional compactor [Rajski 2005], and q-compactor [Han 2003] [Han 2005a,b].

Because q-compactor is simple, this section uses it to introduce the conceptual architecture of a mixed time and space compactor. Figure 2.51 shows an example of a q-compactor assuming the inputs are coming from scan chain outputs. The spatial part of the q-compactor consists of single-output XOR networks (called *spread networks*) connected to the flip-flops by means of additional two-input XOR gates interspersed between successive storage elements. As the figure shows, every error in a scan cell can reach storage elements and then outputs in several possible ways. The spread network that determines this property is defined in terms of *spread polynomials* indicating how particular scan chains are connected to the register flip-flops.

Different from a conventional MISR, the q-compactor presented in Figure 2.51 does not have a feedback path; consequently, any error or X injected into the compactor is shifted out after at most five cycles. The shifted-out data will be compared with the expected data and then the error will be detected.

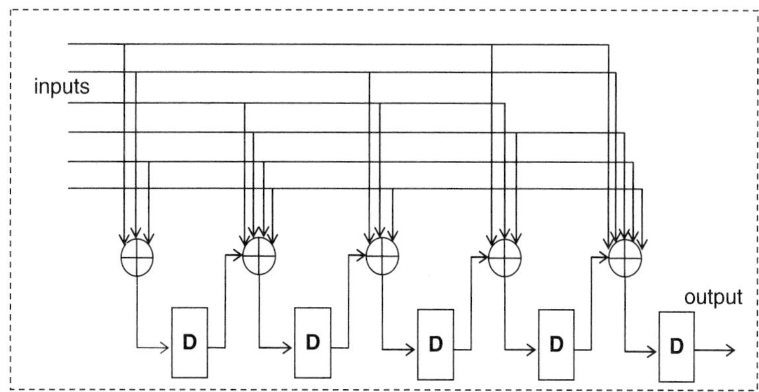

■ **FIGURE 2.51**

An example q-compactor with single output.

Example 2.3

Figure 2.51 shows an example of a q-compactor with six inputs, one output, and five storage elements—five per output. For the sake of simplicity, the injector network is shown in a linear form rather than as a balanced tree.

2.4.3 Low-Power Test Compression Architectures

The bandwidth-matching low-power scan design given in Figure 2.9 is also applicable for test compression. The general UltraScan architecture shown in Figure 2.52

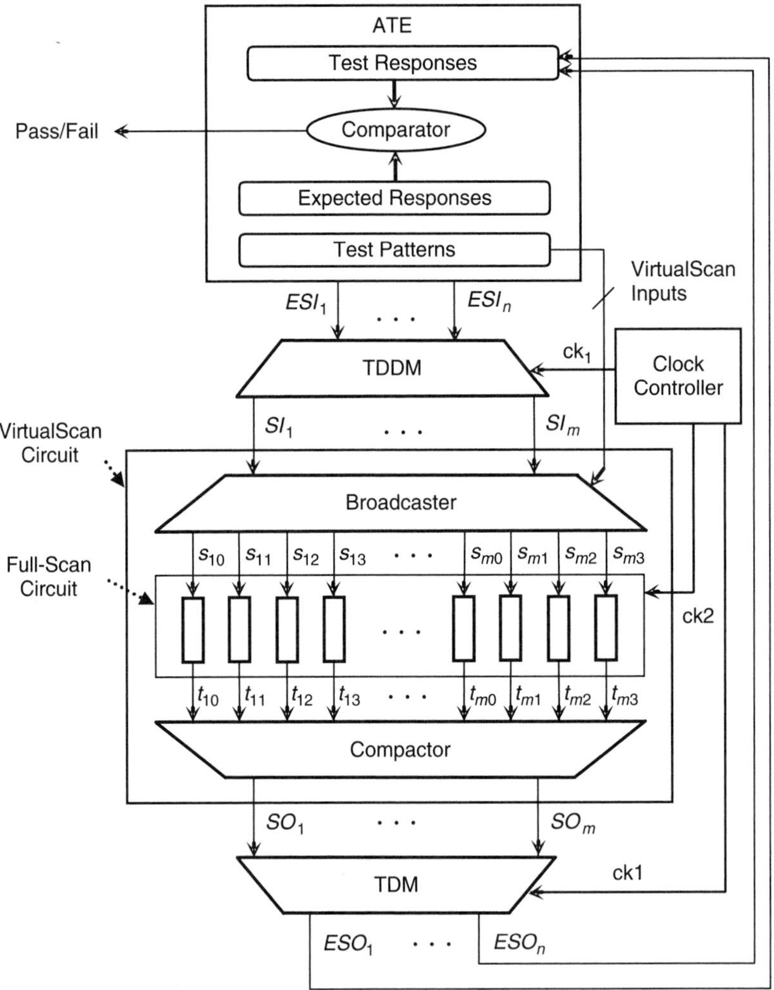

■ **FIGURE 2.52**

UltraScan architecture.

uses a *time-division demultiplexer* and a *time-division multiplexer* (TDDM/TDM) pair, as well as a clock controller to create the UltraScan circuit [Wang 2005b]. The TDDM is typically a serial-input/parallel-output shift register, whereas the TDM is a parallel-input/serial-output shift register. The clock controller is used to derive the scan shift clock, ck2, by dividing the high-speed clock, ck1, by the demultiplexing ratio. The broadcaster can be a general decompressor using any linear-decompression-based scheme or broadcast-scan-based scheme.

In this UltraScan circuit, assume that eight high-speed input pads are used as external scan input ports, which are connected to the inputs of the TDDM circuit. The TDDM circuit uses a high-speed clock, provided externally or generated internally using a *phase-locked loop* (PLL), to demultiplex the high-speed compressed stimuli into compressed stimuli operating at a slower data rate for scan shift. Similarly, the TDM circuit will use the same high-speed clock to capture and shift out the test responses to high-speed output pads for comparison. Assume the demultiplexing ratio, the ratio between the high-speed data rate and the low-speed data rate, is 10. This means that designers can split eight original scan chains into 1280 internal scan chains for possible reduction in test data volume by 16X and test application time by 160X. In this example, for a desired scan shift clock frequency of 10 MHz, the external I/O pads are operated at 100 MHz. The TDDM/TDM circuit will not compress test data volume but only reduce test application time or test pin count by additional 10X. For low-power applications, however, it is possible to use UltraScan as a **low-power test compression architecture** to reduce shift power dissipation. In these cases, one can reduce shift power dissipation by 10X, by slowing down the shift clock frequency to 1 MHz and operating the high-speed I/O pads at 10 MHz. Although shift power dissipation has been reduced by 10X, reduction in test application time can now reach a maximum of 16X only.

2.4.4 Industry Practices

Several test compression products and solutions have been introduced by some of the major DFT vendors in the CAD industry. These products differ significantly with regard to technology, design overhead, design rules, and the ease of use and implementation. A few second-generation products have also been introduced by a few of the vendors. This section summarizes a few of the products introduced by companies such as Cadence [Cadence 2007], Mentor Graphics [Mentor 2007], SynTest [SynTest 2007], Synopsys [Synopsys 2007], and LogicVision [LogicVision 2007].

Current industry solutions can be grouped under two main categories for stimulus decompression. The first category uses *linear-decompression-based schemes*, whereas the second category employs *broadcast-scan-based schemes*. The main difference between the two categories is the manner in which the ATPG engine is used. The first category includes products such as ETCompression [LogicVision 2007] from LogicVision, TestKompress [Rajski 2004] from Mentor Graphics, XOR Compression [Cadence 2007] from Cadence, and SOCBIST [Wohl 2003b] from Synopsys. The second category includes products such as OPMISR+ [Cadence 2007] from Cadence, VirtualScan [Wang 2004] and UltraScan [Wang 2005b] from SynTest, and DFT MAX [Sitchinava 2004] from Synopsys.

For designs using *linear-decompression-based schemes*, test compression is achieved in two distinct steps. During the first step, conventional ATPG is used to generate sparse ATPG patterns (called test cubes), in which *dynamic compaction* is performed in a nonaggressive manner, while leaving unspecified bit locations in each test cube as "*X*." This is accomplished by not aggressively performing the *random fill* operation on the test cubes, which is used to increase coverage of individual patterns and hence reduce the total pattern count. During the second step, a system of linear equations describing the hardware mapping from the external scan input ports to the internal scan chain inputs are solved in order to map each test cube into a compressed stimulus that can be applied externally. If a mapping is not found, a new attempt at generating a new test cube is required.

For designs using *broadcast-scan-based schemes*, only a single step is required to perform test compression. This is achieved by embedding the constraints introduced by the decompressor as part of the ATPG tool, such that the tool operates with much more restricted constraints. Hence, whereas in conventional ATPG, each individual scan cell can be set to 0 or 1 independently, for *broadcast-scan-based-schemes* the values to which related scan cells can be set are constrained. Thus, a limitation of this solution is that in some cases, the constraints among scan cells can preclude some faults from being tested. These faults are typically tested as part of a later top-up ATPG process if required, similar to using *linear-decompression-based schemes*.

On the response compaction side, industry solutions have utilized either space compactors such as XOR networks, or time compactors such as MISRs, to compact the test responses. Currently, space compactors have a higher acceptance rate in the industry because they do not involve the process of guaranteeing that no unknown (*X*) values are generated in the circuit under test.

A summary of the different compression architectures used in the commercial products is shown in Table 2.2. Six products from five DFT companies are included. Since June 2006, Cadence added XOR Compression as an alternative to the OPMISR+ product described in [Wang 2006a].

TABLE 2.2 ■ Summary of Industry Practices for Test Compression

Industry Practices	Stimulus Decompressor	Response Compactor
XOR Compression or OPMISR+	Combinational XOR network or Fanout Network	XOR network with or without MISR
TestKompress	Ring generator	XOR network
VirtualScan	Combinational logic network	XOR network
DFT MAX	Combinational MUX network	XOR network
ETCompression	(Reseeding) PRPG	MISR
UltraScan	TDDM	TDM

TABLE 2.3 ■ Summary of Industry Practices for At-Speed Delay Fault Testing

Industry Practices	Skewed-Load	Double-Capture
XOR Compression or OPMISR+	√	√
TestKompress	√	√
VirtualScan	√	√
DFT MAX	√	√
ETCompression	√	Through service
UltraScan	√	√

It is evident that the solutions offered by the current EDA DFT vendors are quite diverse with regard to stimulus decompression and response compaction. For stimulus decompression, OPMISR+, VirtualScan, and DFT MAX are broadcast-scan-based, whereas TestKompress and ETCompression are linear-decompression-based. For response compaction, OPMISR+ and ETCompression can include MISRs, whereas other solutions purely adopt (X-tolerant) XOR networks. The UltraScan TDDM/TDM architecture can be implemented on top of any test compression solution to further reduce test application time and test pin count. What is common is that all six products provide their own diagnostic solutions.

Generally speaking, any modern ATPG compression program supports at-speed clocking schemes used in its corresponding at-speed scan architecture. For at-speed delay fault testing, ETCompression currently uses a **skewed-load based at-speed test compression architecture** for ATPG. The product can also support the double-capture clocking scheme through service. All other ATPG compression products—including OPMISR+, TestKompress, VirtualScan, DFT MAX, and UltraScan—support the **hybrid at-speed test compression architecture** by using both skewed-load (a.k.a. launch-on-shift) and double-capture (a.k.a. launch-on-capture). In addition, almost every product supports inter-clock-domain delay fault testing for synchronous clock domains. A few on-chip clock controllers for detecting these inter-clock-domain delay faults at-speed have been proposed in [Beck 2005], [Furukawa 2006], [Nadeau-Dostie 2006], and [Nadeau-Dostie 2007].

The clocking schemes used in these commercial products are summarized in Table 2.3. It should be noted that compression schemes may be limited in effectiveness if there are a large number of unknown response values, which can be exacerbated during at-speed testing when many paths do not make the timing being used.

2.5 RANDOM-ACCESS SCAN DESIGN

Our discussions in previous sections mainly focus on *serial scan design* that requires shifting data into and out of a scan cell through adjacent scan cells. Although serial scan design has been one of the most successful DFT techniques in use that has minimum routing overhead, one inherent drawback with this architecture is its test power dissipation. Test power consists of shift power and capture power. Because of the serial shift nature, excessive heat can cumulate and damage the circuit under test. Excessive dynamic power during capture can also cause IR drop and induce yield loss. In addition, any fault present in a scan chain makes fault diagnosis difficult, because the fault can mask out all scan cells in the same scan chain. When scan chain faults are combined with combinational logic faults, the fault diagnosis process can even become more complex.

All of these problems result from the underlying architecture used for serial scan design. ***Random-access scan*** (RAS) [Ando 1980] offers a promising solution. Rather than using various hardware and software approaches to reduce test power dissipation in serial scan design [Girard 2002], random-access scan attempts to alleviate these problems by making each scan cell randomly and uniquely addressable, similar to storage cells in a ***random-access memory*** (RAM). As its name implies, because each scan cell is randomly and uniquely addressable, random-access scan design can reduce shift power dissipation with an increase in routing overhead. In addition, because there are no scan chains, scan chain diagnosis will be no longer an issue. One can simply apply combinational logic diagnosis techniques for locating faults within the combinational logic [Wang 2006a]. What has to be explored next is whether random-access scan could further reduce capture power dissipation.

In this section, we first introduce the basic concepts on random-access scan design. Next, RAS architectures along with their associated scan cell designs to reduce routing overhead are presented. As these RAS architectures do not specifically target test cost reduction, we then examine test compression RAS architectures to further reduce test application time and test data volume. At-speed RAS architectures are finally discussed.

2.5.1 Random-Access Scan Architectures

Traditional RAS design [Ando 1980] is illustrated in Figure 2.53. All scan cells are organized into a two-dimensional array, where they can be accessed individually for observing (reading) or updating (writing) in any order. This full-random access capability is achieved by decoding a full address with a row (X) decoder and a column (Y) decoder. A $\lceil \log_2 n \rceil$-bit address shift register, where n is the total number of scan cells, is used to specify which scan cell to access. A scan-in port *SI* is connected to all scan cells and a scan-out port *SO* is used to observe the state of each scan cell.

Therefore, the RAS design can access any selected scan cell without changing the states of other scan cells. This significantly reduces shift power dissipation, because there is no need to shift data into and out of the selected scan cell through scan chains; data in each scan cell can be directly observed and updated through the *SO* and *SI* ports, respectively. As opposed to serial scan design, however, there is no guarantee that the RAS design can further reduce the test application time or test data volume if a large number of scan cells has to be updated for each test vector or the addresses of scan cells to be accessed consecutively have little overlap.

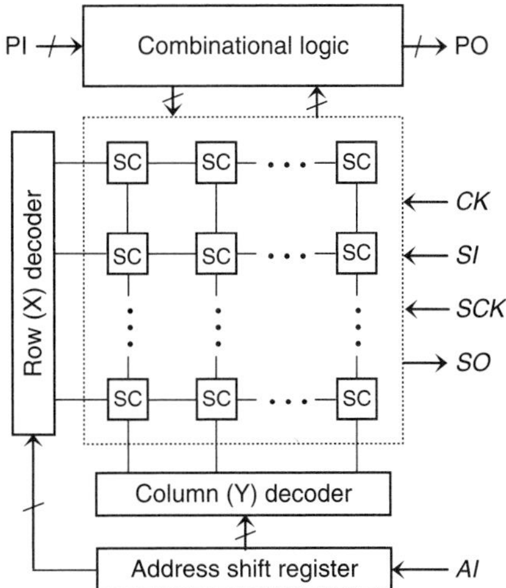

■ **FIGURE 2.53**

Traditional random-access scan architecture.

Although RAS design can easily reduce shift power dissipation and simplify fault diagnosis, a major disadvantage of using this architecture is its high area and routing overhead, which has unfortunately hindered the approach from being considered for practical applications since it was invented in the 1980s. Only until recently, since silicon gates are no longer expensive for nanometer VLSI designs, has the RAS design started to regain its momentum.

A traditional RAS scan cell design proposed in [Wagner 1984] is shown in Figure 2.54a. An additional multiplexer is placed at the *SI* port of the muxed-D scan cell to either update data from the external *SI* port or keep its current state. This is controlled by the address select signal *AS*. Each scan cell output Q is directly fed to a ***multiple-input signature register*** (MISR) for output response compaction. As

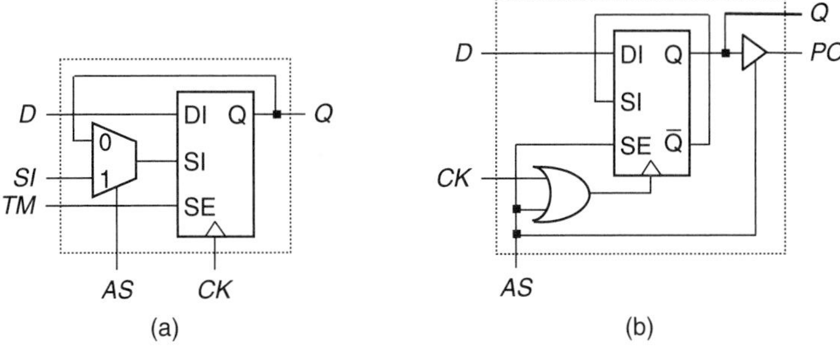

■ **FIGURE 2.54**

Traditional random-access scan cell designs: (a) traditional RAS scan cell design and (b) toggle scan cell design.

it is required to broadcast the external *SI* port to all scan cells and connect each scan cell output to the MISR, routing becomes a serious problem. A toggle scan cell design is proposed in [Mudlapur 2005] and illustrated in Figure 2.54b that eliminates the external *SI* port and connects selected scan cell outputs to a bus lead to an external *SO* port. Because this scheme eliminates the global *SI* port, a clear mechanism is required to reset all scan cells before testing. This introduces additional area and routing overhead.

2.5.1.1 Progressive Random-Access Scan Design

A ***progressive random-access scan*** (PRAS) design [Baik 2005a] was proposed in an attempt to alleviate the problems associated with the traditional serial scan design. The PRAS scan cell, as shown in Figure 2.55a, has a structure similar to that of a ***static random access memory*** (SRAM) cell or a grid-addressable latch [Susheel 2002], which has significantly smaller area and routing overhead than the traditional scan cell design [Ando 1980]. In normal mode, all horizontal row enable *RE* signals are set to 0, forcing each scan cell to act as a normal *D* flip-flop. In test mode, to capture the test response from *D*, the *RE* signal is set to 0 and a pulse is applied on clock Φ, which causes the value on *D* to be loaded into the scan cell. To read out the stored value of the scan cell, clock Φ is held at 1, the *RE* signal for the selected scan cell is set to 1, and the content of the scan cell is read out through the bidirectional scan data signals *SD* and $\overline{SD}$. To write or update a scan value into the scan cell, clock Φ is held at 1, the *RE* signal for the selected scan cell is set to 1, and the scan value and its complement are applied on SD and $\overline{SD}$, respectively.

The PRAS architecture is shown in Figure 2.55b, where rows are enabled in a fixed order, one at a time, by rotating a 1 in the row enable shift register. That is, it is only necessary to supply a column address to specify which scan cell in an enabled row to access. The length of the column address, which is $\lceil \log_2 m \rceil$ for a circuit with m columns, is considerably shorter than a full (row and column) address; therefore, the column address is provided in parallel in one clock cycle instead of providing a full address in multiple clock cycles. This reduces test application time. To minimize the need to shift out test responses, the scan cell outputs are compressed with a *multiple-input signature register* (MISR).

The test procedure of the PRAS design is shown in Figure 2.55c. For each test vector, the test stimulus application and test response compression are conducted in an interleaving manner when the test mode signal TM is enabled. That is, all scan cells in a row are first read into the MISR for compression simultaneously, and then each scan cell in the row is checked and updated if necessary. Repeating this operation for all rows compresses the test response to the previous test vector into the MISR and sets the next test vector to all scan cells. Next, TM is disabled and the normal clock is applied to conduct test response acquisition. The figure shows that the smaller the number of scan cells to be updated for each row, the shorter the test application time. This can be achieved by reducing the Hamming distance between the next test vector and the test response to the previous test vector. Possible solutions include test vector reordering and test vector modification [Baik 2004; 2005a,b; 2006] [Le 2007].

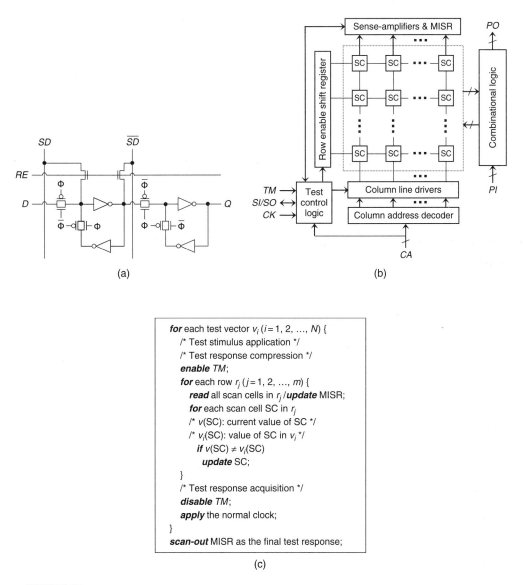

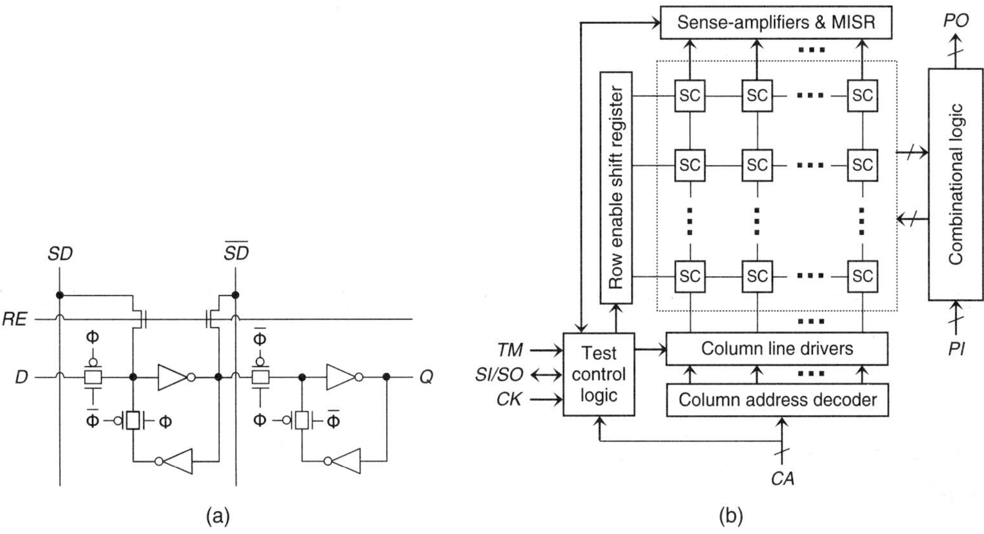


```
for each test vector vi (i = 1, 2, ..., N) {
  /* Test stimulus application */
  /* Test response compression */
  enable TM;
  for each row rj (j = 1, 2, ..., m) {
    read all scan cells in rj /update MISR;
    for each scan cell SC in rj
    /* v(SC): current value of SC */
    /* vi(SC): value of SC in vi */
      if v(SC) ≠ vi(SC)
        update SC;
  }
  /* Test response acquisition */
  disable TM;
  apply the normal clock;
}
scan-out MISR as the final test response;
```

(c)

FIGURE 2.55

Progressive random-access scan design: (a) PRAS scan cell design, (b) PRAS architecture, and (c) PRAS test procedure.

2.5.1.2 Shift-Addressable Random-Access Scan Design

The PRAS design has demonstrated that it can significantly reduce shift power dissipation by 100X and reduce routing overhead to within 10%. One difficulty is the control complexity of the test control logic for updating the selected scan cells one at a time. In fact, when a PRAS design contains 100 or more rows (scan cells), reduction in shift power dissipation could also possibly reach 100X, even if all columns would have to be updated simultaneously [Wang 2006c].

The ***shift-addressable random-access scan*** (STAR) architecture proposed in [Wang 2006c] uses only one row (*X*) decoder and supports two or more *SI* and *SO* ports. All rows are enabled (selected) in a fixed order one at a time by rotating a 1 in the row enable shift register. When a row is enabled, all columns (or scan cells) associated with the enabled row are selected at the same time; therefore, there is no need to provide a column address. This reduces the test application time as opposed to traditional RAS designs, which require a column address to write a selected scan cell one at a time [Ando 1980] [Baik 2005a] [Mudlapur 2005] [Hu 2006]. The STAR architecture and its associated test procedure are shown in Figure 2.56. The STAR architecture can use any RAS scan cell design as proposed in [Wagner 1984], [Baik 2005a], or [Mudlapur 2005].

It has been reported in [Baik 2005a] and [Mudlapur 2005] that RAS design can easily provide a 100X reduction in shift power dissipation. Because each scan cell is updated when needed, a 2X to 3X reduction in test data volume and test application time is also achieved. These results indicated that RAS design achieved a significant reduction in shift power dissipation, as well as a good reduction in test data volume and test application time. Whether or not RAS design can further reduce capture power dissipation remains a research topic.

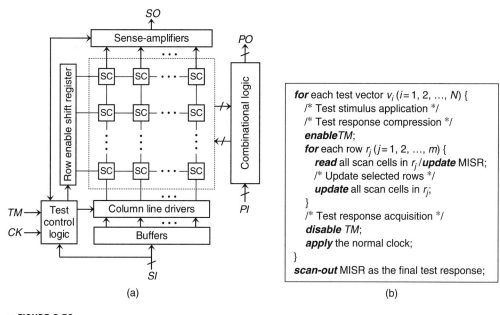

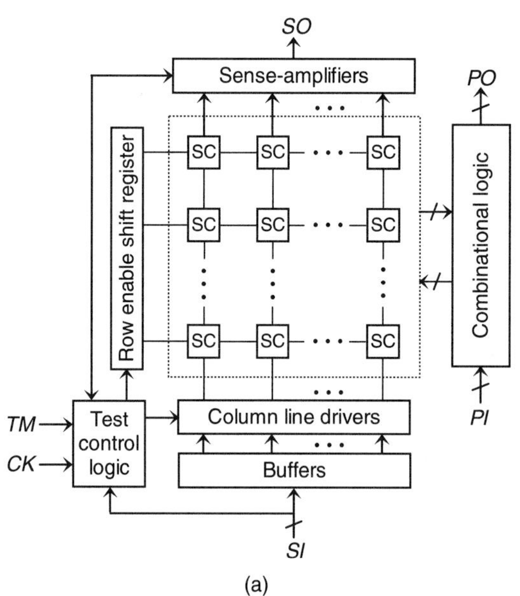

for each test vector v_i ($i = 1, 2, \ldots, N$) {
/* Test stimulus application */
/* Test response compression */
enable *TM*;
for each row r_j ($j = 1, 2, \ldots, m$) {
read all scan cells in r_j / ***update*** MISR;
/* Update selected rows */
update all scan cells in r_j;
}
/* Test response acquisition */
disable *TM*;
apply the normal clock;
}
scan-out MISR as the final test response;

(b)

■ **FIGURE 2.56**

Test procedure for a shift-addressable random-access scan (STAR) design: (a) STAR architecture and (b) STAR test procedure.

2.5.2 Test Compression RAS Architectures

Although RAS design has proven to be effective in reducing shift power dissipation at the cost of an increased area and routing overhead, reduction in test data volume and test application time is not significant. Since 2000, many test compression schemes have been developed to drastically reduce test data volume and test application time [Wang 2006a]. Even though these schemes are not aimed at reducing test power and mainly target serial scan design, they are applicable for use in RAS design.

All of these test compression schemes require that the design contain many short scan chains and data in the scan cells located on the same row be shifted in and out of the scan cells simultaneously within one shift clock cycle. Because most RAS architectures adopt the traditional RAS design architecture given in [Ando 1980] that updates the states of scan cells one at a time, they could substantially increase test application time.

The STAR architecture shown in Figure 2.56a overcomes the problem by allowing all scan cells on the same row to be accessed simultaneously. A general test compression RAS architecture based on the STAR architecture, called **STAR compression architecture** [Wang 2006c], is shown in Figure 2.57.

A decompressor is used to decompress the ATE-supplied stimuli, and a compactor is used to compact the test responses. In principle, the decompressor can be a pure buffer network as used in broadcast scan [Lee 1999] or Illinois scan [Hamzaoglu 1999], a MUX network as proposed in reconfigurable broadcast scan [Pandey 2002]

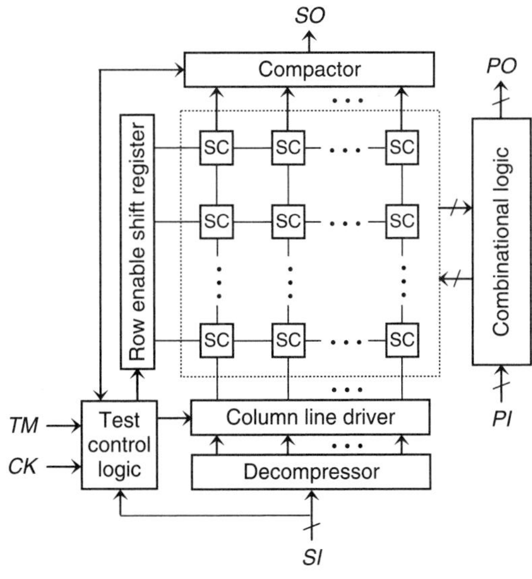

■ **FIGURE 2.57**

STAR compression architecture.

[Sitchinava 2004], a broadcaster as practiced in virtual scan [Wang 2004], a linear decompressor as used in [Wohl 2003b] and [Rajski 2004], or a coding-based decompressor [Hu 2005]. The compactor can be a MISR, an XOR network, or an X-tolerant XOR network [Mitra 2004].

One important feature in the STAR compression architecture is its ability to reconfigure the RAS scan cells into a serial scan mode. The purpose is to uncover faults, which go undetected because of the decompression-compression process. Unlike serial scan design where multiplexers are inserted to merge two or more short scan chains into a long scan chain, the reconfiguration in RAS design is accomplished by adding a multiplexer at the scan input of each column (short scan chain) and an AND gate at the scan output of each column (short scan chain). The multiplexer allows transmitting scan-in stimulus from one column to another column, whereas the AND gate enables or disables the scan-out test response on the column to be fed to the compactor in serial scan mode. One or more additional pins may be required to support the reconfiguration. Figure 2.58 shows the reconfigured STAR compression architecture. This architecture is also helpful for fault diagnosis, failure analysis, and yield enhancement.

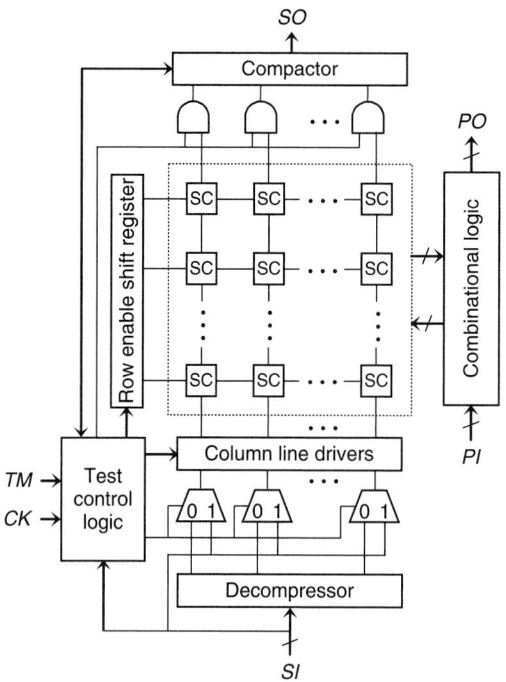

■ **FIGURE 2.58**

Reconfigured STAR compression architecture.

2.5.3 At-Speed RAS Architectures

In addition to the major advantages of providing significant shift power reduction and facilitating fault diagnosis, RAS design offers an additional benefit for at-speed delay fault testing. Typically, the launch-on-shift (also known as skewed-load) or launch-on-capture (also known as double-capture) capture-clocking scheme is employed for at-speed testing of path-delay faults and transition faults in serial scan design. Testing for a delay fault requires applying a pair of test vectors in an at-speed fashion. Either scheme requires generating a logic value transition at a signal line or at the source of a path in order to be able to capture the circuit response to this transition at the circuit's operating frequency.

In random-access design, these delay tests can be easily generated and applied using an **enhanced-scan** scheme [Malaiya 1983] [Glover 1988] [Dervisoglu 1991] [Kuppuswamy 2004] [Le 2007]. Rather than generating a functionally dependent pair of vectors, a **single-input-change** pair of vectors can be easily generated by a combinational ATPG. [Gharaybeh 1997] showed that any testable path can be tested by a single-input-change vector pair. Hence, an **enhanced-scan based at-speed RAS architecture** allows RAS design to maximize the delay fault detection capability. This is in sharp contrast with using launch-on-shift or launch-on-capture in serial scan design, which relies on scan chains to shift in the first initialization vector. In enhanced scan, the second test vector can be applied by simply flipping the state of the selected scan cell. Moreover, no additional hardware is required, as opposed to applying the enhanced-scan scheme to serial scan design [Dervisoglu 1991] [Wang 2006a] [Le 2007].

Although enhanced scan offers many benefits as stated here, the vector count could be a problem. The reason is that a design can contain millions of delay faults and hundreds of thousands of scan cells. Generating single-input-change vector pairs may not yield a sufficient compacted test set. For RAS designs containing many noninteracting clock domains, the enhanced-scan scheme fails to generate a single vector set to test these clock domains simultaneously.

One approach to overcome the long vector count problem is to use the enhanced-scan cell that adds a latch to a storage element as shown in Figure 2.7 [Dervisoglu 1991] [Kuppuswamy 2004]. This allows the application of two independent test vectors in two successive clock cycles. The drawback is that this enhanced-scan based at-speed RAS architecture adds more hardware overhead to the RAS design.

Another approach is to employ the conventional launch-on-capture scheme. The **launch-on-capture based at-speed RAS architecture** allows applications of multiple transitions on the initialization vector, thereby reducing the vector count. Because RAS design does not contain scan chains, the launch-on-shift clocking scheme is not applicable for RAS design. One promising **hybrid at-speed RAS architecture** would be to first support launch-on-capture and then supplement it with enhanced scan when required to maximize the delay fault coverage. To ease silicon debug and failure analysis, it may be advantageous to use a **faster-than-at-speed RAS architecture**, which applies delay tests at faster than the operating speed to the clock domain being tested. This can catch small delay defects that escape traditional transition fault tests [Kruseman 2004] [Amodeo 2005].

2.6 CONCLUDING REMARKS

Scan and *logic built-in self-test* (BIST) are currently the two most widely used *design-for-testability* (DFT) techniques for ensuring circuit testability and product quality. For completeness, we first cover a number of fundamental scan and logic BIST architectures in use today [Wang 2006a]. Because a scan design can now contain tens to hundreds of millions of transistors and test set with 100% single stuck-at fault coverage using scan ATPG can no longer guarantee adequate product quality, we have seen *at-speed delay testing* and *test compression* rapidly become a requirement for 90-nanometer designs and below. Many physical failures have manifested themselves as delay faults, thus requiring at-speed delay test patterns for detection of these faults [Ferhani 2006]. As the need for additional test sets to detect manufacturing faults grows, test compression is becoming crucial for reducing the explosive test data volume and long test application time problems.

Because scan ATPG assumes a single-fault model, physical failures that cannot be modeled as single faults for ATPG can potentially escape detection [Gizopoulos 2006]. To detect these physical failures, logic BIST is of growing importance in VLSI manufacturing, when combined with its major advantages of performing on-chip self-test and in-system remote diagnosis. We anticipate that for VLSI designs at 65 nanometers and below, logic BIST and *low-power testing* will gain more industry acceptance. Although the STUMPS-based architecture [Bardell 1982] is the most popular logic BIST architecture now practiced for scan-based designs, the efforts required to implement the BIST circuitry and the loss of the fault coverage for using pseudo-random patterns have prevented the BIST architecture from being widely used across all industries.

As the semiconductor manufacturing technology moves into the nanometer design era, it remains to be seen how the CBILBO-based architecture proposed in [Wang 1986], which can always guarantee 100% single stuck-at fault coverage and has the ability of running 10 times more BIST patterns than the STUMPS-based architecture, will perform. Challenges lie ahead with regard to whether or not pseudo-exhaustive testing will become a preferred BIST pattern generation technique and *random-access scan* will be a promising DFT technique for test power reduction.

2.7 EXERCISES

2.1 **(Muxed-D Scan Cell)** Show a possible CMOS implementation of the muxed-D scan cell shown in Figure 2.3a.

2.2 **(Low-Power Muxed-D Scan Cell)** Design a low-power version of the muxed-D scan cell given in Figure 2.3a by adding gated-clock logic, which includes a lock-up latch to control the clock port.

2.3 **(At-Speed Scan)** Assume that a scan design contains three clock domains running at 100 MHz, 200 MHz, and 400 MHz, respectively. In addition, assume that the clock skew between any two clock domains is manageable. List all possible at-speed scan ATPG methods, and compare their advantages and disadvantages in terms of fault coverage and test pattern count.

2.4 **(At-Speed Scan)** Describe two major capture-clocking schemes for at-speed scan testing, and compare their advantages and disadvantages. Also discuss what will happen if three or more captures are used.

2.5 **(BIST Pattern Generation)** Implement a period-8 in-circuit *test pattern generator* (TPG) using a binary counter. Compare its advantages and disadvantages with using a Johnson counter (twisted-ring counter).

2.6 **(BIST Pattern Generation)** Implement a period-31 in-circuit *test pattern generator* (TPG) using a modular *linear feedback shift register* (LFSR) with *characteristic polynomial* $f(x) = 1 + x^2 + x^5$. Convert the modular LFSR into a muxed-D scan design with minimum area overhead.

2.7 **(BIST Pattern Generation)** Implement a period-31 in-circuit *test pattern generator* (TPG) using a five-stage *cellular automaton* (CA) with *construction rule* = 11001, *where* "0" denotes a *rule* 90 cell and "1" denotes a *rule* 150 cell. Convert the CA into an LSSD design with minimum area overhead.

2.8 **(Cellular Automata)** Derive a construction rule for a cellular automaton of length 54, and then derive construction rules up to length 300 in order to match the list of primitive polynomials up to degree 300 reported in [Bardell 1987].

2.9 **(Test Point Insertion)** For the circuit shown in Figure 2.22, calculate the detection probabilities, before and after test point insertion, for a stuck-at-0 fault present at input X_3 and a stuck-at-1 fault present at input X_6 simultaneously.

2.10 **(BIST Response Compaction)** Discuss in detail what errors can be and cannot be detected by a MISR.

2.11 **(STUMPS *Versus* CBILBO)** Compare the performance of a STUMPS design and a CBILBO design. Assume that both designs operate at 400 MHz and that the circuit under test (CUT) has 100 scan chains each having 1000 scan cells. Calculate the test time required to test each design when 100,000 test patterns are to be applied. In general, the scan shift frequency is much slower than a circuit's operating speed. Assuming the scan shift frequency is 50 MHz, calculate the test time for the STUMPS design again. Explain further why the STUMPS-based architecture is gaining more industry acceptance than the CBILBO-based architecture.

2.12 **(Scan *Versus* Logic BIST *Versus* Test Compression)** Compare the advantages and disadvantages of a scan design, a logic BIST design, and a test compression design, in terms of fault coverage, test application time, test data volume, and area overhead.

2.13 **(Test Stimulus Compression)** Given a circuit with four scan chains, each having five scan cells, and with a set of test cubes listed as follows:

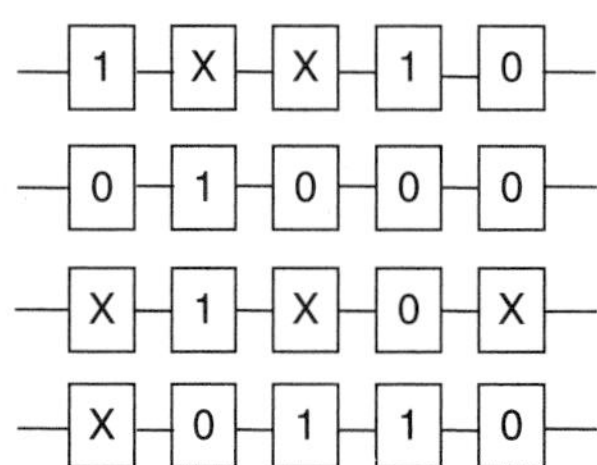

a. Design the multiple-input broadcast scan decompressor that fulfills the test cube requirements.

b. Explain the compression ratio.

c. The assignment of X's will affect the compression performance dramatically. Give one X-assignment example that will unfortunately lead to no compression with this multiple-input broadcast scan decompressor.

2.14 **(Test Stimulus Compression)** Derive mathematical expressions for the following in terms of the number of tester channels, c, and the expansion ratio, k.

a. The probability of encoding a scan slice containing 2 specified bits with Illinois scan.

b. The probability of encoding a scan slice containing 3 specified bits where each scan chain is driven by the XOR of a unique combination of 2 tester channels such that there are a total of $C_2^n = n(n-1)/2$ scan chains.

2.15 **(Test Stimulus Compression)** For the sequential linear decompressor shown in Figure 2.39 whose corresponding system of linear equations is shown in Figure 2.40, find the compressed stimulus $X_1 - X_{10}$ necessary to encode the following test cube: $< Z_1, \ldots, Z_{12} > = <$ 0-0-1-0--011 $>$.

2.16 **(Test Stimulus Compression)** For the MUX network shown in Figure 2.44 and then the XOR network shown in Figure 2.45a, find the compressed stimulus at the network inputs necessary to encode the following test cube: <1-0---01>.

2.17 (**Test Response Compaction**) Explain further how many errors and how many unknowns (X's) can be detected or tolerated by the X-tolerant compactor and q-compactor as shown in Figures 2.47 and 2.51, respectively.

2.18 **(Test Response Compaction)** For the X-compact matrix of the X-compactor given as follows:

$$\begin{bmatrix} 0 & 1 & 1 & 1 & 0 \\ 0 & 1 & 0 & 1 & 1 \\ 1 & 1 & 0 & 0 & 1 \\ 1 & 1 & 0 & 1 & 0 \\ 1 & 0 & 1 & 0 & 1 \\ 1 & 0 & 0 & 1 & 1 \\ 1 & 0 & 1 & 1 & 0 \\ 0 & 0 & 1 & 1 & 1 \end{bmatrix}$$

a. What is the compaction ratio?

b. Which outputs after compaction are affected by the second scan chain output?

c. How many errors can be detected by the X-compactor?

2.19 **(Random-Access Scan)** Assume that a sequential circuit with n storage elements has been reconfigured as a scan design as shown in Figure 2.3b and two random-access scan designs as shown in Figures 2.53 and 2.56. In addition, assume that the scan design has m balanced scan chains and that a test vector v_i is currently loaded into the scan cells of the three scan designs. Now consider the application of the next test vector v_{i+1}. Assume that v_{i+1} and the response of v_i are different in d bits. Calculate the number of clock cycles required for applying v_{i+1} to each of the three designs.

2.20 **(A Design Practice)** Write a C/C++ program to find the smallest number of clock groups in clocking grouping.

2.21 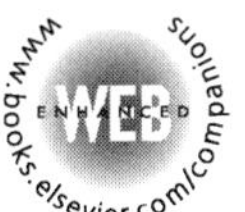

(A Design Practice) Assume that a scan clock and a PLL clock operate at 20 MHz and 50 MHz, respectively. Write an RTL code in Verilog *hardware description language* (HDL) to design the on-chip clock controller shown in Figure 2.14. Revise the RTL code to generate two pairs of staggered double-capture clock pulses each pair to control one clock domain.

2.22 

(A Design Practice) Design a STUMPS-based logic BIST system in Verilog RTL code using staggered double-capture for a circuit with two interactive clock domains, one with 10 inputs and 20 outputs, the other with 16 inputs and 18 outputs. Report the circuit's BIST fault coverage every 10,000 increments up to 100,000 pseudo-random patterns.

2.23 

(A Design Practice) Repeat Exercise 2.19 but to design a CBILBO-based logic BIST system. Compare the observed BIST fault coverage with the BIST fault coverage given in Exercise 2.19, and explain why both methods produce the same or different fault coverage numbers.

2.24 

(A Design Practice) Use the ATPG programs and user's manuals contained on the Companion Web site to generate test patterns for the three largest ISCAS-1985 combinational circuits and record the number of test patterns needed for each circuit. Then combine the three circuits into one circuit by connecting their inputs in such a way that the first inputs of the three circuits are connected to the first shared input of the combined circuit, the second inputs of the three circuits are connected to the second shared input, etc. Use the ATPG tool again to generate test patterns for this combined circuit. Compare the number of test patterns generated for the combined circuit with that generated for each individual circuit.

2.25 

(A Design Practice) Repeat Exercise 2.24, but this time try to use different input connections so as to reduce the number of test patterns for the combined circuit as much as you can. What is the least number of test patterns you can find?

Acknowledgments

The author wishes to thank Professor Xinghao Chen of The City College and Graduate Center of The City University of New York for contributing a portion of the Scan Architectures section; Professor Nur A. Touba of the University of Texas at Austin for contributing a portion of the Coverage-Driven Logic BIST Architectures section; Professor Xiaowei Li of the Chinese Academy of Sciences, Professor Kuen-Jong Lee of National Cheng Kung University, and Professor Nur A. Touba of the University of Texas at Austin for contributing a portion of the Circuits for Test Stimulus Compression and Circuits for Test Response Compaction sections; and Professor Xiaoqing Wen of the Kyushu Institute of Technology and Shianling Wu of SynTest Technologies for contributing a portion of the Random-Access Scan Architectures section. The author also would like to express his gratitude to Claude E. Shannon Professor John P. Hayes of the University of Michigan, Professor Kewal K. Saluja of the University of Wisconsin-Madison, Professor Yinhe Han of the Chinese Academy of Sciences, Dr. Patrick Girard of LIRMM, Dr. Xinli Gu of Cisco Systems, Dr. Rohit Kapur and Khader S. Abdel-Hafez of Synopsys, Dr. Brion Keller of Cadence Design Systems, Anandshankar S. Mudlapur of Intel, Dr. Benoit Nadeau-Dostie of LogicVision, Dr. Peilin Song of IBM, Dr. Erik H. Volkerink of Verigy US, Inc., and Dr. Seongmoon Wang of NEC Labs for reviewing the text and providing valuable comments; and Teresa Chang of SynTest Technologies for drawing most of the figures.

References

R2.0 Books

[Abramovici 1994] M. Abramovici, M. A. Breuer, and A. D. Friedman, *Digital Systems Testing and Testable Design*, IEEE Press, Revised Printing, Piscataway, NJ, 1994.

[Bardell 1987] P. H. Bardell, W. H. McAnney, and J. Savir, *Built-In Test for VLSI: Pseudorandom Techniques*, John Wiley & Sons, Somerset, NJ, 1987.

[Bushnell 2000] M. L. Bushnell and V. D. Agrawal, *Essentials of Electronic Testing for Digital, Memory & Mixed-Signal VLSI Circuits*, Springer, Boston, 2000.

[Crouch 1999] A. Crouch, *Design for Test for Digital IC's and Embedded Core Systems*, Prentice-Hall, Upper Saddle River, NJ, 1999.

[Gizopoulos 2006] D. Gizopoulos, editor, *Advances in Electronic Testing: Challenges and Methodologies Series: Frontiers in Electronic Testing*, Springer, Boston, 2006.

[Golomb 1982] S. W. Golomb, *Shift Register Sequence*, Aegean Park Press, Laguna Hills, CA, 1982.

[Jha 2003] N. Jha and S. Gupta, *Testing of Digital Systems*, Cambridge University Press, London, 2003.

[McCluskey 1986] E. J. McCluskey, *Logic Design Principles: With Emphasis on Testable Semicustom Circuits*, Prentice-Hall, Englewood Cliffs, NJ, 1986.

[Mourad 2000] S. Mourad and Y. Zorian, *Principles of Testing Electronic Systems*, John Wiley & Sons, Somerset, NJ, 2000.

[Nadeau-Dostie 2000] B. Nadeau-Dostie, *Design for At-Speed Test, Diagnosis and Measurement*, Springer, Boston, 2000.

[Rajski 1998a] J. Rajski and J. Tyszer, *Arithmetic Built-In Self-Test for Embedded Systems*, Prentice-Hall, Englewood Cliffs, NJ, 1998.

[Stroud 2002] C. E. Stroud, *A Designer's Guide to Built-In Self-Test*, Springer, Boston, 2002.

[Wang 2006a] L.-T. Wang, C.-W. Wu, and X. Wen, *VLSI Test Principles and Architectures: Design for Testability*, Morgan Kaufmann, San Francisco, 2006.

R2.1 Introduction

[Ando 1980] H. Ando, Testing VLSI with random access scan, in *Proc. COMPCON*, pp. 50–52, February 1980.

[Baik 2005a] D. Baik and K. K. Saluja, Progressive random access scan: A simultaneous solution to test power, test data volume and test time, in *Proc. Int. Test Conf.*, pp. 359–368, November 2005.

[Hu 2006] Y. Hu, C. Li, J. Li, Y. Han, X. Li, W. Wang, H. Li, L.-T. Wang, and X. Wen, Test data compression based on clustered random access scan, in *Proc. Asian Test Conf.*, pp. 231–236, November 2006.

[Mudlapur 2005] A. S. Mudlapur, V. D. Agrawal, and A. D. Singh, A random access scan architecture to reduce hardware overhead, in *Proc. Int. Test Conf.*, Paper 15.1, November 2005.

[SIA 2003] SIA, *The International Technology Roadmap for Semiconductors: 2003 Edition—Design*, pp. 30–36, Semiconductor Industry Association, San Jose, CA (http://public.itrs.net), 2003.

[SIA 2006] SIA, *The International Technology Roadmap for Semiconductors: 2006 Update*, Semiconductor Industry Association, San Jose, CA (http://public.itrs.net), 2006.

[Touba 2006] N. A. Touba, Survey of test vector compression techniques, *IEEE Design & Test of Computers*, 23(4), pp. 294–303, July/August 2006.

R2.2 Scan Design

[Beck 2005] M. Beck, O. Barondeau, M. Kaibel, F. Poehl, X. Lin, and R. Press, Logic design for on-chip test clock generation: Implementation details and impact on delay test quality, in *Proc. Design, Automation and Test in Europe Conf.*, pp. 56–61, March 2005.

[Bonhomme 2001] Y. Bonhomme, P. Girard, L. Guiller, C. Landrault, and S. Pravossoudovitch, A gated clock scheme for low power scan testing of logic ICs or embedded cores, in *Proc. Asian Test Symp.*, pp. 253–258, November 2001.

[Chandra 2001] A. Chandra and K. Chakrabarty, Combining low-power scan testing and test data compression for system-on-a-chip, in *Proc. European Design Automation Conf.*, pp. 166–169, June 2001.

[Cheung 1996] B. Cheung and L.-T. Wang, The seven deadly sins of scan-based designs, *Integrated System Design*, August 1996.

[DasGupta 1982] S. DasGupta, P. Goel, R. G. Walther, and T. W. Williams, A variation of LSSD and its implications on design and test pattern generation in VLSI, in *Proc. Int. Test Conf.*, pp. 63–66, November 1982.

[Dervisoglu 1991] B. I. Dervisoglu and G. E. Strong, Design for testability: Using scan path techniques for path-delay test and measurement, in *Proc. Int. Test Conf.*, pp. 365–374, October 1991.

[Eichelberger 1977] E. B. Eichelberger and T. W. Williams, A logic design structure for LSI testability, in *Proc. Design Automation Conf.*, pp. 462–468, June 1977.

[Furukawa 2006] H. Furukawa, X. Wen, L.-T. Wang, B. Sheu, Z. Jiang, and S. Wu, A novel and practical control scheme for inter-clock at-speed testing, in *Proc. Int. Test Conf.*, Paper 17.2, October 2006.

[Girard 2002] P. Girard, Survey of low-power testing of VLSI circuits, *IEEE Design & Test of Computers*, 19(3), pp. 82–92, May/June 2002.

[Glover 1988] C. T. Glover and M. R. Mercer, A method of delay fault test generation, in *Proc. Design Automation Conf.*, pp. 90–95, June 1988.

[Iyengar 2006] V. Iyengar, T. Yokota, K. Yamada, T. Anemikos, B. Bassett, M. Degregorio, R. Farmer, G. Grise, M. Johnson, D. Milton, M. Taylor, and F. Woytowich, At-speed structural test for high-performance ASICs, in *Proc. Int. Test Conf.*, Paper 2.4, October 2006.

[Khoche 2002] A. Khoche, Test resource partitioning for scan architectures using bandwidth matching, *IEEE Test Resource Partitioning Workshop*, pp. 1.4-1–1.4-8, October 2002.

[Kuppuswamy 2004] R. Kuppuswamy, P. DesRosier, D. Feltham, R. Sheikh, and P. Thadikaran, Full hold-scan systems in microprocessors: Cost/benefit analysis, *Intel Technology J.*, 8(1), pp. 69–78, February 2004.

[Malaiya 1983] Y. K. Malaiya and R. Narayanaswamy, Testing for timing faults in synchronous sequential integrated circuits, in *Proc. Int. Test Conf.*, pp. 560–571, October 1983.

[Nadeau-Dostie 1994] B. Nadeau-Dostie, A. Hassan, D. Burek, and S. Sunter, Multiple Clock Rate Test Apparatus for Testing Digital Systems, U.S. Patent No. 5,349,587, September 20, 1994.

[Rosinger 2004] P. Rosinger, B. M. Al-Hashimi, and N. Nicolici, Scan architecture with mutually exclusive scan segment activation for shift- and capture-power reduction, *IEEE Trans. on Computer-Aided Design*, 23(7), pp. 1142–1153, July 2004.

[Savir 1993] J. Savir and S. Patil, Scan-based transition test, *IEEE Trans. on Computer-Aided Design*, 12(8), pp. 1232–1241, August 1993.

[Savir 1994] J. Savir and S. Patil, Broad-side delay test, *IEEE Trans. on Computer-Aided Design*, 13(8), pp. 1057–1064, August 1994.

[Saxena 2001] J. Saxena, K. M. Butler, and L. Whetsel, A scheme to reduce power consumption during scan testing, in *Proc. Int. Test Conf.*, pp. 670–677, October 2001.

[Wang 2005a] L.-T. Wang, M.-C. Lin, X. Wen, H.-P. Wang, C.-C. Hsu, S.-C. Kao, and F. -S. Hsu, Multiple-Capture DFT System for Scan-Based Integrated Circuits, U.S. Patent No. 6,954,887, October 11, 2005.

[Wang 2007] L.-T. Wang, P.-C. Hsu, and X. Wen, Multiple-Capture DFT System for Detecting or Locating Crossing Clock-Domain Faults during Self-Test or Scan-Test, U.S. Patent Application No. 11,098,703, April. 18, 2007 (allowed).

[Whetsel 1998] L. Whetsel, Core test connectivity, communication, and control, in *Proc. Int. Test Conf.*, pp. 303–312, November 1998.

[Yoshida 2003] T. Yoshida and M. Watari, MD-scan method for low power scan testing, in *Proc. Int. Test Conf.*, pp. 480–487, October 2003.

R2.3 Logic Built-In Self-Test

[Aboulhamid 1983] M. E. Aboulhamid and E. Cerny, A class of test generators for built-in testing, *IEEE Trans. on Computers*, C-32(10), pp. 957–959, October 1983.

[Agarwal 1981] V. K. Agarwal and E. Cerny, Store and generate built-in testing approach, in *Proc. of Fault Tolerant Computing Symp.*, pp. 35–40, June 1981.

[Al-Yamani 2005] A. Al-Yamani, S. Mitra, and E. J. McCluskey, Optimized reseeding by seed ordering and encoding, *IEEE Trans. on Computer-Aided Design*, 24(2), pp. 264–270, February 2005.

[Bardell 1982] P. H. Bardell and W. H. McAnney, Self-testing of multiple logic modules, in *Proc. Int. Test Conf.*, pp. 200–204, November 1982.

[Bershteyn 1993] M. Bershteyn, Calculation of multiple sets of weights for weighted random testing, in *Proc. Int. Test Conf.*, pp. 1031–1040, October 1993.

[Bhawmik 1997] S. Bhawmik, Method and Apparatus for Built-In Self-Test with Multiple Clock Circuits, U.S. Patent No. 5,680,543, October 21, 1997.

[Cadence 2007] Cadence Design Systems, www.cadence.com, 2007.

[Chin 1984] C. K. Chin and E. J. McCluskey, Weighted pattern generation for built-in self-test, Center for Reliable Computing, Technical Report (CRC TR) No. 84-7, Stanford University, August 1984.

[Dandapani 1984] R. Dandapani, J. Patel, and J. Abraham, Design of test pattern generators for built-in test, in *Proc. Int. Test Conf.*, pp. 315–319, October 1984.

[Das 2000] D. Das and N. A. Touba, Reducing test data volume using external/LBIST hybrid test patterns, in *Proc. Int. Test Conf.*, pp. 115–122, October 2000.

[Dorsch 2001] R. Dorsch and H.-J. Wunderlich, Tailoring ATPG for embedded testing, in *Proc. Int. Test Conf.*, pp. 530–537, October 2001.

[Edirisooriya 1992] G. Edirisooriya and J. P. Robinson, Design of low cost ROM based test generators, in *Proc. VLSI Test Symp.*, pp. 61–66, April 1992.

[Foote 1997] T. G. Foote, D. E. Hoffman, W. V. Huott, T. J. Koprowski, B. J. Robbins, and M. P. Kusko, Testing the 400MHz IBM generation-4 CMOS chip, in *Proc. Int. Test Conf.*, pp. 106–114, November 1997.

[Furukawa 2006] H. Furukawa, X. Wen, L.-T. Wang, B. Sheu, Z. Jiang, and S. Wu, A novel and practical control scheme for inter-clock at-speed testing, in *Proc. Int. Test Conf.*, Paper 17.2, October 2006.

[Girard 2001] P. Girard, L. Guiller, C. Landrault, S. Pravossoudovitch, and H. J. Wunderlich, A modified clock scheme for a low power BIST test pattern generator, in *Proc. VLSI Test Symp.*, pp. 306–311, April/May 2001.

[Girard 2002] P. Girard, Survey of low-power testing of VLSI circuits, *IEEE Design & Test of Computers*, 19(3), pp. 82–92, May/June 2002.

[Hellebrand 1995a] S. Hellebrand, J. Rajski, S. Tarnick, S. Venkataramann, and B. Courtois, Generation of vector patterns through reseeding of multiple-polynomial linear feedback shift registers, *IEEE Trans. on Computers*, 44(2), pp. 223–233, February 1995.

[Hellebrand 1995b] S. Hellebrand, B. Reeb, S. Tarnick, and H.-J. Wunderlich, Pattern generation for a deterministic BIST scheme, in *Proc. Int. Conf. on Computer-Aided Design*, pp. 88–94, November 1995.

[Hetherington 1999] G. Hetherington, T. Fryars, N. Tamarapalli, M. Kassab, A. Hassan, and J. Rajski, Logic BIST for large industrial designs: Real issues and case studies, in *Proc. Int. Test Conf.*, pp. 358–367, October 1999.

[Hortensius 1989] P. D. Hortensius, R. D. McLeod, W. Pries, D. M. Miller, and H.C. Card, Cellular automata-based pseudorandom number generators for built-in self-test, *IEEE Trans. on Computer-Aided Design*, 8(8), pp. 842–859, August 1989.

[Ichino 2001] K. Ichino, T. Asakawa, S. Fukumoto, K. Iwasaki, and S. Kajihara, Hybrid BIST using partially rotational scan, in *Proc. Asian Test Symp.*, pp. 379–384, November 2001.

[Iyengar 1989] V. S. Iyengar and D. Brand, Synthesis of pseudo-random pattern testable designs, in *Proc. Int. Test Conf.*, pp. 501–508, August 1989.

[Jas 2004] A. Jas, C. V. Krishna, and N. A. Touba, Weighted pseudo-random hybrid BIST, *IEEE Trans. on Very Large Scale Integration (VLSI) Systems*, 12(12), pp. 1277–1283, December 2004.

[Jervan 2003] G. Jervan, P. Eles, Z. Peng, R. Ubar, and M. Jenihhin, Test time minimization for hybrid BIST of core-based systems, in *Proc. Asian Test Symp.*, pp. 318–323, November 2003.

[Kapur 1994] R. Kapur, S. Patil, T. J. Snethen, and T. W. Williams, Design of an efficient weighted random pattern generation system, in *Proc. Int. Test Conf.*, pp. 491–500, October 1994.

[Kiefer 1998] G. Kiefer and H.-J. Wunderlich, Deterministic BIST with multiple scan chains, in *Proc. Int. Test Conf.*, pp. 1057–1064, October 1998.

[Könemann 1991] B. Koenemann, LFSR-coded test patterns for scan designs, in *Proc. European Test Conf.*, pp. 237–242, April 1991.

[Krishna 2003a] C. V. Krishna and N. A. Touba, Hybrid BIST using an incrementally guided LFSR, in *Proc. Symp. on Defect and Fault Tolerance*, pp. 217–224, November 2003.

[Krishnamurthy 1987] B. Krishnamurthy, A dynamic programming approach to the test point insertion problem, in *Proc. Design Automation Conf.*, pp. 695–704, June 1987.

[Lai 2005] L. Lai, J. H. Patel, T. Rinderknecht, and W.-T. Cheng, Hardware efficient LBIST with complementary weights, in *Proc. Int. Conf. on Computer Design*, pp. 479–481, October 2005.

[Lei 2005] L. Lei and K. Chakrabarty, Hybrid BIST based on repeating sequences and cluster analysis, in *Proc. Design, Automation and Test in Europe Conf.*, pp. 1142–1147, March 2005.

[Liang 2001] H.-G. Liang, S. Hellebrand, and H.-J. Wunderlich, Two-dimensional test data compression for scan-based deterministic BIST, in *Proc. Int. Test Conf.*, pp. 894–902, September 2001.

[LogicVision 2007] LogicVision, www.logicvision.com, 2007.

[Manich 2000] S. Manich, A. Gabarro, M. Lopez, J. Figueras, P. Girard, L. Guiller, C. Landrault, S. Pravossoudovitch, P. Teixeira, and M. Santos, Low power BIST by filtering non-detecting vectors, *JETTA: J. Electron. Test.: Theory and Applications*, 16(3), pp. 193–202, June 2000.

[McCluskey 1985] E. J. McCluskey, Built-in self-test structures, *IEEE Design & Test of Computers*, 2(2), pp. 29–36, April 1985.

[Mentor 2007] Mentor Graphics, www.mentor.com, 2007.

[Nadeau-Dostie 1994] B. Nadeau-Dostie, A. Hassan, D. Burek, and S. Sunter, Multiple Clock Rate Test Apparatus for Testing Digital Systems, U.S. Patent No. 5,349,587, September 20, 1994.

[Nadeau-Dostie 2006] B. Nadeau-Dostie and J.-F. Côté, Clock Controller for At-Speed Testing of Scan Circuits, U.S. Patent No. 7,155,651, December 26, 2006.

[Nadeau-Dostie 2007] B. Nadeau-Dostie, Method and Circuit for At-Speed Testing of Scan Circuits, U.S. Patent No. 7,194,669, March 20, 2007.

[Rajski 1998b] J. Rajski, J. Tyszer, and N. Zacharia, Test data decompression for multiple scan designs with boundary scan, *IEEE Trans. on Computers*, 47(11), pp. 1188–1200, November 1998.

[Rajski 2003] J. Rajski, A. Hassan, R. Thompson, and N. Tamarapalli, Method and Apparatus for At-Speed Testing of Digital Circuits, U.S. Patent Application No. 20030097614, May 22, 2003.

[Savir 1993] J. Savir and S. Patil, Scan-based transition test, *IEEE Trans. on Computer-Aided Design*, 12(8), pp. 1232–1241, August 1993.

[Savir 1994] J. Savir and S. Patil, Broad-side delay test, *IEEE Trans. on Computer-Aided Design*, 13(8), pp. 1057–1064, August 1994.

[Schnurmann 1975] H. D. Schnurmann, E. Lindbloom, and R. G. Carpenter, The weighted random test-pattern generator, *IEEE Trans. on Computers*, 24(7), pp. 695–700, July 1975.

[Seiss 1991] B. H. Seiss, P. Trouborst, and M. Schulz, Test point insertion for scan-based BIST, in *Proc. European Test Conf.*, pp. 253–262, April 2003.

[Sugihara 1998] M. Sugihara, H. Date, and H. Yasuura, A novel test methodology for core-based system LSIs and a testing time minimization problem, *Proc. Int. Test Conf.*, pp. 465–472, October 1998.

[SynTest 2007] SynTest Technologies, www.syntest.com, 2007.

[Tamarapalli 1996] N. Tamarapalli and J. Rajski, Constructive multi-phase test point insertion for scan-based BIST, in *Proc. Int. Test Conf.*, pp. 649–658, October 1996.

[Touba 1995] N. A. Touba and E. J. McCluskey, Transformed pseudo-random patterns for BIST, in *Proc. VLSI Test Symp.*, pp. 410–416, April 1995.

[Touba 1996] N. A. Touba and E. J. McCluskey, Test point insertion based on path tracing, in *Proc. VLSI Test Symp.*, pp. 2–8, April 1996.

[Touba 2001] N. A. Touba and E. J. McCluskey, Bit-fixing in pseudorandom sequences for scan BIST, *IEEE Trans. on Computer-Aided Design*, 20(4), pp. 545–555, April 2001.

[Tsai 1998] H.-C. Tsai, K.-T. Cheng, C.-J. Lin, and S. Bhawmik, Efficient test point selection for scan-based BIST, *IEEE Trans. on Very Large Scale Integration (VLSI) Systems*, 6(4), pp. 667–676, December 1998.

[Tsai 2000] H.-C. Tsai, K.-T. Cheng, and S. Bhawmik, On improving test quality of scan-based BIST, *IEEE Trans. on Computer-Aided Design*, 9(8), pp. 928–938, August 2000.

[Waicukauski 1989] J. A. Waicukauski, E. Lindbloom, E. B. Eichelberger, and O. P. Forenza, WRP: A method for generating weighted random test patterns, *IBM J. Res.Dev.*, 33(2), pp. 149–161, March 1989.

[Wang 1986] L.-T. Wang and E. J. McCluskey, Concurrent built-in logic block observer (CBILBO), in *Proc. Int. Symp. on Circuits and Systems*, 3(3), pp. 1054–1057, May 1986.

[Wang 1999] S. Wang and S. K. Gupta, LT-RTPG: A new test-per-scan BIST TPG for low heat dissipation, in *Proc. Int. Test Conf.*, pp. 85–94, October 1999.

[Wang 2005a] L.-T. Wang, M.-C. Lin, X. Wen, H.-P. Wang, C.-C. Hsu, S.-C. Kao, and F.-S. Hsu, Multiple-Capture DFT System for Scan-Based Integrated Circuits, U.S. Patent No. 6,954,887, October 11, 2005.

[Wang 2005b] L.-T. Wang, X. Wen, P.-C. Hsu, S. Wu, and J. Guo, At-speed logic BIST architecture for multi-clock designs, in *Proc. Int. Conf. on Computer Design: VLSI in Computers & Processors*, pp. 475–478, October 2005.

[Wang 2006b] L.-T. Wang, P.-C. Hsu, S.-C. Kao, M.-C. Lin, H.-P. Wang, H.-J. Chao, and X. Wen, Multiple-Capture DFT System for Detecting or Locating Crossing Clock-Domain Faults during Self-Test or Scan-Test, U.S. Patent No. 7,007,213, February 28, 2006.

[Wen 2006] X. Wen, S. Kajihara, K. Miyase, T. Suzuki, K. K. Saluja, L.-T. Wang, K. S. Abdel-Hafez, and K. Kinoshita, A new ATPG method for efficient capture power reduction during scan testing, in *Proc. VLSI Test Symp.*, pp. 58–63, May 2006.

[Wohl 2003a] P. Wohl, J. Waicukauski, S. Patel, and M. Amin, X-tolerant compression and applications of scan-ATPG patterns in a BIST architecture, in *Proc. Int. Test Conf.*, pp. 727–736, October 2003.

[Wunderlich 1987] H.-J. Wunderlich, Self test using unequiprobable random patterns, in *Proc. Fault-Tolerant Computing Symp.*, pp. 258–263, July 1987.

[Zhang 2000] X. Zhang, W. Shan, and K. Roy, Low-power weighted random pattern testing, *IEEE Trans. on Computer-Aided Design*, 19(11), pp. 1389–1398, November 2000.

R2.4 Test Compression

[Barnhart 2002] C. Barnhart, V. Brunkhorst, F. Distler, O. Farnsworth, A. Ferko, B. Keller, D. Scott, B. Koenemann, and T. Onodera, Extending OPMISR beyond 10x scan test efficiency, *IEEE Design & Test of Computers*, 19(5), pp. 65–73, May/June 2002.

[Bayraktaroglu 2001] I. Bayraktaroglu and A. Orailoglu, Test volume and application time reduction through scan chain concealment, in *Proc. Design Automation Conf.*, pp. 151–155, June 2001.

[Bayraktaroglu 2003] I. Bayraktaroglu and A. Orailoglu, Concurrent application of compaction and compression for test time and data volume reduction in scan designs, *IEEE Trans. on Computers*, 52(11), pp. 1480–1489, November 2003.

[Beck 2005] M. Beck, O. Barondeau, M. Kaibel, F. Poehl, X. Lin, and R. Press, Logic design for on-chip test clock generation: Implementation details and impact on delay test quality, in *Proc. Design, Automation and Test in Europe Conf.*, pp. 56–61, March 2005.

[Cadence 2007] Cadence Design Systems, www.cadence.com, 2007.

[Dutta 2006] A. Dutta and N. A. Touba, Using limited dependence sequential expansion for decompressing test vectors, in *Proc. Int. Test Conf.*, Paper 23.1, October 2006.

[Furukawa 2006] H. Furukawa, X. Wen, L.-T. Wang, B. Sheu, Z. Jiang, and S. Wu, A novel and practical control scheme for inter-clock at-speed testing, in *Proc. Int. Test Conf.*, Paper 17.2, October 2006.

[Hamzaoglu 1999] I. Hamzaoglu and J. H. Patel, Reducing test application time for full scan embedded cores, in *Proc. Fault-Tolerant Computing Symp.*, pp. 260–267, July 1999.

[Han 2003] Y. Han, Y. Xu, A. Chandra, H. Li, and X. Li, Test resource partitioning based on efficient response compaction for test time and tester channels reduction, in *Proc. Asian Test Symp.*, pp. 440–445, November 2003.

[Han 2005a] Y. Han, Y. Hu, H. Li, and X. Li, Theoretic analysis and enhanced X-tolerance of test response compact based on convolutional code, in *Proc. Asia and South Pacific Design Automation Conf.*, pp. 53–58, January 2005.

[Han 2005b] Y. Han, X. Li, H. Li, and A. Chandra, Test resource partitioning based on efficient response compaction for test time and tester channels reduction, *J. Comp. Sci. Tech*, 20(2), pp. 201–210, February 2005.

[Han 2005c] Y. Han, S. Swaminathan, Y. Hu, A. Chandra, and X. Li, Scan data volume reduction using periodically alterable MUXs decompressor, in *Proc. Asian Test Symp.*, pp. 372–377, November 2005.

[Hsu 2001] F. F. Hsu, K. M. Butler, and J. H. Patel, A case study on the implementation of Illinois scan architecture, in *Proc. Int. Test Conf.*, pp. 538–547, October 2001.

[Könemann 1991] B. Koenemann, LFSR-coded test patterns for scan designs, in *Proc. European Test Conf.*, pp. 237–242, April 1991.

[Könemann 2001] B. Koenemann, C. Barnhart, B. Keller, T. Snethen, O. Farnsworth, and D. Wheater, A SmartBIST variant with guaranteed encoding, in *Proc. Asian Test Symp.*, pp. 325–330, November 2001.

[Könemann 2003] B. Koenemann, C. Barnhart, and B. Keller, Real-Time Decoder for Scan Test Patterns, U.S. Patent No. 6,611,933, August 26, 2003.

[Krishna 2001] C. V. Krishna, A. Jas, and N. A. Touba, Test vector encoding using partial LFSR reseeding, in *Proc. Int. Test Conf.*, pp. 885–893, October 2001.

[Krishna 2002] C. V. Krishna, A. Jas, and N. A. Touba, Reducing test data volume using LFSR reseeding with seed compression, in *Proc. Int. Test Conf.*, pp. 321–330, October 2002.

[Krishna 2003b] C. V. Krishna and N. A. Touba, Adjustable width linear combinational scan vector decompression, in *Proc. Int. Conf. on Computer-Aided Design*, pp. 863–866, September 2003.

[Lee 1998] K.-J. Lee, J. J. Chen, and C. H. Huang, Using a single input to support multiple scan chains, in *Proc. Int. Conf. on Computer-Aided Design*, pp. 74–78, November 1998.

[Lee 1999] K.-J. Lee, J. J. Chen, and C. H. Huang, Broadcasting test patterns to multiple circuits, *IEEE Trans. on Computer-Aided Design*, 18(12), pp. 1793–1802, December 1999.

[Li 2004] L. Li and K. Chakrabarty, Test set embedding for deterministic BIST using a reconfigurable interconnection network, *IEEE Trans. on Computer-Aided Design*, 23(9), pp. 1289–1305, September 2004.

[LogicVision 2007] LogicVision, www.logicvision.com, 2007.

[Mentor 2007] Mentor Graphics, www.mentor.com, 2007.

[Mitra 2004] S. Mitra and K. S. Kim, X-compact: An efficient response compaction technique, *IEEE Trans. on Computer-Aided Design*, 23(3), pp. 421–432, March 2004.

[Mitra 2006] S. Mitra and K. S. Kim, XPAND: An efficient test stimulus compression technique, *IEEE Trans. on Computers*, 55(2), pp. 163–173, February 2006.

[Mrugalski 2004] G. Mrugalski, J. Rajski, and J. Tyszer, Ring generators: New devices for embedded test applications, *IEEE Trans. on Computer-Aided Design*, 23(9), pp. 1306–1320, September 2004.

[Nadeau-Dostie 2006] B. Nadeau-Dostie and J.-F. Côté, Clock Controller for At-Speed Testing of Scan Circuits, U.S. Patent No. 7,155,651, December 26, 2006.

[Nadeau-Dostie 2007] B. Nadeau-Dostie, Method and Circuit for At-Speed Testing of Scan Circuits, U.S. Patent No. 7,194,669, March 20, 2007.

[Naruse 2003] M. Naruse, I. Pomeranz, S. M. Reddy, and S. Kundu, On-chip compression of output responses with unknown values using LFSR reseeding, in *Proc. Int. Test Conf.*, pp. 1060–1068, October 2003.

[Pandey 2002] A. R. Pandey and J. H. Patel, Reconfiguration technique for reducing test time and test volume in Illinois scan architecture based designs, in *Proc. VLSI Test Symp.*, pp. 9–15, April 2002.

[Patel 2003] J. H. Patel, S. S. Lumetta, and S. M. Reddy, Application of Saluja-Karpovsky compactors to test responses with many unknowns, in *Proc. VLSI Test Symp.*, pp. 107–112, April 2003.

[Rajski 2004] J. Rajski, J. Tyszer, M. Kassab, and N. Mukherjee, Embedded deterministic test, *IEEE Trans. on Computer-Aided Design*, 23(5), pp. 776–792, May 2004.

[Rajski 2005] J. Rajski, J. Tyszer, C. Wang, and S. M. Reddy, Finite memory test response compactors for embedded test applications, *IEEE Trans. on Computer-Aided Design*, 24(4), pp. 622–634, April 2005.

[Rajski 2006] J. Rajski, J. Tyszer, G. Mruglaski, W.-T. Cheng, N. Mukherjee, and M. Kassab, X-Press compactor for 1000x reduction of test data, in *Proc. Int. Test Conf.*, Paper 18.1, October 2006.

[Saluja 1983] K. K. Saluja and M. Karpovsky, Test compression hardware through data compression in space and time, in *Proc. Int. Test Conf.*, pp. 83–88, October 1983.

[Samaranayake 2003] S. Samaranayake, E. Gizdarski, N. Sitchinava, F. Neuveux, R. Kapur, and T. W. Williams, A reconfigurable shared scan-in architecture, in *Proc. VLSI Test Symp.*, pp. 9–14, April 2003.

[Shah 2004] M. A. Shah and J. H. Patel, Enhancement of the Illinois scan architecture for use with multiple scan inputs, in *Proc. VLSI Symp.*, pp. 167–172, February 2004.

[Sitchinava 2004] N. Sitchinava, S. Samaranayake, R. Kapur, E. Gizdarski, F. Neuveux, and T. W. Williams, Changing the scan enable during shift, in *Proc. VLSI Test Symp.*, pp. 73–78, April 2004.

[Synopsys 2007] Synopsys, www.synopsys.com, 2007.

[SynTest 2007] SynTest Technologies, www.syntest.com, 2007.

[Touba 2006] N. A. Touba, Survey of test vector compression techniques, *IEEE Design & Test of Computers*, 23(4), pp. 294–303, July/August 2006.

[Volkerink 2003] E. H. Volkerink and S. Mitra, Efficient seed utilization for reseeding based compression, in *Proc. VLSI Test Symp.*, pp. 232–237, April 2003.

[Volkerink 2005] E. H. Volkerink and S. Mitra, Response compaction with any number of unknowns using a new LFSR architecture, in *Proc. Design Automation Conf.*, pp. 117–122, June 2005.

[Wang 2002] L.-T. Wang, H.-P. Wang, X. Wen, M.-C. Lin, S.-H. Lin, D.-C. Yeh, S.-W. Tsai, and K. S. Abdel-Hafez, Method and Apparatus for Broadcasting Scan Patterns in a Scan-Based Integrated Circuit, U.S. Patent Application No. 20030154433, January 16, 2002.

[Wang 2004] L.-T. Wang, X. Wen, H. Furukawa, F.-S. Hsu, S.-H. Lin, S.-W. Tsai, K. S. Abdel-Hafez, and S. Wu, VirtualScan: A new compressed scan technology for test cost reduction, in *Proc. Int. Test Conf.*, pp. 916–925, October 2004.

[Wang 2005b] L.-T. Wang, K. S. Abdel-Hafez, X. Wen, B. Sheu, S. Wu, S.-H. Lin, and M.-T. Chang, UltraScan: Using time-division demultiplexing/multiplexing (TDDM/TDM) with VirtualScan for test cost reduction, in *Proc. Int. Test Conf.*, pp. 946–953, November 2005.

[Wohl 2003b] P. Wohl, J. A. Waicukauski, S. Patel, and M. B. Amin, Efficient compression and application of deterministic patterns in a logic BIST architecture, in *Proc. Design Automation Conf.*, pp. 566–569, June 2003.

[Wohl 2004] P. Wohl, J. A. Waicukauski, and S. Patel, Scalable selector architecture for X-tolerant deterministic BIST, in *Proc. Design Automation Conf.*, pp. 934–939, June 2004.

[Wohl 2005] P. Wohl, J. A. Waicukauski, S. Patel, F. DaSilva, T. W. Williams, and R. Kapur, Efficient compression of deterministic patterns into multiple PRPG seeds, in *Proc. Int. Test Conf.*, pp. 916–925, November 2005.

R2.5 Random-Access Scan Design

[Amodeo 2005] M. Amodeo and B. Cory, Defining faster-than-at-speed delay tests, *Cadence Nanometer Test Quarterly*, 2(2), May 2005, [www.cadence.com/newsletters/nanometer_test].

[Ando 1980] H. Ando, Testing VLSI with random access scan, in *Proc. COMPCON*, pp. 50–52, February 1980.

[Baik 2004] D. Baik, S. Kajihara, and K. K. Saluja, Random Access Scan: A solution to test power, test data volume and test time, in *Proc. Int. Conf. on VLSI Design*, pp. 883–888, January 2004.

[Baik 2005a] D. Baik and K. K. Saluja, Progressive random access scan: A simultaneous solution to test power, test data volume and test time, in *Proc. Int. Test Conf.*, pp. 359–368, November 2005.

[Baik 2005b] D. Baik and K. K. Saluja, State-reuse test generation for progressive random access scan: Solution to test power, application time and data size, in *Proc. Asian Test Symp.*, pp. 272–277, November 2005.

[Baik 2006] D. Baik and K. K. Saluja, Test cost reduction using partitioned grid random access scan, in *Proc. Int. Conf. on VLSI Design*, pp. 169–174, January 2006.

[Dervisoglu 1991] B. I. Dervisoglu and G. E. Strong, Design for testability: Using scan path techniques for path-delay test and measurement, in *Proc. Int. Test Conf.*, pp. 365–374, October 1991.

[Gharaybeh 1997] M. A. Gharaybeh, M. L. Bushnell, and V. D. Agrawal, Classification and test generation for path-delay faults using single stuck-at fault tests, *J. Electronic Testing: Theory and Applications*, 11, pp. 55–67, August 1997.

[Girard 2002] P. Girard, Survey of low-power testing of VLSI circuits, *IEEE Design & Test of Computers*, 19(3), pp. 82–92, May/June 2002.

[Glover 1988] C. T. Glover and M. R. Mercer, A method of delay fault test generation, in *Proc. Design Automation Conf.*, pp. 90–95, June 1988.

[Hamzaoglu 1999] I. Hamzaoglu and J. H. Patel, Reducing test application time for full scan embedded cores, in *Proc. Fault-Tolerant Computing Symp.*, pp. 260–267, July 1999.

[Hu 2005] Y. Hu, Y.-H. Han, X. Li, H.-W. Li, and X. Wen, Compression/scan co-design for reducing test data volume, scan-in power dissipation and test application time, in *Proc. Pacific Rim Int. Symp. on Dependable Computing*, pp. 175–182, December 2005.

[Hu 2006] Y. Hu, C. Li, J. Li, Y. Han, X. Li, W. Wang, H. Li, L.-T. Wang, and X. Wen, Test data compression based on clustered random access scan, in *Proc. Asian Test Conf.*, pp. 231–236, November 2006.

[Kruseman 2004] B. Kruseman, A. K. Majhi, G. Gronthoud, and S. Eichenberger, On hazard-free patterns for fine-delay fault testing, in *Proc. Int. Test Conf.*, pp. 213–222, October 2004.

[Kuppuswamy 2004] R. Kuppuswamy, P. DesRosier, D. Feltham, R. Sheikh, and P. Thadikaran, Full hold-scan systems in microprocessors: Cost/benefit analysis, *Intel Technology J.*, 8(1), pp. 69–78, February 2004.

[Le 2007] K. T. Le, D. Baik, and K. K. Saluja, Test time reduction to test for path-delay faults using enhanced random access scan, in *Proc. Int. Conf. on VLSI Design*, pp. 769–774, January 2007.

[Lee 1999] K.-J. Lee, J. J. Chen, and C. H. Huang, Broadcasting Test patterns to multiple circuits, *IEEE Trans. on Computer-Aided Design*, 18(12), pp. 1793–1802, December 1999.

[Malaiya 1983] Y. K. Malaiya and R. Narayanaswamy, Testing for timing faults in synchronous sequential integrated circuits, in *Proc. Int. Test Conf.*, pp. 560–571, October 1983.

[Mitra 2004] S. Mitra and K. S. Kim, X-compact: An efficient response compaction technique, *IEEE Trans. on Computer-Aided Design*, 23(3), pp. 421–432, March 2004.

[Mudlapur 2005] A. S. Mudlapur, V. D. Agrawal, and A. D. Singh, A random access scan architecture to reduce hardware overhead, in *Proc. Int. Test Conf.*, pp. 350–358, November 2005.

[Pandey 2002] A. R. Pandey and J. H. Patel, Reconfiguration technique for reducing test time and test volume in Illinois scan architecture based designs, in *Proc. VLSI Test Symp.*, pp. 9–15, April 2002.

[Rajski 2004] J. Rajski, J. Tyszer, M. Kassab, and N. Mukherje, Embedded deterministic test, *IEEE Trans. on Computer-Aided Design*, 23(5), pp. 776–792, May 2004.

[Sitchinava 2004] N. Sitchinava, S. Samaranayake, R. Kapur, E. Gizdarski, F. Neuveux, and T. W. Williams, Changing the scan enable during shift, in *Proc. VLSI Test Symp.*, pp. 73–78, April 2004.

[Susheel 2002] T. G. Susheel, J. Chandra, T. Ferry, and K. Pierce, ATPG based on a novel grid-addressable latch element, in *Proc. Design Automation Conf.*, pp. 282–286, June 2002.

[Wagner 1984] K. D. Wagner, Design for testability in the AMDAHL 580, in *Proc. COMPCON.*, pp. 384–388, February 1984.

[Wang 2004] L.-T. Wang, X. Wen, H. Furukawa, F.-S. Hsu, S.-H. Lin, S.-W. Tsai, K. S. Abdel-Hafez, and S. Wu, VirtualScan: A new compressed scan technology for test cost reduction, in *Proc. Int. Test Conf.*, pp. 916–925, October 2004.

[Wang 2006c] L.-T. Wang, B. Sheu, Z. Jiang, Z. Wang, and S. Wu, Method and Apparatus for Broadcasting Scan Patterns in a Random-Access Scan Based Integrated Circuit, U.S. Patent Application No. 20030154433, February 8, 2006. (Continuation in Part.)

[Wohl 2003b] P. Wohl, J. A. Waicukauski, S. Patel, and M. B. Amin, Efficient compression and application of deterministic patterns in a logic BIST architecture, in *Proc. Design Automation Conf.*, pp. 566–569, June 2003.

R2.6 Concluding Remarks

[Bardell 1982] P. H. Bardell and W. H. McAnney, Self-testing of multiple logic modules, in *Proc. Int. Test Conf.*, pp. 200–204, November 1982.

[Ferhani 2006] F.-F. Ferhani and E. J. McCluskey, Classifying bad chips and ordering test sets, in *Proc. Int. Test Conf.*, Lecture 1.2, October 2006.

[Wang 1986] L.-T. Wang and E. J. McCluskey, Concurrent built-in logic block observer (CBILBO), in *Proc. Int. Symp. on Circuits and Systems*, 3(3), pp. 1054–1057, May 1986.

CHAPTER 3

FAULT-TOLERANT DESIGN

Nur A. Touba
University of Texas, Austin, Texas

ABOUT THIS CHAPTER

Fault tolerance is the ability of a system to continue error-free operation in the presence of unexpected faults. Faults can be either temporary (because of radiation, noise, ground bounce, etc.) or permanent (because of manufacturing defects, oxide breakdown, electromigration, etc.). As technology continues to scale, circuit behavior is becoming less predicable and more prone to failures, thereby increasing the need for fault-tolerant design. Fault tolerance requires some form of redundancy in time, space, or information. When an error occurs, it either needs to be masked/corrected or the operation needs to be retried if it is a temporary fault. If there is a permanent fault, then retrying an operation will not solve the problem. In that case, sufficient redundancy or spare units are required to continue error-free operation, or the part needs to be repaired or replaced.

This chapter gives an overview of a number of fault-tolerant design schemes suitable for nanometer ***system-on-chip*** (SOC) applications. It starts with an introduction to the basic concepts in fault-tolerant design and the metrics used to specify and evaluate the dependability of the design. Coding theory is next reviewed and some of the commonly used error detecting and correcting codes are described. Then fault-tolerant design schemes using hardware, time, and information redundancy are discussed. This is followed by some examples of various types of fault-tolerant applications used in industry.

3.1 INTRODUCTION

Fault tolerance is the ability of a system to continue error-free operation in the presence of an unexpected fault. Fault tolerance has always been important in mission-critical applications, such as medical, aviation, and banking, where errors can be costly [Lala 2001] [Pradhan 1996] [Siewiorek 1998]. As semiconductor manufacturing technology has scaled to a point where circuit behavior is less predictable and more prone to failures, fault tolerance is becoming important even for mainstream applications in order to keep failure rates within acceptable levels.

Faults can be either permanent or temporary. **Permanent faults** can result from manufacturing defects, early life failures, or wearout failures. Defects introduced during the manufacturing process can cause failures right away or early in the lifetime of the part, which can result in a permanent fault. **Wearout failures** can occur later in the lifetime of a part because of various mechanisms, such as electromigration, hot carriers degradation, and time-dependent dielectric breakdown, which can also lead to a permanent fault. **Temporary faults**, on the other hand, occur because of either transient or intermittent disturbances that are only present for a short period of time. **Transient errors** are *nonrecurring errors* caused by a disturbance external from the fault site such as radiation, noise, or power disturbance. **Intermittent errors** are *recurring errors* caused by marginal design parameters such as timing problems that result from races, hazards, skew, or signal integrity problems caused by crosstalk, ground bounce, and other factors. Wearout failures often first present themselves as intermittent errors. Note that some fault-tolerant design techniques can only tolerate permanent faults, only temporary faults, or a combination of both.

Fault tolerance requires some form of redundancy in time, hardware, or information [Lala 2001] [Siewiorek 1998]. An example of **time redundancy** would be to perform the same operation twice and see whether the same result is obtained both times (if not, then a fault has occurred). This can detect temporary faults but not permanent faults, as a permanent fault would affect both computations. The advantage of time redundancy is that it requires little or no extra hardware. however, the drawback is that it impacts the system or circuit performance. **Hardware redundancy** involves replicating hardware and comparing the outputs from two or more replicated hardware units. This can detect both temporary and permanent faults. The advantage of hardware redundancy is that it allows the redundant computation to be done in parallel and thus has little or no impact on performance. The drawback is the cost in terms of area and power for the redundant hardware. One approach to reduce the hardware requirements is to use **information redundancy** where the outputs are encoded with error detecting or correcting codes [Lin 1983] [Peterson 1972]. Generally, information redundancy using codes requires less hardware to generate the redundant information bits than hardware redundancy that replicates hardware units. The codes can be selected to minimize the redundancy required for the class of faults that need to be tolerated. The drawback of information redundancy is the added complexity in design compared with simply replicating hardware units.

3.2 FUNDAMENTALS OF FAULT TOLERANCE

The failure rate, $\lambda(t)$, for a component varies over time. As was shown in Figure 1.2 in Chapter 1, the graph of the failure rate *versus* time tends to look like a bathtub curve. It is high during the infant mortality period and then becomes relatively constant over the normal lifetime of the part, and finally increases again during the wearout period. The failure rate is typically measured in units of **FITS**, which is the number of failures per 10^9 hours.

A system is constructed from various components. If no fault tolerance is provided in the system, then if any component fails, the whole system will fail. In that case, the failure rate for a system, λ_{sys}, that consists of k components is equal to the sum of the individual failure rates of each of its components ($\lambda_{c,i}$):

$$\lambda_{sys} = \sum_{i=1}^{k} \lambda_{c,i}$$

However, as we will show, fault-tolerant design can be used to construct a more dependable system from less dependable components by using redundancy. Some important measures of dependability that will be defined in this section are reliability, mean time to failure (MTTF), maintainability, and availability.

3.2.1 Reliability

The **reliability** of a component is defined as the probability that the component will perform its required function for a specified period of time. If the component is working at time $t = 0$, then $R(t)$ is the probability that the component is still working at time t. If the failure rate for the component is assumed to be constant, which is generally a good approximation for components that have passed the infant mortality period, then the reliability of a component follows the **exponential failure law** and is given by the following expression:

$$R(t) = e^{-\lambda t}$$

where λ is the failure rate.

Consider the system shown in Figure 3.1, which consists of three components: A, B, and C. All three components must be operational in order for the system to be operational. Hence, the system reliability in this case would be equal to the product of the component reliabilities:

$$R_{sys} = R_A R_B R_C$$

Now consider the system in Figure 3.2 where there are two components B's in parallel. In this case, there is a redundant component B, and the system will be

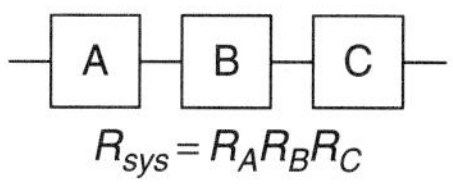

A B C

$R_{sys} = R_A R_B R_C$

■ **FIGURE 3.1**

System reliability for series system.

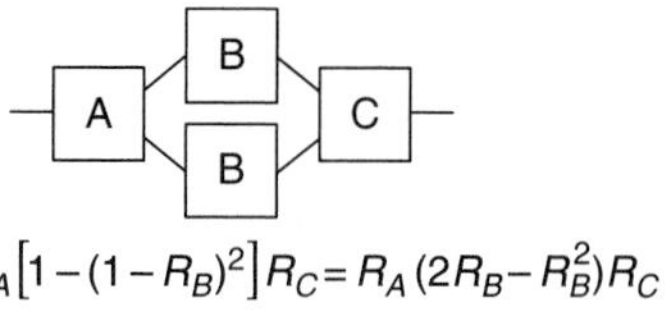

■ **FIGURE 3.2**

System reliability with component B in parallel.

operational if both component A and C are operational and if either of the two component B's is operational. Hence, the system reliability in this case is as shown:

$$R_{sys} = R_A\left[1-(1-R_B)^2\right]R_C = R_A(2R_B - R_B^2)R_C$$

Here the system is able to tolerate one of the component B's failing, and hence its reliability is improved. This simple example illustrates the general principle of how redundancy can be used to improve system reliability. A number of schemes for introducing redundancy in a system to improve its reliability are further described in Section 3.4.

3.2.2 Mean Time to Failure (MTTF)

The average time before a component or system fails is called the ***mean time to failure*** (MTTF) and is given by:

$$MTTF = \int_0^{\infty} R(t)dt$$

In other words, it is equal to the area under the reliability curve when plotted against time. This holds for any failure distribution. The units used for MTTF are typically hours.

For the exponential failure law, MTTF is equal to:

$$MTTF = \int_0^{\infty} e^{-\lambda t}dt = \frac{1}{\lambda}$$

Example 3.1

For the system shown in Figure 3.1, if each of the three components A, B, and C has a failure rate (λ_i) of 10,000 FITS, then the system failure rate (λ_{sys}) is (10,000/10^9) × 3 = 3×10^{-5} failures/hour. Because $MTTF_{sys} = 1/\lambda_{sys}$, the mean time before the system fails is 33,333 hours. Given that $R(t) = e^{-\lambda t}$, the probability that the system has not failed after 1 year is equal to:

$$R(1\ year) = e^{-(3\times 10^{-5})(365\times 24)} = 0.769$$

Example 3.2

For the system shown in Figure 3.2, if each of the three components A, B, and C, has the same failure rate of 10,000 FITS, then they each have the same reliability. Hence, $R_{sys} = R_A(2R_B - R_B^2)R_C = 2R^3 - R^4$. $MTTF_{sys}$ can then be computed as:

$$MTTF_{sys} = \int_0^\infty R_{sys}dt = \int_0^\infty (2R^3 - R^4)dt = \int_0^\infty (2e^{-3\lambda t} - e^{-4\lambda t})dt = \frac{2}{3\lambda} - \frac{1}{4\lambda}$$

$$= \frac{2}{3 \times 10^{-5}} - \frac{1}{4 \times 10^{-5}} = 41{,}666 \text{ hours}$$

The probability that the system has not failed after 1 year is equal to:

$$R(1\ year) = 2e^{-(3\times 10^{-5})(365\times 24)} - e^{-(4\times 10^{-5})(365\times 24)} = 0.834$$

Comparing this result with what was computed in Example 3.1 shows the benefit of adding the redundant B component in terms of improving both system reliability and MTTF.

3.2.3 Maintainability

When a system permanently fails and needs to be manually repaired, the probability that the system will be restored back to operation after a specified period of time is defined as the **maintainability**, $M(t)$, of the system. Assuming the system is failed at time $t = 0$, then $M(t)$ is the probability that it is repaired and operational again after time t. The time for repairing a system can be divided into two periods. The first is the **passive repair time**, which is the time required for a service engineer to travel to the repair site, and the second is the **active repair time**, which is the time required to locate the failing component(s), repair or replace it (them), and verify that the system is operational again. The active repair time can be improved by designing the system so that it is easy to locate the failed components and verify that the system is working correctly again.

The repair rate, μ, measures the rate at which a system is repaired (analogous to the failure rate λ). Maintainability is often modeled as:

$$M(t) = 1 - e^{-\mu t}$$

The mean time to repair, MTTR, is then equal to $1/\mu$.

3.2.4 Availability

Suppose the state of system operation is represented as S, where $S = 0$ means the system is operational, and $S = 1$ means the system is failed. Then S is a function of time t, as illustrated in Figure 3.3.

Suppose the system is operational at $t = 0$, it fails at t_1, and is repaired and brought back to operation at t_2 by some software modification, reset, or hardware

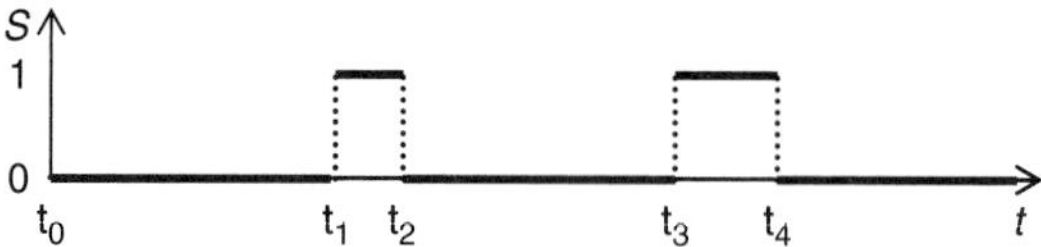

■ **FIGURE 3.3**

System operation and repair.

replacement. Similar failure and repair events happen at t_3 and t_4. The average duration of intervals when the system is operational such as $t_1 - t_0$ and $t_3 - t_2$ is equal to the MTTF. The average duration of intervals when the system is unoperational such as $t_2 - t_1$ and $t_4 - t_3$ is equal to the MTTR.

The fraction of time that a system is operational is called the **system availability** and is given by:

$$system\ availability = \frac{MTTF}{MTTF + MTTR}$$

This formula is widely used in reliability engineering; for example, telephone systems are required to have system availability of 0.9999 (simply called **four nines**), whereas high-reliability systems may require seven nines or more. Achieving such high system availability requires using fault-tolerant design to construct a dependable system from less reliable components.

Component reliability depends on the robustness of the component design as well as the thoroughness of the reliability screens used during manufacturing test. One of the key reliability screens that is used is burn-in where a component is operated at an elevated temperature to age it past its infant mortality period. This helps to eliminate marginal parts so that the surviving parts that are placed in a system will have a lower failure rate. Another useful reliability screen is I_{DDQ} testing, which detects anomalies in quiescent current, which may indicate a marginal part. Unfortunately, burn-in is expensive, and many of the other reliability screens such as I_{DDQ} testing are becoming ineffective for designs manufactured at 90 nm or below. As indicated in [SIA 2004], fundamentally new long-term solutions must be developed for reliability screening and may include significant on-die hardware for stressing or special reliability measurements.

The bottom-line is that component reliability is limited and becoming lower with increasing IC complexity. Component reliability is often not high enough to construct a system that meets availability requirements without the use of fault-tolerant design.

Example 3.3

Consider a system that consists of 100 components, each with a failure rate of 10,000 FITS. When the system fails, it requires an average of 4 hours to repair. What is the system availability? The system failure rate (λ_{sys}) is equal to $(10{,}000/10^9) \times 100 = 10^{-3}$. $MTTF_{sys} = 1/\lambda_{sys}$, so the mean time to failure of the system is 1000 hours and the mean time to repair is 4 hours. Thus, the system availability is equal to $MTTF/(MTTF+MTTR) = 1000/(1000+4) = 0.996$.

3.3 FUNDAMENTALS OF CODING THEORY

Coding data involves using more bits to represent the data than necessary. This information redundancy provides a way to detect and even sometimes correct errors. Errors occur when the value of one or more bits gets flipped because of noise or other disturbances to the signal values. There are many different error-detecting codes [Lin 1983] [Peterson 1972]. They have different properties in terms of the amount of redundancy that they use, the class of errors that they can detect, and how easy it is to encode and decode the data. Some codes are more suitable for certain applications. In this section, some of the most commonly used codes are described and their properties are discussed.

3.3.1 Linear Block Codes

In a **block code**, the data being encoded, called the message, is encoded using n-bit blocks. If each n-bit block can encode m distinct messages, then the **redundancy** of the block code is equal to:

$$redundancy = 1 - \frac{\log_2(m)}{n}$$

If there is no redundancy, then m messages are encoded using $log_2(m)$ bits, which is the minimum number of bits possible. To detect errors, some redundancy is required. In that case, the space of all distinct 2^n blocks is partitioned into codewords and noncodewords, and each of the m messages is assigned to distinct codewords. Errors that cause a codeword to become a noncodeword can be detected. Errors that cause a codeword to become another codeword cannot be detected.

A block code is said to be **separable** if the n-bit blocks can be partitioned into k information bits and $(n-k)$ check bits where the k information bits directly represent 2^k distinct messages. The advantage of a separable code is that the k-bit message can be directly extracted from the codeword without the need for any decoding. The **rate** of a separable block code is defined as k/n.

Example 3.4

An example of a separable block code is a *single-bit parity code* in which one extra check bit is added to the message to indicate whether it contains an odd or even number of 1's. For example, if the message is 101, then the corresponding codeword would contain 3 information bits ($k = 3$), which are the same as the message plus 1 check bit, which is equal to the modulo-2 sum (XOR) of the 3 information bits. Hence, the codeword would have a total of 4 bits (n=4) and be equal to 101**0** where the last 0 is the parity of the 3 information bits. The redundancy and rate of a single-bit parity code is equal to:

$$redundancy = 1 - \frac{\log_2(m)}{n} = 1 - \frac{\log_2(2^k)}{n} = 1 - \frac{k}{n} = \frac{n-k}{n} = \frac{1}{n}$$

$$rate = \frac{k}{n} = \frac{n-1}{n}$$

For the example here with 3 information bits, the redundancy would be 1/4 and the rate would be 3/4. Separable block codes are typically denoted as **(*n*,*k*) block codes**. So in this example, the single-bit parity code with 3 information bits is a (4,3) block code.

Example 3.5

An example of a nonseparable block code is a *one-hot code* in which every codeword contains a single 1 and the rest 0. For example, if $n = 8$, then there are eight codewords (10000000, 01000000, 0010000, 00010000, 00001000, 00000100, 00000010, 00000001). For an n-bit one-hot code, the number of distinct messages that can be represented is n. The redundancy of a one-hot code is equal to:

$$redundancy = 1 - \frac{\log_2(m)}{n} = 1 - \frac{\log_2(n)}{n}$$

If there are eight messages, the redundancy would be 5/8. This is higher than the redundancy for a single-bit parity code with eight messages, which is equal to 1/8 (as shown in Example 3.4). Moreover, the drawback of a nonseparable code is that some decoding logic would be needed to convert the one-hot codeword back to the corresponding message.

Linear block codes are a special class of block codes in which the modulo-2 sum of any two codewords is also a codeword [Lin 1983]. The codewords for a linear block code correspond to the null space of a $(n-k) \times n$ Boolean matrix, which is called the **parity check matrix, $H_{(n-k)\times n}$**. In other words,

$$cH^T = 0$$

for any n-bit codeword c. The corresponding n-bit codeword c is generated for a k-bit message m, by multiplying m with a **generator matrix, $G_{k\times n}$**. In other words,

$$c = mG$$

and thus

$$GH^T = 0$$

A **systematic block code** is one in which the first k-bits correspond to the message and the last $n-k$ bits correspond to the check bits [Lin 1983]. For a systematic code, the generator matrix has the following form: $G = [I_{k\times k} : P_{k\times(n-k)}]$—that is, it consists of a $k \times k$ identity matrix followed by $P_{k\times(n-k)}$. Note that any generator matrix can be put in systematic form by performing Gaussian elimination. If $G = [I_{kxk} : P_{kx(n-k)}]$, then $H = [I_{(n-k)x(n-k)} : P^T_{(n-k)xk}]$.

Example 3.6

For a single-bit parity code, there is one check bit, which is equal to the modulo-2 sum of all the information bits. The G-matrix and H-matrix for a single-bit parity (4,3) code are shown here:

$$G = \begin{bmatrix} 1 & 0 & 0 & 1 \\ 0 & 1 & 0 & 1 \\ 0 & 0 & 1 & 1 \end{bmatrix} \quad H = \begin{bmatrix} 1 & 1 & 1 & 1 \end{bmatrix}$$

For the message $m = [1\ 0\ 1]$. The corresponding codeword c is obtained as follows:

$$c = mG = \begin{bmatrix} 1 & 0 & 1 \end{bmatrix} \begin{bmatrix} 1 & 0 & 0 & 1 \\ 0 & 1 & 0 & 1 \\ 0 & 0 & 1 & 1 \end{bmatrix} = \begin{bmatrix} 1 & 0 & 1 & 0 \end{bmatrix}$$

The codeword c is in the null space of the H-matrix:

$$cH^T = \begin{bmatrix} 1 & 0 & 1 & 0 \end{bmatrix} \begin{bmatrix} 1 \\ 1 \\ 1 \\ 1 \end{bmatrix} = 0$$

Example 3.7

Consider the code corresponding to the following G-matrix and H-matrix:

$$G = \begin{bmatrix} 1 & 0 & 0 & 1 & 0 & 1 \\ 0 & 1 & 0 & 1 & 1 & 1 \\ 0 & 0 & 1 & 0 & 1 & 1 \end{bmatrix} \quad H = \begin{bmatrix} 1 & 1 & 0 & 1 & 0 & 0 \\ 0 & 1 & 1 & 0 & 1 & 0 \\ 1 & 1 & 1 & 0 & 0 & 1 \end{bmatrix}$$

From the dimensions of the matrices it can be determined that it is a (6,3) code (*i.e.*, $n = 6$, $k = 3$, and $n - k = 3$). The matrices are in systematic form, so each of the first three columns of the H-matrix corresponds to the 3 information bits, and each of the last three columns correspond to the 3 check bits. Each check bit is equal to the parity of some of the information bits. Each row in the H-matrix corresponds to a check bit and indicates the set of bits that it is a parity check for. In other words, the first row of the H-matrix indicates that the first check bit (corresponding to the fourth column) is equal to the parity of the first two information bits. The second row of the H-matrix indicates that the second check bit (corresponding to the fifth column) is the parity of the second and third information bits. The last row indicates that the last check bit is the parity of all 3 information bits. For the message $m = [1\ 0\ 1]$, the corresponding codeword, c, would be computed as follows:

$$c = mG = \begin{bmatrix}1 & 0 & 1\end{bmatrix}\begin{bmatrix}1&0&0&1&0&1\\0&1&0&1&1&1\\0&0&1&0&1&1\end{bmatrix} = \begin{bmatrix}1 & 0 & 1 & 1 & 1 & 0\end{bmatrix}$$

The **distance** between two codewords is the number of bits in which they differ. The **distance of a code** is equal to the minimum distance between any two codewords in the code. If $n = k$ (*i.e.*, no redundancy is used), then the distance between any two messages is 1. For single-bit parity, the distance between any two codewords is 2 because all codewords differ in at least 2 bits.

A code with distance d can detect up to $d-1$ errors and can correct up to $\lfloor (d-1)/2 \rfloor$ errors. So a single-bit parity code can detect one-bit errors, but cannot correct them. Consider the codewords 0000 and 0011 (which have even parity), a single-bit error could cause 0000 to become 0010 which is a noncodeword (because it has odd parity) and hence the error would be detected. However, a single-bit error could also cause 0011 to become 0010. As 0010 can be reached because of a single-bit error from more than one codeword, there is no way to correct 0010 because it cannot be determined from which original codeword the error occurred. To correct single-bit errors, the code must have a distance of 3.

Example 3.8

An example of a code with a distance of 3 is a **triplication code** in which each codeword contains three copies of the information bits. For $k = 3$, the triplication code has $n = 9$. The corresponding H-matrix for a (9,3) triplication code is shown here:

$$G = \begin{bmatrix}1&0&0&1&0&0&1&0&0\\0&1&0&0&1&0&0&1&0\\0&0&1&0&0&1&0&0&1\end{bmatrix}$$

$$H = \begin{bmatrix}1&0&0&1&0&0&0&0&0\\0&1&0&0&1&0&0&0&0\\0&0&1&0&0&1&0&0&0\\1&0&0&0&0&0&1&0&0\\0&1&0&0&0&0&0&1&0\\0&0&1&0&0&0&0&0&1\end{bmatrix}$$

Consider the codeword 100100100; if there is a single-bit error in the first bit position, it will result in 000100100 which is a noncodeword. There is no single-bit error from any other codeword that can result in 000100100. Consequently, that noncodeword can be corrected back to 100100100 assuming that only a single-bit error occurred.

A code with distance 3 is a called a **single-error-correcting (SEC) code**. A code with distance 4 is called a **single-error-correcting and double-error-detecting (SEC-DED) code**. A systematic procedure for constructing a SEC code for any value of n was described by [Hamming 1950]. Any H-matrix in which all columns are distinct and no column is all 0 is a SEC code. Based on this, for any value of n, a SEC code can be constructed by simply setting each column equal to the binary representation of the column number starting from 1. The number of rows in the H-matrix (*i.e.*, the number of check bits) will be equal to $\lceil log_2(n+1) \rceil$.

Example 3.9

A SEC Hamming code for $n = 7$ will have $\lceil log_2(n+1) = 3 \rceil$ check bits. The (7,4) code is shown here:

$$H = \begin{bmatrix} 0 & 0 & 0 & 1 & 1 & 1 & 1 \\ 0 & 1 & 1 & 0 & 0 & 1 & 1 \\ 1 & 0 & 1 & 0 & 1 & 0 & 1 \end{bmatrix}$$

It is constructed by making the first column equal to the binary representation of 1, the second column equal to the binary representation of 2, and so forth.

For a Hamming code, correction is done by computing the **syndrome**, ***s***, of the received vector. The syndrome for vector v is computed as $s = Hv^{\mathrm{T}}$. If v is a codeword, then the syndrome will be all 0. However, if v is a noncodeword and only a single-bit error has occurred, then the syndrome will match one of the columns of the H-matrix and hence will contain the binary value of the bit position in error.

Example 3.10

For the (7,4) Hamming code from Example 3.9, suppose the codeword 0110011 had a 1-bit error in the first bit position, which changed it to **1110011**. The syndrome would be equal to:

$$s = vH^T = [1110011] \begin{bmatrix} 0 & 0 & 1 \\ 0 & 1 & 0 \\ 0 & 1 & 1 \\ 1 & 0 & 0 \\ 1 & 0 & 1 \\ 1 & 1 & 0 \\ 1 & 1 & 1 \end{bmatrix} = [001]$$

The syndrome is the binary representation of 1, and hence it indicates that the first bit is in error. By flipping the first bit, the error can be corrected.

An SEC Hamming code can be made into an SEC-DED code by adding a parity check over all bits. This extra parity bit will be 1 for a single-bit error, but will be 0 for a double bit error. This makes it possible to detect double-bit errors so that they are not assumed to be a single-bit error and hence miscorrected.

Example 3.11

The (7,4) SEC code from Example 3.9 can be converted into a (7,3) SEC-DED code by adding a parity check over all bits, which corresponds to adding an all 1 row to the H-matrix as shown here:

$$H = \begin{bmatrix} 0 & 0 & 0 & 1 & 1 & 1 & 1 \\ 0 & 1 & 1 & 0 & 0 & 1 & 1 \\ 1 & 0 & 1 & 0 & 1 & 0 & 1 \\ 1 & 1 & 1 & 1 & 1 & 1 & 1 \end{bmatrix}$$

Suppose the codeword 0110011 has a 2-bit error in the first two bit positions, which changes it to **10**10011. The syndrome is computed as follows:

$$s = vH^T = [1010011] \begin{bmatrix} 0 & 0 & 1 & 1 \\ 0 & 1 & 0 & 1 \\ 0 & 1 & 1 & 1 \\ 1 & 0 & 0 & 1 \\ 1 & 0 & 1 & 1 \\ 1 & 1 & 0 & 1 \\ 1 & 1 & 1 & 1 \end{bmatrix} = [0010]$$

In this case, the syndrome does not match any column in the H-matrix thereby indicating that it is a double-bit error, which cannot be corrected (although it is detected).

Another method for constructing an SEC-DED code was described in [Hsiao 1970]. The **weight** of a column is defined as the number of 1's in the column. Any H-matrix in which all columns are distinct and all columns have odd weight corresponds to an SEC-DED code. Based on these criteria, a Hsiao code for any value of n can be constructed by first using all possible weight-1 columns, then using all possible weight-3 columns, then using all possible weight-5 columns, and so forth up until n columns have been formed. The number of check bits required for an **SEC-DED Hsiao code** is $\lceil log_2(n+1) \rceil$. The advantage of a Hsiao code is that it minimizes the number of 1's in the H-matrix, which results in less hardware and less delay for computing the syndrome. The disadvantage is that correcting an error requires some decoding logic to determine which bit is in error because the syndrome is no longer simply the binary value of the bit position in error (*i.e.*, each column in the H-matrix is not necessarily the binary representation of the column number).

Example 3.12

A (7,3) Hsiao SEC-DED code is constructed as shown here:

$$H = \begin{bmatrix} 0 & 0 & 0 & 1 & 0 & 1 & 1 \\ 0 & 0 & 1 & 0 & 1 & 0 & 1 \\ 0 & 1 & 0 & 0 & 1 & 1 & 0 \\ 1 & 0 & 0 & 0 & 1 & 1 & 1 \end{bmatrix}$$

First the four weight-1 column vectors are used. Then weight-3 column vectors are used. This (7,3) Hsiao code has 13 entries in the H-matrix that are 1's whereas the SEC-DED code in Example 3.11 had 19 entries in the H-matrix that are 1's. Fewer 1's in the H-matrix requires fewer XOR gates being needed to generate the syndrome and consequently less propagation delay. Because the syndrome needs to be generated every time to check for an error, having less delay is beneficial for performance. The drawback is that the correction is a little more complicated because the syndrome is not the binary representation of the bit position in error; however, correction is rarely required and hence has negligible impact on performance. If a single-bit error occurs that results in the syndrome 0100, determining which bit is in error is done by identifying which column it matches in the H-matrix. In this case, 0100^T would match the third column, so the third bit is in error.

3.3.2 Unidirectional Codes

Unidirectional errors are defined as errors in a block of data that only cause 1→0 errors or only cause 0→1 errors, but not both within the same block of data [Lala 1985]. In other words, there are two possibilities: either all the errors in a block of data are 1's that become 0's, or all the errors in a block of data are 0's that become 1's. For example, if the correct codeword is 111000, then examples of unidirectional errors would be **00**1000, **000**000, and 1**0**1000 (all of which are 1→0 errors) or 111**1**0**1**, 111**1**00, and 111**111** (all of which are 0→1 errors). Examples of nonunidirectional errors would be 1**0**100**1**, **0**110**0**1, and **0**11**0**1**1** because both 1→0 errors and 0→1 errors are present at the same time.

Error-detecting codes that are able to detect all unidirectional errors in a codeword (*i.e.*, any number of bits in error provided they are unidirectional) are called **all unidirectional error detecting (AUED) codes**. Single-bit parity can detect all errors where there are an odd number of bits in error, but it cannot detect errors where there is an even number of bits in error, and hence it is not an AUED code. In fact, no linear code is AUED because all linear codes must contain the all 0 codeword (which always exists in the null space of the H-matrix) and hence there will always be some set of 0→1 errors that will make the all 0 codeword equal to some other codeword. Detecting AUED requires a nonlinear code.

3.3.2.1 Two-Rail Codes

A **two-rail code** has one check bit for each information bit which is equal to the complement of the information bit [Lala 1985]. For example, if the message is 101, then the check bits for a two-rail code would be the complement which is 010. The set of codewords for a (6,3) two-rail code where the information bits are the first 3 bits and the check bits are the last 3 bits would be 000111, 001110, 010101, 011100, 100110, 101010, 110001, and 111000.

A two-rail code is AEUD because no unidirectional errors can change the value of both an information bit and its corresponding check bit. 0→1 errors can only affect 0's and 1→0 errors can only affect 1's. Either an information bit will be affected or its corresponding check bit will be affected, but not both because they have complementary values. Hence, no codeword can be changed into another codeword through any set of unidirectional errors.

For a two-rail code, $n = 2k$ because there is one check bit for each information bit; thus it requires 50% redundancy, which is quite high. Lower redundancy AUED codes exist and are described next.

3.3.2.2 Berger Codes

Berger codes [Berger 1961] have the lowest redundancy of any separable code for detecting all possible unidirectional errors. If there are k information bits, then a Berger code requires $log_2\lceil k+1 \rceil$ check bits, which are equal to the binary representation of the number information bits that are 0. For example, if the information bits are 1000101, then there are $log_2\lceil 7+1 \rceil = 3$ check bits, which are equal to 100, which is the binary representation of 4 since four information bits are 0. The set of codewords for a (5,3) Berger code where the 3 information bits come first followed by the two check bits are 00011, 00110, 01010, 01101, 10010, 10101, 11001, and 11100.

Any unidirectional error will be detected by a Berger code. If the unidirectional error contains $1 \rightarrow 0$ errors, then it will increase the number of 0's in the information bits, but can only decrease the binary number represented by the check bits thereby causing a mismatch. If the unidirectional error contains $0 \rightarrow 1$ errors, then it will reduce the number of 0's in the information bits, but it can only increase the binary number represented by the check bits.

Example 3.13

If there are 8 information bits ($k = 8$), then a Berger code requires $log_2\lceil 8+1 \rceil = 4$ check bits (*i.e.*, $n = 12$), so its redundancy is:

$$redundancy = 1 - \frac{\log_2(m)}{n} = 1 - \frac{\log_2(2^k)}{n} = 1 - \frac{8}{12} = \frac{1}{4} = 25\%$$

This is much less than a (16,8) two-rail code, which requires 50% redundancy. The redundancy advantage of the Berger code becomes greater as k increases.

3.3.2.3 Constant Weight Codes

The codes discussed so far have been separable codes. Constant weight codes are **nonseparable codes,** which have lower redundancy than Berger codes for detecting all unidirectional errors. Each codeword in a constant weight code has the same number of 1's. For example, a 2-out-of-3 constant weight code would have three codewords that each have two 1's in them (*i.e.*, 110, 101, and 011). One type of constant weight code is a *one-hot code,* which is a 1-out-of-n code where there are n codewords each having a single 1.

Constant weight codes detect all unidirectional errors because unidirectional errors will either increase the number of 1's or decrease the number of 1's, thus resulting in a noncodeword.

The number of codewords in an m-out-of-n constant weight code is equal to C_m^n. The number of codewords is maximized when m is as close to $n/2$ as possible. Thus, for an n-bit constant weight code, the minimum redundancy code is $(n/2)$-out-of-n if n is even and $(n/2-0.5$ or $n/2+0.5)$-out-of-n if n is odd as that maximizes the number of codewords. For example, if $n = 12$, then the 6-out-of-12 code has minimum redundancy, and if $n = 13$, then either the 6-out-of-13 or the 7-out-of-13 would have the minimum redundancy.

Example 3.14

A 6-out-12 constant weight code has $C_6^{12} = 924$ codewords, so its redundancy is:

$$redundancy = 1 - \frac{\log_2(m)}{n} = 1 - \frac{\log_2(924)}{12} = 17.9\%$$

In comparison, a 12-bit Berger code has 4 check bits and thus has only $2^8 = 256$ codewords and a redundancy equal to:

$$redundancy = 1 - \frac{\log_2(m)}{n} = 1 - \frac{\log_2(2^8)}{12} = 33.3\%$$

A constant weight code requires less redundancy than a Berger code, but the drawback of a constant weight code is that it is nonseparable, so some decoding logic is needed to convert each codeword back to its corresponding binary message. This is not the case for a Berger code where the information bits are separate from the check bits and hence can be extracted without any decoding.

3.3.3 Cyclic Codes

Cyclic codes are a special class of linear codes in which any codeword shifted cyclically (rotating the last bit around to become the first bit) is another codeword. Cyclic codes are used for detecting **burst errors** including those that wrap around from the last to the first bit of the codeword. The length of a burst error is the number of bits between the first error and last error, inclusive. For example, if the original codeword is 00000000 and errors occur in the 3rd, 4th, 5th, and 6th bit positions resulting in 00**1111**00, then it is a burst error of length 4. If the errors occurred only in the 3rd, 4th, and 6th bit positions resulting in 00**1101**00, it would still be a burst error of length 4. Any number of errors can occur between the first error and last error. The reason why burst errors are of particular interest is because multi-bit errors tend to be clustered together. Noise sources tend to affect a contiguous set of bus lines in communication channels. Less redundancy is required to detect burst errors than general multi-bit errors. For example, some distance-2 codes can detect all burst errors of length 4, whereas detecting all possible 4-bit errors requires a distance-5 code.

The most widely used cyclic code is a **cyclic redundancy check (CRC) code**, which uses a binary alphabet and is based on GF(2). A CRC code is an (n,k) block code formed using a **generator polynomial**, $\mathbf{g(x)}$, which is also called the **code generator**. It is a degree $n-k$ polynomial (*i.e.*, its degree is the same as the number of check bits) and has the following form:

$$g(x) = g_{n-k}x^{n-k} + \ldots + g_2x^2 + g_1x + g_0$$

Each coefficient g_i is binary. A codeword, $c(x)$, is formed by Galois (modulo-2) multiplication of the message, $m(x)$, with the generator, $g(x)$:

$$c(x) = m(x)g(x)$$

Example 3.15

The codewords for a (6,4) CRC code generated by $g(x) = x^2 + 1$ are shown in Table 3.1.

TABLE 3.1 ■ Codewords for (6,4) CRC Code Generated by $g(x) = x^2+1$

Message	m(x)	g(x)	c(x)	Codeword
0000	0	x^2+1	0	000000
0001	1	x^2+1	x^2+1	000101
0010	x	x^2+1	x^3+x	001010
0011	$x+1$	x^2+1	x^3+x^2+x+1	001111
0100	x^2	x^2+1	x^4+x^2	010100
0101	x^2+1	x^2+1	x^4+1	010001
0110	x^2+x	x^2+1	$x^4+x^3+x^2+x$	011110
0111	x^2+x+1	x^2+1	x^4+x^3+x+1	011011
1000	x^3	x^2+1	x^5+x^3	101000
1001	x^3+1	x^2+1	$x^5+x^3+x^2+1$	101101
1010	x^3+x	x^2+1	x^5+x	100010
1011	x^3+x+1	x^2+1	x^5+x^2+x+1	100111
1100	x^3+x^2	x^2+1	$x^5+x^4+x^3+x^2$	111100
1101	x^3+x^2+1	x^2+1	$x^5+x^4+x^3+1$	111001
1110	x^3+x^2+x	x^2+1	$x^5+x^4+x^2+x$	110110
1111	x^3+x^2+x+1	x^2+1	x^5+x^4+x+1	110011

A CRC code is a linear block code, so it has a corresponding G-matrix and H-matrix. Each row of the G-matrix is simply a shifted version of the generator polynomial as shown here:

$$G = \begin{bmatrix} g_{n-k} & \cdots & g_1 & g_0 & 0 & 0 & 0 \\ 0 & g_{n-k} & \cdots & g_1 & g_0 & 0 & 0 \\ \cdot & \cdot & \cdot & \cdot & \cdot & \cdot & \cdot \\ 0 & 0 & \cdots & g_{n-k} & \cdots & g_1 & g_0 \end{bmatrix}$$

Example 3.16

A (6,4) CRC code generated by $g(x) = x^2 + 1$ has the following G-matrix:

$$G = \begin{bmatrix} 1 & 0 & 1 & 0 & 0 & 0 \\ 0 & 1 & 0 & 1 & 0 & 0 \\ 0 & 0 & 1 & 0 & 1 & 0 \\ 0 & 0 & 0 & 1 & 0 & 1 \end{bmatrix}$$

The CRC codes shown so far have not been systematic (*i.e.*, the first k bit positions of the codeword do not correspond to the message). To obtain a systematic CRC code, the codewords can be formed as follows:

$$c(x) = m(x)x^{n-k} + r(x)$$

$$r(x) = remainder\ of\ \frac{m(x)x^{n-k}}{g(x)}$$

Note that this involves using Galois (modulo-2) division rather than multiplication. This is nice because Galois division can be performed using a ***linear feedback shift register*** (LFSR), which is a compact circuit.

Example 3.17

To illustrate Galois division, consider encoding $m(x) = x^2 + x$ with $g(x) = x^2 + 1$. This requires dividing $m(x)x^{n-k} = (x^2 + x)(x^2) = x^4 + x^3$ by $g(x)$. Note that subtraction in Galois arithmetic is the same as modulo-2 addition.

```
      111
101 |11000
     101
      110
      101
       110
       101
       [11] remainder
```

The remainder $r(x) = x + 1$, so the codeword is equal to:

$$c(x) = m(x)x^{n-k} + r(x) = (x^2 + x)(x^2) + x + 1 = x^4 + x^3 + x + 1$$

Consider another example of encoding $m(x) = x^2$ with $g(x) = x^2 + 1$. The Galois division is shown next. Notice that in this case, the first bit of the quotient is 1 even though 101 is larger than 100. Each bit of the quotient will be a 1 anytime the high order bit of the remaining dividend is 1 and will be 0 anytime the high order bit of the remaining dividend is 0.

```
      101
101 |10000
     101
      010
      000
       100
       101
       [01] remainder
```

The remainder $r(x) = 1$, so the codeword is equal to:

$$c(x) = m(x)x^{n-k} + r(x) = (x^2)(x^2) + 1 = x^4 + 1$$

Example 3.18

The codewords for a systematic (6,4) CRC code generated by $g(x)=x^2+1$ are shown in Table 3.2.

TABLE 3.2 ■ Codewords for Systematic (6,4) CRC Code Generated by $g(x)=x^2+1$

Message	m(x)	g(x)	r(x)	c(x)	Codeword
0000	0	x^2+1	0	0	000000
0001	1	x^2+1	1	x^2+1	000101
0010	x	x^2+1	x	x^3+x	001010
0011	$x+1$	x^2+1	$x+1$	x^3+x^2+x+1	001111
0100	x^2	x^2+1	1	x^4+1	010001
0101	x^2+1	x^2+1	0	x^4+x^2	010100
0110	x^2+x	x^2+1	$x+1$	x^4+x^3+x+1	011011
0111	x^2+x+1	x^2+1	x	$x^4+x^3+x^2+x$	011110
1000	x^3	x^2+1	x	x^5+x	100010
1001	x^3+1	x^2+1	$x+1$	x^5+x^2+x+1	100111
1010	x^3+x	x^2+1	0	x^5+x^3	101000
1011	x^3+x+1	x^2+1	1	$x^5+x^3+x^2+1$	101101
1100	x^3+x^2	x^2+1	$x+1$	x^5+x^4+x+1	110011
1101	x^3+x^2+1	x^2+1	x	$x^5+x^4+x^2+x$	110110
1110	x^3+x^2+x	x^2+1	1	$x^5+x^4+x^3+1$	111001
1111	x^3+x^2+x+1	x^2+1	0	$x^5+x^4+x^3+x^2$	111100

Generating the check bits for CRC codes can be performed using an LFSR whose **characteristic polynomial**, $f(x)$, corresponds to the generator polynomial for the code. To encode a message, $n-k$ 0's are appended to the end of a message, and then it is shifted into an LFSR to compute the remainder. This is illustrated in the following example.

Example 3.19

To encode the message $m(x)=x^2+x+1$ using a (6,3) CRC code with the generator $g(x)=x^3+x+1$, the following LFSR whose characteristic polynomial, $f(x)$, corresponds to the generator polynomial can be used. The initial state of the LFSR is reset to all 0, and $(n-k=3)$ 0's are appended to the end of the message which is then shifted into the LFSR is shown in Figure 3.4.

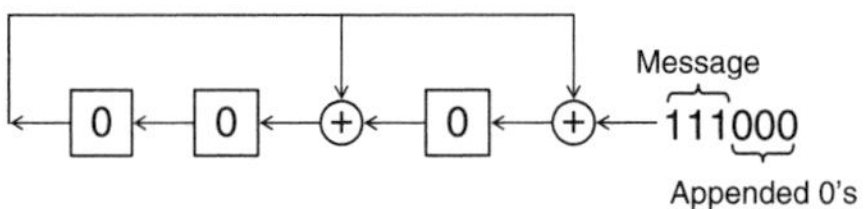

■ **FIGURE 3.4**

Using LFSR to generate check bits for CRC code.

After shifting this in, the final state of the LFSR contains the remainder 010 as shown in Figure 3.5. The three appended 0's are then replaced with the remainder to form the codeword, which is thus equal to 111010.

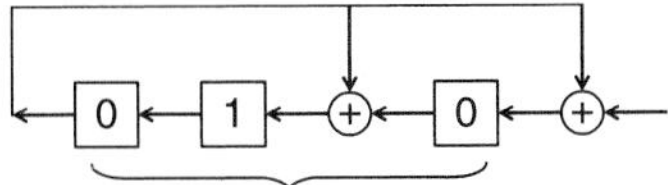

■ **FIGURE 3.5**

Final LFSR state after shifting in message and appended 0's.

The CRC codeword can be transmitted across a communication channel to a receiver. To check if there were any errors in transmission, the receiver can simply shift the codeword into an LFSR with the same characteristic polynomial as was used to generate it (as illustrated in Figure 3.6). The final state of the LFSR will be all 0 if there is no error and will be nonzero if there is an error.

0 0 0 111010

codeword to check

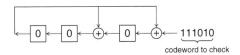

■ **FIGURE 3.6**

Using LFSR to checking CRC codeword.

One key issue for CRC codes is selecting the generator polynomial. A generator polynomial in which the first and last bit of the polynomial is 1 will detect burst errors of length $n-k$ or less. If the generator polynomial is a multiple of $(x+1)$, then it will detect any number of odd errors. Moreover, if the generator polynomial is equal to:

$$g(x) = (x+1)p(x)$$

where $p(x)$ is a **primitive polynomial** of degree $n-k-1$ and $n < 2^{n-k-1}$, then it will detect all single, double, triple, and odd errors.

There are some commonly used CRC generator polynomials. Some of these are shown in Table 3.3 [Tanenbaum 1981].

TABLE 3.3 ■ Some Commonly Used CRC Generator Polynomials

CRC Code	Generator Polynomial
CRC-5 (USB token packets)	x^5+x^2+1
CRC-12 (Telecom systems)	$x^{12}+x^{11}+x^3+x^2+x+1$
CRC-16-CCITT (X25, Bluetooth)	$x^{16}+x^{12}+x^5+1$
CRC-32 (Ethernet)	$x^{32}+x^{26}+x^{23}+x^{22}+x^{16}+x^{12}+x^{11}+x^{10}+x^8+x^7+x^5+x^4+x+1$
CRC-64 (ISO)	$x^{64}+x^4+x^3+x+1$

3.4 FAULT TOLERANCE SCHEMES

As discussed earlier, without using redundancy, the reliability of a system is limited by the reliability of its components. Moreover, the system is unprotected from transient errors. Adding fault tolerance to a design to improve the dependability of a system requires the use of redundancy. In this section, a number of fault tolerance schemes are described. They are organized based on the type of redundancy used: hardware, time, or information [Lala 2001].

3.4.1 Hardware Redundancy

Hardware redundancy involves replicating hardware units, which can be done at any level of design (gate-level, module-level, chip-level, board-level, or system-level). There are three basic forms of hardware redundancy:

1. *Static* (also called *passive*). **Static redundancy** masks faults rather than detects them. It prevents faults from resulting in errors at the output. It achieves fault tolerance without any action from the operator or CPU of the system. The interconnections are fixed.
2. *Dynamic* (also called *active*). **Dynamic redundancy** detects faults and then reconfigures the system to use spare fault-free hardware. It requires the ability to detect, locate, and recover from a fault. The interconnections are reconfigurable.
3. *Hybrid*. **Hybrid redundancy** combines both active and passive approaches. It uses fault masking to prevent errors, but it also uses fault detection to reconfigure the system to replace the failed hardware with a spare.

Schemes using each of the three forms of hardware redundancy are described in this section.

3.4.1.1 Static Redundancy

Static redundancy masks faults so that they do not produce erroneous outputs. For real-time systems where there is no time to reconfigure the system or retry an operation, the ability to mask faults is essential to permit continuous operation. In addition to allowing uninterrupted operation, the other advantage of static redundancy is that it is simple and self-contained without requiring the need to update or roll back the system state.

Triple Modular Redundancy (TMR)

One well-known static redundancy scheme is ***triple modular redundancy*** (TMR). The idea in TMR is to have three copies of a module and then use a **majority voter** to determine the final output (see Figure 3.7). If an error occurs at the output of

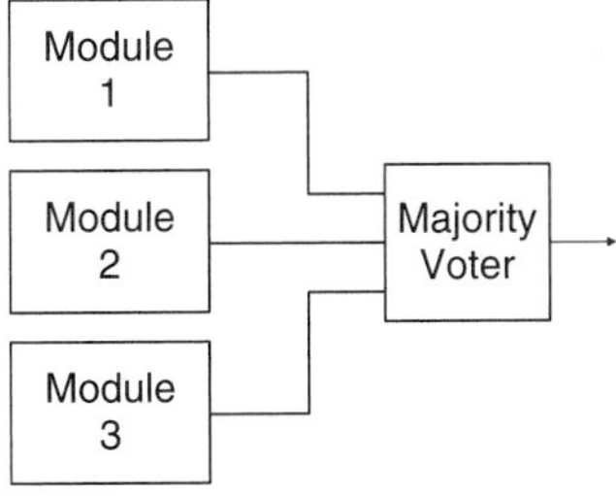

■ **FIGURE 3.7**

Triple modular redundancy (TMR).

one of the modules, the other two error-free modules will outvote it and hence the final output will be correct.

A TMR system will work if the voter works and either all three modules are working or any combination of two modules is working. Thus, if the reliability of each module is R_m and the reliability of the voter is R_v, then the reliability for the TMR system is expressed as follows:

$$R_{TMR} = R_v[R_m^3 + C_2^3 R_m^2(1 - R_m)] = R_v(3R_m^2 - 2R_m^3)$$

The MTTF for a TMR system is equal to:

$$\begin{aligned} MTTF_{TMR} &= \int_0^\infty R_{TMR} dt = \int_0^\infty R_v(3R_m^2 - 2R_m^3)dt = \int_0^\infty e^{-\lambda_v t}(3e^{-2\lambda_m t} - 2e^{-3\lambda_m t})dt \\ &= \frac{3}{2\lambda_m + \lambda_v} - \frac{2}{3\lambda_m + \lambda_v} \end{aligned}$$

Neglecting the failure rate of the voter (which is generally much lower than the module itself), the expression simplifies to:

$$MTTF_{TMR} = \frac{3}{2\lambda_m} - \frac{2}{3\lambda_m} = \left(\frac{5}{6}\right)\left(\frac{1}{\lambda_m}\right) = \frac{5}{6} MTTF_{simplex}$$

$MTTF_{simplex}$ is the MTTF of a system that consists of just the module itself. What this implies is that the MTTF for a TMR system is actually shorter than that for a simplex system. The advantage of TMR is that it can tolerate temporary faults that a simplex system cannot, and it has higher reliability for short mission times. The crossover point can be computed as follows:

$$\begin{aligned} &R_{TMR} = R_{simplex} \\ &3e^{-2\lambda_m t} - 2e^{-3\lambda_m t} = e^{-\lambda_m t} \\ &Solve \Rightarrow t = \frac{\ln 2}{\lambda_m} \approx (0.7)\, MTTF_{simplex} \end{aligned}$$

So R_{TMR} will be greater than $R_{simplex}$ as long as the mission time is shorter than 70% of the MTTF.

Example 3.20

For a component with a failure rate of 1000 FITS, what is its reliability after 2 years?

$$R_{simplex} = e^{-\lambda t} = e^{-(1000\times 10^{-9})(2)(365)(24)} = 0.983$$

If TMR is used, what is the reliability after 2 years (neglecting the failure rate of the voter)?

$$R_{TMR} = 3e^{-2\lambda t} - 2e^{-3\lambda t} = 3e^{-(2)(1000\times 10^{-9})(2)(365)(24)} - 2e^{-(3)(1000\times 10^{-9})(2)(365)(24)} = 0.999$$

Note that the TMR reliability equations are actually somewhat pessimistic because TMR is able to continue error-free operation in the presence of multiple faults as long as they do not cause simultaneous errors at the outputs of two or more modules. For example, the outputs of modules 1 and 2 can be stuck-at-0 and stuck-at-1, respectively, yet the TMR circuit will continue to function properly; this is referred to as compensating module faults [Stroud 1994] [Siewiorek 1998].

One vulnerable part of TMR is the voter. Errors in the voter are not masked. The voter circuit generally is small though, so the probability of errors originating in the voter is fairly minimal. However, the voter circuit can also be replicated; this is referred to as **triplicated TMR** in [Lala 1985].

N-Modular Redundancy (NMR)

TMR is actually one case of a general class of error masking circuits referred to as ***N-modular redundancy*** (NMR). In NMR circuits, N modules can be used along with a majority voter. In this case, the number of failed modules that can be masked is equal to $\lfloor (N-1)/2 \rfloor$. As N increases, the MTTF of the system decreases, but the reliability for short missions increases. However, if the main goal is only to tolerate temporary faults, then typically TMR is sufficient as temporary faults generally cause an error in only one module at a time.

Interwoven Logic

Another method that uses static redundancy to mask errors is **interwoven logic** [Pierce 1965]. Interwoven logic is implemented at the gate level by replacing each gate with 4 gates and using an interconnection pattern that automatically corrects error signals. This is illustrated in Figure 3.8. In this example, the functional design shown on the left is implemented with interwoven logic by replacing each NOR gate with 4 NOR gates shown on the left. So although the original design has a total of 4 NOR gates, the interwoven logic design has a total of 16 NOR gates. The gates are interconnected with a systematic pattern that ensures that any single error will be masked after one or two levels of logic.

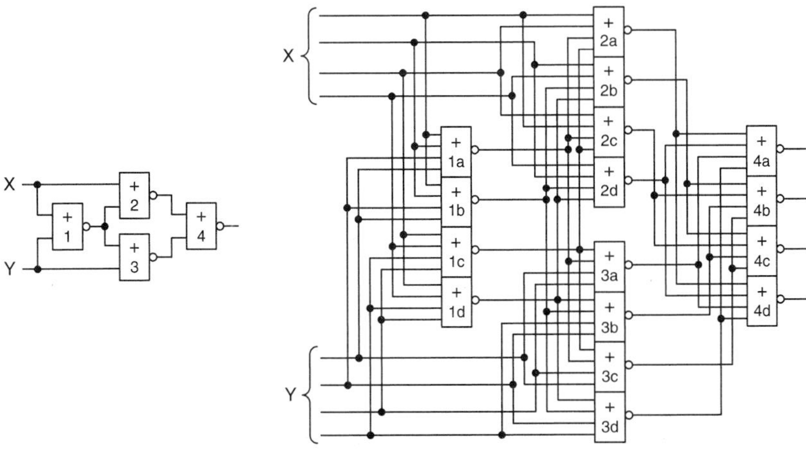

■ **FIGURE 3.8**

Example of interwoven logic for design with four NOR gates.

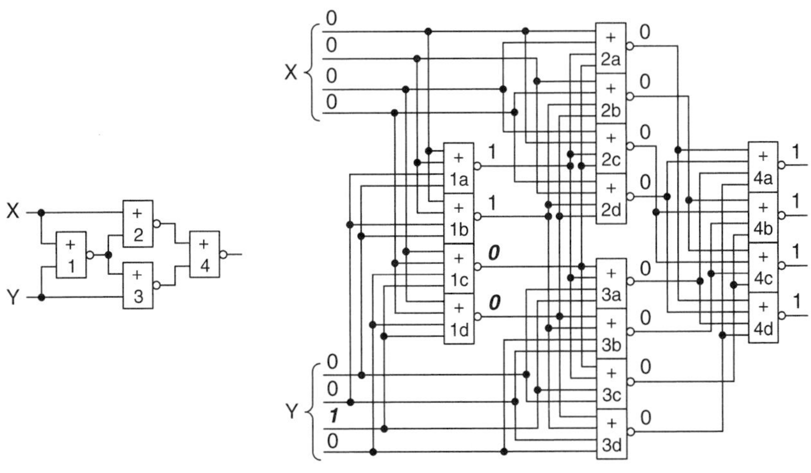

■ **FIGURE 3.9**

Example of error on third *Y* input.

Consider the example shown in Figure 3.9 where the primary inputs *X* and *Y* are both 0. In this case, the output of gates 1 and 4 is 1 and the output of gates 2 and 3 is 0. If there were an error on the third *Y* input that causes it to be 1 instead of a 0, then that error would propagate through the next level of logic and cause the output of gates 1*c* and 1*d* to be a 0 instead of a 1. However, the error would not propagate to the level of logic after that because the output of gates 3*a*, 3*b*, 3*c*, 3*d*, 4*a*, 4*b*, 4*c*, and 4*d* would still be 1 (*i.e.*, the errors would all be masked).

More details about interwoven logic including a systematic design procedure can be found in [Pierce 1965]. Several similar design methodologies for fault masking logic at the gate level have been described in [Tyron 1962], [Schwab 1983], and [Barbour 1989].

Traditionally, interwoven logic has not been as attractive as TMR because it requires a lot of area overhead because of the additional interconnect and gates. However, it has seen some renewed interest by researchers investigating fault-tolerant approaches for emerging nanoelectronic technologies.

3.4.1.2 Dynamic Redundancy

Dynamic redundancy involves detecting a fault, locating the faulty hardware unit, and reconfiguring the system to use a spare hardware unit that is fault-free.

Unpowered (Cold) Spares

One option is to leave the spares unpowered ("cold" spares). The advantage of using this approach is that it extends the lifetime of the spare. If the spare is assumed to not fail until it is powered, and perfect reconfiguration capability is assumed (*i.e.*, failures in reconfiguring to the spare are neglected), then the reliability and MTTF of a module with one unpowered spare is equal to:

$$R_{w/cold\_spare} = (1+\lambda t)e^{-\lambda t}$$

$$MTTF_{w/cold\_spare} = \frac{2}{\lambda}$$

Note that adding one unpowered spare doubles the MTTF. This is because when the original module fails, then the spare powers up and replaces it thereby doubling the total MTTF (because the spare cannot fail until it is powered up). Using N unpowered spares would increase the MTTF by a factor of N. However, keep in mind that this is assuming that faults are always detected and the reconfiguration circuitry never fails.

One drawback of using a cold spare is the extra time required to power and initialize the cold spare when a fault occurs. Another drawback is that the cold spares cannot be used to help in detecting faults. Fault detection in this case requires either using periodic offline testing or using online testing based on time or information redundancy.

Powered (Hot) Spares

Another option is to use powered ("hot" spares). In this case, it is possible to use the spares for online fault detection. One approach is to use **duplicate-and-compare** as illustrated in Figure 3.10. Two copies of the module are compared. If the outputs mismatch at some point, it indicates that one of them is faulty. At that point, a diagnostic procedure must be run to determine whether module A is faulty or module B is faulty. The faulty module is then replaced by a spare module so that

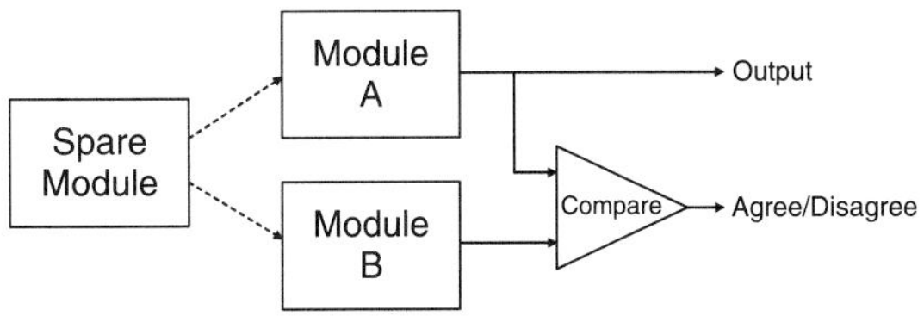

■ **FIGURE 3.10**

Duplicate-and-compare scheme with spares.

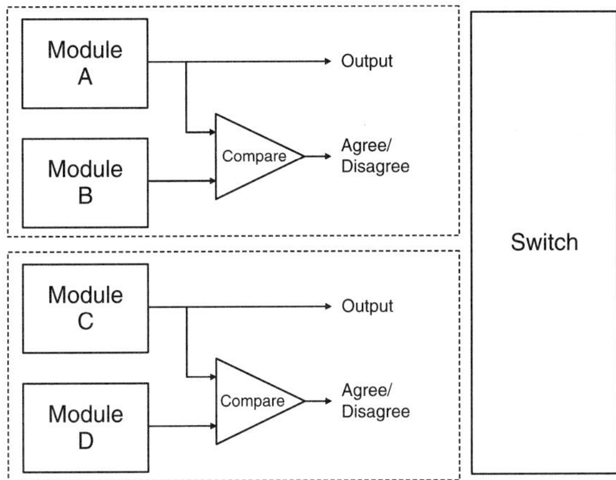

■ **FIGURE 3.11**

Pair-and-a-spare scheme.

the system can keep operating. In the meantime, the system can notify the operator that it has a failed module so that the operator can manually repair or replace the failed module, hopefully before another failure occurs. Any number of spares can be used with this scheme.

One drawback of using a single duplicate-and-compare unit is that a diagnostic procedure has to be executed to determine which module is faulty so that the spare can replace the right one. The system has to halt operation until the diagnostic procedure is complete. To avoid the need for using a diagnostic procedure, a pair of duplicate-and-compare units can be used. This scheme is commonly called "**pair-and-a-spare.**" This is illustrated in Figure 3.11. In this case, if one duplicate-and-compare unit mismatches, then the system switches over and uses the spare duplicate-and-compare unit. Meanwhile, the system can notify the operator that one duplicate-and-compare unit has failed and needs to be repaired or replaced.

TMR/Simplex

As Section 3.4.1.1 demonstrated, the MTTF for a TMR system is lower than for a simplex system. This occurs because once one module fails, then there are two

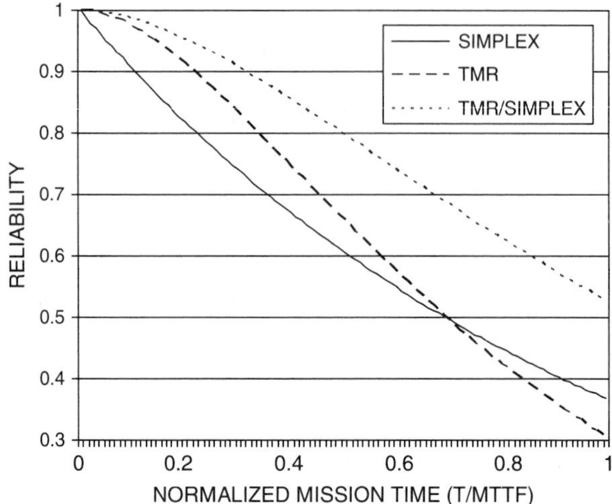

■ **FIGURE 3.12**

Reliability *versus* normalized mission time for simplex, TMR, and TMR/simplex systems.

remaining modules connected to the voter. If either of those modules fails, the whole system fails. The idea in TMR/simplex is that once one of the modules in TMR fails, then the system is reconfigured as a simplex system using one of the remaining good modules. This improves the MTTF while still retaining all the advantages of a normal TMR system while there are three good modules. The graph in Figure 3.12 compares the reliability of TMR alone, simplex alone, and TMR/simplex *versus* the normalized mission time ($time/MTTF_{simplex}$). As the figure shows, TMR is better than simplex up until 0.7 $MTTF_{simplex}$ as demonstrated earlier. However, TMR/simplex is always better than either TMR or simplex alone.

3.4.1.3 Hybrid Redundancy

Hybrid redundancy combines both static and dynamic redundancy. It masks faults like static redundancy while also detecting faults and reconfiguring to use spares like dynamic redundancy.

TMR with Spares

One approach for hybrid redundancy is to use TMR along with spares. If one TMR module fails, then it is replaced with a spare that could be either a hot or cold spare. Although the system has at least three working modules, the TMR will provide fault masking for uninterrupted operation in the presence of a fault.

Self-Purging Redundancy

Another approach for hybrid redundancy is self-purging redundancy [Losq 1976]. It is illustrated in Figure 3.13 using five modules. It uses a **threshold voter** instead of a majority voter. A threshold voter outputs a 1, if the number of its inputs that are 1 is greater than or equal to the threshold value; otherwise it outputs a 0. The threshold voter in the design shown in Figure 3.13 has five inputs and uses a threshold value of 2. This is different from a majority voter, which outputs a 1 only if a majority of the inputs are 1, which in this case would require three inputs to be a 1. The idea in self-purging redundancy is that if only one module fails, then its output will be different from the others. If the faulty module outputs a 1 while all the others output a 0, then the threshold voter will still output a 0 because only one of its inputs is a 1, which is less than the threshold value of 2.

The design of the elementary switch is shown in Figure 3.14. The elementary switch checks if a module's output differs from the output of the threshold voter. If it does differ, then the module is assumed to be faulty and its control flip-flop is reset to 0. This permanently masks the output of the module so that its input to the threshold voter will always be 0, which is a safe value. The self-purging system in Figure 3.13 will continue to operate correctly as modules are masked out as long as at least 2 modules are still working. When it reaches the point where only one module is not being masked then when the correct output should be 1, there is only one module that can generate a 1, which is not enough to reach the threshold and consequently the threshold voter will erroneously output a 0.

In comparing the self-purging system in Figure 3.13 with a 5MR system using a majority voter, the self-purging system is able to tolerate up to three failing modules and still operate correctly, whereas a 5MR system can only tolerate two failing modules. The advantage of the 5MR system is that it can tolerate two

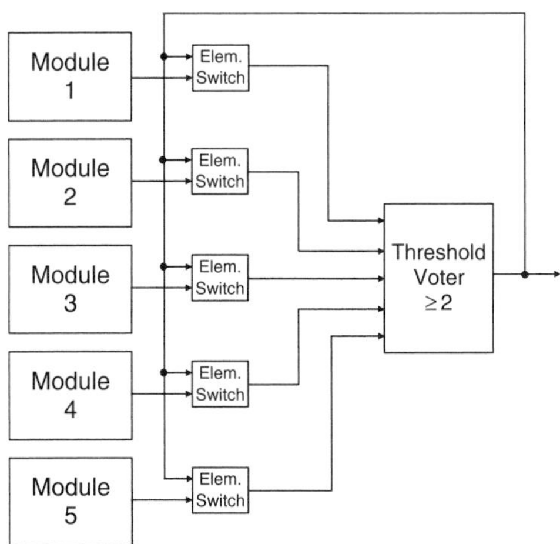

■ FIGURE 3.13

Self-purging redundancy.

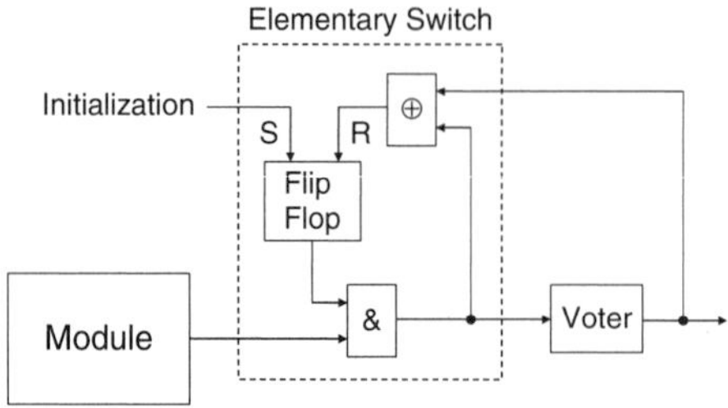

■ **FIGURE 3.14**

Elementary switch.

modules simultaneously failing, whereas a self-purging system with a threshold of two cannot. A self-purging system with a threshold value of T can tolerate up to $T-1$ simultaneous failures.

In comparing the self-purging system in Figure 3.13 with a TMR system having two spares, the fault tolerance capability is the same. The advantage of the self-purging system is that the elementary switches that it uses are much simpler than the reconfiguration circuitry that is required to support TMR with two spares. This is important because it reduces the failure rate of the reconfiguration circuitry, which can be a significant factor in high reliability systems having lots of spares. An advantage of TMR with two spares is that it can support unpowered spares, whereas self-purging redundancy uses powered spares.

3.4.2 Time Redundancy

An alternative to hardware redundancy is to use time redundancy. The advantage of time redundancy is that it requires less hardware, but the drawback is that it can only detect temporary faults. It cannot detect permanent faults. However, temporary faults occur much more frequently than permanent faults, and thus for applications where it is desirable to reduce the error rate without using a lot of additional hardware, time redundancy is an attractive approach. If an error is detected, then the system needs to roll back to a known good state before resuming operation.

3.4.2.1 Repeated Execution

The simplest way to use time redundancy is to simply repeat an operation twice and compare the results. This detects temporary faults, which occur during only one of the executions (but not both) as it will cause a mismatch in the results. Repeated execution can reuse the same hardware unit for both executions and thus requires only one copy of the functional hardware. However, it does require some mechanism for storing the results of the executions and comparing them. In a processor-based system, this can be done by storing the results in memory or on a disk and then comparing the results in software. The main cost of repeated execution is the additional time required for the redundant execution and comparison.

In a processor-based system that can run multiple threads (different streams of instructions run in parallel), multithreaded redundant execution can be used [Mukherjee 2002]. In this case, two copies of a thread are executed concurrently and the results are compared when both complete. This takes advantage of a processor system's built-in capability to exploit available processing resources to reduce execution time for multiple threads. This can significantly reduce the performance penalty.

3.4.2.2 Multiple Sampling of Outputs

At the circuit level, one attractive way to implement time redundancy with low performance impact is to use multiple sampling of the outputs of a circuit. The idea is that in a clock cycle, the output of the functional circuit is sampled once at the end of the normal clock period and then a second time after a delay of Δt. The two sampled results are then compared, and any mismatch indicates an error. This method is able to detect any temporary fault whose duration is less than Δt, because it would only cause an error in one of the sampled results and not the other one. The performance overhead for this scheme depends on the size of Δt relative to the normal clock period, because the time for each clock cycle is increased by that amount. The duration of most common temporary faults is generally relatively short compared with the clock period, so the value of Δt can be selected to obtain good coverage with relatively small impact on performance.

A simple approach for implementing multiple sampling of outputs is to use two latches as illustrated in Figure 3.15. It is also possible to implement multiple sampling of outputs using only combinational logic as described in [Franco 1994]. The approach in [Franco 1994] uses a **stability checker**, which is activated after the normal clock period to monitor an output during the Δt period and give an error indication if the output changes (*i.e.*, transitions) during that time period

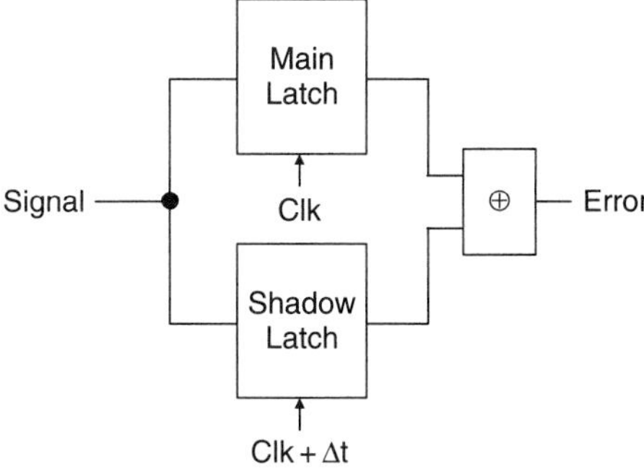

■ FIGURE 3.15

Multiple sampling of outputs.

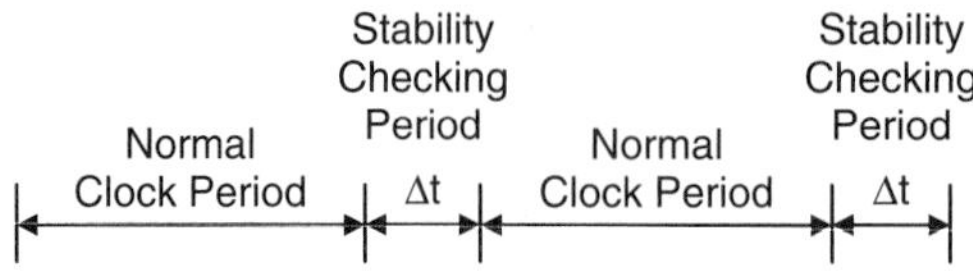

■ **FIGURE 3.16**

Timing diagram for stability checking.

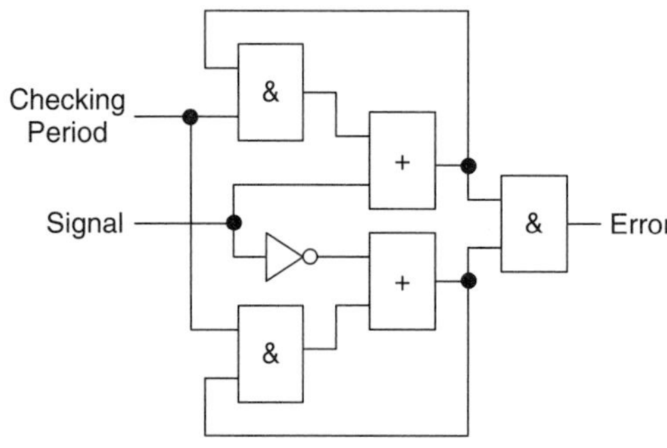

■ **FIGURE 3.17**

Example of a stability checker.

(Figure 3.16 shows a timing diagram). An example of a gate-level stability checker design is shown in Figure 3.17. If the output makes a transition during the stability checking period, then a pulse whose width is equal to the propagation delay of the inverter will be generated on the error indication signal. One stability checker is needed for each output, and the error signals can be logically ORed together. In [Franco 1994], a transistor-level design that integrates the stability checker into a flip-flop is also given, which significantly reduces the overhead. Other implementations of multiple sampling of the outputs can be found in [Metra 1998], [Favalli 2002], [Austin 2004], and [Mitra 2005].

3.4.2.3 Diverse Recomputation

Pure time redundancy approaches where the same hardware is used to repeat a computation the same way cannot detect permanent faults. However, if the computation can be performed differently the second time to utilize the hardware in a different manner, then it can be possible to detect permanent faults if the fault causes the two computations to mismatch. One simple approach used for some arithmetic or logic operations is to shift the operands when performing the computation the second time, which yields a shifted version of the result that can be compared with the result obtained the first time [Patel 1982]. With this approach, a permanent fault in the hardware could be detected if it affected the computations differently. For example, a permanent fault affecting only one bit-slice of the hardware could be detected because it would create an error in different bit positions of the normal and shifted versions of the computation. A more general approach would be to have two different versions of a software routine that perform the same function but do it in a different manner thereby exercising different portions of the hardware [Oh 2002]. By running both versions of the routine and comparing the result, it is possible to detect permanent faults in hardware.

3.4.3 Information Redundancy

Information redundancy has the advantages of hardware redundancy in that it can detect both temporary and permanent faults with minimal impact on performance, yet it can generally be implemented with less hardware overhead than by using multiple copies of the module. Information redundancy is based on using error detecting and correcting codes. Section 3.2 reviewed the fundamentals of coding theory and described a number of commonly used codes. This section describes fault tolerance schemes that use these codes to detect and correct errors.

3.4.3.1 Error Detection

Error-detecting codes can be used to detect errors. If an error is detected, then the system needs to roll back to a previous known error-free state so that the operation can be retried. Rollback requires adding some storage to save a previous state of the system so that the system can be restarted from the saved state. The amount of rollback that is required depends on the latency of the error detection mechanism. If errors are detected in the same clock cycle in which they occur, then **zero-latency error detection** is achieved and the rollback can be implemented by simply preventing the state of the system from updating in that clock cycle. If errors are detected only after an n clock cycle latency, then rollback requires restoring the system to the state it had at least n clock cycles earlier. Often the execution is divided into a set of operations, and before each operation is executed a **checkpoint** is created where the system's state is saved. Then if any error is detected during the execution of an operation, the state of the system is rolled back to the last checkpoint and the operation is retried. If multiple retries do not result in an error-free execution, then operation halts and the system flags that a permanent fault has occurred.

The basic idea in information redundancy is to encode the outputs of a circuit with an error-detecting code and have a checker, which checks if the encoded output is a valid codeword code [Lin 1983] [Peterson 1972]. A noncodeword output indicates an error.

As discussed in Section 3.2, some error-detecting codes are "separable" and some are "nonseparable." A separable code is one in which each codeword is formed by appending check bits to the normal functional output bits. The advantage of a separable code is that the normal functional outputs are immediately available

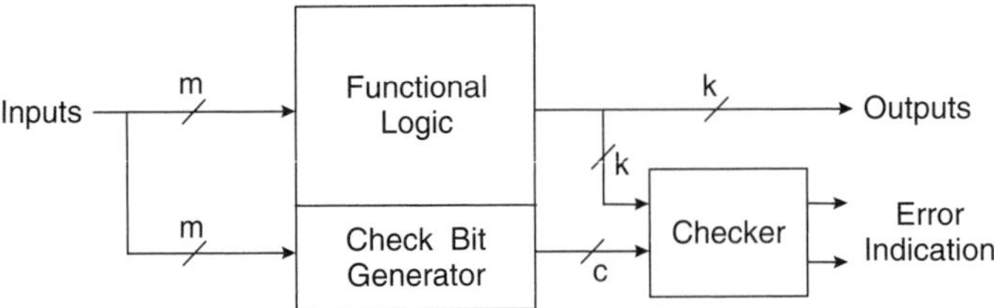

■ **FIGURE 3.18**

Block diagram of error detection using a separable code.

without the need for any decoding. A nonseparable code requires some decoding logic in order to extract the normal functional outputs.

Figure 3.18 shows a block diagram for using a separable error-detecting code. A **check bit generator** is added to generate the check bits. Then there is a checker, which checks if the functional output bits plus the check bits form a valid codeword. If not, then an error indication is given. Typically a **self-checking checker** is used which is capable of detecting errors in the checker itself as well.

A self-checking checker has two outputs, which in the normal error-free case have opposite values, (1,0) or (0,1). If they are equal to each other—that is, (1,1) or (0,0)—then an error is indicated. The reason why a self-checking checker cannot have a single output is that if a stuck-at 0 fault was on its output, then it would permanently indicate no error. By having two outputs which are equal to (1,0) for one or more codeword inputs and (0,1) for one or more codeword inputs, if either output becomes stuck-at 0, then it will be detected by at least one codeword input. In other words, at least one codeword input will cause the output of the checker to become (0,0), which would detect the fault in the checker.

In order for a checker to be **totally self-checking**, it must possess the following three properties: **code disjoint**, **fault secure**, and **self-testing** [Anderson 1971] [Lala 2001]. A checker is said to be **code disjoint** if all codeword inputs get mapped to codeword outputs, and all noncodeword inputs get mapped to noncodeword outputs. This property ensures that any error causing a noncodeword input to the checker will be detected by the checker. A checker is said to be **fault secure** for a fault class F if for all codeword inputs, the checker in the presence of any fault in F will either produce the correct codeword output or will produce a noncodeword output. This property ensures that either the checker will work correctly, or it will give an error indication (noncodeword output). A checker is said to be **self-testing** for a fault class F if for each fault in F there exists at least one codeword input that causes a noncodeword output (*i.e.*, detects the fault by giving an error indication). Assuming that all codeword inputs occur periodically with a higher frequency than faults occur, this property ensures that if a permanent fault occurs in the checker, it will be detected before a second fault can occur. The reason for this property is that if more than one permanent fault occurs, the checker may no longer be fault secure. The self-testing property ensures that the first permanent fault that occurs in the checker will get flagged with an error indication before subsequent faults can occur and thus prevents the circuit from losing its fault secure property. Alternatively, [Smith 1978] showed that if the checker is designed to be **strongly fault secure** such that it remains fault secure in the presence of any number of faults in *F*, then the self-testing property is not necessary.

Schemes using some of the most commonly used error-detecting codes are described next, and the design of self-checking checkers for them is discussed.

Duplicate-and-Compare

The simplest error-detecting code is duplication. A copy of the functional logic is used as the check bit generator, and an equality checker is used to compare the outputs of the duplicate functional logic as illustrated in Figure 3.19. If the outputs mismatch, then the equality checker indicates an error. With this approach, any fault that causes an error in only one module is guaranteed to be detected because its output will mismatch with the duplicate module's error-free output. The only way for an undetected error to occur would be if there was a **common-mode failure**, which caused the exact same error to be produced at the output of both modules [Mitra 2000]. An example of a common-mode failure would be if there was a particular gate in the module that was especially sensitive to power supply noise; then if just the right amount of noise occurred in the power rail, it could cause an error at that one gate in both copies of the module, thereby creating the exact same error at the output of both modules and thus go undetected. One way to reduce the chance of a common-mode failure is to use **design diversity** where the duplicated modules are each designed differently [Avizienis 1984] [Mitra 2002]. The two modules implement the same function, but they are designed differently. This reduces the likelihood of the two modules producing identical errors in the presence of coupled disturbances.

Note that with a duplicate-and-compare scheme, no fault on the primary input stems or in any circuitry driving the primary inputs can be detected. Only faults that occur after the primary inputs stems can be detected, because those faults will

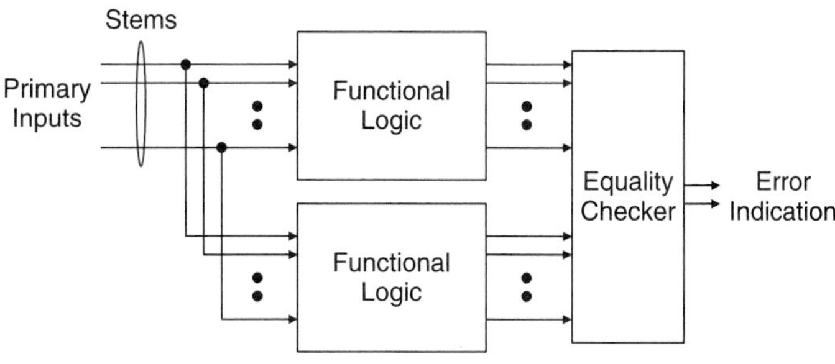

■ **FIGURE 3.19**

Duplicate-and-compare scheme.

only affect one of the modules but not both. Faults on the primary input stems fan out to both copies of the module and thus would produce identical errors at the outputs of both modules and thus go undetected. To detect faults on the primary input stems or in the circuitry driving the primary input stems, a separate checker is required for that portion of the design.

The advantage of a duplicate-and-compare scheme is that the design is simple and modular, and it provides high coverage missing only common-mode failures. The drawback of duplicate-and-compare is the large hardware overhead required. Greater than 100% hardware overhead is needed for adding the extra copy of the module plus the equality checker.

Single-Bit Parity Code

On the other end of the spectrum from duplication is single-bit parity. It is the simplest error-detecting code, which uses a single check bit to indicate the parity of the functional output bits. The parity check bit is equal to the XOR of all the functional output bits and indicates if an even or odd number of functional output bits is equal to 1. When using single-bit parity, the check bit generator predicts what the parity should be, and then the checker determines whether the functional outputs plus the parity bit form a valid codeword. If not, then an error is indicated.

The checker for single-bit parity is simple. It consists of just an XOR tree. To make the checker totally self-checking, the final gate is removed so that there are two outputs that have opposite values in the fault-free case (see Figure 3.20). The functional circuit has six outputs, and the parity prediction logic predicts the parity bit so that the 7-bit codeword (six outputs plus parity bit) has odd parity (*i.e.*, there is an odd number of 1's in each codeword). This ensures that the two error indication outputs will have opposite values for valid codeword inputs. If a noncodeword having even parity is applied to the checker, the two error indications outputs will have equal value, thereby indicating an error.

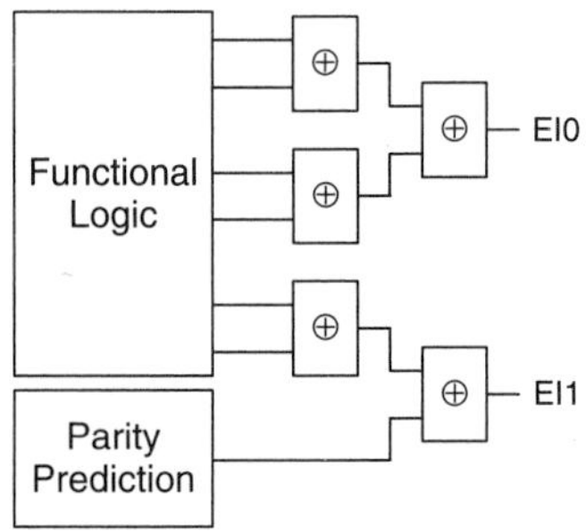

■ FIGURE 3.20

Example of a totally self-checking parity checker.

If there is a fault in the parity prediction logic, it will be detected because the parity will be incorrect. If there is a fault in the checker, it will be detected because the checker is self-checking. If there is a fault in the functional logic that causes an error on an odd number of functional outputs, it will be detected. However, if a fault in the functional logic causes an error on an even number of functional outputs, then it will not be detected because the parity will remain unchanged.

One way to ensure that no even bit errors occur is to design the functional logic such that each output is generated with its own independent cone of logic (*i.e.*, no gate has a structural path to more than one output). In this case, a single point fault could only cause a single-bit error, which is guaranteed to be detected by a single-bit parity code. However, using independent cones of logic with no logic sharing between the outputs generally results in large area, so this approach is typically only feasible for small circuits.

For large circuits that have a lot of logic sharing between the outputs, typically somewhere around 75% of the errors will be odd errors. The reason that it is not evenly distributed around 50% is that single-bit errors typically occur with much higher frequency than multiple bit errors, which skews the distribution toward odd errors. If the error coverage with single-bit parity is not sufficient, then a code with multiple check bits can be used.

Parity-Check Codes

Single-bit parity is actually a special case of the more general class of parity-check codes. In parity-check codes, each check bit is a parity check for some group of the outputs called a *parity group*. A parity-check code can be represented by a parity-check matrix, $\boldsymbol{H}$, in which there is one row for each parity group and each column corresponds to a functional output or a check bit. An entry in the matrix is a 1 if the parity group corresponding to the row contains the output or check bit corresponding to the column. Figure 3.21 shows an example of a parity-check matrix, $\boldsymbol{H}$, for a circuit with 6 outputs (Z_1–Z_6) and 3 check bits (c_1–c_3). There are three parity groups each containing one check bit and a subset of the outputs. For example, check bit c_1 is a parity check for outputs Z_1, Z_4, and Z_5.

	Z_1	Z_2	Z_3	Z_4	Z_5	Z_6	c_1	c_2	c_3
Parity Group 1	1	0	0	1	1	0	1	0	0
Parity Group 2	0	1	1	0	0	0	0	1	0
Parity Group 3	0	0	0	0	0	1	0	0	1

FIGURE 3.21

Example of a parity-check matrix (***H*** matrix).

With a parity-check code, an *error combination* (*i.e.*, a particular combination of functional output bits in error) will be detected provided that at least one parity group has an odd number of bits in error. By increasing the number of check bits (*i.e.*, having more parity groups), the number of error combinations that can be detected will increase. If there are c check bits and k functional outputs, then the number of possible parity-check codes is equal to 2^{ck} because each entry in the parity-check matrix can be either a 1 or 0. Selecting the best parity-check code to use for a particular functional circuit depends on the structure of the circuit and the set of modeled faults because that determines which error combinations can occur. For example, if there is no shared logic between the cones of logic driving outputs Z_1 and Z_2 (*i.e.*, there is no gate that has a structural path to both Z_1 and Z_2), then it is not possible to have an error combination that includes both Z_1 and Z_2 because of a single point fault. Consequently, Z_1 and Z_2 could be in the same parity group with no risk of an even bit error because Z_1 and Z_2 cannot both have an error at the same time. Given a functional circuit, it can be analyzed to determine what error combinations are possible based on the sensitized paths in the circuit, and a parity-check code with the minimum number of check bits required to achieve 100% coverage can be selected [Fujiwara 1987]. If it is possible to synthesize the functional circuit or to modify or redesign it, then the structure-constrained logic synthesis procedure described in [Touba 1997] can be used to constrain the fanout in the functional circuit to simplify the parity-check code that is needed to achieve 100% coverage. This is done by intelligently factoring the logic in a way that minimizes the number of ways that errors can propagate to the outputs.

The checker for a parity-check code can be constructed by using a self-checking single-bit parity checker for each of the parity groups and then combining the two-rail outputs of each self-checking parity checker using a self-checking two-rail checker. A **two-rail checker** takes two pairs of two-rail inputs and combines them into a single pair of two-rail outputs where if either or both of the two sets of two-rail inputs is a noncodeword, then the output of the two-rail checker will be a noncodeword input. The design of a total self-checking two-rail checker is shown in Figure 3.22 where the two two-rail input pairs (A0, A1) and (B0, B1) are combined into a single two-rail output pair (C0, C1). Figure 3.23 shows the self-checking checker for the party-check code shown in Figure 3.21.

Berger Codes

As discussed in Section 3.2, Berger codes have the lowest redundancy of any separable code for detecting all possible unidirectional errors. If there are k functional output bits, then a Berger code requires $log_2\lceil k+1\rceil$ check bits.

The only way to get a nonunidirectional error for a single point fault is if there are one or more paths from the fault site to a primary output which have an even number of inversions and one or more paths from the fault site to a primary output that has an odd number of inversions. It was shown in [Jha 1993] that if a circuit is synthesized using only algebraic transformations, then it is possible to transform the circuit so that it is *inverter-free* (*i.e.*, it has inverters only at the primary inputs), and hence all single point faults will cause only unidirectional errors. Thus, for an

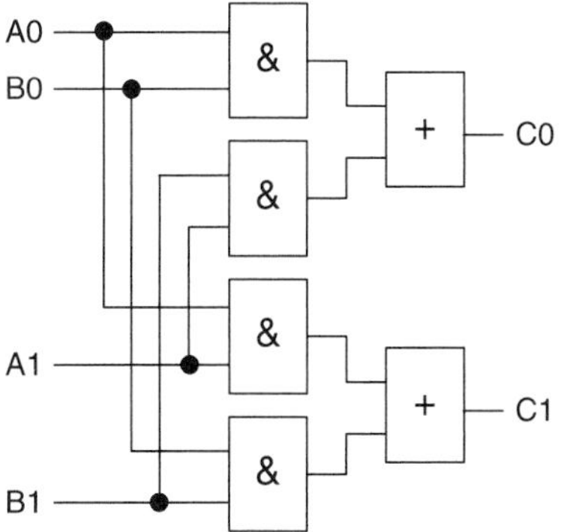

■ **FIGURE 3.22**

A totally self-checking two-rail checker.

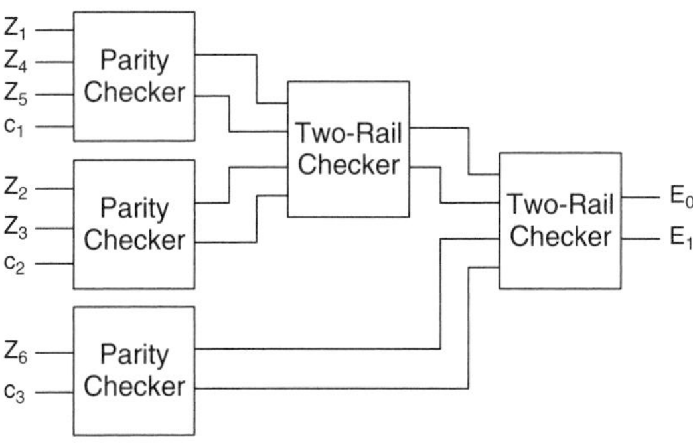

■ **FIGURE 3.23**

A self-checking checker for parity-check code in Figure 3.21.

inverter-free circuit, a Berger code is guaranteed to give 100% coverage for single point faults.

An ALU design using a Berger code was described in [Lo 1992]. Self-checking checker designs for Berger codes can be found in [Piestrak 1987], [Lo 1988], and [Kavousianos 1998].

Constant Weight Codes

The codes discussed so far are all separable codes. As discussed in Section 3.2, constant weight codes are nonseparable codes that have lower redundancy than Berger codes for detecting all unidirectional errors. The drawback of constant weight codes is that some decoding logic is needed to convert each codeword back to its corresponding binary value. However, one application where constant weight codes are useful is for state encoding of a *finite state machine* (FSM). The states in an FSM can be encoded with a constant weight code, and consequently there is no need for decoding when generating the next state, so the next state logic can be checked with a constant weight code checker, which is much simpler than a Berger code checker.

Self-checking checker designs for constant weight codes can be found in [Marouf 1978], [Gaitanis 1983], and [Nanya 1983].

3.4.3.2 Error Correction

Information redundancy can also be used to mask (*i.e.*, correct) errors by using an *error-correcting code* (ECC). For logic circuits, using ECC is not as attractive as using TMR because the logic for predicting the check bits is generally complex, and the number of output bits that can be corrected is limited. However, ECC is commonly used for memories [Peterson 1972]. Errors do not propagate in memories as they do in logic circuits, so *single-error-correction and double-error-detection* (SEC-DED) codes are usually sufficient.

Memories are dense and prone to errors especially because of ***single-event upsets*** (SEUs) from radiation. However, schemes that only detect errors are generally not attractive as there may be a long latency from when the memory was written until the point at which an error is detected, thereby making rollback infeasible. Hence, ECC codes are needed. Each word in the memory is encoded by adding extra check bits that are stored together with the data. SEC-DED codes are commonly used. For a 32-bit word, SEC-DED codes require 7 check bits thereby requiring that the memory store 39 bits per word. For a 64-bit word, 8 check bits are required and thus the memory must store 72 bits per word. So using SEC-DED ECC increases the size of the memory by 7/32 = 21.9% for 32-bit words and 8/64 = 12.5% for 64-bit words. SEC-DED codes are able to correct single errors and detect double bit errors in either the data bits or the check bits.

The hardware required for implementing memory with SEC-DED ECC is shown in Figure 3.24. When the memory is written to, the check bits are computed for

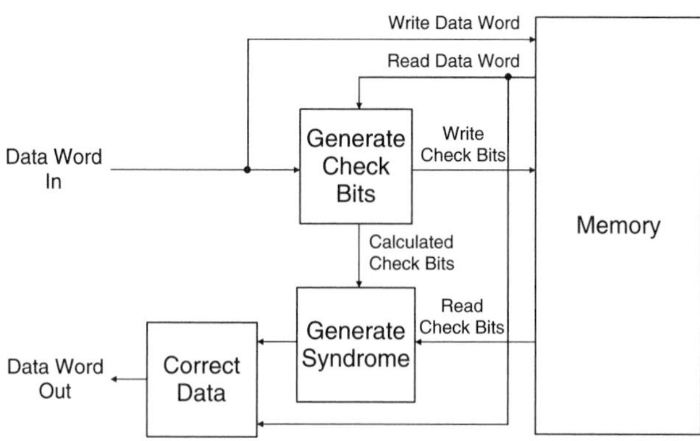

■ **FIGURE 3.24**

Memory ECC architecture.

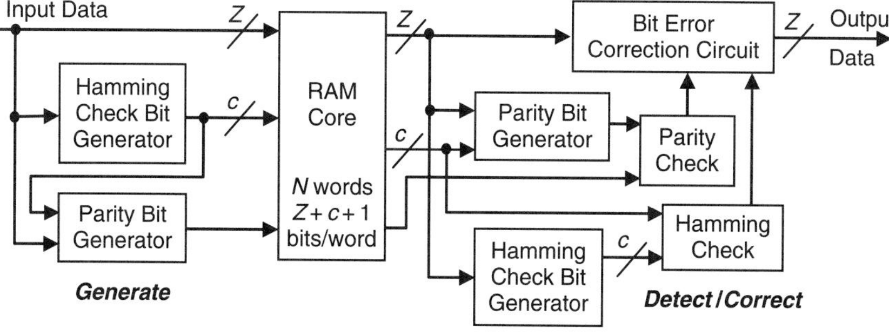

■ **FIGURE 3.25**

Hamming code generation/regeneration for ECC RAM.

the data word. This is done using the "generate check bits" block. The check bits are then stored along with the data bits in the memory. When the memory is read, the check bits for the data word are computed again using the generate check bits block. This is then XORed with the check bits read from the memory to generate the **syndrome**. The syndrome indicates which check bits computed from the data word mismatch with the check bits read from the memory. As explained in Section 3.3, if the syndrome is all zero, then no error has occurred and the data word can be used as is. If the syndrome is nonzero, then an error has been detected. If the parity of the syndrome is odd, then a single-bit error has occurred and the value of the syndrome indicates which bit is in error. That bit is then flipped to correct it. Generally the corrected word is written back to the memory to reduce the risk of a second error occurring in that word, which would make it uncorrectable the next time it is read. If the syndrome has even parity, then a double bit error has occurred, and it cannot be corrected. However, the error is flagged so that the system can take appropriate action.

Figure 3.25 shows the application of a Hamming code to construct memory ECC that supports pipelined RAM access. In this example, the Hamming code generation circuitry is inserted at the input to the RAM and Hamming code detection/correction circuitry is added at the output of the RAM. Here the Hamming code generation circuitry is not shared by both write and read ports as was the case in Figure 3.24. The Hamming check bits are generated by the exclusive-OR of a particular subset of data bits associated with each individual Hamming check bit as illustrated by the Hamming parity-check matrix example in Figure 3.26 [Lala 1985]. The Hamming check bits are then stored in the RAM along with the data bits. At the output of the RAM, the Hamming check bits are regenerated for the data bits. The regenerated Hamming check bits are compared to the stored Hamming check bits to determine the syndrome which points to the bit to be inverted to provide SEC. The Hamming code that is shown in Figure 3.26 has a distance of 3 such that an additional parity bit can be included to obtain a distance of 4 and to provide DED. This additional bit gives the parity across the entire codeword (data

$$
\begin{array}{r}
\\
\text{Parity Group 1} \\
\text{Parity Group 2} \\
\text{Parity Group 3} \\
\text{Parity Group 4}
\end{array}
\begin{array}{c}
\begin{array}{cccccccccccc} Z_1 & Z_2 & Z_3 & Z_4 & Z_5 & Z_6 & Z_7 & Z_8 & c_1 & c_2 & c_3 & c_4 \end{array} \\
\begin{bmatrix}
1 & 1 & 0 & 1 & 1 & 0 & 1 & 0 & 1 & 0 & 0 & 0 \\
1 & 0 & 1 & 1 & 0 & 1 & 1 & 0 & 0 & 1 & 0 & 0 \\
0 & 1 & 1 & 1 & 0 & 0 & 0 & 1 & 0 & 0 & 1 & 0 \\
0 & 0 & 0 & 0 & 1 & 1 & 1 & 1 & 0 & 0 & 0 & 1
\end{bmatrix}
\end{array}
$$

■ **FIGURE 3.26**

Example 8-bit Hamming code parity-check matrix.

TABLE 3.4 ■ Error Indications for ECC RAM

Error Type	Condition
No bit error	Hamming check bits match, no parity error
Single-bit correctable error	Hamming check bits mismatch, parity error
Double-bit error detection	Hamming check bits mismatch, no parity error

bits plus Hamming bits) and is used to determine the occurrence of SEC/DED as summarized in Table 3.4.

Memory SEC-DED ECC is generally effective because memory bit flips tend to be independent and uniformly distributed. If a bit flip occurs, the next time the memory word is read, it will be corrected and written back. The main risk is that if a memory word is not accessed for a long period of time, then the chance for multiple bit errors to accumulate is increased. One way to avoid this is to use **memory scrubbing**, in which every location in the memory is read on a periodic basis. This can be implemented by having a memory controller that, during idle periods, cycles through the memory reading every location and correcting any single-bit errors that have occurred in a word. This reduces the time period in which multiple bit errors can accumulate in a word, thereby reducing the chance of an uncorrectable or undetectable error.

Another issue is that some SEUs can affect more than one memory cell, resulting in ***multiple-bit upsets*** (MBUs). MBUs typically result in errors in adjacent memory cells. To prevent MBUs from resulting in multiple bit errors in a single word, **memory interleaving** is typically used. In memory interleaving, the memory has more columns than the number of bits in a single word, and the columns corresponding to each word are interleaved. For example, if the memory has four times as many columns as the width of a word, then every fourth column starting with column 0 is assigned to one word, and every fourth column starting with column 1 is assigned to another word, and so forth. This is illustrated in Figure 3.27. Thus, an MBU affecting up to four adjacent memory cells will cause single-bit errors in separate words, each of which can be corrected with an SEC code.

Note that memory ECC also helps for permanent faults because faulty memory cells can be tolerated up to the limit of the correcting capability of the code.

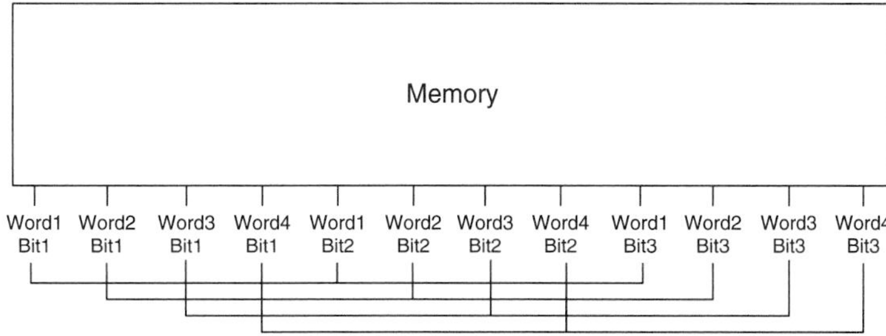

■ **FIGURE 3.27**

Bit interleaving with four words.

However, the existence of faulty memory cells increases the vulnerability to additional temporary faults as there is less redundancy remaining to correct them. ECC can be used to identify faulty memory cells by logging the errors that occur and looking for any memory cell that has an unusually high frequency of errors.

3.5 INDUSTRY PRACTICES

The dependability requirements for industrial designs vary greatly depending on the application. Table 3.5 characterizes different types of applications in terms of their dependability requirements and the fault tolerance schemes that are used.

- **Long-life systems.** Some systems are difficult or expensive to repair (*e.g.*, those used in space or implanted in a human). Consequently, dynamic redundancy is typically used to maximize their MTTF so they will have a long lifetime without the need for repair. Such systems carry spares so that when a module fails, it can be replaced by a spare.
- **Reliable real-time systems.** Other systems can be repaired, but they need to be able to tolerate failures when they occur without any delay. Such real-time systems (*e.g.*, aircraft) cannot afford the time required to diagnose a fault and switch to a spare. These systems need to mask faults, which is typically done using TMR. At some safe point in time (*e.g.*, after the aircraft has landed), a failed module can be repaired so that the TMR fault masking capability is restored before the next mission.
- **High-availability systems.** For some systems, downtime is costly and may result in losing business or alienating customers. Examples include a reservation system, stock exchange, or telephone systems. In these systems, the goal is to have high availability. To accomplish this objective, it is important to avoid any single point of failure that could take the system down. Everything is typically replicated including the power subsystem and cables along with

TABLE 3.5 ■ Use of Fault Tolerance in Different Types of Systems

Type	Issues	Goal	Examples	Techniques
Long-life systems	Difficult or expensive to repair	Maximize MTTF	Satellites, spacecraft, implanted biomedical	Dynamic redundancy
Reliable real-time systems	Error or delay catastrophic	Fault masking capability	Aircraft, nuclear power plant, air bag electronics, radar	TMR
High-availability systems	Downtime very costly	High availability	Reservation system, stock exchange, telephone systems	No single point of failure; self-checking pairs; fault isolation
High-integrity systems	Data corruption very costly	High data integrity	Banking, transaction processing, database	Checkpointing, time redundancy; ECC; redundant disks
Mainstream low-cost systems	Reasonable level of failures acceptable	Meet failure rate expectations at low cost	Consumer electronics, personal computers	Often none; memory ECC; bus parity; changing as technology scales

all boards and buses. Self-checking pairs are often used where if one pair fails, the other pair replaces it. The system can then notify a maintenance engineer, who can then repair or replace the failed pair. Good fault isolation is important in high-availability systems to make them fast and easy to repair.

- **High-integrity systems.** In some systems, (*e.g.*, banking) data corruption is costly. It is important in these systems to maintain high data integrity. Such systems typically use checkpointing, in which the state of the system is saved before performing an operation so that if a fault occurs during the operation, then the operation can be rolled back. Time redundancy is often used to perform the operation multiple times to ensure that it is correct before committing it to the database. Using ECC and having redundant disks protect the database itself. Note that many high-integrity systems are also high-availability systems and vice versa.

- **Mainstream low-cost systems.** In the applications mentioned previously, dependability is the main concern, and the user is willing to pay for a highly redundant system to obtain it. However, for most applications, the primary concern is low cost. In such systems, the user is typically willing to accept some reasonable level of failures. For example, in a personal computer, users typically would not be willing to pay twice the price for a redundant system if they can get a reasonably reliable system that does not incorporate fault tolerance. Traditionally, most mainstream low-cost systems have not used any fault tolerance. However, this is changing as technology scales and failure rates increase. Personal computers, for example, now typically use memory ECC because it provides a significant improvement in failure rates at low cost. Including a parity line on a bus is another commonly used low-cost technique. As technology continues to scale, soft error rates are becoming a major concern and meeting users' expectations for reasonable failure rates will likely require incorporating some form of fault tolerance. We examine this subject in greater detail in Chapter 8.

3.6 CONCLUDING REMARKS

There are many different fault-tolerant design schemes. Choosing the scheme to use in a particular design depends on what types of faults need to be tolerated (temporary or permanent, single or multiple point failures, etc.) and the constraints on area, performance, and power. As technology continues to scale and circuits become increasingly prone to failure, achieving sufficient fault tolerance will be a major design issue.

Note that this chapter is introductory in that it only discusses fundamental fault tolerance schemes that are applicable for coping with various types of faults in the nanometer design era. Further details can be found in subsequent chapters. In particular, Chapter 8 extensively covers various sources of permanent faults and temporary faults (soft errors) and goes into more detail on techniques for coping with them, and Chapter 12 discusses implementing fault tolerance in *field programmable gate arrays* (FPGAs).

3.7 EXERCISES

3.1 **(Reliability)** Assume that the failures of three computers—A, B, and C—are independent, with constant failure rates $\lambda = 1/1000, \lambda = 1/1200$, and $\lambda = 1/900$ failures per hour, respectively.

a. What is the probability that at least one system fails in a 4-week period?

b. What is the probability that all three systems fail in a 4-week period?

3.2 **(Mean Time to Failure)** There is one light bulb in your apartment and it burns out. You are only going to be living in the apartment for another 30 weeks. You go to the store to buy a light bulb and there are two choices. One has a MTTF of 100 weeks and costs \$1, and the other has a MTTF of 300 weeks and cost \$1.50.

a. If you buy the cheaper bulb, what is the probability that you will have to buy another before you move out?

b. If you buy the more expensive bulb, what is the probability that you will have to buy another before you move out?

c. Which bulb would you buy?

3.3 **(System Reliability)** The system in Figure 3.28 functions correctly if all components along at least one path from input to output are functioning. All components failures can be assumed to be independent.

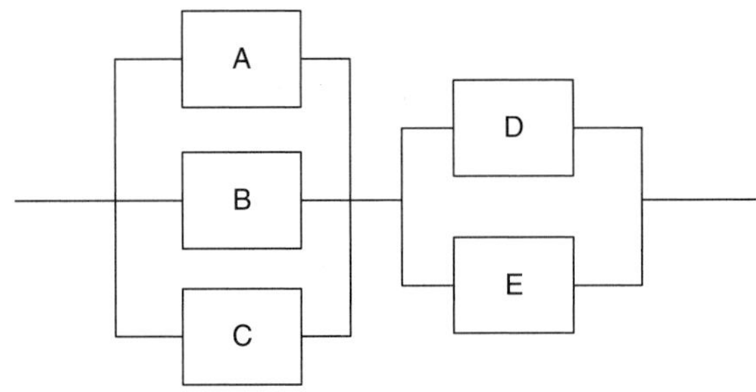

■ **FIGURE 3.28**

System reliability diagram.

a. Express the system reliability as a function of the component reliabilities. Each component has a reliability R_m.

b. What is the MTTF of the system if the MTTF of each component is 1 year and each component has a constant failure rate?

3.4 **(Availability)** Assume you own a computer with a failure rate of one failure per 100 weeks. Computer service company ABC promises to provide repairs fast enough that your MTTR will be 2 weeks. Computer service company XYZ promises to provide repairs fast enough that your availability will be 99%. If both companies charge the same, which would you contract with? Explain.

3.5 **(Linear Code)** For the following generating matrix:

$$G = \begin{bmatrix} 1 & 0 & 0 & 1 & 1 \\ 0 & 1 & 0 & 0 & 1 \\ 0 & 0 & 1 & 1 & 0 \end{bmatrix}$$

a. List all valid codewords.

b. Derive the H-matrix.

3.6 **(Linear Code)** Write the generator matrix for the following codes:

a. Single-bit even parity for 4 message bits.

b. Duplication code having 3 message bits and 3 check bits.

3.7 **(Code Redundancy)** If the number of information bits is 16 ($k = 16$), calculate the redundancy of the following codes:

a. Berger code.

b. 9-out-of-19 constant weight code.

c. Code in which the information bits are partitioned into two blocks of 8 bits each, and each block is encoded with a 6-out-of-12 constant weight code.

3.8 **(CRC Code)** For a (6,3) CRC code with $G = x^3 + x^2 + 1$:

a. Write out all the codewords.

b. Derive the G-matrix for this code.

3.9 **(CRC Code)** For a systematic (8,4) CRC code with $G = x^4 + x^3 + x^2 + 1$:

a. What are the check bits for the message $x^3 + x^2 + 1$?

b. What are the check bits for the message $x + 1$?

3.10 **(ECC)** For the following SEC Hamming code, correct the single-bit error in each of the following words:

$$H = \begin{bmatrix} 1 & 1 & 0 & 1 & 1 & 0 & 1 & 0 & 1 & 0 & 0 & 0 \\ 1 & 0 & 1 & 1 & 0 & 1 & 1 & 0 & 0 & 1 & 0 & 0 \\ 0 & 1 & 1 & 1 & 0 & 0 & 0 & 1 & 0 & 0 & 1 & 0 \\ 0 & 0 & 0 & 0 & 1 & 1 & 1 & 1 & 0 & 0 & 0 & 1 \end{bmatrix}$$

a. 110001111110

b. 101101110100

3.11 **(ECC)** Construct an H-matrix for the (14,9) odd-column-weight (Hsiao) code. Try to keep the row weights as equal as possible.

3.12 **(TMR)** For a TMR system, suppose the three modules have the same function but are designed differently such that the failure rate for each of the three modules is 1000 FITS, 2000 FITS, and 3000 FITS, respectively.

a. Compute the reliability of the TMR system after 2 years.

b. Compute the MTTF of the TMR system.

3.13 **(NMR)** For an NMR system with five modules each with a failure rate of λ:

a. Derive an expression for the reliability.

b. Derive an expression for the MTTF.

Acknowledgments

The author wishes to thank Avijit Dutta of the University of Texas, Austin, Professor Kartik Mohanram of Rice University, and Professor Subhasish Mitra of Stanford University for their helpful comments. The author also wishes to thank Professor Charles Stroud of Auburn University for contributing material on ECC RAMs.

References

R3.0 Books

[Lala 1985] P. Lala, *Fault Tolerant and Fault Testable Hardware Design*, Prentice Hall, London, 1985.

[Lala 2001] P. K. Lala, *Self-Checking and Fault-Tolerant Digital Design*, Morgan Kaufmann, San Francisco, 2001.

[Lin 1983] S. Lin and D. J. Costello, *Error Control Coding*, Prentice Hall, Englewood Cliffs, NJ, 1983.

[Peterson 1972] W. W. Peterson and E. J. Weldon, Jr., *Error-Correcting Codes*, Second Edition, MIT Press, Cambridge, MA, 1972.

[Pierce 1965] W. H. Pierce, *Fault-Tolerant Computer Design*, Academic Press, New York, 1965.

[Pradhan 1996] D. K. Pradhan, *Fault-Tolerant Computer System Design*, Prentice Hall, Upper Saddle River, NJ, 1996.

[Siewiorek 1998] D. P. Siewiorek and R. S. Swarz, *The Theory and Practice of Reliable System Design*, Third Edition, AK Peters, Natick, MA, 1998.

[Tanenbaum 1981] A. S. Tanenbaum, *Computer Networks*, Prentice Hall, Englewood Cliffs, NJ, 1981.

[Tyron 1962] J. C. Tyron, Quadded logic, *Redundancy Techniques for Computing Systems*, pp. 205–228, Spartan, Washington, DC, 1962.

R3.2 Fundamentals of Fault Tolerance

[SIA 2004] SIA, *The International Technology Roadmap for Semiconductors: 2004 Update*, Semiconductor Industry Association, San Jose, CA (http://public.itrs.net), 2004.

R3.3 Fundamentals of Coding Theory

[Berger 1961] J. M. Berger, A note on error detecting codes for asymmetric channels, *Information Control*, 4(3), pp. 68–73, March 1961.

[Hamming 1950] R. W. Hamming, Error detecting and error correcting codes, *Bell System Tech. J.*, 29(2), pp. 147–160, April 1950.

[Hsiao 1970] M. Y. Hsiao, A class of optimal minimum odd-weight-column SEC-DED codes, *IBM J. Res. Dev.*, 14(4), pp. 395–401, July 1970.

R3.4 Fault Tolerance Schemes

[Anderson 1971] D. A. Anderson, Design of self-checking digital networks using coding techniques, *Tech. Report R-527*, Coordinated Science Lab., Univ. of Illinois, Urbana-Champagne, September 1971.

[Austin 2004] T. Austin, D. Blaauw, T. Mudge, and K. Flautner, Making typical silicon matter with Razor, *IEEE Computer*, 37(3), pp. 57–65, March 2004.

[Avizienis 1984] A. Avizienis and J. P. J. Kelly, Fault tolerance by design diversity: Concepts and experiments, *IEEE Computer*, 17(8), pp. 67–80, August 1984.

[Barbour 1989] A. E. Barbour and A. S. Wojcik, A general, constructive approach to fault-tolerant design using redundancy, *IEEE Trans. on Computers*, 38(1), pp. 15–20, January 1989.

[Favalli 2002] M. Favalli, and C. Metra, Online testing approach for very deep-submircon Ics, *IEEE Design & Test of Computers*, 19(2), pp. 16–23, March 2002.

[Franco 1994] P. Franco and E. J. McCluskey, On-line delay testing of digital circuits, in *Proc. VLSI Test Symp.*, pp. 164–173, April 1994.

[Fujiwara 1987] E. Fujiwara and K. Matsouka, A self-checking generalized prediction checker and its use for built-in testing, *IEEE Trans. on Computers*, C-36(1), pp. 86–93, January 1987.

[Gaitanis 1983] N. Gaitanis and C. Halatsis, A new design method for M-out-of-N codes, *IEEE Trans. on Computers*, C-32(3), pp. 273–283, March 1983.

[Jha 1993] N. K. Jha and S. Wang, Design and synthesis of self-checking VLSI circuits, *IEEE Trans. on Computer-Aided Design*, 12, pp. 878–887, June 1993.

[Kavousianos 1998] X. Kavousianos and D. Nikolos, Novel single and double output TSC Berger code checkers, in *Proc. VLSI Test Symp.*, pp. 348–353, April 1998.

[Lo 1988] J.-C. Lo and S. Thanawastien, The design of fast totally self-checking berger code checkers based on Berger code partitioning, in *Proc. Int. Symp. on Fault-Tolerant Computing*, pp. 226–231, June 1988.

[Lo 1992] J.-C. Lo, S. Thanawastein, and M. Nicolaidis, An SFS Berger check prediction ALU and its application to self-checking processor designs, *IEEE Trans. on Computer-Aided Design*, 11(4), pp. 525–540, April 1992.

[Losq 1976] J. Losq, A highly efficient redundancy scheme: Self-purging redundancy, *IEEE Trans. on Computers*, C-25(6), pp. 569–578, June 1976.

[Marouf 1978] M. A. Marouf and A. D. Friedman, Efficient design of self-checking checker for M-out-of-N code, *IEEE Trans. on Computers*, C-27(6), pp. 482–490, June 1978.

[Metra 1998] C. Metra, M. Favalli, and B. Ricco, On-line detection of logic errors due to crosstalk, delay, and transient faults, in *Proc. Int. Test Conf.*, pp. 524–533, October 1998.

[Mitra 2000] S. Mitra, N. R. Saxean, and E. J. McCluskey, Common-mode failures in redundant VLSI systems: A survey, *IEEE Trans. on Reliability*, 49(3), pp. 285–295, September 2000.

[Mitra 2002] S. Mitra, N. R. Saxena and E. J. McCluskey, A design diversity metric and analysis of redundant systems, *IEEE Trans. on Computers*, 51(5), pp. 498–510, May 2002.

[Mitra 2005] S. Mitra, N. Seifert, M. Zhang, Q. Shi, and K. S. Kim, Robust system design with built-in soft-error resilience, *IEEE Computer*, 38(2), pp. 43–52, February 2005.

[Mukherjee 2002] S. S. Mukherjee, M. Kontz, and S. Reinhardt, Detailed design and evaluation of redundant multithreading alternatives, in *Proc. Int. Symp. on Computer Architecture*, pp. 99–110, May 2002.

[Nanya 1983] T. Nanya and Y. Tohma, A 3-level realization of totally self-checking checkers for M-out-of-N codes, in *Proc. Int. Symp. on Fault-Tolerant Computing*, pp. 173–176, June 1983.

[Oh 2002] N. Oh, S. Mitra, and E. J. McCluskey, ED^4I: Error detection by diverse data and duplicated instructions, *IEEE Trans. on Computers*, 51(2), pp. 180–199, February 2002.

[Patel 1982] J. H. Patel and L. Y. Fung, Concurrent error detection in ALU's by recomputing with shifted operands, *IEEE Trans. on Computers*, C-31(7), pp. 589–595, July 1982.

[Piestrak 1987] S. J. Piestrak, Design of fast self-checking checkers for a class of Berger codes, *IEEE Trans. on Computers*, C-36(5), pp. 629–634, May 1987.

[Schwab 1983] T. E. Schwab and S. S. Yau, An algebraic model of fault-masking logic circuits, *IEEE Trans. on Computers*, C-32(9), pp. 809–825, September 1983.

[Smith 1978] J. E. Smith and G. Metze, Strongly fault secure logic networks, *IEEE Trans. on Computers*, C-27(6), pp. 491–499, June 1978.

[Stroud 1994] C.E. Stroud, Reliability of majority voting based VLSI fault-tolerant circuits, *IEEE Trans. on Very Large Scale Integration (VLSI) Systems*, 2(4), pp. 516–521, December 1994.

[Touba 1997] N. A. Touba and E. J. McCluskey, Logic synthesis of multilevel circuits with concurrent error detection, *IEEE Trans. on Computer-Aided Design*, 16(7), pp. 145–155, July 1997.

CHAPTER 4

System/Network-on-Chip Test Architectures

Chunsheng Liu
University of Nebraska-Lincoln, Omaha, Nebraska

Krishnendu Chakrabarty
Duke University, Durham, North Carolina

Wen-Ben Jone
University of Cincinnati, Cincinnati, Ohio

ABOUT THIS CHAPTER

The popularity of ***system-on-chip*** (SOC) integrated circuits has led to an unprecedented increase in test costs. This increase can be attributed to the difficulty of test access to embedded cores, long test development and test application times, and high test data volumes. The on-chip network is a natural evolution of traditional interconnects such as shared bus. An SOC whose interconnection is implemented by an on-chip network is called a ***network-on-chip*** (NOC) system. Testing an NOC is challenging, and new test methods are required.

This chapter presents techniques that facilitate low-cost modular testing of SOCs. Topics discussed here include techniques for wrapper design, test access mechanism optimization, test scheduling, and applications to mixed-signal and hierarchical SOCs. Recent work on the testing of embedded cores with multiple clock domains and wafer sort of core-based SOCs are also discussed. Together, these techniques offer SOC integrators with the necessary means to manage test complexity and reduce test cost. The discussion is then extended to the testing of NOC-based designs. Topics discussed here include reuse of on-chip network for core testing, test scheduling, test access methods and interface, efficient reuse of the network, power-aware and thermal-aware testing, and on-chip network (including interconnects, routers, and network interface) testing. Finally, two case studies, one for SOC testing and the other for NOC testing, are presented based on industrial chips developed by Philips.

4.1 INTRODUCTION

Shrinking process technologies and increasing design sizes have led to billion-transistor ***system-on-chip*** (SOC) integrated circuits. SOCs are crafted by system designers who purchase ***intellectual property*** (IP) circuits, known as embedded cores, from core vendors and integrate them into large designs. Embedded cores are complex, predesigned, and preverified circuits that can be purchased off the shelf and reused in designs. Although SOCs have become popular as a means to integrate complex functionality into designs in a relatively short amount of time, there remain several roadblocks to rapid and efficient system integration. Primary among these is the lack of ***design-for-testability*** (DFT) tools upon which core design and system development can be based. Importing core designs from different IP sources and stitching them into designs often entails cumbersome format translation. A number of SOC and core development working groups have been formed, notable among these being the ***virtual socket interface alliance*** (VSIA) [VSIA 2007]. The IEEE 1500 standard has also been introduced to facilitate SOC testing [IEEE 1500-2005].

The VSIA was formed in September 1996 with the goal of establishing a unifying vision for the SOC industry and the technical standards required to facilitate system integration. VSIA specifies interface standards, which will allow cores to fit quickly into "virtual sockets" on the SOC, at both the architectural level and the physical level [VSIA 2007]. This will allow core vendors to produce cores with a uniform set of interface features, rather than having to support different sets of features for each customer. SOC integration is in turn simplified because cores may be imported and plugged into standardized "sockets" on SOCs with relative ease. The IEEE 1500 standard includes a standardized ***core test language*** (CTL) and a test wrapper interface from cores to on-chip test access mechanisms.

An SOC test is essentially a composite test comprised of the individual tests for each core, the ***user defined logic*** (UDL) tests, and interconnect tests. Each individual core or UDL test may involve surrounding components and may imply operational constraints (*e.g.*, safe mode, low power mode, bypass mode), which necessitate special isolation modes.

SOC test development is especially challenging for several reasons. Embedded cores represent intellectual property, and core vendors are reluctant to divulge structural information about their cores to users. Thus, users cannot access core netlists and insert *design-for-testability* (DFT) hardware that can ease test application from the surrounding logic. Instead, the core vendor provides a set of test patterns that guarantees a specific fault coverage. These test patterns must be applied to the cores in a given order, using a specific clocking strategy. Care must often be taken to ensure that undesirable patterns and clock skews are not introduced into these test streams. Furthermore, cores are often embedded in several layers of user-designed or other core-based logic and are not always directly accessible from chip I/Os. Propagating test stimuli to core inputs may therefore require dedicated test transport mechanisms. Moreover, it is necessary to translate test data at the inputs and outputs of the embedded-core into a format or sequence suitable for application to the core.

A conceptual architecture for testing embedded core-based SOCs is shown in Figure 4.1 [Zorian 1999]. It consists of three structural elements:

1. **Test pattern source and sink.** The test pattern source generates the test stimuli for the embedded cores, and the test pattern sink compares the response(s) to the expected response(s).
2. **Test access mechanism (TAM).** The TAM transports test patterns. It is used for the on-chip transport of test stimuli from the test pattern source to the core under test, and for the transport of test responses from the core under test to a test pattern sink.
3. **Core test wrapper.** The core test wrapper forms the interface between the embedded core and its environment. It connects the terminals of the embedded core to the rest of the integrated circuit and to the TAM.

Once a suitable test data transport mechanism and test translation mechanism have been designed, the next major challenge confronting the system integrator is test scheduling. This refers to the order in which the various core tests and tests for user-designed interface logic are applied. A combination of BIST and external testing is often used to achieve high fault coverage [Sugihara 1998] [Chakrabarty 2000a], and tests generated by different sources may therefore be applied in parallel, provided resource conflicts do not arise. Effective test scheduling for SOCs is challenging because it must address several conflicting goals: (1) SOC test time minimization, (2) resource conflicts between cores arising from the use of shared TAMs and on-chip BIST engines, (3) precedence constraints among tests, and (4) power constraints.

Finally, analog and mixed-signal cores are increasingly being integrated onto SOCs with digital cores. Testing mixed-signal cores is challenging because their failure mechanisms and testing requirements are not as well modeled as they are for digital cores. It is difficult to partition and test analog cores, because they may be prone to crosstalk across partitions. Capacitance loading and complex timing issues further exacerbate the mixed-signal test problem.

Section 4.2 presents a survey of modular test methods for SOCs that enhance the utilization of test resources, such as test data and test hardware, to reduce test cost such as test time.

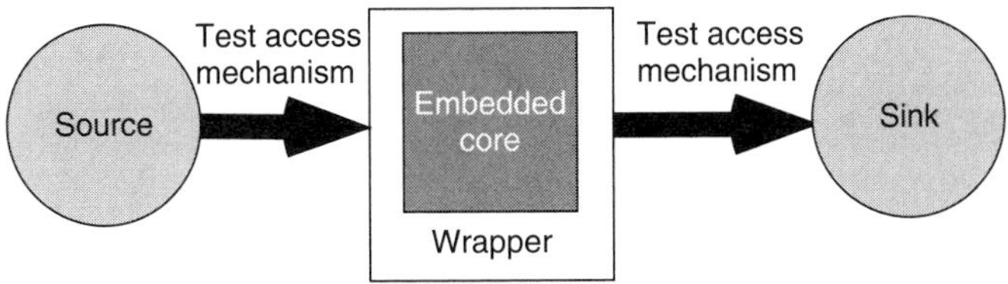

■ **FIGURE 4.1**

Overview of the three elements in an embedded-core test approach: (1) test pattern source and sink, (2) test access mechanism, and (3) core test wrapper [Zorian 1999].

Current core-based SOCs represent the success of the reuse paradigm in the industry. The design effort of a complex and mixed system can be strongly reduced by the reuse of predesigned functional blocks. However, for future SOCs with a large number of cores and increased interconnection delay, the implementation of an efficient and effective communication architecture among the blocks is becoming the new bottleneck in the performance of SOCs. It has been shown that the conventional point-to-point or bus-based communication architectures can no longer meet the system requirements in terms of bandwidth, latency, and power consumption [Vermeulen 2003] [Zeferino 2003].

Some works [Daly 2001] [Benini 2002] have proposed the use of integrated switching networks as an alternative approach to interconnect cores in SOCs. Such networks rely on a scalable and reusable communication platform, the so-called network-on-chip (NOC) system, to meet the two major requirements of current systems: reusability and scalable bandwidth. A conceptual architecture of an NOC system based on a 2-D mesh network is shown in Figure 4.2. Cores are connected to the network by routers or switches. Data are organized in the form of packets and transported through the interconnection links. Various network topologies and routing algorithms can be adopted to meet the requirements on performance, overhead, power consumption, among others.

Reusing the on-chip network as a test access mechanism for embedded cores in an SOC has been proposed in [Cota 2003]. Further results presented in [Cota 2004] show that test time can be reduced by network reuse even under power constraints, while other cost factors such as pin count and area overhead are strongly minimized. The reuse method assumes that most or all the embedded cores are connected or accessible through the on-chip communication network. This easier access to the cores and high parallelism of communications make the network a cost-effective test access mechanism (TAM).

The main advantage of on-chip network reuse is the availability of several accesses to each core, depending on the number of system input and output ports used during testing. Therefore, more cores can be tested in parallel as more access paths are available. In [Cota 2003], the idea of network parallelism is explored so that all

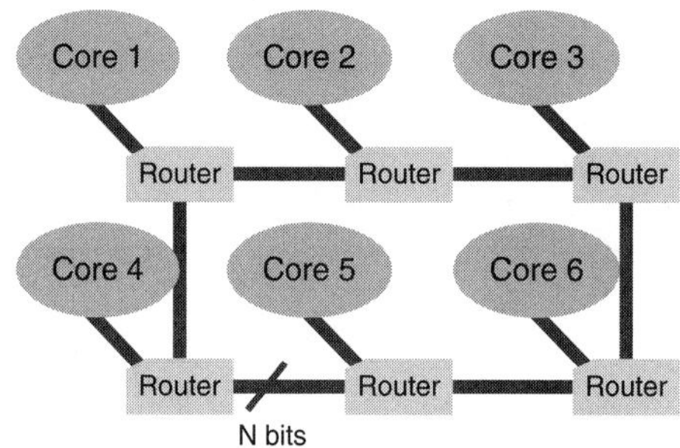

■ FIGURE 4.2

Conceptual architecture of a 2-D mesh NoC system.

available communication resources (channels, routers, and interfaces) are used in parallel to transmit test data. Test data are organized into test packets, which are scheduled so that the network usage is maximized to reduce the system test time.

For the rest of this chapter, we use the term *NOC* to denote an on-chip interconnection network (in general a packet-switching network) consisting of switches/routers, channels and other network components. And we use the term *NOC-based system* or *NOC-based design* to denote the entire SOC system consisting of an NOC and the embedded cores.

Section 4.3 presents advances in cost-efficient testing of embedded cores and interconnection fabrics in NOC-based systems.

4.2 SYSTEM-ON-CHIP (SOC) TESTING

This section first explains the importance of modular testing of SOC through the design of test wrappers and *test access mechanisms* (TAMs) that are used to transport test data. Optimizations of wrapper and TAM designs are thoroughly reviewed and discussed. Test scheduling with power and resource constraints is also elaborated for the cost-efficient testing of SOCs. Because modern SOCs contain mixed-signal circuits, analog wrapper design and its TAM support are also included. Finally, advanced topics such as hierarchical core-based testing and wafer-sort optimization are discussed.

4.2.1 Modular Testing of SOCs

Modular testing of embedded cores in a system-on-a-chip (SOC) is being increasingly advocated to simplify test access and test application [Zorian 1999]. To facilitate modular test, an embedded core must be isolated from surrounding logic, and test access must be provided from the I/O pins of the SOC. Test wrappers are used to isolate the core, whereas test access mechanisms (TAMs) transport test patterns and test responses between SOCs pins and core I/Os [Zorian 1999].

Effective modular testing requires efficient management of the test resources for core-based SOCs. This involves the design of core test wrappers and TAMs, the assignment of test pattern bits to *automatic test equipment* (ATE) channels, the scheduling of core tests, and the assignment of ATE channels to SOCs. The challenges involved in the optimization of SOC test resources for modular testing can be divided into three broad categories:

1. **Wrapper/TAM co-optimization.** Test wrapper design and TAM optimization are of critical importance during system integration because they directly impact hardware overhead, test time, and tester data volume. The issues involved in wrapper/TAM design include wrapper optimization, core assignment to TAM wires, sizing of the TAMs, and routing of TAM wires. As shown in [Chakrabarty 2001], [Iyengar 2002d], and [Marinissen 2000], most of these problems are *NP*-hard. Figures 4.3a and b illustrate the position of TAM design and test scheduling in the SOC DFT and test generation flows.

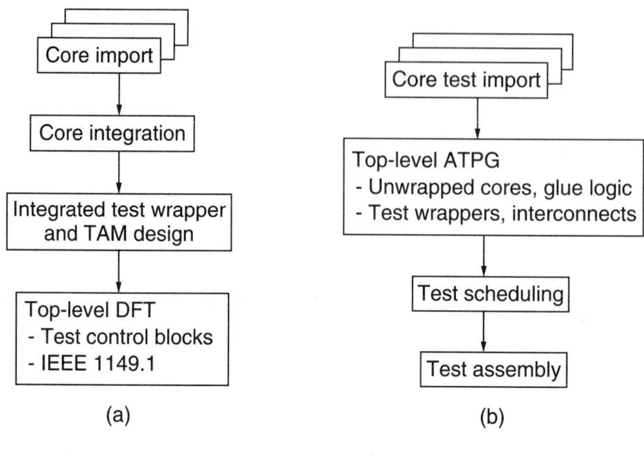

■ **FIGURE 4.3**

The (a) DFT generation flow and (b) test generation flow for SOCs [Iyengar 2002c].

2. **Constraint-driven test scheduling.** The primary objective of test scheduling is to minimize test time while addressing one or more of the following issues: (1) resource conflicts between cores arising from the use of shared TAMs and BIST resources, (2) precedence constraints among tests, and (3) power dissipation constraints. Furthermore, test time can often be decreased further through the selective use of test preemption [Iyengar 2002a]. As discussed in [Chakrabarty 2000a] and [Iyengar 2002a], most problems related to test scheduling for SOCs are also *NP*-hard.

3. **Minimizing ATE reload under memory depth constraints.** Given test data for the individual cores, the entire test suite for the SOC must be made to fit in a minimum number of ATE memory loads (preferably one memory load). This is important because, whereas the time required to apply digital vectors is relatively small, the time required to load several gigabytes of data to the ATE memory from workstations is significant [Barnhart 2001] [Marinissen 2001]. Therefore, to avoid splitting the test into multiple ATE load–apply sessions, the number of bits required to be stored on any ATE channel must not exceed the limit on the channel's memory depth.

In addition, the rising cost of ATE for SOC devices is a major concern [SIA 2005]. Because of the growing demand for pin counts, speed, accuracy, and vector memory, the cost of high-end ATE for full-pin, at-speed functional test, is predicted to be excessively high [SIA 2005]. As a result, the use of low-cost ATE that will perform structural rather than at-speed functional testing is increasingly being advocated for reducing test costs. Multisite testing, in which multiple SOCs are tested in parallel on the same ATE, can significantly increase the efficiency of ATE usage, as well as reduce test time for an entire production batch of SOCs. The use of low-cost ATE and multisite testing involves test data volume reduction and test pin count (TAM width) reduction, such that multiple SOC test suites can fit in ATE memory in a single test session [Marinissen 2001] [Volkerink 2002].

As a result of the intractability of the problems involved in test planning, test engineers adopted a series of simple ad hoc solutions in the past [Marinissen 2001]. For example, the problem of TAM width optimization is often simplified by stipulating that each core on the SOC have the same number of internal scan chains, say W; thus, a TAM of width W bits is laid out and cores are simply daisy-chained to the TAM. However, with the growing size of SOC test suites and the rising cost of ATE, the application of more aggressive test resource optimization techniques that enable effective modular test of highly complex next-generation SOCs using current-generation ATE is critical.

4.2.2 Wrapper Design and Optimization

A core test wrapper is a layer of logic that surrounds the core and forms the interface between the core and its SOC environment. Wrapper design is related to the well-known problems of circuit partitioning and module isolation and is therefore a more general test problem than its current instance (related to SOC test using TAMs). For example, earlier proposed forms of circuit isolation (precursors of test wrappers) include boundary scan and the *built-in logic block observer* (BILBO) [Abramovici 1994].

The test wrapper and TAM model of SOC test architecture was presented in [Zorian 1999]. In this paper, three mandatory wrapper operation modes listed were (1) normal operation, (2) core-internal tests, and (3) core-external tests. Apart from the three mandatory modes, two optional modes are "core bypass" and "detach."

Two proposals for test wrappers have been the "test collar" [Varma 1998] and TestShell [Marinissen 1998]. The test collar was designed to complement the test bus architecture [Varma 1998] and the TestShell was proposed as the wrapper to be used with the TestRail architecture [Marinissen 1998]. In [Varma 1998], three different test collar types were described: combinational, latched, and registered. For example, a simple combinational test collar cell consisting of a 2-to-1 multiplexer can be used for high-speed signals at input ports during parallel, at-speed test. The TestShell described in [Marinissen 1998] is used to isolate the core and perform TAM width adaptation. It has four primary modes of operation: function mode, IP test mode, interconnect test mode, and bypass mode. These modes are controlled using a test control mechanism that receives two types of control signals: pseudo-static signals (that retain their values for the duration of a test) and dynamic control signals (that can change values during a test pattern).

An important function of the wrapper is to adapt the TAM width to the core's I/O terminals and internal scan chains. This is done by partitioning the set of core-internal scan chains and concatenating them into longer wrapper scan chains, equal in number to the TAM wires. Each TAM wire can now directly scan test patterns into a single wrapper scan chain. TAM width adaptation directly affects core test time and has been the main focus of research in wrapper optimization. Note that to avoid problems related to clock skew, either internal scan chains in different clock domains must not be placed on the same wrapper scan chain or anti-skew (lock-up) latches must be placed between scan flip-flops belonging to different clock domains.

The issue of designing balanced scan chains within the wrapper was addressed in [Chakrabarty 2000b] (see Figure 4.4). The first techniques to optimize wrappers for test time reduction were presented in [Marinissen 2000]. To solve the problem, the authors proposed two polynomial-time algorithms that yield near-optimal results. The *largest processing time* (LPT) algorithm is taken from the multiprocessor scheduling literature and solves the wrapper design problem in short computation times. At the expense of a slight increase in computation time, the COMBINE algorithm yields even better results. It uses LPT as a start solution, followed by a linear search over the wrapper scan chain length with the First Fit Decreasing heuristic.

To perform wrapper optimization, the authors in [Iyengar 2002d] proposed *design wrapper*, an algorithm based on the *best fit decreasing* heuristic for the bin packing problem. The algorithm has two priorities: (1) minimizing core test time and (2) minimizing the TAM width required for the test wrapper. These priorities are achieved by balancing the lengths of the wrapper scan chains designed and identifying the number of wrapper scan chains that actually need to be created to minimize test time. Priority (2) is addressed by the algorithm, as it has a built-in reluctance to create a new wrapper scan chain, while assigning core-internal scan chains to the existing wrapper scan chains [Iyengar 2002d].

Wrapper design and optimization continue to attract considerable attention. Work in this area has focused on "light wrappers"—that is, the reduction of the

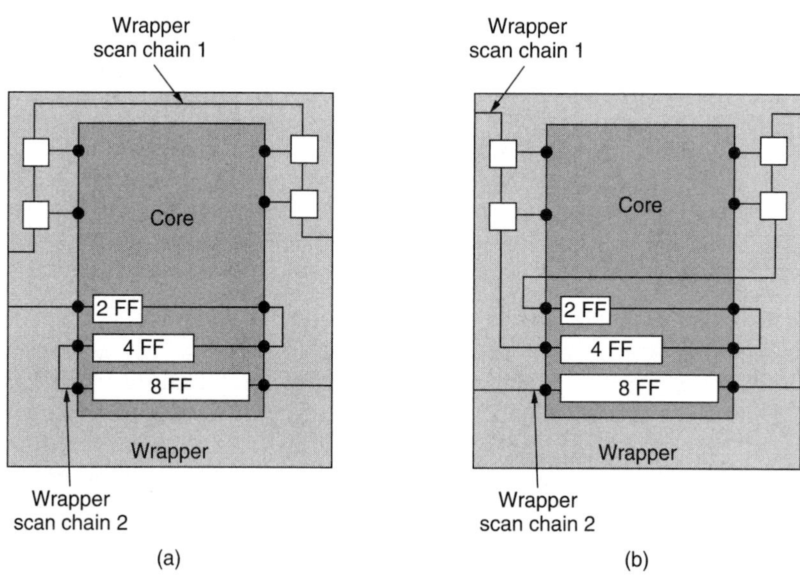

FIGURE 4.4

Wrapper chains: (a) unbalanced and (b) balanced.

number of register cells [Xu 2003]—and the design of wrappers for cores and SOCs with multiple clock domains [Xu 2004].

4.2.3 TAM Design and Optimization

Many different TAM designs have been proposed in the literature. TAMs have been designed based on direct access to cores multiplexed onto the existing SOC pins [Immaneni 1990], reusing the on-chip system bus [Harrod 1999], searching transparent paths through or around neighboring modules [Ghosh 1998] [Nourani 2000] [Chakrabarty 2003], and 1-bit boundary scan rings around cores [Touba 1997] [Whetsel 1997].

The most popular appear to be the dedicated, scalable TAMs such as test bus [Varma 1998] and TestRail [Marinissen 1998]. Despite the fact that their dedicated wiring adds to the area costs of the SOC, their flexible nature and guaranteed test access have proven successful. Three basic types of such scalable TAMs have been described in [Aerts 1998] (Figure 4.5): (1) the *multiplexing* architecture, (2) the *daisy-chain* architecture, and (3) the *distribution* architecture. In the multiplexing and daisy-chain architectures, all cores get access to the total available TAM width, while in the distribution architecture, the total available TAM width is distributed over the cores.

In the multiplexing architecture, only one core wrapper can be accessed at a time. Consequently, this architecture only supports serial schedules, in which the cores are tested one after the other. An even more serious drawback of this architecture is that testing the circuitry and wiring in between cores is difficult; interconnect test requires simultaneous access to multiple wrappers. The other two basic architectures do not have these restrictions; they allow for both serial as well as parallel test schedules, and they also support interconnect testing.

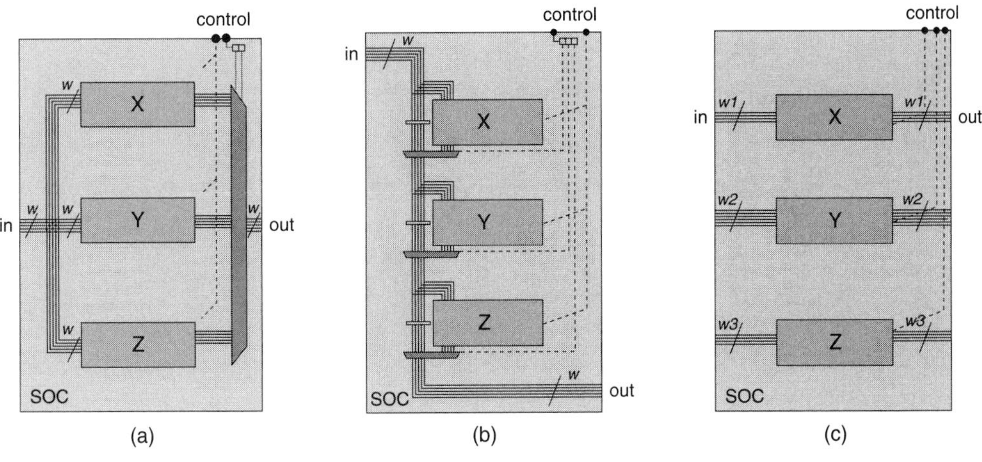

■ **FIGURE 4.5**

The (a) multiplexing, (b) daisy-chain, and (c) distribution architectures [Aerts 1998, Iyengar 2002c].

The *test bus* architecture [Varma 1998] (see Figure 4.6a) is a combination of the multiplexing and distribution architectures. A single test bus is in essence the same as what is described by the multiplexing architecture; cores connected to the same test bus can only be tested sequentially. The test bus architecture allows for multiple test buses on one SOC that operate independently, as in the distribution architecture. Cores connected to the same test bus suffer from the same drawback as in the multiplexing architecture (*i.e.*, their wrappers cannot be accessed simultaneously), making core-external testing difficult or impossible.

The *TestRail* architecture [Marinissen 1998] (see Figure 4.6b) is a combination of the daisy-chain and distribution architectures. A single TestRail is in essence the same as what is described by the daisy-chain architecture: scan-testable cores connected to the same TestRail can be tested simultaneously as well as sequentially. A TestRail architecture allows for multiple TestRails on one SOC, which operate independently, as in the distribution architecture. The TestRail architecture supports serial and parallel test schedules, as well as hybrid combinations of those.

In most TAM architectures, the cores assigned to a TAM are connected to *all* wires of that TAM. This is referred to this as *fixed-width* TAMs. A generalization of this design is one in which the cores assigned to a TAM each connect to a (possibly different) subset of the TAM wires [Iyengar 2003c]. The core–TAM assignments are made at the granularity of TAM wires, instead of considering the entire TAM bundle as one inseparable entity. These are referred to as *flexible-width* TAMs. This concept can be applied to both test bus and TestRail architectures. Figure 4.6c shows an example of a flexible-width test bus architecture.

Most SOC test architecture optimization algorithms proposed have concentrated on fixed-width test bus architectures and assume cores with fixed-length scan chains. In [Chakrabarty 2001], the author described a test bus architecture optimization approach that minimizes test time using ***integer linear programming*** (ILP). ILP is replaced by a genetic algorithm in [Ebadi 2001]. In [Iyengar 2002b],

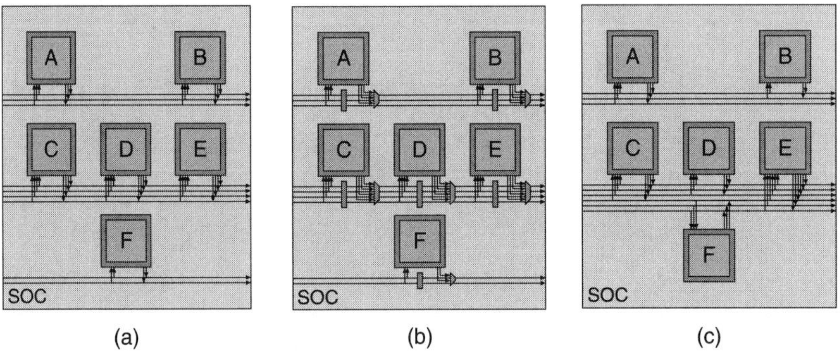

■ **FIGURE 4.6**

The (a) fixed-width test bus architecture, (b) fixed-width TestRail architecture, and (c) flexible-width test bus architecture [Iyengar 2003c].

the authors extend the optimization criteria of [Chakrabarty 2001] with place-and-route and power constraints, again using ILP. In [Huang 2001] and [Huang 2002], test bus architecture optimization is mapped to the well-known problem of two-dimensional bin packing, and a best fit algorithm is used to solve it. Wrapper design and TAM design both influence the SOC test time, hence their optimization needs to be carried out in conjunction in order to achieve the best results. The authors in [Iyengar 2002d] were the first to formulate the problem of integrated wrapper/TAM design; despite its *NP*-hard character, it is addressed using ILP and exhaustive enumeration. In [Iyengar 2003b], the authors presented efficient heuristics for the same problem.

Idle bits exist in test schedules when parts of the test wrapper and TAM are underutilized, leading to idle time in the test delivery architecture. In [Marinissen 2002a], the authors first formulated the test time minimization problem both for cores with fixed-length scan chains as well as for cores with flexible-length scan chains. Next, they presented lower bounds on the test time for the test bus and TestRail architectures and then examined three main reasons for underutilization of TAM bandwidth, leading to idle bits in the test schedule and test times higher than the lower bound [Marinissen 2002a]. The problem of reducing the amount of idle test data was also addressed in [Gonciari 2003].

The optimization of a flexible-width multiplexing architecture (*i.e.*, for one TAM only) was proposed in [Iyengar 2002d]. This work again assumes cores with fixed-length scan chains. The paper describes a heuristic algorithm for the co-optimization of wrappers and test buses based on rectangle packing. In [Iyengar 2002a], the same authors extended this work by including precedence, concurrency, and power constraints while allowing a user-defined subset of the core tests to be preempted.

Fixed-width TestRail architecture optimization was investigated in [Goel 2002]. Heuristic algorithms have been developed for the co-optimization of wrappers and TestRails. The algorithms work both for cores with fixed-length and flexible-length scan chains. TR-ARCHITECT, the tool presented in [Goel 2002], is currently in actual industrial use.

4.2.4 Test Scheduling

Test scheduling for SOCs involving multiple test resources and cores with multiple tests is especially challenging, and even simple test scheduling problems for SOCs have been shown to be *NP*-hard [Chakrabarty 2000a]. In [Sugihara 1998], a method for selecting tests from a set of external and BIST tests (that run at different clock speeds) was presented. Test scheduling was formulated as a combinatorial optimization problem. Reordering tests to maximize defect detection early in the schedule was explored in [Jiang 1999]. The entire test suite was first applied to a small sample population of ICs. The fault coverage obtained per test was then used to arrange tests that contribute to high fault coverage earlier in the schedule. The authors used a polynomial-time algorithm to reorder tests based on the defect data as well as execution time of the tests [Jiang 1999]. A test scheduling technique based on the defect probabilities of the cores has been reported [Larsson 2004].

Macro testing is a modular testing approach for SOC cores in which a test is broken down into a *test protocol* and list of test patterns [Beenker 1995]. A test protocol is defined at the terminals of a macro and describes the necessary and sufficient conditions to test the macro [Marinissen 2002b]. The test protocols are expanded from the macro level to the SOC pins and can be either applied sequentially to the SOC or scheduled to increase parallelism. In [Marinissen 2002b], a heuristic scheduling algorithm based on pair-wise composition of test protocols was presented. The algorithm determines the start times for the expanded test protocols in the schedule, such that no resource conflicts occur and test time is minimized [Marinissen 2002b].

SOCs in test mode can dissipate up to twice the amount of power they do in normal mode, because cores that do not normally operate in parallel may be tested concurrently [Zorian 1993]. *Power-constrained* test scheduling is therefore essential to limit the amount of concurrency during test application to ensure that the maximum power budget of the SOC is not exceeded. In [Chou 1997], a method based on approximate vertex cover of a resource-constrained test compatibility graph was presented. In [Muresan 2000], the use of list scheduling and tree-growing algorithms for power-constrained scheduling was discussed. The authors presented a greedy algorithm to overlay tests such that the power constraint is not violated. A constant additive model is employed for power estimation during scheduling [Muresan 2000]. The issue of reorganizing scan chains to tradeoff test time with power consumption was investigated in [Larsson 2001b]. The authors presented an optimal algorithm to parallelize tests under power and resource constraints. The design of test wrappers to allow for multiple scan chain configurations within a core was also studied.

In [Iyengar 2002a], an integrated approach to test scheduling was presented. Optimal test schedules with precedence constraints were obtained for reasonably sized SOCs. For precedence-based scheduling of large SOCs, a heuristic algorithm was developed. The proposed approach also includes an algorithm to obtain preemptive test schedules in $O(n^3)$ time, where n is the number of tests [Iyengar 2002a]. Parameters that allow only a certain number of preemptions per test can be used to prevent excessive BIST and sequential circuit test preemptions. Finally, a new power-constrained scheduling technique was presented, whereby power constraints can be easily embedded in the scheduling framework in combination with precedence constraints, thus delivering an integrated approach to the SOC test scheduling problem.

Integrated TAM Optimization and Test Scheduling

Both TAM optimization and test scheduling significantly influence the test time, test data volume, and test cost for SOCs. Furthermore, TAMs and test schedules are closely related. For example, an effective schedule developed for a particular TAM architecture may be inefficient or even infeasible for a different TAM architecture.

Integrated methods that perform TAM design and test scheduling *in conjunction* are therefore required to achieve low-cost, high-quality test.

In [Larsson 2001a], the authors presented an integrated approach to test scheduling, TAM design, test set selection, and TAM routing. The SOC test architecture was represented by a set of functions involving test generators, response evaluators, cores, test sets, power and resource constraints, and start and end times in the test schedule modeled as Boolean and integral values [Larsson 2001a]. A polynomial-time algorithm was used to solve these equations and determine the test resource placement, TAM design and routing, and test schedule, such that the specified constraints are met.

The mapping between core I/Os and SOC pins during the test schedule was investigated in [Huang 2001]. TAM design and test scheduling was modeled as two-dimensional bin-packing, in which each core test is represented by a rectangle. The height of each rectangle corresponds to the test time, the width corresponds to the core I/Os, and the weight corresponds to the power consumption during test. The objective is to pack the rectangles into a bin of fixed width (SOC pins) such that the bin height (total test time) is minimized while power constraints are met. A heuristic method based on the best fit algorithm was presented to solve the problem [Huang 2001]. The authors next formulated constraint-driven pin mapping and test scheduling as the chromatic number problem from graph theory and as a dependency matrix partitioning problem [Huang 2002]. Both problem formulations are *NP*-hard. A heuristic algorithm based on clique partitioning was proposed to solve the problem.

The problem of TAM design and test scheduling with the objective of minimizing the *average* test time was formulated in [Koranne 2002a]. The problem was reduced to one of minimum-weight perfect bipartite graph matching, and a polynomial-time optimal algorithm was presented. A test planning flow was also presented.

The power-constrained TAM design and test scheduling problem was studied in [Zhao 2003]. By formulating into a graph theoretic problem, the test scheduling problem was reduced to provide test access to core level test terminals from system level pins and efficiently route wrapper-configured cores on TAMs to reduce the total test application time. The seed sets were effectively selected from the conflict graph to initiate scheduling and the compatible test sets were derived from the power-constrained test compatible graph to facilitate test concurrency and dynamic TAM width distribution.

In [Iyengar 2003c], a new approach for wrapper/TAM co-optimization and constraint-driven test scheduling using rectangle packing was described. Flexible-width TAMs that are allowed to fork and merge were designed. Rectangle packing was used to develop test schedules that incorporate precedence and power constraints, while allowing the SOC integrator to designate a group of tests as preemptable. Finally, the relationship between TAM width and tester data volume was studied to identify an effective TAM width for the SOC.

The work reported in [Iyengar 2003c] was extended in [Iyengar 2003a] to address the minimization of ATE buffer reloads and include multisite test. The ATE is assumed to contain a pool of memory distributed over several channels, such that the memory depth assigned to each channel does not exceed a maximum limit. Furthermore, the sum of the memory depth over all channels equals the total pool of ATE memory. Idle bits appear on ATE channels whenever there is idle time on a TAM wire. These bit positions are filled with don't-cares if they appear between useful test bits; however, if they appear only at the end of the useful bits, they are not required to be stored in the ATE.

The SOC test resource optimization problem for multisite test was stated as follows. Given the test set parameters for each core, and a limit on the maximum memory depth per ATE channel, determine the wrapper/TAM architecture and test schedule for the SOC, such that (1) the memory depth required on any channel is less than the maximum limit, (2) the number of TAM wires is minimized, and (3) the idle bits appear only at the end of each channel. A rectangle packing algorithm was developed to solve this problem.

A new method for representing SOC test schedules using k-tuples was discussed in [Koranne 2002b]. The authors presented a p-admissible model for test schedules that is amenable to several solution methods such as local search, two-exchange, simulated annealing and genetic algorithms that cannot be used in a rectangle-representation environment. The proposed approach provides a compact, standardized representation of test schedules. This facilitates fast and efficient evaluation of SoC test automation solutions to reduce test costs.

Finally, work on TAM optimization has focused on the use of ATEs with port scalability features [Sehgal 2003a, 2004a, 2004c]. To address the test requirements of SOCs, automatic test equipment (ATE) vendors have announced a new class of testers that can simultaneously drive different channels at different data rates. Examples include the Agilent 93000 series tester based on port scalability and the test processor-per-pin architecture [Agilent] and the Tiger system from Teradyne [Teradyne 2007] in which the data rate can be increased through software for selected pin groups to match SOC test requirements. However, the number of tester channels with high data rates may be constrained in practice because of ATE resource limitations, the power rating of the SOC, and scan frequency limits for the embedded cores. Optimization techniques have been developed to ensure that the high data rate tester channels are efficiently used during SOC testing [Sehgal 2004a].

The availability of dual-speed ATEs was also exploited in [Sehgal 2003a] and [Sehgal 2004c], where a technique was presented to match ATE channels with high data rates to core scan chain frequencies using virtual TAMs. A *virtual TAM* is an on-chip test data transport mechanism that does not directly correspond to a particular ATE channel. Virtual TAMs operate at scan-chain frequencies; however, they interface with the higher frequency ATE channels using bandwidth matching. Moreover, because the virtual TAM width is not limited by the ATE pin count, a larger number of TAM wires can be used on the SOC, thereby leading to lower test times. A drawback of virtual TAMs, however, is the need for additional TAM wires on the SOC as well as frequency division hardware for bandwidth matching. In [Sehgal 2004a], the hardware overhead is reduced through the use of a smaller number of on-chip TAM wires; ATE channels with high data rates directly drive SOC TAM wires, without requiring frequency division hardware.

4.2.5 Modular Testing of Mixed-Signal SOCs

Prior research on modular testing of SOCs has focused almost exclusively on the digital cores in an SOC. However, most SOCs in use today are mixed-signal circuits containing both digital and analog cores [Liu 1998] [Kundert 2000] [Yamamoto 2001]. Increasing pressure on consumer products for small form factors and extended battery life is driving single chip integration and blurring the lines between analog and digital design types. As indicated in the 2005 International Technology Roadmap for Semiconductors [SIA 2005], the combination of these circuits on a single die compounds the test complexities and challenges for devices that fall in an increasing commodity market. Therefore, an effective modular test methodology should be capable of handling both digital and analog cores, and it should reduce test cost by enabling test reuse for reusable embedded modules.

In traditional mixed-signal SOC testing, tests for analog cores are applied either from chip pins through direct test access methods, such as via multiplexing or through a dedicated analog test bus [Sunter 1996] [Cron 1997], which requires the use of expensive mixed-signal testers. For mid- to low-frequency analog applications, the data are often digitized at the tester, where it is affordable to incorporate high-quality data converters. In most mixed-signal ICs, analog circuitry accounts for only a small part of the total silicon ("big-D/small-A"). However, the total production testing cost is dominated by analog testing costs. This is because expensive mixed-signal testers are employed for extended periods of time resulting in high overall test costs. A natural solution to this problem is to implement the data converters on-chip. Because most SOC applications do not push the operational frequency limits, the design of such data converters on-chip appears to be feasible. Until recently, such an approach has not been deemed desirable because of its high hardware overhead. However, as the cost of on-chip silicon is decreasing and the functionality and the number of cores in a typical SOC are increasing, the addition of data converters on-chip for testing analog cores now promises to be cost-efficient. These data converters eliminate the need for expensive mixed-signal test equipment.

Results have been reported on the optimization of a unified test access architecture that is used for both digital and analog cores [Sehgal 2003b]. Instead of treating the digital and analog portions separately, a global test resource optimization problem is formulated for the entire SOC. Each analog core is wrapped by a DAC-ADC pair and a digital configuration circuit. Results show that for "big-D/small-A" SOCs, the test time and test cost can be reduced considerably if the analog cores are wrapped and the test access and test scheduling problems for the analog and digital cores are tackled in a unified manner.

Each analog core is provided a test wrapper where the test information includes only digital test patterns, clock frequency, the test configuration, and pass/fail criteria. This analog test wrapper converts the analog core to a virtual digital core with strictly sequential test patterns, which are the digitized analog signals. To utilize test resources efficiently, the analog wrapper needs to provide sufficient flexibility in terms of required resources with respect to all the test needs of the analog core. One way to achieve this uniform test access scheme for analog cores is to provide an on-chip ADC-DAC pair that can serve as an interface between each analog core and the digital surroundings, as shown in Figure 4.7.

Analog test signals are expressed in terms of a signal shape, such as sinusoidal or pulse, and signal attributes, such as frequency, amplitude, and precision. The core vendor provides these tests to the system integrator. In the case of analog testers, these signals are digitized at the high precision ADCs and DACs of the tester. In the case of on-chip digitization, the analog wrapper needs to include the lowest cost data converters that can still provide the required frequency and accuracy for applying the core tests. Thus, on-chip conversion of each analog test to digital patterns imposes requirements on the frequency and resolution of the data converters of the analog wrapper. These converters need to be designed to accommodate all the test requirements of the analog core.

Analog tests may also have a high variance in terms of their frequency and test time requirements. Whereas tests involving low-frequency signals require low bandwidth and high test times, tests involving high-frequency signals require high bandwidth and low test time. Keeping the bandwidth assigned to the analog core constant results in underutilization of the precious test resources. The variance of analog test needs have to be fully exploited in order to achieve an efficient test plan. Thus, the analog test wrapper has to be designed to accommodate multiple configurations with varying bandwidth and frequency requirements.

Figure 4.8 shows the block diagram of an analog wrapper that can accommodate all the abovementioned requirements. The figure highlights the control and clock signals generated by the test control circuit. The registers at each end of the data converters are written and read in a semiserial fashion depending on the frequency requirement of each test. For example, for a digital TAM clock of 50MHz, 12-bit DAC and ADC resolution, and an analog test requirement of 8MHz sampling frequency, the input and output registers can be updated with a serial-to-parallel ratio of 6. Thus, the bandwidth requirement of this particular test is only 2 bits. The digital test control circuit selects the configuration for each test. This configuration includes the divide ratio of the digital TAM clock, the serial-to-parallel conversion rate of the input and output registers of the data converters, and the test modes.

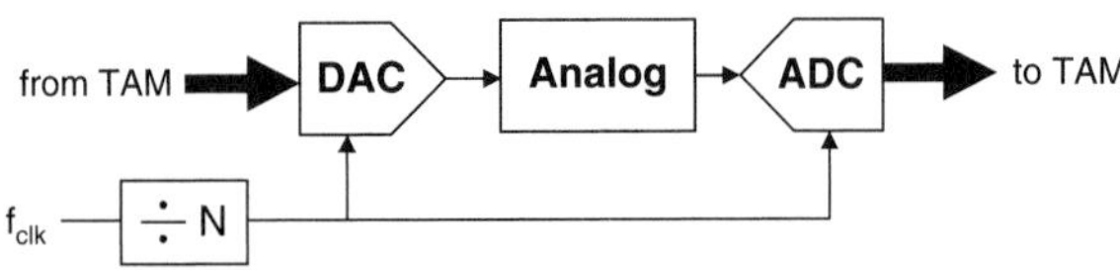

FIGURE 4.7

On-chip digitization of analog test data for uniform test access [Sehgal 2003b].

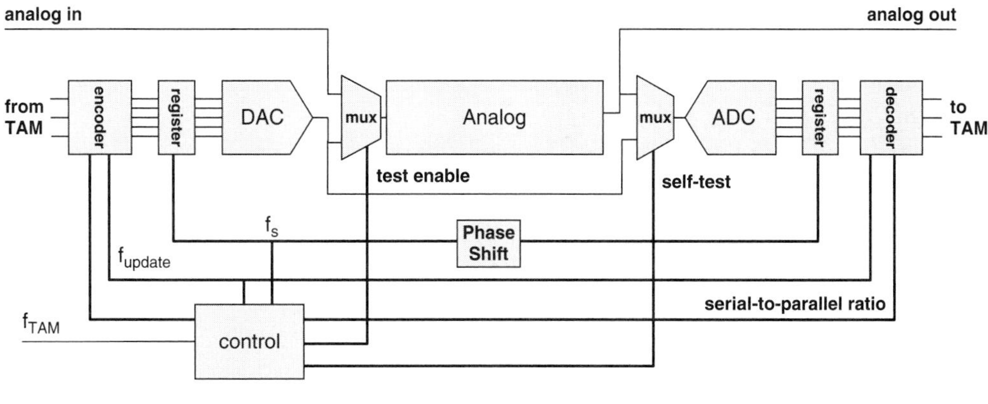

■ **FIGURE 4.8**

Block diagram of the analog test wrapper [Sehgal 2003b].

Analog Test Wrapper Modes

In the normal mode of operation, the analog test wrapper is completely bypassed; the analog circuit operates on its analog input/output pins. During testing, the analog wrapper has two modes, a *self-test* mode and a *core-test* mode. Before running any tests on the analog core, the wrapper data converters have to be characterized for their conversion parameters, such as the nonlinearity and the offset voltage. The self-test mode is enabled through the analog multiplexer at the input of the wrapper ADC, as shown in Figure 4.8. The parameters of the DAC-ADC pair are determined in this mode and are used to calibrate the measurement results. Once the self-test of the test wrapper is complete, core testing can be enabled by turning off the *self-test* bits.

For each analog test, the encoder has to be set to the corresponding serial-to-parallel ***conversion ratio*** (*cr*), where it shifts the data from the corresponding TAM inputs into the register of the ADC. Similarly, the decoder shifts data out of the DAC register. The update frequency of the input and output registers, $f_{update} = f_s \times cr$, is always less than the TAM clock rate, f_{TAM}. For example, if the test bandwidth requirement is 2 bits and the resolution of the data converters is 12 bits, the input and output registers of the data converters are clocked at a rate six times less than the clock of the encoder, and the input data are shifted into the encoder and out of the decoder at a 2-bits/cycle rate. The complexity of the encoder and the decoder depends on the number of distinct bandwidth and TAM assignments (the number of possible test configurations). For example, for a 12-bit resolution, the bandwidth assignments may include 1, 2, 3, 4, 6, and 12 bits, where in each case the data may come from distinct TAMs. Clearly, to limit the complexity of the encoder-decoder pair, the number of such distinct assignments has to be limited. This requirement can be imposed in the test scheduling optimization algorithm.

The analog test wrapper transparently converts the analog test data to the digital domain through efficient utilization of the resources; thus this obviates the need for analog testers. The processing of the collected data can be done in the tester by adding appropriate algorithms, such as the FFT algorithm. Further details and experimental results can be found in [Sehgal 2003b], [Sehgal 2005], and [Sehgal 2006b].

4.2.6 Modular Testing of Hierarchical SOCs

A hierarchical *system-on-chip* (SOC) is designed by integrating heterogeneous technology cores at several layers of hierarchy [SIA 2005]. The ability to reuse embedded cores in a hierarchical manner implies that today's SOC is tomorrow's embedded core [Gallagher 2001]. Two broad design transfer models are emerging in hierarchical SOC design flows.

1. **Noninteractive.** The noninteractive design transfer and hand-off model is one in which there is limited communication between the core vendor and the SOC integrator. The hard cores are taken off-the-shelf and integrated into designs as optimized layouts.
2. **Interactive.** The interactive design transfer model is typical of larger companies where the business units producing *intellectual property* (IP) cores may be part of the same organization as the business unit responsible for system integration. Here, there is a certain amount of communication between the core vendor and core user during system integration. The communication of the core user's requirements to the core vendor can play a role in determining the core specifications.

Hierarchical SOCs offer reduced cost and rapid system implementation; however, they pose difficult test challenges. Most TAM design methods assume that the SOC hierarchy is flattened for the purpose of test. However, this assumption is often unrealistic in practice, especially when older-generation SOCs are used as hard cores in new SOC designs. In such cases, the core vendor may have already designed a TAM within the "megacore" that is provided as an optimized and technology-mapped layout to the SOC integrator.

A *megacore* is defined as a design that contains nonmergeable embedded cores. To ensure effective testing of an SOC based on megacores, the top-level TAM must communicate with lower level TAMs within megacores. Moreover, the system-level test architecture must be able to reuse the existing test architecture within cores; redesign of core test structures must be kept to a minimum, and it must be consistent with the design transfer model between the core designer and the core user [Parulkar 2002].

A TAM design methodology that closely follows the design transfer model in use is necessary because if the core vendor has implemented "hard" (*i.e.*, nonalterable) TAMs within megacores, the SOC integrator must take into account these lower-level TAM widths while optimizing the widths and core assignment for higher-level TAMs. On the other hand, if the core vendor designs TAMs within megacores in consultation with the SOC integrator, the system designer's TAM optimization method must be flexible enough to include parameters for lower-level cores.

Finally, multilevel TAM design for SOCs that include reused cores at multiple levels is needed to exploit "TAM reuse" and "wrapper reuse" in the test development process.

It is only recently that the problem of designing test wrappers and TAMs of multilevel TAMs for the "cores within cores" design paradigm has been investigated [Iyengar 2003a] [Chakrabarty 2005]. Two design flows have been considered for the scenario in which megacores are wrapped by the core vendor before delivery. In an alternative scenario, the megacores can be delivered to the system integrator in an unwrapped fashion, and the system integrator appropriately designs the megacore wrappers and the SOC-level TAM architecture to minimize the overall test time.

Figure 4.9 illustrates a megacore that contains four embedded cores and additional logic external to the embedded cores. The core vendor for this megacore core has wrapped the four embedded cores, and implemented a TAM architecture to access the embedded cores. The TAM architecture consists of two test buses of widths 3 bits and 2 bits, respectively, that are used to access the four embedded cores. It is assumed here that the TAM inputs and outputs are not multiplexed with the functional pins. Next, Figure 4.10 shows how a two-part wrapper (wrapper 1 and wrapper 2) for the megacore can be designed not only to drive the TAM wires

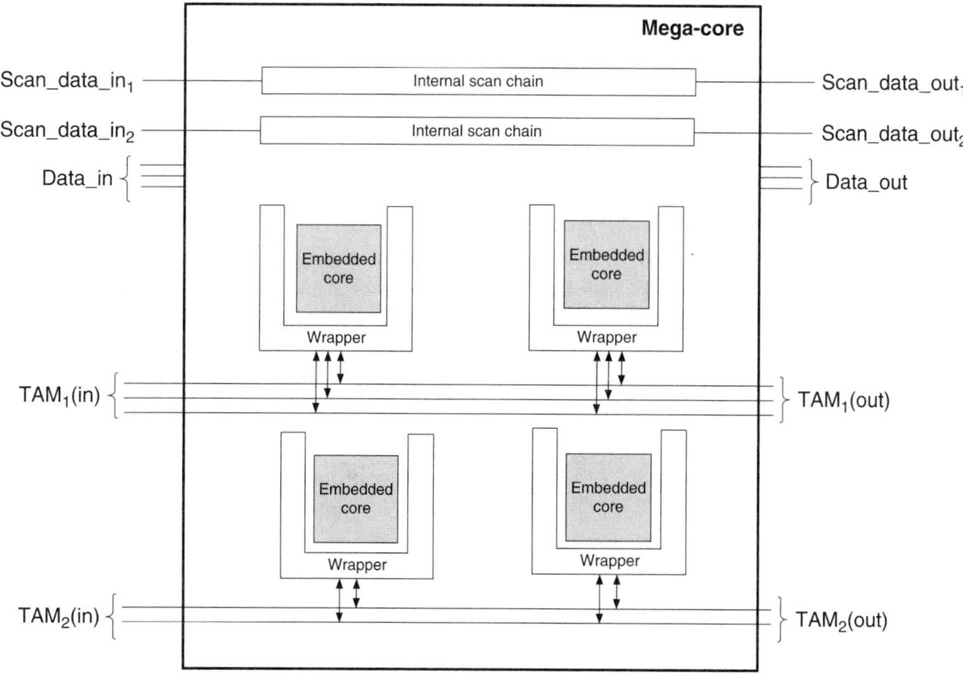

■ **FIGURE 4.9**

An illustration of a megacore with a predesigned TAM architecture.

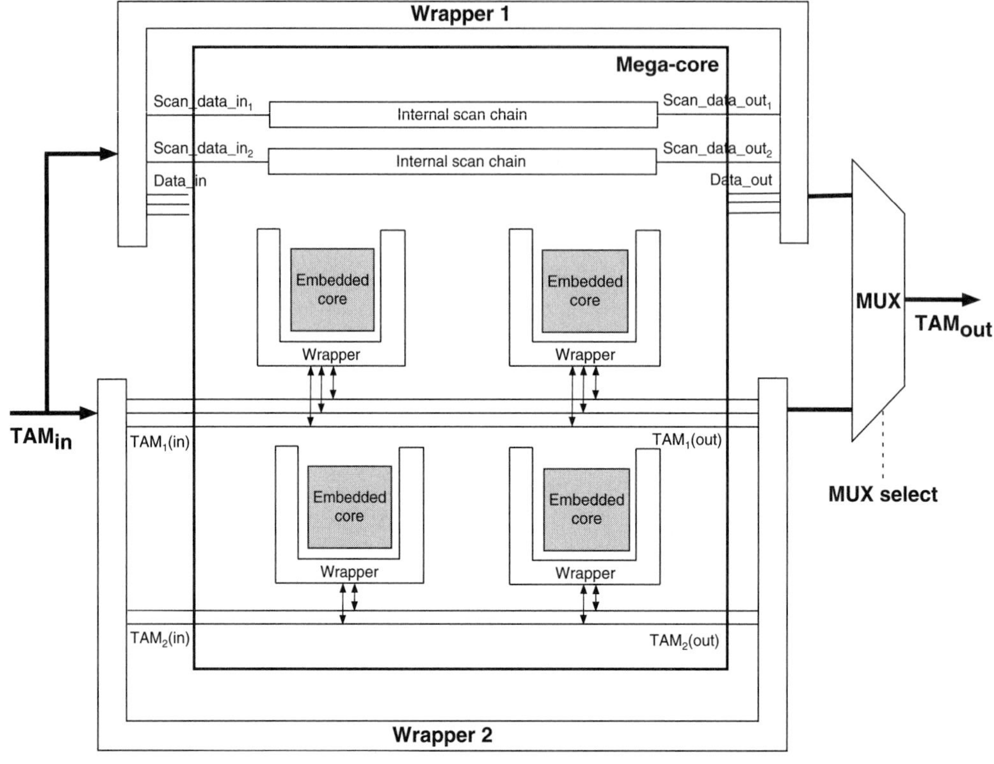

■ **FIGURE 4.10**

An illustration of a two-part wrapper for the megacore that is used to drive the TAMs in the megacore and to test the logic external to the embedded cores.

within the megacore but also to test the logic that is external to the embedded cores. In this design, the TAM inputs for wrapper 1 and wrapper 2 are multiplexed in time such that the embedded cores within the megacore are tested before the logic external to them, or vice versa. Test generation for the top-level logic is done by the megacore vendor with the wrappers for the embedded cores in functional mode. During the testing of the top-level logic in the megacore using wrapper 1, the wrappers for the embedded cores must therefore be placed in functional mode to ensure that the top-level logic can be tested completely through the megacore I/Os and scan terminals.

Megacores may be supplied by core vendors in varying degrees of readiness for test integration. For example, the IEEE 1500 standard on embedded core test defines two compliance levels for core delivery: 1500-wrapped and 1500-unwrapped [IEEE 1500-2005]. Here we describe three other scenarios, based in part on the 1500 compliance levels. These scenarios refer to the roles played by the system integrator and the core vendor in the design of the TAM and the wrapper for the megacore. For each scenario, the design transfer model refers to the type of information about the megacore that is provided by the core vendor to the system integrator. The term *wrapped* is used to denote a core for which a wrapper has been predesigned, as in [Marinissen 2002b]. The term *TAM-ed* is used to denote a megacore that contains an internal TAM structure.

1. **Scenario 1.** Not TAM-ed and not wrapped: In this scenario, the system integrator must design a wrapper for the megacore as well as TAMs within the megacore. The megacores are therefore delivered either as soft cores or before final netlist and layout optimization such that TAMs can be inserted within the megacores.
2. **Scenario 2.** TAM-ed and wrapped: In this scenario, we consider TAM-ed megacores for which wrappers have been designed by the core vendor. This scenario is especially suitable for a megacore that was an SOC in an earlier generation. It is assumed that the core vendors wrap such megacores before design transfer, and test data for the megacore cannot be further serialized or parallelized by the SOC integrator. This implies that the system integrator has less flexibility in top-level TAM partitioning and core assignment. At the system level, only structures that facilitate normal/test operation, interconnect test, and bypass are created. This scenario includes both the interactive and noninteractive design transfer models.
3. **Scenario 3.** TAM-ed but not wrapped: In this scenario, the megacore contains lower-level TAMs, but it is not delivered in a wrapped form; therefore, the system integrator must design a wrapper for the megacore. To design a wrapper as sketched in Figure 4.10, the core vendor must provide information about the number of functional I/Os, the number and lengths of top-level scan chains in the megacore, the number of TAM partitions and the size of each partition, and the test time for each TAM partition. Compared to the noninteractive design transfer model in scenario 2, the system integrator in this case has greater flexibility in top-level TAM partitioning and core assignment. Compared to the interactive design transfer model in scenario 2, the system integrator here has less influence on the TAM design for a megacore; however, this loss of flexibility is somewhat offset by the added freedom of being able to design the megacore wrapper. Width adaptation can be carried out in the wrapper for the megacore such that a narrow TAM at the SOC-level can be used to access a megacore that has a wider internal TAM.

Optimization techniques for these scenarios are described in detail in [Iyengar 2003a], [Chakrabarty 2005], [Sehgal 2004b], and [Sehgal 2006a]. As hierarchical SOCs become more widespread, it is expected that more research effort will be devoted to this topic.

4.2.7 Wafer-Sort Optimization for Core-Based SOCs

Product cost is a major driver in the consumer electronics market, which is characterized by low profit margins and the use of SOC designs. Packaging has been recognized as a significant contributor to the product cost for such SoCs [Kahng 2003]. To reduce packaging cost and the test cost for packaged chips, the semiconductor industry uses wafer-level testing (wafer sort) to screen defective dies [Maxwell 2003]. However, because test time is a major practical constraint for wafer sort, even more so than for package test, not all the scan-based digital tests can be applied to the die under test. An optimal test-length selection technique for wafer-level testing of core-based SOC has been developed [Bahukudumbi 2006]. This technique, which is based on a combination of statistical yield modeling and integer linear programming, allows us to determine the number of patterns to use for each embedded core during wafer sort such that the probability of screening defective dies is maximized for a given upper limit on the SOC test time.

One SOC test scheduling method attempted to minimize the average test time for a packaged SOC, assuming an abort-on-first fail strategy [Larsson 2004] [Ingelsson 2005]. The key idea in this work is to use defect probabilities for the embedded cores to guide the test scheduling procedure. These defect probabilities are used to determine the order in which the embedded cores in the SOC are tested, as well as to identify the subsets of cores that are tested concurrently. The defect probabilities for the cores were assumed in [Larsson 2004] to be either known a priori or obtained by binning the failure information for each individual core over the product cycle [Ingelsson 2005]. In practice, however, short product cycles make defect estimation based on failure binning difficult. Moreover, defect probabilities for a given technology node are not necessarily the same for the next (smaller) technology node. Therefore, a yield modeling technique is needed to accurately estimate these defect probabilities.

In [Bahukudumbi 2006], the researchers show how statistical yield modeling for defect-tolerant circuits can be used to estimate defect probabilities for embedded cores in an SOC. The test-length selection problem for wafer-level testing of core-based SOC is next formulated. Then, integer linear programming is used to obtain optimal solutions for the test-length selection problem. The application of this approach to reduced pin-count testing is presented in [Bahukudumbi 2007].

4.3 NETWORK-ON-CHIP (NOC) TESTING

Testing an NOC-based system includes testing of embedded cores and testing of the on-chip network. The former is similar to conventional SOC testing, the latter tests the NOC itself including interconnects, switches/routers, input/output ports, and other mechanisms other than the cores. Because of the excessive routing overhead, all tests in the NOC domain should be done in a cost-efficient manner, which can be done through the reuse of the NOC as a *test access mechanism* (TAM).

4.3.1 NOC Architectures

A typical packet-switching network model called ***system-on-chip interconnection network*** (SOCIN) [Zeferino 2003] that implements on a *two-dimensional* (2-D) mesh topology is shown in Figure 4.11. Here the d695 circuit from the ITC-2002 SOC

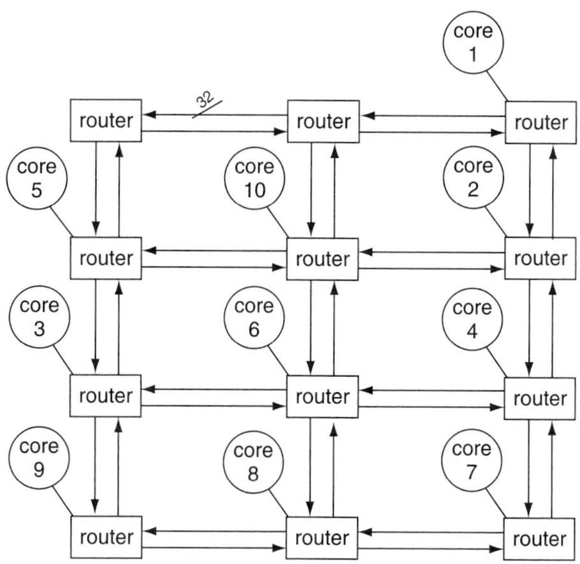

FIGURE 4.11

System d695 implemented in the SOCIN NOC.

benchmarks [ITC 2002] is stitched in the network for illustration. The bidirectional communication channels are defined to be 32-bits wide on each direction, and the packets have unlimited length. It uses credit-based flow-control and XY routing—a deadlock-free, deterministic and source-based approach, in which a packet is first routed on the X direction and then on the Y direction before reaching its destination. Switching is based on the wormhole approach, where a packet is broken up into flits (flow control units, the smallest unit over which the flow control is performed), and the flits follow the header in a pipelined fashion. The flit size equals the channel width. This platform is assumed in the rest of this section for illustration.

NOCs typically use the message-passing communication model, and the processing cores attached to the network communicate by sending and receiving request and response messages. To be routed by the network, a message is composed of a header, a payload, and a trailer. The header and the trailer frame the packet, and the payload carries the data being transferred. The header also carries the information needed to establish the path between the sender and the receiver. Depending on the network implementation, messages can be split into packets, which have the same format of a message and are individually routed. Packet-based networks present better resource utilization, as packets are short and reserve only a small number of channels during their transportation.

Besides its topology, an NOC can be described by the approaches used to implement the mechanisms for flow-control, routing, arbitration, switching, and buffering, as follows. The flow control deals with data traffic on the channels and inside the routers. Routing is the mechanism that defines the path a message takes from a sender to a receiver. The arbitration makes a scheduling when two or more messages request the same resource. Switching is the mechanism that takes an incoming message of a router and puts it in an output port of the router. Finally, buffering is the strategy used to store messages when a requested output channel is busy. Current embedded cores usually need to use wrappers to adapt their interfaces and protocols to the ones of the target NOC. Such wrappers pack and unpack data exchanged by the processing cores.

4.3.2 Testing of Embedded Cores

Testing of the embedded cores in NOC-based systems poses considerable challenges. In a traditional SOC, test data are transported through a dedicated TAM. This strategy, however, could lead to difficulty of routing in an NOC-based system, because the network (routers, channels, etc.) has already imposed significant routing overhead. Therefore, many current approaches for testing NOC-based systems rely on the reuse of the existing on-chip communication infrastructure as a TAM [Cota 2003, Cota 2004] [Liu 2004].

4.3.2.1 Reuse of On-Chip Network for Testing

To reuse the on-chip network as an access mechanism, the test vectors and test responses of each core are first organized into sets of packets that can be transmitted through the network. To keep the original wrapper design of each core unchanged to minimize the test application time and cost, the test packets are defined in such a way that each flit arriving from the network can be unpacked in one cycle. Each bit of a packet flit fills exactly one bit of a scan chain of the core. Functional inputs and outputs of the core, as well as the internal scan chains, are concatenated into wrapper (external) scan chains of similar length such that the channel width is enough to transport one bit for each wrapper scan chain. Control information, such as scan shift and capture signals, is also delivered in packets, either in the test header (to be interpreted by the wrapper) or as specific bits in the payload (for direct connection to the target pins). This concept will not considerably affect the core wrapper design, which can still follow the conventional SOC wrapper design methodology. A wrapper configuration during test and normal operation is depicted in Figure 4.12 [Cota 2003].

Test packets can be injected from an external tester and routed to the core under test, and corresponding test responses will be assembled into packets and routed back to the tester. This requires some test interfaces, which can be either dedicated I/O ports or reused embedded cores. Figure 4.13 [Liu 2004] illustrates the scenario using the NOC-based system shown in Figure 4.11. Note that there are two input ports and two output ports, each associated with a core by reusing the core's input/output or wrapper (if the core is wrapped) infrastructure. In a BIST environment, test data packets can be generated on-chip, and responses can also be analyzed via MISRs or comparators. In practice, each core can have several BIST and external test sets simultaneously.

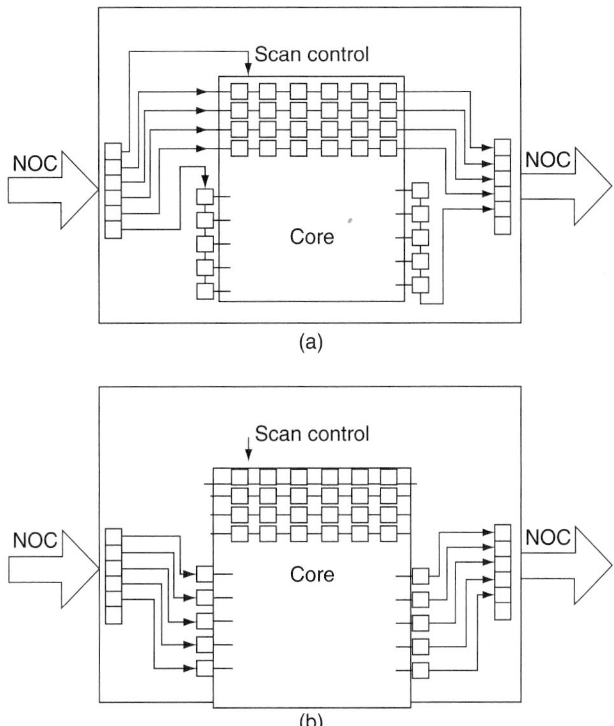

■ **FIGURE 4.12**

Wrapper configurations of cores in NOC-based system [Cota 2003]: (a) test mode and (b) function mode.

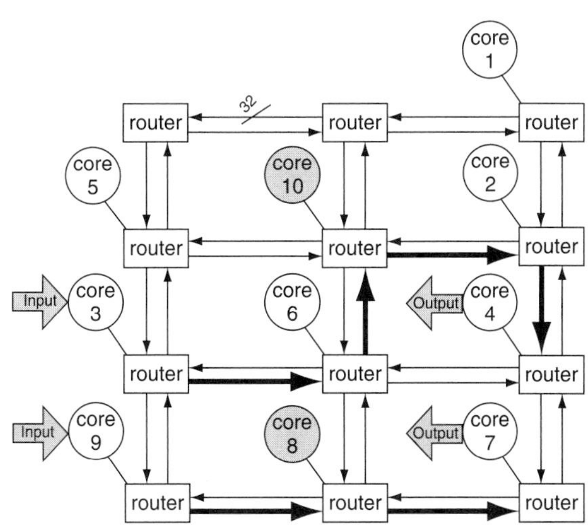

■ **FIGURE 4.13**

System d695 with test ports and routing paths [Liu 2004].

4.3.2.2 Test Scheduling

Test scheduling is one of the major problems in embedded core testing. A general form $\boldsymbol{S}$ of this problem can be formulated as follows: for an NOC-based system $\mathcal{N}$, a maximal number $\mathcal{P}$ of input/output ports, a set $\mathcal{T}$ of test sets (each core may have multiple test sets, either deterministic or random, or functional test for processor cores), a set $\boldsymbol{C}$ of constraints, and a set $\mathcal{R}$ of test resources (dedicated TAM, BIST engines, etc.), determine a selection of input/output ports, an assignment of test resources to cores, and a schedule of test sets such that the optimization objective(s) is (are) minimized and all the constraints are met. Note that the objective can be any cost factor such as test time and hardware overhead. A basic subset of this problem $\boldsymbol{S}$ is problem $\boldsymbol{S}_0$, where only core tests are optimized under some constraints. It was proved as $\mathcal{NP}$-complete in [Liu 2004]. $\boldsymbol{S}$ and other subset of $\boldsymbol{S}$ can also be similarly proved as $\mathcal{NP}$-complete.

In an NOC-based system, test scheduling can be done in a manner of either preemptive or nonpreemptive. As in a general SOC, an optimized schedule should maximize the test data parallelism. In an NOC-based system, this is done through the exploration of network parallelism so that all available communication resources (channels, routers, and interfaces) are used in parallel to transmit test data. In preemptive test scheduling, test data are transformed into test packets, which are transmitted through the network in such a way that one test packet contains one test vector or one test response. Because each test vector or test response can be scheduled individually, the network parallelism has privilege over the core test pipeline, and the test of a core can be interrupted. As a result, the pipeline of the core's scan-in and scan-out operations cannot be maintained.

Preemptive testing is not always desirable in practice, especially for BIST and sequential circuit testing [Iyengar 2002a]. In addition, it is always desirable that the test pipeline of a core is not interrupted—that is, the nth test vector will be shifted into the scan chains as the $(n-1)^{th}$ test response is shifted out, such that test time is minimized. However, in the case of preemption, the test pipeline has to be halted if either the test vector or test response packet cannot be scheduled because of the unavailability of test resources (*i.e.*, channels and input/output ports). This will not only increase the complexity of wrapper control but also cause potential increase on test time.

A nonpreemptive schedule maintains the test pipeline so that the wrapper can remain unchanged and the test time can be potentially reduced. In this approach, the scheduler will assign each core a routing path, including an input port, an output port, and the corresponding channels that transport test vectors from the input to the core and the test responses from the core to the output in the form of packets. Once the core is scheduled on this path, all resources (input, output, channels) on the path will be reserved for the test of this core until the entire test is completed. Test vectors will be routed to the core and test responses to the output in a pipelined fashion. Therefore, in this nonpreemptive schedule, the test of a core is identical to a normal test and the flow control becomes similar to circuit switching. Note that in Figure 4.13, cores 8 and 10 are scheduled on two I/O pairs in a nonpreemptive manner. It has been shown that usually nonpreemptive scheduling can yield shorter test time compared to preemptive scheduling [Liu 2004]. This is because the test pipeline can be maintained, scan-in and scan-out can be overlapped, and, hence, test time is reduced. It can also avoid the possibility of resource conflict. The complexity of a nonpreemptive scheduling algorithm is much lower than that of a preemptive scheduling algorithm because the minimum manageable unit in scheduling is a set of test packets in the former, instead of a single packet in the latter.

In practice, it is more realistic to have both preemptive and nonpreemptive test configurations in testing a system. It is also necessary to consider various constraints such as power, precedence, and multiple test sets such as external test and BIST. A preemptive schedule can be useful under these requirements. For instance, excessive power dissipation and inadequate heat convection/conduction can cause some cores to be significantly hotter than others, so called *hot spots*. Applying the entire test suite continuously can lead to dangerous temperature on these cores. In this case, the test suite can be split into several test sessions (or even single test vector in the extreme case) that can be scheduled individually in a preemptive manner. Sufficient time can be allowed between test sessions for hot spots to be cooled down.

4.3.2.3 Test Access Methods and Test Interface

Test access and test interface design in an NOC-based system need techniques different from those in conventional SOCs. A typical instance is the multiprocessor system discussed in [Aktouf 2002]. Each node in such an NOC-based system can contain a processor core and the corresponding router or switch, buffers, etc. Testing of a processor-based system usually mandates both deterministic test and functional test. Most functional test approaches such as software-based BIST [Chen 2003] can be applied. However, all test patterns and test responses must be organized into sets of packets, and all necessary network interfaces need to be added. For deterministic tests, nodes can be organized into groups, and each group can be tested using a conventional boundary scan approach. Figure 4.14 shows the test configuration in this scheme when nodes are organized into 1×1 and 2×2 groups, respectively. Nodes in the same group share the test infrastructure for boundary

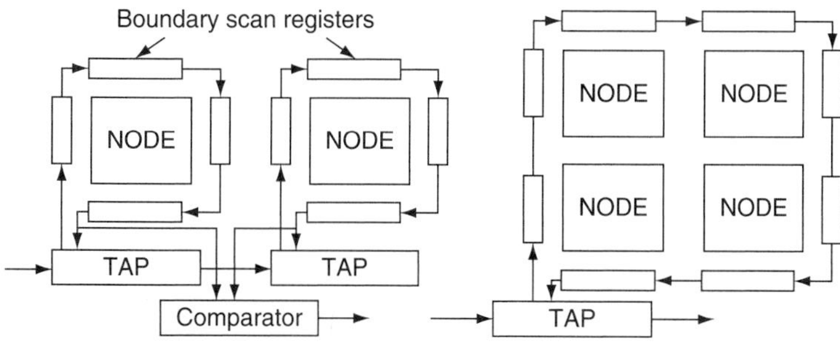

FIGURE 4.14

Testing identical nodes in NOC-based multiprocessor system using boundary scan in group of 1×1 (*left*) and 2×2 (*right*) [Aktouf 2002].

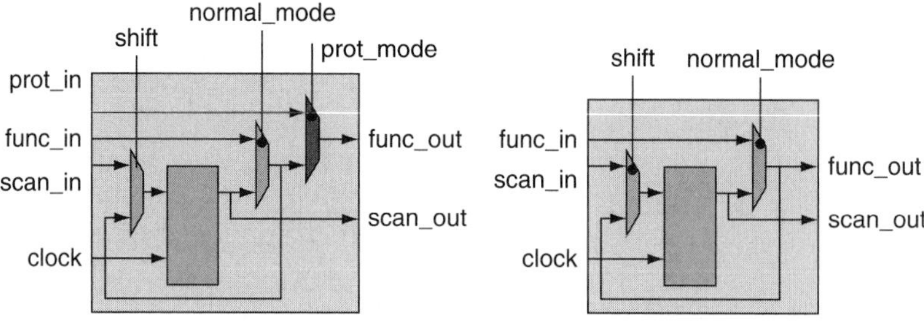

■ **FIGURE 4.15**

Standard 1500 wrapper cell (*right*) and modified wrapper cell for NOC-based system (*left*) [Amory 2006].

scan (***test access port*** [TAP] controller, I/O cells, etc.) and are tested in parallel. Because nodes are identical, each bit of test data is broadcasted to all nodes in the group. Test responses from all nodes in the group can be processed on-chip by feeding to a comparator, as shown in Figure 4.14.

To reuse the network to transport test data, a test interface has to be established to handle both functional protocol from network and test application to the core. A wrapper is therefore needed for each core as an interface. Because the core and the network may use difference protocols, the wrapper must be extended to incorporate the standard IEEE 1500 test modes. This includes modifications on test wires, wrapper cells, and test control logic. The TAM port in the 1500 wrapper should be replaced by a port connecting to the network. The control logic should include the process of network protocols. Further, wrapper cell should be correspondingly modified to implement the protocol. An instance of such a wrapper cell, as well as a standard 1500 wrapper cell, is shown in Figure 4.15 [Amory 2006]. Note that both cells need functional/scan input/output terminals and a few control terminals for the MUXs. When compared with the traditional 1500 wrapper cell, the modified cell has an additional MUX. Terminal prot_in receives the required values for protocol operation from the control logic. In test mode, terminal prot_mode is asserted to "1" to ensure that test signal does not interfere with the functional protocol. The actual protocol is implemented in the control logic. Other modes in the 1500 wrapper can be maintained.

4.3.2.4 Efficient Reuse of Network

Test efficiency is also critical in future massive NOC-based systems containing a large number of cores. One challenge in this reuse-based approach is that the channel width is determined by the system performance in design process and hence cannot be optimized for test purpose. This can be illustrated through an example shown in Figure 4.16 [Liu 2005]. Core test wrapper is usually designed through the use of balanced wrapper scan chains. Figure 4.16a shows a core with two internal scan chains of lengths 4 and 8, two primary inputs and two primary outputs. In Figure 4.16b, two balanced wrapper scan chains are designed for the

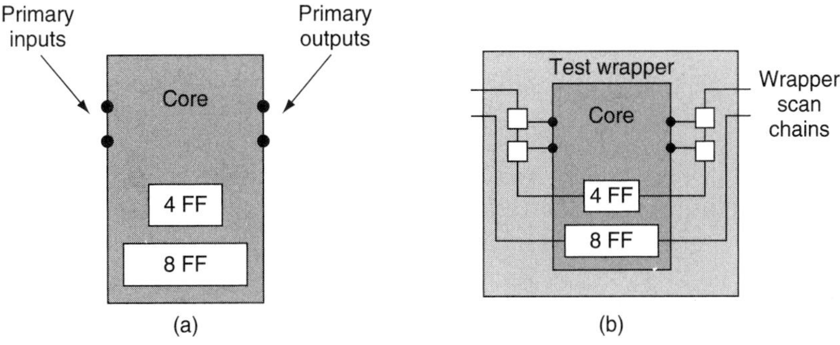

■ **FIGURE 4.16**

An example of (a) an unwrapped core and (b) a balanced wrapper scan chain design.

core. Note that eight test cycles are needed to scan in a test vector. If this core is used in an NOC with channel width of 2, each test vector can be loaded using eight payload flits (in eight clock cycles) with 2 bits for each flit. However, in Figure 4.16b, if we increase the number of wrapper scan chains from two to four, the longest wrapper scan chain remains eight, and the test time is unchanged. Therefore, two wrapper scan chains are sufficient for minimizing the test time.

However, in the context of network reuse in NOC test, the available TAM or channel width for wrapper scan chain design is already determined by the bandwidth requirements of cores in mission mode, not for test mode. If the channel width is predesigned to be 4, then half of the channel wires will be idle during core test. When legacy cores with test wrappers designed for minimal dedicated TAM widths are integrated into NOC, a significant part of the network channel width can therefore remain idle during test. Table 4.1 [Liu 2005] illustrates how NOC channels can remain idle during test of some cores in the d695 benchmark shown in Figure 4.11, by presenting statistics of test data of the 10 cores. Column 2 lists the

TABLE 4.1 ■ Test Data Statistics for Cores in d695 of Figure 4.11 [Liu 2005]

Cores	Number of Test Patterns	Channel Width = 16		Channel Width = 32	
		Flits/Packet	Test Cycles	Flits/Packet	Test Cycles
1	24	2	38	1	25
2	146	13	1029	7	588
3	**150**	**32**	**2507**	**32**	**2507**
4	**210**	**54**	**5829**	**54**	**5829**
5	220	109	12192	55	6206
6	468	50	11978	41	9869
7	190	43	4219	34	3359
8	**194**	**46**	**4605**	**46**	**4605**
9	24	128	1659	64	836
10	136	109	7586	55	3863

total number of test data packets (test vectors and the corresponding test responses) for each core. Note that test data are organized in packet(s) in both preemptive and nonpreemptive scheduling because of the nature of network. Columns 3 and 4 list the number of payload flits per packet and the total number of test cycles needed to test each core using a channel width of 16. Columns 5 and 6 list the corresponding numbers of flits and test cycles when the channel width equals 32. Test time is calculated assuming full-scan with balanced scan chain design and full test pipeline. Note that when the channel width increases from 16 to 32, test times of cores 3, 4 and 8 do not decrease. Hence, channel width of 16 is already adequate for optimized test time for these three cores. Using channel width of 32 will cause idle channel wires during test data transportation.

To fully utilize the channel width to reduce test time, the on-chip clocking scheme presented in [Gallagher 2001] can be used to provide multiple test clocks to different cores such that test data throughput can be maximized [Liu 2005]. This can be done through a combination of on-chip clocking and parallel-serial conversion. Let the channel width for the NOC be w and the number of predesigned wrapper scan chains for a legacy core in the system be w', where $w' < w$. Further, let $n = \lfloor w/w' \rfloor$. The channel width w can be used to transport n flits in parallel to the wrapper, and the wrapper can serially scan in each flit to the core. To synchronize the core test wrapper operation with test data transportation on the network channel, the frequency of the on-chip clock supplied to the wrapper must be n times the frequency of the slower tester clock supplied to the network. This fast clock is generated by an on-chip ***phase-locked loop*** (PLL) [Gallagher 2001]. Additionally, a multiplexer controlled by the on-chip clock is used to select between the flits on the network channel during one tester clock cycle.

A possible test architecture using on-chip clocking is illustrated in Figure 4.17 with a simple instance where $n = 2$. Test data for two flits at a time are presented to

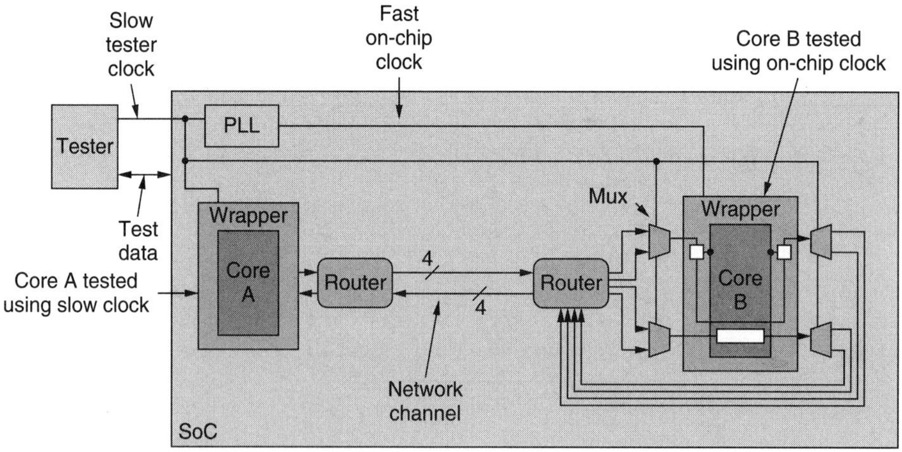

■ FIGURE 4.17

Test architecture using on-chip clocking [Liu 2005].

the core test wrapper on the network channels. These test data remain stable on the network channel for the period of the slow tester clock. Test data are loaded into the wrapper scan chains one flit at a time every on-chip clock cycle. No changes are required to the core test wrapper, thereby protecting the core vendor's IP as well as easing the core's integration into the system.

Another method of fully utilizing the channel width to transport test data can be accomplished by multiple data flit formats [Li 2006] through the use of dynamically reconfigurable data flit format to adapt the number of unfilled wrapper scan chains. As shown in Figure 4.18, when both shorter scan chains are filled, the data flit can be reformatted to send more test data for both longer scan chains by data flit reformatting. A wrapper scan chain configuration method was proposed in [Li 2006] to minimize the waste in NOC channel capacities. The basic idea is to organize the internal scan chains and functional I/Os of a core into *factorial wrapper scan chain groups* (FSCGs) with balanced-length chains within each group as shown in Figure 4.19. For example, if the channel (flit) width equals 32, then FSCG1 and FSCG2 should contain 2 wrapper scan chains in each group, FSCG3 (FSCG4) should contain 4 (8) wrapper scan chains. Therefore, the first data flit format involves 16 wrapper scan chains (from scan chain 1 to 16), and each wrapper scan chain gets two bits in each data flit. When wrapper scan chains in FSCG4 are filled, the second data flit format involving 8 wrapper scan chains (from scan chain 1 to 8) starts, and each wrapper scan chain gets 4 bits in each data flit. This avoids the waste of channel capacity significantly. Among 40 SOC benchmark cores [ITC 2002], 20 of them can be solved by using equal-length configuration methods for traditional SOC testing without any waste, but the remaining 20 have significant waste in channel capacities. For these 20 cores, the proposed wrapper scan chain configuration can reduce the channel capacity waste to zero for 14 cores, and have slight improvement for the other 6 cores. Only two or three data flit formats are required for all of these cases, so the hardware overhead is quite small. Note that

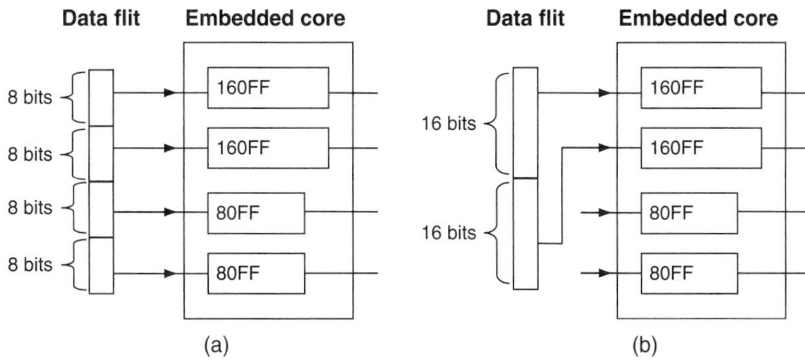

■ **FIGURE 4.18**

(a) Data flit format with each flit containing 8 bits/chain; (b) data flit format with each flit containing 16 bits/chain [Li 2006].

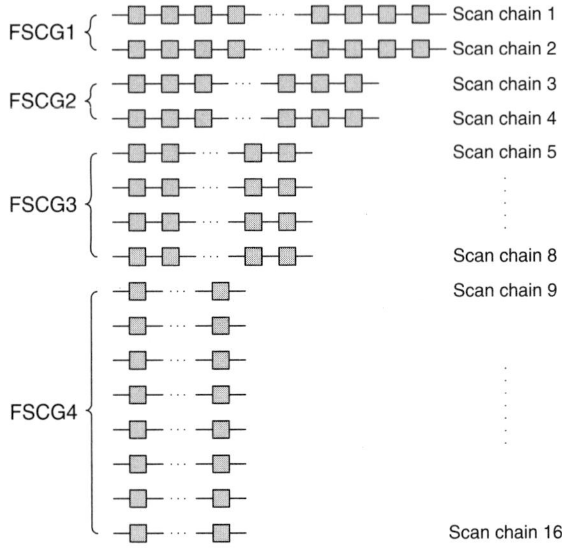

FIGURE 4.19

Factorial scan chain groups with channel width equal 32 [Li 2006].

the wrapper scan chain configuration can be applied only when the core wrapper can be redesigned.

4.3.2.5 Power-Aware and Thermal-Aware Testing

On-chip power and thermal management is critical for future NOC-based systems. High power consumption has been a critical problem in system testing; this is exacerbated when faster test clocks are used. Therefore, test scheduling needs to be *power-aware*. This is usually done by setting power constraints during test scheduling [Cota 2004] [Liu 2004]. However, stringent power constraints can cause failure of test scheduling for some high power-consumption cores. In this case, test clocks need to be slowed down by scaling down the tester clock. A frequency divider can generate a slower test clock. If the slower clock rate is a factor $1/n$ of the tester clock rate, then no change is needed for the core test wrapper. Similar to the virtual channel routing method in [Duato 1997], each NOC channel can be viewed as n virtual channels, and each core using a slower clock will occupy only one of them. Therefore, time-division scheduling of test packets is required. A conceptual timing diagram on a specific channel, where $n = 3$, is shown in Figure 4.20 [Liu 2005]. The figure shows that during one test clock cycle of core A, three test packets are routed through the channel to cores A, B, and C, respectively.

The on-chip variable clocking scheme can also be used for removing the hot spots on chip by assigning different test clocks to cores during test scheduling. This is specifically suitable for NOC-based systems because here cores are globally asynchronous, and they communicate by sending and receiving messages in the

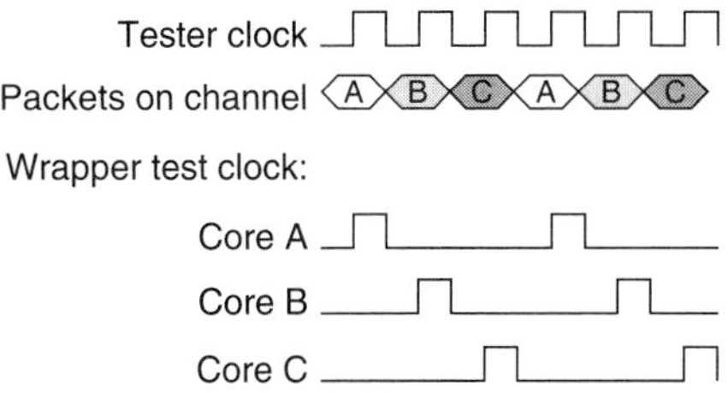

■ **FIGURE 4.20**

Slower on-chip clock using time-division scheduling [Liu 2005].

form of packets via network. Each core can receive several test clocks generated by an on-chip clock generator. During the test application process, a core can vary its power dissipation, and hence temperature, by choosing a different clock based on the test control information carried in test packets. Slower clocks are used to reduce temperature while faster clocks are used to reduce test time. This dynamic clock scaling scheme can not only guarantee thermal safety but also achieve thermal balance and optimize test time. Hot spots are efficiently removed and test time can be reduced. Test scheduling using this method can be found in [Liu 2006].

4.3.3 Testing of On-Chip Networks

An NOC usually consists of routers, network interfaces (between routers and embedded cores), and interconnects (channels). Although numerous works have been done (as discussed in Section 4.3.2) on core testing by transporting test data through the on-chip network fabric, works on testing on-chip networks have been limited [Amory 2005] [Grecu 2006] [Stewart 2006]. Unless the on-chip network of an NOC-based system has been thoroughly tested, it cannot be used to support the testing of embedded cores.

4.3.3.1 Testing of Interconnect Infrastructures

Interconnect testing has been discussed in many works [Singh 2002], and most of them can be directly applied to the NOC domain. A major difference in interconnect testing between SOC and NOC is that test patterns can be delivered using packets in the latter case. Hardware overhead can be greatly reduced if the test patterns and responses are transported using the existing on-chip network.

Based on the well-known *maximal aggressor fault* (MAF) model [Cuviello 1999], an efficient built-in self-test methodology has been proposed for testing interconnects of an on-chip network in [Grecu 2006]. For a set of interconnects with N wires, the MAF model assumes the worst-case situation with one victim and (N-1) aggressors. The MAF model consists of six crosstalk errors: rising/falling delay, positive/negative glitch, and rising/falling speed-up. Though it has been found that MAF may not always generate the worst case [Attarha 2001], the test coverage is generally high. Using this model, a total of 6N faults are to be tested, and 6N two-vector test patterns

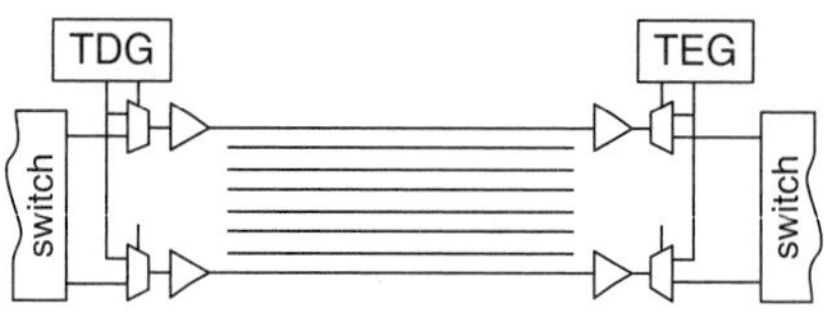

■ **FIGURE 4.21**

Test data generator (TDG) and test error detector (TED) [Grecu 2006].

are required. They also cover traditional faults such as stuck-at, stuck-open, and bridging faults [Grecu 2006]. Both unidirectional and bidirectional transactions are considered.

A self-test structure without considering the special property of NOCs is shown in Figure 4.21. Here, a pair of *test data generator* (TDG) and *test error detector* (TED) are inserted to generate all MAF test patterns and observe the test responses. As shown in Figure 4.21, test patterns are launched before the line drivers and sampled after the receiver buffers. This tests the crosstalk effects that are highly dependent on the driver size and load. The self-test circuit also allows one clock cycle delay for data to travel from transmitters to receivers. Multiple clock-cycle delays are also allowed if the interconnects are pipelined. Detailed design of TDG and TED can be found in [Grecu 2006]. This self-test structure can be inserted to interconnects between each pair of routers (called *point-to-point* MAF self-test), and all interconnects can be tested in parallel as long as the power consumption is not exceeded.

By taking advantage of the on-chip network, the MAF test patterns can be *broadcast* to all interconnects in the form of test packets with only one TDG. Note that one TED is still required for interconnects between each pair of routers as shown in Figure 4.22. Here, the test packets are broadcasted in a *unicast* manner

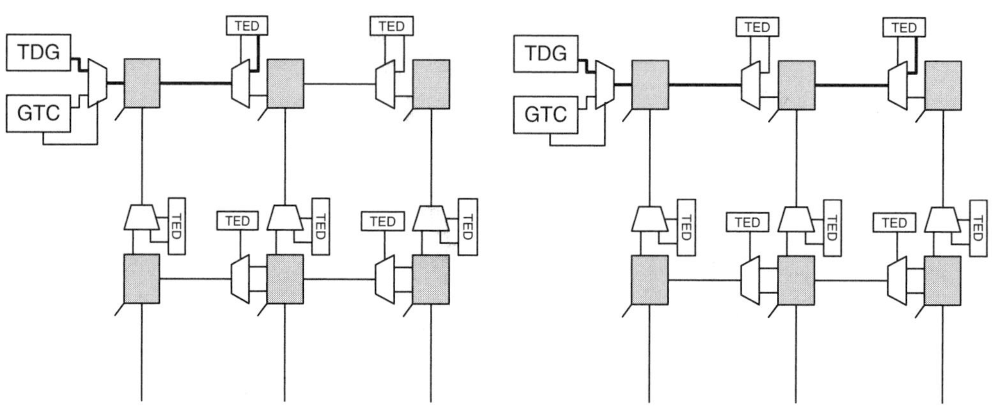

■ **FIGURE 4.22**

Interleaved unicast MAF test [Grecu 2006].

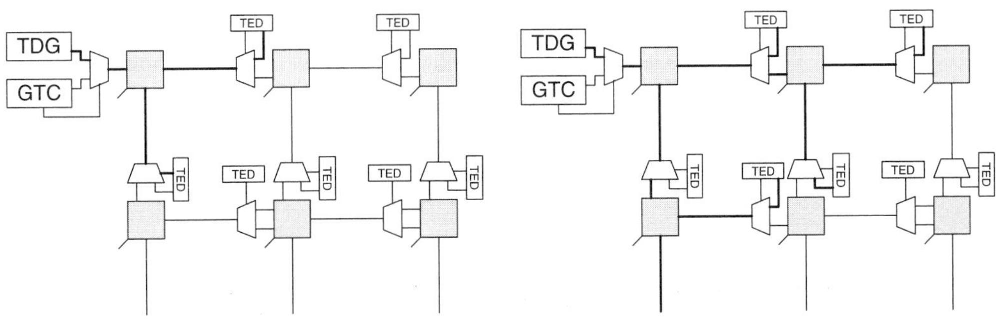

■ **FIGURE 4.23**

Interleaved multicast MAF test [Grecu 2006].

(*i.e.*, interconnects between a pair of routers are tested for each test packet broadcasting). A *global test controller* (GTC) is also designed to inject test patterns for testing routers. A more powerful test packet broadcasting is *multicast* MAF self-test as shown in Figure 4.23. The major difference between unicast and multicast is that in the latter, test packets are broadcast to interconnects of different pairs of routers to achieve the maximum test parallelism. A detailed design of test packets, TDG, TED, and multidestination broadcasting to support the interconnect BIST infrastructure can be found in [Grecu 2006]. This proposed interconnect test approach was validated using a 64-router NOC. The results demonstrate that the point-to-point (unicast) test method has the smallest (largest) test application time but the largest (smallest) hardware overhead. The multicast method gives a good compromise between test application time and hardware overhead [Grecu 2006].

4.3.3.2 Testing of Routers

Routers are used to implement the functions of flow control, routing, switching, and buffering of packets for an on-chip network. Figure 4.24 shows a typical organization of a router, whereas Figure 4.25 shows a typical structure of an NOC-based design [Amory 2005]. As shown in Figure 4.25, router testing can be considered the same way as testing a sequential circuit. However, a special property of an on-chip network is its regularity. Based on this property, the idea of test pattern broadcasting discussed in Section 4.3.3.1 can be applied to reduce the test application time. In [Amory 2005], an efficient test method has been developed based on partial scan and an IEEE 1500-compliant test wrapper by taking advantage of the NOC regularity.

In [Amory 2005], router testing has been dealt with in three parts: the testing of each router, the testing of all routers (without considering network interfaces and interconnects), and the testing of wrapper design. Testing a router consists of testing the control logic (routing, arbitration, and flow control modules) and input *first-in first-out buffers* (FIFOs) as shown in Figure 4.24. Control logic testing can be done using traditional sequential circuit testing methods such as scan. A smart

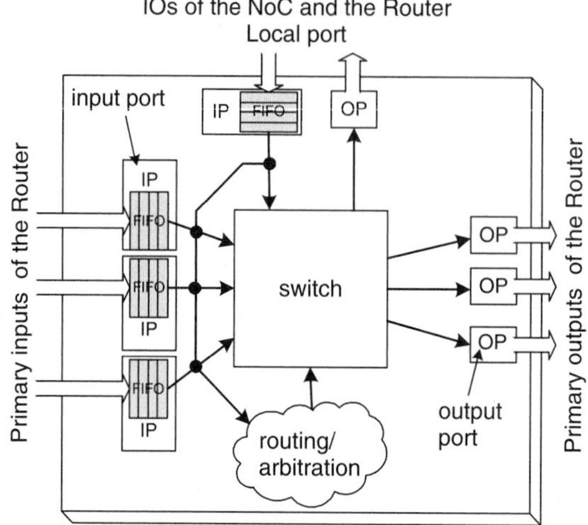

■ **FIGURE 4.24**

A typical organization of a router [Amory 2005].

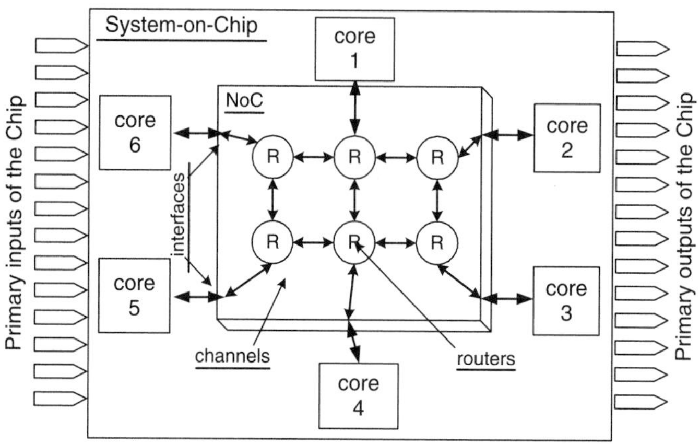

■ **FIGURE 4.25**

An NOC-based system [Amory 2005].

way to test each FIFO is to configure the first register of the FIFO as part of a scan chain, and other registers can be tested through this scan chain. Because FIFOs are generally not deep, this method proves to be very efficient [Amory 2005]. Because routers are identical, all can be tested in parallel by test pattern broadcasting as shown in Figure 4.26. Comparator implemented by XOR gates can be used to evaluate the output responses. The comparator logic also supports diagnosis.

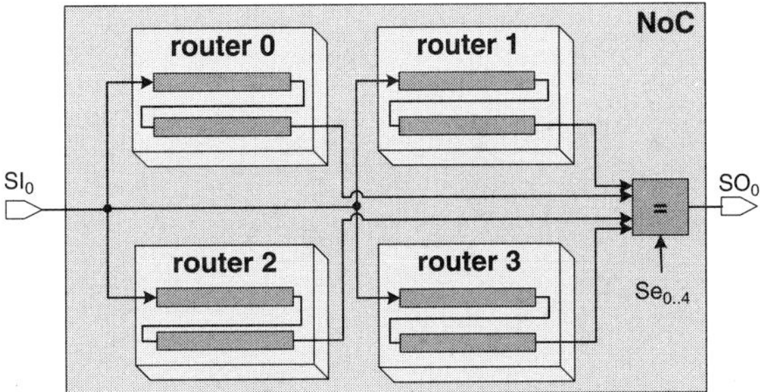

■ **FIGURE 4.26**

Testing multiple identical routers [Amory 2005].

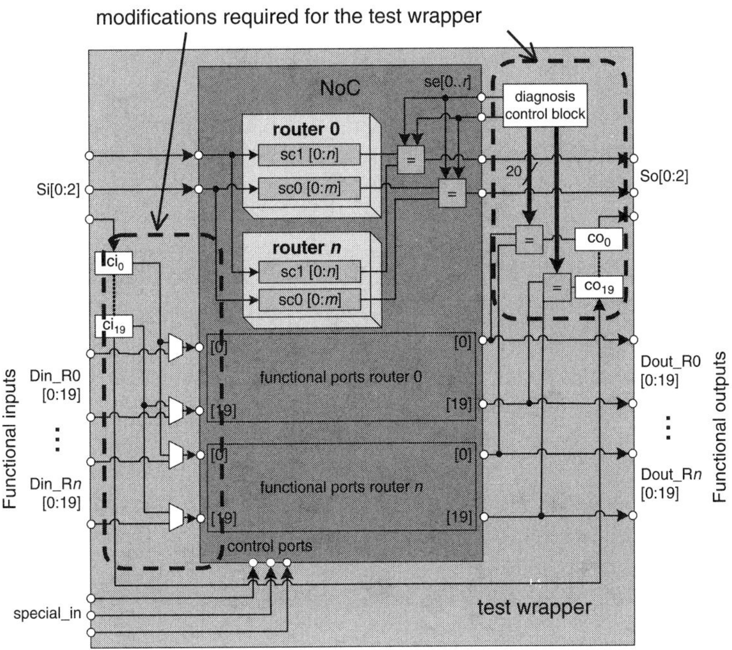

■ **FIGURE 4.27**

Test wrapper design [Amory 2005].

To support the proposed test strategy, an IEEE-1500 compliant test wrapper is designed to support test pattern broadcasting and test response comparison as shown in Figure 4.27. For example, all SC1 scan chains of these routers share the same set of test patterns. Similarly, all Din[0] (*i.e.*, Din-R0[0], …, Din-Rn[0]) data inputs of these routers share the same set of test patterns. As Figure 4.27 shows, this wrapper also supports test response comparison for scan chains and data outputs. Finally, the diagnosis control block can activate diagnosis. Simulation results demonstrate that the proposed router test method achieves the goals of small hardware overhead (about 8.5% relative to router hardware), small number of test patterns (several hundreds) because of test pattern broadcasting, and small amount of test application time (several tens of thousands test cycles) using multiple, balanced scan chains and test pattern broadcasting. Most important, the method is scalable—that is, the test volume and test time increase at a much lower rate than the increase in the NOC size. More details for Figure 4.27 can be found in [Amory 2005].

4.3.3.3 Testing of Network Interfaces and Integrated System Testing

A ***network interface*** (NI) is used to receive data bits from its corresponding IP core (router), packetize (de-packetize) the bits, and perform clock domain conversions between the router and the IP core. NIs might be the most difficult to test components in an on-chip network, because clock domain conversion introduces nondeterministic device behavior, which is detrimental to conventional stored response testing. New structural test solutions must be developed to deal with NI testing. In [Stewart 2006], based on the architectures of AEthereal [Goossens 2005] and Nostrum [Wiklund 2002], functional testing has been used to detect faults in NIs, routers, and the corresponding interconnects. The following discussion is mainly based on the work in [Stewart 2006] using AEthereal as the target NOC architecture.

The AEthereal NI in [Stewart 2006] is outlined in Figure 4.28. The NI faults in AEthereal are represented with the four-tuple NI(c1, c2, o1 and o2), where c1 is the identification of the NI under test, and c2 indicates whether the NI works as a source (S) or destination (D) during testing. O1 is an optional field, which represents the transmission mode, BE (best effort) or GT (time guarantee) of the NI, and o2 is another optional field, which represents the connection type U (unicast),

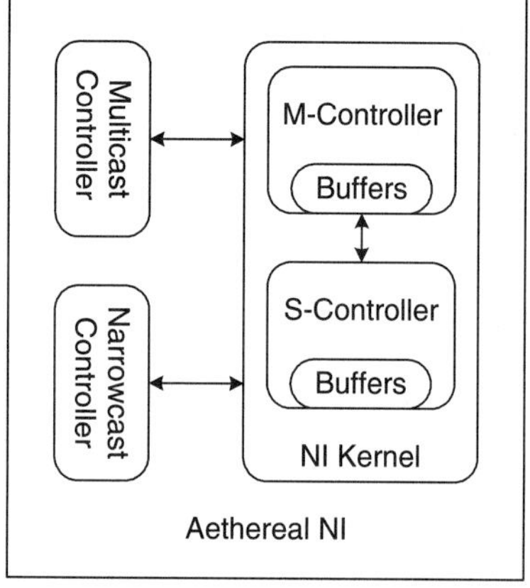

FIGURE 4.28

The AEthereal NI [Stewart 2006].

N (narrowcast) or M (multicast) of the NI. Details of transmission modes and transmission connections of AEthereal can be found in [Radulescu 2005]. Therefore, each NI must be tested based on all different combinations of these tuples. For example, each NI must be tested as a source (master) and as a destination (slave), respectively; in each case, the NI must be tested with both BE and GT transmission modes. Two additional tests are required to test narrowcast (N) and multicast (M) for the NI. Consequently, a total of six faults must be dealt with for thoroughly testing each NI. Note that unicast (U) is not required to be added to the last two tests, because it has been applied during the first four tests. By following the same process, 10 functional faults can be identified for each router. Test patterns must be generated to detect all 6 (10) faults for each NI (router).

It is important to develop an efficient method that can generate test patterns shared for NI faults and router faults. A test scheduling method is also proposed in [Stewart 2006] to mix the testing of router functional faults (ten faults for each router) and NI functional faults (six faults for each NI) such that the total test application time can be minimized. Initially, a preprocessing step is used to broadcast data packets (GT data and BE data) from I/O pins to the local memory of each core. During the test phase, an instruction packet is sent from the input port of the NOC to the source router by the GT transmission mode. The instruction packet contains information of the destination core, the transmission path, and the time at which the test pattern application should take place. After the test pattern is applied, the destination node generates a signature packet indicating whether a fault is detected [Stewart 2006].

NI is indeed a complex device that implements the functions of arbitration between channels, transaction ordering, end-to-end flow control, packetization and de-packetization, and a link protocol with the router. Moreover, it is necessary to design the NI to adapt to existing on-chip communication protocols. Functional testing for NI is not sufficient and efficient structural test methods must be investigated. Testing an on-chip network piece by piece is inadequate. Similarly, testing an NOC-based system by separating core testing from on-chip network testing is inadequate as well. Interactions between cores and the on-chip network must be tested using extensive functional testing. The interactions among the on-chip network components (routers, interconnects, and NIs) must be thoroughly tested by functional testing as well. Reliable NOC design can be found in [Tamhankar 2005] [Pande 2006].

4.4 DESIGN AND TEST PRACTICE: CASE STUDIES

Although multicore platforms are now being designed by a number of semiconductor companies (AMD, Intel, Sun), and incorporated in products such as servers and desktops, not much has been published in the open literature about testing and design-for-testability techniques adopted for these platforms. The literature on the testing of industrial SOCs and NOCs has largely been limited to publications from Philips. Hence, this chapter presents case studies based on Philips chips. For the SOC testing example, we consider the PNX8550 chip, which is used in the well-known Nexperia Home Platform [Goel 2004]. The example for NOC testing is a companion chip of PNX8550 [Steenhof 2006]. Philips developed both chips for its high-end (digital) TVs.

4.4.1 SOC Testing for PNX8550 System Chip

PNX8550 is a chip that is used in the well-known Nexperia digital video platform (Figure 4.29) developed by Philips [Goel 2004]. This chip is fabricated using a 0.13 μm process, six metal layers, with 1.2V supply voltage. It is packaged using PBGA564 and the die size is $100mm^2$. The entire chip contains 62 logic cores (five are hard cores while the rest are soft cores), 212 memory cores, and 94 clock domains. Five hard cores are: one MIPS CPU, two TriMedia CPUs, a custom analog block containing the PLLs and the ***delay-locked loops*** (DLLs), and a digital to analog converter. The 62 logic cores are partitioned into 13 chiplets (a chiplet is a group of cores placed together), because either they are synchronous to each other or they are not timing critical. Each chiplet is considered as an independent block to be connected to a specific set of TAM wires. As shown in Figure 4.29, PNX8550 contains one or more 32-bit MIPS CPUs (hard core) to control the process of the entire chip, and one or more 32-bit Tri-Media VLIW processors (hard core) for streaming data. Other modules include the MPEG decoder, UART, PIC 2.2 bus interface, etc. CPUs and many modules have access to external memory via a high-speed memory access network. Two device control and status (DCS) networks

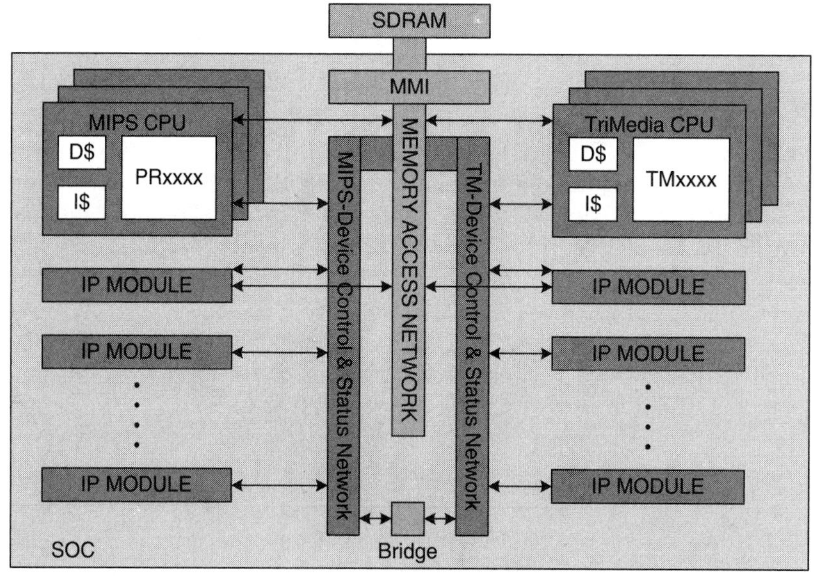

FIGURE 4.29

Nexperia Home Platform [Goel 2004].

enable each processor to control or observe the on-chip modules. A bridge is used to allow both DCS networks to communicate [Goel 2004].

PNX8550 inherits the requirement to have modular test strategy that allows tests to be reused through the use of wrappers (so-called TestShell [Marinissen 2000]) and TAMs (so-called TestRails [Marinissen 2000]). Full scan design is used for logic core testing to achieve 99% of stuck-at fault coverage for all logic cores. Small, embedded memories are also tested using scan chains, though large memories are testing using BIST. The design team of PNX8550 decided to give 140 TAM wires (*i.e.*, 280 chip pins) for core testing. The design issues are how to assign these TAM wires to different cores and how to design the wrapper for each core. Both issues must be considered carefully such that the data volume per test channel (28M) can be met and the overall test cost (mainly test application time) can be minimized. To solve these problems, Philips has developed a tool, called TR-ARCHITECT, to deal with these core-based testing requirements [Goel 2004]. TR-ARCHITECT supports three test architectures: daisy chain (see Figure 4.5b), distribution (see Figure 4.5c), and hybrid (of daisy chain and distribution) as discussed in Section 4.2; and requires two different kinds of inputs: the SOC data file and a list of user options. The SOC data file describes the SOC parameters such as the number of cores in the SOC and the number of test patterns and scan chains in each core. The user options give the test choices such as the number of SOC test pins, type of modules (hard or soft), TAM type (test bus/TestRail), architecture type (daisy chain, distribution, or hybrid), test schedule type (serial or parallel for daisy chain), and external bypass per module (yes/no).

As discussed previously, 140 TAM wires are spent, and 13 chiplets are identified. The distribution of these 140 TAM wires to 13 chiplets is done *manually* based on test data of a predecessor version and experienced engineering judgments. The assignment of TAM wires for a chiplet ranges from 2 (for chiplet UCLOCK) to 21 (for chiplet UQVCP5L). Once this step is done, the next job is to design the test architecture inside each chiplet. Basically, the distribution architecture is chosen for all except two chiplets: UMDCS and UTDCS. For these two chiplets (hybrid test architecture), some TAM wires are shared by two or more cores (using daisy chain), as there are more cores than wires; some cores are connected by the distribution architecture. Test architecture design is trivial, if the chiplet under consideration has only one core (*e.g.*, chiplets UMCU, UQVCP5L, UCLOCK, MIPS, TM1, and TM2). Given the number of TAM wires and the core parameters, the wrapper design method presented in [Marinissen 2000] can be applied to design the core wrapper.

For a chiplet containing multiple cores and using the distribution test architecture, TR-ARCHITECT is used to determine the number of TAM wires assigned to each core. TR-ARCHITECT applies the idea in [Goel 2002] to determine the number of TAM wires assigned to each individual core and to design the wrapper for the core. For the chiplets UMDCS (22 soft cores) and UTDCS (17 soft cores) that have a hybrid test architecture, TR-ARCHITECT can be applied to determine the number of TAM-wire groups, the width assigned to each group, and the assignment of cores to each group [Goel 2002].

TR-ARCHITECT contains four major procedures: Create-Start-Solution, Optimize-BottomUp, Optimize-TopDown, and Reshuffle [Goel 2002]. As it is named, initially, Crate-Start-Solution is used to assign at least one TAM wire for each core. If there are cores left unassigned, they will be added to the least occupied TAMs. However, if there are TAM wires left unassigned, they will be added to the most-occupied TAMs. After the initial assignment, Optimize-BottomUp is used to merge the TAM (may be with several wires) with the shortest test time with another TAM, such that the wire freed up in this process can be used for an overall test time reduction. For example, TAM-1 contains three wires with 500 test cycles for Core-1, TAM-2 contains four wires with 200 test cycles for Core-2, and TAM-3 contains two wires with 100 test cycles for Core-3. Core-1 is then the bottleneck core and the whole system needs to be tested by 500 test cycles. Now, if Core-3 is merged to TAM-2, the overall test time is not increased (200 + 100 = 300 test cycles). Thus, the two wires freed up by TAM-3 can be given to TAM-1. Assume that adding the extra two lines to TAM-1 can greatly reduce the test time of Core-1 from 500 to 350 test cycles (this may not always be the case). Finally, the overall test time can be reduced from 500 to 350 test cycles. The processes of Optimize-TopDown and Reshuffle follow the same idea and can be found in [Goel 2002]. It should be noted that each of the procedures in TR-ARCHITECT requires the information of wrapper design and test time for each assignment of TAM wires, which can be provided by the work in [Marinissen 2000].

As described earlier, PNX8550 spends 140 TAM wires and the distribution of the wires to each chiplet is done manually. This is because TR-ARCHITECT became available halfway of the PNX8550 design process. In this case, the total test time is dominated by chiplet UTDCS with 3,506,193 test cycles. If these 140 TAM wires are distributed to the 13 chiplets by TR-ARCHITECT (instead of manually) and a hybrid test architecture is used, then the overall test time can be reduced by 29% [Goel 2004]. In this case, UTDCS is assigned three more TAM lines (to reduce test time) by TR-ARCHITECT, and the chiplet dominating the test time is changed to UMCU with 2,494,687 test cycles. This demonstrates the effectiveness and efficiency of TR-ARCHITECT. If the designer can further modify and optimize the number and lengths of the core internal scan chains of all cores (except Trimedia and MIPS), the test time can be further decreased by 50%. In this case, the dominating chiplet is TM2 with 1,766,095 test cycles. The test data volume can fit onto the Agilent 93000-P600 test system with 28M deep vector memories. The computing complexity of TR-ARCHITECT is light, and the computing time is negligible. More SOC test strategies can be found in [Vermeulen 2001] for PNX8525, which is a predecessor chip of PNX8550. The work in [Wang 2005] presents a BIST scheme dealing with at-speed testing for multiclock design, based on an SOC chip developed by Samsung.

4.4.2 NOC Testing for a High-End TV System

In Philips, NOC has been proved as a mature technology using an existing SOC architecture for picture improvement in high-end TVs [Steenhof 2006]. Traditional interconnect design of several chips has been replaced by the AEthereal NOC

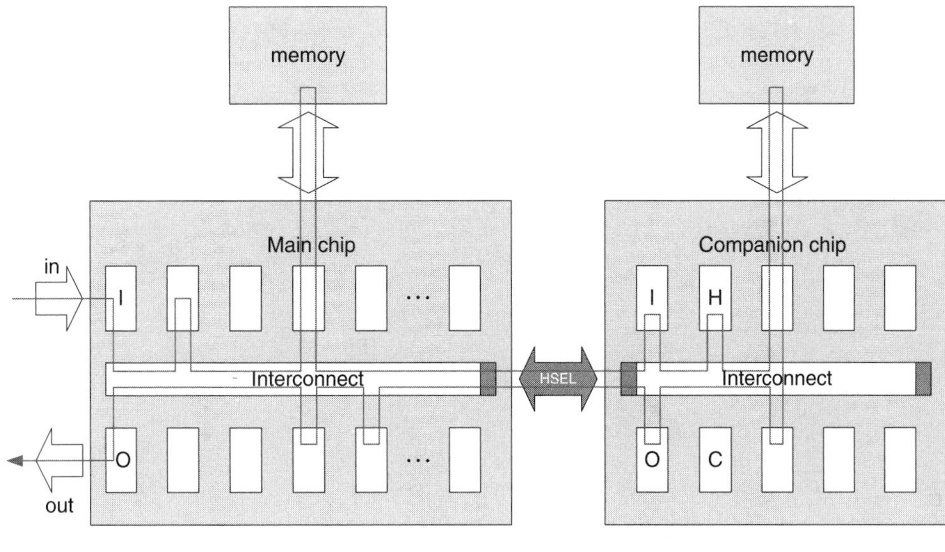

■ **FIGURE 4.30**

The main TV chip and companion chip [Steenhof 2006].

architecture [Goossens 2005]. Figure 4.30 outlines the main TV chip (PNX8550 discussed in Section 4.4.1), a companion chip, and external memories. The *main chip*, the master of the entire system, contains 61 IP blocks with the responsibilities of interacting with users, TV source, TV display, peripherals, and configuration of the companion chip. The *companion chip* contains 9 IP blocks for enhancing the video quality. A companion chip usually implements more advanced technologies that will not be released to competitors. Both main and companion chips, implemented by dedicated interconnect structures, are connected using a *high-speed external link* (HSEL). The dashed lines in Figure 4.30 represent a task that involves 11 IP blocks in the main chip, companion chip, and two memories. The functionality of the whole system usually contains hundreds of tasks. I (O) stands for input (output), H for horizontal scaler, and C for control processor. The method of partitioning a complex system into multiple chips including main and companion chips has many advantages. For example, it can reduce the development risk (because of implementing smaller and less complex chips), manage the different innovation rates in different market segments, and encapsulate differentiating functionality [Steenhof 2006]. To further enhance the flexibility, the dedicated interconnect of the companion chip has been replaced by an NOC structure [Steenhof 2006].

Figure 4.31 shows the detailed diagram of the on-chip network of the companion chip in Figure 4.30. The on-chip network contains routers (R), interconnects, and network interfaces (NIs). Each NI contains one kernel (K), one shell (S) and several ports. The functions of each on-chip network component can be found in [Radulescu 2005]. The numbers of master (M) and slave (S) ports are indicated

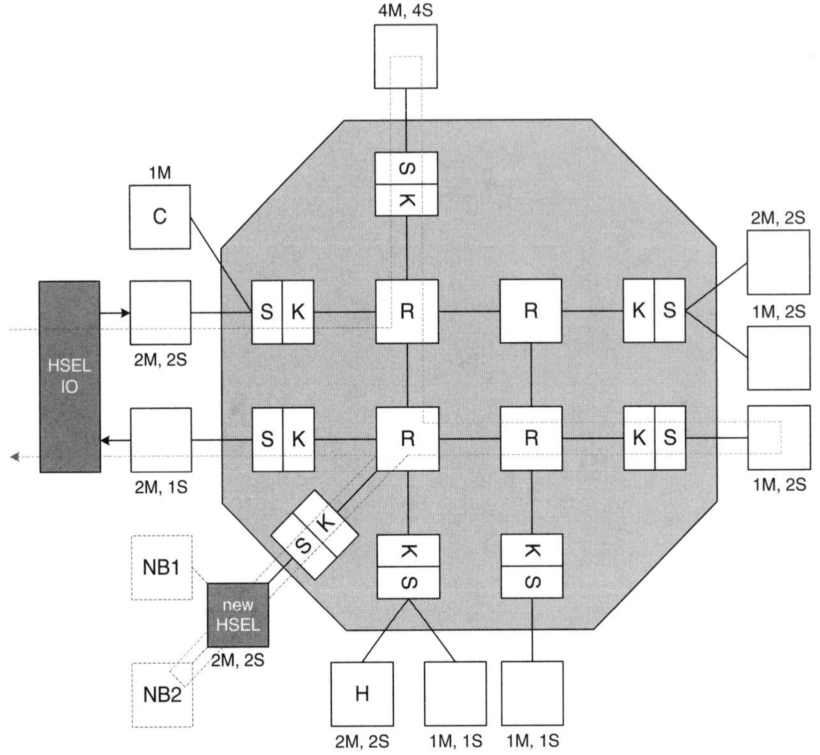

■ **FIGURE 4.31**

On-chip network architecture of the companion chip [Steenhof 2006].

in each NI. For example, the box with 2M, 2S is a NI port with two masters and two slaves. The ports are connected to IPs of microprocessor, DSP, or memory arrays. The HSEL IO is the high-speed external link used to connect the main chip and the companion chip, whereas the new HSEL is used to attach another companion chip (*e.g.*, an FPGA) to this chip. Thus, the on-chip network in the companion chip is basically a 2×2 mesh. The AEthereal NOC is configured at run time with the required task diagram. The NOC structure offers great flexibility and reuse potential of the companion chip with the price of increased area (4%), larger power consumption (12%), and larger latency (10%), which is viewed as tolerable [Steenhof 2006].

Although no test strategy has been proposed in [Steenhof 2006], the test method for Philips's AEthreal NOC architecture has been presented in [Vermeulen 2003]. It has been suggested that the on-chip network shown in Figure 4.31 can be treated as a core, and the knowledge about the NOC can be utilized to modify the standard core-based test approach to obtain a better suited test. For example, to test the NOC-based system in Figure 4.31, all identical blocks (*e.g.*, all routers) can reuse the test data by test broadcasting as described in Section 4.3.3.2. The responses can be compared to each other and any mismatch will be sent off-chip. It is emphasized that *timing test* is very important for the following two reasons: (1) many long wires exist in the NOC-based design, and they can cause crosstalk errors; and (2) all clock boundaries between cores are inside the NI and timing errors can occur easily. Thus, the method on interconnect testing discussed in Section 4.3.3.1 can be applied to (1). However, multiple clock domain testing for the NIs is still under investigation. Once the NOC-based system in Figure 4.31 has been structurally tested, the network can be used to transfer test data for all cores as discussed in Section 4.3.2 in a flexible way. No new TAM wires need to be added to the design, and the NOC is fully reused for testing. The NOC structure also enables parallel testing of multiple cores, if the channel capacity can support their test data transportation under a specific power budget [Vermeulen 2003].

4.5 CONCLUDING REMARKS

Rapid advances in test development techniques are needed to reduce the test cost of million-gate SOC devices. This chapter has described a number of state-of-the-art techniques for reducing test time, thereby decreasing test cost. Modular test techniques for digital, mixed-signal, and hierarchical SOCs must develop further to keep pace with design complexity and integration density. The test data bandwidth needs for analog cores are significantly different from that for digital cores; therefore, unified top-level testing of mixed-signal SOCs remains a major challenge. Most SOCs include embedded cores that operate in multiple clock domains. Because the 1500 standard does not address wrapper design for at-speed testing of such cores, research is needed to develop wrapper design techniques for multifrequency cores. There is also a pressing need for test planning methods that can efficiently schedule tests for these multifrequency cores. The work reported in [Xu 2004] is a promising first step in this direction. Meanwhile, as a result of the revolutionary RF interconnect technology, the integrated wireless test framework for heterogeneous nanometer SoC testing is emerging to address future global routing needs and sustain performance improvement [Zhao 2006].

NOC has become a promising design paradigm for future SOC design, and cost-efficient test methodologies are imminently needed. This chapter has included advances in testing NOC-based systems, including core testing and network testing. The focus has been on how to efficiently utilize the on-chip network as a test access mechanism without compromising fault coverage or test time. Case studies have been provided to demonstrate some initial efforts in testing real-world NOC designs. Research on NOC test is still premature when compared to industrial needs because of the limited support for various network topologies, routing strategies, and other factors. Future research and development are needed to provide an integrated platform and a set of methodologies that are suitable for various networks such that the design and test cost of the overall NOC-based system (both cores and network) can be reduced. Wrapper design techniques for SOC test can also be adopted by NOC-based systems.

4.6 EXERCISES

4.1 **(SOC Test Infrastructure Design)** Consider an embedded core, referred to as *C* in a core-based SOC. *C* has 8 functional inputs a[0:7], 11 functional outputs z[0:10], 9 internal scan chains of lengths 12, 12, 8, 8, 8, 6, 6, 6, and 6 flip-flops, respectively, and a scan enable control signal *SE*. The test wrapper for *C* is to be connected to a 4-bit TAM. Design the wrapper for this core, and present your results in the form of the following table for each wrapper scan chain *n* (1 to 4):

	Wrapper scan chain *n*
Internal scan chains	Which scan chains are included? Provide number of scan elements.
Wrapper input cells	How many wrapper input cells are included?
Wrapper output cells	How many wrapper output cells are included?
Scan-in length	Number of bits?
Scan-out length	Number of bits?

4.2 **(SOC Test Infrastructure Design)** Next consider the same embedded core *C* as in Exercise 4.1. The test wrapper for *C* is to be connected to two TAMs: a 4-bit TAM and a 6-bit TAM. Design a reconfigurable wrapper for this scenario, and compute the savings in test time (as a percentage) if a 6-bit-wide TAM is used.

4.3 **(NOC Infrastructure)** Refer to the ITC-2002 SOC benchmarks [ITC 2002], and verify the data illustrated in Table 4.1 using the methodology introduced in Section 4.2.

4.4 **(NOC Test Scheduling)** Using the system d695 shown in Figure 4.13 and data in Table 4.1, develop a nonpreemptive schedule for cores in system d695. Based on this result, assume all routers are identical and the test time of a router equals the average test time of embedded cores. Calculate the test time for testing all routers.

4.5 **(Integrated NOC Test Scheduling for Test Time Reduction)** In the previous problem, what method can be used to reduce the test time of all routers in the NOC? What method can be used to reduce the overall test time of cores and routers?

4.6 **(NOC Test Using On-Chip Clocking)** Now assume on-chip clocking scheme is used and the operation clock on NOC for testing is CLK. Also assume each core can be tested using CLK/2, CLK, or CLK*2 under certain constraints. Using the data shown in Table 4.1, design the wrapper architecture for each core in d695, and then upgrade your scheduling method for embedded cores (without consideration of routers) by incorporating these multiple clocks. Note that cores tested using slow clocks may share a physical channel in a time-multiplexing manner. Observe the test time, and compare it with the result you obtained in Problem 4.4.

4.7 **(NoC Interconnect Test Application)** Figure 4.22 shows an example for interleaved unicast MAF test of an NOC. Assume interconnects between each pair of routers are unidirectional (*i.e.*, the wires for data transmission from router R1 to router R2 are different from those from router R2 to router R1). Find test configurations (paths) to apply the MAF test to all wires without any wire traveled redundantly. Show your answer by using a mesh with size MxN where $M = 3$, $N = 4$; $M = 4$, $N = 4$, and $M = 5$, $N = 4$. Note that M is the number of routers in the x direction, and N is the number of routers in the y direction.

To further simplify the analysis, you can replace each (bidirectional) interconnect in Figure 4.22 with two unidirectional interconnects, and assume there are only two unidirectional interconnects between each pair of routers.

Acknowledgments

The authors wish to thank Erik Jan Marinissen of NXP Semiconductors; Dr. Anuja Sehgal of AMD; Dr. Vikram Iyengar of IBM, Dr. Ming Li of Siemens, Ltd., China; Ke Li of University of Cincinnati; C. Grecu of University of British Columbia; Professor Erika Cota and Alexandre de Morais Amory of UFRGS, Brazil; and Professor Dan Zhao of University of Louisiana for their help and contribution during the preparation of this chapter. Thanks are also due to Professor Partha Pande of Washington State University; Dr. Ming Li of Siemens, Ltd., China; Professor Erika Cota of UFRGS, Brazil; Professor Tomokazu Yoneda of Nara Institute of Science and Technology, Japan; and Professor Erik Larsson of Linköping University, Sweden, for reviewing the text and providing valuable comments.

References

R4.0 Books

[Abramovici 1994] M. Abramovici, M. A. Breuer, and A. D. Friedman, *Digital Systems Testing and Testable Design*, IEEE Press, Revised Printing, Piscataway, NJ, 1994.

[Beenker 1995] F. Beenker, B. Bennetts, and L. Thijssen, *Testability Concepts for Digital Ics: The Macro Test Approach,* Frontiers in Electronic Testing, Vol. 3, Kluwer Academic, Boston, 1995.

[Duato 1997] J. Duato, S. Yalamanchili, and L. M. Ni, *Interconnection Networks: An Engineering Approach,* IEEE Computer Society Press, Los Alamitos, CA, 1997.

[Singh 2002] R. Singh, editor, *Signal Integrity Effects in Custom IC and ASIC Designs*, John Wiley & Sons, Hoboken, NJ, 2002.

R4.1 Introduction

[Benini 2002] L. Benini and G. De Micheli, Networks on chips: A new SOC paradigm, *IEEE Computer*, 35(1), pp. 70–78, January 2002.

[Chakrabarty 2000a] K. Chakrabarty, Test scheduling for core-based systems using mixed-integer linear programming, *IEEE Trans. on Computer-Aided Design*, 19(10), pp. 1163–1174, October 2000.

[Cota 2003] E. Cota, M. Kreutz, C. A. Zeferino, L. Carro, M. Lubaszewski, A. Susin, The impact of NOC reuse on the testing of core-based systems, in *Proc. VLSI Test Symp.*, pp. 128–133, April 2003.

[Cota 2004] E. Cota, L. Carro, and M. Lubaszewski, Reusing an on-chip network for the test of core-based systems, *ACM Trans. on Design Automation of Electronic Systems*, 9(4), pp. 471–499, October 2004.

[Daly 2001] W. J. Dally and B. Towles, Route packets, not wires: On-chip interconnection networks, in *Proc. ACM/IEEE Design Automation Conf.*, pp. 684–689, June 2001.

[Sugihara 1998] M. Sugihara, H. Date, and H. Yasuura, A novel test methodology for core-based system LSIs and a testing time minimization problem, in *Proc. Int. Test Conf.*, pp. 465–472, October 1998.

[Vermeulen 2003] B. Vermeulen, J. Dielissen, K. Goznes, and C. Ciordas, Bringing communication networks on a chip: Test and verification implications, *IEEE Communication Magazine*, pp. 74–81, September 2003.

[VSIA 2007] Virtual Socket Interface Alliance (www.vsi.org), 2007.

[Zeferino 2003] C. Zeferino and A. Susin, SoCIN: A parametric and scalable network-on-chip, in *Proc. Symp. on Integrated Circuits and Systems Design*, pp. 121–126, May 2003.

[Zorian 1999] Y. Zorian, E. J. Marinissen, and S. Dey, Testing embedded-core-based system chips, *IEEE Computer*, 32(6), pp. 52–60, June 1999.

R4.2 System-on-Chip (SOC) Testing

[Aerts 1998] J. Aerts and E. J. Marinissen, Scan chain design for test time reduction in core-based ICs, in *Proc. Int. Test Conf.*, pp. 448–457, October 1998.

[Agilent] Agilent Technologies, Winning in the SOC market, available online at http://cp.literature.agilent.com/litweb/pdf/5988-7344EN.pdf.

[Bahukudumbi 2006] S. Bahukudumbi and K. Chakrabarty, Defect-oriented and time-constrained wafer-level test length selection for core-based SOCs, in *Proc. Int. Test Conf.*, Paper 19.1, October 2006.

[Bahukudumbi 2007] S. Bahukudumbi and K. Chakrabarty, Test-length selection, reduced pin-count testing, and TAM optimization for wafer-level testing of core-based digital SoCs, in *Proc. IEEE Int. Conf. on VLSI Design*, pp. 459–464, January 2007.

[Barnhart 2001] C. Barnhart, V. Brunkhorst, F. Distler, O. Farnsworth, B. Keller, and B. Koenemann, OPMISR: The foundation for compressed ATPG vectors, in *Proc. Int. Test Conf.*, pp. 748–757, October 2001.

[Chakrabarty 2000a] K. Chakrabarty, Test scheduling for core-based systems using mixed-integer linear programming, *IEEE Trans. on Computer-Aided Design*, 19(10), pp. 1163–1174, October 2000.

[Chakrabarty 2000b] T. J. Chakrabarty, S. Bhawmik, and C.-H. Chiang, Test access methodology for system-on-chip testing, in *Proc. Int. Workshop on Testing Embedded Core-Based System-Chips*, pp. 1.1-1–1.1-7, May 2000.

[Chakrabarty 2001] K. Chakrabarty, Optimal test access architectures for system-on-a-chip, *ACM Trans. on Design Automation of Electronic Systems*, 6(1), pp. 26–49, January 2001.

[Chakrabarty 2003] K. Chakrabarty, A synthesis-for-transparency approach for hierarchical and system-on-a-chip test, *IEEE Trans. on VLSI Systems*, 11(4), pp. 167–179, April 2003.

[Chakrabarty 2005] K. Chakrabarty, V. Iyengar, and M. Krasniewski, Test planning for modular testing of hierarchical SOCs, *IEEE Trans. on Computer-Aided Design*, 24(3), pp. 435–448, March 2005.

[Chou 1997] R. M. Chou, K. K. Saluja, and V. D. Agrawal, Scheduling tests for VLSI systems under power constraints, *IEEE Trans. on VLSI Systems*, 5(2), pp. 175–184, June 1997.

[Cron 1997] A. Cron, IEEE P1149.4: Almost a standard, in *Proc. Int. Test Conf.*, pp. 174–182, November 1997.

[Ebadi 2001] Z. S. Ebadi and A. Ivanov, Design of an optimal test access architecture using a genetic algorithm, in *Proc. Asian Test Symp.*, pp. 205–210, November 2001.

[Gallagher 2001] P. Gallagher, V. Chickermane, S. Gregor, and T. S. Pierre, A building block BIST methodology for SOC designs: a case study, in *Proc. Int. Test Conf.*, pp. 111–120, October 2001.

[Ghosh 1998] I. Ghosh, S. Dey, and N. K. Jha, A fast and low cost testing technique for core-based system-on-chip, in *Proc. ACM/IEEE Design Automation Conf.*, pp. 542–547, June 1998.

[Goel 2002] S. K. Goel and E. J. Marinissen, Effective and efficient test architecture design for SOCs, in *Proc. Int. Test Conf.*, pp. 529–538, October 2002.

[Gonciari 2003] P. T. Gonciari, B. Al-Hashimi, and N. Nicolici, Addressing useless test data in core-based system-on-a-chip test, *IEEE Trans. on Computer-Aided Design*, 22(11), pp. 1568–1590, November 2003.

[Harrod 1999] P. Harrod, Testing re-usable IP: A case study, in *Proc. Int. Test Conf.*, pp. 493–498, September 1999.

[Huang 2001] Y. Huang, W.-T. Cheng, C.-C. Tsai, N. Mukherjee, O. Samman, Y. Zaidan, and S. M. Reddy, Resource allocation and test scheduling for concurrent test of core-based SOC design, in *Proc. Asian Test Symp.*, pp. 265–270, November 2001.

[Huang 2002] Y. Huang, W.-T. Cheng, C.-C. Tsai, N. Mukherjee, O. Samman, Y. Zaidan, and S. M. Reddy, On concurrent test of core-based SOC design, *J. Electronic Testing: Theory and Applications*, 18(4–5), pp. 401–414, August/October 2002.

[IEEE 1500-2005] IEEE Std. 1500-2005, *IEEE Standard for Embedded Core Test*, IEEE Press, New York, 2005.

[Immaneni 1990] V. Immaneni and S. Raman, Direct access test scheme: Design of block and core cells for embedded ASICs, in *Proc. Int. Test Conf.*, pp. 488–492, September 1990.

[Ingelsson 2005] U. Ingelsson, S. K. Goel, E. Larsson, and E. J. Marinissen, Test scheduling for modular SOCs in an abort-on-fail environment, in *Proc. European Test Symp.*, pp. 8–13, May 2004.

[Iyengar 2002a] V. Iyengar and K. Chakrabarty, System-on-a-chip test scheduling with precedence relationships, preemption, and power constraints, *IEEE Trans. on Computer-Aided Design*, 21(9), pp. 1088–1094, September 2002.

[Iyengar 2002b] V. Iyengar and K. Chakrabarty, Test bus sizing for system-on-a-chip, *IEEE Trans. on Computers*, 51(5), pp. 449–459, May 2002.

[Iyengar 2002c] V. Iyengar, K. Chakrabarty, and E. J. Marinissen, Recent advances in TAM optimization, test scheduling, and test resource management for modular testing of core-based SOCs, in *Proc. IEEE Asian Test Symp.*, pp. 320–325, November 2002.

[Iyengar 2002d] V. Iyengar, K. Chakrabarty, and E. J. Marinissen, Test wrapper and test access mechanism co-optimization for system-on-chip, *J. Electronic Testing: Theory and Applications*, 18(2), pp. 213–230, April 2002.

[Iyengar 2003a] V. Iyengar, K. Chakrabarty, M. D. Krasniewski, and G. N. Kumar, Design and optimization of multi-level tam architectures for hierarchical SOCs, in *Proc. VLSI Test Symp.*, pp. 299–304, April 2003.

[Iyengar 2003b] V. Iyengar, K. Chakrabarty, and E. J. Marinissen, Efficient test access mechanism optimization for system-on-chip, *IEEE Trans. on Computer-Aided Design*, 22(5), pp. 635–643, May 2003.

[Iyengar 2003c] V. Iyengar, K. Chakrabarty, and E. J. Marinissen, Test Access Mechanism Optimization, Test Scheduling and Tester Data Volume Reduction for System-on-Chip, *IEEE Trans. on Computers*, 52(12), pp. 1619–1632, December 2003.

[Jiang 1999] W. Jiang and B. Vinnakota, Defect-oriented test scheduling, in *Proc. VLSI Test Symp.*, pp. 433–438, April 1999.

[Kahng 2003] A. B. Kahng, The road ahead: The significance of packaging, *IEEE Design & Test of Computers*, 20(6), pp. 1619–1632, December 2003.

[Koranne 2002a] S. Koranne, On test scheduling for core-based SOCs, in *Proc. Int. Conf. on VLSI Design*, pp. 504–510, January 2002.

[Koranne 2002b] S. Koranne and V. Iyengar, On the use of K-tuples for SoC test schedule representation, in *Proc. Int. Test Conf.*, pp. 539–548, October 2002.

[Kundert 2000] H. Kundert, K. Chang, D. Jefferies, G. Lamant, E. Malavasi, and F. Sendig, Design of mixed-signal systems-on-a-chip, *IEEE Trans. on Computer-Aided Design*, 19(12), pp. 1561–1571, December 2000.

[Larsson 2001a] E. Larsson and Z. Peng, An integrated system-on-chip test framework, in *Proc. ACM/IEEE Design, Automation and Test in Europe Conf.*, pp. 138–144, March 2001.

[Larsson 2001b] E. Larsson and Z. Peng, Test scheduling and scan-chain division under power constraint, in *Proc. Asian Test Symp.*, pp. 259–264, November 2001.

[Larsson 2004] E. Larsson, J. Pouget, and Z. Peng, Defect-aware SoC test scheduling, in *Proc. VLSI Test Symp.*, pp. 359–364, April 2004.

[Liu 1998] E. Liu, C. Wong, Q. Shami, S. Mohapatra, R. Landy, P. Sheldon, and G. Woodward, Complete mixed-signal building blocks for single-chip GSM baseband processing, in *Proc. IEEE Custom Integrated Circuits Conf.*, pp. 11–14, May 1998.

[Marinissen 1998] E. J. Marinissen, R. Arendsen, G. Bos, H. Dingemanse, M. Lousberg, and C. Wouters, A structured and scalable mechanism for test access to embedded reusable cores, in *Proc. Int. Test Conf.*, pp. 284–293, October 1998.

[Marinissen 2000] E. J. Marinissen, S. K. Goel, and M. Lousberg, Wrapper design for embedded core test, in *Proc. Int. Test Conf.*, pp. 911–920, October 2000.

[Marinissen 2001] E. J. Marinissen and H. Vranken, On the role of DfT in IC-ATE matching, *IEEE Int. Workshop on Test Resource Partitioning*, November 2001.

[Marinissen 2002a] E. J. Marinissen and S. K. Goel, Analysis of test bandwidth utilization in test bus and TestRail architectures in SOCs, *Digest of papers, IEEE Int. Workshop on Design and Diagnostics of Electronic Circuits and Systems*, pp. 52–60, April 2002.

[Marinissen 2002b] E. J. Marinissen, P. Kapur, M. Lousberg, T. McLaurin, M. Ricchetti, and Y. Zorian, On IEEE P1500's standard for embedded core test, *J. Electronic Testing: Theory and Applications*, 18(4), pp. 365–383, August/October 2002.

[Maxwell 2003] P. C. Maxwell, Wafer-package test mix for optimal defect detection and test time savings, *IEEE Design & Test of Computers*, 20(5), pp. 84–89, September 2003.

[Muresan 2000] V. Muresan, X. Wang, and M. Vladutiu, A comparison of classical scheduling approaches in power-constrained block-test scheduling, in *Proc. Int. Test Conf.*, pp. 882–891, October 2000.

[Nourani 2000] M. Nourani and C. Papachristou, An ILP formulation to optimize test access mechanism in system-on-chip testing, in *Proc. Int. Test Conf.*, pp. 902–910, October 2000.

[Parulkar 2002] I. Parulkar, T. Ziaja, R. Pendurkar, A. D'Souza, and A. Majumdar, A scalable, low cost design-for-test architecture for UltraSPARC^{TM} chip multi-processors, in *Proc. Int. Test Conf.*, pp. 726–735, October 2002.

[Sehgal 2003a] A. Sehgal, V. Iyengar, M. D. Krasniewski, and K. Chakrabarty, Test cost reduction for SOCs using virtual TAMs and lagrange multipliers, in *Proc. ACM/IEEE Design Automation Conf.*, pp. 738–743, June 2003.

[Sehgal 2003b] A. Sehgal, S. Ozev, and K. Chakrabarty, TAM optimization for mixed-signal SOCs using test wrappers for analog cores, in *Proc. IEEE Int. Conf. on Computer-Aided Design*, pp. 95–99, November 2003.

[Sehgal 2004a] A. Sehgal and K. Chakrabarty, Efficient modular testing of SOCs Using dual-speed TAM architectures, in *Proc. ACM/IEEE Design, Automation and Test in Europe Conf.*, pp. 422–427, February 2004.

[Sehgal 2004b] A. Sehgal, S. K. Goel, E. J. Marinissen, and K. Chakrabarty, IEEE P1500-compliant test wrapper design for hierarchical cores, in *Proc. Int. Test Conf.*, pp. 1203–1212, October 2004.

[Sehgal 2004c] A. Sehgal, V. Iyengar, and K. Chakrabarty, SOC test planning using virtual test access architectures, *IEEE Trans. on VLSI Systems*, 12(12), pp. 1263–1276, December 2004.

[Sehgal 2005] A. Sehgal, F. Liu, S. Ozev, and K. Chakrabarty, Test planning for mixed-signal SoCs with wrapped analog cores, in *Proc. Design, Automation and Test in Europe Conf.*, pp. 50–55, February 2005.

[Sehgal 2006a] A. Sehgal, S. K. Goel, E. J. Marinissen, and K. Chakrabarty, Hierarchy-aware and area-efficient test infrastructure design for core-based system chips, in *Proc. Design, Automation and Test in Europe Conf.*, pp. 285–290, March 2006.

[Sehgal 2006b] A. Sehgal, S. Ozev, and K. Chakrabarty, Test Infrastructure design for mixed-signal SOCs with wrapped analog cores, *IEEE Trans. on VLSI Systems*, 14(3), pp. 292–304, March 2006.

[SIA 2005] SIA, *The International Technology Roadmap for Semiconductors: 2005 Edition*, Semiconductor Industry Association, San Jose, CA (http://public.itrs.net), 2005.

[Sugihara 1998] M. Sugihara, H. Date, and H. Yasuura, A novel test methodology for core-based system LSIs and a testing time minimization problem, in *Proc. Int. Test Conf.*, pp. 465–472, October 1998.

[Sunter 1996] S. K. Sunter, Cost/benefit analysis of the P1149.4 mixed-signal test bus, *IEE Proceedings-Circuits, Devices and Systems*, 143(6), pp. 393–398, December 1996.

[Teradyne 2007] Teradyne Technologies, Tiger: Advanced digital with silicon germanium technology (www.teradyne.com/tiger/digital.html), 2007.

[Touba 1997] N. A. Touba and B. Pouya, Using partial isolation rings to test core-based designs, *IEEE Design & Test of Computers*, 14(4), pp. 52–59, October/December 1997.

[Varma 1998] P. Varma and S. Bhatia, A structured test re-use methodology for core-based system chips, in *Proc. Int. Test Conf.*, pp. 294–302, October 1998.

[Volkerink 2002] E. Volkerink, A. Khoche, J. Rivoir, and K. D. Hilliges, Test economics for multi-site test with modern cost reduction techniques, in *Proc. VLSI Test Symp.*, pp. 411–416, May 2002.

[Whetsel 1997] L. Whetsel, An IEEE 1149.1 based test access architecture for ICs with embedded cores, in *Proc. Int. Test Conf.*, pp. 69–78, November 1997.

[Xu 2003] Q. Xu and N. Nicolici, On reducing wrapper boundary register cells in modular SOC testing, in *Proc. Int. Test Conf.*, pp. 622–631, September 2003.

[Xu 2004] Q. Xu and N. Nicolici, Wrapper design for testing IP cores with multiple clock domains, in *Proc. Design, Automation and Test in Europe Conf.*, pp. 416–421, February 2004.

[Yamamoto 2001] T. Yamamoto, S.-I. Gotoh, T. Takahashi, K. Irie, K. Ohshima, and N. Mimura, A mixed-signal 0.18-μm CMOS SoC for DVD systems with 432-MSample/s PRML read channel and 16-Mb embedded DRAM, *IEEE J. of Solid-State Circuits*, 36(11), pp. 1785–1794, November 2001.

[Zhao 2003] D. Zhao and S. Upadhyaya, Power constrained test scheduling with dynamically varied TAM, in *Proc. VLSI Test Symp.*, pp. 273–278, April 2003.

[Zorian 1993] Y. Zorian, A distributed BIST control scheme for complex VLSI devices, in *Proc. VLSI Test Symp.*, pp. 6–11, April 1993.

[Zorian 1999] Y. Zorian, E. J. Marinissen, and S. Dey, Testing embedded-core-based system chips, *IEEE Computer*, 32(6), pp. 52–60, June 1999.

R4.3 Network-on-Chip (NOC) Testing

[Aktouf 2002] C. Aktouf, A complete strategy for testing an on-chip multiprocessor architecture, *IEEE Design & Test of Computers*, 19(1), pp. 18–28, January/February 2002.

[Amory 2005] A. M. Amory, E. Briao, E. Cota, M. Lubaszewski, and F. G. Moraes, A scalable test strategy for network-on-chip routers, in *Proc. Int. Test Conf.*, Paper 25.1, November 2005.

[Amory 2006] A. M. Amory, K. Goossens, E. J. Marinissen, M. Lubaszewski, and F. Moraes, Wrapper design for the reuse of networks-on-chip as test access mechanism, in *Proc. European Test Symp.*, pp. 213–218, May 2006.

[Attarha 2001] A. Attarha and M. Nourani, Testing interconnects for noise and skew in giga-hertz SoCs, in *Proc. Int. Test Conf.*, pp. 304–314, October 2001.

[Chen 2003] L. Chen, S. Ravi, A. Raghunathan, and S. Dey, Scalable software-based self test methodology for programmable processors, in *Proc. ACM/IEEE Design Automation Conf.*, pp. 548–553, June 2003.

[Cota 2003] E. Cota, M. Kreutz, C. A. Zeferino, L. Carro, M. Lubaszewski, A. Susin, The impact of NOC reuse on the testing of core-based systems, in *Proc. VLSI Test Symp.*, pp. 128–133, April 2003.

[Cota 2004] E. Cota, L. Carro, and M. Lubaszewski, Reusing an On-Chip Network for the Test of Core-Based Systems, *ACM Trans. on Design Automation of Electronic Systems*, 9(4), pp. 471–499, October 2004.

[Cuviello 1999] M. Culiello, S. Dey, X. Bai, and Y. Zhao, Fault modeling and simulation for crosstalk in system-on-chip interconnects, in *Proc. ACM/IEEE Int. Conf. on Computer-Aided Design*, pp. 297–303, November 1999.

[Gallagher 2001] P. Gallagher, V. Chickermane, S. Gregor, and T. S. Pierre, A Building Block BIST methodology for SOC designs: A case study, in *Proc. Int. Test Conf.*, pp. 111–120, October 2001.

[Grecu 2006] C. Grecu, P. Pande, A. Ivanov, and R. Saleh, BIST for network-on-chip interconnect infrastructures, in *Proc. IEEE VLSI Test Symp.*, pp. 30–35, April 2006.

[Goossens 2005] K. Goossens, J. Dielissen, and A. Radulescu, AEthereal network on chip: Concepts, architectures, and implementations, *IEEE Design & Test*, 22(5), pp. 414–421, September/October 2005.

[ITC 2002] ITC, *Int. Test Conf.*, SoC Test Benchmarks, available online at www.extra.research.philips.com/itc02socbenchm, 2002.

[Li 2006] M. Li, W.-B. Jone, and Q. Zeng, An efficient wrapper scan chain configuration method for network-on-chip testing, in *Proc. IEEE Computer Society Annual Symp. on VLSI*, pp. 147–152, March 2006.

[Liu 2004] C. Liu, E. Cota, H. Sharif and D. K. Pradhan, Test Scheduling for network-on-chip with BIST and precedence constraints, in *Proc. Int. Test Conf.*, pp. 1369–1378, October 2004.

[Liu 2005] C. Liu, V. Iyengar, J. Shi, and E. Cota, Power-aware test scheduling in network-on-chip using variable-rate on-chip clocking, in *Proc. VLSI Test Symp.*, pp. 349–354, May 2005.

[Liu 2006] C. Liu, V. Iyengar and D. K. Pradhan, Thermal-aware testing of network-on-chip using multiple clocking, in *Proc. VLSI Test Symp.*, pp. 46–51, April 2006.

[Iyengar 2002a] V. Iyengar and K. Chakrabarty, System-on-a-chip test scheduling with precedence relationships, preemption, and power constraints, *IEEE Trans. on Computer-Aided Design*, 21(9), pp. 1088–1094, September 2002.

[Pande 2006] P. P. Pande, A. Gangguly, B. Freero, B. Belzer, and C. Gredu, Design of low-power & reliable networks on chip through joint crosstalk avoidance and forward error correction code, in *Proc. IEEE Int. Symp. on Defect and Fault Tolerance in VLSI Systems*, pp. 466–476, October 2006.

[Radulescu 2005] A. Radulescu, J. Dielissen, S. G. Pestana, O. P. Gangwal, E. Rijpkema, P. Wielage, and K. Goossens, An efficient on-chip in offering guaranteed services, shared-memory abstraction, and flexible network configuration, *IEEE Trans. on Computer-Aided Design*, 25(1), pp. 4–17, January 2005.

[Stewart 2006] K. Stewart and S. Tragoudas, Interconnect testing for network on chips, in *Proc. IEEE VLSI Test Symp.*, pp. 100–105, April 2006.

[Tamhankar 2005] R. R. Tamhankar, S. Murali, and G. De Micheli, Performance driven reliable design for networks on chips, in *Proc. Asia and South Pacific Design Automation Conf.*, 2, pp. 749–754, January 2005.

[Wiklund 2002] D. Wiklund and D. Liu, Design of a system-on-chip switched network and its design support, in *Proc. Int. Conf. on Communications, Circuits, and Systems*, pp. 1279–1283, June 2002.

[Zeferino 2003] C. Zeferino and A. Susin, SoCIN: A parametric and scalable network-on-chip, in *Proc. Symp. on Integrated Circuits and Systems Design*, pp. 121–126, May 2003.

R4.4 Design and Test Practice: Case Studies

[Goel 2002] S. K. Goel and E. J. Marinissen, Effective and efficient test architecture design for SOCs, in *Proc. Int. Test Conf.*, pp. 529–538, October 2002.

[Goel 2004] S. K. Goel, K. Chiu, E. J. Marinissen, T. Nguyen, and S. Oostdijk, Test infracture design for the Nexperia home platform PNX8550 system chip, in *Proc. Design, Automation and Test in Europe Conf.*, pp. 108–113, February 2004.

[Goossens 2005] K. Goossens, J. Dielissen, and A. Radulescu, AEthereal network on chip: Concepts, architectures, and implementations, *IEEE Design & Test*, 22(5), pp. 414–421, September/October 2005.

[Marinissen 2000] E. J. Marinissen, S. K. Goel, and M. Lousberg, Wrapper design for embedded core test, in *Proc. Int. Test Conf.*, pp. 911–920, October 2000.

[Radulescu 2005] A. Radulescu, J. Dielissen, S. G. Pestana, O. P. Gangwal, E. Rijpkema, P. Wielage, and K. Goossens, An efficient on-chip NI offering guaranteed services, shared-memory abstraction, and flexible network configuration, *IEEE Trans. on Computer-Aided Design,* 25(1), pp. 4–17, January 2005.

[Steenhof 2006] F. Steenhof, H. Duque, B. Nilsson, K. Goossens, and P. Llopis, Network on chips for high-end consumer-electronics TV system architectures, in *Proc. Design, Automation and Test in Europe Conf.*, pp. 124–129, March 2006.

[Vermeulen 2001] B. Vermeulen, S. Oostdijk, and F. Bouwman, Test and debug strategy of the PNX8525 Nexperia digital video platform system chip, in *Proc. Int. Test Conf.*, pp. 121–130, October 2001.

[Vermeulen 2003] B. Vermeulen, J. Dielissen, K. Goznes, and C. Ciordas, Bringing communication networks on a chip: Test and verification implications, *IEEE Communication Magazine*, pp. 74–81, September 2003.

[Wang 2005] L.-T. Wang, X. Wen, P.-C. Hsu, S. Wu, and J. Guo, At-speed logic BIST architecture for multi-clock designs, in *Proc. IEEE Int. Conf. on Computer Design*, pp. 474–478, October 2005.

R4.5 Concluding Remarks

[Xu 2004] Q. Xu and N. Nicolici, Wrapper design for testing IP cores with multiple clock domains, in *Proc. Design, Automation and Test in Europe Conf.*, pp. 416–421, February 2004.

[Zhao 2006] D. Zhao, S. Upadhyaya, and Martin Margala, Design of a wireless test control network with radio-on-chip technology for nanometer systems-on-chip, *IEEE Trans. on Computer-Aided Design*, 25(6), pp. 1411–1418, June 2006.

CHAPTER 5

SIP TEST ARCHITECTURES

Philippe Cauvet
NXP Semiconductors, Caen, France

Michel Renovell
LIRMM–CNRS/University of Montpellier, Montpellier, France

Serge Bernard
LIRMM–CNRS/University of Montpellier, Montpellier, France

ABOUT THIS CHAPTER

Since the 1980s, we have seen the emergence and rapid growth of ***system-on-chip*** (SOC) applications. The same trend has happened in ***system-in-package*** (SIP) applications since the mid-1990s. The development of SIP technology has benefited from the SOC technology; however, this emerging SIP technology presents specific test challenges because of its complex design and test processes. Indeed, one major difference between an SOC and an SIP is that an SOC contains only one die on a packaged chip whereas an SIP is an assembled system composed of a number of individual dies on a packaged chip. Each die in the SIP can also use a different process technology, such as silicon or *gallium-arsenide* (GaAs), which includes a ***radiofrequency*** (RF) or ***microelectromechanical systems*** (MEMS) components. This fundamental difference implies that to test an SIP, each bare die in the SIP must be tested first before the bare die is packaged in the SIP. Then, a functional system test or embedded component test at the system level can be performed. The passing bare dies are often called ***known-good-die*** (KGD).

In this chapter, we first discuss the basic SIP concepts, explore SIP technology's difference from the SOC technology, and show some SIP examples. We highlight the specific challenges from the testing point of view and derive the assembled yield and defect level for the packaged SIP. Next, various bare-die test techniques to find known-good-dies are described, along with their limitations. Finally, we present two techniques to test the SIP at the system level: the functional system test and the embedded component test. The functional system test aims to check the functions of all dies per their specifications at the same time, whereas the embedded component test tries to detect each faulty die individually. We conclude the chapter with a brief discussion on future SIP design and test challenges.

5.1 INTRODUCTION

Systems are moving to higher and higher levels of complexity, integrating ever more complex functions under ever more stringent economical constraints. With advances in semiconductor manufacturing technology, the SOC technology has appeared as a viable solution to reduce device cost through higher levels of integration. Thanks to the exploitation of the so-called reuse concept, the SOC development time remains at a reasonable limit despite the increasing complexity. However, the push toward more functionality in a single box requires the integration of heterogeneous devices that cannot be intrinsically achieved in single-technology SOC. In this context, SIP clearly appears as the only viable solution to integrate more functions in an equal or smaller volume.

Cellular handset designs have been one good example of this global trend. Today, a new cellular design must support multiband and multimode features in addition to Bluetooth™ networking, ***global positioning systems*** (GPSs), and ***wireless local area networks*** (WLANs), not to mention user applications such as games, audio, and video. This triggers the evolution of the SIP technology that has become one viable solution to solve these highly complex design integration problems in an economical way.

5.1.1 SIP Definition

The SIP concept started with the development of the ***multichip module*** (MCM) in the 1990s. MCM was the pioneer technology to integrate several active dies on one common substrate. A typical integration at that time included memories and a processor, but it was not possible to build a complete system in one package because of limited integration capabilities. With its associated system integration benefits, the SIP technology has since been adopted rapidly as a packaging alternative to save product cost. In mobile phone applications, the system integrators often face short product life cycles. As a result, they came to the conclusion that integrating existing and available ***integrated circuits*** (ICs) into an SIP is much easier than creating new SOC designs from scratch or even by reusing existing ***intellectual property*** (IP) cores.

The International Technology Roadmap for Semiconductors published by the Semiconductor Industry Association defines a *system-in-package* (SIP) as any combination of semiconductors, passives, and interconnects integrated into a single package [SIA 2005]. The definition does not limit SIP to any single technology or integration; it clearly indicates that an SIP can combine different die technologies and applications with active and passive components to form a complete system or subsystem. Consequently, an SIP usually includes logic and memory components, but it may also include analog and RF components. These various components are interconnected by wire-bond, flip-chip, stacked-die technology, or any combination of the above.

■ **FIGURE 5.1**

Multiple dies, components, and interconnections on an SIP for GSM application.

The differences between SOC and SIP can be better outlined as follows:

- The SOC is created from a single piece of substrate (*i.e.*, a *single* die); the single die is fabricated in a *single* process technology with a *single* level of interconnections from the die to the package pads or to the interposer. The SOC uses a *single* interconnection technology, such as wire-bond or flip-chip.
- The SIP is created from several different dies (*i.e.*, *multiple* parts); these dies can come from a broad mix of *multiple* process technologies, such as CMOS, GaAs, or BiCMOS, with *multiple* levels of interconnections from die/component to die/component, or from die/component to the package pads or to the interposer. The SIP includes *multiple* interconnection technologies, such as wire-bond, flip-chip, soldering, and gluing.

In short, we can say that for an SOC everything is single while for an SIP everything is multiple. Figure 5.1 shows an example SIP in a ***global system for mobile communications*** (GSM) application where multiple dies, components, and leadframe connections are embedded in a single package.

5.1.2 SIP Examples

In recent years, many types of SIP have been developed. They differ in the type of carrier or interposer to be used for holding the bare die (or component) and the type of interconnections to be used for connecting components. The carrier or interposer can be a leadframe, an organic laminate, or silicon based. Passive components such as resistors, inductors, and capacitors can be embedded as part of the substrate construction, or they can be soldered or epoxy attached on the surface. Active components such as microprocessors, memories, RF transceivers, and mixed-signal components can be distributed on the carrier surface. Another possibility is to stack components on top of each other, called **stacked dies**. The *three-dimensional* (3-D) stacked-die assembly can drastically reduce the overall size of the packaged system resulting in low package cost and increased component density that is critical in many applications.

The component interconnections are made from the component to the carrier and from die to die. They can use any type of interconnect technology: wire bonding, flip-chip, or soldering. The final packaged SIP looks like any conventional ceramic-type package, ***ball grid array*** (BGA), ***land grid array*** (LGA), or leadframe-based package. Figures 5.2 and 5.3 illustrate some examples of carriers and stacked dies. In Figure 5.2a, two dies are glued and interconnected on a leadframe. Figure 5.2b shows an SIP made of active dies and discrete components soldered on a laminate substrate (a micro printed circuit board) and encapsulated in a low-cost package. An example of silicon-based SIP is shown in Figure 5.2c, where three active dies

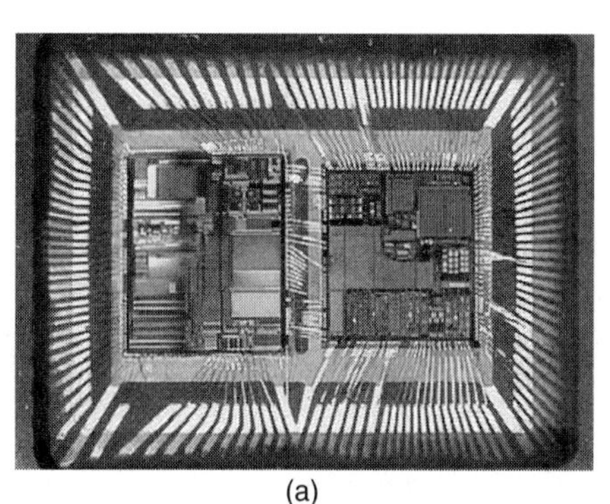
(a)

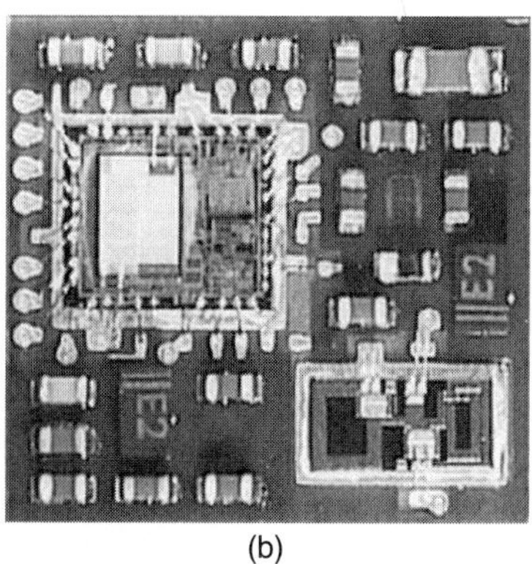

(b)

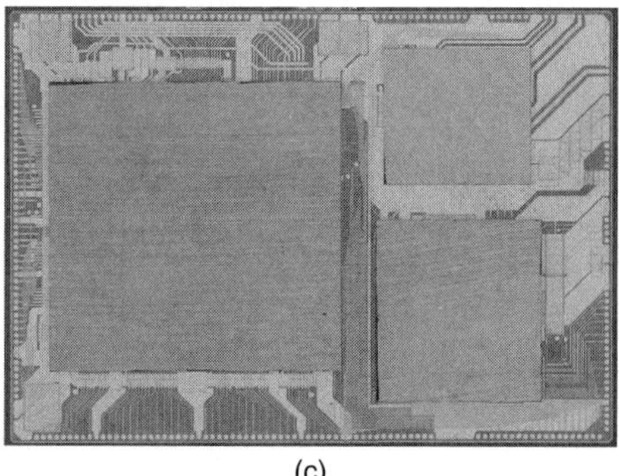
(c)

FIGURE 5.2

Examples of carrier style: (a) leadframe, (b) laminate, and (c) silicon based.

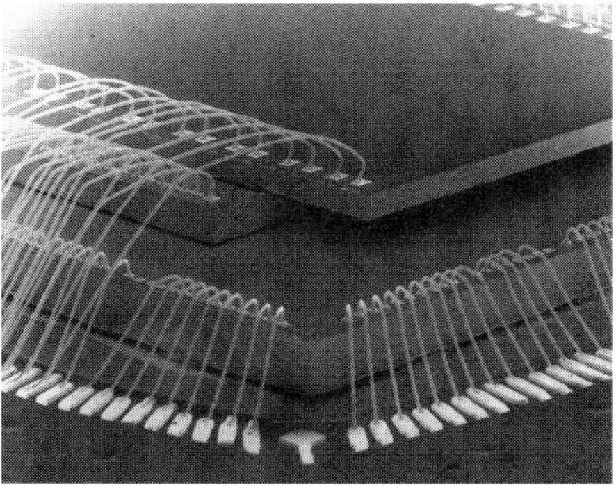

■ **FIGURE 5.3**

Example of stacked components. (Courtesy of [DPC 2007].)

are flipped and soldered on a passive silicon substrate. These three technologies are planar, whereas Figure 5.3 illustrates an SIP using the 3-D stacked die concept.

SIP products may also be classified according to their development flow:

- SIP development using off-the-shelf components: the individual dies were designed independently, thus no specific DFT for SIP test has been introduced, which in turn may cause difficult test problems.
- SIP development with application specific components: in this case, test requirements can be considered in the SIP design and test strategy, which may simplify the SIP test significantly.

The SIP offers a unique advantage over the SOC in its ability to integrate any type of semiconductor technology and passive components into a single package. Currently, for wireless systems requiring the use of digital and analog components, the SIP is becoming a favorite choice. The SIP is also a good candidate for the integration of MEMS with circuitry to provide a fully functional system, as opposed to acting as a pure sensor or actuator. The MEMS-integrated SIP can handle sensors for physical entities such as light, temperature, blood/air pressure, or magnetic fields for automotive, biological/chemical, consumer, communication, medical, and RF applications, to name a few.

The SIP actually offers even more advantages including the following:

- Combining different die technologies (Si, GaAs, SiGe, etc.)
- Combining different die geometries (250 nm, 90 nm, etc.)
- Including other technologies (MEMS, optical, vision, etc.)
- Including other components (antennas, resonators, connectors, etc.)
- Increasing circuit density and reducing printed circuit board area

- Reducing design effort by using existing ICs
- Minimizing risks by using proven existing blocks
- Improving performance

In the early years, an SIP was only developed and used for high-volume production because of packaging cost concerns. The reduction of packaging cost through the utilization of proven low-cost packaging platform technologies has opened a broad range of applications. However, two points still require particular attention when developing or manufacturing an SIP:

- Usually an SIP has a longer supply chain and may introduce additional planning risks and dependencies because of external suppliers for the carrier and the higher complexity of the assembly process.
- The design and layout of carriers are often at the leading edge of substrate fabrication technology, which may result in low yield of the carrier production and a higher carrier cost.

5.1.3 Yield and Quality Challenges

The fabrication and test flow of a standard SIP is represented in Figure 5.4. The SIP under consideration is basically composed of *n* different dies. Figure 5.4 concentrates on die fabrication (Figure 5.4a) and SIP assembly (Figure 5.4b); passive

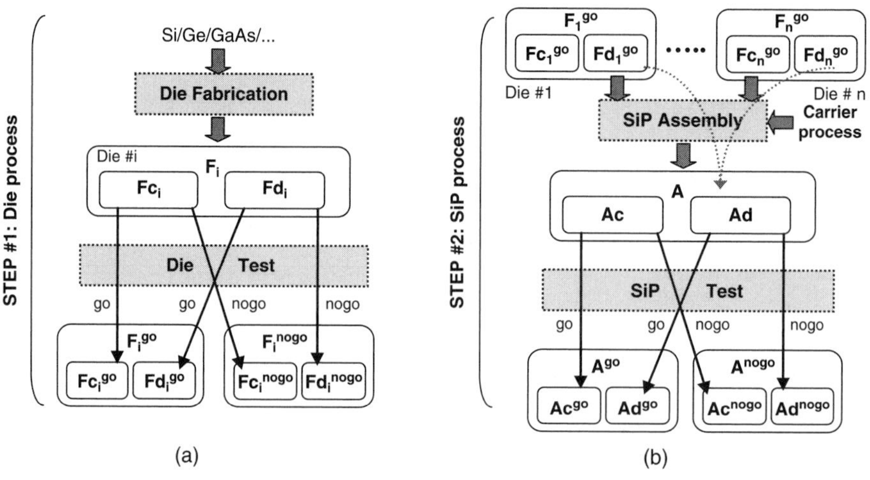

FIGURE 5.4

SIP creation and test: (a) fabrication of die #i and (b) assembly of the SIP.

components and carrier process are omitted. This flow can be seen as being composed of two fundamental steps referred to as the die process and the SIP process.

The die process represents the flow for all dies. Starting from a given material (silicon or other), a number of dies, given by F_i, are fabricated using a technological process. Among these F_i fabricated dies, a number of dies, Fc_i, are correct while a number of dies, Fd_i, are defective. Testing is then used to screen defective dies (F_i^{nogo}) from correct dies (F_i^{go}) where the terms "go" and "nogo" indicate that the die has passed or failed the screening test(s), respectively. Because testing is not a perfect process, we can classify the testing results of the dies into four possible categories [Bushnell 2000]:

1. A number of correct dies (Fc_i^{go}) that are declared "go"
2. A number of defective dies (Fd_i^{go}) that are declared "go"
3. A number of correct dies (Fc_i^{nogo}) that are declared "nogo"
4. A number of defective dies (Fd_i^{nogo}) that are declared "nogo"

Clearly, the correct decision has been made in categories 1 and 4, whereas categories 2 and 3 correspond to wrong decisions. Category 2 is usually characterized by defect level DL_i, which is defined as the number of defective parts declared "go" over the total number of parts declared "go":

$$DL_i = Fd_i^{go}/F_i^{go} = Fd_i^{go}/(Fc_i^{go} + Fd_i^{go})$$

This defect level can be viewed as a metric of test quality. A high-quality test should minimize the defect level: $Fd_i^{go} \to 0 \Rightarrow DL_i \to 0$. The defect level is a measure of the number of dies remaining in the "go" category that is actually defective and will not function correctly in the application. In the context of the SIP, this concept of defect level is of great importance since all dies in the "go" category (*i.e.*, $Fc_i^{go} + Fd_i^{go}$) are used during the second step of the SIP process as illustrated in Figure 5.4b. In this step, an SIP is assembled from a set of incoming dies (die #1 to die #n) and possibly some passive components, using a carrier or interposer. At the end of this assembly process, a total number of SIPs, A, is assembled, among which Ac SIPs are correct and Ad SIPs are defective. Testing is used once again to screen defective SIPs from correct SIPs.

The resulting SIP can be a very sophisticated and expensive product because a single SIP may include many dies and passive components. Furthermore, SIP assembly can be a very complex and expensive process, where many different interconnect techniques may be used for wire bonding and flip-chip connections. In addition to these costs, it is important to note that typically a defective SIP cannot be repaired. These factors combined lead to the following observation:

The SIP process is economically viable only if the associated yield Y_{SIP} of the packaged SIP is high enough.

Referring to Figure 5.4b, the production yield of the packaged SIP can be defined as the number of correct SIPs, Ac, over the total number of assembled SIPs, A:

$$Y_{SIP} = Ac/A = (A - Ad)/A$$

To optimize the assembly yield, Y_{SIP}, it is important to minimize the number of defective parts, *Ad*. This requires a detailed analysis of all possible root causes for defective SIPs. A defective SIP can originate from the following:

- Defective substrate
- Incorrect placement or mounting of dies and components
- Defective soldering or wire-bonding of dies and components
- Stress on dies during assembly
- Defective dies, etc.

The first three cases are under the control and the responsibility of the SIP assembler. The substrate used as a carrier is generally made with a mature technology with a high level of quality, whereas the assembly process and the quality of the mounted dies are particularly critical. Indeed, SIPs are only viable if the quality of the assembly process is sufficiently high. In the IC fabrication context, yields (Y_{ic}) of around 75% are quite common. As a matter of fact, the SIP assembly context is totally different because the yield associated with the assembly process, Y_a, should be much higher. For example, viable yields for SIP assembly are typically around 99%.

The fourth case corresponds to a situation where a die that is correct before assembly sustains some form of flaw. The die is stressed by the assembly process and becomes defective.

The last case is illustrated in Figure 5.4b with curved arrows and is totally independent of the assembly process. It is clear that any defective die belonging to the Fd_i^{go} category will contribute to the number, *Ad*, of defective SIPs. This important point demonstrates that the defect level DL_i of each die process has a direct impact on the yield of the SIP assembly process. This is not a new point; in the early 1990s, after passing the first euphoria for MCMs, it was discovered that the yield of assembled modules was related to the number of devices in the module and the defect levels associated with each device. This simple relationship holds for any assembly process of components that must all function correctly (assuming no redundancy) to provide the correct system function [DPC 2007].

For a given die #i, the probability P_i of being defect free is related to its defect level:

$$P_i = 1 - DL_i$$

Considering an SIP assembly process including different dies, the overall yield Y_{SIP} can be ascertained by:

$$Y_{SIP} = 100\ [P_1 \times P_2 \times \cdots \times P_n] \times P_s \times P_{int}{}^{Q} \times P_w$$

where P_i is the probability that die #i is defect-free, P_s is the probability of substrate being defect-free, P_{int} is the probability of die interconnect being defect-free, Q is the

TABLE 5.1 ■ Assembly Yield Example

	DL_i (ppm)	P_i (%)
Substrate	600	99.94
Die 1	4100	99.59
Die 2	35500	97.45
Die 3	1200	99.88
Y_{SIP}		**96.87**

quantity of die interconnect, and P_w is the probability of placement and mounting being defect-free.

The preceding equation demonstrates the cumulative effect of the different defect levels DL_i expressed in ***parts per million*** (ppm). The resulting yield is related to the multiplication of the different defect levels. Table 5.1 gives a simplified example where passive components, interconnect, and mounting problems are omitted. The example only considers problems related to the substrate and three dies.

From this example, the overall yield Y_{SIP} will always be lower than the probabilities of individual dies being defect-free. In other words, we could say that the assembly process accumulates all the problems of the individual dies. Another important feature is the sensitivity of this phenomenon. Indeed, to consider a high ppm for only one die, let's assume that die 1 has a defect level of 20000 ppm. In this case, the resulting yield falls from $Y_{SIP} = 96.87$ to $Y_{SIP} = 95.33$.

An acceptable assembly yield requires that every die in the SIP exhibits a very low defect level. In other words, only high-quality components are used in the SIP assembly process. In Figure 5.4a, the die test process is responsible for this high quality level. In practice, this quality level is not easy to guarantee because tests are applied to bare dies and not to packaged dies. A test applied to a bare die is usually called *wafer test,* whereas a test applied to packaged IC is called *final test.* It is well-known that many factors limit the efficiency of tests applied to bare dies including limited contact and probe speed, among others. Despite these constraints, bare dies used in the SIP assembly process should exhibit the same, or better, quality level than a packaged IC. As noted earlier, this is referred to as the *known-good-die* (KGD) concept, which can be stated as follows:

KGD: A bare die with the same, or better, quality after wafer test than its packaged and "final tested" equivalent.

5.1.4 Test Strategy

We also have to consider the SIP test process in Figure 5.4b, which is responsible for the final delivery quality. As already discussed for the die process, testing is not a perfect process and, as a result, the SIP test process may exhibit some defect level or yield loss. Consequently the *test* yield Y_{SIP}^{test} of the packaged SIP is not exactly the same as the *production* yield Y_{SIP} defined earlier. The test yield is defined as follows:

$$Y_{SIP}{}^{test} = A^{go}/A = (A - A^{nogo})/A$$

Under the assumption that only high-quality components (KGD) are used in the assembly process, the SIP test process must focus on the faults that originate from the assembly process itself. Under ideal conditions, an SIP test strategy includes the following:

- A precondition stating that all the components used for the assembly have a KGD quality
- A test of the placement and connections
- A global functional test for critical parameters

The first test concentrates on the faults that result from the assembly itself. The second test is more oriented to system aspects and concentrates on parameters that were not testable at wafer test of the separate components. Under ideal conditions (*i.e.*, using KGD components), it is expected that failing SIPs will most frequently be encountered during placement and connections test. In practice, the situation is far different from the ideal one:

- The status of the KGD is not always perfect.
- The use of mixed component technologies and different connection technologies gives rise to a wide spectrum of interconnect faults.
- The functionality may include mixed signal domains such as digital, analog, and RF devices.

As a consequence, the ideal strategy described earlier represents only a global guideline, and more realistic approaches must be used. In particular, most of the testing performed for the dies at wafer level must be repeated because a given die may become defective because of assembly stress. In some cases, additional tests are required to detect defects that passed the wafer test. For example, RF devices cannot be fully tested at wafer test and require at-frequency functional test after mounting.

Testing dies is not trivial because of the limited test access. The total number of pads on each of the different mounted dies is usually much higher than the number of pins of the final SIP package. This means that many test signals, such as scan-in and scan-out signals, multiplexed on die pads may not be connected to package pins. In this case, the recommended solution is the widely adopted IEEE standard for boundary scan. Because of the limited access to internal dies, ***design for testability*** (DFT) and ***built-in self-test*** (BIST) connected to the boundary scan ***test access port*** (TAP) are also recommended whenever possible.

Because of these practical points, testing an SIP is a combination of functional test at the system level with structural test, commonly referred to as *defect-based test*, at the die and interconnect level. The majority of the challenges lie in the testing of mixed signal and RF blocks.

5.2 BARE DIE TEST

Two major factors have to be considered by SIP manufacturers in bare die testing: test escapes of the die (defective die declared "go") and system level infant mortality in the final package. The number of test escapes depends on the yield of the lot, ***automated test equipment*** (ATE) inaccuracy, and insufficient fault coverage obtained by the set of tests. The number of test escapes is unpredictable; therefore, meeting the KGD target for high-volume markets represents a big challenge for the industry. For stand-alone devices, good test coverage is often achieved thanks to the final package test, where high pin count, high speed, and high frequencies can be handled more easily than at wafer level [Mann 2004]. For bare die, on the other hand, probing technology is crucial.

5.2.1 Mechanical Probing Techniques

The probe card technology has not evolved at the same speed as IC technology. Until recently, most products did not require sophisticated probe cards. The traditional technology, based on tungsten needles shaped in a cantilever profile as illustrated in Figure 5.5, still represents 85% of the probe card market [Mann 2004].

The vertical probe from 1977 was developed to fulfill the requirements for array configurations, anticipating the increasingly high number of contact pads in circuits. However, the development of complex devices, combined with the expansion of multisite (parallel) testing, pushed the fabrication of probe cards toward the limits of the technology, one of the key issues being co-planarity. Meanwhile, the size and the pitch of the pads has regularly decreased, forcing probe cards toward novel, but expensive, technologies [SIA 2004]. On top of those requirements, the growth of chip-scale and wafer-level packages put further demands on probes, which contact solder bumps instead of planar pads.

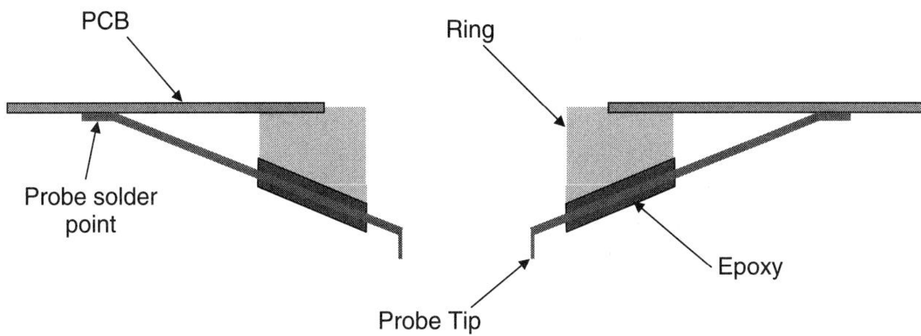

■ FIGURE 5.5

Principle of cantilever probe card using epoxy.

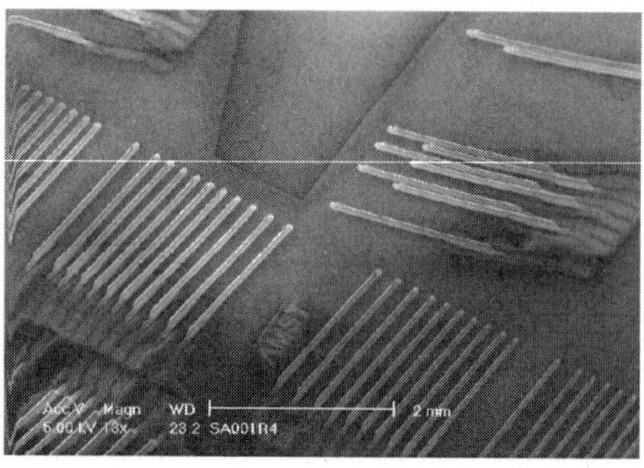

■ **FIGURE 5.6**

View of probe tips in MEMS technology. (Courtesy of Advanced Micro Silicon Technology Company.)

Fortunately, some solutions have been able to push the mechanical limits forward. Among them, MEMS-based implementations of probe cards are now starting to replace the traditional macroscopic technologies. An example of a MEMS-based probe card is shown in Figure 5.6, which represents a silicon micro probe with Cr/Au metal wiring. The cantilevers are kept on the wafer and aligned, via an intermediate ***printed circuit board*** (PCB), with a holder to make the complete probe card. Pitches of 35 μm can be achieved. There are many examples of either industrial products or research demonstrations that show the growing interest in the MEMS-based technologies [Cooke 2005].

Another solution for KGD wafer testing is also emerging, which consists of replacing the traditional probe card with a noncontact interface [Sellathamby 2005]. This technology uses short-range, near field communications to transfer data at gigabit per second rates between the probe card and the ***device under test*** (DUT) on a wafer. The principle is depicted in Figures 5.7 and 5.8. The test system consists of

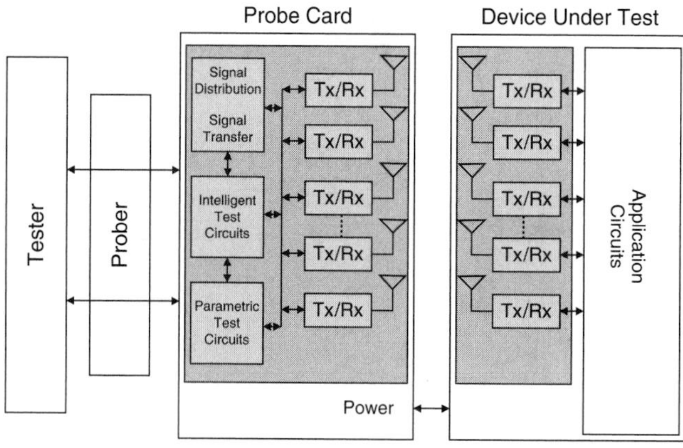

■ **FIGURE 5.7**

Generic principle of noncontact testing. (Courtesy of Scanimetrics.)

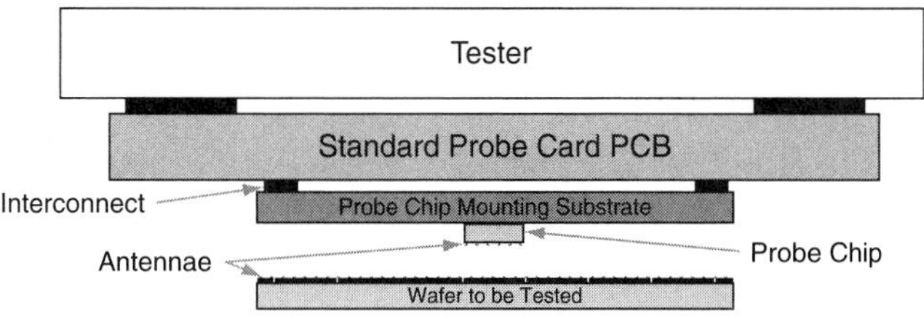

■ **FIGURE 5.8**

Cross-sectional view of the noncontact test solution. (Courtesy of Scanimetrics.)

a probe chip with micro antennas and transceivers. Each antenna and transceiver probes one ***input/output*** (I/O) on the DUT with each I/O site on the DUT being connected to a single antenna and transceiver circuit. The antennas and transceiver circuits are designed into the DUT and operate at the same carrier frequency. This novel technology offers many advantages, including the following:

- High reliability because no damage is caused by scratching the pad
- High density because of the reduction of the size and pitch of the die pads
- High throughput because it enables a massively parallel test

5.2.2 Electrical Probing Techniques

From an electrical perspective, it is fair to say that KGD cannot be achieved for most of the chips with RF interfaces. Thus far, analog ICs are mainly tested against their specified parameters. This strategy has proved to be effective, but it places a lot of requirements on the test environment including ATE, test board, and probe card. A lot of progress has been made by ATE vendors, in test board technology, and more recently in the probe cards. Testing RF and high-speed mixed-signal ICs represents a big challenge because the propagation of the signal along the path may be disturbed by parasitic elements. Moreover, the emergence of new communication protocols beyond 3GHz (*i.e., ultrawideband* [UWB]) adds many constraints to the wafer probing of these devices. A correct propagation is achieved when the following occurs:

1. The integrity of the signal properties (frequency, phase and power) is maintained from the source to the load.
2. The load does not "see" any parasitic signal.

Many elements must be controlled to fulfill these requirements. First, impedance matching will impact the path loss. The serial inductance and the shunt capacitance will create the conditions for proper oscillations and requested frequency bandwidths. The crosstalk will depend on the mutual inductance, the quality of shielding, and filtering. Finally, the noise level can be kept low by adequate decoupling.

All of these effects can be reduced thanks to short, impedance-matched connections between the source and the load to minimize the parasitic elements of the RF connections and related effects. Figure 5.9 represents the cross-section of a typical cantilever probe card for testing RF dies. Components, such as ***surface-mounted devices*** (SMDs), are placed in the RF path to ensure a proper impedance matching (Z_{in}) to optimize the test environment. Mass production implies thousands of touchdowns for the probe tips, thus the drift of the electrical characteristics must be minimal. Finally, despite a high degree of purity that can be reached, the test environment may differ slightly from the application. This is particularly the case with a bare die compared to either a stand-alone IC or an assembled SIP. Therefore, a good correlation is needed between the test and application environments.

The "membrane" technology (see Figure 5.10) can solve many problems by reducing the distance between the pads and the tuning components. A set of microstrip transmission lines is designed on a flexible dielectric material to connect the test electronics from the ATE to the DUT. A conductor on one side of the dielectric membrane and a thin metal film on the opposite side, which acts as a common ground plane, form each microstrip line. The width of the trace depends on the line impedance needed to match the device technology. The die is contacted by an array of microtips formed at the end of the transmission lines through via holes in the membrane (see Figure 5.10a) [Leslie 1989], [Leung 1995], [Wartenberg 2006]. Figure 5.10b shows the structure of the membrane probe. The membrane is mounted on a PCB carrier that interfaces with the test board.

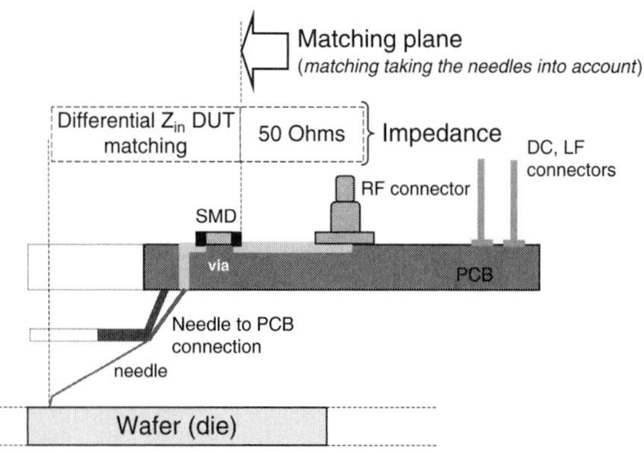

■ **FIGURE 5.9**

RF probe card based on cantilever technology.

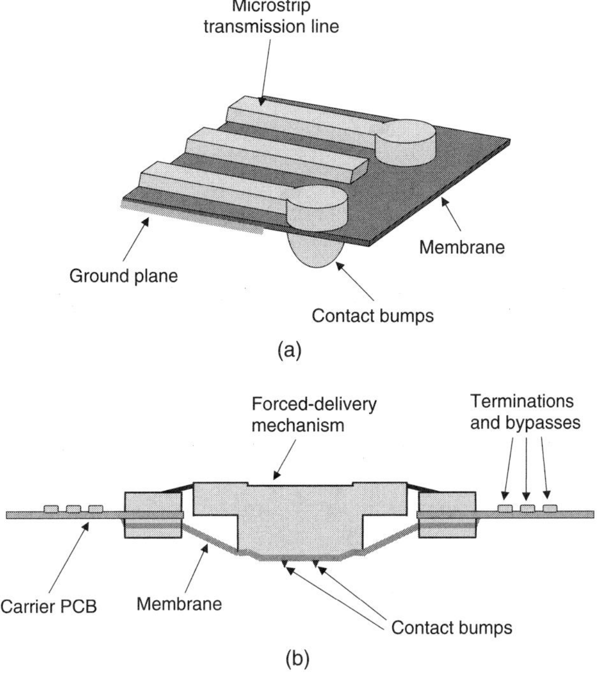

■ **FIGURE 5.10**

Membrane probe card [Leslie 1989]: (a) membrane technology and (b) structure of the probe card.

This technology offers a number of significant advantages for high-performance wafer test including the following:

- High-frequency signals can stimulate the DUT with high fidelity because the membrane probe provides a wideband interface with low crosstalk between the ATE and the DUT.
- Photo-lithographic techniques enable an accurate positioning of the conductor traces and contact pads.
- Contact reliability is much higher than with cantilever probe cards, as is final package test thanks to the absence of scratching.

Although this technology offers better signal integrity and better prediction thanks to modeling capabilities, the correlation issues are still not fully solved because the test environment cannot be exactly identical to that of the application.

An effective probing setup is required in order to achieve full test coverage (*i.e.*, a test at "full specs"). Testing at "full specs" means that the DUT operates under worst-case conditions. In the silicon-based SIP technology [Smolders 2004], the active dies are designed to fulfill their respective performance requirements in

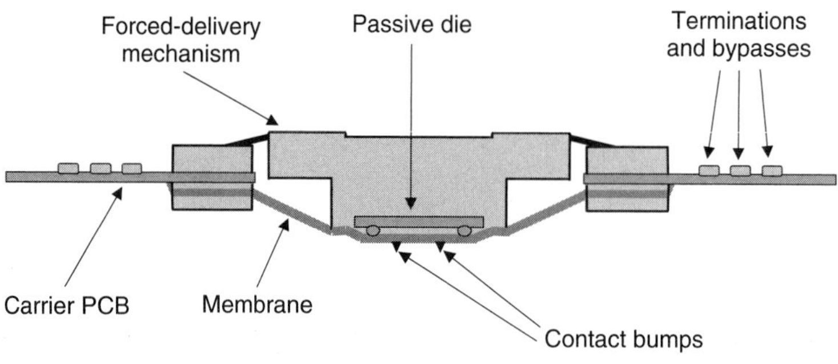

■ **FIGURE 5.11**

Direct-die concept.

a well-defined environment. In some cases, the active die requires its associated passive die to be functional. The passive die has to be closely connected (~10 μm) to ensure nominal performances of the active die under test. These conditions cannot be achieved with traditional probe technologies. A solution has been proposed, namely the direct-die-contacting concept, which intends to test the active die (wafer) through a reference passive die held on the probing fixture. The DUT is contacted with short connections through a passive die, which is the same as, or very close to, the application passive die (see Figure 5.11).

Another approach to KGD for RF/analog products relies on alternative test methods. A representative example is given by a technique that consists of ramping the power supply and observing the corresponding quiescent current signatures [Pineda 2003]. With all transistors being forced into various regions of operation, the detection of faults is done for multiple supply voltages. This method of structural testing exhibits fault coverage results comparable to functional RF tests. Given that no RF signal needs to be generated or measured, probing becomes much less critical, and the reproducibility of the results is potentially better; therefore, the test escapes can be also lowered. This generic approach is illustrated in Figure 5.12.

Other approaches propose to reuse some low-speed or digital internal resources of the DUT, and to add some DFT features to get rid of RF signals outside of the DUT. For example, in [Akbay 2004], the authors apply an alternate test that automatically extracts features from the component response to predict RF specifications. In [Halder 2001], a test methodology is proposed that makes use of a voltage comparator and a simple waveform generator. A digitized response is then shifted out and externally analyzed. The combination of such methods in conjunction with some structural testing techniques will help reach high defect coverage for analog, mixed-signal, and RF devices.

5.2.3 Reliability Screens

In principle, burn-in testing is done by applying abnormally high voltage and elevated temperatures to the device, usually above the limits specified in the data sheet

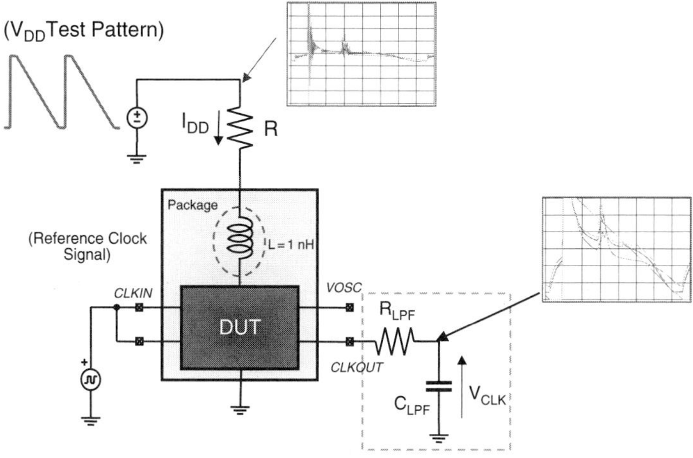

■ **FIGURE 5.12**

Generic approach of DC signature method [Pineda 2003].

for normal operation. Burn-in testing becomes more efficient when the device can be excited by some stimuli (*i.e.*, clock signals). However, manufacturers of high-volume ICs for low-cost consumer and mobile markets are interested in developing novel reliability screens that can be applied at the wafer level and that may fulfill the targets without burn-in testing. Test equipment vendors propose various probe models with high and low temperature capabilities. By applying high or low ambient temperatures during the wafer test, some of the infantile defects can be detected. However, this solution is not cost-effective, especially for very low temperatures (below $-30°$ C), because of the price of the equipment, ramp-up and ramp-down times and possible disturbances during the test that can degrade the throughput time (*i.e.*, air leaks). Diverse alternate methods have been developed and published. One of the most used techniques consists of applying a high-voltage stress to the device before the normal test sequence. A high-voltage stress breaks down weak oxides, thereby creating failure modes that can be detected by I_{DDQ}, structural, or functional tests. The voltage level is significantly higher than the operating voltage [Barrette 1996] [Kawahara 1996]. Other potential failure mechanisms can be detected by low-voltage testing. Bridging faults due to resistive via contacts are detected during structural or functional testing at lower operating voltage.

The I_{DDQ} test for CMOS devices is one of the most meaningful techniques for test engineers. It is a measurement of the supply current at the V_{DD} node while the device is in a quiescent state. In [Arnold 1998], the authors discuss the benefits of I_{DDQ} test and propose implementation of ***built-in current sensors*** (BICS) in each die for quiescent current testing. This paper also explains how a statistical approach can efficiently be used to achieve KGD. The technique consists of calculating test limits for parametric tests based on the quartile values (the **Tukey method**) of the distribution of a given population, which could be one wafer, a sublot, a full lot, or a group of lots. Another statistical method consists of detecting failure clusters on the wafers and declaring all the passing neighbors as defective [Singh 1997]. These statistical techniques are based on outlier identification. Outliers can be distinguished from a variety of patterns: vector to vector, test to test, or die to die. It is worth noting that statistical methods have been used extensively for several years in the test of automotive ICs. In summary, screening methods are not unique, and the trend is to couple them in order to achieve a reliability level that fulfills the requirements for KGD.

5.3 FUNCTIONAL SYSTEM TEST

System test at the SIP level can be considered as (1) functional test of the whole system or subsystem and (2) access methods implemented in the SIP components to enable functional, parametric, or structural tests. These access methods enable testing of the SIP once all the dies are assembled.

The first method is the functional system test in which the system is tested against the application specifications and the functionality is checked. The biggest advantage of this test method is the good correlation at the system level between the measurement results of the SIP supplier and those of the SIP customer (the end-integrator). However, some drawbacks may be as follows:

1. There is a complex test setup with expensive instruments (*i.e.*, combination of RF sources and analyzers with digital channels).
2. There are long test times because of a large number of settings in the SIP (*i.e.*, application software can take seconds to be completed).
3. Testing full paths makes diagnostics difficult (with no intermediate test points).

Several efforts have been made to improve the functional test, especially with the growth of the mobile applications, where SIP technologies represent a big market share. The proposed solutions deal with either DFT, ***digital signal processing*** (DSP) techniques, BIST capabilities, or a combination of these.

5.3.1 Path-Based Testing

Despite the fact that all SIPs are not for application in wireless communications, the following subsections describe techniques developed in this field. Nevertheless, most of the techniques presented can be adapted to other types of SIPs. We consider a system made of a ***transmitter*** (Tx) and a ***receiver*** (Rx), with DFT and BIST features for test embedded inside. In case of a transmitter or a receiver-only system (*i.e.*, a TV set, a Set Top Box, etc.), the same solutions can be applied, provided that the missing part is on the test board instead of the DUT. A block diagram of a typical system with an RF transceiver is shown in Figure 5.13. The partitioning of the functional blocks may depend on the application and the technologies. In this didactical example, we consider an SIP made of three dies: digital plus mixed-signal circuitry, an RF transceiver including a ***low-noise amplifier*** (LNA), and finally

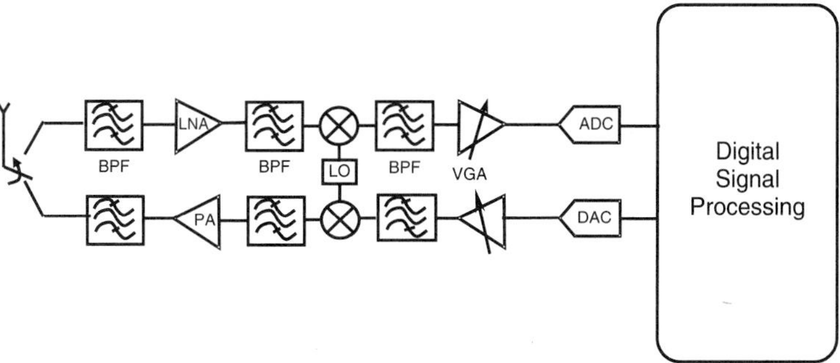

FIGURE 5.13

A typical transceiver system.

a ***power amplifier*** (PA). Other elements, such as switches or ***band-pass filters*** (BPFs), can also be placed on the substrate.

In this architecture, two paths are considered: the transmitter path and the receiver path. For transmission operation, the DSP generates a periodic bit sequence, which is modulated using a digital modulation type, such as ***wideband code-division multiple access*** (WCDMA) or GSM. The modulated sequences are up-converted in frequency and amplified. The receive subsystem down-converts the RF signal into the base-band, where an ***analog-to-digital converter*** (ADC) converts the base-band signal into digital words that are processed by the DSP.

To measure the system performance, the usual strategy consists of splitting the test in two paths, the receiver and the transmitter paths, respectively. In practice, a receiver is tested using sources able to generate digitally modulated RF signals with a spectral purity better than the DUT, and a transmitter is tested using a demodulator and digitizer that have a sufficient noise floor and resolution compared to the DUT.

At the system level, the quality of a receiver is given by its ***bit error rate*** (BER) performance. A basic BER test, applied to a receiver, for example, consists of the comparison of the demodulated binary string out of the system under test to the original data (which were modulated to produce the RF signal). The BER is determined by calculating the ratio of the total number of errors and the total number of bits checked during the comparison. In practice, such a test requires a lot of data to achieve the target accuracy, thus the test time becomes unacceptable. This problem may be overcome by varying the test conditions. Let us assume that a sampled voltage value corresponding to a digital bit follows a Gaussian distribution such that the likelihood of a bit-error, p_e, is given by:

$$p_e = 0.5 \cdot erfc\left(\sqrt{E_b/N_o}\right)$$

where N_o is the noise power spectral density, and E_b is the energy of the received bit, expressed by:

$$E_b = C/f_b$$

where C represents the power of the carrier, and f_b is the data rate.

From these equations, it can be observed that varying the ***signal-to-noise ratio*** (SNR) of the RF signal will influence the BER, in theory. However, attention must be paid to the nonlinear behavior of the receiving system with respect to amplitude variations. In practice, the ***intermodulation distortion*** (IMD) and the crest factor (the peak amplitude of a waveform divided by the root-mean-square value) will influence the BER. In [Halder 2005], the authors have proposed a low-cost solution for production testing of BER for a wireless receiver, based on statistical regression models that map the results of the AC tests to the expected BER value. BER testing is performed under various conditions, which guarantee several parameters such as sensitivity, co-channel, and ***adjacent channel power ratio*** (ACPR).

The transmitter Tx channel is usually tested at the system level by measuring the ***error vector magnitude*** (EVM). The EVM test (see Figure 5.14) is based on a time-domain analysis of a modulated signal represented by an I/Q diagram. Assuming that $v(t)$ represents the transmitted signal at a carrier frequency w_c, then the following relationship may be established:

$$v(t) = I(t)\cos(w_c t) + Q(t)\sin(w_c t)$$

where $I(t)$ and $Q(t)$ are the data signals that can be evaluated using the constellation diagram. For clarity in Figure 5.14, we represent one symbol only. The EVM parameter is expressed as follows:

$$EVM = \sqrt{(I - I_{ref})^2 + (Q - Q_{ref})^2}$$

In a transmission system, the data are collected just at the ADC outputs, where a tradeoff must be considered among the amount of data to be collected and transferred, accuracy, and test time. In [Halder 2006], a new test methodology for wireless transmitters is described in which a multitone stimulus is applied at the base band of the transmitter under test. The method enables multiple parameters to be tested in parallel, reducing the overall test time. An original method is proposed

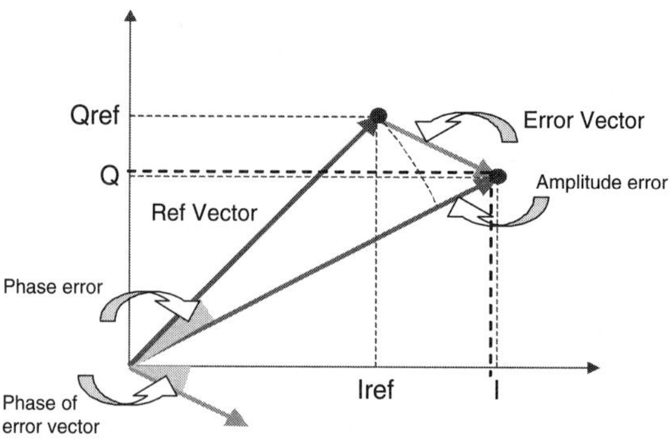

■ FIGURE 5.14

Definition of error vector magnitude.

in [Ozev 2004], where the signals propagated through the analog paths are used to test the digital circuitry. Although this methodology was developed for SOC testing, it can be easily applied to SIP testing. Although it represents an attractive solution for Tx testing, EVM alone cannot detect all defective dies and systems. In [Acar 2006], the authors proposed enhanced EVM measurements in conjunction with a set of simple path measurements (input-output impedances) to provide high fault coverage.

5.3.2 Loopback Techniques: DFT and DSP

Loopback techniques are increasingly proposed in literature. Most of these techniques are combined with alternate test methods to reduce the test time and to be more predictable. Practical techniques may depend on the radio architecture of the system under test: time-division *versus* frequency-division duplex, half *versus* full duplex, shared *versus* separate Tx and Rx local oscillators, same *versus* different Tx and Rx modulation, etc. However, a general approach can be considered without considering the radio architecture. This section describes two generic concepts, external and internal, respectively.

One solution consists of creating the loop between the output of the PA of the transmitter and the input of the LNA of the receiver. Such a configuration is described in [Srinivasan 2006] and [Yoon 2005], where the authors propose an external loopback circuit. As illustrated in Figure 5.15, an attenuator is connected to the PA output, the frequency of the attenuated signal is then divided, and an offset mixer and band-pass filter are subsequently connected to the input of the receiver path. Creating a loop between the Tx and the Rx requires such a loopback circuit to overcome the limitations of the method applied to a TRx (such as in GSM or WLAN), where the shared VCO, the modulation cancellation, and the half duplex architecture affect synchronization between the Tx and Rx signals.

In [Lupea 2003], some blocks and intermediate loopbacks are added inside the RF transceiver at the low-pass filter, ADC, detector, etc. Creating the loop in the front-end IC was also investigated in [Dabrowski, 2003], as shown in Figure 5.16.

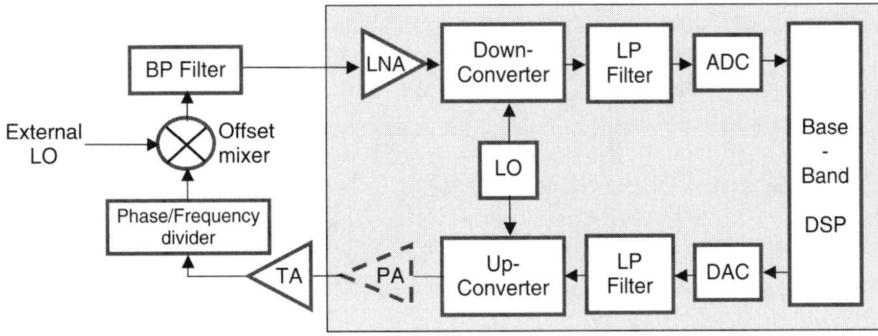

■ **FIGURE 5.15**

External loopback principle.

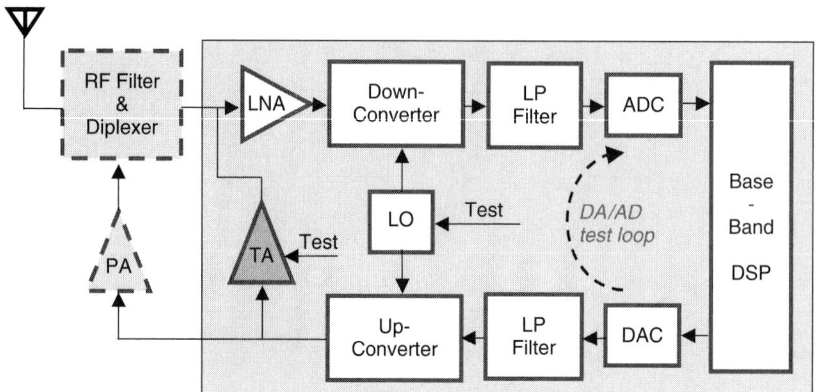

■ **FIGURE 5.16**

Internal loopback using a test attenuator [Dabrowski 2003].

The author proposes connecting the output of the up-converter of the Tx to the input of the LNA of the Rx through a so-called ***test attenuator*** (TA), a complementary BIST sharing its circuitry with on-chip resources. Consequently, the total area overhead for test is negligible. This TA and associated BIST are used so that EVM errors can be detected. These errors are caused by the faults in RF blocks that degrade the gain or the noise figure of the chain.

The faults close to the Rx analog output can be detected by reducing the TA gain or by running a complementary test for gain that is insensitive to fault location. The ***third-order intercept point*** (IP3) test was enhanced using statistical analysis. It should be noted that the path-based and the loopback techniques are still emerging because of limitations from both hardware and processing perspectives, leading to long test times. In [Dabrowski 2004] and [Battacharya 2005], de-embedding techniques for the BER test are proposed to reduce the test time of the system. In [Srinivasan 2006], a computation of the specifications from the measurements is performed in order to eliminate the need for standard specification test. This so-called **alternate diagnosis** approach is based on statistical correlation between the measurements and specifications enabling a drastic reduction of the test time. In Section 5.4.4, we describe another processing method applied to data converters that may also be considered as a preliminary solution to a full system test.

5.4 TEST OF EMBEDDED COMPONENTS

As discussed in Section 5.1, testing bare dies after assembly is a critical phase to achieve an economically viable SIP and to give some diagnostic capabilities. The test consists of two complementary steps:

1. Structural testing of interconnections between dies
2. Structural or functional testing of dies themselves

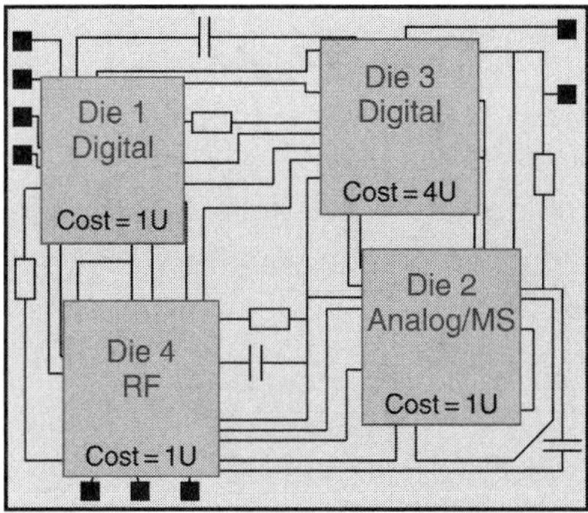

■ **FIGURE 5.17**

Conceptual view of an example of SIP.

The main challenge is accessing the dies from the primary I/O of the SIP. The total number of effective pins of the embedded dies is generally much higher than the number of I/O for the package. Moreover, in contrast to SOC where it is possible to add some DFT between IP to improve controllability and observability, the only available active circuitry for testing in the SIP is that made up of the connected active dies. Consequently, improving testability places requirements on the bare dies used for the SIP and the definition of a specific SIP ***test access port*** (TAP). In the next sections, we illustrate SIP TAP constraints, the test of interconnections, and the test of dies with a didactic example of the SIP shown in Figure 5.17.

In this example, we consider the assembly of an SIP consisting of four active dies soldered onto a passive substrate: the first die is a cheap digital core (*i.e.*, a microcontroller), the second one is a complex mixed-signal die (*i.e.*, a video channel decoder), the third die is an expensive digital block (*i.e.*, a complex decoder, with large embedded memory), and the fourth one is an RF die (*i.e.*, a transceiver). This configuration is quite realistic and allows one to consider various testing issues.

5.4.1 SIP Test Access Port

The SIP imposes some specific constraints on the TAP. This SIP TAP must afford several features, mainly:

- Access for die and interconnection tests
- SIP test enabling at system level as it would be for an SOC
- Additional recursive test procedures during the assembly phase

The IEEE 1500 standard was developed to manage the test of a complex SOC at system level and to give access to and control of each embedded core. Unfortunately, this approach is unsuitable for the SIP, as the standard is based on a specific global ***test access mechanism*** (TAM) embedded into the SOC but outside of the cores. In many SIPs, the only active circuitry is in the dies themselves, which are equivalent to cores in an SOC. Thus, a more viable approach to give the best accessibility to embedded active dies consists of using boundary scan compliant bare dies. In other words, digital active dies should be IEEE 1149.1 compliant, and mixed-signal or analog active dies in the SIP should be IEEE 1149.4 compliant. Figures 5.18 and 5.19 summarize the implications on the internal architecture of each die.

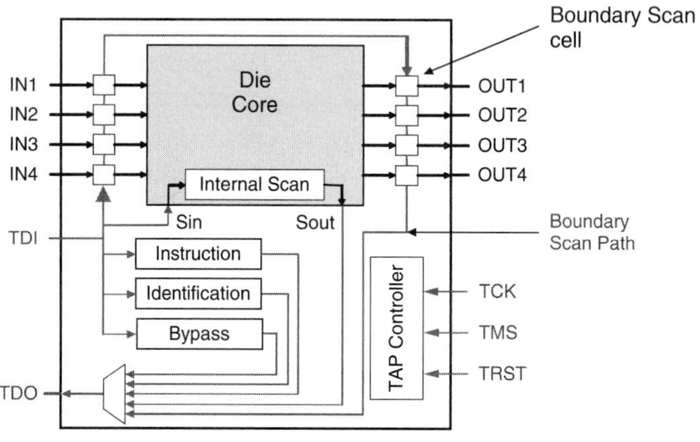

■ **FIGURE 5.18**

Compliant digital die with IEEE 1149.1 [Landrault 2004].

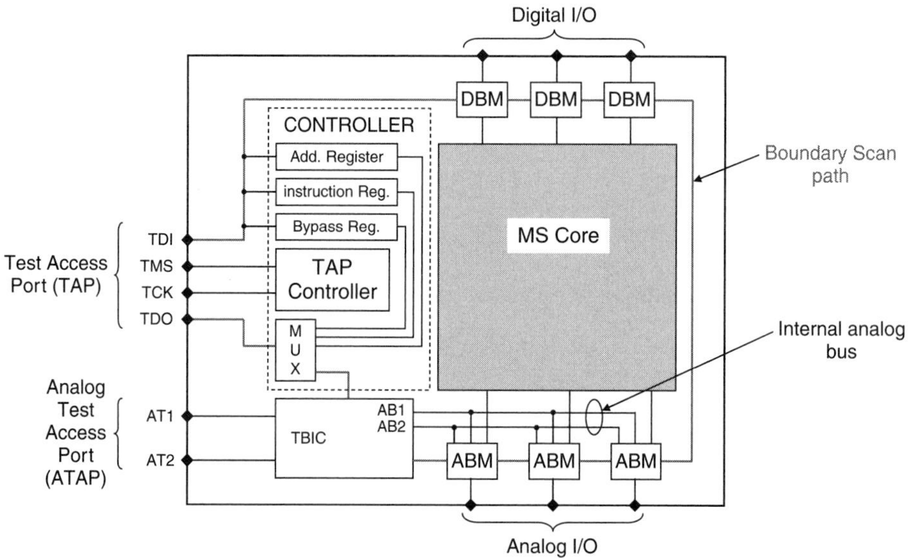

■ **FIGURE 5.19**

Compliant mixed-signal die with IEEE 1149.4 [Landrault 2004].

Around the internal die core, the required circuitry consists of at least two registers (***instruction*** and ***bypass registers***), and some required digital boundary scan cells (also referred to as ***digital boundary-scan modules* [DBMs]**) on the digital I/O and a TAP controller. The extensions for mixed-signal circuits include a ***test bus interface circuit*** (TBIC) and ***analog boundary modules*** (ABM) on the analog pins. Consequently, at the top level, the SIP TAP must be able to manage the local digital and analog TAPs by controlling four or five digital signals and two additional analog signals including the following:

- TCK: the test clock
- TMS: the test mode select signal
- TDI: the test data input pin
- TDO: the test data output
- TRST*: the test reset signal (optional in the IEEE 1149.1 standard)
- AT1 and AT2: the analog signals used in the IEEE 1149.4 standard

In addition to direct accessibility and controllability of dies, these boundary scan signals will control DFT or BIST circuitry embedded on each die and connected to the TAP of each die. For the consumer or integrator to perform system level test, an SIP has to be equivalent to an SOC in final application when the packaged system is soldered on the PCB. This assumption involves two constraints: how to distribute and manage the local TAPs at the die level and determining the identity of the top level TAP (specifically, the SIP ID code). The latter constraint comes from the specificity of the SIP and the significant impact on the assembly phase of active dies on the substrate. Because KGD is sometimes not achievable, and because the assembly process may introduce additional failures, intermediate tests after every die soldering might be required. This strategy combined with a die assembly ordering (from the least to the most expensive dies) allows one to optimize the overall SIP cost. During these incremental tests, the SIP TAP controller must manage boundary scan resources for interconnection and die tests even while some dies are missing.

Taking all the requirements into consideration, the SIP TAP controller must have two configurations: one during the incremental test and the other for the end-user test. Following the ordered assembly strategy, the first die will integrate the SIP TAP controller and, as a result, the ID code of the SIP will be the ID code of this first die. Figure 5.20 shows a conceptual view of the two configurations if the four dies in our example are boundary scan compliant.

The "star" configuration (see Figure 5.20a) attempts to facilitate incremental testing during the assembly. Obviously, the link between the dies (the daisy chain) is broken during intermediate testing because all dies are not yet soldered onto the substrate. This configuration requires as many TMS control signals as there are dies in the SIP. The "ring" configuration (see Figure 5.20b) is designed such that the end-user cannot detect the presence of several dies, either for identification

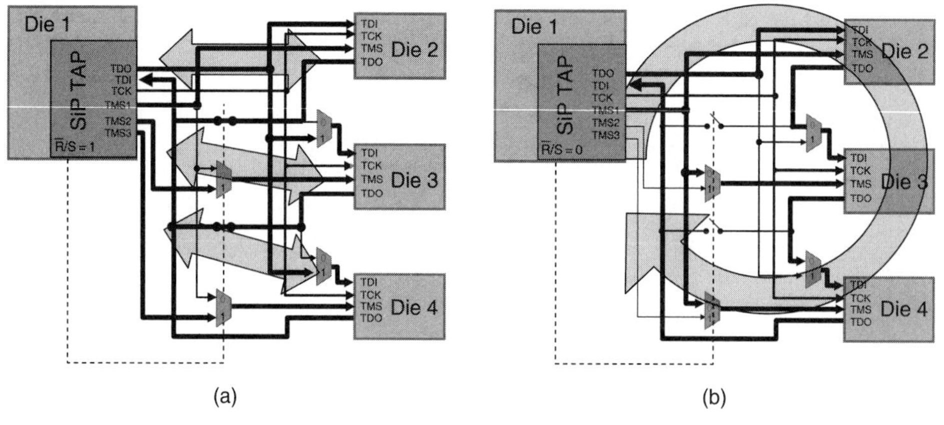

■ **FIGURE 5.20**

Possible TAP configurations: (a) star configuration for the intermediate test and (b) ring configuration for the end-user test.

(there is only one ID code) or for the boundary scan test. Only one TMS control signal is required in this configuration. Thus far, no SIP TAP standard exists, but architectures have been proposed based on the IEEE 1149.1 standard [De Jong 2006] and the IEEE 1500 standard [Appello 2006].

5.4.2 Interconnections

There are two types of interconnections: interconnection between dies and interconnection between die and SIP package pads. The test method for interconnections is equivalent in both cases, but the access issues are obviously different.

The techniques used to test the analog interconnections are based on both structural and functional approaches. The basic principle consists of forcing a specific current (voltage) and measuring its associated voltage (current). In fact, passive components on an SIP are placed on the analog interconnections between analog I/O. Therefore, testing of analog interconnections must also embed impedance measurements.

For testing digital interconnections, the IEEE 1149.1 boundary scan circuitry is used, if available, with classical stuck-at structural tests where the interconnection test is performed through boundary scan in *external test mode* (*EXTEST*). The fundamental principle consists of applying a digital value onto a die output (*i.e.*, at the interconnection start-point) and evaluating the response to this stimulus at the input(s) of the other die(s) (*i.e.*, at the interconnection end-point(s)). To illustrate the test procedure, consider the test of interconnections between die 1 and die 3 in the SIP example as shown in Figure 5.21, where die 4 is not yet soldered on the substrate. Table 5.2 shows an example of a two-vector sequence with the instructions sent to each die.

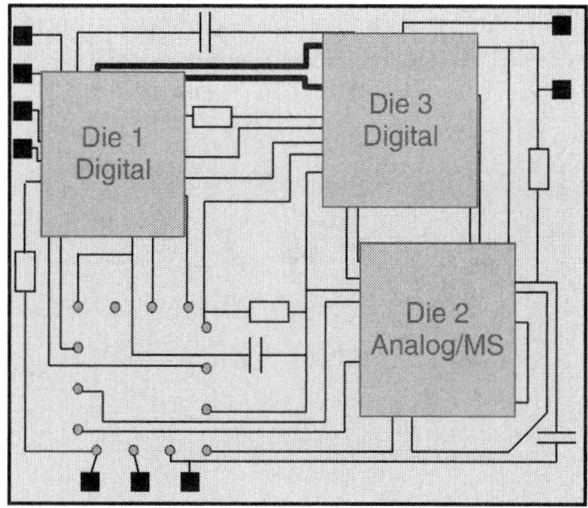

■ **FIGURE 5.21**

Test of interconnections.

TABLE 5.2 ■ Example of Test Procedure

Step	Instruction Die 1	Instruction Die 3	Test Vector
1	*Reset*	*Reset*	
2	*PRELOAD*	*PRELOAD*	
3	*EXTEST*	*EXTEST*	Vector #1
4	*EXTEST*	*EXTEST*	Vector #2
5	*Reset*	*Reset*	

After initialization, the first vector is loaded in the DBMs of die 1 by the *PRELOAD* command on the instruction register. Next, the *EXTEST* instruction is used to apply the preloaded value on the interconnection wires and the value obtained at the other end of the wires (on the DBMs of die 3) are extracted with the subsequent EXTEST instruction. The second test vector is processed in a similar way such that only two vectors are required to test all the possible stuck-at and bridging faults for two adjacent wires. Considering all possible bridging faults affecting k wires, it has been shown that a minimum of $log_2(2k+2)$ vectors are required.

If no boundary scan capability is available on a die, the interconnections to be tested should be directly accessible from packaged pads. Implications of this additional DFT on the signal integrity should be analyzed and minimized as much as possible.

5.4.3 Digital and Memory Dies

Similar to an SOC with internal IP, we face the problem of accessing the inputs and the outputs of internal dies. In fact, an SIP with four times less external pads than internal pins of embedded dies is common. Consequently, we again rely on boundary scan capabilities for testing the internal dies. By activating the bypass function in dies, it is possible to reduce the length of the scan chain. However, "at-speed testing" requires using techniques such as compression, DFT, and BIST. Unfortunately, in many SIPs, no additional active silicon is available and, as a result, no additional circuitry can be implemented such that either BIST or DFT should already exist in the die itself.

If additional DFT is required to test a specific die, this DFT may be integrated on another die. Obviously, there is little chance that this DFT facility will be available on the hardware of one of the other dies because the design of each die is typically independent. Therefore, the only solution is to use the software or programmable capabilities available on the other digital dies to implement a configurable DFT. The assumption of sufficient programmable facilities on the SIP is often realistic with the incorporation of programmable digital dies, such as DSP, microprocessors or microcontrollers, ***field programmable gate arrays*** (FPGAs), etc. Another alternative is to use a transparent mode of the other dies to directly control and observe from the primary I/O of the package as illustrated in Figure 5.22. This concept looks simple, but, in fact, this parallel and at-speed connection through other dies is not necessarily an easy task.

Obviously, we might find an SIP implementation where none of these techniques can be applied. In this case, the only solution to access the specific internal pin is to add direct physical connections to SIP I/O pins while attempting to meet all the associated requirements in terms of signal integrity. In the specific case of a memory die, the access problem is critical because these embedded memories are generally already packaged. These ***package-on-package*** (POP) or ***package-in-package*** (PIP) configurations may have no BIST capabilities, thus the BIST has to be implemented in another digital core for application to the embedded memory.

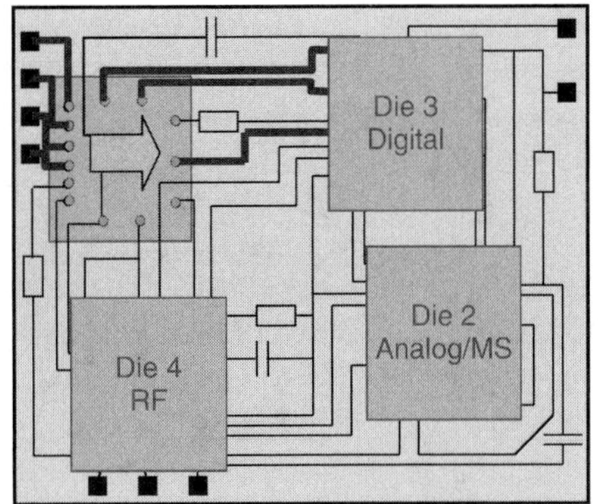

■ FIGURE 5.22

Transparent mode principle.

5.4.4 Analog and RF Components

For the test of analog, mixed-signal, or RF dies, the two most significant challenges are as follows:

1. Cost reduction of the required test equipment
2. Testing of embedded dies because of the difficulty in accessing these dies after SIP assembly

5.4.4.1 Test Equipment Issues

The main advantage of SIP over SOC is the ability to assemble various and heterogeneous dies of different types and technologies into the same package. However, from the point of view of a test engineer, this assembly possibility can be a testing nightmare because the test equipment has to be able to address the whole set of testing requirements in all domains: digital, RF, analog, etc. Using ATE with expensive analog, mixed-signal, and RF options may result in unacceptable test costs. Moreover, analog, mixed-signal, and RF circuits need long functional test procedures, which inevitably impact the test time and cost.

The functional tests are required to achieve a satisfactory test quality and to give diagnostic capabilities at the die level. Even if all the tests previously performed at the wafer level for each die are not necessarily required after assembly, the price of the test equipment and the long test sequences usually make the test cost prohibitive. As a result, specific approaches must be considered to reduce the test time and test equipment cost.

A common approach is to move some or all the tester functions onto the chip itself. Based on this idea, several BIST techniques have been proposed where signals are internally generated or analyzed [Ohletz 1991] [Toner 1993] [Sunter 1997] [Azais 2000, 2001]. However, the generation of pure analog stimuli or accurate analog signal processing to evaluate the system response remains the main roadblock.

Another proposed approach is based on indirect test techniques to achieve a better fault coverage at wafer test and to come closer to the KGD status [Pineda 2003] as described in Section 5.2. The fundamental idea is to replace the difficult direct measurements by easier indirect measurements, provided that a correlation exists between what is measured and what is expected from the direct measurements.

Other techniques consist of transforming the signal to be measured into a signal that is easier to be measured by ATE. For example, timing measurement is easier for ATE than a precise analog level evaluation and a solution is to convert an analog signal on-chip to a proportional timing delay. Another possible solution consists of using DFT techniques to internally transform the analog signals to digital signals that are controllable and observable from the chip I/Os [Ohletz 1991] [Nagi 1994]. As a result, only digital signals are externally handled by less-expensive digital test equipment (a low-cost tester, for example). These techniques are limited by the accuracy of the conversion of the analog signal. A similar approach attempts to avoid the problem of conversion accuracy by assuming several ***digital-to-analog converters*** (DACs), and ADCs are already available to obtain a fully digital test

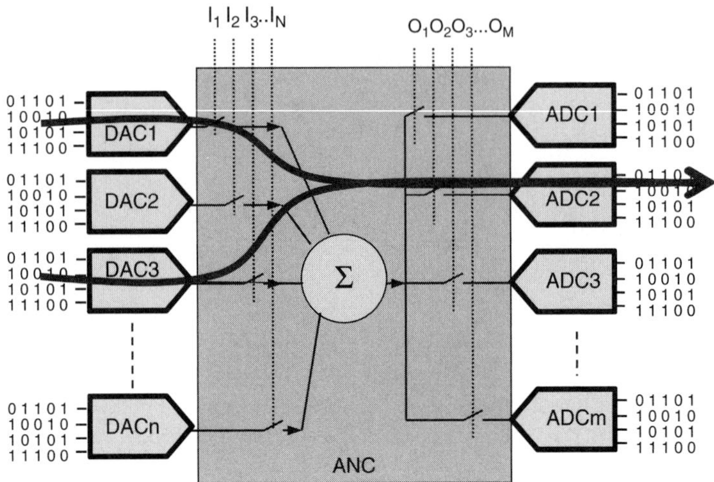

■ **FIGURE 5.23**

DFT principle of the ANC technique.

[Kerzerho 2006] as illustrated in Figure 5.23. Note that this assumption is realistic for the majority of mixed-signal circuits used as active dies in some SIPs. In creating a so-called ***analog network of converters*** (ANC) by adding some analog DFT on the analog side of the converters, new paths can be used to test the DACs and ADCs as well as the analog blocks with fully digital inputs and outputs. The main difficulty is to discriminate the influence of each element (each converter and analog block in a same path) on the digital signature. The proposed solution takes advantage of multiconfiguration interconnections between the converters.

5.4.4.2 Test of Analog, Mixed-Signal, and RF Dies

Access is also critical for testing analog, mixed-signal, and RF dies. The two types of signals to be controlled or observed include analog signals as well as digital and clock signals for mixed-signal dies. It is common in SIP implementations that some of the analog signals are not connected to the external pins of the SIP, and, as a result, there is no direct access to these signals. Fortunately, for static or low-frequency signals, access to internal analog nodes is possible using the IEEE 1149.4 standard where two pads AT1 and AT2 are used to transmit stimuli and receive responses. According to the standard, the maximum frequency must be lower than 10 kHz and the resolution less than or equal to 16 bits. If the IEEE 1149.4 circuitry is not available on the die, or if the required signal frequency or the resolution is too high, a possible solution is to add some internal access nodes in the SIP. This can introduce disturbances (load modifications, parasitic effects, etc.) in critical signal paths and decrease the system performance. For critical signals, the only viable solution is to have BIST or DFT integrated into the die to preserve the signal integrity.

The digital signals for mixed-signal dies, which are generally used to control the analog portion of the die, seldom have direct access from SIP pins. However, these digital signals are often controllable through another fully digital die. For characterization of mixed-signal die, these digital control signals must be operated at-speed, so that the IEEE 1149.1 standard does not fulfill the specifications. Including a transparent mode in digital dies represents one unique solution. As illustrated in Figure 5.22 for digital die testing, this mode allows a direct transmission of digital signals from the primary inputs of the SIP to mixed-signal digital pins of the embedded die. The concept of transparent mode may seem simple, but routing digital signals simultaneously, while preserving an efficient control, is more complex than in a stand-alone configuration of the mixed-signal die. For instance, the delay introduced by the bypass configuration of the digital die is not accurately known and controllable. Furthermore, the required synchronization between digital signals, clocks, and analog signals is a challenge. The control of clock signals is actually a problem by itself because of jitter effect. Another problem is how to choose, a priori, the best candidates for transparent mode if we do not know which pins are likely to be available on the SIP package.

RF dies would seem to represent the worst-case scenario because of the high-frequencies and low analog levels. But, actually, RF signals are rarely only internal to the SIP, which makes direct access likely in most applications.

5.4.5 MEMS

Microelectromechanical systems (MEMS) correspond to the extreme cases of heterogeneous systems. A typical MEMS device can be an accelerometer, pressure sensor, temperature or humidity sensor, microfluidic system, or bioMEMS device, among others [Bao 2000]. The first problem for MEMS testing begins with the required test equipment. MEMS devices are generally dedicated to generate or monitor nonelectrical signals. Consequently, test equipment should allow generation and measurement using sound, light, pressure, motion, or even fluidics. Because of their price, the difficulty to implement them, and the long associated test time, the use this type of equipment for production test (especially at the wafer level) is rarely an option [Mir 2004]. In a production test environment, only fully electrical signals are actually viable. In this context, two approaches are likely:

1. Perform an indirect structural or functional test on an electrical signal that would be an image of the physical signal associated with the MEMS under test.
2. Implement some DFT circuitry allowing one to convert the physical signal associated with the MEMS to an electrical signal. This approach is used on the most famous MEMS device, the accelerometer from Analog Devices [Analog 2007], with a network of capacitors to convert the electrical stimulus to a force equivalent to a 10 g acceleration [Allen 1989].

Another major challenge in MEMS testing is due to the significant package influence. MEMS characteristics depend on the properties and the quality of the package used. For example, the frequency of a mechanical resonator is directly linked to the humidity (moisture) inside the package. Therefore, we have a paradox with expensive MEMS packaging—because the package is required, it is difficult to detect defective MEMS before packaging. As the result, the cost of MEMS testing can be prohibitive because of the price of rejected packaged devices [Charlot 2001].

For MEMS integration into an SIP, the classical problems of MEMS testing are exacerbated. Indeed, the presence of additional active dies close to the MEMS might disturb and modify the MEMS quality observed at the system level. Moreover, because several MEMS can be mounted in the same SIP, the test needs to manage both stimulus generation and response analysis including various types of nonelectrical signals. As a result, the alternative techniques with only electrical signals are the only viable options.

From the package standpoint, the SIP concept poses new challenges. For monolithic MEMS in CMOS technology, direct integration of the bare MEMS onto the passive substrate is conceivable. For more complex MEMS, the bare die can be flipped onto the passive substrate. The new challenges then involve achieving a perfect etching and sealing of the cavity and guaranteeing the cavity quality during the life of the system. In this context, one solution is to add additional simple MEMS into the cavity to monitor the cavity characteristics as illustrated in Figure 5.24.

Considering access to MEMS in the SIP, for both smart MEMS composed of significant digital processing and for simple analog sensor, the problem is equivalent to digital and mixed-signal die. As a result, the solutions are thus similar to those described earlier according to the nature of the electric signal to be accessed.

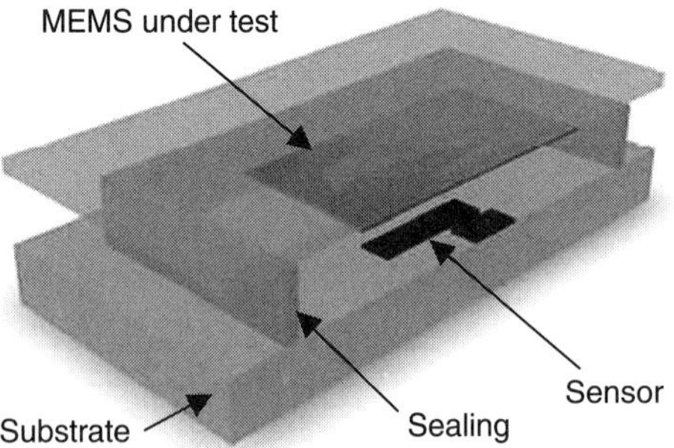

■ FIGURE 5.24

Cavity monitoring via additional sensor.

5.5 CONCLUDING REMARKS

A ***system-in-package*** (SIP) is a packaged device composed of two or more embedded bare dies and, in most cases, passive components. The SIP technology has found many applications, in particular, in the customer electronics industry, such as cellular handsets. An SIP provides a system or subsystem solution in a single package, using various types of carriers and interconnect technologies.

Testing these complex SIP devices first requires extensive testing of the bare dies to reach the desired ***known-good-die*** (KGD) quality. If the KGD quality cannot be guaranteed, then each embedded component (bare die) must be tested within the SIP. After packaging, the assembled SIP must be tested from a functional point of view.

In this chapter, the problems of testing bare dies were described and analyzed. We illustrated the need for implementing IEEE 1149.1 boundary scan in the assembled dies to test die-to-die and pad-to-die interconnects. In addition, we discussed test limitations when a bare die is an analog, mixed-signal, RF, or MEMS component. The mechanical and electrical limitations of probing technologies were also described, along with a number of means to improve reliability. For example, we described a path-based technique using a *transmitter* (Tx) and *receiver* (Rx) pair to perform functional system testing, where tests are conducted to measure the responses of the SIP against its functional specifications using ***bit error rate*** (BER) and ***error vector magnitude*** (EVM). To reduce test cost, a loopback technique was then described that creates a loop between the Tx and Rx paths during testing. Both techniques make use of the bidirectional architecture of communication-oriented circuits.

In its current application fields of predilection, the SIP is moving toward ever more sophisticated packaging technologies, which will require new test solutions. The trend toward more functionality combined with more communication features for emergent applications, such as health care, smart lighting, or ambient computing, drives the integration of a large variety of sensors and actuators. Consequently, heterogeneous SIP implementations will be developed, posing many new test challenges.

5.6 EXERCISES

5.1 **(SOC *versus* SIP)** Explain the major differences and test challenges between a *system-on-chip* (SOC) and a *system-in-package* (SIP).

5.2 **(Yield and Profit Using an SIP)** Assume that a *system-in-package* (SIP) contains four active dies as follows:
Die 1 is a digital core with a price of 50 cents and an estimated defect level at 1000 PPM.
Die 2 is complex mixed-signal die; the price of this die is $1, and the defect level is estimated at 100 PPM.

Die 3 is a RF die; its price is only $1, but the defect level is 1000 PPM because of the inefficiency of wafer testing.
Die 4 is a $20 processor with a low defect level estimated at 10 PPM.
The price of package and passive substrate is $5.

a. Estimate the final yield of this SIP if no additional failure appears during the assembly process.

b. Estimate the direct profit of using incremental testing during the assembly phase if the cost of handling (ATE, time, etc.) is omitted.

5.3 **(Known-Good-Die)** List the three most significant limiting factors that must be addressed in order to achieve a *known-good-die* (KGD) quality level.

5.4 **(Functional System Testing)** Explain the differences between path-based test and loopback test techniques.

5.5 **(Functional System Testing)** Give the two main approaches to solve the test equipment issues.

5.6 **(SIP-TAP)** List and justify three major features when a *test access port* (TAP) is used on the SIP (SIP-TAP).

5.7 **(MEMS Testing)** List the most critical factors for MEMS testing in the context of SIP.

Acknowledgments

The authors wish to thank Peter O'Neill of Avago Technologies, Herbert Eichinger of Infineon Technologies, Dr. Florence Azais of LIRMM, Professor Sule Ozev of Duke University, and Professor Charles E. Stroud of Auburn University for reviewing the text and providing valuable comments. They also would like to thank Dr. Christian Landrault of LIRMM and Frans de Jong of NXP Semiconductors for their invaluable technical and editorial advice.

References

R5.0 Books

[Bao 2000] M.-H. Bao, Volume 8: Micro mechanical transducers, pressure sensors, accelerometers and gyroscopes, pp. 362–365, in *Handbook of Sensors and Actuators,* S. Middelhoek, editor, Morgan Kaufmann, San Francisco, CA, 2000.

[Bushnell 2000] M. L. Bushnell and V. D. Agrawal, *Essentials of Electronic Testing for Digital, Memory & Mixed-Signal VLSI Circuits*, Springer, New York, 2000.

[Landrault 2004] C. Landrault, F. Azais, S. Bernard, Y. Bertrand, M.L. Flottes, P. Girard, L. Latorre, S. Pravossoudovitch, M. Renovell, and B. Rouzeyre, *Test de Circuits et de Systèmes Intégrés*, Hermes Science, Paris, 2004.

R5.1 Introduction

[DPC 2007] Die Product Consortium, www.dieproducts.org.

[SIA 2005] SIA, *The International Technology Roadmap for Semiconductors: 2005 Edition—Assembly & Packaging,* Semiconductor Industry Association, San Jose, CA (www.itrs.net/Links/2005ITRS/AP2005.pdf), 2005.

R5.2 Bare Die Test

[Akbay 2004] S. S. Akbay and A. Chatterjee, Feature extraction based built-in alternate test of RF components using a noise reference, in *Proc. IEEE VLSI Test Symp.*, pp. 273–278, April 2004.

[Arnold 1998] R. Arnold, Test methods used to produce highly reliable known good die (KGD), in *Proc. IEEE Int. Conf. on Multichip Modules and High Density Packaging,* pp. 374–382, April 1998.

[Barrette 1996] T. Barrette. V. Bhide, K. De, M. Stover, and E. Sugasawara, Evaluation of early failure screening methods, in *Proc. IEEE Int. Workshop on IDDQ Testing*, pp. 14–17, October 1996.

[Cooke 2005] M. D. Cooke and D. Wood, development of a simple microsystems membrane probe card, in *Proc. IEEE Symp. on Design, Test, Integration and Packaging of MEMS and MOEMS*, pp. 399–404, June 2005.

[Halder 2001] A. Halder and A. Chatterjee, Specification based digital compatible built-in test of embedded analog circuits, in *Proc. IEEE Asian Test Symp.*, pp. 344–349, November 2001.

[Leung 1995] J. Leung, M. Zargari, B. A. Wooley, and S. S. Wong, Active substrate membrane probe card, in *Proc. Int. Electron Devices Meeting,* pp. 709–712, December 1995.

[Mann 2004] W. R. Mann, F. L. Taber, P. W. Seitzer, and J. J. Broz, The leading edge of production wafer probe test technology, in *Proc. IEEE Int. Test Conf.*, pp. 1168–1195, October 2004.

[Pineda 2003] J. Pineda de Gyvez, G. Gronthoud, and R. Amine, Vdd ramp testing for RF circuits, in *Proc. IEEE Int. Test Conf.*, pp. 651–658, September 2003.

[Sellathamby 2005] C. Sellathamby, M. Reja, L. Fu, B. Bai, E. Reid, S. Slupsky, I. Filanovsky, and K. Iniewski, Noncontact wafer probe using wireless probe cards, in *Proc. IEEE Int. Test Conf.*, Paper 18.3, pp. 1–6, November 2005.

[SIA 2004] SIA, *The International Technology Roadmap for Semiconductors: 2004 Update*, Semiconductor Industry Association, San Jose, CA (http://public.itrs.net), 2004.

[Leslie 1989] B. Leslie and F. Matta, Wafer-level testing with a membrane probe, *IEEE Design & Test of Computers*, 6(1), pp. 10–17, February 1989.

[Kawahara 1996] R. Kawahara. O. Nakayama, and T. Kurasawa, The effectiveness of IDDQ and high voltage stress for burn-in elimination, in *Proc. IEEE Int. Workshop on IDDQ Testing*, pp. 9–13, October 1996.

[Singh 1997] A. D. Singh, P. Nigh, and C. M. Krishna, Screening for known good die (KGD) based on defect clustering: An experimental study, in *Proc. IEEE Int. Test Conf*, pp. 362–369, November 1997.

[Smolders 2004] A. Smolders, N. Pulsford, P. Philippe, and V. Straten, RF, SIP: The next wave for wireless system integration, *Digest of Papers, IEEE Radio Frequency Integrated Circuits Symp.*, pp. 233–236, June 2004.

[Wartenberg 2006] S. Wartenberg, Six-gigahertz equivalent circuit model of an RF membrane probe card, *IEEE Trans. on Instrumentation and Measurement,* 55(3), pp. 989–994, June 2006.

R5.3 Functional System Test

[Acar 2006], E. Acar, S. Ozev, and K.B. Redmond, Enhanced error vector magnitude (EVM) measurements for testing WLAN transceivers, *IEEE Int. Conf. on Computer-Aided Design*, pp. 210–216, November 2006.

[Bhattacharya 2005] S. Bhattacharya, R. Senguttuvan, and A. Chatterjee, Production test technique for measuring BER of ultra-wideband (UWB) devices, *IEEE Transactions on Microwave Theory and Techniques*, 53(11), pp. 3774–3481, November 2005.

[Dabrowski 2003] J. Dabrowski, BiST model for IC RF-transceiver front-end, in *Proc. IEEE Int. Symp. on Defect and Fault Tolerance in VLSI Systems*, pp. 295–302, November 2003.

[Dabrowski 2004] J. Dabrowski, and J.G. Bayon, Mixed loopback BIST for RF digital transceivers, *IEEE Int. Symp. on Defect and Fault Tolerance in VLSI Systems*, pp. 220–228, October 2004.

[Halder 2005] A. Halder and A. Chatterjee, Low-cost production test of BER for wireless receivers, in *Proc. IEEE Asian Test Symp.*, pp. 64–69, December 2005.

[Halder 2006] A. Halder and A. Chatterjee, Low-cost production testing of wireless transmitters, in *Proc. Int. Conf. on VLSI Design*, 6, January 2006.

[Lupea 2003] D. Lupea, U. Pursche, and H-J. Jentschel, RF BIST: Loopback spectral signature analysis, in *Proc. IEEE Design, Automation, and Test in Europe*, pp. 478–483, February 2003.

[Ozev 2004] S. Ozev, A. Orailoglu, and I. Bayraktaroglu, Seamless test of digital components in mixed-signal paths, *IEEE Design & Test of Computers*, 21(1), pp. 44–55, January/February 2004.

[Srinivasan 2006] G. Srinivasan, A. Chatterjee, and F. Taenzler, Alternate loop-back diagnostic tests for wafer-level diagnosis of modern wireless transceivers using spectral signatures, in *Proc. IEEE VLSI Test Symp.*, pp. 222–227, April 2006.

[Yoon 2005] J. S. Yoon, and W. R. Eisenstadt, Embedded loopback test for RF, *IEEE Trans. on Instrumentation and Measurement*, 54(5), pp. 1715–1720, October 2005.

R5.4 Test of Embedded Components

[Allen 1989] H. V. Allen, S. C. Terry, and D. W. DeBruin, Accelerometer systems with self-testable features, in *Proc. IEEE Sensors and Actuators*, pp. 153–161, February 1989.

[Analog 2007] Analog Devices, (www.analog.com), 2007.

[Appello 2006] D. Appello, P. Bernardi, M. Grosso, and M. S. Reorda, System-in-package testing: Problems and solutions, *IEEE Design & Test of Computers*, 23(3), pp. 203–211, May/June 2006.

[Azais 2000] F. Azais, S. Bernard, Y. Bertrand, and M. Renovell, Towards an ADC BIST scheme using the histogram test technique, in *Proc. IEEE European Test Workshop*, pp. 53–58, May 2000.

[Azais 2001] F. Azais, S. Bernard, Y. Bertrand, and M. Renovell, Implementation of a linear histogram BIST for ADCs, in *Proc. IEEE Design, Automation and Test in Europe*, pp. 590–595, March 2001.

[Charlot 2001] B. Charlot, S. Mir, F. Parrain, and B. Courtois, Electrically induced stimuli for MEMS self-test, in *Proc. IEEE VLSI Test Symp.*, pp. 60–66, April/May 2001.

[De Jong 2006] F. de Jong and A. Biewenga, SIP-TAP: JTAG for SIP, in *Proc. IEEE Int. Test Conf.*, pp. 389–395, October 2006.

[Kerzerho 2006] V. Kerzerho, P. Cauvet, S. Bernard, F. Azais, M. Comte, and M. Renovell, Analog network of converters: A DFT technique to test a complete set of ADCs and DACs embedded in a complex SIP or SoC, in *Proc. IEEE European Test Symp.*, pp. 159–164, May 2006.

[Mir 2004] S. Mir, L. Rufer, and B. Courtois, On-chip testing of embedded silicon transducers, in *Proc. IEEE Int. Conf. on VLSI Design*, pp. 463–472, October 2004.

[Nagi 1994] N. Nagi, A. Chatterjee, and J. Abraham, A signature analyzer for analog and mixed-signal circuits, in *Proc. IEEE Int. Conf. on Computer Design*, pp. 284–287, October 1994.

[Ohletz 1991] M. J. Ohletz, Hybrid built-in self-test (HBIST) for mixed analog/digital integrated circuits, in *Proc. IEEE European Test Conf.*, pp. 307–316, May 1991.

[Pineda 2003] J. Pineda de Gyvez, G. Gronthoud, and R. Amine, Vdd ramp testing for RF circuits, in *Proc. IEEE Int. Test Conf.*, pp. 651–658, September 2003.

[Sunter 1997] S. Sunter and N. Nagi, A simplified polynomial-fitting algorithm for DAC and ADC BIST, in *Proc. IEEE Int. Test Conf.*, pp. 389–395, November 1997.

[Toner 1993] M. Toner and G. Roberts, A BIST scheme for an SNR test of a sigma-delta ADC, in *Proc. IEEE Int. Test Conf.*, pp. 805–814, October 1993.

CHAPTER 6

DELAY TESTING

Duncan M. (Hank) Walker
Texas A&M University, College Station, Texas

Michael S. Hsiao
Virginia Tech, Blacksburg, Virginia

ABOUT THIS CHAPTER

Delay testing is used to verify that a circuit meets its timing specifications. This chapter introduces delay testing of digital logic in synchronous circuits and then focuses on delay tests that target the circuit structure. Approaches for applying delay tests in scan designs are first discussed, including issues in clocking. Delay fault models are then described along with their sensitization criteria. The chapter includes a discussion of recent research in defect-based delay fault models.

The second half of the chapter discusses delay fault simulation and automatic test pattern generation. Delay fault simulation is used to grade the coverage of functional test patterns, as well as to drop faults that are fortuitously detected while targeting other faults.

Through this chapter, the reader will learn about the major delay fault modeling, simulation, and test generation techniques and their application to modern digital circuits. This background will be valuable for selecting the delay test methodology that best meets the design needs.

6.1 INTRODUCTION

Timing is one of the specifications that must be verified during digital circuit testing. This type of test is referred to as a **delay test**. Sometimes delay testing is referred to as **AC testing** to distinguish it from the DC test conditions of stuck-at faults. Delay testing is becoming increasingly important, as timing margins are reduced to maximize performance while minimizing power dissipation.

There are three basic approaches to delay testing: random, functional, and structural. Random patterns applied at the rated clock speed will detect most delay defects that cause circuit failure. The advantages of random pattern testing are that test generation is not required; the patterns can be generated on-chip with ***built-in self-test*** (BIST) hardware (as discussed in Chapter 2); the patterns are applied one after the other, as in normal operation; and the same hardware targets all types of defects, not just delay defects. The disadvantages of random pattern testing are that it provides poor coverage of the longest circuit paths and may produce circuit activity much higher than normal. This higher activity generates higher-than-normal chip temperatures and power supply noise, causing the circuit to operate slower than normal. This may cause chips to be incorrectly rejected.

Functional tests applied at the rated clock speed provide the most accurate delay testing, because the chip is tested in the same way it is used in normal operation. Instruction set processors such as microprocessors, digital signal processors, and microcontrollers can be tested at low cost by loading instruction sequences into on-chip cache or memory and executing them. The primary disadvantages of functional testing are that it is difficult to write high-coverage test sequences, and designs without large, on-chip instruction memories require an expensive, full-featured tester to apply the functional patterns.

Structural testing uses knowledge of the circuit structure and the corresponding delay fault models to generate delay test patterns. The advantages of structural testing are that the test pattern generation process is automated, and it can achieve high fault coverage. It also makes it considerably easier to diagnose delay faults. The main disadvantages are that simplifying assumptions must be made in the delay fault model to make the problem tractable, the design must include ***design for testability*** (DFT) features to achieve high coverage, and the delay test application is very different from normal operation. A structurally testable path may not be functionally feasible. Therefore, care must be taken to avoid over-testing the false long paths.

This chapter focuses on structural delay testing of synchronous digital circuits. We consider circuits represented by gate-level primitives, including storage elements, such as flip-flops and latches. The circuit may also contain **black boxes**—that is, undefined modules. These most commonly represent embedded memory arrays or analog blocks.

The Huffman model of a synchronous sequential circuit is shown in Figure 6.1. The circuit consists of combinational logic and flip-flops *synchronized* by a common clock. The inputs to the combinational logic consist of ***primary inputs*** (PI),$x_1, x_2, \ldots, x_n$, and the flip-flop outputs $y_1, y_2, \ldots, y_l$, (also called ***pseudo primary inputs*** [PPI]). Its outputs consist of ***primary outputs*** (PO),$z_1, z_2, \ldots, z_m$, and the flip-flop inputs $Y_1, Y_2, \ldots, Y_l$ (also called the ***pseudo primary outputs*** [PPO]).

Some delay fault models require knowledge of circuit delays in order to test them. Most delay test approaches use no more than the rising and falling delays from each gate input to output, with the interconnect delay lumped into the gate delay. This can be split into the **gate transport delay** and **interconnect propagation delay** to compute the delay to each fanout separately. To analyze glitch behavior during testing, the **inertial delay** must also be considered. The inertial delay defines the minimum pulse width that propagates through a gate. Process variation can be

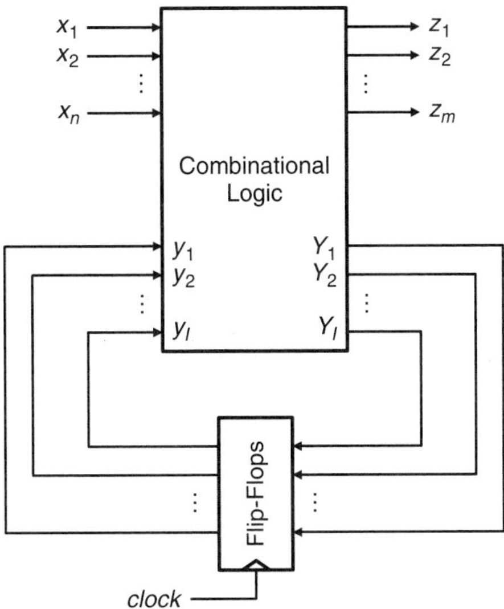

■ **FIGURE 6.1**

Huffman model of sequential digital circuits.

modeled using the **min-max delay** model, but the correlation between gate delays is usually not available.

Delay testing requires that transitions be **launched** into the circuit from the PIs and PPIs, and the outputs **captured** at the POs and PPOs at the specified time. The outputs are compared to the correct results to detect any delay faults. Launching transitions into the circuit requires a **two-pattern test**. The first vector initializes the circuit and is termed the **initialization vector**. The second vector launches the transitions and is termed the **test vector**.

6.2 DELAY TEST APPLICATION

In this section, we discuss how delay tests are applied to synchronous sequential circuits. As with testing for stuck-at faults, automatic generation of delay tests for large sequential circuits is impractical. DFT circuitry must be inserted to enhance the controllability and observability of the circuit. As discussed in Chapter 2, this is primarily done by connecting most latches and flip-flops together into scan chains. These scan chains provide direct access to the PPIs and PPOs. They are used to initialize the circuit for the delay test and to read out the results. The difference from stuck-at testing is that delay testing requires a two-pattern test. This presents a problem, because in normal operation, the flip-flops or latches only store one value. There are two common approaches to applying the two patterns: enhanced scan and muxed-D scan.

6.2.1 Enhanced Scan

In **enhanced scan**, each storage element contains the corresponding bits of both the initialization and test vector. This can be implemented in several ways. In a flip-flop design, the flip-flops can be connected in a scan chain, and their outputs connected to holding latches. After the initialization vector has been scanned into the flip-flops, the vector is transferred to the holding latches, initializing the circuit under test. The test vector is then scanned in and transferred to the holding latches to apply the test to the circuit. This approach has the advantage that the initialization and test vectors are independent of one another. A second advantage is that the circuit PPIs are fixed while the test vector is scanned in, drastically reducing test power. Test power is discussed in Chapter 7. Because the initialization and test vectors are independent, a third advantage of enhanced scan is increased delay fault coverage. The disadvantage of enhanced scan is the extra area, power, and delay associated with the holding latch. The area and delay overhead of the holding latch preclude the use of enhanced scan in most designs.

6.2.2 Muxed-D Scan

Most designs implement scan by using **muxed-D flip-flops**, as discussed in Chapter 2. That is, the D (data) input of a D-type flip-flop is fed by a 2:1 multiplexer, with one multiplexer input fed by the circuit logic (the PPO) and the other fed by the previous flip-flop in the scan chain. The **scan enable** (SE) signal controls the multiplexer. The advantage of the muxed-D scan design is that the system clock is used to clock the flip-flops in both normal and test modes, and the design has low area and delay overhead. The disadvantage of the muxed-D design approach is that PPIs change with each shift of the scan chain, resulting in high power dissipation during scan-in or scan-out. This is discussed further in Chapter 7.

6.2.3 Scan Clocking

Delay test pattern application in scan designs uses a **variable clock** approach. The test pattern is slowly shifted into the scan chains, fast system clock cycles are used to apply the test, and then slow shift cycles are used to shift out the result (and shift in the next test pattern). This approach makes each test pattern independent. The slow shift speed avoids critical timing along the serial data path of the scan chain and reduces the power dissipated during scan shifting.

For enhanced scan, the initialization vector is scanned in and then loaded into the hold latches. The test vector is then scanned in and launched when the hold latches are loaded, and then it is captured with the system clock. The disadvantage of this approach is that the hold latch enable must be able to operate at system speed, so that the latch enable to system clock timing is accurate. In contrast, scan testing using muxed-D flip-flops uses the system clock to launch and capture patterns, so there are no timing correlation issues related to the clock tree.

The muxed-D scan design has only one bit stored in each PPI. Because the flip-flop input is only connected to the PPOs of the circuit under test and the neighboring scan cell, this constrains the relationship between the initialization and test vectors. The initialization vector is scanned in, so it does not have any constraints. The test vector, however, can only be derived in two ways. The first is to have the test vector be a 1-bit shift from the initialization vector. This is generated by shifting the scan chain by 1 bit. This 1-bit shift to launch the transition is known as ***launch-on-shift*** (LOS), launch-off-shift, or skewed load [Patel 1992] [Savir 1992]. The procedure for applying a LOS test is as follows:

1. The circuit is set to scan mode. The initialization vector is scanned in, and values are set on PIs.
2. The test vector is obtained by shifting the scan chain by 1 bit (applying one system clock cycle). Usually the PIs do not change values because of the timing constraints of low-cost testers.
3. The circuit is set to normal operation by flipping the scan enable and pulsing the system clock at the rated speed to capture the circuit PPO values in the flip-flops. The values on POs are captured if necessary.
4. The circuit is set to scan mode and the results are shifted out. This step is overlapped with step 1.

The advantage of LOS is that a combinational ATPG can enforce the pattern constraint with only minor modifications; fast test generation methodologies for combinational circuits can be applied without many modifications; and the constraint can be met for most faults, minimizing test patterns and test generation time. The primary disadvantage of LOS is that in a muxed-D scan implementation, after the shift has occurred to launch the transition, the scan enable must be flipped at the rated system clock speed, so that the PPOs can capture the test result on the next system clock cycle. This requires designing the scan enable to have the same timing performance as the system clock. This requires too much power, area, and design effort for most designs. An alternative is to have separate scan enables for the launch and capture flip-flops, so that the launch SE can remain in scan mode while the capture SE is already in capture mode. Multiple scan enables are sometimes used to reduce test power dissipation (see Chapter 7). This places constraints on the scan design, because the inputs and outputs of a combinational logic block must be controlled by separate scan enables. ***Level sensitive scan design*** (LSSD) based scan can use LOS because it uses separate clocks for the L1 and L2 latches [Savir 1992]. Another disadvantage of LOS is that some false long paths may be exercised. Because the test vector is the shifted version of the initialization vector, such a state transition may not be possible in the functional mode, thereby allowing nonfunctional behavior.

If LOS cannot be used, the alternative is to derive the test vector from the PPOs feeding the flip-flops. The PPO values in turn are generated by the initialization vector. This is known as ***launch-on-capture*** (LOC), launch-off-capture, functional

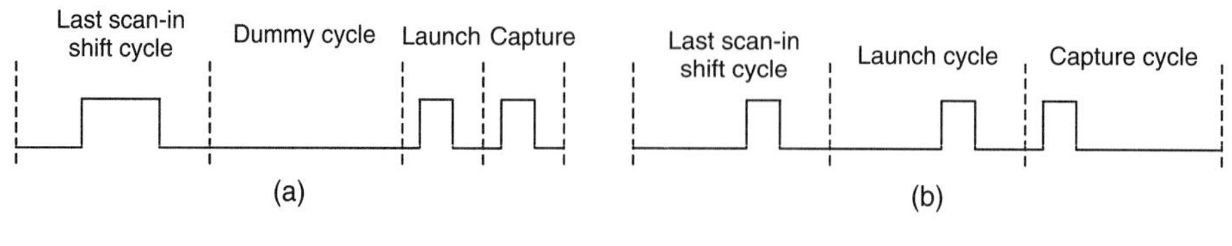

■ **FIGURE 6.2**

Launch-on-capture clock waveforms.

justification, **double-capture** [Wang 2006a], or **broad-side test** [Savir 1994]. The procedure for applying a LOC test is as follows:

1. Same as LOS step 1.
2. The scan enable is set to normal operation. Dummy cycles are inserted as needed to give the SE time to flip. Figure 6.2a shows the clock waveform with one dummy cycle. Figure 6.2b shows the clock waveform if no dummy cycle is needed.
3. The system clock is pulsed twice at the rated clock speed. At the first clock pulse, the test vector is derived from the initialization vector. At the second clock pulse, the results are captured. The values on POs are captured if necessary.
4. Same as LOS step 4.

The advantages of the LOC approach are that the SE has no timing constraint, and the test vector is a legal state of the circuit, assuming the initialization vector is a legal state. This reduces the testing of **false paths**, which can lead to yield loss. This is discussed further in Section 6.7. The primary disadvantage of LOC is the need to justify the test vector back one time frame to determine the initialization vector, increasing test generation time, and pattern count. This constraint also makes some faults untestable. This was listed earlier as an advantage, but a sequentially redundant fault may present a reliability hazard. If such redundant faults occur in online error-checking hardware, they must be tested to verify the correct function of this hardware.

6.2.4 Faster-Than-At-Speed Testing

The preceding discussion assumes that the timing between the launch and capture of a test pattern uses the rated system clock period. The drawback is that some test generation methods, particularly for transition faults, typically use a breadth-first search algorithm that tends to select short paths through each fault site. As a result, when tested at rated speed, many paths have considerable timing slack, and so only relatively large delay defects can be detected. One solution is to generate one or more longest paths through each fault site [Majhi 2000] [Sharma 2002] [Qiu 2003a], as discussed in Section 6.6. However, the increased path length increases test data volume [Qiu 2004a]. Rather than maximizing the length of the tested paths, the alternative is to shrink the capture clock timing to minimize the slack for each pattern [Mao 1990]. However, separate timing for each pattern drastically increases test data volume. An alternative is to group patterns into sets of almost equal-length paths [Kruseman 2004]. The test engineer must trade off between the number of groups and test data volume. Because the chip is being tested at faster than its rated speed, logic transitions may still be occurring when the capture clock is applied. Those flip-flops fed by paths that exceed the cycle time or containing hazards must be masked off. The faults that are not detected because of masking are targeted by patterns run at the next lower clock speed [Barnhart 2004]. Applying transition fault patterns at faster than the rated speed has been shown to catch small delay defects that escape traditional transition fault tests [Kruseman 2004] [Amodeo 2005].

Faster-than-at-speed testing can be applied whenever test patterns propagate on paths with timing slack. There are two primary concerns when using this test approach. The first is the accuracy of the timing models. If paths are slower than predicted, a faster-than-at-speed test can reject chips that pass at the rated speed. Timing margin must be added statically or dynamically, reducing the effectiveness at screening small delay defects. The second problem is the additional power supply noise introduced by operating at faster than rated speed. The power grid has less time to recover from the launch, so the circuit may operate more slowly than it would if clocked at the rated speed.

6.3 DELAY FAULT MODELS

A **delay defect** is a defect that causes an extra delay in the circuit. An example is a spot defect that causes a resistive short or open. To make the delay testing problem tractable, delay defect behavior must be abstracted to a **delay fault**. A large number of delay fault models have been proposed. In the following sections, we discuss some popular delay fault models.

6.3.1 Transition Fault Model

The most commonly used delay fault model is the ***transition fault (TF) model*** [Barzilai 1983]. In this model, a gate input or output is assumed to have a ***slow-to-rise*** (STR) or ***slow-to-fall*** (STF) delay fault. For every stuck-at fault in a circuit, there is a corresponding transition fault (TF). The delay of the TF is assumed large enough that any path through it will be detected as slow—that is, the delay increase because of the TF must exceed the slack of the shortest path through that line.

The primary advantage of the TF model is that test generation does not need to consider circuit timing. A stuck-at fault test generator can be modified to meet the additional requirements for generating TF tests [Waicukauski 1987]. Because a stuck-at fault can be considered a very slow TF, a TF test set will detect all the corresponding stuck-at faults. The TF model has more constraints than the stuck-at fault model, so the TF coverage is normally lower than stuck-at fault coverage. Top-off vectors can be generated to test the stuck-at faults not detected by the TF test set.

The primary disadvantage of the TF model is that its resolution is limited by the difference in delay between the longest and shortest path through the fault site. Stuck-at and TF test generators normally select the easiest path to test, which is the shortest path, because this has the fewest necessary assignments. This problem can be avoided by propagating along low-slack paths [Lin 2006a]. Another major problem with the TF model is that TF tests often only propagate a glitch from the fault site, which may not be detected unless the delay fault is large [Lin 2005a].

6.3.2 Inline-Delay Fault Model

Production experience shows that many delay defects are due to resistive interconnect vias, which cause both the rising and falling transitions through that line to be slow [Benware 2003]. This can be modeled by the **inline-delay fault**. This fault is analogous to the TF, except that only one of the STR or STF faults must be detected at each fault site. A test set for inline-delay faults is smaller than a test set for TFs.

Even though resistive vias are more common in newer technologies, misshapen transistors, resistive contacts, and resistive shorts can still cause STR or STF faults without the opposite slow transition. Therefore, high delay fault coverage still requires a TF test set. If a technology contains many inline-delay faults, test generation can first be done for them, and then top-off TF patterns can be generated. If testing stops on the first failure, this combination of patterns will reduce average test time without loss of TF coverage.

6.3.3 Gate-Delay Fault Model

A spot defect that causes a small delay fault for a particular transition on a gate input or output can be modeled as a **gate-delay fault**, also termed a **local delay fault** [Carter 1987]. Here "small" is a delay that is larger than the minimum slack but smaller than the maximum slack for that line. Testing such faults requires testing a long or longest path with the appropriate transition through the fault site. The quality of a test set is defined by how close the minimum detected delay fault sizes are to the minimum detectable fault sizes [Iyengar 1988].

The **line delay fault model** [Majhi 2000] extends the gate-delay fault model to test a rising or falling delay fault on each line in the circuit. Detecting the smallest delay defect that can cause failure requires propagating along the longest sensitizable path through that line.

6.3.4 Path-Delay Fault Model

The **path-delay fault model** [Smith 1985] models the distributed delay on a path. A path is a sequence of gates from PIs to POs, with a transition on each gate output along the path. The gate input on the path is referred to as the **on-input**, or **on-path input**, whereas the other inputs are **side inputs** or off-inputs. If the circuit contains a path that is slow for a rising or falling transition, then it contains a **path-delay fault**. Unless explicitly mentioned, we refer to the transition direction at the input to the path. The path-delay fault model assumes that any path can have any delay, so fault coverage is defined as the fraction of all paths (or all sensitizable paths) tested. The path-delay fault model is more general than the fault models discussed above, because a path-delay fault test will detect the corresponding transition, inline, or gate-delay faults, as well as distributed delay resulting from process variation *(global delay faults)* or supply noise.

The primary drawback of the path-delay fault model is that the number of paths can be exponential in the size of the circuit, so computing and achieving high coverage is difficult. For example, ISCAS-1985 benchmark circuit c6288, a 16-bit multiplier, has close to 10^{20} paths. Almost all of the long paths in this circuit are false paths [Qiu 2003c]. Coverage estimates can be improved by eliminating untestable paths [Cheng 1993] [Lam 1993], but these methods are expensive. Path-delay testing is primarily used to test a set of longest (or *critical*) paths provided by static timing analysis.

A fault model intermediate between gate and path delay is the **segment delay fault model**. It assumes that a segment along a path is slow [Heragu 1996]. Because the segment length is bounded, the number of segment faults is linear in the circuit size. The **propagation delay fault model** combines the transition fault and path-delay fault models [Lin 2005a]. It assumes that the sum of the local delay and the distributed delay of the fault propagation path causes a failure, with at least one robust propagation path.

6.3.5 Defect-Based Delay Fault Models

The models described here do not adequately describe the delay defects that are more common in advanced technologies. As a result, such models cannot achieve the desired product quality. In practice, delay faults can be a combination of local and global disturbances. Delay faults caused by local disturbances are termed **local delay faults**, and those caused by global disturbances are termed **global delay faults** [Luong 1996] or distributed path-delay faults [Sivaraman 1996]. Resistive shorts and opens and capacitive crosstalk cause local delay faults, whereas power supply noise, intradie temperature distributions, and process variation cause global delay faults. The ***combined delay fault*** (CDF) **model** was developed to incorporate all of these effects [Qiu 2003b, 2004b]. Resistive short, open, and capacitive crosstalk locations can be extracted from the layout design, and are roughly linear in circuit size [Stanojevic 2001]. The circuit-level delay behavior of shorts, opens, and crosstalk can be abstracted to rules for test generation [Chen 1999] [Krstic 2001] [Li 2003]. Linear approximations [Lu 2005] can be used for nonlinear process correlation effects [Luong 1996]. A Monte Carlo approach can identify potentially longest paths under process variation [Liou 2002a] [Krstic 2003]. The statistical delay quality model incorporates process variation and defect characteristics into the coverage model [Sato 2005].

Rather than directly targeting a delay fault model, defect-based delay tests can attempt to maximize observations of each fault site. For example, the *N-detect stuck-at test* [Ma 1995] can be extended to target transition faults [Pomeranz 1999]. A related approach is to generate transition tests to each reachable output of a transition fault site, that is, ***transition fault propagation to all reachable outputs*** (TARO). The TARO test was shown to achieve high defect coverage [Tseng 2001] [McCluskey 2004] [Park 2005]. Similarly, the DOREME [Dworak 2004] test approach that increases the excitation and observation of fault sites increases the detection of delay defects.

Technology scaling has led to increased power supply noise, which can cause the chip to incorrectly fail at rated speed during testing [Saxena 2003]. Test generation must take power supply noise into account [Krstic 1999] [Liou 2003]. The primary challenge is accurate supply noise (delay) estimation at low cost [Wang 2006b]. An alternative is to use on-chip test structures to calibrate the supply noise [Rearick 2005]. See Section 7.3 for a further discussion of noise during testing.

The CDF coverage metric [Qiu 2004b] was developed to accurately reflect test quality, considering realistic delay defects. In the CDF coverage metric, the correlation in path delays is used to substantially reduce the pessimism of the path-delay fault model while avoiding the optimism of the transition and gate-delay fault models. The CDF coverage metric exploits structural and spatial correlation between paths [Luong 1996] [Lu 2005]. Two paths have *structural correlation* when they share a common path segment. For example, in Figure 6.3, path *a-d-e* and *b-d-e* are structurally correlated because they share segment *d-e*. Two paths can also have *spatial correlation* because the path delays are functions of the manufacturing process parameters, which are spatially correlated. For two paths that are physically close to each other, the delay correlation is high because the paths have similar process parameters.

Figure 6.4 shows the delay space [Luong 1996] for two paths, assuming the path delay is a one-dimensional function of process parameters. The delay space is the bounded region in which the probable delay value combinations are represented. It is assumed that each path has min-max delays. If the two paths have *no correlation*, the delay value combination can be anywhere within the rectangle. If they are *perfectly correlated*, the delay space shrinks to a line, which means if path 1 has max (min) delay under a combination of certain process parameters, path 2 also reaches its max (min) delay under the same combination of process parameters. In reality, the correlation is somewhere in between, and the realistic delay space is the shaded area. Using correlation information, the delays on the untested paths

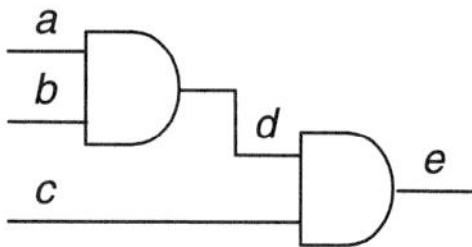

■ FIGURE 6.3

Example of structurally correlated paths.

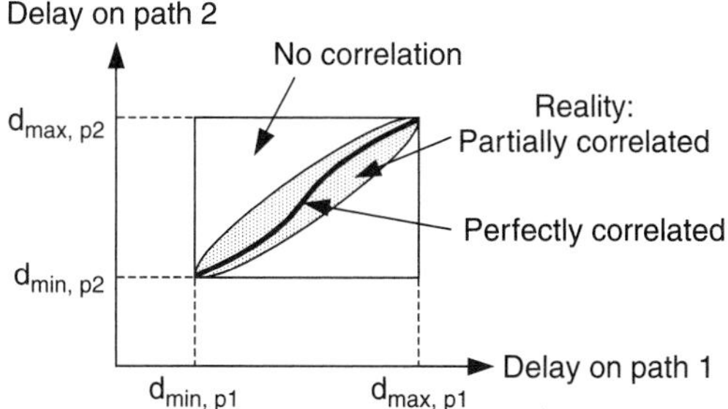

■ **FIGURE 6.4**

Delay spaces for different path correlations.

can be predicted by the delays on the tested paths [Brockman 1989], reducing the paths that must be tested.

In terms of probability, the coverage for test set t is as follows:

$$P(t \text{ detects delay fault} \mid \text{chip has a delay fault}) \tag{6.1}$$

Under the CDF model, a single local delay fault on a line is assumed, termed the fault site. Any path through the site is subject to process variation. A delay fault may be caused by a local delay fault, a global delay fault, or a combination of the two.

CDF detection is probabilistic, instead of deterministic. For example, suppose there are two paths, P_1 and P_2, through a fault site, and the local extra delay is not large enough for either path to be definitely slow. Figure 6.5 shows the delay space [Sivaraman 1996] for this fault. t_{max} is the maximum specified delay of the circuit. The circuit has some probability that path 1 or 2 is slow. If test set t_1 tests path 1 only and test set t_2 tests path 2 only, then both t_1 and t_2 are required to guarantee detection of the fault.

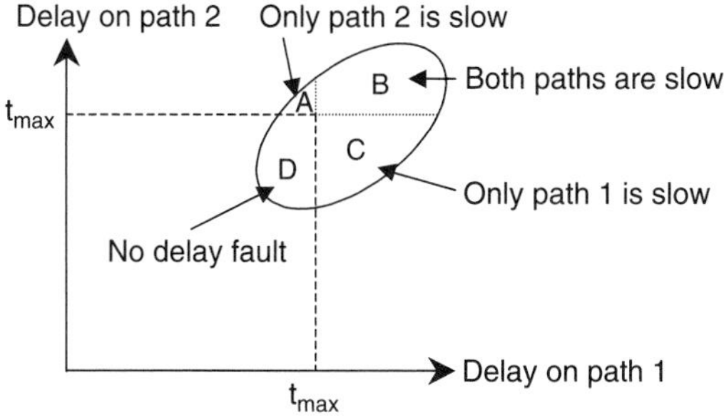

■ **FIGURE 6.5**

Delay space of a fault.

Using this probability model, Formula (6.1) can be translated into Formula (6.2) to compute the ***detection probability*** (DP) for fault site i with local extra delay Δ (the size of the local delay fault):

$$\mathrm{DP}_{i,\Delta}(t) = \mathrm{P}\ (\text{at least one tested path through } i \text{ is slow} \mid \text{at least one path through } i \text{ is slow}) \quad (6.2)$$

In the example shown in Figure 6.5, if test set t tests path 1 only, the DP is area(B∪C)/area(A∪B∪C); if t tests path 2 only, the DP is area(B∪A)/area(A∪B∪C); and if t tests both paths, the DP is 100%.

The preceding analysis is for a fixed local extra delay Δ. For fault site i with an arbitrary Δ, the DP for site i is computed as:

$$\mathrm{DP}_i(t) = \int_{\Delta>\Delta_{0,i}} \mathrm{DP}_{i,\Delta}(t) \cdot \mathrm{p}_i(\Delta)\mathrm{d}\Delta \quad (6.3)$$

where $\Delta_{0,i}$ is the minimum slack at the fault site and $p_i(\Delta)$ is the probability density function (PDF) of Δ at fault site i, based on the PDF of resistive shorts and opens. The overall fault coverage for test set t is:

$$\mathrm{FC}(t) = \sum_i \mathrm{DP}_i(t) \cdot w_i \times 100\% \quad (6.4)$$

where w_i is the weight for fault site $i\,(\sum_i w_i = 1)$. Fault sites might be weighted based on the probability of a delay fault occurrence, such as the critical area.

Based on Formula (6.2), if the path delays are correlated, the DP computation is dependent on the delay space. For example, in Figure 6.5, the areas of A, B, and C change if the delay space changes. Because accurate spatial delay correlations are usually not known, coverage can be computed assuming no correlation and 100% correlation. This provides lower and upper bounds on the coverage.

The CDF coverage metric (Formulae 6.2–6.4) is inexpensive, because only a small subset of paths must be considered. For example, Figure 6.6 shows the delays of four paths through a fault site, each having a delay distribution because of process variation. Suppose path P_1 is tested by t, and the longest testable path P_0 is not tested. When $\Delta_0 < \Delta < \Delta_1$, $\mathrm{DP}_{i,\Delta}(t)$ is 0; when $\Delta > \Delta_2$, $\mathrm{DP}_{i,\Delta}(t)$ is 100%, because the tested path P_1 is definitely slow; when $\Delta_1 < \Delta < \Delta_2$, $\mathrm{DP}_{i,\Delta}(t)$ increases from 0 to 100%

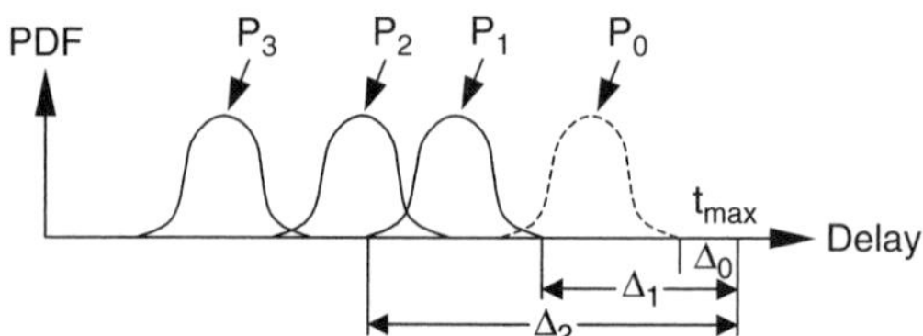

■ **FIGURE 6.6**

Fault coverage computation.

as Δ increases. Thus, the fault coverage computation is required only in this interval. The main cost is computing the fault efficiency, which is the number of tested faults over the number of testable faults. This requires checking the sensitization for all the paths whose length is within the interval where $DP_{i,\Delta}(t)$ rises from 0 to 100%. A lower bound on coverage can be quickly computed by assuming all structural paths are testable. Experiments on ISCAS-1985 circuits show the error of this approximation is $<4\%$ [Qiu 2004b].

The CDF fault coverage metric suggests a test strategy:

1. Apply transition fault tests to detect large local delay faults. In practice, most local delay faults are large.
2. Test one of the longest paths through each gate or line at the rated clock speed to eliminate or reduce the 0-DP area between Δ_0 and Δ_1 in Figure 6.6, because this is the second largest source of coverage loss.
3. Test more potentially longest paths (such as P_2 in Figure 6.6, if P_0 does not exist) to increase the DP between Δ_1 and Δ_2.

Figure 6.7 shows the conceptual relationship between fault coverage and the percentage of tested paths, sorted in decreasing order of length. If there is no local delay fault, the fault coverage increases quickly and reaches 100% after all potentially critical paths are tested. The curves for the CDF coverage metric have "jumps," where the first path through a fault site is tested. The traditional path-delay fault coverage assumes the percentage of testable paths tested.

Simulation experiments with the CDF coverage metric on ISCAS-1985 benchmark circuits [Qiu 2004b] show that a TF test set achieves reasonably high coverage (98.96%), but lower than what can be achieved by a ***K longest paths per gate*** (KLPG) test set (see Section 6.6.4), or a combination of TF and a set of critical paths. Figure 6.8 shows the fault efficiency for c7552, assuming the number of longest rising and falling paths per line increases from 1 to 5. As the figure shows, only a small number of longest paths are needed through each fault site to achieve high

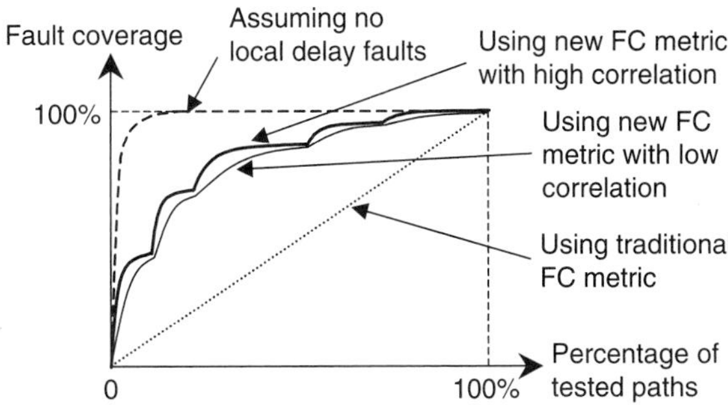

■ **FIGURE 6.7**

Fault coverage versus percentage of tested paths.

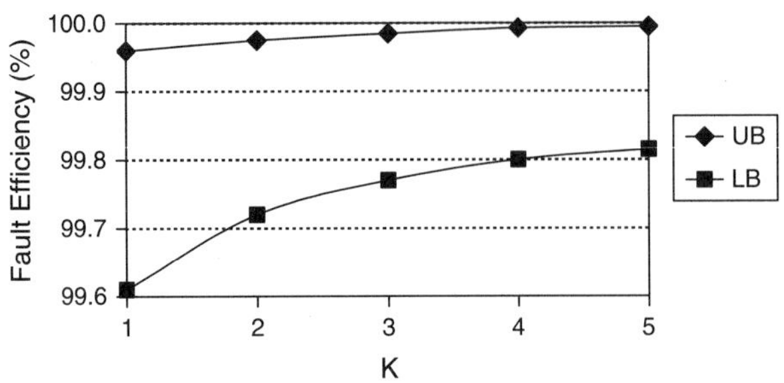

FIGURE 6.8

Fault efficiency for c7552, assuming the K longest rising and falling paths through each gate are tested ($K = 1$ to $K = 5$). UB is the upper bound (no intradie process variation), and LB is the lower bound (random intradie process variation).

fault efficiency. In addition, the benefit of testing one longest rising and one longest falling path (the fault efficiency increase from the transition fault test to the $K = 1$ test) is more significant than testing more long paths (increasing K from 1 to 5).

6.4 DELAY TEST SENSITIZATION

A path is said to be **testable** if a rising/falling transition can propagate from the primary input to the primary output associated with the path, under certain sensitization criteria [Lin 1987] [McGeer 1989] [Benkoski 1990] [Chang 1993]. If a path is not testable, it is called an **untestable** or **false path** [Liou 2002a, 2002b]. For example, in Figure 6.9, path *a-c-d* is a false path under the single-path sensitization criterion, because to propagate a transition through the AND gate requires line b to be logic 1 and to propagate the transition through the OR gate requires line b to be logic 0. We use the terms "untestable" and "false" interchangeably.

Tests that target paths, including gate-delay, line-delay, path-delay, and KLPG tests, have differing quality, depending on the path **sensitization criteria** that they meet. These tests are usually classified as **robust** or **nonrobust** [Smith 1985] [Lin 1987] [Bushnell 2000] or **functionally sensitizable** [Cheng 1996a]. Robust and nonrobust tests propagate a transition along the target path, while functionally sensitizable tests propagate on multiple paths. An overview of the classification schemes in the literature can be found in [Krstic 1998] [Jha 2003].

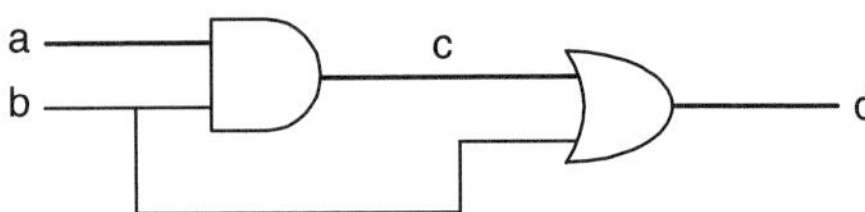

FIGURE 6.9

A circuit with a false path *a-c-d*.

A robust test is one that will detect the delay fault independent of the delays in the rest of the circuit. Detection by a nonrobust test depends on circuit delays. A test along a functionally sensitizable path requires increased delay on several paths for detection of the delay fault. Consider, for example, a path with a rising transition into an OR gate. If the side input of the OR gate is a stable 0, then the transition will pass through the OR gate, independent of the delays on the path. This is the robust test condition. If the side input has a falling transition that arrives before the on-path transition, then the OR gate will **glitch** low (have a transient pulse), with a wider glitch for increasing delay at the on-input to the OR gate. If the side input transition is late, the glitch will shrink or disappear. This is the nonrobust condition. Reconvergence from multiple paths of different lengths can cause multiple glitches or **hazards**. In both the robust and nonrobust cases, the path is **statically sensitized**, because the side inputs along the path have noncontrolling values in the test vector. In functionally sensitizable paths, the test vector has at least one controlling value on a side input along the path. Detection of a delay fault on such paths requires propagation of the delay fault on multiple paths. For example, if the side input of the OR gate has a rising transition, then both the on-input and side input must be slow for the rising transition on the output to be slow.

One challenge in determining path sensitization is the presence of uncontrollable **unknown** (X) values in the circuit. These are produced by uninitialized flip-flops or "black boxes," most commonly embedded memory arrays. For example, suppose a 2:1 multiplexer has an X on its control line and transitions on both of its inputs. Then the path that propagates through the multiplexer is unknown. All that is known is that a path at least as long as the shorter of the two paths through the multiplexer will propagate.

6.5 DELAY FAULT SIMULATION

Delay fault simulation is used to fault grade existing test sets, such as functional tests, and to drop faults that are fortuitously detected during ATPG.

6.5.1 Transition Fault Simulation

Transition fault simulation is similar to stuck-at fault simulation, except for the additional criterion that the fault is sensitized by the appropriate transition [Waicukauski 1987]. Good machine simulation determines the transitions at each fault site. For example, a rising transition sensitizes an STR fault. Each sensitized fault is then propagated to see if the fault effect reaches a PO or PPO.

6.5.2 Gate/Line Delay Fault Simulation

Gate or line delay fault simulation must account for circuit delays, in order to determine the smallest detected delay fault [Carter 1987] [Pramanick 1988]. The method in [Carter 1987] assumes that an **earliest arrival time** and **latest stabilization**

time is known for each signal waveform when transitioning from its **initial value** to its **final value**. The signal has an unknown value during this time interval. First good machine simulation is performed to compute these values for all signals. Propagating the earliest arrival times and latest stabilization times is based on the controlling values at each gate. For example, the latest stabilization time for a signal transitioning from 0 to 1 on an AND gate input determines the latest stabilization time for the output to transition from 0 to 1. For each pair of input vectors, fault simulation consists of injecting an extra delay at each fault site, determining if this extra delay propagates in terms of new arrival and stabilization times, and then whether it is detected at the POs and PPOs at the observation time. This process is repeated for all fault sites and patterns. Fault detection may be uncertain, if the observation time is between the earliest arrival and latest stabilization times.

6.5.3 Path-Delay Fault Simulation

A large number of path-delay fault simulators have been developed, for example [Smith 1985] [Schulz 1989] [Fink 1992] [Kagaris 1997]. These can be classified into **enumerative** and **nonenumerative** methods. Enumerative methods use a list of paths and compare the paths sensitized by a given vector pair to the list. The value algebra used for determining signal propagation must be sufficient to handle the sensitization criteria. For example, [Smith 1985] used a six-valued algebra to handle hazards when computing robust test coverage. Because the number of paths in most circuits is very large, nonenumerative methods were developed to compute fault coverage without an explicit list of paths. Such techniques allow for a compact graph representation of all path-delay faults in the circuit, called the ***path status graph*** **(PSG)**. The PSG is modified dynamically during test generation or fault simulation to account for the set of faults that remain to be targeted.

6.5.4 Defect-Based Delay Fault Model Simulation

CodSim is a CDF fault simulator for combinational circuits [Qiu 2003b]. It incorporates resistive shorts and opens, capacitive coupling, and process variation, and considers robust and nonrobust propagation on paths through each fault site. CodSim computes the detection probability for each fault site, for test set t. If the DP is above the specified threshold, the fault site is dropped. The DPs for all fault sites are then used to compute the overall fault coverage. The three phases of the fault simulation algorithm are as follows:

1. For each test pattern, run good machine timing simulation and identify the robust/nonrobust propagation paths from each line to primary outputs.
2. Check the validation of the nonrobustly sensitized paths through a line by introducing a local delay fault at that line and running fault simulation.
3. Run fault simulation considering capacitive coupling to the selected long paths for each fault site.

After good machine simulation, the initial and final logic values and the nominal transition time of the last event for each line are known, as shown in Figure 6.10. The numbers next to the transition symbols indicate the transition time, assuming a unit gate delay model. S1 or S0 indicates a stable 1 or 0.

In phase 2, the robust/nonrobust propagation paths through each line are identified. A line's robust propagation paths can be computed using its immediate fanout lines' robust propagation paths. In Figure 6.10, suppose line d has a robust propagation path P_1 with length 6, and line e has path P_2 with length 7. The robust propagation paths for line b are computed by checking the final logic values on the side inputs of gate G_1 and G_2. Then two paths are identified: b-P_1 with length 7 and b-P_2 with length 8. Because the propagation paths are robust, the extra delay Δ on a line must be detected if $t_{trans} + \Delta + l_{prop} > t_{max}$, where t_{trans} and l_{prop} are the transition time and propagation path length associated with that line, respectively. Because t_{trans} and l_{prop} are statistical values, the computed Δ is also statistical. For resistive shorts, the opposite logic value on the other shorted line must be checked.

The nonrobust propagation paths can be identified in a similar way. The difference is that if there is no transition on a line, the nonrobust propagation paths from that line must also be computed, such as line g in Figure 6.11. The reason is that a local delay fault on line d may generate a glitch on g, and the computation of the nonrobust propagation paths from d uses g's propagation paths. The time complexity of phase 1 is $O(V \cdot C)$, for V vectors and C lines in the circuit.

Fault detection through a nonrobust propagation path is dependent on the delays on the side inputs to the path. In Figure 6.11, an extra delay on line b does not affect the transition time on line h, even though line b has a nonrobust propagation path. Therefore, the validation of these paths must be checked (phase 2). After phase 1, each line has a few nonrobust propagation paths. The validation check can be

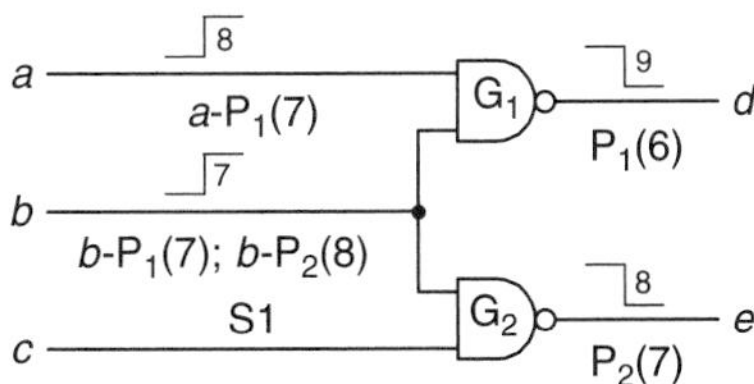

■ FIGURE 6.10

Robust propagation path identification.

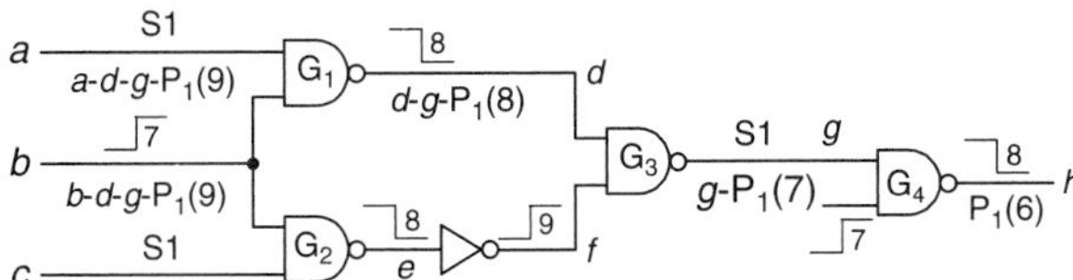

■ FIGURE 6.11

Nonrobust propagation path identification.

performed by introducing an extra delay Δ on the line, where $\Delta = t_{max} - t_{trans} - l_{prop}$, and running fault simulation for the test pattern which sensitizes this path to check if the slow signal can be detected at any PO or PPO. This procedure starts with the smallest Δ. If a small Δ can be detected, the validation check for the paths that can only detect large Δ is not necessary. Experiments show that normally only a few paths must be checked for each line.

Checking only robust and nonrobust paths may miss some functional sensitizable paths [Cheng 1996b]. However, because these paths always appear in pairs and the shorter path determines the delay, in most cases they do not contribute to the fault coverage. Thus, these paths are not checked unless there is no long robust or nonrobust propagation path through the fault site.

In phase 3, the extra delay due to coupling is computed. Similar to phase 2, phase 3 introduces an extra delay Δ at the fault site, where Δ can cause one propagation path (either robust or nonrobust) to be slow, and runs the fault simulation considering coupling for the vector sensitizing the path. The delay Δ is introduced because the coupling alignment is dependent on Δ. Because of the interdependence of aggressor and victim signal timing, iterative simulation is used. The phase 3 analysis is only performed for long paths.

Simulation experiments on ISCAS-1985 circuits with resistive opens show that for 10,000 random patterns applied at rated speed, the CDF and transition fault coverages are within 1%, as each fault site has a high probability of having many long paths tested. The fault efficiency of a KLPG-5 test (see Section 6.6.4) is 99.9%+. Simulation of resistive shorts shows fault coverage of approximately 90%. The reduced coverage is because resistive shorts on fault sites with large slack cannot be detected until the short resistance is so low that it nearly causes a stuck-at fault [Li 2003]. Simulation of KLPG-5 tests shows much higher resistive short coverage, because the longest paths are tested.

6.6 DELAY FAULT TEST GENERATION

6.6.1 Transition/Inline Fault ATPG

Transition fault ATPG is based on modified stuck-at fault ATPG [Waicukauski 1987]. The initialization vector sets the initial value, and the test vector launches the transition and propagates it to a PO or PPO. For example, for a STR fault on line g, a logic 0 is set on g by the initialization vector, and in the test vector the value on g is set to 1, with a propagation path sensitized. These conditions can result in test invalidation due to hazards [Lin 2005a]. Such hazards can be reduced by attempting to generate ***single input change*** (SIC) patterns, in which the test vector differs from the initialization vector by only a single bit.

A simple ATPG model is created by introducing a seven-valued algebra, with logic values representing two time frames of a sequential circuit. The seven values are S1, S0, R1, F0, U1, U0, and X, denoting steady 1 across both time frames, steady 0 across both time frames, a rising transition across the two frames, a falling transition across the two frames, don't care in the first frame and logic 1 in the second frame, don't care in the first frame and logic 0 in the second frame, and don't care in both frames, respectively. Boolean operators can operate on this seven-valued logic, and an ATPG can be developed to generate test vectors for each target transition fault.

One may also employ a stuck-at ATPG to generate tests for transition faults. By noting that a transition fault can be mapped to two stuck-at faults, vectors from a stuck-at suite can be used to construct the transition test set. For example, consider a fault x STR. To detect this fault, signal x must be set to logic 0 in the first time frame, and in the second vector, launch a transition by setting $x = 1$ and simultaneously propagating its fault effect to a PO or PPO. In other words, the first vector excites x stuck-at 1, and the second vector detects x stuck-at 0. As the two vectors may be uncorrelated, this technique works well under the enhanced-scan architecture. A test set can thus be constructed by matching stuck-at vectors. A graphical representation of all the test vectors and stuck-at faults can permit finding a minimum length test set. A concept of **transition test chains** was proposed in [Liu 2002, 2005] [Wang 2006a] to compute short chains that maximize detection of transition faults.

Despite the work that has gone into transition fault test generation, a complete transition fault test set may not detect all critical paths, because the ATPG decision process often favors those easier paths to launch or propagate the fault. The easier paths are generally the shorter paths. Conversely, a test set that exercises the longest paths in the circuit may not detect all transition faults, because a transition fault may not be robustly or nonrobustly launched or propagated [Gupta 2004]. Therefore, there has been effort to generate test sets that also cover long paths, such as [Sharma 2002], [Qiu 2003a], and [Gupta 2004]. Among these approaches, some try to generate a set of longest paths such that every gate is covered. Subsequently, a test set that tests these paths would also achieve full transition fault coverage. Alternatively, in [Gupta 2004], the aim was to launch the transition as late as possible, while at the same time propagating it through the longest propagation path. This technique thus elevates the quality of generated transition tests so that long paths are exercised.

A transition fault that is testable under one application scheme (*e.g.*, LOC) may become untestable under another test application. Consider the circuit fragment shown in Figure 6.12. Suppose a, b, and c are outputs of scan flip-flops. The STF transition on line d would be untestable under the LOS test application. This is because to detect this STF, $ab = 00$ and $c = 0$ in the test (second) vector are needed for the transition to propagate through the AND gate. In other words, the test

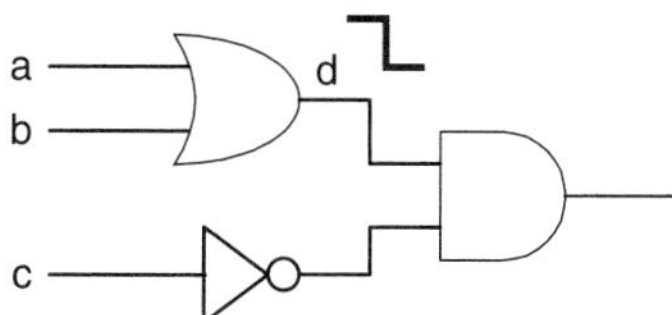

FIGURE 6.12

Launch-on-shift untestable transition fault.

vector must be $abc = 000$. However, under LOS, this vector is a shifted version of the initialization vector. For the test vector to be $abc = 000$, the initialization vector must be $abc = 00X$. However, this initialization vector cannot set $d = 1$. Consequently, this STF fault is untestable under LOS.

Now, as the previous example illustrated, a LOS untestable fault may be testable under other methods of test application. Interestingly, however, if a transition fault is LOC untestable, it is guaranteed to be functionally untestable. This is because under LOC, the two vectors are functionally correlated in that the second vector is derived from the first via clocking the circuit. Because the starting state of a LOC test is fully controllable, any fault that is untestable would also be functionally untestable as no initial state would be able to both launch the transition and observe its effect. Note that, however, a **functionally untestable transition fault** may be detected under LOS, because the two vectors are not functionally correlated [Liu 2003].

6.6.2 Gate-Delay Fault ATPG

Gate delay fault ATPG attempts to generate a long path through every gate input and output for both rising and falling transitions. This overlaps with the goal of many path delay subset ATPG approaches [Sharma 2002], described later. Here we focus on extensions to transition fault ATPG to improve the detection of small delay defects. One approach is to make random decisions when selecting the propagation path from the fault site. This should result in a longer path length than the normal approach of using breadth-first search, which tends to produce the shortest path. However, this approach had little benefit in a recent experiment [Qiu 2006].

A second approach is to use ***static timing analysis*** (STA) to compute the timing slack of each line in the circuit. The ATPG then attempts to propagate through the fault site on one or more low-slack paths [Lin 2006a]. Because the least slack path may be false, the ATPG must consider other low-slack paths. When the paths through a fault site are of similar length, it does not matter which path is chosen, particularly if none of the paths is critical [Qiu 2004b]. When there is a significant difference in path length, the ATPG should attempt to propagate on the longest path. A low-cost approach to achieve this goal is to take the difference between the maximum and minimum structural path lengths in the fanin and fanout cones of a fault site [Kajihara 2006]. The cone with the larger difference is given search priority. On average, this results in longer path lengths than a standard TF ATPG algorithm.

6.6.3 Path-Delay Fault ATPG

Test generation for path-delay faults can be categorized into enumerative and nonenumerative approaches. Enumerative methods generate a list of long structural paths and check their testability [Li 1989] [Majhi 2000] [Murakami 2000] [Shao 2002]. This approach works well when most paths are testable, but it is inefficient if many long paths are untestable. Rather than testing all paths, the globally longest paths can be targeted [Lin 1987].

To increase the efficiency, NEST [Pomeranz 1995] generates paths in a nonenumerative way, but it is only effective in highly testable circuits. DYNAMITE [Fuchs 1991] improved the efficiency for poorly testable circuits, but it has excessive memory consumption for highly testable circuits. RESIST [Fuchs 1994] exploits the fact that many paths in a circuit have common subpaths and sensitizes those subpaths only once, which reduces repeated work and identifies large sets of untestable paths. RESIST can handle circuits such as ISCAS-1985 benchmark c6288 with an exponential number of paths. However, both the enumerative and nonenumerative algorithms are too slow to handle industrial circuits.

The RESIST algorithm was extended to find a set of longest testable paths that cover every gate [Sharma 2002]. This test set targets line delay faults. This method takes advantage of the relations between the longest paths through different gates and guarantees their testability. However, this work assumes a unit delay model and fails on c6288.

6.6.4 K Longest Paths per Gate (KLPG) ATPG

The CDF fault coverage metric suggests that testing multiple paths through each fault site will increase fault coverage. The ***K longest paths per gate*** (KLPG) test generation algorithm [Qiu 2003a, 2004a] was developed to test the K longest sensitizable paths through each gate or line for both STR and STF faults at the fault site. Figure 6.13 illustrates the KLPG path generation algorithm. A *launch point*

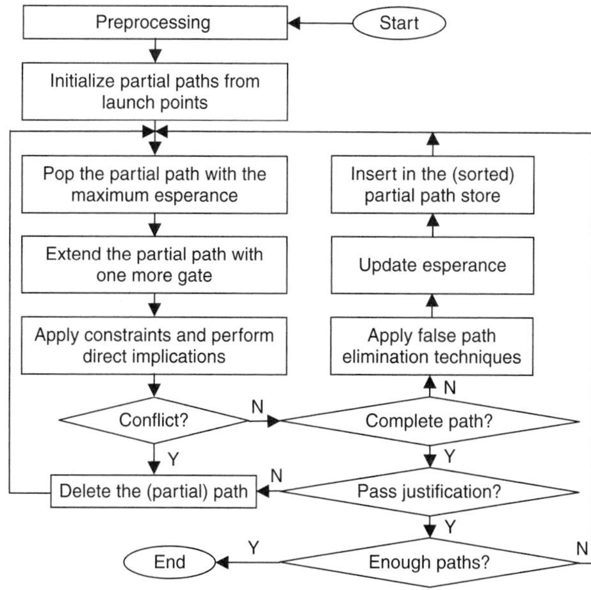

■ **FIGURE 6.13**

KLPG path generation algorithm.

(of a path) is a PI or PPI, and a *capture point* is a PO or PPO. In the preprocessing phase, STA computes the maximum delay from each gate to capture points, without considering any logic constraint. This value is termed the *PERT delay* or *STA delay*. In the path generation phase, *partial paths* are initialized from launch points. A partial path is a path that originates from a launch point but has not reached any capture point. A value called *esperance* [Benkoski 1990] is associated with a partial path. The esperance is the sum of the delay of the partial path and the PERT delay from its last node to a capture point. In other words, the esperance of a partial path is the upper bound of its delay when it becomes a *complete path* that reaches a capture point.

In each iteration of the path generation algorithm, the partial path with the maximum esperance value is extended by adding one gate. If the last gate of the partial path has more than one fanout, the partial path splits. Then the constraints to propagate the transition on the added gate, such as noncontrolling side input values required under the robust or nonrobust sensitization criterion, are applied. Direct implications are then used to propagate the constraints throughout the circuit. If there are any conflicts, the search space containing the partial path is trimmed off. If the partial path does not reach a capture point, some false path elimination techniques [Qiu 2003a] are applied to prevent it from growing to a false path. Then, its esperance is updated and the partial path is inserted back into the partial path store. If a partial path becomes a complete path, final justification is performed to find a vector. This process repeats until enough longest testable paths are generated. Because the longest path through a fault site is likely the longest path through other fault sites along the path, fault dropping is performed when a new path is generated.

An iteration of the path generation begins by popping the max esperance partial path from the path store. The partial path is extended by adding a fanout gate along the max esperance path. For example, in Figure 6.14, the partial path $g_0 \ldots g_i$ is extended by adding gate g_j because extending to g_j could potentially preserve

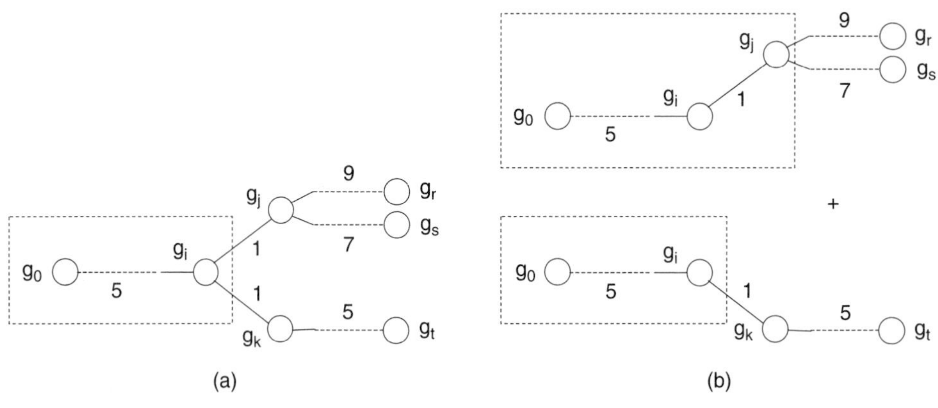

■ **FIGURE 6.14**

Extending a partial path: (a) before extension and (b) after extension.

the max esperance. If the partial path has more than one extendable fanout, it must be copied, its esperance updated and pushed into the path store. For example (Figure 6.14), because gate g_i has two fanouts, and extending the partial path to g_j may result in false paths later, the partial path $g_0 \ldots g_i$ must be saved because extending it to gate g_k may get a longer testable path. Because fanout g_j has been tried, the min-max esperance in the copy becomes 11/11.

After the partial path is extended ($g_0 \ldots g_i g_j$ in Figure 6.14) the constraints to propagate the transition on the added gate (g_j) are applied. Under the nonrobust sensitization criterion, noncontrolling final values on the side inputs are required. Under the robust sensitization criterion, in addition to noncontrolling final values, the side inputs must remain noncontrolling if the on-path input has a transition to the controlling value. Then direct implications are used to propagate the constraints throughout the circuit and discard false partial paths. For example (Figure 6.14), if extending partial path $g_0 \ldots g_i$ to gate g_j results in a conflict (see Figure 6.15), both path $g_0 \ldots g_r$ and $g_0 \ldots g_s$ are false, and the partial path is deleted from the path store. Experimental results show that most false paths can be eliminated by direct implications [Benkoski 1990] [Qiu 2003a].

If the launch-on-shift approach is used, the logic values on neighboring scan flip-flops are dependent on each other. For example, in Figure 6.16, the logic value of cell A in the initialization vector is the same as that of cell B in the test vector.

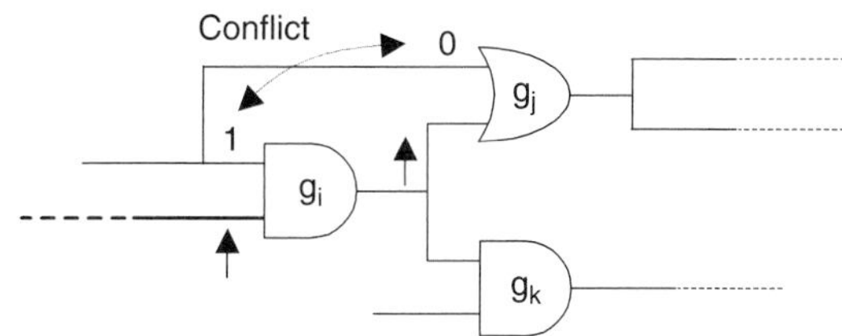

■ **FIGURE 6.15**

Conflict after applying direct implications.

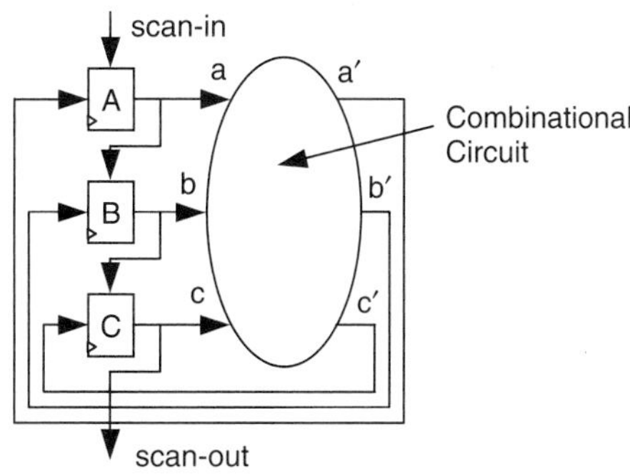

■ **FIGURE 6.16**

Implications on scan flip-flops.

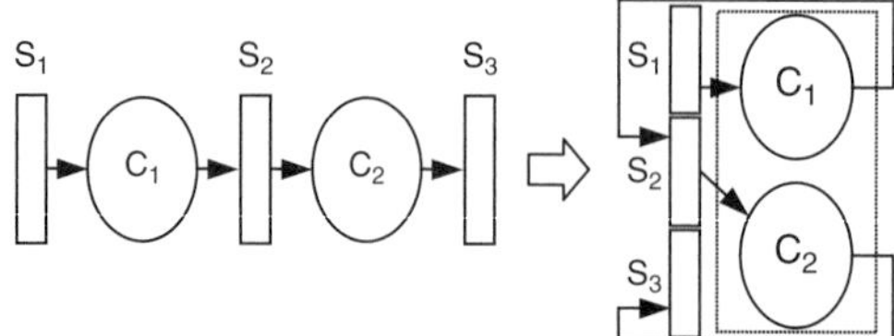

FIGURE 6.17

A pipeline circuit structure.

The relation between cell *B* and *C* is the same. Therefore, if there is a rising transition assigned to cell *B*, direct implications will try to assign logic 1 to cell *A* in the initialization vector and logic 0 to cell *C* in the test vector and propagate the new assignments throughout the circuit. If there are any conflicts, the partial path is a sequential false path under the launch-on-shift constraints. This algorithm does not consider modification of the scan chain design to reduce the dependence, such as inserting dummy cells between the scan flip-flops.

If LOC is used, dependence exists between the two vectors. Even if the circuit had a pipeline structure, in which the two vectors are independent, the structure can also be seen as the general structure shown in Figure 6.16. The conversion is shown in Figure 6.17. Thus, the test vector is the output of the combinational circuit, derived from the initialization vector, excluding the PIs and POs. In other words, $V_2 = C(V_1)$, where V_1 and V_2 are the two vectors and C is the logic of the combinational circuit. For example, if it is assumed that a testable path has a rising transition launched from cell *A* and a rising transition captured on cell *B*, in Figure 6.16, then for the initialization vector, output a' must be logic 1 (then it becomes the value for input a in the second vector); and for the test vector, input b must be logic 0 because it is derived from the initialization vector. Then more direct implications can be performed from a' and b.

Most designs include non-scan flip-flops. Even after simulation of the test setup sequence and scan procedure, some flip-flops remain uninitialized. In addition, embedded memories and "black boxes" are assumed uncontrollable, so their outputs have unknown values. A seven-valued algebra is used to distinguish between controllable and uncontrollable unknown values. The seven values are logic 0/1, x (unknown/unassigned), u (uncontrollable), $0/u$ (0 or uncontrollable), $1/u$ (1 or uncontrollable), and x/u (unknown or uncontrollable). At the beginning of test generation, the lines from the uncontrollable memories are u and all the other lines are x. The x values may be assigned a known value during test generation. In a two-input AND gate, if one input is x and the other is u, the output is $0/u$ because if the input with x is assigned a logic 0, the output becomes 0, but if this input is assigned a logic 1, the output becomes uncontrollable. Figure 6.18 shows two examples, assuming M_1 is a non-scan memory cell and M_2 is a scan flip-flop. The logic values assigned by simulation are shown. If the conventional three-value algebra is used, all the lines are assigned x's.

In Figure 6.18a, line n_3 can never be logic 1, so all paths through n_4 are false. Using a three-value algebra, test generation may have to check all paths through n_4.

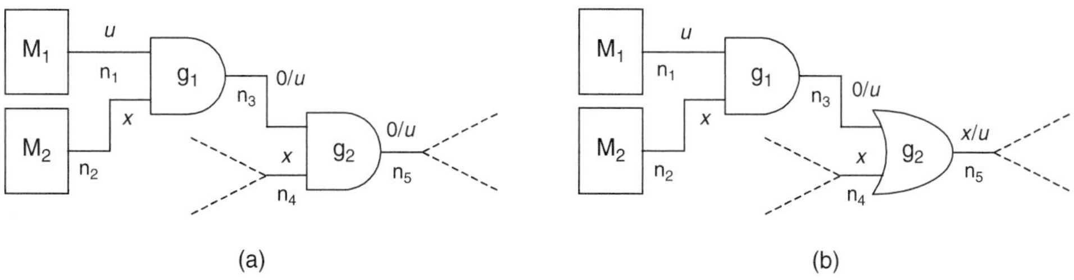

■ **FIGURE 6.18**

Application of seven-value algebra in path-delay test: (a) trimming the search space and (b) enabling additional search.

Moreover, it can be learned that n_5 can never be logic 1, so its faults are untestable. Because all the paths through n_4 contain n_5, the faults on n_4 are also untestable. Figure 6.18b shows how this algebra permits additional searching. When a partial path reaches gate g_2, the value of the test vector on side input n_3 is set to logic 0 and direct implications performed. The value on n_2 is set to 0 and further direct implications can be performed from M_2. In the conventional three-value logic, direct implications stop at gate g_1.

Final justification is used to find a delay test pattern when a complete path is found, using the corresponding LOS or LOC constraints. For LOC, one time frame expansion is used, as shown in Figure 6.19. The initialization vector V_1 can be generated within one time frame, but since the test vector $V_2 = C(V_1')$, the goal is to find a satisfying V_1'. Because V_1 and V_1' are identical excluding the "don't care" bits, in the justification process, there must be no conflicts between V_1 and V_1'.

A KLPG-1 test set is constructed as follows. For each fault, a test is first generated using the longest robustly testable path (as this is the most reliable test, even though it may not be the longest path). If there is no robust test, the longest restricted nonrobustly testable path is selected. The restrictions are that the test must have the required transition at the fault site, and the local delay fault must be detected at a capture point, if there are no other delay faults. In other words, the test is also a TF test.

For example, path n_1-n_2-n_3-n_5 is a nonrobustly testable path in Figure 6.20. It is a valid nonrobust test for lines n_2 and n_3. However, it is not a valid nonrobust test

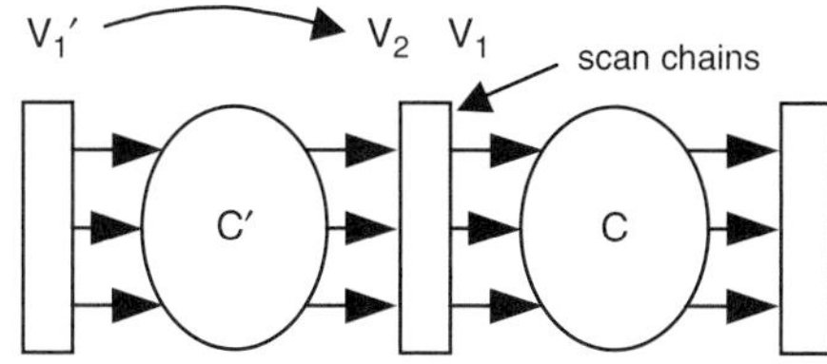

■ **FIGURE 6.19**

Time frame expansion for final justification using launch-on-capture.

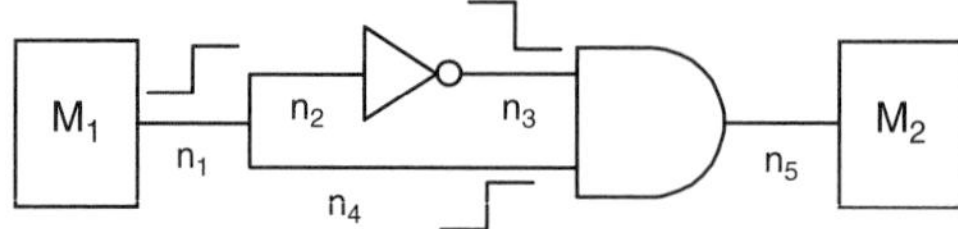

FIGURE 6.20

Restricted nonrobust test.

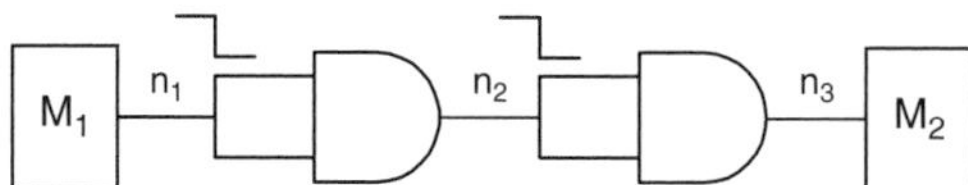

FIGURE 6.21

Transition fault test.

for line n_5 because the glitch (or transition) may not happen if the delay of path n_1-n_4 is greater than the delay of path n_1-n_2-n_3. Similarly, it is not a valid nonrobust test for line n_1 because the slow-to-rise fault may not be detected even if there are no other delay faults.

If a fault has no robust or nonrobust test, a transition fault test that can detect a small local delay fault is generated. In KLPG test generation, this case usually happens when the local delay fault can only be activated or propagated through multiple (functionally sensitizable) paths, such as the slow-to-fall fault on line n_2 in Figure 6.21. The length of the shortest paths in the activating or propagating path set determines the test quality. The longer the shortest path, the smaller the local delay fault that can be detected. A similar situation occurs when an uncontrollable value selects between two propagation paths. Only the shorter of the two paths can be assumed.

Experimental results with KLPG test patterns show that the execution time and number of paths rise slowly with K, since many gates share the same paths. More faults are detected with LOS than LOC, with each having unique detects [Qiu 2004a]. The LOS robust path lengths are close to those of enhanced scan, while LOC path lengths are often shorter. This indicates that LOS and enhanced scan sensitize false paths. Silicon results with KLPG-1 test patterns have shown that they detect delay faults that escape TF tests [Qiu 2006].

6.7 PSEUDO-FUNCTIONAL TESTING TO AVOID OVER-TESTING

Design for testability (DFT) is a popular technique used to increase the testability of a design, and it has been effective for structural testing, as discussed in the earlier sections of this chapter. It also allows the underlying test generator to focus only on generating a two-vector delay test that assumes the initial state can be loaded via the scan structure. However, unlike stuck-at faults, when testing for delay, one must keep in mind those **functionally infeasible paths**. Incidental activation of those functionally infeasible paths may **over-test** the chip. In addition, if a chip fails only the particular test that exercises such paths, it may be considered a bad chip when in reality it is fully functional.

Clearly, a delay fault that cannot be activated in the combinational portion of the circuit is functionally infeasible. On the other hand, some delay faults that can be activated under scan may not be functionally feasible. This is because the scanned-in state of the two-vector pattern may not be functionally reachable. In LOC, if the scanned-in starting state is unreachable, the LOC test may inadvertently exercise some functional false paths. Likewise, in LOS, the transition that is launched and propagated may not map to a functionally realizable state transition.

Having established the preceding discussion, the main hurdle in using scan tests for at-speed testing is that of yield loss [Rearick 2001] [Krstic 2003] [Lin 2005b]. That is, a timing failure caused by a functionally infeasible path may not be a true failure and will result in throwing good parts away. To address this problem, **pseudo-functional tests** have been proposed. In pseudo-functional scan testing, any test that is scanned in should conform "closely" to a functionally reachable state. The intuition behind this is that if the scanned state is known to be a functionally reachable state, then the chip will operate close to the functional mode during the application of the target test. Subsequently, the yield loss problems can be considerably reduced.

The following discussion gives a broad overview of various techniques that have been proposed to generate such pseudo-functional tests. Pseudo-functional tests can be computed by first extracting a set of constraints that specifies the illegal states [Lin 2005b] [Zhang 2005] [Lin 2006b] [Syal 2006]. With these constraints fed to the underlying ATPG, the test generator generates suitable tests that attempt to avoid the supplied constraints, which specify the illegal states. Alternatively, one may start with an initial set of unconstrained test patterns that has been generated by an ATPG without any constraints. Then, the unconstrained test set is modified to transform them to pseudo-functional scan tests by incorporating the knowledge of a subset of reachable states [Pomeranz 2004]. For instance, one may first compute a subset of reachable states via random simulation from a known valid state and collect all the states reached from the simulation. Any valid state identified is guaranteed to be reachable. Consequently, the pseudo-patterns generated by adhering to this information are guaranteed to be functionally valid states. When any pattern generated cannot strictly adhere to the valid state information, either a validity check of the modified state is invoked, or the initial state portion of the pregenerated pattern can be modified to adhere to one of the known valid states.

Therefore, identification of unreachable/illegal states (the converse of the reachable states) is a fundamental problem in the context of pseudo-functional testing. Sequential ATPG has been dealing with this problem since its conception. Given a state, it tries to determine if the state can be justified from an all-unknown state. When the target state is found to be unjustifiable, DUST [Gouders 1993] and DAT [Konijnenburg 1999] take this known illegal state and attempt to eliminate one specified assignment at a time, try to justify it, and end up with as large an illegal state space as possible. In other words, suppose the initial illegal state is 0X01X1. If after this process of elimination, state 0XX1XX is also illegal, we have obtained

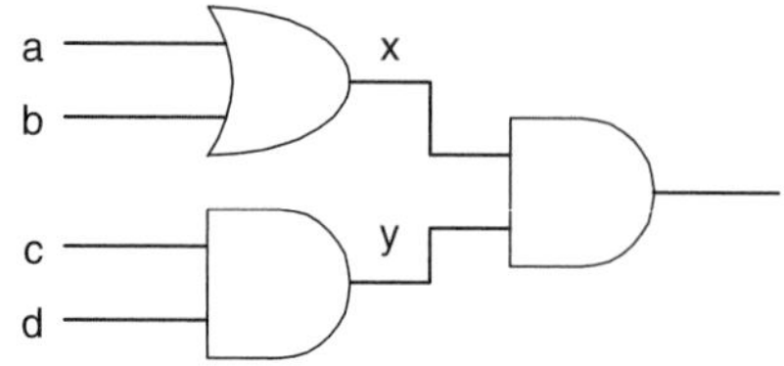

FIGURE 6.22

Circuit fragment for constraint expression.

a larger illegal state space. Methods to merge these cubes to form even larger cubes are discussed in [Konijnenburg 1999].

In addition, succinct representation of such illegal states is essential to avoid memory explosion and enhance search efficiency. One representation is by expressing constraints as Boolean expressions over nets in the design [Liu 2003]. All such constraints are referred to as **functional constraints**. Another is the use of ***binary decision diagrams*** (BDDs). The constraints can be explicitly represented as a set of all computed illegal states. Consider three illegal states in a circuit with six flip-flops s1 through s6. The first illegal state is s1s2s3s4s5s6 = 111XXX. The illegal state subspace can be represented simply as s1 · s2 · s3. Negating this conjunction gives us the clause $w = (\neg s1 + \neg s2 + \neg s3)$. This clause essentially restricts any solution to fall within the subspace expressed by w. For example, when s1s2s3 = 111, w would evaluate to 0, indicating that an illegal state space has been entered. The clauses for the other two illegal states can be obtained in a similar manner. Finally, the constrained formula is formed simply by the conjunction of all the constrained clauses. Thus, at any time, none of the constraint clauses must evaluate to 0 to ensure that the search remains *outside* of any of the illegal state spaces.

Alternatively, the illegal state spaces can be represented as Boolean expressions of illegal internal signal combinations [Syal 2006]. For example, suppose we have a fragment of the circuit shown in Figure 6.22. A single constraint $(\neg x + y)$ can represent several illegal states, namely $abcd = \{1000, 0100, 1100, 1001, 0101, 1101, 1010, 0110, 1110\}$. When the signals $abcd$ are assigned any of these values, the constraint $(\neg x + y)$ will be violated.

6.7.1 Computing Constraints

Logic implications capture Boolean dependencies between various nets in a circuit. Logic implications have been used in ATPG engines in the past. For details of computing static implications, refer to [Zhao 1997] [Wang 2006a]. In the rest of the discussion, the following notations will be used:

Impl[g, v, t]: Set of implications resulting from assigning logic v to gate g in time frame t.

[g, v, t] $\rightarrow$ [h, w, t_1]: The assignment $g = v$ in time frame t implies $h = w$ in time frame t_1.

There are relationships between nets that are valid in the same time frame but can only be learned through "sequential learning" (*i.e.*, Boolean learning across time frames). Such relationships are missed if only one time frame is analyzed. Furthermore, not all sequential relationships can be learned by time frame expansion [Syal 2006]. During ATPG, all combinational constraints will be automatically factored in the process and need not be stored. On the other hand, sequential relations are of great help to the ATPG to avoid those undesirable state spaces.

6.7.1.1 Pair-Wise Constraints

The first mechanism to identify functional constraints is by using the logic implications directly. For each gate g in the circuit, the sets Impl[g, 1, 0] and Impl[g, 0, 0] over the time window $-t$ to t (t is user specified) are first computed and stored. All relationships for [g, 1, 0] and [g, 0, 0] that exist in time frame 0 are recorded. Let these relationships be denoted by sets $S_{seq}{}^{g1}$ and $S_{seq}{}^{g0}$. Note that some of these relationships are learned through sequential reasoning, whereas some relationships may be purely combinational.

To identify useful **pair-wise constraints** that denote illegal states (*i.e.*, functional constraints), combinational relationships must be filtered out. This can be performed by using a single time frame analysis. For each gate g in this single-frame circuit, Impl[g, 1, 0] and Impl[g, 0, 0] are also computed. These combinational implications are recorded as $S_{comb}{}^{g1}$ and $S_{comb}{}^{g0}$. Next, set difference operations are performed for each gate g: $S_{seq}{}^{g1} - S_{comb}{}^{g1}$ and $S_{seq}{}^{g0} - S_{comb}{}^{g0}$. This operation ensures that all combinational relationships that do not represent any illegal state(s) are pruned from the list of functional constraints. Note that the remaining sequential constraints are those that still reside in the same time frame. For instance, $x = 1$ implies $y = 1$. However, because this constraint is sequential in nature, there must exist a starting state such that when $x = 1$, y can be made to equal 0. By adding this constraint $(\neg x + y)$, the set of states that violate this implication is thus captured [Syal 2006].

6.7.1.2 Multiliteral Constraints

Although the pair-wise sequential constraints may help to capture a large set of illegal states, they capture relationships only between a pair of gates. These may be insufficient to capture illegal states that require an analysis on more than two gates. To cover such illegal states, an approach to identify sequential relationships between multiple nets (**multinode constraints**) was proposed [Syal 2006].

The concept of multinode implications is rather simple. Starting with $(A = 0,\ B = 1,\ C = 0)$, if via simulation an implication $(D = 1)$ is obtained, then the multinode implication would simply be $[(A = 0,\ B = 1,\ C = 0) \rightarrow (D = 1)]$.

Let A_i represent an assignment of the form $[g_i, v_i, t]$, where g_i is a gate in the circuit, $v_i \in \{0, 1\}$, and t is the time frame of this assignment. In other words, each A_i denotes a value assignment to a variable in the net-list. The form of pair-wise implication is $A_i \rightarrow A_j$ whereas the form of multinode implication is $(A_1 \Lambda A_2 \Lambda \ldots \Lambda A_m) \rightarrow A_{m+1}$.

The element on the left-hand side of the arrow is called the **implicate**. The element on the right-hand side of the arrow is called the **implicant**.

Because the implicate of a multinode constraint is a conjunction of terms, the probability of the implicate evaluating to true reduces with the number of terms. Thus, the usefulness of a multinode constraint decreases with its size. Thus, multinode constraints are often restricted to a small size. Similar to pair-wise constraints, multinode constraints are also restricted to reside within the same time frame.

Consider the circuit fragment illustrated in Figure 6.23. Two important implications can be computed for gate *D*, namely (1) $[D, 1, 0] \rightarrow [F, 0, 1]$ and (2) $([A, 1, 0] \wedge [B, 1, 0]) \rightarrow [D, 1, 0]$. By the transitive property of implications, a third relationship can be learned in a straightforward manner: (3) $([A, 1, 0] \wedge [B, 1, 0]) \rightarrow [F, 0, 1]$. Note that implication (i) is a pair-wise relation while relationships (2) and (3) are multinode constraints. Such relationships can be derived in linear time simply by analyzing every two-input gate.

However, deriving constraints this way is not entirely desirable. Specifically, constraint (2) is a combinational relationship and, as stated previously, is useless as a functional constraint because it does not block any states that could violate this constraint. Constraint (3) involves gates across time frames and is thus not considered for representing multinode constraints.

As a result, a different approach is needed. Assume that the following sequential relationships have been identified: $[g_i, v_i, 1] \rightarrow [A, 1, 0]$ and $[g_j, v_j, 1] \rightarrow [B, 1, 0]$, then a new multinode constraint can be deduced: $([g_i, v_i, 1] \wedge [g_j, v_j, 1]) \rightarrow [F, 0, 1]$. Note that this multinode constraint has all the desirable characteristics, namely that it is not a combinational constraint and all the involved gates reside in the same time frame. In addition, the constraint can be learned by analyzing relationships around every single gate, retaining the linear complexity.

The preceding discussion begs the following question: How can one efficiently discover the gates g_i that hold the relations $[g_i, v_i, 1] \rightarrow [A, 1, 0]$ and $[g_j, v_j, 1] \rightarrow [B, 1, 0]$? In essence, the task is to identify the implicates of [A, 1, 0] as well as for [B, 1, 0].

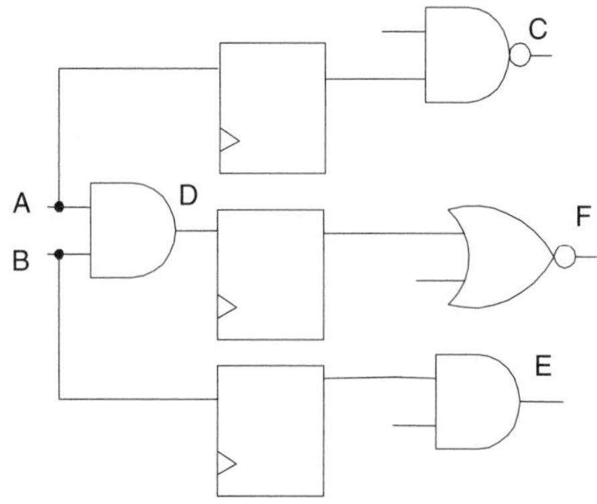

■ **FIGURE 6.23**

Fragment of sequential circuit.

At first glance, this task seems daunting. However, the contrapositive law allows for an efficient computation of such implicates. Recall that the contrapositive law states that if $x \rightarrow y$, then $\neg y \rightarrow \neg x$. In other words, it is sufficient to compute what [A, 0, 0] implies, then by the contrapositive law, the implicates can be obtained. For example, if [A, 0, 0] $\rightarrow$ [$g_i, \neg v_i$}, 1] and [B, 0, 0] $\rightarrow$ [$g_j, \neg v_j$, 1], then by the contrapositive law, [g_i, v_i, 1] $\rightarrow$ [A, 1, 0] and [g_j, v_j, 1] $\rightarrow$ [B, 1, 0]. These implications are exactly what are needed to form the desired multinode constraint discussed earlier. Consider the circuit of Figure 6.22 again. Because it is known that [A, 0, 0] $\rightarrow$ [C, 1, 1] and [B, 0, 0] $\rightarrow$ [E, 0, 1], by contrapositive law, [C, 0, 1] $\rightarrow$ [A, 1, 0] and [E, 1, 1] $\rightarrow$ [B, 1, 0] are obtained. So the multinode constraint can be constructed to be [C, 0, 1] Λ [E, 1, 1] $\rightarrow$ [F, 0, 1]. Again, this constraint can be generalized (([C, 0, t] Λ [E, 1, t]) $\rightarrow$ [F, 0, t]; for all values of t) and it can be used as a constraint during constrained ATPG.

Finally, note that for the AND gate $D, D = 1$ is used to search for the multinode constraint. If it were an OR gate, the opposing value would have been chosen. In essence, the particular value assignment is chosen in order to form a conjunction of values necessary to form a multinode constraint. In the AND gate case—that is, ([A, 1, 0] Λ [B, 1, 0]) $\rightarrow$ [D, 1, 0].

The complexity for computing the prescribed multinode constraints is thus linear with respect to the number of gates in the circuit. Both the pair-wise and multinode constraints are computed and stored, which are then used to constrain the ATPG process.

6.7.2 Constrained ATPG

Suppose we are given a set of functionally untestable delay faults, F_u, and we want to make sure that the generated vectors will not incidentally detect any fault in F_u. A naive approach is to fault simulate the faults in F_u whenever a vector is obtained, and the ATPG engine would backtrack if some faults in F_u are incidentally detected and continue to search for a different pattern. However, this naive approach can be computationally expensive. To reduce the expense, instead of focusing on F_u, each fault in F_u is projected onto the state space S, to identify subspaces that will detect them. The projected state space S essentially constitutes a portion of the illegal states. This is because any state s that can detect any fault f in F_u is an unreachable state, because fault f would otherwise be functionally detectable. Subsequently, the ATPG needs to use the projected state space to avoid searching in the identified illegal subspaces. The projected state space, S, can be represented as a constraint formula, as explained earlier in Section 6.7.

Given a constraint formula that represents the set of illegal states, during the ATPG process, we must make sure that no clause in the constraint formula ever evaluates to false (*i.e.*, all literals in a clause simultaneously evaluate to false). The set of constraints also enables speed up of the ATPG by avoiding futile search spaces. Whenever a decision is made on a state variable s_i, we apply this decision assignment to all the constrained clauses in the formula that contain s_i. Application of this assignment may result in some unit clauses (a unit clause is an unsatisfied clause with exactly one remaining unassigned literal left). This remaining literal is called an **implication**. The implied variable automatically becomes the next decision variable. For example, consider a constraint clause $(\neg s1 + \neg s2 + \neg s3)$. Suppose $s1$ has been decided or implied to be 1, and the current decision makes $s2 = 1$. At this time, this clause yields an implication, $s3 = 0$. With $s1 = s2 = 1$, the only way not to cause any conflict is to have $s3 = 0$. We also check whether there is conflict (where one clause evaluates to false). If there is a conflict, the ATPG backtracks immediately [Liu 2003].

6.8 CONCLUDING REMARKS

Delay testing is becoming increasingly critical with rising clock frequencies and reduced timing margins. Structural delay testing is applied primarily using scan chain structures and slow-fast-slow clocking, using the transition fault model. Increasing quality demands have made the transition fault model inadequate. The transition fault model has been extended to target small delay defects by incorporating timing information. The path-delay fault model has also been made practical for small delay defects by generating path-delay tests that cover all transition fault sites. These tests run the risk of over-testing chips by detecting sequentially untestable faults. Pseudo-functional test can be used to avoid such over-testing.

In this chapter we described the common delay test approaches, with a focus on structural delay testing using two-pattern tests, where transitions are launched into the circuit and then the results captured within the rated clock cycle time. The use of design-for-test features such as scan chains provides the controllability and observability to make structural testing practical. This is combined with a slow scan in of the initialization vector, then application of the test vector via launch-on-shift (LOS) or launch-on-capture (LOC), and then capture of the results. With knowledge of the expected path delays, faster-than-at-speed testing can be used to increase detection of small delay defects.

We described the traditional delay fault models, including the transition fault, gate-delay fault, and path-delay fault, as well as newer models, including inline-delay fault, propagation delay fault, segment delay fault, and defect-based delay fault models. The quality of tests, in terms of test sensitization, was also discussed. Sensitization is particularly important when testing for small delay defects.

We discussed fault simulation and test generation for transition, gate/inline-delay path-delay faults and defect-based delay faults. Transition fault simulation and test generation is widely used, because it is a minor modification of stuck-at fault simulation and test generation. Defect-based delay fault simulation and test generation combines some aspects of both gate-delay and path-delay fault simulation and test generation. We discussed in more detail KLPG test generation, which illustrates the issues in both gate-delay and path-delay test generation. A common problem in structural test is that the initialization vector may not be a legal state of the circuit. As a result, functionally infeasible paths may be tested. If these tests fail, a chip that passes functional test will be rejected. The solution is to generate pseudo-functional tests, which have a high probability of starting from a functionally reachable state.

An increasing challenge in delay testing is to achieve high coverage of small delay defects and good correlation to functional test results. A major part of this challenge is to account for supply noise during delay testing, as discussed in Chapter 7. A further challenge is achieving this coverage in the presence of process variation, which increases the number of potentially longest paths in the circuit. The greatest challenge is incorporating these new goals into testing while avoiding an increase in test time.

6.9 EXERCISES

6.1 **(Untestable Transition Faults)** Consider the following circuit fragment. Assume the three inputs to the circuit are outputs of flip-flops.

a. Compute the set of untestable transition faults with launch-on-capture.

b. Compute the set of untestable transition faults with launch-on-shift.

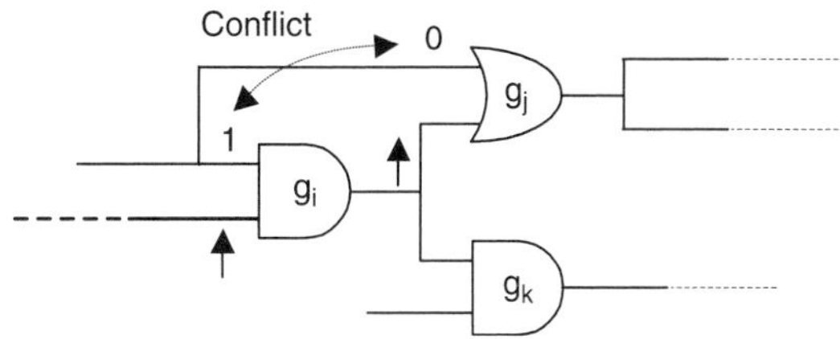

6.2 **(Pseudo-functional Testing)** Consider the following circuit fragment, and assume the three inputs to the circuit are outputs of flip-flops. Compute the set of unreachable states represented by the constraint ($g_j = 0$, $g_k = 1$).

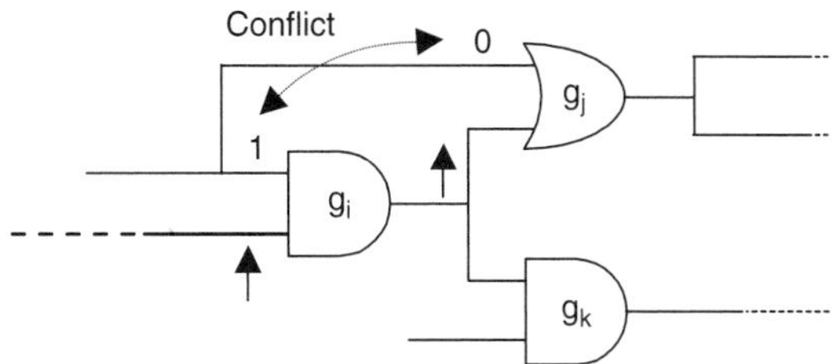

6.3 **(Pseudo-functional Testing)** Consider the following circuit fragment, where a, b, c, and d are outputs of flip-flops. Let the set of unreachable states be $abcd$ = {0000, 0001, 0010, 0100, 0101, 0110}. Derive a succinct constraint/expression for representing this set of unreachable states in terms of a,b,c,d,x, and y.

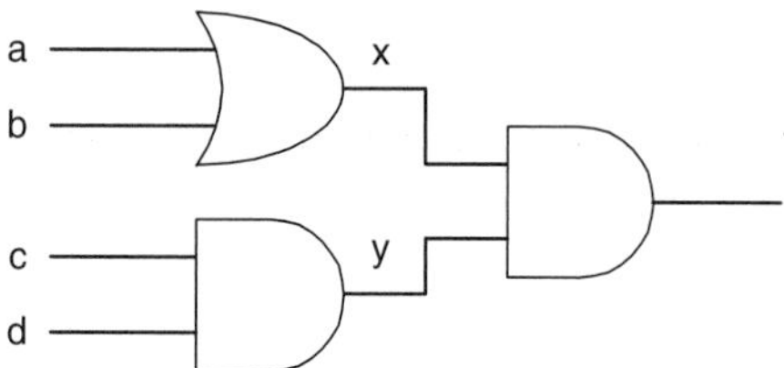

6.4 **(Transition Fault Testing)** For the circuit shown here, find the following tests, if they exist.

a. Find a test for H s@1.

b. Find a test for transition fault D slow to rise.

c. Find a test for transition fault F slow to fall.

d. To ensure test quality in part c, we want to make A a static 0. Please explain why.

e. For path C-F-J-L, find all its static, robust, and nonrobust path delay tests.

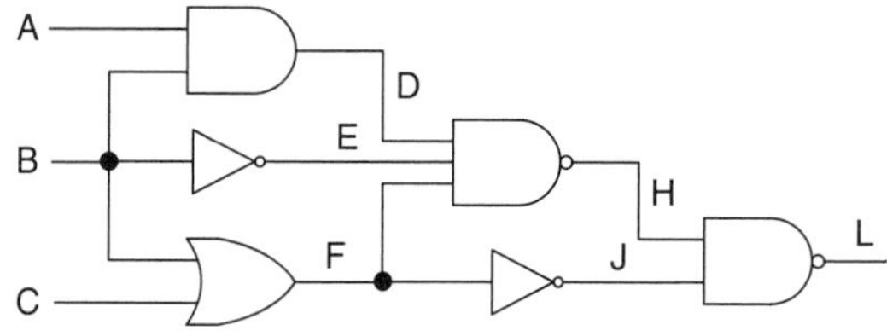

6.5 **(Circuit Delays)** For the circuit shown here and the given input transitions on inputs A and B, draw the waveforms for C, D, and E, for the specified delays.

a. Zero delay for each gate.

b. Unit delay for each gate.

c. Rising delay of 5, and falling delay of 7 for each gate.

d. Rising/falling delay of 5 for every gate, with inertial delay of 3.

e. Rising/falling delay (min, max) = (5, 7) for every gate.

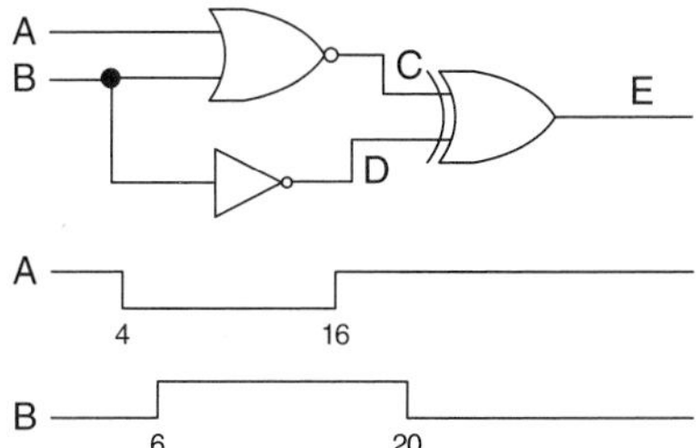

6.6 (Fault Count) For the two circuits shown here, determine the following.

a. The total number of transition delay faults.

b. The total number of path delay faults.

c. Longest path (assume unit gate delay).

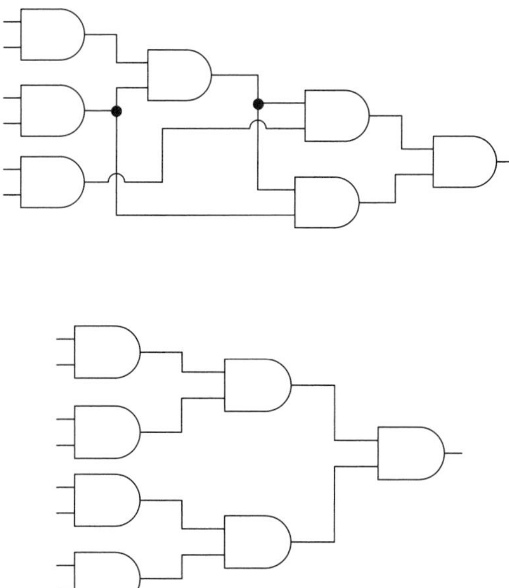

6.7 (Scan Test Time) Given a combinational circuit with eight scan inputs and three scan outputs, determine the total time to apply one LOS pattern and one LOC pattern. Assume the scan clock frequency is 20 MHz, and system clock frequency is 100 MHz.

6.8 **(Robust Path Delay Test)** For the circuit shown here, verify that all path-delay faults are robustly testable.

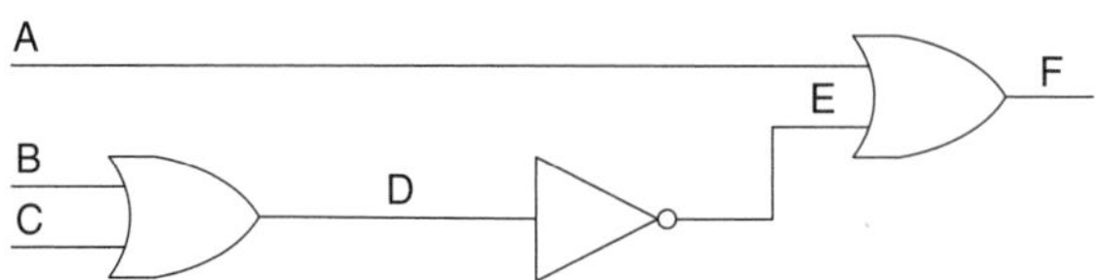

6.9 **(Path Delay Test)** Consider path P1: *B-D-F-G-I* and path P2: *B-D-E-H-I* in the following circuit.

a. Derive a robust test for path-delay fault ↑P2, where ↑ means a rising transition on the input to the path.

b. Can you derive a robust test for path-delay fault ↑P1? Why or why not?

c. Derive a nonrobust test for path-delay fault ↑P1.

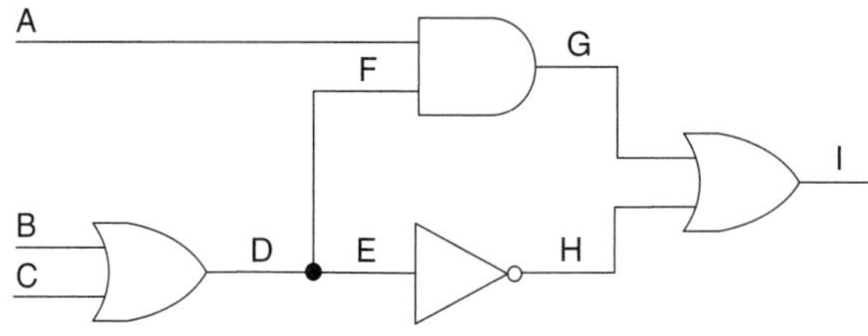

6.10 **(LOS and LOC)** In the timing diagram that follows, the launch clock cycle is labeled *L* and the capture clock cycle is labeled *C*. Draw the corresponding waveform of the scan enable signal for both the LOS and LOC test application methods.

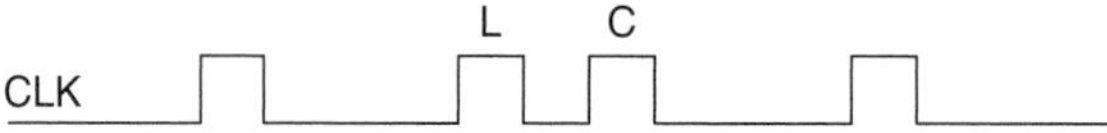

6.11 **(Path Sensitization)** What is the difference between robust test and nonrobust test?

6.12 **(Off-Path Signals)** Specify the off-path signal states for delay testing of a two-input XNOR gate.

6.13 **(Path Delay Test)** Consider the following circuit. Is the path delay fault {↓,*b-c-e-g-h-i*} robustly testable, nonrobustly testable, or only functionally sensitizable? How about path delay fault {↑,*a-c-f-h-i*}? Count the total number of robustly testable path delay faults in the circuit.

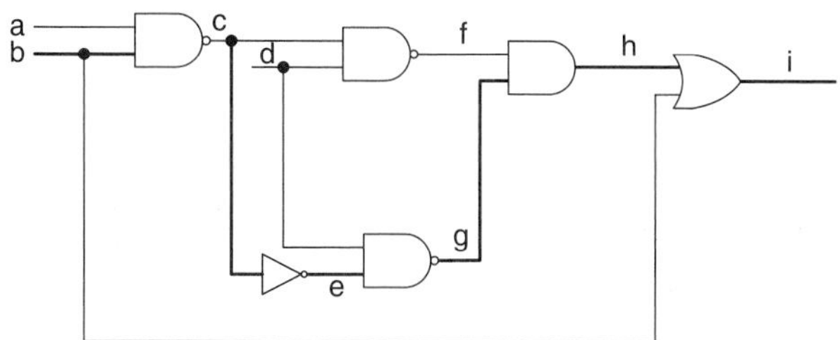

6.14 **(Sequential Path-Delay Fault Test)** In the following circuit, determine whether there are robust path-delay fault tests for faults {↑, *h-f-g*} and {↓, *h-f-g*}.

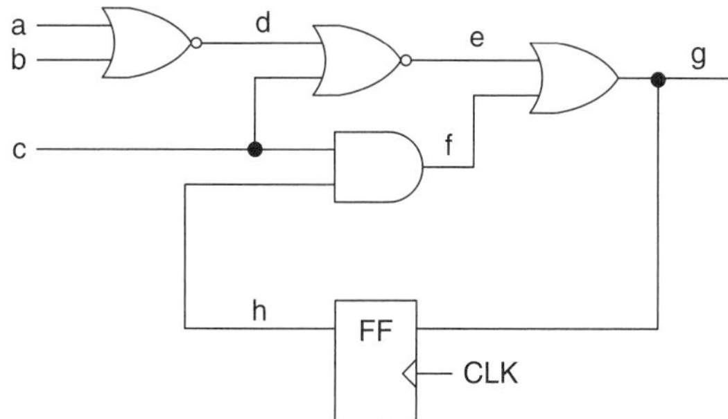

6.15  **(A Design Practice)** Use the ATPG programs and user's manuals contained on the Companion Web site to generate transition fault test patterns for the combinational (full-scan) versions of the three largest ISCAS89 circuits and record the number of test patterns needed for each circuit.

6.16  **(A Design Practice)** Follow the instructions for Exercise 6.15, but generate transition fault test patterns when the SDF file for the combinational circuit is available. Compare the delay size generated for each pattern.

Acknowledgments

The authors wish to thank Shrirang Yardi and Karthik Channakeshava of Virginia Tech for reviewing the text. The authors would also like to thank Jing Wang, Zheng Wang, and Lei Wu of Texas A&M University for reviewing the text and providing exercises.

References

R6.0 Books

[Bushnell 2000] M. Bushnell and V. Agrawal, *Essentials of Electronic Testing for Digital, Memory & Mixed-Signal VLSI Circuits*, Springer Science, New York, 2000.

[Jha 2003] N. Jha and S. Gupta, *Testing of Digital Systems*, Cambridge University Press, Cambridge, United Kingdom, 2003.

[Krstic 1998] A. Krstic and K.-T. Cheng, *Delay Fault Testing for VLSI Circuits*, Kluwer Academic Publishers, Boston, 1998.

[Wang 2006a] L.-T. Wang, C.-W. Wu, and X. Wen, editors, *VLSI Test Principles and Architectures: Design for Testability*, Morgan Kaufmann, San Francisco, 2006.

R6.2 Delay Test Application

[Amodeo 2005] M. Amodeo and B. Cory, Defining faster-than-at-speed delay tests, *Cadence Nanometer Test Q.*, 2(2), Paper 1, May 2005.

[Barnhart 2004] C. Barnhart, What is true-time delay test?, *Cadence Nanometer Test Q.*, 1(2), pp. 2–7, 2004.

[Kruseman 2004] B. Kruseman, A. Majhi, G. Gronthoud, and S. Eichenberger, On hazard-free patterns for fine-delay fault testing, in *Proc. IEEE Int. Test Conf.*, pp. 213–222, October 2004.

[Majhi 2000] A. Majhi, V. Agrawal, J. Jacob, and L. Patnaik, Line coverage of path delay faults, *IEEE Trans. on VLSI Systems*, 8(5), pp. 610–613, October 2000.

[Mao 1990] W. Mao and M. Ciletti, A variable observation time method for testing delay faults, in *Proc. ACM/IEEE Design Automation Conf.*, pp. 728–731, June 1990.

[Patel 1992] S. Patel and J. Savir, Skewed-load transition test: Part II, in *Proc. IEEE Int. Test Conf.*, pp. 714–722, September 1992.

[Qiu 2003a] W. Qiu and D. Walker, An efficient algorithm for finding the K longest testable paths through each gate in a combinational circuit, in *Proc. IEEE Int. Test Conf.*, pp. 592–601, September 2003.

[Qiu 2004a] W. Qiu, J. Wang, D. Walker, D. Reddy, X. Lu, Z. Li, W. Shi, and H. Balachandran, K longest paths per gate (KLPG) test generation for scan-based sequential circuits, in *Proc. IEEE Int. Test Conf.*, pp. 223–231, October 2004.

[Savir 1992] J. Savir, Skewed-load transition test: Part I, calculus, in *Proc. IEEE Int. Test Conf.*, pp. 705–713, September 1992.

[Savir 1994] J. Savir and S. Patel, Broad-side delay test, *IEEE Trans. on Computer-Aided Design*, 13(8), pp. 1057–1064, August 1994.

[Sharma 2002] M. Sharma and J. Patel, Finding a small set of longest testable paths that cover every gate, in *Proc. IEEE Int. Test Conf.*, pp. 974–982, October 2002.

R6.3 Delay Fault Models

[Barzilai 1983] Z. Barzilai and B. Rosen, Comparison of AC self-testing procedures, in *Proc. IEEE Int. Test Conf.*, pp. 89–94, October 1983.

[Benware 2003] B. Benware, R. Madge, C. Lu, and R. Daasch, Effectiveness comparisons of outlier screening methods for frequency dependent defects on complex ASICs, in *Proc. IEEE VLSI Test Symp.*, pp. 39–46, April/May 2003.

[Brockman 1989] J. Brockman and S. Director, Predictive subset testing: Optimizing IC parametric performance testing for quality, cost, and yield, *IEEE Trans. on Semiconductor Manufacturing*, 2(3), pp. 104–113, August 1989.

[Carter 1987] J. Carter, V. Iyengar, and B. Rosen, Efficient test coverage determination for delay faults, in *Proc. IEEE Int. Test Conf.*, pp. 418–427, September 1987.

[Chen 1999] W. Chen, S. Gupta, and M. Breuer, Test generation for crosstalk-induced delay in integrated circuits, in *Proc. IEEE Int. Test Conf.*, pp. 191–200, September 1999.

[Cheng 1993] K.-T. Cheng and H.-C. Chen, Delay testing for non-robust untestable circuits, in *Proc. IEEE Int. Test Conf.*, pp. 954–961, October 1993.

[Dworak 2004] J. Dworak, B. Cobb, J. Wingfield, and M. Mercer, Balanced excitation and its effect on the fortuitous detection of dynamic defects, in *Proc. Design, Automation, and Test in Europe Conf.*, pp. 1066–1071, February 2004.

[Heragu 1996] K. Heragu, J. Patel, and V. Agrawal, Segment delay faults: A new fault model, in *Proc. IEEE VLSI Test Symp.*, pp. 32–39, May 1996.

[Iyengar 1988] V. Iyengar, B. Rosen, and I. Spillinger, Delay test generation 1: Concepts and coverage metrics, in *Proc. IEEE Int. Test Conf.*, pp. 857–866, September 1988.

[Krstic 1999] A. Krstic, Y. Jiang, and K.-T. Cheng, Delay testing considering power supply noise effects, in *Proc. IEEE Int. Test Conf.*, pp. 181–190, September 1999.

[Krstic 2001] A. Krstic, J.-J. Liou, Y. Jiang, and K.-T. Cheng, Delay testing considering crosstalk-induced effects, in *Proc. IEEE Int. Test Conf.*, pp. 558–567, October 2001.

[Krstic 2003] A. Krstic, L.-C. Wang, K.-T. Cheng, J. Liou, and M. Abadir, Delay defect diagnosis based upon statistical timing models: The first step, in *Proc. IEEE Design, Automation and Test in Europe Conf.*, pp. 328–333, March 2003.

[Lam 1993] W. Lam, A. Saldanha, R. Brayton, and A. Sangiovanni-Vincentelli, Delay fault coverage and performance trade-offs, in *Proc. ACM/IEEE Design Automation Conf.*, pp. 446–452, June 1993.

[Li 2003] Z. Li, X. Lu, W. Qiu, W. Shi, and D. Walker, A circuit level fault model for resistive bridges, *ACM Trans. on Design Automation of Electronic Systems (TODAES)*, 8(4), pp. 546–559, October 2003.

[Lin 2005a] X. Lin and J. Rajski, Propagation delay fault: A new fault model to test delay faults, in *Proc. IEEE Asian and South Pacific Design Automation Conf.*, pp. 178–183, January 2005.

[Lin 2006a] X. Lin, K.-H. Tsai, C. Wang, M. Kassab, J. Rajski, T. Kobayashi, R. Klingenberg, Y. Sato, H. Shuji, and A. Takashi, Timing-aware ATPG for high quality at-speed testing of small delay defects, in *Proc. IEEE Asian Test Symp.*, pp. 139–146, November 2006.

[Liou 2002a] J. Liou. A. Krstic, L.-C. Wang, and K.-T. Cheng, False-path-aware statistical timing analysis and efficient path selection for delay testing and timing validation, in *Proc. ACM/IEEE Design Automation Conf.*, pp. 566–569, June 2002.

[Liou 2003] J. Liou, A. Krstic, Y. Jiang, and K.-T. Cheng, Modeling, testing and analysis for delay defects and noise effects in deep submicron devices, *IEEE Trans. on Computer-Aided Design*, 22(6), pp. 756–769, June 2003.

[Lu 2005] X. Lu, Z. Li, W. Qiu, D. Walker, and W. Shi, Longest path selection for delay test under process variation, *IEEE Trans. on Computer-Aided Design*, 24(12), pp. 1924–1929, December 2005.

[Luong 1996] G. Luong and D. Walker, Test generation for global delay faults, in *Proc. IEEE Int. Test Conf.*, pp. 433–442, October 1996.

[Ma 1995] S. Ma, P. Franco, and E. McCluskey, An experimental chip to evaluate test techniques: Experimental results, in *Proc. IEEE Int. Test Conf.*, 663–672, October 1995.

[Majhi 2000] A. Majhi, V. Agrawal, J. Jacob, and L. Patnaik, Line coverage of path delay faults, *IEEE Trans. on VLSI Systems*, 8(5), pp. 610–613, October 2000.

[McCluskey 2004] E. McCluskey, A. Al-Yamani, J. Li, C.-W. Tseng, E. Volkerink, F.-F. Ferhani, E. Li, and S. Mitra, ELF-Murphy data on defects and test sets, in *Proc. IEEE VLSI Test Symp.*, pp. 16–22, April 2004.

[Park 2005] I. Park, A. Al-Yamani and E. McCluskey, Effective TARO pattern generation, in *Proc. IEEE VLSI Test Symp.*, 161–166, May 2005.

[Pomeranz 1999] I. Pomeranz and S. Reddy, On N-detection test sets and variable N-detection test sets for transition faults, in *Proc. IEEE VLSI Test Symp.*, pp. 173–179, April 1999.

[Qiu 2003b] W. Qiu, X. Lu, Z. Li, D. Walker, and W. Shi, CodSim: A combined delay fault simulator, in *Proc. IEEE Int. Symp. on Defect and Fault Tolerance in VLSI Systems*, pp. 79–86, November 2003.

[Qiu 2003c] W. Qiu and D. Walker, Testing the path delay faults for ISCAS85 circuit c6288, in *Proc. IEEE Int. Workshop on Microprocessor Test and Verification*, pp. 38–43, May 2003.

[Qiu 2004b] W. Qiu, X. Lu, J. Wang, Z. Li, D. Walker, and W. Shi, A statistical fault coverage metric for realistic path delay faults, in *Proc. IEEE VLSI Test Symp.*, pp. 37–42, April 2004.

[Rearick 2005] J. Rearick and R. Rodgers, Calibrating clock stretch during AC-scan test, in *Proc. IEEE Int. Test Conf.*, Paper 11.3, November 2005.

[Sato 2005] Y. Sato, S. Hamada, T. Maeda, A. Takatori, and S. Kajihara, Evaluation of the statistical delay quality model, in *Proc. IEEE Asian and South Pacific Design Automation Conf.*, pp. 305–310, January 2005.

[Saxena 2003] J. Saxena, K. Butler, V. Jayaram, S. Kundu, N. Arvind, P. Sreeprakash, and M. Hachinger, A case study of IR-drop in structured at-speed testing, in *Proc. IEEE Int. Test Conf.*, pp. 1098–1104, September 2003.

[Sivaraman 1996] M. Sivaraman and A. Strojwas, Delay fault coverage: A realistic metric and an estimation technique for distributed path delay faults, in *Proc. IEEE/ACM Int. Conf. on Computer Aided Design*, pp. 494–501, November 1996.

[Smith 1985] G. Smith, Model for delay faults based upon paths, in *Proc. IEEE Int. Test Conf.*, pp. 342–349, October 1985.

[Stanojevic 2001] Z. Stanojevic and D. Walker, FedEx: A fast bridging fault extractor, in *Proc. IEEE Int. Test Conf.*, pp. 696–703, October 2001.

[Tseng 2001] C.-W. Tseng and E. McCluskey, Multiple-output propagation transition fault test, in *Proc. IEEE Int. Test Conf.*, pp. 258–366, October 2001.

[Waicukauski 1987] J. Waicukauski, E. Lindbloom, B. Rosen, and V. Iyengar, Transition fault simulation, *IEEE Design & Test of Computers*, 4(5), pp. 32–38, April 1987.

[Wang 2006b] J. Wang, D. Walker, A. Majhi, B. Kruseman, G. Gronthoud, L. Elvira Villagra, P. van de Wiel, and S. Eichenberger, Power supply noise in delay testing, in *Proc. IEEE Int. Test Conf.*, Paper 17.3, October 2006.

R6.4 Delay Test Sensitization

[Benkoski 1990] J. Benkoski, E. Meersch, L. Claesen, and H. Man, Timing verification using statically sensitizable paths, *IEEE Trans. on Computer-Aided Design*, 9(10), pp. 1073–1084, October 1990.

[Chang 1993] H. Chang and J. Abraham, VIPER: An efficient vigorously sensitizable path extractor, in *Proc. ACM/IEEE Design Automation Conf.*, pp. 112–117, June 1993.

[Cheng 1996a] K.-T. Cheng and H.-C. Chen, Classification and identification of non-robust untestable path delay faults, *IEEE Trans. on Computer-Aided Design*, 15(8), pp. 845–853, August 1996.

[Lin 1987] C. Lin and S. Reddy, On delay fault testing in logic circuits, *IEEE Trans. on Computer-Aided Design*, 6(5), pp. 694–703, September 1987.

[Liou 2002a] J. Liou, A. Krstic, L.-C. Wang, and K.-T. Cheng, False-path-aware statistical timing analysis and efficient path selection for delay testing and timing validation, in *Proc. ACM/IEEE Design Automation Conf.*, pp. 566–569, June 2002.

[Liou 2002b] J. Liou, L.-C. Wang, and K.-T. Cheng, On theoretical and practical considerations of path selection for delay fault testing, in *Proc. IEEE/ACM Int. Conf. on Computer Aided Design*, pp. 94–100, November 2002.

[McGeer 1989] P. McGeer and R. Brayton, Efficient algorithms for computing the longest viable path in a combinational network, in *Proc. ACM/IEEE Design Automation Conf.*, pp. 561–567, June 1989.

[Smith 1985] G. Smith, Model for delay faults based upon paths, in *Proc. IEEE Int. Test Conf.*, pp. 342–349, October 1985.

R6.5 Delay Fault Simulation

[Carter 1987] J. Carter, V. Iyengar, and B. Rosen, Efficient test coverage determination for delay faults, in *Proc. IEEE Int. Test Conf.*, pp. 418–427, September 1987.

[Cheng 1996b] K.-T. Cheng, A. Krstic, and H.-C. Chen, Generation of high quality tests for robustly untestable path delay faults, *IEEE Trans. on Computers*, 45(12), pp. 1379–1396, December 1996.

[Fink 1992] F. Fink, K. Fuchs, and M. Schulz, Robust and nonrobust path delay fault simulation by parallel processing of patterns, *IEEE Trans. on Computers*, 41(12), pp. 1527–1536, December 1992.

[Kagaris 1997] D. Kagaris, S. Tragoudas, and D. Karayiannis, Improved non-enumerative path delay fault coverage estimation based on optimal polynomial time algorithms, *IEEE Trans. on Computer-Aided Design*, 16(3), pp. 309–315, March 1997.

[Li 2003] Z. Li, X. Lu, W. Qiu, W. Shi, and D. Walker, A circuit level fault model for resistive bridges, *ACM Trans. on Design Automation of Electronic Systems (TODAES)*, 8(4), pp. 546–559, October 2003.

[Pramanick 1988] A. Pramanick and S. Reddy, On the detection of delay faults, in *Proc. IEEE Int. Test Conf.*, pp. 845–856, September 1988.

[Qiu 2003b] W. Qiu, X. Lu, Z. Li, D. Walker, and W. Shi, CodSim: A combined delay fault simulator, in *Proc. IEEE Int. Symp. on Defect and Fault Tolerance in VLSI Systems*, pp. 79–86, November 2003.

[Schulz 1989] M. Schulz, F. Fink, and K. Fuchs, Parallel pattern fault simulation of path delay faults, in *Proc. ACM/IEEE Design Automation Conf.*, pp. 357–363, June 1989.

[Smith 1985] G. Smith, Model for delay faults based upon paths, in *Proc. IEEE Int. Test Conf.*, pp. 342–349, October 1985.

[Waicukauski 1987] J. Waicukauski, E. Lindbloom, B. Rosen, and V. Iyengar, Transition fault stimulation, *IEEE Design & Test of Computers*, 4(2), pp. 32–38, April 1987.

R6.6 Delay Fault Test Generation

[Benkoski 1990] J. Benkoski, E. Meersch, L. Claesen, and H. Man, Timing verification using statically sensitizable paths, *IEEE Trans. on Computer-Aided Design*, 9(10), pp. 1073–1084, 1990.

[Fuchs 1991] K. Fuchs, F. Fink, and M. Schulz, DYNAMITE: An efficient automatic test pattern generation system for path delay faults, *IEEE Trans. on Computer-Aided Design*, 10(10), pp. 1323–1355, October 1991.

[Fuchs 1994] K. Fuchs, M. Pabst, and T. Rossel, RESIST: A recursive test pattern generation algorithm for path delay faults considering various test classes, *IEEE Trans. on Computer-Aided Design*, 13(12), pp. 1550–1562, December 1994.

[Gupta 2004] P. Gupta and M. Hsiao, ALAPTF: A new transition fault model and the ATPG algorithm, in *Proc. IEEE Int. Test Conf.*, pp. 1053–1060, October 2004.

[Kajihara 2006] S. Kajihara, S. Morishima, A. Takuma, X. Wen, T. Maeda, S Hamada, and Y. Sato, A framework of high-quality transition fault ATPG for scan circuits, in *Proc. IEEE Int. Test Conf.*, Paper 2.1, October 2006.

[Li 1989] W. Li, S. Reddy, and S. Sahni, On path selection in combinational logic circuits, *IEEE Trans. on Computer-Aided Design*, 8(1), pp. 56–63, January 1989.

[Lin 1987] C. Lin and S. Reddy, On default testing in logic circuits, *IEEE Trans. on Computer-Aided Design*, 6(5), pp. 694–703, September 1987.

[Lin 2005a] X. Lin and J. Rajski, Propagation delay fault: A new fault model to test delay faults, in *Proc. IEEE Asian and South Pacific Design Automation Conf.*, pp. 178–183, January 2005.

[Lin 2006a] X. Lin, K.-H. Tsai, C. Wang, M. Kassab, J. Rajski, T. Kobayashi, R. Klingenberg, Y. Sato, H. Shuji, and A. Takashi, Timing-aware ATPG for high quality at-speed testing of small delay defects, in *Proc. IEEE Asian Test Symp.*, pp. 139–146, November 2006.

[Liu 2002] X. Liu, M. Hsiao, S. Chakravarty, and P. Thadikaran, Novel ATPG algorithms for transition faults, in *Proc. IEEE European Test Workshop*, pp. 47–52, May 2002.

[Liu 2003] X. Liu and M. Hsiao, Constrained ATPG for broadside transition testing, in *Proc. IEEE Int. Symp. on Defect and Fault Tolerance in VLSI Systems*, pp. 175–182, November 2003.

[Liu 2005] X. Liu and M. Hsiao, A novel transition fault ATPG to reduce yield loss, *IEEE Design & Test of Computers*, 22(6), pp. 576–584, November/December 2005.

[Majhi 2000] A. Majhi, V. Agrawal, J. Jacob, and L. Patnaik, Line coverage of path delay faults, *IEEE Trans. on VLSI Systems*, 8(5), pp. 610–613, October 2000.

[Murakami 2000] A. Murakami, S. Kajihara, T. Sasao, I. Pomeranz, and S. Reddy, Selection of potentially testable path delay faults for test generation, in *Proc. IEEE Int. Test Conf.*, pp. 376–384, October 2000.

[Pomeranz 1995] I. Pomeranz, S. Reddy, and P. Uppaluri, NEST: A nonenumerative test generation method for path delay faults in combinational circuits, *IEEE Trans. on Computer-Aided Design*, 14(12), pp. 1505–1515, December 1995.

[Qiu 2003a] W. Qiu and D. Walker, An efficient algorithm for finding the K longest testable paths through each gate in a combinational circuit, in *Proc. IEEE Int. Test Conf.*, pp. 592–601, September 2003.

[Qiu 2004a] W. Qiu, J. Wang, D. Walker, D. Reddy, X. Lu, Z. Li, W. Shi, and H. Balachandran, K longest paths per gate (KLPG) test generation for scan-based sequential circuits, in *Proc. IEEE Int. Test Conf.*, pp. 223–231, October 2004.

[Qiu 2004b] W. Qiu, X. Lu, J. Wang, Z. Li, D. Walker, and W. Shi, A statistical fault coverage metric for realistic path delay faults, in *Proc. IEEE VLSI Test Symp.*, pp. 37–42, April 2004.

[Qiu 2006] W. Qiu, D. Walker, N. Simpson, D. Reddy, and A. Moore, Delay tests on silicon, in *Proc. IEEE Int. Test Conf.*, Paper 11.3, October 2006.

[Shao 2002] Y. Shao, S. Reddy, I. Pomeranz, and S. Kajihara, On selecting testable paths in scan designs, in *Proc. IEEE European Test Workshop*, pp. 53–58, May 2002.

[Sharma 2002] M. Sharma and J. Patel, Finding a small set of longest testable paths that cover every gate, in *Proc. IEEE Int. Test Conf.*, pp. 974–982, October 2002.

[Waicukauski 1987] J. Waicukauski, E. Lindbloom, B. Rosen, and V. Iyengar, Transition fault simulation, *IEEE Design & Test of Computers*, 4(5), pp. 32–38, April 1987.

R6.7 Pseudo-Functional Testing to Avoid Over-Testing

[Gouders 1993] N. Gouders and R. Kaibel, Advanced techniques for sequential test generation, in *Proc. IEEE European Test Conf.*, pp. 293–300, April 1993.

[Konijnenburg 1999] M. Konijnenburg, J. Van Der Linden, and A. Van de Goor, Illegal state space identification for sequential test generation, in *Proc. IEEE Design and Test in Europe*, pp. 741–746, March 1999.

[Krstic 2003] A. Krstic, J. Liou, and K.-T. Cheng, On structural versus functional tests for delay faults, in *Proc. IEEE Int. Symp. on Quality Electronic Design*, pp. 453–469, March 2003.

[Lin 2005b] Y.-C. Lin, F. Lu, K. Yang, and K.-T. Cheng, Constraint extraction for pseudo-functional scan-based delay testing, in *Proc. IEEE Asian and South Pacific Design Automation Conf.*, pp. 166–171, January 2005.

[Lin 2006b] Y.-C. Lin, F. Lu, and K-T. Cheng, Pseudo-functional testing, *IEEE Trans. on Computer-Aided Design*, 25(8), pp. 1535–1546, August 2006.

[Liu 2003] X. Liu and M. Hsiao, Constrained ATPG for broadside transition testing, in *Proc. IEEE Int. Symp. on Defect and Fault Tolerance in VLSI Systems*, pp. 175–182, November 2003.

[Pomeranz 2004] I. Pomeranz, On the generation of scan-based test sets with reachable states for testing under functional operations conditions, in *Proc. ACM/IEEE Design Automation Conf.*, pp. 928–933, June 2004.

[Rearick 2001] J. Rearick, Too much delay fault coverage is a bad thing, in *Proc. IEEE Int. Test Conf.*, pp. 624–633, October 2001.

[Syal 2006] M. Syal, K. Chandrasekar, V. Vimjam, M. Hsiao, Y.-S. Chang, and S. Chakravarty, A study of implication based pseudo functional testing, in *Proc. IEEE Int. Test Conf.*, Paper 24.3, October 2006.

[Zhang 2005] Z. Zhang, S. Reddy, and I. Pomeranz, On generating pseudo-functional delay fault tests for scan designs, in *Proc. IEEE Int. Symp. on Defect and Fault Tolerance in VLSI Systems*, pp. 398–405, October 2005.

[Zhao 1997] J. Zhao, E. Rudnick, and J. Patel, Static logic implication with application to redundancy identification, in *Proc. IEEE VLSI Test Symp.*, pp. 288–293, April 1997.

CHAPTER 7

Low-Power Testing

Patrick Girard
LIRMM/CNRS, Montpellier, France

Xiaoqing Wen
Kyushu Institute of Technology, Fukuoka, Japan

Nur A. Touba
University of Texas, Austin, Texas

ABOUT THIS CHAPTER

Power dissipation has become a major design objective in many application areas, such as wireless communications and high-performance computing, thus leading to the production of numerous low-power designs. At the same time, power dissipation is also becoming a critical parameter during manufacturing test, as the design can consume much more power during testing than during the functional mode of operation. Because test throughput and manufacturing yield are often affected by **test power**, dedicated test methodologies have emerged since the mid-1990s.

In this chapter, we discuss issues arising from excessive power consumption during test application as well as provide structural and algorithmic solutions that can alleviate the low-power test problems. We first review some basic elements of power modeling and related terminologies. After discussing test power issues, promising low-power test techniques to deal with nanometer ***system-on-chip*** (SOC) designs are presented. These techniques can be broadly classified into those that apply during **scan** testing and those that apply during ***built-in self-test*** (BIST). A few of them are also applicable to test compression circuits or memory designs.

In the literature, techniques that reduce power consumption during test application are generally referred to as *power-conscious testing*, *power-aware testing*, *power-constrained testing*, or *low-power testing*. These terms are used interchangeably throughout the chapter whenever fit.

7.1 INTRODUCTION

With the advance in semiconductor manufacturing technology, a ***very-large-scale-integrated*** (VLSI) device can now contain tens to hundreds of millions of transistors. Because this trend is predicted to continue at least into the 2010s per Moore's law [Moore 1965], severe challenges are imposed on tools and methodologies used to design and test complex VLSI circuits. Addressing these design and test challenges in an efficient way is becoming increasingly difficult [SIA 2005].

Testing currently ranks among the most important issues in the development process of an integrated circuit. The issues that center on testing are manufacturing yield, product quality, and test cost. To address these test issues, ***design-for-testability*** (DFT) techniques [Bushnell 2000] [Jha 2003] [Wang 2006] have become widely used in industry since the 1990s. Traditionally, these techniques are mainly employed to improve the circuit's fault coverage, test application time, and test development efforts. The advances in low-power design techniques and deep-submicron manufacturing technologies, however, have spurred the rapid growth of electronic products into consumer markets using laptop computers, cellular phones, audio and video-based multimedia products, energy-efficient desktop computers, etc. These new products make power management a critical issue that needs to be considered not only during circuit design but also during test development [Crouch 1999] [De Colle 2005].

The main motivation for considering power consumption during testing is that, generally, a circuit consumes much more power in test mode than in normal mode [Zorian 1993] [Rajski 1998] [Girard 2000] [Pouya 2000] [Bushnell 2000] [SIA 2001] [Saxena 2003] [Nicolici 2003]. In [Zorian 1993], the author showed that test power can be more than twice the power consumed in normal functional mode. There are several explanations for this increase in **test power**. First, modern automatic test pattern generation (ATPG) tools tend to generate test patterns with a high toggle rate in order to reduce pattern count and thus test application time. Thus, the node switching activity of the device in test mode is often several times higher than that in normal mode. Second, parallel testing (*e.g.*, testing a few memories in parallel) is often used to reduce test application time, particularly for ***system-on-chip*** (SOC) devices. This parallelism inevitably increases power dissipation during testing. Third, the DFT circuitry inserted in the circuit to alleviate test issues is often idle during normal operation but may be used intensively in test mode. This surplus of active elements during testing again induces an increase of power dissipation. Finally, this elevated test power can come from the lack of correlation between consecutive test patterns, while the correlation between successive functional input vectors applied to a given circuit during normal operation is generally high [Wang 1997].

For instance, in a speech signal processing circuit, the input vectors behave in a predictable manner, with the least significant bits more likely to change than the most significant bits. Similarly, in high-speed circuits that process digital audio and video signals, the inputs to most of those modules change relatively slowly over time. The low-power designers often take advantage of this fact when they determine the thermal and electrical limits of the circuit and system packaging requirements. In contrast, there is no definite correlation between successive test patterns generated by an ATPG tool during scan testing or produced by a ***pseudo-random pattern generator*** (PRPG) during logic BIST. As power dissipation in CMOS circuits is proportional to switching activity, this excessive switching activity during testing can cause catastrophic problems, such as instantaneous circuit damage, test-induced yield loss because of noise phenomena, reduced reliability, product cost increase, or reduced autonomy for battery-operated devices.

To reduce this increased power consumption during test application, the industry generally resorts to ad hoc solutions [Monzel 1997]. These solutions include the following:

- *Oversizing power and ground rails to allow higher current densities in the circuit under test.* This allows additional power to be supplied to the circuit to satisfy the increase in switching activity that occurs during testing. However, this solution raises several problems. By increasing the power available for the circuit, the amount of energy (heat) that needs to be dissipated also increases, which in turn leads to additional problems related to the thermal constraints of the circuit (these problems are discussed in Section 7.3). It is possible to avoid these problems by using packages with higher thermal capabilities or by using higher performance cooling systems. However, the impact on the final product cost may prevent the use of these solutions. Another problem is that this solution affects the entire design and may require an early estimation of the power consumption during testing. As test data are generally not available in the early stages of the design process, this solution may not be satisfactory in all cases.
- *Testing with a reduced operating frequency.* This solution does not require additional hardware, but it increases the test application time and may lead to a loss of defect coverage as timing-related faults may escape detection. In effect, this solution reduces power consumption at the expense of longer test time and does not reduce the total energy consumed during testing.
- *Partitioning of the circuit under test with appropriate test planning.* This solution, although effective from a power-reduction point of view, increases test time because it reduces test concurrency. Moreover, it generally requires circuit design modifications (often with additional multiplexers), thus impacting final product cost and circuit performance.

Considering the problems associated with these ad hoc approaches and the need to provide an adequate remedy to the problems, numerous solutions have been proposed to cope with test power problems during testing. These solutions can be classified based on whether they apply during scan testing or whether they apply during logic BIST. A few of them can also be used with memory designs or can be used in conjunction with test compression. These solutions are explained in detail in the next sections.

7.2 ENERGY AND POWER MODELING

A logical step before discussing promising low-power test solutions is to correctly define the terminology and the associated energy and power models. In this section, we first review the electronic basics related to power consumption and power dissipation and then discuss terminology and test-power modeling.

7.2.1 Basics of Circuit Theory

Consider the generic representation of a ***complementary metal-oxide semiconductor*** (CMOS) logic gate shown in Figure 7.1. The load capacitance C_L of the output node, representing the input capacitance of the next logic stage as well as interconnect and diffusion capacitances, is connected to the supply voltage V_{dd} through a pull-up block composed of ***positive metal-oxide semiconductor*** (PMOS) transistors and to the ground through a pull-down block composed of ***negative metal-oxide semiconductor*** (NMOS) transistors.

A switching on the gate output corresponds to the charge or discharge of the load capacitance C_L. In the process of charging the output (from 0 to 1), a charge $Q = C_L \cdot V_{dd}$ is delivered to the load. The power supply must supply this charge at voltage V_{dd}, so the energy supplied is $Q \cdot V_{dd} = C_L \cdot V_{dd}^2$. However, the energy stored on a capacitance C_L charged to V_{dd} is only half of this (*i.e.*, $\frac{1}{2} \cdot C_L \cdot V_{dd}^2$) [Athas 1994]. In accordance with the **energy conservation principle**, the other half must be dissipated by the PMOS transistors in the pull-up network. Similarly, when the inputs change again causing the output to discharge (from 1 to 0), all the energy stored on the capacitance C_L is inevitably dissipated in the pull-down network, because no energy can enter the ground rail ($Q \cdot V_{gnd} = Q \cdot 0 = 0$). In both cases, the energy is dissipated as heat.

There are three components to the power consumed by the logic gate: (1) the **dynamic power**, because of the charge of capacitance C_L; (2) the **short-circuit power**, because of the short circuit between power and ground during switching; and (3) the **leakage power**. The main component is the dynamic power, which still represents a significant fraction of the total power consumption despite the proportional increase of the other two components with technology improvements. This dynamic power consumption occurs during the charge of the load capacitance C_L (transition from 0 to 1 on the gate output) as a current I flows between power and ground through the capacitance. The dynamic power consumed during the time interval [0,*T*] is therefore $P_{dyn} = V_{dd} \cdot I = V_{dd} \cdot Q \cdot 1/T$, where $Q = C_L \cdot V_{dd}$.

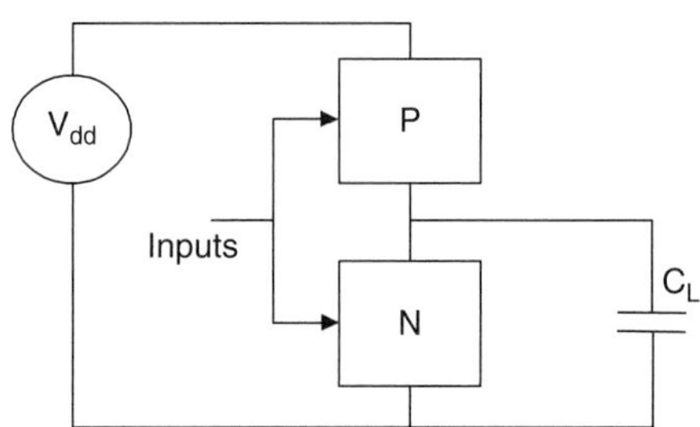

FIGURE 7.1

Generic representation of a CMOS logic gate.

As several transitions may occur during the time interval [0,T], the dynamic power consumption can be expressed as follows:

$$P_{dyn} = C_L \cdot V_{dd}^2 \cdot N_{0\to1} \cdot 1/T$$

where $N_{0\to1}$ represents the number of rising transitions at the gate output during the time interval [0,T]. Without loss of generality, it can be assumed that the number of rising transitions is equal to half of the total number of N transitions at the gate output. The dynamic power consumption of the logic gate during the time interval [0,T] can finally be expressed as follows:

$$P_{dyn} = ½ \cdot C_L \cdot V_{dd}^2 \cdot N \cdot 1/T$$

This analysis shows that **dynamic power consumption** occurs during the charge of node output capacitance, whereas **power dissipation**, which is related to **energy dissipation**, occurs during the charge or discharge of each node. Because power dissipated by N rising or falling transitions during the time interval [0,T] is $E_n/T = ½ \cdot C_L \cdot V_{dd}{}^2 \cdot N \cdot 1/T$, which is the same as power consumption, the terms *power dissipation* and *power consumption* are used without distinction throughout this chapter.

7.2.2 Terminology

We use the same terminology as defined in [Weste 1993] to denote power consumption measures used for low-power testing:

- **Energy** represents the total switching activity generated during the application of the complete test sequence. An energy increase during testing has impact on the battery lifetime of battery-operated devices, particularly those equipped with online test facilities or those submitted to test procedures during power-up (such as cellular phones).
- **Average power** corresponds to the ratio between the total energy dissipated during testing and the test time. Elevated average power during testing adds to the thermal load that must be vented away from the device under test (temperature increase). It may cause structural damage to the silicon (hot spots), or it may lead to phenomena that alter the circuit reliability.
- **Instantaneous power** corresponds to the power consumed at any given instant during testing. Usually, it is defined as the power consumed right after the application of a synchronizing clock signal. Elevated instantaneous power may cause a supply voltage drop and alter the correct behavior of the circuit.
- **Peak power** corresponds to the highest value of instantaneous power measured during testing. The peak power generally determines the thermal and electrical limits of the circuit and the system packaging requirements. If the peak power exceeds a certain limit, the circuit may be subjected to structural

degradation and, in some cases, be destroyed. From a theoretical point of view, the peak power is defined from the values of instantaneous power measured in short time intervals (*i.e.*, the system clock period). In practice, the time window for the definition of peak power is related to the thermal capacity of the chip, and restricting this window within just one clock period is not realistic enough. For example, if the circuit has a peak power consumption during only one cycle but it has power consumption within the limit of thermal capacity of the chip for all other cycles, the circuit may not be damaged because the energy consumed may not be enough to elevate chip temperature over the thermal capacity limit of the chip (unless the peak power consumption is far higher than normal power consumption). To damage the circuit, high power consumption should last for several successive clock cycles to consume enough energy to elevate chip temperature over the limit [Shi 2004]. On the other hand, high peak power in only one clock cycle can be an issue if it results in a significant ground bounce or an IR-drop phenomenon that causes a memory element to lose its state and the test procedure to unnecessarily fail. This problem will be discussed further in Section 7.3.2.

7.2.3 Test-Power Modeling and Evaluation

As mentioned previously, most power dissipated in a CMOS circuit comes from the charge and discharge of capacitances during switching. To explain this power dissipation during testing, let us consider a circuit composed of N nodes and a test sequence of length L applied to the circuit inputs. The **average energy** consumed at node i per switching is $\frac{1}{2} \cdot C_i \cdot V_{dd}^2$, where C_i is the equivalent output capacitance at node i and V_{dd} the power supply voltage [Cirit 1987]. A good approximation of the energy consumed at node i in a time interval t is $\frac{1}{2} \cdot C_i \cdot S_i \cdot V_{dd}^2$, where S_i is the average number of transitions during this interval (also called the **switching activity factor** at node i). Furthermore, nodes connected to more than one logic gate in the circuit are nodes with a higher output capacitance. Based on this fact, and in a first approximation, it can be stated that output capacitance C_i is proportional to the fanout at node i, denoted as F_i [Wang 1995]. Therefore, an estimation of the energy E_i consumed at node i during the time interval t is given by:

$$E_i = \frac{1}{2} \cdot S_i \cdot F_i \cdot C_0 \cdot V_{dd}^2$$

where C_0 is the minimum output capacitance of the circuit. According to this expression, energy consumption at the logic level is a function of the fanout F_i and the switching activity factor S_i. The fanout F_i is defined by circuit topology, and the activity factor S_i can be estimated by a logic simulator. The product $F_i \cdot S_i$ is named ***weighted switching activity*** (WSA) at node i and represents the only variable part in the energy consumed at node i during test application.

According to the preceding formulation, the energy consumed in the circuit after application of a pair of successive input vectors (V_{k-1}, V_k) can be expressed by:

$$E_{Vk} = \frac{1}{2} \cdot C_0 \cdot V_{dd}^2 \cdot \sum_i S_i(k) \cdot F_i$$

where i ranges across all the nodes of the circuit, and $S_i(k)$ is the number of transitions provoked by V_k at node i. Now, let us consider the complete test sequence of length L required to achieve the target fault coverage. The **total energy** consumed in the circuit after the application of the complete test sequence is given here, where k ranges across all the vectors of the test sequence:

$$E_{total} = ½ \cdot C_0 \cdot V_{dd}^2 \cdot \sum_k \sum_i S_i(k) \cdot F_i$$

By definition, power is given by the ratio between energy and time. The instantaneous power is generally calculated as the amount of power required during a small instant of time t_{small} such as the portion of a clock cycle immediately following the system clock rising or falling edge. Consequently, the instantaneous power dissipated in the circuit after the application of a test vector V_k can be expressed by:

$$P_{inst}(V_k) = E_{Vk}/t_{small}$$

The peak power corresponds to the maximum value of instantaneous power measured during testing. Therefore, it can be expressed in terms of the highest energy consumed during a small instant of time during the test session:

$$P_{peak} = Max_k P_{inst}(V_k) = Max_k(E_{Vk}/t_{small})$$

Finally, the average power consumed during the test session can be calculated from the total energy and the test time. Considering that the test time is given by the product $L \cdot T$, where T corresponds to the nominal clock period of the circuit, the average power can be expressed as follows:

$$P_{average} = E_{total}/(L \cdot T)$$

The preceding expressions of power and energy, although based on a simplified model, are accurate enough for the intended purpose of power analysis during testing. According to these expressions, and assuming a given CMOS technology and a supply voltage for the considered circuit, it appears that the switching activity factor S_i is the only parameter that has impact on the energy, peak power, and average power. This explains why most of the methods proposed so far for reducing power or energy during testing are based on a reduction of the switching activity factor.

7.3 TEST POWER ISSUES

When verifying the correct functions of high-density systems such as an SOC, test procedures and test techniques have to satisfy all power constraints defined in the design phase. In other words, these procedures and techniques must be so that the power consumed during testing remains comparable to that consumed during the functional mode. Ignoring these constraints can expose the circuit to various problems, such as premature destruction, noise phenomena that can lead to yield loss, reduced reliability, product cost increase, reduced autonomy (for battery-operated devices), etc. This section lists a few of these important problems.

7.3.1 Thermal Effects

The heat produced during the operation of a circuit is proportional to the dissipated power. This heat is produced by the collision of carriers with the conductor molecular structure (a friction phenomenon called the **Joule effect**) and is responsible for the temperature increase observed during operation [Altet 2002]. Therefore, there is a relationship between die temperature and power dissipation. It can be formulated from the **laws of thermodynamics** as follows [Weste 1993]:

$$T_{die} = T_{air} + \theta \times P_d$$

where T_{die} is the *die temperature*, T_{air} is the temperature of surrounding air, θ is the *package thermal impedance* expressed in °C/Watt, and P_d is the *average power* dissipated by the circuit. From this expression, it is clear that an excessive power dissipated during testing will increase the circuit temperature well beyond the value measured (or calculated) during the functional mode [SIA 2003]. If the temperature is too high, even during the short duration of a test session, it can result in irreversible structural degradations. Some of these degradations, such as **hot spots**, appear during test data application and may lead to premature destruction of the circuit [Pouya 2000]. Others, which are accelerated gradually over time (*ageing*), may affect circuit performance or cause functional failures after a given lifetime [Hertwig 1998] [Shi 2004]. In this case, the main mechanisms leading to these structural degradations are *corrosion* (oxidizing of conductors), *electromigration* (molecular migration of the conductor structure toward the electronic flow), *hot-carrier-induced* defects, or *dielectric breakdown* (loss of insulation of the dielectric barrier) [Altet 2002]. These types of degradations have a big impact on long-term circuit reliability.

7.3.2 Noise Phenomena

These types of problems can occur when testing the circuit at the wafer level (for *characterization testing* or *verification testing*). For this type of test, the power must be supplied to the circuit through probes, which typically have higher inductances than the power and ground pins of the package planned for circuit encapsulation. If the switching activity during testing is equal to or higher than the switching activity during functional mode, the **power supply noise** (which is given by $L[di/dt]$ where L is the inductance of a power line and di/dt represents the magnitude of the variation of the current flowing through this line) will increase [Wang 1997]. This excessive noise can erroneously change the logic state of some circuit nodes at a given instant, causing some good dies to fail the test, thus leading to an unnecessary loss of yield. To avoid such phenomena, it is important to reduce test power.

Comparable inductive phenomena, known as **ground bounce** or **voltage surge/droop**, may occur during testing of the packaged circuit (*production testing*). Actually, wire/substrate inductances or package lead inductances associated with power or ground rails appear in circuits designed with deep submicron technologies. When high switching currents occur in the circuit under test, caused by high switching activity, voltage glitches can be observed at the nodes of these inductances [Jiang 2000]. These voltage glitches are proportional to both the inductance value and the magnitude of the variation of the current flowing through this inductance. In some cases, these voltage glitches may change the rise/fall times of some signals in the circuit (timing performance degradation). In other cases, they can erroneously change the logic state of some circuit nodes or flip-flops and cause some good dies to fail the test thus leading to yield loss [Chang 1997]. Once again, high switching rates, elevated operating frequencies, and short rise/fall times of internal signals are the primary causes of these phenomena, worsened by the increased susceptibility of today's circuits to these noise phenomena.

Similarly, **IR-drop** and crosstalk effects are noise phenomena that may show up as an error in test mode but not in functional mode. IR-drop refers to the amount of decrease (increase) in the power (ground) rail voltage and is linked to the existence of a non-negligible resistance between the rail and each node in the circuit under test. **Crosstalk** refers to capacitive coupling between neighboring nets within an IC. With high peak current demands during testing, the voltages at some gates in the circuit are reduced. This causes these gates to exhibit higher delays, possibly leading to test fails and yield loss [Butler 2004]. These phenomena have been widely reported in the literature, in particular when at-speed transition delay testing is performed [Shi 2004]. Typical examples of voltage drop sensitive applications are gigabit switches containing millions of logic gates.

7.3.3 Miscellaneous Issues

The **cost** constraints of consumer electronic products typically require the use of plastic packages for integrated circuit packaging. This type of package, although quite cheap, is not always able to dissipate high levels of heat. However, the use of packages with higher thermal capacities, such as ceramic or organic packages, which would allow removal of the excessive heat dissipated during testing, would significantly increase the final product cost. Similarly, the use of special cooling systems that could be used for venting away the excess of heat generated during testing, such as a radiator or fan, would also have a negative impact on the product cost. Moreover, in portable systems, where weight and size are important, these solutions are completely out of the question. Thus, it is important to reduce test power to avoid cost increases in these types of products.

Embedded electronic systems powered by batteries are employed in various types of applications (computing, aerospace, avionics, telephony, automotive, military, etc.). In mission-critical and safety-critical applications (*e.g.*, avionics and aerospace), these systems are equipped with BIST features to periodically check that the circuits are functioning correctly by taking advantage of idle periods in the system operation [Nicolaidis 1998]. For applications such as telephony, power-up self-test procedures are used to check the system integrity and alert the user when problems occur. In this case, test resources are also embedded in the system to facilitate such operations. For all these applications, **autonomy** is a critical issue that needs to be addressed during testing by minimizing the switching activity. As the main issue here is the amount of energy used, it is also possible to minimize the impact of testing by reducing the length of the test sequences used.

Finally, another reason why it is important to reduce power consumption during testing is the need for applying at-speed tests. In the past, tests were typically applied at rates much lower than the functional clock rate of the circuit. The main goal was to screen static faults (such as stuck-at faults). Thus, the excess of switching activity generated during testing was compensated by the reduction of the test clock frequency. Timing defects are becoming prevalent because of the use of nanometer process technology. This makes it essential to test for delay faults to ascertain circuit performance. Therefore, tests have to be applied at-speed, and it is no longer practical to reduce the test clock frequency [Krstic 1998]. Minimizing switching activity during testing for reducing power consumption thus becomes imperative.

7.4 LOW-POWER SCAN TESTING

In the context of scan testing, the problem of excessive power during testing is much more severe than in functional mode. This is mostly because the application of each test pattern in a scan design requires a number of shift clock cycles that contribute to an unnecessary increase of switching activity [Bushnell 2000] [Wang 2006]. A study reported in [Saxena 2003] shows that while 10% to 20% of the memory elements (D flip-flops and D latches) in a digital circuit change state during one clock cycle in functional mode, 35% to 40% of these memory elements when reconfigured as scan cells can switch state during scan testing. In the worst case, all scan cells can switch state. Another report [Shi 2004] further indicates that the average power during scan testing can be 3 times the power consumed during normal functional operation, and the peak power can be 30 times what it is in normal functional operation. In this section, we discuss various low-power scan test techniques to reduce excessive test power.

7.4.1 Basics of Scan Testing

Scan design requires the reconfiguration of memory elements (often D flip-flops) into scan cells and then stitching them together to form scan chains [Bushnell 2000] [Jha 2003] [Wang 2006]. During **slow-speed scan testing**, each scan test pattern first must be shifted into the scan chains. This requires setting the scan cells to shift mode and applying a number of load/unload (shift) clock cycles. Scan shifting is generally done at slow speed to avoid high power dissipation. A capture clock cycle is then applied to capture the test response of the design into scan cells. This

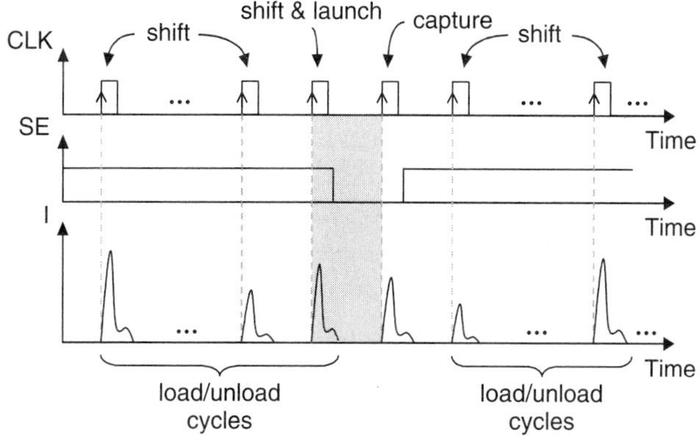

■ **FIGURE 7.2**

Slow-speed scan testing with associated current waveform.

requires setting the scan cells to normal/capture mode. A scan enable signal (*SE*) is typically used for this setting. When *SE* is set to 1, the scan design is in shift mode; when *SE* is set to 0, the circuit is switched to normal/capture mode.

Since the early 1990s, this slow-speed scan test technique has been widely used in industry to test stuck-at faults and bridging faults. The shift and capture operations for a typical scan design with the associated current waveform for each clock cycle are shown in Figure 7.2. This current waveform varies cycle by cycle because the current is proportional to the number of 0-to-1 and 1-to-0 transitions on the scan cells, which in turn produce switching in the ***circuit under test*** (CUT).

The problem of excessive power during (slow-speed) scan testing can be split into two subproblems: excessive power during the shift operation (called **excessive shift power**) and excessive power during the capture operation (called **excessive capture power**) [Girard 2002]. The latter is intended to address clock skew problems that may arise when many flip-flops change their output values simultaneously after capture operation. Since the mid-1990s, numerous techniques have been proposed to reduce shift power, capture power, or both at the same time during slow-speed scan testing. These low-power scan test techniques are introduced in detail in the following subsections.

In the meantime, as feature size shrinks into the ***deep-submicron*** (DSM) scale and circuit speed starts to operate at the GHz range, we have seen more and more chips fail because of timing-related defects. As a result, **at-speed scan testing**, which captures the test response of the scan design at the rated clock speed, is becoming mandatory to ensure high product quality.

There are two types of at-speed scan test schemes: ***launch-on-shift*** (LOS) and ***launch-on-capture*** (LOC) [Wang 2006]. Figure 7.3 shows the clock diagram of both schemes. Using the launch-on-shift scheme, vector V_1 (after the next to last shift) is loaded to the scan cells for initialization, and a second vector V_2 (after the last shift) is then shifted into the scan cells to launch a transition on selected scan cells in shift mode. In this case, V_2 is a 1-bit shift of the first vector V_1. One capture clock

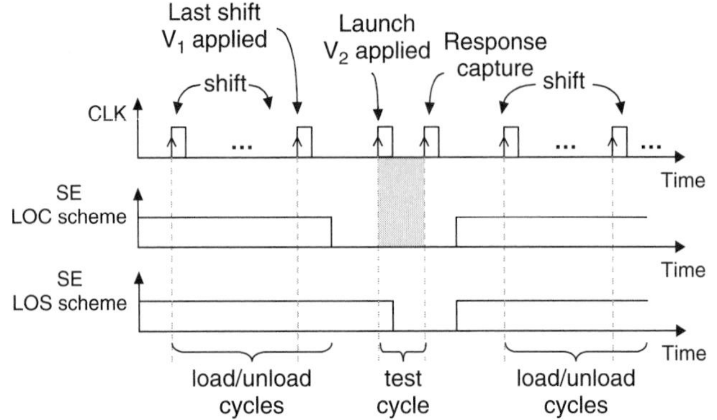

FIGURE 7.3

At-speed scan testing with LOS and LOC test schemes.

cycle is then applied at-speed to the design in normal/capture mode to capture the test response. On the other hand, using the launch-on-capture scheme, two capture clock cycles (launch and capture pulses as shown in Figure 7.3) are applied at speed in normal/capture mode to capture the final test response to scan cells. In this case, V_2 is the circuit's response to V_1, which is then captured to the scan cells in capture mode again. Experiments have shown that a LOS test can have higher delay fault coverage than a LOC test [Xu 2006]. However, because LOS requires an at-speed scan enable signal (*SE*), it is more difficult to lay out the scan design. Moreover, LOS suffers from an overkill issue, which could reject good chips as more false paths can be activated than using the LOC scheme.

The applicability of at-speed scan testing is also further challenged by **test-induced yield loss**, an emerging test problem that occurs when good chips fail only during testing. The main source of test-induced yield loss is **excessive IR-drop** caused by the high switching activity generated in the CUT between the launch of the test stimulus and the capture of the corresponding test response [Butler 2004]. Excessive IR-drop, during the short period between launch and capture (called the **test cycle**), may lead to a situation where gates in the circuit exhibit higher delays so an erroneous response may be captured to the scan cells at the end of the test cycle (the response capture edge). This makes at-speed scan testing especially vulnerable to IR-drop. A few solutions, based on power-aware ATPG or *X*-filling and presented in Section 7.4.2, have been proposed to avoid IR-drop-induced yield loss.

7.4.2 ATPG and X-Filling Techniques

In conventional scan ATPG, each don't care bit (*X*) in a **test cube** is filled with 0 or 1 randomly; the resulting fully specified test cube (called a scan test pattern or simply test pattern) is then fault-graded to confirm the detection of all targeted faults and additional faults. Although state-of-the-art dynamic and static test pattern compaction techniques have been extensively used to reduce pattern count in scan ATPG, the number of don't care bits in a given test cube remains high [Wohl 2003] [Hiraide 2003]. This provides a great opportunity that can be exploited for power minimization during scan testing.

The first set of techniques to reduce test power is to use novel low-power ATPG algorithms for generating low-power test patterns that still meet the original ATPG objectives (maximum fault coverage and minimum pattern length with reasonable run time). The authors in [Wang 1994] enhanced the ***path-oriented decision-making*** (PODEM) algorithm by assigning don't care bits present at the CUT inputs in a clever manner to minimize the number of transitions between two consecutive test patterns. This reduces both average and peak power dissipation during shift operations. In [Wang 1997], the authors further extend the ATPG approach proposed in [Wang 1994] to full-scan sequential circuits by exploiting all don't cares that occur during scan shifting, test application, and response capture to minimize shift power and capture power simultaneously.

Another low-power ATPG method for efficient capture power reduction during scan testing [Wen 2006] tries to achieve two goals: the primary goal being the detection of targeted faults and the secondary goal being the minimization of the difference between before-capture and after-capture output values of scan cells. This is achieved by introducing the concept of a capture conflict (*C-conflict*) in addition to the conventional detection conflict (*D-conflict*). A C-conflict occurs when a difference between the before-capture and after-capture output values of a scan cell is created by logic value assignment during ATPG. A C-conflict, in the same manner as a D-conflict, may be avoided through the backtrack operation. However, backtracking for a C-conflict may make fault detection impossible. In this case, the backtracking for the C-conflict is reversed, and the transition at the scan cell is tolerated because the primary goal is fault detection.

The second set of techniques to reduce test power is to use **power-aware *X*-filling** heuristics that do not modify the overall ATPG process. Given a set of deterministic test cubes, the main goal of these techniques is to assign values to the don't care bits of each test cube so that the number of transitions in the scan cells is minimized. By reducing the number of transitions in the scan cells during scan shifting, the overall switching activity in the CUT is also reduced; power consumption during testing is thus minimized. Most of the time, the *X*'s are assigned with the help of the following classical *nonrandom filling heuristics*:

- ***Minimum transition filling*** (MT-filling), also called **adjacent filling.** All don't care bits in a test cube are set to the value of the last encountered care bit. That is, when applying MT-filling, the most recent care bit value is used to fill successive *X* values until a care bit is reached.
- **0-filling.** All don't care bits in a pattern are set to 0.
- **1-filling.** All don't care bits in a pattern are set to 1.

MT-filling results in the fewest number of transitions in the scan chains, which generally corresponds to the lowest switching activity in the overall circuit, and is thus the preferred approach. Consider the test cube <0*XXX*1*XX*0*XX*0*XX*>. By applying the above three nonrandom filling heuristics, the resulting patterns become as follows:

- 0000111000000 with MT-filling heuristics
- 0000100000000 with 0-filling heuristics
- 0111111011011 with 1-filling heuristics

These classical nonrandom filling heuristics (among a few others) have been evaluated [Butler 2004] to measure the reduction in average power consumption during scan shifting (load/unload cycles). These heuristics have also been evaluated to measure the reduction in peak power consumption with respect to a random filling of don't care bits [Badereddine 2006]. Complete results on benchmark circuits have shown that both average and peak power consumption during testing can be efficiently minimized with the MT-filling heuristics.

In the context of at-speed scan testing, a few *X*-filling solutions have also been described to reduce power during the test cycle and thus avoid IR-drop-induced yield loss [Wen 2005a, 2005b] [Remersaro 2006]. These solutions have been developed to provide power-aware LOC delay tests. The basic idea is to minimize the bit differences between V_1, the initialization vector, and V_2, the sensitizing vector (which in this case is equal to the output response of V_1), while maintaining the original transition fault coverage.

Compared to other solutions, *X*-filling techniques have the advantage of being applicable at the end of the design process (without imposing any impact on the design flow) and thus do not require any modification of the circuit and hence do not incur any area overhead. These methods reduce test power consumption sometimes at the expense of an increase in the pattern count because they may not be as effective in detecting additional faults as random filling, thereby requiring incrementally more patterns to achieve the target fault coverage.

7.4.3 Low-Power Test Vector Compaction

Static compaction involves minimizing the number of test cubes generated by an ATPG tool by merging test cubes that are compatible in all bit positions (*i.e.*, they have no conflicting bit position where one test cube has a specified "1" and another has a specified "0"). Conventional approaches for static compaction merge test cubes in an arbitrary order until no more merging is possible. In [Sankaralingam 2000], the authors showed that by carefully selecting the order in which test cubes are merged, the number of transitions can be minimized. To measure the number of transitions that result when shifting a scan vector into a scan chain, a **weighted transition metric** can be used. To illustrate the weighted transition metric, consider the example given in Figure 7.4. It has two transitions. When this vector is scanned into the CUT, transition 1 passes through the entire scan chain. This transition dissipates power at every scan cell in the scan chain. On the other hand,

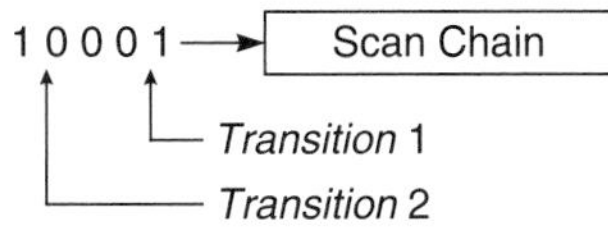

■ **FIGURE 7.4**

Transitions in scan vector.

transition 2 dissipates power only at the first scan cell during scan in. The number of scan cell transitions caused by a transition in a scan vector being scanned in depends on its position in the scan vector. In this example where there are five scan cells, a transition in position 1 (which is where transition 1 is) would be weighted four times more than a transition in position 4 (which is where transition 2 is). The weight assigned to a transition is the difference between the size of the scan chain and the position in the vector in which the transition occurs. Hence, the power dissipated when applying two vectors can be compared by counting the number of weighted transitions in each vector. The number of weighted transitions is given by:

$$\text{weighted\_transitions} = \Sigma\ (\text{size\_of\_scan\_chain} - \text{position\_of\_transition})$$

Using this metric, a greedy heuristic procedure is given in [Sankaralingam 2000] for merging test cubes in a way that minimizes the number of transitions. Significant reductions in average and peak power consumption can be obtained by using this approach.

7.4.4 Shift Control Techniques

Several techniques have also been proposed to reduce or cancel the switching activity in the CUT during scan shifting. The authors in [Huang 1999] try to find an input vector, called a *control vector,* such that when this vector is applied to the primary inputs of the CUT during scan shifting, the switching activity in the combinational part of the CUT is minimized. To determine this input control vector, a modified version of the D-algorithm is used. The method has achieved some reasonable reduction in average power consumption.

Another technique proposed in [Hertwig 1998] is to modify each scan cell in the scan chain so as to block transitions at the scan cell outputs during scan shifting and thereby prevent all switching activity in the combinational portion of the CUT. The scan cell modification consists of adding a NOR gate and an additional fanout to the output of each scan cell (see Figure 7.5). During scan shifting, the NOR gate prevents data in the scan cell from propagating to the combinational part of the CUT. This technique obviously is effective in test power reduction; however, it requires a significant area overhead and may degrade the circuit performance.

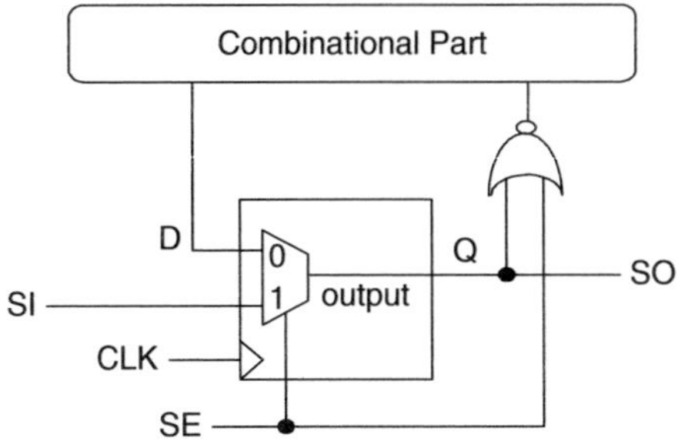

FIGURE 7.5

Scan cell modification.

7.4.5 Scan Cell Ordering

In scan design, switching activity, and hence power dissipation, can be further reduced by changing the order of the scan cells in each scan chain. Consider a scan chain composed of four ordered scan cells (FF_1-FF_2-FF_3-FF_4). Assume that the test vector $V = $ <0101> is to be loaded in the scan chain and the initial state of the four scan cells is <0000> in this scan chain. The total number of transitions generated in the scan chain by the loading of vector V will be equal to 10 (see Figure 7.6a). Now, suppose that the order of the scan cells in the scan chain is changed to FF_2-FF_4-FF_1-FF_3; the total number of transitions in the scan chain in this case becomes 2 (see Figure 7.6b). Of course, changing the order of the scan cells in the scan chain implies a change of the bit order in each test vector.

A first scan-cell ordering technique was proposed in [Dabholkar 1998], which uses two heuristics to find the best ordering of scan cells. The first heuristic performs a random search where the scan cells are randomly permuted a predefined number of times and the entire deterministic test sequence is simulated to measure the switching activity. The second heuristic uses a **simulated annealing** algorithm that

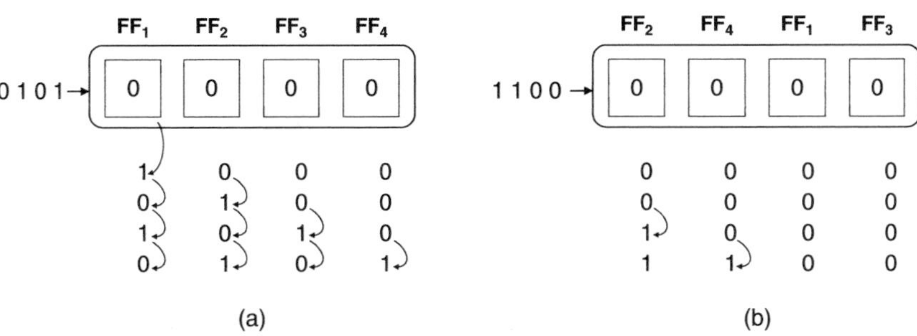

FIGURE 7.6

Impact of scan cell reordering on switching activity.

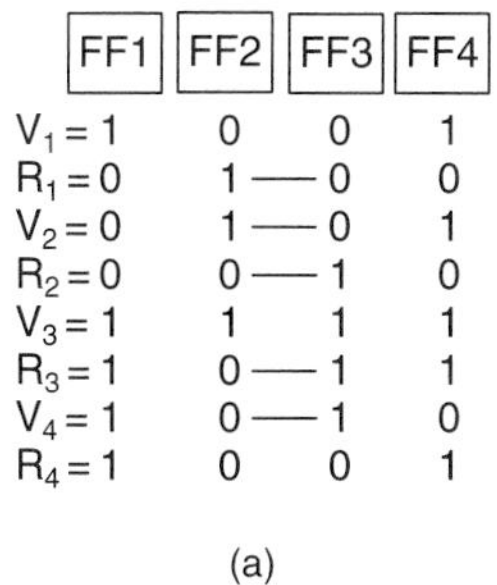

FF1
FF2
FF3
FF4
V1 = 1 0 0 1
R1 = 0 1 0 0
V2 = 0 1 0 1
R2 = 0 0 1 0
V3 = 1 1 1 1
R3 = 1 0 1 1
V4 = 1 0 1 0
R4 = 1 0 0 1
(a)

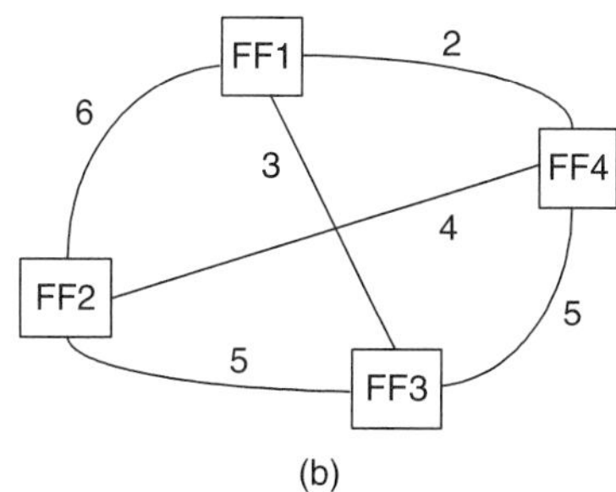

■ **FIGURE 7.7**

An example test sequence and the corresponding weighted graph.

explores the whole space of solutions to search for a global optimum of the cost function. Both heuristics have met with limited success.

Another solution given in [Bonhomme 2002] starts from a set of scan cells and the deterministic test sequence generated to test the corresponding scan-based circuit. The method first constructs a complete undirected graph in which each vertex represents a scan cell and each edge represents a possible connection between two scan cells in the scan chain (see Figure 7.7b). The weight on each edge of the graph represents the total number of bit differences between two scan cells for the corresponding test sequence. In Figure 7.7a, V_i is a test vector, R_i is the corresponding output response, and there are 5 bit differences between scan cells *FF2* and *FF3* in this example. This weight reflects the number of transitions that may be generated in the corresponding portion of the scan chain by connecting these two scan cells.

From this weighted graph, the problem then amounts to finding a Hamiltonian path of minimum cost in the graph. The cost of a path is obtained by summing the weights on edges belonging to this path. This problem is equivalent to the well-known traveling salesman problem, which is known to be *nondeterministic polynomial-time hard* (NP-hard) and for which different polynomial-time approximation algorithms can be used. The solution implemented in [Bonhomme 2002] uses a greedy algorithm to find the scan cell ordering that minimizes the occurrence of transitions in the scan chain during scan-in and scan-out operations. The heuristic procedure can be exploited by any layout synthesis program during scan-cell placement and routing.

Scan-cell ordering has many advantages: (1) it does not require additional hardware, (2) the fault coverage and test time are left unchanged, and (3) the impact on the design flow is low, and (4) test power can be significantly reduced. The only drawback is that power-driven stitching of the scan cells may result in a longer interconnect between the scan cells and potential routing congestion problems during scan routing (see Figure 7.8a).

To provide better scan routing between scan cells after scan reordering for low power, the authors in [Bonhomme 2003] proposed to partition the circuit in clusters (by using geographical criteria) and then reorder the scan cells within each cluster

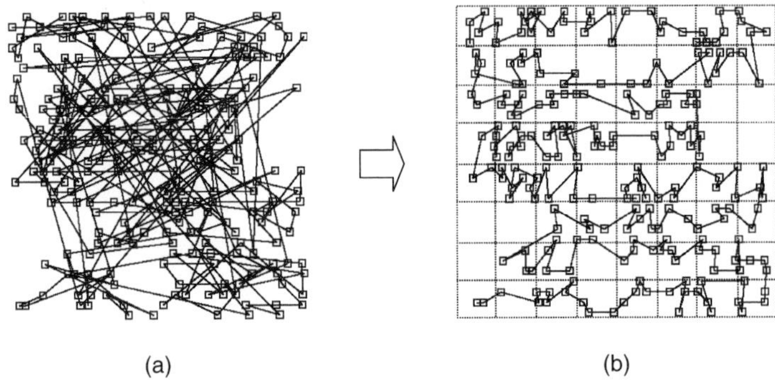

FIGURE 7.8

An example of power-driven scan chain routing.

to reduce the switching activity. The clusters are then stitched together using the nearest neighbor criteria. This technique offers a good tradeoff between test power reduction and scan chain length (see Figure 7.8b), and it is applicable to circuits with multiple scan chains and clock domains.

7.4.6 Scan Architecture Modification

This section presents techniques that involve modifying the scan architecture by inserting new elements. A first solution involves partitioning the scan chain(s) into N segments and having only one segment at a time active when loading and unloading test data [Whetsel 2000]. An on-chip test module that contains a counter activates one segment at a time when it receives the scan enable signal from the ATE. When one segment has been completely loaded/unloaded, then the next segment is activated. This technique reduces the average power dissipated by a factor of N, without any change in the test time, the test sequence, or the fault coverage. Nevertheless, the power dissipation in the clock tree feeding the circuit, which represents a significant part of the total power dissipated during testing, is not decreased. To address this, an alternative is to have separate clock trees for each scan segment so that the activation of the scan segments can be controlled by gating the clock trees rather than the scan enable signals [Saxena 2001]. In this way, the average test power is reduced in both the circuit and the clock tree without changing the overall principle of the method (see Figure 7.9).

The same approach can be used to reduce peak power consumption during both shift operation and capture operation [Rosinger 2004]. A dedicated scan architecture with mutually exclusive scan segment activation is used for this purpose, and a high reduction in peak power consumption can be obtained.

Two other possible techniques based on scan architecture modification can be used [Sinanoglu 2002] [Lee 2000]. The first one [Sinanoglu 2002] consists of inserting logic elements (XOR gates) between the scan cells so as to minimize the occurrence of transitions in the scan chain (and hence in the CUT) during shift operations.

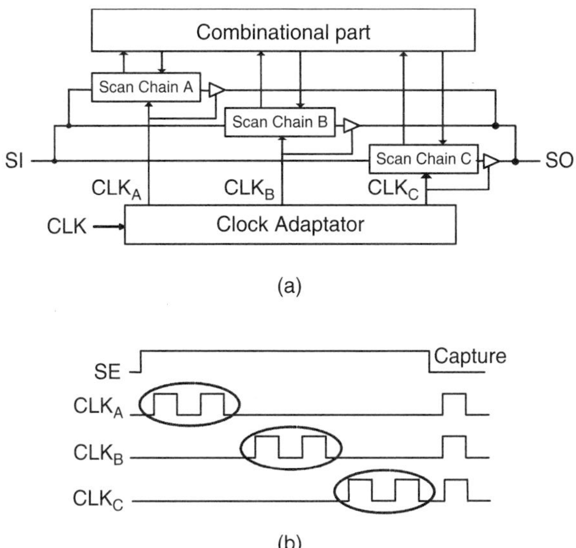

■ **FIGURE 7.9**

Scan chain segmentation.

Adding logic elements in the scan chains transforms the logic values that need to be shifted in. By doing this intelligently, it is possible to transform the scan vectors so that they contain fewer transitions. The second technique [Lee 2000] is applicable to circuits with multiple scan chains and consists of inserting buffers of different sizes at each scan chain input to create a slight temporal skew between the scan chains during the shift operation. When a scan chain shifts, it creates transitions in the CUT, which results in a current spike. By creating temporal skew between the shifting of the scan chains, the current spikes for each scan chain are spread out over time so that they do not occur simultaneously, thereby allowing significant reductions in peak power during testing. The technique requires some changes in the scan structure as well as in the scan controller.

Another interesting and original approach [Huang 2001] is based on a novel scan architecture, called a token scan architecture, that uses the concept of a token ring—a well-known structure in the field of communication networks—to reduce the shift power in both the scan chain and the combinational logic. Basically, this approach starts from a multi-phase technique, which is applied to scan-based circuits using the architecture shown in Figure 7.10a. The scan-in wire *SI* is broadcasted to all scan cells, but only one scan cell is activated at a time. For a scan chain with N scan cells, an N-phase nonoverlapping clocking scheme is applied with one clock for each scan cell as shown in Figure 7.10a. Because only one scan cell is activated at a time, a high reduction of data transitions can be achieved during scan shifting.

However, the multi-phase technique may have two problems because of the discrete multi-phase generator. First, because the scan cells are usually distributed over the chip, the N multi-phase clock routes will require large area overhead.

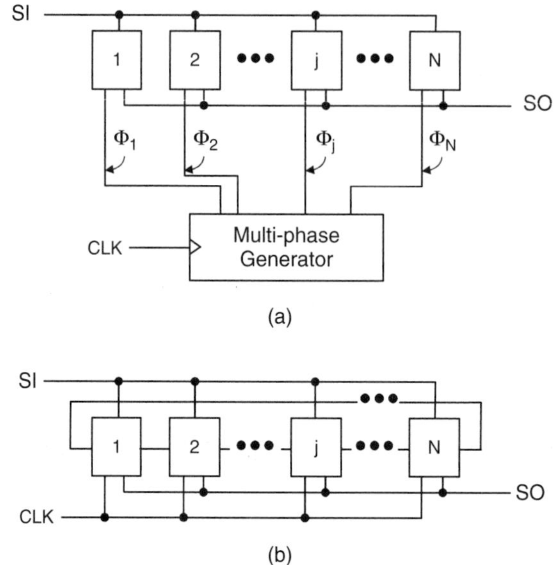

■ **FIGURE 7.10**

The token scan architecture.

Second, an interphase skew may occur because of the different lengths of the N clock routes, which may make multi-phase clocks overlap and cause a data error from *SI* or a bus contention at *SO*. To overcome these problems, a token ring-like structure [Huang 2001] can be used to embed the multi-phase clock generator into the scan cell. As Figure 7.10b shows, a token bit is rotated in the scan chain, and only the scan cell receiving the token can be activated. The N phase wires are thus reduced to a single phase clock wire. The interphase skew resulting from the different delays of the N phase wires can also be eliminated. This solution requires the use of a new type of scan cells, called token scan cells, to compose each scan chain (see Figure 7.11). A token scan cell consists of a data FF D_1, a token FF D_2, two multiplexers M_1 and M_2, and a switch S. D_1 and M_1 behave as a basic conventional scan cell while D_2 and M_2 serve as a phase generator.

To additionally reduce test power in the clock and scan-in data trees, the authors also proposed a novel clock gating technique that takes the advantage of the regularity and periodicity of the token scan chain. By combining all of these concepts, the token scan architecture can efficiently reduce shift power as well as clock power—power consumed in the clock tree during scan testing. The main drawback of this solution is the significant area overhead.

7.4.7 Scan Clock Splitting

To reduce the power consumption during scan testing, it is also possible to modify the scan clock (*i.e.*, the clock that drives all the scan cells of the chain[s]). The first technique based on scan clock splitting [Bonhomme 2001] involves reducing the

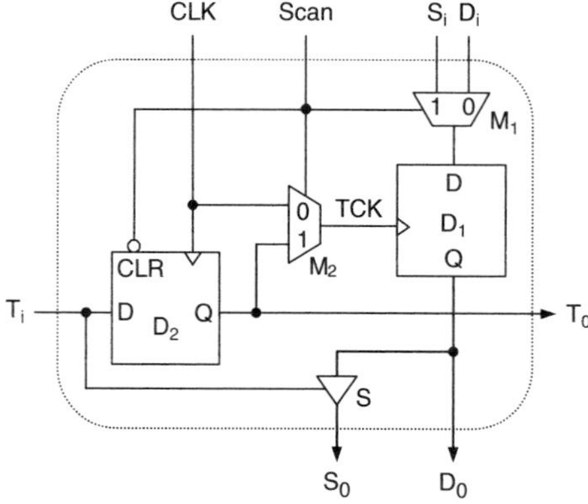

■ **FIGURE 7.11**

The token scan chain.

operating frequency of the scan cells during scan shifting without modifying the total test time. For this purpose, a clock whose speed is half that of the normal (functional) clock is used to activate one half of the scan cells during one clock cycle of the scan operation (see Figure 7.12). During the next clock cycle, the second half of the scan cells in the scan chain(s) is activated by another clock whose speed is also half of the normal speed. The two clocks are synchronous with the system clock and have the same period during shift operation except that they are shifted in time. During capture operation, the two clocks operate as the system clock. The use of such a modified clock scheme lowers the transition density in the CUT, the scan chains, and the clock tree feeding the scan chains during shift operation. Consequently, the switching activity in a time interval (*i.e.*, the average power) as well as the peak power consumption is minimized. Moreover, the total energy consumption is also reduced as the test length with the proposed clock scheme is exactly the same as the test length with a conventional scan design to reach the same stuck-at fault coverage.

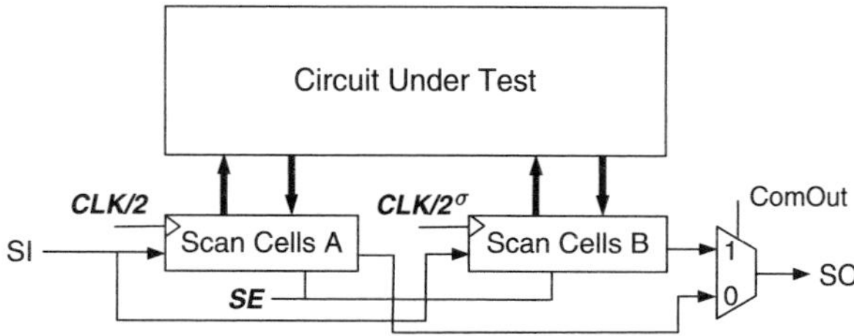

■ **FIGURE 7.12**

Scan clock splitting.

Another technique [Sankaralingam 2003] uses a staggered clock scheme to reduce peak power dissipation during testing. The principle of this approach is similar to the one used in [Lee 2000]. The scan chains of the CUT are grouped together to form *N* groups of scan chains. Then, each clock cycle of the load/unload phase is divided into *N* periods, where each period corresponds to the activation of a given group of scan chains. Staggering the activation of each group in this manner reduces the number of scan cells that are simultaneously switching, thereby greatly lowering the peak power consumption. Note, however, that the total number of transitions generated during testing (*i.e.*, the energy) is unchanged.

7.5 LOW-POWER BUILT-IN SELF-TEST

The logic *built-in self-test* (BIST) is a DFT technique in which a portion of the ***circuit under test*** (CUT) is used to test itself. Because it can provide self-test ability, logic BIST is crucial in many applications, in particular, for safety-critical and mission-critical applications. One major objective of logic BIST is to obtain high fault coverage; however, a major issue is that power consumption during BIST can exceed the power rating of the chip or package. Increased average power can heat the chip, and increased peak power can produce noise-related failures [Bushnell 2000]. In this section, we discuss a number of low-power BIST architectures and methodologies to reduce power consumption.

7.5.1 Basics of Logic BIST

Figure 7.13 illustrates a typical logic BIST system. A **logic BIST controller** is required to control the BIST operation. The ***test pattern generator*** (TPG) automatically generates test patterns that are applied to the inputs of the ***circuit under test*** (CUT), and an ***output response analyzer*** (ORA) is used for compacting the circuit's output responses. In practice, in-circuit TPGs constructed from ***linear feedback shift registers*** (LFSRs) are commonly used for exhaustive, pseudo-exhaustive, or

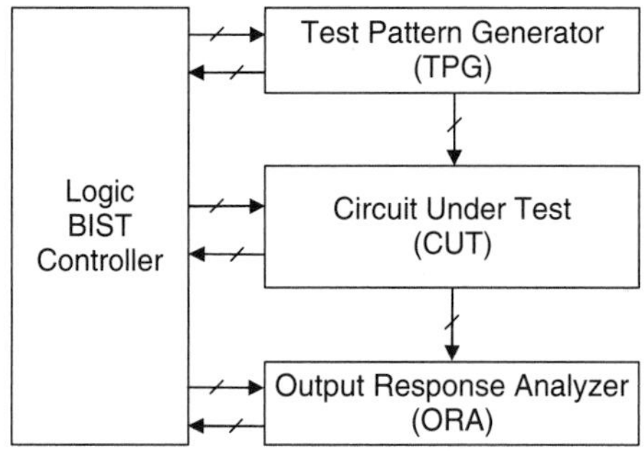

■ **FIGURE 7.13**

A typical logic BIST system.

pseudo-random testing [Wang 2006]. This is mostly because these LFSRs incur little area overhead and can be used as both TPG and ORA.

There are two basic BIST schemes for testing the circuit under test: (1) **test-per-clock BIST** for architectures using register configuration and (2) **test-per-scan BIST** for designs that incorporate scan chains [Wang 2006]. In test-per-clock BIST, test vectors are applied every clock cycle from the TPG, and test responses are captured in the ORA and compared to a reference value. This scheme has the advantage of running much faster in applying tests and yields higher fault coverage than test-per-scan BIST. The drawbacks of this scheme are high area overhead and incompatibility with scan design. In test-per-scan BIST, also called **scan-based BIST**, pseudo-random patterns are first shifted into the scan chains during shift operation; the test responses to these patterns are then captured in scan cells during the capture operation. The captured test responses are shifted out to the ORA for response compaction, while a new test is being shifted in. Clearly, the test-per-scan BIST system will run much slower than the test-per-clock BIST system; however, it takes advantage of the existing scan design, thereby requiring much simpler BIST circuitry, and is thus the industry preferred solution today.

7.5.2 LFSR Tuning

The aim of LFSR tuning is to find a way of decreasing the energy consumed during BIST by appropriately selecting the parameters of the LFSR (*i.e.*, the seed and characteristic polynomial). A preliminary step in this approach is to analyze the impact of these parameters on the switching density generated in the CUT. In [Girard 1999b], a number of experiments on benchmark circuits were conducted where for each circuit, several characteristic polynomials were used for the LFSR, and for each of these polynomials, several seeds were tried. Polynomials were taken from the list of primitive polynomials of an n-stage LFSR (n being the number of primary inputs of the CUT), and seeds were randomly chosen for each selected polynomial. In each experiment, the length of the test sequence required to reach the target fault coverage was determined through fault simulation. Figure 7.14 shows the experimental results for an 8-by-8 multiplier targeting 99% stuck-at fault coverage. Each number on the X-axis corresponds to a particular primitive polynomial of the LFSR, and each dot corresponds to the internal WSA resulting

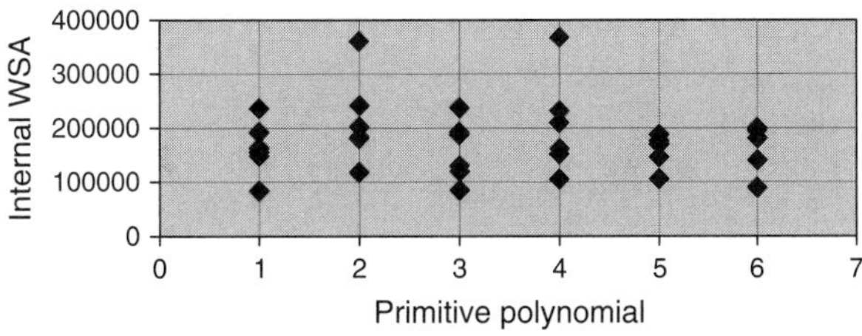

FIGURE 7.14

Impact of LFSR polynomial selection on energy.

from a randomly selected seed for the particular polynomial. Note that the *internal* WSA refers to the **weighted switching activity** of the internal nodes of the CUT.

As the figure shows, the WSA obtained for a given primitive polynomial of the LFSR strongly depends on the seed selected. Indeed, the deviation between best seeds and worst seeds is significant in terms of WSA. On the other hand, the sensitivity of the WSA to a given primitive polynomial is much lower; the value of the minimum WSA is almost the same regardless of which primitive polynomial is used. Therefore, selecting a primitive polynomial to minimize energy dissipation during BIST is not as crucial as selecting a good seed for the LFSR. For a given polynomial and target fault coverage, selecting the best seed of an LFSR for low-power BIST can then easily be done by using a method based on a simulated annealing algorithm [Girard 1999b].

7.5.3 Low-Power Test Pattern Generators

Several approaches have been proposed for designing on-chip test generators that can generate effective test patterns while reducing the transition density in the CUT. A first approach, called ***dual speed LFSRs*** (DS-LFSRs) [Wang 1997], is based on the use of two LFSRs operating at different clock frequencies (see Figure 7.15). Average power during testing is reduced by connecting the CUT inputs with the highest transition densities to the low-speed LFSR while CUT inputs with the lowest activity are connected to the normal speed LFSR. Note that this technique is applicable in a test-per-clock BIST environment. A second approach is also based on a modified LFSR [Girard 2001]. The original LFSR is replaced by two LFSRs that operate out-of-phase at half the clock rate of the original (functional) speed. Compared to the previous approach, the power dissipation is reduced not only in the CUT but also in the clock tree feeding the circuit. Fault coverage and test time are left unchanged. A third solution [Zhang 1999] consists of inserting logic between the LFSR and the CUT to allow the generation of weighted random test patterns that reduce the switching activity in the circuit while maintaining a high fault coverage. This solution uses a genetic algorithm-based search to determine optimal weight sets at primary inputs to minimize energy dissipation. One last approach that can be used [Corno 2000] is based on selecting a **cellular automaton** that generates a

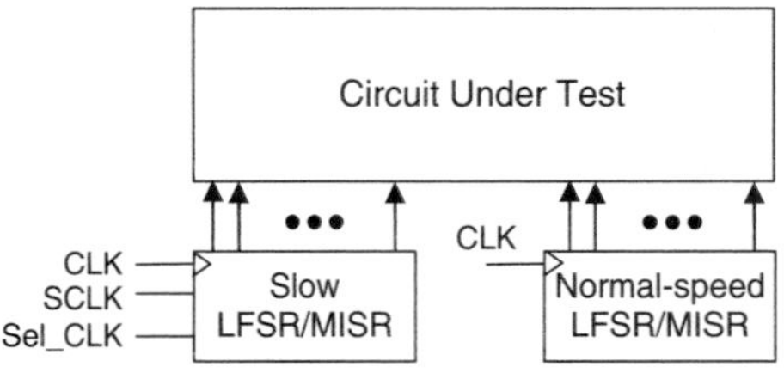

FIGURE 7.15

Low-power BIST with DS-LFSRs.

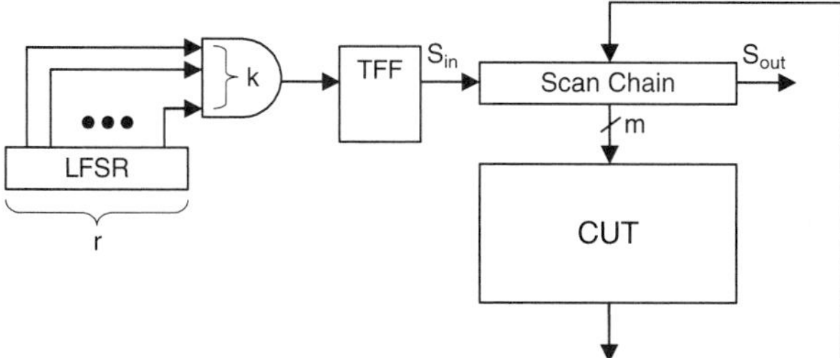

■ **FIGURE 7.16**

The LT-RTPG structure.

test sequence with a low transition density and has a good tradeoff between fault coverage and test time.

For scan-based BIST (test-per-scan BIST), an interesting approach for low-power testing, called ***low transition random test pattern generator*** (LT-RTPG) [Wang 1999], was proposed. It involves inserting an AND gate and a *toggle flip-flop* (TFF) between the LFSR and the input of the scan chain to increase the correlation of neighboring bits in the scan vectors (see Figure 7.16). Because the TFF holds its previous values until it receives a 1 on its input, the same value (0 or 1) is repeatedly scanned into the scan chain until the value at the output of the AND gate becomes 1. Hence, neighboring scan cells are assigned identical values in most test vectors if a large k is used—that is, if the AND gate has many inputs (the probability that the TFF toggles at any time t is given by $1/2^k$). In this manner, the number of transitions generated in the CUT can be significantly reduced. Although the pseudo-random test sequence is modified by this logic, it still provides a good tradeoff between fault coverage and test time.

Interesting low-power test pattern generators that are applicable for data path architectures based on multipliers and accumulators are described in [Gizopoulos 2000]. Two hardware solutions are proposed depending on whether the concern is energy reduction or power reduction. They are both based on the use of Gray counters, which can generate successive test vectors with a Hamming distance of 1. Significant energy and power reductions can be obtained.

7.5.4 Vector Filtering BIST

BIST techniques based on vector filtering to reduce power consumption have also been proposed in the literature [Corno 1999] [Gerstendörfer 1999] [Girard 1999a] [Manich 2000]. These techniques are based on the observation that as self-test progresses, the detection capability of the pseudo-random test vectors generated by an LFSR decreases quickly. Therefore, many of the pseudo-random test vectors do not detect new faults despite consuming a significant amount of energy. For example, only 159 patterns among the 2524 required to reach 99.9% fault coverage actually detect faults in the benchmark circuit *c5315* [Brglez 1985]. In addition, the length of the subsequence of consecutive nondetecting test vectors is often long. For example, the longest subsequence of consecutive nondetecting vectors in the

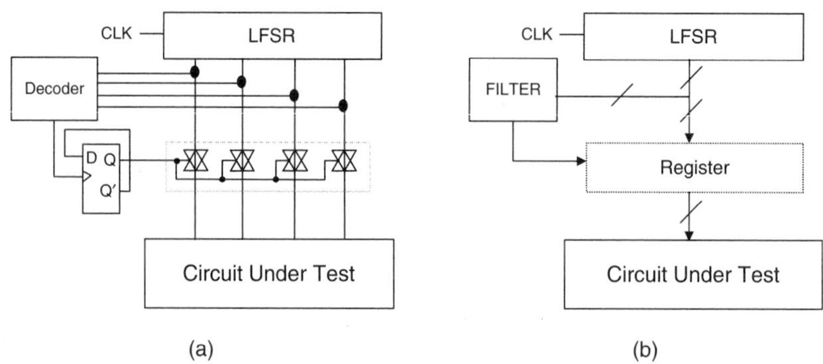

■ **FIGURE 7.17**

Two vector filtering BIST structures.

pseudo-random test sequence generated for the benchmark circuit *s1488* [Brglez 1985] to reach 100% fault coverage contains 509 vectors, whereas the complete test sequence is of length 2931.

Consequently, the main goal of these techniques is to filter test vectors that do not detect additional faults, thus preventing the CUT from being excited by these undesired vectors. For this purpose, a decoder can be used during BIST pattern generation to store the first and last vectors of each subsequence of consecutive nondetecting vectors to be filtered [Girard 1999a]. The output of this decoder provides the logic value 1 after detection of each of these vectors. Then, the vector filtering structure has to allow or prevent application of these test vectors to the circuit inputs. A toggle D flip-flop is used to control the transmission of stimuli from the LFSR to the CUT. The transmission is activated or inhibited by means of a transmission gate network (see Figure 7.17a).

Rather than just filtering the subsequences of nondetecting vectors, it is also possible to filter all vectors that do not detect any faults [Manich 2000]. Using this approach, a register made of latches is used to control the transmission of test vectors from the LFSR to the CUT. A filter module is used to provide the needed control signals to the register (see Figure 7.17b). A similar solution was also proposed in [Corno 1999].

The authors in [Gerstendörfer 1999] exploited the same idea but applied it to scan-based BIST. A gating signal is derived from the decoder/filter and is used to enable or disable the shift clock to the scan chains.

The main advantage of all these techniques is that they allow a significant reduction of energy and average power consumption during testing. The drawbacks are the negative impact on circuit performance and the area overhead, which may be high in some cases.

7.5.5 Circuit Partitioning

Another approach involves partitioning the original circuit into two structural subcircuits so two different BIST sessions can be used to successively test each

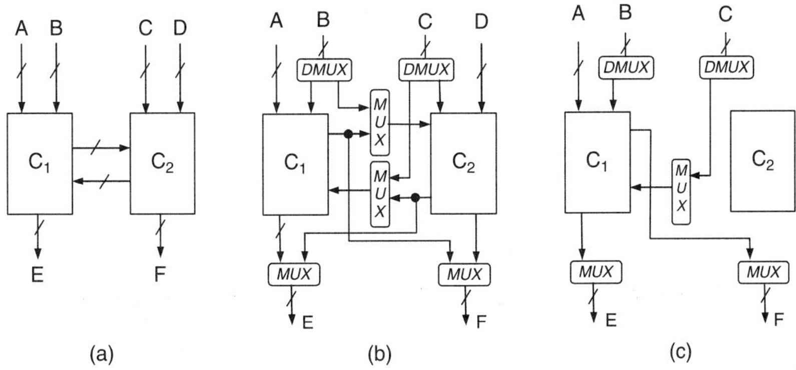

■ **FIGURE 7.18**

Circuit partitioning for low-power testing.

subcircuit. To minimize the area overhead of the resulting BIST scheme, however, the number of connections between the subcircuits of the partition, called the *cut size,* has to be minimal. The basic scheme of partitioning a circuit into two subcircuits is shown in Figure 7.18. In Figure 7.18a, a logic circuit is partitioned into two subcircuits C_1 and C_2. Many such partitions exist for large VLSI circuits. Figure 7.18b depicts how multiplexers are inserted between the two subcircuits. By controlling the multiplexers, all inputs and outputs of each subcircuit can be accessed using primary inputs and primary outputs. For example, to test subcircuit C_1, the multiplexers can be controlled as shown in Figure 7.18c. The demultiplexers (DMUX) on the sets B and C of input signals are added to avoid switching activity in C_2 during the test of C_1.

A circuit partitioning technique for low power testing was proposed in [Girard 1999c]. This technique tries to find an optimal partitioning solution, which is a *NP*-complete problem, by using a simple *graph partitioning* algorithm. An improved version [Girard 2000] uses a circuit partitioning tool based on a *multilevel hypergraph partitioning* algorithm [Karypis 1998]. Traditional partitioning algorithms compute a partition of a graph by directly operating on the original graph. This approach is often too slow and can lead to poor quality partitions in terms of cut size, which is representative of the area overhead of the BIST scheme. The multilevel partitioning algorithm follows a completely different approach. The algorithm successively decreases the size of the graph (or the **hypergraph**) by collapsing vertices and edges, partitions the smallest graph, and then uncoarsens it to construct a partition for the original graph. At each level of the uncoarsening phase, an iterative refinement algorithm is used to improve the partition.

By partitioning the circuit into two subcircuits and testing the subcircuits in successive test sessions, average and peak power consumption are minimized. In addition, this approach reduces the total energy consumed during BIST operation because the test length required for the two subcircuits is usually shorter than that of the original circuit. This is because circuit partitioning increases the controllability and observability of the internal nodes in the CUT. The area overhead with this approach is low. Drawbacks are a slight penalty on circuit performance and a non-negligible impact on routing. The proposed strategy can be applied to scan-based BIST or parallel BIST by adapting the test pattern generation structure.

7.5.6 Power-Aware Test Scheduling

A test scheduling technique for low power consumption [Zorian 1993] considers a set of blocks (memories, logic, analog, test resources, etc.) in an SOC and a specified limit of power dissipation for the SOC during testing. The objective is to find the best combination of blocks to be tested in parallel so the overall test time is minimal and the power limit is satisfied. This technique also takes into account the fact that, to minimize the area overhead associated to BIST, some of the test resources (test pattern generators and output response analyzers) must be shared among the various blocks.

A similar technique [Chou 1994] addresses the *NP*-complete test scheduling problem by using a compatibility graph and heuristic-driven algorithms. The power constraint is established with respect to the peak power consumption of each block. Two different problems are considered depending on the test length of each block: (1) scheduling equal-length tests with power constraints and (2) scheduling unequal-length tests with power constraints. Optimal solutions are sought for both problems. The algorithms consist of four basic steps. First, a test compatibility graph is constructed from a resource graph in which a resource represents either a combinational block or a register block. Second, the test compatibility graph is used to identify a complete set of **time compatible tests** (tests that can be executed concurrently) with power dissipation information associated with each test. Third, from the set of time compatible tests, lists of **power compatible tests** are extracted. Finally, a minimum cover approach is used to find an optimum scheduling for these power compatible tests.

Based on these two basic test scheduling techniques, several solutions have been further proposed for testing SOC designs [Muresan 2000] [Iyengar 2001] [Larsson 2002] [Pouget 2003]. For given power constraints and parameters related to the test organization (fixed, variable, or undefined test sessions with or without precedence constraints) or to the test structure (test bus width, test resources sharing), these solutions allow for optimized overall SOC test time.

Another test scheduling technique [Ubar 2005] has a slightly different objective, as the main focus is on total test energy minimization for SOC testing. This technique assumes a hybrid BIST test architecture, where the test set is composed of core-level locally generated pseudo-random test patterns and additional deterministic test patterns that are generated offline and stored in the system (see Figure 7.19). The exact composition of these patterns defines not only the test length and test memory requirements but also the energy consumption. In general, because a deterministic test pattern is more effective in detecting faults than a pseudo-random pattern, using more deterministic test patterns for a core will lead to a short test sequence with, consequently, less energy. However, the total number of deterministic test patterns is constrained by the test memory requirements, and at the same time, the deterministic test patterns of different cores of a SOC have different energy

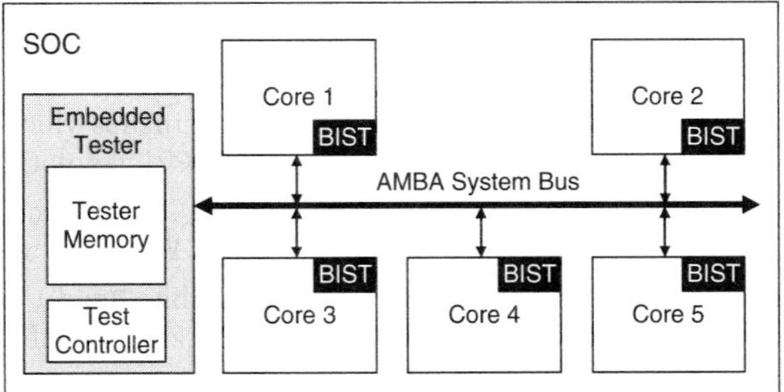

■ **FIGURE 7.19**

AMBA bus-based hybrid BIST architecture.

and fault detection characteristics. A careful tradeoff between the deterministic pattern lengths of the core must therefore be made in order to produce a globally optimal solution. Two heuristics [Ubar 2005] can be proposed to try to minimize the total switching energy without exceeding the assumed test memory constraint. The solutions are obtained by modifying the ratio of pseudo-random and deterministic test patterns for every individual core such that the total energy dissipation is minimized.

Another category of test scheduling, called thermal-aware test scheduling, has been proposed to address the problem of chip overheating during the testing of complex core-based systems. Here, the basic idea is to consider that the spatial distribution of power across the chip is nonuniform, so that imposing a chip-level maximum power constraint during test scheduling (as for system-level BIST solutions described previously) does not necessarily avoid local overheating and hence destructive hot spots. A few solutions have been proposed in this area [Rosinger 2005] [Liu 2005] [He 2006], mainly based on incorporating thermal constraints during test scheduling so as to spread heat more evenly over the chip and reduce hot spots during testing. The proposed approaches facilitate the rapid generation of thermal-safe test schedules without requiring time-consuming thermal simulations.

7.6 LOW-POWER TEST DATA COMPRESSION

Test data volume is now recognized as a major contributor to the cost of manufacturing testing of ICs. High test data volume leads to a high test time and may exceed the limited memory depth of *automatic test equipment* (ATE). Test application time for scan testing can be reduced by using a large number of scan chains. However, the number of ATE channels that can directly drive scan chains is limited because of pin count constraints.

Test data compression is an efficient solution to the problem of increasing test data volume. Test data compression involves encoding a test set to reduce its size.

By using this reduced set of test data, the ATE limitations (*i.e.*, tester storage memory and bandwidth gap between the ATE and the CUT) may be overcome. On the other hand, using compressed test data involves having a small on-chip decoder, which decompresses the data as it is fed into the scan chains during test application.

Despite its ability in reducing test data volume and test application time, test data compression does not solve the problem of excessive test power during scan testing. A case study of a Motorola ColdFire microprocessor core [Pouya 2000] has been used to illustrate the commercial means of reducing test data volume and how they affect test power. To address this issue, several techniques have been proposed to simultaneously reduce test data volume and test power during scan testing of digital ICs. As in [Wang 2006], these low-power test data compression techniques can be classified into three categories: **coding-based** schemes, **linear-decompression-based** schemes, and **broadcast-scan-based** schemes.

7.6.1 Coding-Based Schemes

Code-based schemes use data compression codes to encode the test cubes of a test set. An interesting encoding algorithm that can be used to concurrently reduce scan power dissipation and test data volume during SOC testing is proposed in [Chandra 2001]. In this approach, test cubes generated by ATPG are encoded using **Golomb codes** which are an evolved form of **run-length codes**. All don't care bits of the test cubes are mapped to 0 and Golomb coding is used to encode runs of 0's. More details about Golomb codes can be found in [Wang 2006]. Golomb coding efficiently compresses test data, and the mapping of don't cares to all 0's reduces the number of transitions during scan-in, thus significantly reducing power dissipation (up to 75%). One drawback of Golomb coding is that it is inefficient for runs of 1's. In fact, the test storage can even increase for test cubes that have many runs of 1's. Moreover, implementing this test compression scheme requires a synchronization signal between the ATE and the CUT as the size of the compressed data (*codeword*) is of variable length.

Another method based on an **alternating run-length coding** [Chandra 2002] improves the encoding efficiency of Golomb coding. Whereas a Golomb code only encodes runs of 0's, an alternating run-length code can encode both runs of 0's and runs of 1's. In this case, the drawback is that the coding becomes inefficient when a pattern with short runs of 0's or 1's has to be encoded.

7.6.2 Linear-Decompression-Based Schemes

Another class of low-power test stimulus compression schemes is based on using **linear decompressors** to expand the data coming from the tester to fill the scan chains during test application. Linear decompressors consist only of XOR gates and flip-flops, and they are described in detail in [Wang 2006].

An example of a low-power linear-decompression-based scheme using LFSR reseeding is proposed in [Lee 2004]. The basic idea in LFSR reseeding is to generate deterministic test cubes by expanding seeds. A seed is an initial state of the LFSR that is expanded by running the LFSR in an autonomous mode. Given a deterministic test cube, a corresponding seed can be computed by solving a set of linear equations—one for each specified bit—based on the feedback polynomial of the LFSR. Because typically 1% to 5% of the bits in a test vector are specified, most bits in a test cube do not need to be considered when a seed is computed because they are don't care bits. Therefore, the size of a seed is much smaller than the size of a vector. Consequently, reseeding can significantly reduce test data volume and bandwidth. However, it is not as good for power consumption because the don't care bits in each test cube get filled with random values, thereby resulting in excessive switching activity during scan shifting.

The key idea of the encoding scheme proposed in [Lee 2004] is to take advantage of the fact that the number of transitions in a test cube is always less than its number of specified bits. A transition in a test cube is defined as a specified 0 (1) followed by a specified 1 (0) with possible *X*'s between them (*e.g.*, X10XXX or XX0X1X). Thus, rather than using LFSR reseeding to directly encode the specified bits as in conventional LFSR reseeding, the proposed encoding scheme divides each test cube into blocks and only uses LFSR reseeding to produce the blocks that contain transitions. For the blocks that do not contain transitions, the logic value fed into the scan chain is simply held constant. This approach reduces the number of transitions in the scan chain and hence reduces test power. Despite the area overhead caused by the use of *hold flag* shift registers, this scheme is an efficient solution to tradeoff between test data compression and test power reduction.

7.6.3 Broadcast-Scan-Based Schemes

The third class of low-power test data compression schemes is based on broadcasting the same value to multiple scan chains. An example of a low-power broadcast-scan-based scheme is the **segmented addressable scan** architecture presented in [Al-Yamani 2005]. This architecture involves modifying the **Illinois scan architecture** [Hamzaoglu 1999] in which a given scan chain is split into multiple scan segments, thus allowing the same data to be loaded simultaneously into all segments when compatibility exists. The segmented addressable scan architecture enhances the Illinois scan architecture by avoiding the limitation of needing all segments to be compatible to benefit from the segmentation. In other words, any combination of compatible segments for a given test pattern can be used to load the same data to these segments and hence increase the compression rate. The compatible segments are loaded in parallel using a multiple-hot decoder (see Figure 7.20). Test power is reduced as segments that are incompatible within a given round (*i.e.*, during the time needed to upload a given test pattern) are not clocked. One drawback of this solution is that the multiple-hot decoder is designed with respect to a given test set. This means that the test set has to be known early during the design phase of the circuit and that changing the test set during verification testing or production testing involves changing the design of the circuit.

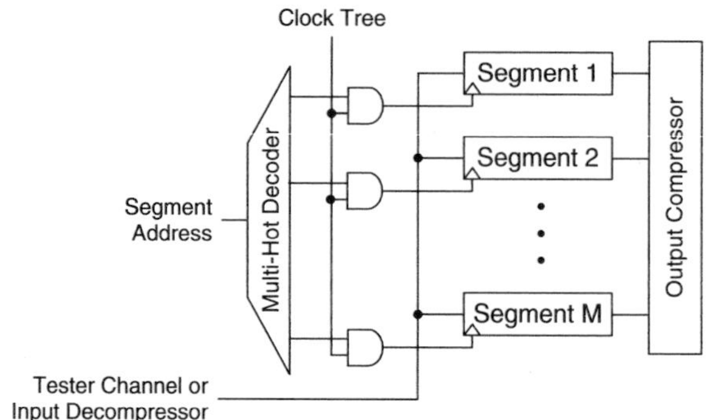

■ **FIGURE 7.20**

The segmented addressable scan architecture.

Another example is the ***progressive random access scan*** (PRAS) architecture proposed in [Baik 2005] that allows individual accessibility to each scan cell. In this architecture, scan cells are configured as an SRAM-like grid structure using specific PRAS scan cells and some additional peripheral and test control logic (see Figure 7.21). Providing such accessibility to every scan cell eliminates unnecessary switching activity during the scan while reducing the test application time and test data volume by updating only a small fraction of scan-cells throughout the test application. Power consumption during testing is drastically reduced—up to 99%. The main drawback of the PRAS architecture is the significant hardware overhead.

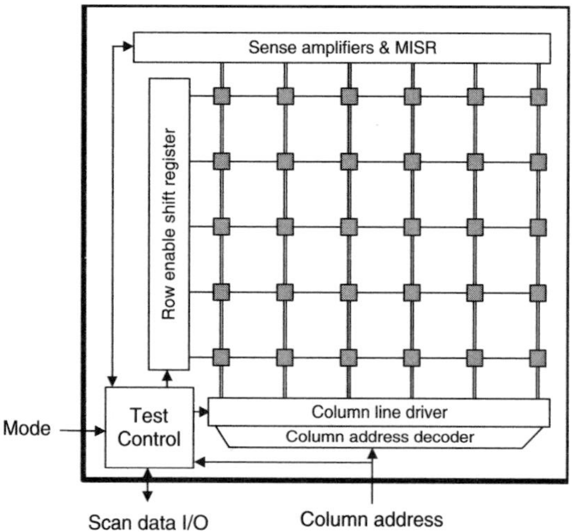

■ **FIGURE 7.21**

Progressive random access scan (PRAS) architecture.

7.7 LOW-POWER RAM TESTING

Although numerous techniques for constraining power dissipation during testing exist, there appear to be only a few solutions that are dedicated to memories. The main motivation for reducing test power in memories can be explained as follows. System memories, or embedded device memories, are divided into banks for increasing access speed and optimizing system cost [Cheung 1996]. During normal system operation, only one memory bank is accessed at any given time. In contrast, concurrent self-testing of all memory modules is highly desirable to reduce test time and simplify BIST control circuitry. However, by concurrently testing several banks of memories, the power dissipation can by far exceed that during normal system operation. For this reason, reducing test power in memories becomes mandatory when concurrent testing is used. Note that this statement also applies when testing memories embedded in an SOC.

A first methodology for low-power test of *random access memories* (RAMs) [Cheung 1996] is based on modifying several common memory tests (Zero-One, Checker Board, March B, Walking-0-1 and SNP 2-Group) in a way that reduces power dissipation during testing. The modified tests are based on the following principle: reorder the original tests to minimize the switching activity on each address line while retaining the fault coverage. The number of signal transitions on each address line depends on the address counting method (*i.e.*, the order in which addresses are enumerated during a read or a write loop of a memory test algorithm) and the address bit position. For example, the LSB (respectively MSB) address line has the largest (respectively smallest) number of transitions when binary address counting is used. Table 7.1 describes the original and low-power test algorithms for two memory tests. The symbol '$\updownarrow$' is used to describe a sequential access to all memory cells in any addressing order (increasing or decreasing). Binary address counting is typically used for such addressing. The low-power tests are described using the symbol $\updownarrow_s$ which represents *single bit change* (SBC) counting. For example, {00, 01, 11, 10} is the counting sequence of a 2-bit SBC code. Finally, W0 (W1) represents writing a 0 (1) to an address location, and R0 (R1) represents reading a 0 (1) from an address location.

Each proposed test has the same fault coverage and time complexity as the original version but reduces power dissipation by a factor of 2 to 16 thanks to a

TABLE 7.1 ■ Original and Low-Power Memory Test Algorithms

	Original Test	Low-Power Test
Zero-One	$\updownarrow$ (W0); $\updownarrow$ (R0); $\updownarrow$ (W1); $\updownarrow$ (R1);	$\updownarrow_s$ (W0, R0, W1, R1);
Checker Board	$\updownarrow$ (W($1_{odd}/0_{even}$)); $\updownarrow$ (R($1_{odd}/0_{even}$)); $\updownarrow$ (W($0_{odd}/1_{even}$)); $\updownarrow$ (R($0_{odd}/1_{even}$));	$\updownarrow_s$ (W($1_{odd}/0_{even}$), R($1_{odd}/0_{even}$), W($0_{odd}/1_{even}$), R($0_{odd}/1_{even}$));

modified addressing sequence. A special design of the BIST circuitry [Cheung 1996] is required to implement the proposed low-power tests.

Another methodology [Dilillo 2006] to minimize test power in SRAM memories is to exploit the predictability of the addressing sequence. It is known that the precharge circuits are the principal contributor to power dissipation in SRAM. It has indeed been shown that it may represent up to 70% of the overall power dissipation of an SRAM memory [Liu 1994]. These circuits have the role of precharging and equalizing the long and high capacitive bit lines. This action is essential to ensure correct memory operation. To reduce the precharge activity during testing, one can use the fact that in functional mode the cells are selected in random sequence, and therefore all precharge circuits need to be always active, while during the test mode the access sequence is known, and hence only the columns that are to be selected need to be precharged [Dilillo 2006]. This low-power test mode can be implemented by using a modified precharge control circuitry and by exploiting the first degree of freedom of March tests, which allows choosing a specific addressing sequence. The modified precharge control logic contains an additional element for each column (see Figure 7.22). This element consists of one multiplexer (two transmission gates and one inverter) and one NAND gate. Signal LP_{test} allows the selection between the functional mode of the memory and the low-power test mode in which the addressing sequence is fixed to "word line after word line" and the precharge activity is restricted to two columns for each clock cycle: the selected column and

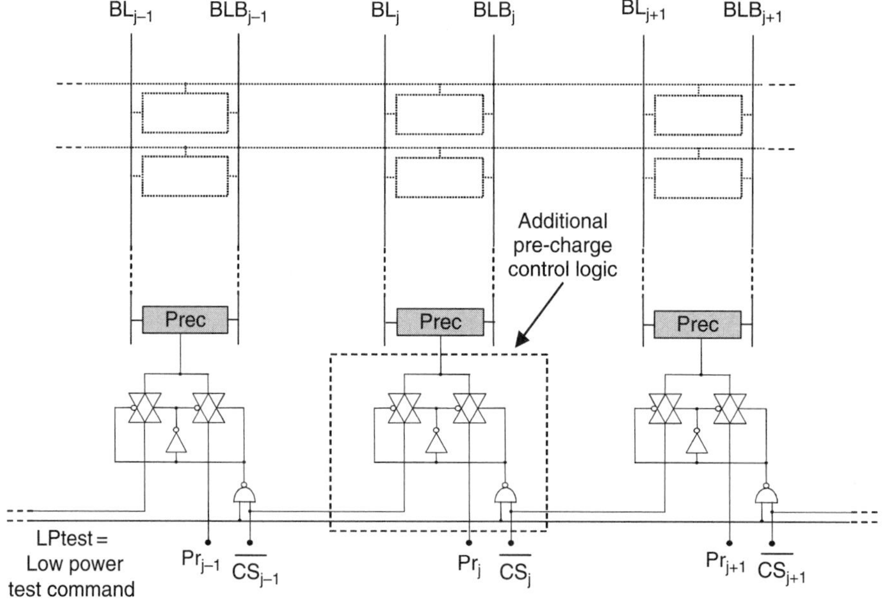

FIGURE 7.22

A precharge control logic for low-power testing.

the following one. Signal Pr_j is the precharge signal originally used, while CS_j' is the complement of the column selection signal. The multiplexer operates the mode selection, whereas the NAND gate forces the functional mode for a given column when it is selected for a read/write operation during testing. When LP_{test} is ON, the signal CS_j' of a column j drives the precharge of the next column $j+1$. Note that the precharge is active with the input signal at 0. Experiments used to validate the proposed method have shown a significant test power reduction (~50%) with negligible impact on area overhead and memory performance.

7.8 CONCLUDING REMARKS

Numerous studies from academia and industry, including the *International Technology Roadmap for Semiconductors* (ITRS) published by the Semiconductor Industry Association (SIA) [SIA 2005, 2006], have shown the need to reduce power consumption during testing of digital and memory designs. This need is triggered by the fact that typically test power can be more than twice the power consumed in normal functional mode.

Because test throughput and manufacturing yield are often affected by test power, various test solutions have been proposed since the mid-1990s. In this chapter, we discussed many low-power test solutions to address the above-mentioned problems. Both structural and algorithmic solutions were described along with their impacts on parameters such as fault coverage, test time, area overhead, circuit performance penalty, and design flow modification. These solutions cover a broad spectrum of testing environments, including scan testing, scan-based BIST, test-per-clock BIST, test compression, and memory testing.

Although solutions presented in this chapter can be used to address most of the problems caused by excessive test power, not all problems have been solved. One concern is when multiple issues arise at the same time when developing low-power test solutions. For example, almost all digital circuits today have scan chains and quite a few require test compression for test data volume reduction along with at-speed testing for screening timing defects. Thus far, few solutions have been proposed to address the problem of low-power scan testing when both test compression and at-speed scan testing are used. Another concern relates to the growing complexity and increasing use of core-based systems. In this case, we are now facing situations where several cores, such as scan cores, memory cores, and logic BIST cores each with embedded at-speed test features, have to be tested in parallel to avoid prohibitive test time. Power-aware or thermal-aware test scheduling is required for these SOC designs so that power and thermal constraints are satisfied while maintaining an optimized test throughput. This complicates the low-power test problems and requires the joint development of core-level and system-level low-power test solutions in the nanometer SOC design era.

Finally, concerns arise from how testing is to be done when new low-power design techniques, such as dynamic power management and multiple-voltage design techniques, are employed. The idea of dynamic power management is to shut down parts of a design when they are idle. Thus far, testing of those designs has been done sequentially (*i.e.*, dealing with power domains one at a time). However, this practice will soon become inadequate because of test time concerns. Similarly, multiple-voltage domains have also been used in designs to reduce power consumption. Among others, the challenges we are facing now include how to build scan chains that span more than one voltage domain, how to cope with physical design constraints when dealing with level-shifters, and how to safely handle the test of such designs. A more future issue relates to asynchronous design, which is now seeing renewed interest as a way to reduce power. Although still far from being practical, asynchronous design will also require new and dedicated low-power test solutions.

7.9 EXERCISES

7.1 **(Test Power)** Provide at least three examples to show why scan test power can be significantly higher than functional power.

7.2 **(Test Power Reduction)** List three ad hoc solutions for reducing power consumption during test application, and show the advantages and disadvantages of each solution.

7.3 **(Terminology)** Explain the difference between dynamic power, short-circuit power, and leakage power.

7.4 **(Terminology)** Explain the difference between energy and power. Also explain what the following terms mean: average power, instantaneous power, and peak power.

7.5 **(Test Power Evaluation)** Show the equations for estimating average power, instantaneous power, and peak power. Explain all the parameters used in the equations.

7.6 **(Noise Phenomena)** Describe three types of circuit noise and their impact on testing.

7.7 **(Terminology)** Explain the following terms:
shift, test, capture cycles in scan testing
slow-speed scan testing and at-speed scan testing
launch-on-shift (LOS) and *launch-on-capture* (LOC)

7.8 **(Test-Induced Yield Loss)** Give at least three reasons why yield loss may occur as a result of scan testing.

7.9 **(*X*-Filling)** Conduct *minimum transition filling* (MT-filling), 0-filling, and 1-filling for the following test cube, and calculate the number of weighted transitions for each resulting fully specified test vector:

Scan-Input-Pin : 1XX0XXXX0XXXXX101XX0

7.10 **(Scan Cell Ordering)** Consider the application of two test vectors to a scan chain composed of four scan cells as follows:

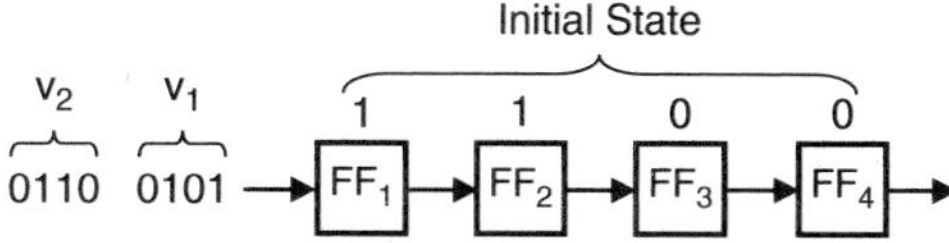

Find the best order of scan cells for reducing the average switching activity measured by the number of weighted transitions.

7.11 **(Token Scan Architecture)** Consider the token scan scheme shown in Figure 7.10. This scheme allows only one scan cell to be activated each time. Show a new token scan scheme that allows exactly two scan cells to be activated each time.

7.12 **(Scan Clock Splitting)** Consider the scan clock splitting scheme shown in Figure 7.12. This scheme divides scan cells into two groups that are not operated simultaneously. Show a new scan clock splitting scheme that divides scan cells into three groups that are not operated simultaneously.

7.13 **(Test Vector Filtering BIST)** Explain the basic idea of test vector filtering BIST, and then show two possible techniques to implement this idea.

7.14 **(Circuit Partitioning)** Study the example (in Figure 7.18) of circuit partitioning for low-power testing in logic BIST. Then show how to partition the following circuit for low-power testing.

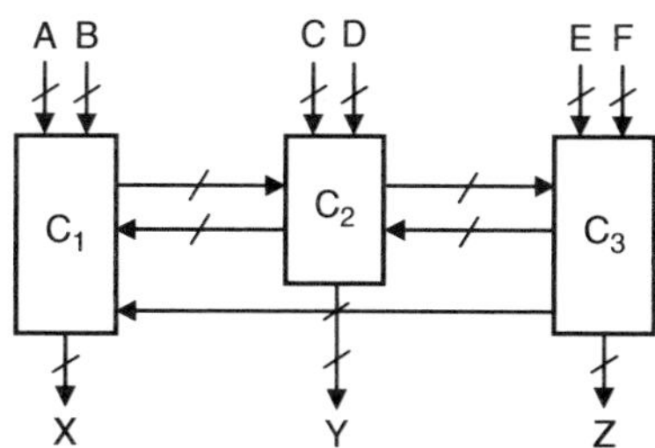

7.15 7.15 **(Low-Power RAM Testing)** Compare the original test algorithms and low-power test algorithms shown in Table 7.1. Discuss the reasons why the low-power test algorithms can reduce test power.

7.16 

(A Design Practice) Use the ATPG programs and user's manuals contained on the Companion Web site to generate test sets for a number of full-scan benchmark circuits, with and without using the low shift power option. Compare the resulting test sets in terms of fault coverage, test data volume, and estimated shift power dissipation.

7.17 

(A Design Practice) Use the ATPG programs and user's manuals contained on the Companion Web site to generate test sets for a number of full-scan benchmark circuits, with and without using the low capture power option. Compare the resulting test sets in terms of fault coverage, test data volume, and estimated capture power dissipation.

7.18 

(A Design Practice) Use the ATPG programs and user's manuals contained on the Companion Web site to generate test sets for a number of full-scan benchmark circuits, with and without using the low shift/capture power option. Compare the resulting test sets in terms of fault coverage, test data volume, and estimated capture power dissipation.

Acknowledgments

The authors wish to thank Prof. Nicola Nicolici of McMaster University, Dr. Mokhtar Hirech of Synopsys, and Dr. Christian Landrault of LIRMM/CNRS for reviewing this chapter and providing helpful comments. The authors also would like to thank Dr. Zhigang Jiang of SynTest Technologies for providing the Design Practices questions in the Exercises section.

References

R7.0 Books

[Altet 2002] J. Altet and A. Rubio, *Thermal Testing of Integrated Circuits*, Springer, Boston, 2002.

[Bushnell 2000] M. L. Bushnell and V. D. Agrawal, *Essentials of Electronic Testing for Digital, Memory & Mixed-Signal VLSI Circuits*, Springer, Boston, 2000.

[Crouch 1999] A. Crouch, *Design-for-Test for Digital IC's and Embedded Core Systems*, Prentice-Hall, Englewood Cliffs, NJ, 1999.

[Jha 2003] N. Jha and S. Gupta, *Testing of Digital Systems*, Cambridge University Press, London, 2003.

[Krstic 1998] A. Krstic and K.-T. Cheng, *Delay Fault Testing for VLSI Circuits*, Springer, Boston, 1998.

[Nicolici 2003] N. Nicolici and B. Al-Hashimi, *Power-Constrained Testing of VLSI Circuits*, Springer, Boston, 2003.

[Rajski 1998] J. Rajski and J. Tyszer, *Arithmetic Built-In Self-Test for Embedded Systems*, Prentice-Hall, Englewood Cliffs, NJ, 1998.

[Wang 2006] L.-T. Wang, C.-W. Wu, and X. Wen, editors, *VLSI Test Principles and Architectures: Design for Testability*, Morgan Kaufmann, San Francisco, 2006.

[Weste 1993] N. H. E. Weste and K. Eshraghian, *Principles of CMOS VLSI Design: A Systems Perspective*, Second Edition, Addison Wesley, Reading, MA, 1993.

R7.1 Introduction

[De Colle 2005] A. De Colle, S. Ramnath, M. Hirech, and S. Chebiyam, Power and design for test: A design automation perspective, *ASP J. Low Power Electronics*, 1(1), pp. 73–84, April 2005.

[Girard 2000] P. Girard, Low power testing of VLSI circuits: Problems and solutions, in *Proc. Int. Symp. on Quality of Electronic Design*, pp. 173–179, March 2000.

[Monzel 1997] J. Monzel, S. Chakravarty, V. D. Agrawal, R. Aitken, J. Braden, J. Figueras, S. Kumar, H.-J. Wunderlich, and Y. Zorian, Power dissipation during testing: Should we worry about it?, *IEEE VLSI Test Symp.*, Panel Session, April 1997.

[Moore 1965] G. Moore, Cramming more components onto integrated circuits, *Electronics*, pp. 114–117, April 19, 1965.

[Pouya 2000] B. Pouya and A. Crouch, Optimization trade-offs for vector volume and test power, in *Proc. Int. Test Conf.*, pp. 873–881, October 2000.

[Saxena 2003] J. Saxena, K. M. Butler, V. B. Jayaram, S. Kundu, N. V. Arvind, P. Sreeprakash, and M. Hachinger, A case study of IR-drop in structured at-speed testing, in *Proc. Int. Test Conf.*, pp. 1098–1104, October 2003.

[SIA 2001] SIA, *The International Technology Roadmap for Semiconductors: 2001 Edition*, Semiconductor Industry Association, San Jose, CA (http://public.itrs.net), 2001.

[SIA 2005] SIA, *The International Technology Roadmap for Semiconductors: 2005 Edition*, Semiconductor Industry Association, San Jose, CA (http://public.itrs.net), 2005.

[Wang 1997] S. Wang and S. K. Gupta, DS-LFSR: A new BIST TPG for low heat dissipation, in *Proc. Int. Test Conf.*, pp. 848–857, November 1997.

[Zorian 1993] Y. Zorian, Testing the monster chip, *IEEE Spectrum*, 36(7), pp. 54–60, July 1999.

R7.2 Energy and Power Modeling

[Athas 1994] W. C. Athas, L. J. Svensson, J. G. Koller, N. Tzartzanis, and E. Ying-Chin Chou, Low-power digital systems based on adiabatic-switching principles, *IEEE Trans. on Very Large Scale Integration Systems*, 2(4), pp. 398–416, December 1994.

[Cirit 1987] M. A. Cirit, Estimating dynamic power consumption of CMOS circuits, in *Proc. Int. Conf. on Computer-Aided Design*, pp. 534–537, November 1987.

[Shi 2004] C. Shi and R. Kapur, How power aware test improves reliability and yield, *IEEDesign.com*, September 15, 2004.

[Wang 1995] C.-Y. Wang and K. Roy, Maximum power estimation for CMOS circuits using deterministic and statistical approaches, in *Proc. VLSI Conf.*, pp. 364–369, January 1995.

R7.3 Test Power Issues

[Butler 2004] K. M. Butler, J. Saxena, T. Fryars, G. Hetherington, A. Jain, and J. Lewis, Minimizing power consumption in scan testing: Pattern generation and DFT techniques, in *Proc. Int. Test Conf.*, pp. 355–364, October 2004.

[Chang 1997] Y.-S. Chang, S. K. Gupta, and M. A. Breuer, Analysis of ground bounce in deep sub-micron circuits, in *Proc. VLSI Test Symp.*, pp. 110–116, May 1997.

[Hertwig 1998] A. Hertwig and H. J. Wunderlich, Low power serial built-in self-test, in *Proc. European Test Workshop*, pp. 49–53, May 1998.

[Jiang 2000] Y. M. Jiang, A. Krstic, and K.-T. Cheng, Estimation for Maximum Instantaneous Current Through Supply Lines For CMOS circuits, *IEEE Trans. on Very Large Scale Integration Systems*, 8(1), pp. 61–73, February 2000.

[Nicolaidis 1998] M. Nicolaidis and Y. Zorian, On-Line testing for VLSI: A compendium of approaches, *JETTA J. Electronic Testing: Theory and Applications*, 12(1–2), pp. 7–20, February/April 1998.

[Pouya 2000] B. Pouya and A. Crouch, Optimization trade-offs for vector volume and test power, in *Proc. Int. Test Conf.*, pp. 873–881, October 2000.

[Shi 2004] C. Shi and R. Kapur, How power aware test improves reliability and yield, *IEEDesign.com*, September 15, 2004.

[SIA 2003] SIA, *The International Technology Roadmap for Semiconductors: 2003 Edition*, Semiconductor Industry Association, San Jose, CA (http://public.itrs.net), 2003.

[Wang 1997] S. Wang and S. K. Gupta, DS-LFSR: A new BIST TPG for low heat dissipation, in *Proc. Int. Test Conf.*, pp. 848–857, November 1997.

R7.4 Low-Power Scan Testing

[Badereddine 2006] N. Badereddine, P. Girard, S. Pravossoudovitch, C. Landrault, A. Virazel, and H. J. Wunderlich, Minimizing peak power consumption during scan testing: Test pattern modification with X filling heuristics, in *Proc. Int. Conf. on Design & Test of Integrated Systems*, pp. 259–264, September 2006.

[Bonhomme 2001] Y. Bonhomme, P. Girard, L. Guiller, C. Landrault, and S. Pravossoudovitch, A gated clock scheme for low power scan testing of logic ICs or embedded cores, in *Proc. Asian Test Symp.*, pp. 253–258, November 2001.

[Bonhomme 2002] Y. Bonhomme, P. Girard, C. Landrault, and S. Pravossoudovitch, Power driven chaining of flip-flops in scan architectures, in *Proc. Int. Test Conf.*, pp. 796–803, October 2002.

[Bonhomme 2003] Y. Bonhomme, P. Girard, L. Guiller, C. Landrault, and S. Pravossoudovitch, Efficient scan chain design for power minimization during scan testing under routing constraint, in *Proc. Int. Test Conf.*, pp. 488–493, October 2003.

[Butler 2004] K. M. Butler, J. Saxena, T. Fryars, G. Hetherington, A. Jain, and J. Lewis, Minimizing power consumption in scan testing: Pattern generation and DFT techniques, in *Proc. Int. Test Conf.*, pp. 355–364, October 2004.

[Dabholkar 1998] V. Dabholkar, S. Chakravarty, I. Pomeranz, and S. M. Reddy, Techniques for reducing power dissipation during test application in full scan circuits, *IEEE Trans. on Computer-Aided Design*, 17(12), pp. 1325–1333, December 1998.

[Girard 2002] P. Girard, "Survey of Low-Power Testing of VLSI Circuits," *IEEE Design & Test of Computers*, 19(3), pp. 82–92, May/June 2002.

[Hertwig 1998] A. Hertwig and H. J. Wunderlich, Low power serial built-in self-test, in *Proc. European Test Workshop*, pp. 49–53, May 1998.

[Hiraide 2003] T. Hiraide, K. O. Boateng, H. Konishi, K. Itaya, M. Emori, H. Yamanaka, and T. Mochiyama, BIST-aided scan test: A new method for test cost reduction, in *Proc. VLSI Test Symp.*, pp. 359–364, May 2003.

[Huang 1999] T.-C. Huang and K.-J. Lee, An input control technique for power reduction in scan circuits during test application, in *Proc. Asian Test Symp.*, pp. 315–320, November 1999.

[Huang 2001] T.-C. Huang and K.-J. Lee, A token scan architecture for low power testing, in *Proc. Int. Test Conf.*, pp. 660–669, October 2001.

[Lee 2000] K.-J. Lee, T.-C. Huang, and J.-J. Chen, Peak-power reduction for multiple-scan circuits during test application, in *Proc. IEEE Asian Test Symp.*, pp. 453–458, December 2000.

[Remersaro 2006] S. Remersaro, X. Lin, Z. Zhang, S. M. Reddy, I. Pomeranz, and J. Rajski, Preferred fill: A scalable method to reduce capture power for scan based designs, in *Proc. Int. Test Conf.*, Paper 32.2, October 2006.

[Rosinger 2004] P. Rosinger, B. Al-Hashimi, and N. Nicolici, Scan architecture with mutually exclusive scan segment activation for shift- and capture-power reduction, *IEEE Trans. on Computer-Aided Design*, 23(7), pp. 1142–1153, July 2004.

[Sankaralingam 2000] R. Sankaralingam, R. Oruganti, and N. A. Touba, Static compaction techniques to control scan vector power dissipation, in *Proc. VLSI Test Symp.*, pp. 35–42, May 2000.

[Sankaralingam 2003] R. Sankaralingam and N. A. Touba, Multi-phase shifting to reducing instantaneous peak power during scan, in *Proc. Latin American Test Workshop*, pp. 78–83, February 2003.

[Saxena 2001] J. Saxena, K. M. Butler, and L. Whetsel, A scheme to reduce power consumption during scan testing, in *Proc. Int. Test Conf.*, pp. 670–677, October 2001.

[Saxena 2003] J. Saxena, K. M. Butler, V. B. Jayaram, S. Kundu, N. V. Arvind, P. Sreeprakash, and M. Hachinger, A case study of ir-drop in structured at-speed testing, in *Proc. Int. Test Conf.*, pp. 1098–1104, October 2003.

[Shi 2004] C. Shi and R. Kapur, How power aware test improves reliability and yield, *IEEDesign.com*, September 15, 2004.

[Sinanoglu 2002] O. Sinanoglu, I. Bayraktaroglu, and A. Orailoglu, Dynamic test data transformations for average and peak power reductions, in *Proc. European Test Workshop*, pp. 113–118, May 2002.

[Wang 1994] S. Wang and S. K. Gupta, ATPG for heat dissipation minimization during test application, in *Proc. Int. Test Conf.*, pp. 250–258, October 1994.

[Wang 1997] S. Wang and S. K. Gupta, ATPG for heat dissipation minimization for scan testing, in *Proc. Design Automation Conf.*, pp. 614–619, June 1997.

[Wen 2005a] X. Wen, Y. Yamashita, S. Morishima, S. Kajihara, L.-T. Wang, K. K. Saluja, and K. Kinoshita, Low-capture-power test generation for scan-based at-speed testing, in *Proc. Int. Test Conf.*, Paper 39.2, November 2005.

[Wen 2005b] X. Wen, T. Suzuki, S. Kajihara, K. Miyase, Y. Minamoto, L.-T. Wang, and K.K. Saluja, Efficient test set modification for capture power reduction, *ASP J. Low Power Electronics*, 1(3), pp. 319–330, December 2005.

[Wen 2006] X. Wen, S. Kajihara, K. Miyase, T. Suzuki, K. K. Saluja,L.-T. Wang, K. S. Abdel-Hafez, and K. Kinoshita, A new ATPG method for efficient capture power reduction during scan testing, in *Proc. VLSI Test Symp.*, pp. 58–63, May 2006.

[Whetsel 2000] L. Whetsel, Adapting scan architectures for low power operation, in *Proc. Int. Test Conf.*, pp. 863–872, October 2000.

[Wohl 2003] P. Wohl, J. A. Waicukauski, S. Patel, and M. B. Amin, Efficient compression and application of deterministic patterns in a logic BIST architecture, in *Proc. Design Automation Conf.*, pp. 566–569, June 2003.

[Xu 2006] G. Xu and A. D. Singh, Low cost launch-on-shift delay test with slow scan enable, in *Proc. European Test Symp.*, Paper 3a-1, May 2006.

R7.5 Low-Power Built-In Self-Test

[Brglez 1985] F. Brglez and H. Fujiwara, A neutral netlist of 10 combinational benchmark circuits and a target translator in Fortran, in *Proc. Int. Symp. on Circuits and Systems*, pp. 663–698, June 1985.

[Chou 1994] R.-M. Chou, K. K. Saluja, and V. D. Agrawal, Power constraint scheduling of tests, in *Proc. Int. Conf. on VLSI Design*, pp. 271–274, January 1994.

[Corno 1999] F. Corno, M. Rebaudengo, M. Sonza Reorda, and M. Violante, A new BIST architecture for low power circuits, in *Proc. European Test Workshop*, pp. 160–164, May 1999.

[Corno 2000] F. Corno, M. Rebaudengo, M. Sonza Reorda, G. Squillero, and M. Violente, Low power BIST via non-linear hybrid cellular automata, in *Proc. VLSI Test Symp.*, pp. 29–34, May 2000.

[Gerstendörfer 1999] S. Gerstendörfer and H. J. Wunderlich, Minimized power consumption for scan-based BIST, in *Proc. Int. Test Conf.*, pp. 77–84, September 1999.

[Girard 1999a] P. Girard, L. Guiller, C. Landrault, and S. Pravossoudovitch, A test vector inhibiting technique for low energy BIST design, in *Proc. VLSI Test Symp.*, pp. 407–412, April 1999.

[Girard 1999b] P. Girard, L. Guiller, C. Landrault, S. Pravossoudovitch, J. Figueras, S. Manich, P. Teixeira, and M. Santos, Low energy BIST design: Impact of the LFSR TPG parameters on the weighted switching activity, in *Proc. Int. Symp. on Circuits and Systems*, CD-ROM Proceedings, June 1999.

[Girard 1999c] P. Girard, L. Guiller, C. Landrault, and S. Pravossoudovitch, Circuit partitioning for low power BIST Design with minimized peak power consumption, in *Proc. Asian Test Symp.*, pp. 89–94, November 1999.

[Girard 2000] P. Girard, L. Guiller, C. Landrault, and S. Pravossoudovitch, Low power BIST design by hypergraph partitioning: Methodology and architectures, in *Proc. Int. Test Conf.*, pp. 652–661, October 2000.

[Girard 2001] P. Girard, L. Guiller, C. Landrault, S. Pravossoudovitch, and H. J. Wunderlich, A modified clock scheme for a low power BIST test pattern generator, in *Proc. VLSI Test Symp.*, pp. 306–311, May 2001.

[Gizopoulos 2000] D. Gizopoulos, N. Kranitis, A. Paschalis, M. Psarakis, and Y. Zorian, "Low Power/Energy BIST Scheme for Datapaths," in *Proc. VLSI Test Symp.*, pp. 23–28, May 2000.

[He 2006] Z. He, Z. Peng, P. Eles, P. Rosinger, and B. Al-Hashimi, Thermal-aware SOC test scheduling with test set partitioning and interleaving, in *Proc. Int. Symp. on Defect and Fault Tolerance in VLSI Systems*, pp. 477–485, October 2006.

[Iyengar 2001] V. Iyengar, and K. Chakrabarty, Precedence-based, preemptive, and power-constrained test scheduling for system-on-a-chip, in *Proc. VLSI Test Symp.*, pp. 42–47, May 2001.

[Karypis 1998] G. Karypis, R. Aggarwal, V. Kumar, and S. Shekhar, Multilevel hypergraph partitioning: Applications in VLSI domain, Technical Report, Department of Computer Science, University of Minnesota, (www.cs.umn.edu/~karypis/metis), November 1998.

[Larsson 2002] E. Larsson and H. Fujiwara, Power-constrained preemptive TAM scheduling, in *Proc. European Test Workshop*, pp. 119–126, May 2002.

[Liu 2005] C. Liu, K. Veeraraghavant, and V. Iyengar, Thermal-aware test scheduling and hot spot temperature minimization for core-based systems, in *Proc. Int. Symp. on Defect and Fault Tolerance in VLSI Systems*, pp. 552–562, October 2005.

[Manich 2000] S. Manich, A. Gabarro, M. Lopez, J. Figueras, P. Girard, L. Guiller, C. Landrault, S. Pravossoudovitch, P. Teixeira, and M. Santos, Low power BIST by filtering non-detecting vectors, *JETTA J. Electronic Testing: Theory and Applications*, 16(3), pp. 193–202, June 2000.

[Muresan 2000] V. Muresan, X. Wang, and M. Vladutiu, A comparison of classical scheduling approaches in power-constrained block-test scheduling, in *Proc. Int. Test Conf.*, pp. 882–891, October 2000.

[Pouget 2003] J. Pouget, E. Larsson, Z. Peng, M.L. Flottes, and B. Rouzeyre, An efficient approach to SOC wrapper design, TAM configuration and test scheduling, in *Proc. European Test Workshop*, pp. 117–122, May 2003.

[Rosinger 2005] P. Rosinger, B. Al-Hashimi, and K. Chakrabarty, Rapid generation of thermal-safe test schedules, in *Proc. Design, Automation and Test in Europe*, pp. 840–845, March 2005.

[Ubar 2005] R. Ubar, T. Shchenova, G. Jervan, and Z. Peng, Energy minimization for hybrid BIST in a system-on-chip test environment, in *Proc. European Test Symp.*, pp. 2–7, May 2005.

[Wang 1997] S. Wang and S. K. Gupta, DS-LFSR: A new BIST TPG for low heat dissipation, in *Proc. Int. Test Conf.*, pp. 848–857, November 1997.

[Wang 1999] S. Wang and S. K. Gupta, LT-RTPG: A new test-per-scan BIST TPG for low heat dissipation, in *Proc. Int. Test Conf.*, pp. 85–94, September 1999.

[Zhang 1999] X. Zhang, K. Roy, and S. Bhawmik, POWERTEST: A tool for energy conscious weighted random pattern testing, in *Proc. Int. Conf. on VLSI Design*, pp. 416–422, January 1999.

[Zorian 1993] Y. Zorian, Testing the monster chip, *IEEE Spectrum*, 36(7), pp. 54–60, July 1999.

R7.6 Low-Power Test Data Compression

[Al-Yamani 2005] A. Al-Yamani, E. Chmelar, and M. Grinchuck, Segmented addressable scan architecture, in *Proc. VLSI Test Symp.*, pp. 405–411, May 2005.

[Baik 2005] D. H. Baik and K. K. Saluja, Progressive random access scan: A simultaneous solution to test power, test data volume and test time, in *Proc. Int. Test Conf.*, paper 15.2, November 2005.

[Chandra 2001] A. Chandra and K. Chakrabarty, Combining low-power scan testing and test data compression for system-on-a-chip, in *Proc. Design Automation Conf.*, pp. 166–169, June 2001.

[Chandra 2002] A. Chandra and K. Chakrabarty, Reduction of SOC test data volume, scan power and testing time using alternating run-length codes, in *Proc. Design Automation Conf.*, pp. 673–678, June 2002.

[Hamzaoglu 1999] I. Hamzaoglu and J. Patel, Reducing test application time for full scan embedded cores, in *Proc. Int. Symp. on Fault Tolerant Computing*, pp. 260–267, June 1999.

[Lee 2004] J. Lee and N. A. Touba, Low power test data compression based on LFSR reseeding, in *Proc. Int. Conf. on Computer Design*, pp. 180–185, October 2004.

[Pouya 2000] B. Pouya and A. Crouch, Optimization trade-offs for vector volume and test power, in *Proc. Int. Test Conf.*, pp. 873–881, October 2000.

R7.7 Low-Power RAM Testing

[Cheung 1996] H. Cheung and S. Gupta, A BIST methodology for comprehensive testing of RAM with reduced heat dissipation, in *Proc. Int. Test Conf.*, pp. 22–32, October 1996.

[Dilillo 2006] L. Dilillo, P. Rosinger, P. Girard, and B. M. Al-Hashimi, Minimizing test power in SRAM through pre-charge activity reduction, in *Proc. Design, Automation and Test in Europe*, pp. 1159–1165, March 2006.

[Liu 1994] D. Liu and C. Svensson, Power consumption estimation in CMOS VLSI chips, *IEEE J. Solid-State Circuits*, 29(6), pp. 663–670, June 1994.

R7.8 Concluding Remarks

[SIA 2005] SIA, *The International Technology Roadmap for Semiconductors:* 2005 *Edition*, Semiconductor Industry Association, San Jose, CA (http://public.itrs.net), 2005.

[SIA 2006] SIA, *The International Technology Roadmap for Semiconductors:* 2006 *Update*, Semiconductor Industry Association, San Jose, CA (http://public.itrs.net), 2006.

CHAPTER 8

Coping with Physical Failures, Soft Errors, and Reliability Issues

Laung-Terng (L.-T.) Wang
SynTest Technologies, Inc., Sunnyvale, California

Mehrdad Nourani
University of Texas at Dallas, Richardson, Texas

T. M. Mak
Intel Corporation, Santa Clara, California

ABOUT THIS CHAPTER

Physical failures caused by manufacturing defects and process variations, as well as soft errors induced by alpha-particle radiation, have been identified as the main source of faults attributed to chip or system failure. Today, the semiconductor industry relies heavily on two test technologies: scan and *built-in self-test* (BIST). Existing scan implementations may no longer be sufficient as scaling introduces new failure mechanisms that exceed the ability to capture by single-fault-model-based tests. BIST will also become problematic if it does not achieve sufficient fault coverage in reasonable time. Faced with significant test problems in the nanometer design era, it is imperative that we seek viable test solutions now to complement the conventional scan and BIST techniques.

In this chapter, we focus on test techniques to cope with physical failures for digital logic circuits. Techniques for improving process yield, silicon debug, and system diagnosis, along with test methods and DFT architectures for testing *field programmable gate array* (FPGA), *microelectromechanical systems* (MEMS), and *analog and mixed-signal* (AMS) circuits are covered in subsequent chapters. In this chapter, we first discuss test techniques to deal with signal integrity problems. We then describe test techniques to screen manufacturing defects and process variations. Finally, we present a number of promising online error-resilient architectures and schemes to cope with soft errors as well as for defect and error tolerance.

8.1 INTRODUCTION

Since the early 1980s, *complementary metal oxide semiconductor* (CMOS) process has become the dominant manufacturing technology. At the introduction rate of a new process technology (node) roughly every 2 years, which is a reflection of the Moore's law [Moore 1965], new defect mechanisms have initially caused low manufacturing yield, elevated infant mortality rates, and high defect levels. As CMOS scaling continues (down to a feature size of 65 nanometers and below), additional defect mechanisms caused by new manufacturing defects (such as defects as a result of optical effects) and process variations continue to create various failure mechanisms. To meet yield, reliability, and quality goals (referred to as ***defective parts per million*** [DPM]), these defects must be screened during manufacturing or be tolerated during system operation.

Defects are physical phenomena that occur during manufacturing and can cause functional or timing failures. Examples of defects are missing conducting material or extra insulating material (possibly causing **opens**), the presence of extra conducting material, or missing insulating material between two signal lines (possibly causing **shorts**), among others. A defect does not always manifest itself as a single isolated problem such as an open or a short. When a circuit parameter is out of given specifications, it can also cause a failure or become susceptible to other problems (temperature effects, **crosstalk**, leakage power, etc.). New manufacturing processes, such as a changeover from aluminum metallization to copper metallization and from SiO_2 to low-K interlayer dielectric [Tyagi 2000], have created new defect and fault mechanisms. These defect mechanisms include copper-related defects (an effect of dual damascene copper deposition process), optical induced defects (an effect of undercorrection or overcorrection), and design-related defects (low threshold voltage and multiple voltages in low power designs) [Aitken 2004] [Guardiani 2004]. These defect mechanisms along with their associated potential defects and failure mechanisms are extensively discussed in [Gizopoulos 2006]. The more recently announced changeover of the gate stack, from polysilicon/SiO_2 to high-K/metal gate dielectric, will certainly bring forth new defect mechanisms of their own [Chau 2004].

Broadly speaking, defects can be *random* or *systematic,* and they can be *functional* or *parametric*. **Random defects** are caused by manufacturing imperfections and occur in random places. Most of these random defects are relatively easy to find except for a few that may require a lot of test time to find. In certain cases, 50% of test time for some devices can go after 0.5% of the defects [Nigh 2007, personal communication]. **Systematic defects** are caused by process or manufacturing variations (sometimes the result of lithography, planarity, film thickness, etc.). A systematic defect can also include a temporal systematic component (*e.g.*, every 10th wafer). At 65 nm and below, these systematic variations are the greatest cause of catastrophic chip failures and electrical issues related to timing, signal integrity, and leakage power. Reference [Clear Shape 2007] indicates that at 65 nm, systematic variations of 3 nm on a transistor gate can cause a 20% variation in delay and have a 2× impact on **leakage power**. For some devices, leakage power can vary by 10× within the process window [Nigh 2007, personal communication]. Leakage power is gradually dominating process variations, in becoming a major yield detractor as process technology continues to scale down.

The traditional treatment of defects focuses more on *functional random (spot) defects*, which lead to existing yield models. Growing process variations and other uncertainty issues require looking at the other types of defects. In a narrow sense, defects are caused by process variations or random localized manufacturing imperfection [Sengupta 1999].

Process variations, such as transistor channel length variation, transistor threshold voltage variation, metal interconnect thickness variation, and interlayer dielectric thickness variation, have a big impact on device speed characteristics. In general, the effect of process variation shows up first in the most critical paths in the design, those with maximum and minimum delays. For instance, a shorter channel length or lower threshold voltage can result in potentially faster device speed but with a significantly higher leakage current. A thinner gate oxide can result in potentially increased device speed at the expense of a significantly increased gate tunneling current and reliability concerns.

Random imperfection, such as resistive shorts between metal lines, resistive opens on metal lines, improper via formations, and shallow trench isolation defects, is yet another source of defects, called random defects. Based on the electrical characteristics of the defect and neighboring parasitic, the defect may result in a functional (hard) or parametric (delay or marginal) failure. For instance, a missing contact could cause an open fault. A resistive short between two metal lines or an extra contact could cause a bridging fault. A resistive open on a metal line or an improper via formation could cause a delay fault.

Recall that defect level (DL) is a function of process yield (Y) and fault coverage (FC) [Williams 1981]. The authors in [McCluskey 1988] further showed that:

$$DL = 1 - Y^{(1-FC)}$$

This indicates that in order to reduce defect level to meet a given DPM goal, one can improve the fault coverage of the chips (devices) under test, the process yield, or both at the same time. In reality, not all chips passing manufacturing tests would function correctly in the field. Reports have shown that (1) chips can be exposed to *alpha-particle radiation* and (2) *nonrecurring transient errors* caused by single-event upsets, called **soft errors**, can occur [May 1979] [Baumann 2005]. The chips can also be exposed to noises, such as power supply noise or signal integrity, and cause unrecoverable errors [Dally 1998].

For nanometer *system-on-chip* (SOC) designs, there is also a growing concern as to whether one can find defect-free or error-free dies [Breuer 2004a]. Advanced design and test technologies are eminent now in order to meet yield and DPM goals and ensure that defective chips function harmlessly in the field.

There are two fundamentally complementary test technologies that can be taken to meet our goals, similar to those approaches used to improve the **reliability** of computer systems: *design for testability* (DFT) [Williams 1983] [Abramovici 1994] [Bushnell 2000] [Mourad 2000] [Jha 2003] [Wang 2006] and fault tolerance

[Siewiorek 1998] [Lala 2001]. The fault tolerance approach aims at preventing the chip (computer system) from malfunction despite the presence of physical failures (errors), whereas design for testability uses design techniques to reduce defect levels or the probability of chip (system) failures as a result of manufacturing faults by screening those defective devices more effectively during manufacturing test.

In the following subsections, we first discuss promising test techniques to deal with signal integrity problems induced by physical failures. We next describe promising schemes to screen manufacturing defects and process variations and to improve manufacturing yield. We then discuss a few online test architectures designed to protect soft errors induced by radiation or other reliability concerns. Finally, promising schemes for defect and error tolerance are presented to ensure that defective chips can still function in nanometer designs.

8.2 SIGNAL INTEGRITY

Signal integrity is the ability of a signal to generate correct responses in a circuit. Informally speaking, signal integrity indicates how clean or distorted a signal is. A signal with good integrity stays within *safe* (acceptable) margins for its voltage amplitude and transition time. For example, an input signal to a flip-flop with good integrity arrives on time to satisfy the setup/hold time requirements and does not have large undershoots that may cause erroneous logic readout or large overshoots that affect the transistor's lifetime.

Leaving the safe margins may not only cause failure in a system (*e.g.*, unexpected ringing) but also shorten the system's lifetime. The latter occurs because of the ***time-dependent dielectric breakdown*** (TDDB) [Hunter 1999] or injection of high-energy electrons and holes (also called **hot carriers**) into the gate oxide. Such phenomena ultimately cause permanent degradation of *metal oxide semiconductor* (MOS) transistors. To quantify these, systematic methods can be employed to perform the lifetime analysis and measure the performance degradation of logic gates under stress (*e.g.*, repeated overshoots) [Fang 1998].

8.2.1 Basic Concept of Integrity Loss

Signal integrity depends on many internal (*e.g.*, interconnects, data, characteristics of transistors, power supply noise, process variations) and external (*e.g.*, environmental noise, interactions with other systems) factors. By using accurate simulation in the design phase, one can apply conservative techniques (*e.g.*, stretched sizing/spacing, shielding) to minimize the effect of integrity loss. There are interdependencies among these parameters, which can result in performance degradation or permanent/intermittent failures. Because of the uncertainty caused by these interdependent parameters, it is impossible (with our current state of knowledge) to have a guaranteed remedy at the design phase. Thus, testing future *very-large-scale integration* (VLSI) chips for signal integrity seems to be inevitable.

True characteristics of a signal are reflected in its waveform. In practice, digital electronic components can tolerate certain levels of voltage swing and transition/

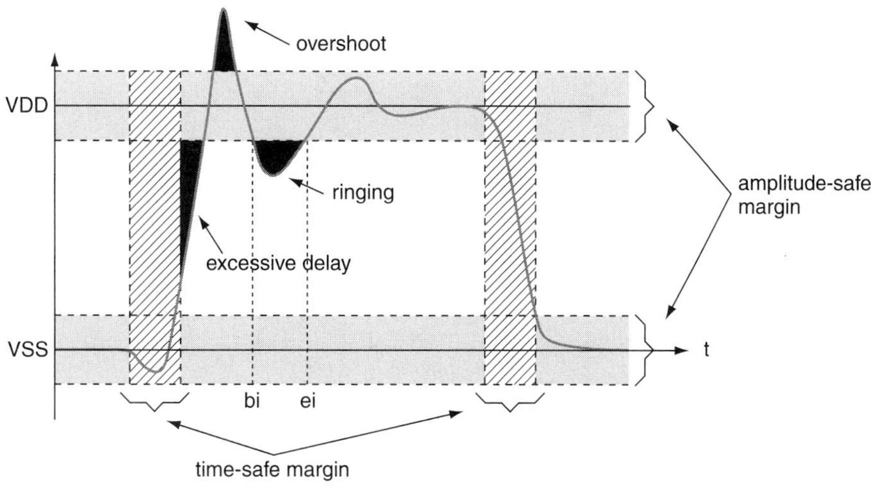

■ **FIGURE 8.1**

The concept of signal integrity loss.

propagation delay. Any portion of a signal that exceeds these levels represents ***integrity loss*** (IL). This concept has graphically been shown in Figure 8.1, in which the horizontal and vertical shaded strips correspond to the amplitude- and time-safe margins, respectively. The black areas illustrate the time frames in which the signal has left the safe margin and thus integrity loss has occurred.

Any portion of a signal $f(t)$ that exceeds the safe margins contributes to the integrity loss metric, which can be quantified as:

$$IL = \sum_{i} \left(\int_{b_i}^{e_i} |V_i - f(t)| \cdot dt \right)$$

where V_i is one of the acceptable amplitude levels (*i.e.*, a border of safe margin) and $[b_i, e_i]$ is a time frame during which integrity loss occurs.

Figure 8.1 and the preceding formula show the basic concept of integrity loss. Not all signals experience the same fluctuations. The presence or the level of integrity loss (overshoot, ringing, and delay) depends on the technology (*e.g.*, interconnect parasitic *R*/*L*/*C* values), process variations (*e.g.*, changes of threshold voltage, oxide thickness), and the application (*e.g.*, on-chip *versus printed circuit board* [PCB] wiring). For example, overshoots and ringing are more commonly found in PCBs and chip packaging where inductance (*L*) of wires is not negligible. Delays (transitions or settling), on the other hand, are more important for on-chip interconnects because of the larger effect of wire's parasitic resistance (*R*) and capacitance (*C*).

With today's computing power, the computationally intensive analysis/simulation recommended by this model would not be practical for real-world circuits; yet it implies three main requirements in testing VLSI chips for signal integrity: (1) the target source (location) selection to stimulate or sample/monitor IL, (2) integrity loss sensors/monitors, and (3) readout circuitry to deliver IL information to an observation/analysis point. Almost all solutions presented in the literature so far point to the necessity of a combination of these three requirements. Next, we briefly discuss these requirements and some of the techniques described in the literature.

8.2.2 Sources of Integrity Loss

To have a practical evaluation of integrity loss, we need to decide where and what to look at. Various sources of signal integrity loss in VLSI chips have been identified. The most important ones are the following:

- *Interconnects*, which contribute to *crosstalk* (signal distortion caused by cross-coupling effects among signals), *overshoot* (signal rising momentarily above the power supply voltage), and *electromagnetic interference* (resulting from the antenna properties) [Bai 2000] [Chen 2001] [Nourani 2002].
- *Power supply noise*, whose large fluctuations, mainly the result of simultaneous switchings, affect the functionality of some gates and eventually may lead to failure [Senthinatharr 1994] [Zhao 2000a].
- *Process variations*, which are deviations of parameters from their desired values because of the imperfect nature of the fabrication process. Sources of process variations include random dopant fluctuation, annealing effects, lithographic limitations, among others [Borkar 2004a].

The pattern generation mechanism depends on the source (location) of IL that is the target of testing. For interconnects and power supply noise, deterministic test pattern generation methods are often used. For testing IL caused by process variation, however, pattern generation is replaced by sensing or monitoring devices.

8.2.2.1 Interconnects

The ***maximum aggressor*** (MA) fault model [Cuviello 1999] is a simplified model that many researchers use mainly for crosstalk analysis and testing. This model, shown in Figure 8.2, assumes the signal traveling on a victim line V may be affected by signals/transitions on other aggressor line(s) A in its neighborhood. The coupling can be represented by a generic coupling component Z. In general, the result could be noise and excessive delay that may lead to functional error and performance degradation, respectively. There is, however, controversy as to what patterns trigger maximal integrity loss. Specifically, in the traditional MA model that takes only coupling C into account, all aggressors make the same simultaneous transition in the same direction, whereas the victim line is kept quiescent for maximal ringing (see pattern pair 1) or makes an opposite transition for maximal delay (see pattern pair 2).

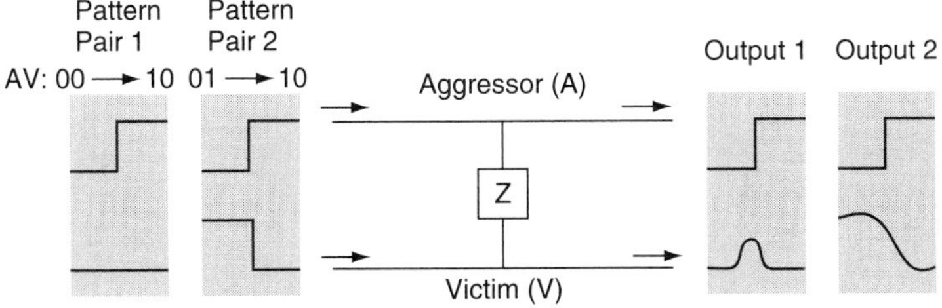

■ **FIGURE 8.2**

Signal integrity fault model using concept of aggressor/victim lines.

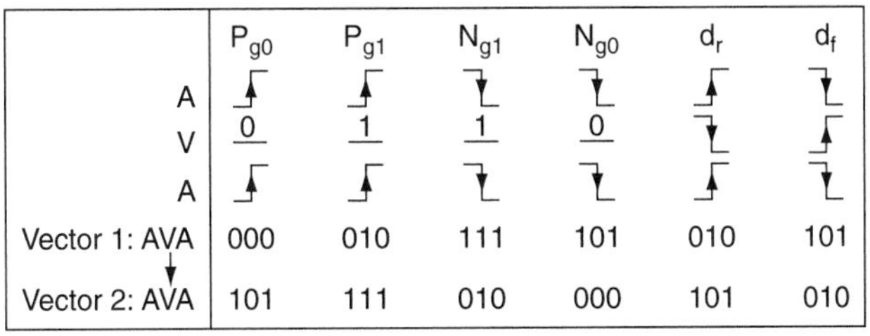

	P_{g0}	P_{g1}	N_{g1}	N_{g0}	d_r	d_f
A						
V	0	1	1	0		
A						
Vector 1: AVA	000	010	111	101	010	101
Vector 2: AVA	101	111	010	000	101	010

■ **FIGURE 8.3**

The MA fault model and test patterns.

Figure 8.3 pictures six cases (patterns) for three line interconnects where the middle one is the victim based on the MA fault model. Test patterns for signal integrity are vector pairs. For example, when the victim line is kept quiescent at 0 or 1 (see columns 1 through 4), four possible transitions on the aggressors are examined.

When mutual inductance comes into play, some researchers have shown that the MA model may not reflect the worst case and have presented other test generation approaches (using pseudo-random, weighted pseudo-random, or deterministic patterns) to stimulate the maximal integrity loss [Chen 1998, 1999] [Attarha 2002].

As reported in [Naffziger 1999], a device failed when the nearest aggressor lines change in one direction and the other aggressors change in the opposite direction. The MA fault model does not cover these and many similar scenarios. Exhaustive testing covers all situations, but it is time consuming because of the huge number of test patterns. In [Tehranipour 2004], the authors proposed a *multiple transition* (MT) fault model assuming a single victim, a limited number of aggressors, a *full* transition on victim, and *multiple* transitions on aggressors. In the MT model, all possible transitions on the victim and aggressors are applied, whereas in the MA model only a subset of these transitions is generated. Briefly, the MA-pattern set is a subset of the MT-pattern set.

8.2.2.2 Power Supply Noise

Power supply is distributed in a VLSI chip through wires that contain parasitic R/L/C elements. Hence, drawing current from a power source produces voltage fluctuations. Noise is mainly created in the power supply lines because of the resistive and inductive parasitic elements. Specifically, the inductive noise, also known as *di/dt noise*, is caused by the instantaneous change in current drawn from the power supply. Inductive noise becomes significant in high-frequency designs because of the sharp rise and fall times. High-frequency switching causes the current to be drawn (from the power supply) for very short duration (usually in hundreds of picoseconds) causing very high *di/dt*. Resistive noise (IR drop) is dependent on the current drawn from the power supply. Hence, the ***power supply noise***, shown as *PSN(t)* or simply *PSN*, is collectively given by:

$$PSN(t) = L \cdot \frac{di(t)}{dt} + R \cdot i(t)$$

For an n-input circuit, there are $2^n(2^n - 1) \approx 2^{2n}$ possible pattern pairs that may cause internal transitions. Because simulating all possible pairs is unrealistic, it is essential to be able to select a small set of pattern pairs without exhaustive simulation. One approach is to use random-pattern-based simulation. Unfortunately, because of the random nature of these patterns, such approaches cannot guarantee to create maximum PSN in a reasonable amount of simulation time. Researchers have, therefore, applied deterministic or probabilistic heuristics to find such pattern sets with no guarantee of optimality.

Authors in [Lee 2001] have presented the generation and characterization of three types of noise induced by electrostatic discharge in power supply systems. These three types are I/O protection induced signal loss, latent damage after electrostatic discharge stress, and power/ground coupling noise. To speed up the power supply noise analysis and test generation process, some works exploited the concept of random search. For example, the authors in [Jiang 1997] and [Bai 2001] used a genetic algorithm (with random basis) to stimulate the worst-case PSN. Some researchers, such as [Zhao 2000b], precharacterize cells using transistor-level simulators and annotate the information into the PSN analysis phase. A technique for vector generation for power supply noise estimation and verification is described in [Jiang 2001]. The authors used a genetic algorithm to derive a set of patterns producing high power supply noise. A pattern generation method to minimize the PSN effects during testing is presented in [Kristic 2001]. In [Nourani 2005], the authors identified three design metrics (level, fanin, and fanout) that have the maximum effects on PSN. A greedy algorithm and a conventional fault simulator are then used to quickly construct pattern pairs that simulate the worst-case PSN based on circuit topology, regardless of whether the circuit is in functional mode or in test mode.

8.2.2.3 Process Variations

As device technology progresses toward 45 nm and below, the fidelity of the process parameter modeling becomes questionable. For every process, there is some level of uncertainty in its device parameters because of limitations imposed by the laws of physics, imperfect tools, and properties of materials not fully comprehended [Visweswariah 2003]. The deviation of parameters from desired values because of the limited controllability of a process is called ***process variation*** (PV). Sources of process variation include random dopant fluctuation, annealing effects, and lithographic limitations. Typical variations are 10% to 30% across wafers and 5% to 20% across dies, and they may change the behavior of devices and interconnects [Borkar 2004a].

Researchers have explored various ways of analyzing and dealing with process variation. The solutions are broadly labeled ***design for manufacturability*** (DFM) techniques. DFM struggles to quantify the impact of PV on circuits and systems. This role has made DFM techniques of high interest to semiconductor and manufacturing companies [Nassif 2004]. There are tens of factors that affect or contribute to process variation, including interconnects, thermal effects, and gate capacitance [Dryden 2005]. Process variation has been shown to potentially cause a 40% to 60% variation for *effective gate length* L_{eff}, a 10% to 20% fluctuation in V_t and T_{ox} that may lead to malfunction and ultimately failure [Borkar 2004a].

The PV monitoring/analysis approaches can be classified using different criteria. From a source point of view, variation can be intradie (within-die) or interdie. The latter can be classified as a die-to-die, center-to-edge (in a wafer), wafer-to-wafer, lot-to-lot, or fab-to-fab variation. From a methodology point of view, the solutions are classified into two broad classes: statistical and systematic. Examples of statistical approaches are PV modeling [Sato 1998], analyzing the impact of parameter fluctuations on critical-path delay [Bowman 2000] [Agarwal 2003], mapping statistical variations into an analytical model [Cao 2005], and addressing the effect of PV on crosstalk delay and noise [Narasimha 2006].

In systematic approaches, because of the complexity of parameters, almost all researchers traced a limited number of PV metrics and their effects on design characteristics. [Orshansky 2000] explored the impact of the gate length on performance. The authors in [Chen 2005] proposed a current monitor component to design PV-tolerant circuits, and [Azizi 2005] analyzed the effect of voltage scaling on making designs more resistant to process variations. Other works that focused on the effect of PV on key design characteristics include [Mehrotra 1998], which studied the effect of manufacturing variation on microprocessor interconnects; [Ghanta 2005], which showed the effect of PV on the power grid; [Agarwal 2005], which presented the failure analysis of memories by considering the effect of process variation; and [Ding 2005], in which the authors investigated the effect of PV on soft error vulnerability.

Because of the nature of parameters (*e.g.*, V_t, T_{ox}, and L_{eff}) affected by process variation, tracing and pinpointing each variation for any realistic circuit is not a viable option. Hence, it becomes necessary to abstract and limit the problem to a specific domain to look at it collectively. Such abstraction can be clearly seen in prior works that consider process variation for clock distribution [Bowman 2001], delay test [Lu 2004], defect detection [Acharyya 2005], PV monitoring techniques [Kim 2005], reliability analysis [Borkar 2004b], and yield prediction [Jess 2003].

The authors in [Mohanty 2007] considered simultaneous variation of V_t, T_{ox}, and V_{dd} for the transistor's current characterization and optimization.

Tracing process variations for individual parameters is not possible because of the size, complexity, and unpredictability of playing factors. Like in conventional testing approaches (*e.g.*, stuck-at fault, path-delay fault), we need a simplified model for PV-faults to be able to devise and apply a PV test methodology. In spite of its simplicity, the model should be generic in concept, straightforward in measurement, and practical in application. A single **PV-fault model** is defined in [Nourani 2006]. This model assumes that there exists only one grid (unit area) in the layout of the circuit under test, where a sensor planted in that region can generate a faulty metric z_f instead of a fault-free metric z such that $\Delta z = |z_f - z|$ is measurable (in terms of delay, frequency, etc.).

8.2.3 Integrity Loss Sensors/Monitors

The integrity loss metric depends on the signal's waveform (see Figure 8.1), which is subject to change as it travels through a circuit. For an accurate and uniform measurement, the integrity loss needs to be captured or sampled right after creation. This will be practical only by limiting the observation sites and designing cost-effective sensors and readout circuitry. Various types of on-chip sensors, potentially useful for integrity loss detection, have been reported in the literature. This section explores a few of the on-die environmental sensors that can be used in BIST or scan-based architectures for targeting integrity loss and process variations.

8.2.3.1 Current Sensor

Current sensors are often used to detect the completion of asynchronous circuits [Lampinen 2002] [Chen 2005]. Figure 8.4 shows a conventional current sensor. The supply current of a logic circuit block is mirrored through a current mirror transistor pair (*M*0 and *M*1) to a bias-generation circuit. The bias-generation circuit

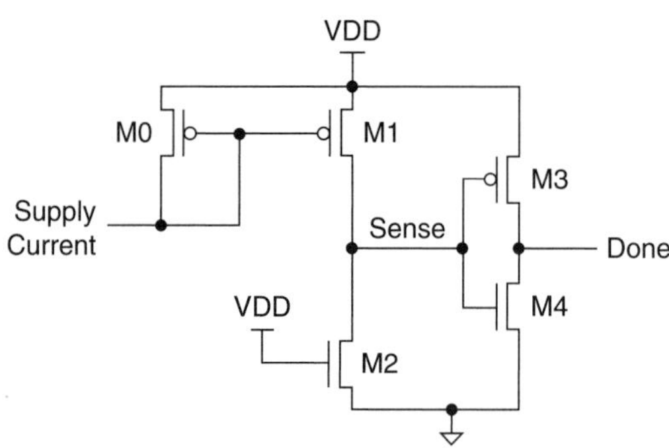

FIGURE 8.4

Conventional current sensor.

contains an *N-channel metal oxide semiconductor* (NMOS) biased as a resistor (*M2*). If the supply current is high, the voltage drop across *M2* is high, which generates *Done* = 0 indicating the job is not completed. When the circuit operation is completed, only the leakage current flows through the circuit. This makes the voltage drop across *M2* quite small, thus producing *Done* = 1.

In general, designing and fine-tuning the current sensors are challenging tasks. Yet, different versions of current sensors have been used for various monitoring and testing applications. For example, the authors in [Yang 2001] proposed boundary scan combined with ***transition power supply current*** (I_{DDT}) for testing interconnect buses; the authors in [Chen 2005] proposed a ***leakage canceling current sensor*** (LCCS). The sensor was then recommended for self-timed logic to design process-variation tolerant systems. The self-timed systems can accept input data synchronously and supply their outputs asynchronously (*i.e.*, by a "Done" signal generated by the current sensor).

8.2.3.2 Power Supply Noise Monitor

A PSN monitor was presented in [Vazquez 2004] by which the authors claimed to detect high-resolution (100 ps) PSN at the power/ground lines. The schematic of this circuit is shown in Figure 8.5. Briefly, the three inverters work as a delay line, whose delay depends on its effective supply voltage. The charge supplied to C_x (voltage V_x) is proportional to the propagation delay of the inverter block. The V_x voltage at the end of sampling period (controlled by the NOR gate) depends on the supply voltage. Thus, the voltage V_x depends on the power/ground bounce: the higher the PSN is, the longer the propagation delay and the higher the voltage V_x will be.

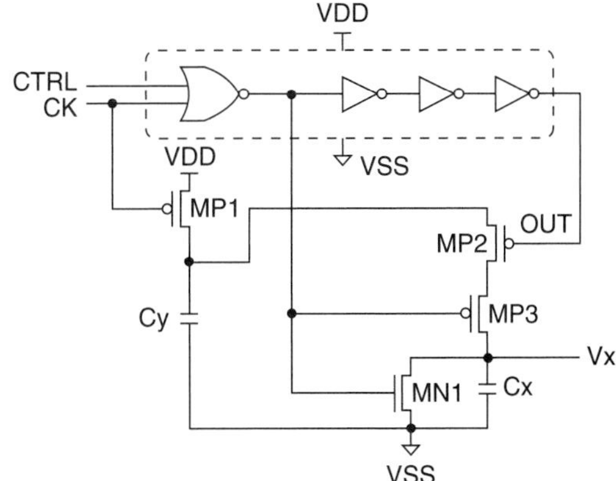

FIGURE 8.5

Power supply noise monitor.

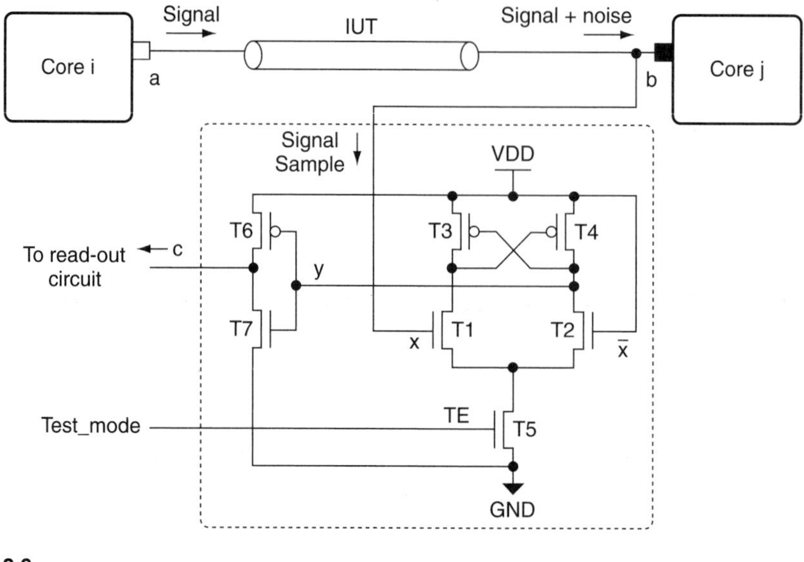

■ **FIGURE 8.6**

Noise detector (ND) sensor using a cross-coupled PMOS amplifier.

8.2.3.3 Noise Detector (ND) Sensor

A modified cross-coupled *P-channel metal oxide semiconductor* (PMOS) differential sense amplifier is designed to detect integrity loss (noise) relative to voltage violations [Nourani 2002]. Figure 8.6 shows a *noise detector* (ND) sensor, which sits physically near the receiving *Core j* for sampling the actual signal plus noise transmitted through *Core i*. *TE* is connected to *test mode* to create a permanent current source in the test mode, and input $\bar{x}$ is connected to V_{DD} to define the threshold level for sensing V_b, (*i.e.*, the voltage received in x). By adjusting the size of the PMOS transistors (*i.e.*, W and L), the current through transistors T_1 and T_2 and the threshold voltages to turn the transistors on or off can be tuned. A designer uses this tuning technique to set the high and low noise threshold levels (V_{Hthr} and V_{Hmin}) in the ND sensor. Each time when noise occurs (*i.e.*, $V_b > V_{Hthr}$), the ND sensor generates a 0 signal that remains unchanged until V_b drops below V_{Hmin}. The ND sensor shows a *hysteresis* (Schmitt-trigger) property, which implies a (temporary) storage behavior. This property helps to detect the violation of two threshold voltages (V_{Hthr} and V_{Hmin}) with the same ND sensor.

8.2.3.4 Integrity Loss Sensor (ILS)

The ***integrity loss sensor*** (ILS) is a **delay violation sensor** shown in Figure 8.7 [Tehranipour 2004]. The ILS sensor consists of two parts: the sensor and the detector (XNOR gate). An ***acceptable delay region*** (*ADR*) is defined as the time interval from the triggering clock edge during which all output transitions must occur. The test clock *TCK* is used to create a delayed signal b, and together they determine the

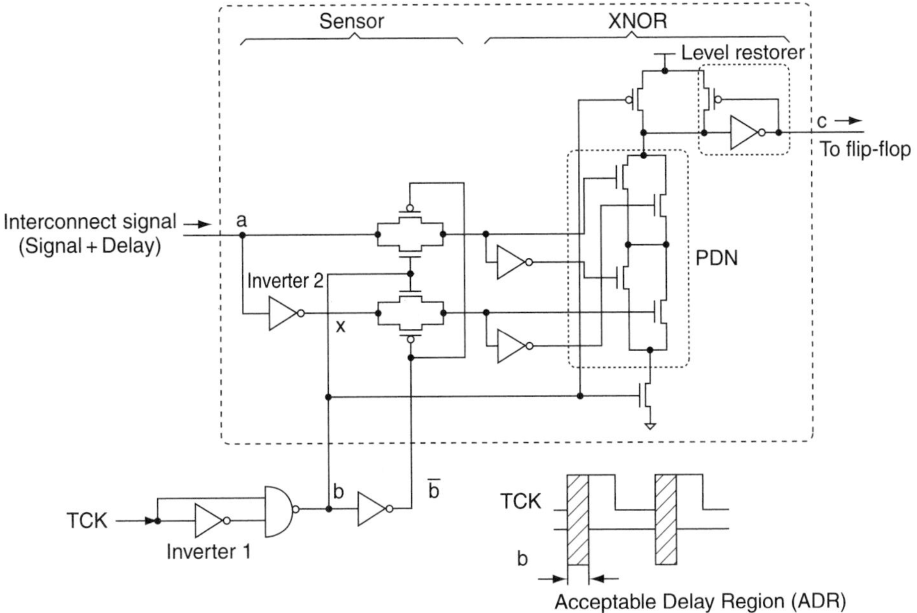

■ **FIGURE 8.7**

Integrity loss sensor (ILS).

ADR window. The input interconnect signal a is in the acceptable delay period if its transition occurs during the period when b is at logic 0. Any transition that occurs during the period when b is at logic 1 is passed through the transmission gates to the detector composed of an XNOR gate. The XNOR gate is implemented using dynamic precharged logic. Output c becomes 1 when a signal transition occurs during $b = 1$ and remains unchanged till $b = 0$, the next precharge cycle. Output c is used to trigger a flip-flop. The minimum detectable delay can be decreased by increasing the delay of inverter 2 (T_{inv2}) or decreasing the delay of inverter 1 (T_{inv1}). T_{inv2} can be decreased until the duration is enough to precharge the dynamic logic.

8.2.3.5 Jitter Monitor

Jitter is often defined as the time deviation of a signal from its ideal location in time. Jitter characterization is important for ***phase-locked loops*** (PLLs) and other circuits with time-sensitive outputs. Jitter measurement of a data signal with sub-gate resolution can be done using two delay lines feeding a series of D latches as shown in Figure 8.8. Such a structure is known as a ***Vernier delay line*** (VDL) [Gorbics 1997]. Assuming the clock signal is jitter-free, the propagation delay of the clock and data paths differ by $\Delta T = T_d - T_c$ (*e.g.*, the time difference between rising edges). The time difference decreases by ΔT after each stage, and the phase relationship between these two rising edges is recorded by a D latch in each stage. A counter reads the output of the D latch and counts the number of times the data

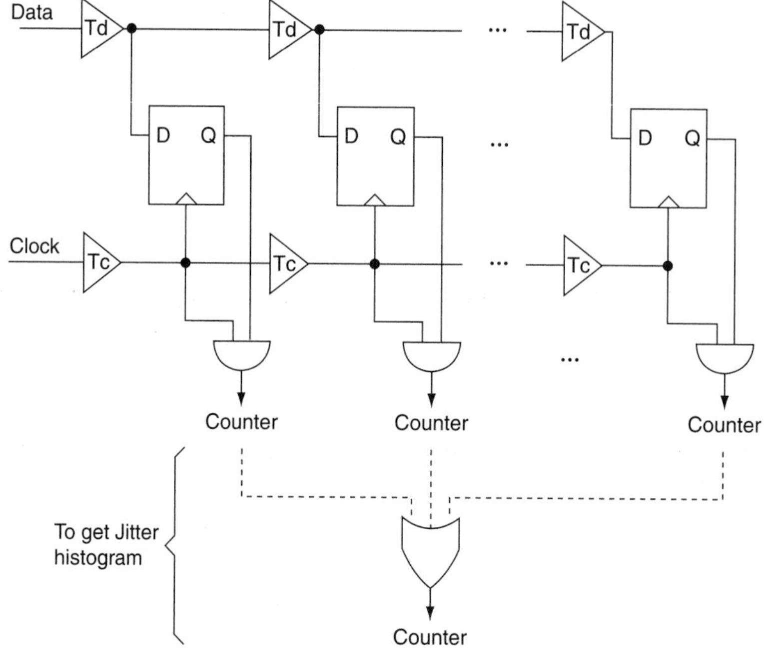

■ FIGURE 8.8

Jitter monitor using VDL.

signal leads the clock signal with a delay difference that depends on the position of D latch in the chain. Alternatively, the histogram of the jitter (*i.e.*, the jitter's probability density function) can be directly derived by ORing the outputs of all D latches and counting the number of 1's over the time period of the clock. The accuracy of jitter measurement using VDL depends on the matching of delay elements between stages [Chan 2001].

8.2.3.6 Process Variation Sensor

Using ***ring oscillators*** (ROs) to probe on-die process variation is a long-standing practice. The oscillators inserted into the system are affected by *process variation* (PV) along with the rest of the system. The variation of delay caused by PV-faults in any of the inverters in the loop results in deviation in the frequency of the oscillator, which can be detected. Several foundries have already used ring oscillators on wafers to monitor process variations. This often serves as a benchmark performance measure (sanity check) [MOSIS 2007]. Conventionally, several in-wafer ROs are placed on dicing lines (scribe lines) for process parameter monitoring. Unfortunately, this is insufficient for evaluating variations on each die, as there is a growing demand for sensors and methodologies that allow process variations to be evaluated with precision.

Ring oscillators are implemented by cascading an odd number of inverters to form a loop. By using an odd-numbered loop, we can assure that the output of the last inverter is the opposite (inverse) of the previous input to the first inverter, thus preventing it from stabilizing to a steady state. The oscillation frequency of a ring oscillator is given as the reciprocal of the total delay of the inverters. That is $f_{RO} = 1/(N_{inv}T_{inv})$, where N_{inv} is an odd number of inverters and T_{inv} is the delay of one inverter. Using standard CMOS inverters, the following equation can be written [Rabaey 1996]:

$$f_{RO} \approx \frac{1}{N_{inv} \cdot V_{dd} \cdot C_{Load}} \left(\frac{\mu \varepsilon W^2}{2T_{ox}} \right) (V_{GS} - V_t)^2 \left(1 + \frac{K}{L_{eff}} V_{DS} \right)$$

This equation is a first-order approximation of the relationship among the current, load capacitance, and frequency of an oscillator. Being an approximation, the formula cannot accurately capture the complex relationships among the factors. Yet, in general, process variation collectively causes a *measurable frequency shift* (Δf) in the output of a ring oscillator. For example, simulation results using a commercially available SPICE circuit simulator at a TSMC 180-nm process node for a 41-stage ring oscillator shows a shift in frequency of 16 MHz and 6 MHz for 10% variation of *threshold voltage* (V_t) and *transistor oxide thickness* (T_{ox}), respectively [Nourani 2006].

8.2.4 Readout Architectures

Both popular test methodologies (*i.e.*, BIST and scan) can be used to coordinate the activities among IL sensors and the readout mechanism in a signal integrity test session. We now briefly address these two basic architectures.

8.2.4.1 BIST-Based Architecture

In a logic BIST environment, a *test pattern generator* (TPG) is used to generate pseudo-random patterns for detecting manufacturing faults in a *circuit under test* (CUT). An *output response analyzer* (ORA) is used to compact the test responses of the CUT and form a signature. Under the control and coordination of the BIST controller, the final signature is then compared against an embedded *golden signature* to determine pass/fail of the CUT. Figure 8.9 shows a typical logic BIST architecture to test SOC interconnects for integrity. The integrity loss monitoring cell (IL sensor) can be any of those sensors discussed earlier such as ND or ILS. The TPG and ORA are located on the two sides of the *interconnect under test* (IUT). The IUTs can be long interconnects or those suspicious of having noise/delay violations as a result of environmental factors, (crosstalk, electromagnetic effects, environmental noise, etc.).

The rationale in using pseudo-random patterns for integrity testing is the fact that finding patterns that are guaranteed to create the worst-case scenarios for integrity loss (*e.g.*, noise and delay) is prohibitively expensive with the current state

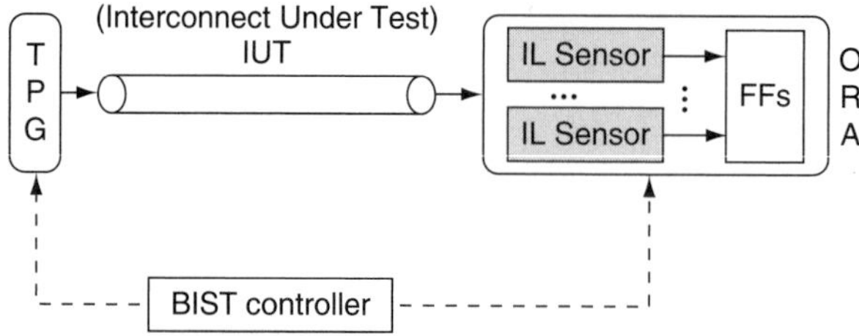

■ **FIGURE 8.9**

Typical logic BIST architecture for integrity testing.

of knowledge. This is mainly because of the complexity of the distributed *resistance-inductance-capacitance* (RLC) interconnect model, parasitic values, and too many influential factors.

Detecting signals that leave the noise-safe and time-safe regions is a crucial step in IL monitoring and testing. Various IL sensors may be needed per interconnect to detect noise (crossing the threshold supply voltage V_{Hthr} and the minimum supply voltage V_{Hmin}) and delay violations. The test architecture used to read out the information stored in these cells is based on a DFT decision, which depends on the overall SOC test methodology, testing objective, and cost consideration. Figure 8.10 shows one such test architecture given in [Nourani 2002]. IL sensors are pairs of *ND* and *ILS* cells, which, in coordination with scan cells, record the occurrence of any noise or delay violation. The results are scanned out via *Sout* to the *scan-out chain* for analysis. In test mode, the *flag* signal is first transmitted through the multiplexer to the test controller. When a noise or delay violation (low integrity signal) occurs (*flag* = 1), the contents of all scan cells are then scanned out through *Sout* for further reliability and diagnosis analysis. Suppose an n-bit interconnect is under test for m cycles (*i.e.*, with m pseudo-random test patterns). The pessimistic worst-case

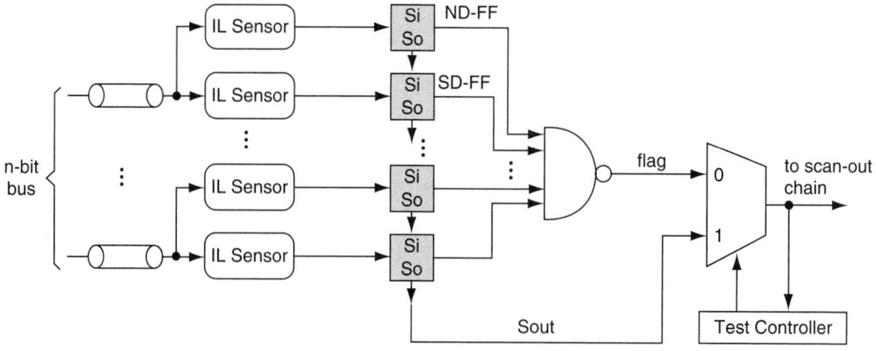

■ **FIGURE 8.10**

The readout circuitry.

scenario in terms of test time is a case in which all lines are subject to noise in all m test cycles. This situation requires overall m and mn cycles for response capture and readout, respectively. In practice, a much shorter time (*e.g.*, kn, where $k << m$) is sufficient, as the presence of defects or environmental factors causing an unacceptable level of noise/delay (integrity loss) is limited.

8.2.4.2 Scan-Based Architecture

The IEEE 1149.1 boundary-scan test standard [IEEE 1149.1-2001], also known as ***Joint Test Action Group*** (JTAG) standard, has been widely accepted and practiced in the electronics industry for testing interconnects between devices and providing external access to a device under test or diagnosis. The standard provides excellent test and diagnosis capabilities for devices mounted on a *printed-circuit board* (PCB) or embedded in a system with low complexity, but it was not intended to address high-speed testing and signal integrity loss.

To address signal integrity loss, in [Whetsel 1996], the author proposed a method to simplify the development of a mixed-signal test standard by adding the analog interconnect test to 1149.1. The IEEE 1149.4 mixed-signal test bus standard [IEEE 1149.4-1999] was then developed to allow access to the analog pins of a mixed-signal device. In addition to the ability to test interconnects using digital patterns, the 1149.4 standard includes the ability to measure actual passive components, such as resistors and capacitors; however, it cannot support high-frequency phenomena, such as crosstalk on interconnects. To deal with high-speed testing, the IEEE 1149.6 standard provides a solution for testing AC-coupled interconnects between integrated circuits on PCBs or systems [IEEE 1149.6-2003]. Various issues on the extended JTAG architecture to test SOC interconnects for signal integrity are reported in [Ahmed 2003], [Tehranipour 2003a], and [Tehranipour 2003b] where ***maximum aggressor*** (MA) and ***multiple transition*** (MT) fault models are employed.

Integrating pseudo-random pattern generators and IL sensors within scan test architecture is a relatively straightforward task. To activate IL sensors, a separate test mode is needed. For example, the authors in [Tehranipour 2004] proposed to modify the *boundary-scan cells* (BSCs) for testing integrity loss on interconnects. At the driving side of an interconnect, a modified BSC that generates test patterns, called *pattern generation BSC* (PGBSC), is used. At the receiving side of the interconnect, the authors proposed to use an *observation BSC* (OBSC) that includes an *integrity loss sensor*.

Figure 8.11 shows the overall test architecture with n interconnects between core i and core j in a two-core SOC. The five standard JTAG interfaces (*TDI, TCK, TMS, TRST,* and *TDO*) are still used without any modification. Two new instructions (called G-SITEST and O-SITEST) have been defined for signal integrity test, one to activate PGBSCs to generate test patterns and the other to read out the test results. The cells at the output pins of core i are changed to PGBSCs, and the cells at the input pins of core j are changed to the OBSCs. The remaining cells are standard BSCs, which are present in the scan chain during signal integrity test mode. In the case of bidirectional interconnects, boundary-scan cells used at

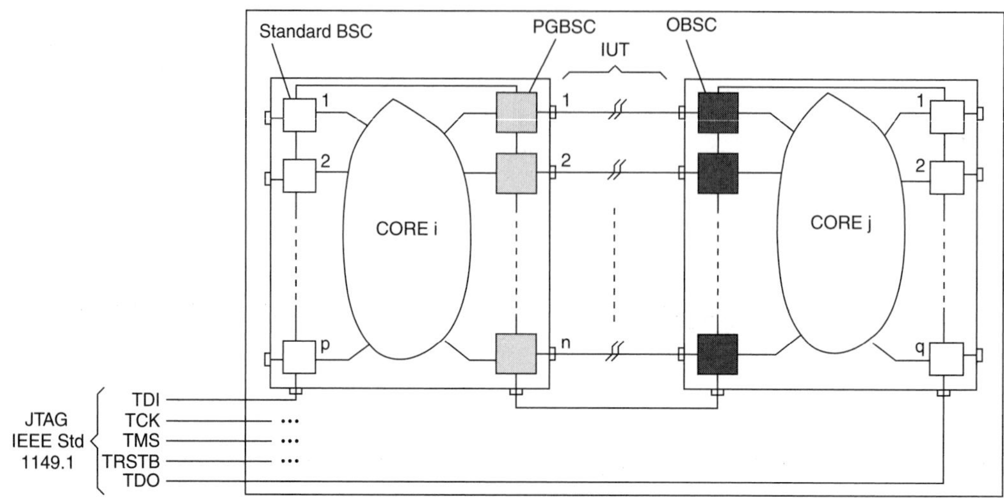

■ **FIGURE 8.11**

Test architecture.

both ends are a combination of PGBSC and OBSC to test the interconnects in both directions. The IL sensing part of OBSCs does not need any special control and automatically captures the occurrence of integrity loss. After all patterns are generated and applied, signal integrity information stored in the IL sensing scan cell is scanned out to determine which interconnect has a problem.

8.2.4.3 PV-Test Architecture

In PV testing, because of the self-generating and on-spot nature of process variation, no fault stimulation is needed. However, the output of a sensor needs to be carried out to an observation point for analysis. Researchers have observed that the intradie variations are significantly more difficult to predict and deal with than die-to-die variations on a wafer [Bowman 2000] [Nassif 2004]. On-chip ROs with counters, embedded in a test chip, were presented in [Hatzilambrou 1996] to detect process variation by measuring the RO's frequency shifts. This frequency variation was then used to *grade* the die performance or the performance of individual cores. This approach is not intended as a pass/fail test but instead as a **grading test**.

There are a number of techniques on PV probing and monitoring. Two examples were described in [Ukei 2001] and [Samaan 2003]. The *monitor test element group* (TEG) proposed in [Ukei 2001] consists of a ring oscillator and a control circuitry. Five TEGs are arranged in the four corners and the middle of a die, and their signals are reported one by one for process variation and manufacturing yield analysis. In [Samaan 2003], ROs are disposed over an integrated circuit chip depending on available layout space. To record its oscillation frequency, only one RO is allowed to operate at any one time. An analog-frequency wire is used to deliver test data to the counting and monitoring units for analysis. In [Bhushan 2006], the authors employed ring oscillators to evaluate the effect of process variations on key transistor parameters like switching delay and active/leakage power. These metrics reflect the average behavior of a few hundred MOSFETs embedded within each RO test structure. The authors experimented with IBM's 90-nm and 65-nm circuits and showed that their test mechanism can be useful for manufacturing test.

In [Nourani 2006], a distributed network of several (extendable to a large number of) ring oscillators per die was presented. The methodology targets detecting those process variation changes that collectively cause a *measurable frequency shift* (Δf) in the output of a ring oscillator planted in that region. These measurements can be further used to identify the problematic region(s) of the die and even grade the quality of the die or assembled chip. This approach carefully chooses three parameters of architecture, namely, types, numbers, and positions of ring oscillators. The basic architecture is shown in Figure 8.12.

The layout under PV test may be a full die or portion of die such as a sensitive core. In this figure, each RO symbolically occupies one or more regions out of $9 \times 9 = 81$ regions. Practically, the placement/layout generation tools position ROs automatically. The concept of *fault sampling* was used to stay practical while collecting a good estimate of coverage. The authors assumed that the area of each RO is a multiple of a grid area and the size of a die is equivalent to N_p grids. Applying the theory of sampling and the tradeoff between accuracy *versus* number of samples, the authors randomly chose N_s grids (where PV-faults occur) out of N_p and devised sensors to collect data on process variations.

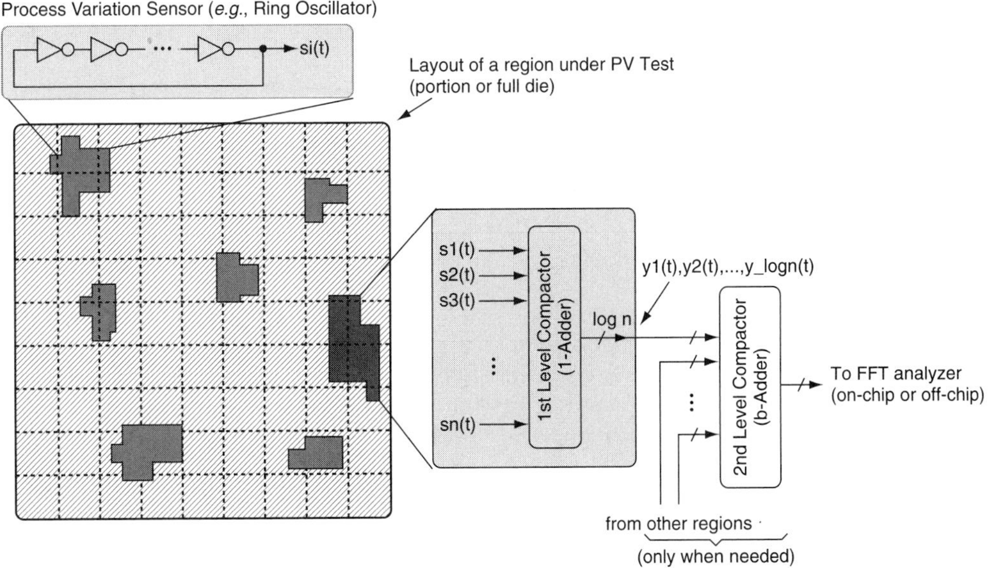

FIGURE 8.12

Basic concept of PV test architecture with ROs and compactor(s).

8.3 MANUFACTURING DEFECTS, PROCESS VARIATIONS, AND RELIABILITY

Aside from noise-induced errors such as power supply noise and signal integrity, manufacturing faults caused by manufacturing defects and process variations can severely impact device (process) yield and DPM levels. An internal document from a foundry reports that when **defect avoidance** and **defect tolerance** schemes were employed for one 50-mm^2 design at a 130-nm process node on an 8-inch fabrication line, the defect density was reduced from 0.2 to 0.1 defects per square inch, the device yield was increased from 85% to 92%, and there were 40 more good dies on one wafer (490 *versus* 450).

Fault-model-based (structural) tests, such as stuck-at tests and transition tests, have become the requirement for improving a device's fault coverage during manufacturing test. Studies have shown that stuck-at tests with 100% single stuck-at fault coverage could not guarantee perfect product quality (*i.e.*, no **test escape**) [McCluskey 2000] [Li 2002] [McCluskey 2004]. An investigation by [Ferhani 2006] further revealed that only 6% of the 483 defective ELF18 chips contained defects that acted as single-stuck-at faults, whereas 18% of the 205 defective ELF35 chips and 35% of the 116 defective Murphy chips acted as single-stuck-at faults. The remaining defects were (1) timing-dependent, (2) sequence-dependent, or (3) attributed to timing-independent, non-single-stuck-at faults, such as multiple stuck-at faults or nonfeedback bridging faults. A timing-dependent defect is sequence dependent because timing dependence implies that a transition arrives either earlier or later than expected; these transitions are created by a sequence of values applied to the circuit inputs that form a test for the defect.

Possible causes of timing-dependent defects are resistive opens, connections that have significantly higher resistance than intended, or transistors with lower drive than intended [McCluskey 2004]. Possible causes of sequence-dependent defects are (1) a defect that acts like a stuck-open fault [Li 2002] or (2) one that causes a feedback bridging fault [Franco 1995].

8.3.1 Fault Detection

To detect these manufacturing faults caused by manufacturing defects and process variations, one common approach is to generate multiple test sets each targeting a different fault model. These **fault-model-based tests** are commonly referred to as **structural tests**. Stuck-at tests and delay tests (including transition tests and path-delay tests) belong to this category. There are also structural tests, which are modeless [Boppana 1999]; they were first used for fault diagnosis but later for ATPG. Because these structural tests cannot provide sufficient defect coverage, a conventional approach has been to supplement structural tests with **functional tests** running at the circuit's rated speed. As we move toward the nanometer design era, meeting the stringent DPM goal is becoming a serious problem. This has prompted the need to generate **defect-based tests** by enumerating likely defect sites (failures) from the layout based on physical characteristics [Sengupta 1999]

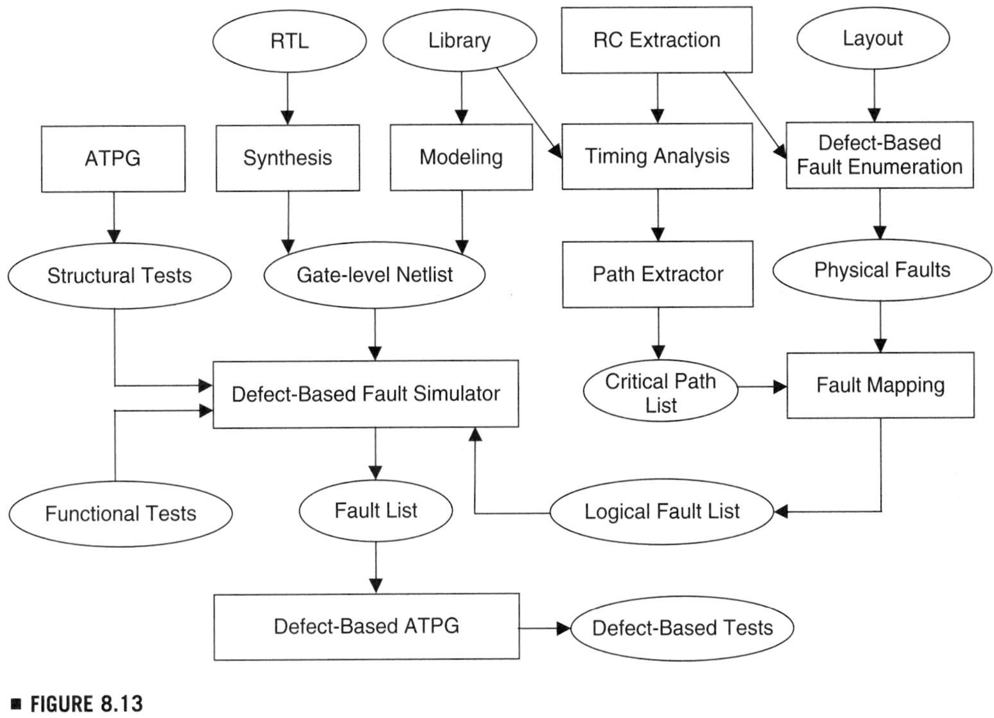

■ **FIGURE 8.13**

A defect-based test architecture.

[Segura 2002] [Gizopoulos 2006]. The physical characteristics of defects were studied to find better tests [Hawkins 1994] or understand yield learning [Maly 2003].

Figure 8.13 shows a defect-based test architecture [Sengupta 1999]. Structural tests are first generated from ATPG based on conventional fault models, such as stuck-at faults and transition faults. These structural tests are then combined with functional tests, and the resulting tests are fault-graded using a *defect-based fault simulator* for a given logical fault list, such as small delay defects and bridging faults, extracted from the physical layout. The undetected fault list is then sent to a *defect-based ATPG* for generating additional defect-based tests to meet the product's fault coverage and DPM goals.

8.3.1.1 Structural Tests

Structural tests are fault-model-based tests that usually include stuck-at tests for detecting stuck-at faults and delay tests for detecting transition faults and path-delay faults. In the 1980s and 1990s, the most commonly used structural tests are stuck-at tests. Because stuck-at tests have difficulty in meeting a product's DPM goal, functional tests are often used to supplement these structural tests during manufacturing test. An experiment conducted in [Maxwell 1991] has shown that a structural test with 92% single stuck-at fault coverage had lower overall defect coverage than a combined structural and functional test with only 82% single stuck-at fault coverage. More recent experiments further confirm that even a structural test with 100% stuck-at fault coverage is inadequate to screen most manufacturing faults [Ferhani 2006].

To further improve the circuit's defect coverage, at-speed delay tests have been used to supplement stuck-at tests since the 1990s when process node started to move to 180 nanometers and below [Foote 1997]. One study on a 733-MHz PowerPC microprocessor designed at IBM showed that if at-speed delay tests were removed from the test program, the *escape rate* would rise nearly 3% [Gatej 2002]. As a result, *at-speed delay testing* has become mandatory [Iyengar 2006] [Tendolkar 2006] [Vo 2006] for designs manufactured at 90 nanometers and below. These at-speed delay tests can come from scan or BIST [Wang 2005, 2006].

Modern ATPG and logic BIST programs can also take test power consumption into consideration when generating **power-aware structural tests**. These power-aware structural tests can avoid excessive heat during shift operation and IR-drop-induced yield loss during capture operation [Girard 2002] [Nicolici 2003] [Butler 2004] [Wen 2005] [Remersaro 2006] [Wen 2006]. These techniques have been extensively discussed in Chapter 7.

8.3.1.2 Defect-Based Tests

Defect-based tests include tests that are generated to target specific manufacturing faults arising from imprecise process technologies such as process variation and lithography. These defect-based test methods have been found crucial in screening additional physical failures during manufacturing test for designs at 130-nm process node or below. To increase manufacturing yield and meet a stringent DPM goal, these defect-based tests must supplement structural tests.

Small Delay Defect Tests

Small delay defect tests are delay tests that take timing delay associated with the fault sites and propagation paths from the layout into consideration. Although it is more accurate to compute path delay from the layout, because of process variation, the critical path in one chip may differ from another chip. So one can just approximate the longest paths without layout and target the set of longest paths rather than one critical path. These small delay defect tests by targeting the set of longest paths rather than one critical path for each delay defect are intended to catch small delay defects that escape traditional transition fault tests [Park 1988] [Williams 1991]. This is because shrinking feature size, growing circuit scale, increasing clock speed, and decreasing power supply voltage have made small delay defects the dominant failure mechanism in the nanometer design era [Mitra 2004].

One approach to test small delay defects is to group these structural transition tests into sets of almost equal-length paths [Kruseman 2004] and then test each group at faster than its rated speed. Those flip-flops fed by paths that exceed the cycle time or containing hazards must be masked off. The faults that are not detected because of masking are targeted by patterns run at the next lower clock speed [Barnhart 2004]. Applying transition tests at faster than the rated speed has been shown to catch small delay defects that escape traditional transition fault tests [Kruseman 2004] [Amodeo 2005]. The drawback of this approach is that circuit design may limit how much faster than the rated speed tests may be safely generated. In addition, both hazard [Kruseman 2004] and IR drop [Ahmed 2005] issues need to be considered when applying this approach.

By targeting or favoring the longest delay paths, the authors in [Sato 2005], [Hamada 2006], and [Qiu 2006] have reported that transition fault tests simply running at the circuit's rated speed can improve the quality of these transition tests and detect small delay defects. When the launch-on-capture clocking scheme is used, it is important to stretch the double-capture clock with an on-chip programmable clock controller that contains delay lines, in order to provide true at-speed delay tests [Rearick 2005].

The ***statistical delay quality model*** (SDQM) [Sato 2005] has been proposed for the evaluation of delay test quality. The SDQM is generated by first assuming a delay defect distribution that is based on the actual defect probability in a fabrication process and then investigating the sensitized transition paths and calculating their delay lengths. Detectable delay defect sizes are defined as the difference between the test timing and the path lengths. Finally, the probability of detecting small delay defects is calculated by multiplying the distribution probability for each defect. The calculated value is called the ***statistical delay quality level*** (SDQL).

Bridging Defect Tests

It is also important to further supplement the defect-based delay tests and the traditional structural tests with **bridging defect tests** (often called *bridging tests* only). One of the most common defect types in CMOS designs is the interconnect bridge. Bridges (shorts) involving many nodes are typically catastrophic, so although they are vital to yield calculation, they are less important for fault modeling. In [Aitken 2004], the author indicated that a **bridging fault** between two small NAND gates using a particular 130-nm process can result in (1) a **dominance fault** where strongest node always wins, (2) a **voting fault** where winner depends on relative drive strengths of opposite values, (3) an **analog fault** including the **Byzantine Generals** behavior (where downstream gates interpret an intermediate shorted voltage differently, some as 0, some as 1), and (4) a **delay** fault. The bridge behavior depends on the resistance of the defect, when running at different voltages and temperatures (see Tables 8.1 and 8.2). Thus, *multiple voltage bridge tests* are required to detect delay-independent bridging faults, and delay tests running at different speeds shall be used to detect short-induced delay faults [Aitken 2004]. As the number of potential bridges is astronomical, it is more realistic to enumerate likely bridging fault sites (*physical bridging faults*) from the layout [Stroud 2000] [Stanojevic 2001] and map them to *logical bridging faults (realistic bridging faults)* for fault simulation, scan ATPG, or fault diagnosis [Stanojevic 2000] [Zou 2005] [Ko 2006] [Wang 2006].

TABLE 8.1 ■ Example Bridge Behavior at 1.2 V, 25° C

Bridge Resistance	Behavior (1.2 V, 25° C)
<3000 ohms	Dominance or voting fault
3000–3700 ohms	Analog or Byzantine fault
>3700 ohms	Delay fault

TABLE 8.2 ■ Example Bridge Behavior at 1.32 V, 25° C

Bridge Resistance	Behavior (1.32 V, 25° C)
<2500 ohms	Dominance or voting fault
2500–3000 ohms	Analog or Byzantine fault
>3000 ohms	Delay fault

N-Detect Tests

It has been reported in [Ma 1995] that *N-detect stuck-at tests* that detect every stuck-at fault multiple (*N*) times are better at closing DPM holes than tests that detect each fault only once. This approach, called ***N*-detect**, works because each fault is generally targeted in several different ways, increasing the probability to activate a particular defect when the observation path to the fault site opens up. *N-detect at-speed tests* can also be used [Pomeranz 1999], but a promising study shows that by generating *transition tests* one for each reachable output for a given transition fault, ***transition fault propagation to all reachable outputs*** (TARO) was able to detect faults that other tests could not on a test chip [Tseng 2001] [McCluskey 2004] [Park 2005]. TARO can be a good candidate for tests that require extreme thoroughness, such as sample-based quality assurance tests, and for logic diagnosis when a much better resolution is required.

On the other hand, *gate exhaustive tests* that apply all possible input combinations to each gate and observe test responses of the gate at a scan cell or primary output have also shown effectiveness in detecting more defective chips than *single stuck-at tests* [McCluskey 1993] [Cho 2005]. The authors in [Ferhani 2006] further demonstrated that *N*-detect and gate exhaustive test sets have higher TARO coverage than the transition and single stuck-at test sets; however, they cannot have higher transition coverage than the *transition test set* because the transition test set has been generated to have the highest possible transition coverage. Therefore, TARO seems to be a better metric than the other three in detecting timing-dependent and sequence-dependent defective chips, though it may require more test patterns.

I_{DDQ} *Tests*

Normally, the leakage current of CMOS circuits under a quiescent state is small and negligible. When a fault, such as a transistor stuck short or a bridging fault, occurs and causes a conducting path from power to ground, it may draw an excessive supply current [Bushnell 2000] [Jha 2003]. Additionally, a weak CMOS chip can contain flaws or defects that do not cause functional failures in normal operation mode but degrade the chip's performance, reduce noise margins, or draw excessive supply current. Such flaws or defects must be also screened. Using I_{DDQ} tests by monitoring the **quiescent power supply current** (I_{DDQ}), it was shown to have detected many types of these defects, including some timing-dependent defects [Levi 1981] [Hawkins, 1986] [Nigh 1998, 2000]. Thus, in addition to detecting leakage current related failures [SIA 2005], I_{DDQ} tests can also be used for reliability screening. Effective methods to screen I_{DDQ} outliers at the wafer or lot level further include using **current ratios** [Maxwell 2000] and the **statistical outlier screening** method [Daasch 2001].

I_{DDQ} testing became an accepted test method for the IC industry in the 1980s. The small geometry sizes of today's devices, however, have made normal fault-free I_{DDQ} quite large because of the collective leakage currents of millions of transistors on a chip. This makes the detection of the additional I_{DDQ} current difficult; hence, I_{DDQ} testing is becoming ineffective and has caused many companies to abandon or rely less on I_{DDQ} tests [Williams 1996a,1996b].

A similar approach is **transient power supply current** (I_{DDT}) testing. When a CMOS circuit switches states, a momentary path is established between the supply lines V_{DD} and V_{SS} that results in a dynamic current I_{DDT}. The I_{DDT} waveform exhibits a spike every time the circuit switches with the magnitude and frequency components of the waveform dependent on the switching activity; therefore, it is possible to differentiate between fault-free and faulty circuits by observing either the magnitude or the frequency spectrum of I_{DDT} waveforms. Monitoring the I_{DDT} of a CMOS circuit may also provide additional diagnostic information about possible defects unmatched by I_{DDQ} and other test techniques [Min 1998]; however, I_{DDT} testing suffers many of the same problems as I_{DDQ} testing as the number of transistors in VLSI devices continues to grow.

MinV$_{DD}$ Tests

Similar to I_{DDQ} testing, which can be used to screen outliers for VLSI designs, a traditional **minimum supply voltage (MinV$_{DD}$) testing** can also be used to detect manufacturing faults. In general, a datasheet V_{DD} specification is in the range of $\pm 10\%$ of V_{DD} to allow a certain level of power supply tolerance. The device is supposed to still run at the rated speed with the raised/lowered V_{DD}. This minV$_{DD}$ test technique will try to find the lowest operating voltage when the device operates functionally, usually below the rated -10% V_{DD}. Similar to I_{DDQ} testing, this **MinV$_{DD}$ test** technique can also be used for reliability screening. The observation is that as we push the device to operate at the edge with barely enough voltage, marginal devices will show up as failures, not meeting the rated speed. In MinV$_{DD}$ testing, a minimum pass/fail V_{DD} level is measured, usually by a binary search method over a large sample of dies, and the resulting voltage level is used to test each die by applying a full vector set of *MinV$_{DD}$ tests*. This screening method has shown to be effective while it may miss some **MinV$_{DD}$ outliers**.

To further look for these MinV_{DD} outliers, it may be more practical to use the **statistical outlier screening** method [Madge 2002]. The statistical outlier screening method, referred to as **feed-forward voltage testing** in [Madge 2002], consists of three steps:

1. The first step is to use a *reduced vector set* (RVS), typically 3% to 15% of the *full vector set* (FVS), to search for a minimum passing voltage (the intrinsic defect-free MinV_{DD}) for each die. The reduced vector set can be composed of only stuck-at tests, delay tests, memory BIST tests, logic BIST tests, functional tests, or any combination of the above. A binary search algorithm for reducing test time during wafer sort is then employed to find and record the MinV_{DD} by which the die will function correctly at the nominal test frequency.
2. The second step is to use the *full vector set* (FVS) to test each die at this recorded minimum passing voltage (MinV_{DD}) plus a small voltage guard-band. This **guard-band** is user definable and depends on the level of screening required to meet the product's fault coverage and DPM goals. This calculated voltage is referred to as the *feed-forward* $MinV_{DD}$, and the bad dies screened are called **feed-forward outliers**.
3. The last step is to "post-process" the RVS binary search data by using complementary ***statistical post-processing*** (SPP) methods, namely **delta** $\mathbf{V_{DD}}$ and ***nearest neighbor residual*** (NNR), to screen **SPP outliers** that are not screened by feed-forward but are statistical Min_{VDD} outliers identified using the RVS binary search data. Delta V_{DD} uses the MinV_{DD} data collected from different intradie tests (in memory cores, scan cores, functional blocks, etc.) and calculates the delta between the values for each test. The differences between vector values on the same die are then compared against an expected difference to get an **intradie residual**, which is then measured against a *threshold* to determine a downgrade. The threshold value is adaptive (*i.e.*, data-based) and varies from die to die because of the gradual across-wafer variation of the intrinsic MinV_{DD} data. NNR computes an **interdie residual** based on the vector value averages of the surrounding sites. The interdie residual is then compared to an adaptive threshold value to flag outlier dies. The concept of SPP outlier screening was mainly borrowed from I_{DDQ} testing where **delta** $\mathbf{I_{DDQ}}$ and *NNR* have been shown to be effective in screening I_{DDQ} outliers [Gattiker 1996] [Powell 2000] [Daasch 2001].

The statistical outlier screening or **adaptive** $\mathbf{MinV_{DD}}$ method is most effective when there is a good correlation between RVS MinV_{DD} and FVS MinV_{DD}. The authors applied the method to four ASIC chips designed at LSI Logic using a guard-band of 50 mV at a 180-nm process node [Madge 2002]. Experimental results indicated that MinV_{DD} yield fallout was 0.2%–0.8% depending on product complexity, and around 20% of these outlier dies showed positive or negative V_{DD} shifting during burn-in. Resistive vias and tungsten stringers were identified as the source of defects during failure analysis.

VLV Tests

As opposed to MinV$_{DD}$ testing, which performs testing at a datasheet rated speed and minimum voltage, **very-low-voltage (VLV) testing** is conducted at a test speed that is below anything guaranteed in the device datasheet (*e.g.*, 600 MHz for a 2-GHz device). The test voltage can be at or below the MinV$_{DD}$ level. VLV testing was first proposed in [Hao 1993] as an alternative to burn-in to detect **delay flaws**. A delay flaw (*nonoperational delay fault*) is a defect that causes a local timing failure but the failure is not severe enough to cause circuit malfunction. An example failure is NMOS gate oxide shorts, which can be modeled as resistive shorts [Hao 1993]. A **delay fault** (*operational delay fault*) is a timing failure that causes the circuit to malfunction at its rated speed, although it is functional when operated at a lower speed. Example timing failures include high resistance interconnects, via defects, and tunneling opens that can only be detected by at-speed tests [Chang 1996b].

The VLV test technique is based on the observation that delay can increase substantially as VDD lowers or the driving strength of a gate (transistor) weakens to a certain level. The authors in [Chang 1996a] and [Chang 1996b] indicated that by setting the supply voltage at a low supply voltage between 2 and 2.5 times the threshold voltage (V_t) of the CMOS transistors, VLV testing can detect many types of delay flaws at the transistor level, including (1) transmission gate opens, (2) threshold voltage shifts, (3) diminished-drive gates, (4) gate oxide shorts, (5) metal shorts, and (6) defective interconnect buffers. The authors in [Ali 2006] further claimed that testing at the lowest operating voltages is only required for certain types of design flaws such as transmission gate opens and bridging faults; weak resistive opens that cause delay faults are best tested at higher operating voltages. To guarantee the quality of ***dynamic voltage scaling*** (DVS) systems (see Section 8.3.4), it is necessary to select a number of voltage-specific delay tests.

Although these design flaws do not do any harm to the normal circuit operation, they may cause circuit malfunction (intermittent or early-life failures) if the supply voltage changes during operations because of IR drop or simultaneous switching noise. The test speed for VLV testing can be determined using the methods presented in [Chang 1996a] to achieve high design flaw coverage. One potential problem with this setting is that the VLV limit may have hit the operating voltage in 90-nm designs [Gizopoulos 2006].

Though improved methods could be further explored to determine the VLV setting and test speed, the author in [Roehr 2006] showed that by (1) empirically setting the test speed to 12.5 MHz, (2) using MinV$_{DD}$ testing to search for a VLV setting at 0.9 V, and (3) applying a new **VLV ratio** (VLVR) at 10% on each die, during burn-in for initial product qualification of the first two 130-nm *low power* (LP) CMOS products, the intermittent failures caused by an interplay of several factors in different wafer fabrication lots were screened. The VLV test at 0.9 V running at 12.5 MHz was able to first eliminate the "*tail*" dies (the MinV$_{DD}$ outliers beyond 0.9 V). The VLV ratio that is defined as (Max-Min)/Min of the two MinV$_{DD}$ values for two VLV tests (in two logic blocks: Registers and SysMem) on the same die and set to 10% was then able to eliminate the "*flip*" dies (the intermittent failures).

The differences between MinV$_{DD}$ testing and VLV testing will slowly disappear as CMOS scaling continues. The typical operating V$_{DD}$ will continue to trend down as technology scales (to ensure device reliability). With V_{DD} well below 1 V for some of the mobile products at 65 nm, there is simply not much room to lower V_{DD}, when instrumentation errors and test interface limitations come into play. Moreover, as vendors seek to offer ever lower power products, a reduced-frequency part with lower voltage is already part of the product portfolio. Thus, we have seen vendors start to screen their low-power products with lower voltage at reduced frequencies. VLV testing becomes mainstream product testing but it does not leave much for reliability screen.

8.3.1.3 Functional Tests

Functional testing, once the sole test method that allows for testing actual functional paths at-speed, has begun to regain its acceptance in the industry. Microprocessor testing is one particular example [Sengupta 1999]. To meet the aggressive DPM goal, *functional tests* must be added to supplement *structural tests* (*at-speed tests*, *stuck-at tests*, and *bridging tests*).

Traditionally, functional tests are manually generated. It is a resource-intensive process, and fault coverage can only be assessed through fault simulation (again, a tedious process). Usually, fault coverage with functional tests cannot be improved upon easily, because the circuit could be complex and few commercial tools are available to help with this manual test generation process.

Efforts have been made to utilize validation techniques to tackle this problem. One of the common postsilicon validation techniques is that of random test generation with random instruction and data [Shen 1998]. With the proper test templates, these procedures can permute instruction sequences, addressing modes, data types, etc., to allow a more diverse set of tests to find potential design errors. These types of tests also help in detecting more manufacturing faults if deployed during manufacturing test. The limitation with doing this in manufacturing has been the immense amount of tester storage required to hold all of these randomly generated test vectors. The authors in [Parvathala 2002] came up with an idea to utilize the large cache that is a common feature for modern-day microprocessors. They shoehorned a version of this random test generator into the cache of a microprocessor. The area overhead required to load the test generator into the cache and run the code without invalidating the cache is considered very small as it piggybacks on other memory test (DFT) architectures. By running this test generator on-chip repeatedly, an infinite number of new test vectors can be generated with minimal additional test storage since we only need to load up the cache memory just once. Fault detection is through the periodic signature compression of the memory maps (the cached memory), further reducing bandwidth requirements. The additional benefit of this mode of testing is that the test can be run on the processor core at-speed, thus detecting delay faults as well as signal integrity faults. The authors in [Parvathala 2002] provided data showing that this method is effective in catching additional faulty chips and has improved DPM. The only limitation to this test method is that portions of the device dealing with external bus transactions and the cache miss related logic cannot be covered.

Another advancement came in the area of actually generating specific instruction sequences targeting manufacturing faults. [Tupuri 1997] and [Tupuri 2002] described such methods to reuse deterministic ATPG for targeted faults within an embedded module while extracting the input/output constraints for delivering those tests to the targeted module. The authors in [Chen 2003] used basic test templates and learned the characteristics (how they can reach control states and how they can propagate key signals, etc.) of these instruction sequences. Once a sufficient number of test templates are learned, this test generator is able to piece together these instruction sequences to activate the nodes (to be tested) and enable the propagation logic to allow them to be observable. Although these are promising early results, these techniques have to be proven on more complicated test cases. More functional test methods can also be found in Chapter 11.

An optimal test ordering of these tests plays an important role in minimizing the overall test time while maintaining the fault coverage. The study in [Butler 2000] revealed that functional tests should be placed early in the test flow before the use of transition tests, stuck-at tests, and I_{DDQ} tests. The authors in [Pouya 2000] and [Ferhani 2006] have also indicated that an optimal test ordering to have the shortest test time is to apply *path-delay tests*, *transition tests*, and *stuck-at tests*, in that order. However, finding an optimal test sequence is not a simple process, and the shortest test time is not always the best when tests are required for yield, defect, and performance learning. In that case, an ideal optimal test sequence should mean to trade off test time with test data collection and to pick the best point.

The key issue is thus how to generate these *manufacturing tests* (made up of *structural tests, defect-based tests,* and *functional tests*) in a timely manner to best meet time-to-market, DPM, and test budget goals all at the same time. In manufacturing processes at the 65-nm node and below, the test data volume can become so considerable that the cost of testing will soar. Therefore, we expect test compression to become indispensable. Similarly, functional BIST techniques like the cache-resident functional test generator described earlier may become more of a necessity. Active research will be more directed toward reducing scan ATPG time and test power, physical fault modeling, speedup of concurrent fault simulation, coverage enhancement of logic BIST, and logic ***built-in self-repair*** (BISR) [SIA 2005, 2006].

8.3.2 Reliability Stress

Some manufacturing defects (such as bridges and opens) are not necessarily screenable because of the defect mechanisms that make them too high or too low ohmic (depending on whether it is a bridge or open) to exhibit a distinguishable behavior from a fault-free circuit. However, they may degrade with normal life and cause the circuit to fail in time. These are categorized as *infant mortality* failures (see the bathtub curve in Figure 1.2 of Chapter 1). It is highly desirable that these failures be screened during manufacturing test before the devices are shipped out.

The reliability screen methods described previously (by either monitoring the supply current or lowering the supply voltage) are better at detecting infant mortality failures. Little evidence, however, shows that they can replace the costly reliability stress or burn-in that potentially reduce the normal life of the device.

The most common method to screen infant mortality is to **burn-in** the devices, followed by subsequent test screening. Burn-in essentially is a stress process by which devices are aged with proper excitation mechanisms, such as elevated voltage, temperature, and humidity, while the internal nodes are biased alternately (*e.g.*, node toggles with some kind of built-in tests or externally supplied tests). The basic acceleration mechanism is described with the following **Arrhenius equation**:

$$ttf = C \cdot e^{E_A/kT}$$

where *ttf* is time to failure (hours), C is a constant (hours), E_A is the activation energy (eV), k is the Boltzman's constant (8.616×10^{-5} eV/°K), and T is an absolute temperature (°K).

Essentially, we are using up some lifetime of the product with burn-in so when the product is eventually in the customer's hands, it has reached the normal life portion of its lifetime, with a low failure rate for the next 7 to 10 years. Some levels of defects may still exist, but they can be covered under the warranty (repair or replacement of products) offered by most manufacturers.

There are limits as to how much we can accelerate the defects. For example, if we elevate the voltage during burn-in, it also produces a higher than normal electric field across the drain and source of the MOS transistor. This higher than normal field increases the number of hot electrons that will then degrade the gate oxide and cause a more than normal shift of the threshold voltage (V_t). Stress-induced leakage currents and failures in CMOS devices have also been found in [Lee 2002] and [Pacha 2004]. This can essentially hasten the wearout process and reduce the normal life of the device. So it is a delicate balance between trading off infant mortality *versus* normal life. Increasing the temperature also has its limit, as a higher temperature (and voltage too) will increase static power consumption dramatically, requiring chips to be cooled during burn-in. This seems to be contradictory between refrigerated cooling and burn-in, but it is true for chips with hundreds of millions of transistors. In the extreme situation, if heat is not removed fast enough, this will cause a phenomenon called **thermal runaway**. Increased leakage causes more power consumption and more power consumption causes more thermal buildup, and the cycle continues until something is melted.

There are also alternate stress methods, such as **elevated voltage stress** or SHOVE as described in [Chang 1997]. This method is essentially a short-term burn-in, with stress time in fraction of a second to a few seconds. It has a certain degree of effectiveness for defects that will degrade quickly with elevated voltage, such as oxide thinning, which occurs when the oxide thickness of a transistor is less than expected, and via defects [Chang 1997]. In addition, it can avoid the wearout effect, as the time is so short, but by the same token, it may not be effective in screening all the defects that require both temperature and time to reveal themselves. Nevertheless, it is a common stress method used in some manufacturing test flows. This method may be a complement to burn-in or will lower the failure rate enough to avoid burn-in altogether. Again, there is a limit to how high a voltage one can apply. A voltage that is too high will also lead to (gate) oxide breakdown, which may cause irreversible damage if the current is not limited. This voltage is typically determined by the process technology of each process generation.

8.3.3 Redundancy and Memory Repair

General memory redundancy scheme uses spare rows, columns, or blocks. These schemes are good for repair of bad cells or sense amps, etc., resulting from manufacturing defects. However, if the defect rate is higher, setting aside redundant elements will have a diminishing return because the redundant elements themselves can also be faulty [Spica 2004]. The yield of additional redundant elements is subjected to the same exponential diminishing yield inversely proportional to the area.

Because *error-correcting code* (ECC) is also generally applied to detect and repair transient errors (*e.g.*, soft error) for high reliability memories as well as on-chip caches, one can also consider them to be a repair mechanism. ECC operates by creating a syndrome (check bits) from the data word, when the data are first written. These check bits are stored together with the data themselves. When the data are read, the syndrome is again calculated and compared with the stored syndrome (detail of the mathematics involved is not contained here). To minimize the performance impact for these calculations, sometimes the syndrome will not be calculated directly (as in the flow-through architecture) during the read but will be calculated in a parallel path in the data pipeline. If the stored syndrome matches with the recomputed syndrome, the data flow will not be interrupted. If there is a mismatch, the erroneous bit position can be identified from the new syndrome (*versus* the stored one) and the bit is flipped (if it is not a 1, then it must be a 0, the beauty of a binary system). It will take extra logic processing (maybe an extra clock cycle) to identify this bit position, so once the error is detected, the data flow has to be halted while the corrected data are generated and sent forth for further consumption. Theoretically, one can employ ECC to correct hard errors as well. If ECC (such as Hamming code) is used to correct hard errors, it will lose its primary function to protect against transient errors. Because transient errors are expected to occur infrequently, ECC can be the first line of defense (always detect and correct, no matter what the cause is). The system has to react to the situation (that a hard error has developed) and apply a remedy so that the reliability of the system will not degrade further. The chosen correction scheme has to deal with a convolution of the failure distributions to guarantee that it will be sufficient to meet the short-term (manufacturing) yield goals, the infant mortality goals, as well as reliability goals. The sum of the aforementioned distributions now must be considered for an overall error correction scheme.

The Pellston technology is one such scheme [Wuu 2005]. As described previously, this technology essentially makes use of ECC to protect large on-chip caches. Once an ECC event occurs, it issues a ***machine check architecture*** (MCA) interrupt and immediately branches to a handler that will pick up the erroneous location. This handler then writes the corrected data into the location and reads them back again. If the error persists, it sets a “not to use” bit for that particular cache line. This bit is similar to the ***modified/exclusive/shared/invalid*** (MESI) cache coherency protocol bits, and this cache line will not be used to cache any new data. This scheme again assumes that the amount of field failures (as a result of infant mortality, etc.) will be low. The loss of a few cache lines will have minimal impact on system-level performance. The small loss in performance is more than made up for with the increase in reliability against any soft error as well as hard (infant mortality) errors that might develop with the usage of the system.

There is also research that describes more elaborate schemes for dealing with increasing field failures that may be brought about by scaling to ever smaller devices. The authors in [Bhattacharjee 2004] described a scheme whereby the actual erroneous cache lines are mapped to spare cache lines through a mapping table. Memory testing during initial system boot time provides the failing address/location. Conceivably, this can also be coupled with the ECC mechanism and essentially ECC checking performs a test every time a cache line is read. Yet another scheme that advocates the recycling of erroneous bits was described in [Agarwal 2005]. The authors envision that these erroneous cells may not really be hard defects; they may fail to operate properly with power (or other) constraints (such as lowered V_{DD} in a mobile application). On-chip self-testing will identify the bad cache lines, and the memory addressing will be reconfigured so that the erroneous locations are used. Once the operation environment is changed again (*e.g.*, V_{DD} has resumed to normal), the on-chip self-test circuit will provide new failure map information to allow reconfiguration to a fuller-size cache. Innovations continue in this area to deal with the ever-complex failure mechanisms brought about as CMOS scaling continues.

8.3.4 Process Sensors and Adaptive Design

Traditionally, process sensors are used only by the process and yield engineers to monitor the process variation and yield analysis. These are generally test structures (transistors of various sizes, via chains, contact chains, etc.) put on the scribe lines (the four sides of the die). Generally, not all test structures are measured; only sample test data (usually called **e-test data**) are taken from each wafer and stored into a database (for later analysis). After dies are assembled into packages, these structures disappear with the wafer saw process (to extract dies). So additional test data will not be measurable in case one would like to investigate the causes of failure with a particular packaged chip. They can only extract the data from the databases.

With deep submicron scaling, **on-die process variation** slowly becomes a more significant cause of variation. Data have shown that the on-die process variation can be as much as the whole process variation. So monitoring the test structure on the scribe line would be of limited value. Embedding additional process sensors on-chip (the actual product) is essential to the understanding of on-die process variation. **On-die sensors** have the additional advantage that they are still available with the separated die and after package assembly. With the appropriate board-level hookup, one can even extract data when the die is running in the end-user application, so it is very powerful.

8.3.4.1 Process Variation Sensor

Ring oscillators have long served this purpose. With an odd number of inverter stages, a ring oscillator will oscillate naturally, and its frequency can be measured easily with a counter clocked appropriately. Because many factors can affect the frequency of the ring oscillator, it is generally difficult to de-convolute the cause(s) of the varied frequency. The authors in [Samaan 2003], [Stinson 2003], [Nassif 2004], and [Krishnamoorthy 2006] have proposed the use of multiple ring oscillators of varying design parameters. With many of these embedded ring oscillators, de-convolution of the process variation, temperature, and even voltage is possible, providing an immense source of information during manufacturing and online operational feedback. Because these sensors are so helpful, hundreds of them can be sprinkled onto different areas of the die to provide useful information about on-die process variation, local temperature (hot spots), and local power grid fluctuation, serving multiple purposes with a single set of sensors. The only requirement is that one has to know how to gain access to these sensors, and analysis must be done to de-convolute the underlying changes, which may be difficult to be carried out with on-die resources alone.

Another type of process variation sensor utilizes an analog circuit, which is sensitive to process parameters; in fact, almost all analog circuits are sensitive to different process parameters [Cherubal 2001]. It is not critical to pick an ideal circuit, as the analog circuit may have multiple specifications and these specifications may in turn be sensitive to different process parameters. These specifications can be measured using any on-die or instrumentation-based test methods. Once enough data are collected on a large sample of sensors (in a die or even from various locations of the die for across-die process variation) and with appropriate statistical analysis techniques such as ***principal component analysis*** (PCA), the authors claimed that they can de-convolute the various device/process parameters to which the circuit is sensitive. The technique is similar to solving the variables with many equations and many unknowns. In this case, one must be careful about choosing the analog circuit itself. If the circuit is not sensitive to some types of process parameters, the analysis will not reveal any such variation at all. Because the analysis is essentially based on statistical methods, a large sample of data is needed. Unlike ring oscillators, the analog process variation sensor also does not report the process variation at the specific spot on the die (unless enough data are collected from that very spot). It is also unlikely that one can extract and analyze these data in real time.

8.3.4.2 Thermal Sensor

This leads us to the discussion of more mission-critical sensors, such as **thermal sensors**. As high performance drives higher-frequency operations and a higher level of power consumption, today's processors and high-end SOC designs could generate lots of heat, which needs to be dissipated properly. If, for whatever reason, the heat sink is not properly mounted or the cooling fan (a mechanical device) fails to turn, the device can accumulate enough heat to go into **thermal runaway**. In an extreme situation, it may cause irreversible damage to the chip. Thus, thermal sensors are extremely important for protecting the chip as well as the rest of the system (power supply, socket, motherboard, etc.). As a result, the industry has started putting thermal sensors onto the heat sink or the motherboard itself. Putting thermal sensors on the motherboard or even inside packages may not have the fast response time needed (thermal conduction is a relatively slow process as opposed to electronic processes) to control power consumption. Often, **hot spots** are local (such as in the *floating point unit* [FPU] or *integer execution unit* [IEU]), and the device can heat up and cool off quickly with code changes. Power control has to happen really fast; otherwise, the accumulated heat may lead to circuit failure or, worst yet, may melt down as a result of thermal runaway. Designing conservatively so heat will never be a problem can be a solution, but that leaves performance on the table. There is also the inevitable accident that the heat sink may fall off or the fans may stop working. All of these require on-chip thermal sensors to act as the last defense to prevent system crash or permanent damage to the chip. Therefore, these are mission-critical sensors.

An example of a thermal sensor is illustrated in Figure 8.14 [Pham 2006]. It is essentially a diode coupled with a current source. The diode current (and voltage) is a function of temperature, so the trip point can be set to detect the reach of a particular temperature. With multiple sensors, one can construct a system whereby different temperatures trigger different events—for example, slowing the clock down with the first trigger; if the second trigger happens, a more drastic measure has to be taken, such as stopping all clocks; and the last sensor will shut down the system power supply to protect itself.

Diode does have its issues (same for any fabricated device), as the diode voltage varies with process variation and not just with its design parameters. In applications where absolute temperature measurement is needed, calibration is necessary. It can be accomplished at manufacturing time and with correction factors burned into on-chip fuses. For protection mechanisms like the one described earlier, calibration may be optional.

Thermal protection actually brought us to an important system feature that has made a significant impact on the world of testing. The chips of today are increasingly adaptive because of power savings, maximizing performance within a given

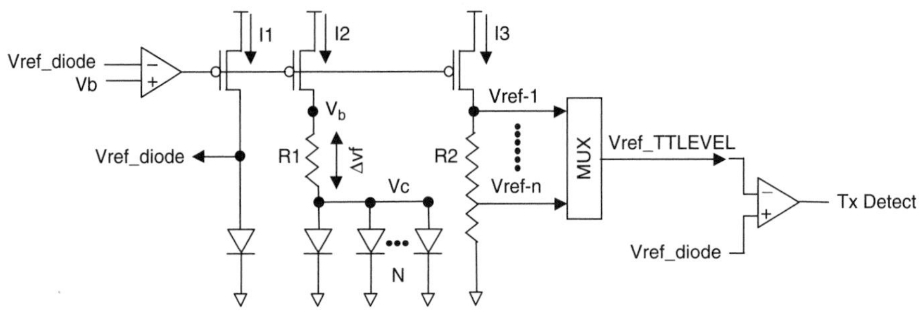

■ **FIGURE 8.14**

Thermal sensor example.

power envelop, multiple clock domains with variable frequencies, high-speed IO interfaces, etc. Using the thermal diode as an example, each diode is slightly different from another because of fabrication variation. Each chip also may consume power differently (varying load capacitance and leakage). So the trip point from one die to the next can be very different. When they trip can also be different, even though both of them run the same patterns. Although this poses few issues with system-level operation (nobody would figure out down to the millisecond when the trigger event occurs), this is totally unacceptable in the digital testing paradigm.

The digital testing paradigm rests on the principle of stored stimuli and stored responses. *Automatic test equipment* (ATE) essentially stored all test patterns during logic or RTL simulation, and the ***device under test*** (DUT) is expected to perform exactly as simulated. Any bits coming off in an earlier or later cycle cause an error and the device is discarded. ATEs simply do not have the intelligence to figure out that the clock has changed to a different frequency, especially when the trip point can vary so much in time because of all the variances. Similar nondeterminism happens with all the conditions mentioned earlier, so our digital chips are increasingly difficult to test. One can implement DFT to turn off this nondeterminism (if one knows about it), but then we are either not testing a particular chip feature or will pose much greater design effort to turn off a natural feature of the chip. The saving grace is that structural tests are not impacted (because logic is tested from latch to latch anyway with structural testing). However, we are losing coverage as increasingly we have to turn an adaptive chip into a nonadaptive chip.

8.3.4.3 Dynamic Voltage Scaling

Another adaptive feature for power management is ***dynamic voltage scaling*** (DVS) [Flautner 2001]. As an efficient power reduction technique, DVS has been implemented in several commercial embedded microprocessors such as Transmeta Crusoe [Transmeta 2002], Intel Xscale [Intel 2003], and ARM IEM [ARM 2007]. DVS exploits the fact that the clock frequency (f) of a processor changes proportionally with the supply voltage (V), whereas the dynamic energy (P) is proportional to the *square* of the processor's supply voltage ($P \propto V^2 f$). Thus, to save power, one can simply lower the clock frequency of the processor. If the frequency is lowered, then the voltage supply also will be lowered, resulting in a *cubic* power reduction. When the system goes into a period of low activity (*e.g.*, a user is looking at the screen while thinking), it will signal the circuit to go into a low power state whereby frequency and voltage are scaled down. Once there is a need to process more data (*e.g.*, a key is pressed), the system will resume to higher voltage and frequency, making the system appear to be responsive all the time. An example of the DVS scheme is shown in Figure 8.15.

Additional adaptability features may fit into this category. One example is the use of **sleep transistors** [Tschanz 2003] (which essentially scales the voltage close to 0 for specific circuits at a specific time) and **dynamic body biasing** (which varies the back bias of the transistor to increase or decrease the threshold voltage) to save power. The function of the back bias is to affect the threshold voltage V_t of the

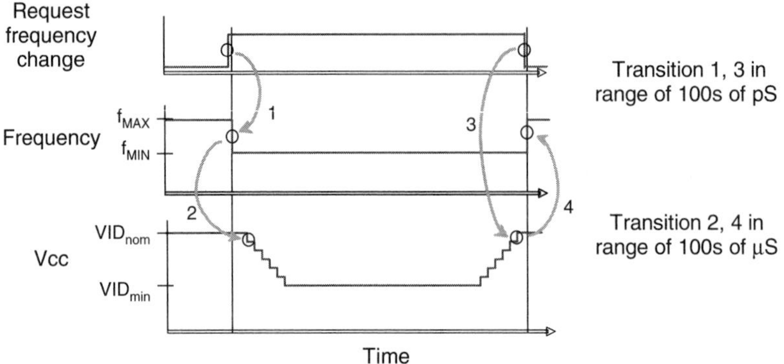

■ **FIGURE 8.15**

Dynamic voltage scaling scheme.

transistor. It can be forward biased and it will lower V_t and speed up the transistor, but at the expense of leaking much more current. It can also be reversed biased, and it will increase V_t and lower the leakage current, but it will slow down the transistor. The other example is the **adaptive test** method [Edmondson 2006] for smart binning (not necessarily just speed), whereby devices are tested for power level before the optimal voltage/frequency classification of the chip is made. This is done through a ***voltage identifier*** (voltage ID), which is supplied by the chip to the ***voltage regulator module*** (VRM) on the motherboard during power up. The core frequency is determined by an internal multiplier, which is programmed during test through a fuse (after determining how much power it consumes). The voltage ID is also fused at the same time. Both uses of sleep transistors and adaptive testing are unconventional ways of adaptations to cope with process variations.

8.4 SOFT ERRORS

Soft errors are transient *single-event upsets* (SEUs) caused by various types of radiation. Cosmic radiation has long been regarded as the major source of soft errors, especially in memories [May 1979], and chips used in space applications typically use parity or *error-correcting code* (ECC) for soft error protection. As circuit features begin to shrink into the nanometer ranges, error-causing activation energies are reduced. As a result, terrestrial radiation, such as alpha particles from the packaging materials of a chip, is also beginning to cause soft errors more frequently. This has created reliability concerns, especially for microprocessors, network processors, high-end routers, and network storage components.

In this section, we first illustrate the sources of soft errors and the ***soft error rate*** (SER) trends. Following a discussion of general fault tolerance schemes for soft error protection, we then discuss **DIVA** [Austin 1999] and **Razor** [Ernst 2003] [Ernst 2004], two representative error-resilient processor microarchitectures, as well as three soft error mitigation methods through ***built-in soft-error resilience*** (BISER) [Mitra 2005] and circuit-level modifications [Almukhaizim 2006] [Zhou 2006]. DIVA and Razor are mainly used for high-performance processor designs. BISER and circuit-level modification methods, however, are applicable to any design for soft error protection.

8.4.1 Sources of Soft Errors and SER Trends

Soft errors are the result of *transients* that are induced in the circuit when a radiation particle strikes. This radiation can range from cosmic origin (when stars are formed and die) or from everyday material (*e.g.*, lead isotopes) [Ziegler 2004]. When high-energy cosmic rays reach into our atmosphere, they collide and strip off air molecules and send off neutrons. These neutrons continue their journey and penetrate through most types of matter (so shielding is largely out of the question). As these neutrons transverse through silicon, they ionize the silicon lattice and leave a trail of holes and electrons behind. These will then be moved by the electric field of surrounding diffusion and wells. As holes and electrons recombine, they charge or discharge the node appropriately. From a circuit standpoint, we simply have a glitch.

Radioactive isotopes emit alpha particles as the radioactive decay process occurs. These alpha particles are larger and heavier, so they will not have deep penetration; however, because they may exist (in a stray amount) in packaging material (*e.g.*, ceramic or solder), they are located close to the die and can lead to a relatively high error rate (in fact, early soft errors were first discovered in radioactive elements in packaging material [May 1979]). If such a glitch is induced in a memory element, its state can be reversed. As an example, let us examine a SRAM cell that has two back-to-back inverter pairs, as shown in Figure 8.16.

When the select transistors are off, this cell holds the state in a stable configuration. If a glitch is introduced at the drain of the PMOS or the source of the NMOS, the other inverter can pick up the glitch and the state of the cell is reversed. A similar problem can occur for all storage elements, such as D latches and D flip-flops (see Figure 8.17). If a glitch strikes the combinational logic elements, the resulting glitch is evaluated and passed on by the succeeding logic.

Soft errors can happen to all memory and storage elements. Sometimes, they can be benign (*e.g.*, the memory elements are not used in the application); other times, they can cause a system crash or, even worse, a ***silent data corruption*** (SDC) if they are undetected. That is why we have to devise online detection (or fault tolerance and error mitigation) mechanisms to protect against such transients and cope with soft errors. These kinds of detection and fault tolerance mechanisms are further discussed in the following section. Unlike defects or other fault types, a soft error is transient and is induced by a glitch at one time at one location; it is not repeatable; thus the term "soft error." This property is well utilized in the solution space.

Logic circuits are less susceptible to these glitches than memories for the following reasons [Mohanram 2003]. First, the glitch must be of sufficient strength to propagate from the location of the strike, through each stage of the logic circuit, until it reaches an output; otherwise, the glitch is attenuated and the transient error

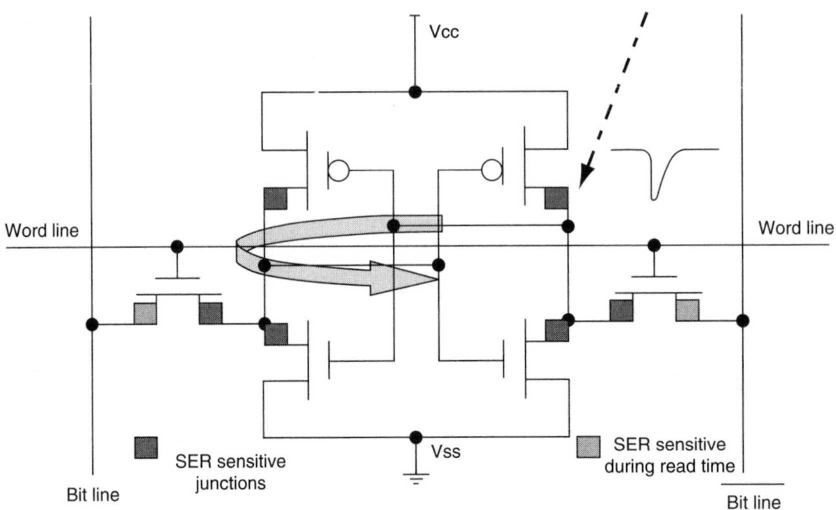

■ **FIGURE 8.16**

Induced soft error on a SRAM cell.

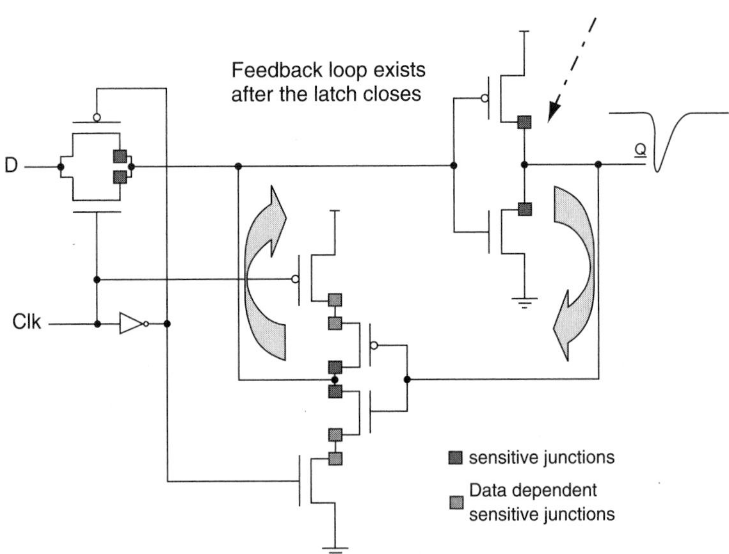

■ **FIGURE 8.17**

Induced soft error on a D latch.

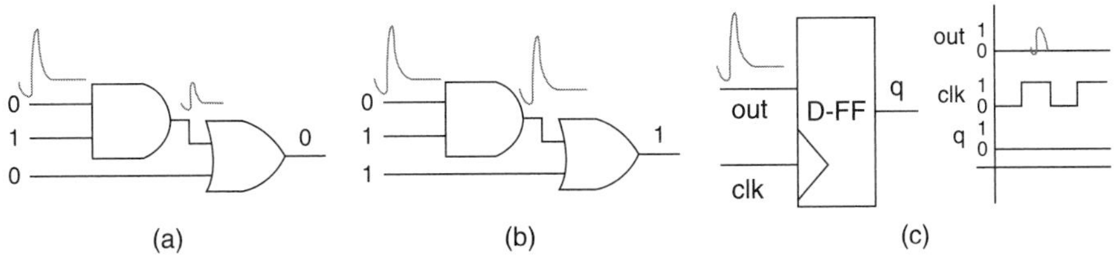

FIGURE 8.18

Masking factors of soft errors in combinational logic: (a) electrical masking, (b) logical masking, and (c) latching window masking.

is electrically masked. Second, the glitch needs to have a functionally sensitized path to a latch; otherwise, the glitch is logically masked. Finally, the timing of the glitch must be such that the glitch arrives at a latch during its latching window; otherwise, the glitch is latching-window masked (an exception is a domino-type circuit, where its logic states are held at every gate by a feedback element). These three masking factors are illustrated in Figure 8.18; historically they have made soft errors a non-issue for combinational logic circuits. Technology trends, however, are rapidly reducing the effect of these masking factors. For example, reduced logic depth implies a higher probability of having a sensitized path to the output and less attenuation of the glitch; lower supply voltage leads to less noise margin and, hence, smaller glitches may produce an error; higher clock frequency leads to a higher probability that a glitch will be latched, etc. These masking factors are judiciously utilized by circuit-level soft error mitigation techniques that reduce the susceptibility of logic circuits to soft errors.

Because the physics of soft errors involves node charging and discharging, the amount of stored charge at a given node determines how sensitive it is to a particle strike. The charge (Q) is represented by the following relationship with voltage (V) and capacitance (C):

$$Q = CV.$$

As processing technology scales, capacitance for a given node decreases. This is good for performance but bad for soft error. Because of the *hot electron* type of degradation, reliability requirements also force V_{DD} to be lowered. This compounded effect causes a decrease in stored charge and increases the ***soft error rate*** (SER). With scaling, we also get more transistors (roughly 2×) per chip, resulting in increase in the soft error rate [Baumann 2005]. The saving grace is that, because the transistor junction area is also scaled, the ability for the node to collect stray charge is also reduced; however, this is not sufficient to slow the increase in soft error vulnerability. The **Moore's law** prediction of doubling transistors with every process generation [Moore 1965] effectively doubles the soft error rate, so not only should SRAM cells (such as caches and registers) be protected, but protection on storage elements as well as against glitches creeping through the combinational logic is also important. These areas are now all hot research topics.

The implication of soft errors with regard to chip testing varies. From the surface, soft errors really cannot be tested. Even good circuits are susceptible to soft errors, so there is nothing to screen for. Soft errors are also not easily exercisable with electrical test stimulus. The natural occurrence of radiation also does not usually happen during the short test time of component testing. What really requires attention is an online detection scheme or a fault tolerance scheme.

Often the three types of redundancy—hardware (spatial), time (temporal), and information—involve extra circuit elements (refer to Chapter 3 for more detailed description). At a minimum, there is a *self-checking equality checker*, which indicates whether there is any error. With *information redundancy*, there are extra check bits (or code bits). With *hardware redundancy*, there is even duplicate circuitry. Each redundancy circuit has to be tested to make sure that the redundancy scheme can detect soft errors and, when found, can signal that there is an error or simply correct the error.

As it is difficult to test every redundancy circuitry and they are probably not testable without appropriate DFT means, special attention must be given to such redundancy circuitry so it is accounted for in the overall test strategy. If a redundancy scheme is capable of correcting errors by itself, then even manufacturing faults can hide behind the redundancy scheme, as output results are always correct. The undiscovered defects will consume the correction capability, and any subsequent soft error hit to that functional circuitry or redundancy circuitry may cause an unrecoverable error.

8.4.2 Coping with Soft Errors

As chips are susceptible to soft errors, the growing circuit sizes of these chips have drastically increased their *soft error rate* (SER). To cope with these soft errors, many soft error protection schemes targeting chip designs have been proposed.

8.4.2.1 Fault Tolerance

One approach to improve the reliability of a chip is to remove the source of soft errors. Because the early discovery of soft error was the result of contaminated packaging material, the solution was simple: eliminate the radiation contaminant from packaging material [May 1979]. However, trace amounts of radioactive isotopes do exist in common processing and packaging materials—such as the boron in ***borophosphosilicate glass*** (BPSG) and ceramics and the lead in solder—and their removal is costly. Their radioactive decay still leads to some level of soft error rate. Because alpha-particle radiation (see the previous section) can be stopped on the surface of the die, a die coating (epoxy resins that are deposited on the surface of the die) was introduced and used for some period of time. Die coating is only effective if the radiation comes from the outside and is of limited value if the alpha emitter is among the materials that transistors or interconnects are made of. As we move from wire bonding packaging to flip-chip solder (tin/lead) joint (or C4) packaging, solders are never far away from the surface of the die and die coatings would have no effect at all. Today, the primary alpha-particle source is solders (lead radiation isotopes). The careful selection of raw material has resulted in low-alpha solder. As a result of environmental and health concerns over lead, tin/lead solder is being phased out in packaging material, which will reduce alpha-induced soft errors. However, the other source of radiation, *high-energy neutrons*, cannot be stopped by anything associated with packaging.

All of these preventive measures combined with advances in manufacturing process technology have improved the system-level reliability substantially. In the past, soft errors were not critical for most computer systems for terrestrial applications. Thus, traditionally, only high-reliability applications, especially those deployed in the financial transaction, transportation, and aerospace/defense industries, have required fault tolerance to prevent the systems from crashes and silent data corruption errors.

As these measures are not effective enough to prevent soft errors from happening, traditional fault tolerance schemes commonly used for high-reliability applications have started to emerge. There are three fundamental fault tolerance schemes that can be used to protect such systems or devices from hard errors or soft errors: (1) hardware (special) redundancy, (2) time (temporal) redundancy, and (3) information redundancy [Pradhan 1996] [Siewiorek 1998] [Lala 2001]:

1. **Hardware (spatial) redundancy** relies on the assumption that defects and radiation particles will only hit on a specific device and not another device (at least not simultaneously). So having a duplicate circuitry of the functional circuitry and with their outputs compared using a *self-checking equality checker* (checking circuitry), mismatches will point to an error (hard or soft error) (see Figure 3.19 in Chapter 3). This can happen at the circuit level (*e.g.*, one adder is compared with another adder while both are fed the same data) or at the system level (*e.g.*, a processor's front side bus is compared with another one on the same bus while executing the same codes). Because the computation occurs in parallel, there is little or no penalty on the overall system performance, but there has to be hardware duplication and a *self-checking equality checker*, resulting in higher hardware costs and a much higher level of power consumption.

2. **Time (temporal) redundancy** relies on the assumption that even if functional circuitry receives a radiation strike it is unlikely that the strike will happen on the same circuitry again at a slightly later time, so the scheme will not require duplicate circuitry. In this case, the same computation is repeated on the same functional circuitry for a second time, and the results of the first computation are not committed without comparison to the second computation. This obviously has the benefit of not requiring additional hardware, but the software must be coded to execute the program twice, which means saving the results of the first computation on memory or disk. One serious problem with this scheme is that it cannot detect any hard error; the same erroneous result will happen when recomputed. Therefore, time redundancy may give a false sense of security with regard to any hard error that results from physical failures.

3. **Information redundancy** uses ***error-detecting code*** (EDC) or ***error-correcting code*** (ECC) to represent information contents [Peterson 1972]. Some of these coding properties are maintained even after computation, so by checking these codes before and after, one can determine if a hard or soft error has occurred. Parity is one such code. Parity represents whether the number of ones in a computer word is odd or even. Normally, this parity is computed when the information is generated and stored in the memory system. Upon reading of the word, parity again is recalculated and compared against the earlier stored parity bit. A mismatch identifies that an error has occurred. One major benefit of using parity code is that a single parity bit can detect any odd number of bit errors (caused by soft errors and hard errors) in each computer word; however, there is always the danger that a single radiation strike could affect more than a single bit, and when an even number of bits are flipped, the errors escape detection (because parity only counts odd or even). In this case, more sophisticated codes (such as Hamming code) can be used [Peterson 1972]. This code allows detection of 2-bit errors as well as correction of single-bit error. This, in general, is referred to as the *error-correcting code*. It requires the storage of more check bits (codes) and a computation unit that does the check code generation. It is important to note that properties like parity or ECC are often embedded for arithmetic operations (such as add/subtract/multiply) during normal operation. Thus, they are also used for arithmetic computation protection. Because additional information is stored, it can protect against both hard and soft errors.

Having detection capability is only half the story. After an error is detected, some recovery actions have to be taken. The most common action is for the operating system to stop the application, generate the necessary error message/log, and close the application. This action has varying degrees of system integrity implications, as partially computed results may have been written to disk already. Of course, this is better than simply letting the system crash, but we need better schemes that can provide more integrity. **Checkpointing** and **rollback** constitute one such scheme. Checkpointing is essentially taking a snapshot of the system states, which when revoked (rollback) will cause the system to restart from that point without rebooting or terminating the application. However, one has to ensure that no error has occurred (and the system is intact) before the checkpoint. One also has to consider where (or how regular) checkpoints are done to optimize for performance (because of the checkpointing process) as well as make sure that system states are not contaminated before the checkpoint.

Thus, having explained the basic principle, what are some of the common fault tolerance schemes used in high reliability systems? As mentioned before, duplicate and compare is one method that is commonly used in mainframes and high-end servers [Spainhower 1999] [Bartlett 2004]. For systems that cannot fail, a more secure system is ***triple modular redundancy*** (TMR) [Sklaroff 1976] [Siewiorek 1998] [Lala 2001]. Consider the TMR example shown in Figure 8.19. Here we have three pieces of computing units (modules), and their results are constantly compared. Because we have three results, the two matched results outvote the

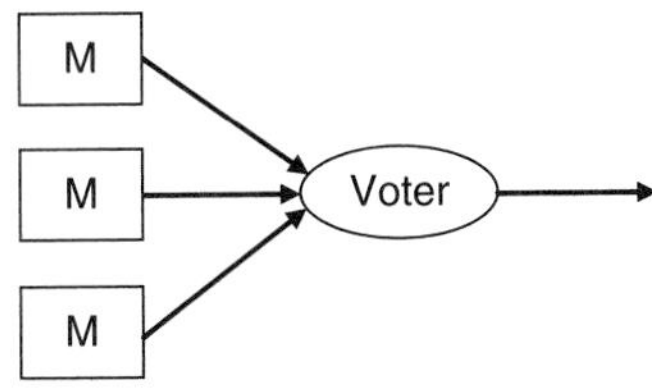

■ **FIGURE 8.19**

A TMR example.

mismatched result and the deemed correct result will be sent on. The aerospace industry particularly favors this approach (for obvious reasons). More recently, because a ***central processing unit*** (CPU) is capable of running multiple threads (different streams of instructions that can be run in parallel on a CPU), one can also send a redundant thread through another path (virtual compute units) and check the results before retiring the thread/instructions. This is called ***redundant multithreading*** (RMT) [Mukherjee 2002].

The incorporation of fault tolerance in a compute system did not begin with the processor. It began with the circuit that had the highest transistor density and the part of the system that had the most transistors—the main memory. The protection scheme is the use of ECC, which is quickly followed by ***redundant array of inexpensive disk*** (RAID). Even though the disk system does not necessarily have the highest number of transistors, disk drives are susceptible to mechanical failures; hence, it is essential to protect the data that it holds. Because of the need to have signals routed over long wires or traces, buses and backplanes that interconnect various subsystems are protected with parity. The networking communication protocols also contain error codes, such as ***cyclic redundancy check*** (CRC) and checksum (the sum of all the binary numbers in a particular packet of data). In the early 1990s, the importance of fault tolerance to CPUs became apparent, as on-chip cache memories have become large enough to warrant their own ECC protection. On some high-end server CPUs, register files are protected with parity, and duplicated execution blocks help to identify errors [Spainhower 1999]. So the last holdout seems to be flip-flops and combinational logic. Researchers have come up with hardened latch/flops [Calin 1997], where circuit design has decreased the internal nodes' vulnerability to a radiation strike. More recently, enhancing the storage elements or scan cells to fill the role of duplication (as the states are already duplicated) has been suggested [Franco 1994] [Austin 1999] [Ernst 2003] [Mitra 2005] [Tamhankar 2005]. These on-chip fault tolerance schemes are referred to as **error resilience**.

As we move toward the nanotechnology era, more and more system-level functions (in the form of IP cores) will be integrated on a single piece of silicon (or package). This trend has substantially exposed the nanometer SOC design to ever more manufacturing faults and soft errors; therefore, it is becoming more and more important to embed online error detection or correction schemes in these chips as the distinction between computer systems and ***systems-on-chip*** (SOCs) increasingly narrows in nanometer designs.

8.4.2.2 Error-Resilient Microarchitectures

Because hardware redundancy requires one or more redundant modules for error detection or correction, the hardware overhead using this fault tolerance scheme is a concern for use in chips. At the same time, while information redundancy can significantly reduce hardware overhead for array logic such as ***programmable logic array*** (PLA) and memory, this fault tolerance scheme is not applicable for random logic. This has left time redundancy as a viable solution to providing online error detection or correction for random logic on-chip.

These on-chip fault tolerance schemes for error detection or correction, referred to as error resilience, were first proposed for processor designs. While the major purpose of these error-resilient processor microarchitectures is to achieve maximum processor performance and power savings using *dynamic voltage scaling* (DVS), the online error detection and correction circuits used therein are applicable for soft error protection.

In this section, we only discuss two representative error-resilient processor microarchitectures: **DIVA** [Austin 1999] and **Razor** [Ernst 2003] [Ernst 2004]. Both DIVA and Razor adopt the fault tolerance scheme using spatial redundancy. DIVA uses a simpler **DIVA checker** to verify the recomputed results before commit, while Razor uses a **Razor flip-flop** similar to the **stability checker** (see Figure 3.17 of Chapter 3) proposed in [Franco 1994] to check whether each main flip-flop during normal operation functions correctly. For more information on other DVS schemes that are also applicable for soft error protection, refer to [Ernst 2003] and [Ernst 2004]. Chapter 3 of this book also provides a good reference to the basics of these fault tolerance schemes. For a more specific explanation, other implementations of the error-resilient microarchitectures can be found in [Patel 1982], [Metra 1998], [Oh 2002], [Favalli 2002], and [Tamhankar 2005].

DIVA

One error-resilient processor microarchitecture is the ***dynamic implementation verification architecture*** (DIVA) [Austin 1999]. As illustrated in Figure 8.20, DIVA uses a smaller and simpler shadow processor (*DIVA checker*) that computes concurrently as the main processor (*DIVA core*). Instead of using two large complex cores as in the case of using the ***double modular redundancy*** (DMR) scheme, the *DIVA checker* largely depends on the *DIVA core* to do the computation while it verifies the correctness of the core processor's computation with simpler hardware (to save on silicon cost).

The DIVA core constitutes the entire microprocessor design except the retirement stage. The main processor core fetches, decodes, and executes instructions, holding their speculative results in the *reorder buffer* (ROB). The *DIVA checker* contains a functional checker stage (CHK) that verifies the correctness of all core computations, only permitting correct results to pass through to the commit stage (CT)

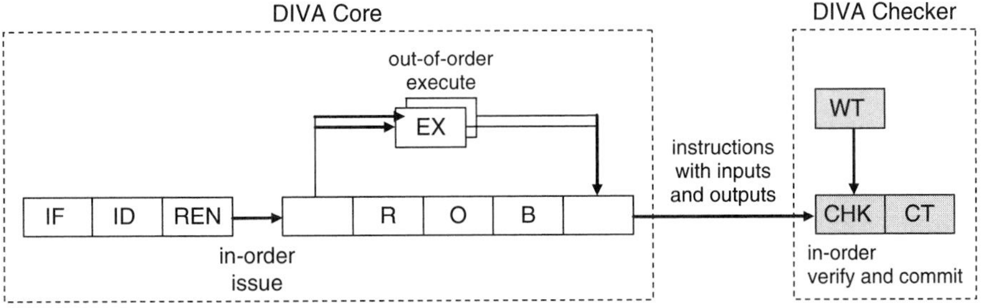

■ **FIGURE 8.20**

DIVA architecture.

where they are written to architected storage. A *watchdog timer* (WT) is added to detect faults that can lock up the core processor or put it into a *deadlock* or *livelock* state where no instructions attempt to retire. If an error is detected during any core computation, then the *DIVA checker* will fix the errant computation, flush the processor pipeline, and restart the processor at the next instruction.

This **dynamic verification** scheme, a microarchitecture-based technique that can significantly reduce the burden of correctness in microprocessor designs, is only applicable to the processor world where these microarchitectures make sense. This implies the protection is only limited to portions of the design which are not shared; because the whole bus unit, instruction decoding, and the backend of the pipeline are shared, they are not protected. Like any DMR scheme, this technique is good for detecting an error (be it permanent or SER); recovery is yet another issue. Similarly to DMR, it can be coupled with **checkpointing** and **rollback** as a reasonable *error recovery* scheme to achieve fault tolerance.

Razor

With increasing clock frequencies and silicon integration that come with scaling, power-aware computing has become a critical concern in the design of high-performance computing. In addition, the energy consumption and dissipation issues also spread to embedded processors and SOCs. One approach for power conservation is to run the circuit at the lowest V_{DD} where it still executes correctly. Instead of having a fixed V_{DD}, the voltage is adjusted dynamically, hence *dynamic voltage scaling* (DVS). DVS is one of the more effective and widely used methods for **power-aware computing** [Pering 1998]. To obtain maximum power savings from DVS, it is essential to scale the supply voltage as low as possible while ensuring correct operation of the design. This critical supply voltage is difficult to be set correctly all the time considering a wide diversity of process and environmental variations, where the design might not operate correctly.

The authors in [Austin 1999] attempted to address the power conservation issues by using a self-tuned clock and voltage scheme in which dynamic verification is used to reclaim frequency and voltage margins. The **Razor** scheme proposed in

[Ernst 2003] and [Ernst 2004], on the other hand, is based on the use of *in situ timing error detection and correction* to permit increased energy reduction as voltage margins are completely eliminated. The key idea of Razor is to tune the supply voltage by monitoring the error during circuit operation, thereby eliminating the need for voltage margins and exploiting the data dependence of circuit delay. Similar to DIVA, this is accomplished with a shadow unit, but this shadow unit has been pushed all the way down into a **Razor flip-flop.** This Razor flip-flop, shown in Figure 8.21a, double-samples pipeline stage values, one with a fast clock, *clk*, and the other with a time-borrowing delayed clock, *clk_del*. It includes a *main flip-flop* controlled by the fast clock and a *shadow latch* controlled by the delayed clock. The value of the shadow latch is assumed to be correct under any operating voltage. A *metastability-tolerant comparator* then validates the values stored in the main flip-flop.

Actually, using a delay checker for on-chip error detection was first proposed in [Franco 1994], long before power became a significant issue in chip design. The stability checker proposed in that paper was designed to deal with delay degradation (reliability). The delayed data are sampled and checked against the primary data. A transistor-level design that integrates the stability checker into a flip-flop is also given, which significantly reduces the overhead. The paper also discussed the possibility of moving the checking ahead of the primary latch to detect oncoming timing errors and changing the dynamic frequency of the circuit to cope with the CUT where field repair is not possible. However, the earlier paper does not address (1) power saving when power was not yet an issue and (2) methods for dealing with an error when it occurs. These are crucial elements in all time redundancy (delay detection) schemes. Razor documents the whole error handling part in detail.

A reduced overhead Razor flip-flop with the metastability detection circuit is illustrated in Figure 8.21b. The operating voltage is tuned to the level so that the worst-case delay is guaranteed to meet the shadow latch setup time, even though the main flip-flop could fail. By comparing the values latched by the main flip-flop and the shadow latch during each clock cycle, a delay error in the main flip-flop can be detected. The value stored in the shadow latch is then used to correct the error.

The operation of the Razor flip-flop is illustrated in Figure 8.21c. In clock cycle 1, the combinational logic *L*1 meets the setup time by the rising edge of both clocks so both the main flip-flop and the shadow latch will latch the correct data. In this case, the error signal at the output of the XOR gate remains logic 0, and the operation of the pipeline is unaltered. In clock cycle 2, we show an example of the operation when the combinational logic *L*1 exceeds the intended setup time of the main flip-flop but does not exceed the worst-case setup time of the shadow latch during subcritical voltage scaling. In this case, the data are not correctly latched by the main flip-flop but are successfully latched by the shadow latch because the shadow latch is controlled by a delayed clock. This means, in order to guarantee that the shadow latch will always latch the input data correctly, the allowable operating voltage must be constrained at design time so under worst-case conditions, the logic delay in *L*1 will never exceed the setup time of the shadow latch. By comparing the valid data of the shadow latch with the incorrect data latched by the main flip-flop, an error signal, *Error_L*, is then generated in clock cycle 3. In clock cycle 4, the valid

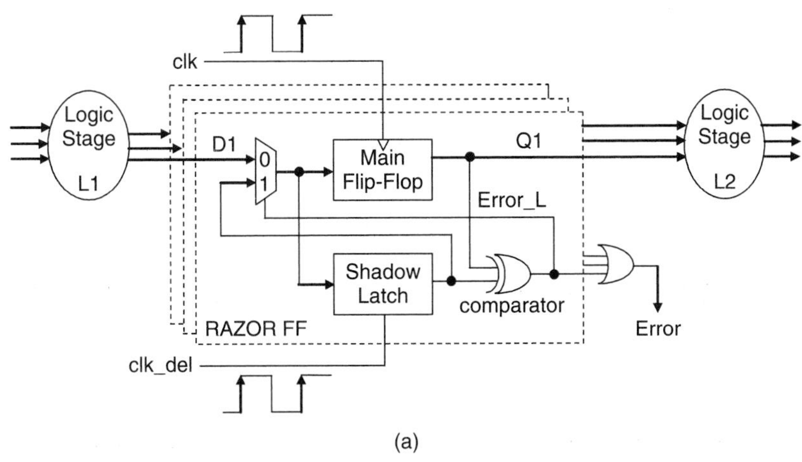

(a)

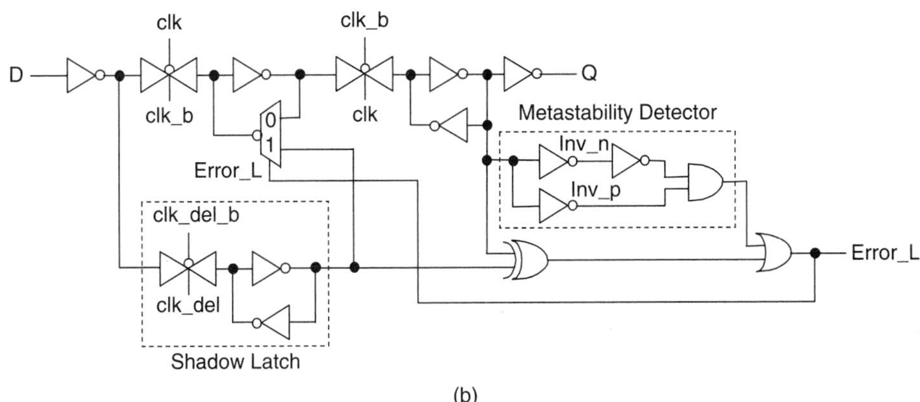

(b)

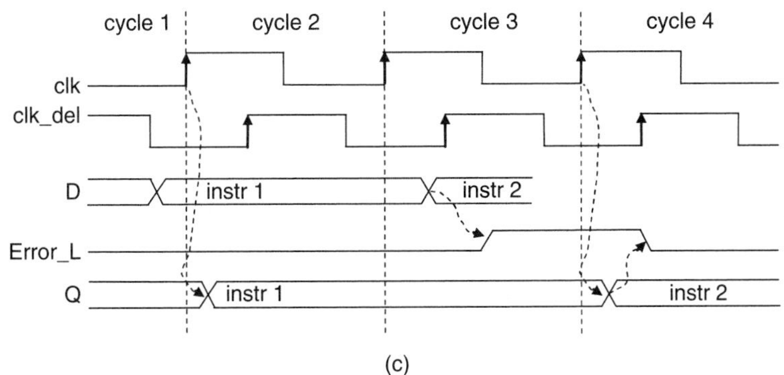

(c)

■ **FIGURE 8.21**

Razor flip-flop: (a) schematic of the Razor flip-flop, (b) reduced overhead Razor flip-flop with metastability detection circuit, and (c) waveform.

data in the shadow latch are restored into the main flip-flop and become available to the next pipeline stage *L*2. These local error signals are OR'ed together to ensure that all contents of the main flip-flops are restored even when only one of the Razor flip-flops generates an error.

Having said that, it is still a delicate balance that we have to guarantee the shadow latch latches the correct data. If both of them are latched incorrectly as a result of unpredictable timing mismatches caused by whatever reasons, the corrupted system will remain undetected and becomes a ***silent data corruption*** (SDC). This also involves a complicated pipeline reloading mechanism (see Figure 8.21), and it is unclear what the implication would be for applications of a nonpipelined (finite-state machine) type. The voltage adjustment mechanism also has to be designed such that it can be changed reasonably fast (within a couple of clock cycles); otherwise the whole system will keep circulating until the main flip-flop matches the shadow latch. This can be a problem for large circuits where a power-adjusting circuit would not be able to react fast enough to the heavy load.

8.4.2.3 Soft Error Mitigation

The objective of soft error mitigation techniques is to provide partial immunity of a design to potential soft errors while significantly minimizing the required cost over fault tolerance schemes. Thus, these methods are highly suitable for mainstream applications where the SER is reduced in a cost-effective manner without providing complete tolerance capabilities. We review three soft error mitigation methods through *built-in soft-error resilience* (BISER) [Mitra 2005] and circuit-level modifications [Zhou 2006] [Almukhaizim 2006].

Built-In Soft-Error Resilience

The ***built-in soft-error resilience*** (BISER) proposed in [Mitra 2005] can be used to allow scan design to protect a device from soft errors during normal system operation. BISER is based on the observation that soft errors either (1) occur in memories and storage elements and manifest themselves by flipping their stored states or (2) result in a transient fault in a combinational gate, as caused by an ion striking a transistor within the combinational gate, and can be captured by a memory or storage element [Nicolaidis 1999]. Data from [Mitra 2005] show that combinational gates and storage elements contribute to a total of 60% of the ***soft error rate*** (SER) of a design manufactured using current state-of-the-art technology *versus* 40% for memories. Hence, it is no longer enough to consider soft error protection only for memories without considering any soft error protection for storage elements as well.

Figure 8.22 shows the BISER scan cell design [Mitra 2005] that reduces the impact of soft errors affecting storage elements by more than 20 times. This scan cell consists of a system flip-flop and a scan portion, each comprising a one-port D latch and a two-port D latch, a C-element, and a bus keeper. This scan cell supports two operation modes: system mode and test mode.

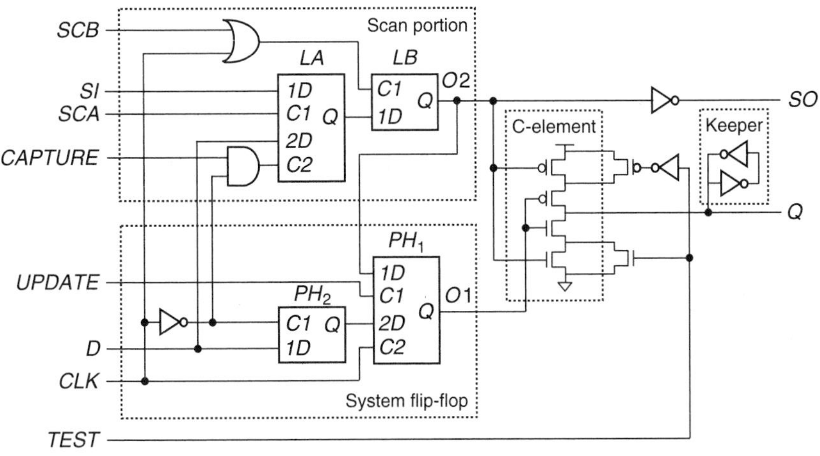

■ **FIGURE 8.22**

Built-in soft-error resilience (BISER) scan cell.

In test mode, *TEST* is set to 1, and the C-element acts as an inverter. During the shift operation, a test vector is shifted into latches *LA* and *LB* by alternately applying clocks *SCA* and *SCB* while keeping *CAPTURE* and *CLK* at 0. Then, the *UPDATE* clock is applied to move the content of *LB* to PH_1. As a result, a test vector is written into the system flip-flop. During the capture operation, *CAPTURE* is first set to 1, and then the functional clock *CLK* is applied which captures the circuit response to the test vector into the system flip-flop and the scan portion simultaneously. The circuit response is then shifted out by alternately applying clocks *SCA* and *SCB* again.

In system mode, *TEST* is set to 0, and the C-element acts as a hold-state comparator. The function of the C-element is shown in Table 8.3. When inputs O_1 and O_2 are unequal, the output of the C-element keeps its pervious value. During this mode, a 0 is applied to the *SCA*, *SCB*, and *UPDATE* signals, and a 1 is applied to the *CAPTURE* signal. This converts the scan portion into a master-slave flip-flop that operates as a shadow of the system flip-flop. That is, whenever the functional clock *CLK* is applied, the same logic value is captured into both the system flip-flop and the scan portion. When *CLK* is 0, the outputs of latches PH_1 and *LB* hold their

TABLE 8.3 ■ C-Element Truth Table

O_1	O_2	Q
0	0	1
1	1	0
0	1	Previous value retained
1	0	Previous value retained

previous logic values. If a soft error occurs either at PH_1 or at LB, O_1 and O_2 will have different logic values. When CLK is 1, the outputs of latches PH_2 and LA hold their previous logic values, and the logic values drive O_1 and O_2, respectively. If a soft error occurs either at PH_2 or at LA, O_1 and O_2 will have different logic values. In both cases, unless such a soft error occurs after the correct logic value passes through the C-element and reaches the keeper, the soft error will not propagate to the output Q, and the keeper will retain the correct logic value at Q.

The beauty of this scheme is that the BISER scan design has **self-correction** capability. Each BISER scan cell can still function as a normal scan cell in test mode. Once the chip is in the final application, it can be configured in self-checking mode (system mode) and no errors will propagate any further than the C-element. There are no new routing and new control signals to be added other than the existing scan control signals.

The only shortcoming, however, is that this will incur more power and area to some degree. The scan portion of the scan cell is much weaker because scan routing does not have to go far and each scan chain can run much slower (to ease design complexity) than the functional logic. For the scan cell to latch the same data, it has to be sized up appropriately so that it can run at more or less the same speed as the normal system flip-flop. This will increase loading and power to some degree (up to 2×). However, as not all system flip-flops need to be protected before we can achieve a 20× reduction of the SER, the area and power penalty for the whole chip is in the 3% to 5% range, much smaller than any conventional SER detection mechanism like DMR (where it is at least 100% more) [Mitra 2005].

Another important attribute of this scheme is that it is applicable for any latch-based or flip-flop-based logic design. Other architectural-level solutions, such as checkpointing and rollback, are good for processor designs but are totally nonrelevant for nonprocessor types of logic design.

Circuit-Level Approaches

Circuit-level approaches attempt to increase the ability of logic circuits to mask glitches by increasing the soft error masking factors in the circuit. While the reduction in the ability of a circuit to mask soft errors via latching-window masking is a consequence of the rapid increase in the operating frequency of logic circuits, the electrical and logical masking factors can be improved in a cost-effective manner for a targeted design. We describe two techniques that successfully mitigate soft errors in logic circuits by improving the electrical masking and logical masking factors of a design using gate resizing [Zhou 2006] and netlist transformations [Almukhaizim 2006], respectively.

Gate resizing for soft error mitigation [Zhou 2006] is based on physical-level design modifications, wherein individual transistor characteristics are perturbed to reduce the sensitivity of gates to glitches. Specifically, a select set of gates are resized, by altering the *width/length* (W/L) ratios of transistors in these gates, in order to increase their immunity to glitches and, by extension, reduce the SER of the logic circuit. Figure 8.23 illustrates the effect of gate resizing on the amplitude and width of a 0-to-1 transient at the output of a gate. As illustrated in the figure,

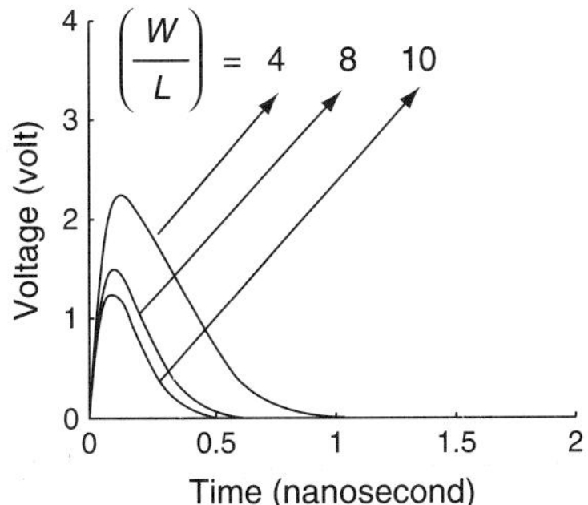

■ **FIGURE 8.23**

Effect of gate resizing on the amplitude/width of SETs [Zhou 2006].

the magnitude and duration of the transient diminish rapidly as the size of the transistor(s) that are collecting this charge. Because there may be multiple transistor diffusions (capacitors) on any given node, the collected charge tends to redistribute among the nodes once it is collected at any given junction. It also follows the RC discharge curve, as the wires connecting these diffusion areas are resistive. Thus, transistors in highly susceptible gates can be resized to disperse the injected charge as quickly as it is collected; hence, the transient does not achieve sufficient magnitude and duration to propagate to the fanout of the gate. SER estimation and assessment of the susceptibility of logic gates to soft errors is performed either through fault injection and simulation [Zhou 2006], wherein the soft error masking factors are evaluated separately, or through symbolic representation [Miskov-Zivanov 2006], wherein all the masking factors are evaluated in a unified approach. In both cases, gate resizing is performed for the most susceptible logic gates (*i.e.*, the ones that contribute the most to the SER of the design). Results in [Zhou 2006] indicate that a 10-fold reduction in the SER of logic circuits implemented using present-day technologies is achievable at an overhead of roughly 30% in area and power consumption.

More recent data reported in [Seifert 2006] indicate that SER may be flattened or even declined with scaling, and it points to the **collection efficiency** of the smaller diffusion area. This is still somewhat controversial as the same effect has not been reported by the majority of the researchers. However, if this theory is proved true, the gate resizing method would have to be considered carefully as it may increase the collection area and hence increase SER.

Netlist transformation for soft error mitigation [Almukhaizim 2006] is based on logic-level design modifications, wherein the logic circuit is modified while preserving the functionality of the netlist, to reduce the probability of sensitizing glitches to an output of the circuit. This has to take into account the vulnerability of all gates in the circuit, minimizing the overall vulnerability of the circuit and not just shifting the SER of one path to another. Design modification is performed using an ATPG-based rewiring method to generate functionally equivalent yet structurally different gate-level circuit implementations. Together with a SER estimation method, the

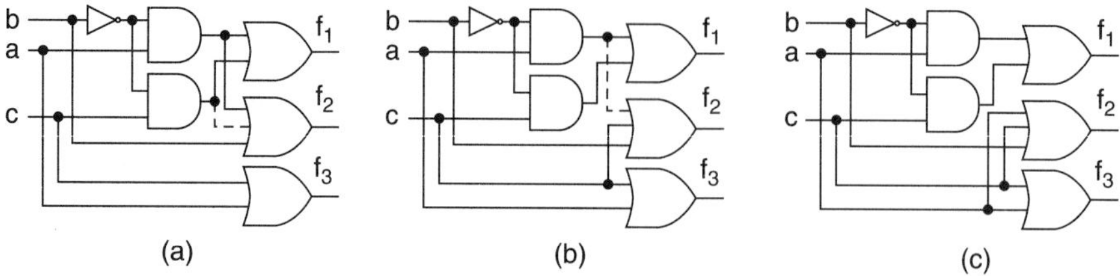

■ **FIGURE 8.24**

Example of rewiring to reduce the soft error failure rate: (a) original circuit, (b) circuit after first rewiring, and (c) circuit after second rewiring.

design is iteratively modified through the selection of rewiring operations that minimize the overall SER of the circuit.

Consider, for example, the logic circuit in Figure 8.24. If rewiring is performed on the dashed wire in the circuit in Figure 8.24a, the dashed wire is replaced with input *c* in the circuit in Figure 8.24b. Performing a second rewiring on the dashed wire in the circuit in Figure 8.24b generates the circuit in Figure 8.24c. The soft error failure rate computed using SERA [Zhang 2006] indicates that the circuit in Figure 8.24c is improved by 5.00% over the circuit in Figure 8.24b and by 9.50% over the original circuit in Figure 8.24a. Because soft errors are mitigated at the logic level, the method proposed in [Almukhaizim 2006] is technology independent and, thus, enables design modifications for SER reduction that are equally effective, independent of the technology to which the circuit will be eventually mapped. Moreover, the mechanisms through which soft errors are mitigated at the logic level are orthogonal to those at the physical level; hence, not only does it provide a better starting point, it may also be applied synergistically with the current state of the art in gate resizing-based soft error mitigation techniques. Results in [Almukhaizim 2006] indicate that it is often possible to reduce the SER of a logic circuit without incurring any overhead in terms of area, delay, or power consumption.

8.5 DEFECT AND ERROR TOLERANCE

A couple of tolerance terminologies have surfaced: **defect tolerance** [Koren 1998] and **error tolerance** [Breuer 2004a, 2004b]. Defect tolerance requires inserting redundancy circuitry in a circuit under test so the circuit can continue correct operation in the presence of defects. An example is adding self-diagnosis and self-repair circuitry or *error-correcting code* (ECC) to ensure correct memory operation. Error tolerance, on the other hand, allows the circuit to continue *acceptable* operation in the presence of errors. An example is the MPEG player where a pixel may be faulty but the player can continue acceptable operation.

In the nanometer design era, the *International Technology Roadmap for Semiconductors* (ITRS) has indicated that the efforts in fabricating a defect-free chip can be extremely expensive [SIA 2006]. Consider random spot defects. Assume a design consists of N submodules each having n unique positions where a defect would cause it to fail its tests. There are D defects uniformly distributed over the submodule such that each defect lands at a unique position. In addition, assume that the number of defects in any submodule is independent of the number of defects in other submodules. The authors in [Breuer 2004a] conducted an analysis of *defect probability* and showed that the probability that an arbitrary position on a submodule is associated with a defect is $p = D/(nN)$, and the probability of having d defects in a given submodule is:

$$P(d) = C(n,d)p^d(1-p)^{n-d}$$

where $C(n,d) = n!/(d!(n-d)!)$. Because $P(d)$ is binomially distributed, the *average number of defects* $E(d)$ in an arbitrary submodule is:

$$E(d) = \lambda = np = D/N$$

For large n and small p, the binomial distribution can be approximated by a Poisson distribution:

$$P(d) = e^{-\lambda}(\lambda^d/d!)$$

Where λ is the failure rate. Assume a submodule is equally likely to be defect-free or defective. We have:

$$P(d=0) = e^{-\lambda}(\lambda^0/0!) = 0.5$$

Thus, $\lambda = 0.693$. Table 8.4 shows some numerical results taken from [Breuer 2004a]. The table shows that for a submodule yield (Y) of 50%, the probability of having exactly one defect at one submodule is 0.35. Even if the submodule yield reaches 80%, the probability of having one or two defects on the submodules is still as high as 20% ($= 0.18 + 0.02$). This suggests that effective yield can increase significantly if the system can accept some defective submodules.

TABLE 8.4 ■ Probability of Having Exact *d* Defects at a Submodule as a Function of Yield (*Y*) for Various Values of Failure Rate λ

d	$\lambda =$ 0.105	$\lambda =$ 0.223	$\lambda =$ 0.357	$\lambda =$ 0.511	$\lambda =$ 0.693	$\lambda =$ 0.916	$\lambda =$ 1.204	$\lambda =$ 1.609	$\lambda =$ 2.303
	Y =	Y =	Y =	Y =	Y =	Y =	Y =	Y =	Y =
0	0.90	0.80	0.70	0.60	0.50	0.40	0.30	0.20	0.10
1	0.09	0.18	0.25	0.31	0.35	0.37	0.36	0.32	0.23
2		0.02	0.04	0.08	0.12	0.17	0.22	0.26	0.27
3			0.01	0.01	0.03	0.05	0.09	0.14	0.20
4						0.01	0.03	0.06	0.12
5							0.01	0.02	0.05
6									0.02
7									0.01

8.5.1 Defect Tolerance

Defect tolerance [Koren 1998] is not new and used to be called *redundancy repair*. A typical defect-tolerant design is shown in Figure 8.25 where two spares (identical modules) are used and a switch is used to select one module. This is in contrast to the TMR system where a voter is used to vote on the majority of the three identical modules (see Figure 8.19). Defect tolerance allows increased process yield (the percentage of good manufactured parts). In the late 1980s, redundancy techniques were used in the manufacture of DRAMs. By using spare rows, columns, or blocks, defective elements can be identified during the manufacturing test process, and fuses are blown to map the spare resources to replace those that are defective. The use of these techniques becomes mandatory as DRAMs scale to the gigabit level. We would not be able to buy and sell DRAMs at the price level we enjoy today without the redundancy repair process.

A similar technique is also used in the manufacture of hard disk drives. During the drive test process, defective sectors are identified, and a map containing those defective sectors is stored permanently on the drive control electronics. These defective sectors are mapped so the drive will not use them to store data. In both situations, spare elements replace the defective elements. Other circuits that have regular structures also can benefit from these defect tolerance techniques, such as FPGAs, cache memories, and processors. For field programmable gate arrays, testing can identify bad cells or routing resources; thus, the mapping and routing tools can work around those obstacles during the mapping process. For processors, it is possible to sell the product with the fewer features (*e.g.*, minus the ***floating-point unit*** [FPU]) upon detection of a fault during manufacturing test; however, defect tolerance has its limit. Not only can the regular circuit elements become faulty because of defects, but the spare elements themselves can also be defective. As the percentage of spare elements increases, therefore, they will occupy more area of the die and the larger die area will result in even more defects, affecting both the normal circuit elements and the spare elements. Therefore, there is a point at which the law of diminishing returns begins to set in [Koren 1990] [Hirase 2001].

In chip design, defect tolerance can also be achieved by using **defect avoidance** in a circuit implementation to improve process yield. These defect avoidance techniques are generally referred to as ***design for manufacturability*** (DFM) or ***design for yield*** (DFY). Layout as well as circuit design methods (*e.g.*, double vias) are

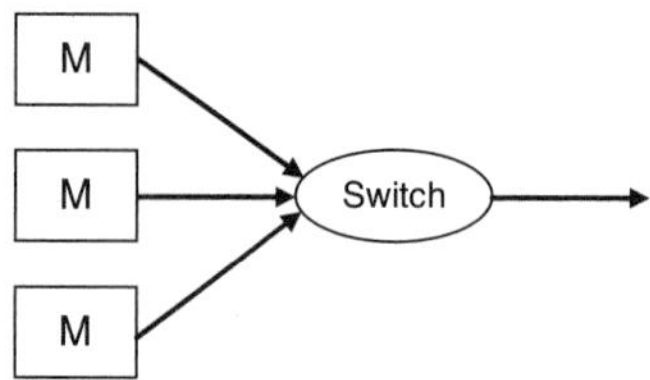

■ FIGURE 8.25

A defect-tolerant design with two spares.

commonly used to reduce the sensitivity of the circuit to fabrication defects and process variations. These DFM and DFY techniques are discussed extensively in Chapter 9.

8.5.2 Error Tolerance

Error tolerance is a different concept. Conventional wisdom would seem to suggest that if an error is injected and trapped in the logic, it will not perform to its intended functionality; however, some logic functionality defies that conventional wisdom. An example is the processing or storage of any kind of multimedia data (*e.g.*, video, pictures, or music). Compression techniques (*e.g.*, JPEG, MPEG, MP3) are generally used for these types of data, and these compression algorithms are *lossy* in nature—that is, some details of the raw data are lost in the compression process. A stuck bit in the least significant portion of the data word may or may not be distinguishable from artifacts with the compression process [Breuer 2004a, 2004b]. Also, these kinds of data appeal to the senses, and our senses are usually not keen enough to spot minute variances (unless one observes them with an expert's eyes or ears). This sort of error tolerance is application-specific, and general-purpose machines that are tolerant of all kinds of errors have not yet been designed. For example, if an error occurs at the most significant bits of the compressed data or within the control logic instead of within the data, the data processing can still lead to incorrect processing and may yield an unacceptable picture or sound.

In essence, the concept of error tolerance is also not new. For instance, microprocessors with different speeds have been sold at different prices in the marketplace, even though they were produced from the same production line. RAMs or flash memories that are fault-free can be sold to any customer, whereas those that are slightly defective might go to a manufacturer of digital answering machines or video games. The main objective of error tolerance is to increase the effective yield of a process by identifying defective but acceptable chips. This lies in the development of an accurate method to estimate **error rate** with a specified level of confidence and an effective method to predict yield increase based on the results of *error-rate estimation* [Lee 2005] [Hsieh 2006].

In [Lee 2005], the authors proposed a **fault-oriented test methodology** to enhance effective yield based on error-rate analysis. The main idea is illustrated in Figure 8.26, where a set of fault models is assumed. Before actually carrying

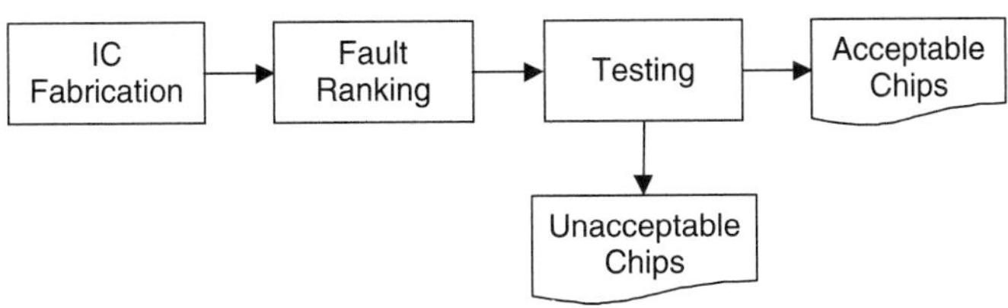

FIGURE 8.26

Fault-oriented test methodology.

out production testing, the *error rate* of each modeled fault is estimated, and a set of *acceptable faults* is identified based on their error rates. These acceptable faults are then excluded from ATPG. Because chips containing only acceptable faults can now pass through the manufacturing test floor, the effective yield of these chips increases.

Instead of using fault models to estimate error rates, the authors in [Hsieh 2006] proposed an **error-oriented test methodology**, where focus is on errors produced by defective chips rather than on modeled faults. Because no fault model is required, the proposed test methodology can be applied to any circuit without knowing its detailed structure. In addition, because the entire circuit is processed at one time, the estimation process can be greatly simplified and becomes easier to carry out. The proposed error-oriented test methodology is illustrated in Figure 8.27. First, a sampling-based method is used to estimate the error rates of these chips. The estimated results are then used to determine the acceptability of the faulty chips. With this test methodology, defective chips with high error rates are considered unacceptable and will be rejected, whereas chips with low error rates can be placed into different acceptable classes depending on their error rates, thus increasing the effective yield of the manufactured chips.

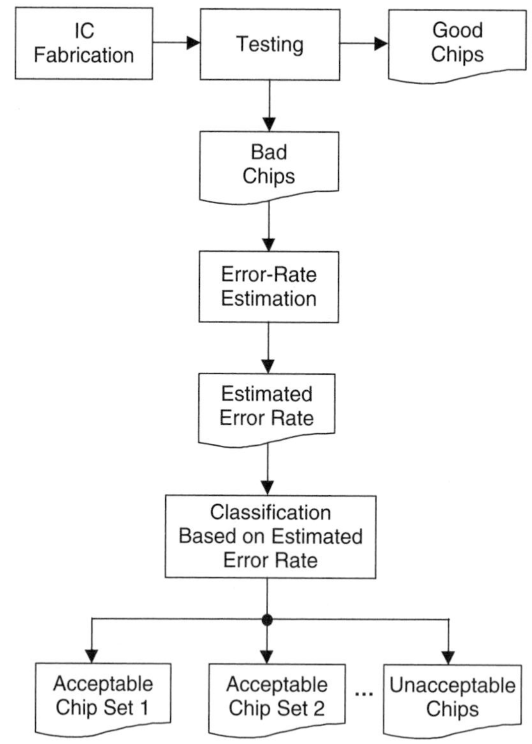

FIGURE 8.27

Error-oriented test methodology.

As for the trend, error tolerance is application-oriented; it is especially useful for audio/video or multimedia applications. The measures for error tolerance may also include factors other than error rate [Breuer 2004a]. Error tolerance may also play its important role in *analog and mixed-signal* (AMS) circuits because these circuits can be naturally graded based on their performance.

8.6 CONCLUDING REMARKS

Advances in semiconductor manufacturing technology have allowed the integration of a billion transistors in a nanometer design. The nanometer design has faced many test challenges [SIA 2005, 2006]. Growing complexity and defect mechanisms of fabricating the nanometer design have made the circuit even more vulnerable than earlier generations to physical failures caused during manufacturing and susceptible to radiation during system applications. In this chapter, we first presented several promising techniques for testing nanometer designs that can cope with physical failures caused by signal integrity, defects, and process variations during manufacturing. These test techniques are developed for reliability screen or DPM reduction. A few include on-chip hardware for stressing or special reliability measurements. These defect avoidance techniques are generally referred to as ***design for manufacturability*** (DFM).

With the process node now scaling down to 65 nanometers and below, DFM alone is not enough. Many new defect mechanisms, such as copper-related defects and defects as a result of optical effects, become much more likely [Gizopoulos 2006]. Any single-event upsets can increase logic and memory soft error rates. It is thus becoming crucial to ensure that the chips can still function at the end system in the presence of these defects and soft errors, especially when the chips are to be installed into airplanes, pacemakers, or cars for safety-critical concerns. This chapter then further covered a number of error-resilient and defect-tolerant designs embedded on-chip to tolerate soft errors and defects. Although these schemes may require additional area overhead, they pave the way to develop more advanced error-resilient and defect-tolerant scan and logic BIST architectures to cope with the physical failures of the nanometer age. These error resilience and defect tolerance schemes are now referred to as ***design for reliability*** (DFR).

8.7 EXERCISES

8.1 **(Signal Integrity)** Figure 8.10 shows one readout circuit for reading the signal integrity loss information through a scan chain. How do you change this architecture to achieve the following?

a. Less integrity test overhead.

b. More accuracy in terms of pinpointing the problematic wire/bus.

In each case, draw your architecture and estimate the duration of the signal integrity test session.

8.2 **(Signal Integrity)** For the equation given in Section 8.2.3 that approximates the frequency of a ring oscillator, do the following:

a. Use a partial derivative to analytically determine the sensitivity of f_{RO} with respect to V_t and T_{ox}.

b. Use the library and technology parameters available to your SPICE tool, replace the subset of parameters needed, and find the frequency shift for a 15% increase of V_t and a 20% reduction in T_{ox}.

c. Run SPICE for the same variations, and compare the frequency shift with the analytical results found in part (b).

8.3 **(Memory Repair)** What is the minimum cost of using Hamming ECC (a single-bit error correction and double-bit error detection code) to protect a 64-bit-wide memory word (*e.g.*, the number of check bits over the number of 64-bit word)?

8.4 **(Redundancy)** What is the marginal increase of redundancy if the number of redundant elements (rows, columns, or blocks) increases by 2×? (*Hint:* You have to look at the yield decrease because the area has increased by 2×.)

8.5 **(Adaptive Design)** What other adaptive mechanisms are available that can reduce power consumption besides the frequency and voltage knobs mentioned in Section 8.3.4?

8.6 **(Soft Errors**) Consider the original circuit in Figure 8.24a and the final circuit in Figure 8.24c. Verify that the probability of sensitizing an error at the inputs of the gate driving f_2 in the final circuit is higher than the probability of sensitizing an error at the inputs of the gate driving f_2 in the original circuit. What is the improvement?

8.7 **(Fault Tolerance)** Calculate the reliability R of the defect-tolerant design with two spares shown in Figure 8.25. Assume that the probability of the module M nonoperational is p, where $0 < p < 1$. Then, calculate the reliability of the TMR system given in Figure 8.19, and plot a chart to explain which system will yield a more reliable operation.

8.8 **(Defect and Error Tolerance)** Refer to Table 8.4. Assume that there are 10 identical modules in a design. Each module has 10,000 unique positions where a defect can exist and make the module behave imperfectly, and 20 defects are uniformly distributed over the module. Let $P(d)$ be the *probability mass function* of the number of defects d on an arbitrary module, where d is a random variable. Calculate the probability, p, that an arbitrary position on a module is associated with a defect, $P(d = 1)$, and the probability of having zero defect, $P(d = 0)$, in a given module. What's the yield of the design?

8.9 **(Defect and Error Tolerance)** Repeat Exercise 8.8. Because $P(d)$ is binomially distributed, calculate the average number of defects, $E(d)$, in an arbitrary module. For large n and small p, we can approximate the *binomial* *distribution P(d)* by a *Poisson distribution* with failure rate λ. Assume that the design is equally likely to be defect-free or defective. Calculate the failure rate λ and the probability of having one defect, $P(d = 1)$, in a given module. Explain the difference between the approximated probability and the probability derived in Exercise 8.8.

8.10 

(A Design Practice) Write a C/C++ program or use the fault simulator on the Web site to estimate the transient failure probability by applying 1000 random patterns to a 4-bit carry-ripple adder implemented in combinational logic gates. Assume soft errors are solely caused by single-event upsets, which can randomly cause one internal net to flip its value from 0 to 1 or 1 to 0 during one clock cycle.

Acknowledgments

The authors wish to thank Professor Yiorgos Makris of Yale University for contributing the Circuit-Level Approaches in the Soft Errors section; Professor Mohammad Tehranipoor of University of Connecticut and Professor Saraju P. Mohanty of University of North Texas for reviewing the Signal Integrity section; Dr. Jonathan T.-Y. Chang of Intel, François-Fabien Ferhani of Stanford University, Professor James C.-M. Li of National Taiwan University, Praveen K. Parvathala and Michael Spica of Intel, Professor Michael S. Hsiao of Virginia Tech, Phil Nigh of IBM, and Dr. Brion Keller of Cadence Design Systems for reviewing the Structural Tests and Defect-Based Tests sections; Praveen K. Parvathala and Dr. Li Chen of Intel for reviewing the Functional Tests section; Dr. Shih-Lien Lu of Intel for reviewing the Process Sensors and Adaptive Design section; Professor Subhasish Mitra of Stanford University for reviewing the Soft Errors section; Professor Kuen-Jong Lee of National Cheng Kung University and Michael Spica of Intel for proofreading the Defect and Error Tolerance section; Professor Shi-Yu Huang of National Tsing Hua University and Dr. Srikanth Venkataraman of Intel for providing helpful comments; as well as Teresa Chang of SynTest Technologies for drawing most of the figures.

References

R8.0 Books

[Abramovici 1994] M. Abramovici, M. A. Breuer, and A. D. Friedman, *Digital Systems Testing and Testable Design*, Revised Printing, IEEE Press, Piscataway, NJ, 1994.

[Bushnell 2000] M. L. Bushnell and V. D. Agrawal, *Essentials of Electronic Testing for Digital, Memory & Mixed-Signal VLSI Circuits*, Springer, Boston, 2000.

[Dally 1998] W. J. Dally and J. W. Poulton, *Digital Systems Engineering*, Cambridge University Press, London, 1998.

[Gizopoulos 2006] D. Gizopoulos, editor, *Advances in Electronic Testing: Challenges and Methodologies Series: Frontiers in Electronic Testing,* Springer, Boston, 2006.

[Jha 2003] N. Jha and S. Gupta, *Testing of Digital Systems*, Cambridge University Press, London, 2003.

[Lala 2001] P. K. Lala, *Self-Checking and Fault-Tolerant Digital Design*, Morgan Kaufmann, San Francisco, 2001.

[Mourad 2000] S. Mourad and Y. Zorian, *Principles of Testing Electronic Systems*, John Wiley & Sons, Somerset, NJ, 2000.

[Nicolici 2003] N. Nicolici and B. Al-Hashimi, *Power-Constrained Testing of VLSI Circuits*, Kluwer Academic, Norwell, MA, 2003.

[Peterson 1972] W. W. Peterson and E. J. Weldon, Jr., *Error-Correcting Codes*, MIT Press, Cambridge, MA, 1972.

[Pradhan 1996] D. K. Pradhan, *Fault-Tolerant Computer System Design*, Prentice Hall, Upper Saddle River, NJ, 1996.

[Rabaey 1996] J. Rabaey, *Digital Integrated Circuits*, Prentice Hall, Upper Saddle River, NJ, 1996.

[Senthinatharr 1994] R. Senthinatharr and J. L. Prince, *Simultaneous Switching Noise of CMOS Devices and Systems*, Kluwer Academic, Norwell, MA, 1994.

[Siewiorek 1998] D. Siewiorek and R. S. Swarz, *Reliable Computer Systems: Design and Evaluation, Third Edition*, AK Peters, Wellesley MA, 1998.

[Wang 2006] L.-T. Wang, C.-W. Wu, and X. Wen, editors, *VLSI Test Principles and Architectures: Design for Testability*, Morgan Kaufmann, San Francisco, 2006.

[Ziegler 2004] J. F. Ziegler and H. Puchner, *SER-History, Trends and Challenges, A Guide for Designing with Memory ICs*, Cypress Semiconductor, San Jose, CA, 2004.

R8.1 Introduction

[Aitken 2004] R. Aitken, New defect behavior at 130nm and beyond, *Informal Proc. European Test Symp.*, pp. 279–284, May 2004.

[Baumann 2005] R. Baumann, Soft errors in advanced computer systems, *IEEE Design & Test of Computers*, pp. 258–266, May/June 2005.

[Breuer 2004a] M. Breuer, S. Gupta, and T. M. Mak, Defect and error tolerance in the presence of massive numbers of defects, *IEEE Design & Test of Computers*, pp. 216–227, May/June 2004.

[Chau 2004] R. Chau, S. Datta, M. Doczy, B. Doyle, J. Kavalieros, and M. Metz, High-k/metal-gate stack and its MOSFET characteristics, *IEEE Electron Device Letters*, 25(6), pp. 408–410, June 2004.

[Clear Shape 2007] Clear Shape Technologies, Inc., www.clearshape.com, 2007.

[Guardiani 2004] C. Guardiani, N. Dragone, and P. McNamara, Proactive design for manufacturing (DFM) for nanometer SoC design, in *Proc. Custom Integrated Circuits Conf.*, pp. 309–316, October 2004.

[May 1979] T. C. May and M. H. Woods, Alpha-particle-induced soft errors in dynamic memories, *IEEE Trans. on Electron Devices*, ED-26(1), pp. 2–9, January 1979.

[McCluskey 1988] E. J. McCluskey and F. Buelow, IC quality and test transparency, in *Proc. Int. Test Conf.*, pp. 295–301, September 1988.

[Moore 1965] G. Moore, Cramming more components onto integrated circuits, *Electronics*, pp. 114–117, April 19, 1965.

[Sengupta 1999] S. Sengupta, S. Kundu, S. Chakravarty, P. Parvathala, R. Galivanche, G. Kosonocky, M. Rodgers, and T. M. Mak, Defect-based test: A key enabler for successful migration to structural test, *Intel Technology J.*, pp. 1–14, Q1 1999.

[Tyagi 2000] S. Tyagi, M. Alavi, R. Bigwood, T. Bramblett, J. Brandenburg, W. Chen, B. Crew, M. Hussein, P. Jacob, C. Kenyon, C. Lo, B. McIntyre, Z. Ma, P. Moon, P. Nguyen, L. Rumaner, R. Schweinfurth, S. Sivakumar, M. Stettler, S. Thompson, B. Tufts, J. Xu, S. Yang, and M. Bohr, A 130 nm generation logic technology featuring 70 nm transistors, dual Vt transistors and 6 layers of Cu interconnects, *Digest of Papers, IEEE Int. Electron Devices Meeting*, pp. 567–570, December 2000.

[Williams 1981] T. W. Williams and N. C. Brown, Defect level as a function of fault coverage, *IEEE Trans. on Computers*, 30(12), pp. 987–988, December 1981.

[Williams 1983] T. W. Williams and K. P. Parker, Design for testability: A survey, in *Proc. of the IEEE*, 71(1), pp. 98–112, January 1983.

R8.2 Signal Integrity

[Acharyya 2005] D. Acharyya and J. Plusquellic, Hardware results demonstrating defect detection using power supply signal measurements, in *Proc. VLSI Test Symp.*, pp. 433–438, May 2005.

[Agarwal 2003] A. Agarwal, D. Blaauw, and V. Zolotov, Statistical timing analysis for intra-die process variations with spatial correlations, in *Proc. Int. Conf. on Computer-Aided Design*, pp. 900–907, November 2003.

[Agarwal 2005] A. Agarwal, B. Paul, S. Mukhopadhyay, and K. Roy, Process variation in embedded memories: Failure analysis and variation aware architecture, *IEEE J. of Solid-State Circuits*, 40(9), pp. 1804–1814, September 2005.

[Ahmed 2003] N. Ahmed, M. H. Tehranipour, and M. Nourani, Extending JTAG for testing signal integrity in SoCs, in *Proc. Design, Automation, and Test in Europe*, pp. 218–223, March 2003.

[Attarha 2002] A. Attarha and M. Nourani, Test pattern generation for signal integrity faults on long interconnects, in *Proc. VLSI Test Symp.*, pp. 336–341, April 2002.

[Azizi 2005] N. Azizi, M. Khellah, V. De, and F. Najm, Variations-aware low-power design with voltage scaling, in *Proc. Design Automation Conf.*, pp. 529–534, June 2005.

[Bai 2000] X. Bai, S. Dey, and J. Rajski, Self-test methodology for at-speed test of crosstalk in chip interconnects, in *Proc. Design Automation Conf.*, pp. 619–624, June 2000.

[Bai 2001] G. Bai, S. Bobba, and I. Haji, Maximum power supply noise estimation in VLSI circuits using multimodal genetic algorithms, in *Proc. Int. Conf. on Electronics, Circuits and Systems*, 3, pp. 1437–1440, September 2001.

[Bhushan 2006] M. Bhushan, A. Gattiker, M. Ketchen, and K. Das, Ring oscillators for CMOS process tuning and variability control, *IEEE Trans. on Semiconductor Manufacturing*, 19(1), pp. 10–18, February 2006.

[Borkar 2004a] S. Borkar, Microarchitecture and design challenges for gigascale integration, *Keynote Speech in Micro 2004 Conf.*, December 2004.

[Borkar 2004b] S. Borkar, T. Karnik, and V. De, Design and reliability challenges in nanometer technologies, in *Proc. Design Automation Conf.*, p. 75, December 2004.

[Bowman 2000] K. Bowman, X. Tang, J. Eble, and J. Meindl, Impact of extrinsic and intrinsic parameter fluctuations on CMOS, *IEEE J. of Solid State Circuits*, 35(8), pp. 1186–1193, August 2000.

[Bowman 2001] K. Bowman and J. Meindl, Impact of within-die parameter fluctuations on future maximum clock frequency distributions, in *Proc. Custom Integrated Circuits Conf.*, pp. 229–232, May 2001.

[Cao 2005] Y. Cao and L. Clark, Mapping statistical process variations toward circuit performance variability: An analytical modeling approach, in *Proc. Design Automation Conf.*, pp. 658–663, June 2005.

[Chan 2001] A. Chan and G. Roberts, A synthesizable, fast and high-resolution timing measurement device using a component-invariant vernier delay line, in *Proc. Int. Test Conf.*, pp. 858–867, October 2001.

[Chen 1998] W. Chen, S. Gupta, and M. A. Breuer, Test generation in VLSI circuits for crosstalk noise, in *Proc. Int. Test Conf.*, pp. 641–650, October 1998.

[Chen 1999] W. Chen, S. Gupta, and M. A. Breuer, Test generation for crosstalk-induced delay in integrated circuits, in *Proc. Int. Test Conf.*, pp. 191–200, October 1999.

[Chen 2001] L. Chen, X. Bai, and S. Dey, Testing for interconnect crosstalk defects using on-chip embedded processor cores, in *Proc. Design Automation Conf.*, pp. 317–322, June 2001.

[Chen 2005] Q. Chen, S. Mukhopadhyay, H. Mahmoodi, and K. Roy, Process variation tolerant online monitor for robust systems, in *Proc. IEEE Int. On-Line Testing Symp.*, pp. 171–176, July 2005.

[Cuviello 1999] M. Cuviello, S. Dey, X. Bai, and Y. Zhao, Fault modeling and simulation for crosstalk in system-on-chip interconnects, in *Proc. Int. Conf. on Computer-Aided Design*, pp. 297–303, November 1999.

[Ding 2005] Q. Ding, R. Luo, and Y. Xie, Impact of process variation on soft error vulnerability for nanometer VLSI circuits, in *Proc. Int. Conf. on ASIC*, pp. 1117–1121, October 2005.

[Dryden 2005] C. Dryden, Survey of design and process failure modes for high-speed SerDes in nanometer CMOS, in *Proc. VLSI Test Symp.*, pp. 285–291, May 2005.

[Fang 1998] P. Fang, J. Tao, J. Chen, and C. Hu, Design in hot carrier reliability for high performance logic applications, in *Proc. Custom Integrated Circuits Conf.*, pp. 25.1.1–25.1.7, October 1998.

[Ghanta 2005] P. Ghanta, S. Vrudhula, R. Panda, and J. Wang, Stochastic power grid analysis considering process variations, in *Proc. Design, Automation, and Test in Europe Conf.*, pp. 964–969, March 2005.

[Gorbics 1997] M. Gorbics, J. Kelly, K. Roberts, and R. Summer, A high resolution multihit time to digital converter integrated circuits, *IEEE Trans. on Nuclear Science,* 44(3), pp. 379–384, June 1997.

[Hatzilambrou 1996] M. Hatzilambrou, A. Neureuther, and C. Spanos, Ring oscillator sensitivity to spatial process variation, *Int. Workshop on Statistical Metrology,* 1996.

[Hunter 1999] W. Hunter, The statistical dependence of oxide failure rates on vdd and tox variations with applications to process design, circuit design and end use, in *Proc. Reliability Physics Symp.*, pp. 72–81, March 1999.

[IEEE 1149.1-2001] IEEE Std. 1149.1-2001, *IEEE Standard Test Access Port and Boundary Scan Architecture*, Institute of Electrical and Electronics Engineers, New York, 2001.

[IEEE 1149.4-1999] IEEE Std. 1149.4-1999, *IEEE Standard for a Mixed Signal Test Bus*, Institute of Electrical and Electronics Engineers, New York, 1999.

[IEEE 1149.6-2003] IEEE Std. 1149.6-2003 *IEEE Standard for Boundary-Scan Testing of Advanced Digital Networks*, Institute of Electrical and Electronics Engineers, New York, 2003.

[Jess 2003] J. Jess, K. Kalafala, S. R. Naidu, R. Otten, and C. Visweswariah, Statistical timing for parametric yield prediction of digital integrated circuits, in *Proc. Design Automation Conf.*, pp. 343–347, June 2003.

[Jiang 1997] Y.-M. Jiang, K.-T. Cheng, and A. Krstic, Estimation of maximum power and instantaneous current using a genetic algorithm, in *Proc. Custom Integrated Circuits Conf.*, pp. 135–138, October 1997.

[Jiang 2001] Y. Jiang and K.-T. Cheng, Vector generation for power supply noise estimation and verification of deep submicron designs, *IEEE Trans. on VLSI*, 9(2), pp. 329–340, April 2001.

[Kim 2005] C. Kim, K. Roy, S. Hsu, R. Krishnamurthy, and S. Borkar, An on-die CMOS leakage current sensor for measuring process variation in sub-90nm generations, in *Proc. Int. Conf. on Integrated Circuit Design and Technology*, pp. 221–222, May 2005.

[Kristic 2001] A. Kristic, Y.-M. Jiang, and K.-T. Cheng, Pattern generation for delay testing and dynamic timing analysis considering power-supply noise effects, *IEEE Trans. on Computer-Aided Design*, 20(3), pp. 416–425, March 2001.

[Lampinen 2002] H. Lampinen and O. Vainio, Current-sensing completion detection method for standard cell based digital system design, in Proc. *Int. Symp. on Circuits and Systems*, 5, pp. 117–120, May 2002.

[Lee 2001] J. Lee, Y. Huh, P. Bendix, and S. Kang, Understanding and addressing the noise induced by electrostatic discharge in multiple power supply systems, in *Proc. Int. Conf. on Computer Design*, pp. 406–411, September 2001.

[Lu 2004] X. Lu, Z. Li, W Qui, D. Walker, and W. Shi, Longest path selection for delay test under process variation, in *Proc. Asia and South Pacific Design Automation Conf.*, pp. 98–103, January 2004.

[Mehrotra 1998] V. Mehrotra, S. Nassif, D. Boning, and J. Chung, Modeling the effects of manufacturing variation on high-speed microprocessor interconnect performance, in *Proc. Int. Electron Device Meeting*, pp. 28.6.1–28.6.4, December 1998.

[Mohanty 2007] S. P. Mohanty, E. Kougianos, and R. N. Mahapatra, A comparative analysis of gate leakage and performance of high-K nanoscale CMOS logic gates, in *Proc. ACM/IEEE Int. Workshop on Logic and Synthesis*, May 2007.

[MOSIS 2007] MOSIS Service, www.mosis.org/Technical/process-monitor.html, 2007.

[Naffziger 1999] S. Naffziger, Design methodologies for interconnects in GHz$^+$ ICs, tutorial lecture in *Proc. Int. Solid-State Circuits Conf.*, February 1999.

[Narasimha 2006] U. Narasimha, B. Abraham, and NS Nagaraj, Statistical analysis of capacitance coupling effects on delay and noise, in *Proc. Int. Symp. on Quality Electronic Design*, pp. 795–800, March 2006.

[Nassif 2004] S. Nassif, D. Boning, and N. Hakim, The care and feeding of your statistical static timer, in *Proc. Int. Conf. on Computer-Aided Design*, pp. 138–139, November 2004.

[Nourani 2002] M. Nourani and A. Attarha, Signal integrity: Fault modeling and testing in high-speed SoCs, *J. of Electronic Testing: Theory and Applications*, 18(4–5), pp. 539–554, August/October 2002.

[Nourani 2005] M. Nourani and A. Radhakrishnan, Power-supply noise in SoCs: ATPG, estimation and control, in *Proc. Int. Test Conf.*, Paper 22.1, November 2005.

[Nourani 2006] M. Nourani and A. Radhakrishnan, Modeling and testing process variation in nanometer CMOS, in *Proc. Int. Test Conf.*, Paper 7.2, October 2006.

[Orshansky 2000] M. Orshansky, L. Milor, P. Chen, K. Keutzer, and C. Hu, Impact of systematic spatial intra-chip gate length variability on performance of high-speed digital circuits, in *Proc. Int. Conf. on Computer-Aided Design*, pp. 62–67, November 2000.

[Samaan 2003] S. Samaan, Parameter variation probing technique, U.S. Patent No. 6,535,013, March 18, 2003.

[Sato 1998] H. Sato, H. Kunitomo, K. Tsuneno, K. Mori, and H. Masuda, Accurate statistical process variation analysis for 0.25-μm CMOS with advanced TCAD methodology, *IEEE Trans. on Semiconductor Manufacturing*, 11(4), pp. 575–582, November 1998.

[Tehranipour 2003a] M. H. Tehranipour, N. Ahmed, and M. Nourani, Testing SoC interconnects for signal integrity using boundary scan, in *Proc. VLSI Test Symp.*, pp. 163–168, April 2003.

[Tehranipour 2003b] M. Tehranipour, N. Ahmed, and M. Nourani, Multiple transition model and enhanced boundary scan architecture to test interconnects for signal integrity, in *Proc. Int. Conf. on Computer Design*, pp. 554–559, October 2003.

[Tehranipour 2004] M. Tehranipour, N. Ahmed, and M. Nourani, Testing SoC interconnects for signal integrity using extended JTAG architecture, *IEEE Trans. on Computer-Aided Design*, 23(5), pp. 800–811, May 2004.

[Ukei 2001] T. Ukei and H. Aoyagi, Monitor TEG test circuit, U.S. Patent No. 6,239,603, May 29, 2001.

[Vazquez 2004] J. Vazquez and J. Gyvez, Power supply noise monitor for signal integrity faults, in *Proc. Design, Automation and Test in Europe Conf.*, pp. 1406–1407, February 2004.

[Visweswariah 2003] C. Visweswariah, Death, taxes and failing chips, in *Proc. Design Automation Conf.*, pp. 343–347, June 2003.

[Whetsel 1996] L. Whetsel, Proposal to simplify development of a mixed signal test standard, in *Proc. Int. Test Conf.*, pp. 400–409, October 1996.

[Yang 2001] S. Yang, C. Papachristou, and M. Tabib-Azar, improving bus test via *IDDT* and boundary scan, in *Proc. Design Automation Conf.*, pp. 307–312, June 2001.

[Zhao 2000a] S. Zhao and K. Roy, Estimation of switching noise on power supply lines in deep sub-micron CMOS circuits, in *Proc. Int. Conf. on VLSI Design*, pp. 168–173, January 2000.

[Zhao 2000b] S. Zhao, K. Roy, and C. Koh, Estimation of inductive and resistive switching noise on power-supply network in deep submicron CMOS circuits, in *Proc. Int. Conf. on Computer Design*, pp. 65–72, September 2000.

R8.3 Manufacturing Defects, Process Variations, and Reliability

[Agarwal 2005] A. Agarwal, B. C. Paul, S. Mukhopadhyay, and K. Roy, Process variation in embedded memories: Failure analysis and variation aware architecture, *IEEE J. of Solid-State Circuits*, 40(9), pp. 1804–1814, September 2005.

[Ahmed 2005] N. Ahmed, M. Tehranipoor, and V. Jayaram, A novel framework for faster-than-at-speed delay test considering IR-Drop effects, in *Proc. Int. Conf. on Computer-Aided Design*, pp. 439–444, November 2005.

[Aitken 2004] R. Aitken, New defect behavior at 130nm and beyond, *Informal Proc. European Test Symp.*, pp. 279–284, May 2004.

[Ali 2006] N. B. Z. Ali, M. Zwolinski, B. M. Al-Hashimi, and P. Harrod, Dynamic voltage scaling aware delay fault testing, in *Proc. IEEE European Test Symp.*, pp. 15–20, May 2006.

[Amodeo 2005] M. Amodeo and B. Cory, Defining faster-than-at-speed delay tests, *Cadence Nanometer Test Quarterly*, 2(2), www.cadence.com/newsletters/nanometer_test, May 2005.

[ARM 2007] ARM Ltd., 1176JZ(F)-S Documentation, 2007, www.arm.com/products/CPUs/ARM1176.html.

[Barnhart 2004] C. F. Barnhart, What is true-time delay test?, *Cadence Nanometer Test Quarterly*, 1(2), pp. 2–7, June 2004.

[Bhattacharjee 2004] S. Bhattacharjee and D. K. Pradhan, LPRAM: A novel low-power high-performance RAM design with testability and scalability, *IEEE Trans. on Computer-Aided Design*, 23(5), pp. 637–651, May 2004.

[Boppana 1999] V. Boppana, R. Mukherjee, J. Jain, M. Fujita, and P. Bollineni, Multiple error diagnosis based on xlists, in *Proc. Design Automation Conf.*, pp. 660–665, June 1999.

[Butler 2000] K. M. Butler and J. Saxena, An empirical study on the effects of test type ordering on overall test efficiency, in *Proc. IEEE Int. Test Conf.*, pp. 408–416, October 2000.

[Butler 2004] K. M. Butler, J. Saxena, T. Fryars, G. Hetherington, A. Jain, and J. Levis, Minimizing power consumption in scan testing: Pattern generation and DFT techniques, in *Proc. Int. Test Conf.*, pp. 355–364, October 2004.

[Chang 1996a] J. T.-Y. Chang and E. J. McCluskey, Quantitative analysis of very-low-voltage testing, in *Proc. VLSI Test Symp.*, pp. 332–337, April 1996.

[Chang 1996b] J. T.-Y. Chang and E. J. McCluskey, Detecting delay flaws by very-low-voltage testing, in *Proc. Int. Test Conf.*, pp. 367–376, October 1996.

[Chang 1997] J. T.-Y. Chang and E. J. McCluskey, SHOrt voltage elevation (SHOVE) test for weak CMOS ICs, in *Proc. VLSI Test Symp.*, pp. 446–451, April 1997.

[Chen 2003] L. Chen, S. Ravi, A. Raghunathan, and S. Dey, A scalable software-based self-test methodology for programmable processors, in *Proc. Design Automation Conf.*, pp. 548–553, June 2003.

[Cherubal 2001] S. Cherubal and A. Chatterjee, Test generation based diagnosis of device parameters for analog circuits, in *Proc. Design, Automation, and Test Conf. in Europe*, pp. 596–602, March 2001.

[Cho 2005] K. Y. Cho, S. Mitra, and E. J. McCluskey, Gate exhaustive testing, in *Proc. Int. Test Conf.*, pp. 771–777, November 2005.

[Daasch 2001] R. Daasch, K. Cota, J. McNames, and R. Madge, Neighbor selection for variance reduction in I_{DDQ} and other parametric data, in *Proc. Int. Test Conf.*, pp. 1240–1248, October 2001.

[Edmondson 2006] R. Edmondson, G. Iovino, and R. Kacprowicz, Optimizing the cost of test at Intel using per device data, in *Proc. Int. Test Conf.*, Paper 1.1, October 2006.

[Ferhani 2006] F.-F. Ferhani and E. J. McCluskey, Classifying bad chips and ordering test sets, in *Proc. Int. Test Conf.*, Lecture 1.2, October 2006.

[Flautner 2001] K. Flautner, S. Reinhardt, and T. Mudge, Automatic performance setting for dynamic voltage scaling, in *Proc. Int. Conf. on Mobile Computing and Networking*, pp. 260–271, May 2001.

[Foote 1997] T. G. Foote, D. E. Hoffman, W. V. Huott, T. J. Koprowski, B. J. Robbins, and M. P. Kusko, Testing the 400MHz IBM generation-4 CMOS chip, in *Proc. Int. Test Conf.*, pp. 106–114, November 1997.

[Franco 1995] P. Franco, W. D. Farwell, R. L. Stokes, and E. J. McCluskey, An experimental chip to evaluate test techniques chip and experiment design, in *Proc. Int. Test Conf.*, pp. 653–662, October 1995.

[Gatej 2002] J. Gatej, L. Song, C. Pyron, R. Raina, and T. Munns, Evaluating ATE features in terms of test escape rates and other cost of test culprits, in *Proc. Int. Test Conf.*, pp. 1040–1048, October 2002.

[Gattiker 1996] A. Gattiker and W. Maly, Current signatures, in *Proc. IEEE VLSI Test Symp.*, pp. 112–117, April 1996.

[Girard 2002] P. Girard, survey of low-power testing of VLSI circuits, *IEEE Design & Test of Computers*, 19(3), pp. 82–92, May/June 2002.

[Hamada 2006] S. Hamada, T. Maeda, A. Takatori, Y. Noduyama, and Y. Sato, Recognition of sensitized longest paths in transition-delay test, in *Proc. Int. Test Conf.*, Paper 11.1, October 2006.

[Hao 1993] H. Hao and E. J. McCluskey, Very-low-voltage testing for weak CMOS logic IC's, in *Proc. Int. Test Conf.*, pp. 275–284, October 1993.

[Hawkins 1985] C. F. Hawkins and J. M. Soden, Electrical characteristics and testing considerations for gate oxide shorts in CMOS ICs, in *Proc. Int. Test Conf.*, pp. 544–555, November 1985.

[Hawkins 1986] C. F. Hawkins and J. M. Soden, Reliability and electrical properties of gate oxide shorts in CMOS ICs, in *Proc. Int. Test Conf.*, pp. 443–451, September 1986.

[Hawkins 1994] C. F. Hawkins, J. M. Soden, A. W. Righter, and F. J. Ferguson, Defect classes: An overdue paradigm for CMOS IC testing, in *Proc. Int. Test Conf.*, pp. 413–425, October 1994.

[Intel 2003] Intel Corp., *Intel Xscale Core Developer's Manual*, 2003, http://developer.intel.com/design/intelxscale.

[Iyengar 2006] V. Iyengar, T. Yokota, K. Yamada, T. Anemikos, B. Bassett, M. Degregorio, R. Farmer, G. Grise, M. Johnson, D. Milton, M. Taylor, and F. Woytowich, At-speed structural test for high-performance ASICs, in *Proc. Int. Test Conf.*, Paper 2.4, October 2006.

[Ko 2006] L.-Y. Ko, S.-Y. Huang, J.-L. Chiou, and H.-C. Cheng, Modeling and testing of intra-cell bridging defects using butterfly structure, in *Proc. VLSI Design, Automation, and Testing*, pp. 159–162, April 2006.

[Krishnamoorthy 2006] A. Krishnamoorthy, and A. M. Detofsky, Mapping variations in local temperature and local power supply voltage that are present during operation of an integrated circuit, U.S. Patent No. 7,071,723, July 4, 2006.

[Kruseman 2004] B. Kruseman, A. K. Majhi, G. Gronthoud, and S. Eichenberger, On hazard-free patterns for fine-delay fault testing, in *Proc. IEEE Int. Test Conf.*, pp. 213–222, October 2004.

[Lee 2002] J.-H. Lee, S.-H. Park, K.-M. Lee, K.-S. Youn, Y.-J. Park, C.-J. Choi, T.-Y. Seong, and H.-D. Lee, A study of stress-induced p^+/n salicided junction leakage failure and optimized process conditions for sub-0.15 μm CMOS technology, *IEEE Trans. on Electron Devices*, 49(11), pp. 1985–1992, November 2002.

[Levi 1981] M. W. Levi, CMOS is most testable, in *Proc. Int. Test Conf.*, pp. 217–220, October 1981.

[Li 2002] J. C.-M. Li and E. J. McCluskey, Diagnosis of sequence-dependent chips, in *Proc. VLSI Test Symp.*, pp. 187–192, April 2002.

[Ma 1995] S. C. Ma, P. Franco, and E. J. McCluskey, An experimental chip to evaluate test techniques: Experimental results, in *Proc. Int. Test Conf.*, pp. 663–672, October 1995.

[Madge 2002] R. Madge, B. H. Goh, V. Rajagopalan, C. Macchietto, R. Daasch, C. Schuermyer, C. Taylor, and D. Turner, Screening MinVDD outliers using feed-forward voltage testing, in *Proc. Int. Test Conf.*, pp. 673–682, October 2002.

[Maly 2003] W. Maly, A. Gattiker, T. Zanon, T. Vogels, R. D. Blanton, and T. Storey, Deformations of IC structure in test and yield learning, in *Proc. Int. Test Conf.*, pp. 856–865, September 2003.

[Maxwell 1991] P. Maxwell, R. C. Aitken, V. Johansen, and I. Chiang, The effect of different test sets on quality level prediction: When is 80% better than 90%, in *Proc. Int. Test Conf.*, pp. 358–364, October 1991.

[Maxwell 2000] P. Maxwell, P. O'Neill, R. Aitken, R. Dudley, N. Jaarsma, M. Quach, and D. Wiseman, Current ratios: A self-scaling technique for production IDDQ testing, in *Proc. Int. Test Conf.*, pp. 1148–1156, October 2000.

[McCluskey 1993] E. J. McCluskey, Quality and single-stuck faults, in *Proc. Int. Test Conf.*, p. 597, October 1993.

[McCluskey 2000] E. J. McCluskey and C.-W. Tseng, Stuck-fault tests vs. actual defects, in *Proc. Int. Test Conf.*, pp. 336–343, October 2000.

[McCluskey 2004] E. J. McCluskey, A. Al-Yamani, J. C.-M. Li, C.-W. Tseng, E. Volkerink, F.-F. Ferhani, E. Li, and S. Mitra, ELF-Murphy data on defects and test sets, in *Proc. VLSI Test Symp.*, pp. 16–22, April 2004.

[Min 1998] Y. Min and Z. Li, IDDT testing versus IDDQ testing, *J. of Electronic Testing: Theory and Applications*, 13(1), pp. 51–55, January 1998.

[Mitra 2004] S. Mitra, E. Volkerink, E. J. McCluskey, and S. Eichenberger, Delay defect screening using process monitor structures, in *Proc. VLSI Test Symp.*, pp. 43–52, April 2004.

[Nassif 2004] S. Nassif, D. Boning, and N. Hakim, The care and feeding of your statistical static timer, in *Proc. Int. Conf. on Computer-Aided Design*, pp. 138–139, November 2004.

[Nigh 1998] P. Nigh, D. Vallett, A. Patel, J. Wright, F. Motika, D. Forlenza, R. Kurtulik, and W. Chong, Failure analysis of timing and I_{DDQ}-only failures from the SEMATECH test methods experiment, in *Proc. Int. Test Conf.*, pp. 43–52, October 1998.

[Nigh 2000] P. Nigh and A. Gattiker, Test method evaluation experiments and data, in *Proc. Int. Test Conf.*, pp. 454–463, October 2000.

[Pacha 2004] C. Pacha, M. Bach, K. von Arnim, R. Brederlow, D. Schmitt-Landsiedel, P. Seegebrecht, J. Berthold, and R. Thewes, Impact of STI-induced stress, inverse narrow width effect, and statistical V_{TH} variations on leakage currents in 120 nm CMOS, in *Proc. European Solid-State Research Conf.*, pp. 397–400, September 2004.

[Park 1988] E. S. Park, M. R. Mercer, and T. W. Williams, Statistical delay fault coverage and defect level for delay faults, in *Proc. Int. Test Conf.*, pp. 492–499, September 1988.

[Park 2005] I. Park, A. Al-Yamani, and E. J. McCluskey, Effective TARO pattern generation, in *Proc. VLSI Test Symp.*, pp. 161–166, May 2005.

[Parvathala 2002] P. Parvathala, K. Maneparambil, and W. Lindsay, FRITS: A microprocessor functional BIST method, in *Proc. Int. Test Conf.*, pp. 590–598, October 2002.

[Pham 2006] D. C. Pham, T. Aipperspach, D. Boerstler, M. Bolliger, R. Chaudhry, D. Cox, P. Harvey, P. M. Harvey, H. P. Hofstee, C. Johns, J. Kahle, A. Kameyama, J. Keaty, Y. Masubuchi, M. Pham, J. Pille, S. Posluszny, M. Riley, D. L. Stasiak, M. Suzuoki, O. Takahashi, J. Warnock, S. Weitzel, D. Wendel, and K. Yazawa, Overview of the architecture, circuit design, and physical implementation of a first-generation cell processor, *IEEE J. of Solid-State Circuits*, 41(1), pp. 179–196, January 2006.

[Pomeranz 1999] I. Pomeranz and S. M. Reddy, On N-detection test sets and variable N-detection test sets for transition faults, in *Proc. VLSI Test Symp.*, pp. 173–179, April 1999.

[Pouya 2000] B. Pouya and A. L. Crouch, Optimization trade-offs for vector volume and test power, in *Proc. Int. Test Conf.*, pp. 873–881, October 2000.

[Powell 2000] T. J. Powell, J. Pair, M. St. Jones, and D. Counce, Delta I_{DDQ} for testing reliability, in *Proc. VLSI Test Symp.*, pp. 439–443, April 2000.

[Qiu 2006] W. Qiu, D. M. H. Walker, N. Simpson, D. Reddy, and A. Moore, Comparison of delay tests on silicon, in *Proc. Int. Test Conf.*, Paper 11.3, November 2005.

[Rearick 2005] J. Rearick and R. Rodgers, Calibrating clock stretch during AC-scan test, in *Proc. Int. Test Conf.*, Paper 11.3, November 2005.

[Remersaro 2006] S. Remersaro, X. Lin, Z. Zhang, S. M. Reddy, I. Pomeranz, and J. Rajski, Preferred fill: A scalable method to reduce capture power for scan based designs, in *Proc. Int. Test Conf.*, Paper 32.2, October 2006.

[Roehr 2006] J. L. Roehr, Very-low voltage (VLV) and VLV ratio (VLVR) testing for quality, reliability, and outlier detection, in *Proc. Int. Test Conf.*, Paper 31.1, October 2006.

[Samaan 2003] S. Samaan, Parameter variation probing technique, U.S. Patent No. 6,535,013, March 18, 2003.

[Sato 2005] Y. Sato, S. Hamada, T. Maeda, A. Takatori, Y. Nozuyama, and S. Kajihara, Invisible delay quality: SDQM model lights up what could not be seen, in *Proc. Int. Test Conf.*, Paper 47.1, November 2005.

[Segura 2002] J. Segura, A. Keshavarzi, J. M. Soden , and C. F. Hawkins, Parametric failures in CMOS Ics: A defect-based analysis, in *Proc. Int. Test Conf.*, pp. 90–99, October 2002.

[Sengupta 1999] S. Sengupta, S. Kundu, S. Chakravarty, P. Parvathala, R. Galivanche, G. Kosonocky, M. Rodgers, and T. M. Mak, Defect-based test: A key enabler for successful migration to structural test, *Intel Technology J.*, pp. 1–14, Q1 1999.

[Shen 1998] J. Shen and J. A. Abraham, Native mode functional test generation for processors with applications to self test and design validation, in *Proc. Int. Test Conf.*, pp. 990–999, October 1998.

[SIA 2005] SIA, *The International Technology Roadmap for Semiconductors: 2005 Edition*, Semiconductor Industry Association, San Jose, CA (http://public.itrs.net), 2005.

[SIA 2006] SIA, *The International Technology Roadmap for Semiconductors: 2006 Update*, Semiconductor Industry Association, San Jose, CA (http://public.itrs.net), 2006.

[Spica 2004] M. Spica and T. M. Mak, Do we need anything more than single bit error correction (ECC)?, in *Proc. Int. Workshop on Memory Technology, Design, and Testing*, pp. 111–116, August 2004.

[Stanojevic 2000] Z. Stanojevic, H. Balachandran, D. M. H. Walker, F. Lakhani, and S. Jandhyala, Defect localization using physical design and electrical test information, in *Proc. IEEE/SEMI Advanced Semiconductor Manufacturing Conf.*, pp. 108–114, September 2000.

[Stanojevic 2001] Z. Stanojevic and D. M. H. Walker, FedEx: A fast bridging fault extractor, in *Proc. Int. Test Conf.*, pp. 696–703, October 2001.

[Stinson 2003] J. Stinson and E. A. De La Iglesia, Process parameter extraction, U.S. Patent No. 6,553,545, April 22, 2003.

[Stroud 2000] C. E. Stroud, J. M. Emmert, J. R. Bailey, K. S. Chhor, and D. Nikolic, Bridging fault extraction from physical design data for manufacturing test development, in *Proc. Int. Test Conf.*, pp. 760–769, October 2000.

[Tendolkar 2006] N. Tendolkar, D. Belete, B. Schwarz, B. Podnar, A. Gupta, S. Karako, W.-T. Cheng, A. Babin, K.-H. Tsai, N. Tamarapalli, and G. Aldrich, Improving transition fault test pattern quality through at-speed diagnosis, in *Proc. Int. Test Conf.*, Paper 11.2, October 2006.

[Transmeta 2002] Transmeta Corp., *Crusoe Processor Documentation*, www.transmeta.com, 2002.

[Tschanz 2003] J. W. Tschanz, S. G. Narendra, Y. Ye, B. A. Bloechel, S. Borkar, and V. De, Dynamic sleep transistor and body bias for active leakage power control of microprocessors, *IEEE J. of Solid-State Circuits*, 38(11), pp. 1838–1845, November 2003.

[Tseng 2001] C.-W. Tseng and E. J. McCluskey, Multiple-output propagation transition fault test, in *Proc. Int. Test Symp.*, pp. 358–366, October 2001.

[Tupuri 1997] R. S. Tupuri and J. A. Abraham, A novel functional test generation method for processors using commercial ATPG, in *Proc. Int. Test Conf.*, pp. 743–752, November 1997.

[Tupuri 2002] R. S. Tupuri, A. Krishnamachary, and J. A. Abraham, Test generation for gigahertz processors using an automatic functional constraint extractor, in *Proc. Design Automation Conf.*, pp. 647–652, June 1999.

[Vo 2006] T. Vo, Z. Wang, T. Eaton, P. Ghosh, H. Li, Y. Lee, W. Wang, R, Fang, D. Singletary, and X. Gu, Design for board and system level structural test and diagnosis, in *Proc. Int. Test Conf.*, Paper 14.1, October 2006.

[Wang 2005] L.-T. Wang, X. Wen, P.-C. Hsu, S. Wu, and J. Guo, At-speed logic BIST architecture for multi-clock designs, in *Proc. Int. Conf. on Computer Design*, pp. 475–478, October 2005.

[Wen 2005] X. Wen, Y. Yamashita, S. Morishima, S. Kajihara, L.-T. Wang, K. K. Saluja, and K. Kinoshita, Low-capture-power test generation for scan-based at-speed testing, in *Proc. Int. Test Conf.*, Paper 39.2, November 2005.

[Wen 2006] X. Wen, S. Kajihara, K. Miyase, T. Suzuki, K. K. Saluja, L.-T. Wang, K. S. Abdel-Hafez, and K. Kinoshita, A new ATPG method for efficient capture power reduction during scan testing, in *Proc. VLSI Test Symp.*, pp. 58–63, May 2006.

[Williams 1991] T. W. Williams, B. Underwood, and M. R. Mercer, The interdependence between delay-optimization of synthesized networks and testing, in *Proc. Design Automation Conf.*, pp. 87–92, June 1991.

[Williams 1996a] T. W. Williams, R. Kapur, M. R. Mercer, R. H. Dennard, and W. Maly, Iddq testing for high performance CMOS: The next ten years, in *Proc. European Design and Test Conference*, pp. 578–583, March 1996.

[Williams 1996b] T. W. Williams, R. H. Dennard, R. Kapur, M. R. Mercer, and W. Maly, Iddq test: Sensitivity analysis to scaling, in *Proc. Int. Test Conf.*, pp. 786–792, October 1996.

[Wuu 2005] J. Wuu, D. Weiss, C. Morganti, and M. Dreesen, The asynchronous 24MB on-chip level-3 cache for a dual-core Itanium-family processor, in *Proc. IEEE Int. Solid-State Circuits Conf.*, 1, pp. 488–612, February 2005.

[Zou 2005] W. Zou, W.-T. Cheng, and S. M. Reddy, Bridge defect diagnosis with physical information, in *Proc. Asian Test Symp.*, pp. 248–253, December 2005.

R8.4 Soft Errors

[Almukhaizim 2006] S. Almukhaizim, Y. Makris, Y.-S. Yang, and A. Veneris, Seamless integration of SER in rewiring based design space exploration, in *Proc. Int. Test Conf.*, Paper 29.3, September 2006.

[Austin 1999] T. M. Austin, DIVA: A reliable substrate for deep submicron microarchitecture design, in *Proc. Int. Symp. on Microarchitecture*, pp. 196–207, November 1999.

[Bartlett 2004] W. Bartlett and L. Spainhower, Commercial fault tolerance: A tale of two systems, *IEEE Trans. on Dependable and Secure Computing*, 1(1), pp. 87–96, January 2004.

[Baumann 2005] R. Baumann, Soft errors in advanced computer systems, *IEEE Design & Test of Computers*, pp. 258–266, May/June 2005.

[Calin 1997] T. Calin, R. Velazco, M. Nicolaidis, S. Moss, S. D. LaLumondiere, V. T. Tran, R. Koga, and K. Clark, Topology-related upset mechanisms in design hardened storage cells, in *Proc. European Conf. on Radiation and Its Effects on Components and Systems, (RADECS)*, pp. 484–488, September 1997.

[Ernst 2003] D. Ernst, N. S. Kim, S. Das, S. Pant, R. Rao, T. Pham, C. Ziesler, D. Blaauw, T. Austin, K. Flautner, and T. Mudge, Razor: A low-power pipeline based on circuit-level timing speculation, in *Proc. Int. Symp. on Microarchitecture*, pp. 7–18, December 2003.

[Ernst 2004] D. Ernst, S. Das, S. Lee, D. Blaauw, T. Austin, T. Mudge, N. S. Kim, and K. Flautner, RAZOR: Circuit-level correction of timing errors for low-power operation, *IEEE Micro*, 24(6), pp. 10–20, November 2004.

[Favalli 2002] M. Favalli, and C. Metra, Online testing approach for very deep-submircon ICs, *IEEE Design & Test of Computers*, 19(2), pp. 16–23, March 2002.

[Franco 1994] P. Franco and E. J. McCluskey, On-line delay testing of digital circuits, in *Proc. VLSI Test Symp.*, pp. 164–173, April 1994.

[May 1979] T. C. May and M. H. Woods, Alpha-particle-induced soft errors in dynamic memories, *IEEE Trans. on Electron Devices*, ED-26(1), pp. 2–9, January 1979.

[Metra 1998] C. Metra, M. Favalli, and B. Ricco, On-line detection of logic errors due to crosstalk, delay, and transient faults, in *Proc. Int. Test Conf.*, pp. 524–533, October 1998.

[Miskov-Zivanov 2006] M. Miskov-Zivanov and D. Marculescu, Circuit reliability analysis using symbolic techniques, *IEEE Trans. on Computer-Aided Design*, 25(12), pp. 2638–2649, December 2006.

[Mitra 2005] S. Mitra, N. Seifert, M. Zhang, Q. Shi, and K. S. Kim, Robust system design with built-in soft-error resilience, *IEEE Computer*, 38(2), pp. 43–52, February 2005.

[Mohanram 2003] K. Mohanram and N. A. Touba, Cost-effective approach for reducing soft error failure rate in logic circuits, in *Proc. Int. Test Conf.*, pp. 893–901, September 2003.

[Moore 1965] G. Moore, Cramming more components onto integrated circuits, *Electronics*, pp. 114–117, April 19, 1965.

[Mukherjee 2002] S. S. Mukherjee, M. Kontz, and S. Reinhardt, Detailed design and evaluation of redundant multithreading alternatives, in *Proc. Int. Symp. on Computer Architecture*, pp. 99–110, May 2002.

[Nicolaidis 1999] M. Nicolaidis, Time redundancy based soft-error tolerance to rescue nanometer technologies, in *Proc. VLSI Test Symp.*, pp. 86–94, April 1999.

[Oh 2002] N. Oh, S. Mitra, and E. J. McCluskey, ED^4I: Error detection by diverse data and duplicated instructions, *IEEE Trans. on Computers*, 51(2), pp. 180–199, February 2002.

[Patel 1982] J. H. Patel and L. Y. Fung, Concurrent error detection in ALU's by recomputing with shifted operands, *IEEE Trans. on Computers*, C-31(7), pp. 589–595, July 1982.

[Pering 1998] T. Pering, T. Burd, and R. Brodersen. The simulation and evaluation of dynamic voltage scaling algorithms, in *Proc. Int. Symp. on Low Power Electronics and Design*, pp. 76–81, June 1998.

[Seifert 2006] N. Seifert, P. Slankard, M. Kirsch, B. Narasimham, V. Zia, C. Brookreson, A. Vo, S. Mitra, B. Gill, and J. Maiz, Radiation-induced soft error rates of advanced CMOS bulk devices, in *Proc. IEEE Int. Reliability Physics Symp.*, pp. 217–225, March 2006.

[Sklaroff 1976] J. R. Sklaroff, Redundancy management technique for space shuttle computers, *IBM J. of Research and Development*, 20(1), pp. 5–19, January 1976.

[Spainhower 1999] L. Spainhower and T. A. Gregg, IBM S/390 parallel enterprise server G5 fault tolerance: A historical perspective, *IBM J. Research and Development*, 43(5/6), pp. 863–873, September/November 1999.

[Tamhankar 2005] R. Tamhankar, S. Murali, and G. De Micheli, Performance driven reliable link for networks on chip, in *Proc. Asian and South Pacific Design Automation Conf.*, pp. 749–754, January 2005.

[Zhang 2006] M. Zhang and N. R. Shanbhag, A soft error rate analysis (SERA) methodology, in *Proc. Int. Conf. on Computer-Aided Design*, pp. 111–118, November 2004.

[Zhou 2006] Q. Zhou and K. Mohanram, Gate sizing to radiation harden combinational logic, *IEEE Trans. on Computer-Aided Design*, 25(1), pp. 155–166, January 2006.

R8.5 Defect and Error Tolerance

[Breuer 2004a] M. Breuer, S. Gupta, and T. M. Mak, Defect and error tolerance in the presence of massive numbers of defects, *IEEE Design & Test of Computers*, pp. 216–227, May/ June 2004.

[Breuer 2004b] M. Breuer, Intelligible test sequence to support error-tolerance, *Asian Test Symp.*, pp. 386–393, November 2004.

[Hirase 2001] J. Hirase, Yield increase of VLSI after redundancy-repairing, in *Proc. Asian Test Symp.*, pp. 353–358, November 2001.

[Hsieh 2006] T.-Y. Hsieh, K.-J. Lee, and M. A. Breuer, An error-oriented test methodology to improve yield with error-tolerance, in *Proc. VLSI Test Symp.*, pp. 130–135, May 2006.

[Koren 1990] I. Koren and A. D. Singh, Fault tolerance in VLSI Circuits, *IEEE Computer*, 23(7), pp. 73–83, July 1990.

[Koren 1998] I. Koren and Z. Koren, Defect tolerance in VLSI circuits: Techniques and yield analysis, *Proceedings of the IEEE*, 86(9), pp. 1819–1838, September 1998.

[Lee 2005] K.-J. Lee, T.-Y. Hsieh, and M. A. Breuer, A novel test methodology based on error-rate to support error-tolerance, in *Proc. Int. Test Conf.*, pp. 1136–1144, November 2005.

[SIA 2006] SIA, *The International Technology Roadmap for Semiconductors: 2006 Update*, Semiconductor Industry Association, San Jose, CA (http://public.itrs.net), 2006.

R8.6 Concluding Remarks

[SIA 2005] SIA, *The International Technology Roadmap for Semiconductors: 2005 Edition*, Semiconductor Industry Association, San Jose, CA (http://public.itrs.net), 2005.

[SIA 2006] SIA, *The International Technology Roadmap for Semiconductors: 2006 Update*, Semiconductor Industry Association, San Jose, CA (http://public.itrs.net), 2006.

CHAPTER 9

Design for Manufacturability and Yield

Robert C. Aitken
ARM, Sunnyvale, California

ABOUT THIS CHAPTER

Design for manufacturability (DFM) in the context of integrated circuit design is a broad topic that covers a number of activities, all loosely connected by their intention to improve yield. Clearly, no one sets out to design against manufacturability. Instead, many design methods have tended to ignore manufacturability and treat yield as a problem to be solved by the wafer *fabrication facility* (fab) or foundry. This has happened both in vertically oriented organizations and in fabless environments, but it has been worse in the latter. In some ways, this has been a direct result of progress in the semiconductor industry. Future progress now depends on reversing this trend.

In this chapter, we first provide some background information and historical context for DFM and show how it fits into the modern semiconductor industry. Next, we introduce and define the concept of yield, introduce a variety of yield models, and show how these models can be used to quantify the benefits of repairable circuit elements. This is followed by an introduction to the basic concepts of photolithography. All these concepts are then combined, and we show how they interact for both DFM and ***design for yield*** (DFY). We also show their relationship with ***design for testability*** (DFT). Following this discussion, we introduce process variation and its relationship to DFM, DFY, and DFT. The combination of these topics is known as DFX, where "X" stands for any of the aforementioned letters, but it might also be thought of as ***design for excellence***. We discuss metrics for quantifying various aspects of DFX, and, finally, we present future directions and conclusions.

9.1 INTRODUCTION

In 1965, Gordon Moore published his famous observation in *Electronics* magazine [Moore 1965]. He predicted that technology innovation would allow the number of transistors to double in a given silicon area every year (he revised the scale to every 2 years in 1975) [Hiremane 2005]. For more than 40 years, the semiconductor industry has been able to track this prediction. In 2006, chips with hundreds of millions of transistors were produced. Moore's prediction has been so prescient that it is now referred to as **Moore's law.**

When Moore made his prediction, he was director of research and development at Fairchild Semiconductor. His biography lists him as "one of the new breed of electronic engineers, schooled in the physical sciences rather than in electronics." In 1965, when a chip consisted of roughly 50 transistors and only a small number of chips were in production, design and manufacturing were difficult to separate. The entire chip life cycle would fit within a single organization. This remained true through the 1980s. When a company wanted chips for a product, it was often economical to build a fab to go with it. Large companies often ended up with multiple fabs working on multiple product lines (*e.g.*, calculators and computers at Hewlett-Packard, with fabs in Corvallis, Oregon for the former and Fort Collins, Colorado for the latter).

As the ***integrated circuit*** (IC) business evolved, it made sense to specialize. When the number of transistors climbed into the hundreds and thousands, digital design split between circuit design (schematics) and transistor design (artwork layout), and manufacturing with its own specialties (mask making, transistor fabrication {front end}, metal fabrication {back end}, and test). Links between design and manufacturing were formalized: design rules to make layout manufacturable and associated ***design rule checkers*** (DRCs) [Lindsay 1976], plus ***design for testability*** (DFT) rules and methods [Eichelberger 1977]. Eventually, the IC design and manufacturing process stratified into a large number of layers, approximated by Figure 9.1, where it has remained relatively stable. Economic pressure and the benefits of specialization has led to disaggregation of the industry and the introduction of companies specializing in one or more layers, including entire new industries such as wafer foundries and ***intellectual property*** (IP) companies. Note that mask making, foundry equipment, etc., are not represented in this example, but they could be.

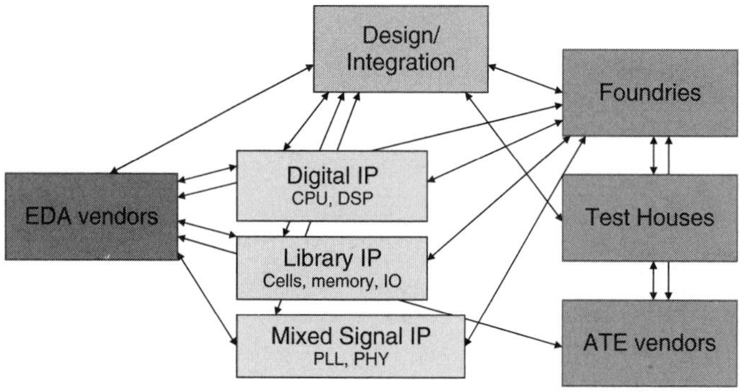

■ FIGURE 9.1

Disaggregated supply chain example.

The existence of this layered structure has allowed organizations to concentrate on areas where they can add the most value, whereas the introduction of standard approaches and hand-offs at layer boundaries have allowed for increasing levels of abstraction with the understanding that the underlying components are sound. For example, a *register-transfer-level* (RTL) designer can synthesize an RTL design to a standard cell library with confidence that each of the functions in the library will behave in silicon precisely as its model predicts it will. The library designer is able to guarantee this result because each cell follows the design rules specified by the foundry and is simulated using models extracted from actual silicon. Each step in the chain is supported by standard methods (synthesis, characterization, design rule checking) and by commercial tools.

Notice that these examples all relate to digital design. For analog circuits, abstraction has been more difficult to achieve. It has been and continues to be common practice to tune circuits through an iterated process of schematic design, layout, extraction, and simulation. Often, the tuning continues with multiple iterations on silicon as well. Variation in circuit behavior across the manufacturing process has been modeled using **Monte Carlo** methods [Kennedy 1964] [Balaban 1975], and designs are crafted to be "centered" within a manufacturing process window [Director 1977] (see Figure 9.2). As a result, analog design is tightly linked to manufacturability. An odd side effect, however, is that yield of analog designs is less precisely defined than for digital designs. As will be discussed later, a simple practical definition of yield is "fraction of parts passing applied tests," and pass/fail criteria are usually easier to specify for a digital design than its analog counterpart.

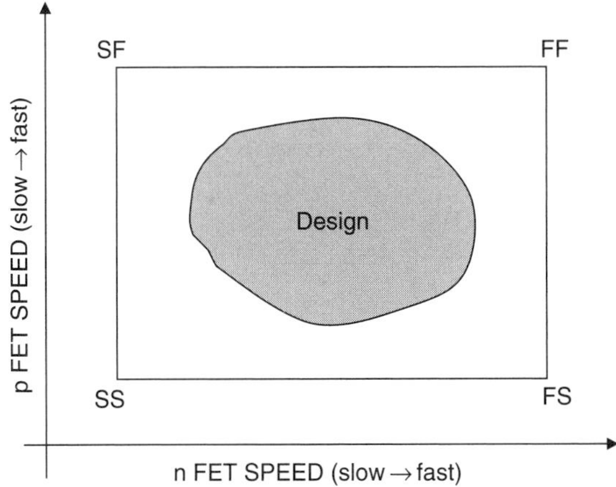

FIGURE 9.2

Conventional (corner-based) design centering.

9.2 YIELD

In its most basic form, yield for a semiconductor manufacturing process can be expressed as:

$$Yield = \frac{good\ chips}{total\ chips}$$

"Total chips" is easy to calculate (although care needs to be taken to include or exclude certain classes of chips—those from zero yielding wafers, for example). "Good chips," on the other hand, is much more difficult to define and measure. The ideal definition of a good chip is one that works in a customer's application, but two problems exist with this definition: (1) it cannot be determined until it is too late, and (2) the requirements may differ between customers and over time. Consider two users of a cell phone as an example. The first uses it to store a couple dozen phone numbers and make phone calls a few times per day, whereas the second adds custom applications, uses it for text messaging, Web access, e-mail, and music video downloads, as well as near constant phone calls. The second user would be far more likely to encounter an inherent flaw than will the first user. There is also a time-based reliability component to "good," based on expected product lifetime, but we will ignore this factor in the following discussion.

The definition of "good" is ambiguous; so is the measurement of "good." For example, a chip with higher-than-average leakage could be the result of an unintended resistive short or of faster-than-average transistors. Similarly, there are specified *versus* observed quantities. Suppose a design is specified to operate above 0.8 V, but the vast majority of manufactured chips are able to operate down to 0.65 V. When a chip is found that fails at 0.75 V, is it good? It meets the spec, but is clearly different from all the other chips. Work in **statistical post processing** of test results suggests that using "different" as a predictor of "bad" works well in improving shipped quality and future reliability (*e.g.*, the authors in [Madge 2002] showed a 30% to 60% reduction in ***defective parts per million*** [DPM] levels by applying such techniques). This emphasizes that the nature and quality of tests is a vital factor in determining both yield and quality. The ambiguous nature of test can be used to expand the definition of measured yield:

$$Measured\ Yield = \frac{good\ parts + test\ escapes - false\ rejects}{all\ parts}$$

"Test escapes" are also known as **type II failures**—chips that pass the tests but will fail in the application. "False rejects" are also known as **type I failures**—chips that fail the test but would work in the application. Relative costs vary, but a common rule of thumb is that it is worth having 10 false rejects to avoid 1 test escape. This discussion shows that test and yield are tightly coupled and that failing to consider both can lead to suboptimal results.

9.3 COMPONENTS OF YIELD

Chips fail for a number of reasons, and these can be grouped to form a classification of yield types:

1. The first class is known as **systematic yield**. Systematic yield loss affects the process as a whole and must be dealt with by the fab. Examples include equipment-related problems, material quality and delivery, managing contamination sources, etc. These operations are managed as an aggregate across all products being run on a wafer line. Additional details are available in the books [Nishi 2000] and [May 2006].
2. The second class is **parametric yield**. Example parameters include ***critical lithography dimensions*** (CD), device mobility, threshold voltage, oxide thickness, metal resistance, etc. These also affect the process as a whole, but they can be accommodated by a mixture of fab adjustments and design tolerance. The important feature of this class of yield is that the **parametric variations** are inherent in the process rather than induced by the design.
3. The third class is **defect-related yield**. Defects are imperfections in a fabricated chip, including particle-induced shorts, cracks, poorly formed vias, etc. These defects tend to fall in random locations and be somewhat unpredictable in their properties. The detection and avoidance of defects is the subject of much interest in test, and solutions must be applied throughout the design process. Segura and Hawkins have produced a good overview of the topic of defects [Segura 2004].
4. The fourth class is **design-related yield**. This category is similar to parametric yield but emphasizes the deterministic aspect of parametric variation, particularly those variations or deformations that can be predicted based on design layout and margining techniques. These will be discussed in detail in the remainder of this chapter.
5. The final category is **test-related yield**. This category explicitly includes the false rejects mentioned earlier, as well as quantification of those based on design and test decisions. As an example, if a part's spec requires that it operate at 1.0 V+/−10%, with minimum operating voltage of 900 mV and tester power supply accuracy of 20 mV, then low-voltage test should be applied at 880 mV to guarantee that no more than 900 mV is applied. This in turn implies that sometimes the test will be applied at 860 mV, so the design must operate at this low voltage. Yield differences between 860 mV and 900 mV operation are an example of test-related yield. Additional discussion is beyond the scope of this chapter but has been discussed elsewhere [Aitken 2006a].

9.3.1 Yield Models

It is desirable to trade off various design approaches in terms of their ability to improve yield. This in turn requires an ability to predict yield—to model expected yield based on some criteria. The simplest such model is the **Poisson model**. This model assumes that defects are rare events that occur randomly and independently and that as such they can be modeled as following a **Poisson distribution**. The resulting Poisson yield equation [Nishi 2000] takes two parameters, die area A and a global defect density D_0:

$$Y = e^{-AD_0}$$

For many applications the Poisson model is adequate, especially those involving changes in yield, but it has been shown to be pessimistic for absolute chip yield prediction. One primary reason is that defect density is not uniform and fixed but rather is itself a distribution. Ideally a **Gaussian distribution** would be used, but the mathematics becomes complex. Instead, approximating a Gaussian defect density distribution by a triangular shape leads to the **Murphy model** [Nishi 2000]:

$$Y = \left(\frac{1 - e^{-AD}}{AD}\right)^2$$

Another approach, common among foundries, is to assume a **uniform distribution** by layer, and to assign a complexity factor to each layer. The cumulative result of these complexity factors becomes "n" in the **Bose-Einstein model** [Nishi 2000]:

$$Y = \frac{1}{(1 + AD)^n}$$

In many foundries, D is given in defects per square inch per layer. For a complexity factor of 12, a value of $D = 0.5$ in the Bose-Einstein model is equivalent to D_0 of 1.12 defects per square centimeter in the Poisson model (the unit change is also typical). Published values for D range from 0.16 to 0.28 per square inch for a commercial 90-nm process as of March 2006 [Sydow 2006].

Another useful model is the **negative binomial model**, which takes into account the tendency of defects to cluster. This model was advocated extensively by Stapper at IBM [Stapper 1989]:

$$Y = \frac{1}{\left(1 + \frac{AD}{\alpha}\right)^{\alpha}}$$

The clustering parameter, α, has been found to lie between 2 and 3 in many processes. As α approaches infinity, the model reduces to the Poisson model. Additional discussion of yield models is available in Chapter 26 of [Nishi 2000].

9.3.2 Yield and Repair

The chances of more than a single defect affecting a given instance of a small memory are sufficiently small that they can be ignored. As a result, the Poisson yield model can be used to estimate ΔY, the change in yield expected across area at global defect density when single defects are repaired with probability *P(repair)*:

$$\begin{aligned}\Delta Y &= P(repair)\cdot(1 - base\_yield)\\ &= P(repair)\cdot(1 - e^{-AD})\end{aligned}$$

D can be calculated "bottom up," based on layer defect densities for each layer used in the memory (*e.g.*, product of all individual step yields up to metal 3), or "top down" by fitting observed yield to a Poisson equation. In either case, it is usually recommended to use different values for logic and *static random-access memory* (SRAM) and to account for the added complexity of SRAM by increasing the value *D* for logic circuits by a process-dependent factor of 1.5 to 3.

For a single instance, the **yield improvement** is small, but the yield improvement for each instance is independent, giving an overall value, for *n* instances, of the following:

$$\begin{aligned}\Delta Y_n &= repair\_yield^n - base\_yield^n\\ &= \left(e^{-(A+A_r)D} + P(repair)\cdot\left(1 - e^{-AD}\right)\right)^n - e^{-nAD}\end{aligned}$$

Note that **repair improvement** applies only to the base area, not the additional area of the test overhead (A_r). Figure 9.3 shows yield improvement plotted against test overhead and defect density for a set of 250 small memories of 16k bits each (128 × 128) in a 90-nm process. A conservative value of 0.5 was assumed for *P(repair)*. Higher or lower values change the scale of yield improvement, but not the shape of the curve.

For large memories, test overhead, including redundant memory cells, remapping logic, fuses, and test controllers is often less than 5% of total area, but for smaller memories this percentage is higher. Figure 9.3 shows that both test overhead and defect density strongly influence yield improvement. Thus, it is vital when designing repair circuitry for small memories to keep overhead low. It is also clear that the need for redundancy is higher when the process is immature and defect densities are higher than it is later in the process life cycle when defect densities are low.

To give some idea of the magnitude of costs involved, if wafer cost is $2500 and there are 100 good dies per wafer (often referred to as **net die per wafer**), a 5% increase in the number of good dies per wafer decreases the die cost from $25 to $23.80. Over a high volume of chips, the savings can be substantial. If redundancy is already in place for larger memories on a chip, the added external cost (*e.g.*, laser fusing) is virtually zero. For a production volume of 10 million chips, the savings can amount to $12 million. A key part of achieving good yield is photolithography, which is discussed in the next section.

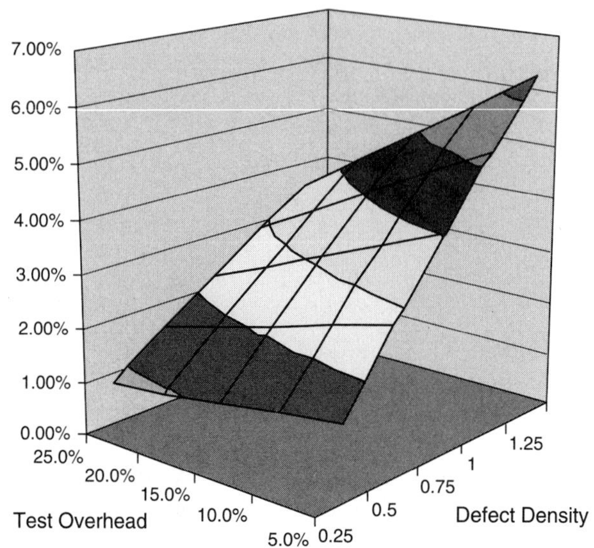

■ **FIGURE 9.3**

Yield improvement *versus* test overhead and defect density.

9.4 PHOTOLITHOGRAPHY

Photolithography is a key challenge in the production of ICs and has a significant influence on both manufacturability and yield. This section outlines the basics of lithography and shows how its complexities directly influence both ***design for manufacturability*** (DFM) and ***design for yield*** (DFY). Figure 9.4 represents a basic lithography system. (The discussion below follows [Liebmann 2003].) A plane of light of wavelength λ passes through gaps in a photomask set at pitch P apart to create two coherent individual light sources, and they interfere with each other, causing diffraction nodes at any angle where an integral multiple m of the wavelength exists across the difference in path length between the paths, as shown in the figure:

$$\sin\theta = \frac{m\lambda}{P}$$

A lens can use these diffraction nodes for image formation provided that its ***numerical aperture*** (NA) contains at least the first diffracted order, giving a minimum value for P:

$$P_{\min} = \frac{\lambda}{NA}$$

The minimum feature size, R, is half of P_{min}. The wavelength used for printing has not kept pace with feature size, shrinking from 365 nm for technology defined

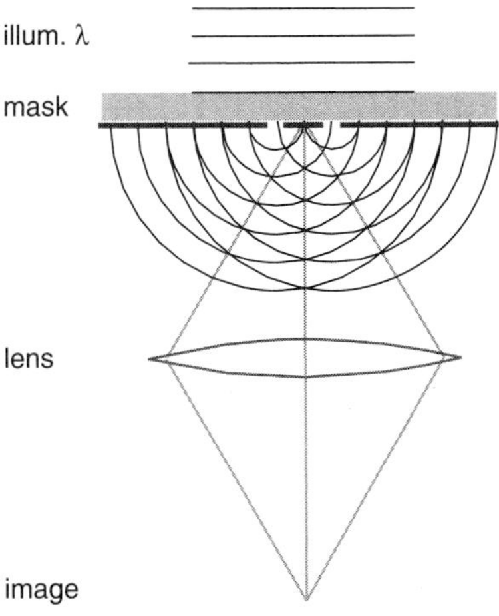

■ **FIGURE 9.4**

Lithography basics.
(Courtesy of L. Liebmann.)

by line widths of 1 μm through 0.35 μm, then 248 nm for 0.25 μm through 0.13 μm, and over recent times to 193 nm for 90 nm and smaller technologies. Only certain wavelengths are available for use in lithography, because each requires a specialized light source (ArF laser for 193 nm, KrF for 248 nm, moving to F_2 for 157 nm). The high cost of switching light sources (which requires a complete replacement of all lithography equipment) has led to work in resolution enhancement with existing technology. One approach is **off-axis imaging**, which allows features to be printed at half the size of conventional lithography as shown in Figure 9.5.

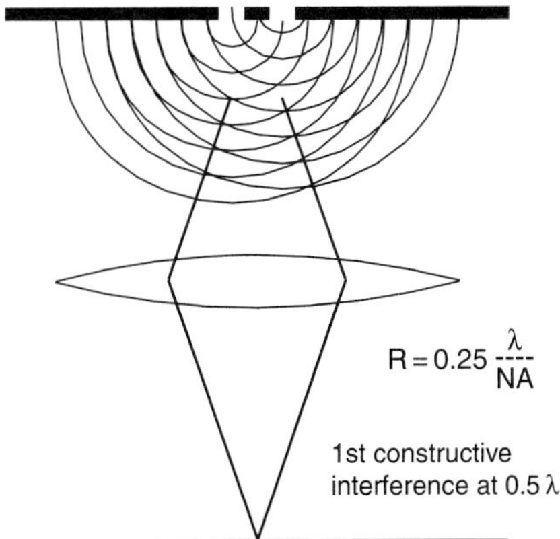

■ **FIGURE 9.5**

Off-axis illumination.
(Courtesy of L. Liebmann.)

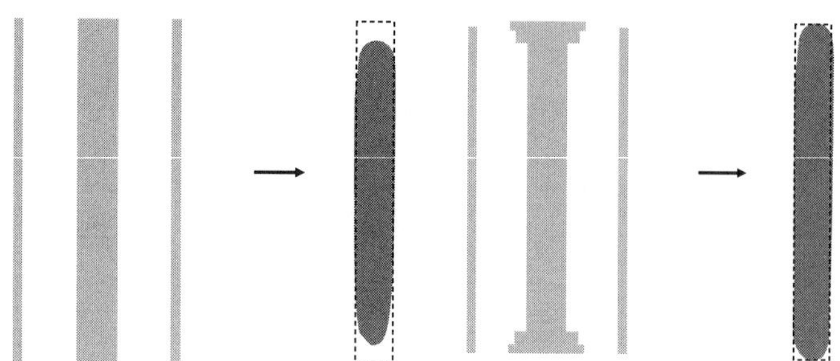

■ **FIGURE 9.6**

Scatter bars and line end features.

Printing of fine-geometry features can be enhanced by adding ***subresolution assist features*** (SRAF), which are features in the mask that do not print themselves but are added to improve the printing of the desired features by creating interference patterns. A common example is **scatter bars**, as shown in Figure 9.6.

The SRAF scatter bars improve printing of the desired line but are insufficient to prevent the line-end shortening, as shown by the first part of the diagram. To help correct line-end shortening, ***optical proximity correction*** (OPC) is used to add serif-like features that force the line end to be printed where it is intended. Current generation processes sometimes use multiple sets of scatter bars. It is clear that the addition of SRAF scatter bars and OPC places restrictions on the allowed layout. For example, scatter bars cannot be placed on top of other features, or one another, and this gives rise to layout restrictions such as **forbidden pitches**. In some cases certain layout features will be problematic without substantial relaxation of pitch, such as that shown in Figure 9.7.

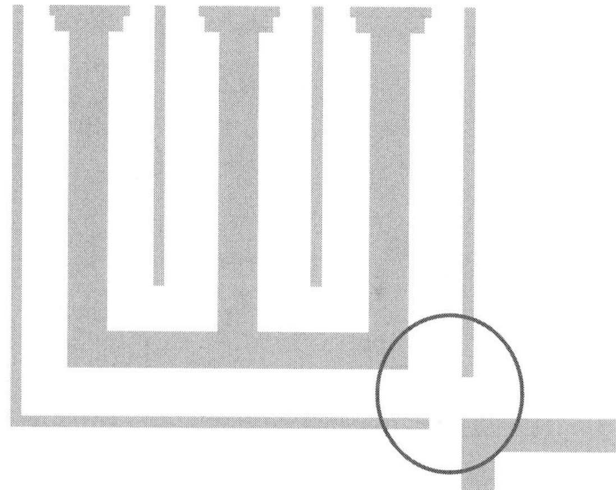

■ **FIGURE 9.7**

Resolution enhancement challenge.

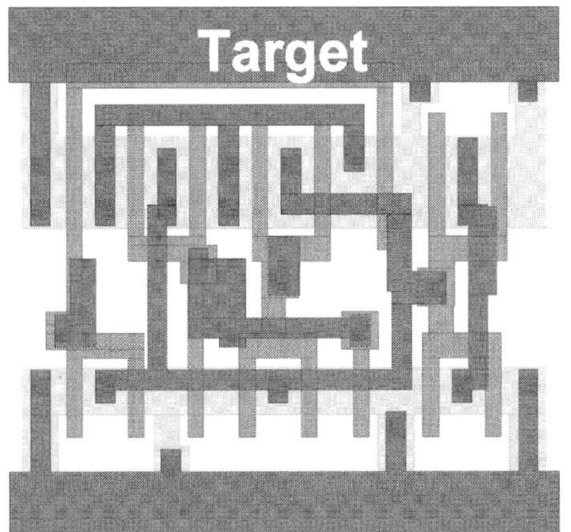

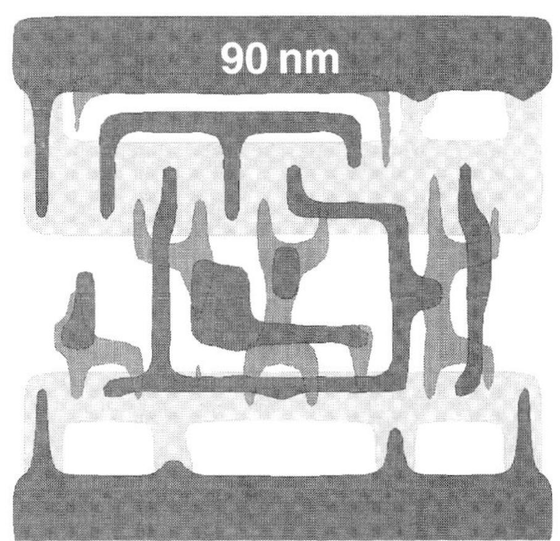

■ **FIGURE 9.8**

Target and printed layouts without RET at 90 nm.
(Courtesy of J. M. Brunet.)

The situation shown in Figure 9.7 can be described in conventional DRC terms, but it is cumbersome. It is representative of a class of geometries that are easily described in terms of shape but are harder to describe with distances, diagonals, and other DRC staples. The DRC of the future is likely to be shape-based as a result.

SRAF scatter bars and OPC are two techniques that can be grouped under the heading of ***resolution enhancement technology*** (RET). Figure 9.8 from [Brunet 2006, personal communication] dramatically illustrates just how important RET is to modern-day chip making: without it, nothing would ever work!

These and other optical complexities have led some processes to force unidirectional criticality (*i.e.*, all transistor gates will have the same orientation). That way, dimensions in the critical direction can be ensured with RET, whereas those in the noncritical direction will suffer if shape-based problems occur in SRAF generation. Many innovative approaches are being proposed to address this major change in layout design methodology. For a good overview, see [Liebmann 2003].

9.5 DFM AND DFY

Design for manufacturability (DFM) and *design for yield* (DFY) mean many things, depending on the context. For the purposes of this chapter, DFM will refer to layout design changes made to improve any aspect of manufacturability, from mask making through lithography and ***chemical-mechanical processing*** (CMP). DFY will refer to techniques specifically targeted to improving manufacturing yield.

The DFM challenge can be phrased simplistically as follows: uniformity is good for manufacturability, but nonuniformity is the source of value in a design. *Field programmable gate arrays* (FPGAs) solve this problem through programmability of regular structures. Memories repeat bit cells and other common layouts over and over. Standard cells push the bounds of regularity but must retain enough uniformity to be manufacturable. Design rules enforce significant uniformity, but their binary nature (pass/fail) prevents them from accurately conveying the tradeoffs that exist in manufacturability.

In general, the changes involved in making a layout DFM-compliant are subtle. The standard tradeoff applied is to make changes that do not increase the area of the cell. Consider, for example, the two layouts shown in Figures 9.9 and 9.10. The single contact at *A* is made more manufacturable by doubling it and adding additional metal overlap. The contact at *B* cannot be doubled without increasing cell area. Instead, additional metal overlap is added. Contact *C* is already doubled, but additional overlap will help its manufacturability as well. *D* is an example of a small metal jog whose removal will simplify mask making and improve lithography. In each case, the effects on yield and manufacturability are minor but will add up across a die.

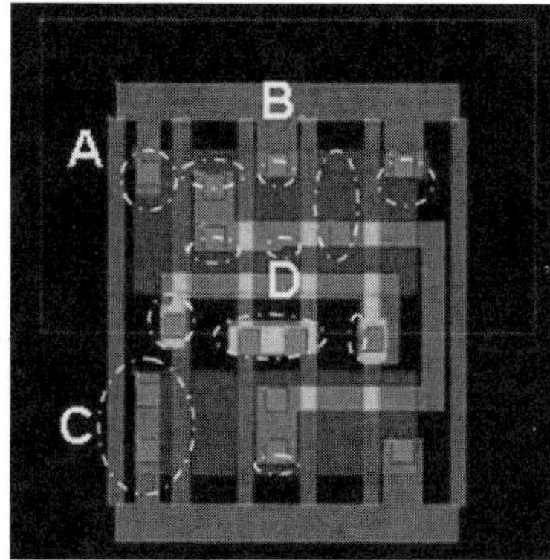

■ FIGURE 9.9

Pre-DFM layout.

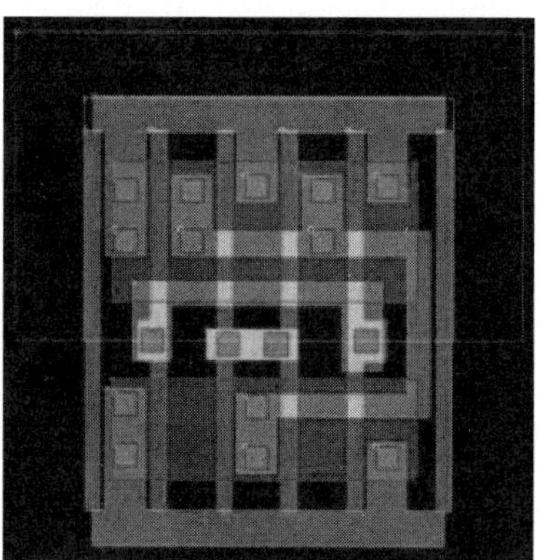

■ FIGURE 9.10

Post-DFM layout.

9.5.1 Photolithography

As explained previously, current IC manufacturing processes at current technology nodes make use of light with a wavelength of 248 nm or 193 nm for photolithography; both wavelengths are in the ***ultraviolet*** (UV) region of the spectrum, specifically the UVC region, which applies to wavelengths below about 400 nm. This means that subwavelength features are now the rule rather than the exception and require *resolution enhancement technology* (RET) for successful printing. The basic optical issues in photolithography were discussed in Section 9.4.

The complex optics needed for printing result in some interesting challenges. For example, the resolution enhancement for dense lines and isolated lines is different, and this can lead to gaps where neither approach works, leading to unprintable lines as shown in Figure 9.11. At 0.13 μm, an intermediate density could not be printed. At finer geometries, there can be multiple regions of such forbidden pitches.

Another issue that arises in photolithography is the concept of shape-based rules. We touched on this subject earlier in the discussion of Figure 9.7. We present another example here. Figure 9.12 shows an example piece of layout (polysilicon gate crossing a diffusion region). At 0.18 μm, the shape prints well enough, so no adjustment on gate width or length dimension is needed, provided that gate width is maintained (compliance with rule A). At 45 nm, the object will print correctly (where correctly in this case refers to a change in effective W of 5% or less) provided that a set of ratios of properties of A, B, and C are maintained (as shown in the upper right part of the diagram; constant $k = 2$ in this example). This property is difficult to code in rule-based DRC, so a set of rules has been developed limiting A, B, and C that will suffice in most cases. Now consider two assignments of values for A, B, and C, both of which are legal according to the rule-based DRC. In the

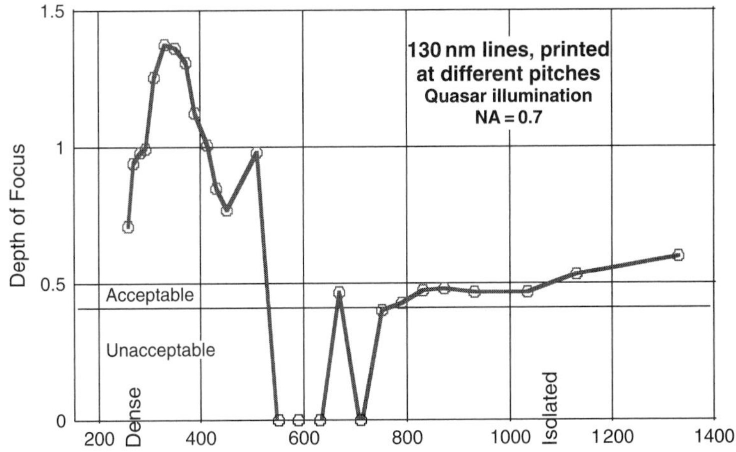

■ **FIGURE 9.11**

Forbidden pitches at 0.13-μm technology because of lithography effects. (Courtesy of D. Pan; Graph from [Socha 2000].)

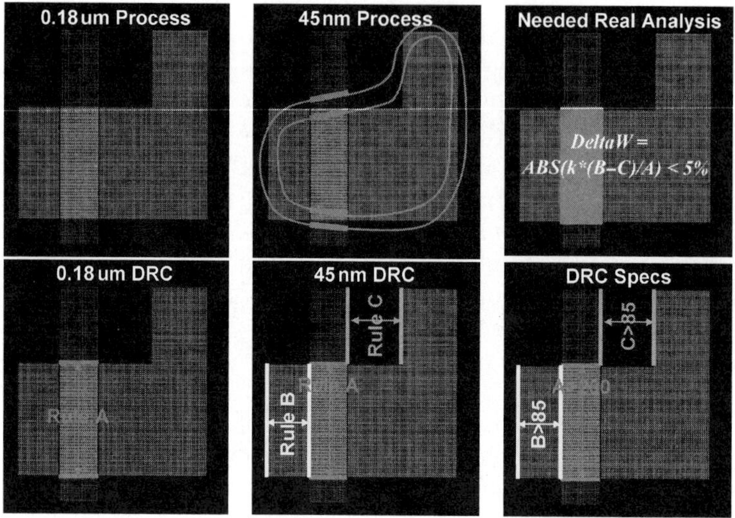

■ **FIGURE 9.12**

Shape-based DRC example. (Courtesy of Mentor Graphics.)

first case, $A = 200$, $B = 85$, $C = 90$. Delta $W = 5\%$, so the underlying "real" rule is met. Now change C from 90 to 95. Delta W goes to 10%, which could well cause a problem in the implemented circuit. This is somewhat counterintuitive, because the spacing is larger than before. An additional rule could be developed to outlaw this particular case, or extraction could be modified to take the change in W into consideration, but the situation could have been avoided in the first place if the DRC were targeted at shapes and not linear dimensions. Future generations of DRC tools will follow this approach in order to reduce complexity while simultaneously improving the ability to identify actual problem regions.

As noted previously, the more uniform an area is, the simpler it is to print. Consider, for example, Figure 9.13, which shows the polysilicon layer for a group of bit cells in an SRAM array. Notice that all gates are aligned in the same direction. The uniform density and repeated shapes allow shape-dependent variation to be effectively controlled, and this enables consistent lithography. Compare this with the polysilicon layer of an older (pre-DFM) standard cell, as shown in Figure 9.14. Some gates are horizontal and some are vertical. The density varies substantially within the cell, and there are various **jogs** and complex routes. Over the past few process generations, standard cell polysilicon has become increasingly uniform, primarily to improve printability. This has led to some loss of flexibility in layout (the design in Figure 9.14 is no longer allowed) but with the benefit of improved lithography.

To assist layout engineers, tools have been developed to identify lithography problem areas. These tools present a challenge for information exchange in a disaggregated supply chain. Detailed process information is required and must be

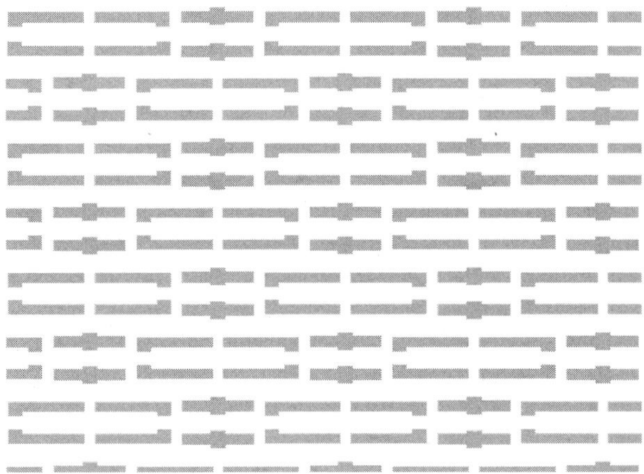

■ **FIGURE 9.13**

Uniform polysilicon layout of SRAM bit cells.

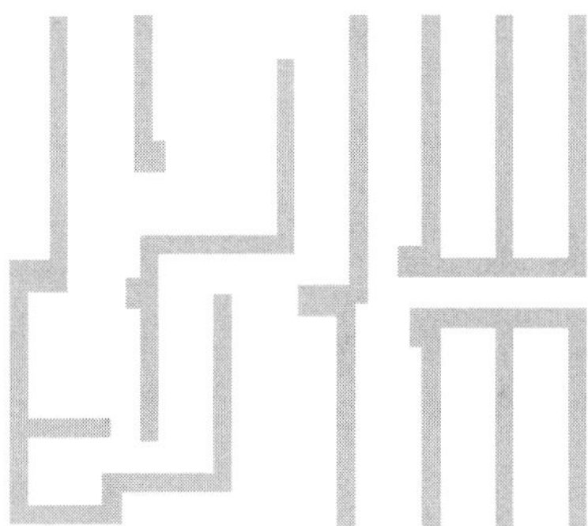

■ **FIGURE 9.14**

Diverse polysilicon layout of a pre-DFM standard cell.

applied to detailed transistor-level layout, but this needs to be done in a timely fashion (see [Aitken 2003] for additional discussion on information sharing issues). Between the 90-nm and 65-nm technology nodes, several approaches have been tried, but the industry is now converging on a method whereby foundries can encrypt some details of their processes, and tools are able to generalize some others based on their knowledge of semiconductor processes. This allows layout designers to quickly identify problem areas (**hot spots**) at a level the foundry deems important without either long turnaround times or risk of *intellectual property* (IP) loss that could happen when layouts leave an organization. Figure 9.15 presents an example of a hot spot (in this case, a potential bridging location) identified by a tool. This particular hot spot is formed because of the difficulty of applying OPC to the metal region because of the proximity of corners of the two shapes involved, as well as the empty area under the right-hand shape.

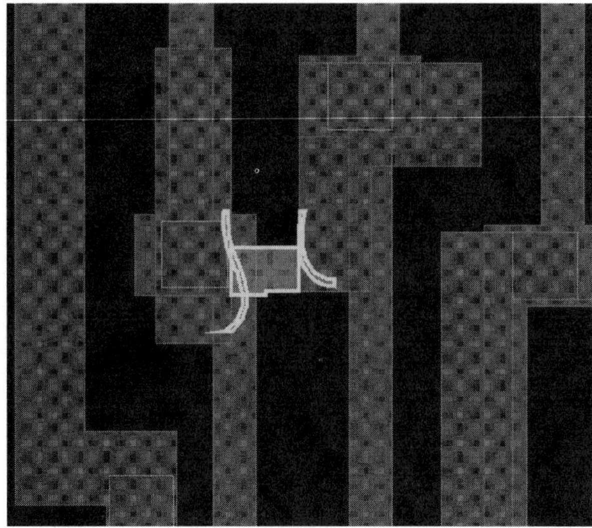

■ **FIGURE 9.15**

Example hot spot identified in 65-nm layout.

The optical region of influence for photolithography is about 1 μm. This means that patterns up to 1 μm away may affect the printability of a given shape. This figure is not reducing quickly as geometries shrink, and in some cases may actually be increasing. At 0.25-μm technology, for example, 1 μm encompasses two transistor gates and their intervening space. At 65 nm, it can cover five or more gates. In fact, the relative size of the region of influence has increased to the point where it exceeds the size of individual standard cells. This has required significant uniformity in layout rules and has led some to propose that designs be composed of larger entities (so-called **bricks**) where uniformity rules could be relaxed inside in favor of tradeoffs for performance or density [Pileggi 2003]. In the extreme case, this might result in FPGA-like layouts, which are notoriously inefficient in power and area but which are very regular for printability. The brick approach has been designed to mitigate these costs, but it may require different synthesis approaches than those needed for present-day standard cells.

Not all lithography issues relate directly to printing. The production of masks is itself a complex process, and some rules are present to reduce the cost of making masks. Also, once lithography itself is complete (the patterns have been printed on a silicon wafer), etch issues come into play. These tend to be at the wafer level and largely result from the chemical challenges associated with thin films. For example, the outer edges of a wafer spin much faster than the center. This is just one of many such phenomena that must be taken into consideration by process engineers. Other examples include mechanical issues (stress), structural issues (contact overlap of gate, leading to salicide formation challenges), and electrical issues (variations in R or C that are strongly influenced by manufactured line width or thickness).

Additional physical effects that must be considered include *chemical mechanical polishing* (CMP) and physical/chemical interactions such as **ion implant**.

Not surprisingly, photolithography remains the biggest challenge in current processes and is thus the target of most DFM/DFY. One technique being explored at the present time to improve DFM/DFY is **immersion lithography**, and this is expected to be used in most 45-nm processes. Immersion lithography takes advantage of the fact that a liquid can have a refractive index greater than 1, and thus improve resolution over air or a vacuum. The maximum expected numerical aperture for water-based immersion is 1.35, *versus* 1.0 for dry lithography. More esoteric materials can give the *numerical aperture* (NA) values of up to 1.65 [Hand 2005]. In addition, lithography using *extreme ultraviolet* (EUV) light, probably at 13.5-nm wavelength, has been proposed for patterning instead of the current 193-nm or 157-nm wavelength.

Other techniques being considered include multiple exposure methods, where different features are patterned separately. One interesting variant has been proposed in [Fritze 2005], where a completely regular pattern is generated initially by using maskless interference patterns and is then selectively "erased" where necessary to generate particular features using **lower resolution mask-based lithography**.

Lithography challenges become design challenges in two primary areas: (1) physical IP design (library design) and (2) during routing. DFM can be considered as part of library design. Certain structures are inherently vulnerable to optical effects and therefore allowance should be made in their design to ensure that subsequent OPC will be able to treat them correctly. An example is line-end shortening of polysilicon as shown in Figure 9.6. The presence of a nearby structure can prevent OPC correction, resulting in an incorrectly printed object. Optical issues undoubtedly pose a significant challenge in IP development for current process generations.

9.5.2 Critical Area

Critical area is a common metric used to evaluate the susceptibility of a given layout to defects [Shen 1985]. The critical area for bridging between wires is the area where the center of a particle of given radius could land and cause a short circuit between two wires (such short circuits are often termed simply "shorts"). In Figure 9.16, the critical area for a 0.5-micron particle is 0.2 square microns, as shown.

Critical area by itself is not a particularly significant metric. It needs to be combined with a defect distribution in order to quantify the relative importance of specific defects.

Critical area is a monotonically increasing function of particle size, because bigger particles cause more defects. Particle size itself is usually modeled as an **inverse cube distribution**, where a particle of size 2X is 1/8 as likely as a particle of size X. Critical area is affected by layout features such as layer density and complexity.

Figure 9.17 shows a defect size distribution for defect sizes ranging from 0.12 μm to 0.5 μm. The "minimum spacing" DRC rule is set at a value where the foundry

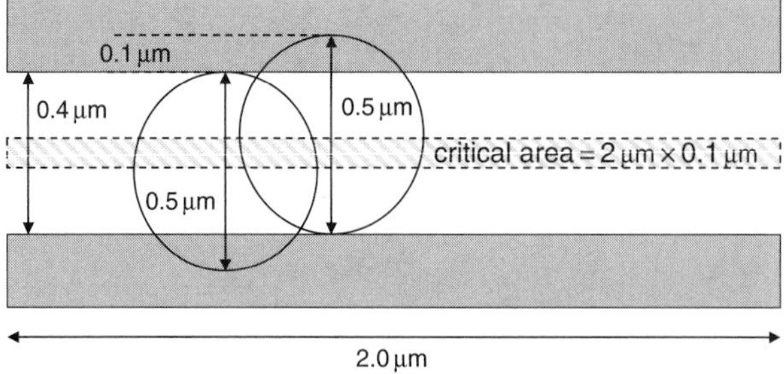

■ **FIGURE 9.16**

Calculating critical area.

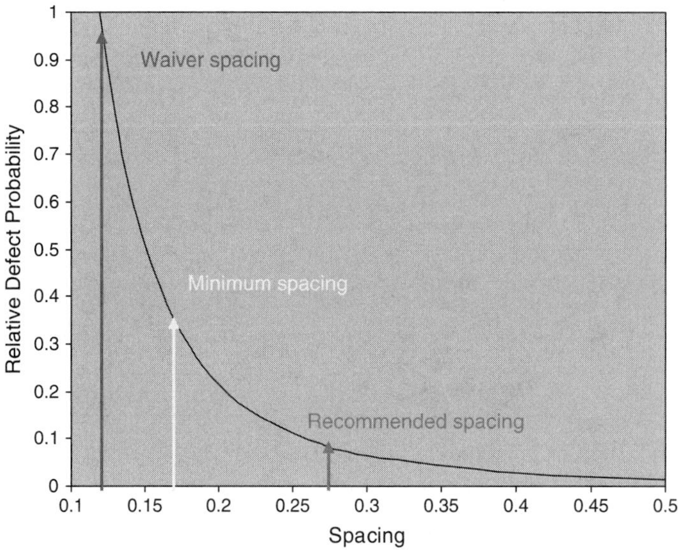

■ **FIGURE 9.17**

Defect size distribution and defect density (0.13-μm process).

believes that good yield can be achieved across a wide variety of designs. At a "recommended spacing" that is significantly higher than the minimum spacing, defect probability is reduced by a factor of about 4. Under some circumstances where a significant area or performance benefit can be achieved, process control or design requirements justify a much tighter "waiver spacing" where defect probability is higher by a factor of nearly 3.

Combining the critical area distribution (which progresses monotonically from 0 to 100% as defect size is increased) with a defect size distribution as shown in

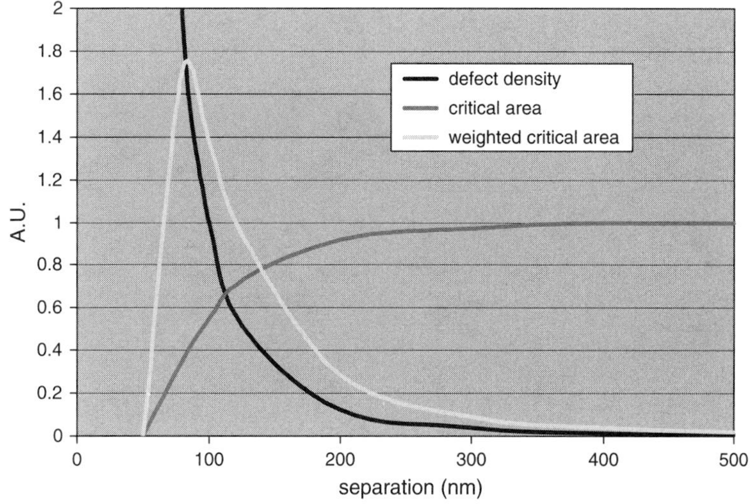

■ **FIGURE 9.18**

Weighted critical area.

Figure 9.17 leads to the concept of **weighted critical area**, where critical area at certain defect sizes is significantly more important than at others. This is shown in Figure 9.18. Weighted critical area peaks at a value slightly above the minimum DRC separation and then tapers off quickly, such that at double the minimum separation it becomes less important.

From a design standpoint, the simplest approach to improve critical area is to increase wire spacing, but as the preceding section described, this can sometimes lead to lithography problems because of forbidden pitches. In addition, because improving critical area leads to an overall area increase, it may not always be clear that this tradeoff is a good one. Defect density in a fab will change over time, but area cannot be recovered once it has been lost.

The "critical area" concept can also be extended to cover missing material (by measuring the area susceptible to a circular open of a given fixed size), contact or via failure rates, and overlap issues.

As an alternative to weighted critical area, sometimes a single defect size value can be chosen, such as the 50th percentile layer defect size (the size where half the expected defect particles are larger and half are smaller) and critical area calculated from that.

9.5.3 Yield Variation over Time

Making changes for DFM reasons that increase area or lower performance cannot be done without accurate metrics to quantify the tradeoffs. A yield metric must be able to adapt to changing process conditions in order to be useful over time.

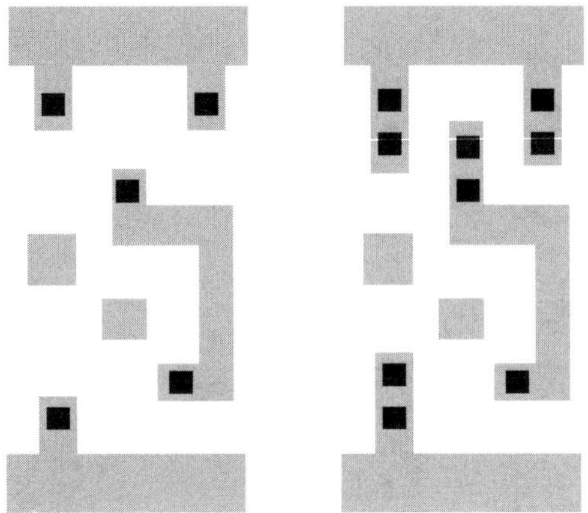

■ FIGURE 9.19

Sample layouts optimized for shorts *(left)* and for contact failures *(right).*

One major difficulty is that a manufacturing process is not static: as problems are found, they are fixed; process recipes are altered; equipment ages or is replaced. Consider the following example. Suppose two metal-1 layouts are available for a certain cell, shown in Figure 9.19. If the target process is highly susceptible to metal-1 shorts, the ideal layout will be the one on the left (L). On the other hand, if contacts are more of a problem in this particular process at this time, then the better layout will be the one on the right (R).

Consider the case where initial process analysis shows that shorts are the bigger problem, and layout L is selected as part of a cell library because its yield, YL, will be better than that of layout R, YR ($YL > YR$). If, after some time, process engineers identify the issue that causes the shorts, then although both YL and YR will improve, YR could potentially become greater than YL ($YR > YL$). A single yield number associated with either would be inadequate. Recharacterization of the library might help a subsequent design, but it is too late for one that has already gone into production. Some process changes are predictable in advance (*e.g.*, a steady decline in defect density over a product lifetime), whereas others are not (*e.g.*, unpredictable changes caused by moving to a different equipment set).

One way of looking at yield variation over time in more detail is to consider graphs such as the one shown in Figure 9.20, which are commonly used to discuss the relative weighting of **feature-limited yield** (systematic effects) *versus* **defect-limited yield** (random effects) [Guardiani 2005]. The graph conveys an important point, namely that feature-limited yield is more important in recent processes than it has been historically, but it is also somewhat misleading in that it implies that these numbers are fixed in time.

In reality, however, yield is not fixed in time. Process engineers expend significant effort identifying systematic and random problems and working to fix them [Wong 2005] [May 2006]. Feature-limited yield, for example, can be addressed either by

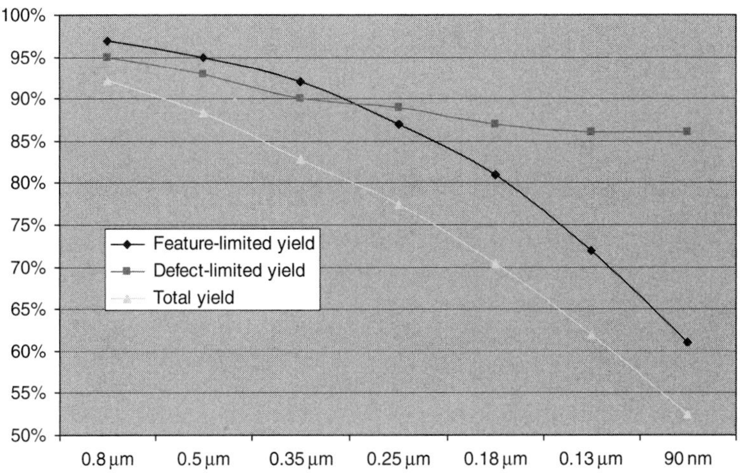

■ **FIGURE 9.20**

Common representation of random *versus* systematic yield by process generation.

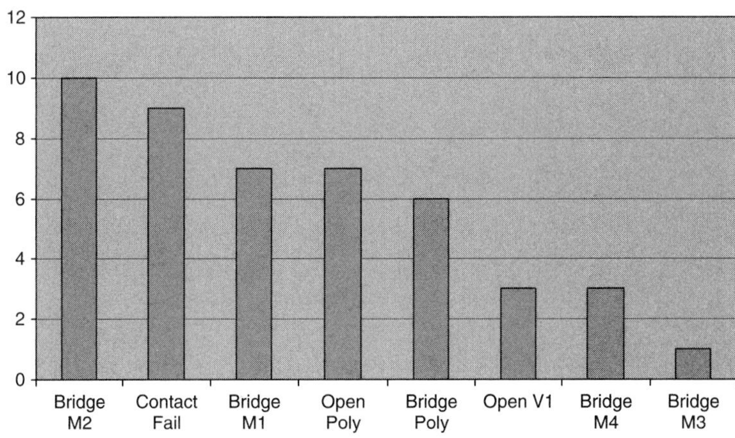

■ **FIGURE 9.21**

Example Pareto chart.

modifying the process to make the feature print (through RET changes or recipe changes) or by making the feature illegal by forcing a DRC change. In both cases, yield loss caused by the individual feature will decline. Similarly, random yield can be improved with reduction in defect density, because of a cleaner process or equipment change for example. Again, yield will improve over time. Process engineers often use a **Pareto chart** (a histogram, sorted by occurrence) to identify the most immediate problem, as shown in Figure 9. 21. Once this problem has been tackled, then the next most significant one may be addressed, etc.

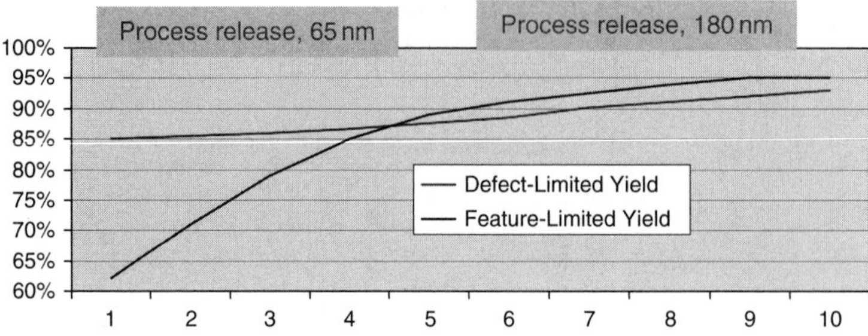

■ **FIGURE 9.22**

Yield variation over time (arbitrary time units in *X*-axis).

As a result, both random and systematic yield improve over time. Early in a process life cycle, most yield loss is systematic, but eventually, systematic yield loss is reduced to the point at which random yield loss becomes more important. The difference between current processes and their ancestors is that economic pressure is driving process release earlier in the life cycle. Figure 9.22 shows this trend. Thus, the curve in Figure 9.20 is representative at process release time, but over time all processes will tend toward being limited by defects rather than systematic effects.

9.5.4 DFT and DFM/DFY

Just as design and manufacturing have long been considered independently, the interaction between testability and yield has been neglected. Test and design have evolved over the history of integrated circuits as separate topics, though of mutual interest. *Design for testability* (DFT) is concerned about structures and methods to improve detection of defective parts. Defects cause faults, which are deterministic, discrete changes in circuit behavior. Design is not concerned with faults. Rather, design worries about parametric variation, manifesting itself as design margin. This has led to the development of DFM, which is concerned with structures and methods to improve manufacturability and yield. This state of affairs is illustrated in Figure 9.23. Test is concerned with distribution 2, while design is concerned with distribution 1.

It has always been known that this separation is artificial, but it has remained because it is convenient. The first signs of a breakdown came in I_{DDQ} test (which tests parts by measuring their leakage current). Results such as those in [Josephson 1995] and [Nigh 1997] showed that there was no readily identifiable breakpoint between good and bad parts, leading to questions about the usefulness of I_{DDQ} test [Williams 1996].

Rather than being an anomaly, **leakage variability**, as highlighted by I_{DDQ} test, has proven to be an indicator of the future. Measurement criteria such as circuit delay and minimum operating voltage are now suffering from the same inability to distinguish parts with defects from inherent variability. While it may be argued

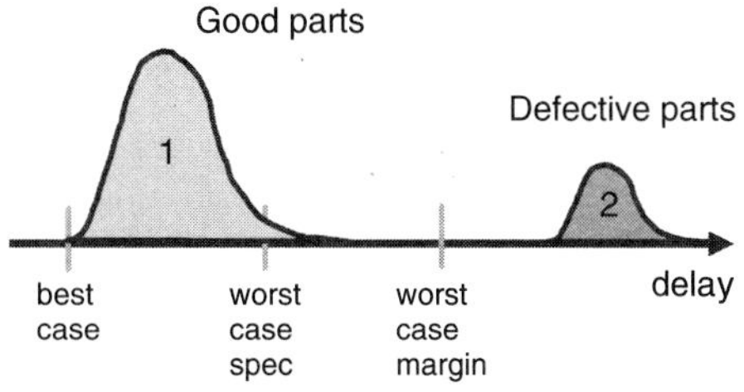

■ FIGURE 9.23

Classic separation between design margin and defects.

that there is no need to discriminate as both are "bad," it is clear that choices made in these matters will directly affect yield and quality.

9.6 VARIABILITY

Historically, testing has attempted to distinguish between "defects" (random events that cause silicon failure) and "design errors" (problems with a design that prevent it from functioning correctly). Process variation (sometimes referred to as parametric variation) occupies a middle ground between these two: design margin will account for some amount of variation, but beyond that a circuit will fail to operate. Had additional margin been provided, the design would have worked, so it may be considered to be a design error, but the excess variation can equally well be thought of as a process defect.

9.6.1 Sources of Variability

Variability occurs throughout a process, lot-to-lot, wafer-to-wafer, die-to-die, and intradie [Campbell 2001] [Segura 2004]. Three primary sources of variation are considered here: (1) transistor length variation, (2) gate dielectric thickness variation, and (3) ion implant variation.

Length variation across a chip tends to be continuous: a map of variation *versus* X,Y location will show a continuous pattern. Transistors with shorter channel lengths will be faster and leakier, while those with longer channels will be slower but less leaky. In most cases, transistor models will not account for the X,Y dependency of this variation, but will instead assume that all variation occurs within the boundaries of the device models as supplied to the designers (the familiar fast and slow silicon corners). Design margin is needed to account for this variability.

Gate dielectric thickness directly affects the electric field of a transistor, so variation can significantly change its behavior. Wafer-to-wafer variation can be caused by changes in temperature and time in the furnace during the dielectric formation process. Because gate dielectric thickness is so vital, great care is taken to ensure uniformity, and localized variation is minimal; even so, **dielectric variation** does

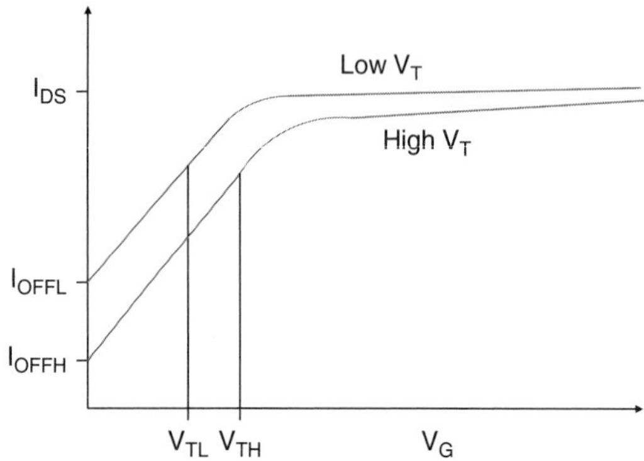

■ **FIGURE 9.24**

Leakage and delay variation with threshold voltage.

contribute to leakage variation, especially at low temperatures where gate leakage is more dominant.

Unlike length and gate dielectric variation, **ion implant variation** is localized. The distribution of implanted ions is roughly Gaussian and depends on the ion's energy. For lighter ions such as **Boron**, lower implant energy results in higher variability [Campbell 2001]. Bombarding ions also scatter beyond masked areas, resulting in changes to diffused areas and deflecting off of nearby structures, thereby creating higher variability near well boundaries. Implant variation between transistors leads to threshold voltage variation, which in turn causes changes to both leakage and delay as shown in Figure 9.24. As a result, leakage variation caused by implant variation can be assumed to be independent on a per-cell basis.

Small variations in threshold voltage lead to substantial changes in leakage, but relatively minor changes in performance. For example, an 85 mV decrease in V_T might lead to 10% improvement in drive current, at the cost of a 10× increase in leakage in a typical 0.13-μm process.

9.6.2 Deterministic *versus* Random Variability

Variability can be considered at several levels of IC production: lot-to-lot, wafer-to-wafer within a lot, die-to-die within a wafer, and within a die. The first three are often grouped together as **interdie variation**, whereas the last is referred to as **intradie variation**. Process engineers spend significant time analyzing sources of variability. They check for variation in localized test structures (*e.g.*, ring oscillators or specially designed transistors) and compare properties visually. Examples of silicon properties that might vary include transistor line width, roughness of line edges, and the thickness of layers such as metal, dielectric, and polysilicon.

Figure 9.25 shows an example of a property varying on a reticle-by-reticle basis. In this case, the property (*e.g.*, line thickness) is at its strongest in the upper left die of each 2 × 2 reticle and at its weakest in the lower right die. Figure 9.26, on

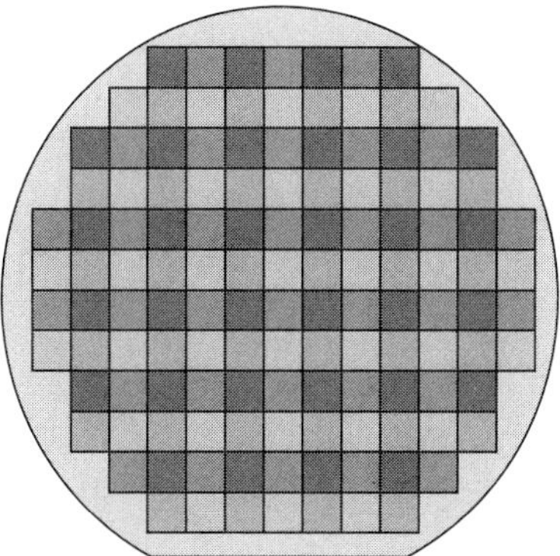

■ **FIGURE 9.25**

Reticle-based variability (2 × 2 die per reticle).

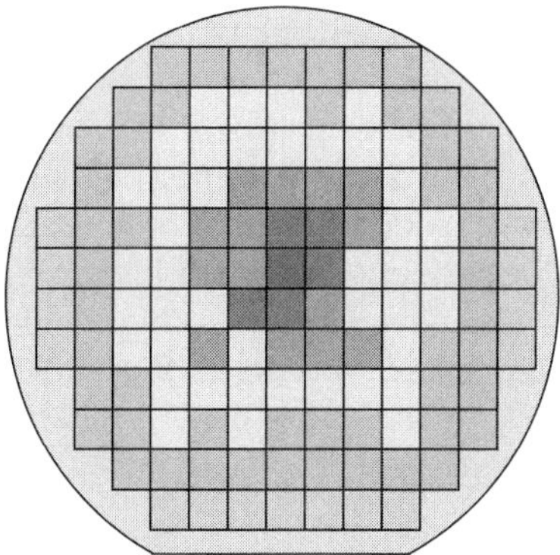

■ **FIGURE 9.26**

Wafer-based variability example (property varies radially from center outward).

the other hand, shows a property that varies on a wafer-by-wafer basis, so that it is strongest in the center of the wafer and then progressively weaker as it works outward. Notice that the variation is not absolutely equal in all directions. Other factors (such as measurement accuracy and random effects) can also influence these results.

Wafer-level variability is not always perfectly centered. It may originate in one region of the die, as shown in Figure 9.27, or follow stripes or bands, or any of a myriad of other possibilities depending on the physical and chemical phenomena underlying the variability. See [Campbell 2001] for additional discussion.

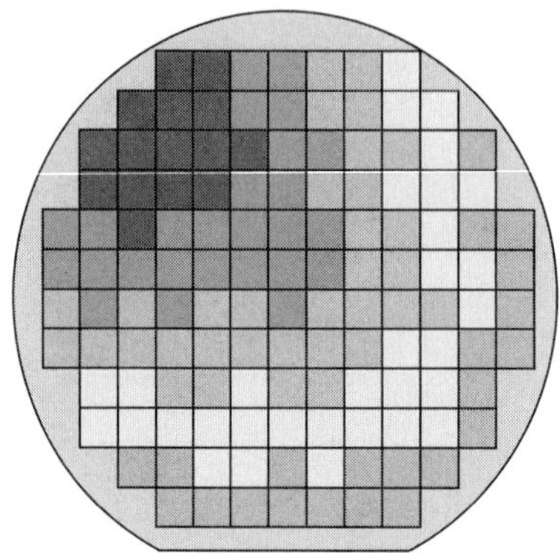

■ **FIGURE 9.27**

Wafer-based variability example (property varies radially from one region outward).

9.6.3 Variability *versus* Defectivity

As discussed previously, variability at some point can be so extreme that it becomes **defectivity** (a term meaning "manifestation of defects"). For example, if via resistance becomes too high, current passing through wires will simply fail to function at the circuit's rated speeds. There is no predefined point where a given property passes from variation to defect; the choice is fab dependent, design dependent, and in some cases application dependent. When deciding whether variation is caused by a defect or the result of simple variation, "neighborhood" context is sometimes used. This is shown in Figure 9.28, which modifies Figure 9.27 slightly to show

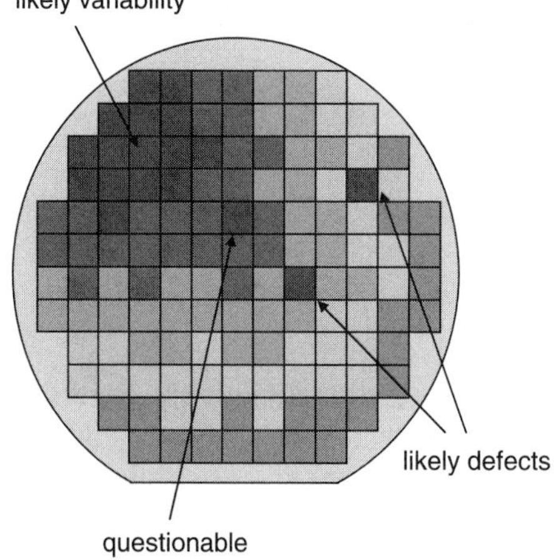

■ **FIGURE 9.28**

Defects *versus* variability example.

how some extreme measures are more likely than others to be the result of defects. For additional information, see [Madge 2002] and [Maly 2003].

9.6.4 Putting It All Together

New issues in manufacturing are requiring visibility earlier in the design cycle. It is no longer sufficient to assume that adjustments for optical effects can be made after tapeout on DRC clean layouts. Neither can designers assume that the physical effects of implant variation are negligible in memory design nor that the measurement environment and equipment have no influence in design timing. As a result, designers must consider a multiplicity of factors as part of optimization.

For physical IP designers, this means developing methodologies and metrics to take into account yield, manufacturability, and test, to ensure that tradeoffs can be made correctly. For example, doubling contacts improves yield caused by contact failures but increases critical area, thereby reducing yield caused by metal shorts, and the extra capacitance added will lower performance and may even increase sensitivity to variability in the transistors. Effective design in these circumstances requires the ability to quantify and measure all these criteria.

In addition to increased cooperation between manufacturing and design, ***design for excellence*** (DFX), which includes DFM, DFY, and DFT, requires increased understanding at higher system levels too. This ranges from simple concepts (some transistors are involved in critical paths, some are not) to more complex interactions between variability, performance targets, power consumption, and more. For example, consider the microarchitectural approach used in Razor [Ernst 2004], where the processor is designed to operate with transient errors that are corrected by execution rollback. This allows the circuits to operate at a lower power level or higher frequency than they might otherwise, but it cannot be achieved without significant interaction between the physical IP and the processor architecture.

9.7 METRICS FOR DFX

This discussion follows what was presented in [Aitken 2006b]. Each of the classic standard cell properties has associated metrics. Area is the simplest and is usually measured in square microns. Timing was initially expressed as a simple delay, typically the propagation delay of the cell in nanoseconds. As technology advanced, this simple delay measurement became inadequate. Delay was next calculated for multiple process corners (slow, typical, and fast), to account for process spread. As wire delay has become a more important component of overall delay, more complex delay models have been developed, including equation-based timing, nonlinear delay tables, current source modeling, etc. It is not uncommon for an inverter to have several hundred delay measurements associated with it. Similarly, power metrics have evolved from a single number representing dynamic current to include leakage, load-dependent power, state-dependent leakage, etc.

These metrics share two important qualities: (1) they move over time toward increasing accuracy, and (2) there is an agreed-upon procedure for calculating them (de facto standard). Both qualities are needed to assure working silicon across multiple designs and designers. The actual delay of a single standard cell is rarely measured for a chip. Instead, it is the cumulative delay effect of critical paths made up of many of these cells connected together that matters. Similarly, the power associated with a given cell can be identified only with special hardware, but cumulative power consumption of all cells is what matters for battery life, heat dissipation, etc. If the low-level metrics were unreliable, cumulative calculations would be meaningless.

The process of calculating metrics for standard cells is known as **characterization**. In general, SPICE is used as the "gold standard" for characterization, because it has proven over time to describe cell behaviors, especially delay, accurately enough to enable high-volume manufacturing. Table 9.1 summarizes the relationship between the important standard cell properties and their associated metrics. For each property, the metric used to calculate that property is listed, along with the level of accuracy that successful manufacturing requires. Also included are the accuracy that is available in state-of-the-art technology and an "implications" column, which comments on the issues facing efforts to match accuracy to requirements.

The last row in Table 9.1 is for yield and manufacturability. The entries are vague because there is much that is not universally agreed upon for yield and standard cells, including a suitable metric (percentage or number of good dies per wafer, across lots or per lot, etc.), an objective means of calculating the metric, and even the data that would be used in the calculation.

9.7.1 The Ideal Case

Ideally, a yield number could be associated with each standard cell and a synthesis tool could use this yield number, along with power, performance, and area to produce a circuit optimal for the designer's needs. Several efforts have been made in this area (*e.g.*, [Guardiani 2004] and [Nardi 2004]), and it is certainly a worthy goal to strive for.

The major difficulty with the ideal case, as noted earlier, is that a manufacturing process is not static (see Figure 9.22). As problems are found, they are fixed. A single yield number associated with any specific problem would be inadequate. Recharacterization of the library might help a subsequent design, but it is too late for one that has already gone to silicon. Some process changes are predictable in advance (*e.g.*, steady decline in defect density over a product lifetime), whereas others are not (*e.g.*, changes caused by moving to a different equipment set).

Observation 1

A yield metric must be able to adapt to changing process conditions in order to be useful over time.

A second issue is objectivity. Before embarking on a design, it is common for designers to evaluate several libraries and to select the one(s) that provide the best results for the target design. As we have shown, standard library metrics

TABLE 9.1 ■ Standard Cell Metrics

Property	Metric	Accuracy Needed	Accuracy Available	Implications
Area	Square microns	High	High	Self-evident to calculate; chip area is a property of library routability
Power	Dynamic and leakage current	Medium	Medium to high; highly dependent on the SPICE model; challenging for leakage because of process variation	Except in special cases, only the cumulative effect of thousands of cells is measurable, not individual values
Performance	Delay	High	High, but requires complex analysis at the cell, block, and chip level; dependence on variation not fully understood	Performance depends on a small number of cells (typically 10–100) forming a set of critical and near-critical paths, so accuracy is vital
Manufac-turability	Yield	Depends	Depends	Failure of a single cell can cause a chip to fail, but failure rates depend on factors that are difficult to characterize; catastrophic failure can be predicted with some accuracy, but interacting failure modes are difficult to model, let alone predict

have objective definitions: given a SPICE model and set of assumptions (*e.g.*, delay of a transition is the time taken for a signal to move from 10% to 90% of the power rail voltage), then the simulated delay of a cell should be the same for a given output load, regardless of who calculates it. Yield, on the other hand, is inherently statistical. The yield of a cell depends on its surrounding context, including neighboring cells, position on a die, position in a reticle, and position on a wafer. Exact values for each of these are unlikely, so approximations must be used, based on some test data. Different organizations may have different access to data, may have used different methods to collect it, and may have collected it at different times. Comparisons can therefore be challenging. For example, critical area is a relatively objective metric, but to be converted to yield it must include a failure rate, and this is subject to all the difficulties just mentioned as well as commercial sensitivity.

Observation 2

A yield metric must have an objective definition before it can be used for comparison.

9.7.2 Potential DFY Metrics

Despite the challenges, there are a number of potential yield metrics that can be useful in *design for yield* (DFY).

9.7.2.1 Critical Area

Critical area (see Section 9.5.2) is one such metric used to evaluate the susceptibility of a given layout to defects. Figures 9.29 and 9.30 show two partial metal-1 layouts for a three-input NOR gate. The critical area for shorts is much smaller for layout 1 (Figure 9.29) than that of layout 2 (Figure 9.30), showing that even for simple functions critical area can be an important criterion.

Combining critical area with a particle distribution leads to *weighted critical area*, as shown in Figure 9.18. An example is given in Figure 9.31. Similar to other outlier-based analyses [Daasch 2001], the extreme cells at either end of the curve should be subjected to additional analysis—those in a high critical area to improve yield and those in a low critical area to assess layout effectiveness.

9.7.2.2 RET-Based Metrics

Defining a metric for *resolution enhancement technology* (RET) is complicated by several factors. First, recipes are foundry proprietary. Changes in yield and

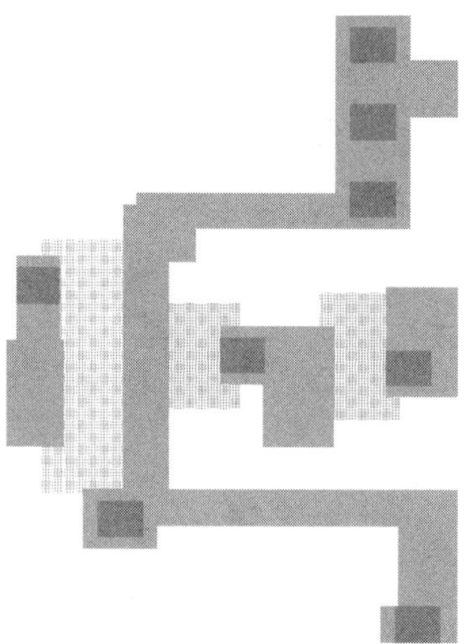

■ FIGURE 9.29

Layout 1 for a three-input NOR gate.

■ **FIGURE 9.30**

Layout 2 for a three-input NOR gate.

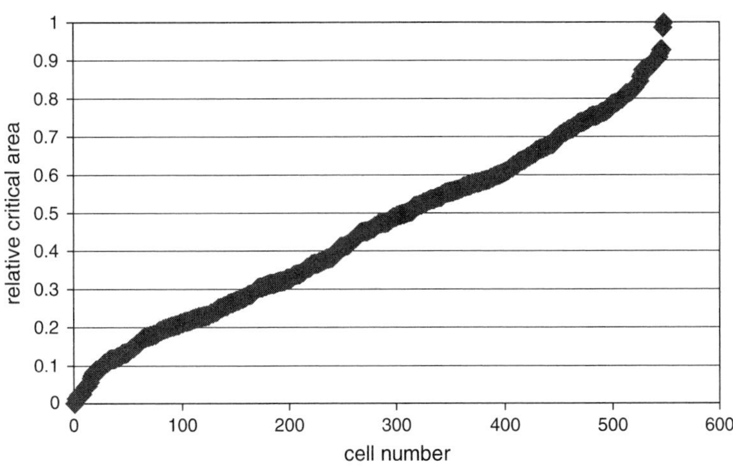

■ **FIGURE 9.31**

Relative critical areas for metal-1 shorts.

manufacturability give foundries an edge over their competitors, so RET recipes are jealously guarded. Current generation tools use data encryption to help with this issue. Second, *optical proximity correction* (OPC) and ***phase shift mask*** (PSM) rules change frequently. Even if it were possible to update library artwork for every revision, this would disrupt user design flows and thus be unacceptable. Finally, there is a data volume issue: post-OPC layouts contain significantly more shapes and are thus significantly larger than pre-OPC data. For users already burdened by huge tape-out file sizes, including OPC information with a library is unacceptable. Still, there is some hope for such metrics, and several commercial vendors have proposed them.

9.7.2.3 Example DRC-Based Metrics for DFM

Because DFM is complicated, foundries have developed special rules and recommendations for DRC. They fall into several major categories, including the following:

1. **Improved printability.** These include line end rules, regularity requirements, diffusion shape near gate rules, and contact overlap rules, among others.
2. **Reduced mask complexity**. These include rules about "jogs," or small changes in dimensions, structures, which could confuse line end algorithms, and space needed for phase-shift mask features.
3. **Reduced critical area.** These include relaxed spacing, increased line thickness, etc.
4. ***Chemical-mechanical processing* (CMP) rules.** These include density fill rules, as well as layer relationship rules.
5. **Performance related rules.** These include extra source width to reduce current crowding, symmetrical contacts around the gate, etc. The collective result of these rules is to ensure that transistors perform as intended.

Sometimes a rule serves multiple purposes, and sometimes the purposes conflict. For example, increasing contact overlap improves yield but also increases critical area for shorts. To allow numerical treatment of rules, a weighted approach is desirable. Each rule can be given a certain weight, and relative compliance can be scored. An example is given in Table 9.2 for four simplified polysilicon rules and a simple scoring system: 0% for meeting minimum value, 50% credit for an intermediate value, 100% for a recommended value; and the inverse of these values for "negative" rules, of which "avoid jogs" is an example—any noncomplying structures are subtracted from the score. Note that the rule values are artificial and are not meant to represent any actual process.

TABLE 9.2 ■ Rules Example

Rule		Weight	Scoring		
			0%	50%	100%
1.	Increase line end	0.4	0.05	0.1	0.15
2.	Avoid jogs in polysilicon	−0.3	Jog > 0.1	Jog > 0.05	Jog < 0.05
3.	Reduce critical area for polysilicon gates	0.2	0.15 spacing	0.2 spacing	0.25 spacing
4.	Maximize contact overlap	0.1	0.05 on two sides	0.05 on three sides	0.05 on four sides

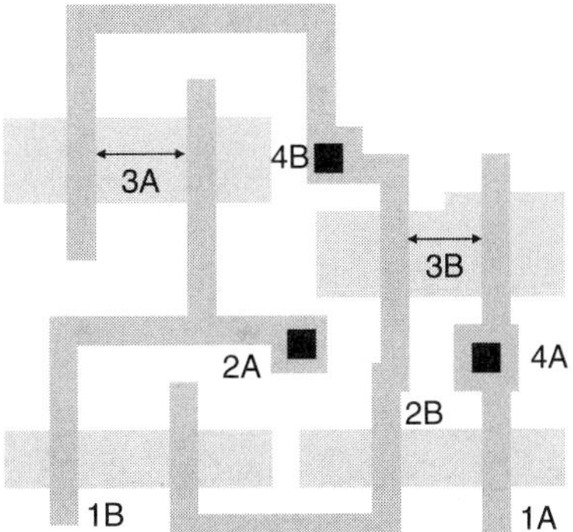

■ **FIGURE 9.32**

Example layout 1.

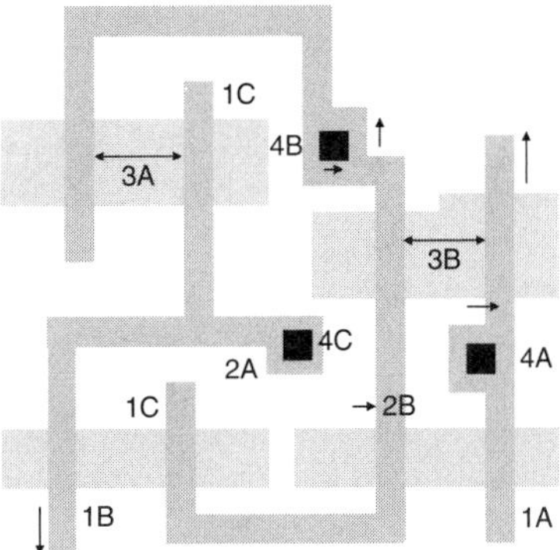

■ **FIGURE 9.33**

Example layout 2.

Figure 9.32 shows a sample layout, together with areas that comply with the minimum rule (*e.g.*, 1B is a minimum line end, 4B is a minimal contact) and the recommended rule (1A meets the recommended value, 4A is an optimal contact). In scoring this cell, there are 6 line ends, 2 minimum (0%), 1 intermediate (50%), and 3 recommended (100%), for a total of 3.5 out of 6. There are 4 small jogs, for a score of −4. Gate spacing is scored at 2.5 out of 4 (two recommended, one intermediate), with contacts at 1 out of 3. Weighting these values gives a total of 0.8 out of a possible total of 3.5. Minor changes to the cell layout, as shown in Figure 9.33, increase the weighted total to 2.65, much closer to the ideal. None of these changes increased cell area. Improving some values further would require an area increase to avoid violating other rules. These are indicated by "C." The results are summarized in Table 9.3.

Building a fractional compliance metric such as the one described here is a straightforward process using the scripting capability of modern DRC tools. Similar

TABLE 9.3 ■ Metrics for Example Layouts

		Layout 1		Layout 2		Ideal Layout	
	Weight	**Raw**	**Weighted**	**Raw**	**Weighted**	**Raw**	**Weighted**
Rule 1	0.4	3.5	1.4	4.5	1.8	6	2.4
Rule 2	0.3	−4	−1.2	0	0	0	0
Rule 3	0.2	2.5	0.5	3	0.6	4	0.8
Rule 4	0.1	1	0.1	2.5	0.25	3	0.3
Total	1		0.8		2.65		3.5

scripting methods have been shown in [Pleskacz 1999] to calculate critical area. The challenge is to determine an effective set of weights for the various rules. Additionally, there are some DFM requirements that cannot readily be expressed in either rule or metric format, and these still require hand analysis. For example, a DRC-clean structure might be tolerable in a rarely used cell but be flagged as a potential yield limiter in a frequently used cell.

9.8 CONCLUDING REMARKS

As ***complementary metal oxide semiconductor*** (CMOS) processes evolve, *design for manufacturability* (DFM) and *design for yield* (DFY) will become increasingly important, and their interaction with *design for testability* (DFT) will also become critical. The combination of these and other effects is classified as *design for excellence* (DFX) in this chapter. A key aspect of DFX will be the ability to cope with ever increasing amounts of variability that will be encountered with reduced feature sizes. For example, the number of dopant atoms in a minimum-sized channel at 65-nm process node is on the order of 200, gate oxides are four to five molecules thick, and resolution enhancement continues to push physical limits. Innovative approaches continue to be developed to counter this complexity. Some of the more recent promising approaches include the use of regular fabrics to simplify processing [Pileggi 2003], radically restricted layout practices [Liebmann 2003], and statistical approaches to timing [Visweswariah 2004].

This chapter has attempted to bring forth several important issues regarding design for manufacturability and yield enhancement. Among these are the following:

- The economics of CMOS scaling and Moore's law have driven the IC industry to a disaggregated supply chain model, with increasing consolidation in the various specialty areas.
- "Yield" has no single objective definition but is measured and modeled in different ways depending on the situation.
- Overall yield is composed of five subcategories: systematic yield, parametric yield, defect-related yield, design-related yield, and test-related yield.

- The possibility of repair introduces additional complexity to yield calculations, but with the benefit of recovering parts that would otherwise be unusable.
- Photolithography has moved well into the realm of subwavelength features and now requires significant resolution enhancement technology and subresolution enhancement features.
- Uniformity and regularity are key to successful DFM.
- A variety of useful metrics can be developed for DFM, including DRC-derived metrics, shape-based metrics, CMP-based metrics, and weighted critical area.
- Manufacturing processes are dynamic, so yield enhancement must evolve with the process and is subject to change over time.
- As relative variation increases, DFM, DFY, and DFT are becoming increasingly intertwined.
- Variability contains deterministic and random components. Control and measurement can quantify and limit the deterministic portion but can never eliminate the random portion.

The demise of Moore's law and CMOS scaling has been predicted for more than 20 years, but innovation has so far trumped the detractors. The effective use of design in order to improve manufacturability and yield has always been a competitive advantage for those who have tried it and will continue to be a vital component of IC design and manufacturing. As DFX evolves, it will absorb new areas, including variability, reliability, and power management.

9.9 EXERCISES

9.1 **(Introduction)** Redraw Figure 9.1 to include semiconductor equipment vendors and their relationship to foundries. What other industries could form part of this ecosystem?

9.2 **(Yield)** Defect density in a Poisson process drops from 1 per cm^2 to 0.5 per cm^2.

a. How does yield change for a chip of area 1 cm^2?

b. How does yield change for a chip of area 0.25 cm^2?

9.3 **(Photolithography)** A lens has a numerical aperture of 0.5 for light with a wavelength of 193 nm.

a. What is the minimum feature size that can be printed with the lens using conventional imaging?

b. What is the minimum value if off-axis imaging is used?

c. Immersion lithography has the effect of increasing the numerical aperture by the refractive index of the immersion medium. Assume that air has refractive index 1 and water has refractive index 1.44. What is the theoretical minimum feature size for water-based immersion lithography using off-axis imaging and the previously described lens?

9.4 **(Photolithography)** Uniformity in layout is desirable for lithography purposes because it enables better printing of a reduced number of shapes. It is also helpful for ***chemical-mechanical polishing*** (CMP) because it allows for consistent etch rates. Which of the following layouts is better for CMP and why?

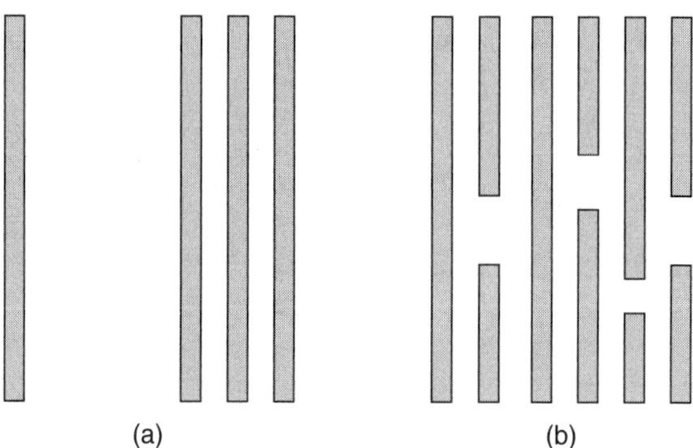

■ **FIGURE 9.34**

Metrics for Example Layouts

9.5 **(Critical Area)** Which of the preceding layouts has lower critical area for shorts because of particles of a size equal to twice the wire width?

9.6 **(Critical Area)** Two well-known methods for calculating critical area are "dot throwing," where random dots are inserted in a layout to determine if they join two lines, and "polygon expansion," where rectangle boundaries are enlarged and then subsequently analyzed to see if shorts are created. Which approach is likely to arrive at an answer faster for a large (*e.g.*, chip size) layout?

9.7 **(Variability)** Why are the outlying red dies in Figure 9.28 more likely to be caused by defects than those in the main group? Are the ones in the main group known to be defect free?

9.8 **(DFX)** How would the "ideal" layout described in Table 9.3 differ from that shown in Figure 9.33?

Acknowledgments

The author wishes to thank Dr. Lars Liebmann of IBM, Dr. Jean-Marie Brunet of Mentor Graphics, and Professor David Pan of the University of Texas at Austin for their contributions to this chapter, as well as Dr. Anne Gattiker of IBM, Dr. Anurag Mittal of ARM, and Dr. Harry Oldham of ARM for their valuable review comments.

References

R9.0 Books

[Aitken 2006a] R. C. Aitken, Defect-oriented testing, Chapter 1, in *Advances in Electronic Testing: Challenges and Methodologies*, D. Gizopoulos, editor, Springer Science, New York, 2006.

[Campbell 2001] S. A. Campbell, *The Science and Engineering of Microelectronic Fabrication*, Oxford University Press, Oxford, United Kingdom, 2001.

[May 2006] G. S. May and C. J. Spanos, *Fundamentals of Semiconductor Manufacturing and Process Control*, IEEE Press (Wiley), Hoboken, NJ, 2006.

[Nishi 2000] Y. Nishi and R. Doering, editors, *Handbook of Semiconductor Manufacturing Technology*, Marcel Dekker, New York, 2000.

[Segura 2004] J. Segura and C. F. Hawkins, *CMOS Electronics: How It Works, How It Fails*, IEEE Press (Wiley), Hoboken, NJ, 2004.

[Wong 2005] B. P. Wong, A. Mittal, Y. Cao, and G. Starr, *Nano-CMOS Circuit and Physical Design*, IEEE Press (Wiley), Hoboken, NJ, 2005.

R9.1 Introduction

[Balaban 1975] P. Balaban and J. J. Golembeski, Statistical analysis for practical circuit design, *IEEE Trans. on Circuits and Systems*, CAS-22(2), pp. 100–108, February 1975.

[Director 1977] S. W. Director and G. D. Hachtel, The simplicial approximation approach to design centering, *IEEE Trans. on Circuits and Systems*, CAS-24(7), pp. 363–372, July 1977.

[Eichelberger 1977] E. B. Eichelberger and T. W. Williams, A logic design structure for LSI testability, in *Proc. ACM/IEEE Design Automation Conf.*, pp. 462–468, June 1977.

[Hiremane 2005] R. Hiremane, From Moore's law to Intel innovation: Prediction to reality, *Technology@Intel Magazine*, 3(3), www.intel.com/technology/magazine/archive/silicon.htm#2005, April 2005.

[Kennedy 1964] D. P. Kennedy, P. C. Murley, and R. R. O'Brien, A statistical approach to the design of diffused junction transistors, *IBM J. Research and Development*, 8(5), pp. 482–495, November 1964.

[Lindsay 1976] B. W. Lindsay and B. T. Preas, Design rule checking and analysis of IC mask designs, in *Proc. ACM/IEEE Design Automation Conf.*, pp. 301–308, June 1976.

[Moore 1965] G. E. Moore, Cramming more components onto integrated circuits, *Electronics*, pp. 114–117, April 19, 1965.

R9.2 Yield

[Madge 2002] R. Madge, M. Rehani, K. Cota, and W. R. Daasch, Statistical post-processing at wafersort: An alternative to burn-in and a manufacturable solution to test limit setting for sub-micron technologies, in *Proc. IEEE VLSI Test Symp.*, pp. 69–74, April 2002.

R9.3 Components of Yield

[Stapper 1989] C. H. Stapper, Large-error fault clusters and fault tolerance in VLSI circuits: A review, *IBM J. Research and Development*, 33(2), pp. 162–173, March 1989.

[Sydow 2006] M. Sydow, Compare logic-array to ASIC-chip cost per good die, *Chip Design*, pp. 17–19, February/March 2006.

R9.4 Photolithography

[Liebmann 2003] L. Liebmann, Layout impact of resolution enhancement techniques: Impediment or opportunity?, in *Proc. IEEE Int. Symp. on Physical Design*, pp. 110–117, April 2003.

R9.5 DFM and DFY

[Aitken 2003] R. Aitken, Applying defect-based test to embedded memories in a COT model, in *Proc. IEEE Int. Workshop on Memory Technology Design and Test*, pp. 72–77, July 2003.

[Fritze 2005] M. Fritze, B. Tyrrell, T. Fedynyshyn, M. Rothschild, and P. Brooker, High-throughput hybrid optical maskless lithography: All-optical 32-nm node imaging, *Emerging Lithographic Technologies IX, Proc. SPIE Vol. 5751*, pp. 1058–1068, May 2005.

[Guardiani 2005] C. Guardiani, M. Bertoletti, C. Dolainsky, N. Dragone, M. Malcotti, and P. McNamara, An effective DFM strategy requires accurate process and IP pre-characterization, in *Proc. ACM/IEEE Design Automation Conf.*, pp. 760–761, June 2005.

[Hand 2005] A. Hand, High-index fluids look to 2nd-generation immersion, *Semiconductor International*, (4), April 1, 2005.

[Josephson 1995] D. Josephson, M. Storey, and D. Dixon, Microprocessor IDDQ testing: A case study, *IEEE Design & Test of Computers*, 12(2), pp. 42–52, Summer 1995.

[Nigh 1997] P. Nigh, W. Needham, K. Butler, P. Maxwell, and R. Aitken, An experimental study comparing the relative effectiveness of functional, scan, IDDq and Delay Fault Testing, in *Proc. IEEE VLSI Test Symp.*, pp. 459–464, April 1997.

[Pileggi 2003] L. Pileggi, H. Schmit, A. J. Strojwas, P. Gopalakrishnan, V. Kheterpal, A. Koorapaty, C. Patel, V. Rovner, and K. Y. Tong, Exploring regular fabrics to optimize the performance-cost trade-off, in *Proc. ACM/IEEE Design Automation Conf.*, pp. 782–787, June 2003.

[Shen 1985] J. P. Shen, W. Maly, and F. J. Ferguson, Inductive fault analysis of MOS integrated circuits, *IEEE Design & Test of Computers*, 2(6), pp. 13–36, December 1985.

[Socha 2000] R. Socha, M. Dusa, L. Capodieci, J. Finders, J. Chen, D. Flagello, and K. Cummings, Forbidden pitches for 130 nm lithography and below, *Optical Microlithography XIII, Proc. SPIE Vol. 4000*, pp. 1140–1155, July 2000.

[Williams 1996] T. W. Williams, R. Dennard, R. Kapur, M. R. Mercer, and W. Maly, Iddq test: Sensitivity analysis of scaling, in *Proc. IEEE Int. Test Conf.*, pp. 786–792, October 1996.

R9.6 Variability

[Ernst 2004] D. Ernst, S. Das, S. Lee, D. Blaauw, T. Austin, T. Mudge, N. Kim, and K. Flautner, Razor: Circuit-level correction of timing errors for low-power operation, *IEEE Micro*, 24(6), pp. 10–20, November/December 2004.

[Madge 2002] R. Madge, M. Rehani, K. Cota, and W. R. Daasch, Statistical post-processing at wafersort: An alternative to burn-in and a manufacturable solution to test limit setting for sub-micron technologies, in *Proc. IEEE VLSI Test Symp.*, pp. 69–74, April 2002.

[Maly 2003] W. Maly, A. Gattiker, T. Zanon, T. Vogels, R. D. Blanton, and T. Storey, Deformations of IC structure in test and yield learning, in *Proc. IEEE Int. Test Conf.*, pp. 856–865, October 2003.

R9.7 Metrics for DFX

[Aitken 2006b] R. Aitken, DFM metrics for standard cells, in *Proc. IEEE Int. Symp. on Quality in Elect. Des.*, pp. 491–496, March 2006.

[Daasch 2001] R. Daasch, K. Cota, J. McNames, and R. Madge, Neighbor selection for variance reduction in IDDQ and other parametric data, in *Proc. IEEE Int. Test Conf.*, pp. 92–100, October 2001.

[Guardiani 2004] C. Guardiani, N. Dragone, and P. McNamara, Proactive design for manufacturing (DFM) for nanometer SoC designs, in *Proc. IEEE Custom Integrated Circuits Conf.*, pp. 309–316, September 2004.

[Nardi 2004] A. Nardi and A. Sangiovanni-Vincentelli, Synthesis for manufacturability: A sanity check, in *Proc. IEEE Design, Automation and Test in Europe*, pp. 796–801, February 2004.

[Pleskacz 1999] W. A. Pleskacz, C. H. Ouyang, and W. Maly, A DRC-based algorithm for extraction of critical areas for opens in large VLSI circuits, *IEEE Trans. on Computer-Aided Design*, 18(2), pp. 151–162, February 1999.

R9.8 Concluding Remarks

[Liebmann 2003] L. Liebmann, Layout impact of resolution enhancement techniques: Impediment or opportunity?, in *Proc. IEEE Int. Symp. on Physical Design*, pp. 110–117, April 2003.

[Pileggi 2003] L. Pileggi, H. Schmit, A. J. Strojwas, P. Gopalakrishnan, V. Kheterpal, A. Koorapaty, C. Patel, V. Rovner, and K. Y. Tong, Exploring regular fabrics to optimize the performance-cost trade-off, in *Proc. ACM/IEEE Design Automation Conf.*, pp., 782–787, June 2003.

[Visweswariah 2004] C. Visweswariah, K. Ravindran, K. Kalafala, S. G. Walker, and S. Narayan, First-order incremental block-based statistical timing analysis, in *Proc. ACM/IEEE Design Automation Conf.*, pp. 331–336, June 2004.

CHAPTER 10

DESIGN FOR DEBUG AND DIAGNOSIS

T. M. Mak
Intel Corporation, Santa Clara, California

Srikanth Venkataraman
Intel Corporation, Hillsboro, Oregon

ABOUT THIS CHAPTER

Designers have to be prepared for the scenario where chips do not work as intended or do not meet performance expectations after they are fabricated. Product yield engineers need to know what has caused their product yield to be below expectation. Reliability engineers need to know what circuits or elements have failed from various stresses and which ones customers have returned. It is necessary to debug and diagnose chip failures to find the root cause of these issues, and this is best facilitated if the design has accommodated features for debug and diagnosis.

This chapter focuses on the design features at the architectural, logic, circuit, and layout level that are needed to facilitate silicon debug and defect diagnosis of integrated circuits. These design features are generally referred to as ***design for debug and diagnosis*** (DFD). We explain how these DFD features are used effectively in a debug or diagnosis environment for applications ranging from design validation, low-yield analysis, and all the way to field failure analysis. Common probing techniques used to access internal signal values (logic and timing) and alter the execution and timing behavior of an integrated circuit in a controlled manner are covered. We also describe circuit editing techniques and tools to enable probing, validate hypotheses, and confirm root-cause design fixes. These DFD features are categorized into ***logic DFD*** (LDFD) features, which enable the nonintrusive isolation of failures, and ***physical DFD*** (PDFD) features, which enable physically intrusive analysis.

10.1 INTRODUCTION

Few chips ever designed function or meet their performance goal the first time. Many fabricated chips may also fail to function because of defects, circuit/process sensitivity, infant mortality, or wearout throughout their product lifetime. When fabricated ***integrated circuits*** (ICs) of a new design do not function correctly or fail to meet their performance goal, the silicon debug process starts immediately to identify the root cause of the failure. For manufacturing defect-induced failures, the cause of failure may not be important as long as the failure rate is comparable to the process norm and is acceptable. However, if the yield is low or if the chip performance varies with process variation, the diagnosis process is utilized to identify the root cause so that corrective actions (with process and/or design changes) can be taken to improve yield. Failing to do so will reduce the profitability of the product or result in the product being unable to meet the volume demands of the market because of availability shortfall. When chips fail in the field and the field failure rate is high and above acceptable levels, customers will typically initiate action focused on the quality of the incoming product. Customers usually send these chips back, and the chips then have to be analyzed for the root cause of the failure. Following this step, corrective actions have to be put into place. If the cause of the problem is not understood and the customer is not convinced, a *customer lines-down* situation or even *product recall* may result, which can cripple the product and cause a loss of market share relative to competitors.

In all of the above scenarios, speedy and accurate debug and diagnosis is needed. This chapter focuses on the changes that can make a design more debug-able and diagnosable—hence, the acronym DFD, which stands for **design for debug and diagnosis**.

10.1.1 What Are Debug and Diagnosis?

Webster's dictionary defines **diagnosis** as the investigation or analysis of the cause or nature of a condition, situation, or problem. In the context of ***very-large-scale integrated*** (VLSI) circuits and systems, diagnosis is a general term that applies to various phases of the design and manufacturing process that involves isolating failures, faults, or defects. The term **debug** refers to the process of isolating bugs or errors that cause a design to behave differently from its intended behavior under its specified operating conditions. Debug and diagnosis techniques try to address the following questions:

- What was wrong with this device?
- Which chip was bad on this board?
- Why did this system crash?
- Why did the simulation of the ***arithmetic logic unit*** (ALU) show that $2 + 2 = 5$?
- Why was this signal late?

The aim of debug and diagnosis is to locate the root cause of the device bugs or failures.

10.1.2 Where Is Diagnosis Used?

Electronic systems are diagnosed at different stages of design and manufacturing of the constituent components of the system for several different objectives. However, the common aim of any *diagnosis* procedure is to locate the root cause of the device failures. Depending on whether the device is a VLSI chip, *multichip module* (MCM), a board, or a system, diagnosis is performed with different objectives (see Figure 10.1). In the case of MCMs, boards, or systems, diagnosis is intended to identify and subsequently replace the faulty subcircuit (a chip on a board or a board in a system) or to reconfigure the circuit around the failure. In the case of chips or integrated circuits, diagnosis is performed to improve the manufacturing process. Diagnosis followed by failure analysis is vital to IC design and manufacturing. Diagnosis identifies root cause defects for yield enhancement, finds design flaws that hinder circuit operation, and isolates reliability problems that could lead to early product failure. The primary focus of this chapter is on debug and diagnosis usage during the manufacturing and postsilicon phase of digital ICs or VLSI chips.

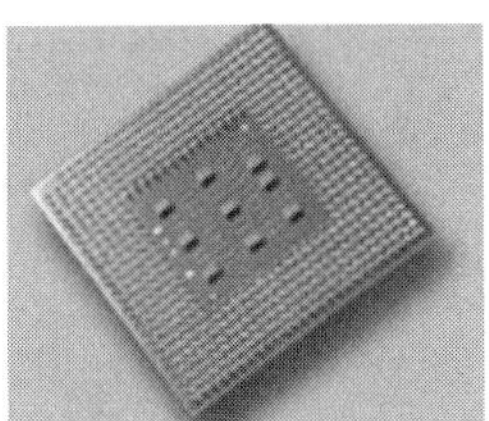

■ **FIGURE 10.1**

Diagnosis at the chip, board, and system level.

10.1.3 IC-Level Debug and Diagnosis

Figure 10.2 shows the typical life cycle of an integrated circuit or VLSI chip, from the requirements stage to the end of life. Three major phases of the life cycle are

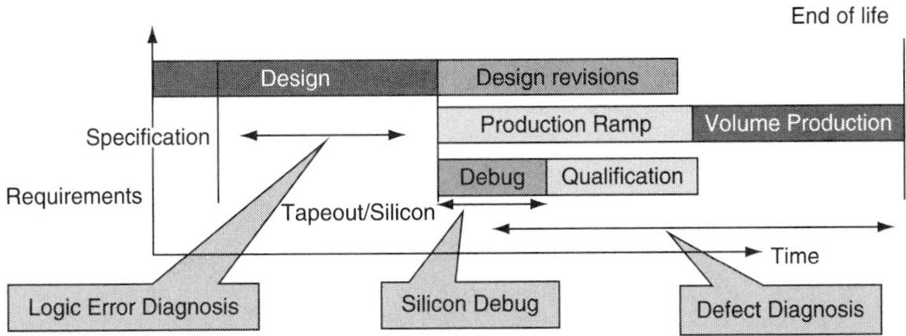

■ **FIGURE 10.2**

Debug and diagnosis applications across an IC life cycle.

shown: (1) **design** along with validation and verification, (2) **ramp-up** to production with design revisions or steppings, and (3) **volume production.** There are three major applications of diagnosis and debug techniques and technologies at the IC level to root-cause issues that are either design or manufacturing process related: (1) **design error diagnosis** during the design phase, (2) ***silicon debug*** during the ramp-up to production, and (3) **defect diagnosis** through production ramp and volume production. The focus of this chapter is on silicon debug and defect diagnosis, and they are described in more detail in the following sections.

10.1.4 Silicon Debug *versus* Defect Diagnosis

Silicon debug is the first and necessary step in the first/initial silicon stage of a product life cycle for any complex VLSI chip. Silicon debug starts with the arrival of first silicon and continues until volume production. Once first silicon arrives, these early chips are sent to the design team for validation, test, and debugging of bugs not uncovered during the presilicon design process. Many bugs and errors, such as logic errors, timing errors, and physical design errors, may be introduced during the design process. These bugs and errors should have been uncovered during the design verification and validation processes that include simulation, timing verification, logic verification, and design rule checking. However, very often bugs and errors still escape these checkpoints and cause silicon not to behave as designed. These errors are often caused by limitations and flaws in the circuit models, simulations, or verification. Problems could range from logic or functional bugs, to circuit sensitivities or marginalities, to timing or critical speed-path issues [van Rootselaar 1999] [Josephson 2001]. Verification tests performed during the design phase may not cover all corner cases encountered in a real application leading to logic or functional bugs. Static timing analysis and dynamic timing simulations may not accurately and precisely predict every critical path. This results in timing or speed paths being encountered in silicon that were not factored into design. As an interesting side note, one of the first examples of "debugging" was in 1945 when a computer failure was traced to a moth that was caught in a relay between contacts [Gizopoulos 2006].

Manufacturing defects are physical imperfections in the manufactured chip. The process of locating manufacturing defects is called defect diagnosis, fault diagnosis, or fault isolation. A key application of defect diagnosis is supporting low-yield analysis and yield enhancement activities.

One application of defect diagnosis to manufacturing yield enhancement is shown in Figure 10.3. Initially the effort is mainly focused on memories, test chips, and inline monitors. The regular and repetitive structure of a memory makes it easier to diagnose or to isolate its faults. Test chips and *static random-access memories* (SRAMs) are used to bring up a new process, and the large embedded memories in products are usually used to maintain or improve yield [Gangatirkar 1982] [Hammond 1991] [Segal 2001]. However, because of the differences in the possible defect types between logic and memories (layout topology, number of layers, and transistor density/count), memories may not capture all manufacturing process issues, leading to the use of logic diagnosis in yield learning in the intermediate and

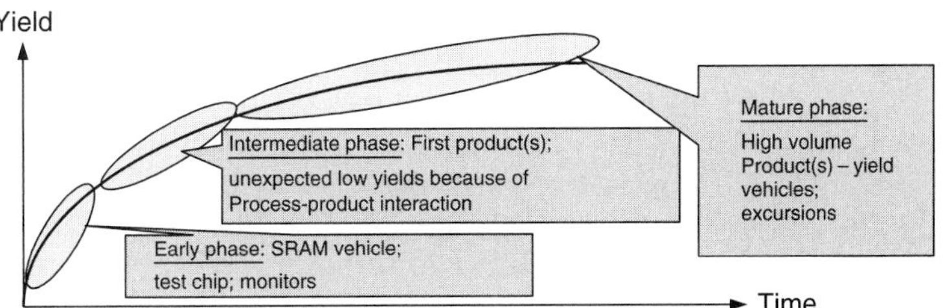

■ **FIGURE 10.3**

Defect diagnosis applications to manufacturing yield.

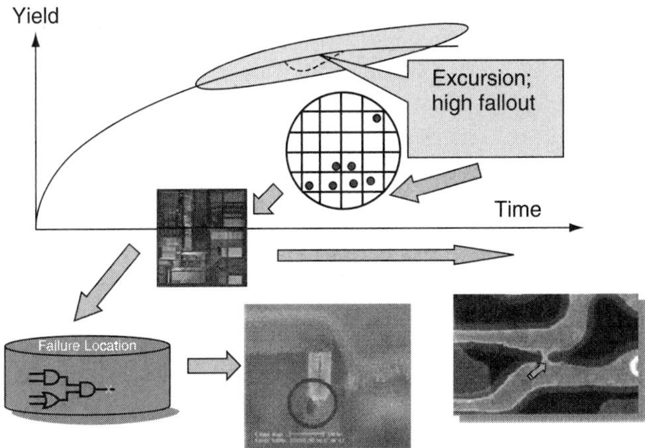

■ **FIGURE 10.4**

Defect diagnosis applications to manufacturing excursions.

mature phases of the manufacturing process. A second application of defect diagnosis is dealing with manufacturing excursions or low-yield situations as shown in Figure 10.4. Diagnosis is performed on low-yield chips to isolate the root cause of failure (process abnormality leading to high defect density, process variation, design and process sensitivities, etc.) and follow up with appropriate corrective actions. Other applications of defect diagnosis include analysis of failures from the qualification of the product, which includes reliability testing (both infant mortality and wearout) and failures from the field or customer returns.

10.1.5 Design for Debug and Diagnosis

Debug and diagnosis require a high degree of **observability**. The ability to observe erroneous events close to when they happen is important. In addition,

controllability is needed to further validate what the cause of the problem is and to narrow down the circuits that manifest the symptoms.

Many ***design for testability*** (DFT) features (*e.g.*, scan), which enable both controllability and observability, can be reused for debug and diagnosis [Gu 2002]. There are also specific DFD features (*e.g.*, clock controllability or reconfigurable logic) that are developed or tailored for debug and diagnosis.

The DFD features can be broadly bucketed into **logic DFD structures,** which are added to the design to extract logic information and to manipulate the operation of a chip in a physically nonintrusive manner. There are several physical tools to extract logic and timing information from a chip in a minimally physically intrusive manner. Circuit or ***focused ion beam*** (FIB) editing is performed to either enable probing or make corrections to the circuit to verify root-cause fixes. DFD structures to enable probing and circuit editing are called **physical DFD structures** [Livengood 1999].

Because probing is also needed to confirm the hypothesis of the cause of failures, we will touch on physical debug tools, and we will also cover some layout related DFD features that enable probing and circuit editing. **Speed debug** is an extension of **logic debug**. Throughout the description of these techniques, emphasis is placed on debugging for speed-related problems in addition to logic failures. We have decided not to include memory-related diagnosis techniques that are common for yield analysis purposes. However, references are provided should the reader want to study this area [Gangatirkar 1982] [Hammond 1991] [Segal 2001].

10.2 LOGIC DESIGN FOR DEBUG AND DIAGNOSIS (DFD) STRUCTURES

Logic design for debug and diagnosis (DFD) *structures* are added to the design to extract logic information and to manipulate the operation of a chip in a physically nonintrusive manner. This section explores key logic DFD features, including scan, observation-only scan, observation points and multiplexers, array dumps, clock control, partitioning and core isolation, as well as reconfigurable logic.

10.2.1 Scan

Scan is a widely used DFT feature. Scan is also useful for debug and diagnosis as a DFD feature (see Chapter 7 of [Wang 2006]). Although scan is typically utilized to enable *automatic test pattern generation* (ATPG), the increased observability of the design because of the insertion of scan also enables debug and diagnosis. For more information on scan and other DFT features, please refer to DFT textbooks [Abramovici 1994] [Bushnell 2000] [Jha 2003] [Wang 2006].

For ATPG patterns generated automatically and applied using scan, automated analysis of the failures can be performed using diagnosis tools that provide fault candidates for further validation [Waicukauski 1989] [Venkataraman 2001] [Guo 2006]. Modern-day DFT tool vendors also provide diagnosis tools along with their ATPG offerings.

Scan also helps in functional-pattern-based debug and diagnosis. There are two commonly used types of scan design: (1) **muxed-scan** or **clocked-scan** for flip-flop–based scan designs [Wang 2006] and (2) ***level-sensitive scan design*** (LSSD) for latch-based scan designs [Eichelberger 1977]. In both types of scan design, a functional test can be executed up to a certain clock cycle and then stopped. Once the system flip-flops and latches are reconfigured as scan cells, the functional state of these system flip-flops and latches in the chip can be shifted out for analysis during the shift operation. This process has to be coupled with *register-transfer level* (RTL) or logic simulation to compare against the expected state. For most scan designs (using muxed-scan, clocked-scan, or LSSD-like), unloading the scan cell contents is destructive in that the functional state is destroyed during shift. Hence, functional execution cannot be resumed after the data are unloaded. If a functional state at a further clock cycle needs to be investigated, then the functional pattern would have to be reexecuted and then stopped at that later cycle. This can be time consuming if the time from reset to the failure point is long, which is common in system-level debugging. An enhancement to this scheme is to wrap the scan chain around from scanout to scanin. If properly accounted for, the system flip-flops and latches will contain exactly the same contents as they did before it was shifted. Hence, functional execution can continue until the next clock cycle of interest [Hao 1995] [Levitt 1995]. This approach also requires that the scan shift operation be nondestructive with respect to all other storage elements (flip-flops and latches) in the circuit that are not scanned. Alternately, many designs add observation-only scan cells (described next). It is important to note that scan only provides a means to access the internal state of the storage elements that are scanned and help determine the incorrect state captured during the execution of a failing test. It does not tell exactly what caused the incorrect state. More debugging is necessary from the point of observed failures to determine its root cause.

10.2.2 Observation-Only Scan

Observation-only scan is a specialized version of scan that is typically used only for debugging purposes [Carbine 1993, 1997] [Needham 1995] [Josephson 2001] [Gizopoulos 2006]. This is also called **scanout** or **shadow scan**. Figure 10.5 shows the schematic of a typical scanout cell. The *Sin* port is connected to the *Sout* port of the preceding stage to create a scanout chain, whereas the *Data* input port is connected to the signal of interest. When the *LOAD* signal is enabled (set to logic 1), the signal of interest is latched into a separate **scanout cell** (where a **snapshot** is taken just for a clock cycle of interest) and can be kept there until it is ready to be shifted out when enabling the *SHIFT* signal. The whole captured state from all the signals of interest connected to scanout cells can be shifted out through a special scanout chain for analysis. This type of observation-only scan is part of a separate scan chain that is typically clocked by the same system clock capturing the signals of interest at-speed. From a performance penalty standpoint, observation-only scan merely adds some small capacitance to the signal of interest, and the functional signal does not have to pass through any additional multiplexers. Hence, this type of scan is more acceptable for functional debugging for high-performance circuits and systems.

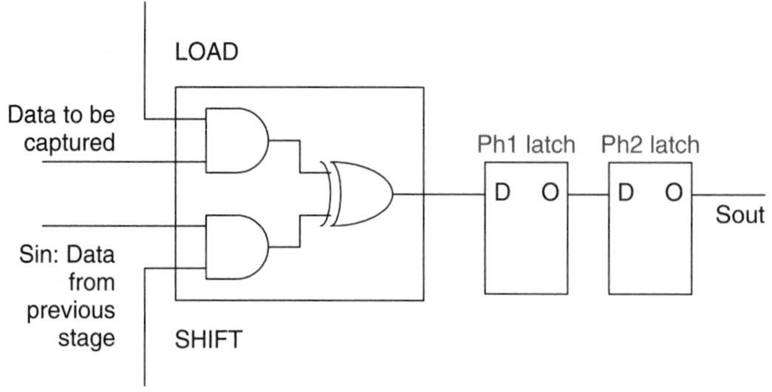

FIGURE 10.5

A scanout cell.

Scanout cells are judiciously placed throughout the design to maximize the observability of key states and signals. Potentially, they can be inserted at any signal line, but that would increase the cost because of the added area. A selective placement of this kind of scanout cells on key signals helps the debugging process tremendously while keeping area overhead low. Because of their unique speed-capturing nature, these cells can also be put on different segments of long signal lines to observe the effects of signal degradation or noise. Very often, the area of the scanout cells can also be hidden (or further reduced) if they are placed under routing channels, where silicon is not utilized. Just like conventional scan design, the placement and routing of these cells can also be automated, so the scheme is perfectly compatible with both custom and ***application-specific integrated circuit*** (ASIC) design styles.

As indicated before, observation-only scan is primarily aimed at speed debugging, because it is running concurrently with the system clock. At the specific clock cycle of interest, one would trigger the *LOAD* signal so the *Data* signal of interest would be captured into the scanout cell. On one hand, the captured state can be kept there until the test is completed and then shifted out for analysis. On the other hand, the captured state can also be shifted out simultaneously while the system is running. Repeated capturing and shifting is possible provided that the time to shift the whole chain is shorter than the time it takes to arrive at the next capture point. Again, RTL or logic simulation is executed to allow the comparison of expected states. Figure 10.6 illustrates how the scanout system is used with an ***automatic test equipment*** (ATE) to allow the capturing of internal state using scanout cells while the chip is being tested with a functional pattern that is being executed on the ATE.

Additionally, scanout systems typically have a **compressed signature** mode built into the cell, whereas the content of an upstream cell is XORed with the content of the current scanout cell. This is accomplished by enabling with *LOAD* and *SHIFT* signals simultaneously as shown in Figure 10.5. The final signature after the execution of a test can be shifted out of the scanout chain. The advantage of a signature *versus* a single snapshot is that it is more efficient to simply check an accumulated signature to see if an error is captured during a time interval starting from one

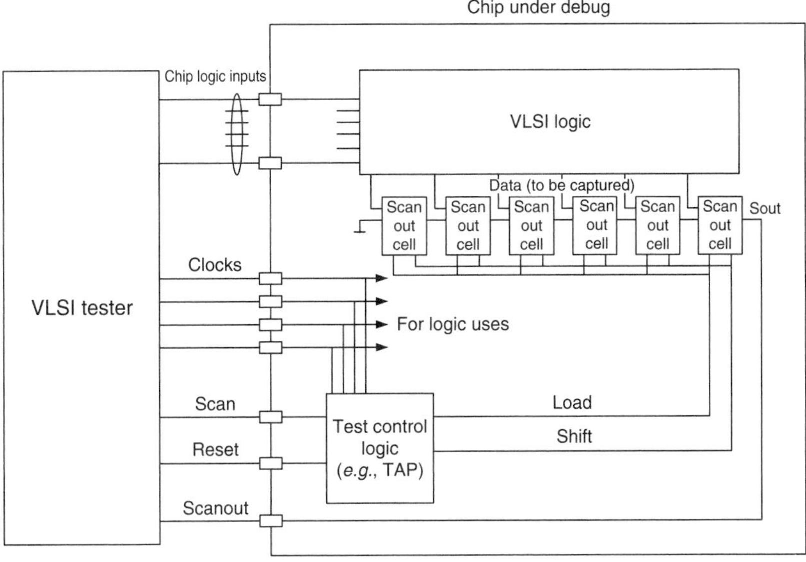

■ **FIGURE 10.6**

Operation of an observation-only scan design.

trigger event, which starts recording to another trigger event, which stops recording. This can narrow down the cycles where errors occurred during the execution of a test. A signature compare operation can also be done with the chip running in a passing condition rather than running in a failing condition. An alternate observation-only cell design was proposed in [Sogomonyan 2001].

10.2.3 Observation Points with Multiplexers

Multiplexers (MUXes) are a common alternate way to observe on-chip signals. Once a set of important signals is identified for observation, multiplexers can be added so that individual signals can be mapped out to a test port (*e.g.*, through the *Test Data Out* [TDO] port defined in the IEEE 1149.1 boundary-scan standard [IEEE 1149.1-2001]). However, the test control signals for the MUXes can add to the overhead, especially when a large set of signals is observed. Also during debugging, the signal that one would like to observe may not be included for observation and would not be available during the debugging process. The authors in [Abramovici 2006] proposed to have a layer of programmable logic that can be programmed to allow specific signals to be observed. Figure 10.7 shows how this can be done by using wrappers around individual blocks of circuits. The MUXes bring various signals of interest to each wrapper and allow signals from various wrappers to be mapped to observation points, thus alleviating the amount of overhead required while at the same time maximizing the likelihood for signals to be observed. However, the overall architecture of placing the programmable logic where it will provide a high degree of observability may still be expensive for certain designs.

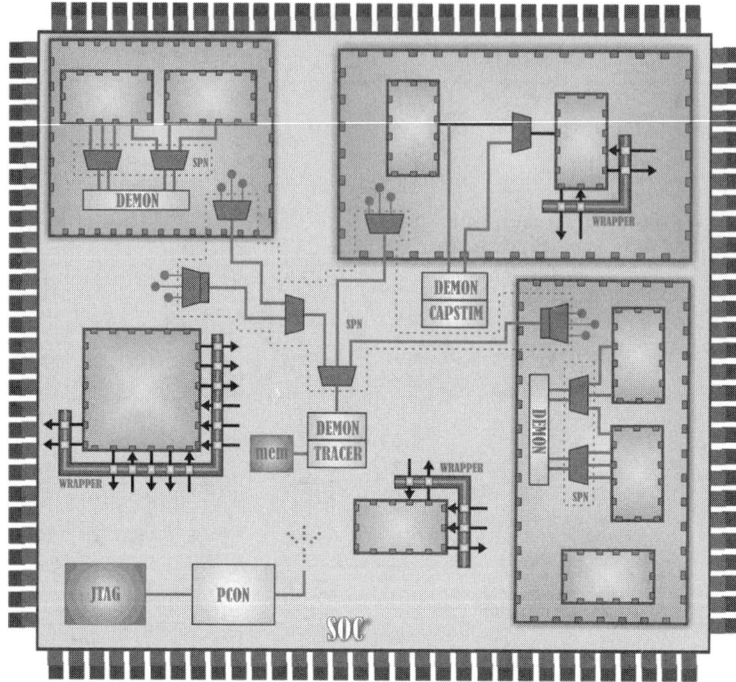

■ **FIGURE 10.7**

Programmable observation of internal signals.

The same programmable logic has an additional function beyond providing observation. Once a hypothesis of the problem is formulated, the erroneous logic can be altered with this programmable logic to provide a soft fix. Because this programming logic is already part of the chip, this repair method is more flexible than the **blue wire** patching method, which we will describe in Section 10.4. Sample or even production units can be supported provided that this logic can be programmed before the chip is initialized for use in the end user system.

10.2.4 Array Dump and Trace Logic Analyzer

In addition to logic states, which can be observed using scan and scanout, other memory states such as those held in embedded arrays (caches, register files, buffers/pointers, etc.) are also important for debug and diagnosis. Because these structures are usually not scanned, scan dump cannot access the information stored in these arrays. So an array dump mechanism is typically designed [Carbine 1997] so that the contents stored in these arrays can be observed after a normal functional operation is stopped with clock control. (Clock control is covered in the next section.) The information can be dumped on an external bus, which can then be captured by the ATE (tester) or a logic analyzer. Alternately, the data may be accessed through other test data pins (*e.g.*, *test access port* [TAP] as defined in the IEEE 1149.1 boundary-scan standard [IEEE 1149.1-2001]).

The inverse of array dump is to use the existing array(s) to store on-chip activities, such as internal bus traffic and specific architectural state changes. Multiplexers or even programmable interfaces can be designed to redirect these types of information to be stored on specific on-chip arrays (*e.g.*, the L2 cache on a microprocessor). This is called *trace logic analyzer* (TLA) in [Pham 2006]. Of course, these arrays have to be resized so that they will not impair existing functionality. Alternately, we can design in dedicated arrays to capture these traces, but that will be too expensive just for debugging purposes.

10.2.5 Clock Control

Clock control is an important DFD feature for speed debugging. Internal observation mechanisms like scan and scanout help obtain information in the spatial domain (where did the error originate?), while clock control supplements it by enabling extraction of information in the time domain (when did the error start to occur?).

Most modern VLSI chips with clock frequencies that run at hundreds of *megahertz* (MHz) to *gigahertz* (GHz) usually have an internally generated clock from a ***phase-locked loop*** (PLL). This generated clock is usually synchronized to an external system clock of much lower frequency. High-frequency oscillators are difficult to build, and high-frequency clocks are difficult to route on a board. Thus, the solution of having an internal PLL multiply to a high-frequency internal clock is common. Moreover, modern-day VLSI chips also typically contain multiple clock domains that drive specialized circuits or external I/O interfaces.

Starting, stopping, and restarting these internal clocks while keeping them synchronized to specific external and internal events is critical for debug [Josephson 2001]. Being able to start and stop the clock is an important first step in the debugging process. Whereas scan or scanout capture can extract internal logic state, the question of when to take this internal observation is answered by clock control. Stopping the clock at a specific clock cycle and synchronizing it to an internal or external event are the most basic of clock control features. If the clock stops at the wrong cycle, the internal observation results will be wrong, and the analysis will be performed down the wrong path.

Although it is possible to stop the clock after an external event, it is definitely preferable to use internal events to control the clock, as an external event may be too imprecise. This happens because the external event is synchronized to a slower external clock, and there may be an offset between the slow external clock and the faster internal clock.

Specific **offset counters** are typically added to make the clock stopping points much more flexible. These offset counters are often part of a more comprehensive set of **debug registers** designed specifically to capture information from or assert control to different parts of the chip, specifically for debugging purposes [Pham 2006]. These are also commonly referred to as **control registers** [Carbine 1997]. For example, one can specify that a scan or scanout capture be taken at 487 clock cycles after an exception event has occurred. The programming of these clock stop event(s) and offset is usually preshifted in through specific scan chains so that they are ready to execute when the debug tests are run.

Another useful function, in addition to starting and stopping the clocks, is to issue a single clock pulse. This is also sometimes called **single stepping**. Single stepping allows one to observe the execution of the circuit in a controlled manner. A scanout or observation-only scan like capability that is nondestructive is needed to complement single stepping with the internal observation of circuit state.

Besides starting and stopping the internal clocks, the capability of stretching a specific clock cycle is useful in the debugging process. When the clock is supplied externally, this is relatively simple. This can be accomplished by swapping in a different set of timing generators on a specific clock from the ATE. However, for internally generated clocks, circuit modification to the clock circuit is needed. Debugging performance problems (for example, "Why won't the chip run any faster than 500 MHz?") requires isolating the offending path or paths in the circuit. By sequentially stretching the clock from the failure point backward, the specific clock cycle where the failure was latched internally can be determined. This will provide clues as to where the problem lies considering the pipelining nature of high-performance systems. Multiple failing paths may exist requiring multiple iterations of stretching clocks to find the root cause in an onion-peeling fashion. Debug flows will be described in more detail in Section 10.6.

Stretching of individual clock phases can also be implemented to pinpoint the path to a particular clock phase. For flip-flop-based designs, the specific phase may not be important, but for latch-based designs, this phase stretching feature is indispensable.

Figure 10.8 illustrates the clock controls for the Pentium 4 microprocessor [Kurd 2001]. An external system clock is fed into individual Core PLL and IO PLL, where the clock is multiplied. A skew detect circuit will make sure that the skew between the 2 PLLs can be monitored for deviation. Both clock systems can be adjusted

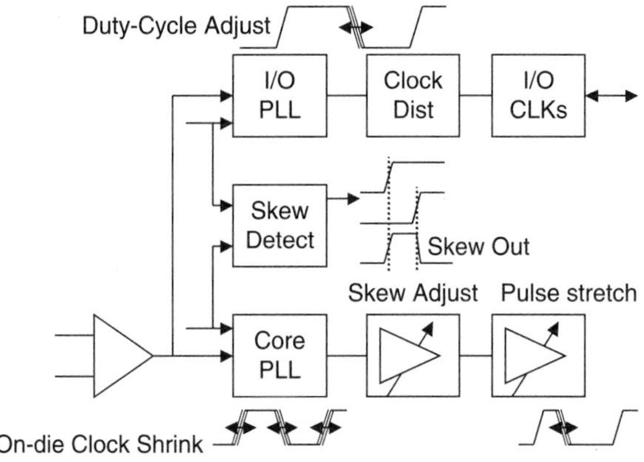

■ **FIGURE 10.8**

Pentium 4's on-die clock control.

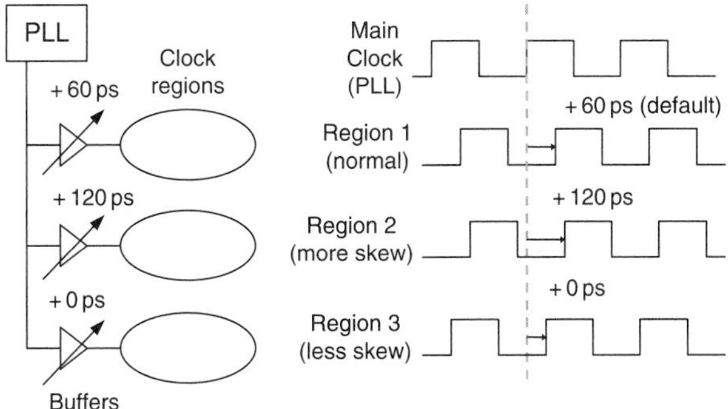

■ **FIGURE 10.9**

Introducing intentional clock skew.

for their duty cycles, while the core clock allows for skew adjustment between its various distribution buffers because this clock is distributed all over the chip and is subjected to on-die process variation. Finally, this core clock can also be adjusted for its duty cycle by stretching either one phase or the other.

Another clock control feature that can help with debugging is the ability to introduce relative skew between different clock domains. As a chip can contain tens of millions to hundreds of millions of transistors (and can include tens of thousands to hundreds of thousands of storage elements), the clock system has to be distributed from the PLL to all these storage elements with an elaborate clock distribution system. The clock distribution system consisting of a series of buffers must have the ability to deskew the various domains so that all the clocks arrive at their destination within a tight tolerance. Without this kind of clock distribution system, much performance could be lost because of the natural skews introduced by on-die process variations [Kurd 2001] [Fetzer 2006].

Because the clock distribution system can be deskewed, intentional skews can also be introduced through additional adjustment above and beyond what is needed for deskewing [Josephson 2001] [Josephson 2004] [Mahoney 2005] [Gizopoulos 2006]. This intentional skew, which is illustrated in Figure 10.9, gives more sampling time for the storage elements using the delayed clocks. Of course, the storage elements using this delayed clock also launch their outputs later, so failures may be observed in their downstream logic as well. This feature provides additional information as to where the failing path may lie. If pulling in the clock edges of a clock region causes the test to pass, then the source or the driving storage elements of the failing path can be pinned to that clock region. Similarly, if pulling in the clock edges of a clock region causes the test to fail, then the destination or the receiving storage elements of the failing path can be pinned to that clock region.

10.2.6 Partitioning, Isolation, and De-featuring

VLSI chips, including ***system-on-chips*** (SoCs) and microprocessors, are complex devices. They typically consist of multiple heterogeneous subfunctional blocks operating together to deliver the final functionality that the chip is supposed to deliver. Debugging such a chip will be difficult if we cannot isolate or narrow down the problem at the macro level before taking internal observation. Partitioning is one mechanism to determine if the problem still exists after some blocks are separated. This could be accomplished by logic partitioning or by rewriting software so as to confine the execution to a much smaller logic unit. For example, multiple execution units on the processor can be disabled so that the parallel execution of instructions can be restricted to a few or even down to one unit. The instructions can then be directed to individual units by subsequently enabling or disabling specific units. Of course, this ability is limited to units that have some level of redundancy built in. This is also called **de-featuring**.

10.2.7 Reconfigurable Logic

As mentioned in Section 10.2.3, the authors in [Abramovici 2006] proposed placing a programmable logic (also called **reconfigurable fabric**) between blocks or interspersing them into the regular logic fabric to aid both debugging and making a fix (see Figure 10.7). Patching can be done by re-programming the logic to replace or change existing logic functionality. This can be seen as halfway to fully programmable logic. However, it does provide benefits that *focused ion beam* (FIB) patching cannot achieve—that is, it ensures (1) instantly patched logic and (2) that every chip can be patched because every chip can be reprogrammed. (FIB patching is covered in Section 10.4.) This will allow very fast time-to-market by allowing units to be shipped without having to wait for a new design revision and a new set of masks. Should a bug appear, a software download can provide the fix.

10.3 PROBING TECHNOLOGIES

The logic DFD structures described in the previous section, along with debug and diagnosis techniques to be described later, provide the first level of reasoning and isolation down to the signals and signal paths that failed to capture the correct logic values within the prescribed timing window of a clock cycle under the operating window of supply voltage and temperature. However, little else can be derived about the exact nature of the problem. Was the problem caused by a weakly driven signal somewhere along the combinational path? Did it occur because the signal suffers from coupling noise? Was it caused by a power glitch at the vicinity of the affected logic? Was there charge leakage that is supply voltage sensitive? We need to know what causes the signals or signal paths not to propagate with the correct logic values within the prescribed timing window and under the operating condition window. A fix for the circuit cannot be put into place to solve the issue if the root cause is not verified. Probing technologies (or probing in short) can provide more detailed information to further isolate the issue, come up with root-cause hypotheses, and verify the correct hypothesis.

Probing technologies can be broadly classified as contacting and noncontacting. **Contact probing** is also called **mechanical probing**. **Noncontact probing** can be accomplished using two classes of probing, both of which are minimally invasive. The first class includes techniques and tools that *inject* beams (*e.g.*, laser or ***electron beams*** [e-beams]) and sense the response while the circuit is operational. These can also be categorized as **active** techniques. The behavior of the chip may be intentionally altered by this injection as well in some cases. The second class involves measuring the *emissions* (photon or thermal infrared) during the operation of the circuit. These can be categorized as **passive** techniques. These techniques are described next.

10.3.1 Mechanical Probing

Mechanical probing has been the mainstay of silicon analysis for decades. Essentially, it is a mechanical probe with a tip that can be manipulated to land on metal lines or specific layout pads. At this scale of layout geometry, this is called **pico-probing**. Its predecessor was called **micro-probing**. Pico-probe tips are on the scale of ~1 μm in diameter with the most sophisticated version maybe a quarter of that. Because of the relatively large capacitance that the probe will present to the signal, this is pretty much limited to the probing of strong signals such as clock or power buses with today's technology. Because the pico-probe is usually hooked up to high-frequency, high-impedance oscilloscopes, it serves a function that can never be satisfied with other means—high signal fidelity. This technique potentially gives the most accurate signal extraction, should the additional capacitance not be a problem to the signal being observed.

Figure 10.10 shows a typical mechanical probe station. The platen is for the placement of these probes. The wafer/chip is placed on the circular chuck at the center. The microscope is for the optical viewing of the probe placement. The fine

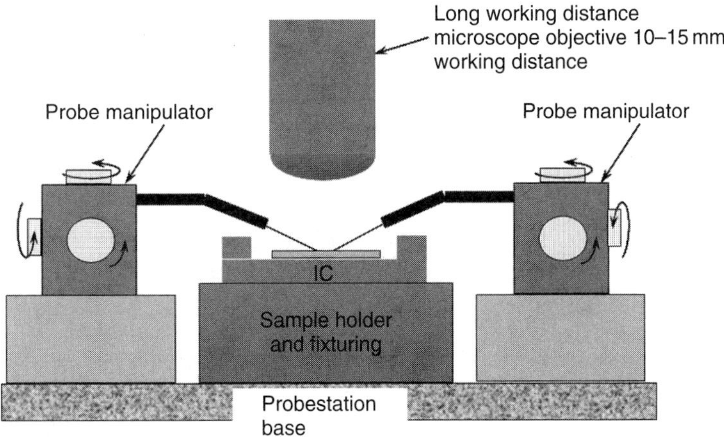

■ **FIGURE 10.10**

Mechanical probe setup.

adjustment (x, y, z) of the probes is provided by the three tuning knobs on the probes at top/side and the back located on the probe manipulator.

10.3.2 Injection-Based Probing

These **signal injection-based probing tools** are also alternately known as **beam-based probing tools**. Internal signals and transistors can be probed by using an *electron beam* (E-beam) or laser to inject a signal and by measuring the characteristics of the returned signal, which is modulated by the actual signal on the probed node while a circuit is in operation under a test. **E-beam probing** was the mainstay for a long time. However, in the current generation of integrated circuits using multiple metal layers and flip-chip packaging, inspection of the circuits from the front side (top side) of the die or wafer using E-beam probing is difficult. Fortunately, silicon is transparent at infrared wavelengths. Thus, by injecting a laser in the infrared spectrum from the backside of the die or wafer, it is possible to probe circuits in a relatively noninvasive manner. This technique is called **laser voltage probing**. In the case of E-beam, there is a voltage contrast effect where the ***secondary electron*** (SE) yield is attenuated by the voltage on the line. For **laser (optical) probing**, the electric fields modify the phase and amplitude of the returning signal.

10.3.2.1 E-beam Probing

E-Beam (electron beam) probing uses a principle that is similar to that used by a ***scanning electron microscope*** (SEM). In a vacuum environment, electron beams are scanned over the area that one wishes to observe. As the beam hits upon the metal lines, *secondary electrons* (SEs) are generated. A sensor pickup (**SE detector**) nearby collects these secondary electrons. For the metals that have a positive charge, the secondary electrons are attracted to fall back onto the metal line itself and result in little collected charge from the sensor. If the line is negatively charged, secondary electrons are expelled, and more of them will be collected by the pickup mechanism (SE detector or **energy analyzer**). Hence, a contrasting image can be developed based on these varying levels of electron pickup. This resulting image is also called **voltage contrast imaging**. Figures 10.11 and 10.12 illustrate this principle.

E-beam probing requires a line of sight to the metal line that is to be probed. It also requires certain clearing from neighboring metal lines so that the reading will not be distorted because of the charge on these lines. Thus, certain layout design rules have to be followed (see Section 10.5). The passivation or interlayer dielectrics (insulation) also have to be stripped off completely so that the electron beams can hit upon the metal line directly and secondary electrons can be scattered freely. For signals carried on lower levels of metals, one can only observe them through the spaces between the upper level metals or by cutting holes through wide fat power buses (so that the underlying layers can be exposed). E-beam probing is almost impossible for signals carried on deeper layers (*e.g.*, M2 [metal-2] or M3 [metal-3] in a six-layer metal system), so this technology is not effective for modern-day multiple-layer metallization chips. By detecting logic states in time,

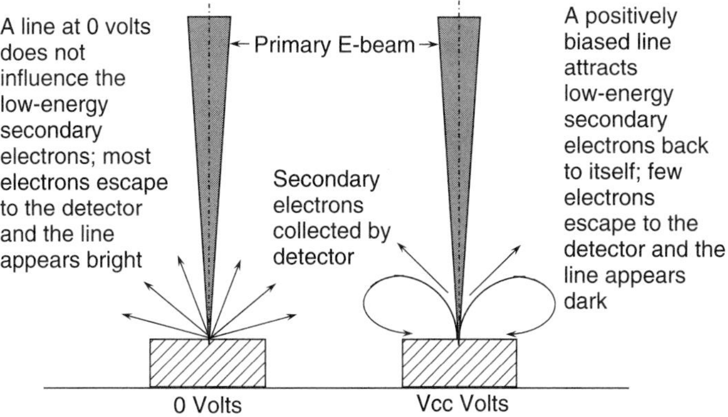
A line at 0 volts does not influence the low-energy secondary electrons; most electrons escape to the detector and the line appears bright
Primary E-beam
Secondary electrons collected by detector
A positively biased line attracts low-energy secondary electrons back to itself; few electrons escape to the detector and the line appears dark
0 Volts
Vcc Volts

■ **FIGURE 10.11**

E-beam probing technique.

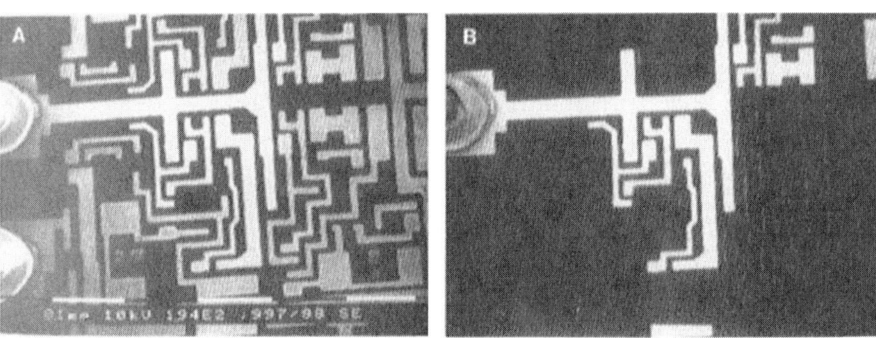
A
B

■ **FIGURE 10.12**

Voltage contrast image (with E-beam probing): normal image *(left)*, low potential (0 V) area highlighted *(right)*.

E-beam probing can be used to recreate the timing waveform on a signal as shown in Figure 10.13.

10.3.2.2 Laser Voltage Probing

E-beam probing as a debug technique was impacted by the introduction of the ***controlled collapse chip carrier*** (C4) flip-chip packaging technology. In flip-chip, to improve performance over bond wires, **solder bumps** replace the conventional bond wires. This can be achieved by forming solder bumps on existing die bond pads and corresponding mirrored pads on the surface of the package. The die is then flipped and placed on top of the corresponding package. Through reheating, the solder on both sides will fuse and form a small solid solder connection, much

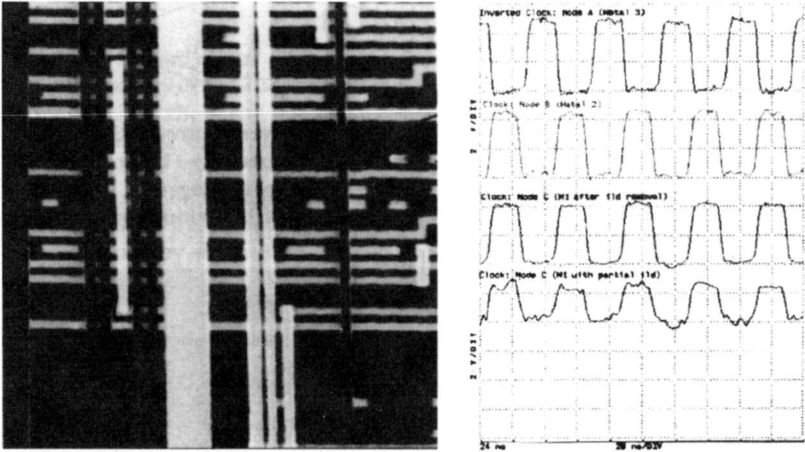

FIGURE 10.13

E-beam probing, which can show logic state map at any clock (*left*) or timing waveform at any node (*right*).

like the much larger surface mounting ***printed-circuit board*** (PCB) assembly, albeit with smaller sized solder bumps instead. Other than lower inductance for each of these bumps, many more such solder bumps can be placed all over the die area, resulting in thousands of parallel paths, which lower the resistance/inductance of the chip's power grid connections. Also, because all the bumps can be fused in one single reheat process, productivity is superior to the serial wire bonding technology.

With this kind of packaging, the surface of the die is out of view. One can only see the backside of silicon as seen in Figure 10.14. The metal wiring side is hidden between the die and the package. Probing from the backside of the silicon poses a new challenge.

With E-beam, it would require thinning and etching the die from the backside to expose to M1 for E-beam probing as shown on Figure 10.15. However, such removal of material would require the M1 exposure to be reasonably far away from active silicon so as not to impact device and circuit performance. This would require debug design rules that are not density friendly.

Additionally, each M1 (metal-1) exposure would have to be done using a direct write etching technique, such as **FIB nano-machining**, to expose the lines without damaging adjacent transistors. Such techniques are time consuming and are not conducive to in situ signal probing.

A solution to the problem of probing from the backside came from material characteristics. The technique is called *laser voltage probing* (LVP) [Paniccia 1998] [Yee 1999]. It was noticed that *infrared* (IR) light can be observed (*i.e.*, transmitted) from silicon, as illustrated in Figure 10.16, as a result of hole electron recombination. Because IR can be emitted and observed through silicon, it can also be transmitted into and through the silicon. Not only is silicon transparent to IR emission, but the

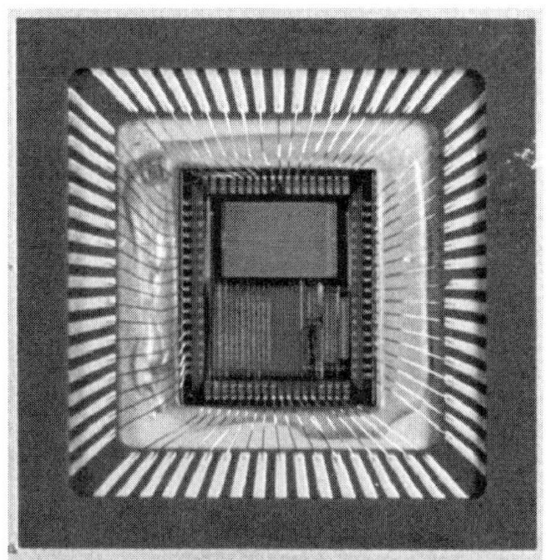

■ **FIGURE 10.14**

Wirebond packaging *(left)* and flip-chip packaging *(right)*.

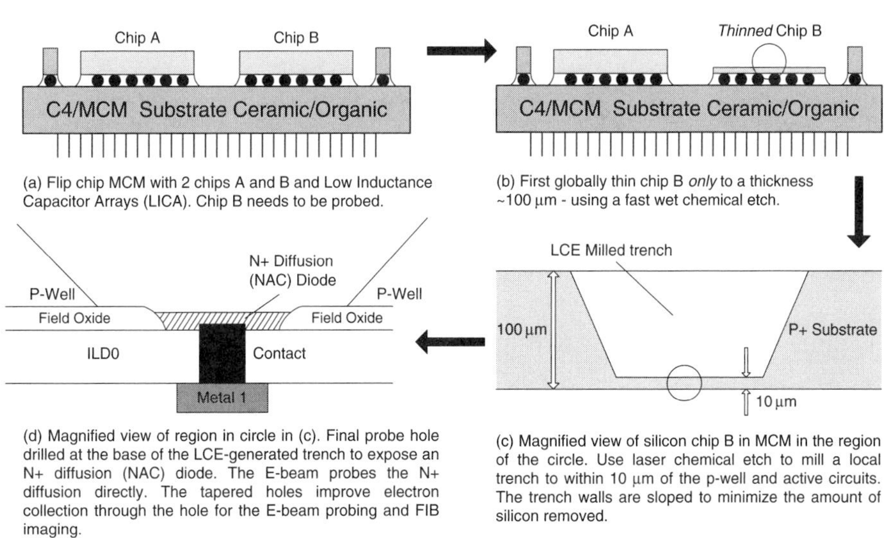

Chip A
Chip B
C4/MCM Substrate Ceramic/Organic
(a) Flip chip MCM with 2 chips A and B and Low Inductance Capacitor Arrays (LICA). Chip B needs to be probed.
Chip A
Thinned Chip B
C4/MCM Substrate Ceramic/Organic
(b) First globally thin chip B *only* to a thickness ~100 μm - using a fast wet chemical etch.
LCE Milled trench
100 μm
P+ Substrate
10 μm
(c) Magnified view of silicon chip B in MCM in the region of the circle. Use laser chemical etch to mill a local trench to within 10 μm of the p-well and active circuits. The trench walls are sloped to minimize the amount of silicon removed.
N+ Diffusion
(NAC) Diode
P-Well
P-Well
Field Oxide
Field Oxide
ILD0
Contact
Metal 1
(d) Magnified view of region in circle in (c). Final probe hole drilled at the base of the LCE-generated trench to expose an N+ diffusion (NAC) diode. The E-beam probes the N+ diffusion directly. The tapered holes improve electron collection through the hole for the E-beam probing and FIB imaging.

■ **FIGURE 10.15**

E-beam probing scheme for backside probing.

infrared light energy also can be reflected off a charged object. The phase change of this reflected light can be analyzed and a voltage contrast image reconstructed. However, light absorption by silicon does require the silicon to be thinned down to 50 μm or so, but that can be supported through general silicon planarization

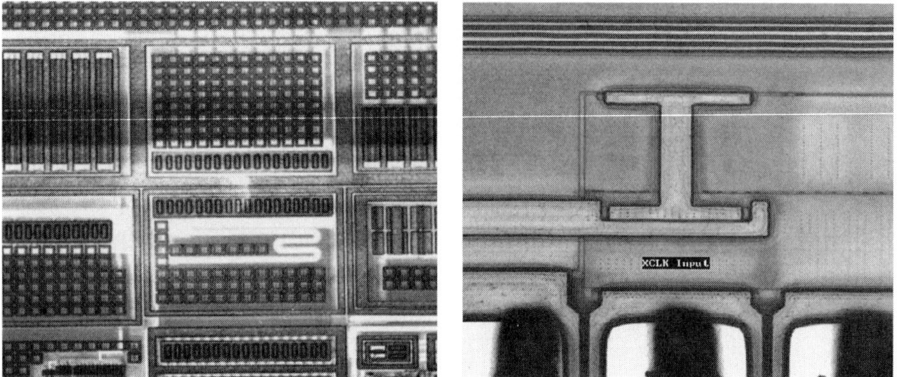

■ **FIGURE 10.16**

Backside view *(left)* using IR *versus* frontside view *(right)* at the same location of the die.

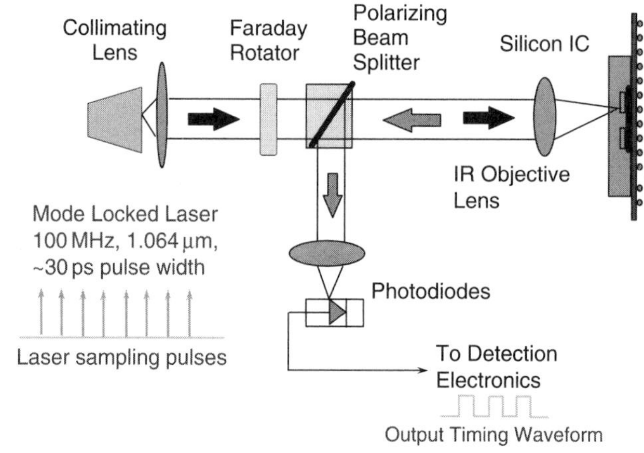

■ **FIGURE 10.17**

IR prober setup.

technology (*e.g.*, ***chemical-mechanical polishing*** [CMP]) as well as a *focused ion beam* (FIB).

Figure 10.17 illustrates how a system can be set up to allow a laser pulse to be focused on the device junctions and the reflected energy to be observed with sensitive photodiodes and further processed to extract voltage information. Figure 10.18 illustrates in more detail how the field around a junction will modulate the reflected energy. This variation of reflected energy provides one with the ability to figure out the voltage transitions happening at the junction.

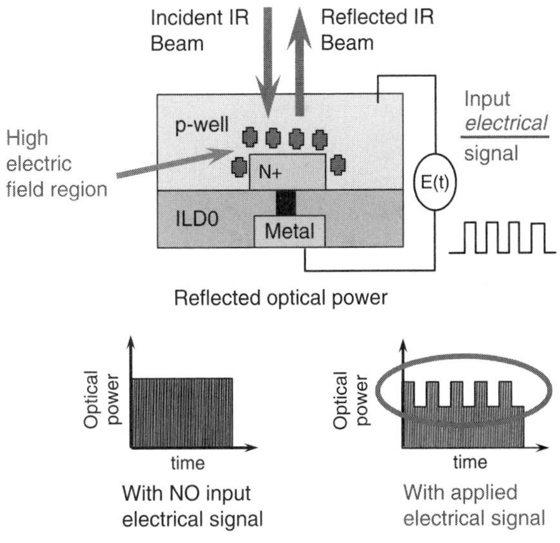

■ FIGURE 10.18

Laser voltage probing (LVP) theory of operation.

10.3.3 Emission-Based Probing

As mentioned earlier, in the current generation of integrated circuits using multiple metal layers and flip-chip packaging, inspection of the circuits using E-beam probing from the front side (top side) of the die or wafer is difficult. Fortunately, silicon is transparent at infrared wavelengths. Thus, by sensing infrared emission from the backside of the die or wafer, it is possible to examine circuits in a noninvasive manner.

10.3.3.1 Infrared Emission Microscopy (IREM)

Complementary metal oxide semiconductor (CMOS) failures often result in unwanted or unexpected electron-hole recombination. This electron-hole recombination is accompanied by weak photon emission. Light is also emitted by intraband carrier transition in high field situations. Sensitive and long-exposure cameras mounted in dark boxes can register these emissions and the associated emission sites. These emissions can be overlaid on a reference image to determine the location of the lighted object. This technique is also known as ***emission microscopy*** (EMMI) [Uraoka 1992].

A sensor array collects near-IR radiation in the 800- to 2500-nm wavelength range. Combined with advanced camera optics and precise stage movement, this emission detection method allows super-accurate pinpointing of the location of emissions. The tester is docked to the ***infrared emission microscopy*** (IREM), and failing patterns are applied to the circuit under test. Emissions are observed from the backside through a thinned substrate [Liu 1997] [Chew 1999]. A comparison can be made between passing and failing units. An abnormal emission seen on the failing unit but not seen on the passing unit indicates that a defect is likely associated with the abnormal emission, and localization techniques such as those aligned to a layout database can be used to determine the location of the emission [Loh 1999].

The die substrate is thinned and ***antireflective-coating*** (AR-coating) is applied to the backside of the die to allow a more uninterrupted light path when the die is observed through the backside. Strong abnormal emission can indicate a variety of circuit or process defects, as shown in Figure 10.19. Saturated transistors, devices with contention (because of improper control), and high-leakage devices give out strong emission.

In addition to observing abnormal emissions, it is also possible to map the normal emissions from transistor diffusions to the logic state as shown in Figure 10.20. N-diffusion emissions can be mapped to logic state 1 and P-diffusion emissions mapped to logic state 0. This is called **logic state imaging** [Bockelman 2002]. Emissions are observed on the N and P diffusions of inverters. If the input to the inverter is logic 0, the N-diffusion region emits; if the input to the inverter is logic 1, the P-diffusion region emits. The emission mechanism being observed is soft avalanche breakdown, in which a high concentration of electron-hole plasma in the drain region recombines in part through indirect radiative recombination [Bockelman 2002].

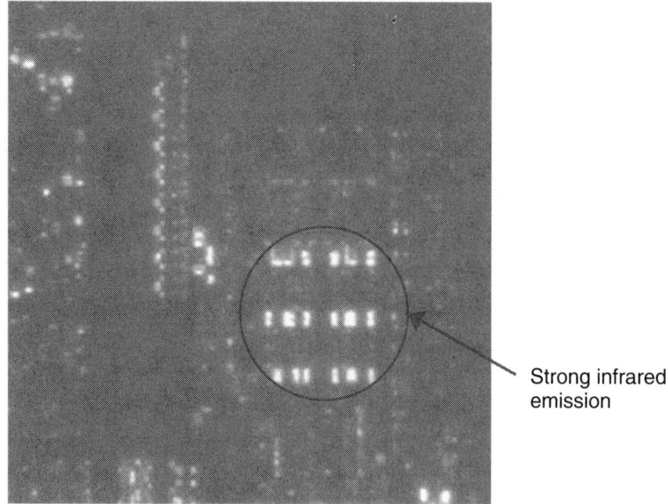

■ FIGURE 10.19

Emissions observed on an IREM.

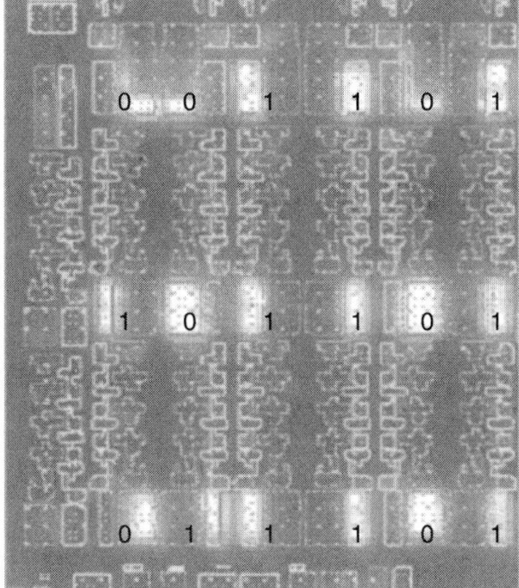

■ **FIGURE 10.20**

Emissions observed on an IREM mapped to logic state.

10.3.3.2 Picosecond Imaging Circuit Analysis (PICA)

When electrons speed through the drain-source region, they emit **optical energy** (mainly in the infrared band) in the process. This emission occurs during switching [Kash 1997] [Knebel 1998] [Sanda 1999], and by detecting this optical energy in real time it is possible to identify when a transistor switches. With a high-speed simultaneous spatial detection system using an image sensing array, it is possible to detect how the transistors of a local area switched in sequence. After the data are captured, they can be analyzed by replaying them at a much slower speed for more detailed examination. This process is termed as ***picosecond imaging circuit analysis*** (PICA). For example, if a clock distribution system is observed, the light pulses start appearing at the first stage of the clock driver tree and then spread to different areas of the chip as successive stages of the clock tree fire. Then the process repeats itself with subsequent clock switching. The optical energy from N-transistors is much stronger than those of the P-transistors, marking the relative switching between the pulldown and pullup network. This is useful in identifying relative timing problems, especially when the circuits are nearby and within the field of the detection system.

Because this is optical detection only, it will not disturb the dynamics of the switching event around the device junctions and will preserve the timing behavior of the signals. An additional advantage is that this light detection is possible from both the front side as well as the backside of silicon, making it more versatile for various packaging technologies.

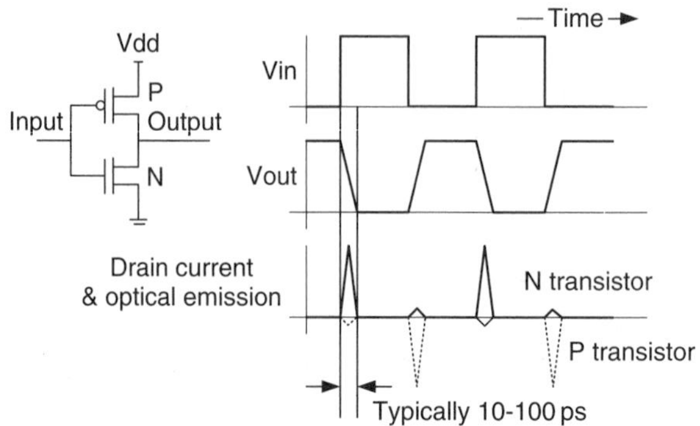

■ **FIGURE 10.21**

Optical pulses as revealed by PICA.

In Figure 10.21, the respective N/P-transistors switch with the input signal. The switching energy of the N-transistor is much stronger than that of the P-transistor making the detection of individual edges possible.

10.3.3.3 Time Resolved Emissions (TRE)

The use of an imaging array in PICA produces pictures of how the emission occurs over an area of the circuit. However, the use of an **array imaging sensor** is slow and

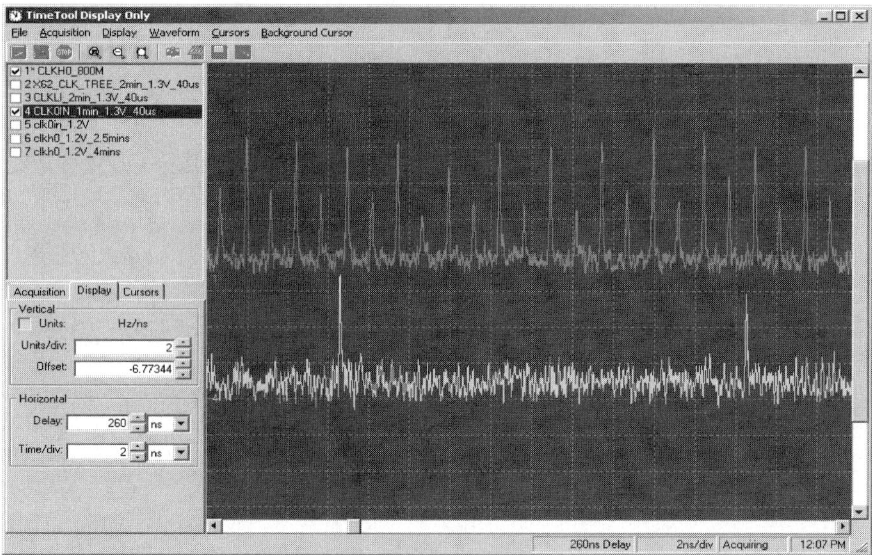

■ **FIGURE 10.22**

Switching emissions observed on a TRE.

requires longtime sampling of the signals. For repetitive signals such as a clock, this is not an issue. However, it is ineffective for small emission from weak transistors with low signal switching rates. Moreover, technology scaling and the associated voltage scaling also reduce emission and shift the emission spectrum to those of longer wavelengths. Many authors [Bodoh 2002] [Varner 2002] [Vickers 2002] have improved upon the technique with a more sensitive single element photon counting detector to detect the emission from a specific node in question; they have named the technique ***time resolved emissions*** (TRE). A high-speed low-noise InGaAs avalanche photodiode with below-breakdown bias and gated operation is typically used. Figure 10.22 shows the waveform caused by the switching activity of a signal using TRE.

10.4 CIRCUIT EDITING

Circuit editing is performed to either enable probing or make corrections to the circuit to verify root-cause fixes. **Circuit editing** is also called **silicon microsurgery** (or, more recently, **nanosurgery**). This involves either removing or adding material. Removal operations typically include cross-sections for observations, trenching to access and probe signals, and cutting signals. Deposition of short wires to create new connections can also be made. This is achieved through a tool called a **focused ion beam** or FIB. FIB along with DFD structures to enable circuit editing as well as layout-database-driven navigation systems to enable circuit editing are described next.

10.4.1 Focused Ion Beam

A *focused ion beam* (FIB) itself is not a probing tool but rather the enabler to mechanical or optical probing and, more importantly, it offers a means to do some patching at the silicon level to confirm a hypothesis of what an error is or how an error can be fixed.

A FIB used for circuit rewiring combines a high-energy particle beam (typically gallium) with locally introduced gas species to enable a desired ion beam–assisted chemical etch or ion-beam–induced ***chemical vapor deposition*** (CVD) as shown in Figure 10.23. This capability is analogous to a direct-write back end fab, where metal traces and devices can be cut or trimmed and rewired using dielectric and metal deposition.

With the right chemistry, an ion beam can also react in a chemical atmosphere to result in solid formation, resulting in deposited material on the surface of the die. The deposited material is not in a crystalline structure and will result in a resistive line structure, playing the role of the "blue wires" for patching up circuitry on silicon (*i.e.*, rerouting circuits).

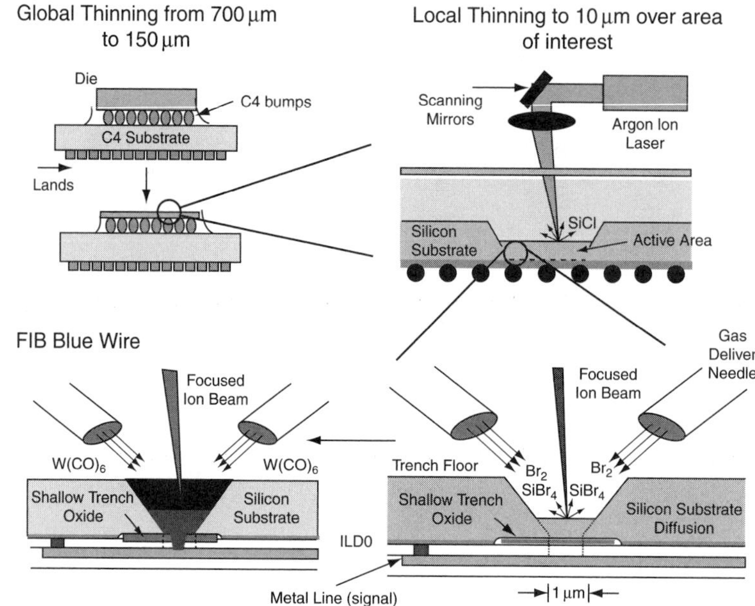

■ **FIGURE 10.23**

Focused ion beam.

10.4.2 Layout-Database-Driven Navigation System

To navigate around the die, one needs to make use of a layout-database-driven navigation system to position an E-beam, IR beam, or FIB to the location of the devices, wires, or circuits that one would like to observe or repair. Because of the complexity of the chips, manual navigation is almost impossible.

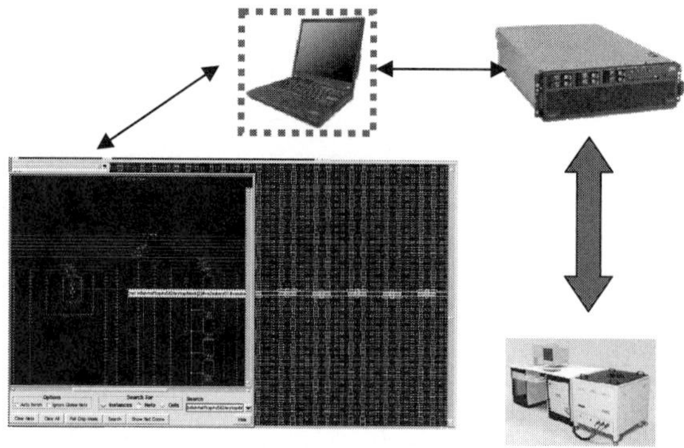

■ **FIGURE 10.24**

A layout-database-driven navigation system.

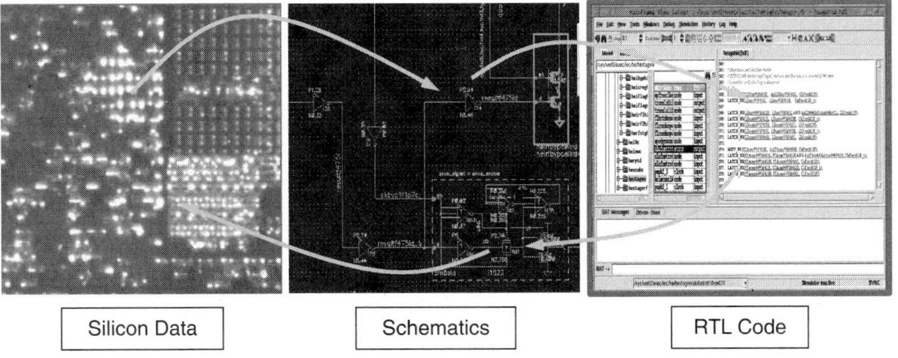

■ **FIGURE 10.25**

Mapping of probed topological information to schematics and then to high-level logic models.

Once certain reference coordinates of the die can be located, all other device or wire locations can be accessed through an automated system so that one can call up any circuit or signal with ease. The reverse process is also possible. If a signal is traced to a different part of the layout, one can also look up the database to find out what the design layout should look like. Figure 10.24 shows a layout-database-driven navigation system. Signals can be looked up in a schematic viewer and mapped to their corresponding layout polygons (see Figure 10.25). These polygons and their coordinates are communicated to the FIB stage driver, which moves the FIB to the right location on the silicon.

10.4.3 Spare Gates and Spare Wires

Based on the data collected from logic DFD and probing, a hypothesis of what the problem is may be formulated. However, the hypothesis needs confirmation. Further, other problems may lurk beneath the surface, hidden by the problem being debugged. One way to answer these questions would be to implement the circuit with logic fixes and tape out a new chip. However, this would push out any further validation by several months and can cost more than a million dollars in mask and other silicon fabrication expenses. Validating that a hypothesis is correct and continuing further validation requires patching up the circuit and reexecuting all the validation tests. With a FIB, we can cut and repatch wires, but what about transistors? How can we extend the patching concept to circuits in silicon, where adding transistors is impossible, unless we have the transistors already in place? This is where spare gates and spare wires (also called bonus gates and bonus wires) [Livengood 1999] come in handy. If spare transistors or even gates are placed in the layout at strategic locations or even randomly where space is available, one may find the replacement transistors or gates nearby and patch the circuit up so as to perform a fix. Even though it is tedious to cut and reconnect a bunch of wires using FIB, it is still far faster than the redesign-tapeout-fabrication route, and it costs much less. This kind of patching is indispensable because time-to-market or

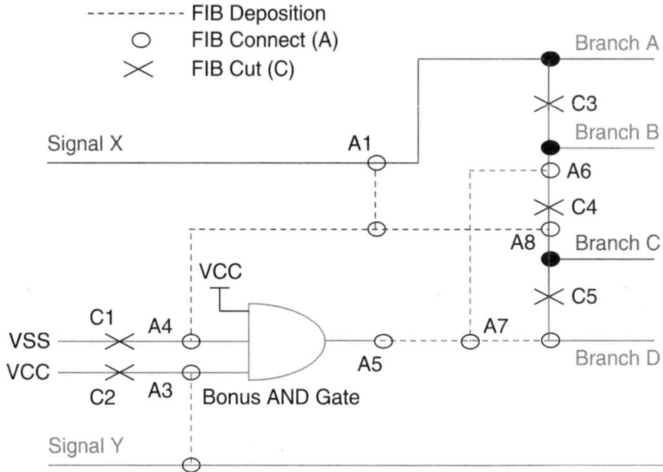

■ **FIGURE 10.26**

Sample spare gate and spare wire usage with FIB patching.

time-to-manufacturing is critical for any product. Ultimately a new tapeout is still needed to get robust fixes and correct silicon, but a temporary fix verifies that the hypothesis is correct and makes sure that the initial bug did not mask any problems.

Figure 10.26 illustrates a preplanned layout with a bonus AND gate on a circuit diagram. The inputs are normally shorted to Vcc and Vss so that it is dormant. If this gate is needed, then the inputs are cut from the tie-downs, and the gate is then connected to the respective signals (X and Y). The output of the gate can be hooked up to several routing wires and will be isolated with more cuts to drive the desired signal.

10.5 PHYSICAL DFD STRUCTURES

The probing technologies outlined in the previous section cannot be successful without the appropriate support from the layout design. We have collectively termed these layout changes ***physical DFD*** (PDFD). The term "physical" also refers to "physical design," which is another name for layout design. Physical DFD are DFD features implemented on the layout design to enable probing and circuit editing.

10.5.1 Physical DFD for Pico-Probing

To facilitate probing, one should designate the placement of well-thought-out locations with plenty of open space around them to ease probing and to avoid accidental shorting. To facilitate pico-probing, the probe pads have to be reasonably large so that landing the probe does not present a problem. Care must also be taken not to place active signal wires underneath it to avoid leakage, should the pressure of the tip crack the interlayer dielectric. To help with planarization, buffer (dummy) metals should be placed under its footprint.

10.5.2 Physical DFD for E-Beam

Specific design rules have to be developed to space out the metal lines so that they do not interfere with the signal pickup of the subsequent lower metal layers. This is because the lower layer metal lines are only visible through the gaps between metal lines on the upper layers. For fat metal buses (*e.g.*, power buses), specific design rules also have to be developed so that holes of reasonable size can be dug to open up for lower-level signal lines to be observed.

Figure 10.27 shows all the features that are needed for exposing signals from various layers to the E-beam. This is a five-layer metallization. Because the top

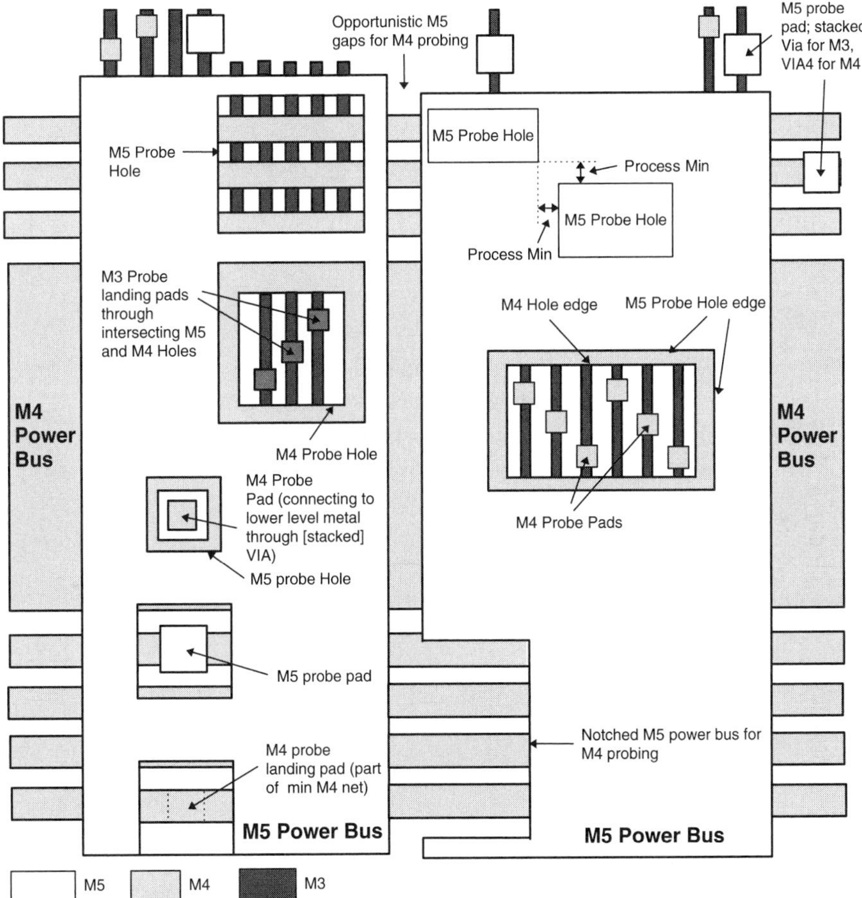

■ **FIGURE 10.27**

Sample E-beam layout openings illustration.

layers may have wide power buses, the cutting of holes or notching is necessary to expose signals at the subsequent layers. It is also preferable to widen the signal lines designated as probe pads where the signal is to be probed to allow maximum reflectivity of the electrons.

10.5.3 Physical DFD for FIB and Probing

Because the front side of the die is not available (or not visible), there is no geometry or marking on the backside to indicate relative positioning. Even for probing, there is a need to locate where we want to thin out silicon for probing so as to not miss the probe point when thinning out by using a FIB. Specific markers formed by diffusion have to be placed on the layout at the four corners of the die as shown in Figure 10.28. These are infrared visible so that they can be used to orient the die precisely.

In general, the rules are roughly driven by the FIB precision control. Spacing between the cut point and the surrounding geometries has to be allocated generously so that it will not cut into other undesirable or sensitive areas. Sometimes it may be preferable to plan and design the cut sites to facilitate this.

During probing, because infrared energy can disturb the nodes under observation, it is preferable to have a specific test structure away from the transistor junction as a probe point. The plasma protection diode or an additional reverse biased diode at the input of the gate is typically used. Although it will add some small loading, it is not to the degree that it will interfere with the switching properties of the transistor.

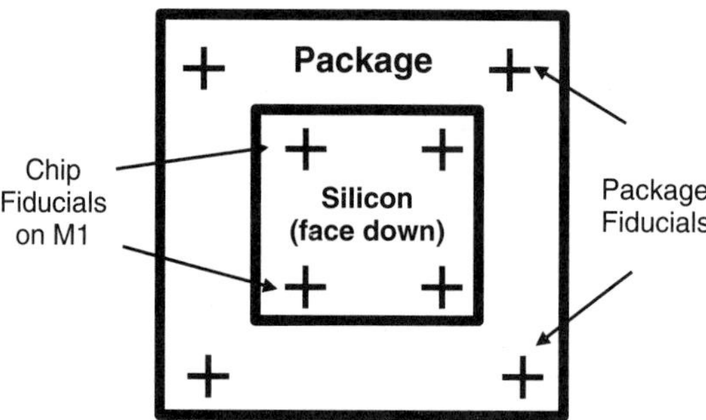

■ **FIGURE 10.28**

Flip-chip markers on silicon for infrared probing alignment.

10.6 DIAGNOSIS AND DEBUG PROCESS

Figure 10.29 shows a generic diagnosis flow. The diagnosis process involves starting with the test results, which capture all the observed failures and mapping the test results to defects or errors, which are then used as a starting point for

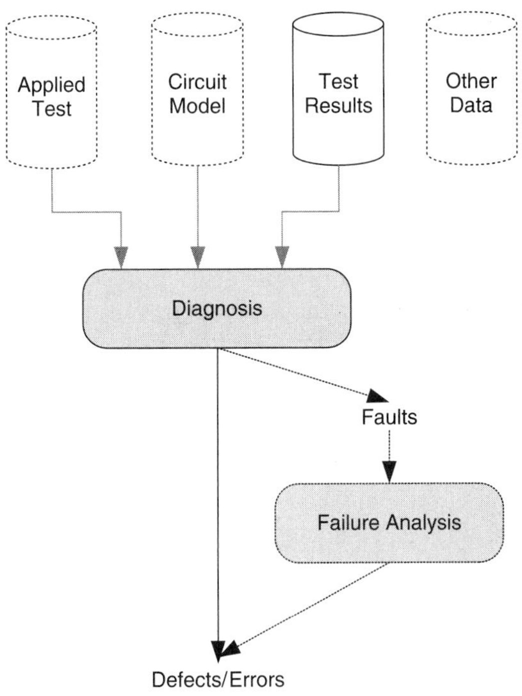

■ **FIGURE 10.29**

A generic diagnosis flow.

repair (replacement or redesign) or finding the root cause for process improvement depending on the goals of the diagnosis process. Depending on the objectives of the diagnosis flow and the type of systems under diagnosis, the defects or errors may include defective components (for example, a faulty IC on a board, a faulty cell in a ***random-access memory*** [RAM], a faulty column of cells in a ***field programmable gate array*** [FPGA], or a faulty board in a system), defective interconnections (shorts or opens), logic implementation errors, timing errors, or IC manufacturing defects.

The diagnosis process would typically use information about the circuit model and applied tests along with other data that may be available. Fault models [Aitken 1995] [Wang 2006] are also typically used as a means to arrive at the final defects or errors for consideration as shown in Figure 10.29.

Defects are fabrication errors caused by problems in the manufacturing process or human errors. They may also be physical failures that are caused by wearout or environmental factors. Manufacturing defects are unpredictable in both location and effect, and the processes that cause them are continuous over a wide range of variables (*e.g.*, a short between two adjacent wires may occur anywhere along their length with a range of possible resistances, capacitances, etc.). Ideally, the test process would take every defect into account, develop a test for each, and apply these tests during manufacturing testing. However, because the space of all possible defects is continuous and unpredictable, there is no way to apply a finite number of tests that are guaranteed to detect everything in that space. The complexity of defect behavior makes a strict defect-based test approach impossible; some simplification is necessary.

Defects themselves could be approximated. This approach is usually taken both during testing and during diagnosis [Abramovici 1994] [Wang 2006]. The infinite defect space is approximated by a finite set of faults. A fault is a deterministic, discrete change in circuit behavior. It is important to stress that the fault is an approximation of defective behavior, not a true representation. By this definition, a fault can never be "found" during diagnosis. Fault models are used as a means to arrive at the final defects or errors for consideration. Figure 10.29 illustrates this process. Faults are often thought of as localized within a circuit (*e.g.*, a particular gate is broken), but they may also be thought of as transformations that change the Boolean function implemented by a circuit. Many fault models are timing independent, whereas a few include timing behavior explicitly. The most commonly used fault models include *stuck-at faults*, *bridging faults*, *delay faults* (including *path-delay faults* and *transition faults)*, and *functional faults*. The choice of fault model depends on its intended use (test generation, manufacturing quality prediction, defect diagnosis, characterization for defect tolerance, etc.).

Fault models are an integral part of the fault diagnosis process and thus help find the root cause of the defective device under consideration. In its most basic form, a fault model is used to predict the behavior of faulty circuits, compare these predictions to the actual observed behavior of defective chips, and identify the predicted behavior that most closely matches the observations. An analogy to this process is a detective story. A fault can be likened to a criminal or suspect that exists in the circuit model. The fault may be **permanent**, **intermittent** (with alibis most of the time), or **transient** (hit-and-run) depending on the nature of the fault model. An *error* (or *fault effect*) is created by applying stimulus to *activate* the fault (provoke the criminal) at the fault site. The fault effect may *propagate* through the circuit and be detected. A *detection* implies that the fault effect has propagated to an observation point and can be observed as an error or failure. Figure 10.30 illustrates fault effect propagation. The goal of the process is to enable further analysis by identifying promising locations for further study.

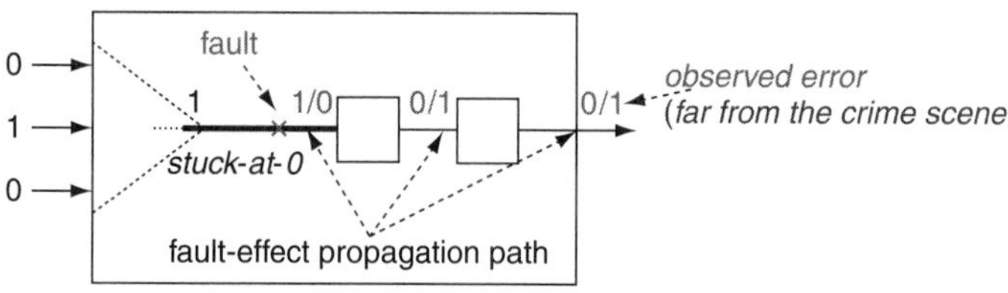

FIGURE 10.30

A fault with its propagated fault effect leading to an observed error.

10.6.1 Diagnosis Techniques and Strategies

The diagnosis process can employ a variety of methodologies. Some of the common types of diagnoses are enumerated:

- **One-step or nonadaptive**. The test sequence is fixed and does not change during the diagnosis process.
- **Multistep or adaptive**. Tests applied in a subsequent step of diagnosis depend on results of the previous steps.
- **Static**. Diagnostic information is precomputed for all possible faults before testing.
- **Dynamic**. Some diagnostic information is computed during diagnosis based on the actual ***device under test*** (DUT) response.
- **Cause-effect**. Computes faulty responses for a set of faults.
- **Effect-cause**. Analyzes the actual DUT response to determine compatible faults.
- **One-step**. Without replacement.
- **Multistep**. Alternate retest and replacement steps.

In line with the detective story, the common elements of most diagnosis methods include the following:

1. Be prepared.
2. Assume, suspect, and exclude.
3. Track them down.

Being prepared involves storing information that may potentially be useful for analysis later. In criminal investigation, this may involve fingerprinting the entire population, communities, or just those individuals with a prior criminal record. This is the strategy employed by cause-effect fault dictionaries [Abramovici 1994] [Wang 2006]. However, the questions of the feasibility of fingerprinting the entire population or determining which communities to fingerprint are tedious at best. The challenges with this criminal investigation strategy also have an analogy in diagnosis. Which faults or faults models (stuck faults, bridging faults, delay faults, etc.) should be considered? Is it feasible to store responses of all faults?

How do we process all the prestored information during analysis? Questions in criminal investigation may include finding matches for fingerprints found at the crime site, what happens when the criminal has not been previously fingerprinted, or what happens when several criminals committed the crime and their fingerprints are mixed.

Analogous situations occur in diagnosis: using a single-stuck-at fault dictionary when in reality the defect is a short that behaves close to a bridging fault, or using a single-stuck-at fault dictionary when the defect behaves like multiple stuck-at faults. Dealing with these situations requires diagnostic models and algorithms that can deal with partial matches. Several diagnostic models and algorithms have been developed and successfully employed [Waicukauski 1989] [Venkataraman 2001] [Guo 2006]. More diagnosis approaches can also be found in [Wang 2006].

The "assume, suspect, and exclude" strategy used in criminal investigation applies equally well to the diagnosis problem. Make some initial assumptions about the criminals (faults in the case of diagnosis). Do they act isolated (single faults) or in gangs (multiple faults)? Are they always bad (permanent faults) or only sometimes (intermittent faults)? Where are they coming from (stuck-at faults, bridging faults, functional faults, etc.)? The next step involves rounding up the plausible suspects and then reducing the set of suspects. In the words of Sherlock Holmes [Doyle 1905]: "One should always look for a possible alternative and provide against it. It is the first rule of criminal investigation." In criminal investigation, this involves checking existing data for alibis. In diagnosis, this may require posttest fault simulation on the data collected from the on-chip logic DFD and clock control features discussed in Section 10.2.5. One could also perform experiments to provide alibis (adaptive diagnosis). Quoting Sherlock Holmes again: "When you have eliminated the impossible, whatever remains, however improbable, must be the truth." This step is performed using the on-chip DFD features and collecting failing information on a tester. This leads to the final step. When there are no more suspects, change the initial assumptions and restart. In diagnosis, this may involve starting over with a new fault model.

The final strategy involves tracking down the suspects or backtracing in diagnosis. "There is no branch of detective science which is so important and so much neglected as the art of tracing footsteps" [Doyle 1888]. Tracking down or backtracing involves starting from the available acts (observed errors in the device), tracing back through the city (tracing through a circuit) and in time (tracing through previous test vectors), and following paths that lead to the criminal(s) (actual fault[s]). This step uses the physical DFD structures on-chip and the physical tools described earlier to retrieve information from inside the chip.

10.6.2 Silicon Debug Process and Flow

The silicon debug process starts with the identification of bugs, which can be classified as logic (functional) or electrical. The identification of functional or logic bugs occurs during functional validation on a tester or system platform using architectural and design validation tests. Code generation using pseudo-random instructions and data is also typically used for microprocessors. Operation is performed at safe operating points (nominal voltage, temperature, and frequency) to ensure that only logic bugs are exposed.

Electrical characterization is performed to expose circuit sensitivities or marginalities, and timing or critical speed-path issues. Testing is performed to the extremes of the operation region [Josephson 2001] by varying frequency, voltage, and temperature to extremes. Characterizing with respect to process variation is performed with skewed wafer lots with the process parameters varied.

Debug can start initially on a *tester* or a *system environment*. The tester environment is easier to handle and more flexible [van Rootselaar 1999] [Josephson 2001] as there is full control of all pins on the chips and full access to all DFD features described earlier. Patterns can be looped (to allow for signal sampling) to perform probing using the physical tools described in previous sections. The operation of the chip in this environment is fully deterministic.

In contrast, the *system environment* is cheaper and it is easier to generate and apply system-level functional patterns. Furthermore, many issues are only uncovered when running real programs (OS, drivers, and applications). However, nondeterminism because of system events such as interrupts and memory refreshes need to be factored in. It is also harder to control the chip in a system environment because of the lack of access to all the debugging features. The typical approach is to quickly locate sections of failing tests on a system and try to replicate the problem on a tester for further debugging [Holbrook 1994] [Hao 1995] [Carbine 1997] [van Rootselaar 1999] [Josephson 2001].

A typical debug flow consists of three steps:

1. Find a pattern that excites the bug.
2. Find the root cause of the bug.
3. Determine a fix to correct the bug.

The first step is finding a test pattern that excites the bug. This may be found during system validation, which was described earlier or in a customer sighting. Very often the pattern is already used on the tester or pops out of the ***high-volume manufacturing*** (HVM) test environment because of process shifts. It is also possible to craft special tests using sequential ATPG and using existing scan to generate patterns for embedded arrays and queues [Kwon 1998].

Finding the root cause of the bug requires three steps. The first step involves using the logic DFD features to extract data from the chip. Next, simulation tools and deductive reasoning are used to arrive at hypotheses. This process is described in more detail next. Finally, probing using the physical DFD features and probing tools is employed to confirm a hypothesis while eliminating others. FIB edits and modified tests are then used to verify the hypothesis.

10.6.3 Debug Techniques and Methodology

The debug techniques used to arrive at hypotheses for the observed failures (bugs) involve using the logic DFD features to perform two operations:

1. Controlled operation of the chip
2. Getting internal access (to signals and memories) in the chip

Controlled operation of the chip is needed to stop the chip close to the point of the first internal error and place it in a test mode. This is accomplished by (1) trigger mechanisms programmed to events, (2) clock control mechanisms, and (3) clock manipulation.

Trigger mechanisms programmed to events [Carbine 1997] [van Rootselaar 1999] [Josephson 2001] involve using control registers [Carbine 1997], probe modes in the system to monitor states, and matchers and brakes [van Rootselaar 1999] to monitor for specific events and stop their occurrence. Mechanisms for stopping on trigger events are also called **breakpoints**.

Clock control mechanisms are used to step one clock at a time or step through a specified number of clocks. **Clock manipulation** involves skipping cycles or phases, moving clock edges by stretching or shrinking specific clock cycles, and skewing a clock region relative to other clock regions as described in Section 10.2.5.

Internal access (to signals and memories) in the chip typically involves taking scan snapshots (see Section 10.1) and dumps. These may be destructive or nondestructive to the functional state depending on the scan style (destructive or observation-only scan). **Sample on the fly** (scanout) involves capturing the state and shifting it out while the chip is still in operation. This can be accomplished with the scanout or observation-only scan structure described in Section 10.2. It is possible to restart after reloading the scan state and initializing array states. Of course, absolute full-scan and full array initialization (to restore all the states) is required to accomplish this end. Freezing arrays and dumping observation-only registers are other observation mechanisms that are typically used.

A typical debug methodology involves the following steps:

- Using a trigger to get close to the cycle of interest
- Sampling on the fly to narrow the range of clock cycles
- Using clock manipulation or scan dumps to isolate the cycle of interest
- Taking a complete internal observation at the offending cycle
- Using simulation and deductive reasoning to isolate the offending circuitry

10.7 CONCLUDING REMARKS

Silicon debug and diagnosis is a complex field that employs a wide set of technologies (architecture, logic, circuit and layout design, simulation at different levels, instrumentation, optics, metrology, and even chemistries of various sorts). Logical deduction and the process of elimination are of paramount importance and central to debug and diagnosis. ***Design for debug and diagnosis*** (DFD) features are critical to enable an efficient debug and diagnosis process.

Debug and diagnosis requires bringing together people with various skills, being open minded, and welcoming or embracing new technologies to keep ahead of challenges. As technology scales and designs grow in complexity, the debug and diagnosis processes are constantly challenged. Although much progress has been made, many challenges lie ahead.

The operating voltage continues to scale down to keep reliability in check and for power reduction. It has been close to 1 V. Further scaling will make circuits operate in the subthreshold region; this will further reduce the emissions from the electron-hole recombination process. Lower voltage combined with smaller capacitance also leads to ever-smaller charge stored at any node. This smaller charge will also lead to its sensitivity to any kind of injected energy, be it optical or electromagnetic in nature. We are getting to the point where the **Heisenberg Uncertainty Principle** applies—the very act of observing something changes its nature.

Power dissipation and in particular leakage power has become an important issue for deep submicron integrated circuits, with leakage power contributing more than 30% of the total power dissipation. Thermal density and heat removal need to be carefully considered in the physical debug and probing processes. In addition, designs are becoming more power-aware and adaptive or dynamic control of voltage and frequency to regulate power is being adopted. Debugging failure in the presence of such adaptive or dynamic control is a major challenge.

Because *automatic test equipment* (ATE) or testers are barely able to keep up with the high pin count and high frequencies of device operation in functional mode, more on-chip DFD features are expected to be needed in the future. More DFD features that monitor a device actively (*e.g.*, on-chip power droop, device variation across the die, and on-chip temperature) will be needed. On-chip DFD features that facilitate in situ system debug will become increasingly important [Abramovici 2006].

Tools to enable automated diagnosis continue to evolve while tools for debug are still in a relatively infant state. Tools that identify sensitive areas of the design, add DFD features in an automated manner, generate tests for validation, debug and diagnosis, and automate the debug process need to advance.

Technologies presented in this chapter will need to continue to advance to keep up with CMOS scaling. Following Moore's law, chip complexity will double every 2 years, while we expect ever fewer design resources and ever faster time-to-market and time-to-profitability. Effective and efficient debug and diagnosis are certainly critical to the success of products. Better debug and diagnosis capabilities and faster fixes to problems are constant imperatives. It is expected and hoped that many of the readers will contribute solutions to this challenge.

10.8 EXERCISES

10.1 **(Debug *versus* Diagnosis)** What are the differences between silicon debug and low-yield analysis? Give at least three attributes.

10.2 **(Logic DFD Structures)** What combination of logic *design for debug and diagnosis* (DFD) features is typically needed to identify slow circuit paths?

10.3 **(Logic DFD Structures)** The purpose of introducing intentional clock skew to a clock domain is to allow more time for the signal to arrive at the storage element (flip-flop or latch) in that clock domain. What is the potential undesirable side effect of introducing this clock skew?

10.4 **(Logic DFD Structures)** What are the advantages of using observation-only scan over using a typical scan design for debug purposes?

10.5 **(Probing Technologies)** Which set of probing tools allows nonintrusive observations of a signal without disturbing the original signal?

10.6 **(Probing Technologies)** Which set of probing tools allows the injection of a signal that overrides what is already there?

10.7 **(Probing Technologies)** Optical probing tools are mostly noninvasive. What are their limitations? What signals really require mechanical types of probing?

10.8 **(Circuit Editing)** In Figure 10.26, which signals are connected to the inputs and outputs of the AND gate shown in the figure after all the new patches and cuts are made?

10.9 **(Physical DFD Structures)** Why do we need physical *design for debug and diagnosis* (DFD) for *focused ion beam* (FIB)? Give at least two reasons.

10.10 **(Physical DFD Structures)** For a chip with multiple metal layers, what physical DFD features are needed to use E-beam probing to observe signals at lower-level metal lines? Give at least three specific features.

10.11 **(Physical DFD Structures)** For backside *laser voltage probing* (LVP), what are the relevant physical DFD features for a successful probing? Give at least three specific features.

Acknowledgments

The authors wish to thank Dr. Rick Livengood of Intel, Professor Irith Pomeranz of Purdue University, and Dr. Franco Stellari of IBM for providing helpful feedback and comments.

References

R10.0 Books

[Abramovici 1994] M. Abramovici, M. A. Breuer, and A. D. Friedman, *Digital Systems Testing and Testable Design*, Revised Printing, IEEE Press, Piscataway, NJ, 1994.

[Bushnell 2000] M. L. Bushnell and V. D. Agrawal, *Essentials of Electronic Testing for Digital, Memory & Mixed-Signal VLSI Circuits*, Springer, Boston, 2000.

[Gizopoulos 2006] D. Gizopoulos, editor, *Advances in Electronic Testing: Challenges and Methodologies Series: Frontiers in Electronic Testing,* Springer, Boston, 2006.

[Jha 2003] N. Jha and S. Gupta, *Testing of Digital Systems*, Cambridge University Press, London, United Kingdom, 2003.

[Wang 2006] L.-T. Wang, C.-W. Wu, and X. Wen, editors, *VLSI Test Principles and Architectures: Design for Testability*, Morgan Kaufmann, San Francisco, 2006.

R10.1 Introduction

[Gangartikar 1982] P. Gangartikar, R. Presson, and L. Rosner, Test/characterization procedures for high density silicon RAMs, in *Proc. Int. Solid-State Circuits Conf.*, pp. 62–63, May 1982.

[Hammond 1991] J. Hammond and G. Sery, Knowledge-based electrical monitor approach using very large array yield structures to delineate defects during process development and production yield improvement, in *Proc. Int. Workshop on Defect and Fault Tolerance in VLSI*, pp. 67–80, November 1991.

[Josephson 2001] D. Josephson, S. Poehlman, and V. Govan, Debug methodology for the McKinley processor, in *Proc. Int. Test Conf.*, pp. 451–460, October 2001.

[Livengood 1999] R. H. Livengood and D. Medeiros, Design for (physical) debug for silicon microsurgery and probing of flip-chip packaged integrated circuits, in *Proc. Int. Test Conf.*, pp. 877–882, September 1999.

[Segal 2001] J. Segal, A. Jee, D. Lepejian, and B. Chu, Using electrical bitmap results from embedded memory to enhance yield, *IEEE Des. & Test of Computers*, pp. 28–39, May/June 2001.

[van Rootselaar 1999] G. J. van Rootselaar, and B. Vermeulen, Silicon debug: Scan chains alone are not enough, in *Proc. Int. Test Conf.*, pp. 892–902, September 1999.

R10.2 Logic Design for Debug and Diagnosis (DFD) Structures

[Abramovici 2006] M. Abramovici, P. Bradley, K. Dwarakanath, P. Levin, G. Memmi, and D. Miller, A reconfigurable design-for-debug infrastructure for SoCs, in *Proc. ACM/IEEE Des. Automation Conf.*, pp. 2–12, July 2006.

[Carbine 1993] A. Carbine, Scan mechanism for monitoring the state of internal signals of a vlsi microprocessor chip, U.S. Patent No. 5,253,255, October 12, 1993.

[Carbine 1997] A. Carbine and D. Feltham, Pentium pro processor design for test and debug, in *Proc. Int. Test Conf.*, pp. 294–303, November 1997.

[Eichelberger 1977] E. B. Eichelberger and T. W. Williams, A logic design structure for LSI testability, in *Proc. Des. Automation Conf.*, pp. 462–468, June 1977.

[Fetzer 2006] E. S. Fetzer, Using adaptive circuits to mitigate process variations in a microprocessor design, *IEEE Des. & Test of Computers*, 23(6), pp. 476–483, June 2006.

[Gu 2002] X. Gu, W. Wang, K. Li, H. Kim, and S. S. Chung, Re-using DFT logic for functional and silicon debugging test, in *Proc. Int. Test Conf.*, pp. 648–656, October 2002.

[Guo 2006] R. Guo and S. Venkataraman, An algorithmic technique for diagnosis of faulty scan chains, *IEEE Tran. Comput.-Aided Des.*, 25(9), pp. 1861–1868, September 2006.

[Hao 1995] H. Hao and R. Avra, Structured design for debug: The SuperSPARC-II methodology and implementation, in *Proc. Int. Test Conf.*, pp. 175–183, October 1995.

[Holbrook 1994] K. Holbrook, S. Joshi, S. Mitra, J. Petolino, R. Raman, and M. Wong, MicroSPARC: A case-study of scan based debug, in *Proc. Int. Test Conf.*, pp. 70–75, October 1994.

[IEEE 1149.1-2001] IEEE Std. 1149.1-2001, *IEEE Standard Test Access Port and Boundary Scan Architecture*, IEEE Press, New York, 2001.

[Josephson 2001] D. Josephson, S. Poehlman, and V. Govan, Debug methodology for the McKinley processor, in *Proc. Int. Test Conf.*, pp. 451–460, October 2001.

[Josephson 2004] D. Josephson and B. Gottlieb, The crazy mixed up world of silicon debug [IC validation], in *Proc. IEEE Custom Integrated Circuits Conf.*, pp. 665–670, October 2004.

[Kurd 2001] N. A. Kurd, J. S. Barkarullah, R. O. Dizon, T. D. Fletcher, and P. D. Madland, A multigigahertz clocking scheme for the Pentium 4 microprocessor, *IEEE J. Solid-State Circuits*, 36(11), pp. 1647–1653, November 2001.

[Levitt 1995] M. E. Levitt, S. Nori, S. Narayanan, G. P. Grewal, L. Youngs, A. Jones, G. Billus, and S. Paramanandam, Testability, debuggability, and manufacturability of the UltraSPARC-I microprocessor, in *Proc. Int. Test Conf.*, pp. 157–166, October 1995.

[Mahoney 2005] P. Mahoney, E. Fetzer, B. Doyle, and S. Naffziger, Clock distribution on a dual-core, multi-threaded Itanium family processor, *Digest of Papers, IEEE Int. Solid-State Circuits Conf.*, pp. 292–599, February 2005.

[Needham 1995] W. Needham and N. Gollakota, DFT strategy for Intel microprocessors, in *Proc. Int. Test Conf.*, pp. 157–166, October 1995.

[Paniccia 1998] M. Paniccia, T. Eiles, V. R. M. Rao, and M. Y. Wai, Novel optical probing technique for flip chip packaged microprocessors, in *Proc. Int. Test Conf.*, pp. 740–747, October 1998.

[Pham 2006] D. C. Pham, T. Aipperspach, D. Boerstler, M. Bolliger, R. Chaudhry, D. Cox, P. Harvey, P. M. Harvey, H. P. Hofstee, C. Johns, J. Kahle, A. Kameyama, J. Keaty, Y. Masubuchi, M. Pham, J. Pille, S. Posluszny, M. Riley, D. L. Stasiak, M. Suzuoki, O. Takahashi, J. Warnock, S. Weitzel, D. Wendel, and K. Yazawa, Overview of the architecture, circuit design, and physical implementation of a first-generation cell processor, *IEEE J. Solid-State Circuits*, 41(1), pp. 179–196, January 2006.

[Sogomonyan 2001] E. S. Sogomonyan, A. Morosov, M. Gossel, A. Singh, and J. Rzeha, Early error detection in systems-on-chip for fault-tolerance and at-speed debugging, in *Proc. VLSI Test Symp.*, pp. 184–189, April 2001.

[van Rootselaar 1999] G. J. van Rootselaar and B. Vermeulen, Silicon debug: Scan chains alone are not enough, in *Proc. Int. Test Conf.*, pp. 892–902, September 1999.

[Venkataraman 2001] S. Venkataraman and S. B. Drummonds, Poirot: Applications of a logic fault diagnosis tool, *IEEE Design & Test of Computers*, 18(1), pp. 19–30, January/February 2001.

[Waicukauski 1989] J. A. Waicukauski and E. Lindbloom, Failure diagnosis of structured VLSI, *IEEE Design & Test of Computers*, 6(4), pp. 49–60, August 1989.

[Yee 1999] W. M. Yee, M. Paniccia, T. Eiles, and V. R. M. Rao, Laser voltage probe (LVP): A novel optical probing technology for flip-chip packaged microprocessors, in *Proc. Int. Symp. on Physical and Failure Analysis of Integrated Circuits*, pp. 15–20, July 1999.

R10.3 Probing Technologies

[Bockelman 2002] D. R. Bockelman, S. Chen, and B. Obradovic, Infrared emission-based static logic state imaging on advanced silicon technologies, in *Proc. Int. Symp. for Testing and Failure Analysis*, pp. 531–537, November 2002.

[Bodoh 2002] D. Bodoh, K. Dickson, R. Wang, T. Cheng, N. Pakdaman, J. Vickers, D. Cotton and B. Lee., Defect localization using time-resolved photon emission on SOI devices that fail scan tests, in *Proc. Int. Symp. for Testing and Failure Analysis (ISTFA)*, pp. 655–661, November 2002.

[Chew 1999] Y. Y. Chew, K. H. Siek, and W. M. Yee, Novel backside sample preparation process for advanced CMOS integrated circuit failure analysis, in *Proc. Int. Symp. on Physical and Failure Analysis of Integrated Circuits*, pp. 119–122, July 1999.

[Kash 1997] J. A. Kash and J. C. Tsang, Optical imaging of picosecond switching in CMOS circuits, in *Proc. Conf. on Lasers and Electro-Optics*, 11, pp. 298–299, May 1997.

[Knebel 1998] D. Knebel, P. Sanda, M. McManus, J. A. Kash, J. C. Tsang, D. Vallett, L. Huisman, P. Nigh, R. Rizzolo, P. Song, and F. Motika, Diagnosis and characterization of timing-related defects by time-dependent light emission, in *Proc. Int. Test Conf.*, pp. 733–739, October 1998.

[Liu 1997] C. H. Liu, C. H. Peng, and C. C. Hsu, Identification of process defects using backside emission microscopy, in *Proc. Int. Symp. on Physical and Failure Analysis of Integrated Circuits*, pp. 230–233, July 1997.

[Loh 1999] T. H. Loh, W. M. Yee, and Y. Y. Chew, Characterization and application of highly sensitive infra-red emission microscopy for microprocessor backside failure analysis, *Proc Int. Symp. on Physical and Failure Analysis of Integrated Circuits*, pp. 108–112, July 1999.

[Sanda 1999] P. N. Sanda, D. R. Knebel, J. A. Kash, H. F. Casal, J. C. Tsang, E. Seewann, and M. Papermaster, Picosecond imaging circuit analysis of the POWER3 clock distribution, in *Proc. IEEE Int. Solid-State Circuits Conf.*, pp. 372–373, February 1999.

[Uraoka 1992] Y. Uraoka, T. Maeda, I. Miyanaga, and K. Tsuji, New failure analysis technique of ULSIs using photon emission method, in *Proc. IEEE Conf. on Microelectronic Test Structures*, pp. 100–105, March 1992.

[Varner 2002] E. B. Varner, C.-L. Young, H. M. Ng, S. P. Maher, T. M. Eiles, and B. Lee, Single element time resolved emission probing for practical microprocessor diagnostic applications, in *Proc. Int. Symp. for Testing and Failure Analysis*, pp. 741–746, November 2002.

[Vickers 2002] J. Vickers, N. Pakdaman, and S. Kasapi, Prospects of time-resolved photon emission as a debug tool, in *Proc. Int. Symp. for Testing and Failure Analysis*, pp. 645–653, November 2002.

R10.4 Circuit Editing

[Livengood 1999] R. H. Livengood and D. Medeiros, Design for (physical) debug for silicon microsurgery and probing of flip-chip packaged integrated circuits, in *Proc. Int. Test Conf.*, pp. 877–882, September 1999.

R10.6 Diagnosis and Debug Process

[Aitken 1995] R. C. Aitken, Finding defects with fault models, in *Proc. Int. Test Conf.*, pp. 498–505, October 1995.

[Carbine 1997] A. Carbine and D. Feltham, Pentium pro processor design for test and debug, in *Proc. Int. Test Conf.*, pp. 294–303, November 1997.

[Doyle 1888] A. C. Doyle, *A Study in Scarlet*, Part 2, Chapter 7, Harper & Brothers, New York, 1888.

[Doyle 1905] A. C. Doyle, *The Adventure of Black Peter, The Return of Sherlock Holmes*, McClure, Phillips, & Co., New York, 1905.

[Guo 2006] R. Guo and S. Venkataraman, An algorithmic technique for diagnosis of faulty scan chains, *IEEE Trans. Comput.-Aided Des.*, 25(9), pp. 1861–1868, September 2006.

[Hao 1995] H. Hao and R. Avra, Structured design for debug: The SuperSPARC-II methodology and implementation, in *Proc. Int. Test Conf.*, pp. 175–183, October 1995.

[Holbrook 1994] K. Holbrook, S. Joshi, S. Mitra, J. Petolino, R. Raman, and M. Wong, MicroSPARC: A case-study of scan based debug, in *Proc. Int. Test Conf.*, pp. 70–75, October 1994.

[Josephson 2001] D. Josephson, S. Poehlman, and V. Govan, Debug methodology for the McKinley processor, in *Proc. Int. Test Conf.*, pp. 451–460, October 2001.

[Kwon 1998] Y.-J. Kwon, B. Mathew, and H. Hao, FakeFault: A silicon debug software tool for microprocessor embedded memory arrays, in *Proc. Int. Test Conf.*, pp. 727–732, October 1998.

[van Rootselaar 1999] G. J. van Rootselaar and B. Vermeulen, Silicon debug: Scan chains alone are not enough, in *Proc. Int. Test Conf.*, pp. 892–902, September 1999.

[Venkataraman 2001] S. Venkataraman and S. B. Drummonds, Poirot: Applications of a logic fault diagnosis tool, *IEEE Des. & Test of Computers*, 18(1), pp. 19–30, January/February 2001.

[Waicukauski 1989] J. A. Waicukauski and E. Lindbloom, Failure diagnosis of structured VLSI, *IEEE Des. & Test of Computers*, 6(4), pp. 49–60, August 1989.

R10.7 Concluding Remarks

[Abramovici 2006] M. Abramovici, P. Bradley, K. Dwarakanath, P. Levin, G. Memmi, and D. Miller, A reconfigurable design-for-debug infrastructure for SoCs, in *Proc. ACM/IEEE Des. Automation Conf.*, pp. 2–12, July 2006.

CHAPTER 11

Software-Based Self-Testing

Jiun-Lang Huang
National Taiwan University, Taipei, Taiwan

Kwang-Ting (Tim) Cheng
University of California, Santa Barbara, California

ABOUT THIS CHAPTER

With the advances in semiconductor manufacturing technology, the concept of ***system-on-chip*** (SOC) has become a reality. An SOC device can contain a large number of complex, heterogeneous components including digital, analog, mixed-signal, ***radiofrequency*** (RF), micromechanical, and other systems on a single piece of silicon. The increasing heterogeneity and programmability of the SOC, along with the ever-increasing frequency and technology changes, however, are posing serious challenges to manufacturing test and on-chip self-test. Scan testing is the most commonly used ***design-for-testability*** (DFT) technique to address the fault coverage and test cost concerns. The problem is lacking self-test ability in the field. Hardware-based structural self-test techniques, such as logic ***built-in self-test*** (BIST), offer a feasible solution. Many solutions have been developed over the years to increase a circuit's fault coverage while reducing area overhead and development time. However, structural BIST usually places the circuit in specific self-test mode. Like scan testing, it may also cause excessive test power consumption, overtesting, and yield loss.

Software-based self-testing (SBST) has attracted much attention to address those problems caused by structural BIST. The idea is to utilize on-chip programmable resources such as embedded processors for on-chip functional (rather than structural) test pattern generation, test data transportation, response analysis, and even diagnosis. In this chapter, we present a variety of methods for SBST. We start by discussing processor self-test techniques followed by a brief description of a processor self-diagnosis method. We continue with a discussion of methods for self-testing global interconnects as well as other nonprogrammable SOC cores. We also describe **instruction-level DFT** methods based on the insertion of test instructions into the data stream. These methods are intended to increase the circuit's fault coverage and reduce its test application time and program size. Finally, we summarize methodologies for ***digital signal processing-based*** (DSP-based) self-test of analog and mixed-signal components.

11.1 INTRODUCTION

There is little doubt that SOC has become the logical solution for modern chip manufacturing because of the stringent demands for devices offering short time-to-market, rich functionalities, high portability, and low power consumption that are ultimately placed on designers. A typical SOC device contains a large number of complex, heterogeneous components including digital, analog, mixed-signal, RF, micromechanical, and other systems on a single piece of silicon. As the trend of limited accessibility to individual components, increased operating frequencies, and shrunken feature sizes continues, testing will face a whole new set of challenges.

The cost of silicon manufacturing *versus* that of testing given in the ***International Technology Roadmap for Semiconductors*** (ITRS) [SIA 1997, 1999] is illustrated in Figure 11.1 where the top and bottom curves show the fabrication and test capital per transistor, respectively. The trend clearly shows that unless some

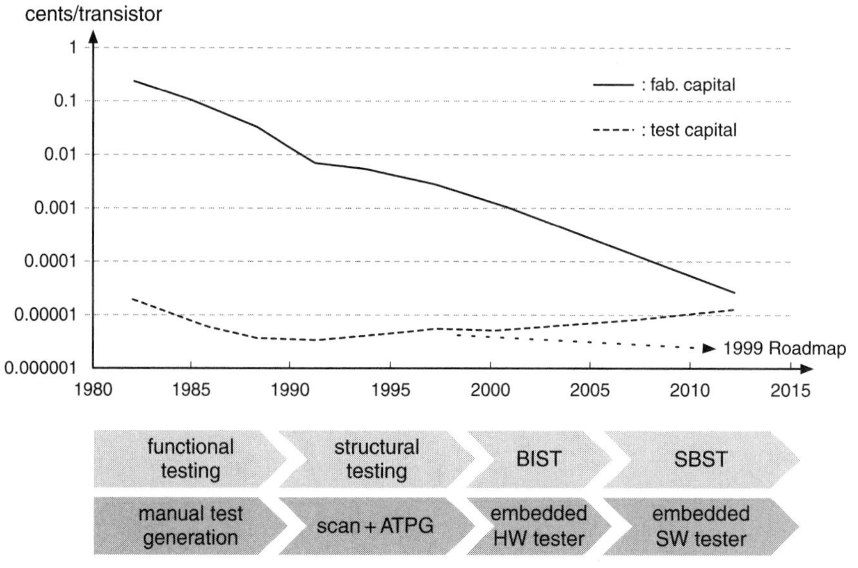

■ **FIGURE 11.1**

Fabrication *versus* test capital (based on 1997 SIA and 1999 ITRS roadmap data).

fundamental changes are made in the test process, it may eventually cost more to test a chip than to manufacture it [SIA 2003].

Also depicted in Figure 11.1 is the test paradigm shift. The difficulties in generating functional test patterns to reduce a chip's defect level and test cost contribute to this trend. DFT and BIST have been regarded as the solutions for changing the direction of the test cost trend in Figure 11.1. BIST empowers the ***integrated circuit*** (IC) by enabling at-speed test signals to be analyzed on-chip using an **embedded hardware tester**. By implementing BIST, not only is the need for high-speed external testers eliminated, but also greater testing accuracy is achieved. Existing BIST techniques are based on structural BIST. Although the most commonly adopted scan-based BIST techniques—[Cheng 1995] [Lin 1995] [Tsai 1998] [Wang 2006]—offer good test quality, the required circuitry to realize the *embedded hardware tester*—including full scan, ***linear-feedback shift register*** (LFSR), ***multiple-input signature register*** (MISR), and BIST controllers—incur nontrivial area, performance, and design time overheads. Furthermore, structural BIST suffers the problem of elevated test power consumption that is inherent in structural testing because test patterns are less correlated spatially or temporally than functional patterns, resulting in higher switching activity. Another serious limitation of existing structural BIST is the complexity associated with applying at-speed patterns to timing related faults. This includes various complex timing issues related to multiple clock domains, multiple frequencies, and test clock skews that must be resolved for such testing to be effective.

This chapter discusses the concept of SBST, a promising solution for alleviating the problems inherent in external testers and structural BIST, and outlines the enabling techniques. In the new SBST paradigm, memory blocks that facilitate SBST are tested first. Then the on-chip programmable components such as processors and **digital signal processors** are self-tested. Finally, these programmable components are configured as an **embedded software tester** to test on-chip global interconnects and other nonprogrammable components. The SBST paradigm is sometimes referred to as **functional self-testing**, **instruction-based self-testing**, or **processor-based self-testing**.

11.2 SOFTWARE-BASED SELF-TESTING PARADIGM

The SBST concept is depicted in Figure 11.2 using a bus-based SOC. In this illustration, the ***central processing unit*** (CPU) accesses the system memory via a shared bus, and all the ***intellectual property*** (IP) cores are connected to the system bus via a ***virtual component interface*** (VCI) [VCI 2000]. Here, the VCI simply acts as the standard communication interface between the core and the shared bus. To support the self-test methodology, each core is surrounded by a **test wrapper** that contains the test support logic needed to control scan chain shifting as well as buffers to store scan data and support at-speed testing.

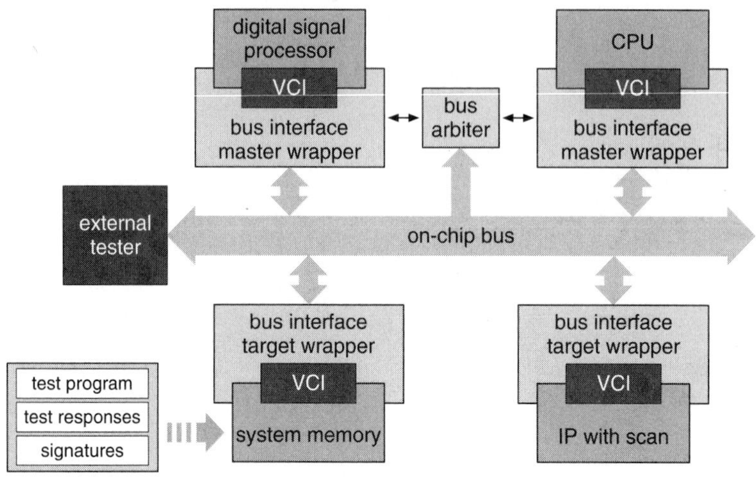

■ **FIGURE 11.2**

A software-based self-testable SOC.

11.2.1 Self-Test Flow

The overall SBST methodology consists of the following steps:

Step 1: Memory self-testing. The memory block (either the system or processor cache memory) that stores the test programs, test responses, and signatures is tested and repaired if necessary. A good reference for learning more about memory BIST and ***built-in self-repair*** (BISR) techniques can be found in [Wang 2006].

Step 2: Processor self-testing. During processor self-testing, the external tester first loads the memory with the test program and the signatures. Then the processor tests itself by executing the test program, aiming at the fault models of interest. The test program responses are written to the memory and later compared to the stored signatures to make the pass/fail decision.

Step 3: Global interconnect testing. To validate the global interconnect functionality, the embedded processor runs the corresponding test programs. Predetermined patterns that activate the defects of interest are transmitted between pairs of cores among which data or address transmission exists. The responses captured in the destination cores are then compared to the stored signatures.

Step 4: Testing nonprogrammable cores. When testing the remaining non-self-testable cores, the embedded processor controls the test pattern generation, test data transportation, and response analysis programs. For analog/mixed-signal cores, processor and DSP cores may be employed to perform the required pre- and postprocessing DSP procedures.

11.2.2 Comparison with Structural BIST

The fundamental difference between SBST and structural BIST is that the former handles testing as a system application whereas the latter places the system in nonfunctional test mode. Handling testing as an application has the following advantages:

1. The need for DFT circuitry is minimized. In structural BIST, one may have to add wrapper cells as well as necessary logic to control or observe the system's external ***input/output*** (I/O) ports that are not connected to low-cost testers. In SBST, additional DFT techniques can be employed if SBST alone does not achieve the desired fault coverage or diagnosis resolution.
2. The performance requirement for the external tester is reduced as all high-speed transactions occur on-chip and the tester's main responsibility is to upload the test program to the system memory and download test responses after self-testing.
3. Performing test pattern application and response capture on-chip achieves greater accuracy than that obtainable with a tester, which reduces the yield loss owing to tester inaccuracy.
4. Because the system is operated in functional mode while executing the test programs, excessive test power consumption and potential overkill problems inherent in structural BIST are eliminated.

One major concern of SBST is its fault detection efficiency. In a programmable component like a processor or DSP core, some structural faults that cannot be detected using instructions do not necessarily map to low fault coverage. If a fault can be detected only by test patterns that cannot be realized by any instruction sequence, then the fault is redundant by definition in the normal operating mode of the programmable component. Thus, there is no need to test for this type of fault during manufacturing test, even though we may still want to detect and locate these faults during silicon debug and diagnosis for manufacturing process improvement.

For an IP core that is not self-testable, one may rely on structural BIST techniques to reach the desired level of fault coverage. In this case, using the processor as the test pattern generator and output response analyzer (*i.e.*, an *embedded software tester*) gives one the flexibility of combining multiple test strategies to achieve the desired fault coverage—one just has to alter the corresponding programs or parameters without any hardware modification. In [Hellebrand 1996], the authors discuss mixed-mode pattern generation for random and deterministic patterns using embedded processors. After identifying the best pattern generation scheme—including random and deterministic techniques to meet the desired fault coverage goal, test time budget, or available memory storage—the test program is synthesized accordingly without the need to alter any BIST hardware.

11.3 PROCESSOR FUNCTIONAL FAULT SELF-TESTING

The complexity of processors, together with the limited accessibility of their internal logic, makes them extremely difficult to test. From the point of view of SBST, the situation is even worse for users or test engineers because the processors' internal design details are often unavailable and difficult to comprehend.

To resolve these problems, **processor functional-level fault models** and their corresponding test generation methods have been proposed in [Thatte 1980] and [Brahame 1984]. Because only knowledge of the processor's instruction set and the functions it performs is needed, these techniques can provide a testing solution for general-purpose processors. In the following subsections, we describe the fault models and test generation procedures for functional faults associated with register decoding, instruction decoding and control, data storage and transfer, and data manipulation as presented in [Thatte 1980].

11.3.1 Processor Model

To facilitate functional-level test generation, the processor is modeled at the ***register-transfer level*** (RTL) by a ***system graph*** (*S*-graph), which is based on its instruction set and the functions it performs. In the *S*-graph, each register that can be explicitly modified by an instruction is represented by a node. Two additional nodes, IN and OUT, are also added to the graph to cover the main memory and I/O devices. Nodes in the *S*-graph are connected by directed edges. There exists a labeled directed edge from node *A* to node *B* if data flow occurs from *A* to *B* during the execution of any instruction. An example *S*-graph is illustrated in Figure 11.3. For

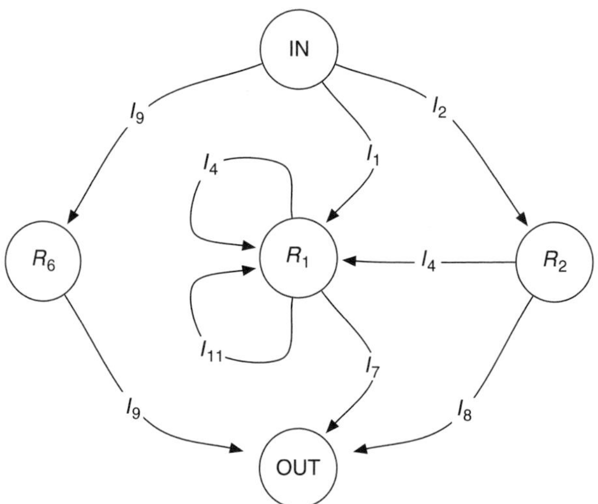

FIGURE 11.3

An example *S*-graph.

TABLE 11.1 ■ Summary of the Registers and Instructions in Figure 11.3

R_1:	Accumulator (ACC)
R_2:	General purpose register
R_6:	Program counter (PC)
I_1:	Load R_1 from the main memory using immediate addressing. (T)
I_2:	Load R_2 from the main memory using immediate addressing. (T)
I_4:	Add the contents of R_1 and R_2 and store the results in R_1. (M)
I_7:	Store R_1 into the main memory using implied addressing. (T)
I_8:	Store R_2 into the main memory using implied addressing. (T)
I_9:	Jump instruction. (B)
I_{11}:	Left shift R_1 by one bit. (M)

convenience, only a subset of instructions and those registers in the processor that are directly involved in carrying out these instructions are shown and summarized in Table 11.1.

As in [Flynn 1974], the instructions are classified as transfer (class T), manipulation (class M), or branch (class B). In Table 11.1, I_4 and I_{11} are class M instructions, I_9 is a class B instruction, and the others are class T instructions. Note that multiple edges may be associated with one instruction (*e.g.*, I_4 and I_9). However, an instruction is allowed to have multiple **destination registers** only if it involves a data transfer between the main memory or an I/O device and if it registers this transfer during its execution.[1] Also, it is assumed that any register can be written (implicitly or explicitly) as well as read (implicitly or explicitly) using a sequence of either class T or B instructions.

After constructing the *S*-graph, one assigns integer labels to the nodes by the node labeling algorithm in Figure 11.4. Regarding the *S*-graph, the node label indicates the shortest distance of a node to the OUT node. On a processor, the node label corresponds to the minimum number of class T or B instructions that need to be executed in order to read the contents of the register represented by that node. The nodes of the *S*-graph in Figure 11.3 are labeled as follows: the OUT node is labeled 0, and R_1, R_2, and R_6 are labeled 1.

The edges in the *S*-graph are then assigned labels in the following way. First, for any instruction that implicitly or explicitly reads a register to the OUT node during its execution, all the edges that are involved in its execution are labeled 1, such as the edges of I_7, I_8, and I_9 for example. In this way, the edges of class B instructions will all be assigned a label of 1. The edges of the remaining instructions are labeled as follows: if the destination register of the instruction is labeled K, all the edges

[1] The *destination registers* of an instruction are defined to be the set of registers that are changed by that instruction during its execution. For example, the destination registers of instruction I_9 in Figure 11.3 are $\{R_6, \text{OUT}\}$.

```
assign label 0 to the OUT node;
K←0;
while there exist unlabeled nodes
        assign K+1 to unlabeled nodes whose contents
        can be transferred to any register(s) labeled K
        by executing a single class T or B instruction;
        K←K+1;
end while
```

■ **FIGURE 11.4**

The node labeling algorithm.

of that instruction will be labeled $K+1$. For the S-graph in Figure 11.3, the edges of instructions I_7, I_8, and I_9 are labeled 1, and those of I_1, I_2, I_4, and I_{11} are labeled as 2.

The purpose of the node and edge labels is to facilitate the test generation process. Tests are generated in a way such that the knowledge gained from the correct execution of tests used to check the decoding of registers and instructions with lower labels is utilized in generating tests to check the decoding of registers and instructions with higher labels.

11.3.2 Functional-Level Fault Models

Processor functional-level fault models are developed at a higher level of abstraction independent of the details of their implementation. The functional-level fault models described include **register decoding fault**, **instruction decoding and control fault**, **data storage fault**, **data transfer fault**, and **data manipulation fault**.

Register decoding fault model. Under the *register decoding fault*, the decoded address of the register(s) is incorrect. As a result, the wrong register(s) may be accessed, or else no register is accessed at all. The resultant outcome retrieved when no or multiple registers are accessed is technology dependent.

Instruction decoding and control fault model. When there is a fault in the instruction decoding and control function, the processor may execute a wrong instruction, execute some other instructions (including the original one), or execute no instruction at all.

Data storage fault model. Single stuck-at faults may occur in any number of cells in any number of registers.

Data transfer fault model. The possible faults include a line in the data transfer path stuck at 1 or 0; or two lines in the data transfer path are coupled.

Data manipulation fault model. No specific fault models are proposed for data manipulation units. Instead, it is assumed that the required test patterns can be derived according to the implementation.

Processor functional-level fault model. The processor may possess any number of faults that belong to the same fault model.

11.3.3 Test Generation Procedures

This section describes the test generation procedures for the modeled functional faults. Note that the *S*-graph as well as the node and edge labels facilitate the test generation process.

11.3.3.1 Test Generation for Register Decoding Fault

The goal is to validate that the **register mapping function**, which is $f_D : \mathcal{R} \rightarrow \mathcal{R}$, where $\mathcal{R}$ is the set of all registers, is correct. That is, $f_D(R_i) = R_i$ for all registers.

The test generation flow is illustrated in Figure 11.5. At the beginning of the process, the first-in first-out queue Q is initialized with the set of all registers such that registers with smaller labels are in the front of Q. A, the set of processed registers, consists of the first element of Q. Then, test generation is performed, one register at a time. Each iteration consists of the write and read phases. In the write phase, all the registers in A are written with ONE (all ones), and the first register in Q, denoted by R_{next}, is written ZERO (all zeros). The needed write operation for each register is the shortest sequence of class T or class B instructions that are necessary to write the target register. In the read phase, registers in A are read in the order of ascending labels. Then, the content of R_{next} is read out. Similarly, the needed read operation for each register denotes the shortest sequence of class T or class B instructions that are necessary to read the target register.

The test generation algorithm in Figure 11.5 assures that all the registers have disjoint image sets under the *register mapping function* f_D, thus establishing that f_D has a one-to-one correspondence. The generated test program is capable of finding any detectable fault in the fault model for the register decoding function. One

```
Q ← sort (ℛ);
R_next ← dequeue (Q);
A ← {R_next};
while Q ≠ φ
    foreach R_i ∈ A
        append write (R_i, ONE) to test program;
    end foreach
    R_next ← dequeue (Q);
    append write (R_next, ZERO) to test program;
    Q' ← sort (A);
    while Q' ≠ φ
        append read (dequeue (Q')) to test program;
    end while
    append read (R_next) to test program;
    A ← A ∪ {R_next};
end while
repeat 1–16 with complementary data;
```

FIGURE 11.5

Test generation for register decoding faults.

example of a possible undetectable fault is the concurrent occurrence of $f_D(R_i) = R_j$ and $f_D(R_j) = R_i$.

11.3.3.2 Test Generation for Instruction Decoding and Control Fault

Let I_j be the instruction to be executed. The faults in which no instruction is executed at all, the wrong instruction is executed such as I_k instead of I_j, or some other instruction I_k is also executed are each denoted by $f(I_j/\phi)$, $f(I_j/I_k)$, and $f(I_j/I_j + I_k)$, respectively. To simplify the test generation task for instruction decoding and control functions, it is assumed that the labels of class M instructions are no greater than 2 and that all class B instructions have the label 1. Only class T instructions can have labels greater than 2.

To ensure fault coverage, the order in which faults are detected is crucial. Figure 11.6 depicts the order in which the tests are applied. Note how the tests are applied in such a way that the knowledge gained from testing instructions of lower labels is utilized for testing instructions with higher labels.

Consider the instruction I_j, label$(I_j) = 2$. To detect the $f(I_j/\phi)$ fault, one first writes O_1 to I_j's *destination register* R_d using one class T or class B instruction. Then, proper operand(s) are written to I_j's **source registers**,[2] such that when I_j is executed, it produces O_2 ($O_2 \neq O_1$) in its *destination register* R_d. I_j is then executed and R_d is read where the expected output is O_2.

To detect the $f(I_j/I_k)$ faults, consider the case when (1) label$(I_j) =$ label$(I_k) = K \geq 3$ and (2) I_j and I_k have the same *destination register* R_d. According to the previous assumptions, both I_j and I_k are class T instructions, and they each have only one *destination register*. The test procedure first writes O_1 and O_2 ($O_1 \neq O_2$) to the *source registers* of I_j and I_k, respectively. Then, I_j is executed, and R_d is read for K times with the expected output being O_1. Finally, I_k is executed and R_d is read with the expected output being O_2. It is interesting to note that I_j is executed K times. If O_2 is really stored in the *source register* of I_k at the beginning of the procedure,

```
1. K←1;
2. for K=1 to K_max
3.     apply tests to detect f(I_j/φ), f(I_j/I_k), and f(I_j/I_j + I_k),
       where label (I_j) = label (I_k) = K;
4.     apply tests to detect f(I_j/I_j + I_k), where 1 ≤ label (I_j) ≤ K,
       label (I_k) = K+1, and K < K_max;
5.     apply tests to detect f(I_j/I_j + I_k), where K+1 ≤ label (I_j) ≤ K_max,
       and label (I_k) = K;
6. end for
```

■ **FIGURE 11.6**

The order of test generation for instruction decoding and control function.

[2] The set of *source registers* for an instruction is defined to be the set of registers that provide the operands for that instruction during its execution.

$f(I_j/I_k)$ will be detected the first time I_j is executed and R_d read out. However, because of the faults involved in instructions used to write O_2 in I_k's *source register*, O_1 may have been stored in I_k's *source register*. In such a case, $f(I_j/I_k)$ will not be detected. In the worst case, I_j has to be executed K times to guarantee the detection of $f(I_j/I_k)$.

As for $f(I_j/I_j+I_k)$, an example when $1 \leq \text{label}(I_j) \leq K$, $\text{label}(I_k) = K+1$, and $K \geq 2$ is the result. Note that I_k is a class T instruction, and the *destination registers* of I_j and I_k, denoted by R_j and R_k, respectively, are different. When the label of I_k's source register is less than K, different operands O_1 and O_2 are first written to I_k's source and destination registers, respectively. Then, I_k's source register is read, and the expected output is O_1. Finally, I_j is executed and I_k's destination register is read. The expected output is O_2.

11.3.3.3 Test Generation for Data Transfer and Storage Function

Depending on the class to which the involved instruction belongs (*i.e.*, class T, B, or M), different test generation procedures are applied.

Consider a sequence of class T instructions $I_{j1}, I_{j2}, \ldots, I_{jk}$ of which the associated edges form a directed path from the IN node to the OUT node in the S-graph. All such paths should be tested. Testing the data transfer faults of the instructions in this path starts by executing I_{j1} with operand O_1. Then $I_{j2}, I_{j3}, \ldots, I_{jk}$ are executed, and the expected output is O_1. Let the width of the data transfer path be w; the procedure is repeated for the following O_1 configurations: $\underbrace{11\cdots1}_{w}, \underbrace{11\cdots1}_{w/2}\underbrace{00\cdots0}_{w/2}, \ldots, \underbrace{1010\cdots10}_{w}$.

For class M instructions, use instruction I_4, which adds the contents of R_1 and R_2 together and stores the sum in R_1, as an example. (The involved edges and nodes of I_4 can be found in Figure 11.3.) The test of I_4 consists of testing the paths from the ***arithmetic logic unit*** (ALU) to R_1, and from R_1 and R_2 to ALU. For the former, I_1 and I_2 are utilized to load R_1 and R_2 with O_1 and all zeros. Then, I_4 is executed, followed by I_7. The result is read and stored in R_1. Assuming that the processor is an 8-bit processor, to fully test the path, the procedure is repeated for the following O_1 configurations (complemented and uncomplemented): 1111 1111, 1111 0000, 1100 1100, and 1010 1010. For the paths from R_1 to ALU, R_1 is loaded with O_1, and R_2 with all zeros. Then, I_4 and I_7 are executed and the expected output is O_1. The procedure is repeated for the following O_1 configurations: 0000 0001, 0000 0010, ..., 1000 0000. Testing the path from R_2 to ALU is similar and not repeated here.

Finally, the transfer paths and registers involved in class B instructions must be tested. Take the JUMP instruction I_9 for example. Assume that the address bus width is 16. Then, I_9 should be executed with the following jump addresses (in both complemented and uncomplemented forms):

0000 0000 0000 0000, 0000 0000 1111 1111,
0000 1111 0000 1111, 0011 0011 0011 0011,
0101 0101 0101 0101

11.3.3.4 Test Generation for Data Manipulation Function

No specific fault model is proposed for the data manipulation functions. Instead, given the test patterns for a data manipulation unit (ALU, shifter, etc.), the desired operands can be delivered to its input(s) and the results can be delivered to the output(s) using class T instructions.

11.3.3.5 Test Generation Complexity

The complexity of the test sequences in terms of the number of instructions used depends on the numbers of registers and instructions, denoted by n_R and n_I, respectively. For *register decoding faults*, the complexity is found to be $O\left(n_R^3\right)$. For *instruction decoding and control faults*, the complexity is $O\left(n_I^2\right)$ if the instruction labels do not exceed two.

The technique was applied to an 8-bit microprocessor with 2200 single stuck-at faults simulated. About 90% of the faults are detected by the test sequences for register decoding, data storage, data transfer, and data manipulation functions. The number of instructions for these sequences was about 1000. The test sequences for the faults that caused simultaneous execution of multiple instructions consist of about 8000 instructions. Many of these faults were subtle and required very elaborate test sequences to detect them. The remaining faults (about 4%) could not be detected with valid instructions; thus, for this particular processor, the fault coverage was excellent.

11.4 PROCESSOR STRUCTURAL FAULT SELF-TESTING

In this section, recent advances in processor SBST techniques that target structural faults, including stuck-at and delay faults, will be discussed.

11.4.1 Test Flow

In general, the structural fault oriented processor SBST techniques [Lai 2000a, 2000b] [Chen 2001] [Kranitis 2003] [Chen 2003] [Bai 2003] [Paschalis 2005] [Krannitis 2005] [Psarakis 2006] consist of two phases: the test preparation phase and the self-testing phase.

11.4.1.1 Test Preparation

In the test preparation phase, instruction sequences that deliver structural test patterns to the inputs of the processor component under test and transport the output responses to observable outputs are generated.

One challenge for processor component test generation is the **instruction-imposed I/O constraints**. For a processor component, the **input constraints** define the input space of the component allowed or realizable by processor instructions. A fault is undetectable if none of its test patterns is in the input space. The **output**

constraints, on the other hand, define the subset of component outputs observable by instructions. A fault is undetected at the chip level if its resulting errors fail to propagate to any observable outputs. Without incorporating these *instruction-imposed I/O constraints*, the following component test generation may produce test patterns that cannot be delivered by processor instructions.

In [Lai 2000a, 2000b], [Chen 2001, 2003], [Bai 2003], the extracted component I/O constraints are expressed in the form of Boolean expressions or ***hardware description language*** (HDL) descriptions and are fed to ***automatic test pattern generation*** (ATPG) for **constrained component test generation**. Next, the **test program synthesis** procedure maps the constrained test patterns to processor instructions. In addition to the test application instruction sequence, test supporting instruction sequences may be added (in front of or after the test application sequence) to set up the required processor state (*e.g.*, register values) and to transport the test responses to main memory.

It is interesting to note that the SBST approach offers great flexibility in test pattern generation and response analysis. For example, depending on which method is more efficient for a particular case, the test patterns may be loaded directly to the data memory or generated on-chip using a test pattern generation program (*e.g.*, a software-based LFSR). Similarly, the captured responses may be compressed on-chip using a software version MISR.

11.4.1.2 Self-Testing

The processor self-testing setup is illustrated in Figure 11.7. Because on-chip system memory (or cache memory) is utilized to store the test program(s) and responses, it has to be tested with standard techniques such as memory BIST [Wang 2006] and repaired if necessary to ensure that it is functioning. Then, a low-cost external tester can be used to load the test program(s) and the test data to the on-chip memory.

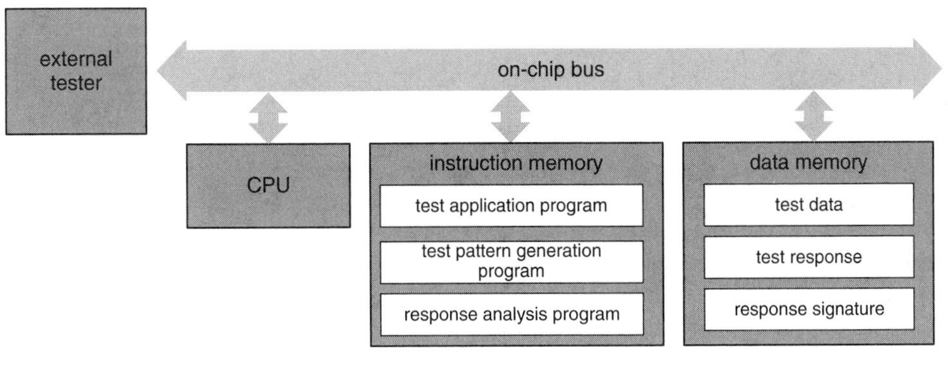

■ **FIGURE 11.7**

Processor self-testing setup.

To apply the structural tests, the processor is set up to properly execute the loaded test program. Finally, the test signatures are downloaded to the external tester for pass/fail decision or diagnosis.

In the following subsections, we describe the SBST methods for processor stuck-at faults [Chen 2001, 2003] and path delay faults [Lai 2000a, 2000b].

11.4.2 Stuck-At Fault Testing

The technique reported in [Chen 2001] targets processor stuck-at faults. Details of its test preparation step and the results of a case study will be illustrated.

11.4.2.1 Instruction-Imposed I/O Constraint Extraction

To reduce the test generation complexity, constrained test generation is performed for subcomponents instead of the full processor. To facilitate *constrained component test generation*, the *instruction-imposed I/O constraints* are extracted first. The constraints can be divided into *input constraint*s and *output constraint*s, which are determined by the instructions for controlling the component inputs and the instructions for observing the component outputs. Furthermore, a constraint may be a **spatial constraint** or a **temporal constraint**.

Take the PARWAN processor's shifter unit (SHU) in Figure 11.8a as an example. The instructions that control the SHU inputs include `lda` (load accumulator), `and`, `add`, `sub`, `asl` (arithmetic shift left), and `asr` (arithmetic shift right), and the corresponding *spatial constraints* on its inputs are listed in Table 11.2. For instance, if the executed instruction is `sub`, then both the *asl* and *asr* control inputs must be 0, the input flag v is set to 1 if flags c and s (the most significant bit of the data input, *i.e.*, *data_in*[7]) are different, and the z flag is set to 1 if the data input is 0 (*i.e.*, $data\_in[i] = 0, i = 0...7$). As for *data_in* and the carry flag c, there is no spatial constraint with respect to the `sub` instruction.

The *temporal constraints* on SHU, on the other hand, are imposed by the sequence of instructions that applies tests to SHU (Figure 11.8b). The sequence consists of

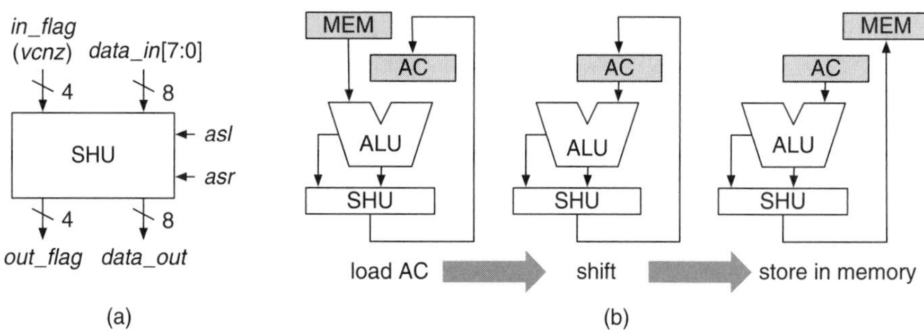

■ **FIGURE 11.8**

The SHU I/O and test application sequence: (a) SHU I/O and (b) SHU test application sequence.

TABLE 11.2 ■ Spatial Constraints at SHU Inputs

	Control Inputs		Flag Inputs				Data Inputs	
	asl	*asr*	*v*	*c*	*z*	*n*	*s* (*data_in*[7])	*data_in*[6 : 0]
`lda`	0	0	0	0	*data_in* ≡ 0	*s*	*X*	*X*
`and`	0	0	0	0	*data_in* ≡ 0	*s*	*X*	*X*
`add`	0	0	$c \oplus s$	*X*	*data_in* ≡ 0	*s*	*X*	*X*
`sub`	0	0	$c \oplus s$	*X*	*data_in* ≡ 0	*s*	*X*	*X*
`asl`	1	0	0	0	*data_in* ≡ 0	*s*	*X*	*X*
`asr`	0	1	0	0	*data_in* ≡ 0	*s*	*X*	*X*

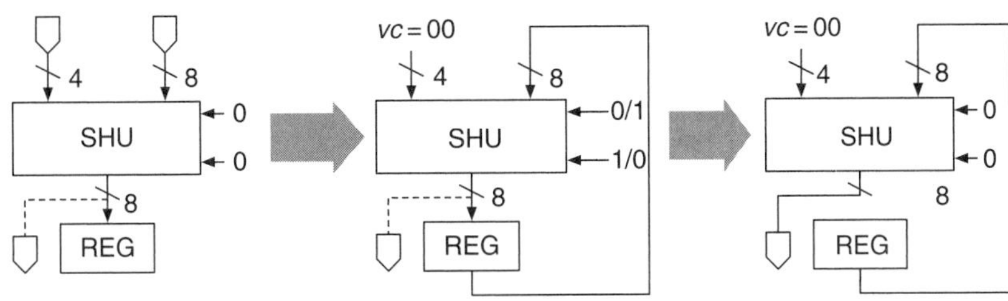

■ **FIGURE 11.9**

The SHU temporal constraint model.

three steps: (1) loading data from memory (MEM) into the accumulator (AC), (2) shifting the data stored in AC and storing the result in AC, and (3) storing the result in memory for later analysis. The corresponding **temporal constraint model** is illustrated in Figure 11.9 and summarized as follows:

1. The SHU inputs are connected to the primary inputs only in the first phase.
2. The SHU data outputs are connected to the primary outputs only in the third phase.
3. The shifting signals, *asl* and *asr*, are set to 0 in the first and third phases.
4. The *v* and *c* flags are set to 0 in the second and third phases because neither the shift nor the store instruction can set them to 1.

11.4.2.2 Constrained Component Test Generation

Once the component spatial and temporal constraints are derived, the *constrained component test generation* algorithm in Figure 11.10 is utilized to generate component structural test patterns.

In the initialization step (line 1), I is initialized to be the set of instructions for controlling the component under test C (*i.e.*, I_C) and F the fault list of C (*i.e.*, F_C).

```
1.  I ← I_C; F ← F_C; V_C ← φ; T_C ← φ;
2.  while I ≠ φ and F ≠ φ
3.      pick i from I;
4.      if not V_i,C ⊆ V_C
5.          (T_i,C, F_det) ← constrainedTG (F, V_i,C);
6.          T_C ← T_C ∪ T_i,C;
7.          F ← F − F_det;
8.          V_C ← V_C ∪ V_i,C;
9.      end if
10.     I ← I − {i};
11. end while
```

■ **FIGURE 11.10**

The constrained component test generation algorithm.

V_C and T_C, the covered input space and generated tests up to now, are initialized to the empty set ϕ. The test generation process then repeats until all the instructions are processed or the fault list is empty (line 2). In each iteration, an instruction i is selected for test generation (line 3). If the input space associated with i, denoted by $V_{i,C}$, is covered by previous instructions, i is skipped (line 4). Otherwise, constrained test pattern generation is performed. In line 5, $T_{i,C}$ is the set of generated test patterns and F_{det} is the set of newly detected faults by $T_{i,C}$. In lines 6 to 8, the test set, the remaining fault set, and the covered input space are updated.

The resulting test set T_C has two important properties. First, if the tests generated under the constraints imposed by any single instruction i achieve the maximum possible fault coverage in the functional mode allowed by i, T_C can achieve the maximum possible fault coverage in the functional mode allowed by I_C. That is, T_C detects any faults detectable by V_C. Second, any test vector in T_C can be realized by at least one instruction in I_C.

In practice, the algorithm faces some challenges. First, determination of I_C is nontrivial. One may utilize simulation-based approaches to identify the set of instructions that affect the inputs of C. Also, the instructions in I_C may be ordered according to the simplicity of *instruction-imposed I/O constraints*—instructions with simpler constraints first. Second, determining whether $V_{i,C} \subseteq V_C$ is true is a co-NP-complete problem. This step can be relaxed to screen out only the instructions that obviously cover the same input space as any previously processed instruction.

To better illustrate the *constrained component test generation* process, the component test preparation results for the PARWAN processor [Navabi 1997] ALU are shown in Table 11.3. The ALU contains two 8-bit data inputs (*in_1* and *in_2*) and one 3-bit control input (*alu_code*). Input *in_1* is connected to the data bus between the memory and the processor, whereas *in_2* is connected to the output of the accumulator. In Table 11.3, column 1 lists the instructions that control the ALU inputs. In columns 2 to 4, the input constraints imposed by the instructions as well as the generated test patterns are shown. The constrained inputs are expressed in the form of fixed values (*e.g.*, the *alu_code* field and the all Z's in *in_1* and *in_2*). The unconstrained inputs, on the other hand, are to be generated by a software LFSR

TABLE 11.3 ■ Component Tests for PARWAN ALU

		Test Pattern	
	alu_code	*in_1*	*in_2*
lda	100	(11111111,01100011,82)	*ZZZZZZZZ*
sta	110	*ZZZZZZZZ*	(11111111,01100011,82)
cma	001	*ZZZZZZZZ*	(11111111,01100011,35)
and	000	(11111111,01100011,98): odd	(11111111,01100011,98): even
sub	111	(11111111,01100011,24): odd	(11111111,01100011,24): even
add	101	(11111111,01100011,26): odd	(11111111,01100011,26): even

procedure; therefore, they are expressed by a self-test signature (S, C, N), where S and C are the seed and configuration of the pseudo-random pattern generator, and N is the number of pseudo random patterns used.

In [Kranitis 2003], [Paschalis 2005], and [Kranitis 2005], a different approach to component test generation is adopted based on the following observations:

1. The functional components such as ALU, multiplexers, registers, and register files should be given the highest priority for test development because their size dominates the processor area and thus they have the largest contribution to the overall processor fault coverage.
2. The majority of these functional components have a regular or semiregular structure. They can be efficiently tested with small and regular test sets that are independent of the gate-level implementation.

Thus, instead of using gate-level ATPG, a component test library of test algorithms that generate small deterministic test sets and provide very high fault coverage for most types and architectures of functional components is developed. Compact loop-based test routines are utilized to deliver these tests to the functional components and the test responses to observable outputs or registers.

11.4.2.3 Test Program Synthesis

Because the component tests are developed under the processor *instruction-imposed I/O constraints*, it will always be possible to find instructions for applying the component tests. On the output end, however, special care must be taken when collecting the component test responses. Inasmuch as data outputs and status outputs have different observability, they should be treated differently during response collection. In general, although there are no instructions for storing the status outputs of a component directly to memory, an image of the status outputs can be created in memory using conditional instructions. Following the PARWAN ALU example, which has an 8-bit data output (*data_out*) and a 4-bit status output (*out_flag* = *vczn*), the test program that observes the ALU status outputs after executing the add instruction is shown in Figure 11.11.

```
            lda   addr(y)      //load  AC
            add   addr(x)
            sta   data_out     //store  AC
            lda   11111111
            brav  ifv          //branch if overflow
            and   11110111
label ifv   brac  ifc          //branch if carry
            and   11111011
label ifc   braz  ifz          //branch if zero
            and   11111101
label ifz   bran  ifn          //branch if negative
            and   11111110
label ifn   sta   flag_out
```

■ **FIGURE 11.11**

Test program for observing ALU status outputs.

11.4.2.4 Processor Self-Testing

The processor self-testing flow is illustrated in Figure 11.12. First, the unconstrained test patterns are generated on-chip using the self-test signatures and the test generation program, which is a software version LFSR. Because all the self-test signatures are the same except in the *N* field, the program constructs and stores a shared array of test patterns.

Once the test patterns are ready, the test application program is executed and the test responses stored. If desired, the captured responses may be analyzed or compressed (*e.g.*, using a software MISR before being delivered to the external tester).

The preceding technique was applied to the PARWAN processor. Although component tests are generated only for a subset of components (ALU, SHU, and program

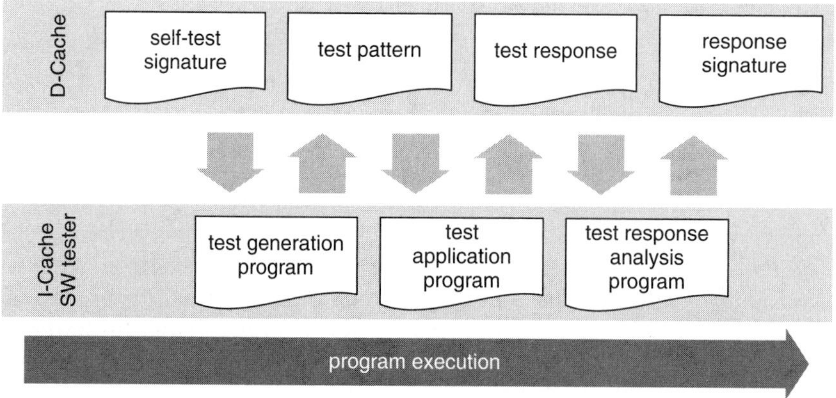

■ **FIGURE 11.12**

Microprocessor self-testing flow.

counter [PC]) that are easily accessible through instructions, other components, such as the instruction decoder, are expected to be tested intensively during the application of the self-test program. The overall fault coverage was 91.42%.

11.4.3 Test Program Synthesis Using Virtual Constraint Circuits (VCCs)

One major challenge of the *test program synthesis* process is to efficiently extract the *instruction imposed I/O constraints*.

Tupuri et al. proposed an approach for generating functional tests for processors by using a gate-level sequential ATPG tool [Tupuri 1997]. It attempts to generate tests for all detectable stuck-at faults under the functional constraints, and then applies these functional test vectors at the system's operational speed. The key idea of this approach lies in the synthesized logic embodying the functional constraints, also known as ***virtual constraint circuits*** (VCCs). The extracted functional constraints are described in HDL and synthesized into a gate-level network. Then a commercial ATPG is used to generate module-level vectors with such constraint circuitry imposed. These module-level vectors are translated to processor-level functional vectors and fault simulated to verify the fault coverage.

Based on the VCC concept but with a different utilization, [Chen 2003] performed module-level test generation such that the generated component test patterns can be directly plugged into the settable fields (*e.g.*, the operand and the source/destination register fields) in **test program templates**. This utilization simplifies the automated generation of test programs for embedded processors. Figure 11.13 illustrates the overall *test program synthesis* process proposed in [Chen 2003], in which the final self-test program can be synthesized directly from (1) a simulatable HDL processor design at RTL level and (2) the **instruction set architecture** specification of the embedded processor. The goals and details of each step are discussed next.

```
M ← partitioning ();
T ← extractTemplate ();
foreach m ∈ M
    T_m ← rankTemplate ();
    while T_m ≠ φ or fault coverage not accceptable
        t ← highest ranked template in T_m;
        F ← deriveMappingFunction (t,m);
        generateVCC ();
        P_m,t ← constrainedTG ();
        TP_m,t ← synthesizeTestProgram (m,t);
        processor-level fault simulation;
        T_m ← T_m − {t};
    end while
end foreach
```

■ **FIGURE 11.13**

VCC-based test program synthesis flow.

Processor partitioning. The first step (line 1) involves partitioning the processor into a collection of ***modules-under-test*** (MUTs), denoted by M. The test program for each MUT will be synthesized separately.

Test template construction. This step (line 2) systematically constructs a comprehensive set of *test program templates*. The *test program templates* can be classified into single-instruction and multi-instruction templates. A **single-instruction template** is built around one key instruction, whereas a **multi-instruction template** includes additional supporting instructions, for example, to trigger pipeline forwarding. To exhaust all possibilities in generating test program templates would be impossible, but generating a wide variety of templates is necessary to achieve high fault coverage.

Test template ranking. Templates are ranked according to a controllability/observability-based testability metric through simulation (line 4). Templates at the top of T_m have high controllability (meaning that it is easy to set specific values at the inputs of the MUT) or high observability (meaning that it is easy to propagate the values at the outputs of the MUT to data registers or to observation points, which can be mapped onto and stored in the memory).

Mapping function derivation. For each MUT, both **input mapping functions** and **output mapping functions** are derived in this step (line 7). The *input mapping function* models the circuits between the instruction template's settable fields, including operands, source registers, or destination registers, and the inputs of the MUT. It is derived by simulating a number of instances of template t to obtain traces followed by regression analysis to construct the mapping function between the settable fields and the MUT inputs. The *output mapping function* models the circuit between the outputs of the MUT and the system's observation points. It is derived by injecting the unknown X value at the outputs of the MUT for simulation, followed by observing the propagation of the X values to the specified template destinations.

VCC generation. The derived mapping functions are synthesized into VCCs (line 8). As will be explained later, the utilization of VCCs not only enforces the *instruction-imposed I/O constraints*, but also facilitates the translation from module-level test patterns to instruction-level test programs.

Module-level test generation. Module-level test generation is performed for the composite circuit of the MUT sandwiched between the input and output VCCs (line 9). An illustration of the composite circuit is shown in Figure 11.14. During the constrained test generation, the test generator sees the circuit including MUT m and the two VCCs (*i.e.*, the shaded area in Figure 11.14). Note that faults within the VCCs will be eliminated from the fault list and will not be considered for test generation. With this composite model, the pattern generator can generate patterns with values directly specified at the settable fields in instruction template t.

Test program synthesis. The test program for MUT m is synthesized using the module-level test patterns generated in the previous step (line 10). Note that the module-level test patterns assign values in some of the settable fields of each instruction template t. The other settable fields without value assignments would be filled with random values. The test program is then synthesized by

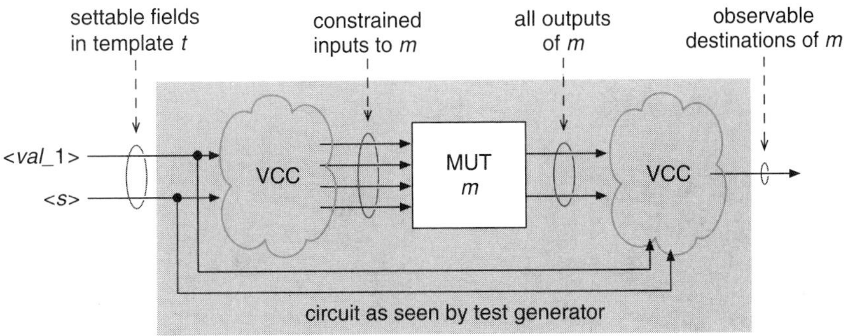

■ **FIGURE 11.14**

Constrained module-level test generation using VCCs.

converting the values of each settable field into its corresponding position in the instruction template t. An example of the target program synthesis flow is given in Figure 11.15. In step 1, the values assigned to the settable fields by the generated test patterns $P_{m,t}$ are identified. Then, in step 2, pseudo-random patterns are assigned to the other settable fields in t. In step 3, t is analyzed to identify the positions of the settable fields (nop stands for the "no operation" instruction). Finally, in step 4, the test program $TP_{m,t}$ is generated by filling the values assigned to the settable fields in their corresponding placeholders in t.

Processor-level fault simulation. Processor-level fault simulation is performed on the synthesized test program segment to identify the set of newly detected faults (line 11). The achieved fault coverage is updated accordingly.

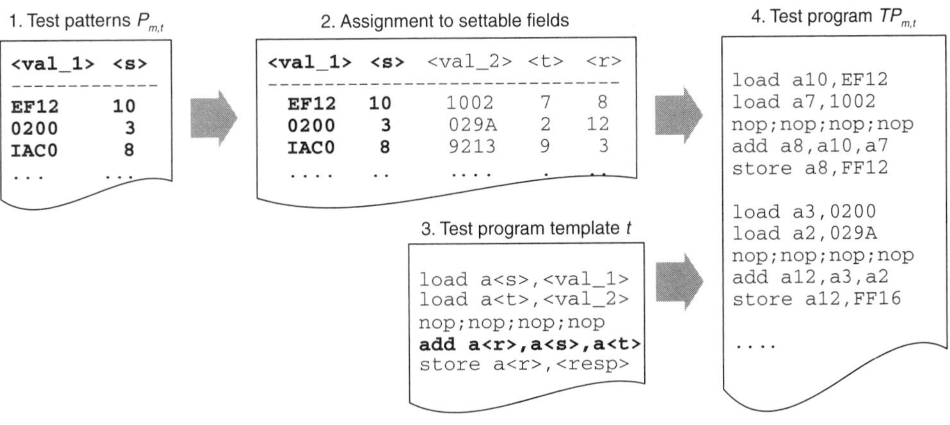

■ **FIGURE 11.15**

A test program synthesis example.

11.4.4 Delay Fault Testing

To ensure that the processor meets its performance specifications requires the application of delay tests. These tests should be applied at-speed and contain two-vector patterns, applied to the combinational portion of the circuit under test, to activate and propagate the fault effects to registers or other observation points. Compared to structural BIST, which needs to resolve various complex timing issues such as multiple clock domains, multiple frequency domains, and test clock skews, processor delay fault self-testing using instruction sequence is a more natural application of at-speed tests.

As in the case of stuck-at faults, not all delay faults in the microprocessor can be tested in the functional mode (*i.e.*, by any instruction sequence). This is simply because no instruction sequence can produce the desired test sequence that can sensitize the path and capture the fault effect into the destination output or flip-flop at-speed. A fault is said to be **functionally testable** if there exists a functional test for that fault; otherwise, it is **functionally untestable**. In practice, one may apply the path classification algorithm [Lai 2000a] to identify a tight superset of the set of *functionally testable* paths in a microprocessor. Then *test program synthesis* algorithms [Lai 2000b] [Singh 2006] can be applied to generate test programs for these *functionally testable* paths.

11.4.4.1 Functionally Untestable Delay Faults

To illustrate the concept of *functionally untestable* delay faults, consider the datapath of the PARWAN processor (Figure 11.16), which contains an 8-bit ALU, an accumulator (AC), and an ***instruction register*** (IR). The data inputs of the ALU, *A*7–*A*0 and *B*7–*B*0, are connected to the internal data bus and AC, respectively. The control inputs of the ALU are *S*2–*S*0, which instruct the ALU to perform the desired arithmetic or logic operation. The outputs of the ALU are connected to the inputs of AC and the inputs of IR.

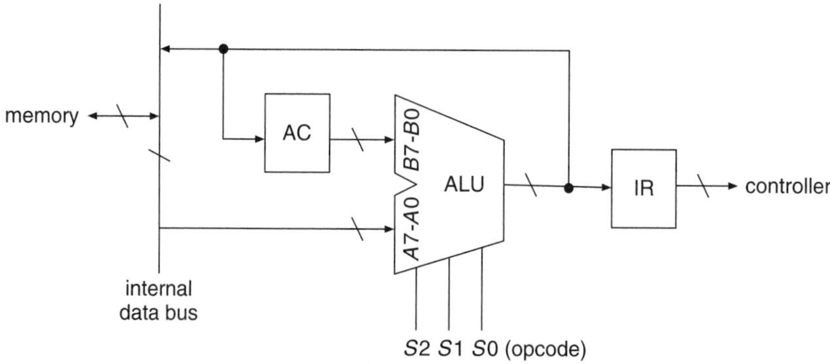

■ **FIGURE 11.16**

Datapath of the PARWAN processor.

Assuming the use of **enhanced scan**, paths that start from $A7$–$A0$, $B7$–$B0$, or $S2$–$S0$ and end at inputs of IR or AC are **structurally testable** if we can find a vector pair to test them. However, some of the paths may be *functionally untestable*. For example, it can be shown that for all possible instruction sequences, whenever a rising transition occurs on signal $S1$ at the beginning of a clock cycle, AC and IR can never be enabled at the end of that cycle. Therefore, paths that start at $S1$ and end at the inputs of IR or AC are *functionally untestable* because delay fault effects on them can never be captured by IR or AC immediately after the vector pair has been applied.

11.4.4.2 Constraint Extraction

Different methodologies are applied to extract the constraints associated with the datapath and the control logic. For datapath logic, all the vector pairs that can be applied to the datapath at speed are symbolically simulated. Constraints are then extracted from the simulation results.

For example, the symbolic simulation results of the instruction sequence, `NOP` followed by `add`, applied to the datapath in Figure 11.16 are listed in rows 2 and 3 of Table 11.4. In the first cycle, the ALU data inputs at $A7$–$A0$ and $B7$–$B0$ are V_{1A} and V_{1B}, respectively, and the ALU executes the `NOP` operation because its control inputs $S2$–$S0$ are set to 100. Note that neither IR nor AC is in the latching mode in this cycle. In the second cycle, the ALU executes an addition and AC will latch the result. The constraints extracted from the simulation results are listed in the bottom row of Table 11.4. Because inputs $A7$–$A0$ change from V_{1A} to V_{2A}, they can be assigned arbitrary vector pairs (*i.e.*, there is no *temporal constraint*, which is denoted by all X's). Inputs $B7$–$B0$, on the other hand, must remain the same in both cycles; therefore, the associated constraint is that they have to be constant, denoted by all C's. The constraint on the control inputs $S2$–$S0$ is apparent. The symbols O, Z, and R here denote 11, 00, and 01, respectively. Because AC is latching the ALU addition result in the second cycle, it is labeled "care" to indicate that it stores cared output.

Note that there exist covering relationships among the extracted constraints. A constraint α covers another constraint β if it is possible to set α exactly the same as β by assigning any of C, Z, O, F, or R to the X terms in α. In such a case, the constraint β can be removed. For the DLX [Gumm 1995] processor, after the reduction process, only 24 constraints remain to be considered.

The controller constraints can be easily identified from the controller's RTL description, including the **input transition constraint**, the **input spatial constraint**, and the *output constraint*. The *input transition constraint* specifies the necessary conditions on the control inputs to those registers that connect to the

TABLE 11.4 ■ Datapath Constraints for the `NOP; add;` Sequence

	$A7$–$A0$	$B7$–$B0$	$S2$–$S0$	IR	AC
Cycle 1	V_{1A}	V_{1B}	100 (`NOP`)		
Cycle 2	V_{2A}	V_{1B}	101 (`add`)		Latch
Constraint	$X \cdots X$	$C \cdots C$	OZR		Care

controller's inputs. The registers must be set to be in the latching or reset mode for transitions to occur at the control inputs. An *input spatial constraint* specifies all the legitimate input patterns to the controller. An *output constraint* records the necessary signal assignments for an output to be latched by a register.

11.4.4.3 Test Program Generation

The flow of delay fault test program generation in [Lai 2000b] is shown in Figure 11.17. In step 1, given the *instruction set architecture* and the microarchitecture of the processor, the spatial and temporal constraints, between and at the registers and control signals, are first extracted. In step 2, the path classification algorithm, extended from [Cheng 1996] [Krstic 1996], implicitly enumerates and examines all paths and path segments with the extracted constraints imposed. If a path cannot be sensitized with the imposed extracted constraints, the path is *functionally untestable* and thus is eliminated from the fault universe. Identifying the *functionally untestable* faults helps reduce the computation effort of the subsequent test generation process. As the preliminary experimental results shown in [Lai 2000a] indicate, a nontrivial percentage of the paths in simple processors (such as PARWAN and DLX) are *functionally untestable* but *structurally testable*.

In step 3, constrained delay fault test generation is performed for a set of long paths selected from the *functionally testable* paths. A gate-level ATPG for path delay faults is extended to incorporate the extracted constraints into the test generation process, where it is used to generate test vectors for each target path delay fault. If the test is successfully generated, it not only sensitizes the path but also meets the extracted constraints. Therefore, it is most likely to be deliverable by processor instruction sequences. (If the complete set of constraints has been extracted, the delivery by instructions could be guaranteed.)

Finally, in the *test program synthesis* process that follows, the test vectors specifying the bit values at internal flip-flops are first mapped back to word-level values in registers and values at control signals. These mapped value requirements are then justified at the instruction level. A predefined propagating routine is used to propagate to the memory the fault effects captured in the registers or flip-flops of the path delay fault. This routine compresses the contents of some or all registers in the processor, generates a signature, and stores it in memory. The procedure is repeated until all target faults have been processed. The test program, which is generated offline, will be used to test the processor at-speed.

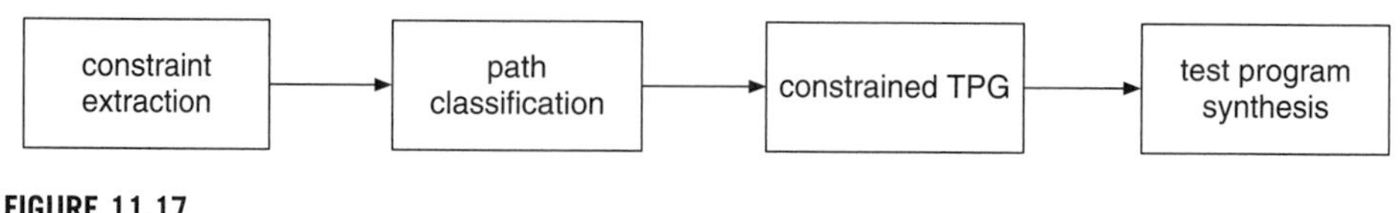

■ **FIGURE 11.17**

Delay fault test program generation flow.

The test program generation algorithm was applied to PARWAN and DLX processors. On the average, 5.3 and 5.9 instructions were needed to deliver a test vector, and the achieved fault coverage for testable path delay faults was 99.8% for PARWAN and 96.3% for DLX.

11.4.5 Functional Random Instruction Testing

One of the major challenges of the SBST methodology is the functional constraint extraction process. The process is time-consuming, usually manually done or partially automated, and in general, only a subset of the functional constraints is extracted, which complicates the succeeding *test program synthesis* process.

In [Parvathala 2002], the authors proposed a technique called ***functional random instruction testing at speed*** (FRITS). FRITS is basically software BIST and is applicable to devices like microprocessors that have an extensive instruction set that can be utilized to realize the software that enables the functional BIST. The FRITS tests (kernels) are different from normal functional test sequences. Once these kernels are loaded into the processor cache, they repeatedly generate and execute pseudo-random or directed sequences of machine codes.

An example FRITS kernel execution sequence is shown in Figure 11.18. After proper initialization, the kernel generates a sequence of machine instructions and the data to be used by these instructions. Note that these instructions as well as the data, not the FRITS kernel, constitute the functional test. The kernel then branches to execute the generated functional test. The test responses, including the

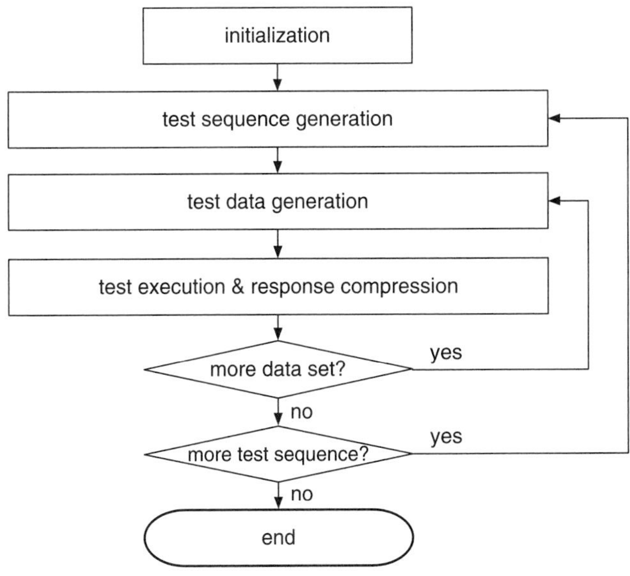

FIGURE 11.18

The FRITS kernel execution flow.

register files and memory locations that are modified by the test, are compressed. To enhance the fault coverage, one may choose to use multiple test data sets for the same random instruction sequence. The test generation and execution process continues until the desired number of test sequences has been generated and executed.

The FRITS technique was applied to a series of Intel microprocessors. Although it is not the primary test generation and test method used on these processors, it provides significant complementary coverage over existing structural and functional test content on these processors.

11.5 PROCESSOR SELF-DIAGNOSIS

Besides enabling at-speed self-test with low-cost testers, SBST eliminates the use of scan chains and the associated test overhead, making it an attractive solution for testing high-end microprocessors. However, the elimination of scan chains poses a significant challenge for accurate fault diagnosis. Deterministic methods for generating diagnostic tests are available for combinational circuits [Grüning 1991], but sequential circuits are much too complex to be handled by the same approach. There have been several proposals related to generating diagnostic tests for sequential circuits by modifying existing detection tests [Pomeranz 2000] [Yu 2000]. However, the success of these methods depends on the success of the sequential test generation techniques. As a result, the existence of scan chains is still crucial for sequential circuit diagnosis [Venkataraman 2001].

Though current sequential ATPG techniques are not sufficiently practical for handling large sequential circuits, SBST methods are capable of successfully generating tests for a particular type of sequential circuits—microprocessors. If properly modified, these tests might possibly achieve high diagnostic capabilities. Functional information (*instruction set architecture* and microarchitecture) can be used to guide and facilitate diagnosis.

11.5.1 Challenges to SBST-Based Processor Diagnosis

Because diagnostic programs rely on instructions to detect and distinguish between faults, SBST-based microprocessor diagnosis may suffer low diagnostic resolution for the following reasons:

1. Faults that are located on the functionality critical nodes in the processor, such as in the instruction decode unit and tristate buffers that control the buses, tend to fail all diagnostic test programs.

2. The test program that targets one module could also activate a large number of faults in other modules.

3. Some faults cannot be detected by SBST at all.

To achieve a high diagnostic resolution, a great number of carefully constructed **diagnostic test programs** are generated. Each program is designed to cover as few faults as possible, while the union of all test programs covers as many faults as possible. The *diagnostic test program* construction principles are as follows:

1. Reduce the variety of instructions in each test program. If possible, use one type of instruction only.
2. Reduce the number of instructions in each test program. Each test program should contain only the essential instructions needed for test data transportation to and from the target internal module.
3. Create multiple copies of the same test program. Each instance is designed to observe the test response on different outputs from the target module. This allows the user to differentiate faults that will cause errors to propagate to different nodes.

11.5.2 Diagnostic Test Program Generation

Initial investigations for the diagnostic potential of SBST were reported in [Chen 2002] and [Bernardi 2006]. Whereas the former attempted to generate test programs that were geared toward diagnosis, the latter adopted an evolutionary approach to refine existing postproduction SBST programs to achieve the desired diagnosis resolution.

Figure 11.19 illustrates the proposed algorithm flow in [Bernardi 2006]. The process starts from a set of test programs for postproduction testing. These test programs could be hand-written by the designers or test engineers to cover the corner cases, or by automatic test program synthesis approaches, and their diagnosis capability is enhanced by the following **sporing**, **sifting**, and **evolutionary improvement** processes.

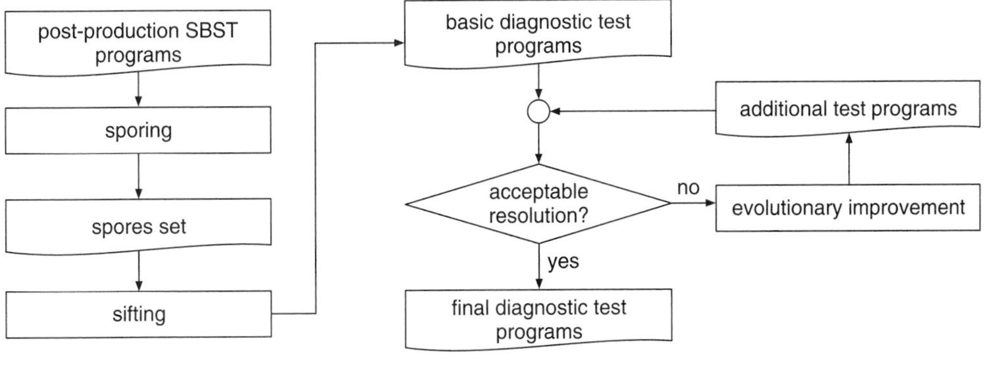

FIGURE 11.19

Refining postproduction test programs for diagnosis.

Sporing. In the *sporing* process, each program belonging to the initial set of SBST programs is fragmented into a huge number of **spores**. Each *spore* represents a completely independent program that is able to excite some processor function, observe the response, and possibly signal the occurrence of faults. A first fault simulation is performed to determine the fault coverage figure for each generated *spore*.

Sifting. The *sifting* process is intended to reduce the number of diagnostic test programs obtained after the *sporing* process. First, each *spore* is assigned a fitness value that indicates its diagnosis capability. The fitness value depends on both the number of faults the *spore* can detect and the number of spores that detect these faults. Then, only the *spores* that possess high fitness values and detect faults not covered by other spores are retained in the basic diagnostic test program set.

Evolutionary improvement. When there are still unsatisfactorily large **equivalence classes**,[3] the *evolutionary improvement* process is applied to generate new test programs that are able to split them.

The diagnosis technique was applied to the Intel i8051 processor core. The processor core's parallel outputs are fed to a 24-bit MISR, and the stored signature is read out at the end of each test program execution. Table 11.5 summarizes the experimental results. The original postproduction test program set consists of eight programs, and the test set size is 4 KB. The percentages of uniquely classified faults $D(1)$ and correctly classified faults $D(10)$[4] are 11.56% and 32.90%, respectively. After the *sporing* process, about 60,000 test programs were generated. This test set is reduced to 7231 test programs after the *sifting* process. The resulting basic diagnosis program set is able to uniquely classify 35.70% and correctly classify 58.02% of faults. Thirty-five new test programs were added in the *evolutionary improvement* process. The percentages of uniquely and correctly classified faults by the final test set are enhanced to 61.93% and 84.30%, respectively.

TABLE 11.5 ■ Summary of the Diagnostic Test Program Generation Results

	Postproduction Test Set	Basic Test Set	Final Test Set
Number of Programs	8	7231	7266
Test Set Size (KB)	4	165	177
$D(1)$ (%)	11.56	35.70	61.93
$D(10)$ (%)	32.90	58.02	84.30

[3] An equivalence class is a set of faults that cause exactly the same faulty behavior for each applied pattern.

[4] $D(n)$ is defined as the percentage of faults that are classified into *equivalence classes* of size n or less by the diagnostic test program set. $D(10)$ is regarded as the percentage of correctly classified faults because exact analysis of fault equivalence cannot be performed for medium or large sequential circuits.

11.6 TESTING GLOBAL INTERCONNECT

In SOC designs, a device must be capable of performing core-to-core communications across long interconnects. As we find ways to decrease gate delay, the performance of interconnects is becoming increasingly important for achieving high overall performance [SIA 2003]. Increases in **cross-coupling capacitance** and **mutual inductance** mean that signals on neighboring wires may interfere with each other, thus causing excessive delay or loss of **signal integrity**. Although many techniques have been proposed to reduce crosstalk, owing to the limited design margins and unpredictable process variations, crosstalk must also be addressed during manufacturing test.

Because of their impact on circuit timing, testing for crosstalk effects may have to be conducted at the rated speed of the circuit under test. At-speed testing of GHz systems, however, may require prohibitively expensive high-speed testers. With external testing, hardware access mechanisms are required for applying tests to interconnects deeply embedded in the system. This may lead to unacceptable costs such as in overhead needed for area or performance.

In [Bai 2000], a BIST technique in which an SOC tests its own interconnects for crosstalk defects using on-chip hardware pattern generators and error detectors has been proposed. Although the amount of area overhead may be amortized for large systems, for small systems, the amount of relative area overhead may be unacceptable. Because this method falls into the category of structural BIST techniques, utilizing this particular technique may cause overtesting and yield loss as not all test patterns generated are valid when the system is operated in normal mode.

For SOCs with embedded processors, utilizing the processor itself to execute interconnect self-test programs is a viable solution because in such SOCs most of the system-level interconnects, such as on-chip buses, are accessible to the embedded processor core(s). During interconnect self-test program execution, test vector pairs can be applied to the appropriate bus in normal functional mode of the system. In the presence of crosstalk-induced glitches or delay effects, the second vector in the vector pair becomes distorted at the receiver end of the bus. The processor, however, can store this error effect in memory as a test response. This can be unloaded later by an external tester and used for off-chip analysis. In this section, the ***maximum aggressor*** (MA) bus fault model is introduced first. Then two software-based interconnect self-test techniques [Lai 2001] [Chen 2001] for MA faults are described.

11.6.1 Maximum Aggressor (MA) Fault Model

The **MA fault model** proposed in [Cuviello 1999] is suitable for modeling crosstalk defects on interconnects. It abstracts the crosstalk defects on global interconnects by using a linear number of faults.

The *MA fault model* defines faults based on the resulting crosstalk error effects, including positive glitch (g_p), negative glitch (g_n), rising delay (d_r), and falling delay (d_f). For a set of N interconnects, the *MA fault model* considers the collective aggressor effects on a given victim line Y_i, whereas all other $N-1$ wires act as

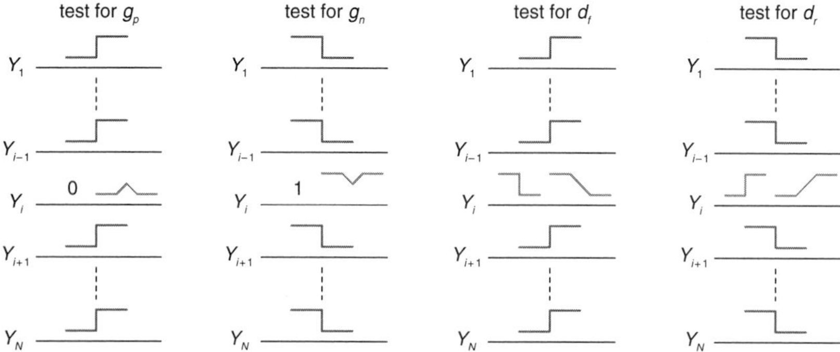

■ **FIGURE 11.20**

Maximum aggressor tests for victim line Y_i.

aggressors. The required transitions on the aggressor/victim lines to excite the four error types are shown in Figure 11.20. The test for positive glitch (g_p) on a victim line Y_i, as shown in the first column, would require that line Y_i hold a constant 0 value, for example, while the other $N-1$ aggressor lines have a rising transition. Under this condition, the victim line Y_i would have a positive glitch created by the crosstalk effect. If excessive, the glitch would result in errors. These patterns, collectively called **MA tests**, excite the worst-case crosstalk effects on the victim line Y_i. For a set of N interconnects, there are $4N$ MA faults, requiring $4N$ *MA tests*. It has been shown that these $4N$ faults cover all crosstalk defects on any of the N interconnects.

11.6.2 Processor-Based Address and Data Bus Testing

In a core-based SOC, the address, data, and control buses are the main types of global interconnects with which the embedded processors communicate with memory and other cores of the SOC via memory-mapped I/O. The proposed technique in [Chen 2001] concentrates on testing the data and address buses in a processor-based SOC. The crosstalk effects on the interconnects are modeled using the *MA fault model*.

11.6.2.1 Data Bus Testing

For a bidirectional bus, such as a data bus, crosstalk effects vary as the bus is driven from different sources. This requires crosstalk tests to be conducted in each direction [Bai 2000]. However, to apply a pair of vectors (v_1, v_2) in a particular bus direction, the direction of v_1 is irrelevant, as long as the logic value at the bus is held at v_1. Only v_2 needs to be applied in the specified bus direction. This is because the signal transition triggering the crosstalk effect takes place only when v_2 is being applied to the bus.

To apply a test vector pair (v_1, v_2) to the data bus from an SOC core to the processor, the processor first exchanges data v_1 with the core. The direction of this data exchange is irrelevant. If the core being tested is the memory, for example, the processor may either read v_1 from the memory or write v_1 to the memory. The processor then requests data v_2 from the core. This might be a memory-read if the core being tested is memory. When the data v_2 is obtained, the processor writes v_2 to memory for later analysis. To apply a test vector pair (v_1, v_2) to the data bus from the processor core to an SOC core, the processor first exchanges data v_1 with the core. Then the processor sends data v_2 to the core or executes a memory-write if the core being tested is memory. If the core is memory, v_2 can be directly stored to an appropriate address for later analysis; otherwise, the processor must execute additional instructions to retrieve v_2 from the core and store it to memory.

11.6.2.2 Address Bus Testing

To apply a test vector pair (v_1, v_2) to the address bus, which is a unidirectional bus from the processor to an SOC core, the processor first requests data from two addresses (v_1 and v_2) in consecutive cycles. In the case of a nonmemory core, because the processor addresses the core via memory-mapped I/O, v_2 must be the address corresponding to the core. If v_2 is distorted by crosstalk, the processor would be receiving data from a wrong address, v'_2, which may be a physical memory address or an address corresponding to a different core. By keeping different data at v_2 and v'_2 (*i.e.*, $\text{mem}[v_2] \neq \text{mem}[v'_2]$), the processor is able to observe the error and store it in memory for analysis.

Figure 11.21 illustrates the address bus testing process. Where the processor is communicating with a memory core, to apply test (0001, 1110) in the address bus from the processor to the memory core for example, the processor first reads data from address 0001 and then from address 1110. In the system with the faulty address bus, the second address may become 1111. If different data are stored at addresses 1110 and 1111, say, mem[1110] = 0100 and mem[1111] = 1001, the processor would receive a faulty value from memory. This might be 1001 instead of 0100. This error response can be stored in memory for future analysis.

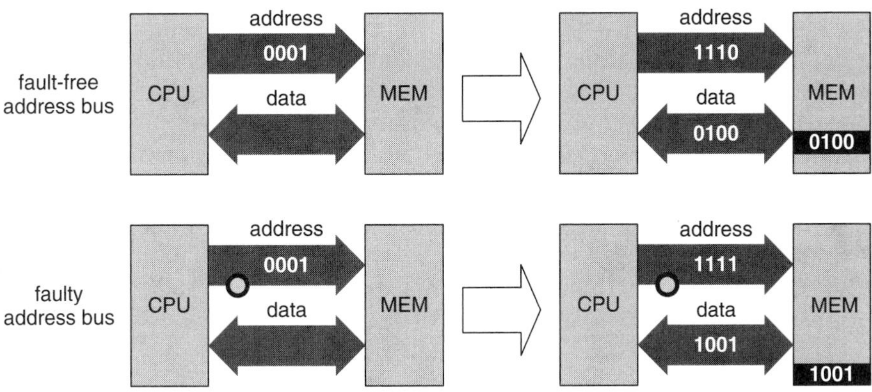

FIGURE 11.21

Address bus testing.

The feasibility of this method has been demonstrated by applying it to test the interconnects of a processor-memory system, and the defect coverage was evaluated using a system-level crosstalk-defect simulation method.

11.6.3 Processor-Based Functional MA Testing

Even though the *MA tests* have been proven to cover all physical defects related to crosstalk between interconnects, Lai et al. observe that many of these types of defects can never occur during normal system operations because of system constraints [Lai 2001]. Therefore, testing buses using *MA tests* might screen out chips that are functionally operational under any pattern produced under normal system operation.

To resolve the overtesting issue associated with the *MA fault model*, ***functionally maximal aggressor*** (FMA) tests that meet the system constraints and can be conducted while operating in the functional mode were proposed [Lai 2001]. These tests provide complete coverage of all crosstalk-induced logical and delay faults that can cause errors during operation in the functional mode.

Given the timing diagrams of all bus operations, the spatial and temporal constraints imposed on the buses can be extracted and **FMA tests** can be generated. A covering relationship between vectors, extracted from the timing diagrams of the bus commands, is used during the FMA test generation process. Because the resulting FMA tests are highly regular, they can be clustered into a few groups. The tests in each group are highly similar except for the victim lines. Similar to a March-test sequence, which is an algorithm commonly used for testing memory, the tests in each group can be synthesized by a software routine. The test program is highly modularized and very small. Experimental results have shown that a test program as small as 3000 to 5000 bytes can detect all crosstalk defects on the bus from the processor core to the target core.

The synthesized test program is applied to the bus from the processor core, and the input buffers of the destination core capture the responses at the other end of the bus. The processor core should read back such responses to determine whether any faults occur on the bus. However, because the processor core cannot read the input buffers of a nonmemory core, a DFT scheme is suggested to allow the processor core to directly observe the input buffers. The DFT circuitry consists of bypass logic added to each I/O core to improve its testability. With DFT support on the target I/O core, the test generation procedure first synthesizes instructions to set the target core to the bypass mode, and then it continues synthesizing instructions for the *FMA tests*. The test generation procedure does not depend on the functionality of the target core.

11.7 TESTING NONPROGRAMMABLE CORES

Testing nonprogrammable SOC cores is a complex problem with many unresolved issues [Huang 2001]. Standards such as IEEE 1500 were created with the intent of relieving these core test problems; however, the test provisions within the standard do not reduce the complexity of the test generation and response analysis problems. Furthermore, the requirement of at-speed testing is not addressed.

A self-testing approach was proposed in [Huang 2001]. In it, the embedded processor running the test program serves as an *embedded software tester* that performs test pattern generation, test pattern application, response capture, and response analysis. The advantages of this approach are as follows:

1. The need for dedicated test circuitry as found in traditional BIST techniques (*i.e., embedded hardware tester*) is eliminated.
2. Tremendous flexibility in the type and quality of patterns that can be generated is provided. One simply has to use a different test pattern generation software procedure or modify the program parameters.
3. The approach is scalable to large IP cores with available structural netlists.
4. Patterns are delivered at the SOC operation speed. Hence, it supports delay testing.

To facilitate core testing using the *embedded software tester*, a *test wrapper* is placed around each core to support pattern delivery. It contains the test support logic needed to control scan chain shifting, buffers to store scan data, buffers to support at-speed test, and so on.

The test flow based on the *embedded software tester* methodology is illustrated in Figure 11.22. It is divided into a preprocessing and a core test phase.

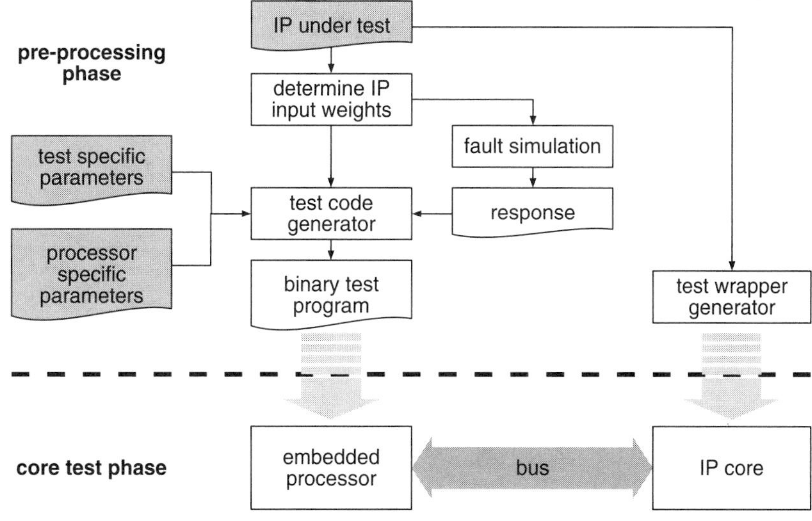

■ **FIGURE 11.22**

The test flow of nonprogrammable cores.

11.7.1 Preprocessing Phase

In the preprocessing phase, a *test wrapper* is automatically inserted around the IP core under test. The *test wrapper* is configured to meet the specific testing needs for the IP core. The IP core is then subjected to fault simulation by using different sets of patterns. Either weighted random patterns generated with multiple weight sets or multiple capture cycles [Tsai 1998] after each scan sequence could be used.

Next, a high-level test program is generated that synchronizes tasks including software pattern generation, starting the test, applying the test, and analyzing the test response. The program can also synchronize testing multiple cores in parallel. The test program is then compiled to generate a processor specific binary code.

11.7.2 Core Test Phase

In the core test phase, the test program is run on the processor core to test various IP cores. A test packet is sent to the IP core *test wrapper* informing it about the test application scheme (single- or multiple-capture cycle). Data packets are then sent to load the scan buffers and the I/O buffers. The test wrapper applies the required number of scan shifts and captures the test response for a preprogrammed number of functional cycles. Test results are stored in the I/O buffers and the scan buffers and then are read by the processor core.

11.8 INSTRUCTION-LEVEL DFT

Self-testing manufacturing defects in an SOC by running test programs using a programmable core has several potential benefits, including the ability to conduct at-speed testing, low-DFT overhead because dedicated test circuitry use is eliminated, and better power and thermal management during testing. Such a self-test strategy might require a lengthy test program, yet it still may not achieve sufficiently high fault coverage. One solution is to apply a DFT methodology, which is based on adding test instructions to the processor core. This methodology is called *instruction-level DFT*.

11.8.1 Instruction-Level DFT Concept

Instruction-level DFT inserts test circuitry into the design in the form of test instructions. It should be a less intrusive testing approach than using gate-level DFT techniques, which attempt to create a separate test mode somewhat orthogonal to the functional mode. This instructional-level methodology is also more attractive for applying at-speed tests and for power/thermal management during testing, as compared with the existing logic BIST approaches.

When adding new instructions, existing hardware should be “reused” as much as possible to reduce area overhead. If the test instructions are carefully designed such that their microinstructions reuse the datapath for the functional instructions and do not require a new datapath, then the controller overhead should be relatively low. In general, adding extra buses or registers to implement new instructions is unnecessary and avoidable. Also, in most cases, a new instruction can be added by introducing new control signals to the datapath rather than by adding hardware.

In [Shen 1998, Lai 2001], the authors propose *instruction-level DFT* methods to address the fault coverage. The approach in [Shen 1998] adds instructions for testing the exceptions such as microprocessor interrupts and reset. With these new instructions, the test program can achieve a fault coverage close to 90% for stuck-at faults. However, this approach cannot achieve a higher coverage because the test program is synthesized based on a random approach and cannot effectively control or observe some internal registers that have low testability.

The DFT methodology proposed in [Lai 2001], on the other hand, systematically adds test instructions to an on-chip processor core to improve its self-testability, to reduce the size of the self-test program, and to reduce the test application runtime. The experimental results of two processors (PARWAN and DLX) show that test instructions can reduce the program size and program runtime by about 20% at the cost of about a 1.6% increase in area overhead. The following discussion elaborates on the *instruction-level DFT* techniques presented in [Lai 2001], including **testability instructions** and **test optimization instructions**.

11.8.2 Testability Instructions

DFT instructions of this type are added to enhance the processor's testability, including controllability and observability of registers and the processor I/O. To determine which instructions to add, the testability of the processor is analyzed first.

A register's testability can be determined based on the availability of data movement instructions between the memory and the register that is targeted for testing. A **fully controllable register** is one for which there exists a sequence of data movement instructions that can move the desired data from memory to that register. Similarly, a **fully observable register** is one for which there exists a sequence of data movement instructions to propagate the register data to memory. Given the microarchitecture of a processor core, it is possible to identify the set of *fully controllable registers* and *fully observable registers*. For registers that are not fully controllable/observable, new instructions can be added to improve their accessibility. In [Lai 2001], four instructions are added to enhance the register accessibility in Figure 11.23:

`s2r` (move *SR* to *Rn*). This instruction is intended to improve the observability of the status register *SR*. It moves the data from the status register to any general-purpose register *Rn*. Note that data in *SR* are propagated through an existing data path from *SR* to ALU, to register *C*, and, finally, to the target register *Rn*.

`r2s` (move *A* to *SR*). This instruction aims to improve the controllability of *SR*. It moves the data from a general-purpose register *A* to *SR*. Again, an existing path (from *A*, to ALU, to *SR*) is utilized.

Read exception signals from *Rn*. This instruction allows the processor to set the values of the exception signals from *Rn*, rather than from external devices.

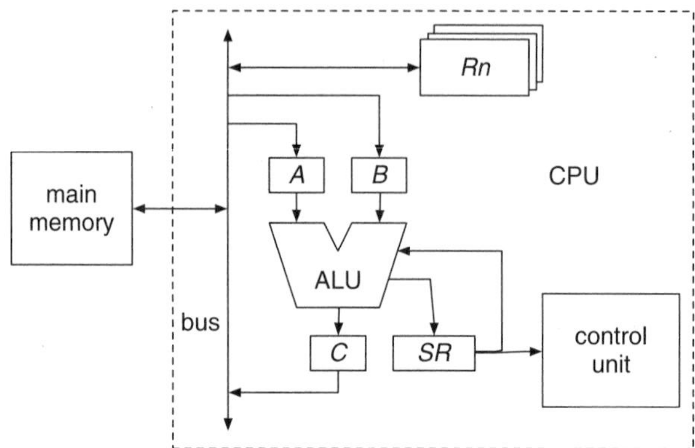

■ **FIGURE 11.23**

An example processor to demonstrate instruction-level DFT.

To enable this instruction to be used, extra hardware is added as shown in Figure 11.24a. Without the loss of generality, *R*27 is selected as the source register of the controller exception signals. By setting the 1-bit register *T*, the processor can switch the exception signal source between *R*27 and external devices using the two multiplexers (MUX). Based on this scheme, the added instruction simply needs to be able to set the value of *T*.

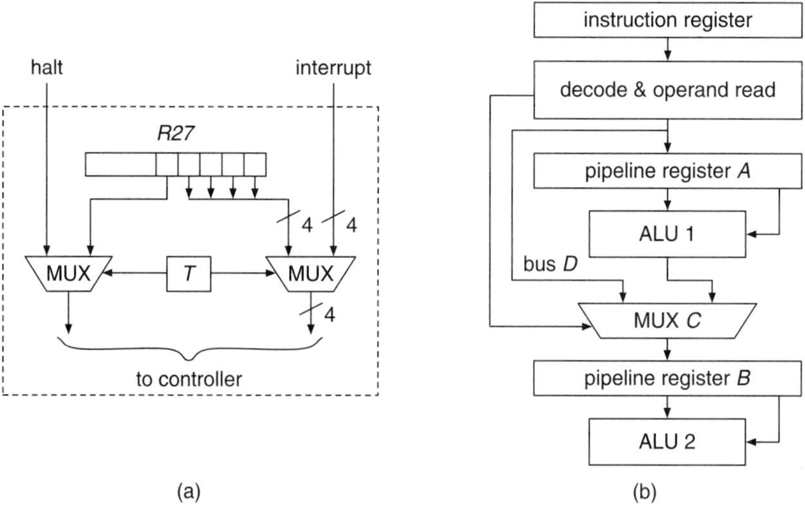

■ **FIGURE 11.24**

DFT instructions for testability enhancement: (a) DFT for exception signals and (b) DFT for pipeline registers.

Pipeline register access. Instructions can be added in pipeline designs to manage the difficult-to-control registers buried deeply in the pipeline. To this end, extra hardware is added. An example of such a pipeline DFT is depicted in Figure 11.24b. To enhance the controllability of pipeline register B, we can add a test instruction, an extra bus (bus D), a multiplexer (MUX C), and a MUX control signal to enable loading data directly from a general-purpose register to the register B. When the test instruction is decoded and its operands become available on bus D, the test instruction will enable MUX C to select bus D as the signal source for B.

11.8.3 Test Optimization Instructions

The *test optimization instructions* aim to optimize the test program in terms of its size and application time. The need to implement such DFT instructions is based on the observation that, in the synthesized self-test program, some code segments (called **hot segments**) appear repeatedly. Therefore, the addition of test instructions that reduce the size of *hot segments* will help to lower the test program size. In addition to program size reduction, DFT instructions may be added to speed up the processes of test vector preparation, response transportation, and response analysis. In [Lai 2001], the authors proposed two *test optimization instructions*:

`load2` (consecutive load to R_i and R_j). This instruction can read two (or more) consecutive words from a memory address, which is stored in another register R_k, and load them to registers R_i and R_j. Whereas a consecutive load needs three words in memory (one for the instruction itself and two for the operands), two load instructions require four words (two for the instruction and two for the operands). Thus, inclusion of the `load2` instruction reduces the test program size.

`xor_all` (signature computation). This instruction performs a sequence of exclusive-OR (XOR) operations on the processor register files (Figure 11.23) and stores the final result in register C. Although replacing a sequence of XOR instructions in the response analysis subroutine with `xor_all` helps reduce the test program run time, it does not significantly reduce its size because there is only one copy of the signature analysis subroutine in the program.

It is interesting to note that although adding test instructions to the programmable core does not improve the testability of other nonprogrammable cores on the SOC, the programs for testing the nonprogrammable cores can also be optimized with the added *test optimization instructions*, such as the `load2` instruction.

11.9 DSP-BASED ANALOG/MIXED-SIGNAL COMPONENT TESTING

For mixed-signal systems that integrate both analog and digital functional blocks onto the same chip, testing of analog/mixed-signal modules has become a production testing bottleneck. Because most analog/mixed-signal circuits are functionally tested, their testing necessitates the use of expensive automatic test equipment for analog stimulus generation and response acquisition. One promising solution is BIST. It utilizes on-chip resources that are shared with either functional blocks or dedicated BIST circuitry to perform on-chip stimulus generation and response acquisition. Under the BIST approach, the demands on the external test equipment are less stringent. Furthermore, stimulus generation and response acquisition are less vulnerable to environmental noise and less limited by I/O pin bandwidth than external tester based testing.

With the advances in CMOS technology, **DSP-based BIST** becomes a viable solution for analog/mixed-signal components—the signal processing needed to make a pass/fail decision can be implemented in the digital domain with digital resources. In *DSP-based BIST* schemes [Toner 1995] [Pan 1995], on-chip ***digital-to-analog converters*** (DACs) and ***analog-to-digital converters*** (ADCs) are used for stimulus generation and response acquisition, and DSP resources (such as processor or DSP cores) are used for the required signal synthesis and response analysis. The *DSP-based BIST* scheme is attractive because of its flexibility—various tests, such as AC, DC, and transient tests, can be performed by modifying the software routines without needing to alter the hardware. However, on-chip ADCs and DACs are not always available in analog/mixed-signal SOC devices.

In [Huang 2000], the authors proposed to use the 1-bit first-order **delta-sigma modulation ADC** as a dedicated BIST module for on-chip response acquisition, in case an on-chip ADC is unavailable. Owing to its oversampling nature, the *delta-sigma modulation ADC* can tolerate relatively high process variations and match inaccuracy without causing functional failure. It is therefore particularly suitable for VLSI implementation. This solution is suitable for low to medium frequency applications such as audio signal. Figure 11.25 illustrates the overall **delta-sigma modulation-based BIST architecture**. It employs the **delta-sigma modulation** technique for both stimulus generation [Dufort 1997] and response analysis. The test process consists of test pattern generation, stimulus application, and response digitization.

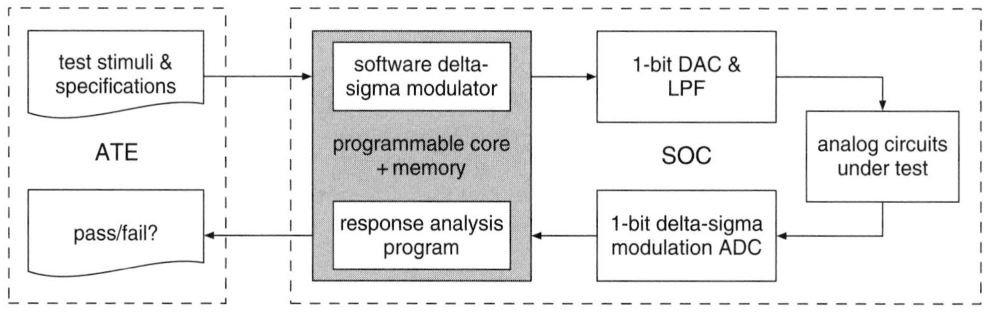

■ **FIGURE 11.25**

DSP-based self-test for analog/mixed-signal cores.

Test pattern generation. A software *delta-sigma modulation ADC* routine is executed to convert the desired test stimulus (*e.g.*, a sinusoidal of 1 V amplitude at 500 KHz) to 1-bit digital stream. For periodic test stimuli, a segment from the *delta-sigma modulation ADC* output bit stream that contains an integer number of signal periods is stored in on-chip memory.

Stimulus application. To transform the stored 1-bit stream segment to the specified analog test stimulus, the stored pattern is repeatedly applied to the 1-bit DAC, which translates the digital 1's and 0's to two discrete analog levels. The succeeding low-pass filter then removes the out-of-band high-frequency **modulation noise** and restores the original analog waveform.

Response digitization. The 1-bit *delta-sigma modulation ADC* is dedicated to converting the analog component output response into a 1-bit stream that will be stored in on-chip memory. Here, the first-order *delta-sigma modulation ADC* is utilized because it is more stable and has a larger input dynamic range than higher-order *delta-sigma modulation ADCs*. However, it is not quite practical for high-resolution applications because a rather high oversampling rate will be needed, and it suffers from ***intermodulation distortion*** (IMD). Compared to the first-order configuration, the second-order configuration has a smaller dynamic range but is more suitable for high-resolution applications.

Response analysis. The stored test responses are then analyzed by software DSP routines (*e.g.*, decimation filter and FFT) to derive the desired performance specifications. Note that, the software part of this technique (that is, the software *delta-sigma modulation ADC* and the response analysis routines) can be performed by on-chip DSP or processor cores, but only if abundant on-chip digital programmable resources are available (as indicated in Figure 11.25) or by external digital test equipment.

11.10 CONCLUDING REMARKS

Embedded ***software-based self-testing*** (SBST) has the potential to alleviate many problems associated with current scan test and structural BIST techniques. These problems include excessive test power consumption, overtesting, and yield loss. This chapter has summarized the recently proposed techniques on this subject. On-chip programmable resources such as embedded processors are used to test processor cores, global interconnects, nonprogrammable cores, and analog/mixed-signal components. The major challenge in using these techniques is extracting functional constraints imposed by the processor instruction set. These extracted constraints are crucial during test program synthesis to ensure that the derived tests are delivered using processor instruction sequences.

Future research in this area must address the problem of automating constraint extraction to make the SBST methodology fully automatic for use in testing general embedded processors. Also, the SBST paradigm should be further generalized for analog/mixed-signal components through the integration of DSP-based test techniques, delta-sigma modulation principles, and some low-cost analog/mixed-signal DFT methods.

11.11 EXERCISES

11.1 **(Functional Fault Testing)** Assign labels to the nodes and edges shown in the following *S*-graph. The instructions and registers are summarized in Table 11.6.

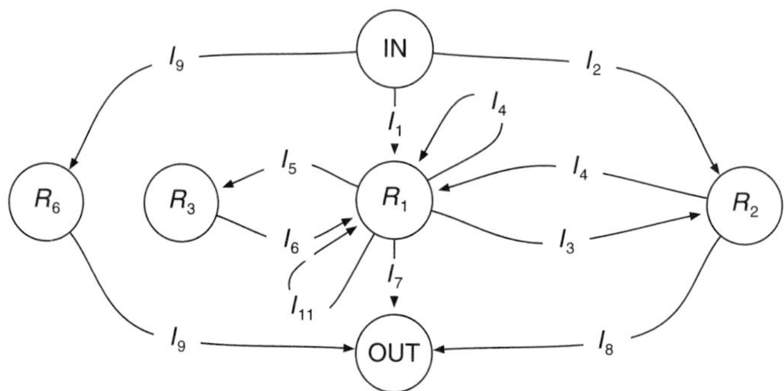

S-graph for Exercise 11.1.

TABLE 11.6 ■ Summary of the Registers and Instructions for Exercise 11.1

R_1: Accumulator (ACC)
R_2: General purpose register
R_3: Scratch-pad register
R_6: Program counter (PC)

I_1: Load R_1 from the main memory using immediate addressing. (T)
I_2: Load R_2 from the main memory using immediate addressing. (T)
I_3: Transfer the contents of R_1 to R_2. (T)
I_4: Add the contents of R_1 and R_2 and store the results in R_1. (M)
I_5: Transfer the contents of R_1 to R_3. (T)
I_6: Transfer the contents of R_3 to R_1. (T)
I_7: Store R_1 into the main memory using implied addressing. (T)
I_8: Store R_2 into the main memory using implied addressing. (T)
I_9: Jump instruction. (B)
I_{11}: Left shift R_1 by one bit. (M)

11.2 **(Functional Fault Test Generation)** Generate tests to detect the register decoding faults of the simplified processor described in Figure 11.3 and Table 11.1.

11.3 **(Structural Fault Testing)** Consider the test program in Figure 11.11. Where will the status flags be stored? How do you interpret the stored results?

11.4 **(Software LFSR)** Using bit-wise shift and logic operations, write a program that generates m state transitions of an LFSR of degree n.

11.5 **(Data Bus Self-Testing)** Assume that the data bus is 4 bits wide. What vector pairs are needed to detect all the MA faults?

11.6 **(Address Bus Self-Testing)** Assume that both the address and data buses are 4 bits wide. What data should you store in the memory in order to test the address bus MA faults by reading the memory contents?

11.7 **(Instruction-Level DFT)** Consider the example processor in Figure 11.22. The `r2s` (move *SR* to *Rn*) DFT instruction is realized by propagating the contents of *SR* to *Rn* through ALU. Another possible way to realize `r2s` is to connect *SR* to the bus directly. What are the advantages and disadvantages of using this second approach?

Acknowledgments

The authors wish to thank Dr. Jayanta Bhadra of Freescale, Professor Dimitris Gizopoulos of University of Piraeus (Greece), and Professor Charles E. Stroud of Auburn University for reviewing the text and providing helpful comments.

References

R11.0 Books

[Navabi 1997] Z. Navabi, *VHDL: Analysis and Modeling of Digital Systems*, Second Edition, McGraw-Hill, New York, 1997.

[Wang 2006] L.-T. Wang, C.-W. Wu, and X. Wen, editors, *VLSI Test Principles and Architectures: Design for Testability*, Morgan Kaufmann, San Francisco, 2006.

R11.1 Introduction

[Cheng 1995] K.-T. Cheng and C.-J. Lin, Timing driven test point insertion for full-scan and partial-scan BIST, in *Proc. Int. Test Conf.*, pp. 506–514, October 1995.

[Lin 1995] C.-J. Lin, Y. Zorian, and S. Bhawmik, Integration of partial scan and built-in self-test, *J. Electronic Testing: Theory and Applications*, 7(1–2), pp. 125–137, August 1995.

[SIA 1997] Semiconductor Industry Association, *The National Technology Roadmap for Semiconductors, 1997.*

[SIA 1999] SIA, *The International Technology Roadmap for Semiconductors: 1999 Edition*, Semiconductor Industry Association, San Jose, CA (http://public.itrs.net), 1999.

[SIA 2003] SIA, *The International Technology Roadmap for Semiconductors: 2003 Edition*, Semiconductor Industry Association, San Jose, CA (http://public.itrs.net), 2003.

[Tsai 1998] H.-C. Tsai, S. Bhawmik, and K.-T. Cheng, An almost full-scan BIST solution—higher fault coverage and shorter test application time, in *Proc. Int. Test Conf.*, pp. 1065–1073, October 1998.

R11.2 Software-Based Self-Testing Paradigm

[Hellebrand 1996] S. Hellebrand, H.-J. Wunderlish, and A. Hertwig, Mixed-mode BIST using embedded processors, in *Proc. Int. Test Conf.*, pp. 195–204, October 1996.

[VCI 2000] On-Chip Bus Development Working Group, *Virtual Component Interface Standard (OCB 2 1.0)*, March 2000.

R11.3 Processor Functional Fault Self-Testing

[Brahme 1984] D. Brahme and J. A. Abraham, Functional testing of microprocessors, *IEEE Trans. Comput.*, C-33(6), pp. 475–485, June 1984.

[Flynn 1974] M. J. Flynn, Trends and problems in computer organizations, in *IFIP Congress*, pp. 3–10, August 1974.

[Thatte 1980] S. M. Thatte and J. A. Abraham, Test generation for microprocessors, *IEEE Trans. Comput.*, C-29(6), pp. 429–441, June 1980.

R11.4 Processor Structural Fault Self-Testing

[Bai 2003] X. Bai, L. Chen, and S. Dey, Software-based self-test methodology for crosstalk faults in processors, in *High-Level Design Validation and Test Workshop*, pp. 11–16, November 2003.

[Chen 2001] L.-Chen and S. Dey, Software-based self-testing methodology for processor cores, *IEEE Trans. Comput.-Aided Des.*, 20(3), pp. 369–380, March 2001.

[Chen 2003] L. Chen, S. Ravi, A. Raghunathan, and S. Dey, A scalable software-based self-test methodology for programmable processors, in *Proc. Des. Automation Conf.*, pp. 548–553, June 2003.

[Cheng 1996] K.-T. Cheng and H.-C. Chen, Classification and identification of nonrobustly untestable path delay faults, *IEEE Trans. Comput.-Aided Des. Integrated Circuits and Systems*, 15(8), pp. 845–853, August 1996.

[Gumm 1995] M. Gumm, *VLSI Design Course: VHDL-Modeling and Synthesis of the DLXS RISC Processor*, University of Stuttgart, Germany, December 1995.

[Kranitis 2003] N. Kranitis, A. Paschalis, D. Gizopoulos, and Y. Zorian, Instructional-based self-testing of processor cores, *J. Electronics Testing: Theory and Applications*, 19(2), pp. 103–112, April 2003.

[Kranitis 2005] N. Kranitis, A. Paschalis, D. Gizopoulos, and G. Xenoulis, Software-based self-testing of embedded processors, *IEEE Trans. Comput.*, 54(4), pp. 461–475, April 2005.

[Krstic 1996] A. Krstic, S. T. Chakradhar, and K.-T. Cheng, Testable path delay fault cover for sequential circuits, in *Proc. European Des. Automation Conf.*, pp. 220–226, September 1996.

[Lai 2000a] W.-C. Lai, A. Krstic, and K.-T. Cheng, On testing the path delay faults of a microprocessor using its instruction set, in *Proc. VLSI Test Symp.*, pp. 15–20, April 2000.

[Lai 2000b] W.-C. Lai, A. Krstic, and K.-T. Cheng, Test program synthesis for path delay faults in microprocessor cores, in *Proc. Int. Test Conf.*, pp. 1080–1089, October 2000.

[Parvathala 2002] P. Parvathala, K. Maneparambil, and W. Lindsay, FRITS: A microprocessor functional BIST method, in *Proc. Int. Test Conf.*, pp. 590–598, October 2002.

[Paschalis 2005] A. Paschalis and D. Gizopoulos, Effective software-based self-test strategies for on-line periodic testing of embedded processors, *IEEE Trans. Comput.-Aided Des. of Integrated Circuits and Systems*, 24(1), pp. 88–99, January 2005.

[Psarakis 2006] M. Psarakis, D. Gizopoulos, M. Hatzimihail, A. Paschalis, A. Raghunathan, and S. Ravi, Systematic software-based self-test for pipelined processors, in *Proc. Des. Automation Conf.*, pp. 393–398, July 2006.

[Singh 2006] V. Singh, M. Inoue, K. K. Saluja, and H. Fujiwara, Instruction-based self-testing of delay faults in pipelined processors, *IEEE Trans. Very Large Integration (VLSI) Systems*, 14(11), pp. 1203–1215, November 2006.

[Tupuri 1997] R. S. Tupuri and J. A. Abraham, A novel functional test generation method for processors using commercial ATPG, in *Proc. Int. Test Conf.*, pp. 743–752, November 1997.

R11.5 Processor Self-Diagnosis

[Bernardi 2006] P. Bernardi, E. Sánchez, M. Schillaci, G. Squillero, and M. S. Reorda, An effective technique for minimizing the cost of processor software-based diagnosis in SoCs, in *Proc. Des., Automation and Test in Europe*, pp. 412–417, March 2006.

[Chen 2002] L. Chen and S. Dey, Software-based diagnosis for processors, in *Proc. Des. Automation Conf.*, pp. 259–262, June 2002.

[Grüning 1991] T. Grüning, U. Mahlstedt, and H. Koopmeiners, DIATEST: A fast diagnostic test pattern generator for combinational circuits, in *Proc. Int. Conf. on Comput.-Aided Des.*, pp. 194–197, November 1991.

[Pomeranz 2000] I. Pomeranz and S. M. Reddy, A diagnostic test generation procedure based on test elimination by vector omission for synchronous sequential circuits, *IEEE Trans. Comput.-Aided Des.*, 19(5), pp. 589–600, May 2000.

[Venkataraman 2001] S. Venkataraman and S. B. Drummonds, POIROT: Applications of a logic fault diagnosis tool, *IEEE Des. & Test of Comput.*, 18(1), pp. 19–30, January/February 2001.

[Yu 2000] X. Yu, J. Wu, and E. M. Rudnick, Diagnostic test generation for sequential circuits, in *Proc. Int. Conf. on Comput.-Aided Des.*, pp. 225–234, October 2000.

R11.6 Testing Global Interconnect

[Bai 2000] X. Bai, S. Dey, and J. Rajski, Self-test methodology for at-speed test of crosstalk in chip interconnects, in *Proc. Des. Automation Conf.*, pp. 619–624, June 2000.

[Chen 2001] L. Chen, X. Bai, and S. Dey, Testing for interconnect crosstalk defects using on-chip embedded processor cores, in *Proc. Des. Automation Conf.*, pp. 317–320, June 2001.

[Cuviello 1999] M. Cuviello, S. Dey, X. Bai, and Y. Zhao, Fault modeling and simulation for crosstalk in system-on-chip interconnects, in *Proc. Int. Conf. on Comput.-Aided Des.*, pp. 297–303, November 1999.

[Lai 2001] W.-C. Lai, J.-R. Huang, and K.-T. Cheng, Embedded-software-based approach to testing crosstalk-induced faults at on-chip buses, in *Proc. VLSI Test Symp.*, pp. 204–209, April 2001.

[SIA 2003] SIA, *The International Technology Roadmap for Semiconductors: 2003 Edition*, Semiconductor Industry Association, San Jose, CA (http://public.itrs.net), 2003.

R11.7 Testing Nonprogrammable Cores

[Huang 2001] J.-R. Huang, M.-K. Iyer, and K.-T. Cheng, A self-test methodology for IP cores in bus-based programmable SoCs, in *Proc. VLSI Test Symp.*, pp. 198–203, April 2001.

[Tsai 1998] H.-C. Tsai, S. Bhawmik, and K.-T. Cheng, An almost full-scan BIST solution—higher fault coverage and shorter test application time, in *Proc. Int. Test Conf.*, pp. 1065–1073, October 1998.

R11.8 Instruction-Level DFT

[Lai 2001] W.-C. Lai and K.-T. Cheng, Instruction-level DFT for testing processor and IP cores in system-on-a-chip, in *Proc. Des. Automation Conf.*, pp. 59–64, June 2001.

[Shen 1998] J. Shen and J. A. Abraham, Native mode functional test generation for processors with applications to self test and design validation, in *Proc. Int. Test Conf.*, pp. 990–999, October 1998.

R11.9 DSP-Based Analog/Mixed-Signal Component Testing

[Dufort 1997] B. Dufort and G. W. Roberts, Signal generation using periodic single and multi-bit sigma-delta modulated streams, in *Proc. Int. Test Conf.*, pp. 396–405, November 1997.

[Huang 2000] J.-L. Huang and K. T. Cheng, A sigma-delta modulation based BIST scheme for mixed-signal circuits, in *Proc. Asia and South Pacific Des. Automation Conf.*, pp. 605–610, January 2000.

[Pan 1995] C. Y. Pan and K. T. Cheng, Pseudo-random testing and signature analysis for mixed-signal circuits, in *Proc. Int. Conf. on Comput.-Aided Des.*, pp. 102–107, November 1995.

[Toner 1995] M. F. Toner and G. W. Roberts, A BIST scheme for a SNR, gain tracking, and frequency response test of a sigma-delta ADC, *IEEE Trans. Circuit and Systems II: Analog and Digital Signal Processing*, 42(1), pp. 1–15, January 1995.

CHAPTER 12

FIELD PROGRAMMABLE GATE ARRAY TESTING

Charles E. Stroud
Auburn University, Auburn, Alabama

ABOUT THIS CHAPTER

Since the mid-1980s, ***field programmable gate arrays*** (FPGAs) have become a dominant implementation medium for digital systems. During that time, FPGAs have grown in complexity as their ability to implement system functions has increased from a few thousand logic gates to tens of millions of logic gates. The largest FPGAs currently available exceed a billion transistors. The ability to program, or **configure**, FPGAs to perform virtually any digital logic function provides an attractive choice, not only for rapid prototyping of digital systems but also for low-to-moderate volume or fast time-to-market systems. The ability to **reconfigure** FPGAs to perform different digital logic functions without modifying the physical system facilitates implementation of advanced system applications such as adaptive computing and fault tolerance.

The chapter begins with an overview of the typical architecture and configuration features of FPGAs as well as the testing challenges posed by these complex devices. We follow this discussion with a review of the various approaches to testing FPGAs that have been proposed and implemented. The remaining sections discuss testing and diagnosis of various resources incorporated in FPGAs, including programmable logic and routing resources as well as specialized cores such as memories and ***digital signal processors*** (DSPs). The chapter concludes with a discussion of one of the new frontiers in FPGA testing in which embedded processor cores within the FPGA are used to test and diagnose the FPGA itself.

12.1 OVERVIEW OF FPGAS

FPGAs come in a variety of sizes and features, which continue to change with advances in integrated circuit fabrication technology and system applications. As a result, it is difficult to write a comprehensive treatise on FPGAs because the material tends to become obsolete shortly after publication. Although a number of books ([Trimberger 1994] [Oldfield 1995] [Salcic 1997]) and papers ([Brown 1996a, 1996b]) have been written about FPGA architectures and operation, a wealth of current information can be found in the data sheets and application notes provided by the FPGA manufacturers such as [Altera 2005], [Atmel 2005], [Cypress 2005], [Lattice 2006], [Xilinx 2006a], and [Xilinx 2006b]. The following subsections provide an overview of FPGAs based on the typical architectures, features, and capabilities that are currently available.

12.1.1 Architecture

FPGAs generally consist of a two-dimensional array of ***programmable logic blocks*** (PLBs) interconnected by a programmable routing network with programmable ***input/output*** (I/O) cells at the periphery of the device, as illustrated in Figure 12.1. Each PLB usually consists of one or more lookup tables (LUTs) and flip-flops, as illustrated in Figure 12.2. A LUT typically has four inputs and is used to implement combinational logic by storing the truth table for any combinational logic equation of up to four input variables. It should be noted, however, that some FPGAs have LUTs with as few as three inputs or as many as six inputs. The LUT in some FPGAs can also be configured to function as a small ***random access memory*** (RAM) or shift register. The flip-flops in the PLB are used to implement sequential logic and often include programmable clock enable and set/reset capabilities. The flip-flops in some FPGAs can also be configured to operate as level sensitive latches. Additional logic besides LUTs and flip-flops is usually incorporated in the PLB; some examples include multiplexers for combining LUTs to construct large combinational logic functions via the Shannon expansion theorem, fast carry logic to implement adders/subtractors or counters, and logic to implement special functions like array multipliers.

The programmable interconnect network consists of horizontal and vertical wire segments of varying lengths along with programmable switches, referred to as ***programmable interconnect points*** (PIPs) or ***configurable interconnect points*** (CIPs). The lengths of the wire segments are usually referred to by the number of PLBs they span where typical lengths include one, two, four, six, and eight

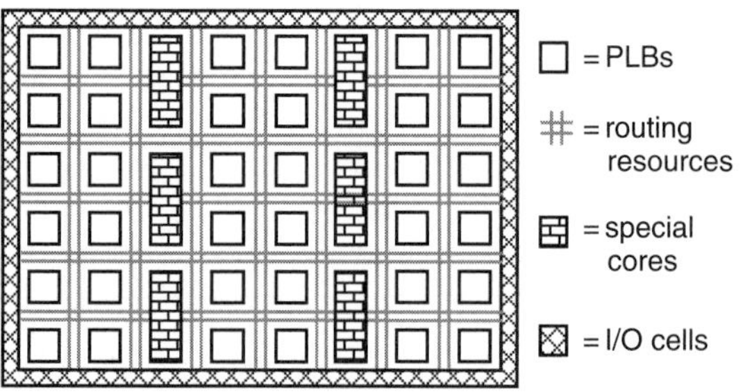

■ FIGURE 12.1

Typical FPGA architecture.

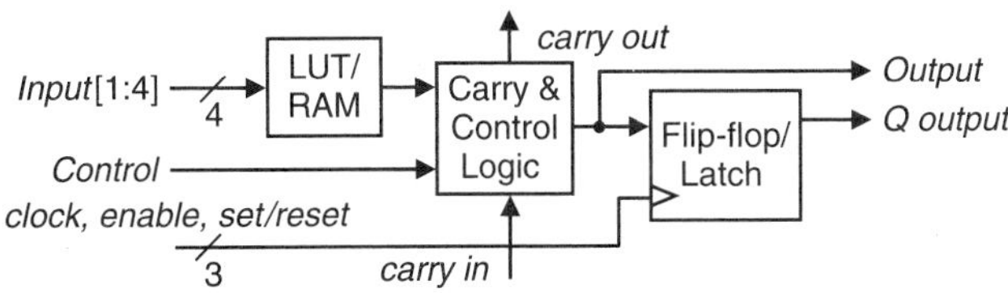

■ **FIGURE 12.2**

Basic PLB architecture.

PLBs. There are also long lines that span one quarter, one half, and the full array of PLBs. The wire segments and their associated PIPs can be classified as local and global routing resources. Local routing resources are used to interconnect internal portions of the PLB, directly adjacent PLBs, or to connect the PLB to the global routing resources. Global routing resources interconnect PLBs with other nonadjacent PLBs, I/O cells, and specialized cores.

The PIPs are used to connect or disconnect the wire segments to form the signal nets for the system application. The basic PIP consists of a transmission gate controlled by a configuration memory bit as illustrated in Figure 12.3a. When the PIP is activated (turned on), the two wire segments are connected to provide a continuous signal path. When the PIP is deactivated (turned off), the two wire segments are isolated and can provide paths for two independent signals. There are several types of PIP structures with the simplest being the break-point PIP (see Figure 12.3b), which connects/disconnects two vertical or two horizontal wire segments. The cross-point PIP (see Figure 12.3c) connects a vertical wire segment and a horizontal wire segment such that the signal path can turn a corner and fanout on both vertical and horizontal routing resources. Multiple PIPs and their associated wire segments can be arranged to form more complicated and sophisticated programmable routing structures. For example, six break-point PIPs can be arranged as illustrated in Figure 12.3d to form a compound cross-point PIP where two independent signal paths can be routed straight through (one vertical and the other horizontal) or turn corners. The programmable routing structure most frequently encountered in current FPGAs uses N break-point PIPs to form a

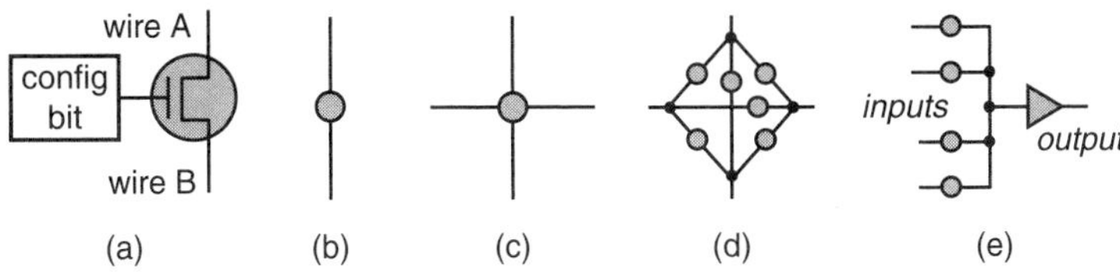

■ **FIGURE 12.3**

Programmable interconnect points: (a) basic PIP structure, (b) break-point PIP, (c) cross-point PIP, (d) compound cross-point PIP, and (e) multiplexer PIP.

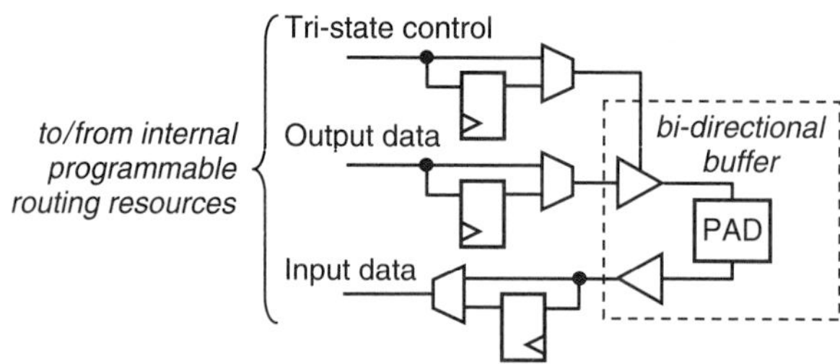

■ **FIGURE 12.4**

Typical I/O cell structure.

multiplexer PIP (see Figure 12.3e) where one of the N inputs can be selected and routed to the multiplexer output. The output of most multiplexer PIPs is buffered to prevent signal degradation on long and/or heavily loaded signal nets.

A simplified programmable I/O cell is illustrated in Figure 12.4 and consists of a bidirectional buffer with tristate control to facilitate user specification of input, output, and bidirectional ports for any given system application. Multiplexers controlled by configuration memory bits are used to select registered or nonregistered signals to or from the pad. These registers can be flip-flops, latches, or flip-flops that can also be configured as a latch. They are used to meet critical system timing specifications including setup time, hold time, and clock-to-output delay. Multiple flip-flops and latches are incorporated for the input and the output portion of the I/O cell to support ***double-data-rate*** (DDR) and/or serial-to-parallel and parallel-to-serial conversions for high-speed data interfaces (SERDES). Programmable clock enable and set/reset signals, under the control of configuration memory bits, are often associated with each register. In addition, the bidirectional buffer typically has programmable current drive and voltage level capabilities for various I/O voltage standards along with programmable pullup/pulldown features and programmable delay characteristics. In some FPGAs, two I/O cells can be combined to support differential pair I/O standards.

A trend in FPGAs is to include cores for specialized functions such as single-port and dual-port RAMs, ***first-in first-out*** (FIFO) memories, multipliers, and DSPs. Within any given FPGA, all memory cores are usually of the same size in terms of the total number of memory bits, but each memory core in the array is individually programmable. For example, RAM cores can be individually configured to select the mode of operation (single-port, dual-port, FIFO, etc.), architecture in terms of number of address locations *versus* number of data bits per address location, active edge of clocks, and active level of control signals (clock enables, set/reset, etc.), as well as options for registered input and output signals. The types and quantities of programmable resources are summarized in Table 12.1 and represent the typical small and large FPGAs available.

Another trend is the incorporation of one or more embedded processor cores. These processor cores range in size from 8-bit to 64-bit architectures and can

TABLE 12.1 ■ Typical Quantities of Programmable Resources in Current FPGAs

	FPGA Resource	Small FPGA	Large FPGA
Logic	PLBs per FPGA	256	25,920
	LUTs and flip-flops per PLB	1	8
Routing	Wire segments per PLB	45	406
	PIPs per PLB	140	4100
Specialized Cores	Bits per memory core	128	36,864
	Memory cores per FPGA	16	576
	DSP cores	0	512
Other	Input/output cells	62	1200
	Configuration memory bits	42,104	79,704,832

operate at clock frequencies as high as 450 MHz. In some devices, the program and data memories are dedicated to the processor core, whereas in other devices the RAM cores within the FPGA are used to construct the program and data memories. In the latter case, programmable logic and routing resources are also required to implement the interface between the processor and the RAM cores used for program and data memories. An alternative to the dedicated embedded processor core, referred to as a **hard core**, is the synthesis of **soft core** processors into the programmable logic, routing, and memory resources of the FPGA. In the case of a soft core, the processor architecture is described in a hardware description language, such as Verilog or VHDL, and the normal system design flow is used for integration and synthesis of the processor core with the system application. Soft core processors typically run at lower maximum clock frequencies than hard core processors with dedicated and optimized resources. However, greater versatility and flexibility are associated with the soft core in terms of the number of processor cores that can be incorporated in a single FPGA and the ability to incorporate the soft processor core in almost any FPGA.

With the incorporation of embedded cores such as memories, DSPs, and microprocessors, FPGAs more closely resemble ***system-on-chip*** (SOC) implementations. At the same time, some SOCs incorporate embedded FPGA cores; these devices are often referred to as *configurable SOCs*. ***Complex programmable logic devices*** (CPLDs) are similar to FPGAs in that they contain programmable logic and routing resources as well as specialized cores such as RAMs and FIFOs. The primary difference is that CPLDs typically use ***programmable logic arrays*** (PLAs) for implementing combinational logic functions instead of the LUTs typically found in FPGAs. As a result, a PLB in a CPLD tends to have more flip-flops and combinational logic resources than a PLB in an FPGA. However, *FPGA* has become the most frequently used term to refer to all programmable logic devices including CPLDs. Therefore, the term *FPGA* will be used in this chapter, but it should be noted the techniques presented are also applicable to CPLDs and embedded FPGA cores in SOC implementations.

12.1.2 Configuration

The system function performed by an FPGA is controlled by an underlying programming mechanism. Early programming technologies included fuses, which later changed to antifuses, as well as floating-gate technologies typically used in erasable ***programmable read-only memories*** (PROMs) and, more recently, in flash memories [Trimberger 1994]. The advantage of these programming technologies is their nonvolatility such that the configuration data for the system function are not lost when the power to the device is turned off. The disadvantage of fuse/antifuse technologies is that they are ***one-time-programmable*** (OTP). The disadvantage of floating-gate technologies is that, with the exception of flash memory, they are not ***in-system reconfigurable*** (ISR) and in some cases are not ***in-system programmable*** (ISP).

The most frequently used programming technology in current FPGAs is a static RAM. Although RAM-based configuration memories are ISR, they are also volatile. For a RAM-based FPGA to operate in a system, the configuration memory must first be written with the data that will define the operation of that FPGA in the system. This process is frequently referred to as **downloading** the configuration data to the FPGA, but is also called *configuring* or *programming* the FPGA. The configuration data include the contents of the LUTs, which specify the combinational logic functions performed as well as the various bits used to control the mode of operation of flip-flops/latches for sequential logic functions. A large portion of the configuration memory is used to control the PIPs that are used to route signals for interconnection of the system logic functions. Other configuration memory bits control the modes of operation of I/O cells and specialized cores such as RAMs, FIFOs, and DSPs. The system function can be changed at any time by rewriting the configuration memory with new data, referred to as **reconfiguration**. The time required to completely configure or reconfigure an FPGA can be long, particularly for FPGAs with large configuration memories. Alternatively, **partial reconfiguration** can be used to reconfigure only the portions of the FPGA that change from one system function to the next. Another feature supported by many current FPGAs is **dynamic partial reconfiguration**, also referred to as **run-time reconfiguration**, where a portion of the FPGA can be reconfigured while the remainder of the FPGA continues to perform normal system operation. Dynamic partial reconfiguration also prevents loss of data contained in system function memory elements (flip-flops, RAMs, etc.) during the reconfiguration process.

There are two basic modes for configuration of FPGAs, typically called master and slave modes. The configuration mode is controlled by dedicated mode input pins to the FPGA that can be hardwired on a ***printed circuit board*** (PCB). In master mode, the FPGA configures itself by reading the configuration data from a PROM residing on the PCB with the FPGA. Configuration using master mode is usually performed during power up of the PCB. In slave mode, configuration of the FPGA is controlled and performed by an external device, such as a processor. As a result, the slave mode is used for partial and dynamic partial reconfiguration of the FPGA. Most FPGAs support the ability to configure multiple FPGAs from a single PROM by daisy-chaining the FPGAs, as illustrated in Figure 12.5. Here the

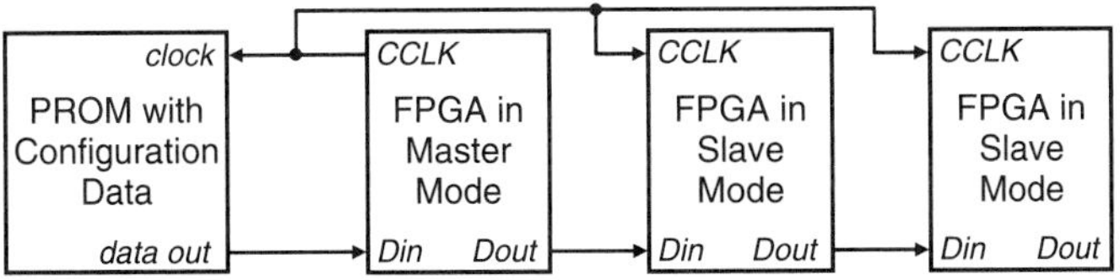

■ **FIGURE 12.5**

Master and slave mode FPGA configuration in a daisy chain.

first FPGA in the chain operates in master mode to supply the ***configuration clock*** (CCLK) to read the configuration data from the PROM, while the remaining FPGAs in the chain operate in slave mode. As each FPGA in the daisy chain completes its configuration process, subsequent configuration data from the PROM are passed on to the next FPGA in the daisy chain by connecting the configuration data input (Din) to the data output (Dout). Many FPGAs provide serial and parallel configuration interfaces for both master and slave configuration modes. The parallel interface is used to speed up the configuration process by reducing the download time via parallel data transfer. The most common parallel interfaces use 8-bit or 32-bit configuration data buses. This facilitates configuration of FPGAs operating in slave mode by a processor, which can also perform reconfiguration or dynamic partial reconfiguration of the FPGAs once the system is operational. To avoid a large number of pins dedicated to a parallel configuration interface, these parallel data bus pins can be optionally reused for normal system function signals once the device is programmed. The IEEE standard **boundary scan** interface, in conjunction with associated instructions to access the configuration memory, can also be used for slave serial configuration in many FPGAs.

Partial reconfiguration data are generated by comparing the new configuration data to be downloaded with the reference configuration data for the implementation residing in the FPGA. The partial reconfiguration data contain the configuration data for only those portions of the FPGA that change between these two configurations along with the appropriate commands to the configuration registers to perform partial reconfiguration. The smallest portion of an FPGA that can be reconfigured is a function of the organization of the configuration memory design. In some FPGAs, for example, a single PLB, or a portion thereof, can be reconfigured without affecting any other PLB in the array; this is referred to as a **PLB addressable** configuration memory. However, most FPGAs have a **frame addressable** configuration memory where a single frame spans multiple PLBs in the array, usually an entire column of PLBs. As a result, the frame size is often a function of the array size and increases for larger FPGAs. Multiple frames, typically 10 to 20, are needed to completely specify the configuration data for a PLB or column of PLBs and the associated routing resources. Therefore, the size of the partial reconfiguration data to be downloaded is a function of the frame size and the number of frames different between the configuration of the reference design residing in the FPGA and the new design to be downloaded.

The ability to perform partial reconfiguration requires an increase in complexity of the configuration interface circuitry compared to a full reconfiguration. For example, the user must have the ability to specify the address of the frame to be written as well as the configuration data to be written into that frame address. As a result, the configuration interface circuitry must support addressing and accessing the ***frame address register*** (FAR) and ***frame data register*** (FDR) as well as other registers such as command and control registers. The command and control registers are provided to facilitate specialized features for not only the partial reconfiguration process but also for the operation of the FPGA once it is configured. An example feature is **multiple frame write** capability where the configuration data contained in the FDR can be written to multiple frames to reduce configuration time, because the size of the frame address is typically much smaller than the frame data. This is particularly useful in the case of very regular designs with repeated identically configured circuit functions. Another example feature is security for **configuration memory readback** where the configuration memory can be read to verify proper download of the configuration data. Once verified, configuration memory readback access can be disabled for intellectual property protection of the design configured in the FPGA. As a result of this type of security feature, different *boundary scan* instructions are needed for write access and read access of the configuration memory. Some FPGAs allow encryption of the configuration data as an additional security feature.

In addition to configuration memory readback for verifying proper download, many FPGAs support partial configuration memory readback. The FAR is used to specify the address of a specific frame, or the starting address for a set of consecutive frames, to be read. In some FPGAs, the contents of memory elements, such as flip-flops or RAMs, can be read during partial configuration memory readback in addition to the configuration data contained in the frame.

A recent and exciting trend in FPGAs is the ability of an embedded processor core, either *hard core* or *soft core*, to read or write the FPGA configuration memory. As a result, the embedded processor core is capable of on-chip *dynamic partial reconfiguration* of portions of the FPGA. Hence, configuration data can be algorithmically generated from within the FPGA itself instead of traditional downloading. This presents opportunities for many new types of system applications in FPGAs, as well as new types of testing solutions, as will be discussed later.

12.1.3 The Testing Problem

The programmability of an FPGA poses a number of challenges when it comes to complete and comprehensive testing of the FPGA itself. First, a large number of configurations must be downloaded into the FPGA to test the programmable resources in their various modes of operation. The size of the configuration memory is an important factor in testing FPGAs, because the total test time is usually dominated by the time required to download configuration data. Fortunately, *partial reconfiguration* can reduce the total time associated with downloading these test configurations by writing only the portions of configuration memory that change from one test configuration to the next. The FPGA testing problem is further complicated by the growing size of FPGAs in terms of the PLB array, frequently changing architectures, and the introduction of specialized embedded cores such as memories and DSPs. If the FPGA can be completely tested and determined to be fault-free, the intended system function can be programmed onto the FPGA with a high probability of proper operation. When faults are detected, the system function can be reconfigured to avoid the faulty resources if they can be diagnosed (identified and located). During manufacturing testing, the ability to locate and identify faults is also important for yield enhancement. In addition, the programmability of FPGAs has resulted in the ability to use faulty devices for system-specific configurations that do not use the faulty resources [Abramovici 2004] [Toth 2004]. Therefore, diagnosis of the faulty resources is an important aspect of FPGA testing in order to take full advantage of the fault- or defect-tolerant potential of these devices.

FPGA testing must consider a variety of fault models. In addition to classical stuck-at faults in the programmable logic resources, the configuration memory bits that define the logic functions performed by these resources must also be tested for stuck-at-0 and stuck-at-1 faults [Abramovici 2001]. For complete testing, the logic resources must be tested in different modes of operation, which in turn requires multiple reconfigurations of the logic resources under test in the FPGA. The number of configuration memory bits associated with the programmable routing resources is typically three to four times the number of configuration memory bits associated with the programmable logic resources. As a result, the programmable interconnect network poses a bigger testing challenge than the logic resources. The fault models used for testing the routing resources include shorts (bridging faults) and opens in the wire segments, wire segments stuck-at-1 and stuck-at-0, as well as PIPs stuck-on and stuck-off along with their controlling configuration memory bits stuck-at-1 and stuck-at-0. Whereas the PIP stuck-off fault can be detected by a simple continuity test, stuck-on faults are similar to bridging faults and require opposite logic values be applied to the wire segments on both sides of the PIP while monitoring both wire segments in order to detect the stuck-on fault [Stroud 2002b]. More recent issues in FPGA testing include delay fault testing [Chmelar 2003]. Testing for delay faults in FPGAs is important because the transmission gates used to construct the PIPs in the interconnect network are particularly susceptible to defects that affect the delay.

FPGA testing is further complicated by the capabilities, or lack thereof, of the ***computer-aided design*** (CAD) tools associated with FPGAs. The primary goal of these CAD tools is the synthesis and implementation of system applications into the FPGA. Because the system functions are usually described in VHDL or Verilog, the synthesis tools attempt to optimize the implementation to obtain the most efficient implementation in the FPGA in terms of area and performance. As a result, CAD tools attempt, during the optimization process, to eliminate unused logic and interconnections in the final implementation. Unfortunately, this optimization is often at odds with the goals of testing. For example, consider the multiplexer in Figure 12.6a and its associated gate-level implementation. To detect the stuck-at-1 (sa1) fault denoted by the "X" in the figure, we must apply the logic values shown next to each input; in other words, we must select the *A* input while applying

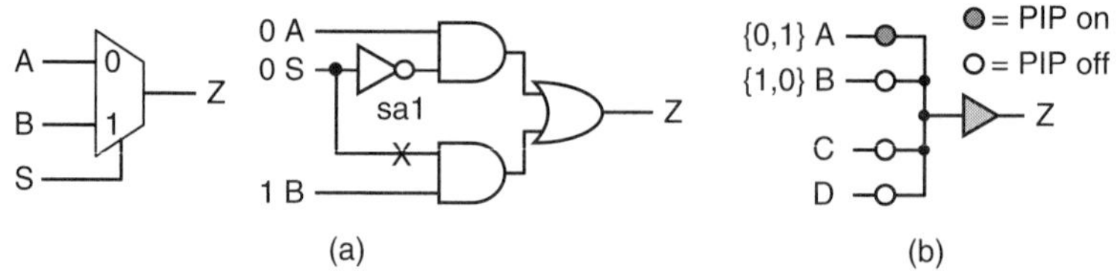

■ FIGURE 12.6

Programmable resource testing problem example: (a) gate-level multiplexer implementation and (b) multiplexer PIP.

a logic 1 to the unselected B input. This is not a problem in a normal circuit, but when the select signal S is sourced by a configuration memory bit, the logic sourcing signals on the unselected inputs is considered to be unused by FPGA CAD tools and will be eliminated, not permitting sourcing of the required test logic values. A more frequent problem exists at the inputs to the multiplexer PIPs that are used extensively in recent FPGAs to construct the programmable interconnect network. Each PIP in the multiplexer shown in Figure 12.6b must be tested for both stuck-on and stuck-off faults. The stuck-off fault can be detected by activating each PIP in turn and passing both logic 0 and logic 1 values through the multiplexer (as shown for input A). To test the stuck-on fault, however, we must apply the opposite logic values, with respect to the logic values being passed through the activated PIP, to one or more of the deactivated PIPs (as shown for input B) to detect the short circuit caused by the PIP stuck-on fault. Therefore, we need a minimum of one test configuration for each input to the multiplexer and, for each of these configurations, we must apply opposite logic values to the activated PIP and at least one of the deactivated PIPs. However, synthesis tools typically require a signal to have both source and sink in order for the signal path to be routed through the programmable interconnect network, which prevents routing a signal to an unselected input of a multiplexer, for example. This problem has been referred to as the "invisible logic problem" and must be overcome to completely test an FPGA [Stroud 1996]. The solution requires using FPGA vendor-specific design tools, intermediate netlist languages, and techniques that are not used in a normal design flow but nevertheless provide the control needed to establish proper testing conditions.

12.2 TESTING APPROACHES

A taxonomy of FPGA testing techniques is summarized in Table 12.2 and includes a number of general types and categories. For example, the generation and application of input test stimuli as well as the analysis of output responses can be performed external or internal to the FPGA. In ***built-in self-test*** (BIST) approaches, both test pattern generation and output response analysis are performed internal to the FPGA, whereas either one or both functions are performed external to the

TABLE 12.2 ■ FPGA Testing Taxonomy

<table>
<tr><th>Test Approach Attribute</th><th colspan="5">Classification</th></tr>
<tr><td>Test pattern application and output response analysis</td><td colspan="3">Internal (BIST)</td><td colspan="2">External</td></tr>
<tr><td>System-level testing</td><td colspan="3">Offline</td><td colspan="2">Online</td></tr>
<tr><td>System application</td><td colspan="3">Independent</td><td colspan="2">Dependent</td></tr>
<tr><td rowspan="2">Target programmable resources</td><td colspan="3">Logic</td><td colspan="2">Routing</td></tr>
<tr><td>PLBs</td><td>I/O cells</td><td>Cores</td><td>Local</td><td>Global</td></tr>
</table>

FPGA in **external testing** approaches. For system-level testing, another classification is based on whether the testing is **online** while the FPGA is also performing its normal system functions, or **offline** where the system is not operational during testing of the FPGA. A testing approach can be further classified based on whether its goal is to test only those resources to be used by the intended system function, referred to as **application-dependent testing**, or all resources regardless of the system function to be implemented in the FPGA, called **application-independent testing**. FPGA manufacturing testing is usually external and, by definition, offline and application independent. On the other hand, a BIST approach could be internal, online, and application dependent.

Different techniques are typically used to test programmable logic and routing resources as a result of the different architectures and fault models associated with the targeted resources. However, the routing resources must be used to test the logic resources and vice versa. Therefore, although this final classification may not necessarily be mutually exclusive, it does provide a general overall classification for test approaches. Programmable logic resources can be subdivided into PLBs, I/O cells, and other specialized cores such as memories and DSPs. Programmable interconnect resources can be subdivided into local and global routing resources, where local routing resources are dedicated to a given logic resource, or pair of adjacent resources, with global routing resources providing the interconnection of nonadjacent resources. The following subsections discuss the first three classifications, and the subsequent section provides a more detailed discussion of testing logic and routing resources in the context of BIST as a representative approach to testing these resources. An earlier classification and survey of FPGA test and diagnostic techniques can be found in [Doumar 2003].

12.2.1 External Testing and Built-In Self-Test

In external testing techniques, the FPGA is programmed for a given test configuration with the application of input test stimuli and the monitoring of output responses performed by external sources such as a test machine [Huang 1998]

[Renovell 1998]. Once the FPGA has been tested for a given test configuration, the process is repeated for subsequent test configurations until the targeted resources within the FPGA have been completely tested. As a result, external test approaches are typically used for manufacturing testing only. For FPGAs with boundary scan that support **INTEST** capabilities, the input test stimuli can be applied, and output responses can be monitored, via the *boundary scan* interface. As a result, internal testing of the FPGA via boundary scan *INTEST* could be performed during system-level testing. Otherwise, the FPGA I/O pins must be used, resulting in package-dependent testing. It should be noted, however, that few FPGA manufacturers support the *INTEST* feature.

The basic idea of BIST for FPGAs is to configure some of the programmable logic and routing resources as ***test pattern generators*** (TPGs) and ***output response analyzers*** (ORAs). These TPGs and ORAs constitute the BIST circuitry used to detect faults in PLBs, routing resources, and specialized cores such as RAMs and DSPs. This facilitates both manufacturing and the system-level use of the BIST configurations. Different approaches are typically used for testing PLBs, routing resources, and embedded cores, as will be discussed later in the chapter.

12.2.2 Online and Offline Testing

BIST approaches for FPGAs have been used at the system level for both online and offline testing. Online testing uses dynamic partial reconfiguration to detect and diagnose faults concurrently with normal system operation [Abramovici 2004] [Verma 2004] [Emmert 2007]. This is accomplished by designating portions of the FPGA as self-test areas while the remaining portions of the FPGA implement the working system function. All of the logic resources within the self-test area can be tested without affecting the operation of the system function because no system function logic is located in the self-test area. Although some of the routing resources within the self-test area must be reserved in order to interconnect the portions of the system function residing on either side of the self-test area, the remaining routing resources can be tested. For example, the self-test area illustrated in Figure 12.7 runs vertically through the array such that the vertical routing resources can be tested while the horizontal routing resources are used to interconnect the system

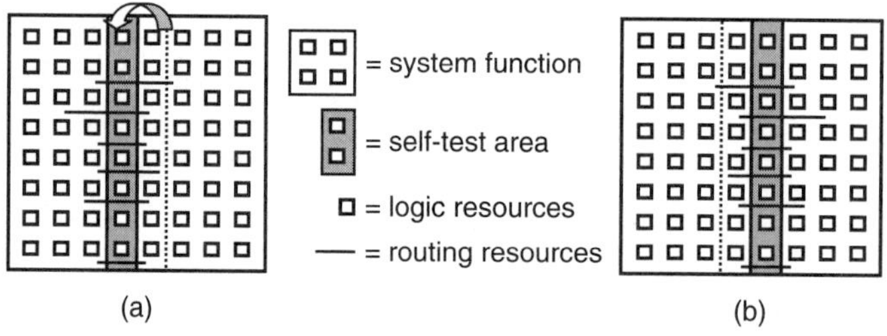

■ **FIGURE 12.7**

Online testing: (a) before relocation and (b) after relocation.

function, illustrated by the horizontal lines crossing the self-test area. To test the horizontal routing resources, a horizontal self-test area must be included in which the vertical routing resources are used to interconnect the system function.

When the logic and routing resources in the self-test area have been tested and found to be fault-free, a portion of the system function is relocated to the self-test area, as illustrated in Figure 12.7a, such that the vacated region now functions as the new self-test area, as illustrated in Figure 12.7b. As a result, the vertical and horizontal self-test areas repeatedly move through the FPGA to test all programmable resources. When faults are detected, the detected faults are diagnosed (located and identified) so that the portion of the system function to be relocated into the self-test area can be reconfigured to use spare resources and avoid the identified faulty resources. One goal of online testing approaches is to minimize the latency between the occurrence of a fault and its subsequent detection and diagnosis. Another goal is to minimize the size of the self-test area as well as the number of resources reserved for spares in order to maximize the resources available for the system function.

Offline testing, on the other hand, requires that the system, or a portion of the system, be taken out of service in order to perform the test. As a result, offline testing is typically performed periodically during low-demand periods of system operation or on demand when an error is detected during system operation via concurrent error detection circuits, such as parity, checksum, or cyclic redundancy code checks. One goal of offline testing is to maximize the quantity of resources under test during each test configuration to minimize the test time and the associated time that the system is out of service. Once the FPGA has been tested and determined to be fault-free, the system function can be configured in the FPGA. Otherwise, when faults are detected, diagnostic configurations or procedures can be applied such that the system function can be reconfigured to avoid the identified faults before returning the system to service. An advantage of offline testing is that there is no penalty in terms of the resources available for system function implementation because the FPGA is completely reconfigured before and after testing.

12.2.3 Application Dependent and Independent Testing

Because test time is usually dominated by the number of times the FPGA must be reconfigured and the associated download time, application-dependent testing approaches seek to test only those resources the intended system function will use [Tahoori 2004] [Tahoori 2007]. One approach to application-dependent testing is to include ***design for testability*** (DFT) circuitry in the system function to be implemented by the FPGA. These DFT approaches include scan design or BIST techniques typically used in ***application-specific integrated circuits*** (ASICs) such that only the logic and routing resources used by the system function are tested. The primary disadvantage of this approach is the additional area overhead and performance penalties traditionally associated with DFT circuitry. Although a few FPGA architectures have incorporated scan chains, or dynamically controlled multiplexers at the inputs to flip-flops that can be used for scan chains, this is not the case in the majority of commercially available FPGAs. As a result, there can be a significant increase in the number of PLBs required to implement the system function when DFT circuitry is incorporated in the design, as much as 50% in some examples [Renovell 2001]. For example, adding a multiplexer to the input of a flip-flop for implementing scan design requires two LUT inputs (one input for the scan input data and one input for the scan mode control) in addition to the inputs needed for the system function. For an FPGA with four-input LUTs, this doubles the number of LUTs that drive flip-flops and that implement system functions requiring more than two inputs.

There is an alternative application-dependent testing approach for FPGAs that support reading the contents of memory elements during partial configuration memory readback as well as the ability to set logic values in memory elements during partial reconfiguration. For FPGAs that support both of these features, test patterns specific to the system function application programmed in the FPGA can be written into the memory elements via the partial reconfiguration process. The FPGA is then clocked once in its system mode of operation. Partial configuration memory readback is then used to retrieve the output responses of the combinational logic from the memory elements. This process is repeated until all necessary test patterns have been applied and output responses have been retrieved. Because the configuration memory bits for the memory elements may be located in frames with other configuration data for the system function, such as LUT contents and routing, a read-modify-write operation on the configuration memory can be used to modify the contents of the memory elements without disturbing (or needing to know and store) the contents of those other configuration memory bits. As a result, this approach is similar to scan design but without any additional area overhead or performance penalties to the system function.

Application-independent testing approaches, on the other hand, seek to test all programmable resources in the FPGA, independent of the system application to be configured in the FPGA. Once the programmable resources have been completely tested, the FPGA can be programmed with the intended system function without any overhead or performance penalties resulting from the DFT circuitry. Because application-independent techniques seek to test all resources in the FPGA to ensure that they are fault-free, regardless of any given system function, the FPGA must be reconfigured many times to test all of the resources. Both external testing and BIST approaches have been developed for application-independent testing. Because of the regularity of the FPGA structure, ***iterative logic array*** (ILA) testing approaches have been applied to facilitate scaling the test architecture and constant test time independent of array size [Huang 1998] [Toutounchi 2004]. Application-independent external testing [Chmelar 2003] and BIST approaches [Abramovici 2003] have been developed to detect delay faults in FPGAs.

12.3 BIST OF PROGRAMMABLE RESOURCES

This section examines various BIST approaches for testing logic and routing resources. Other than the incorporation of TPGs and ORAs, the test configurations for the resources under test are, for the most part, independent of whether

TABLE 12.3 ■ FPGA Test Configurations

FPGA		Number of Test Configurations			
Vendor	**Series**	**PLBs**	**Routing**	**Cores**	**Reference**
Lattice	ORCA2C	9	27	0	[Abramovici 2001]
	ORCA2CA	14	41	0	[Stroud 2002b]
Atmel	AT40K/AT94K	4	56	3	[Sunwoo 2005]
Cypress	Delta39K	20	419	11	[Stroud 2000]
Xilinx	4000E/Spartan	12	128	0	[Stroud 2003]
	4000XL/XLA	12	206	0	
	Virtex/Spartan-II	12	283	11	[Dhingra 2005]
	Virtex-4	15	?	15	[Milton 2006]

the test approach is application-independent external testing or BIST. The testing process consists of (1) configuring the FPGA to test some specific resource (either logic or routing), (2) applying test patterns and analyzing the output responses, (3) reconfiguring to test the target resource in a different mode of operation, and (4) repeating these steps until the target resource is completely tested. There is a significant difference in the number of test configurations needed to completely test logic resources *versus* routing resources, as can be seen in Table 12.3 for some commercially available FPGAs. The number of test configurations needed to completely test routing resources ranges from 3 to more than 23 times the number needed to completely test PLBs. This is primarily because routing resources constitute approximately 80% of the configuration memory bits in most FPGAs, but there are other architectural characteristics that affect the total number of test configurations required [Stroud 2003].

12.3.1 Logic Resources

Early FPGAs consisted only of an array of PLBs and programmable routing resources. As a result, the first testing approaches, including BIST, for logic resources were specifically for PLBs. Over time, however, I/O cells became more complex, and specialized cores with various programmable options were incorporated in the array. Therefore, BIST approaches originally developed for PLBs have been modified and extended to testing I/O cells and specialized cores.

Three BIST approaches for testing logic resources are illustrated in Figure 12.8. In general, multiple identically configured TPGs are implemented to supply test patterns to alternate programmable logic resources under test, which will be referred to as ***blocks under test*** (BUTs). The outputs of the identically configured BUTs are then monitored by one or more comparison-based ORAs. The BUTs are repeatedly

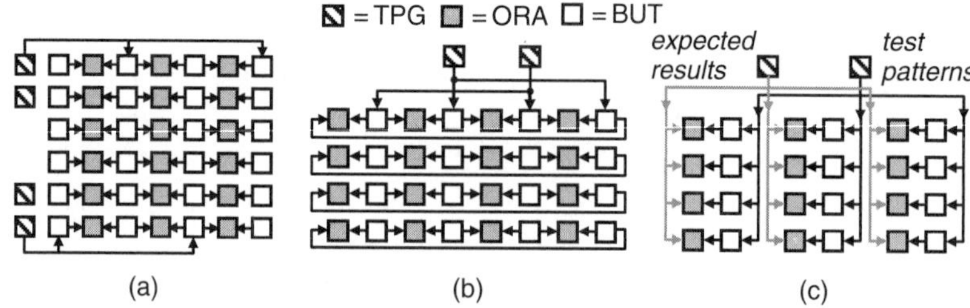

■ **FIGURE 12.8**

FPGA logic resource BIST architectures: (a) basic comparison, (b) circular comparison, and (c) expected results.

reconfigured in their various modes of operation until they are completely tested. A set of test configurations that completely test a set of BUTs in all of their modes of operation is referred to as a **test session**.

Multiple TPGs are used to prevent faults from escaping detection because of a faulty TPG or faulty routing resources interconnecting the TPG and BUTs. With multiple TPGs, a faulty TPG would produce different test patterns from those of the fault-free TPG; therefore, the output responses of the BUTs driven by the faulty TPG would also be different and would be detected by the ORAs. Another important use of multiple TPGs is to control the loading on the TPG outputs, because the number of resources under test can be quite large, as the data in Table 12.1 show. The loading can best be controlled by creating the complete BIST architecture in a small portion of the array and repeating this structure to fill the array [Abramovici 2001]. The use of multiple TPGs also requires a mechanism for synchronizing the TPGs, such as a common reset signal, so that the TPGs apply the same test patterns to their respective BUTs.

In the basic comparison architecture shown in Figure 12.8a, the BUTs in the middle of the array are monitored by two ORAs and compared with the outputs of two different BUTs. Along the edges of the array, the BUTs are monitored by only one ORA. In the circular comparison architecture of Figure 12.8b, all BUTs are monitored by two ORAs and compared to two different BUTs. It has been shown that there are a few pathological cases of multiple faults that can escape detection using the basic comparison architecture [Abramovici 2001], and there are fewer pathological cases with the circular comparison architecture. A final architecture that has been used with success in some cases is the expected results architecture, illustrated in Figure 12.8c, where the ORAs compare the outputs of the BUTs with expected results generated by the TPGs. This architecture is most effective when the expected results can be efficiently generated by the TPG; for example, when testing RAM cores, the expected results are known as part of the test algorithm. However, it is important to have different TPGs generating the test patterns and the expected results to avoid faulty BUTs escaping detection as a result of a faulty TPG. A solution is to have multiple identical TPGs that generate both test patterns and expected results, but to connect the expected results produced by one TPG to

ORAs that compare the output responses from BUTs driven by test patterns from a different TPG.

All of these architectures can be oriented along the columns of the array as shown in Figure 12.8 or can be oriented along the rows of the array by rotating the architectures by 90 degrees. Control of the physical layout is necessary to achieve the desired orientation of the BIST architecture. In most cases, this control cannot be achieved through the normal synthesis design flow and requires the use of FPGA vendor-specific design tools and techniques. The orientation of the architecture is important when using partial reconfiguration to reconfigure the BUTs in their various modes of operation. For example, when the frames of the configuration memory are aligned along the columns of the array, the column oriented architectures of Figure 12.8 give optimum results because the configurations of the TPGs, ORAs, TPG-to-BUT routing, and BUT-to-ORA routing remain unchanged and only the frames containing the BUTs must be written to reconfigure the BUTs for the next test configuration. This minimizes the amount of configuration data that must be downloaded to the FPGA for each test configuration and the total test time.

The programmable resources used to implement the TPGs and ORAs for BIST of the various logic resources are summarized in Table 12.4. The actual TPG design is a function of the logic resources being tested. When implemented in PLBs, a lower bound on the number of PLBs required to implement a TPG, T_{PLB}, is given by:

$$T_{PLB} = B_{IN} \div N_{FF} \tag{12.1}$$

where B_{IN} is the number of inputs to a BUT and N_{FF} is the number of flip-flops in a PLB. However, a TPG can also be implemented by storing test patterns (and expected results) in a RAM core that is then used in a ROM mode of operation with a counter to sequence through the addresses of the ROM; this counter can be implemented in PLBs or a DSP core. Another technique for implementing a TPG is to use a DSP core by itself. Although the accumulator of a DSP can easily be used to implement a counter, another approach is to add a large prime number constant to the accumulator to produce exhaustive test patterns with more transitions in the most significant bits than is the case with a counter [Rajski 1998].

ORAs are most efficiently implemented using PLBs because of the large number of ORAs typically required. The total number of ORAs, N_{ORA}, needed is given by:

$$N_{ORA} = N_{BUT} \times B_{OUT} \tag{12.2}$$

TABLE 12.4 ■ Programmable Resources Used for BIST of Logic Resources

Resource Under Test	TPGs	ORAs
PLBs	PLBs *or* DSP cores	PLBs
LUT RAMs	PLBs *or* DSP and RAM cores	PLBs
I/O cells	PLBs *or* DSP and RAM cores	PLBs
Cores (memories, DSPs, etc.)	PLBs	PLBs

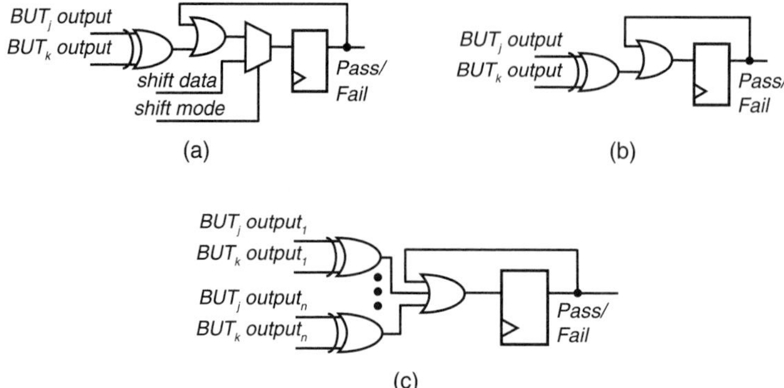

FIGURE 12.9

Comparison-based ORA designs: (a) ORA with shift register retrieval, (b) ORA for configuration readback, and (c) ORA for comparing multiple outputs.

where N_{BUT} is the total number of BUTs and B_{OUT} is the number of outputs per BUT. The ORA can be one of several types of implementations, some of which are summarized in Figure 12.9. In all cases, a comparator is used to detect any mismatches in the output responses of the BUTs caused by faults. Any mismatch is then latched via the flip-flop in conjunction with the feedback and OR gate and held until the end of the BIST sequence where a logic 1 indicates that a mismatch was encountered. The primary ORA design issue is how the contents of the ORAs will be retrieved at the end of the BIST sequence. In the ORA design shown in Figure 12.9a, a shift register mode of operation is incorporated to allow the contents of the ORAs to be shifted out at the end of the BIST sequence. This requires the additional multiplexer functionality, which adds two inputs to each ORA for a total of five inputs including the feedback. Because most FPGAs provide only three-input or four-input LUTs, the ORA with shift register retrieval requires two LUTs and one flip-flop. By using configuration memory readback to retrieve the ORA contents (in those FPGAs that support reading contents of memory elements during configuration readback), the shift register mode can be eliminated as illustrated in the ORA design shown in Figure 12.9b, which requires only one three-input LUT and one flip-flop. As a result, the number of PLBs required to implement ORAs, O_{PLB}, is given by:

$$O_{PLB} = N_{ORA} \div N_{FF} \tag{12.3}$$

Reading back the full configuration memory to obtain the contents of the ORA flip-flops is inefficient in terms of test time. On the other hand, the technique is very effective in FPGAs that support partial configuration memory readback and concentrate the configuration memory bits that contain the contents of the flip-flops in a minimum number of frames. The orientation of the BIST architecture (row or column) with the frames of the configuration memory helps to minimize the time spent retrieving BIST results because only those frames that contain ORA flip-flop contents must be read. BIST results can be retrieved at the end of each BIST sequence or, using dynamic partial reconfiguration, at the end of a test session. Whereas the latter approach reduces overall test time, the tradeoff is some minor loss in diagnostic resolution; the faulty resource(s) can still be identified, but the failing mode of operation of the faulty resource cannot. A final ORA design is illustrated in Figure 12.9c, which compares multiple sets of outputs from the BUTs. This design is effective in FPGAs with LUTs having a large number of inputs, but is also useful for BUTs with a large number of outputs.

12.3.1.1 Programmable Logic Blocks

One of the first BIST architectures for testing PLBs is illustrated in Figure 12.10a where the TPGs are constructed from PLBs [Abramovici 2001]. Both counters and linear feedback shift registers (LFSRs) have been used as TPG functions. A more recent approach uses DSP cores in an accumulator mode of operation with the addition of a large prime constant as the TPG [Milton 2006]. This allows the column of PLBs normally used for the TPGs to implement ORAs instead. Hence, the circular comparison architecture of Figure 12.10b can be implemented. In both PLB BIST architectures, only half of the PLBs can be BUTs at any given time because the other PLBs function as ORAs and TPGs in the case of the basic comparison architecture. As a result, two test sessions are required to completely test all PLBs in the array. Note that the sets of PLB test configurations in Table 12.3 would be applied twice, once for each test session.

After the initial download for a given test session, the BUTs are repeatedly reconfigured in their various modes of operation until they are completely tested. Therefore, it is important to align the BUTs with the orientation of the frames in the configuration memory to minimize the number of frames that must be written to reconfigure the BUTs for subsequent test configurations. This also aligns the ORAs with the frame for retrieving the pass/fail contents of the ORAs after the completion of each BIST sequence, or at the end of each test session, via partial

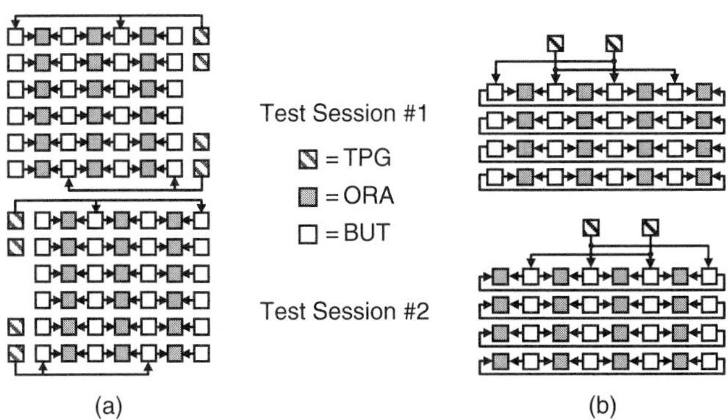

■ **FIGURE 12.10**

PLB BIST architecture: (a) basic comparison and (b) circular comparison.

TABLE 12.5 ■ Test Time Speed-up and Memory Reduction

Download and Readback Method	Test Time Speedup		Memory Reduction	
	Virtex	Virtex-4	Virtex	Virtex-4
Full reconfiguration with full configuration memory readback	1	1	1	1
Partial reconfiguration with partial configuration memory readback	4.6	8.9	3.2	5.3
Dynamic partial reconfiguration with partial configuration memory readback at end of session	5.1	12.9		

configuration memory readback. The resultant test time speedup is illustrated in Table 12.5 for Xilinx Virtex and Virtex-4 FPGAs [Dhingra 2006]. The time required for full reconfiguration and full configuration memory readback at the end of each test configuration is used to normalize the data. The test time speedup obtained for Virtex-4 is better than that obtained in Virtex, by about 2.5 times, the result of better frame organization and multiple frame write capabilities. The table also gives the reduction in the memory storage requirements needed to store the BIST configuration data files when using partial reconfiguration.

The actual test configurations are developed based on the specific PLB architecture. As an example of this process, consider the simple PLB architecture of Figure 12.11, which consists of two three-input LUTs with common inputs, a flip-flop, and four multiplexers. The four multiplexers respectively provide the ability to combine the two LUTs to form a single logic function of four inputs, synchronous

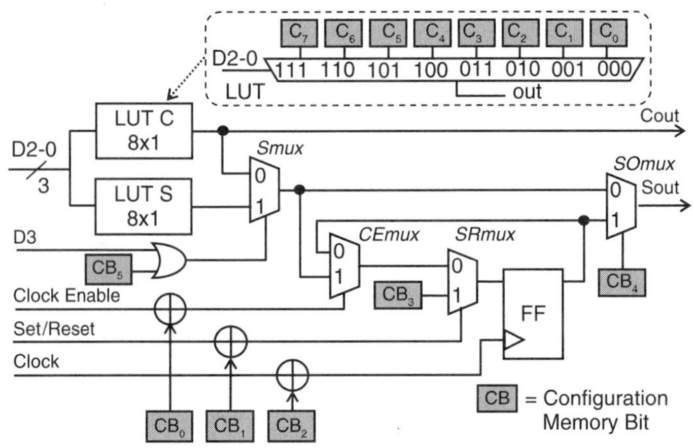

■ **FIGURE 12.11**

Example PLB architecture for test configuration development.

clock enable, synchronous programmable set or reset, and the ability to bypass the flip-flop for a combinational logic function only. Sixteen configuration memory bits define the contents of the two LUTs (C_7–C_0 for LUT C and S_7–S_0 for LUT S) to establish the truth tables for their respective combinational logic functions. For example, LUT C can implement the carry function of a full adder while LUT S implements the sum function. An additional six configuration memory bits (CB_5–CB_0) control functionality of the rest of the PLB. CB_4 controls the output multiplexer to select or bypass the flip-flop for sequential or combinational logic functions, respectively. CB_3 configures the flip-flop to be either set or reset when the set/reset input is active. A logic 1 in CB_5 disables the $D3$ input and forces *Smux* to select the output of LUT S, as would be the case for the full adder implementation example. The final three bits (CB_2–CB_0) drive exclusive-OR functions to define the active levels for the clock enable and set/reset inputs as well as the active edge of the clock, where logic 0's would result in active high control signals and a rising edge clock while logic 1's would invert the control signals and active clock edge.

The primary goal in test configuration development is to minimize the number of test configurations to be downloaded. In this example, a minimum of three test configurations is needed to completely test the PLB for all stuck-at faults. Three possible test configurations are summarized in Table 12.6 in terms of the logic values of the configuration bits as well as the individual and cumulative fault coverage (for 174 total collapsed stuck-at gate-level faults) obtained for each configuration. The input test patterns include all 2^6 possible input values produced either from a counter or an LFSR. Two configurations (configuration 1 and configuration 2) are needed to test the LUTs where opposite logic values are loaded in each LUT for each configuration. In this example, the LUT configuration data implement the truth tables for *exclusive-OR* (XOR) and *exclusive-NOR* (XNOR) functions, which are a good choice for testing LUTs. Note that opposite truth tables are loaded in the two LUTs for each configuration in order to apply opposite logic values to the inputs of the multiplexer controlled by the $D3$ input. Two configurations (configuration 1 and configuration 2) are needed, in which the output multiplexer selects the flip-flop to test the active levels of the control signals to the flip-flops, active edge of the clock, and the programmable set/reset. The last configuration tests the flip-flop bypass.

TABLE 12.6 ■ Test Configurations for Example PLB Architecture

Configuration Bits	Configuration 1	Configuration 2	Configuration 3
C_7-C_0LUT C	01101001 = XNOR	10010110 = XOR	10010110 = XOR
S_7-S_0LUT S	10010110 = XOR	01101001 = XNOR	01101001 = XNOR
CB_5-CB_0	010000	011111	100000
Individual FC	149/174 = 85.6%	149/174 = 85.6%	108/174 = 62.1%
Cumulative FC	149/174 = 85.6%	170/174 = 97.7%	174/174 = 100%

12.3.1.2 Input/Output Cells

The I/O cells have a number of programmable features that not only facilitate user control and specification of the input, output, and bidirectional ports for any given system application but also meet critical system attributes such as timing requirements, voltage standards, and drive capabilities. In newer FPGAs, an increasing amount of programmable logic resources has been incorporated in I/O cells to support system application requirements such as high-speed data transfer. The typical range of programmable resources varies from two flip-flops and 4 multiplexers with 3 I/O standards and 4 delay values in simpler FPGAs to 10 flip-flops and 33 multiplexers with 69 I/O standards and 64 delay values in some of the more complex FPGAs. Hence, the number of test configurations for all of the modes of operation can be large.

Boundary scan, and more specifically the **EXTEST** feature, has traditionally been used to test input and output buffers, tristate control, and pads along with interconnections between devices on a PCB. However, the *EXTEST* capability can only test the bidirectional buffer portion of the programmable I/O cells of an FPGA; it cannot test other programmable logic resources in the I/O cells, such as the flip-flops used for registered inputs and outputs or the programmable routing resources connecting I/O cells to the FPGA core. Although the *boundary scan INTEST* feature could be used to test these programmable logic and interconnect resources, few FPGAs provide *INTEST* capability as it is not a required feature of the IEEE boundary scan standard.

Using the bidirectional capabilities of the I/O cell, the logic BIST architecture can be applied to the I/O cells as illustrated in Figure 12.12 [Vemula 2006]. Test patterns from two identically configured TPGs are applied to the output portion of the I/O cells with the return path through the input portion of the I/O cell to the ORAs. The output responses of identically configured I/O cells under test are compared by ORAs implemented in the PLBs of the FPGA core. The I/O cells under test and the ORAs are arranged to provide a circular comparison such that every I/O cell is observed by two ORAs and compared with two different I/O cells under test. The basic comparison and expected results architectures can also be used for testing I/O cells.

The I/O cells under test are repeatedly reconfigured in their various modes of operation. Because there are more PLBs than I/O cells, all I/O cells can usually be tested in a single test session. This BIST approach also facilitates testing programmable

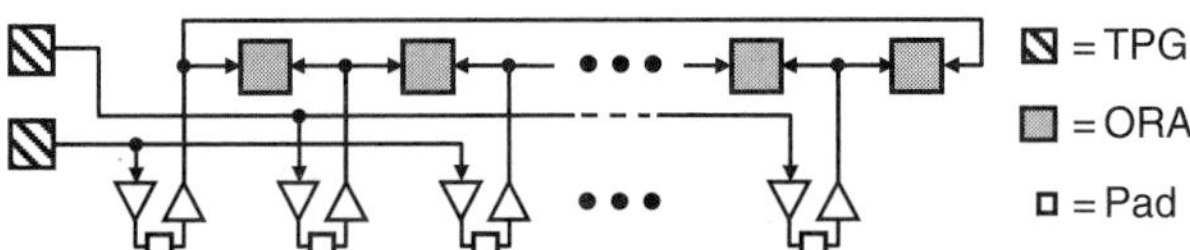

■ **FIGURE 12.12**

I/O cell BIST architecture.

interconnect resources associated with the I/O cells. However, there are some testing limitations because the BIST approach requires that the I/O cells be configured in the bidirectional mode of operation. For example, not all I/O voltage standards can be used in the bidirectional mode and, as a result, cannot be tested with this BIST approach. Furthermore, the potential effects of the bidirectional mode must be considered for the system-level application of BIST for I/O cells as other components may be driving the nets. Differential I/O can be tested using this approach in those FPGAs where the pair of I/O cells can be configured as bidirectional while in the differential mode.

12.3.1.3 Specialized Cores

Current FPGAs typically incorporate multiple regular structure cores such as memories (RAMs and FIFOs), multipliers, and DSPs. To maximize the efficiency of these devices in a large variety of system applications, these regular structure cores are also programmable. For example, RAM cores can be programmed for a range of different address and data bus widths, synchronous and asynchronous write and read modes, as well as single and dual port operation. These cores must be tested in their various modes of operation. All three BIST architectures for programmable logic resources illustrated in Figure 12.8 have been used successfully to test specialized cores embedded in FPGAs. However, the TPGs for testing these cores are often more complex than those for testing PLBs or I/O cells because specific, complex test algorithms are needed.

One of the most efficient RAM test algorithms currently in use, in terms of test time and fault detection capability, is the March LR algorithm [Van de Goor 1996] summarized in Table 12.7. This algorithm has a test time on the order of 16N, where N is the number of address locations. In addition to classical stuck-at faults, this algorithm is capable of detecting pattern sensitivity faults, intraword coupling faults, and bridging faults in the RAM. For word-oriented memories, a ***background data sequence*** (BDS) must be added to detect these faults within each word of the memory [Hamdioui 2004]. The March LR with BDS given in Table 12.7 is for a RAM with 2-bit words, but in general the number of background data sequences, N_{BDS}, is given by:

$$N_{BDS} = \lceil \log_2 K \rceil + 1 \tag{12.4}$$

where K is the number of bits per word. When the RAM core can be configured in a 1-bit/word mode of operation, the March LR without BDS will test for pattern sensitivity and coupling faults. However, applying March LR with BDS to the largest data width access to the RAM will also detect bridging faults in the parallel data bus into and out of the RAM. Furthermore, some FPGA RAM cores include additional space for parity that cannot be accessed in the 1-bit/word mode such that the March LR with BDS must be applied to those modes that give access to the additional memory storage for parity. As a result, it is most efficient in terms of test time to apply March LR with BDS to the largest data width mode of the RAM. Once the memory core has been tested for pattern sensitivity, transition, and coupling

TABLE 12.7 ■ RAM Test Algorithms

Test Algorithm	March Test Sequence
March DPR	↕(w0:-); ↑(r0, w1, r1:↓r); ↓(r1, w0, r0:↑r); ↕(r0:-);
MATS+	↕(w0); ↑(r0, w1); ↓(r1, w0)
March X	↕(w0); ↑(r0, w1); ↓(r1, w0); ↑(r0)
March Y	↕(w0); ↑(r0, w1, r1); ↓(r1, w0, r0); ↑(r0)
March LR w/o BDS	↕(w0); ↓(r0, w1); ↑(r1, w0, r0, r0, w1); ↑(r1, w0); ↑(r0, w1, r1, r1, w0); ↑(r0)
March LR with BDS	↕(w00); ↓(r00, w11); ↑(r11, w00, r00, r00, w11); ↑(r11, w00); ↑(r00, w11, r11, r11, w00); ↑(r00, w01, w10, r10); ↑(r10, w01, r01); ↑(r01)
March S2pf-	↕(w0:n); ↑(r0:r0, r0:-, w1:r0); ↑(r1:r1, r1:-, w0:r1); ↓(r0:r0, r0:-, w1:r0); ↓(r1:r1, r1:-, w0:r1); ↓(r0);
March D2pf	↕(w0:n); $\uparrow_{c=0}^{C-1} (\uparrow_{r=0}^{R-1}(w1_{r,c}{:}r0_{r+1,c},\ r1_{r,c}{:}w1_{r-1,c},\ w0_{r,c}{:}r1_{r-1,c},\ r0_{r,c}{:}w0_{r+1,c}))$; $\uparrow_{c=0}^{C-1} (\uparrow_{r=0}^{R-1}(w1_{r,c}{:}r0_{r,c-1},\ r1_{r,c}{:}w1_{r,c-1},\ w0_{r,c}{:}r1_{r,c+1},\ r0_{r,c}{:}w0_{r,c+1}))$;
Notation:	w0 = write 0 (or all 0's) r1 = read 1 (or all 1's)
portA:portB	↑ = address up ↓ = address down ↕ = address either way

faults, a simpler test algorithm—such as MATS+ (a $5N$ test), March X (a $6N$ test), or March Y (an $8N$ test) from Table 12.7—can be applied for the other address and data width modes of operation to test the programmable address decoding circuitry. Usually, the number of test configurations needed to test the different address and data width options is sufficient to also test all other programmable features such as initialization values, active level for control signals, active edge of clocks, and synchronous/asynchronous operation.

For RAM cores that support true dual-port operation, with two separate address and data ports that can access any location in the memory, the March LR test algorithm is applied to each port in turn to test the single port operation followed by the March S2pf- and D2pf algorithms, summarized in Table 12.7, to test the dual port operation [Hamdioui 2002]. Some RAMs are said to have a "dual-port" mode of operation that is not truly dual port; these RAM simply provide one port for write operations and another port for read operations. This is typical in LUTs that can function as a small RAM. For this case, a simple test algorithm developed for testing these so-called dual-port RAMs is given in Table 12.7 and denoted March DPR [Stroud 2003].

Some FPGAs also incorporate FIFO memories or, more often, RAMs that can also be configured in FIFO modes of operation. In the latter case, it is best to initially test the memory in RAM mode of operation using March LR with BDS to detect pattern sensitivity and coupling faults in the memory core. Subsequent configurations can test the FIFO modes of operation as well as the full and empty flags associated with FIFO operation. Many FPGA FIFOs support programmable "almost full" and "almost empty" flags, and these must be tested as well. The main problem with programmable "almost" full and empty flags is that the FIFO must be reconfigured many times just to set the values at which these flags will go active. The following March X algorithm for testing the FIFO addressing logic and associated flags is based on the algorithm in [Atmel 1999].

March X FIFO Test Algorithm (assuming *N* address locations):

Step 1. Reset the FIFO, check that Empty flag is active.

Step 2. Repeat *N* times: write FIFO with all 0's, check that Empty flag goes inactive after the first write cycle, Full flag goes active after last write cycle, and that Almost Empty flag goes inactive and Almost Full flag goes active at the appropriate points in the sequence. (Partial reconfiguration can be used at this point to set the next value of the "almost" full flag before proceeding with additional write cycles; repeat this process for each "almost" full flag value to be tested while proceeding through the write sequence.) Perform one additional write if the FIFO has a Write Error signal to indicate an attempted write when the FIFO is full.

Step 3. Repeat *N* times: read FIFO expecting all 0's and write FIFO with all 1's, check that Full flag toggles after each read and write cycle.

Step 4. Repeat *N* times: read FIFO expecting all 1's and write FIFO with all 0's, check that Full flag toggles after each read and write cycle.

Step 5. Repeat *N* times: read FIFO expecting all 0's, check that Full flag goes inactive after first read cycle, Empty flag goes active after last read cycle, and that Almost Empty flag goes active and Almost Full flag goes inactive at the appropriate points in the read sequence. (Partial reconfiguration can be used to set the next value of the "almost" empty flag before proceeding with additional read cycles; repeat this process for each "almost" empty flag value to be tested while proceeding through the read sequence.) Perform one additional read if FIFO has a Read Error signal to indicate an attempted read when the FIFO is empty.

More recently, FPGAs have incorporated ***error correcting code*** (ECC) RAMs to tolerate errors, including ***single event upsets*** (SEUs), by adding Hamming code generation circuitry at the input to the RAM along with detection and correction circuitry at the output of the RAM as discussed in Chapter 3. Hamming code bits are generated over the incoming data and then stored in the RAM with the data bits. At the output of the RAM, the Hamming code is regenerated for the data bits as discussed in Chapter 3 (Section 3.4.3.2) and illustrated in Figure 3.25 and compared to the stored Hamming code to identify single bit errors for ***single-error-correction*** (SEC). A parity bit across the entire code word (data bits plus Hamming bits) is

TABLE 12.8 ■ Fault Coverage for Hamming Code Circuits in 64-bit ECC RAM

Circuit	Vectors	# Vectors	Fault Coverage		
			Pin	Gate	Cum.
ECC Generate	All 0's; walk 1-thru-0's	65	100%	87.7%	87.7%
	All 1's	1	50%	26.5%	93.9%
	Walk two 1's-thru-0's	2016	99.9%	99.6%	100%
ECC Detect and Correct	Output of ECC generate vectors	2082	56%	58.4%	58.4%
	Init: Walk 1-thru-0's; all 1's; all Hamming values w/ data = 0's;	321	100%	95.2%	98.1%
	Init: Walk two 1's-thru-0's	2556	73.5%	71.9%	100%

Note: ***Init*** indicates that vectors are initialized in ECC RAM during download.

included to provide ***double error detection*** (DED) as summarized in Table 3.4. ECC RAMs pose the interesting problem of detecting faults in a circuit designed to be fault-tolerant.

Any parity tree can be completely tested for all gate-level stuck-at faults in only four vectors if one knows the connections of the exclusive-OR gates that form the tree [Mourad 1989]. Unfortunately, only the manufacturer may know the connections for the various parity trees in the generate and detect/correct circuits. Regardless, complete testing can be obtained as shown in Table 12.8 for different sets of test vectors that target the generate and detect/correct circuits in an ECC RAM with 64 data bits, seven Hamming bits for SEC, and an overall parity bit for DED. Pin fault coverage considers only the inputs and output of the exclusive-OR gate, whereas gate-level fault coverage considers the internal gate structure (two NOR gates and an AND gate) of the exclusive-OR. The cumulative gate-level fault coverage (denoted "Cum." in the table) is given for the progression through the various sets of test vectors.

The first set of vectors for the ECC generate circuit includes the all 0's test pattern and walking a logic 1 through a field of 0's for a total of 65 vectors. This small set of vectors will detect 100% of the stuck-at pin faults for exclusive-OR gates regardless of the connections in the actual parity trees [Jone 1994]. This is because a logic 1 is propagated through every path from the inputs to the output (this detects all single stuck-at-0's), whereas the all 0's pattern detects all single stuck-at-1 faults. However, when testing pin faults of an exclusive-OR gate, only three combinations {01, 10, and either 00 or 11} are needed to detect each input and the output stuck-at-0 and stuck-at-1 [Mourad 1989]. In actual gate-level implementations of an exclusive-OR function, all four vectors are needed for complete testing; this can be seen in Table 12.8, where only 87.7% of the gate-level stuck-at faults were detected with the same set of 65 vectors. The all 1's vector can be easily applied to the exclusive-OR gates at the inputs to the parity tree, which increases the fault coverage to 93.9% with a total of 66 vectors, but the remaining gates in the parity tree see all 0's from the outputs of the first level exclusive-OR gates. One way to apply the {11} vector to the input of all exclusive-OR gates independent of the internal connections is to generate all possible combinations of two logic 1's in a field of 0's, which, when combined with the other 66 vectors, gives 100% cumulative gate-level stuck-at fault coverage [Jone 1994], as Table 12.8 shows. This combined set of test patterns also detects all multiple stuck-at faults and bridging faults in the parity-based ECC generate circuit.

To detect faults in the ECC detect/correct circuit, error conditions must be created to decode and correct the errors. The input values to this circuit are the data stored in the RAM, which include Hamming bits produced by the generate circuit. As a result, these data values do not include error conditions unless there is a fault in the RAM or in the ECC generate circuit. This result can be seen from the data in Table 12.8, where applying the outputs of the complete set of vectors to the generate circuit obtains only 56% pin fault coverage and 58.4% gate-level fault coverage. The only way to create error conditions is to initialize the ECC RAM (when downloading the configuration) with data and Hamming values that apply error conditions to the ECC detect/correct circuit. By including the all 0's, all 1's, and walking a logic 1 through a field of 0's in the initialization to the ECC RAM during download, 100% pin fault coverage and 98.1% gate-level fault coverage can be obtained. During test application, the ECC RAM is first read to apply these initialized values to the ECC detect/correct circuit. Next, the input vectors are applied to the ECC generate circuit, whose outputs are stored in the RAM and then read for application to the ECC detect/correct circuit. To detect the remaining faults in the ECC detect/correct circuit, independent of the internal connections in the parity trees, we can once again generate every combination of two logic 1's in a field of logic 0's. Unfortunately, these vectors can only be applied by initializing the ECC RAM during configuration of the FPGA such that, depending on the size of the RAM, additional downloads may be required for the complete set of required initialization values. Fortunately, because the TPGs, ORAs, and routing remain constant, a partial reconfiguration will reduce the test time.

Similarly, DSP and multiplier cores are tested algorithmically because the number of inputs makes exhaustive testing impractical. However, a number of test algorithms have been developed for multipliers [Gizopoulos 1998] and multiplier/accumulators [Rajski 1998] that are fundamental components of most DSP cores in current FPGAs. Multiple test configurations are needed to test the associated programmable modes for active edges of clocks, active levels of clock enables and resets, registered inputs and outputs, number of pipelining stages, and cascaded modes of interconnection [Stroud 2005a].

12.3.1.4 Diagnosis

The three BIST architectures for testing programmable logic resources in Figure 12.8 also provide good diagnostic resolution for the identification of faulty resources based on the BIST results. Diagnosis is trivial for the expected results architecture of Figure 12.8c because the failing ORAs indications correlate to the faulty resources, without any additional processing of the BIST results. The basic comparison and circular comparison architectures, on the other hand, require a diagnostic procedure to process the BIST results and identify the faulty resources. A relatively straightforward, tabular diagnostic procedure was developed for the basic comparison architecture of Figure 12.8a [Abramovici 2001]. In that approach, only one ORA monitors the PLBs along the edge of the array. As a result, the diagnostic resolution is lower along the edges of the array, making unique diagnosis of a set of faulty PLBs difficult and sometimes requiring additional BIST configurations. The diagnostic procedure was later extended for use in online BIST of PLBs where a sequence of separate test configurations with two BUTs and one ORA resulted in a small circular array of four BUTs and four ORAs when superimposed [Abramovici 2004]. When the diagnostic procedure was later extended for use when testing the RAMs and multipliers in FPGAs, the circular comparison of Figure 12.8b was found to improve diagnostic resolution over that of the comparison-based BIST architecture of Figure 12.8a as there are no edges where BUTs are observed by only one ORA [Stroud 2005a] [Stroud 2005b].

An important assumption in the original diagnostic procedure was that there are no more than two consecutive BUTs with equivalent faults, where consecutive BUTs in this context means that they are monitored by a common set of ORAs. However, the diagnostic procedure was extended to indicate when this situation may have occurred in the circular comparison architecture in order to reconfigure the BIST architecture to obtain a unique diagnosis. This enhanced tabular diagnostic procedure can be applied to both the basic comparison and circular comparison architectures and is applied as follows (refer to example A in Table 12.9 while reading the procedure to observe the application of steps 1 to 3):

TABLE 12.9 ■ Diagnostic Procedure Examples

	Example A						Example B						Example C					
Arch	*Basic*			*Circular*			*Basic*			*Circular*			*Basic*			*Circular*		
Step	*1*	*2*	*3*	*1*	*2*	*3*	*1*	*2*	*3*	*1*	*2*	*3*	*1*	*2*	*3*	*1*	*2*	*3*
B_1		0	0		0	0		0	0		0	0			1		0	0
O_{12}	0	0	0	0	0	0	0	0	0	0	0	0	1	1	1	1	1	1
B_2		0	0		0	0		0	0		0	0		0	0		0	0
O_{23}	0	0	0	0	0	0	0	0	0	0	0	0	0	0	0	0	0	0
B_3		0	0		0	0		0	0		0	0		0	0		0	0
O_{34}	1	1	1	1	1	1	1	1	1	1	1	1	0	0	0	0	0	0
B_4			1			1			1			1		0	0		0	0
O_{45}	0	0	0	0	0	0	1	1	1	1	1	1	1	1	1	1	1	1
B_5			1			1			?			1			1		0	0
O_{56}	1	1	1	1	1	1	1	1	1	1	1	1	0	0	0	0	0	0
B_6			?		0	0			?		0	0			1		0	0
O_{61}				0	0	0				0	0	0				0	0	0

Logic Resource Diagnostic Procedure

Step 1. Record the ORA results and set the faulty/fault-free status of all BUTs to unknown, as indicated by an empty entry in the table.

Step 2. For every set of two or more consecutive ORAs with 0's (*i.e.*, column 1, Basic example A), enter a 0 for all BUTs observed by these ORAs to indicate that those BUTs are fault-free (*i.e.*, column 2, Basic example A).

Step 3. For every adjacent 0 and 1 followed by an empty entry (*i.e.*, column 2, Basic example A), enter a 1 to indicate that the BUT is faulty (*i.e.*, column 3, Basic example A). This step is recursively applied while such entries exist.

Step 4. If an ORA indicates a failure but both BUTs monitored by the ORA are determined to be fault-free, this is referred to as an ORA inconsistency [Abramovici 2001]. In this case, one of the following three conditions exist: (a) there is a fault in the routing resources between one of the BUTs and the ORA, (b) the ORA is faulty, or (c) there are more than two consecutive BUTs with equivalent faults (for circular comparison only). Condition A or condition B exists if there is only one ORA inconsistency. However, if there are multiple ORA inconsistencies in the circular comparison architecture, then condition C may exist. In the latter case, reconfigure the circular comparison order, and repeat the test and diagnostic procedure.

Step 5. If all BUTs have been marked as faulty or fault-free, then a unique diagnosis has been obtained; otherwise, any BUT that remains marked as unknown may be faulty. In the latter case, reconfigure the circular comparison order to compare different BUTs, or rotate the basic comparison architecture by 90°; then repeat the test and diagnostic procedure.

Three examples of the application of this diagnostic procedure are illustrated in Table 12.9, where O_{ij} denotes an ORA comparing the outputs of BUTs B_i and B_j. As the examples for the basic comparison architecture show, the BUTs with unknown status (indicated by a "?" in the table) are located near the edges of the array where diagnostic resolution is lower because the BUTs are observed by only one ORA. In example A, it was determined that BUTs B_4 and B_5 are faulty and have equivalent faults as a result of their common ORA indicating no mismatch. In example B, it can be determined that at least one of the two BUTs, B_5 and B_6, is faulty. These ambiguities in the diagnosis can be removed by rotating the BIST architecture by 90° where rows of ORAs are comparing rows of BUTs, such that the sets of BUTs being compared are orthogonal, and reapplying the diagnostic procedure to the new BIST results [Abramovici 2001]. This improves diagnostic resolution at the cost of doubling the test time, but a unique diagnosis can be obtained for almost any combination of faulty BUTs. The circular comparison architecture, on the other hand, results in fewer ambiguities leading to a unique diagnosis in more cases. Example C illustrates the reason for the assumption of no more than two consecutive BUTs with equivalent faults when applying this diagnostic procedure to the basic comparison architecture; otherwise, the diagnosis is incorrect. The two

ORA inconsistencies observed in the circular comparison architecture, on the other hand, indicate that there may be more than two consecutive BUTs with equivalent faults and that the circular comparison order should be modified and the test and diagnosis repeated.

The simplicity of this tabular diagnostic procedure facilitates straightforward implementation in, and execution by, an embedded processor core [Stroud 2005b]. Although the new enhancements to the diagnostic procedure in steps 4 and 5 remove the previous assumption of no more than two consecutive BUTs with equivalent faults for the circular comparison architecture, it should be noted that finding a minimum set of circular comparison configurations that guarantee unique diagnosis remains an open problem. A unique diagnosis of the faulty BUT(s) provides sufficient information for reconfiguration of the system function to avoid the fault in coarse-grain fault-tolerant applications. However, finer diagnostic resolution can be obtained with additional test configurations to determine if the fault will not affect operation so that the system configuration can continue to use the faulty resource [Abramovici 2004].

12.3.2 Interconnect Resources

Because the programmable routing network in the FPGA is used to interconnect the TPGs, BUTs, and ORAs when testing logic resources, many faults in routing resources used by those test configurations are detected. In some cases, an interconnect fault can result in an ORA inconsistency during the diagnostic procedure. However, because the programmable interconnect accounts for a large area of the FPGA and a large portion of the configuration memory, it is difficult to determine the thoroughness and quality of interconnect fault detection during BIST for programmable logic resources. Therefore, dedicated programmable interconnect network test configurations are used to ensure effective testing of the routing resources as well as to target the fault models specific to these resources. Similar to BIST for logic resources, BIST for routing resources generally consists of configuring some of the logic resources as TPGs and ORAs while configuring the routing resources (wire segments and PIPs) as wires under test. However, unlike the arrays of identical programmable logic resources, the programmable interconnect network in an FPGA consists of different types of PIPs and wire segments of various lengths associated with global and local, as well as vertical and horizontal, routing resources. Because of the large number of wire segments and PIPs, only a small portion of the routing resources can be under test in any given test configuration. As a result, the sets of wires under test are repeatedly reconfigured to test the various routing resources in the FPGA. The total number of test configurations required to completely test the routing resources depends on the complexity of the interconnect network as well as the PLB architecture used to construct the TPGs and ORAs. Depending on the complexity of the interconnect network, the number of test configurations for routing resources is generally an order of magnitude larger than the number needed to test the logic resources, as illustrated in Table 12.3.

Two general BIST approaches have proven to be effective in testing the programmable interconnect resources in FPGAs including the wire segments, PIPs, and

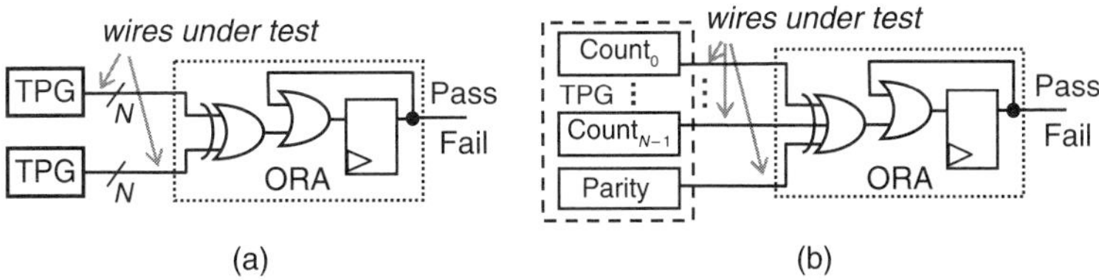

■ **FIGURE 12.13**

FPGA routing BIST architectures: (a) comparison based and (b) parity based.

configuration memory bits that control the PIPs. The first BIST approach for routing resources was comparison based, as illustrated in Figure 12.13a [Stroud 1998]. In this approach, one or more TPGs source exhaustive test patterns over two sets of N wires under test that are monitored at the destination end by comparison-based ORAs to detect any mismatches caused by faults [Stroud 2002b]. Once again, the use of multiple TPGs reduces the probability of faults in the PLBs used to construct the TPGs from causing faults in the interconnect network to escape detection. One potential problem with this approach is that equivalent faults in the two sets of wires under test can escape detection. This problem can be overcome if each set of wires under test is tested more than once and compared to different sets of wires under test. However, this increases the number of test configurations that must be downloaded and applied.

The other BIST approach is parity based as illustrated in Figure 12.13b. In this approach, the TPG sources exhaustive test patterns over one set of N wires under test and produces a parity bit that is also sent to the ORA. The ORA performs a parity check function to detect faults [Sun 2000]. One problem with the original approach was that the parity bit was assumed to be routed over fault-free interconnect resources. However, the faulty/fault-free status of the routing resources is unknown at the start of testing. This approach was later modified to send the parity bit over a wire under test for a total of $N+1$ wires under test and to incorporate multiple types of TPGs and ORAs (for example, countup with even parity and countdown with odd parity) to produce opposite logic values needed to detect PIP stuck-on faults and bridging faults [Sunwoo 2005].

The detailed implementation of these routing BIST approaches varies considerably depending on the types of PIPs, wire segments, and faults targeted for testing. For example, when testing global routing resources, multiple ORAs can be located along the set of wires under test, as illustrated in Figure 12.14a, to test the global-to-local routing resource connections along with the global interconnect. When testing local routing resources, on the other hand, wires under test may be routed through PLBs (through the LUT bypassing the flip-flop) or to adjacent PLBs, as illustrated in Figure 12.14b and Figure 12.14c, respectively. These basic implementations must be reconfigured a number of times to test all of the routing resources along the potential paths. In addition, because of the directional nature of multiplexer PIPs used in most current FPGA interconnect networks, these routing BIST architectures must be flipped about the vertical axis to test routing resources with signal

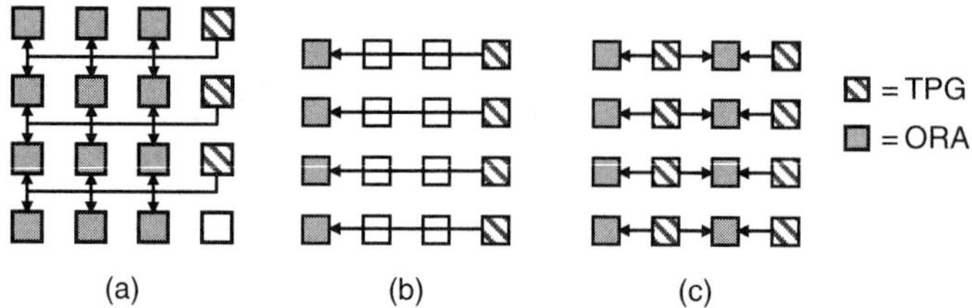

■ **FIGURE 12.14**

Example routing BIST implementations: (a) global routing, (b) local routing, and (c) local routing adjacent PLBs.

flow in the opposite direction. Finally, these routing BIST architectures must also be rotated to test the vertical routing resources. Hence, the total number of routing BIST configurations tends to be large. Although partial reconfiguration can reduce the total time required to download the complete set of test configurations, the total time to test the programmable interconnect resources remains a dominant factor in FPGA testing.

By filling the FPGA with many small routing BIST circuits consisting of independent TPGs, ORAs, and sets of wires under test in the FPGA, the diagnostic resolution based on the BIST results can be improved because an ORA indicating the presence of a fault also identifies the self-test area containing the fault. For example, consider the local routing BIST implementations illustrated in Figure 12.14b and Figure 12.14c. A failing ORA indication in the PLB feed-through implementation isolates the fault within the horizontal routing resources being tested along that row of PLBs. In the adjacent PLBs implementation of Figure 12.14c, a failing ORA indication isolates the fault to the neighboring PLBs. For manufacturing yield enhancement, this resolution may be sufficient for failure mode analysis or for reconfiguring around the faulty area in coarse-grain fault-tolerant system applications. However, for fine-grain fault-tolerant system applications, better diagnostic resolution requires additional test configurations to subdivide the set of wires under test for identification of the faulty wire under test [Stroud 2002b]. Additional test configurations can then be applied to reroute the suspected faulty wire under test for identification of the specific faulty wire segment or PIP.

To illustrate the development of BIST configurations for the programmable routing resources, assume that each example PLB of Figure 12.11 has an associated set of multiplexer PIPs, as illustrated in Figure 12.15. The PLB and its routing resources form a unit cell, which is repeated in an $N \times M$ array to form the core of an FPGA. In this example, there are two input signals sourced from the unit cell above (or north) denoted *NI0* and *NI1*. Similarly there are two input signals from each of the other three directions (east, west, and south), denoted *EI0*, *EI1*, *WI0*, *WI1*, *SI0*, and *SI1*, respectively. As a result, there are a total of 8 input signals from adjacent PLBs to the set of multiplexer PIPs. When combined with the two outputs (*Sout* and *Cout*) of the PLB, 10 possible signals can be selected for inputs (*D0* through *D3*) to the PLB, as illustrated in Figure 12.15a, or selected for outputs to be routed to adjacent unit cells, as illustrated in Figure 12.15b. Each multiplexer PIP has 8 inputs and is

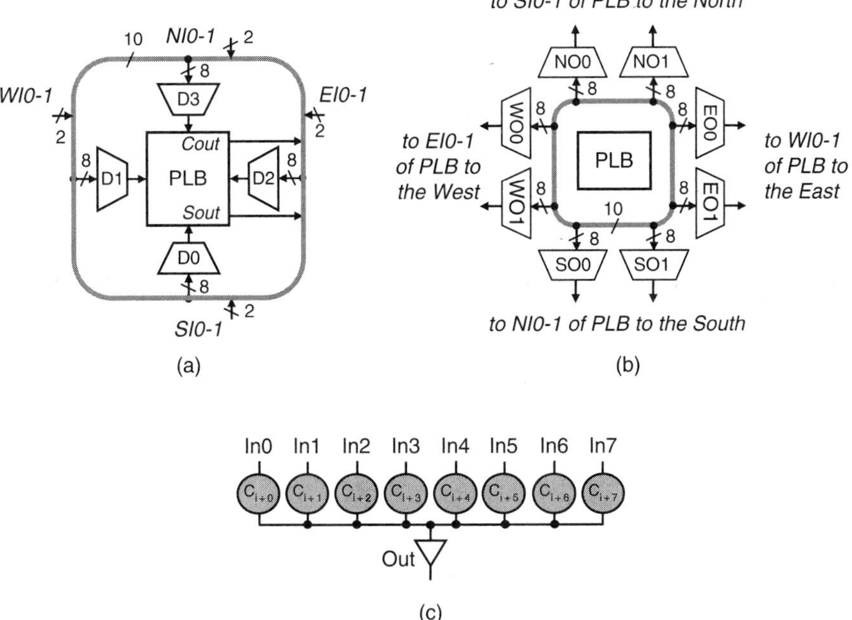

■ **FIGURE 12.15**

Programmable routing resources for simple FPGA example: (a) routing resources for PLB inputs, (b) routing resources for PLB outputs, and (c) multiplexer PIP for PLB routing resources.

constructed as a nondecoded multiplexer with eight associated configuration bits as illustrated in Figure 12.15c. The configuration bits for each of the 12 multiplexer PIPs that form the programmable routing resources in a unit cell are summarized in Table 12.10 in terms of the input that would be selected for a given multiplexer when the associated configuration bit is activated (set to a logic 1); at most, 1 configuration bit will be activated for a given multiplexer PIP. Note that the configuration bit values specified in Table 12.10 are with respect to the specified ordering of the configuration bits given in the table; for example, *CB7* is the most significant bit (MSB) and *CB0* is the least significant bit (LSB) for the multiplexer producing *D0*. The routing resources for a unit cell require 96 configuration bits compared to only 22 for the PLB, which conforms with the fact that routing resources account for about 80% of the total configuration bits in a typical FPGA.

To provide feedback for sequential logic functions such as counters, the *Sout* output of the PLB is also an input to the input multiplexer PIPs *D0* through *D3*. The output multiplexer PIPs provide the selection of signals to be sent to each of the four adjacent unit cells such that there are two output signals per side of the unit cell, denoted *NO0* and *NO1* for the north side, *EO0* and *EO1* for the east side, *WO0* and *WO1* for the west side, and *SO0* and *SO1* for the south side of the PLB. Each output multiplexer can select either of the two outputs of the PLB *(Cout* or *Sout)* or six of the eight signals coming into the unit cell. In the latter case, those

TABLE 12.10 ■ Configuration Bits for Multiplexer PIPs of Simple FPGA

MUX Out	Config Bits	MUX PIP Inputs MSB-Activated Configuration Bits LSB							
D0	CB7-0	NI0	NI1	EI0	EI1	WI0	WI1	SI0	Sout
D1	CB15-8	NI0	NI1	EI0	EI1	WI0	Sout	SI0	SI1
D2	CB23-16	NI0	NI1	EI0	Sout	WI0	WI1	SI0	SI1
D3	CB31-24	NI0	Sout	EI0	EI1	WI0	WI1	SI0	SI1
NO0	CB39-32	Cout	Sout	EI0	EI1	WI0	WI1	SI0	SI1
NO1	CB47-40	Cout	Sout	EI0	EI1	WI0	WI1	SI0	SI1
EO0	CB55-48	NI0	NI1	Cout	Sout	WI0	WI1	SI0	SI1
EO1	CB63-56	NI0	NI1	Cout	Sout	WI0	WI1	SI0	SI1
WO0	CB71-64	NI0	NI1	EI0	EI1	Cout	Sout	SI0	SI1
WO1	CB79-72	NI0	NI1	EI0	EI1	Cout	Sout	SI0	SI1
SO0	CB87-80	NI0	NI1	EI0	EI1	WI0	WI1	Cout	Sout
SO1	CB95-88	NI0	NI1	EI0	EI1	WI0	WI1	Cout	Sout

signals would pass through this unit cell on to an adjacent unit cell, which provides the primary mechanism for routing between nonadjacent cells in our simple FPGA.

Any given multiplexer PIP in this example architecture will require a minimum of eight test configurations. For complete testing with eight configurations, each configuration must test a different PIP for a stuck-off fault and at least one other PIP for a stuck-on fault. The PIP stuck-off fault is tested by a simple continuity test in which both logic 0 and logic 1 are passed through the activated PIP in the signal path from TPG to ORA. To test for an unselected input PIP stuck-on, we must apply opposite logic values to those being passed through the activated PIP for that multiplexer such that if the PIP were stuck-on, the opposite logic value would affect the logic value at the input to the buffer. If only one PIP is tested for the stuck-on fault during each configuration, then a different PIP must be tested during each of the eight test configurations. One difficulty in developing these test configurations is routing the appropriate opposite logic values to the unselected multiplexer input whose PIP is to be tested for the stuck-on fault.

Theoretically, a minimum of eight test configurations would be required to test all of the multiplexer PIPs concurrently, but this is not possible in practice. For example, because any given PLB can function as TPG or ORA, separate sets of test configurations must be applied to test the multiplexers at the inputs to the PLB while functioning as an ORA and to test the output multiplexers when the PLB functions as a TPG. Furthermore, the interconnection pattern (as summarized in Table 12.10 for our simple FPGA) prevents testing all of the multiplexer PIPs concurrently. For example, assume we implement the comparison-based routing BIST architecture illustrated in Figure 12.13a using the test configuration implementation in Figure 12.14c. In this case, only two of the input multiplexers will be tested during any given configuration. The PLBs functioning as ORAs could implement the logic equation $(D0 \oplus D3) + D1$ in the combined four-input LUT. As a result, only input multiplexers for *D0* and *D3* are tested, while the multiplexer for *D1* selects the feedback from *Sout* and the multiplexer for *D2* is unused. Assuming the TPGs to the left and right of the ORA are producing opposite logic values on their two outputs, then four test configurations can be implemented in this arrangement with the following sequence of inputs selected for the two multiplexer PIPs under test: *D0* = {*EI0*, *EI1*, *WI0*, *WI1*} and *D3* = {*WI0*, *WI1*, *EI0*, *EI1*}. In this case, we will have tested four of the eight total inputs for both stuck-on and stuck-off faults in these two multiplexers. By rotating the architecture, we can test the north and south inputs to these multiplexers in an additional four configurations. A similar set of eight configurations would be needed to test the multiplexers for inputs *D1* and *D2*. Note that the *Sout* input to these four multiplexers cannot be tested when the PLB is an ORA, but instead it can be tested by passing signals through the LUTs using the PLB feed-through implementation illustrated in Figure 12.14b.

12.4 EMBEDDED PROCESSOR-BASED TESTING

The ability of embedded processor cores (either hard or soft cores) to write and read the configuration memory via an ***internal configuration access port*** (ICAP) facilitates reconfiguration from within the FPGA. If the processor and program memory are dedicated resources with access to the configuration memory that does not require any additional programmable resources, then the programmable resources of the FPGA can be tested in one test session [Sunwoo 2005]. Otherwise, two test sessions must be applied to the FPGA, as illustrated in Figure 12.16. In each test session, the resources in half of the array are under test, while the other half of the array implements the processor core, TPGs, and interface circuitry to the ICAP [Milton 2006]. Once half of the array is tested, the positions are swapped through a full reconfiguration of the FPGA from external sources, and the other half of the array is tested. Although this doubles the number of test configurations, the overall test time is reduced because internal reconfiguration by the processor core is performed at a much higher clock rate than downloading (usually by a factor of 5 to 25). In addition, the embedded processor core has parallel access to the configuration memory (typically 8, 16, or 32 bits).

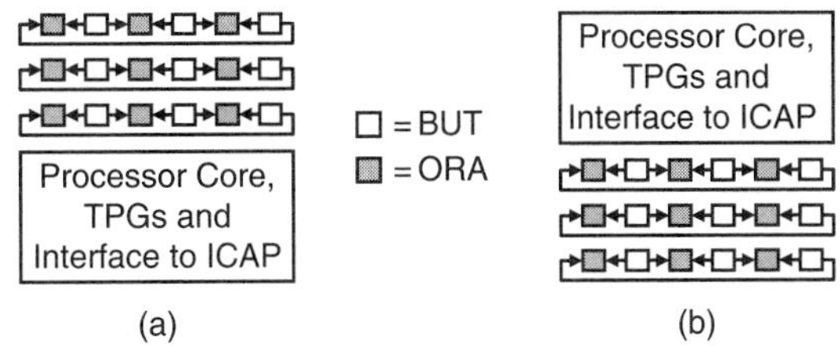

■ **FIGURE 12.16**

Soft core embedded processor-based BIST for FPGAs: (a) session 1 and (b) session 2.

Locating the TPGs in the half of the array with the processor core facilitates implementation of the circular comparison BIST architecture for improved diagnostic resolution. The processor core can also perform the TPG function in some cases, but this introduces the problems associated with a single TPG such as excessive loading and faulty resources under test escaping detection when the TPG is faulty. The latter problem raises the issue of what happens when the processor core or the programmable resources used to construct the processor core are faulty. Although this remains an open problem in embedded processor based BIST of FPGAs, one approach is to emulate faults by manipulating configuration memory bits in the BUTs to create a set of expected failures. This provides a sanity check that the embedded processor is working.

Two approaches have been developed for implementing BIST using embedded processors. In one approach, the BIST configurations are algorithmically generated from within the embedded processor. In this approach, a relatively small program is stored in the program memory of the embedded processor core, which is then used to generate the complete initial BIST configuration as well as to reconfigure the resources under test for the subsequent test configurations [Sunwoo 2005]. In the other approach, the initial BIST configuration for the resources to be tested is downloaded into the FPGA along with the program for reconfiguration of the resources under test for subsequent BIST configurations [Milton 2006]. In this second approach, there is a download for each test session associated with the resources to be tested. Although the first approach gives a better test time speedup, the development effort and size of the reconfiguration program is larger than the second approach. Both of these approaches can be used to test the programmable logic and routing resources as well as other embedded cores such as memories and DSPs. In both approaches, the embedded processor controls and executes the BIST sequence and retrieves the BIST results from the ORAs for each BIST configuration and can also perform on-chip diagnosis in addition to performing reconfiguration of the FPGA for subsequent BIST configurations in the test session [Stroud 2005b].

Experimental results from an actual implementation of an embedded processor-based BIST approach are summarized in Table 12.11. In this BIST approach, the processor generates the initial BIST configurations internally in addition to executing the BIST sequence, retrieving BIST results, and reconfiguring the FPGA for subsequent BIST configurations. In this implementation, the processor was a hard core with dedicated program memory such that all resources in the FPGA could be tested in parallel. Table 12.11 compares the time for external download and execution of a complete set of BIST configurations (see the "External" column) with embedded processor-generated BIST configurations (see the "Processor" column) [Sunwoo 2005]. It should be noted that the download time for the processor-based BIST approach is the time required to load the program memory and that the execution time includes the time required to reconfigure the FPGA and execute the BIST sequence.

To algorithmically and efficiently reconfigure the FPGA resources under test from the embedded processor core, the BIST architecture, including the TPG and ORA routing to the resources under test, should be constant for all BIST configurations

TABLE 12.11 ■ Test Time Speedup with Embedded Processor-Based BIST

Resource	Function	External	Processor	Speedup
PLB BIST	**Download**	7.680 sec	0.101 sec	76.0
	Execution	0.016 sec	0.085 sec	0.2
	Total time	7.696 sec	0.186 sec	41.4
Routing BIST	**Download**	20.064 sec	0.110 sec	182.4
	Execution	0.026 sec	0.343 sec	0.075
	Total time	20.090 sec	0.453 sec	44.3
Total Test Time		27.786 sec	0.639 sec	43.5

associated with that particular test session. This not only reduces the time required to reconfigure, but more importantly it reduces the size of program memory needed to store the program for BIST reconfiguration. This is critical, because the program memory is usually limited to either the size of the dedicated program memory or half of the RAMs cores in the FPGA. Otherwise, the BIST architecture must be confined to a smaller portion of the FPGA (one fourth of the array, for example), which increases the total number of BIST sessions and the resultant test time.

12.5 CONCLUDING REMARKS

With the incorporation of embedded cores including RAMs, DSPs, and processors, FPGAs more closely resemble SOC implementations. At the same time, more SOCs are incorporating embedded FPGA cores. The programmability of FPGAs facilitates the implementation of a wide range of applications and, as a result, presents a number of testing solutions as well as a number of testing challenges. For example, FPGAs can be reprogrammed during system-level offline testing to test other components and functions on a PCB [Stroud 2002a]. Similarly, the PLBs and routing resources of an FPGA core can be reprogrammed to test the other embedded cores within SOCs such as RAM and DSP cores [Abramovici 2002]. With algorithmic generation, execution, and diagnosis of BIST configurations from an embedded processor core, a single program can be stored and used for manufacturing testing or incorporated into the system for on-demand BIST and diagnosis of the FPGA core for fault-tolerant applications. Therefore, FPGA testing techniques are becoming increasingly important for a broader range of system applications. FPGA testing challenges continue to increase with the introduction of new cores and architectures. On the other hand, these testing challenges in conjunction with the programmability of FPGAs provide an excellent platform for research and development of new SOC test architectures, strategies, and methodologies, such as silicon debug and diagnosis [Abramovici 2006].

12.6 EXERCISES

12.1 **(Configuration)** Determine the configuration bits needed to implement the comparison-based ORA shown in Figure 12.9b into the PLB shown in Figure 12.11.

12.2 **(Test Pattern Generation)** What is the minimum number of PLBs needed to implement one TPG to test the PLB of Figure 12.11?

12.3 **(Test Pattern Generation)** What is the number of loads on each output of each TPG used to test PLBs assuming the $N \times N$ array of PLBs shown in Figure 12.11 and the basic comparison BIST architecture shown in Figure 12.8a with two TPGs constructed from PLBs?

12.4 **(Output Response Analysis)** How many PLBs are need to implement a complete set of circular comparison ORAs to test a total of N BUTs using the example PLB shown in Figure 12.11?

12.5 **(Programmable Logic Blocks)** How many test configurations are needed to completely test the example PLB shown here? Specify the test configurations in terms of the values for 25 configuration bits CB_0—CB_{24} for each test configuration. Assume that exhaustive test patterns are applied to inputs A-D and reset during each test configuration.

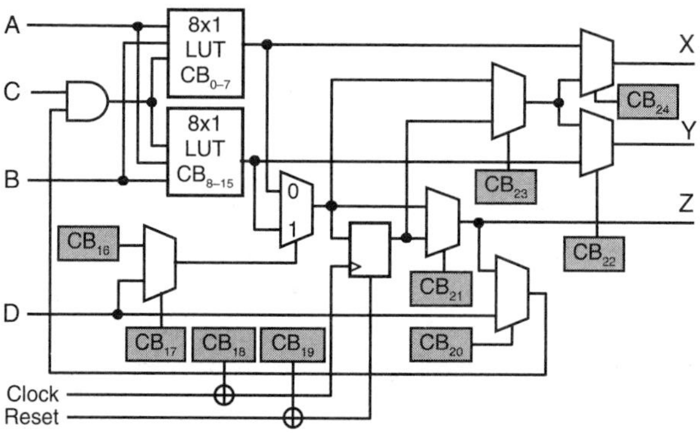

12.6 **(Diagnosis)** Given the following set of ORA results from a circular comparison of 10 BUTs (B_0–B_9), determine the faulty BUTs:

ORA	O_{01}	O_{12}	O_{23}	O_{34}	O_{45}	O_{56}	O_{67}	O_{78}	O_{89}	O_{90}
Results	1	0	0	1	1	0	1	0	0	1

12.7 **(Routing Resources)** Determine the number of test configurations in terms of the configuration bits C_0–C_7 and associated input test vectors to test the two stages of multiplexer PIPs shown here:

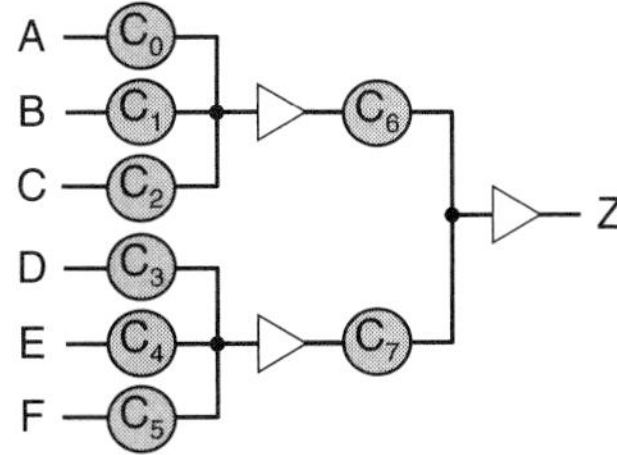

Acknowledgments

The author wishes to acknowledge the assistance of Mustafa Ali, Bobby Dixon, Lee Lerner, and Daniel Milton of the Auburn University Department of Electrical and Computer Engineering Built-In Self-Test Laboratory during the preparation of this chapter. In addition, the author would like to acknowledge Professor Wen-Ben Jone of the Department of Electrical and Computer Engineering of the University of Cincinnati, Cincinnati, Ohio, for his excellent comments and ideas for the enhancement of this chapter.

References

R12.0 Books

[Hamdioui 2004] S. Hamdioui, *Testing Static Random Access Memories*, Springer, Boston, 2004.

[Oldfield 1995] J. Oldfield and R. Dorf, *Field Programmable Gate Arrays*, John Wiley & Sons, New York, 1995.

[Rajski 1998] J. Rajski and J. Tyszer, *Arithmetic Built-In Self-Test for Embedded Systems*, Prentice Hall PTR, Upper Saddle River, NJ, 2002.

[Salcic 1997] Z. Salcic and A. Smailagic, *Digital Systems Design and Prototyping Using Field Programmable Logic*, Kluwer Academic, Boston, 1997.

[Stroud 2002a] C. E. Stroud, *A Designer's Guide to Built-In Self-Test*, Springer, Boston, 2002.

[Trimberger 1994] S. Trimberger, *Field Programmable Gate Array Technology*, Kluwer Academic, Boston, 1994.

R12.1 Overview of FPGAs

[Abramovici 2001] M. Abramovici and C. E. Stroud, BIST-based test and diagnosis of FPGA logic blocks, *IEEE Trans. VLSI Systems*, 9(1), pp. 159–172, February 2001.

[Abramovici 2004] M. Abramovici, C. E. Stroud, and J. Emmert, On-line built-in self-test and diagnosis of FPGA logic resources, *IEEE Trans. VLSI Systems*, 12(12), pp. 1284–1294, December 2004.

[Altera 2005] Altera Corp., Stratix II device handbook, v1–3.0, May 2005.

[Atmel 2005] Atmel Corp., AT94K series field programmable system level integrated circuit, DS1138, June 2005.

[Brown 1996a] S. Brown, FPGA architectural research: A survey, *Design & Test of Computers*, 13(4), pp. 9–15, Winter 1996.

[Brown 1996b] S. Brown and J. Rose, FPGA and CPLD architectures: A tutorial, *Design & Test of Computers*, 13(2), 42–57, Summer 1996.

[Chmelar 2003] E. Chmelar, FPGA interconnect delay fault testing, in *Proc. Int. Test Conf.*, pp. 1239–1247, October 2003.

[Cypress 2005] Cypress Semiconductor Corp., Delta39K ISR CPLD family, DS38-03039 rev1, March 2005.

[Lattice 2006] Lattice Semiconductor Corp., Lattice SC family data sheet, DS1004 v1.2, June 2006.

[Stroud 1996] C. E. Stroud, V. Konala, P. Chen, and M. Abramovici, Built-in self-test of logic blocks in FPGAs, in *Proc. VLSI Test Symp.*, pp. 387–392, May 1996.

[Stroud 2002b] C. E. Stroud, J. Nall, M. Lashinsky, and M. Abramovici, BIST-based diagnosis of FPGA interconnect, in *Proc. Int. Test Conf.*, pp. 618–627, October 2002.

[Toth 2004] F. Toth, The easy path to cost reduction, *Xcell J.*, (48), 96–98, Spring 2004.

[Xilinx 2006a] Xilinx Inc., Virtex-4 user guide, UG070 v1.5, March 2006.

[Xilinx 2006b] Xilinx Inc., Virtex-5 user guide, UG190 v1.1, May 2006.

R12.2 Testing Approaches

[Abramovici 2003] M. Abramovici and C. E. Stroud, BIST-based delay fault testing in field programmable gate arrays, *J. Electronic Testing: Theory & Applications*, 19(5), pp. 549–558, May 2003.

[Abramovici 2004] M. Abramovici, C. E. Stroud, and J. Emmert, On-line built-in self-test and diagnosis of FPGA logic resources, *IEEE Trans. VLSI Systems*, 12(12), pp. 1284–1294, December 2004.

[Chmelar 2003] E. Chmelar, FPGA interconnect delay fault testing, in *Proc. Int. Test Conf.*, pp. 1239–1247, October 2003.

[Doumar 2003] A. Doumar and H. Ito, Detecting, diagnosing, and tolerating faults in SRAM-based field programmable gate arrays: A survey, *IEEE Trans. VLSI Systems*, 11(3), pp. 386–405, June 2003.

[Emmert 2007] J. Emmert, C. E. Stroud, and M. Abramovici, On-line fault-tolerance of FPGA logic resources, *IEEE Trans. VLSI Systems*, 15(2), pp. 216–226, February 2007.

[Huang 1998] W. Huang, F. Meyer, X. Chen, and F. Lombardi, Testing configurable LUT-based FPGAs, *IEEE Trans. VLSI Systems*, 6(2), pp. 276–283, April 1998.

[Renovell 1998] M. Renovell, J. Portal, J. Figueras, and Y. Zorian, Testing the interconnect of RAM-based FPGAs, *IEEE Design & Test of Computers*, 15(1), pp. 45–50, January/March 1998.

[Renovell 2001] M. Renovell, P. Faure, J. Portal, J. Figueras, and Y. Zorian, IS-FPGA: A new symmetric FPGA architecture with implicit scan, in *Proc. Int. Test Conf.*, pp. 924–931, October 2001.

[Tahoori 2004] M. Tahoori, Application-dependent diagnosis of FPGAs, in *Proc. Int. Test Conf.*, pp. 645–649, October 2004.

[Tahoori 2007] M. Tahoori and S. Mitra, Application-dependent delay testing of FPGAs, *IEEE Computer-Aided Design of Integrated Circuits and Systems*, 26(3), pp. 553–563, March 2007.

[Toutounchi 2004] S. Toutounchi and A. Lai, FPGA test and coverage, in *Proc. Int. Test Conf.*, pp. 599–608, October 2002.

[Verma 2004] V. Verma, S. Dutt, and V. Suthar, Efficient on-line testing of FPGAs with provable diagnosabilities, in *Proc. Design Automation Conf.*, pp. 498–503, June 2004.

R12.3 BIST of Programmable Resources

[Abramovici 2001] M. Abramovici and C. E. Stroud, BIST-based test and diagnosis of FPGA logic blocks, *IEEE Trans. VLSI Systems*, 9(1), pp. 159–172, February 2001.

[Abramovici 2004] M. Abramovici, C. E. Stroud, and J. Emmert, On-line built-in self-test and diagnosis of FPGA logic resources, *IEEE Trans. VLSI Systems*, 12(12), pp. 1284–1294, December 2004.

[Atmel 1999] Atmel Corp., Compiled megacell testing, AN0696C, February 1999.

[Dhingra 2005] S. Dhingra, S. Garimella, A. Newalkar, and C. E. Stroud, Built-in self-test for Virtex and Spartan II FPGAs using partial reconfiguration, in *Proc. North Atlantic Test Workshop*, pp. 7–14, May 2005.

[Dhingra 2006] S. Dhingra, D. Milton, and C. E. Stroud, BIST of logic and memory resources in Virtex-4 FPGAs, in *Proc. North Atlantic Test Workshop*, pp. 19–27, May 2006.

[Gizopoulos 1998] D. Gizopoulos, A. Paschalis, and Y. Zorian, Effective built-in self-test for booth multipliers, *IEEE Design & Test of Computers*, 15(3), pp. 105–111, July/September 1998.

[Hamdioui 2002] S. Hamdioui and A. Van de Goor, Efficient tests for realistic faults in dual-port SRAMs, *IEEE Trans. Computers*, 51(5), pp. 460–473, May 2002.

[Jone 1994] W-B. Jone and C-J. Wu, Multiple fault detection in parity checkers, *IEEE Trans. Computers*, 43(9), pp. 1096–1099, September 1994.

[Milton 2006] D. Milton, S. Dhingra, and C. E. Stroud, Embedded processor based built-in self-test and diagnosis of logic and memory resources in FPGAs, in *Proc. Int. Conf. on Embedded Systems and Applications*, pp. 87–93, June 2006.

[Mourad 1989] S. Mourad and E. McCluskey, Testability of parity checkers, *IEEE Trans. Industrial Electronics*, 36(2), pp. 254–260, May 1989.

[Stroud 1998] C. E. Stroud, S. Wijesuriya, C. Hamilton, and M. Abramovici, Built-in self-test of FPGA interconnect, in *Proc. Int. Test Conf.*, pp. 404–411, October 1998.

[Stroud 2000] C. E. Stroud, J. Bailey, J. Emmert, D. Nickolic, and K. Chhor, Bridging fault extraction from physical design data for manufacturing test development, in *Proc. Int. Test Conf.*, pp. 760–769, October 2000.

[Stroud 2002b] C. E. Stroud, J. Nall, M. Lashinsky, and M. Abramovici, BIST-based diagnosis of FPGA interconnect, in *Proc. Int. Test Conf.*, pp. 618–627, October 2002.

[Stroud 2003] C. E. Stroud, K. Leach, and T. Slaughter, BIST for Xilinx 4000 and Spartan series FPGAs: A case study, in *Proc. Int. Test Conf.*, pp. 1258–1267, October 2003.

[Stroud 2005a] C. E. Stroud and S. Garimella, Built-in self-test and diagnosis of multiple embedded cores in SOCs, in *Proc. Int. Conf. on Embedded Systems and Applications*, pp. 130–136, June 2005.

[Stroud 2005b] C. E. Stroud, S. Garimella, and J. Sunwoo, On-chip BIST-based diagnosis of embedded programmable logic cores in system-on-chip devices, in *Proc. Int. Conf. on Computers and Their Applications*, pp. 308–313, March 2005.

[Sun 2000] X. Sun, J. Xu, B. Chan, and P. Trouborst, Novel technique for BIST of FPGA interconnects, in *Proc. Int. Test Conf.*, pp. 795–803, October 2000.

[Sunwoo 2005] J. Sunwoo and C. E. Stroud, Built-in self-test of configurable cores in SOCs using embedded processor dynamic reconfiguration, in *Proc. Int. System-on-Chip Design Conf.*, pp. 174–177, October 2005.

[Van de Goor 1996] A. Van de Goor, G. Gaydadjiev, V. Jarmolik, and V. Mikitjuk, March LR: A test for realistic linked faults, in *Proc. VLSI Test Symp.*, pp. 272–280, April 1996.

[Vemula 2006] S. Vemula and C. E. Stroud, Built-in self-test for programmable I/O buffers in FPGAs and SOCs, in *Proc. Southeastern Symp. on System Theory*, pp. 534–538, March 2006.

R12.4 Embedded Processor-Based Testing

[Milton 2006] D. Milton, S. Dhingra, and C. E. Stroud, Embedded processor based built-in self-test and diagnosis of logic and memory resources in FPGAs, in *Proc. Int. Conf. on Embedded Systems and Applications*, pp. 87–93, June 2006.

[Stroud 2005b] C. E. Stroud, S. Garimella, and J. Sunwoo, On-chip BIST-based diagnosis of embedded programmable logic cores in system-on-chip devices, in *Proc. Int. Conf. on Computers and Their Applications*, pp. 308–313, March 2005.

[Sunwoo 2005] J. Sunwoo and C. E. Stroud, Built-in self-test of configurable cores in SOCs using embedded processor dynamic reconfiguration, in *Proc. Int. System-on-Chip Design Conf.*, pp. 174–177, October 2005.

R12.5 Concluding Remarks

[Abramovici 2002] M. Abramovici, C. E. Stroud, and J. Emmert, Using embedded FPGAs for SOC yield enhancement, in *Proc. Design Automation Conf.*, pp. 713–724, June 2002.

[Abramovici 2006] M. Abramovici, P. Bradley, K. Dwarakanath, P. Levin, G. Memmi, and D. Miller, A reconfigurable design-for-debug infrastructure for SOCs, in *Proc. Design Automation Conf.*, pp. 7–12, July 2006.

CHAPTER 13

MEMS Testing

Ramesh Ramadoss
Auburn University, Auburn, Alabama

Robert Dean
Auburn University, Auburn, Alabama

Xingguo Xiong
University of Bridgeport, Bridgeport, Connecticut

ABOUT THIS CHAPTER

Since the mid-1970s, ***microelectromechanical systems*** (MEMS) have emerged as a successful technology by utilizing the existing infrastructure of the well-established ***integrated circuit*** (IC) industry. MEMS technology along with ***very-large-scale-integration*** (VLSI) technology has created new opportunities in physical, chemical, and biological sensor and actuator applications. MEMS devices are typically manufactured using VLSI IC process-compatible fabrication techniques; however, the test methods for MEMS significantly differ from those used for VLSI circuits. This chapter focuses on MEMS testing and characterization, a topic of special interest to researchers and practicing engineers.

In this chapter, we begin with an overview of the MEMS field followed by a brief discussion of various considerations for testing MEMS devices at the chip, wafer, and package level. A review of various test methods used for MEMS devices is presented, including electrical, mechanical, and environmental tests. We next discuss experimental test setup and performance characteristics of interest for various MEMS devices such as ***radio-frequency*** (RF) MEMS switches, resonators, optical micromirrors, pressure sensors, microphones, humidity sensors, microfluidic systems, accelerometers, and gyroscopes. Due to the growing importance of microfluidics-based biochips, also referred to as **lab-on-a-chip**, and their potential for replacing cumbersome and expensive laboratory equipment, we include a section on the testing of digital microfluidic biochips. Finally, we present ***design-for-testability*** (DFT) and ***built-in self-test*** (BIST) techniques that have been proposed for testing various MEMS devices, followed by a number of examples with actual implementations of BIST in commercially available accelerometers.

13.1 INTRODUCTION

MEMS devices are miniature electromechanical sensors and actuators fabricated using VLSI processing techniques. Typical sizes for MEMS devices range from nanometers to millimeters (100 nm to 1000 μm). MEMS devices are characteristically low cost, highly functional, small, and light weight. Cost and weight reduction are the result of utilizing the semiconductor batch-processing techniques of photolithography and wafer-scale mass production. MEMS devices can sense, control, and actuate on the microscale and function individually (or in combination with other devices) to generate macroscale effects. MEMS enhances realization of ***system-on-chip*** (SOC) by integration of mixed domain technologies such as electrical, optical, mechanical, thermal, and fluidics. MEMS-based SOC combines the functionality of the IC (an information processor) with information gathering (sensing the environment) and actuation (acting on decisions) capabilities of MEMS.

MEMS devices also enable the miniaturization of sensor systems and the possibility of on-chip integration of VLSI electronics with mechanical sensors and actuators. The technology has proven to be revolutionary in many application arenas, including accelerometers, pressure sensors, displays, inkjet nozzles, optical scanners, interferometers, spectrometers, tilt mirrors, and fluid pumps. MEMS devices have potential applications in many fields including the telecommunications, optics, materials science, robotics, automotive, aerospace, healthcare, and information storage industries. Typical examples for commercial MEMS devices include Analog Devices' ADXL series accelerometers, FreeScale Semiconductor's pressure sensors and accelerometers, Texas Instruments' ***digital light processing*** (DLP) displays, and Knowles Electronics' SiSonic MEMS microphone [Petersen 2005].

Advances in microfluidics technology have led to the emergence of miniaturized biochip devices for biochemical analysis [Fair 2003] [Verpoorte 2003] [Dittrich 2006]. **Microfluidics-based biochips**, also referred to as **lab-on-a-chip**, are replacing cumbersome and expensive laboratory equipment for applications such as high-throughput sequencing, parallel immunoassays, protein crystallization, blood chemistry for clinical diagnostics, and environmental toxicity monitoring. Biochips offer advantages of higher sensitivity, lower cost because of smaller sample and reagent volumes, higher levels of system integration, and less likelihood of human error.

To ensure the testability and reliability of these MEMS-based SOCs, MEMS devices need to be thoroughly tested, particularly when used for safety-critical applications such as in the automotive and healthcare industry. Therefore, there is a pressing need for ***design for testability*** (DFT) and ***built-in self-test*** (BIST) of MEMS. However, MEMS devices have diverse structures and working principles. MEMS signals are essentially analog instead of digital, and multiple energy domains are generally involved in device operations. Furthermore, many MEMS devices contain movable parts and, hence, their defects and failure mechanisms are more complex than VLSI circuits. As a result, DFT and BIST for MEMS are more challenging than their VLSI counterparts [Mir 2006].

13.2 MEMS TESTING CONSIDERATIONS

The majority of MEMS devices are inherently mechanical in nature and, as a result, MEMS devices necessitate special considerations during fabrication processes such as handling, dicing, testing, and packaging. This section presents a brief discussion of various considerations involved in testing MEMS.

MEMS devices require consummate care in handling because the micromechanical parts need to be protected from shock and vibration during transport and packaging. Also, extreme care must be taken to avoid particle contamination at various processing steps involved in MEMS fabrication. This is because dust particles are of the same size as many of the common features found in MEMS devices and can lodge in between moving elements disrupting the intended movement [Strassberg 2001]. As a common practice in MEMS industry, the backside of a fully processed wafer is attached to an adhesive plastic film and then mounted in a rigid frame for dicing at the wafer-processing facility. The dicing saw cuts through the silicon, but not all the way through the film such that the die is firmly held in place. This process also electrically and physically separates the dies from one another and permits automatic handling of entire wafers as well as electrical batch testing of individual dies. During dicing of the silicon wafer, the adhesive backed plastic film traps small particles that become liberated from the wafer. A protective overcoat is often used before release of the actuators at the wafer level to keep unwanted debris from entering the devices during the dicing process. This means that careful consideration must be given to the sequencing of the various processing steps and in particular the testing to be performed at the wafer level before and after dicing. If these particles were later to become trapped between the moving and nonmoving parts of the MEMS devices in the subsequent processing steps, they could prevent the MEMS from functioning properly [Strassberg 2001]. Traditional ICs are protected against debris caused by the dicing process through a passivation layer that is deposited on top of the wafer at the end of the processing sequence. However, most MEMS devices contain moving parts on the wafer surface and typically are not suitable for deposition of a passivation layer. As a result, MEMS devices often require packaging before dicing—that is, 0-level packaging at the wafer level by either wafer-to-wafer bonding or local bonding of minature caps (*e.g.*, Si or glass) over the MEMS structure using a hermetic sealing ring [Wolf 2002].

MEMS test methods and instrumentation vary depending on whether the testing is performed at the wafer level (*i.e.*, unpackaged die) or on packaged devices. Wafer-level testing is carried out using precision-controlled wafer probers that step from die to die on the wafer, making electrical contact using needle probes. Testing unpackaged MEMS dies using probes is typically limited to electrical tests that provide basic information about a subset of parameters of the die but cannot fully analyze the mechanical function of sensors and actuators. For a comprehensive characterization of MEMS devices, on-wafer testing technologies for MEMS devices must include nonelectrical stimulation and detection methods. For example, the automotive industry is driven by the relatively high risk of injuries to humans if air bag or antilock braking system sensors fail, and it therefore requires reliable multistimuli tests, including mechanical shock. It is desirable to include these stimulations during the execution of the electrical test. However, standard test handlers cannot expose MEMS devices to additional stimuli such as magnetic fields, required for Hall sensors, or acceleration to test acceleration sensors [Jaeckel 2002]. There have been considerable efforts in the development of multistimuli test systems that allow stimulation or measurement using sound, light, pressure, motion, or fluidics.

Like their VLSI counterparts, testing MEMS devices at wafer-level before packaging can improve overall yield and reduce production costs [Feuerstein 2004]. For new MEMS devices that are still at the design and development stages, a complete analysis at wafer level and throughout the various packaging stages provides a valuable insight into the design characteristics and enables process monitoring. For mature MEMS device technologies, the production costs can be lowered by identifying ***known good die*** (KGD) before packaging by wafer-level testing. Studies show that the packaging process causes 80% of manufacturing costs. By testing at earlier stages of production, the packaging of bad dies can be eliminated, and thus production costs can be reduced. By testing at several production stages, possible errors can be identified and eliminated, resulting in an optimized process and higher yield [Memunity 2007].

Fully packaged MEMS devices can be tested with the electrical and nonelectrical inputs required for the sensor to function. A variety of environmental test methods commonly used for testing packaged ICs can be directly employed for testing packaged MEMS devices, as will be discussed in the next section. Many standard tests are common to both ICs and MEMS, such as thermal cycling, high temperature storage, thermal shock, and high humidity. However, many MEMS packages need to fulfill additional specifications. For instance, the motion of MEMS inertial sensors or RF MEMS resonators is affected by the environment in which they function and would be influenced by particles, humidity, pressure, and ambient gases. Moreover, many MEMS sensors have additional specifications on resistance to shock and vibration.

13.3 TEST METHODS AND INSTRUMENTATION FOR MEMS

MEMS encompass a wide variety of applications such as inertial sensors (accelerometers and gyroscopes), RF MEMS, optical MEMS, and bio or fluidic MEMS. The test instrumentation required for testing MEMS depends on the specific type of MEMS device and the desired performance characteristics. For example, inertial MEMS sensors require different test instrumentation than RF MEMS. Specifically, inertial MEMS sensors require electromechanical shaker and rate table tests, whereas RF MEMS devices typically require spectrum or network analyzers connected to a probe station. Furthermore, within RF MEMS, depending on the type of device under test, one needs different types of test instrumentation. For example, in ohmic contact RF-MEMS switches, the study of possible contamination and degradation of the ohmic contacts is of importance, whereas in capacitive RF-MEMS switches, one has to be able to study charging and degradation of the dielectric layer used in the switch [Wolf 2002].

In general, MEMS testing can be categorized as (1) functionality and performance testing and (2) reliability/failure testing. In functional testing, the characteristic performance parameters are measured and compared against benchmark specifications to verify the intended operation of the MEMS device. In reliability/failure testing, the performance degradation over sustained operation or shelf life and eventual failure of the device are investigated. Quite often the borderline between functional testing and reliability testing is not always clear. This section discusses the various test methods required for operational tests such as electrical, optical, mechanical, and environmental.

13.3.1 Electrical Test

Electrical tests are one of the most important methods employed to characterize MEMS. A typical electrical test setup consists of a probe station interfaced with the required test instrumentation. A wide range of electrical test equipment used for VLSI testing is commonly used to perform electrical characterization of MEMS devices. Typical electrical test instrumentation includes current, voltage, and resistance measurement systems, capacitance-voltage measurement systems, impedance analyzers for low-frequency characterization, network analyzers for high-frequency characterization, and signal analyzers. Other instrumentation can be added as needed. Probe length and wire types (shielded and unshielded) must also be carefully considered. For instance, resistance measurements must include a means for reducing contact errors. Capacitance measurements need to take into account the stray capacitance in test lines. These considerations are particularly important for MEMS devices where the parameters of interest are likely to be small.

A typical experimental setup used for testing an electrostatically actuated MEMS relay is shown in Figure 13.1. The setup shown consists of an Agilent 33220A function generator, Krohnit 7600 Wideband Amplifier, HP 54501A Oscilloscope, MM8060 Micromanipulator Probe Station, and HP3468A 4-wire Multimeter. The MEMS device operation is monitored using a CRT display interfaced to the microscope through a CCD camera. The output of the function generator is set to a 10 V square wave with a 50% duty cycle. The output of the function generator is input to the amplifier and the output of the amplifier is connected to the DC and ground probes in the probe station. The output of the amplifier is measured using the oscilloscope. The resistance of the MEMS relay is measured using the 4-wire multimeter where the two current source wires are connected to probes P1 and P2 and the two voltage sensing wires are connected to probes P3 and P4. The basic test setup described here can be used to test a variety of actuators including electrostatic, thermal, and piezoelectric.

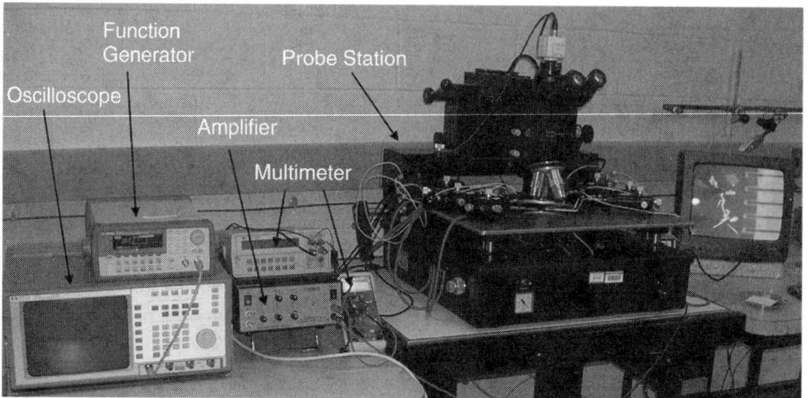

■ **FIGURE 13.1**

Electrical test setup of MEMS using a probe station.

13.3.2 Optical Test Methods

MEMS actuators typically include mechanical motion associated with the electrical signals. In MEMS testing, it is often of interest not only to measure the electrical characteristics of the MEMS but also to monitor its mechanical motion by optical inspection. Optical profilometers, such as an optical microscope, confocal microscope, optical interferometers, and laser Doppler vibrometer, are useful for making static and dynamic measurements of MEMS devices [Lecklider 2006].

An optical microscope equipped with high-resolution objectives and accurate graticule can be used to measure MEMS features in a two-dimensional plane view. High-speed photomicrography is often used in conjunction with an optical microscope to capture video of MEMS devices in motion.

Traditional microscopes focus a magnified area of the specimen to an area of the viewer's eye. A confocal microscope images only a single point at a time rather than an area. Modern confocal microscopes employ low-cost lasers and computers to scan a thin slice through the specimen. By vertically repositioning the specimen, a series of slices are acquired and reassembled into a three-dimensional image.

In optical interferometers, a beam of light is divided into two beams. One beam reflects off the specimen being examined, while the other beam reflects off the reference mirror. The resulting fringe pattern is a function of the specimen height as well as the reference mirror angle. By counting fringes, the height of the specimen can be measured. The optical interferometers can make use of white light (*e.g.*, a sodium lamp) or of coherent monochromatic light (a laser light). Optical interferometers are useful for measuring noncontact three-dimensional profiles of MEMS devices. Examples of optical interferometers include Wyko series manufactured by Veeco Instruments, NewView 6000 series manufactured by Zygo, PhotoMap

3-D profilometers by Fogale nanotech, and the Xi-100 developed by Ambios Technology.

Laser Doppler vibrometry (LDV) is based on the modulation of laser interference fringes caused by motion of the ***device under test*** (DUT). The fringe pattern in a Doppler vibrometer is moving at a rate proportional to the device motion. By measuring the time rate of change in distance between successive fringes, a vibrometer can measure displacement as well as velocity. The direction of motion can be determined by observing the Doppler effect on the modulation frequency. LDV is useful for measuring transient and steady-state responses of MEMS devices. A wide variety of LDVs for MEMS applications are available such as Polytec's MSA-400 Microsystem Analyzer, which uses white light interferometry for static surface topography (see Figure 13.2a), laser Doppler vibrometry for measuring out-of-plane vibrations (see Figure 13.2a), and stroboscopic video microscopy for measuring in-plane motion (see Figure 13.2c and Figure 13.2d) [Polytec 2007].

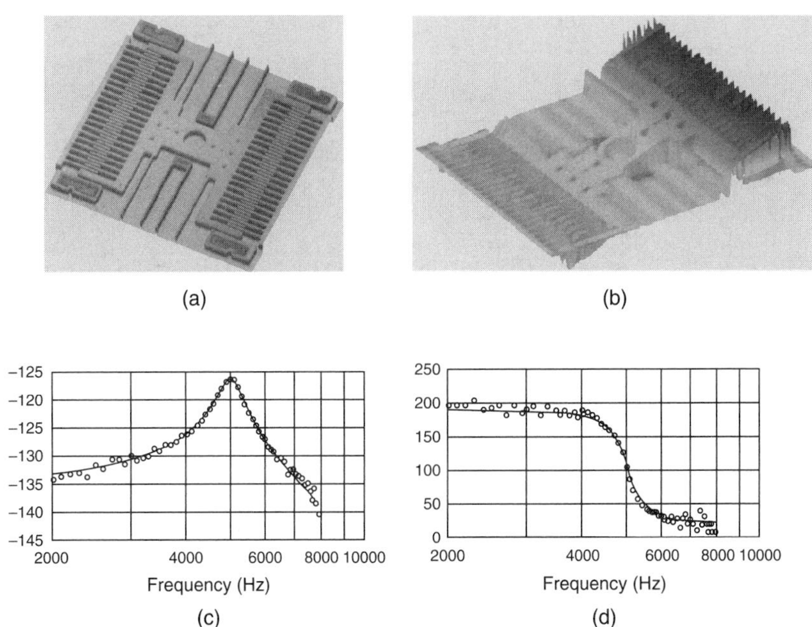

■ **FIGURE 13.2**

Optical profile measurements using Polytec's MSA 400 Microsystem Analyzer for in-plane motion measured using stroboscopic illumination technique: (a) static surface topography measurement using White Light Interferometery, (b) laser-Doppler vibrometry for measuring out-of-plane vibrations, (c) Bode plot for amplitude response, and (d) Bode plot for phase response.
(Courtesy of Polytec.)

13.3.3 Material Property Measurements

Several material properties and processing parameters influence the functionality and reliability of MEMS. Relevant material properties include elastic modulus, Poisson's ratio, fracture toughness and mechanisms, electrical properties (resistivity, migration), interfacial strength, and coefficient of thermal expansion. The specific processing techniques used to fabricate MEMS devices can lead to such outcomes as residual stresses and stress gradients, changes in grain size, surface roughness, doping gradients, stiction, and warping effects. A thorough knowledge and control of the mechanical parameters of all materials utilized for MEMS is important for obtaining good functionality and reliability.

Typically, the structural layers in surface micromachined MEMS devices are fabricated using thin films. The material properties of these films influence the functionality and reliability of the resulting MEMS devices. MEMS often fail when the stresses and stress gradients become too large. MEMS-based test structures such as cantilever beams, clamped-clamped beams, and Guckel rings (see Figure 13.3) are often co-fabricated on the wafer for making stress and strain measurements. Optical profile measurements of these test structures can be used for estimation of the strain gradient, residual strain, and material properties. For example, the curvature of cantilever beams can be used to obtain the stress gradient present in the film. Buckling behavior of fixed-fixed test structures can be used to obtain compressive stresses in the film. A Guckel ring can be used to obtain tensile stress information.

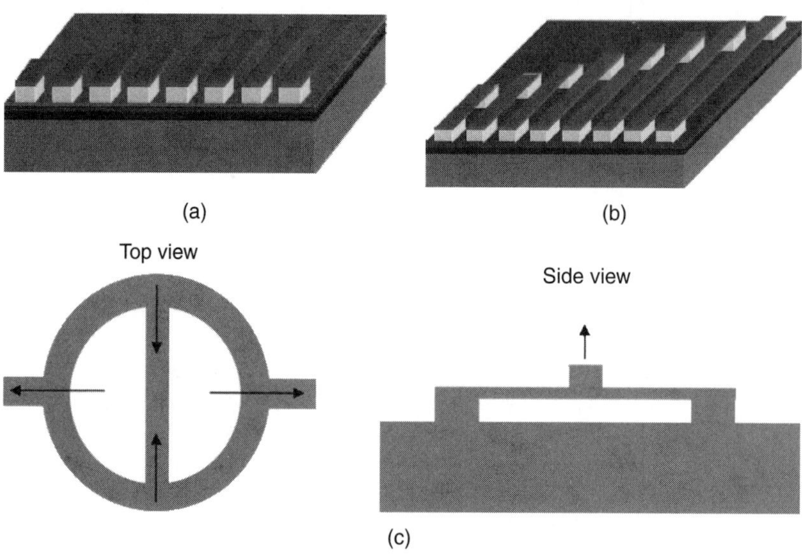

■ **FIGURE 13.3**

MEMS test structures for material and process parameter measurements: (a) cantilever beam, (b) fixed-fixed beam, and (c) Guckel ring.

13.3.4 Failure Modes and Analysis

In addition to many failure modes known from the IC and IC-packaging world, MEMS have specific failure modes, such as fatigue, wear, fracture, and stiction. Several kinds of test structures are commonly used to study materials related reliability issues such as *fatigue*. Typically, samples with a preformed notch are used, such that the growth of a crack during functioning can be studied, either by direct optical observation or by a study of the influence on the Eigen-frequency of a beam or similar structure, for example. Another approach is bending test devices in which the material achieves locally high strain levels during bending tests. Free cantilever beams with different lengths can also be used to study *stiction*. The length of the stuck part depends on the humidity level, the restoring force, and the roughness of the microstructures in contact.

Surface roughness can affect issues such as stiction, wear, contact degradation, and contact resistance. Contact profilometers such as Dektak stylus profilers can be used to measure the surface roughness and the thickness of thin films. Measurements of this sort are usually restricted to special areas on the wafer or before release of the devices. This is because during measurements, the contact type profilometer makes contact with the part being measured and therefore could potentially damage fragile MEMS parts. Contact profilometers are suitable only for static measurements. ***Atomic force microscopy*** (AFM) is a useful tool for measuring surface roughness. It should be pointed out that the roughness of the top surface of a moving MEMS part is not necessarily the same as the roughness of the bottom surface. To measure the bottom side roughness, the moving part can simply be removed destructively or in some cases even cut with a ***focused ion beam*** (FIB) and examined. An AFM can also be used to obtain information on mechanical parameters, contact resistance as a function of force, or even tribological information such as adhesion forces. Also nanoindentor systems are frequently used to study MEMS: they can provide information on the Young's modulus of materials by physically indenting them, and they can also be used to obtain force-displacement curves of moving parts. Noncontact measurements using an optical interferometer can also be employed to determine surface roughness.

Several ***failure analysis*** (FA) techniques that are conventionally used for chips and packages can also be used for MEMS. Especially useful is the ***scanning electron microscope*** (SEM) for inspection and the FIB to make local cross sections. The SEM is a type of electron microscope capable of producing high-resolution surface images of a DUT. The three-dimensional characteristics of SEM images are useful for failure analysis of the surface structure of the DUT. Example SEM pictures of a MEMS relay are presented in Figure 13.4. FIB is a similar technique to the SEM, with the exception that a beam of excited ions is used instead of the conventional electron beam. The ion beam can be used to vaporize material from the DUT surface and is capable of removing layers and machining trenches and troughs to reveal sections or surfaces of interest for imaging and examination. Load-displacement tools such as nanoindentor systems can also be used to obtain information such as Young's modulus of materials and force-displacement characteristics of movable MEMS parts. Additional techniques

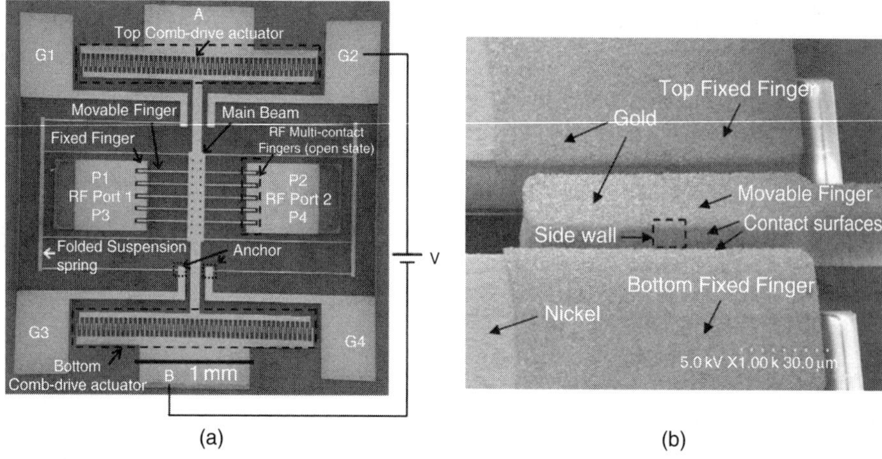

■ **FIGURE 13.4**

SEM pictures of a multicontact MEMS relay using bidirectional comb-drive electrostatic actuator: (a) top view and (b) closeup view of one movable finger and two fixed fingers, from [Almeida 2006].

include ***transmission electron microscopy*** (TEM), ***photon emission microscopy*** (PEM), ***scanning acoustic microscopy*** (SAM), ***infrared inicroscopy*** (In), x-ray, and Raman spectroscopy [Wolf 2002].

13.3.5 Mechanical Test Methods

Mechanical testing consists of subjecting the DUT to some type of external motion, such as translation, vibration, rotation, or mechanical shock, and then detecting the response of the device. Vibration testing is particularly useful for evaluating many types of MEMS devices. Most MEMS devices are composed of movable microstructures that are attached to fixed microstructures by one or more beams. Typically, it is a reasonable assumption to consider the fixed and movable microstructures to be rigid structures and to consider the beams to be flexible structures that elastically deform when an external force is applied to the movable microstructure. The beam or beam structure connecting the microstructures acts as a linear spring, at least for small deflections, and can be modeled by a **system spring constant**, k, where k is a function of the number of beams, the beam geometries, and the material properties of the beams. The movable microstructure is called the **proof mass** and has a mass, m. The fixed microstructure can be referred to as the **frame**. The MEMS device also possesses damping because of internal losses in the beams and external losses such as the interaction of the movable microstructure with the surrounding gas. The composite damping effects are modeled by a single linear damping constant, c. A MEMS device, such as the simple accelerometer illustrated in Figure 13.5, can be modeled by a linear second-order spring-mass-damper system.

In some applications, the input to a MEMS device is a physical displacement to the frame, $x_1(t)$, and the output is the displacement of the proof mass, $x_2(t)$.

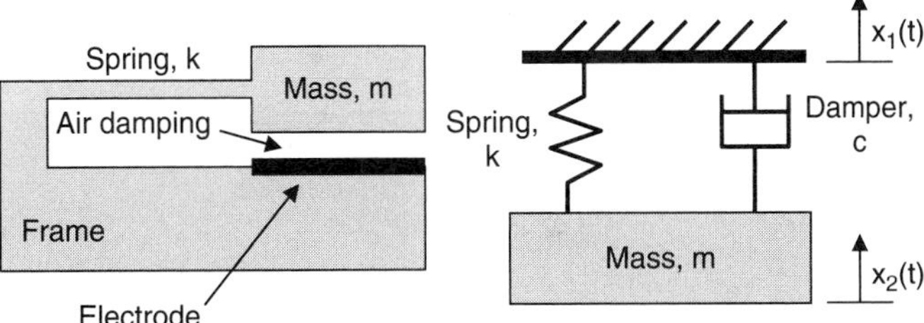

■ **FIGURE 13.5**

Accelerometer modeled as a linear second-order spring-mass-damper system.

However, even in systems where this is not the case, this model is useful for evaluating the motion of movable microstructures because of disturbances to the system resulting from external forces acting on the frame. The governing differential equation describing the system dynamics is:

$$m\ddot{x}_2 + c(\dot{x}_2 - \dot{x}_1) + k(x_2 - x_1) = 0 \tag{13.1}$$

Using a Laplace transformation, the transfer function of the mechanical frequency response can be computed from Equation (13.1) to be:

$$T(s) = \frac{X_2(s)}{X_1(s)} = \frac{\frac{\omega_n}{Q}s + \omega_n^2}{s^2 + \frac{\omega_n}{Q}s + \omega_n^2} \tag{13.2}$$

where ω_n is the **natural frequency** of the second order system model and can be calculated from:

$$\omega_n = \sqrt{k/m} \tag{13.3}$$

and Q is the **system quality factor**, which can be calculated from:

$$Q = \frac{\sqrt{km}}{c} \tag{13.4}$$

The quality factor represents a ratio of the energy stored in a cycle compared to the energy dissipated in a cycle. A high Q system will experience large amplitude motion at its resonant frequency if it is externally excited by a small amplitude signal at that frequency. The quality factor, Q, can be replaced by ζ, the **damping coefficient**, using the following relationship:

$$\zeta = \frac{1}{2Q} \tag{13.5}$$

The magnitude response of Equation (13.2) is called the **transmissibility** [Meirovitch 1986] and is:

$$|T(j\omega)| = \frac{\sqrt{(\frac{\omega_n\omega}{Q})^2 + \omega_n^4}}{\sqrt{(\omega_n^2 - \omega^2)^2 + (\frac{\omega_n\omega}{Q})^2}} \tag{13.6}$$

and the phase response is:

$$\theta = \tan^{-1}\left(\frac{2\zeta(\omega/\omega_n)^3}{1-(\omega/\omega_n)^2 + 2\zeta(\omega/\omega_n)^2}\right) \tag{13.7}$$

Figure 13.6 presents a graph of several plots of the transmissibility for various values of Q where the x-axis variable is represented by ω/ω_n, which is referred to as **normalized frequency**. Therefore, the natural frequency occurs on the plot at the value of ω/ω_n equal to 1. Although the magnitude response in Figure 13.6 is a second-order low-pass filter response, the zero in the numerator of Equation (13.2) results in some interesting properties. When $Q \geq 5$, the magnitude of $T(j\omega)$ at the natural frequency, ω_n, is approximately equal to Q, the system quality factor. Also, the magnitude of $T(j\omega)$ is always greater than 1 at the natural frequency, ω_n, for any value of Q. The stopband attenuation is dependent on the value of Q. Consider the stopband response for three values of Q, 1, 10, and 1000. For Q = 1, the attenuation between $2\omega_n$ and $20\omega_n$ is 21.85 dB, which is approximately the attenuation achieved

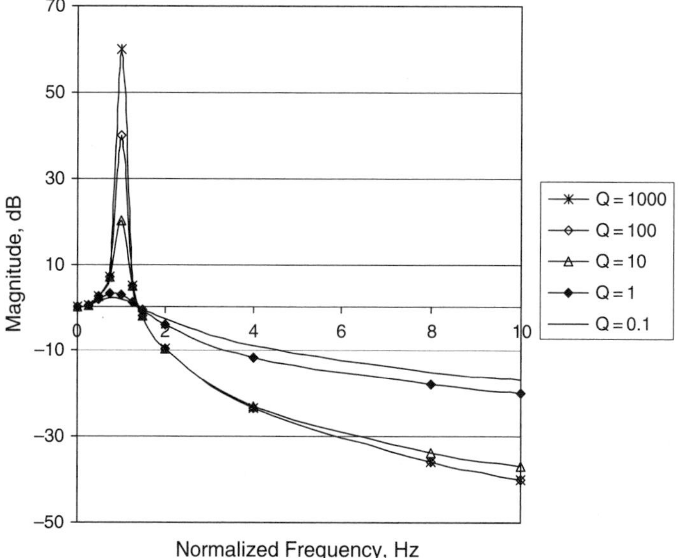

■ **FIGURE 13.6**

Transmissibility *versus* normalized frequency for various values of Q.

in a first-order lowpass system. For $Q = 10$, the attenuation between $2\omega_n$ and $20\omega_n$ is 35.64 dB. For $Q = 1000$, the attenuation between $2\omega_n$ and $20\omega_n$ is 42.48 dB, which is approximately the attenuation achieved in a second-order lowpass system.

Measuring the transmissibility of MEMS devices gives designers or end users an abundance of useful information. In addition to revealing whether the device has a lowpass, highpass, or bandpass frequency response, it reveals the resonant frequency or frequencies. Measurement of the quality factor also reveals both the susceptibility of the device to ringing at its resonant frequency and what the time response will be to an input signal such as a step function. For example, if an end user desires to place several identical devices in close proximity to each other and they each possess a high quality factor, special packaging may be required to prevent the devices from adversely affecting each other through mutual excitation to ring at their resonant frequency, a condition called **intermode coupling**.

An **electromechanical shaker** is a useful piece of test equipment for evaluating the transmissibility of a device. A shaker uses a magnetic field to induce sinusoidal motion in an attached shaker head onto which devices can be attached and vibrated. The magnetic field that induces motion in the shaker head is controlled by an electrical input to the shaker, so that the user can select the amplitude and frequency or bandwidth of the sinusoidal excitation. Several different types of electromechanical shakers are commercially available in different sizes and with different operating characteristics. A photograph of an LDS model V408 electromechanical shaker is presented in Figure 13.7, which can be used to vibrate a MEMS device at a specific amplitude and frequency or over a bandwidth of frequencies. A packaged accelerometer is attached to the shaker head in the photograph. Many shaker heads have a center-tapped hole that can be utilized for mounting the MEMS device. It is relatively easy to attach a ***printed circuit board*** (PCB) to the tapped hole in a shaker head. This is particularly useful if the MEMS device is already

■ **FIGURE 13.7**

Photograph of an LDS model V408 electromechanical shaker with attached accelerometer. (Courtesy of Auburn University.)

packaged and can be soldered to the PCB. The PCB can be designed so that it has a center hole and can be bolted to the shaker head. However, because it is usually desirable to mount the MEMS device directly above the center of the shaker head, one can attach the PCB to the top side of a small aluminum block, either with screws or an adhesive, and then attach the bottom side of the aluminum block to the shaker head using a threaded stud. However, it is important for the PCB to be rigidly attached to the shaker head so that it does not elastically deform during testing and corrupt evaluation of the MEMS device. The PCB approach also allows the use of electrical connections to the MEMS device during testing. Testing of an unpackaged MEMS device on a shaker is more complicated. Plastic fixtures into which the MEMS device can be temporarily clamped can be machined and used to attach the device to the shaker head for testing. Another option is the use of a low temperature adhesive to temporarily bond the die to an aluminum block for attachment to the shaker. Depending on the design of the MEMS device, a cavity may need to be milled into the aluminum block under where the proof mass structure will rest so that it can freely move during testing. After testing is completed, the die can be removed from the aluminum block by heating it on a hot plate.

A convenient method for measuring the mechanical response of an unpackaged MEMS device where the proof mass motion of interest is normal to the surface of the MEMS device is to reflect a laser beam off of the frame using a laser interferometric measurement system, which will yield the motion of the frame, $X_1(f)$, as a function of frequency. A second laser interferometric system can be used to measure the motion of the proof mass, $X_2(f)$, as a function of frequency by reflecting a laser beam off of the proof mass. If $X_1(f)$ and $X_2(f)$ are recorded simultaneously, then the transmissibility of the MEMS device can be determined by computing $X_2(f)/X_1(f)$ and plotting the quotient as a function of frequency. From this plot, the resonant frequency and the quality factor of the MEMS device can be estimated. An example plot of the transmissibility *versus* frequency from the measured frequency response of a MEMS device is presented in Figure 13.8. From examining the data in Figure 13.8, one can determine that the resonant frequency is approximately 1.35 KHz

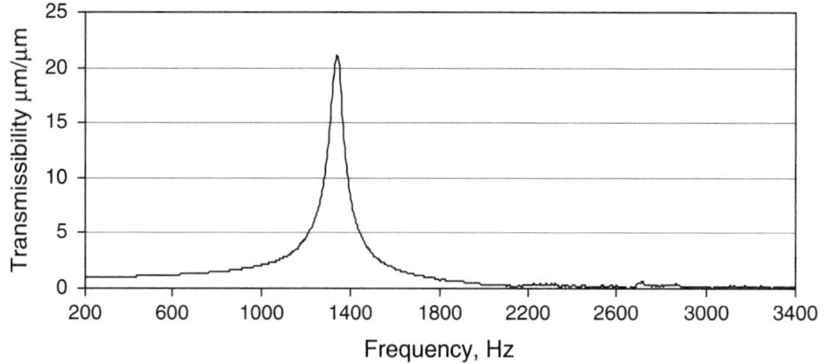

■ **FIGURE 13.8**

Example of the measured mechanical frequency response of the MEMS device.

and the quality factor is approximately 22. Also observe that the MEMS device has a highly underdamped lowpass response. For frequencies significantly less than the resonant frequency, the transmissibility is approximately equal to 1. Because the phase response approaches 0° as the excitation frequency approaches 0 Hz, from Equation (13.7), a steady-state sinusoidal displacement that is input at a frequency significantly less than the resonant frequency will excite almost the same motion in the proof mass as in the frame.

For MEMS devices that have motion parallel to the surface of the device, it is generally not possible to use laser interferometric measurement systems to analyze the response of the proof mass relative to the frame. For evaluating MEMS devices that exhibit this type of motion, the sense mechanism built into the device, such as capacitive displacement or piezoresistive strain sensing, can be used instead. For an accelerometer, the bandwidth of operation is usually DC to some frequency below the resonant frequency of the device. Because the transmissibility of the MEMS device over this bandwidth is approximately 1, as illustrated in Figure 13.8, an electromechanical shaker is of little use in measuring the response of the accelerometer to a specific acceleration level, especially at very low frequencies.

So how can the response of an accelerometer to different acceleration levels be evaluated? Measuring the response of an accelerometer to 1 g, where 1 g is the acceleration that results from gravity on the earth (approximately 9.8 m/s^2), is quite easy to accomplish. If the accelerometer is at rest while oriented such that the positive direction of acceleration measurement is down, the accelerometer is experiencing 1 g of acceleration because of the force of gravity. Likewise, if the accelerometer is rotated 180° so that the positive direction of acceleration measurement is up, the accelerometer is now experiencing −1 g of acceleration from the force of gravity. Also, the accelerometer will experience 0 g of acceleration along its direction of measurement if it is oriented such that the positive direction of acceleration measurement is orthogonal to the force of gravity.

A **rate table** is a test instrument that has a rotating head onto which a MEMS device may be attached and rotated. Consider a round table top with a radius, r, rotating at a constant angular rotation rate, ω, as depicted in Figure 13.9.

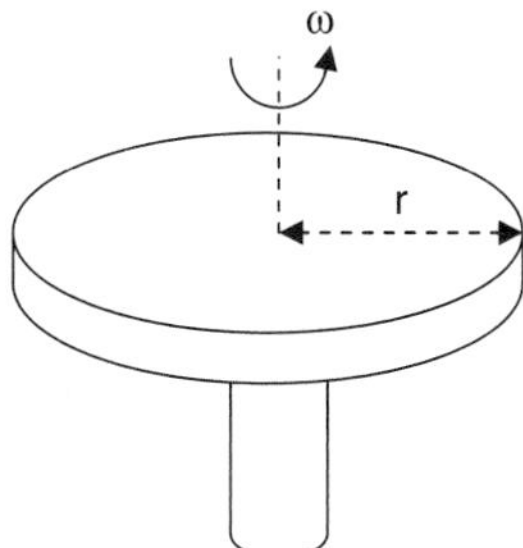

FIGURE 13.9

Illustration of a rate table.

Assume that the axis of rotation is along the direction of gravity. An object with a mass, m, that is held on the edge of the wheel at a distance r from the center of rotation will experience a centripetal force, F_c, equal to:

$$F_c = mr\omega^2 \tag{13.8}$$

The acceleration that the object experiences, a_c, is then equal to:

$$a_c = r\omega^2 \tag{13.9}$$

and is directed radially away from the center of the rotating table. Therefore, a rate table can be utilized to evaluate a device at different acceleration levels, where either the distance that the device is located from the center of rotation or the rotation rate can be varied to adjust the desired acceleration. However, it is usually easier to change the rotation rate than the distance that the device is mounted from the center of rotation. High precision rate tables are commercially available that include closed loop control of the rotation rate or rotation rate profile; of course, these systems can be expensive.

A simple rate table can be constructed fairly inexpensively. A DC motor that is firmly attached to a base along with a variable voltage power supply can be used with a slip ring assembly and a machined plate to realize a simple but useful rate table. The slip ring assembly attaches to the motor shaft and provides electrical contacts to devices mounted on the rate table. Slip ring assemblies with 10 or more electrical feedthroughs can be purchased. The plate can be machined to include mounting holes for the accelerometer and mounting brackets to attach it to the slip ring assembly. To keep the rate table balanced, a dummy accelerometer or a second accelerometer should be attached to the rate table plate at a location opposite to the accelerometer under test. Because the rate table is spinning an object that is temporarily attached to it, appropriate safety measures should be used whenever the rate table is spinning, which, at a minimum, include the use of safety glasses. A safer approach is to place a strong Plexiglas box over the rate table whenever it is in use and to still require the use of safety glasses by all personnel in the vicinity of the unit.

A photograph of a simple rate table, consisting of a DC motor, a motor controller, a 10-pin electrical feedthrough, and a machined plate, is presented in Figure 13.10. Because this simple rate table is operated open loop, a method of measuring the angular rate is needed. One technique is to attach some type of encoder to the motor shaft and then detect the analog or digital output of the encoder. Another technique is to use a laser interferometric measurement system and reflect the laser beam off of a rotating reflector placed on either the motor shaft or the machined plate. A spectrum analyzer is then used to determine the precise rotation rate by computing the frequency response of the output signal from the interferometer.

A rate table is also useful for evaluating the response of a MEMS device to an angular rotation or rotation rate by mounting the device on the rate table at the center of rotation and detecting the output signal from the device. Some commercially available rate tables come with sophisticated software control features that

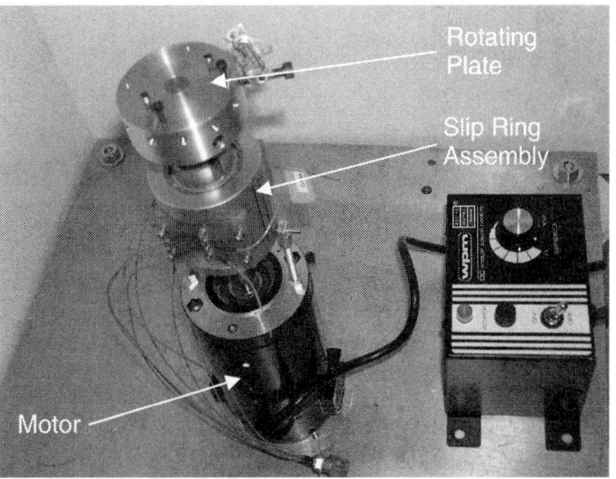

■ **FIGURE 13.10**

Photograph of a simple rate table.
(Courtesy of Auburn University.)

allow complex rate profiles to be performed on the rate table. An example might be a ramp up from 0 *revolutions per minute* (rpm) to 100 rpm over 5 seconds, followed by a steady rotation rate of 100 rpm for 5 seconds, and finally a ramp down to 0 rpm over 5 seconds.

13.3.6 Environmental Testing

Environmental testing can include evaluation of one or more conditions such as temperature, humidity, and atmospheric pressure, to name a few. During thermal evaluation, the device is tested at a temperature different than room temperature or over a time varying temperature range to determine how the temperature affects the performance of the device. Thermal evaluation determines how a device's characteristics change as a function of temperature. Static thermal evaluation consists of holding the device at a constant temperature different than room temperature for some period of time. For temperatures greater than room temperature, this is fairly easy to accomplish. The device can either be heated on a hot plate or in a small oven, called a box oven. A photograph of a box oven is presented in Figure 13.11. Small storage freezers can be purchased that go as low as −40°C, but special equipment must be purchased or constructed if testing at lower temperatures is required. Many types of box ovens, as well as other types of environmental testing chambers, have one or more holes in the back or side panel for electrical feedthroughs so that devices placed in the chamber can be powered up and evaluated using test equipment that resides outside of the chamber. During the test, it is important to use an accurate temperature-measuring device, such as a thermocouple, attached to or close to the MEMS device. Kapton tape, a high temperature polyimide material with attached adhesive, can be used to attach the thermocouple for testing at temperatures up to 400°C. It is important to let the device reach the test temperature before recording the output of the MEMS device.

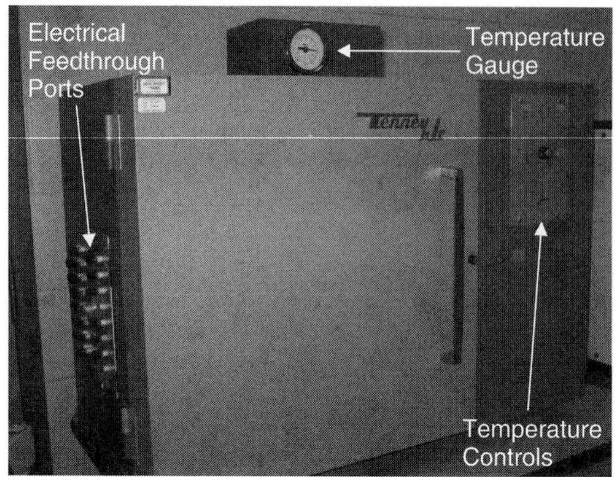

■ **FIGURE 13.11**

Photograph of a box oven.
(Courtesy of Auburn University.)

Another type of thermal evaluation is thermal cycling where a device is placed in a thermal cycling chamber and the temperature is repeatedly cycled between a low temperature and a high temperature. The thermal cycling rate is generally slow (perhaps 30 min) so that the device has ample time to respond to the stimulus. Again, the response of the device during thermal cycling is usually recorded by equipment residing outside of the chamber. The duration of thermal cycling can range from one cycle to several thousand cycles.

High-temperature or low-temperature storage is another thermal evaluation technique, where the unpowered device under evaluation is placed in either a high-temperature oven or a low-temperature freezer for an extended period of time. After the device is removed and brought back to room temperature, it is then powered up and evaluated to see if the thermal storage event had a deleterious effect on the device. Another thermal evaluation technique called thermal shock is sometimes used, where a device is quickly cycled between hot and cold temperatures to stress the device to the point of failure. If the device is cycled while electrically connected, the number of cycles it takes to cause the device to fail can be evaluated. Even if the device is thermally cycled while unpowered, it can be examined after the test to investigate failure mechanisms.

Many types of MEMS sensors have been developed for detecting chemicals such as carbon dioxide, carbon monoxide, hydrogen sulfide, and atmospheric water vapor. Of all of these types of chemical sensors, humidity sensors are among the easiest and probably the safest to test. Although the simplest test would be to have someone breathe on the sensor and then record the output of the device from the stimulus, this is probably not the preferred way to evaluate the sensor. A more quantitative evaluation can be accomplished through utilizing a controlled humidity chamber where a closed-loop system accurately sets the humidity level inside the chamber to the desired level. In addition to humidity, temperature can also usually be controlled inside the chamber. Typically the humidity and temperature can be controlled over a certain range, such as a relative humidity range from 40% to 100% over a temperature range from 20°C to 85°C. Some chambers also allow the humidity and temperature to be cycled, similar to how temperature is cycled in a thermal cycling chamber. Many humidity chambers also have one or more ports for electrical feedthroughs so that the sensor or sensors under evaluation can be powered up and monitored throughout the test. In addition to evaluating the performance of MEMS humidity sensors, the effects of relative humidity levels on other types of MEMS sensors that have direct access to the environment can be evaluated.

For some types of MEMS devices, it is useful to perform the evaluation while changing the pressure of the surrounding atmosphere. Examples include pressure sensors that utilize sealed cavities, resonators, and unpackaged accelerometers. Some pressure sensors utilize an on chip-sealed cavity that is separated from the external environment by a micromachined diaphragm. The pressure difference between the surrounding environment and the sealed cavity determines the deflection of the diaphragm. Micromachined resonators often utilize fluidic damping to control the resonating frequency of the device, and many accelerometers utilize fluidic damping to control the mechanical quality factor of the device. Therefore, evaluating the performance of these types of devices in a chamber where the atmospheric pressure can be controlled is useful when testing and characterizing the devices. A simple low-pressure evaluation system can be assembled using a glass bell jar, a base with electrical feedthroughs, a seal ring with vacuum grease, a pressure gauge, and a mechanical vacuum pump. Mechanical vacuum pumps can typically achieve low vacuums in the range of about 1×10^{-4} Torr (where 760 Torr is standard atmospheric pressure). Turbo and cryopumps can achieve a high vacuum, with pressures approaching 1×10^{-8} Torr. Whereas mechanical pumps can cost as little as a few thousand dollars, turbopumps and cryopumps can easily cost more than $10,000. High-pressure chambers can also be purchased, but pressures less than atmospheric pressure are usually of more interest because MEMS devices are often packaged and sealed at such pressures to obtain the desired fluidic damping levels.

13.4 RF MEMS DEVICES

MEMS employed in ***radio-frequency*** (RF) applications are called RF MEMS. These represent a new class of devices and components that exhibit low insertion loss, high isolation, high Q, small size, and low power consumption, and they enable new system capabilities. The application of MEMS in RF technology can be broadly classified into two categories: active (moving) devices, which involve mechanical motion (RF MEMS switch, RF MEMS capacitors, RF MEMS resonators, etc.) and static (nonmoving) components (micromachined transmission lines, resonators, etc.). This section examines instrumentation and methodologies for testing as well as performing reliability assessments and failure analyses of RF MEMS.

13.4.1 RF MEMS Switches

Peterson was the first to demonstrate microelectromechanical relays [Petersen 1978]. Since then, several MEMS relays and high-frequency MEMS switches have been developed by various researchers and discussed in the literature [De Los Santos 1999] [Rebeiz 2002]. MEMS relays are more preferable than other conventional semiconductor-based switching devices such as field effect transistors because of their low-loss, low power consumption; absence of intermodulation distortion; and broad-band operation from DC to the microwave frequency range. RF MEMS switches can be classified as resistive (or ohmic) contact or capacitive contact type depending on the type of contact used for signal transmission.

An **ohmic contact switch** uses a metal-to-metal contact between the two electrodes for signal transmission. In the ON state, the electrodes are in contact, and in the OFF state, the electrodes are separated by a small gap. Ohmic contact RF MEMS switches can be characterized by measuring the resistance, R, *versus* applied actuation voltage, V. The operating voltage required to obtain electrical continuity can be obtained from measuring the R-V characteristics. RF characteristics of RF MEMS switches are obtained by measuring the S-parameters in both the ON and OFF states of the switch. S-parameters are most commonly used for electrical characterization of devices, components, and networks operating at RF and microwave frequencies. Many electrical properties such as insertion loss, return loss, gain, ***voltage standing wave ratio*** (VSWR), and reflection coefficient can be expressed using S-parameters. The measurement test setup consists of a vector network analyzer interfaced to a probe station. Typically, on-wafer calibration is performed before performing measurements on the switches.

In a **capacitive contact switch**, a thin dielectric layer is present between the two electrodes. ON and OFF states are obtained by changing the capacitance from a high to low value or vice versa in a series or shunt configuration, respectively. Capacitive contact RF MEMS switches can be characterized by measuring the ***capacitance-voltage*** (C-V) characteristics. A C-V meter or an impedance analyzer equipped with a bias-T can be used in conjunction with a probe station to obtain C-V characteristics. The pulldown voltage can be determined from the C-V characteristics. RF characteristics of RF MEMS switches are obtained by measuring the S-parameters using a network analyzer (see Figure 13.13b).

The reliability of MEMS switches has been a major concern that limits the use of MEMS in real-world applications. Ohmic contact MEMS switch reliability issues, such as failure caused by stiction and contact degradation, have been observed to be the key failure modes. *Stiction* is the unintentional adhesion of the movable and fixed parts in MEMS caused by surface adhesion forces. Failure because of stiction is frequently encountered in electrostatically actuated contact type MEMS relays. Typically, MEMS switches are designed for operation at low actuation voltages, which necessitates the design of movable micromechanical parts with low restoring spring forces. Permanent failure because of stiction occurs when the restoring spring force of the movable part is lower than the adhesion forces generated at the contact surfaces. Another common reliability issue in ohmic contact MEMS switches is the increase of resistance over many actuation cycles. Specifically, the

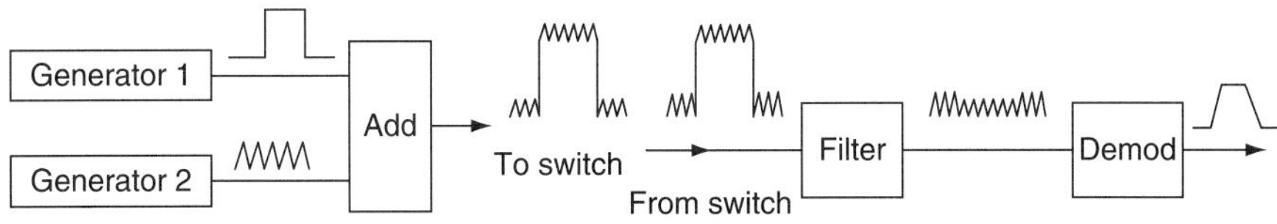

■ **FIGURE 13.12**

Low-frequency electrical test setup for reliability assessment of capacitive RF MEMS switches [van Spengen 2003].

resistance of the MEMS relays gradually increases with actuation cycles and, after several million actuation cycles, leads to an unacceptably high insertion loss. In capacitive contact type MEMS switches, reliability issues such as stiction from charge accumulation in the dielectric layer and capacitance degradation with actuation are commonly encountered failure modes.

The reliability and lifetime testing of RF MEMS switches based on S-parameter measurements is expensive because of the duration of the test time. A low-frequency electrical test setup for reliability testing of RF MEMS switches is shown in Figure 13.12 [van Spengen 2003]. The setup consists of two signal generators, a filter, and a demodulator. The RF MEMS switches are driven by an actuation signal from generator 1. A low frequency RF signal from generator 2 is superimposed on the actuation signal. The combined signal is applied to the RF MEMS switch, which is part of a voltage divider. The carrier voltage is modified by the switching action, resulting in an amplitude modulation of the signal. The modulated signal is detected using a demodulator to obtain switch characteristics such as pull-in voltage, rise time, fall time, and capacitance change for capacitive switches or contact resistance change for ohmic switches. Reliability of switches can be quantified by measuring the drift in any of these parameters.

13.4.2 RF MEMS Resonators

A mechanical filter is composed of multiple coupled lumped mechanical resonators. Mechanical filters transform electrical signals into mechanical energy, perform a filtering function, and then transform the remaining output mechanical energy back into an electrical signal. The characteristics of mechanical filters include low-loss, narrow bandwidth, and high temperature stability. Mechanical filters were first developed in the late 1940s and are widely used in telephone systems. MEMS technology has been applied to the miniaturization of mechanical resonators and filters. Various designs investigated for MEMS-based mechanical resonators include double-folded beam, clamped-clamped beam, free-free beam, wine-glass disk resonator, contour mode disk resonator, and hollow disk ring resonator [Nguyen 2005]. MEMS resonators and filters are characterized by measuring the frequency response characteristics. The performance parameters such as the resonant frequency, Q-factor, and bandwidth are obtained from the measured frequency response characteristics. The equivalent circuit parameters can be extracted from the measured frequency response characteristics. This section discusses instrumentation setup, testing, and characterization of one-port and two-port RF MEMS disk resonators [Clark 2001, 2005] [Hsu 2004].

A MEMS disk resonator in a one-port configuration is shown in Figure 13.13a [Clark 2005]. A typical test setup for testing a one-port contour-mode disk RF MEMS resonator is shown in Figure 13.13b. The required test instrumentation includes a network analyzer, a DC voltage source, a bias-T, and a vacuum chamber. The contour-mode disk resonator consists of a resonating circular disk, two input electrodes, and a bottom output/bias electrode. In this setup, the RF-out port, v_i, of the network analyzer is connected directly to both input electrodes of the resonator inside the vacuum test chamber. The analyzer's RF-in port, v_o, is connected to the resonating disk through the output/bias electrode located under the disk. A bias-T is used to apply a DC bias voltage, V_p, to the disk. The bias-T separates the DC-bias voltage and the RF signal. Before the measurements, a through calibration is performed by replacing the DUT by a short, and the network analyzer is baseline corrected against this measurement. Figure 13.13c shows a transmission spectrum (v_o/v_i) obtained from a one-port measurement of a 156 MHz disk

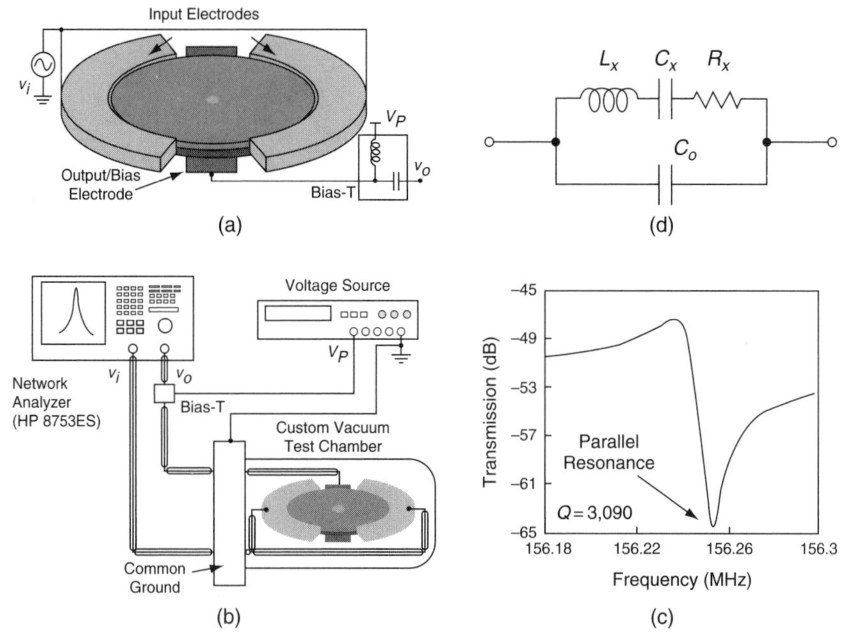

■ **FIGURE 13.13**

RF MEMS disk resonator in one-port configuration [Clark 2005]: (a) schematic of the dark resonator, (b) test setup, (c) measured transmission spectrum, and (d) equivalent circuit model.

resonator. A large mismatch of about 47 dB was measured at the resonant frequency of 156 MHz because of the mismatch between the impedance of the resonator and the 50 Ω impedance of the test equipment. From the measured results, the equivalent circuit model, shown in Figure 13.13d, parameters are extracted and found to be $R_x = 22.287\,\text{k}\Omega$, $L_x = 70.15\,\text{mH}$, $C_x = 14.793\,\text{aF}$, and $C_o = 57.78\,\text{fF}$. This test setup is applicable to a wide variety of one-port resonators.

An RF disk resonator in two-port configuration is illustrated in Figure 13.14a [Clark 2005], and a typical test setup is shown in Figure 13.14b. The RF input signal, v_i, is applied to the input electrode and the RF output signal, v_o, is measured from the output electrode. A DC bias voltage is directly applied to the disk without the need for a bias-T. The equivalent circuit model of the two-port resonator is shown in Figure 13.14d. In the two-port configuration, the input-to-output static capacitance, C_o, is split into half ($C_o/2$) and shunted to ground at the input and output, respectively. The capacitor C_{op} models the feedthrough capacitance between the input and output electrodes. Figure 13.14d shows a transmission spectrum (v_o/v_i) of a 156 MHz resonator obtained via the two-port technique. From the measured results, the equivalent circuit model parameters are extracted and found to be $R_x = 99.9$ kΩ, $L_x = 945.4$ mH, and $C_x = 1.098$ aF. This test setup can also be used for testing other types of two-port resonators.

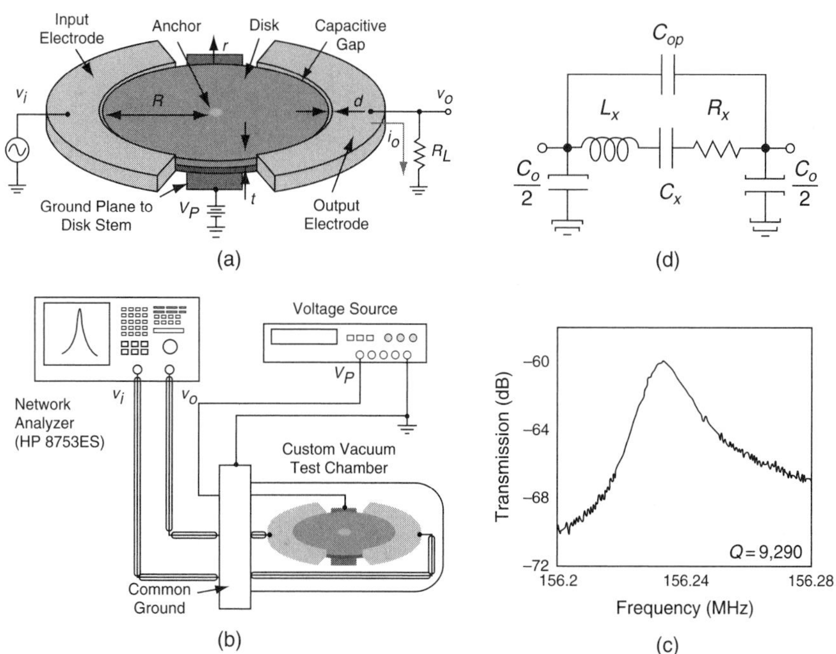

■ **FIGURE 13.14**

RF MEMS disk resonator in one-port configuration [Clark 2005]: (a) schematic of the dark resonator, (b) test setup, (c) measured transmission spectrum, and (d) equivalent circuit model.

13.5 OPTICAL MEMS DEVICES

The integration of microoptics and MEMS has created a new class of devices, termed optical MEMS or *micro-opto-electromechanical systems* (MOEMS). The advantages of optical MEMS devices include high functionality, high performance, and low-cost. This section briefly discusses the testing of piston micromirrors and tilt micromirrors.

A typical piston micromirror consists of a mirror segment supported by four springs and is capable of movement in the direction normal (*i.e.*, vertical) to the mirror surface. Arrays of piston micromirrors are employed in **adaptive optics** to compensate for variable optical aberrations. For a known operating wavelength of incident light, the phase shift obtained from each micromirror depends on the vertical displacement. The two important characteristics of interest are (1) static characteristics (*i.e.*, vertical displacement *versus* applied voltage characteristics) and (2) dynamic characteristics (*i.e.*, transient response). The deflection *versus* applied voltage characteristics can be obtained by measuring the optical profile of the micromirror for various applied voltages. The optical profile of an array of 248 μm × 248 μm gold mirrors on polysilicon plates fabricated using the MUMPS process is shown in Figure 13.15 [Tuantranont 2000]. Dynamic characteristics of piston mirrors can be measured using laser Doppler vibrometers.

A typical tilt micromirror consists of a flat mirror segment supported by two torsional springs. Tilt micromirrors change the angle of reflection of incident light by angular or torsional rotation of micromirror structures. The tilt micromirror is a critical component in an optical scanning system, which is used to deflect the beam of light in different directions and detection of the reflected beam to determine spatial information. The two important characteristics of interest are (1) static characteristics (*i.e.*, tilt angle *versus* applied voltage characteristics) and (2) dynamic characteristics (*i.e.*, transient response). To measure the tilt angle *versus* applied voltage characteristics, a laser beam is directed on the mirror surface while the reflection of the laser beam off the mirror surface is projected onto a screen mounted vertically and parallel to the scanner's chip surface. The screen is

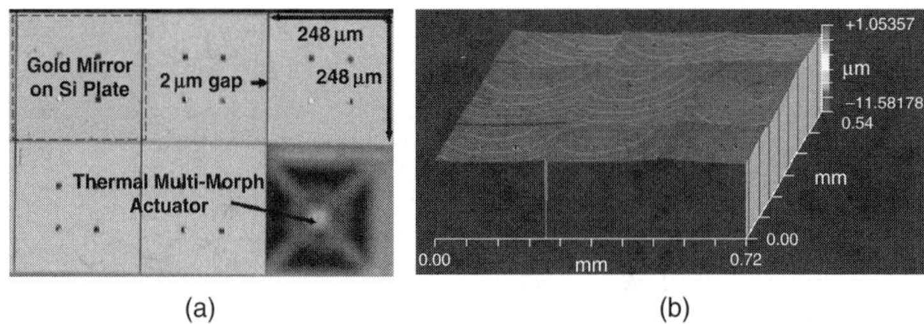

■ **FIGURE 13.15**

Piston micromirror [Tuantranont 2000]: (a) gold micromirrors on polysilicon plates fabricated using MUMPS process and (b) optical profile measured using Zygo interferometer.

spaced from the mirror surface at an arbitrary distance, *d*. In the unactuated state, the mirror displays a circular spot (typically a few millimeters in diameter) on the screen. When the mirror is actuated, the mirror surface is tilted at an angle and therefore the spot location is shifted on the screen. The tilt angle can be calculated using the displacement of the spot on the screen and the distance of separation, *d*, between the mirror and the screen [Motamedi 2005]. The dynamic characteristics of tilt mirrors can be measured using laser Doppler vibrometers.

As an example, dynamic characteristics of an Applied MEMS Durascan two-axis tilt mirror measured using Polytec's Laser Doppler vibrometer are shown in Figure 13.16 [Polytec 2005]. Dynamic parameters such as switching time and settling times can be obtained from these results. The tilt mirror–based ***digital micromirror device*** (DMD) developed by ***Texas Instruments*** (TI) has made tremendous progress in both performance and reliability since it was first invented in 1987. During the development stage, TI spent considerable effort to perform failure analysis of DMDs for various stresses. The key failure mechanisms affecting the reliability of the DMD were identified to be hinge memory, hinge fatigue, stiction, and environmental robustness including shock and vibration failures. After the elimination of various failure modes using clever solutions, DMD has become one of the flagship MEMS technologies. The DMD is now providing high brightness, high contrast, and high reliability in millions of projectors using ***digital light processing*** (DLP) technology [Douglass 2003].

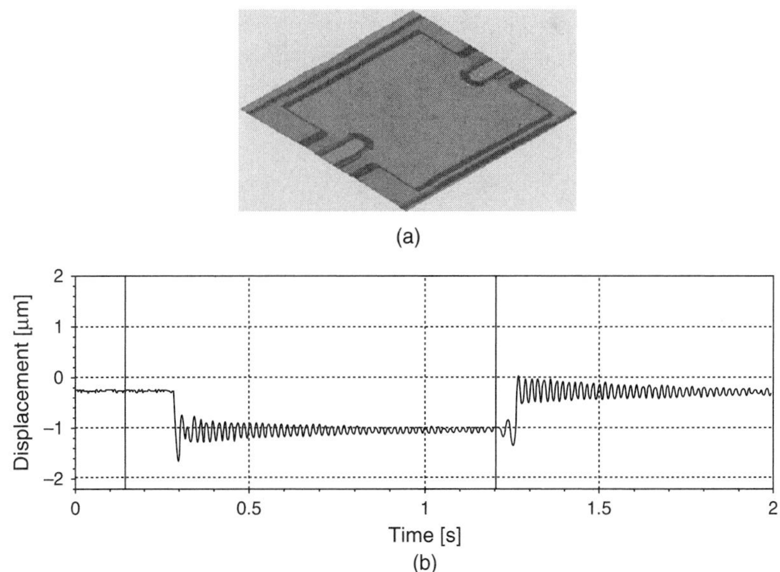

■ **FIGURE 13.16**

Two-axis tilt mirror: (a) Applied MEMS Durascan tilt mirror and (b) two-axis tilt mirror. (Courtesy of Polytec.)

13.6 FLUIDIC MEMS DEVICES

Fluidic MEMS devices are micromachines that respond to or interact with fluids. The fluid can be in liquid or gaseous state. Many liquid microfluidic devices consist of miniature plumbing systems, whereas gaseous microfluidic devices generally possess microstuctures that either affect or are affected by interaction with gasses in the surrounding environment.

Microfabrication techniques can be utilized to realize miniature plumbing systems for transporting and processing fluids, primarily liquids. Standard Si-based MEMS fabrication technologies, such as DRIE etching, anisotropic wet etching, and anodic bonding, as well as non-Si-based fabrication technologies such as ***polymethyl methacrylate*** (PMMA) and ***low temperature co-fired ceramic*** (LTCC), can be utilized to fabricate microchannels and cavities. Using these fabrication processes, microscale flow channels, reaction chambers, heaters, pumps, and valves have been successfully created. Researchers have realized complete microfluidic systems, such as fuel cells [Sakaue 2005], thermal management systems [Shi 2006], tunable die lasers [Bilenberg 2003], and biochemical analysis systems [Choi 2001].

The procedures for testing microfluidic MEMS devices can be quite different from the procedures used to test MEMS devices such as electrostatic actuators, accelerometers, and RF switches. Some types of microfluidic devices utilize a working fluid that is sealed in the device during manufacturing. Testing of these types of devices may be limited to leak testing after manufacture and functional testing. Other types of microfluidic devices only have a fluid inserted into them during actual operation. If this type of device is reusable, then a test fluid can be inserted into the device for evaluation of the device performance. However, if the device is for one-time-only use, then this type of testing is not possible.

A relatively new device that is finding application in liquid microfluidic devices is the flowFET [Kerkhoff 2003b]. This fluidic device is analogous to a FET transistor, except that fluid flow replaces electric current flow. These devices operate using the principle of electro-osmotic flow. In microchannels, the inner surface is naturally electrically charged, which results in the buildup of a thin charged layer in the fluid along the channel walls. By inserting two electrodes, referred to as the Source and Drain electrodes, along one wall of the microchannel and applying a sufficiently large voltage across them, the resulting electric field will move the ions in the fluid, which drags the fluid along with them. A third electrode can be added in between the first two electrodes but on the other side of the microchannel. This electrode acts as the flowFET Gate electrode by controlling the fluidic flow when a suitably large voltage is applied between it and the Source electrode. The achievable fluid velocity is also dependent on the chemical properties of the liquid, such as the PH. The magnitude of the voltage level required to operate the flowFET is typically between 0 V and 100 V, which is compatible with high voltage MOSFET technologies, such as DMOS. Conventional low-voltage electronic technologies such as CMOS can interface to the DMOS transistors that drive the flowFETs. By utilizing this technology, microfluidic devices and systems can be electrically controlled as well as stimulated to implement functional testing algorithms [Kerkhoff 2003a].

For example, by integrating suitable sensors into the microfluidic device or system, the flowFETs could be used to evaluate flow velocity or to test for clogged channels in the device or system.

13.6.1 MEMS Pressure Sensor

Pressure sensors are one of the most successful MEMS devices with a wide range of applications in automotive systems, industrial process control, environmental monitoring, medical diagnostics, and monitoring. A MEMS pressure sensor consists of a mechanical membrane present at the interface between a sealed cavity and the external environment. The pressure difference between the sealed cavity and the surrounding environment causes the diaphragm to deflect. The membrane deflection is converted to a change in the output voltage of the sensor by capacitance or piezoelectric transduction. Pressure sensors are characterized by measuring the output response for various applied pressures. From these results, parameters such as sensitivity (mV/Pa) and linearity can be obtained. Pressure sensors can be tested and characterized using commercially available pressure chambers (see Section 13.3.6). Here, we briefly discuss a custom-made pressure chamber developed for characterizing pressure sensors. The measurement setup for testing capacitive pressure sensors is shown in Figure 13.17 [Palasagaram 2006]. The setup consists of two components: (1) a custom-made pressure chamber, which can withstand large pressures, and (2) the signal conditioning circuitry. The pressure chamber is made of Teflon with dimensions of 9.5" × 8.5" × 3". It has a pressure inlet on one side and a safety valve connected to a hole on the other side. A pressure gauge is used to monitor the pressure inside the chamber. The pressure sensor is placed inside the chamber. When the pressure inside the chamber exceeds the atmospheric

■ **FIGURE 13.17**

Experimental setup used for characterizing pressure sensors [palasagaram 2006].

pressure, the movable diaphragm starts deflecting downward, thereby increasing the capacitance between the top and bottom electrodes. The signal conditioning board (MS3110BDPC from Microsensors Inc.) outputs a voltage corresponding to a change in the sensor capacitance.

13.6.2 MEMS Humidity Sensor

Many types of micromachined devices make use of hermetically sealed structures, either as the package into which a MEMS device is hermetically sealed or as sealed microcavities directly on a substrate. In either case, it is necessary to evaluate the quality of the hermetic seal. A convenient method for doing this is to integrate a MEMS humidity sensor into the sealed cavity, where the cavity is either sealed in high vacuum or backfilled with a dry gas such as nitrogen. The humidity sensor is then monitored to measure the relative humidity inside the sealed cavity in order to detect if moisture has leaked into the cavity. The sealed test specimen can be put through a series of tests, such as thermal cycling or submersion in boiling water, and then the output from the humidity sensor can be probed to see if the internal relative humidity level has increased.

Consider the Hygrometrix HMX2000 MEMS humidity sensor, a commercially available or ***commercial-off-the-shelf*** (COTS) silicon MEMS device. Front and backside photographs of an unpackaged HMX2000 die are presented in Figure 13.18 [Dean 2005a]. The HMX2000 is a micromachined piezoresistive device approximately 2-mm square and consisting of four thin cantilever silicon beams that protrude into a center flow channel. Each beam has been fabricated as a piezoresistor, where the electrical resistance is proportional to the mechanical strain experienced by the beam. Four beams have been fabricated and electrically connected to realize a Wheatstone bridge. The beams are coated with a polymeric material that absorbs or desorbs moisture from the ambient atmosphere in proportion to the amount of moisture present. As moisture is absorbed or desorbed, the volume of the polymeric coating changes, straining the thin cantilevered beams and changing the resistance

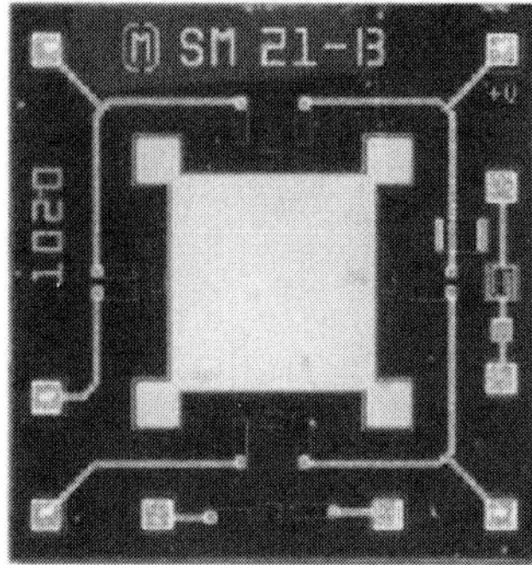

■ FIGURE 13.18

Front and backside photographs of a HMX2000 MEMS humidity sensor [Dean 2005a].

of the piezoresistors. Two of the four beams are constructed to increase in resistance with increasing humidity, whereas the other two decrease. The result is a resistive bridge with a differential resistance proportional to the relative humidity of the ambient atmosphere. Some versions of the device have a solid glass plate attached to the backside of the die, whereas others allow airflow through the device. The HMX2000 sensors can be purchased as bare dies or as packaged devices. Because the HMX2000 does not have on-chip interface electronics, external electronics need to be assembled to utilize the device.

The bare die version of the HMX2000 is small enough to be integrated into most packages and even many sealed micromachined cavities. The sensor requires four electrical traces to measure the relative humidity, and these have to be designed into the sealed cavity or package under investigation. The interface electronics required to operate the HMX2000 can reside on a PCB a short distance away from the sealed package. Because each piezoresistor on the sensor has a nominal resistance of approximately 4.3 kΩ, it is important to keep the cable connecting the sensor to the interface circuit as short as is reasonably possible to minimize noise on the sensor signals from capacitive coupling to the signal traces in the cable.

After the sensor has been sealed inside the cavity, the output of the humidity sensor should be measured, if possible, while the unit is still inside the sealing chamber at the same pressure and gas chemistry as the inside of the sealed cavity, in order to obtain a baseline humidity sensor reading for later comparison. Also, if the temperature of the sealing chamber can be adjusted, it is worthwhile to take a series of sensor readings at various temperatures to characterize the humidity sensor inside the sealed package with regard to temperature. The output of the HMX2000 humidity sensor embedded in a hermetically sealed cavity may change as a function of temperature because of a temperature dependence but also from absorbed moisture being liberated from the inside of the sealed cavity as a result of an increase in temperature. Therefore, it is crucial to characterize the embedded humidity sensor before removing the unit from the sealing chamber.

Once the sealed cavity with the embedded and characterized humidity sensor has been removed from the sealing chamber, a variety of tests can be performed. A temperature sensor should be placed in close proximity to the sealed cavity so that the temperature of the sealed cavity can be recorded at the same time that the output of the humidity sensor is recorded. An example temperature sensor that could be used for this purpose is a thermocouple attached to the outside of the sealed cavity with Kapton tape. Also, the placement of a second HMX2000 humidity sensor, packaged so that it is in direct contact with the environment, will allow the ambient relative humidity level outside of the sealed cavity to be measured. If the sealed cavity leaks, the relative humidity inside the cavity will approach the level of the relative humidity outside the cavity over time, and having both a humidity sensor inside the sealed cavity and one outside the sealed cavity will verify the leak.

The simplest test is to place the sealed cavity at atmospheric conditions for a period of time and record the temperature and humidity sensor output at selected time intervals. If the relative humidity inside the sealed cavity approaches the relative humidity outside of the package during this test, the results indicate that there was a failure in the sealing process. Additional tests can include thermal cycling, high-temperature or low-temperature storage, thermal shock, vibration, mechanical shock, and submersion in boiling water, as well as combinations of these tests. If the relative humidity inside the sealed cavity did not change while the sealed cavity was left at ambient conditions for a reasonably long period of time, but it did change during any of these subsequent tests, it is likely that the subsequent test or tests caused the cavity sealing mechanism to fail, which can indicate either a problem during the sealing process or a possible failure mechanism for the sealing technique.

It should be noted that during any type of testing, it is inadequate to evaluate only a single device. A reasonable number of devices should be evaluated to obtain representative characteristics of the device type. For example, if 5% of the devices in a particular sensor model were faulty, and only five devices were tested, it is possible that all the devices tested would be good devices, leading to the erroneous conclusion that all devices of this model are good. In addition to determining the percentage of good or bad parts of a particular model, most devices have some performance variation differences between otherwise identical units, and evaluating a reasonable number of units yields the statistical performance variations of that particular model.

13.7 DYNAMIC MEMS DEVICES

Dynamic MEMS devices are micromachines that possess one or more members that respond to an applied force by acceleration, resulting in mechanical motion. The applied force could be internally generated, such as the force resulting from a microactuator, or externally generated, such as the force resulting from interaction with the environment. A number of MEMS sensors can be accurately described as dynamic MEMS devices, including microphones, accelerometers, and gyroscopes.

13.7.1 MEMS Microphone

MEMS microphones have been successfully commercialized for use in cell phones, cameras, PDAs, and other high-volume consumer electronics [Loeppert 2006]. The microphones are characterized by measuring sensitivity, frequency response, and noise. The sensitivity (mV/Pa) is obtained by exciting the microphone at a chosen sinusoidal ***sound pressure level*** (SPL) and measuring the output voltage of the microphone for various DC bias voltages. The frequency response is obtained by exciting the microphone with a periodic noise over the desired operating frequency range and measuring the sensitivity of the microphone. The relative gain and resonance frequency can be obtained from the frequency response characteristics. The noise measurements are performed by measuring the frequency response of the microphone in an anechoic chamber.

A typical test setup for acoustical test and characterization of the integrated microphone is illustrated in Figure 13.19 [Pedersen 1998]. The instrumentation includes a signal analyzer and amplifier. The reference microphone, MEMS microphone, and test speaker are located inside the anechoic chamber. The dimensions

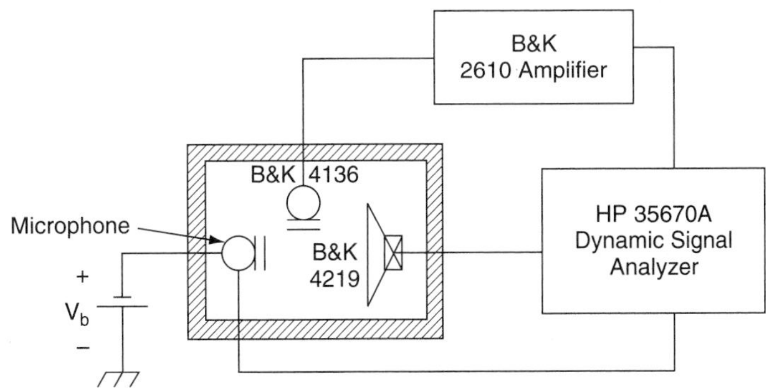

■ **FIGURE 13.19**

Measurement setup for acoustical characterization of MEMS microphone.

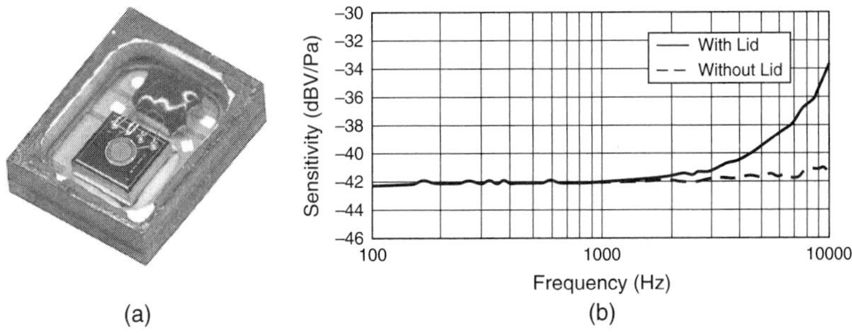

■ **FIGURE 13.20**

Knowles SiSonicTM MEMS microphone [Loeppert 2006]: (a) microphone die with a CMOS die in an acoustic housing and (b) measured frequency response with and without a lid.

of the chamber are chosen such that standing waves are avoided in the frequency range of interest. The inside of the chamber is covered with sound absorbing material to minimize the influence of reflections as well as external noise. This results in an approximate free sound-pressure field. The loudspeaker is driven by a dynamic signal analyzer, which uses a reference microphone in a feedback loop to maintain the output of the loudspeaker at a specified level in the frequency range of interest. The amplifier is used to boost the signal output from the reference microphone. An example frequency response of the Knowles SiSonic MEMS microphone [Loeppert 2006] is shown in Figure 13.20.

13.7.2 MEMS Accelerometer

MEMS accelerometers are one of the most widely used types of MEMS devices that are commercially available. They utilize, in some form, the relative motion between an inertial mass and a surrounding structure to detect or measure an external acceleration that has been applied to the surrounding structure. A common

TABLE 13.1 ■ Angular Rates Required to Achieve the Desired Acceleration Test Plan

Acceleration (Gs)	Angular Rate (rpm)
0	0
0.5	105.69
1.0	149.47
1.5	183.06
2.0	211.38
2.5	236.33
3.0	258.89
3.5	279.63
4.0	298.94
4.5	317.07
5.0	334.23

application for MEMS accelerometers is the detection of an automobile crash in order to engage the inflation of the passenger restraint safety airbags.

Suppose one desires to test a ±5G packaged accelerometer in 0.5G increments. One way to accomplish this is to mount the accelerometer on a small PCB and then attach the PCB onto a small aluminum block that can be fastened to the surface of a rate table where the sensitive axis of the accelerometer is radially aligned toward the center of rotation. The accelerometer power and output signal traces are connected through the rate table slip ring assembly. When the rate table is spinning, the accelerometer will experience acceleration because of the centripetal force. The direction of applied acceleration can be changed by rotating the aluminum block, onto which the accelerometer PCB is mounted, 180° so that the sensitive axis of the accelerometer is radially aligned away from the center of rotation. Suppose that the accelerometer is mounted such that the distance between the proof mass and the center of rotation is 4 cm. Then a table of required rate table angular rates can be generated using Equation (13.9) to accomplish the desired test plan, as illustrated in Table 13.1.

Some accelerometers may have BIST features that allow the user to apply a signal to the device that causes the accelerometer proof mass to be displaced a certain distance to simulate the effect of a particular acceleration on the device. The BIST feature is generally used to verify that a fielded device is still functioning correctly by applying an artificial stimulus and detecting the response of the sensor. For accelerometers with this feature, however, the effects of other types of testing, such as thermal cycling and vibration, along with an applied acceleration can be estimated for the device by utilizing the BIST feature while the other stimulus is being applied to the device.

13.7.3 MEMS Gyroscope

A gyroscope detects the presence of rotational motion about a predefined axis. Most MEMS gyroscopic sensors operate by suspending a miniature proof mass with a microfabricated spring system. Some type of MEMS actuator is used to oscillate the proof mass structure along one axis in a controlled manner. Electrostatic parallel plate actuators, electrostatic comb drive actuators, and piezoelectric actuators have all been used for this purpose. If the sensor experiences a rotation about an orthogonal axis, the resulting Coriolis force causes the proof mass to experience sinusoidal motion along the third orthogonal axis [Bao 2000]. This motion is then detected to yield the angular rate information. Capacitive and piezoresistive measurement techniques have been used in various MEMS gyroscope designs to detect this motion.

A MEMS gyroscopic sensor can be characterized using a rate table by subjecting it to a series of angular rates, recording the output signal and comparing the results to the angular rates. The output signal from a MEMS gyroscope usually has a DC bias offset that needs to be measured and subtracted from the output signal to obtain the rate measurement. Furthermore, the DC offset may be strongly dependent on temperature, and some MEMS gyroscopes integrate a temperature sensor into the device for use in correcting for the temperature induced drift. Another parameter of interest is the noise level on the output signal. This can be investigated by running a series of tests at different fixed angular rates and computing the standard deviation on the recorded data for each angular rate, which is proportional to the noise level. However, it is important that the angular rate is constant, as the gyroscope will detect any "noise" in the rate table angular rate.

Because the microstructure is designed to resonate at a particular frequency and often has a high mechanical quality factor, the sensor can be susceptible to external vibrations present in the operating environment at or near that frequency. Therefore, the resonating frequency is designed to be well above the detection bandwidth of the sensor. An electromechanical shaker can be used to evaluate the susceptibility of a MEMS gyroscope to external mechanical vibrations, one axis at a time. Because different types of MEMS gyroscopic sensors resonate at different frequencies, one will need to obtain the designed internal resonating frequency of a particular MEMS gyroscope so that an electromechanical shaker can be selected that operates at that frequency. Consider the Analog Devices' ADXRS300 MEMS gyroscope that internally oscillates at approximately 14 KHz. To thoroughly evaluate the susceptibility of this MEMS gyroscope to mechanical vibrations, an electromechanical shaker with a bandwidth of 20 KHz is recommended. The packaged ADXRS300 should be mounted on a small PCB with all appropriate interface electronics (power/signal connector, decoupling capacitors, output signal buffer, etc.) and firmly attached to a rigid fixture, such as a small aluminum block, that can be attached to the shaker head. Attaching the PCB to the aluminum block with an adhesive such as glue or an epoxy will minimize vibrational resonances in the PCB that could adversely affect the test. The gyroscope is designed to measure the rotation rate about a particular axis relative to the package, and the device data sheet will indicate the direction of this axis relative to the package. The aluminum block, onto which the PCB containing the MEMS gyroscope is attached, should be machined so that it can be attached to the shaker head in three different orientations so that the gyroscope can be independently shaken along each axis. This allows the susceptibility of the sensor to external vibrations to be evaluated along each axis.

Shaker testing is then performed for each axis, either with a wide bandwidth excitation, a series of several small bandwidth excitations with different center frequencies, or with a swept sinewave excitation covering the desired frequency range. There should be a noticeable increase in the noise floor of the angular rate output signal when the sensor is vibrated close to its resonating frequency. Notice that these tests are performed while the sensor is not experiencing an angular rotation, because the electromechanical shaker cannot rotate and shake the sensor under evaluation. However, consider the Analog Devices' ADXRS300 angular rate sensor, which has a BIST feature that allows two external signals to be applied to the device to cause the internal microstructure to experience the same response as it would to angular rates of +55°/s and −55°/s, respectively [Weinberg 2005]. Therefore, by utilizing these BIST features, the shaker can be used to investigate the effects of external vibrations on the ADXRS300 at angular rates of −55°/s, 0°/s and +55°/s.

Micromachined structures can be utilized to realize miniature passive vibration isolation platforms, consisting of a frame, a spring structure, and a proof mass pad, which can be used to isolate a MEMS gyroscope die from high-frequency vibrations present in the operating environment. A photograph of a micromachined vibration isolation platform that is designed to isolate a MEMS gyroscope from external z-axis high frequency vibrations present in the operating environment is presented in Figure 13.21, where the z-axis is in a direction normal to the plane of the device [Dean 2005b]. The MEMS gyroscope die is attached to the center proof mass pad and wirebonded to pads along the periphery of the center proof mass pad. Eight micromachined springs connect the center proof mass pad to a surrounding frame. Electrical traces run along the top of each spring to connect the wirebond pads on the center proof mass pad to wirebond pads on the frame. The entire device can be mounted inside a standard integrated circuit package, with wirebonds made between the package and the pads on the frame. The frame is attached to a spacer,

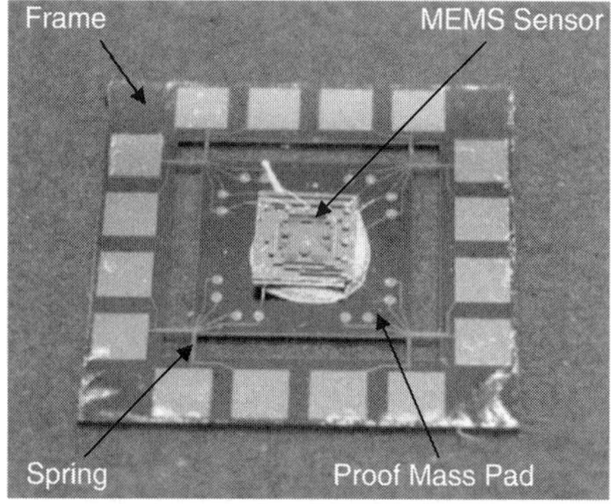

■ FIGURE 13.21

Photograph of a micromachined vibration isolation platform with an attached MEMS gyroscope die [Dean 2005b].

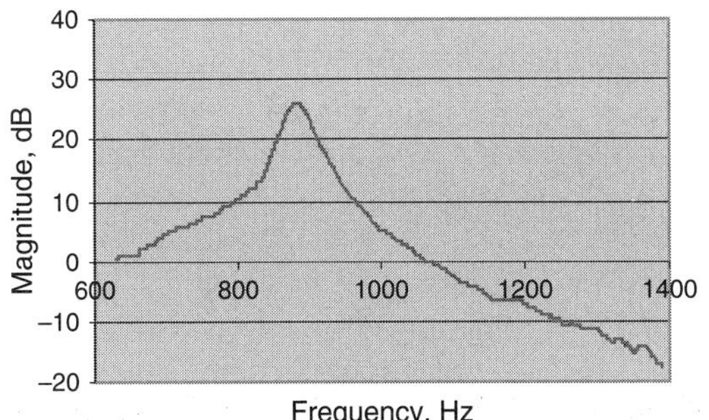

■ **FIGURE 13.22**

Plot of the measured frequency response of the micromachined vibration isolation platform [Dean 2005b].

which is attached to the bottom of the package, so that the proof mass pad can move vertically with respect to the bottom of the package without making contact. The thickness of the spacer is chosen to ensure that contact between the proof mass pad and the bottom of the package does not occur in the intended operating environment, as contact could damage the fragile micromachined device.

To evaluate the isolator, a machined plastic housing is used to hold the isolator frame so that the proof mass pad can move vertically. The plastic housing is attached to the shaker head of an electromagnetic shaker so that the device can be excited with a wide bandwidth vibration excitation. Laser interferometric displacement measurement units are used to measure the motion of the frame and the proof mass pad. Then the relative motion between the frame and the proof mass pad is calculated as a function of vibration frequency to obtain the transmissibility of the device. A plot of the measured transmissibility is presented in Figure 13.22. Observe that the isolator has a low pass response with a resonant frequency of approximately 885 Hz and a quality factor of approximately 20.

13.8 TESTING DIGITAL MICROFLUIDIC BIOCHIPS

Many commercially available biochips are based on continuous fluid, permanently etched microchannels [Schasfoort 1999] [Verpoorte 2003] [Zeng 2004]. Fluid flow in these devices is controlled either using micropumps and microvalves or by electrical methods based on electrokinetics and electroosmosis [2]. **FlowFETs**, where electroosmostic fluid flow is controlled by a gate electrode similar to the behavior of a MOSFET, have also been proposed [Schasfoort 1999].

An alternative category of microfluidic biochips, referred to as **digital microfluidics**, relies on the principle of electrowetting-on-dielectric [Cho 2003] [Fair 2003] [Chatterjee 2006]. Bioassay protocols are scaled down (in terms of liquid volumes and assay times) and run on a microfluidic chip by manipulating discrete droplets of nanoliter volume using a patterned array of electrodes. By reducing the rate of sample and reagent consumption, digital microfluidic biochips enable continuous sampling and analysis for online, real-time, chemical and biological analysis. These systems also have dynamic reconfigurability, whereby microfluidic modules can be relocated to other places on the electrode array, without affecting the functionality, during the concurrent execution of a set of bioassays. Reconfigurability enables the design of multifunctional and "smart" microfluidic biochips that can be used for a wide variety of applications. Moreover, defects can be tolerated through system reconfiguration after testing and fault diagnosis.

As chemists and biologists map more bioassays on a microfluidic platform for concurrent execution, system complexity and integration levels are expected to increase steadily. However, as in the case of ICs, an increase in density and area of microfluidics-based biochips will reduce yield, especially for newer technologies. Moreover, to reduce the cost for disposable devices, device manufacturers are investigating inexpensive processes and materials for low-cost biochip fabrication. As a result, microfluidic biochips are likely to suffer from high defect densities.

Dependability is an important system attribute for biochips that are used for safety-critical applications such as point-of care diagnostics, health assessment and screening for infectious diseases, air-quality monitoring, and food-safety tests, as well as for pharmacological procedures for drug design and discovery that require high precision levels. Some manufacturing defects may be latent and produce errors during field operation. In addition, harsh operational environments and biological samples (*e.g.*, proteins) may introduce physical defects such as particle contamination and residue on surfaces as a result of adsorption. Therefore, biochip platforms must be adequately tested after manufacturing, before the start of a bioassay, and during bioassay execution. Moreover, because disposable biochips are being targeted for a highly competitive and low-cost market segment, test and diagnosis methods should be inexpensive, quick, and effective.

13.8.1 Overview of Digital Microfluidic Biochips

A digital microfluidic biochip utilizes the phenomenon of electrowetting to manipulate and move microliter or nanoliter droplets containing biological samples on a two-dimensional electrode array [Fair 2003]. A unit cell in the array includes a pair of electrodes that acts as two parallel plates. The bottom plate contains a patterned array of individually controlled electrodes, and the top plate is coated with a continuous ground electrode. A droplet rests on a hydrophobic surface over an electrode, as shown in Figure 13.23. It is moved by applying a control voltage to an electrode adjacent to the droplet and, at the same time, deactivating the electrode just under the droplet. This electronic method of wettability control creates interfacial tension gradients that move the droplets to the charged electrode. Using the electrowetting phenomenon, droplets can be moved to any location on a two-dimensional array.

By varying the patterns of control voltage activation, many fluid-handling operations, such as droplet merging, splitting, mixing, and dispensing, can be executed in a similar manner. For example, mixing can be performed by routing two droplets to the same location and then turning them about some pivot points. The digital microfluidic platform offers the additional advantage of flexibility, referred to as reconfigurability, because fluidic operations can be performed anywhere on

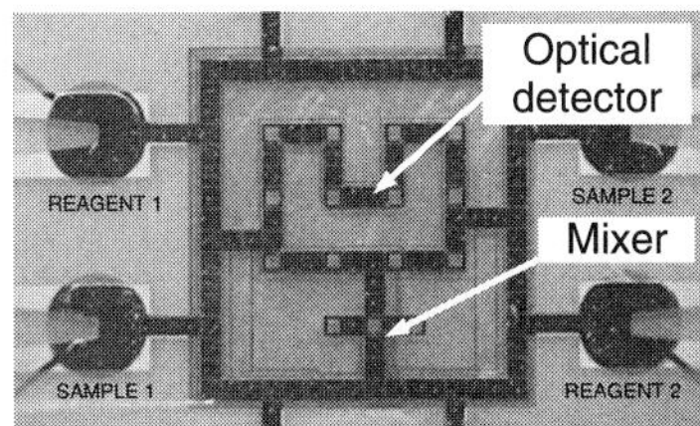

Glass-substrate platform

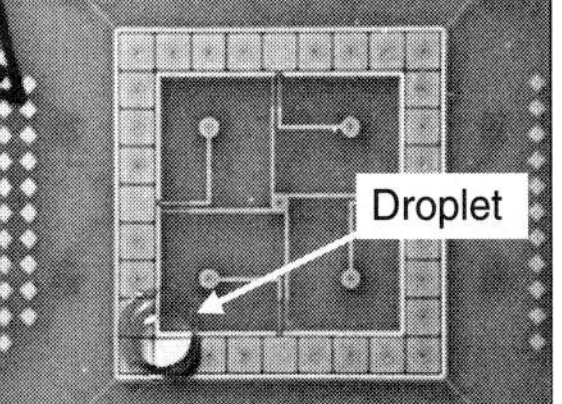

PCB platform

■ **FIGURE 13.23**

Fabricated digital microfluidic arrays.

the array. Droplet routes and operation scheduling results are programmed into a microcontroller that drives electrodes in the array. In addition to electrodes, optical detectors such as LEDs and photodiodes are also integrated in digital microfluidic arrays to monitor colorimetric bioassays [Fair 2003].

13.8.2 Fault Modeling

Like microelectronic circuits, a defective microfluidic biochip is said to have a failure if its operation does not match its specified behavior. To facilitate the detection of defects, fault models that efficiently represent the effect of physical defects at some level of abstraction are required. These models can be used to capture the effect of physical defects that produce incorrect behaviors in the electrical or fluidic domain. As described in [Su 2003], faults in digital microfluidic systems can be classified as being either catastrophic or parametric. Catastrophic faults lead to a complete malfunction of the system, whereas parametric faults cause degradation in the system performance. A parametric fault is detectable only if this deviation exceeds the tolerance in system performance.

Table 13.2 lists some common failure sources, defects, and the corresponding fault models for catastrophic faults in digital microfluidic biochips. Catastrophic faults may be caused by a number of physical defects, including the following:

- *Dielectric breakdown*. The breakdown of the dielectric at high voltage levels creates a short between the droplet and the electrode. When this happens, the droplet undergoes electrolysis, thereby preventing further transportation.
- *Short between the adjacent electrodes*. If a short occurs between two adjacent electrodes, the two electrodes effectively form one longer electrode. When a droplet resides on this electrode, it is no longer large enough to overlap the gap between adjacent electrodes. As a result, the droplet can no longer be actuated.
- *Degradation of the insulator*. This degradation effect is unpredictable and may become apparent gradually during the operation of the microfluidic system. A consequence is that droplets often fragment and their motion is prevented

because of the unwanted variation of surface tension forces along their flow path.

- *Open in the metal connection between the electrode and the control source*. This defect results in a failure in activating the electrode for transport.

Examples of some common parametric faults include the following:

- *Geometrical parameter deviation*. The deviation in insulator thickness, electrode length, and height between parallel plates may exceed their tolerance value.
- *Change in viscosity of droplet and filler medium*. These can occur during operation because of an unexpected biochemical reaction or changes in the operational environment (*e.g.*, temperature variation).

13.8.3 Test Techniques

An excellent survey on the testing of microfluidic biochips is presented in [Kerkhoff 2007]. A unified test methodology for digital microfluidic biochips has been implemented, whereby faults can be detected by controlling and tracking droplet motion electrically [Su 2003, 2004]. Test stimuli droplets containing a conductive fluid (*e.g.*, KCL solution) are dispensed from the droplet source. These droplets are guided through the unit cells following the test plan toward the droplet sink, which is connected to an integrated capacitive detection circuit. Most catastrophic faults result in a complete cessation of droplet transportation. Therefore, we can determine the fault-free or faulty status of the system by simply observing the arrival of test stimuli droplets at selected ports. An efficient test plan ensures that testing does not conflict with the normal bioassay, and it guides test stimuli droplets to cover all the unit cells available for testing. The microfluidic array can be modeled as an undirected graph, and the pathway for the test droplet can be determined by solving the Hamiltonian path problem [Su 2004]. With negligible hardware overhead, this method also offers an opportunity to implement BIST for microfluidic systems and therefore eliminates the need for costly, bulky, and expensive external test equipment. Furthermore, after detection, droplet flow paths for bioassays can be reconfigured dynamically so that faulty unit cells are bypassed without interrupting the normal operation.

Even though most catastrophic faults lead to a complete cessation of droplet transportation, differences exist between their corresponding erroneous behaviors. For instance, to test for the electrode-open fault, it is sufficient to move a test droplet from any adjacent cell to the faulty cell. The droplet will always be stuck during its motion because of the failure in charging the control electrode. On the other hand, if we move a test droplet across the faulty cells affected by an electrode-short fault, the test droplet may or may not be stuck depending on its flow direction. Therefore, to detect such faults, it is not enough to solve only the Hamiltonian path problem. In [Su 2005], the authors describe a solution based on Euler paths for detecting electrode shorts.

TABLE 13.2 ■ List of Catastrophic Defects for Biochips

Cause of Defect	Defect Type	Number of Cells Involved	Fault Model	Observable Error
Excessive actuation voltage applied to an electrode	Dielectric breakdown	1	Droplet-electrode short (short between the droplet and the electrode)	Droplet undergoes electrolysis, which prevents its further transportation
Electrode actuation for excessive duration	Irreversible charge concentration on an electrode	1	Electrode-stuck-on (the electrode remains constantly activated)	Unintentional droplet operations or stuck droplets
Excessive mechanical force applied to the chip	Misalignment of parallel plates (electrodes and ground plane)	1	Pressure gradient (net static pressure in some direction)	Droplet transportation without activation voltage
Coating failure	Nonuniform dielectric layer	1	Dielectric islands (islands of Teflon coating)	Fragmentation of droplets and their motion is prevented
Abnormal metal layer deposition and etch variation during fabrication	Grounding failure	1	Floating droplets (droplet are not anchored)	Failure of droplet transportation
Abnormal metal layer deposition	Broken wire to control source	1	Electrode open (electrode actuation is not possible)	Failure to activate the electrode for droplet transportation
Abnormal metal layer deposition	Metal connection between two adjacent electrodes	2	Electrode short (short between electrodes)	A droplet resides in the middle of the two shorted electrodes, and its transport along one or more directions cannot be achieved
Particle contamination or liquid residue	A particle that connects two adjacent electrodes	2	Electrode short	A droplet resides in the middle of the two shorted electrodes
Protein absorption during a bioassay	Sample residue on electrode surface	1	Resistive open at electrode	Droplet transportation is impeded
			Contamination	Assay results are outside the range of possible outcomes

Despite its effectiveness for detecting electrode shorts, testing based on an Euler path suffers from long test application time. This approach uses only one droplet to traverse the complete microfluidic array, irrespective of the array size. Fault diagnosis is carried out by using multiple test application steps and adaptive Euler paths. Such a diagnosis method is inefficient because defect-free cells are tested multiple times. Moreover, the test method leads to a test plan that is specific to a target biochip. If the array dimensions are changed, the test plan must be completely altered. In addition, to facilitate chip testing in the field, test plans need to be programmed into a microcontroller. However, the hardware implementations of test plans from [Su 2005] are expensive, especially for low-cost, disposable biochips.

More recently, a cost-effective test methodology referred to as "parallel scan-like test" and a rapid diagnosis method based on test outcomes have been proposed for droplet-based microfluidic devices [Xu 2007]. The method is named thus because it manipulates multiple test droplets in parallel to traverse the target microfluidic array, just as test stimuli can be loaded in parallel to multiple scan chains in ICs. This approach allows testing using parallel droplet pathways in both online and offline scenarios. The diagnosis outcome can be used to reconfigure a droplet-based biochip such that faults can be easily avoided.

We first describe the special case of a single test droplet. We determine the pathway for the test droplet, irrespective of the bioassay operation, as shown in Figure 13.24. Starting from the droplet source, the test droplet follows the pathway to traverse every cell in the array, and it finally reaches the sink. During concurrent testing, a test droplet is guided to visit the available cells in accordance with a predetermined path. If the target cell is temporarily unavailable for testing (*i.e.*, it is occupied by a droplet or it is adjacent to active microfluidic modules), the test droplet waits in the current position until the target cell becomes available. The test outcome is read out using a capacitive sensing circuit connected to the electrode for the sink reservoir. This single-droplet, scan-like algorithm is easy to implement. Moreover, the test plan is general, in the sense that it can be applied to any microfluidic array and for various bioassay operations.

Scan-like tests can also be carried out in parallel using multiple droplets. Each column/row in the array is associated with a test droplet and its "target region."

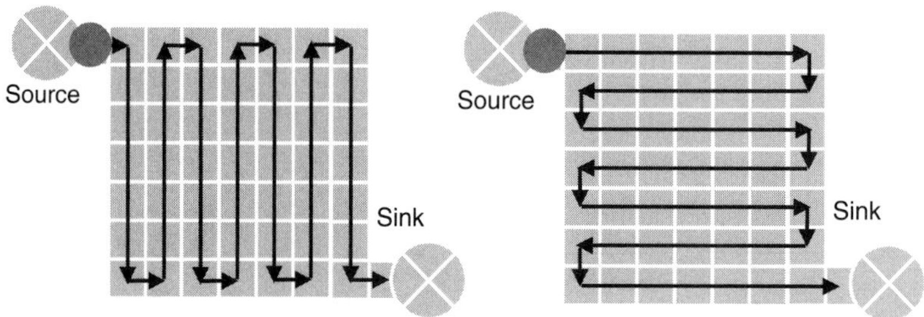

■ **FIGURE 13.24**

Illustration of a single droplet scan-like test using a single droplet.

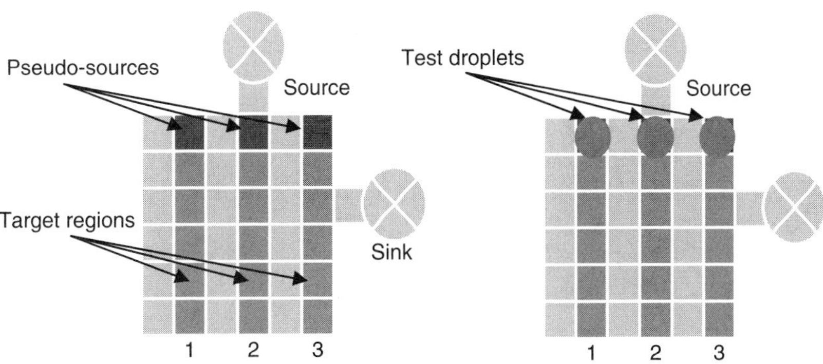

FIGURE 13.25

Example of target regions and pseudo-sources.

A target region for a droplet includes the cells that are traversed by this droplet. Droplets are dispensed from the test droplet source to the start electrodes of their target regions. Because columns/rows are used as target regions, the start electrodes are located on the array boundary, as shown in Figure 13.25. For each target region, the start electrode acts as the test-droplet source for the underlying single-droplet scan-like method. Therefore, start electrodes are referred to as pseudo-sources. Starting from these pseudo-sources, test droplets are routed in parallel (similar to a waterfall in nature) to the electrodes at the other end of the corresponding target regions. These end points are referred to as pseudo-sinks. Finally, the test droplets are routed to the sink reservoir. It is assumed that a microfluidic array has only one source and one sink reservoir to facilitate chip packaging and reduce fabrication cost. Dispensed from the single source, test droplets are aligned one by one and routed in sequence, like components in an assembly line, along the periphery nodes to their pseudo-sources. The reverse process is carried out when the test droplets are routed from the pseudo-sink to the sink reservoir.

The complete parallel scan-like test procedure is as follows:

Step 1. Peripheral test: A test droplet is dispensed from the source. It is routed to traverse all the peripheral electrodes, and the droplet finally returns to the sink.

Step 2. Column test: Two iterations of parallel scan-like test with one column shift are carried out. This step tests every single cell and all "edges" (pairs of adjacent cells) in each column. Therefore, it is referred to as "column test."

Step 3. Row test: Repeat parallel scan-like test (two iterations) for the rows to detect defects involving pairs of adjacent cells in each row. This step is referred to as "row test."

13.8.4 Application to a Fabricated Biochip

The parallel scan-like test method has been applied to a fabricated biochip. The chip under test is a PCB microfluidic platform for DNA sequencing, as shown in

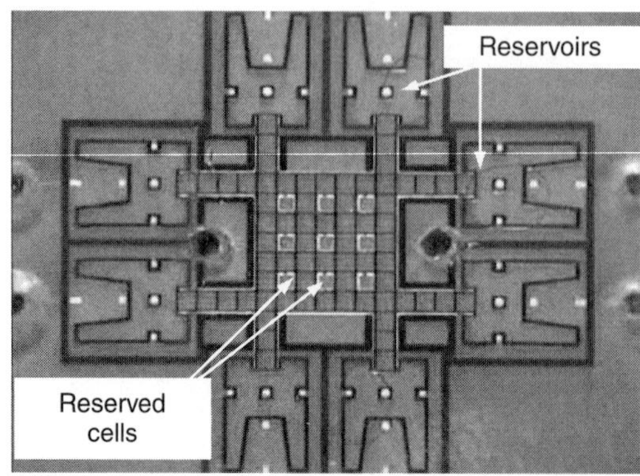

■ **FIGURE 13.26**

Fabricated biochip for DNA sequencing.

Figure 13.26. The platform consists of a 7×7 array, eight reservoirs, and routing electrodes that connect reservoirs to the array. Nine cells are reserved for grounding, and they are not available for droplet transportation.

As a baseline, Euler-path-based testing was applied to this chip [Xu 2007]. The test procedure takes 57 seconds, assuming a (typical) 1-Hz electrode-actuation frequency. Next, a parallel scan-like test (the column-test stage is shown in Figure 13.27) was applied to this chip. Because nine electrodes are not reachable, it is not necessary to test even columns and rows. The test application procedure takes 46 seconds, again for a 1-Hz actuation frequency.

Next we investigate the time needed for fault diagnosis for the two methods. In [Xu 2007], a fabricated chip was used, which was known a priori to contain one defect. The chip with the defect is shown in Figure 13.28. For the Euler-path-based

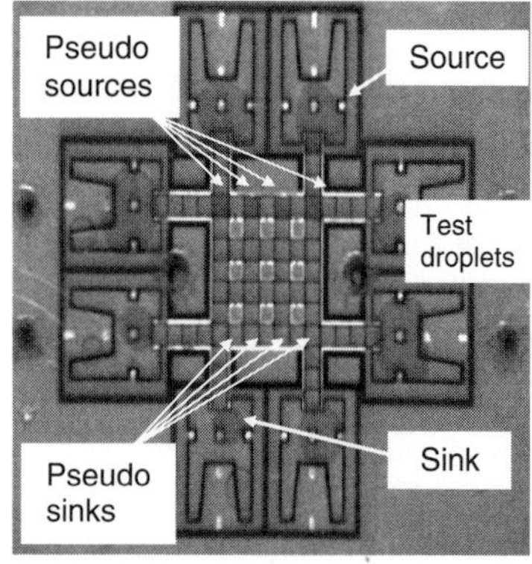

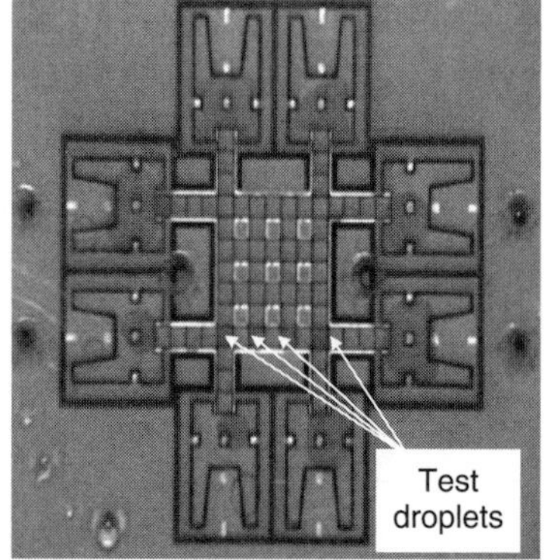

■ **FIGURE 13.27**

Column-test step of a parallel scan-like test.

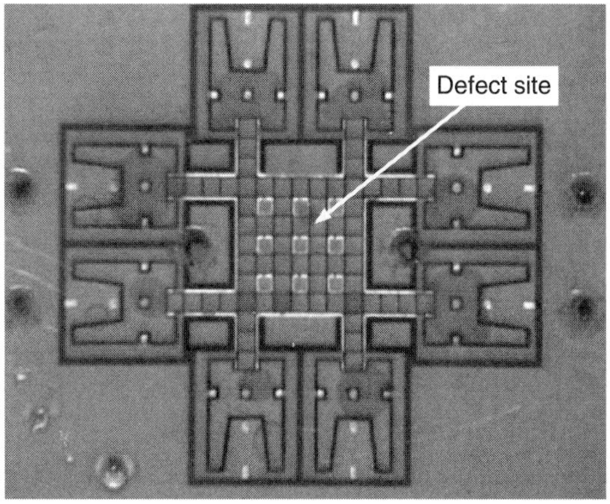

■ **FIGURE 13.28**

Parallel scan-like diagnosis of a single-cell defect.

method, binary search was carried out to locate the defective cell. Seven iterations were needed, and the total diagnosis time was 173 seconds. This value was obtained by summing up the times for the different diagnosis iterations, which are 57, 44, 32, 16, 8, 4, and 2 seconds, respectively. On the other hand, parallel scan-like test can simply determine the defect site from testing readouts. No additional diagnosis steps are needed, and the diagnosis time is the same as the test time (*i.e.*, 44 seconds), which corresponds to a 75% reduction compared to [Su 2005].

13.9 DFT AND BIST FOR MEMS

Because of the diversity of MEMS devices and their working principles, universal *design-for-testability* (DFT) and *built-in self-test* (BIST) solutions for all types of MEMS devices do not exist yet! As a result, different DFT/BIST solutions are required for different types of MEMS. However, these solutions are crucial for mission-critical and safety-critical applications such as in the aerospace, automotive, and healthcare industries. In this section, we briefly review the DFT and BIST approaches that have been proposed and implemented for MEMS. We then present a number of MEMS BIST examples with actual implementations of BIST for accelerometers.

13.9.1 Overview of DFT and BIST Techniques

In [Puers 2002], a built-in shaker realized with electromagnetic microactuator was used to vibrate the accelerometer, and **real acceleration** input test stimuli are generated for self-test of the MEMS accelerometer. Although a certain level of self-test has been achieved, it is often inconvenient to generate real acceleration input test

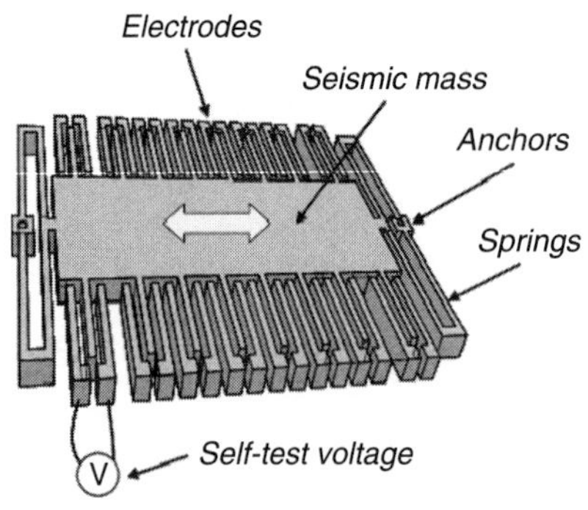

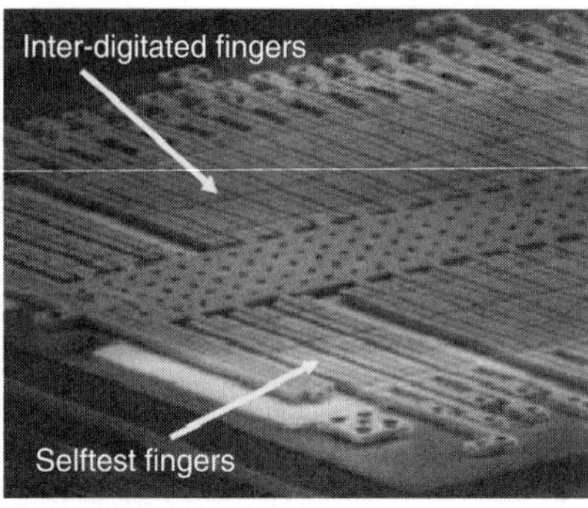

■ **FIGURE 13.29**

BIST of a surface-micromachined comb accelerometer using electrostatic force [Mir 2006].

stimuli for MEMS devices. Generally, alternative test stimuli (electrical voltage, etc.) that are somewhat equivalent, but easier to generate, will be used for **MEMS BIST**. In ADXL series **surface-micromachined comb accelerometers**, electrostatic force generated by a self-test voltage is used to mimic the effect of input acceleration for **in-field BIST** [Analog 2007]. As shown in Figure 13.29, several outer comb finger groups are reserved for BIST implementation [Mir 2006]. In BIST mode, a self-test voltage is applied between left (or right) fixed testing fingers and movable fingers. This will introduce electrostatic force on the movable fingers toward the left (or right) direction, which is somewhat equivalent to the effect of an inertial force caused by input acceleration. The output response of the accelerometer is measured and compared to the good device behavior. If the difference is within a certain tolerance range, the device is considered as good; otherwise, the device is deemed faulty. Voltage-induced electrostatic force is easy to generate, and it is also compatible with BIST circuitry. Hence, it has been widely used for in-field BIST of MEMS accelerometers. In [Zimmermann 1995], **online testing** of surface-micromachined comb accelerometers for automobile airbag application based on electrostatic force activation was reported. Unlike using the conventional offline BIST techniques, the self-test operation of the accelerometers can be performed not only before engine ignition (offline) but also during driving (online). However, none of the preceding MEMS BIST techniques typically can be used to replace the traditional manufacturing test. The reason comes from the fact that considering the fabrication variations, the electrostatic force has to be calibrated first for each individual MEMS device before the device operates in self-test mode, and the calibration process requires the device to be thoroughly tested using external test equipment.

Besides acceleration, other input stimuli can also be induced (or mimicked) on-chip by electrical voltage/current for MEMS BIST. For example, in [Puers 2001], electrically-induced pneumatic actuation is used for the self-test of a **piezoresistive pressure sensor**, as shown in Figure 13.30 [Puers 2001]. During BIST, an electrical pulse voltage is applied to a resistor heater embedded in the cavity of

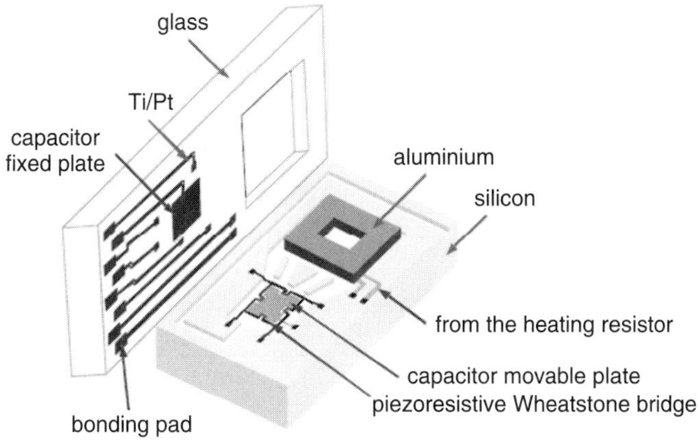

■ **FIGURE 13.30**

Pneumatic actuation for self-test of a piezoresistive pressure sensor [Puers 2001].

the pressure sensor. The air inside the cavity is heated by the induced Joule heat, thereby increasing its pressure. The piezoresistive gauge in the membrane then senses the output response resulting from this input pressure change and compares it with the good device response.

In [Charlot 2001], the authors showed that electrically-induced test stimuli can be used for self-test of parallel-plate-capacitor, piezoresistive-micro-beam, and thermopile based sensors. As an example, electrically-induced resistor heating to mimic the thermal radiation input for BIST of a thermopile-based infrared imager pixel is demonstrated in Figure 13.31 [Mir 2006]. Each pixel of the imager consists of a suspended membrane supported by four beams. The temperature increase caused by the infrared light captured by the membrane is measured by the thermopiles on the support beams. Hence, the infrared image can be generated by an array of such pixels. Self-test is performed on each individual pixel by applying electrical voltage to the heating resistor on the suspended membrane, which will heat up the membrane as incident infrared radiation normally does. In [Cozma Lapadatu 1999], the authors demonstrated a self-test of a pressure sensor utilizing the electrically-induced bimetal effect.

In all of the above cases, electrical signals have been used to induce test stimuli in various energy domains for self-test of MEMS devices. Direct parameters (sensitivity, etc.) are effective to verify the device function. However, they are not always easy to measure. ***Oscillation-based test methodology*** (OTM) measuring indirect parameters was also demonstrated for a MEMS magnetometer [Beroulle 2002]. The electrically-induced Lorentz force in a magnetic field is used as test stimuli. The DUT is reconfigured into an oscillating device with a feedback circuit. Some indirect parameters such as the oscillation frequency and amplitude, which are easier to observe, are measured for testing the MEMS device.

Most MEMS devices have a certain degree of structure symmetry, such as left-right, top-bottom, or rotation symmetry. This has also been utilized to develop another

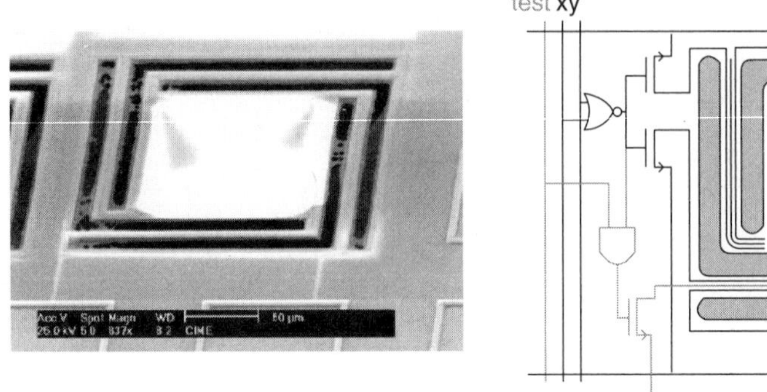

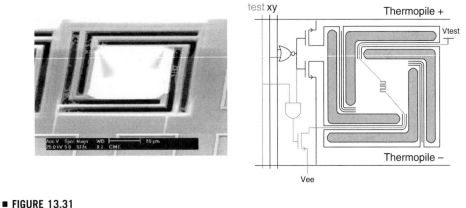

■ **FIGURE 13.31**

Thermal actuation for self-test of an infrared imager pixel [Mir 2006].

structural BIST strategy, called **symmetry testing**. In [Rosing 2000a], **symmetry BIST** for a pressure sensor with internal redundancy was proposed. Based on the left-right symmetry of a device structure, the movable shuttle is activated twice by electrostatic comb driving first toward left and later toward right. The output responses from both activations are stored and compared with each other. Any difference indicates the existence of a local defect leading to a structure asymmetry of the device. In [Deb 2002], symmetry BIST for **CMOS MEMS comb accelerometers** was proposed where the movable shuttle of the accelerometer is divided into two conductors that are physically connected by an insulator layer while electrically insulated from each other. By comparing the voltage outputs from both conductors of the movable shuttle, the BIST technique can effectively detect structure asymmetry caused by local, hard-to-detect defects, such as bridges, finger height mismatch, and local etch variations. In [Deb 2004], this symmetry BIST technique was further extended to a more generalized model so it can be used to characterize a wide range of local manufacturing variations affecting different regions of the device. The proposed symmetry BIST techniques in [Deb 2002] and [Deb 2004] are applicable to MEMS devices in which the movable part is divided into two or more conductors physically connected by insulator layers. In [Xiong 2005a], the authors proposed another symmetry BIST technique that divides fixed instead of movable parts of symmetrical capacitive MEMS devices. This proposed symmetry BIST technique can be applied to MEMS devices such as ADXL accelerometers in which the movable parts are not divided. The preceding symmetry BIST techniques are discussed in detail in the next section.

In [Mir 2006], a **pseudo-random MEMS BIST** technique, illustrated in Figure 13.32, was proposed using bulk-micromachined cantilever. In BIST mode, voltage pulses are applied to a heating resistor on the cantilever. The cantilever deflects because of the induced heat, and the corresponding deflection is measured by a piezoresistor Wheatstone bridge at the anchor. Pseudo-random maximum-length sequences are generated by ***linear feedback shift registers*** (LFSRs).

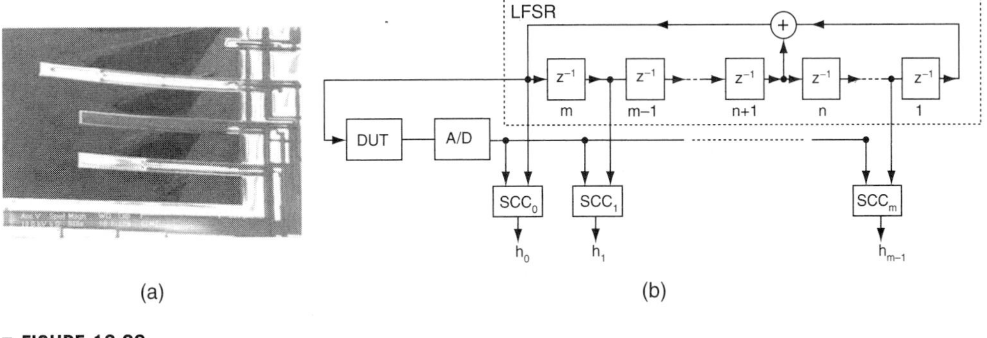

■ **FIGURE 13.32**

Pseudo-random BIST of MEMS cantilevers [Mir 2006]: (a) MEMS cantilevers and (b) BIST architecture.

The output bridge voltage is converted by an A/D converter to digital values, which are then analyzed by a circuit computing the input-output ***cross-correlation function*** (CCF). Each simplified correlation cell in the CCF circuit computes an estimation of an impulse response sample. The estimated samples form the test signature that is compared on-chip with expected values for a Go/No-Go testing. This technique offers an on-chip BIST solution for MEMS devices using electrical pulse-like test signals.

13.9.2 MEMS BIST Examples

Accelerometers represent MEMS devices where BIST has been most widely used in industry. For example, products of the ADXL series from Analog Devices [Analog 2007], such as ADXL190 (single axis accelerometer), ADXL202E (dual-axis accelerometer), and ADXL330 (three-axis accelerometer), all implement BIST. When a voltage V_s activates the self-test pin, an electrostatic force is generated on the movable test fingers of the accelerometer. The force acting on the beam results in an approximately 20% (for ADXL190) of full-scale acceleration input, and a voltage change will be observed on the output pin. Thus, this BIST technique can be used for in-field test where external test equipment is unavailable. Because of its popularity, BIST for accelerometers is used in this section to discuss the basic working principle of MEMS BIST. Other research on BIST for accelerometers is also discussed.

As its counterparts in digital circuit BIST, MEMS BIST also requires circuitry integrated with the accelerometer for test stimulus application and output response analysis. Different BIST methods implemented for accelerometers differ on how the test stimuli are generated, and how the output responses are analyzed. Most BIST designs for accelerometers generate test stimuli using electrostatic [Terry 1989] [Zimmermann 1995] [Charlot 2001] [Deb 2002], thermal [Pourahmadi 1992] [Plaza 1998], piezoelectric [Spineanu 1997], and real acceleration [Puers 2002] inputs. A pseudo-random MEMS BIST method utilizing electrical pulse-like test signals was proposed in [Mir 2004] and [Dhayni 2006]. The test response with respect to the actuation is measured using a sensing circuit and compared with the expected response. This section focuses mainly on *surface-micromachined comb accelerometers,* because this type of accelerometers has been widely used in industry [Analog 2007] and the surface-micromachining process is compatible to the CMOS process for SOC designs (*e.g.*, CMP of French Multiproject Wafer Service) [Castillejo 1998].

A typical surface-micromachined comb accelerometer is shown in Figure 13.33 [Kuehnel 1994]. The comb accelerometer is made of a thin layer of polysilicon on the top of a silicon substrate. The fixed portion of the device includes four anchors and many left and right fixed fingers. The movable portion of the MEMS device includes four tether beams, the central movable mass, and all movable fingers extruding out of the mass. The entire movable portion is floating above the substrate. As Figure 13.33 shows, the central movable mass is connected to the four anchors through four flexible beams. The movable fingers extrude from both sides of the central mass and can move together with it. There is a pair of fixed fingers around the left and right sides of each movable finger, which constitutes a differential capacitance pair c_1 and c_2, as shown in Figure 13.34. In the static state, each movable finger stays in the middle position between the left and right fixed fingers, and the capacitance gaps of both c_1 and c_2 are equal to d_0. If we let C_1 (C_2) represent the sum of all c_1 (c_2) capacitances, then we have:

$$C_1 = C_2 = \frac{n_f \varepsilon_0 (L_f - \Delta)h}{d_0} \tag{13.10}$$

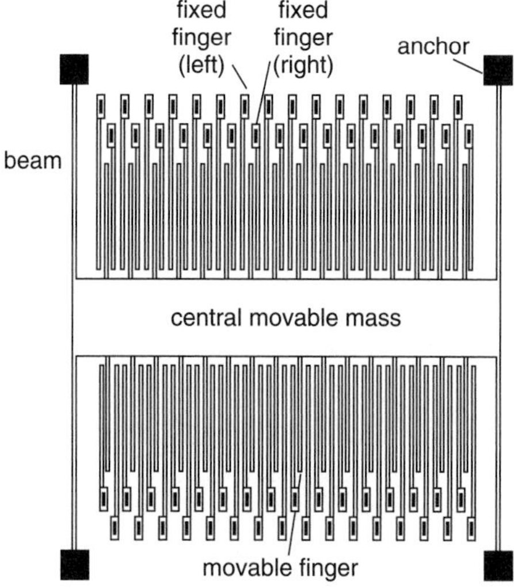

■ FIGURE 13.33

General design of a MEMS comb accelerometer [Kuehnel 1994].

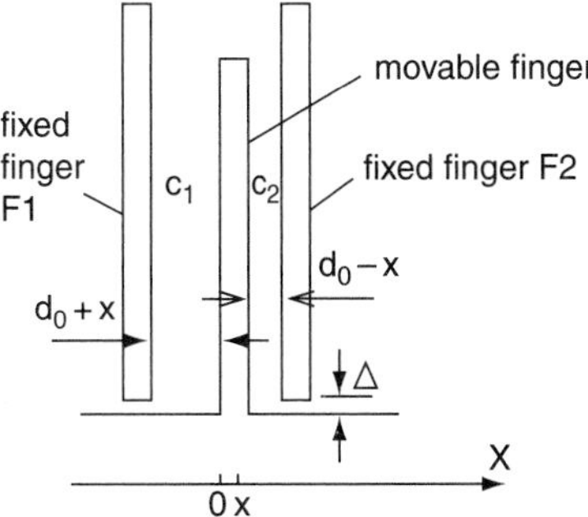

■ **FIGURE 13.34**

The schematic diagram of differential capacitance.

where n_f is the total number of differential capacitance groups, ε_0 is the dielectric constant of air, L_f is the length of each movable finger, Δ is the nonoverlapped length at the root of each movable finger, and h is the thickness of the device.

Assume the mass of both the central movable mass and all the movable fingers is M_s. If there is an acceleration, a, in a direction perpendicular to the beams and parallel to the device plane, the central mass will experience an inertial force. This will result in a certain amount of beam deflection along the direction of the inertial force, hence an equivalent amount of displacement of the central mass and the movable fingers. Thus, each capacitance gap will be changed accordingly, which leads to the change of corresponding capacitances.

As shown in Figure 13.34, the inertial force $F_a = -M_s a$ results in a deflection of the beams and a certain displacement x of movable fingers along the X direction. Assume the total spring constant of four beams as k. Displacement x is given by:

$$x = \frac{F_a}{k} = \frac{-M_s \cdot a}{k} \propto a \tag{13.11}$$

Given $x << d_0$, we have C_1 and C_2 changed to:

$$C_1 = \frac{n_f \varepsilon_0 (L_f - \Delta) h}{(d_0 + x)} \approx \frac{n_f \varepsilon_0 (L_f - \Delta) h}{d_0} \left(1 - \frac{x}{d_0}\right) \tag{13.12}$$

$$C_2 = \frac{n_f \varepsilon_0 (L_f - \Delta) h}{(d_0 - x)} \approx \frac{n_f \varepsilon_0 (L_f - \Delta) h}{d_0} \left(1 + \frac{x}{d_0}\right) \tag{13.13}$$

By sensing the capacitance changes of C_1 and C_2, we get the displacement x and, hence, the experienced acceleration. This is the working principle of a MEMS comb accelerometer.

Before we discuss BIST for accelerometers, fault modeling and simulation should be reviewed. This short survey is mainly based on the work done at Carnegie Mellon University (United States), Lancaster University (United Kingdom), and

TIMA (France). The effect of realistic contaminations on the folded-flexure comb-resonator was studied in [Kolpekwar 1997] by fault simulation. The results demonstrated that realistic contaminations can result in many different fault behaviors. The fault analysis methodology was then developed as a tool called **CARAMEL** (contamination and reliability analysis of microelectromechanical layout) [Kolpekwar 1998a, 1999]. In CARAMEL, a defective MEMS structure is represented by a three-dimensional representation, which is then extracted to mesh netlist for mechanical simulation. A stiction fault model based on the ADXL75 accelerometer was studied in [Kolpekwar 1998b]. A Monte Carlo simulation of a particulate contamination fault model was discussed in [Jiang 2006] based on microresonator. Fault models of vertical stiction, foreign particles, and etch variation for resonators and accelerometers have been investigated in [Deb 2000], and the effects of these faults to resonant frequency have been identified.

The ***failure modes and effect analysis*** (FMEA) approach was proposed in [Rosing 2000a, 2000b]. This technique integrates qualitative failure analysis and quantitative fault simulation to generate a list of realistic faults for MEMS transducers. Industrial failure modes and sensor/actuator are analyzed and simulated by **inductive fault analysis** [Shen 1985] and finite element simulation. Analog and mixed-signals are also simulated using inductive fault analysis (for defect-related faults) and process variation analysis (for parametric faults). The faults are then described by a behavioral model for test purposes. The major faults identified include local defects, global and local parameters out of tolerance, wear, environmental hazards, problems resulting from imperfection in the design process, etc.

Instead of using the inductive fault analysis discussed earlier, in [Castillejo 1998] the fabrication process of MEMS is analyzed in detail to determine realistic defects or failure mechanisms. The failure mechanisms are divided into those that occurred during the CMOS process (*e.g.*, wet oxidation, etching) and those that occurred during the micromachining process. In particular, contaminant particles and oxide residuals left in the fabrication of CMOS can greatly affect the naked silicon exposed for micromachining. Eventually, the defects can be classified into *gauge* (*e.g.*, sensing circuits) faults and *microstructure* faults. Each class can be further divided into *catastrophic* faults (the circuit or structure is totally nonfunctional) and *parametric* faults (machine performance is changed). Gauge faults can be shorts and opens (catastrophic) or changes in width, length, and metal resistivity (parametric). Microstructure faults can be break-around-gauge, stiction, nonreleased microstructure, asymmetrical microstructure (catastrophic), or changes in Young's modulus (parametric). The fault behaviors can be mapped to a physical MEMS device model (based on material and structural properties) for machine behavioral simulation.

A simplified comb accelerometer structure, which can implement BIST functions, is illustrated in Figure 13.35 where, for simplicity, only four groups of driving/sensing fingers are given. M1–M8 are movable fingers, M_s is the central mass, D1–D8 are fixed driving fingers, and S1–S8 are fixed sensing fingers. All beams are connected to the substrate through four anchors. Driving fingers are used to generate test stimuli using electrostatic force, and sensing fingers are used to sense the output voltage change resulting from capacitance change. For test stimulus generation, electrostatic force F_d is used to mimic the effect of inertial force. When voltage

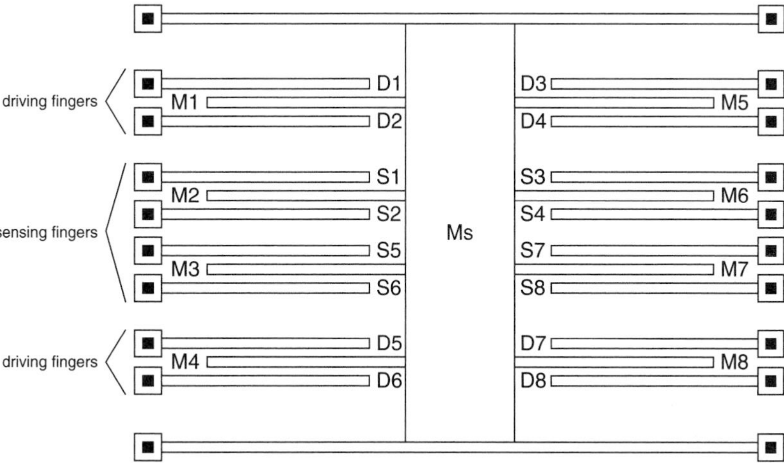

■ **FIGURE 13.35**

Structural diagram of a comb accelerometer [Deb 2002].

V_d is applied to the fixed driving fingers {D1, D3, D5, D7}, and nominal voltage V_{nom} is applied to M_s and {D2, D4, D6, D8}, an electrostatic force F_d toward the top direction will be experienced at the central movable mass M_s. The value of F_d can be represented by the following equation:

$$F_d = \frac{\varepsilon_0 S(V_d - V_{nom})^2}{2d^2} \tag{13.14}$$

where S is the overlap area between {D1, D3, D5, D7} and {M1, M5, M4, M8}, and d is the capacitance gap between them. As a result, the movable mass is activated toward the top with displacement $x = F_d/k$ during self-test.

To sense the displacement x of the movable mass in Figure 13.35 during normal operation, modulation voltage V_{mp} is applied to {S1, S3, S5, S7}, and V_{mn} is applied to {S2, S4, S6, S8}. Here, $V_{mp}(V_{mn})$ can be a sequence of square-wave signals with amplitude V_0 and frequency ω, and it can be represented by $V_0\, sqrt(\omega t)$ $(-V_0\, sqrt(\omega t))$. Because of input acceleration a, the movable mass has a displacement $x = -M_s \cdot a/k$. Assume the sensing capacitance between {S1, S3, S5, S7} ({S2, S4, S6, S8}) and {M2, M6, M3, M7} as C_1 (C_2) separately and the voltage in the movable mass as V_{Ms}. According to charge conservation law, we have:

$$C_1(V_{mp} - V_{Ms}) = C_2(V_{Ms} - V_{mn}) \tag{13.15}$$

Combined with equations for C_1 and C_2 given before, we have:

$$V_{Ms} = \left(\frac{x}{d_0}\right) V_0 sqr(\omega t) \tag{13.16}$$

Measuring the voltage level V_{Ms} in the movable mass, we know the value of displacement x, and hence the acceleration a.

In test mode, a certain test driving voltage V_d is applied to {D1, D3, D5, D7} to activate the device with electrostatic force. A nominal voltage V_{nom} is applied to {D2, D4, D6, D8} and {M1, M5, M4, M8}. Usually, V_{nom} is the time average value of modulation voltage V_{mp}. The modulation voltage V_{mp} is applied to {S1, S3, S5, S7}, whereas V_{mn} is applied to {S2, S4, S6, S8}. Because of electrostatic force F_d, the movable mass has a displacement of $x = -F_d/k$. The output voltage on movable mass M_s is measured for the device sensitivity. This value is compared with the expected good device response with a certain tolerance level to determine whether the device is faulty. Basically, this implements the function of **sensitivity BIST**. For example, in [Zimmermann 1995] and [Analog 2007], a self-test pin is used to send a voltage impulse to generate electrostatic force to actuate the comb accelerometer, and the corresponding voltage change is measured in the output pin for fault detection. Most of the defects that change the sensitivity of a comb accelerometer can be detected by this BIST method.

A fully **differential symmetry BIST** for CMOS MEMS comb accelerometers utilizing device symmetry was introduced in [Deb 2002]. The BIST method can be used for both manufacturing test and in-field test. With this method, the movable mass is physically divided into two (left and right) equal conductors connected by an insulator layer. In this way, the left and right parts of the movable mass/fingers are electrically insulated from each other. Both parts sense capacitance changes separately, without mutual signal interference. In symmetry BIST mode, the movable mass is activated with electrostatic force as in sensitivity BIST. The modulation voltage V_{mp} is applied to {S1, S3, S5, S7}, whereas V_{mn} is applied to {S2, S4, S6, S8}. A differential amplifier senses the difference between voltage V_{s1} from left movable fingers {M2, M3} and voltage V_{s2} from right movable fingers {M6, M7}. Because of the device symmetry, V_{s1} and V_{s2} of a good device should exactly match each other. However, if the left-right symmetry of the device is changed by local defects, then the differential sensing circuit observes the difference between V_{s1} and V_{s2}. Hence, defects causing asymmetry in the device structure (*e.g.*, local and hard-to-detect defects) can be detected. Dummy fingers are also utilized for symmetry BIST of the beams in a similar way. The fully differential scheme of sensing in the BIST can also tolerate any noise common to the left and right sides of the movable mass. Mathematical model analysis for a more generalized differential BIST method to deal with local manufacturing variations can also be found in [Deb 2006].

For comb accelerometers in which the movable mass is not divided (such as Analog Devices' ADXL series accelerometers), the symmetry BIST method needs to be implemented in a different way, as proposed in [Xiong 2005a]. For the comb accelerometer shown in Figure 13.35 in symmetry BIST mode, test driving voltage V_d is applied to {D1, D3, D5, D7}, and V_{nom} is applied to {D2, D4, D6, D8}, {M1, M4, M5, M8}, and {S2, S4, S6, S8}. The movable mass is activated by electrostatic force F_d with displacement $x = F_d/k$. Modulation voltage V_{mp} is applied to {S1, S5}, while V_{mn} is applied to {S3, S7}. Because of device symmetry, the capacitance C_1 between {S1, S5} and {M2, M3} should be always equal to the capacitance C_2 between {S3, S7} and {M6, M7}. Thus, the AC output voltage on the central mass will be held at constant zero. The sensing circuit checks whether the AC output voltage on the central mass is a constant zero to detect any asymmetry caused by local defects. If there is a nonzero AC voltage on the movable electrode M_s (caused by imbalanced V_{mp} and V_{mn}), then it indicates there are local defects that alter the symmetry of the device. Because sensitivity BIST and symmetry BIST each has its own fault coverage, the **dual-mode BIST** in [Xiong 2005a] combines both BIST modes to yield higher fault coverage. The proposed symmetry BIST divides fixed instead of movable capacitance plates. Thus, it can be applied to capacitive MEMS devices in which the movable mass is not divided, such as ADXL series accelerometers, bulk-micromachined MEMS devices, and comb resonators.

In [Natarajan 2006], the authors proposed a technique to extract the mechanical parameters of a comb accelerometer using purely electrical (instead of mechanical) test stimulus. The basic idea is to use a *gradient-search* method to find a single-tone or multitone AC signal (riding on a DC voltage) that will actuate the beams to vary the capacitance of the comb accelerometer. The frequency of beam motion is selected to ensure that the beam is in steady-state oscillation with low vibration amplitude. The effective capacitance of the beam is then measured using a simple op-amp capacitance sensing circuit to predict the mechanical parameters such as mass, damping coefficient, and spring constant using a **regression-based mapping** technique. This method has the potential to eliminate the use of expensive test instrumentation and can also be applied to BIST solutions.

Finally, ***built-in self-repair*** (BISR) may be a possible solution to improve MEMS yield and reliability for safety-critical applications, just like what its counterpart does in VLSI circuits. MEMS devices generally contain movable mechanical parts. The implementation of BISR in MEMS is more challenging than that in VLSI. A BISR technique for comb accelerometers based on *structure modularization* and *redundancy repair* was reported in [Xiong 2005b]. Reliability enhancement of comb accelerometers with BISR can be found in [Xiong 2006]. However, there are still research issues in BISR, such as the signal strength in modularized design and ***built-in self-calibration*** (BISC) of the device after redundancy repair. BISR for MEMS remains a challenging research topic.

13.10 CONCLUDING REMARKS

A majority of *microelectromechanical systems* (MEMS) devices are inherently mechanical in nature and therefore require some special considerations during various manufacturing stages and testing. This chapter discussed some of the important handling considerations during dicing, packaging, and testing. There are a wide variety of test methods, such as electrical, optical, mechanical, and environmental, for characterization of various MEMS devices. This chapter reviewed the instrumentation, typical setup, and important characteristics for testing a wide variety of MEMS devices, including accelerometers, gyroscopes, humidity sensors, RF MEMS, optical MEMS, pressure sensors, and microphones. Due to the growing importance of microfluidics-based biochips, also referred to as **lab-on-a-chip**, and their potential for replacing cumbersome and expensive laboratory equipment, this chapter included a section on the testing of digital microfluidic biochips.

It is primarily the diversity of MEMS devices and their working principles that have prevented the development of universal *design-for-testability* (DFT) and *built-in self-test* (BIST) solutions for all types of MEMS in general. As a result, different DFT/BIST solutions are required for different types of MEMS. The discovery of such solutions needs a thorough understanding of various MEMS defects, fault models, and their associated failure mechanisms. This is further complicated by the variety of physical/chemical stimuli for MEMS devices including acceleration, pressure, heat, and chemical concentration, to name a few. In BIST for MEMS, such stimuli must be generated automatically on-chip. The on-chip generation of nonelectrical test stimuli (such as chemical or fluidic inputs for bioMEMS testing) has also been difficult. This chapter discussed the MEMS DFT and BIST techniques and presented a number of BIST examples that illustrate the variety, diversity, and complexity of MEMS testing, DFT, and BIST. However, MEMS, DFT, and BIST will continue to be important and challenging topics in the future.

13.11 EXERCISES

13.1 **(Mechanical Test Methods)** A certain MEMS device can be modeled as a proof mass attached to a base with a single spring. Testing of the device revealed a low pass response with a natural frequency of 20 KHz and a quality factor of 10. If the proof mass has a mass of 200 mg, compute the system spring constant, k, and the linear damping term, c.

13.2 **(Mechanical Test Methods)** For the MEMS device in Problem 13.1, plot the transmissibility of the device from 100 Hz to 100 KHz using a logarithmic scale for frequency. Repeat for otherwise identical MEMS devices with 100-mg and 400-mg proof masses.

13.3 **(Rate Table)** A certain accelerometer consists of a 10-mg proof mass suspended by a spring system with a system spring constant of 5000 N/m. If the accelerometer is mounted on a rate table 1 cm from the center of rotation and rotated so that the accelerometer experiences a 10-G acceleration, what is the displacement of the proof mass? *Hint:* Use the formula $F = kx$ to determine the displacement. What is the required rotation rate for the accelerometer to experience a 10-G acceleration?

13.4 **(MEMS Accelerometer)** For a ±20 G MEMS accelerometer, develop a test plan for evaluating the performance of the device using a rate table, where the sensor is mounted 2 cm from the center of rotation and is tested in 1-G increments. The test plan should consist of a table of required rate table angular rates.

13.5 **(RF MEMS Resonator)** Let us consider a polysilicon contour-mode disk resonator with a radius R = 17 μm and thickness $h = 2$ μm. Calculate the resonance frequency of the resonator using the following formula:

$$f_o = \frac{\alpha}{R}\sqrt{\frac{E}{\rho}}$$

where $\alpha = 0.342$ for the fundamental mode, Young's modulus $E = 150$ GPa, density $\rho = 2300\,\text{kg/m}^3$ for polysilicon.

13.6 **(MEMS Gyroscope)** A driving signal $x(t) = 10^{-6}\sin(40000\pi t)\hat{x}$ is applied to a gyroscope. Find the amplitude and frequency of the driving signal. If the angular rate is $\vec{\Omega} = 1°/s\ \hat{z}$, find an expression for the Coriolis acceleration. The Coriolis force is given by $\vec{F}_c = 2m\left(\vec{\Omega}\times\vec{v}\right)$, where m is the proof-mass of the gyroscope, $\vec{\Omega}$ is the angular rate of the reference frame, and v is the velocity of the proof-mass.

13.7 **(MEMS Pressure Sensor)** A capacitive pressure sensor fabricated using ***liquid crystal polymer*** (LCP) has a circular diaphragm of radius a = 3 mm. The thickness of the LCP diaphragm is 50 μm, and the spacing between the diaphragm and the substrate is 50 μm. The Young's modulus and Poisson's ratio of the LCP are E = 2.4 GPa and $v = 0.3$, respectively. Find the pressure at which the deflected diaphragm of the pressure sensor contacts the fixed substrate. The maximum deflection of circular diaphragm is given by $w_0 = Pa^4/64D$, where $D = Et^3/12\left(1 - v^2\right)$ is the flexural rigidity, E is the Young's modulus, t is the thickness, and v is the Poisson's ratio of the diaphragm.

13.8 **(MEMS Microphone)** Consider a MEMS microphone with a circular silicon nitride diaphragm of radius R = 2 mm and thickness h = 1 μm. The gap between the diaphragm and the back plate is $z_o = 3$ μm. The diaphragm stress is $\sigma_d = 1.5 \times 10^8\ \text{N/m}^2$. Find the resonance frequency. The resonance frequency of a circular diaphragm is given by $f_o = \alpha/R^2\sqrt{D/\rho h}$, where $D = Eh^3/12\left(1 - v^2\right)$ is the flexural rigidity, h is the thickness of the diaphragm, $\alpha = 10.21$ for the fundamental mode, Young's modulus $E = 250$ GPa, Poisson's ratio $v = 0.23$, and density $\rho = 3187\,\text{kg/m}^3$ for silicon nitride.

13.9 **(BIST)** Read papers and find one reported research work about MEMS BIST other than the examples listed in this chapter. Identify whether electrical signal is used to induce the test stimulus for self-test of the MEMS device. Clearly explain the working principle of the BIST method for the MEMS device in the research work.

13.10 **(BIST)** Consider the MEMS comb accelerometer shown in Figure 13.35, but with 8 driving finger groups and 32 sensing finger groups. Assume the accelerometer works in open-loop mode. The total spring constant of the four beams is $k = 1.72\,N/m$, and the mass of the movable mass M_s is $0.48\,\mu g$. Find the displacement x of the movable mass resulting from a full range acceleration input $a = 50g$ $(1g = 9.8\,m/s^2)$ along the sensitive direction of the accelerometer. Assume the overlap area between each pair of fixed and movable driving fingers is 220 μm × 2 μm and the capacitance gap is 2 μm. Find out the required BIST driving voltage V_d in order to mimic the effect of a 50g acceleration input in sensitivity BIST.

13.11 **(BIST)** Consider the comb accelerometer shown in Figure 13.35. Denote the capacitance between {S1, S5} and {M2, M3} as C_1, and the capacitance between {S3, S7} and {M6, M7} as C_2. When the movable mass experiences no displacement, we have $C_1 = C_2 = 0.3\,pF$. During the symmetry test in [Xiong 2005a], when the movable mass is activated by test-driving voltage for a certain displacement, the capacitances are changed to $C_1 = 0.49\,pF$ and $C_2 = 0.42\,pF$. Assume modulation voltage $V_{mp} = 5\,V \cdot sqr(\omega t)$ is applied to {S1, S5}, and modulation voltage $V_{mn} = -5\,V \cdot sqr(\omega t)$ is applied to {S3, S7} separately. Find out the voltage output in the central movable mass V_{Ms}. Is there any local defect that changes the device left-right symmetry in this case?

Acknowledgments

The authors wish to thank Professor Krishnendu Chakrabarty of Duke University for contributing the Testing Digital Microfluidic Biochips section with assistance from Dr. Fei Su, Tao Xu, William Hwang, Dr. Phil Paik, Dr. Vamsee Pamula, and Professor Richard Fair; and Professor Wen-Ben Jone of University of Cincinnati for contributing the MEMS BIST Examples section. The authors also would like to acknowledge Auburn University for the contribution of numerous photographs of their facilities toward this work. Finally, the authors would like to thank Dr. Phil Reiner of Stanley Associates (Huntsville, Alabama), Professor Ian Papautsky of University of Cincinnati, Derek Strembicke of The AEgis Technologies Group (Huntsville, Alabama), Professor R. D. Shawn Blanton of Carnegie Mellon University, and Dr. Bernard Courtois of CMP (Grenoble, France) for reviewing this chapter.

References

R13.0 Books

[Bao 2000] M.-H. Bao, Volume 8: Micro mechanical transducers, pressure sensors, accelerometers and gyroscopes, pp. 362–365, in *Handbook of Sensors and Actuators*, S. Middelhoek, editor, Morgan Kaufmann, San Francisco, 2000.

[De Los Santos 1999] H. D. Los Santos, *Introduction to Microelectromechanical Microwave Systems*, Artech House, Boston, 1999.

[Meirovitch 1986] L. Meirovitch, *Elements of Vibration Analysis*, McGraw-Hill, New York, 1986.

[Motamedi 2005] M. E. Motamedi, editor, *MOEMS: Micro-Opto-Electro-Mechanical Systems*, SPIE: The International Society for Optical Engineering, Bellingham, WA, 2005.

[Rebeiz 2002] G. M. Rebeiz, *RF MEMS: Theory, Design, and Technology*, John Wiley & Sons, Hoboken, NJ, 2002.

R13.1 Introduction

[Dittrich 2006] P. S. Dittrich, T. Tachikawa, and A. Manz, Micro total analysis systems: Latest advancements and trends, *Analytical Chemistry*, 78(12), pp. 3887–3907, June 2006.

[Fair 2003] R. B. Fair, V. Srinivasan, H. Ren, P. Paik, V. K. Pamula, and M. G. Pollack, Electrowetting-based on-chip sample processing for integrated microfluidics, in *Technical Digest, IEEE Int Electron Devices Meeting*, pp. 32.5.1–32.5.4, December 2003.

[Petersen 2005] K. Petersen, A new age for MEMS, in *Proc. Int. Conf. on Solid-State Sensors, Actuators, and Microsystems*, pp. 1–4, June 2005.

[Verpoorte 2003] E. Verpoorte and N. F. De Rooij, Microfluidics meets MEMS, *Proceedings of the IEEE*, 91(6), pp. 930–953, June 2003.

R13.2 MEMS Testing Considerations

[Feuerstein 2004] D. Feuerstein and F. M. Werner, Testing MEMS at wafer level, *Evaluation Engineering*, September 2004.

[Jaeckel 2002] R. Jaeckel, production handling and testing of MEMS, *Evaluation Engineering*, May 2002.

[Memunity 2007] The MEMS Testing Community, www.memunity.com, 2007.

[Strassberg 2001] D. Strassberg, Testing MEMS: Don't reinvent the wheel but take little on faith, *Electron. Design Magazine*, pp. 46–54, October 2001.

[Wolf 2002] I. D. Wolf, Instrumentation and methodology for MEMS testing, reliability assessment and failure analysis, in *Proc. Int. Conf. on Microelectronics*, pp. 161–167, May 2004.

R13.3 Test Methods and Instrumentation for MEMS

[Almeida 2006] L. Almeida, R. Ramadoss, R. Jackson, K. Ishikawa, and Q. Yu, Study of electrical contact resistance of multi-contact MEMS relay fabricated using MetalMUMPs process, *J. Micromechanics and Microengineering*, 16(6), pp. 1189–1194, July 2006.

[Lecklider 2006] T. Lecklider, Testing written very small, *Evaluation Engineering*, September 2006.

[Polytec 2007] MEMS Geometry and vibrations, *Laser Measurement Systems Application Note VIB-M-05*, Polytec GmbH, www.polytec.com, 2007.

[Wolf 2002] I. D. Wolf, Instrumentation and methodology for MEMS testing, reliability assessment and failure analysis, in *Proc. Int. Conf. on Microelectronics*, pp. 161–167, May 2004.

R13.4 RF MEMS Devices

[Clark 2001] J. R. Clark, W.-T. Hsu, and C. T.-C. Nguyen, Measurement techniques for capacitively-transduced VHF-to-UHF micromechanical resonators, in *Proc. Int. Conf. on Solid-State Sensors & Actuators*, pp. 1118–1121, June 2001.

[Clark 2005] J. R. Clark, W.-T. Hsu, M. A. Abdelmoneum, and C. T.-C. Nguyen, High-Q UHF micromechanical radial-contour mode disk resonators, *J. Microelectromechanical Systems*, 14(6), pp. 1298–1310, December 2005.

[Hsu 2004] W.-T. Hsu, W. S. Best, H. J. De Los Santos, and C. T.-C. Nguyen, Design and fabrication procedure for high Q RF MEMS resonators, *Microwave J.*, 47(2), pp. 60–72, February 2004.

[Nguyen 2005] C. T.-C. Nguyen, RF MEMS in wireless architectures, in *Proc. ACM/IEEE Design Automation Conf.*, pp. 416–420, June 2005.

[Petersen 1978] K. E. Petersen, Micromechanical membrane switches on silicon, *IBM J. Res. Develop.*, 23(4), pp. 376–385, July 1979.

[van Spengen 2003] W. M. van Spengen, R. Puers, R. Mertens, and I. De Wolf, A low frequency electrical test set-up for the reliability assessment of capacitive RF MEMS switches, *J. Micromechanics and Microengineering*, 13(5), pp. 604–612, May 2003.

R13.5 Optical MEMS Devices

[Douglass 2003] M. R. Douglass, DMD reliability: A MEMS success story, in *Proc. Society of Photo-optical Instrumentation Engineers (SPIE), Testing, and Characterization of MEMS/MOEMS II*, 4980, pp. 1–11, January 2003.

[Polytec 2005] Polytec GmbH, MEMS geometry and vibrations, *Laser Measurement Systems Application Note VIB-M-05*, Polytec GmbH, www.polytec.com, 2005.

[Tuantranont 2000] A. Tuantranont, L.-A. Liew, V. M. Bright, J. Zhang, W. Zhang, and Y. C. Lee, Bulk-etched surface micromachined and flip-chip integrated micromirror array for infrared applications in *Proc. IEEE/LEOS Int. Conf. on Optical MEMS*, pp. 71–72, August 2000.

R13.6 Fluidic MEMS Devices

[Bilenberg 2003] B. Bilenberg, B. Helbo, J. P. Kutter, and A. Kristensen, Tunable microfluidic dye laser, in *Proc. Int. Conf. on Solid-State Sensors, Actuators and Microsystems*, pp. 206–209, June 2003.

[Choi 2001] J.-W. Choi, K. W. Oh, J. H. Thomas, W. R. Heineman, H. B. Halsall, J. H. Nevin, A. J. Helmicki, H. Thurman Henderson, and C. H. Ahn, An integrated microfluidic biochemical detection system with magnetic bead-based sampling and analysis capabilities, in *Proc. IEEE Int. Conf. on Micro Electro Mechanical Systems*, pp. 447–450, January 2001.

[Dean 2005a] R. Dean, J. Pack, N. Sanders, and P. Reiner, Micromachined LCP for packaging MEMS sensors, in *Proc. Annual Conf. of IEEE Industrial Electronics Society (IECON 2005)*, pp. 2363–2367, November 2005.

[Kerkhoff 2003a] H. G. Kerkhoff and M. Acar, Testable design and testing of micro-electro-fluidic arrays, in *Proc. IEEE VLSI Test Symp.*, pp. 403–409, April 2003.

[Kerkhoff 2003b] H. G. Kerkhoff and M. Acar, Electronic test solutions for FlowFET fluidic arrays, in *Proc. Symp. on Design, Test, Integration and Packaging of MEMS / MOEMS*, pp. 27–32, May, 2003.

[Palasagaram 2006] J. N. Palasagaram and R. Ramadoss, MEMS capacitive pressure sensor fabricated using printed circuit processing techniques, *IEEE Sensors J.*, 6(6), pp. 1374–1375, December 2006.

[Sakaue 2005] E. Sakaue, Micromachining/nanotechnology in direct methanol fuel cell, in *Proc. IEEE Int. Conf. on Micro Electro Mechanical Systems*, pp. 600–605, January 2005.

[Shi 2006] P. Z. Shi, K. M. Chua, S. C. K. Wong, and Y. M. Tan, Design and performance optimization of miniature heat pipes in LTCC, *J. Physics: Conf. Series*, 34(1), pp. 142–147, April 2006.

R13.7 Dynamic MEMS Devices

[Dean 2005b] R. Dean, G. Flowers, N. Sanders, R. Horvath, M. Kranz, and M. Whitley, Micromachined vibration isolation filters to enhance packaging for mechanically harsh environments, *J. Microelectronics and Electronic Packaging*, 2(4), pp. 223–231, December 2005.

[Loeppert 2006] P. V. Loeppert and S. B. Lee, SiSonic: The first commercialized MEMS microphone, in *Digest of Papers, Solid-State Sensors, Actuators, and Microsystems Workshop*, pp. 27–30, June 2006.

[Pedersen 1998] M. Pedersen, W. Olthuis, and P. Bergveld, High-performance condenser microphone with fully integrated CMOS amplifier and DC-DC voltage converter, *J. Microelectromechanical Systems*, 7(4), pp. 387–394, December 1998.

[Weinberg 2005] H. Weinberg, Using the ADXRS150/ADXRS300 in continuous self-test mode, *Analog Devices Application Note AN-768*, 2005.

R13.8 Testing Digital Microfluidic Biochips

[Chatterjee 2006] D. Chatterjee, B. Hetayothin, A. R. Wheeler, D. King, and R. L. Garrell, Droplet-based microfluidics with nonaqueous solvents and solutions, *Lab on a Chip*, 6(2), pp. 199–206, February 2006.

[Cho 2003] S. K. Cho, H. Moon, and C. J. Kim, Creating, transporting, cutting, and merging liquid droplets by electrowetting-based actuation for digital microfluidic circuits, *J. Microelectromechanical Systems*, 12(1), pp. 70–80, February 2003.

[Fair 2003] R. B. Fair, V. Srinivasan, H. Ren, P. Paik, V. K. Pamula, and M. G. Pollack, Electrowetting-based on-chip sample processing for integrated microfluidics, in *Technical Digest, IEEE Int. Electron Devices Meeting*, pp. 32.5.1–32.5.4, December 2003.

[Kerkhoff 2007] H. G. Kerkhoff, Testing of microelectronic-biofluidic systems, *IEEE Design & Test of Computers*, January/February 2007.

[Schasfoort 1999] R. B. M. Schasfoort, S. Schlautmann, J. Hendrikse, and A. van den Berg, Field-effect flow control for microfabricated fluidic networks, *Science*, 286(5441), pp, 942–945, October 1999.

[Su 2003] F. Su, S. Ozev, and K. Chakrabarty, Testing of droplet-based microfluidic systems, in *Proc. IEEE Int. Test Conf.*, pp. 1192–1200, September/October 2003.

[Su 2004] F. Su, S. Ozev, and K. Chakrabarty, Concurrent testing of droplet-based microfluidic systems for multiplexed biomedical assays, in *Proc. IEEE Int. Test Conf.*, pp. 883–892, October 2004.

[Su 2005] F. Su, W. Hwang, A. Mukherjee, and K. Chakrabarty, Defect-oriented testing and diagnosis of digital microfluidics-based biochips, in *Proc. IEEE Int. Test Conf.*, pp. 487–496, November 2005.

[Verpoorte 2003] E. Verpoorte and N. F. De Rooij, Microfluidics meets MEMS, *Proceedings of the IEEE*, 91(6), pp. 930–953, June 2003.

[Xu 2007] T. Xu and K. Chakrabarty, Parallel scan-like testing and fault diagnosis techniques for digital microfluidic biochips, in *Proc. IEEE European Test Conf.*, pp. 63–68, May 2007.

[Zeng 2004] J. Zeng and T. Korsmeyer, Principles of droplet electrohydrodynamics for lab-on-a-chip, *Lab on a Chip*, pp. 265–277, August 2004.

R13.9 DFT and BIST for MEMS

[Analog 2007] www.analog.com/en/cat/0,2878,764,00.html.

[Beroulle 2002] V. Beroulle, Y. Bertrand, L. Latorre, and P. Nouet, Evaluation of the oscillation-based test methodology for micro-electro-mechanical systems, in *Proc. IEEE VLSI Test Symp.*, pp. 439–444, April 2002.

[Castillejo 1998] A. Castillejo, V. Veychard, S. Mir, J. M. Karam, and B. Courtois, Failure mechanisms and fault classes for CMOS-compatible microelectromechanical systems, in *Proc. IEEE Int. Test Conf.*, pp. 541–550, October 1998.

[Charlot 2001] B. Charlot, S. Mir, F. Parrain, and B. Courtois, Generation of electrically induced stimuli for MEMS self-test, *J. Electronic Testing: Theory and Applications*, 17(6), pp. 459–470, December 2001.

[Cozma Lapadatu 1999] A. Cozma Lapadatu, H. Jakobsen, and R. Puers, A new concept for a self-testable pressure sensor based on the bimetal effect, in *Proc. IEEE Int. Conf. on Solid-State Sensors and Actuators*, pp. 350–353, June 1999.

[Deb 2000] N. Deb and R. D. Blanton, Analysis of failure sources in surface-micromachined MEMS, in *Proc. IEEE Int. Test Conf.*, pp. 739–749, October 2000.

[Deb 2002] N. Deb and R. D. Blanton, Built-in self-test of CMOS-MEMS accelerometers, in *Proc. IEEE Int. Test Conf.*, pp. 1075–1084, October 2002.

[Deb 2004] N. Deb and R. D. Blanton, Multi-modal built-in self-test for symmetric microsystems, in *Proc. IEEE VLSI Test Symp.*, pp. 139–147, April 2004.

[Deb 2006] N. Deb and R. D. Blanton, Built-in self-test of MEMS accelerometers, *J. Microelectromechanical Systems*, 15(1), pp. 52–68, February 2006.

[Dhayni 2006] A. Dhayni, S. Mir, L. Rufer, and A. Bounceur, Pseudorandom functional BIST for linear and nonlinear MEMS, in *Proc. Design, Automation, and Test in Europe*, pp. 1–6, March 2006.

[Jiang 2006] T. Jiang and R. D. Blanton, Inductive fault analysis of surface-micromachined MEMS, *IEEE Trans. Computer-Aided Design*, 25(6), pp. 1104–1116, June 2006.

[Kolpekwar 1997] A. Kolpekwar and R. D. Blanton, Development of a MEMS testing methodology, in *Proc. IEEE Int. Test Conf.*, pp. 923–931, November 1997.

[Kolpekwar 1998a] A. Kolpekwar, C. Kellen, and R. D. Blanton, MEMS fault model generation using CARAMEL, in *Proc. IEEE Int. Test Conf.*, pp. 557–564, October 1998.

[Kolpekwar 1998b] A. Kolpekwar, R. D. Blanton, and D. Woodilla, Failure modes for stiction in surface-micromachined MEMS, in *Proc. IEEE Int. Test Conf*, pp. 551–556, October 1998.

[Kolpekwar 1999] A. Kolpekwar, T. Jiang, and R. D. Blanton, CARAMEL: Contamination and reliability analysis of microelectromechanical layout, *J. Microelectromechanical Systems*, 8(3), pp. 309–318, September 1999.

[Kuehnel 1994] W. Kuehnel and S. Sherman, A surface micromachined silicon accelerometer with on-chip detection circuitry, *Sensors and Actuators A*, 45(1), pp. 7–16, October 1994.

[Mir 2004] S. Mir, L. Rufer, B. Charlot, and B. Courtois, On-chip testing of embedded silicon transducers, in *Proc. IEEE Int. SoC Conf.*, pp. 13–18, September 2004.

[Mir 2006] S. Mir, L. Rufer, and A. Dhayni, Built-in-self-test techniques for MEMS, *Microelectronics J.*, 37(12), pp. 1591–1597, December 2006.

[Natarajan 2006] V. Natarajan, S. Bhattacharya, and A. Chatterjee, Alternate electrical tests for extracting mechanical parameters of MEMS accelerometer sensors, in *Proc. IEEE VLSI Test Symp.*, p. 6, April 2006.

[Plaza 1998] J. A. Plaza, J. Esteve, and E. Lora-Tamayo, Cantilever beam accelerometer with self test system: Simulation, technology and experimental results, in *Proc. IEEE Int. Workshop on Micro Electro Mechanical Systems*, pp. 135–138, January 1998.

[Pourahmadi 1992] F. Pourahmadi, L. Christel and K. Peterson, Silicon accelerometer with thermal self-test mechanism, in *Proc. IEEE Solid-State Sensor and Actuator Workshop*, pp. 122–125, June 1992.

[Puers 2001] R. Puers, S. Reyntjens, and D. De Bruyker, Remote sensors with self-test: New opportunities to improve the performance of physical transducers, *Advanced Engineering Materials*, 3(10), pp. 788–795, October 2001.

[Puers 2002] R. Puers and S. Reyntjens, RASTA: Real-acceleration-for-self-test accelerometer: A new concept for self-testing accelerometers, *Sensors and Actuators A*, 97–98, pp. 359–368, April 2002.

[Rosing 2000a] R. Rosing, A. Lechner, A. Richardson, and A. Dorey, Fault simulation and modeling of microelectromechanical systems, *Computing & Control Engineering J.*, 11(5), pp. 242–250, October 2000.

[Rosing 2000b] R. Rosing, A. M. Richardson, and A. M. Dorey, A fault simulation methodology for MEMS, in *Proc. Design, Automation and Test in Europe Conf.*, pp. 476–483, March 2000.

[Shen 1985] J. P. Shen, W. Maly, and F. J. Ferguson, Inductive fault analysis of MOS integrated circuits, *IEEE Design & Test of Computers*, 2(6), pp. 13–26, December 1985.

[Spineanu 1997] A. Spineanu, P. Benabes, and R. Kielbasa, A digital piezoelectric accelerometer with sigma-delta servo technique, *Sensors and Actuators A*, 60(1–3), pp. 127–133, May 1997.

[Terry 1989] S. C. Terry, H. V. Allen, and D. W. DeBruin, Accelerometer systems with self-testable features, *Sensors and Actuators A*, 20, pp. 153–161, 1989.

[Xiong 2005a] X. Xiong, Y.-L. Wu, and W.-B. Jone, A dual-mode built-in self-test technique for capacitive MEMS devices, *IEEE Trans. Instrumentation and Measurement*, 54(5), pp. 1739–1750, October 2005.

[Xiong 2005b] X. Xiong, Y. Wu, and W.-B. Jone, Design and analysis of self-repairable MEMS accelerometer, in *Proc. IEEE Int. Symp. on Defect and Fault Tolerance in VLSI Systems*, pp. 21–29, October 2005.

[Xiong 2006] X. Xiong, Y. Wu, and W. B. Jone, Reliability analysis of self-repairable MEMS accelerometer, in *Proc. IEEE Int. Symp. on Defect and Fault Tolerance in VLSI Systems*, pp. 236–244, October 2006.

[Zimmermann 1995] L. Zimmermann, J. Ph. Ebersohl, F. Le Hung, J. P. Berry, F. Baillieu, P. Rey, B. Diem, S. Renard, and P. Caillat, Airbag application: A microsystem including a silicon capacitive accelerometer, CMOS switched capacitor electronics and true self-test capability, *Sensors Actuators A*, 46(1–3), pp. 190–195, January/February 1995.

CHAPTER 14

HIGH-SPEED I/O INTERFACES

Mike Peng Li
Wavecrest, San Jose, California

T. M. Mak
Intel Corporation, Santa Clara, California

Kwang-Ting (Tim) Cheng
University of California, Santa Barbara, California

ABOUT THIS CHAPTER

Regardless of how complex a semiconductor device or chip is, the first thing that has to be dealt with is its pins or *input/output* (I/O) interfaces. They can be connected and probed by a myriad set of instruments. From a test perspective, because these pins often are visible, they also carry complex specifications such as drive/sense levels, source/sink capability, leakage, etc. As the signaling rate of a device goes up, more of its signaling characteristics also are specified. This is evident by picking up any datasheet of a chip/component. As processor speed continues to increase, I/O has become the bottleneck, constraining system-level performance. A processor may compute very fast, but if the instructions and data do not reach the processor in time, it simply has to wait. Therefore, improving I/O performance is essential for improving system-level performance. Consequently, in current systems, I/O has become the most important element requiring special attention.

This chapter is devoted to *high-speed parallel/serial I/O link testing* at both chip and system levels. It starts with a discussion of various I/O architectures and then explores various test methodologies for them. In particular, we conduct an extensive overview of the signaling properties of high-speed serial I/O, including jitter, noise, and *bit error rate* (BER). We also present *design-for-testability-assisted* (DFT-assisted) test methods for manufacturing test. Novel DFT approaches for testing the emerging equalization and compensation circuits used in I/O links at signaling rates over 1 GHz are also covered. At the system level, interconnect test methods using the IEEE 1149.1 and 1149.6 boundary-scan standards as well as the *interconnect BIST* (IBIST) method are also included. Finally, we discuss the unique challenges associated with the data rate scaling of I/O interfaces.

14.1 INTRODUCTION

Data communication has been around since the birth of telegraphy, way before any computers arrived on the scene. Early telegraphy primarily used Morse code to send text messages. Because this type of communication had to go hundreds and thousands of miles via a copper medium utilizing the then novel invention of electricity, it was deemed to be more economic to send a signal (coded as a series of short and long pulses) in a serial manner on a pair of wires. The use of the Morse code eventually became standardized, thus signaling the beginning of serial signal engineering.

Serial signaling has evolved to support computer-to-computer communications including **Ethernet** (and its various siblings). For electronic systems that are closer in proximity to each other, such as between boards or between components, designers may take liberty to use more wires or ***printed-circuit board*** (PCB) traces. At this level of communication, it is advantageous to use a less complex signaling scheme at the cost of more wires/board space. The term "bus" was invented. This essentially describes the communication pictorially: a set of data gets on a bus together and travels along together. Upon reaching a destination, they may depart together or continue onto the next stop. The bus can also pick up new data from any bus stop (agent) as long as a simple protocol is followed. A parallel bus is a common communication method, especially for component-to-component (on a board) and between board-to-board in close proximity.

Both parallel and serial communications have coexisted for a long time, and the choice is primarily based on the length of their interconnections. Once communication has to leave a box and goes for any distance greater than tens of meters, serial communication is more preferred. Anything less than that, the simplicity of parallel communication tends to win over.

As the need to move greater and greater amounts of data among big computer networks became more pressing, serial communication links evolved. They are capable of delivering data at a rate of multiple *gigabits per second* (Gbps). Communication standards for technologies capable of transmitting at these data rates are ***synchronous optical network*** (SONET), ***gigabit ethernet*** (GBE), ***fibre channel*** (FC), and those standards created by the ***Optical Internetworking Forum*** (OIF). These communication standards are all serial links that utilize an embedded clock in the transmitting data stream, and the clock signal is recovered at the receive side.

In contrast, chip-to-chip I/O interconnect technology has been largely based on parallel bus technology for the past 40-plus years. The signaling mechanism is mostly based on a global clock, which qualifies the data from the transmit side, and the same global clock also will latch the data at the receive side. Nevertheless, to improve system-level performance, this system clock rate also has been increasing steadily to keep the data rate up. At a data rate of less than 1 Gbps, synchronized **parallel** I/Os with a ***global clock*** (GC) or a ***source synchronous*** (SS) I/O with a strobe is common. These are still expected to be the signaling methods of choice if the data rate is less than 1 Gbps. Special types of buses exist, such as **hypertrans-**

port, where a strobe is sent along with the data on a per link basis. It can deliver data at up to a 5 Gbps rate. The consequence for such architecture is a large pin count for devices with many I/O pins.

Of course, improving system-level performance further is an important goal, and **parallel signaling** begins to show its limitations as transmission speeds increase. At data rates of more than 1 Gbps per wire, the parallel data bus architecture can no longer be sustained for nonforwarding strobe or clock types of architecture because of limitations such as the larger number of I/O pins and channel-to-channel skew. Instead, the distinct properties of serial signaling start to shine as each serial channel is self-timed by using an embedded clock so the channel-to-channel **skew** and mismatch issues associated with synchronized parallel I/O no longer apply. Short distance, chip-to-chip I/O links essentially have adopted the serial communication architectures developed for long-distance network communications, with some technological differences. Digital based ***clock-recovery*** (CR) technologies, for example, are widely used for chip-to-chip I/O links, such as ***phase interpolator*** (PI) or ***oversampling*** (OS), in addition to the conventional analog ***phase-locked loop*** (PLL). To provide for even more data bandwidth, the link can consist of multiple serial channels.

As we move to a serial signaling technology, there are a lot more signaling properties we have to consider. Even the tricky *source synchronous* (SS) timings would seem like child's play compared with asynchronous serial timings. At multiple-Gbps data rates, a digital waveform appears to be an analog waveform at the receiver input because of the frequency-dependent lossy property of the channel or medium (PCB traces, cables, connectors, etc.). Timing edges tend to move around **(jitter)** because of power fluctuations and couplings from neighboring signals. As such, **timing jitter** and **amplitude noise** can be viewed as the consequences of signal waveform degradation, and data might be recognized erroneously and result in bit errors. These errors can be corrected either with error detection and correction protocols or by using a higher transaction-level protocol to request that data be resent. As long as the error rate, or to be more specific, the ***bit error rate*** (BER), is low enough, the higher data rate can compensate for occasional hiccups. Obviously, to achieve a good BER, which is typically 10^{-12} or less for most of the high-speed communication technologies, jitter and noise must satisfy certain limits.

As the data rate keeps increasing, the ***unit interval*** (UI)—the period during which a digital bit can exist—becomes shorter and shorter, and, as such, the system will be more susceptible to failures resulting from jitter and noise. Because of this failure mechanism, ***jitter***, ***noise***, and ***BER*** (JNB) testing becomes necessary for multiple Gbps and GHz devices and systems. Therefore, **JNB testing** needs to be conducted against their corresponding limits so that good interoperability can be verified and ensured.

Because jitter, noise, and BER can accumulate, JNB testing relies on statistical data analysis processes, and all the rules established for those processes should apply. Traditionally, jitter and noise were quantified mostly by either using the statistical range or ***peak-to-peak*** (PK-PK) value, or the ***root-mean-square*** (RMS) of the entire jitter statistical distribution. By the end of 1990, the concept of jitter components such as ***deterministic jitter*** (DJ) and ***random jitter*** (RJ) were first developed to better accommodate the testing needs for mainstream gigabit serial data communications architectures [Wilstrup 1998] [Li 1999] [NCITS 2001]. The motivation for this change was straightforward: neither PK-PK nor RMS can be a good metric for a commonly encountered jitter distribution containing both bounded deterministic (*i.e.*, DJ) and unbounded random (*i.e.*, RJ) processes. Think of deterministic jitter as jitters that are occurring periodically. This could be power noise that happens regularly because of another power hog switching somewhere on the power grid. Random jitter, on the other hand, is a result of various uncorrelated events. This may occur in a particular cycle, but its occurrence pattern is not predictable. Separating jitter into DJ and RJ enables appropriate and accurate statistical metrics to be applied to the corresponding statistical processes. Furthermore, separating jitter into its components also enables developments of diagnostic and debug methods to pin down the exact root causes of the jitter problems, should a system fail. The jitter analysis metrology paradigm shift has triggered many innovations that we will discuss in more detail in the next section.

In addition to obeying statistical rules, JNB testing also follows the rules for serial communication because what matters is the JNB behavior in a system. It does not exist as a pure or unconstrained mathematical or statistical abstraction. Thus, understanding the link architecture, as well as JNB behaviors and characteristics in a link system, is necessary to develop appropriate simulation and measurement methods. An important concept established in testing JNB in a serial link is that the relevant JNB "seen" by a receiver is not necessarily the same raw JNB data as found at a transmitter output or receiver input [Li 2003, 2004]. Thus, to determine receiver jitter and noise responses and JNB properties, knowledge of the receiver architecture is critical. Both the **equalization** and clock recovery circuit are important and relevant to JNB characteristics. Equalization compensates for the frequency-dependent losses (rolling off at the high end) characteristic of the channel or media, and the clock recovery circuit helps to track the data even though it is jittery. For example, a transmitter may have significant jitter but will still work fine in an actual system because of the jitter tracking of the receiver clock recovery circuit. If this transmitter is tested without considering the receiver clock recovery jitter tracking, it would be mistested as a failing part.

At the same time, it is important to test these interfaces on a digital testing platform to keep test costs in check. This is because of the continued commoditization of serial link technology in the mainstream personal computer business and the fact that these interfaces are mostly I/O for large digital circuits (*e.g.*, an I/O for memory or graphics interface hub). Testing the digital circuits on a digital tester while testing the I/O interfaces on an expensive mixed-signal tester is possible, but the high test costs of such an approach certainly limit its use. Many **DFT-assisted test** solutions implemented either on the silicon itself or on the custom loadboard were developed in the early 2000s.

Moreover, in the board test arena, people who have been testing their boards with boundary scan using the IEEE 1149.1 standard [IEEE 1149.1-2001] since the mid-1990s suddenly found that they could not count on the same technique any more as data rates increased. There is an effort to extend the 1149.1 standard to a new 1149.6 standard [IEEE 1149.6-2003] to deal with the AC-coupled nature of these high-speed links. However, because this is still based on an essentially DC testing infrastructure, there is no guarantee that the interfaces are up to the performance and speed requirements of these high-frequency links, even if they are tested to the 1149.6 standard. Several techniques have been in the works, the most visible of which is ***interconnect built-in self-test*** (IBIST) [Nejedlo 2003], which essentially tests the interface at a high data rate or at-speed with worst-case patterns.

14.2 HIGH-SPEED I/O ARCHITECTURES

High-speed I/O architecture is largely driven by the demand of delivering higher data rates. Clock generation and distribution are important considerations in an I/O architecture. On the other hand, architecture determines the testing parameters and functionality requirements, as well as test methodologies. Therefore, it is important to first understand I/O architecture as well as its testing implications and requirements before discussing testing parameters and methodology.

Using a computer I/O bus as an example, at data rates of up to 200 Mbps, the commonly implemented I/O uses the *global clock* (GC) architecture in a parallel way. A global clock is sent to both transmitter and receiver of one link, and the receiver uses the subsequent clock edge of the same clock to drive its data latch. The GC architecture is limited by the skew and propagation delay for clock and data. To increase the data rate to speeds greater than 200 Mbps, *source synchronous* (SS) architecture was developed to deliver the data up to 1 Gbps. In the SS architecture, the transmitter sends a strobe signal along with the data signal. The strobe is used to latch the data at the receiver, and this removes the limitations of both clock skew and propagation delay in the GC. Both GC and SS are synchronized and parallel I/O buses as both transmitter and receiver use the same clock or strobe signal; hence, it is easy to interface with other synchronized systems. As the data rate increases beyond 1 Gbps, the skew between data bits in each parallel channel becomes a key limiting factor. To compensate, an **embedded clock** architecture was developed, which uses a clock embedded in the data stream that is recovered in the receiver. Because there is no clock or strobe sent along with data and data are self-timed, skew and propagation delay are not limitations for the **embedded clock signaling** architecture. We will discuss those architectures in sequence and address what components need to be tested and why.

14.2.1 Global Clock I/O Architectures

The most commonly used GC I/O architecture is shown in Figure 14.1. In this architecture, the signal is launched off one chip with the system clock and received at another chip at the following clock edge. At the sending end, there is a clock to signal delay (T_{co}) specification, and at the receiving end, there is setup (T_{setup}) and hold time (T_{hold}) on either side of the following clock edge.

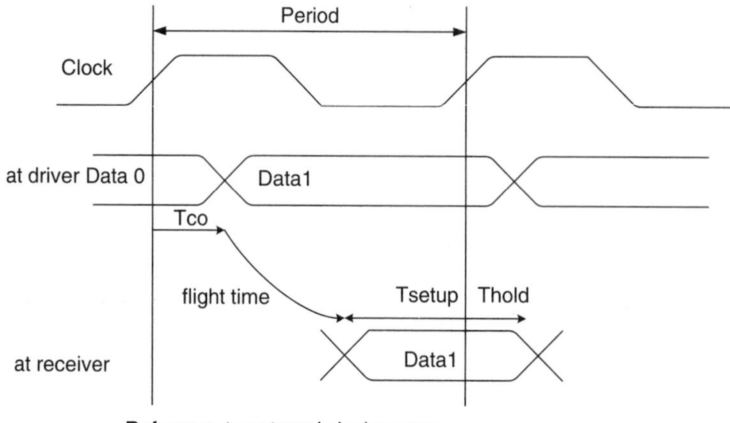

■ **FIGURE 14.1**

Global clock (GC) data and clock timing relationship.

14.2.2 Source Synchronous I/O Architectures

As the signaling rate increases, a problem arises. The clock skew between the sending component and the receiving component (board trace delay A-B shown in Figure 14.2) can cut into the cycle time. To compensate for this clock skew,

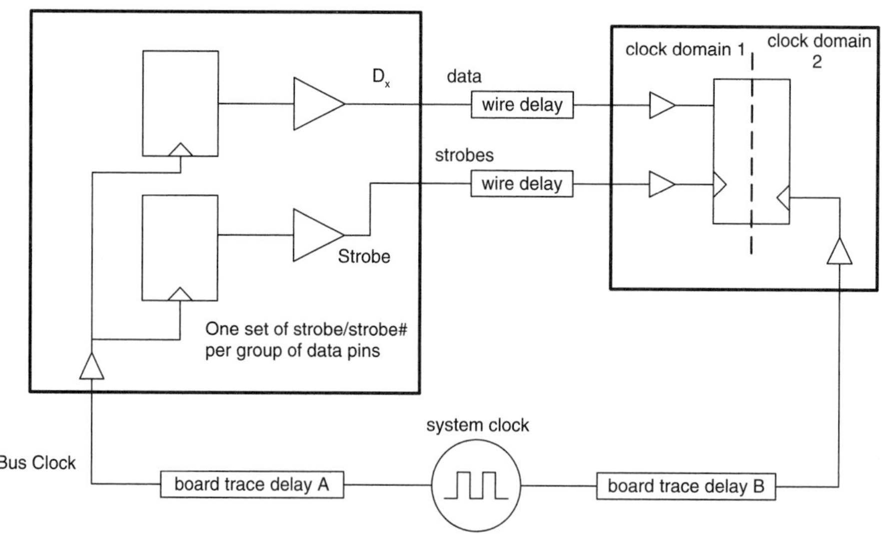

■ **FIGURE 14.2**

Source synchronous (SS) signaling scheme.

the ***source synchronous*** **(SS) architecture** was developed [Ilkbahar 2001]. With this scheme, not only will the sending component send the signal, but also another strobe that is similar to a clock signal goes along with that signal. The receiving component uses this strobe to clock the signal. Hence, system-level clock skew is out of the picture. The designer only cares about the differential skew between the strobe and the signal, with the key timing specifications as the ***time valid before*** (the strobe) and ***time valid after*** (the strobe) (see Figure 14.3). With careful design, (*e.g.*, identically sized drivers and matched layout), this signaling scheme allows the signaling rate to increase gradually from less than 10 to 50 Mbps to today's 1066 Mbps.

As much as this new **SS signaling** scheme has improved system-level performance, another problem starts to show up when the data rate continues to increase. The parallel interface that we use to transfer lots of data becomes a bottleneck itself. The parallel bits of data from the sending/receiving component have to center around the strobes. The skews among these data bits caused by uneven driving speeds and propagation delays between parallel channels become a performance limitation (see Figure 14.3). Additionally, the multiple load nature of a parallel bus creates noises that also affect ***signal integrity*** (SI). Each of these loads can be considered a stub on the transmission line. Because the stub is never perfectly matched to the transmission line characteristics, some of the energy would be reflected and is superimposed on the signal, making it noisier. It is generally believed that beyond 1 Gbps, new signaling technology is required. One may still use SS or similar architectures such as *hypertransport* with higher data rate, but the number of loads has to be decreased correspondingly to minimize disruption to the transmission line characteristics.

Another solution to extend SS is to reduce the number of data bits per strobe or, in other words, provide more strobes for a wide data bus. This will increase the number of board traces or wires and is not economically feasible. This can continue

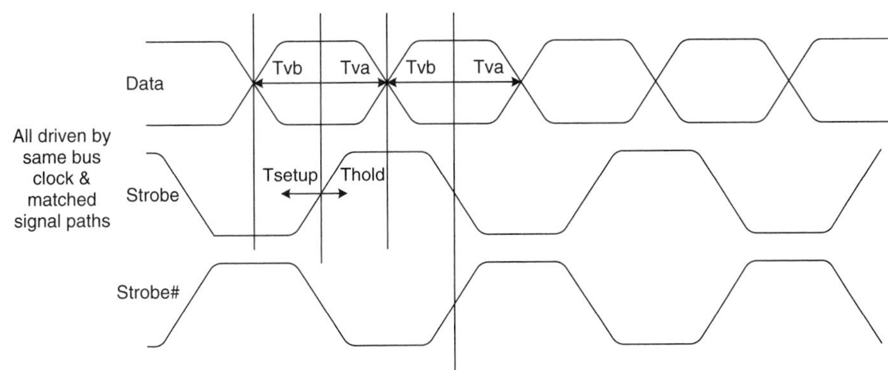

FIGURE 14.3

SS timing definition.

up to the point where we need a strobe with a data bit; then there will not be data skew, but then we also expand the size of the bus to double.

14.2.3 Embedded Clock I/O Architectures

As pointed out earlier, at a data rate beyond 1 Gbps, we have to treat each data channel individually because of the possible skews among data channels. One solution is to apply a serial signaling technique whereby the clock is embedded in its transmitting data stream. At the receive side, this clock is recovered through a CR circuit. PLL circuits are commonly used for clock recovery, but there also exist other digital circuit techniques for this task. Figure 14.4 illustrates the block diagram of a serial link with a CR circuit placed on the receive side.

One may wonder: how does the clock get embedded into the data? The clock is periodic, but data are not. In other words, natural digital data do not resemble the clock. How would the clock recovery scheme work? The trick here is to make the data look more like a clock. By mapping a data word (*e.g.*, 8 bits) onto another codeword (*e.g.*, 10 bits), and only mapping to those with an even number of transitions (*i.e.*, more or less an equal number of 1's and 0's, or DC balanced coding), the data stream will look more like a clock stream and the PLL at the receiver end can periodically synchronize to some of the transitions while it regenerates the clock.

In a serial **embedded clock I/O** link system, jitter is the dominant cause for performance degradation. Therefore, to recover the clock at the receiver, the PLL attempts to track the jitter to mitigate jitter's negative impact on the receiver performance. To characterize the PLL in terms of its performance for tracking jitter, as well as to quantify the jitter "propagation" process from the transmitter to the receiver, some system transfer function concepts, such as ***linear time-invariant*** (LTI) system theory [Papoulis 1977] [Oppenheim 1996], must be incorporated.

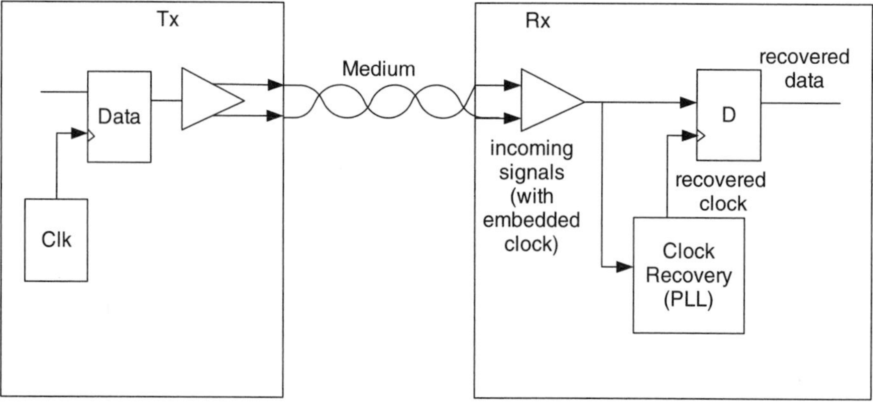

■ **FIGURE 14.4**

Serial link using embedded clock signal scheme.

In an LTI system, its output signal can be determined by the convolution of its input and the impulse response of the system.

In the following, we start with the definition of various jitter components and their interaction hierarchy. Then, we describe how to separate jitter components, an important and practical topic in jitter testing and analysis. Finally, we discuss the extension of the jitter component concept and the separation methods for noise component analysis, and the interactions between jitter, noise, and BER.

14.2.3.1 Jitter Components

A popular and widely used definition for so-called **phase jitter** or **accumulated jitter** states that jitter is the time deviation of an edge transition from its ideal time location [Li 2004] [PCI-SIG 2004]. Phase jitter results from several sources of noise and thus it is desirable to classify jitter into components such that each of them can be attributed to a distinct process or an underlying mechanism causing the jitter. The tree given in Figure 14.5 illustrates the classification scheme of various jitter components and their relationships [Li 2000] [NCITS 2001].

Jitter or ***total jitter*** (TJ) can be separated into two major components: ***deterministic jitter*** (DJ) and ***random jitter*** (RJ). DJ's ***probability density function*** (PDF) is bounded, whereas RJ's PDF is **Gaussian** and **unbounded**. DJ can be further separated into three types: ***data-dependent jitter*** (DDJ), ***periodic jitter*** (PJ), and ***bounded-uncorrelated jitter*** (BUJ). RJ includes ***Gaussian jitter*** (GJ) and ***multiple Gaussian jitter*** (MGJ). There are two types of DDJ: ***duty cycle distortion*** (DCD) and ***intersymbol interference*** (ISI). Figure 14.6 shows the various jitter components and their associated PDFs in the context of an **eye diagram**.

Another means of classifying jitter is based on whether the jitter correlates to the data pattern or not: **correlated jitter** and **uncorrelated jitter**. **Correlated jitter** types include DDJ, DCD, and ISI, and **uncorrelated jitter** types include PJ, BUJ, MGJ, and GJ. This particular means of classifying jitter components offers additional insights to the jitter processes and enables development of better quantification methods.

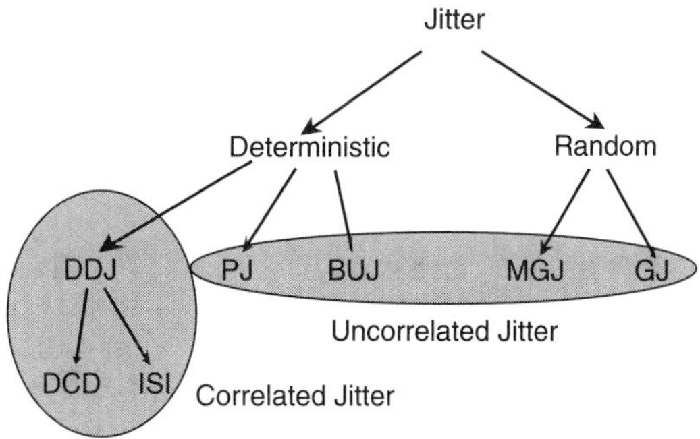

■ **FIGURE 14.5**

Classification of various jitter components.

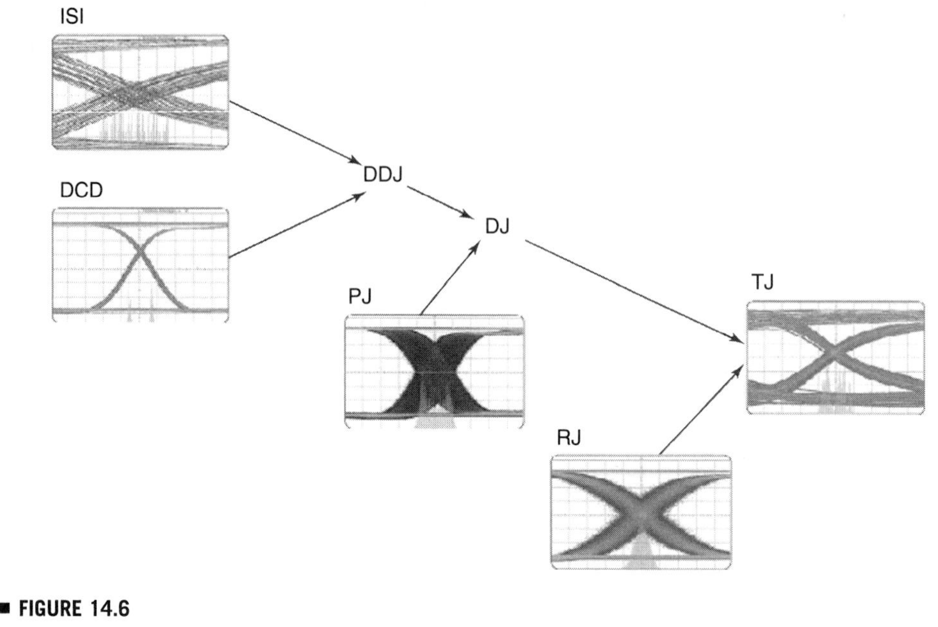

■ **FIGURE 14.6**

Various jitter components and their associated PDFs.

Each jitter component has specific physical mechanisms and root causes associated with it. For DJ components, DCD can be caused by a reference voltage offset or a time delay between the rising and falling clock edges when it is digitally synthesized. ISI results from interaction between successive data bits. These data bits form an analog waveform with rich high-frequency components. As the waveform transverses through lossy and band-limiting materials (*e.g.*, PCB traces), energy at some specific frequencies may be attenuated, resulting in data-dependent variation; hence the term ISI. PJ can be caused by periodic modulations or interferences. BUJ can be caused by crosstalk. RJ can be caused by thermal, flicker, or shot noises. Knowing the amount of each jitter component provides valuable diagnostic information for finding and, in turn, fixing a specific jitter failure problem.

14.2.3.2 Jitter Separation

There are two major approaches to separating individual jitter components. One method is based on the jitter PDF (see Figure 14.7) or ***cumulative distribution function*** (CDF) (see Figure 14.8) measurement. PDF is the normalized histogram of the signal edge times (*i.e.*, how often the signal transitions at a particular time point). CDF, on the other hand, sorts the sampled edge time data in an ascending order to show the distribution profile. Another method is based on the jitter time record. Jitter PDF can be measured by instruments such as the ***sampling oscilloscope*** (SO) or ***time interval analyzer*** (TIA). Jitter CDF (sometimes called BER CDF) can be

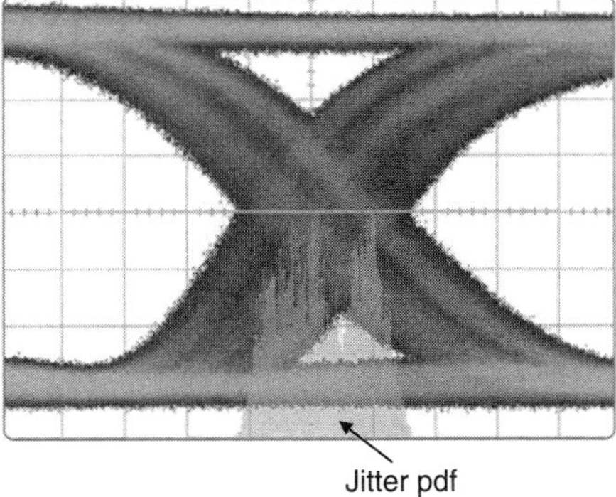

FIGURE 14.7

Jitter probability density function (PDF).

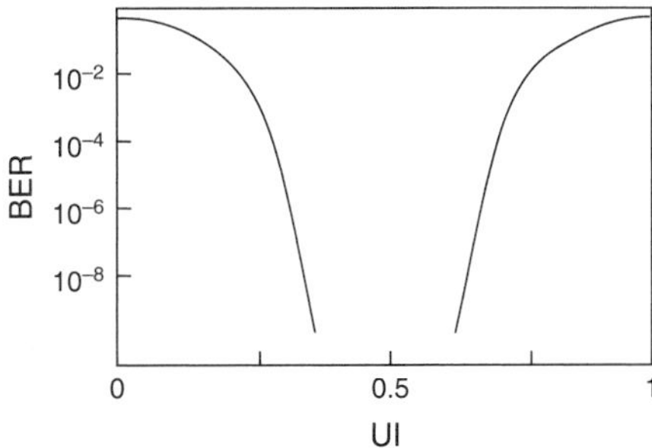

FIGURE 14.8

Jitter cumulative distribution function (CDF).

measured by a ***bit-error-rate tester*** (BERT). The jitter time record can be measured by a TIA or a ***real-time oscilloscope*** (RTO).

Jitter Separation Based on Statistical PDF or CDF

A widely accepted method for conducting PDF-based jitter separation is the **Tailfit** method [Li 1999]. Because the deterministic PDF is bounded, beyond certain jitter limit or the range of DJ, all the PDF-based jitter will be random jitter. Random jitter is naturally modeled by a **Gaussian function** (bell-shaped distribution). Therefore, the tail of a jitter PDF reflects the random jitter process, which, in general should be a Gaussian-type distribution. The Tailfit algorithm identifies a Gaussian curve with a symmetrical tail region to that of the distribution under evaluation. Two Gaussian curves (left and right side of the PDF) are fitted against each of the tail regions of the distribution until optimal matches are found. When an analytical Gaussian model is used to match those measured tail region PDFs through nonlinear fit or other optimization procedures, all the parameters defining a Gaussian distribution such as mean and RMS can be determined. Then the DJ PK-PK can be estimated as the difference between two means, and RJ σ (or RMS) will be the average of two tail σs. The application of the Tailfit method to a total jitter PDF is shown schematically in Figure 14.9.

The Tailfit method can be applied to a BER CDF if it is directly measured via instrument such as a BERT. In general, the mechanism is similar to that used in PDF fitting. However, because the base data are CDF, the model used to fit the tail region of the CDF should be an integrated Gaussian, which represents an **error function** [Hänsel 2004]. Also, the left tail of PDF corresponds to the right tail of the CDF. Similarly, the right tail of PDF corresponds to the left tail of the CDF. Tailfit for BER CDF is shown in Figure 14.10.

It could be hard to extract some second- or third-level DJ components shown in Figure 14.5, such as DCD, ISI, PJ, and DDJ, from the jitter PDF or CDF data. The reason is that the features and characteristics of those jitter components are "washed out" because of the summing and integration in deriving the PDF and CDF distributions.

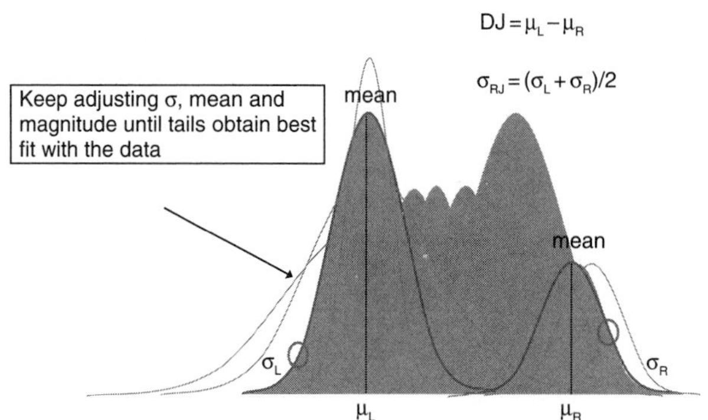

■ **FIGURE 14.9**

Fitting Gaussian distribution curves to a jitter histogram Tailfit.

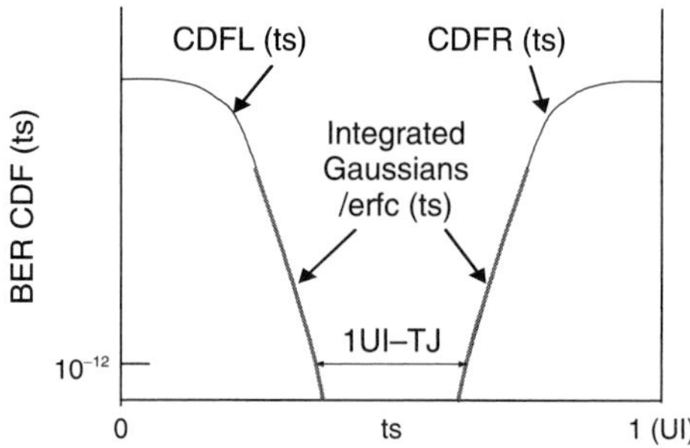

■ **FIGURE 14.10**

Tailfit for BER CDF.

In addition to separating jitter to its DJ and RJ components, Tailfit is used to estimate or extrapolate *total jitter* (TJ) with smaller measurement sample size or time. TJ is defined as the timing closure at a low BER level (10^{-12} or lower for most of the high-speed I/Os) on the BER CDF curve (see Figure 14.10). TJ can be determined based on direct measurement of BER CDF down to 10^{-12}, but such method is time consuming and slow. With the Tailfit method, BER CDF at a low BER level (*e.g.*, 10^{-12} or lower) can be estimated or extrapolated from a higher BER level (*e.g.*, 10^{-6} or higher) with smaller sample size, enabling TJ estimation at the low BER level with fast throughput.

Jitter Separation Based on Frequency Spectrum

If a data stream is repeatedly sampled and captured in the time-domain via instruments such as TIA or RTO, then spectrum analysis can be carried out via time to frequency domain transformations such as ***Fourier transformation*** (FT). The mainstream methods for spectrum analysis of a stochastic process such as RJ include the **autocorrelation** method (for the time domain) and the ***power spectrum density*** (PSD) method (for the frequency domain) [Papoulis 1977] [Oppenheim 1996]. In [Wilstrup 1998], these two methods were first applied for jitter separation in the time and frequency domains, respectively. In these methods, repeated sampling of time deviations from the ideal locations produces a distribution of both DCD and ISI, whereas the variance function in the frequency domain results in a jitter PSD for PJ and RJ, which is *uncorrelated* to the data patterns. An exemplar jitter PSD function having both PJ and RJ is shown in Figure 14.11.

PJ components will be shown as spectral lines in the PSD function, and a **sliding window** technique can be used to identify the magnitude and frequency for each of the PJ components. Then, all PJ spectral lines are removed from the PSD function record. The residues can then be summed over a frequency band and the square root of the sum gives the RMS value of the RJ over that frequency band. Another similar jitter separation method—which employs undersampling, targets production testing, and measures the means of the edge transition times—has been developed by [Cai 2005].

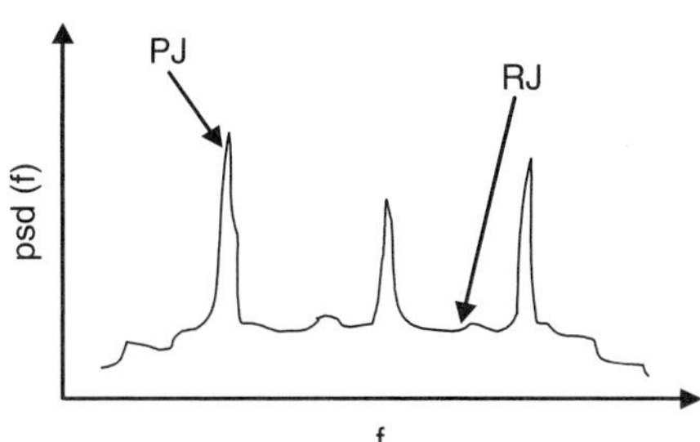

■ FIGURE 14.11

An exemplar plot of jitter power spectrum density (PSD).

A less accurate jitter estimation method using voltage time records can be accomplished by applying direct *Fourier transform* (FT) or *fast Fourier transform* (FFT) operation to convert the interpolated jitter time record (which is not a directly measured jitter time record) to its frequency domain spectrum. The data can then be squared to derive an approximated PSD [Ward 2004]. Unlike the spectrum shown in Figure 14.11, all jitter components, including the DJs, will show up in the PSD derived by this direct FFT approach. The DJs will appear as spectral lines as well. Therefore, if there is a PJ component whose frequency is an integer fraction of the pattern repeating frequency, then PJ and DDJ can no longer be separated. Similarly, this method will also introduce inaccuracy in the RJ estimation [Davenport 1987].

14.2.3.3 Jitter, Noise, and Bit-Error-Rate Interactions

A bit error can be caused by either timing jitter or amplitude noise, or both. An exemplar eye diagram degraded by both timing jitter and amplitude noise is shown in Figure 14.12. If both timing jitter and amplitude noise are considered for estimating the BER, then the BER CDF will become a ***two-dimensional*** (2-D) function. At 0.5 UI of the eye diagram, the noise PDF will contribute to the BER CDF in a manner similar to that of the timing jitter at zero-cross voltage. A 2-D jitter and noise PDF can be measured by a SO or RTO. Although a BERT can measure the 2-D BER CDF, the test time will be very long (several hours, typically) for BER measurement at the 10^{-12} level. Figure 14.12 shows an eye diagram in which the associated jitter and noise PDFs are highlighted.

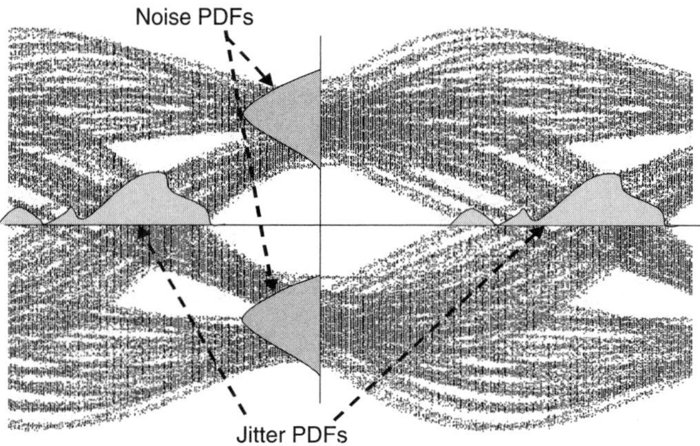

■ **FIGURE 14.12**

A distorted eye diagram caused by timing jitter and amplitude noise.

Receiver Jitter Transfer Function

It is well known that a PLL [Gardner 1979] [Best 1999] has certain frequency response characteristics. Therefore, when a receiver uses a PLL to recover the clock and then to time/retime the received data, the jitter that the receiver "sees" will follow those certain frequency characteristics as well. Any good simulation or test methodology should emulate the actual system/device behavior. In the case of jitter output/receiver jitter input determination, the model setup for both design and test should be such that it determines the jitter specifically as what a receiver sees in the link system [Li 2003, 2004]. A receiver sees jitter on the data from its recovered clock; therefore it is a difference function from clock to data as shown in Figure 14.13.

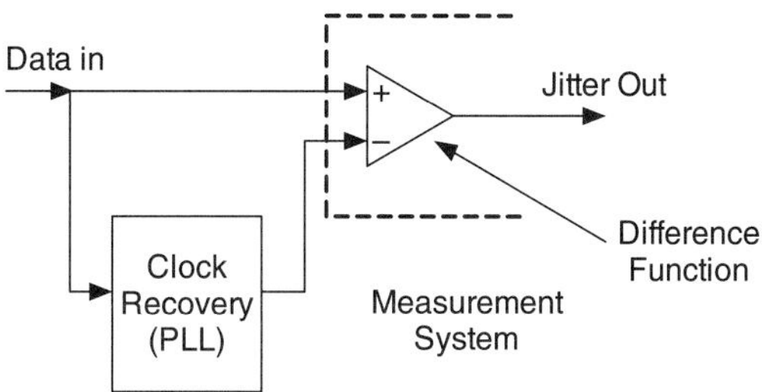

■ FIGURE 14.13

A receiver jitter model.

Because the clock recovery circuits (*e.g.*, PLL) typically have a low-pass transfer function $H_L(s)$, the jitter output will have a high-pass transfer function of $H_H(s)$ as shown in Figure 14.14. This is because $H_H(s) = 1 - H_L(s)$, assuming that there is no phase delay between data and clock.

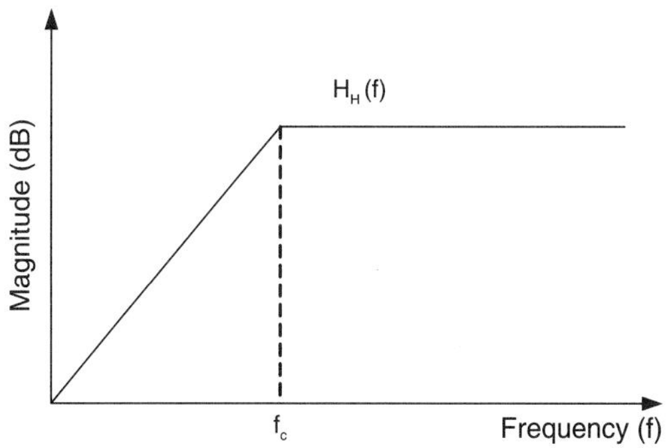

■ FIGURE 14.14

Receiver jitter transfer function.

Receiver Jitter Tolerance Function

The high-pass jitter magnitude transfer function $|H_H(s)|$ shown in Figure 14.15 suggests that a receiver is able to track or attenuate more low-frequency jitter at $f < fc$ than at a higher frequency of $f > fc$. fc is the corner frequency below which

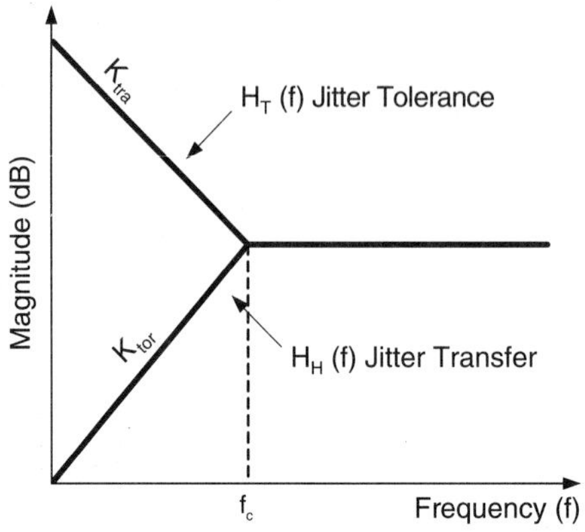

FIGURE 14.15

Receiver jitter tolerance function.

the transfer function magnitude begins to decrease and above which it maintains a constant. This implies that a receiver can tolerate more low-frequency jitter than high-frequency jitter for a given BER performance goal. Therefore, the jitter tolerance function is the "mirror" function of the jitter transfer function around the unity gain, as shown in Figure 14.15. The higher the corner frequency (f_c) is, the better the jitter tolerance capability will be. Similarly, the steeper the jitter transfer function slope is, the better the jitter tolerance capability will be. The corner frequency (f_c) is typically defined as the data rate divided by 1667 for data communication standards such as FC. For example, when the data rate is 2.5 Gbps, the corner frequency *fc* is 1.5 MHz (2.5 GHz/1667), whereas the number 1667 is inherited from the early SONET standard and is related to the PLL phase track speed capability. Note that jitter transfer function determines the receiver jitter tolerance.

Obviously, the slope for the jitter transfer function k_{tra} and the slope k_{tor} for the jitter tolerance function both satisfy the relationship of $k_{tra} = -k_{tor}$. If the CR PLL, as shown in Figure 14.4, is a second-order system, then we will have $k_{tra} = 40$ dB/decade. For a first-order or "golden" PLL, we will have $k_{tra} = 20$ dB/decade. Certainly, a clock and data recovery unit using a higher order PLL will result in a better jitter tolerance capability. For the same corner frequency (f_c), a second-order PLL has better jitter tolerance compared to a first-order PLL.

14.3 TESTING OF I/O INTERFACES

We have introduced the architectures for three I/O links. In this section, we discuss the testing of these interfaces. We start with the test for global clock I/O, and then move to source synchronous I/O, and finally to embedded clock I/O. For embedded clock I/O testing, we further cover testing of the transmitter, receiver, channel, reference clock, and system BER. Tester attributes of functionality, accuracy, and throughput are also discussed.

14.3.1 Testing of Global Clock I/O

Testing of global clock interfaces is relatively straightforward on commercial ***automatic test equipment*** (ATE). The input waveform of both the clock and data inputs can be generated from the tester with the proper format and timing control. Because the clock is preprogrammed (usually set at somewhere within the tester's cycle), all input setup and hold time will simply be timing generator edge placement around that clock edge (see Figure 14.16). The tester can strobe the output signal using either window strobe mode or edge strobe mode. The strobe edge or strobe window will be defined by the delay time specification.

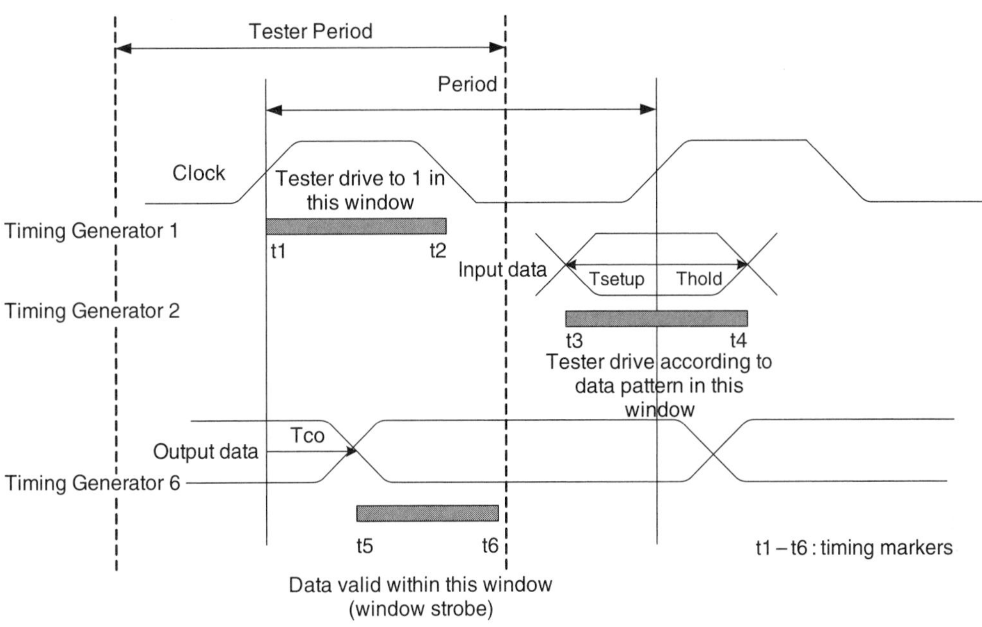

■ **FIGURE 14.16**

Global clock I/O timing tests as implemented on ATE.

14.3.2 Testing of Source Synchronous I/O

The testing of source synchronous interfaces is somewhat more complicated. For testing the input side, not much is changed from that of the global clock situation. One can program where the clocks (or strobes) are and then the other input signals are programmed around it according to the setup and hold time specification.

However, at the output side, the ***device under test*** (DUT) will send out both a strobe and the output data. The tester is supposed to use the strobe to strobe the data and has to ensure that the data have sufficient valid time before and after the strobe; however, the traditional tester architecture is not designed to use an external clock to strobe the data. This earlier architecture of the tester is designed to parallel-process all data coming from the DUT. Because of the limited level of tester pin-electronics integration, there is simply not enough time for the signal coming in through a pin-electronics card to be routed to another card to strobe the data.

Ten years after the introduction of the first *source synchronous* (SS) bus, newer generations of ATE are finally able to take advantage of the pin-electronics integration to route signals between the channels. The tester can now take the clock/strobe signal(s) out of the DUT to strobe the rest of the data pins natively [Sivaram 2004].

With the earlier generations of ATE, the usual way to get around the problem of testing for SS is to program a search routine to find where the output clock edge is and then program the window strobe for the corresponding data to ensure that the data are stable within the valid specification window (see Figure 14.17). This has the disadvantage of a longer test time, as time search is time consuming on the tester (each setting and resetting of the timing generator requires some time to settle once programmed). The tester accuracy also cuts into the measurement twice (first for the clock, and then for the data valid check) and will pose a hefty toll on product margin and yield. Another solution is to use the capture memory of the tester to **undersample** the outputs over a range of cycles to estimate where the data pin timing distribution is.

Both methods may work well for lower data rate (<400 Mbps) but are usually ineffective for higher data rate because of the native **common mode noises**, such as coupling and ground bounce effects, present on the SS bus. If there is a noise source in the DUT, this noise usually will manifest itself as a common mode symptom causing both strobe and data to jitter. Although the SS receiver may interpret the data correctly, external instrumentation using search or undersampling methods can see that the data eye shrinks, causing yield loss or lower bin-split.

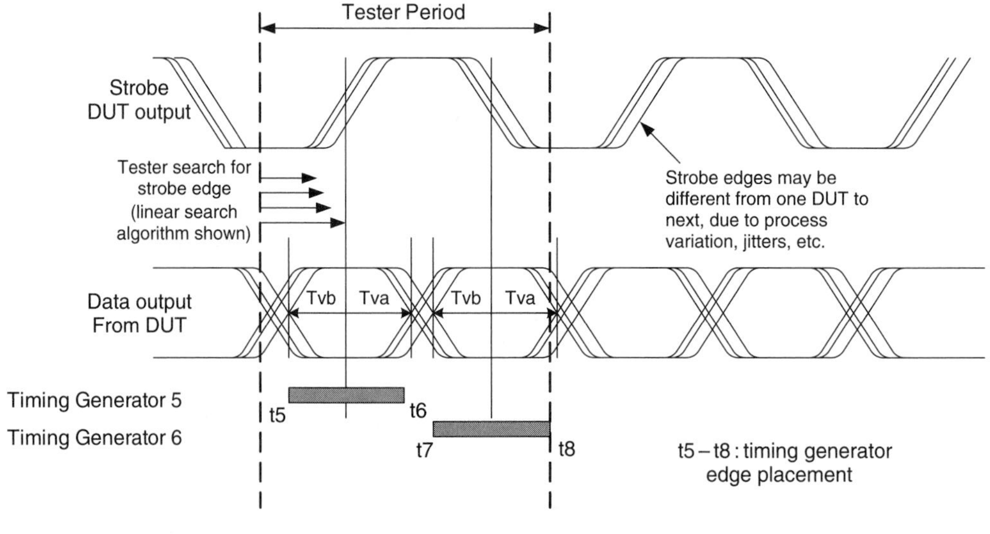

FIGURE 14.17

Source synchronous timing tests as implemented on ATE.

An alternative test method is to add DFT to support high-speed SS I/O testing, which is the subject of Section 14.4.1. Because of the availability of newer generations of ATE that can support native SS, the problem also goes away if there is liberty to upgrade the tester fleet to support high-performance bus signaling.

14.3.3 Testing of Embedded Clock High-Speed Serial I/O

We focus on **JNB tests** for an embedded clock serial I/O, as they are the most challenging tasks because of the complexity in the test method and high precision requirements for the test equipment. We classify JNB tests into three major categories: (1) the JNB output test, (2) the JNB tolerance test, and (3) the link system test. JNB output is typically tested at either the transmitter output pin or the receiver input connector. For most long-haul network devices, a **JNB output test** verifies the performance of either the transmitter or the transmitter plus the medium or channel. For some new Gbps computer I/O standards, JNB testing may also include the reference clock. A **JNB tolerance test** involves setting the worst-case jitter and noise condition(s) at the receiver input pin and measuring the BER at its output. A **JNB system test** checks the overall BER comparing the data bits received by the receiver with the data bits sent by the transmitter.

14.3.3.1 Transmitter

There are two closely related requirements for a JNB output test: (1) a minimum eye-opening at BER $= 10^{-12}$ under a compliance clock recovery jitter transfer function, and (2) worst-case test patterns. The minimum eye-opening defines two important metrics: TJ and ***total noise*** (TN), both correspond to a BER at the 10^{-12} level. A worst-case test pattern for generating worst-case DDJ and crosstalk jitter is needed for compliance test. Jitter components such as DJ and RJ are required for diagnostic testing and may also be required for some types of compliance testing.

Figure 14.18 shows a generic setup for JNB output testing. The critical elements of the clock recovery function (in terms of H[s]) and the data and clock input difference functions need to be in place. The eye diagram and its corresponding PDFs and CDFs measured in this way will have the right transfer function applied. Important jitter components, such as DJ, RJ, and TJ, as well as noise components, such as DN, RN, and TN, will be derived from those PDFs and CDFs with the required statistical confidence level.

Figure 14.19 illustrates the effect of using a recovered clock and a noncompliant clock (*i.e.*, a jitter-free clock whose phase does not track the phase of data) when making a clock-to-data measurement at 2.5 Gbps with a 200-KHz periodic modulation added to the data signal. The measurement was performed with a built-in, programmable hardware clock recovery circuit. The results show that the DJ was 13 ps with the recovered clock (with a corner frequency $f_c = 1.5$ MHz) and 140 ps with a noncompliant clock. Eye diagrams for both are also provided. This example

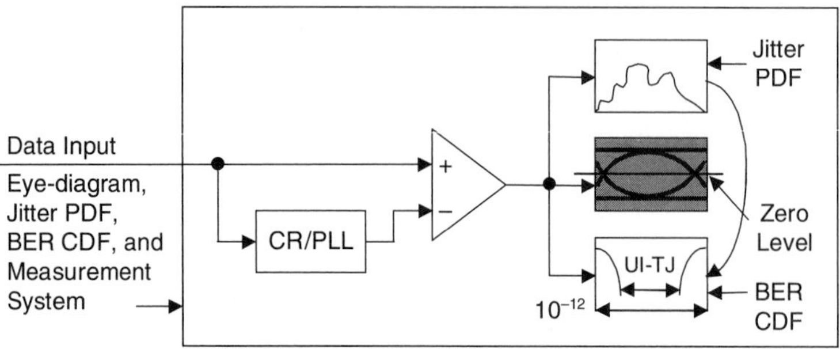

FIGURE 14.18

A test setup for JNB output.

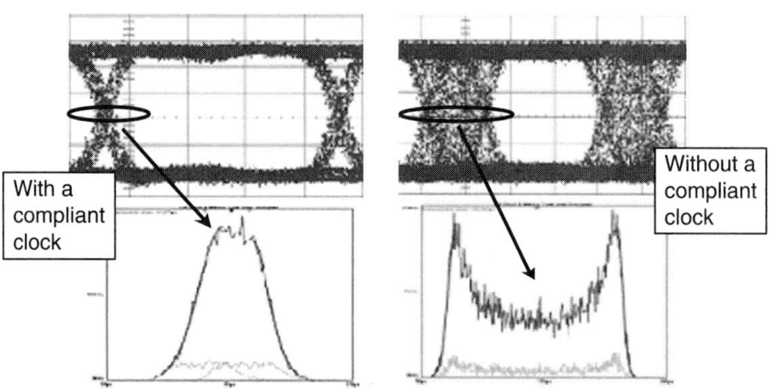

FIGURE 14.19

Eye diagrams based on compliance and noncompliance clocks.

illustrates the importance of incorporating the clock recovery frequency response in testing because the system with a higher corner frequency can tolerate more periodic jitter. Without a correct clock recovery frequency response incorporated in the testing system, good parts could easily be rejected, resulting in serious yield and revenue losses.

Figure 14.20 shows an example of JNB output compliance test results. This example shows a relatively clean eye diagram with sufficient jitter and noise margin with respect to a $\leq 10^{-12}$ BER compliance zone. A good JNB output will have a wide-open eye with large eye-openings in both timing and amplitude axes or, equivalently, small timing jitter and amplitude noise.

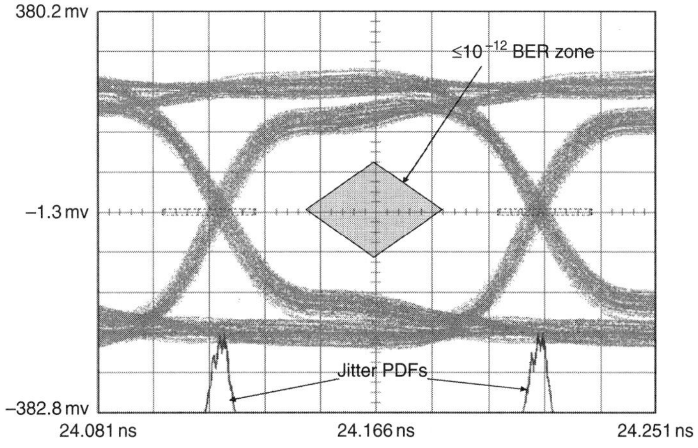

■ FIGURE 14.20

A compliance JNB output test example.

14.3.3.2 Channel or Medium

The channel or medium of a high-speed I/O plays an important role in determining both system and component architectures. Because of the infrastructure legacy limitations and cost constraints, the improvement of channel performance has not been able to keep up with the data rate increase. Most of the improvements were in the architecture of the link as well as in the integrated circuits of transmitters and receivers.

Most channels used for high-speed I/O link are either copper or fiber-optics based. Most computer system channels, for example, are copper traces on PCB with FR-4 material. A copper-based channel typically suffers from frequency-dependent losses such as conductive skin effect and dielectric losses [Johnson 1993].

The effect of the lossy channel to the bit symbol, the waveform, the DDJ, and the ***data dependent noise*** (DDN) can be seen in the eye diagram shown in Figure 14.21. If the bandwidth of the channel is greater than the highest frequency content of the signal, then there will be no waveform distortion nor any DDJ or DDN introduced by the channel.

In theory, the channel can be characterized by a transfer function based on the LTI theorem. In practice, a channel is commonly characterized by the S-parameters that can be measured by a vector network analyzer (VNA) instrument. The S_{21} parameter gives the loss function from port 1 (input) to port 2 (output) and is the same as the channel transfer function $H_{ch}(s)$ in the LTI description of the channel.

For any given reference transmitter and receiver, the channel characteristic requirements can be specified to achieve the overall BER performance goal of the link [PCI-SIG, 2007]. If frequency-dependent loss is the only concern, then whether a channel meets the compliance requirement or not may be judged by comparing the measured S_{21} parameters function with the compliance mask, as shown in Figure 14.22.

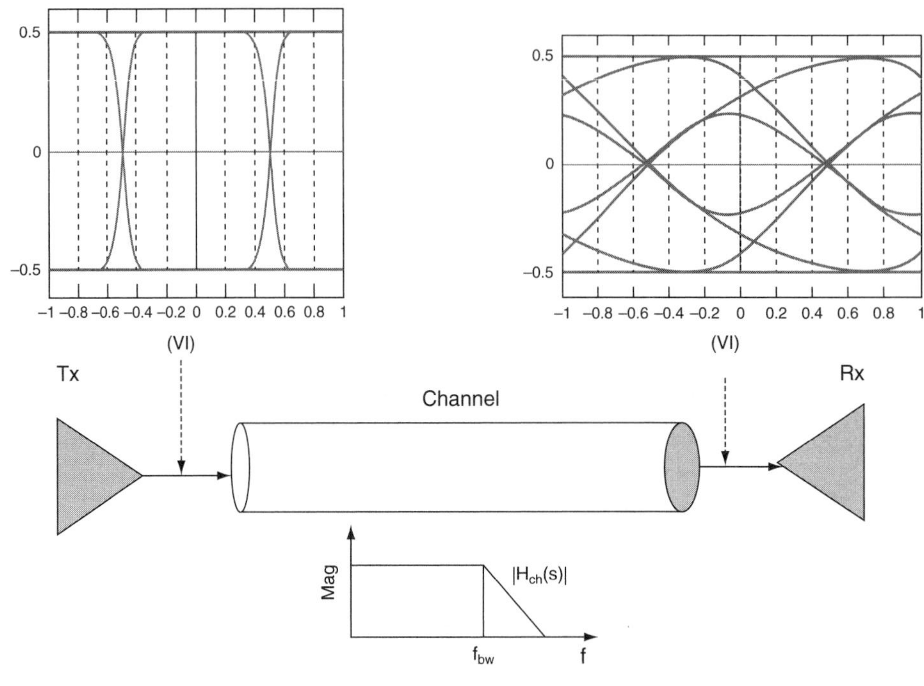

■ **FIGURE 14.21**

Lossy channel effect in terms of eye diagrams.

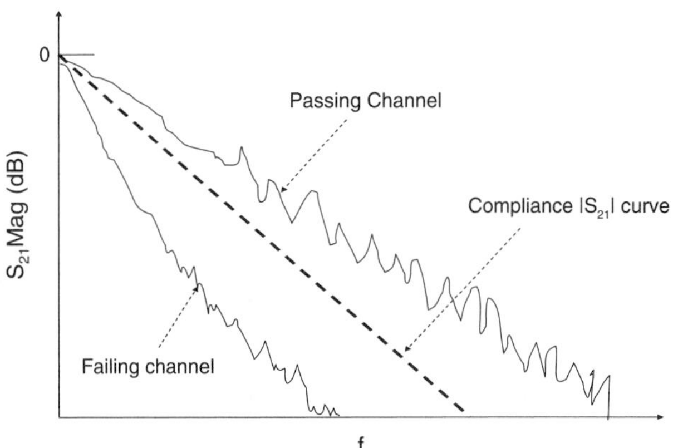

■ **FIGURE 14.22**

S-parameter magnitude function for a channel.

However, a go-no go test method for the channel simply based on S_{21} magnitude can be too coarse because (1) S_{21} phase information is ignored, and (2) crosstalk and reflection-caused ripple may not be well captured and factored in for worst-case scenarios. If the channel is known to have a linear phase or a constant group delay, and has insignificant crosstalk and reflection, then the simple S_{21} magnitude-based channel test method may still work well.

14.3.3.3 Receiver

The ultimate goal for JNB tolerance testing of a receiver is to verify that the receiver can operate at a target BER (*i.e.*, 10^{-12}) when either the input signal or jitter is operating under the worst possible conditions. There are two aspects regarding JNB tolerance testing. The first aspect involves verifying whether the receiver CR tolerance frequency response is indeed better than that of the target threshold. Figure 14.23 shows the concept for testing the receiver CR jitter tolerance frequency response.

The target threshold curve of the frequency response is indicated in the figure. If the measured jitter frequency tolerance curve is above the threshold curve, it indicates a passing result because the receiver clock recovery can tolerate or track more jitter than required. On the other hand, if the measured jitter frequency tolerance curve is below the threshold curve, then it indicates a failing result. Note that all three curves shown in the figure correspond to the same target BER (*e.g.*, 10^{-12}).

The second aspect is to verify the receiver tolerance capability under worst-case signaling, jitter, and noise input conditions. The worst-case conditions critically depend on the link architecture, and the jitter and signaling budgets [Li 2004] [PCI-SIG 2004, 2007]. A worst-case input signaling condition can be intuitively viewed with a worst-case input eye as shown in Figure 14.24.

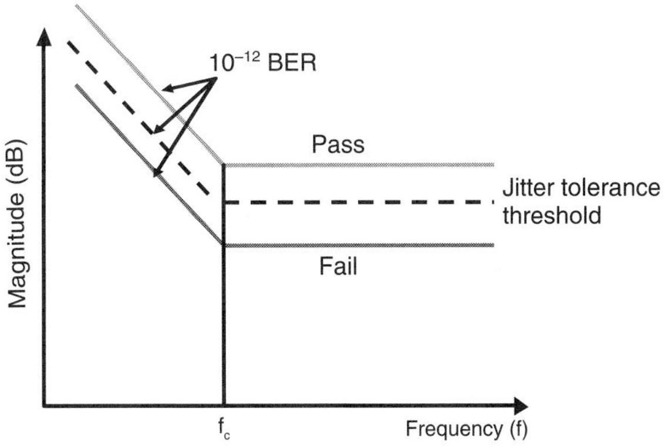

■ FIGURE 14.23

Receiver jitter tolerance threshold for testing.

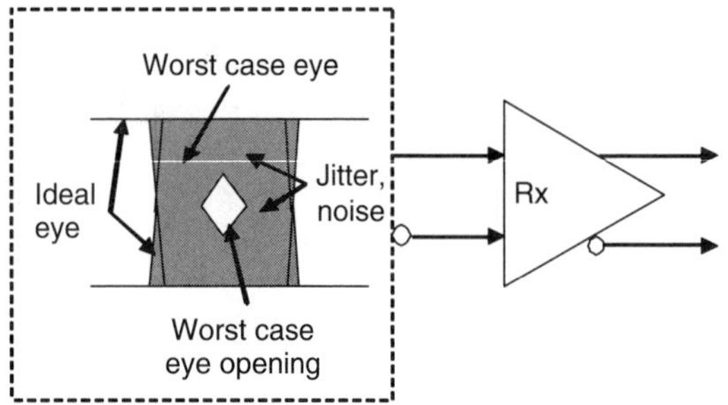

■ **FIGURE 14.24**

A worst-case eye condition for receiver tolerance testing.

Attention must be paid to the frequency content for this worst-case eye diagram because different jitter or noise spectrum content can give rise to a seemingly worst-case eye, yet the stressing level to the receiver and, in turn, the resulting BER can be quite different because of different receiver architectures.

To test for the worst-case jitter and noise condition, all the possible jitter and noise components should be present in the input signal. Figure 14.25 shows a testing setup to stress the worst-case receiver tolerance condition for achieving good test quality. The frequency response tolerance test is covered in this generic setup. Receiver jitter, noise, and signaling test can be exhaustive, given the relatively small number of potential control parameters and the resulting possible combinations. Furthermore, accurately calibrating the signal, jitter, and noise stimulus is important because inaccuracy in the jitter and noise stimulus can cause inaccurate results.

Specialized methodologies for receiver tolerance testing can be found in the literature. For example, in [Yamaguchi 2002], the authors proposed a method for measuring jitter tolerance of a high-speed receiver by utilizing the timing misalignment between the jittered source clock and the recovered clock.

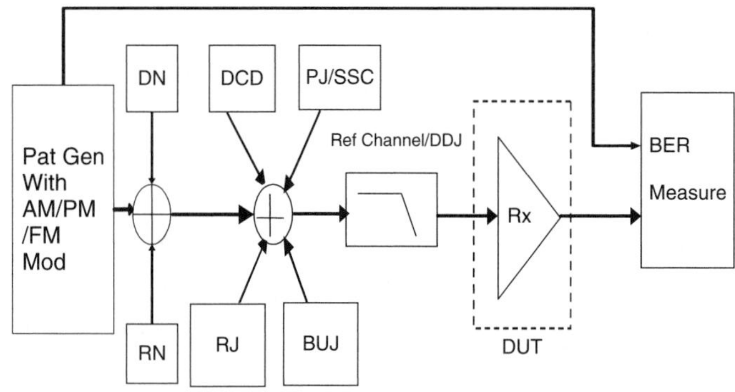

■ **FIGURE 14.25**

A generic setup for receiver jitter and signaling tests.

14.3.3.4 Reference Clock

When testing Tx output or Rx input for a PI-based architecture, as shown in Figure 14.4, the reference clock signal needs to be clean so that there will be no contamination in JNB between the Tx, medium, and reference clock. At the same time, the reference clock itself must be tested according to bandpass filter function [Li 2004] [PCI-SIG 2004]. Figure 14.26 shows a typical transfer function magnitude frequency response. Notice that it is a bandpass function with a peaking. The 3 dB frequencies, peaking, and the overall shape depend on the PLL parameters and the propagation delay between the Tx and Rx route for the reference clock. An example of a reference clock jitter spectrum before and after the filter function that tracks the shape of the transfer function is illustrated in Figure 14.27.

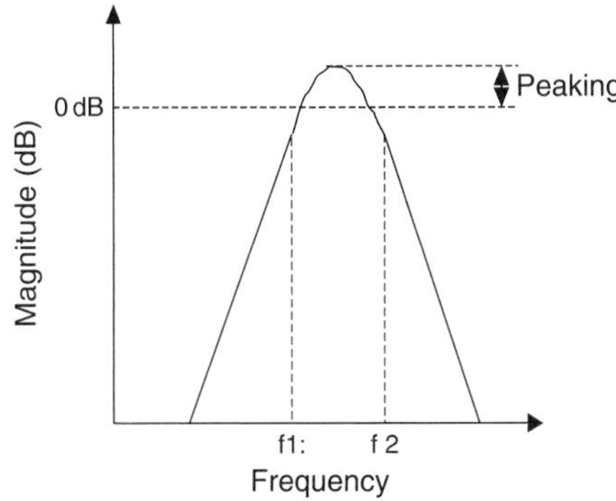

FIGURE 14.26

Jitter transfer function for a reference clock.

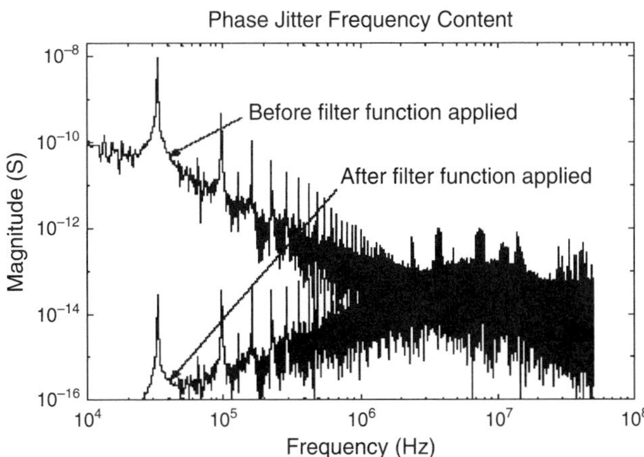

FIGURE 14.27

Reference clock jitter spectra before and after the filter function.

14.3.3.5 System-Level Bit-Error-Rate Estimation

An approach proposed in [Hong 2004] and [Hong 2007] attempts to estimate the BER using the following two sets of parameters: (1) the jitter spectral information, extracted from the signal at the input of the receiver, which includes the RMS value of the RJ and the DJ characteristics, such as frequencies and amplitudes of the PJ components, and (2) the jitter transfer characteristics of the ***clock and data recovery*** (CDR) circuit including both magnitude and phase responses. In principle, if the frequency of input jitter is relatively low, the CDR circuit can track the jitter, and thus insignificant bit errors will occur. However, if the input jitter varies rapidly, the CDR circuit may not track the jitter, and some bit errors will occur. On the other hand, the CDR circuit has an opposite reaction to the internal noise of the CDR circuit. That is, the high-frequency component of the internal noise is transparent (and thus will be directly added) to the recovered clock, instead of being filtered out by the CDR circuit. In addition, the phase response of the CDR circuit, which determines the timing response in clock recovery, has a strong correlation to the BER [Hong 2004]. If the jitter frequency falls into the range where the phase delay is nonzero, the CDR circuit introduces some timing delay to the recovered clock, which will, in turn, contribute to the BER. At a specific frequency range, this timing delay can cause a significant increase of the BER. This work provides insight to the dependency of the BER variations on the jitter spectrum and the jitter transfer characteristics of a CDR circuit. Equations were derived for BER estimation based on the measured parameters mentioned above. The results of Figure 14.28 comparing the estimated and measured BER on a 2.5-Gbps commercial CDR circuit show high accuracy of the estimation technique.

14.3.3.6 Tester Apparatus Considerations

The accuracy of any tester apparatus used will affect the pass/fail results and product yield in testing JNB. This is true for both on-chip and off-chip testing, with a laboratory instrument or a production-oriented ATE system. Receiver emulation with appropriate clock recovery built-in may be achieved using a loopback method for system testing, but it remains a challenging problem for component testing. Also, BER at the level of 10^{-12} or below is not test-time friendly. Test time in the range of hours is required for BER at the level of 10^{-12}. In this section, we discuss challenges related to the tester apparatus, as well as possible solutions.

Hardware Bandwidth and Accuracy

Testing hardware, including both the signal source stimulus and measurement receiver, needs to have little intrinsic DJ, RJ, DN, RN, as well as sufficient bandwidth for testing JNB. A bandwidth of 2.5 to five times the data rate is generally needed to generate an accurate JNB, its waveform, and the associated rise/fall time measurements. Figure 14.29 is an example of such a requirement with the assumption of a 10% total jitter margin for BER at the 10^{-12} level, of which 2% is from DJ and 8% from RJ. Using a 10-Gbps data rate as an example for which

■ **FIGURE 14.28**

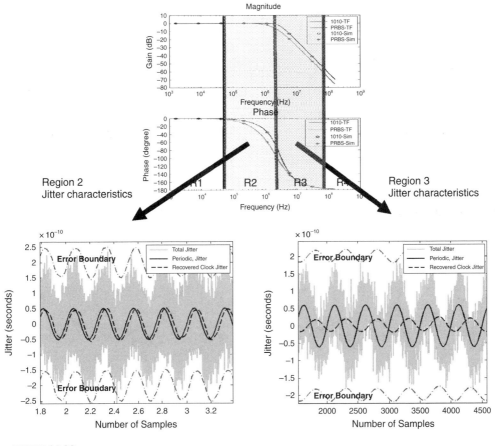

BER estimation considering the CDR and jitter characteristics at the frequency region.

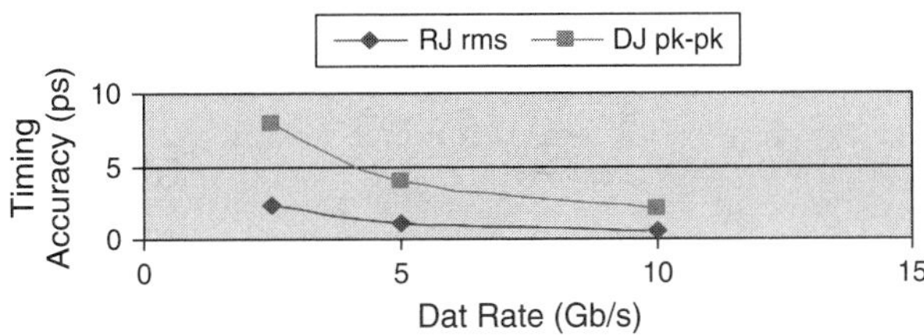

■ **FIGURE 14.29**

DJ and RJ for the tester as a function of data rate.

UI = 100 ps, at 10^{-12} BER, we obtain TJ = 0.1UI = 10 ps, DJ = 0.02UI = 2 ps, and RJ (rms) = (0.08UI)/14 = 0.571ps. The corresponding numbers can be derived for other data rates and the graph shows the upper limits of RJ and DJ as a function of data rate.

"In Situ" Testing to Emulate the Receiver

As described earlier, serial data communication uses the recovered clock as its reference for recovering data at the receiver. This CR unit forms a high-pass frequency response that allows the receiver to track the low-frequency jitter, as well as constitutes the mask for jitter tolerance testing. The jitter transfer function must be incorporated in JNB testing. Otherwise, JNB testing would result in either underestimation or overestimation. Both hardware clock recovery and ***digital-signaling-processing*** (DSP) based soft clock recovery can be used for JNB testing. However, designing a programmable clock recovery receiver for JNB testing, which has the features of sub-ps jitter generation and UI jitter tolerance at low frequency, is a nontrivial task. Soft clock recovery is possible if the real-time record has sufficient resolution and is obtained by a real-time sampling circuit. The real-time record is the key to enable the soft clock recovery, as it captures the phase information of the JNB [Li 2003].

Throughput

Throughput is a critical metric for production testing. To catch one error at 10^{-12} BER for a 1-Gbps link requires about 10^3 seconds. To enhance the statistical confidence, a bit error sample of about 20 or more is needed per Poisson statistical requirement. This means that a good BER measurement with a BERT will take about 2×10^4 seconds to complete, which is too long for any practical production test to accept. Consider another example where an equivalent sampling scope is used for testing jitter via its eye-diagram function. Obtaining a timing jitter histogram with only a few thousand hits will take a few minutes. Although the time needed is much less than that required for taking 10^{12} samples, it is still much longer than the typical IC production test time, which often is less than 10 to 100 ms. Fortunately, there are available model-based extrapolation methods that have demonstrated the capability of estimating the TJ at 10^{-12} BER with good accuracy and high confidence at a total test time around 50 ms [Li 1999] [Hänsel 2004] [Cai 2005]. While achieving high test throughput is always favorable for any test process, this goal is more critical for high-volume and low-cost production test than low-volume laboratory design verification and characterization test.

14.4 DFT-ASSISTED TESTING

Testing methods described in previous sections work well for characterization-level testing, but they have several distinct disadvantages for high volume manufacturing test. For high-volume manufacturing test, typical testing of the I/O interfaces on an

ATE requires it to match the performance of the DUT's I/O. I/O data rates can range from 1 GHz for parallel buses and more than 6 Gbps for serial links. ATE that meet these requirements cost quite a lot; in addition, test time and test programming complexities often mean delayed time-to-market. This combination can result in high product cost that may make the product noncompetitive. This becomes a problem when mass-market personal computers and consumer digital appliances increase their performance bandwidth and both high-speed parallel and serial I/O interfaces become the norm.

These trends and costs have motivated ways to test these products on mainstream test platforms (*e.g.*, digital VLSI tester) with reasonable test time. There have been two trends: (1) making changes to the loadboard, and (2) incorporating I/O-specific DFT into the design. We refer to these test methods as **DFT-assisted testing** in which the DFT circuits are embedded on the loadboard or within the silicon itself. With this approach, no new or special equipment has to be purchased and time-to-market and product cost will fit into the business of mass marketing these technologies into the hands of consumers.

14.4.1 AC Loopback Testing

In the early 1990s, an I/O structural test methodology called **I/O wrap** [Gillis 1998] was developed. More commonly called **I/O loopback**, this method involves applying a transition fault test methodology to I/O circuitry. By tying an output to an input, the output data are launched and latched back into the input buffer on the following clock. As most signal pads are I/O in nature, the I/O wrap (I/O loopback) methodology is convenient. Input-only or output-only pads can be connected with the DFT circuit or wires/relays present on the loadboard. The limitation to this method is that, because the delay path is tested with the clock, the delay cannot be characterized without overstressing the other peripheral circuits. This approach is limited to testing gross delay defects, and timing specifications cannot be measured.

By the early 2000s, a test methodology known as **AC I/O loopback testing** had been proposed (see Figure 14.30). This uses the same loopback principle, but with a twist [Tripp 2004]. Rather than just using the clock to launch and capture the signal, the launch can be carried out by a delayed version of the clock or the capture be accomplished using an early version of the clock (see Figure 14.31). By controlling the delay, the relative delay between the strobes and data can be actually measured without precision timing measurements from ATE.

Essentially, this is transition fault testing of the I/O pair with a tighter clock cycle. Only the I/O circuits are tested in this method, thus preventing false fails from other circuits as in the case of speeding up the I/O wrap. This method works well with the *source synchronous* (SS) scheme, where the strobes are generated by the transmit side. In the SS signaling protocol, the absolute delay of the I/O is not critical; instead, the relative delay of the strobes and any associated data bits are important. These timing delays, denoted as *time valid before* (Tvb) and *time valid after* (Tva), describe the relationships between the strobe and the data bit (see Figure 14.3). So by moving the strobes from their central position to the trailing edges of the data (see Figure 14.32), we are stressing Tva and the setup time of the

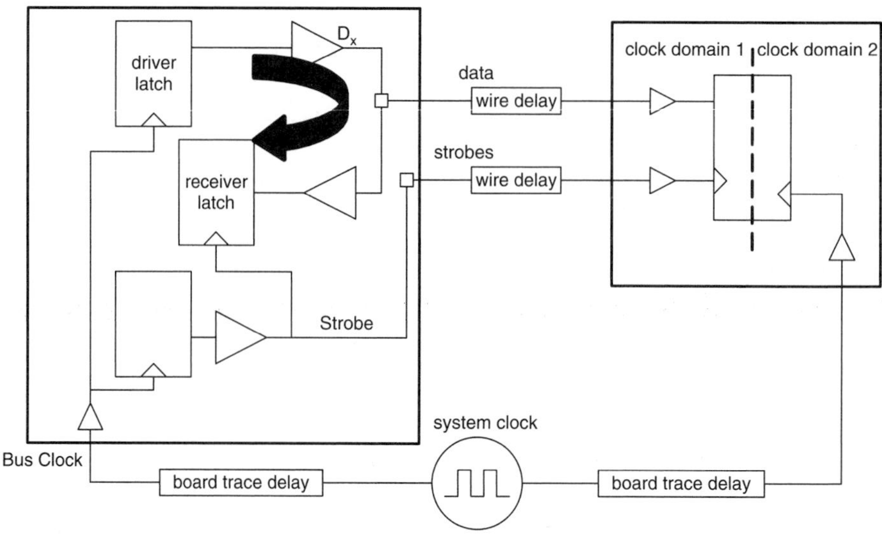

■ **FIGURE 14.30**

Using IO buffers as loopback components for an SS interface in which the strobes that clock the data out to the receive side also strobe the data loopback to itself.

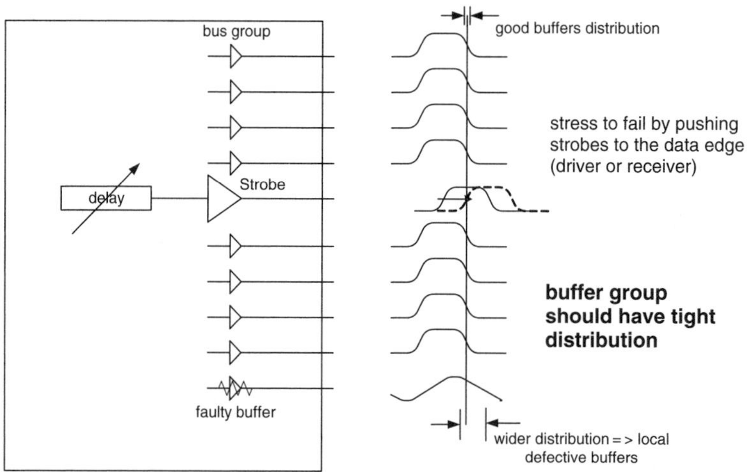

■ **FIGURE 14.31**

AC I/O loopback testing as a defect-based test method.

receiver latch. If we move the strobes toward the leading edge of the data, we are stressing Tvb and the receiving latch's hold time. By stressing this combined timing, we know how much margin there is with the combined pair. If the induced delay to the clock/strobes is calibrated, we can even have more accurate measurements

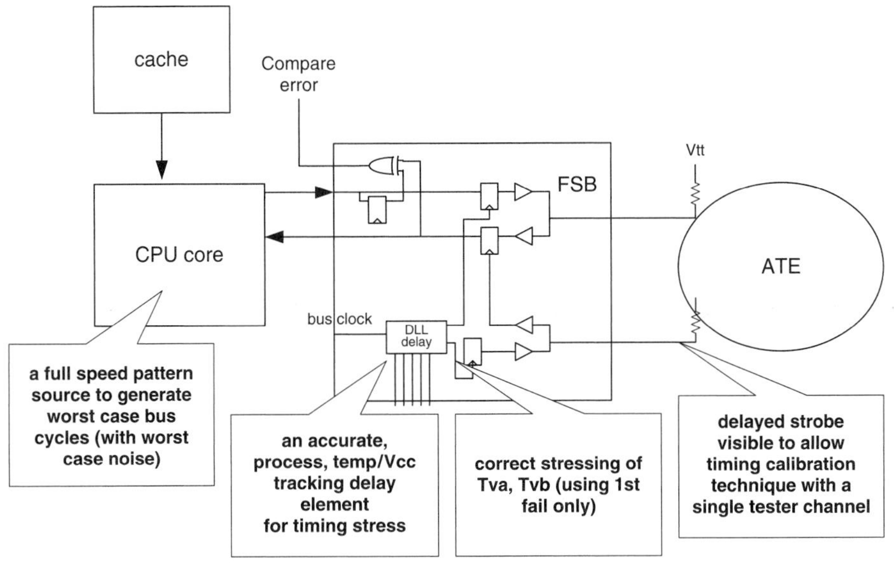

■ **FIGURE 14.32**

Various elements in the AC IO loopback system to facilitate accurate timing test.

(timing margining) using this combined loop time than is possible with external instrumentation (see Figure 14.32). Because the failure mechanisms for signal delay and input setup/hold time are different, the probability of aliasing is low. Slightly different implementations can be applied to *global clock* (GC) interfaces.

Furthermore, because we are not measuring each data bit independently (this is a bus nonetheless), the delays of all of these data bits should be close to one another unless there are defects or local process variations. If a particular data bit is substantially different from the other data bits (see Figure 14.31), we can also conclude that a defect or local process variation exists with that particular bit and declare that to be a failure. This can also be viewed as a **defect-based test method**, especially if no calibration of the induced delay is made. The testing can also be carried out faster when all bits of the bus are compared simultaneously when the loop timing is tightened. A tight distribution of the passing region indicates all the bits are aligned, and a wider distribution of the passing region indicates some bits are slow or mis-aligned; all are causes to fail the chip. If further diagnosis is desired, then we can resume the stressing of individual bits.

14.4.2 High-Speed Serial-Link Loopback Testing

Instrumentation-based testing as explained in the previous section is much more accurate than that of DFT-assisted loopback test. However, it also requires special instruments to be set up, calibrated, and integrated into test platforms. The lengthy setup time, long test pattern sequence application, and data capture, which is typically followed by extensive numerical analysis, often require experienced personnel (*e.g.*, a technician) and longer test time than would be desired in a high-volume manufacturing setting. As serial link technologies are deployed to an increasingly cost-sensitive commodity computing world, it is required to lower test costs.

For high-speed serial links, loopback testing is common, because input and output are conducted on separate channels (serial pair) and there is usually the accompanying transmit and receive channels even in a given component. Loopback testing has been around for as long as serial interfaces have been around.

However, simple loopback testing is just a simple functional test. All the required jitter and noise tests are not possible with this method, and there is no guarantee that the simple loopback test can interoperate with other similar components in an extreme environment. Loopback also makes diagnosis difficult because the defective component cannot be easily located. There is also the nonzero probability of aliasing—that is, a bad receiver, for example, may be covered up with an extremely good transmitter. Thus, within the industry one may also characterize loopback test as just a "system test," as it is commonly used in a system setting to verify that the system is functional.

To improve on simple loopback testing, several approaches have incorporated jitter injection capabilities into the loopback to "stress" the signal and thus test the receiver's jitter rejection or noise-tolerant capabilities [Laquai 2001] [Cai 2002, 2005] [Lin 2005]. Many of these approaches consist of passive *resistance-inductance-capacitance* (RLC) filters, which introduce frequency-dependent delays and losses to the various frequency components (or harmonics) of a high-speed serial signal. When various combinations of 1's and 0's are transmitted (*e.g.*, ISI patterns), the signal will appear as various frequency harmonics (as data are never constant in the frequency domain). As different frequency components are attenuated differently, the resultant timing waveform will be distorted enough to resemble that of a worst-case jittering data eye. The filter can simply be placed on the loadboard between the input and output channels. It is a relatively cheap solution to an expensive test problem. Of course, the filter design has to be coupled with the driver's characteristics (original spectral components), and it is data rate dependent (which determines the base frequency). This method also does not address the need to test the transmitter's drive characteristics. A jittering transmitter signal will add to the distortion of the filter and will likely cause the Tx-Rx pair to fail to communicate reliably, resulting in the rejection of otherwise working devices. However, an extremely good receiver can also potentially hide a marginal transmitter, as the pair is essentially tested together.

Another proposed jitter measuring solution utilizes a **undersampling** method [Huang 2001] [Cai 2005]. The method utilizes the Rx to capture the data sent from its Tx via a loopback on the TIU or on the die itself. Instead of using the recovered clock to strobe the data, an alternate reference clock that has a frequency close enough to the data clock frequency is used to strobe the data. It will result in a much lower data rate (the "beat" f_B of the two frequencies f_D and f_S) as shown in Figure 14.33, where f_D is the frequency of the data, f_S is the frequency of the sampling clock, and f_B is the frequency of the "beat." The jitter that exists on the data channel will dither the data captured. By analyzing the percentage of 1's *versus* 0's, one can deduce the jitter amount. Hardware and methodologies for

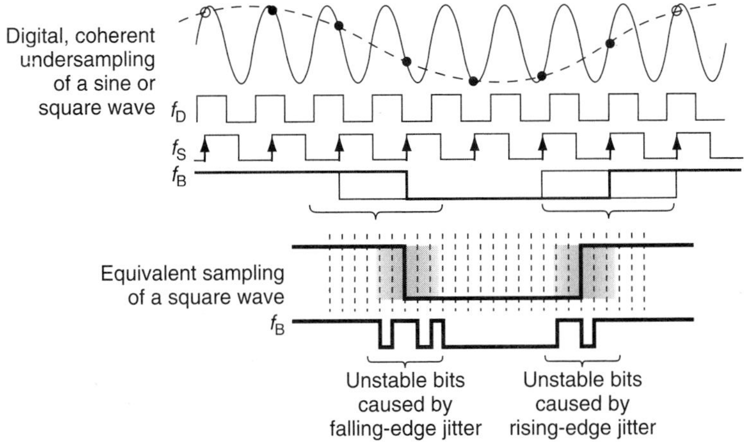

■ **FIGURE 14.33**

Principle of undersampling for jitter measurement.

performing this analysis in real time have been developed [Sunter 2004a, 2004b, 2005]. This can be implemented as a *field programmable gate array* (FPGA) on the loadboard (see Figure 14.34) or even embedded in the silicon itself, making the whole thing close enough to be called **IOBIST**. While undersampling has been proven to work theoretically, the authors have not addressed the issue with practical implementation (*e.g.*, finding the right PLL for the 9.999-GHz sampling clock for a

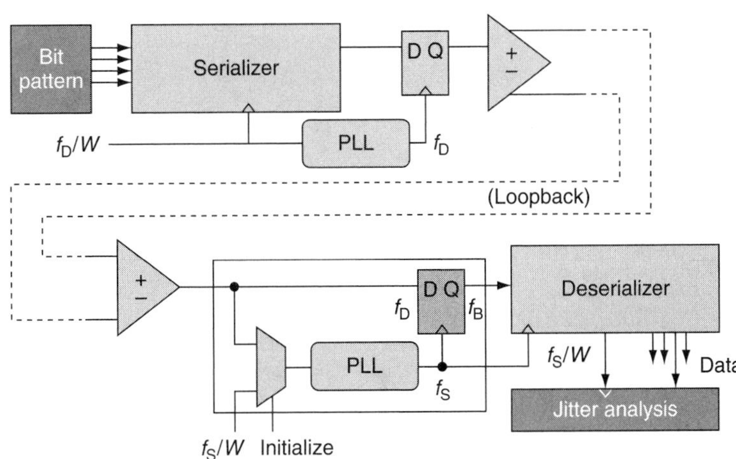

■ **FIGURE 14.34**

Jitter measurement setup where reference clock, pattern generator, and analysis can be off-chip.

10-GHz data frequency or the jitters introduced with routing this external clock to various **serializers/deserializers** on a chip with many high-speed serial channels).

In [Sunter 2005], the authors proposed that by monitoring the placement of the recovered clock *versus* the received data eye, jitter tolerance can be reduced. Although this is one indication that the recovered clock has centered and has the highest probability of correctly strobing the data eye, it cannot predict the dynamic behavior of the CDR circuit well enough for high-frequency jitters (*i.e.*, the data eye and the recovered clock may close down the margins statistically, but the CDR may still be able to recover the data should the CDR circuit be dynamic enough to track the jittering signal). Hence, the scheme may be more conservative and may reject devices that are operational.

The last method as proposed in [Mak 2005] involves assessing data eye via margining as an approach to testing Gbps SerDes. The approach assumes that defective components are relatively rare, and loopback is established with a data pattern streaming through the lookback path. Testing consists of two parts. The first part uses the receiver, which is assumed to be good, to test for the size of a valid data eye. By varying the data strobe point (*i.e.*, by changing the phase of the recovered clock) and the receiver threshold, an eye for the resulting data can be captured [Casper 2003]. A production test can consist of the four points that define the size of the minimum data eye required. The second part of this method is the stress test of the receiver with the transmitter. Again, it is assumed that the transmitter is good. Additional DFT, such as jitter injection into the high-speed clock or introducing small delays into the signal stream (such as those in pre- or de-emphasis circuits) will play the role in jitter injection so that the receiver can be stress tested. It is virtually impossible to generate all kinds of deterministic and random jitter required for the characterization of the serial link with all the necessary built-in DFT, particularly when a new interface is first designed and there is no way of knowing what the receivers are sensitive to; however, it is advisable for these jitter injection circuits to be programmable so that a richer or more diverse set of stimuli can be generated. The proper set of stimuli is defined only after extensive characterization of early samples to minimize both escapes and test time.

Similar techniques have also been reported in [Robertson 2005]. These techniques may not fully test the CDR circuit, as the same clock source may be used to generate the output data stream at the transmit side and drive the receiver circuits. One has to study the actual design to be sure that the jitter generated at the transmit side indeed can stress test the receiver circuits without aliasing.

14.4.3 Testing the Equalizers

A comprehensive test methodology would be more desirable and unavoidable as the frequency/data rate continues to increase. At a higher data rate, signal degradation through the channel will simply close the data eye. To alleviate the problem of signal degradation, modern receiver design employs a means for compensating or reducing the ISI in the received signal. The compensator for the ISI is called the equalizer. Various equalization techniques, which multiply the inverse response of the channel to flatten out the overall frequency response, have been developed to compensate for this channel effect. In addition, the channel characteristics may not be known in advance and might be time variant. To cope with such problems, several adaptation algorithms have also been developed to adjust the overall response depending on the channel conditions. The equalizer can be implemented either in the transmitter or in the receiver. The implementation of the transmitter equalizer is relatively easier than that of the receiver equalizer because the required ***finite impulse response*** (FIR) filter deals with the digital data at the transmit side, rather than the received analog data at the receive side. However, as channel information is not easily available at the transmitter, it is difficult to apply the adaptive technique at the transmitter. More recent publications [Jaussi 2005] tend to use the latter.

The approaches of equalization at the receiver can be divided into two categories: discrete-time equalization and continuous-time equalization. A discrete-time equalizer, which is based on the FIR filter, can take advantage of various digital adaptive algorithms. However, because equalization is based on the samples captured by the receiver's recovered clock, there exists a cross-dependence between the equalizer and the clock recovery circuit. As the data rate increases, the power consumption would increase dramatically as a result of the large number of taps implemented in this type of equalizer. On the other hand, a continuous-time equalizer does not require a sampling clock, and thus the equalizer would work independent of the clock recovery circuit. Continuous-time equalizers have been investigated for low power and high-speed applications, and promising performance has been reported.

A DFT concept for testing digital equalizers is developed in [Lin 2006]. This proposed technique can be applied to a wide range of linear equalizers using various adaptive algorithms. The extra hardware required for the DFT solution needs only a scan chain, which can directly observe the state of the adaptation status of the equalizer, and one simple digital pattern generator. The overhead and the additional design effort of the approach are insignificant. With the proposed DFT solution, the equalizer in the serial-link receiver can be characterized and tested without direct access to the equalizer output. This alleviates the need for observing the equalizer output, which is often infeasible.

A method for testing continuous-time adaptive equalizers has recently been proposed [Hong 2007] which was validated by simulation on a low-power, 20-Gbps continuous-time adaptive passive equalizer. Figure 14.35 shows the architecture of a typical continuous-time adaptive equalizer. The equalizing filter either boosts the high-frequency components or attenuates the low-frequency components of the received input signal to compensate the high-frequency loss resulting from the channel. The adaptive servo loop, which adjusts the compensation gain of the equalizing filter, determines the control voltage by comparing the input and the output signals of the comparator. In practice, it is difficult to design a comparator that can generate a clean waveform for comparison at very high frequencies. Several design approaches for adaptation have been proposed to address this problem. These new methods use the power spectrum derived from the output signal of the equalizing filter for the adaptation, as shown in Figure 14.36. Because the power spectrum of a random signal can be described by a sinc-square function, the high-frequency loss can be detected by comparing the power densities of two different frequency

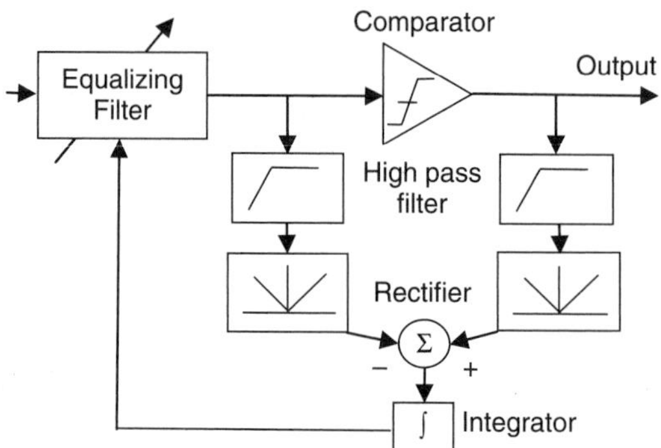

■ **FIGURE 14.35**

Architecture of a conventional continuous-time adaptive equalizer.

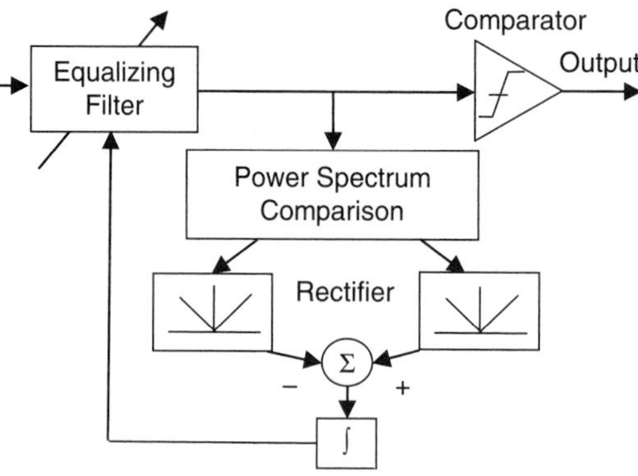

■ **FIGURE 14.36**

Architecture of a modified continuous-time adaptive equalizer.

ranges. Three different methods have been proposed to compare the power spectrum of the random signal: (1) two **band-pass filters** can be used to compare the power of two specific frequencies (see Figure 14.37a); (2) one **low-pass filter** and one **high-pass filter** can be used to compare the power between low-frequency and high-frequency portions of the signal (see Figure 14.37b); and (3) only one low-pass filter is used and the entire signal power is compared to the power of the low-frequency portion of the signal (see Figure 14.37c).

The main idea behind the approach in [Hong 2007] is to directly apply a two-sinusoidal-tone signal as test stimulus at the input of the equalizer, as opposed to a data pattern stressed through the channel, which has been commonly used for validating and characterizing the equalizers. The frequency for one of the two sinusoidal tones, denoted as f_L, falls within the frequency band of the left filter in Figures 14.37a and 14.37b, and the frequency of the other tone, denoted as f_H, falls within the frequency band of the right filter in Figures 14.34a, 14.34b, and 14.34c.

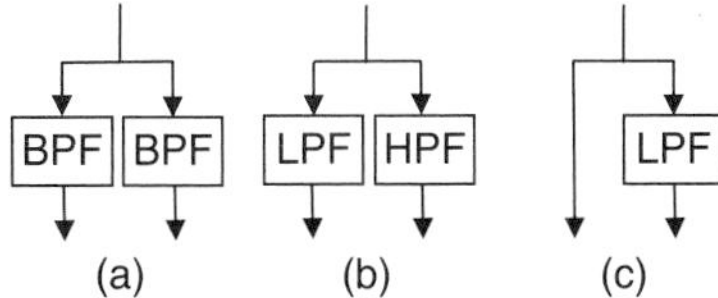

FIGURE 14.37

Three different architectures for power spectrum comparison.

To thoroughly mimic the channel response, this technique repeatedly applies the two-tone signal by gradually varying the magnitude ratio of f_H and f_L. Such test stimuli mimic different relative loss of the high-frequency components caused by the channel.

With the test stimuli, the RMS value at the output of the equalizer is measured for characterization and for fault detection. Based on the principle of how the continuous-time equalizer works, the adaptive servo loop of the equalizer attempts to maintain the ratio of f_L to f_H to the expected level based on the sinc^2 function. Thus, if the test stimulus's f_L and f_H magnitudes are within the range of the equalizer's maximum compensation gain, the RMS value of the equalizer output should be a constant. Therefore, by carefully crafting the two-tone stimuli and inserting an RMS detector on-chip for monitoring the equalizer's response to the two-tone stimuli, this technique can detect defects either in the equalization filter or in the adaptive servo loop that might not be easily detectable by the eye-diagram method.

In addition to the equalizers, other types of compensation elements are used in advanced I/O interfaces. One type of such compensation elements is the crosstalk canceler. Figure 14.38 illustrates an example architecture of an advanced transceiver, which includes a CDR circuit and a ***decision-feedback equalizer*** (DFE) in the receiver, a ***feed-forward equalizer*** (FFE) in the transmitter, and a crosstalk canceler between the transmitter and receiver. Such a transceiver has the capability

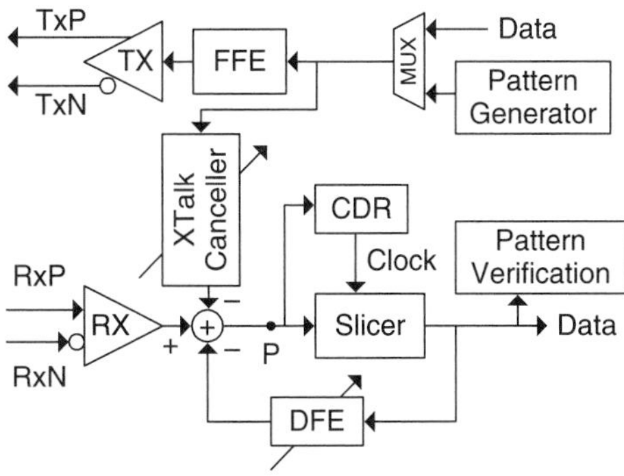

FIGURE 14.38

Architecture of an advanced transceiver with a crosstalk canceler.

to equalize the lossy channels, to remove signal reflections, and to actively cancel the crosstalk noise in the transceiver. For 10-Gbps data rate over backplane channels or other long-range applications, crosstalk that is the dominant noise for microstrip interconnects is becoming a limiting factor for signal integrity. Among various types of crosstalk sources, the ***near-end crosstalk*** (NEXT) is the most severe one [Vrazel 2006]. Active NEXT cancellation using a crosstalk canceler in the transceiver could be an effective solution to addressing the crosstalk problem [Pelard 2004] [Hur 2005].

A DFT solution for such an advanced transceiver is recently proposed in [Lin 2007]. The solution is an enhancement of the technique proposed in [Lin 2006]. The hardware requirement for the DFT solution contains only one shift-register chain and one simple digital pattern generator in the DFE, and three full-swing taps in the crosstalk canceler. With the proposed DFT solution, the transfer characteristic of the CDR in the receiver can be derived, and the DFE can be characterized and tested without direct access to the DFE output. This alleviates the need of external instruments for jitter testing and the need for observing the equalizer output, which are often infeasible in the high-volume production environment.

14.5 SYSTEM-LEVEL INTERCONNECT TESTING

Even though the chips are tested from the chip manufacturers, the chips will then have to be soldered (or socketed) onto the board. Such a board assembly process has its own set of manufacturing defect issues, not to mention that the bare PCB fabrication also can introduce its own defects. Please review Chapter 1.5.7 of the DFT book [Wang 2006]. Many of these defects are not easy for a **bed-of-nails** board tester to find, but the defects can affect system-level operation at the rated speed and worst-case environmental conditions (voltage, noise, etc.). Consequently, board or system manufacturers have to test these interfaces on the board as well as in the system and there is a spectrum of methodologies to choose from. It is, of course, possible to just put together the system and perform an end-user functional test of the whole system. However, the success rate of such a test is not high (because of the many chips mounted on a given board), and, even worse, there is no diagnostic information when the system fails to boot (or startup). Hence, the following methodologies are commonly developed and deployed in various kinds of board assembly and system-level manufacturing.

14.5.1 Interconnect Testing with Boundary Scan

The aim for this test method is to primarily identify board-level manufacturing problems, such as incorrectly placed devices (chips do look alike), rotated devices, bent leads, cold solder joints, improper socketing (if sockets are used), cracked PCB and traces, via problems, etc. The IEEE 1149.1 boundary-scan standard [IEEE 1149.1-2001] is commonly used for interconnect testing of above-mentioned manufacturing problems (see Figure 14.39).

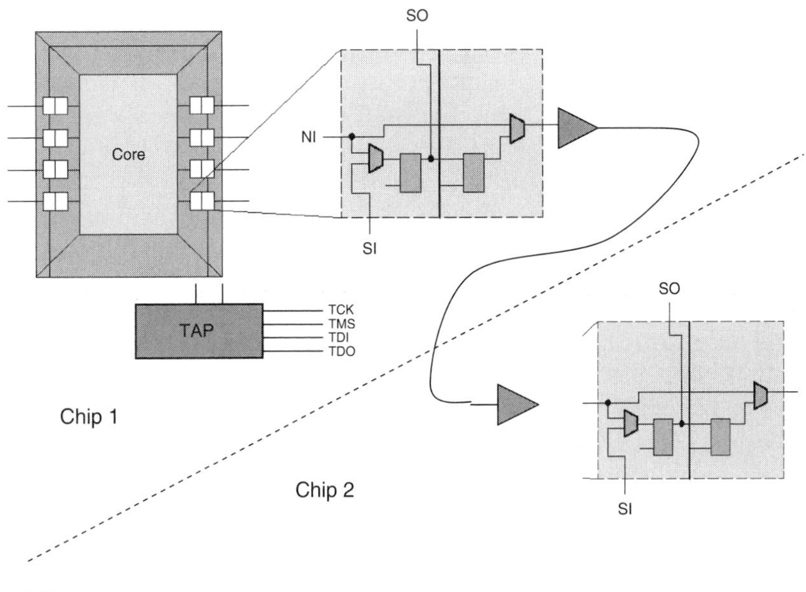

■ **FIGURE 14.39**

P1149.1 system-level interconnect test.

The basic concept of the 1149.1 standard is the addition of a boundary-scan cell for every input or output pad in the chip. Through the *test access port* (TAP), each boundary-scan cell can be set to 1 or 0 independently through the boundary-scan chain. When the net connected to both ends of two chips (chip 1 and chip 2) implemented per the IEEE 1149.1 standard, the data that are driven by one end of the net are captured by the receiving boundary-scan cell and subsequently shifted out through the boundary-scan chain for analysis. This can detect and locate permanent faults, such as stuck-at, open, and short faults on the nets. Each chip often includes a 32-bit device identification (device-ID) register, which stores the vendor's identification code (containing the manufacturer's identity, the part number, and the version number of the chip). The 1149.1 boundary scan allows the designer or user to immediately pinpoint the errors to the defective interconnects or bad chips.

Many interconnect test algorithms and solutions have been proposed [Wang 1992]. For more information on the basics of the IEEE 1149.1 standard, please refer to [Bushnell 2000] and [Wang 2006].

14.5.2 Interconnect Testing with High-Speed Boundary Scan

The IEEE 1149.1 boundary-scan standard has been the mainstay for board-level interconnect testing since the early 1990s. However, the world of chip-level interconnects has changed from tens to hundreds of Mbps to that of multiple Gbps data rate. The signaling technologies for transferring those data have also changed from single-ended signaling to that of differential, clock embedded, and in some situation, even AC-coupled. Although one can design in the 1149.1 ***test access port*** (TAP) controller to run faster for testing these modern-day high-speed (SS) buses and serial links, it is becoming ineffective because of the non-performance-driven nature of the 1149.1 architecture. The state transition through the 1149.1 TAP controller state diagram (see Figure 1.4b in Chapter 1) from Update-DR (data are driven onto the net in one chip) to Capture-DR (data on the net are captured in the other chip) takes 2.5 TCK cycles (see [IEEE 1500-2005]). Moreover, the nature of differential and AC-coupled (there is no DC path) signaling on high-speed serial links also makes it incompatible with the 1149.1 standard.

The IEEE 1149.6 standard [IEEE 1149.6-2003] is an extension of the IEEE 1149.1 standard to deal with this differential, AC-coupled interface. With this standard, independently controllable **digital driver logic** and **digital receiver logic** (with the **analog test receiver**) under the control of the 1149.1 TAP controller are now implemented on both ends of the differential pads of the serial link so interconnect tests can be applied on these differential pads to detect defects that may exist between the differential lines. The analog test receiver is specially designed for edge detection for capacitive coupling. For a more thorough description of the IEEE 1149.6 standard, please refer to [IEEE 1149.6-2003] or Chapter 10 of [Wang 2006]. Because it is an extension of the 1149.1 standard and there is no change to the underlying TAP architecture, 1149.6 still basically runs at a relatively slow data rate. What this test method fails to address is that these new high-speed serial interfaces have to run billions of bits every second. Passing 1149.6 testing alone cannot guarantee the data rate nor transmission reliability, particularly when the system I/O is running at multiple Gbps data rate.

To examine this high-speed signaling issue a bit further, we can look into the test pattern requirement of testing these high-speed serial channels (interconnects). Because of the loss of signal energy in the medium (sockets, PCB traces, or cables), simple 1 and 0 patterns (or clock patterns) do not stress the interface enough. While an extended JTAG architecture to test SOC interconnects (parallel buses) for signal integrity has been reported in [Ahmed 2003], [Tehranipour 2003a], and [Tehranipour 2003b] where ***maximum aggressor*** (MA) and ***multiple transition*** (MT) fault models are employed (see Chapter 8), these methods have limited use for testing high-speed serial channels. Because differential signaling is used, they are resistant to common mode noises, such as coupling and ground bounce effects. The common mode noises are simply canceled out unless they make a difference between both lines. The issue with serial signaling is that the pattern changes so fast that before the pattern has a chance to go from one state to the opposite state fully, it has to go back to the original state again because of fast data changes (see the tightly packed data train in Figure 14.40). This creates data waveforms that are more difficult to differentiate (whether it is a 1 or a 0). Also, a long series of 0 or 1 patterns (the fatter portion of the waveform in Figure 14.40) also makes the signal saturated to one side, and when a single bit of data change comes along the signal will take a longer time to ramp before it has to change again. This makes for the worst-case data eye. This is called ISI. So data pattern is critical for board-level and system-level testing when the media play a major role.

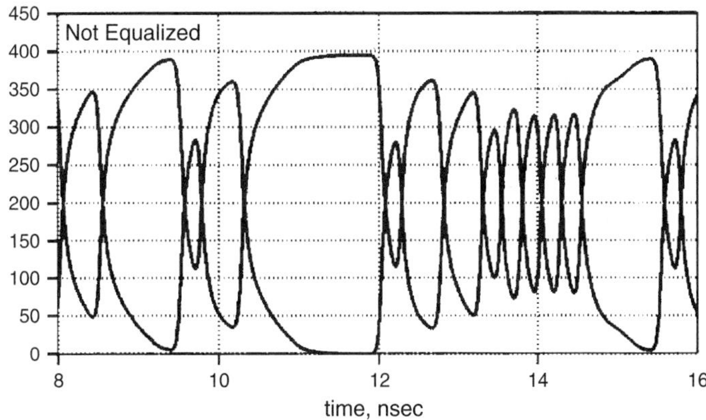

■ **FIGURE 14.40**

Signal effect of intersymbol interference (ISI) on a high-speed serial link.

14.5.3 Interconnect Built-In Self-Test

Interconnect built-in self-test (IBIST) was introduced in [Nejedlo 2003]. It is based on the premise that high-performance interfaces have to be tested at high speed. In a board manufacturing environment, it is desirable to separate board assembly defects without running full board-level functional tests. These types of tests should be carried out before the system attempts to boot/start up to increase the confidence level of the overall board manufacturing process.

Figure 14.41 shows an IBIST example with two components as applied for high-speed serial testing at the board level. The two sides (components) of the I/O interface are named master and slave (the order is not that important as the role of master/slave can be reversed). On the transmit side, IBIST consists of a

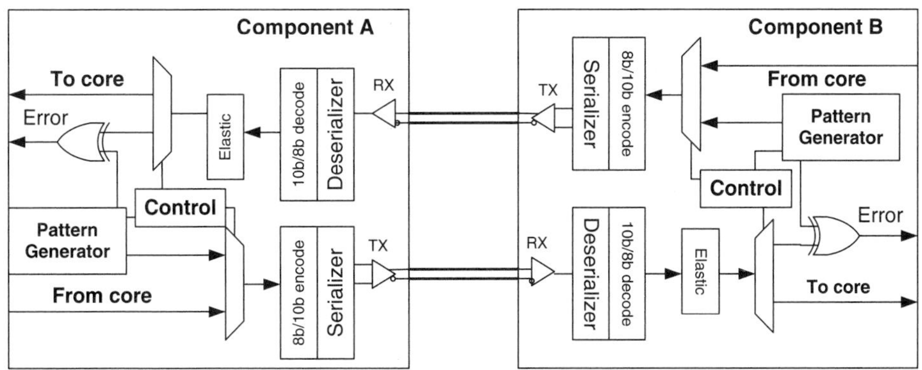

■ **FIGURE 14.41**

BIST as applied for high-speed serial testing at the board level.

programmable pattern generator behind the Tx driver or transmitter of the I/O interface in the high-speed serial link. On the receive side, IBIST consists of logic to route the received data (Rx) back to their own transmitter as well as error checking logic (XOR gates), which can check the pattern transmitted *versus* the received pattern. To further reduce the circuit requirement, the slave can simply implement the internal loopback circuit to support only the bounced back mode. With this reduced logic, it cannot independently send a pattern but will support the bounced back mode only. The independent pattern generator allows the interface to be tested independent of the core logic of the chip that has these high-speed interfaces. With this scheme, the master will take control of the whole test process. The included pattern generator on the master then drives a high-speed pattern to the slave side. Upon receiving it, the slave resends all the data that it has received (in bounce-back mode) immediately (bit-per-bit) through its own transmitter. The master's receiver receives this retransmitted pattern and checks it against the original transmitted pattern. Error checking is done on a symbol-by-symbol basis so that the error can be pinned down to the exact bits and further logic reasoning can be done to figure out what the cause of the error may be. A second stream of data can also be initiated from the slave side to aid with further diagnosis. Control registers keep track of when an error is first detected and can be examined after the test patterns are transmitted and received.

To stress test the system, ample data varieties—from pseudo-random pattern generation to specific preloaded patterns (some specific patterns may be needed for specific jitter and ISI requirements)—can all be supported.

From the manufacturer standpoint, if such a system can be coupled with the electrical AC loopback system (described in an earlier section), in-system margin characterization is also possible, providing a much richer set of characterization data. This simply cannot be done with any aforementioned methods.

14.6 FUTURE CHALLENGES

Because we need to match the data bandwidth of the chip-to-chip connection with that of the core operating speed, increased use of serial link signaling is expected. Soon serial link signaling will replace most buses and perhaps many of the control signals as well. The methodologies for testing all of these types of high-speed serial signaling architectures must consider cost and quality. Increasingly, this points to the increased use of self-testing with DFT support [Kundu 2004], although its accuracy and fault coverage are not at a desirable level [Lin 2003].

To maintain a low BER, lower cost structures, higher channel counts, and more advanced architectures and silicon technologies are expected to be developed for multiple-Gbps I/Os in the near future. In particular, the implementation of even more aggressive equalization methods: transmitter-based, receiver-based, or hybrid, will occur as the data rates increase. Furthermore, to reduce crosstalk, reflection, and lossy-medium-induced jitter, the DFE will be widely implemented in the receiver to reduce the BER; consequently, test methodologies will have to advance to keep pace with future link and silicon architectures and technologies. To achieve an optimized test solution with acceptable accuracy, fault coverage, throughput, and cost, it is anticipated that both on-chip and off-chip test solutions will be necessary; for example, simple logical capabilities such as pattern generation and error detection should all be done internally via DFT/BIST. Other functions, such as jitter and noise generation and calibration, from pico second down to femto second signal and jitter output measurement, and component separation, may likely remain external operations—that is, until and unless better on-chip testing solutions are found. Although, because of its flexibility and accuracy, instrumentation-based testing is more preferred at the post-silicon validation phase, DFT-based test techniques will be more preferred for use in high-volume manufacturing because of their low costs and reasonable accuracy. Of course, DFT-based testing assumes that an interface will have reasonable margins to begin with. If precise testing is needed to test marginal products, direct instrumentation-based testing may not be avoided.

Another complication for testing is that serial signaling utilizes a layered communication protocol stack; the layer that connects to the pins is the ***physical layer*** (PHY). Not only does this layer drive the data and perform the clock/data recovery during receiving, it also has to initialize, detect, and train/retrain the signal between the sending and receiving ends. There is also a need for a link layer where tasks such as error detection and correction and data flow control are handled. In addition, there is a need for a **protocol layer**. This turns the internal data transfer into data packets that can then be handled by the link and PHY layers. A massive increase in logic content will result from the advances in I/O subsystems. To make matters worse, I/O subsystems run on their own clocks, each of which is synchronized to the recovered clock. This creates multiple clock domains on a given chip with multiple I/O channels. Cross-domain asynchronous clocking will result in nondeterministic chip responses and will ultimately lead to mismatches and yield loss [Mak 2004].

14.7 CONCLUDING REMARKS

Any semiconductor device must have I/O interface/interfaces, and an I/O failure will cause the device to fail. At multiple Gbps data rates, I/O testing goes beyond the conventional defect and functional tests, and *jitter noise*, and *bit error rate* (JNB) testing becomes mandatory to ensure the interoperability of the I/O link. In this chapter, we first reviewed JNB from statistical and system frequency response views. Jitter and noise component concepts and separation methods were reviewed, and the relationships between jitter and noise *probability density functions* (PDFs) and *bit-error-rate* (BER) *cumulative density function* (CDF) were also discussed.

Next, we reviewed serial link architectures and associated jitter transfer functions for two commonly implemented link *inputs/outputs* (I/Os), namely the data-driven *phased-locked loop* (PLL) clock recovery receiver and *phase-interpolator* (PI) clock recovery receiver. Jitter tracking and tolerance transfer functions were given for those receivers. Furthermore, the transfer function for the reference clock jitter at the receiver was also discussed.

We then presented the JNB component tests for output and receiving tolerances, as well as system tests including loopback. The output tests included both transmitter and reference clock outputs. We emphasized the importance of employing the clock recovery jitter transfer function in JNB testing. A compliance eye diagram with an eye mask corresponding to BER $= 10^{-12}$ for output testing was introduced, and receiver tolerance testing was also introduced and reviewed. Test hardware bandwidth and noise floor, compliance clock recovery receiver, and test throughput were also discussed. We also discussed a technique for estimating BER based on jitter spectrum information in the data at the receiver's input and the receiver's jitter transfer characteristics, which are easier to measure than taking a direct measurement of the BER.

On-chip built-in *design-for-testability* (DFT) methods were presented. AC loopback tests, high-speed serial loopback, and receiver equalization tests were discussed. For the AC loopback test, both I/O wrap and AC I/O loopback test methods were reviewed, along with their characteristics.

Following the on-chip DFT methods, system interconnect test methods of high-speed boundary scan and *interconnect built-in self-test* (IBIST) were discussed. The IEEE 1149.1 and 1149.6 standards were covered for high-speed boundary scan test method.

As the data rate keeps increasing, JNB testing will encounter some challenging problems. At a 10-Gbps data rate, for example, the jitter noise floor of the testing instrument will need to be much smaller (in the hundreds of fs RMS or less) so that it will not eat away the jitter margin from the *device under test* (DUT). Following the same argument, jitter caused by the test fixture and socket will need to be removed or de-embedded from the testing results. At higher data rates, many equalization techniques will be used to compensate for jitter and noise from transmitter and channel so that an "open eye" can be achieved before the receiver sampling flop to warrant the required BER. In the multiple-lane high-speed I/O topology, as jitter amplification resulting from the mode conversion of the channel crosstalk becomes significant, new tests such as *pulse width shrinkage* (PWS) need to be enforced, posing a new and challenging test requirement. Whereas some novel ideas for testing equalizers have been proposed, some of which were discussed in this chapter, receiver equalization testing is still at its early stage. As data transmission rates continue to increase, many challenges remain for testing these advanced I/O transceivers in the coming years.

14.8 EXERCISES

14.1 **(Jitter)** Why does jitter separate into different components? What are the benefits of the jitter component tree diagram shown in Figure 14.5?

14.2 **(Embedded Clock)** Why does the embedded clock link architecture become the mainstream when data rate reaches 1 Gbps?

14.3 **(Jitter, Noise, and Bit Error Rate [JNB])** Is it possible to establish pass/fail JNB criteria for a link component (transmitter, receiver, channel, or reference clock) without knowing or assuming the properties for the other components and link architecture? Explain your answer.

14.4 **(Jitter, Noise, Bit Error Rate, Transmitter, and Receiver)** What are the major differences of the JNB limits between testing a transmitter and testing a receiver? Why does a standard receiver need to be defined when testing the transmitter, and why does a standard for the rest of the link (transmitter, channel, and reference clock) need to be defined when testing the receiver?

14.5 **(Global Clock Timing Test)** In Figure 14.16, which timing marker needs to be changed/programmed if the minimum delay specification needs to be tested?

14.6 **(Source Synchronous Timing Test)** In Figure 14.17, what other edge search algorithms exist that can be applied to test for the position of the strobe edge?

14.7 **(System Interconnect Test)** Why may random patterns not be sufficient for high-speed serial link testing? Identify at least two reasons.

Acknowledgments

The authors wish to thank Dr. Yi Cai of Agere, Dr. Anne Meixner of Intel, Masashi Shimanuchi of Credence, Dr. Takahiro Yamaguchi of Advantest Japan, William Eklow of Cisco Systems, and Professor Kuen-Jong Lee of National Cheng Kung University for reviewing the chapter and giving critical comments and suggestions.

References

R14.0 Books

[Best 1999] R. E. Best, *Phase-Locked Loops: Design, Simulation, and Applications*, Fourth Edition, McGraw-Hill, New York, 1999.

[Bushnell 2000] M. L. Bushnell and V. D. Agrawal, *Essentials of Electronic Testing for Digital, Memory & Mixed-Signal VLSI Circuits*, Springer Science, New York, 2000.

[Davenport 1987] W. B. Davenport and W. L. Root, *An Introduction to the Theory of Random Signals and Noise*, IEEE Press, New York, 1987.

[Gardner 1979] F. M. Gardner, *Phaselock Techniques*, Second Edition, John Wiley & Sons, New York, 1979.

[Johnson 1993] H. Johnson and M. Graham, *High-Speed Digital Design: A Handbook of Black Magic*, Prentice Hall, Upper Saddle River, NJ, 1993.

[Oppenheim 1996] A. V. Oppenheim, A. S. Willsky, and S. H. Nawab, *Signals & Systems*, Prentice Hall, Upper Saddle River, NJ, 1996.

[Papoulis 1977] A. Papoulis, *Signal Analysis*, McGraw-Hill, New York, 1977.

[Wang 2006] L.-T. Wang, C.-W. Wu, and X. Wen, editors, *VLSI Test Principles and Architectures: Design for Testability*, Morgan Kaufmann, San Francisco, 2006.

R14.1 Introduction

[IEEE 1149.1-2001] IEEE Std. 1149.1-2001, *IEEE Standard Test Access Port and Boundary Scan Architecture*, IEEE Press, New York, 2001.

[IEEE 1149.6-2003] IEEE Std. 1149.6-2003, *IEEE Standard for Boundary Scan Testing of Advanced Digital Networks*, IEEE Press, New York, 2003.

[Li 1999] M. Li, J. Wilstrup, R. Jessen, and D. Petrich, A new method for jitter decomposition through its distribution tail fitting, in *Proc. Int. Test Conf.*, pp. 788–794, October 1999.

[Li 2003] M. Li and J. Wilstrup, Paradigm shift for jitter and noise in design and test >1 Gbps communication systems, in *Proc. Int. Conf. on Computer Design*, pp. 467–472, October 2003.

[Li 2004] M. Li, A. Martwick, G. Talbot, and J. Wilstrup, Transfer functions for the reference clock jitter in a serial link: Theory and applications, in *Proc. Int. Test Conf.*, pp. 1148–1167, October 2004.

[NCITS 2001] National Committee for Information Technology Standardization (NCITS), Working draft for fibre channel: Methodologies for jitter and signal quality specification—MJSQ, Rev. 4, December 2001.

[Nejedlo 2003] J. J. Nejedlo, IBIST (interconnect built-in-self-test) architecture and methodology for PCI express, in *Proc. Int. Test Conf.*, pp. 114–122, September 2003.

[Wilstrup 1998] J. Wilstrup, A method of serial data jitter analysis using one-shot time interval measurements, in *Proc. Int. Test Conf.*, pp. 819–824, October 1998.

R14.2 High-Speed I/O Architectures

[Cai 2005] Y. Cai, A. Bhattacharyya, J. Martone, A. Verma, and W. Burchanowski, Comprehensive production test solution for 1.5 Gbps and 3.0 Gbps serial ATA based on AWG and undersampling techniques, in *Proc. Int. Test Conf.*, Paper 27.1, November 2005.

[Hänsel 2004] G. Hänsel, K. Stieglbauer, G. Schulze, and J. Moreira, Implementation of an economic jitter compliance test for a multi-gigabit device on ATE, in *Proc. Int. Test,* pp. 1303–1312, October 2004.

[Ilkbahar 2001] A. Ilkbahar, S. Venkataraman, and H. Muljono, Itanium™ Processor System Bus Design, *IEEE J. Solid-State Circuits*, 36(10), pp. 1565–1573, October 2001.

[Li 1999] M. Li, J. Wilstrup, R. Jessen, and D. Petrich, A new method for jitter decomposition through its distribution tail fitting, in *Proc. Int. Test Conf.*, pp. 788–794, October 1999.

[Li 2000] M. Li and J. Wilstrup, Signal integrity: How to measure it correctly?, *Proc. DesignCon,* February 2000.

[Li 2003] M. Li and J. Wilstrup, Paradigm shift for jitter and noise in design and test >1 Gbps communication systems, in *Proc. Int. Conf. on Computer Design*, pp. 467–472, October 2003.

[Li 2004] M. Li, A. Martwick, G. Talbot, and J. Wilstrup, Transfer functions for the reference clock jitter in a serial link: Theory and applications, in *Proc. Int. Test Conf.*, pp. 1158–1167, October 2004.

[NCITS 2001] National Committee for Information Technology Standardization (NCITS), Working draft for fibre channel: Methodologies for jitter and signal quality specification—MJSQ, Rev. 4, December 2001.

[PCI-SIG 2004] PCI Express Jitter white paper (I), (www.pcisig.com/specifications/pciexpress/technical_library), 2004.

[Ward 2004] B. A. Ward, K. Tan, and M. L. Guenther, Apparatus and method for spectrum analysis-based serial data jitter measurement, U.S. Patent No. 6,832,172, December 14, 2004.

[Wilstrup 1998] J. Wilstrup, A method of serial data jitter analysis using one-shot time interval measurements, in *Proc. Int. Test Conf.*, pp. 819–824, October 1998.

R14.3 Testing of I/O Interfaces

[Cai 2005] Y. Cai, A. Bhattacharyya, J. Martone, A. Verma, and W. Burchanowski, Comprehensive production test solution for 1.5 Gbps and 3.0 Gbps serial ATA based on AWG and undersampling techniques, in *Proc. Int. Test Conf.*, Paper 27.1, November 2005.

[Hänsel 2004] G. Hänsel, K. Stieglbauer, G. Schulze, and J. Moreira, Implementation of an economic jitter compliance test for a multi-gigabit device on ATE, in *Proc. Int. Test Conf.*, pp. 1303–1312, October 2004.

[Hong 2004] D. Hong, C. K. Ong, and K.-T. Cheng, BER estimation for serial links based on jitter spectrum and clock recovery characteristics, in *Proc. Int. Test Conf.*, pp.1138–1147, October 2004.

[Hong 2007] D. Hong, S. Saberi, K.-T. Cheng, and P. Yue, A two-tone test method for continuous-time adaptive equalizers, in *Proc. IEEE Design, Automation, and Test in Europe*, pp. 1283–1288, April 2007.

[Li 1999] M. Li, J. Wilstrup, R. Jessen, and D. Petrich, A new method for jitter decomposition through its distribution tail fitting, in *Proc. Int. Test Conf.*, pp. 788–794, October 1999.

[Li 2003] M. Li and J. Wilstrup, Paradigm shift for jitter and noise in design and test >1 Gbps communication systems, in *Proc. Int. Conf. on Computer Design*, pp. 467–472, October 2003.

[Li 2004] M. Li, A. Martwick, G. Talbot, and J. Wilstrup, Transfer functions for the reference clock jitter in a serial link: Theory and applications, in *Proc. Int. Test Conf.*, pp. 1158–1167, October 2004.

[PCI-SIG 2004] PCI Express Jitter White Paper (I), www.pcisig.com/specifications/pciexpress/technical_library, 2004.

[PCI-SIG 2007] PCI-SIG, PCI Express 2.0 Base Specification Revision 1.0, 2007.

[Sivaram 2004] A. T. Sivaram, M. Shimanouchi, H. Maassen, R. Jackson, Tester architecture for the source synchronous bus, in *Proc. Int. Test Conf.*, pp. 738–747, October 2004.

[Yamaguchi 2002] T. J. Yamaguchi, M. Soma, H. Musha, L. Malarsie, and M. Ishida, A new method for testing jitter tolerance of SerDes devices using sinusoidal jitter, in *Proc. Int. Test Conf.*, pp. 717–725, October 2002.

R14.4 DFT-Assisted Testing

[Cai 2002] Y. Cai, B. Laquai, and K. Luehman, Jitter testing for gigabit serial communication transceivers, *IEEE Des. Test Comput.*, 19(1), pp. 66–74, January/February 2002.

[Cai 2005] Y. Cai, A. Bhattacharyya, J. Martone, A. Verma, and W. Burchanowski, Comprehensive production test solution for 1.5 Gbps and 3.0 Gbps serial ATA based on AWG and undersampling techniques, in *Proc. Int. Test Conf.*, Paper 27.1, November 2005.

[Casper 2003] B. Casper, A. Martin, J. E. Jaussi, J. Kennedy, and R. Mooney, An 8-Gbps simultaneous bidirectional link with on-die waveform capture, *IEEE J. of Solid-State Circuits*, 38(12), pp. 2111–2120, December 2003.

[Gillis 1998] P. Gillis, F. Woytowich, K. McCauley, and U. Baur, Delay test of chip I/Os using LSSD boundary scan, in *Proc. Int. Test Conf.*, pp. 83–90, October 1998.

[Hong 2007] D. Hong, S. Saberi, K.-T. Cheng, and P. Yue, A two-tone test method for continuous-time adaptive equalizers, in *Proc. IEEE Design, Automation and Test in Europe Conf.*, pp. 1283–1288, April 2007.

[Huang 2001] J.-L. Huang and K.-T. Cheng, An on-chip short-time interval measurement technique for testing high-speed communication links, in *Proc. VLSI Test Symp.*, pp. 380–385, April 2001.

[Hur 2005] Y. Hur, M. Maeng, C. Chun, F. Bien, H. Kim, S. Chandramouli, E. Gebara, and J. Laskar, Equalization and near-end crosstalk (NEXT) noise cancellation for 20-Gbps 4-PAM backplane serial I/O interconnections, *IEEE Trans. Microwave Theory and Techniques*, 53(1), pp. 246–255, January 2005.

[Jaussi 2005] J. E. Jaussi, G. Balamurugan, D. R. Johnson, B. Casper, A. Martin, J. Kennedy, N. Shanbhag, and R. Mooney, 8-Gbps source-synchronous I/O link with adaptive receiver equalization, offset cancellation, and clock deskew, *IEEE J. of Solid-State Circuits*, 40(1), pp. 80–88, January 2005.

[Laquai 2001] B. Laquai and Y. Cai, Testing gigabit multilane SerDes interfaces with passive jitter injection filters, in *Proc. Int. Test Conf.*, pp. 297–304, October 2001.

[Lin 2005] M. Lin, K.-T. Cheng, J. Su, M. C. Sun, J. Chen, and S. Lu, Production-oriented interface testing for PCI express by enhanced loopback technique, in *Proc. Int. Test Conf.*, Paper 27.3, November 2005.

[Lin 2006] M. Lin, and K.-T. Cheng, Testable design for adaptive linear equalizer in high-speed serial links, in *Proc. Int. Test Conf.*, Paper 33.3, October 2006.

[Lin 2007] M. Lin and K.-T. Cheng, Testable design for advanced serial-link transceivers, in *Proc. IEEE Design, Automation and Test in Europe*, pp. 695–700, April 2007.

[Mak 2005] T. M. Mak and M. J. Tripp, Device testing, U.S. Patent No. 6,885,209, April 26, 2005.

[Pelard 2004] C. Pelard, E. Gebara, A. J. Kim, M. G. Vrazel, F. Bien, Y. Hur, M. Maeng, S. Chandramouli, C. Chun, S. Bajekal, S. E. Ralph, B. Schmukler, V. M. Hietala, and J. Laskar, Realization of multigigabit channel equalization and crosstalk cancellation integrated circuits, *IEEE J. Solid-State Circuits*, 39(10), pp. 1659–1670, October 2004.

[Robertson 2005] I. Robertson, G. Hetherington, T. Leslie, I. Parulkar, and R. Lesnikoski, Testing high-speed, large scale implementation of SerDes I/Os on chips used in throughput computing systems, in *Proc. Int. Test Conf.*, Paper 38.1, November 2005.

[Sunter 2004a] S. Sunter and A. Roy, On-chip digital jitter measurement, from megahertz to gigahertz, *IEEE Design & Test of Computers*, 21(4), pp. 314–321, July/August 2004.

[Sunter 2004b] S. Sunter, A. Roy, and J.-F. Cote, An automated, complete, structural test solution for SerDes, in *Proc. Int. Test Conf.*, pp. 95–104, October 2004.

[Sunter 2005] S. Sunter and A. Roy, Structural tests for jitter tolerance in SerDes receivers, in *Proc. Int. Test Conf.*, Paper 9.1, November 2005.

[Tripp 2004] M. Tripp, T. M. Mak, and A. Meixner, Elimination of traditional functional testing of interface timings at Intel, in *Proc. Int. Test Conf.*, pp. 1448–1454, October 2004.

[Vrazel 2006] M. Vrazel and A. J. Kim, Overcoming signal integrity issues with wideband crosstalk cancellation technology, in *Proc. DesignCon*, February 2006.

R14.5 System-Level Interconnect Testing

[Ahmed 2003] N. Ahmed, M. H. Tehranipour, and M. Nourani, Extending JTAG for testing signal integrity in SoCs, in *Proc. Design, Automation, and Test in Europe*, pp. 218–223, March 2003.

[IEEE 1149.1-2001] IEEE Std. 1149.1-2001, *IEEE Standard Test Access Port and Boundary Scan Architecture*, IEEE Press, New York, 2001.

[IEEE 1149.6-2003] IEEE Std. 1149.6-2003, *IEEE Standard for Boundary Scan Testing of Advanced Digital Networks*, IEEE Press, New York, 2003.

[IEEE 1500-2005] IEEE Std. 1500-2005, *IEEE Standard for Embedded Core Test*, IEEE Press, New York, 2005.

[Nejedlo 2003] J. J. Nejedlo, IBIST (interconnect built-in-self-test) architecture and methodology for PCI express, in *Proc. Int. Test Conf.*, pp. 114–122, October 2003.

[Tehranipour 2003a] M. H. Tehranipour, N. Ahmed, and M. Nourani, Testing SoC interconnects for signal integrity using boundary scan, in *Proc. VLSI Test Symp.*, pp. 163–168, April 2003.

[Tehranipour 2003b] M. Tehranipour, N. Ahmed, and M. Nourani, Multiple transition model and enhanced boundary scan architecture to test interconnects for signal integrity, in *Proc. Int. Conf. on Computer Design*, pp. 554–559, October 2003.

[Wang 1992] L.-T. Wang and P. Y.-F. Wu, PATRIOT: A boundary-scan test and diagnosis system, in *Proc. IEEE Int. Conf. on COMPCON Spring*, pp. 436–439, February 1992.

R14.6 Future Challenges

[Kundu 2004] S. Kundu, T. M. Mak, and R. Galivanche, Trends in manufacturing test methods and their implications, in *Proc. Int. Test Conf.*, pp. 679–687, October 2004.

[Lin, 2003] H.-C. Lin, K., Taylor, A. Chong, E. Chan, M. Soma, H. Haggag, J. Huard, and J. Braat, CMOS built-in test architecture for high-speed jitter measurement, in *Proc. Int. Test Conf.*, pp. 67–76, September 2003.

[Mak 2004] T. M. Mak, How do we test for adaptive computing?, *Workshop on Test Resource Partitioning*, pp. 17–21, April 2004.

CHAPTER 15

Analog and Mixed-Signal Test Architectures

F. Foster Dai
Auburn University, Auburn, Alabama

Charles E. Stroud
Auburn University, Auburn, Alabama

ABOUT THIS CHAPTER

The scale of analog integrated circuits (ICs) and the analog portion of mixed-signal ICs is usually relatively small compared to digital IC counterparts—hundreds of devices in an analog circuit compared to millions to hundreds of millions of devices in a digital circuit. However, testing analog circuits poses a number of unique testing problems when compared to digital circuits and, as a result, requires a variety of different and unique test architectures and approaches. For example, there are no analog fault models that have been widely accepted, as is the case in digital testing. As a result, most analog testing tends to be specification-oriented as opposed to defect-oriented approaches typically used when testing digital circuitry. This is due in part to the fact that there is a range of good circuit signal values that result from acceptable component tolerances, environmental variations (including temperature and supply voltage), and noise.

An introduction to analog and mixed-signal testing was given in [Wang 2006], and a number of exceptional books have been dedicated to testing analog and mixed-signal circuits, including [Burns 2000], [Mahoney 1987], [Roberts 1995], and [Vinnakota 1998]. Therefore, this chapter begins with an overview of some of the challenges in analog and mixed-signal testing followed by a discussion of some fundamental analog test techniques and measurements. The chapter then focuses on test architectures that can be included for on-chip test and measurement of the analog portion(s) of ***system-on-chip*** (SOC) implementations. ***Built-in self-test*** (BIST) for digital circuits has been an active area of research and development since the mid-1970s, resulting in a number of good BIST approaches to test the digital portion of mixed-signal systems. Until recently, however, BIST for analog circuitry has received much less attention. As a result, testing the analog portion of mixed-signal integrated circuits and systems has been identified as one of the major challenges for the future, and BIST has been identified as one of the potential solutions to this testing challenge [SIA 2005, 2006].

15.1 INTRODUCTION

There are two basic approaches for testing analog circuitry: *functional testing*, also referred to as *specification-oriented testing*, and *structural testing*, also referred to as *defect-oriented testing*. Specification-oriented testing has long been the preferred approach because of the continuous signals produced by analog circuits, the wide range and variety of analog circuits, and their nonlinear characteristics. Feedback is frequently incorporated to overcome the effect of acceptable component parameter variations caused by the nature of the manufacturing process as well as operational temperature and voltage variations. These factors complicate the process of testing to determine if an analog circuit is faulty or operating within acceptable range. This determination often requires highly accurate (and expensive) test equipment to perform the necessary measurements. Whereas a digital circuit can be decomposed to gates or to transistors for defect-oriented testing, the components of an analog circuit function collectively to perform the overall functionality of a given circuit and, as a result, cannot always be treated as a collection of individual components.

Fault models have been proposed for defect-oriented analog and mixed-signal testing. These fault models can be classified into two categories: *catastrophic faults* (or *hard faults*) and *parametric faults* (or *soft faults*) [Kaminska 1997]. Parametric faults are deviations of component parameters that are outside the acceptable range or specified tolerance limits. A catastrophic fault is similar to digital fault models in that terminals of the component can be stuck-open or stuck-short (with another terminal). Stuck-open faults are hard faults in which the component terminals are not in contact with the rest of the circuit or create a high resistance at the location of the fault in the circuit. A stuck-short fault, on the other hand, is a short between terminals of the component, effectively shorting out the component from the circuit. Alternatively, the terminals of two different components can be shorted, similar to bridging faults in a digital circuit.

The primary purpose of fault models is to facilitate emulation of defects for the evaluation of the fault detection capabilities associated with a given set of test waveforms via fault simulation. Analog fault simulation requires many simulations of the fault-free circuit to establish normal variations in the output waveforms that result from acceptable component parameter variations as well as temperature and voltage variations. A known range of acceptable values for circuit component parameters is necessary to establish the fault-free behavior for a given circuit, which can then be used to determine if a fault can be detected. An analog fault simulator may support Gaussian (normal) or uniform component parameter variations. Standard deviation or the 1σ value is typically used for specifying the normal distribution, which, in turn, assumes the components will vary up to $\pm 3\sigma$ while the analog circuit continues to operate within the system specifications. **Monte Carlo analysis** is typically used to generate different component values for fault-free components within the normal or uniform distributions. During catastrophic and parametric fault simulation, each emulated faulty circuit must also undergo a number of simulations for Monte Carlo analysis. This means that analog fault simulation is more time intensive than digital fault simulation and, as a result, the debate continues between functional (specification-oriented) and structural (defect-oriented) testing approaches.

15.2 ANALOG FUNCTIONAL TESTING

Analog signals are continuous waveforms and, as illustrated in Figure 15.1, the interpretations for a time domain response are different in the analog and digital domains. Whereas a digital designer is primarily concerned with logic ones and zeros, an analog designer cares about the detailed shape of the waveform and its corresponding spectrum. Logic high and low voltages (V_H and V_L) as well as rise and fall times (t_{LH} and t_{HL}) are the major attributes of concern in testing a digital waveform. In an analog waveform, on the other hand, the major testing parameters include the amplitude (V_A), slew rate (SR), overshoot (V_{OV}), settling time (t_{Settle}), bandwidth, phase noise, and timing jitter. As a result, there is a variety of different test and measurement concerns for analog designers and test engineers, as discussed in the following subsections.

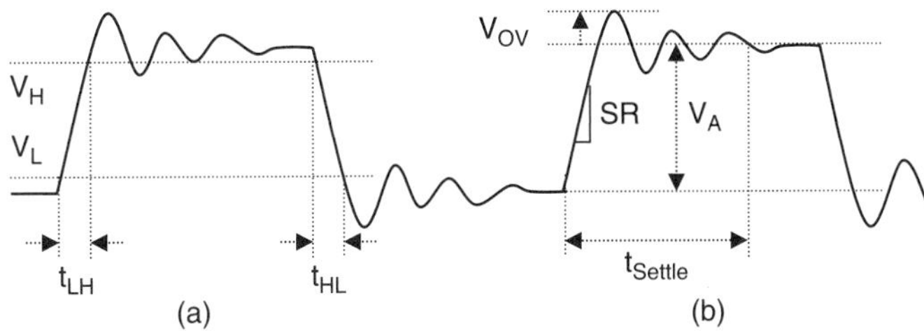

FIGURE 15.1

Digital and analog representations for a time domain response: (a) digital and (b) analog.

15.2.1 Frequency Response Testing

Frequency response measures the frequency dependence of the output signal of an analog ***device under test*** (DUT). As an example, Figure 15.2 illustrates the frequency response of a band-pass filter, which is characterized by its pass band, stop band, and transition band in between. As shown in the figure, the pass band is defined as the frequency range in which a signal can pass with little attenuation. Conversely, a signal falling in the stop band will be significantly attenuated. *Pass band ripple* is the variation of the amplitude response in the pass band. The difference in attenuation from pass band to stop band is defined as the *stop band rejection ratio* and is measured in dB. In the transition bands, the filter rise or rolloff slope is determined by the order of the filter, which follows the 6 dB/octave/order or 20 dB/decade/order rule.

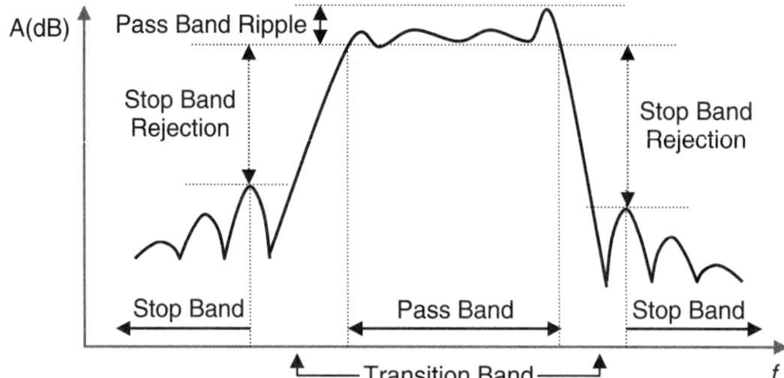

■ **FIGURE 15.2**

Frequency-domain transfer function of a bandpass filter.

In frequency response testing, the DUT is usually driven by a signal generator and its output is analyzed by a spectrum analyzer to obtain the measurement result. To correctly use a spectrum analyzer, let's briefly review its basic operation theory. As shown in Figure 15.3, the spectrum analyzer down-converts the input ***radio frequency*** (RF) signal to baseband frequency through two filters, an analog ***intermediate frequency*** (IF) filter and a digital video filter, whose ***bandwidths*** (BWs) are defined as ***resolution bandwidth*** (RBW) and ***video bandwidth*** (VBW), respectively and can be determined by the analyzer setting. The video filter is used to smooth noise for easier identification and measurement of low-level signals. The IF filter, on the other hand, integrates the input signal and noise power over the RBW. Hence, the phase noise reading on the spectrum analyzer is not the real noise level and is tightly dependent on the RBW. For example, a phase noise reading of −100 dBc with a RBW setting of 10 Hz in a spectrum analyzer would be

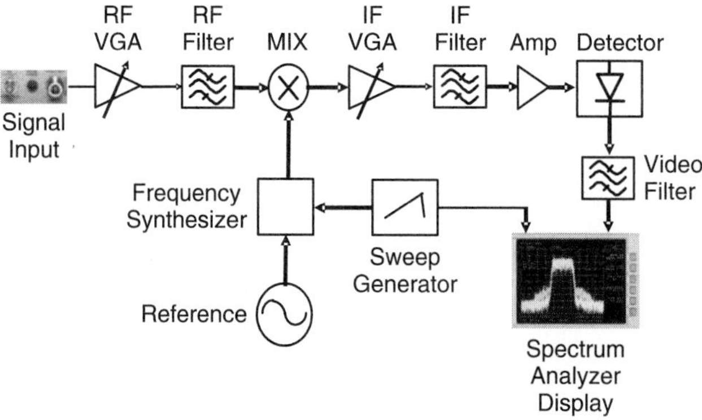

■ **FIGURE 15.3**

Illustration of spectrum analyzer building blocks.

−90 dBc with a RBW setting of 100 Hz. Accordingly, the best sensitivity in a spectrum analyzer is achieved by using the narrowest RBW, minimum RF attenuation, and sufficient video filtering to smooth the noise (normally, VBW is less than 1% of the RBW). Because a spectrum analyzer measures the baseband signal power, the ***power spectral density*** (PSD) reading is normally single sideband. A proper sweeping setting is required to obtain symmetric sideband measurement. The penalty for fast sweeping is an uncalibrated display on sidebands.

15.2.2 Linearity Testing

Linearity is an important performance measure of many analog components such as amplifiers and mixers. The nonlinear characteristics of a circuit result in intermodulation of various components of a signal, referred to as ***intermodulation distortion*** (IMD). IMD is of concern in communication circuits because it can modulate the tone in adjacent bands and cause interference. Theoretically, IMD could be tested using multitone signals. However, only two or three tones are used in real analog testing. Assume two tones $x(t) = A_1\cos\omega_1 t + A_2\cos\omega_2 t$ with frequencies of $f_1 = \omega_1/2\pi$ and $f_2 = \omega_2/2\pi$ are applied to the input of an amplifier with transfer function expressed as $y(t) = \alpha_0 + \alpha_1 x(t) + \alpha_2 x^2(t) + \alpha_3 x^3(t) + \ldots$, where α_j is, in general, independent of time if the system is time invariant. Substituting the two-tone input into the transfer function, we obtain the amplifier output as:

$$
\begin{aligned}
y(t) = {} & \frac{1}{2}\alpha_2({A_1}^2 + {A_2}^2) \\
& + \left[\alpha_1 A_1 + \frac{3}{4}\alpha_3 A_1({A_1}^2 + 2{A_2}^2)\right]\cos\omega_1 t \\
& + \left[\alpha_1 A_2 + \frac{3}{4}\alpha_3 A_2(2{A_1}^2 + {A_2}^2)\right]\cos\omega_2 t \\
& + \frac{1}{2}\alpha_2\left[{A_1}^2\cos 2\omega_1 t + {A_2}^2\cos 2\omega_2 t\right] \\
& + \alpha_2 A_1 A_2\left[\cos(\omega_1+\omega_2)t + \cos(\omega_1-\omega_2)t\right] \\
& + \frac{1}{4}\alpha_3\left[{A_1}^3\cos 3\omega_1 t + {A_2}^3\cos 3\omega_2 t\right] + \frac{3}{4}\alpha_3 \cdot \\
& \left\{\begin{array}{l} {A_1}^2 A_2\left[\cos(2\omega_1+\omega_2)t + \cos(2\omega_1-\omega_2)t\right] \\ + A_1{A_2}^2\left[\cos(2\omega_2+\omega_1)t + \cos(2\omega_2-\omega_1)t\right] \end{array}\right\}
\end{aligned}
\tag{15.1}
$$

Figure 15.4 illustrates the spectrum of a waveform in response to two tones. When the two-tone test signal passes through an amplifier, both fundamental and ***intermodulation*** (IM) terms will be present at the amplifier output. Any linear combinations of f_1 and f_2 may appear because of the IMD. However, the closest intermodulation terms to the fundamental terms are the ***third-order intermodulation*** (IM3) terms with frequencies of $2f_1 - f_2$ and $2f_2 - f_1$. Therefore, the linearity is normally measured using the ***third-order intercept point*** (IP3), where the linear term intercepts the third-order term in a two-tone test as illustrated in Figure 15.5b.

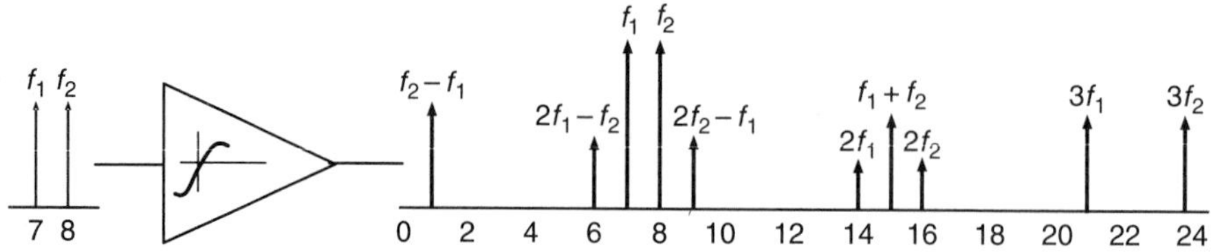

■ **FIGURE 15.4**

Spectrum with noise and distortion.

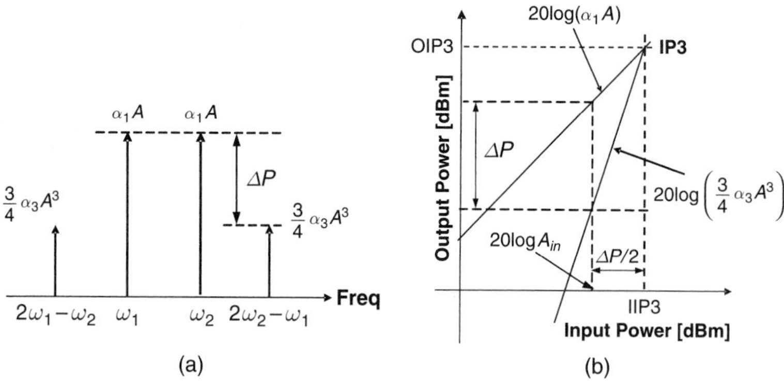

■ **FIGURE 15.5**

Linearity measurement using two-tones test: (a) output spectrum under two-tone test and (b) third-order intercept point (IP3).

According to Equation (15.1) with $A_1 = A_2$, the ***input referred IP3*** (IIP3) and the ***output referred IP3*** (OIP3) can be found as:

$$IIP_3 \approx \sqrt{\frac{4}{3}\left|\frac{\alpha_1}{\alpha_3}\right|}, \quad \text{if} \quad \alpha_1 >> \frac{9}{4}\alpha_3 A^2$$
$$OIP_3 = \alpha_1 IIP_3 \tag{15.2}$$

where the assumption for IIP3 is normally valid when the test tone magnitude is relatively small, such that the amplifier is not desensitized. In the two-tone test, the IIP3 can be found by measuring the difference, ΔP, between fundamental and IM3 terms. As shown in Figure 15.5, the IIP3 is given by:

$$IIP_3[dBm] = \frac{\Delta P[dB]}{2} + P_{in}[dBm] \tag{15.3}$$

where P_{in} is the signal power at the amplifier input. As discussed later in this chapter, there are various ways to generate a two-tone test for IIP3 measurement.

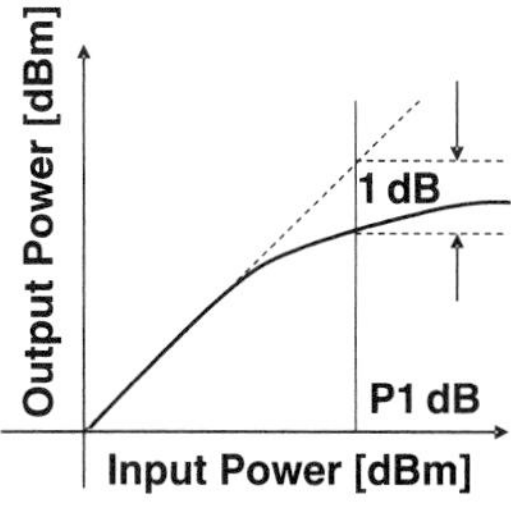

■ **FIGURE 15.6**

Linearity measurement using gain compression.

Another measure of the linearity is the so-called gain compression test. Referring to Figure 15.6, the gain of an amplifier, defined as the ratio of the output power to the input power, remains constant with small input power. When the input power increases, the gain is compressed, deviating from the linear gain curve. Define power unit dBm as $P_{dBm} = 10\log_{10}\left(\frac{Pwatt[mW]}{1mW}\right)$. The ***1-dB compression point*** is defined as the input signal level that causes small-signal gain to drop 1 dB from the linear gain. P1dB is the input power in dBm that corresponds to the 1-dB compression point. It is a measure of the maximum input range. With a single tone input, the P1dB can be found as:

$$P1dB = \sqrt{0.145\left|\frac{\alpha_1}{\alpha_3}\right|} = 0.3808\sqrt{\left|\frac{\alpha_1}{\alpha_3}\right|} \tag{15.4}$$

Considering Equation (15.2), the relationship between the P1dB and the IIP3 can be estimated by:

$$IIP3 = P1dB + 9.66dB \tag{15.5}$$

In other words, IIP3 is about 9.66 dB higher than the P1dB value. Note that Equation (15.5) is valid only for single tone input. For two-tone input, IIP3 is about 14.4 dB higher than the P1dB value.

15.2.3 Signal-to-Noise Ratio Testing

The ***signal-to-noise ratio*** (SNR) reflects how pure a signal is and is defined as the ratio of the signal power to the noise power represented in dB. There are many sources for noise in an electronic system including thermal noise, flicker noise, power supply noise, switching noise, coupling noise, etc. Distortions, such as harmonic distortion and IMD, further degrade the signal's quality. Distortions often result from crossover, clip, saturation, and mismatch of differential signal paths. When a pure sinusoidal waveform is generated with ***additive white Gaussian noise*** (AWGN) and applied to the DUT, a typical output spectrum can be illustrated as in Figure 15.7.

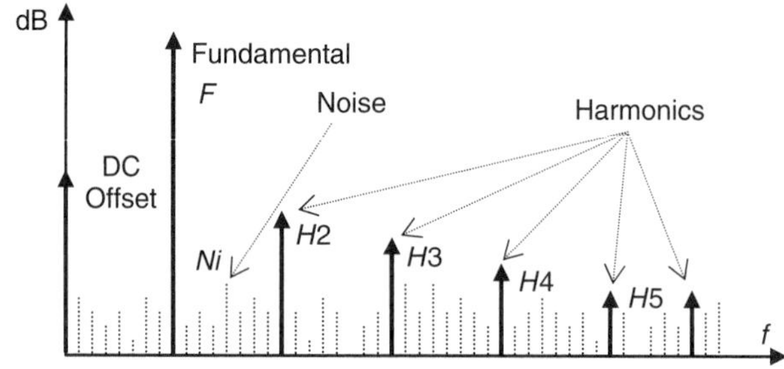

■ **FIGURE 15.7**

Spectrum with noise and distortion.

Frequency components of a sinusoidal waveform are classified into three categories. The first category, F, is the fundamental component—the desired signal. F can be a voltage or a current signal, and F^2 represents the corresponding power for $P = I^2R = V^2/R$. The second category, H_i, is the ith harmonic, and the final category, N_i, represents the ith noise term. The SNR and distortion can be obtained from the following equations:

$$\begin{cases} SNR = 10\log\frac{F^2}{\sum N_i^2} \\ THD = 10\log\frac{F^2}{\sum H_i^2} = 100 \times \frac{F^2}{\sum H_i^2}\% \\ SINAD = 10\log\frac{F^2}{\sum H_i^2 + \sum N_i^2} \end{cases} \tag{15.6}$$

The ***total harmonic distortion*** (THD) is the ratio of signal power to the harmonic power in total and, as a result, harmonic terms H_i are included. THD is represented in either dB or percentage. In the ***signal-to-noise and distortion ratio*** (SINAD), both noise and distortion terms are considered. The *peak harmonic* is also important because it can often be used to locate the source of distortion. For example, if the second harmonic is the peak harmonic, crossover distortion or symmetric nonlinear distortion is the most likely source. On the other hand, clipping and saturation are the most likely sources if the third harmonic is the peak harmonic.

15.2.4 Quantization Noise

In sampled systems such as mixed-signal systems with data converters, quantization noise needs to be carefully analyzed. A quantizer converts the continuous analog signal to a discrete digital signal with a characteristic shown in Figure 15.8, where the output is a function of the input, but has discrete levels. Thus, unless the input happens to be an integer multiple of the quantizer step size Δ, there will always be an error in representing the input. This error e will be bounded over one quantizer level by a value of:

$$-\frac{\Delta}{2} \leq e \leq \frac{\Delta}{2} \tag{15.7}$$

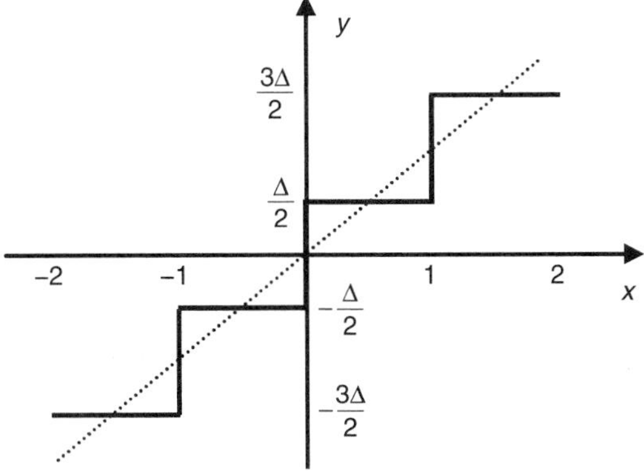

FIGURE 15.8

Transfer function of a multibit quantizer.

Thus, the quantized signal y can be represented by a linear function Gx with an error e as $y = Gx + e$, where the gain G is the slope of the straight line in Figure 15.8. If the input is "random" in nature, then the instantaneous error will also be random. The error is thus uncorrelated from sample to sample and can be treated as noise. Quantization and the resultant quantization noise can be modeled as a linear circuit including an additive error source to represent the quantization noise.

The quantization noise for a random signal can be treated as additive white noise with equal probability of locating anywhere in the range $-\Delta/2$ to $\Delta/2$ and has probability density of:

$$p(e) = \begin{cases} \frac{1}{\Delta} & if \quad -\frac{\Delta}{2} \le e \le \frac{\Delta}{2} \\ 0 & otherwise \end{cases} \tag{15.8}$$

where the normalization factor $1/\Delta$ is needed to guarantee that $\int_{-\Delta/2}^{\Delta/2} p(e)de = 1$. The mean square *rms* error voltage e_{rms} can be found by integrating the square of the error voltage and dividing by the quantization step size, namely:

$$e_{rms}^2 = \int_{-\infty}^{+\infty} p(e)e^2 de = \frac{1}{\Delta}\int_{-\Delta/2}^{+\Delta/2} e^2 de = \frac{\Delta^2}{12} \tag{15.9}$$

Note that frequency spectrums of sampled systems repeat once every sampling frequency. Thus, the spectrum of the quantization noise in a sampled system will be centered around DC and spread out to $f_S/2$, and there will be a copy of the noise spectrum from $f_S/2$ to $3f_S/2$, etc. Considering that all the noise power lies in the range $0 \le f < \infty$, the quantization noise power thus folds into the band from DC to $f_S/2$. Assuming white noise, the power spectral density of the quantization noise is given by $N(f) = \frac{e_{rms}^2}{f_s/2} = 2Te_{rms}^2$, where the sample period $T = 1/f_S$. For a band limited signal with bandwidth over $[0, f_0]$, the quantization noise power that falls into the signal band can thus be found as:

$$N_0 = \int_0^{f_0} e^2(f)df = 2f_0 T e_{rms}^2 = \frac{e_{rms}^2}{OSR} \tag{15.10}$$

where the ***oversampling rate*** (OSR) is defined as the ratio of the sampling frequency f_S to the Nyquist rate $2f_0$, namely, OSR $= f_S/2f_0$. Thus, for the same amount of total quantization noise power, every doubling of the sampling frequency reduces the in-band quantization noise by 3 dB. Oversampling has the advantage that it eases requirements on analog antialiasing filters for an ***analog-to-digital converter*** (ADC) or deglitch filters for a ***digital-to-analog converter*** (DAC). This is because oversampling results in wide transition bands; hence only low-order filters are required. However, the higher sampling rate requires not only faster digital circuits but also much lower signal bandwidth, which also means a lower conversion rate.

In an N-bit sampled system, if the quantizer has 2^N quantization levels equally spaced by Δ, then the maximum peak-to-peak amplitude is given by $v_{max} = (2^N - 1)\cdot\Delta$. If the signal is sinusoidal, the associated signal power can be found by $P = \frac{1}{8}\left(2^N - 1\right)^2 \cdot \Delta^2$. Thus, the SNR caused by the quantization noise power that falls into the signal band becomes:

$$\text{SNR} = 10\log\left(\frac{\frac{1}{8}\left(2^N - 1\right)^2 \Delta^2}{n_0^2}\right) \approx 10\log\left(\frac{3\cdot 2^{2N}\text{OSR}}{2}\right) \tag{15.11}$$

Noting that $\log_{10}(x) = \log_{10}(2)\cdot\log_2(x)$, the preceding expression can be rewritten as:

$$\text{SNR} \approx 6.02\cdot N + 3\cdot\log_2(\text{OSR}) + 1.76 \tag{15.12}$$

Therefore, the SNR improves by 6 dB for every bit added to the quantizer. For the same amount of total quantization noise power, every doubling of the sampling frequency reduces the in-band quantization noise by 3 dB. Hence, doubling the oversampling ratio is equivalent to increasing the quantizer levels by a half bit as far as the quantization noise is concerned.

15.2.5 Phase Noise

The noise performance for an analog circuit is usually classified in terms of phase noise, which is a measure of how much the output diverges from an ideal impulse function in the frequency domain. We are primarily concerned with noise that causes fluctuations in the phase of the output rather than noise that causes amplitude fluctuations in the tone, because the oscillator output typically has limited amplitude. A spectrum analyzer is often used to test the phase noise of an analog signal. To understand the phase noise reading of dBc/Hz on a spectrum analyzer, consider a ***phase-locked loop*** (PLL) as an example. The output signal of a PLL can be described as [Rogers 2006]:

$$v_{\text{out}}(t) = V_o \cos\left(\omega_{\text{LO}} t + \varphi_n(t)\right) \tag{15.13}$$

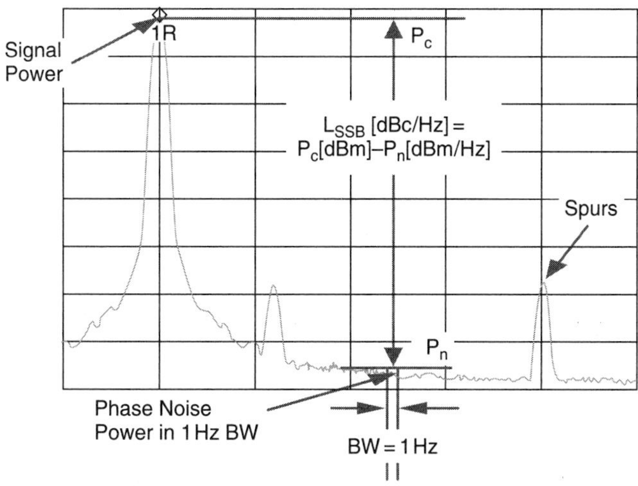

■ **FIGURE 15.9**

Phase noise and spurs shown in a spectrum analyzer.

where $\omega_{LO}t$ is the desired phase of the output and $\varphi_n(t)$ is random phase fluctuations. Phase noise is often quoted in units of dBc/Hz, whereas random phase fluctuation is often quoted in units of rad^2/Hz. The phase fluctuation term $\varphi_n(t)$ may be random phase noise or discrete spurious tones, as shown in Figure 15.9. The discrete spurs at a synthesizer output are most likely due to the fractional-N mechanism and the phase noise in an oscillator, which is mainly due to thermal, flicker, or $1/f$ noise and the finite Q of the oscillator tank.

Assume the phase fluctuation is of a sinusoidal form $\varphi_n(t) = \varphi_p \sin(\omega_m t)$, where φ_p is the peak phase fluctuation and ω_m is the offset frequency from the carrier. Substituting the phase fluctuation expression into Equation (15.13) gives:

$$\begin{aligned} v_{\text{out}}(t) &= V_0 \cos\left\lfloor \omega_{LO}t + \varphi_p \sin(\omega_m t)\right\rfloor \\ &= V_0 \left[\cos(\omega_{LO}t)\cos\left(\varphi_p \sin(\omega_m t)\right) - \sin(\omega_{LO}t)\sin\left(\varphi_p \sin(\omega_m t)\right)\right] \end{aligned} \tag{15.14}$$

Assuming a small phase fluctuation, Equation (15.14) can be simplified as:

$$\begin{aligned} v_{out}(t) &= V_0 \left\lfloor \cos(\omega_{LO}t) - \varphi_p \sin(\omega_m t)\sin(\omega_{LO}t)\right\rfloor \\ &= V_0 \left[\cos(\omega_{LO}t) - \frac{\varphi_p}{2}\left[\cos(\omega_{LO} - \omega_m)t - \cos(\omega_{LO} + \omega_m)t\right]\right] \end{aligned} \tag{15.15}$$

It is now evident that the phase-modulated signal includes the carrier signal tone and two symmetric sidebands at any offset frequency. A spectrum analyzer measures the phase-noise power in dBm/Hz, but in most cases phase noise is normalized by the carrier power as illustrated in Figure 15.9 and reported in units of rad²/Hz or dBc/Hz, namely:

$$\varphi_n{}^2(\Delta\omega) = \frac{Noise\,(\omega_{\mathrm{LO}} + \Delta\omega)}{P_{\mathrm{carrier}}(\omega_{\mathrm{LO}})} \tag{15.16}$$

where *Noise* and P_{carrier} are the PSD of the noise and the carrier signal, respectively. Furthermore, both single sideband and double sideband phase noise can be defined. ***Single sideband*** (SSB) phase noise-to-carrier ratio is defined as the ratio of power in one phase modulation sideband per Hertz bandwidth, at an offset Δω away from the carrier frequency, to the total signal power as:

$$PN_{SSB}\,(\Delta\omega) = 10\log\left[\frac{Noise\,(\omega_{\mathrm{LO}} + \Delta\omega)}{P_{\mathrm{carrier}}(\omega_{\mathrm{LO}})}\right] \tag{15.17}$$

This equation can be rewritten as:

$$PN_{\mathrm{SSB}}\,(\Delta\omega) = 10\log\left[\frac{\frac{1}{2}\left(\frac{V_0\varphi_p}{2}\right)^2}{\frac{1}{2}V_0{}^2}\right] = 10\log\left[\frac{\varphi_p^2}{4}\right] = 10\log\left[\frac{\varphi_{\mathrm{rms}}^2}{2}\right] \tag{15.18}$$

where φ_{rms}^2 is the ***root mean square*** (rms) phase-noise power density in units of [rad²/Hz]. Alternatively, ***double sideband*** (DSB) phase noise-to-carrier ratio is given by:

$$PN_{\mathrm{DSB}}\,(\Delta\omega) = 10\log\left[\frac{Noise\,(\omega_{\mathrm{LO}} + \Delta\omega) + Noise\,(\omega_{\mathrm{LO}} - \Delta\omega)}{P_{\mathrm{carrier}}(\omega_{\mathrm{LO}})}\right] = 10\log\left[\varphi_{\mathrm{rms}}^2\right] \tag{15.19}$$

Note that SSB phase noise is by far the most commonly used measure for phase noise and the subscript SSB is often ignored. From either the SSB or DSB phase noise, the *rms* jitter can be obtained as follows:

$$\varphi_{\mathrm{rms}}(\Delta f) = \frac{180}{\pi}\sqrt{10^{\frac{PN_{\mathrm{DSB}}(\Delta f)}{10}}} = \frac{180\sqrt{2}}{\pi}\sqrt{10^{\frac{PN_{\mathrm{SSB}}(\Delta f)}{10}}}\,\left[\mathrm{deg}/\sqrt{\mathrm{Hz}}\right] \tag{15.20}$$

It is also quite common to quote *rms* integrated jitter which is given by:

$$\mathrm{Jitter}_{\mathrm{rms}} = \sqrt{\int_{\Delta f_1}^{\Delta f_2} \varphi_{\mathrm{rms}}{}^2(f)df} \tag{15.21}$$

The limits of the integration are usually the offsets of the lower and upper frequencies of the bandwidth of the information being transmitted relative to the carrier frequency. In addition, it should be noted that dividing or multiplying a signal in the time domain also divides or multiplies the phase noise. Thus, if a signal is translated in frequency by a factor N, then the phase noise is related by:

$$\begin{aligned} \varphi^2{}_{\text{rms}}(N\omega_{\text{LO}}+\Delta\omega) &= N^2\cdot\varphi^2{}_{\text{rms}}(\omega_{\text{LO}}+\Delta\omega) \\ \varphi^2{}_{\text{rms}}\left(\frac{\omega_{\text{LO}}}{N}+\Delta\omega\right) &= \frac{\varphi^2{}_{\text{rms}}(\omega_{\text{LO}}+\Delta\omega)}{N^2} \end{aligned} \tag{15.22}$$

In the preceding equations, we have assumed that the circuit for frequency translation, a mixer for example, is noiseless. Also, note that the phase noise is scaled by N^2 rather than N because we are dealing with noise power in units of V^2 rather than noise voltage.

15.2.6 Noise in Phase-Locked Loops

A ***phase-locked loop*** (PLL) is the critical component in frequency synthesis and ***clock data recovery*** (CDR). A variety of factors influence the noise and spurious outputs of a PLL. Every component in a PLL contributes a certain amount of noise at the output. As shown in Figure 15.10, the phase noise sources include noise from the reference source θ_{REF}, noise from the phase detector θ_{PD}, noise from the reference divider θ_{R} and from the feedback divider θ_{N}, noise from the low-pass loop filter θ_{LPF}, and noise from the ***voltage controlled oscillator*** (VCO) θ_{VCO}.

In a feedback system, phase noise coming from different sources is independent and has different impact on the system noise performance. The total output noise PSD can be determined by [Rogers 2006]:

$$S_0(s) = \left[\frac{S_{REF}}{R^2}+S_R+S_N+\frac{S_{PD}+S_{LPF}}{k_{PD}^2}\right]\left(\frac{G}{1+G/N}\right)^2+S_{VCO}\left(\frac{1}{1+G/N}\right)^2 \tag{15.23}$$

where S represents the phase noise power spectral density, k_{PD} is the phase detector gain in V/rad, k_{VCO} is the VCO gain in rad/sec/V, G_{LPF} is the low-pass filter

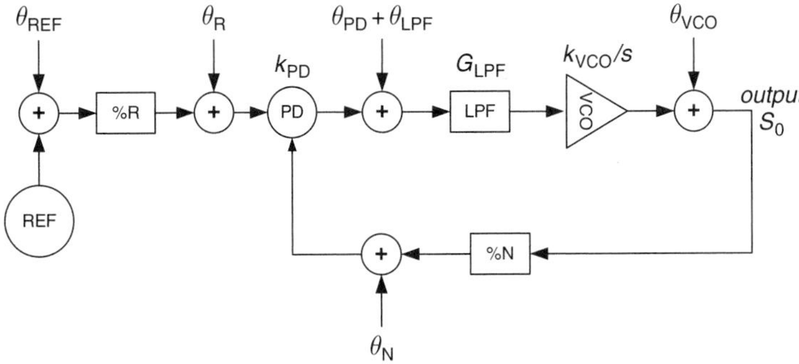

■ **FIGURE 15.10**

Additive noise sources in a phase-locked loop.

transfer function, $G(s) = k_{PD}G_{LPF}k_{VCO}/s$ is the open loop transfer function, and $\left(\frac{G}{1+G/N}\right) = \frac{Nk_{PD}G_{LPF}k_{VCO}}{Ns+k_{PD}G_{LPF}k_{VCO}}$ is the close loop transfer function, which approaches N at DC. Notice the loop has a high-pass effect on the VCO noise because of the term $\left(\frac{1}{1+G/N}\right) = 1 - \frac{1}{N}\left(\frac{G}{1+G/N}\right)$ and has a low-pass effect on other noise sources because of the term $\left(\frac{G}{1+G/N}\right)$. Thus, the in-band phase noise is determined by noise coming from the reference, phase detector, low-pass filter, and frequency dividers. On the other hand, the out-band phase noise is dominated by VCO noise, which determines the interchannel interference.

15.2.6.1 In-Band PLL Phase Noise

The in-band phase noise determines the close-in noise at the synthesized frequency. Suppose the closed loop transfer function has a flat response within the pass band and the loop bandwidth is sufficient to reject the VCO noise; we obtain the phase noise within the loop bandwidth as [Larson 1996]:

$$S_0(s) = \left[\frac{S_{REF}}{R^2} + S_R + S_N + \frac{S_{PD} + S_{LPF}}{k_{PD}^2}\right]\left(\frac{Nk_{PD}G_{LPF}k_{VCO}}{Ns + k_{PD}G_{LPF}k_{VCO}}\right)^2$$
$$\underset{s\to 0}{\rightarrow}\left[\frac{S_{REF}}{R^2} + S_R + S_N + \frac{S_{PD} + S_{LPF}}{k_{PD}^2}\right]N^2 \qquad (15.24)$$

Note that the PLL magnifies the noise from the reference, phase detector, ***low pass filter*** (LPF) and the dividers by the amount of *20logN* dB. The phase detector gain is normally given by $k_{PD} = {}^{V_{PD}}/_{2\pi}$ [V / rad], where V_{PD} is the phase detector output voltage. For V_{PD}= 500 mV, k_{PD}= 0.0796 [V/rad].

Crystal reference noise. Quartz crystal resonators are widely used in frequency control applications because of their unequaled combination of high Q, stability, and small size. The resonators are classified according to "cut," which is the orientation of the quartz wafer with respect to the crystallographic axes of the material. Some examples are AT-, BT-, CT-, DT-, and SC-cut, but they can also be specified by orientation, for example a +5° X-cut. Although a large number of different cuts have been developed, only AT- and SC-cuts are primarily used at frequencies above approximately 1 MHz (others either are used only at low frequencies and applications other than frequency control and selection, or have been made obsolete by later developments). In addition to the thermal noise with its floor around −160 dBm/Hz, $1/f$ noise exists in a crystal oscillator. The total noise power spectral density of a crystal oscillator can be determined by Leeson's formula [Watanabe 2000]:

$$S_{REF}(\Delta f) = 10^{-16\pm 1}\cdot\left[1 + \left(\frac{f_0}{2\Delta f \cdot Q_L}\right)^2\right]\left[1 + \frac{f_c}{|\Delta f|}\right] \qquad (15.25)$$

where f_0 is the oscillator output frequency, Δf is the offset frequency, Q_L is the loaded Q of the resonator, and f_c is the corner frequency between $1/f$ and thermal noise regions, which is normally in the range 1~10 kHz. Because the Q of a crystal resonator is large (*e.g.*, 1,000,000), the reference noise contributes only to the very close-in noise and it quickly reaches the thermal noise floor at the offset frequency around f_c.

Frequency divider noise. Frequency dividers consist of switching logic circuits that are sensitive to the clock timing jitter. The jitter in the time domain further converts to phase noise in the frequency domain. The origin of the time jitter/phase noise encountered with rising and falling edges of a digital divider is due to superimposed spurious signals such as Johnson and flicker noise in semiconductor materials or ambient effects (variation of the triggering level caused by temperature and humidity, etc.). Frequency dividers generate spurious noise especially at high frequency operation. Because of the frequency dependence of the timing jitter, the phase noise presented at the input of a divider will be reduced by factor of N at its output. An empirical formula for noise power spectral density of ***transistor-transistor logic*** (TTL) dividers is [Egan 1990] [Kroupa 2001] given by:

$$S_D(\Delta f) \approx \frac{S_D^{in}}{N^2} + \frac{10^{-14\pm1} + 10^{-27\pm1} f_{do}^2}{\Delta f} + 10^{-16\pm1} + 10^{-22\pm1} f_{do} \tag{15.26}$$

where S_D^{in} is the PSD of the input signal, f_{do} is the divider output frequency, and Δf is the offset frequency. Notice that the third term in Equation (15.26) represents the white thermal noise floor and the second term gives the flicker noise. Timing jitter, resulting from coupling, ambient, and supply variations, causes the last term.

Frequency/phase detector noise. Phase/frequency detectors suffer both flicker and thermal noise. The noise power spectral density of phase/frequency detectors is given by [Kroupa 1982]:

$$S_{PD}(\Delta f) \approx \frac{10^{-14\pm1}}{\Delta f} + 10^{-16\pm1} \tag{15.27}$$

From (15.27), a phase detector generates a white phase noise floor of around −160 dBm/Hz, which is thermal noise dominant, at large offset frequency. For example, passive phase detectors such as balanced mixers have excellent phase noise performance near −165 dBm/Hz.

Low pass loop filter noise. When the loop filter is a passive one (a simple RC lag or lag-lead network, for example), there are two major sources of noise. One source is 1/f noise generated by capacitors and resistors. The other source is thermal noise caused by LPF input resistance and possibly the decoupling resistance separating the varactor circuits from the loop filter. Thus, the noise power spectral density of a low pass filter can also be given by:

$$S_{LPF}(\Delta f) \approx \frac{10^{-14\pm1}}{\Delta f} + 10^{-16\pm1} \tag{15.28}$$

15.2.6.2 Out-Band PLL Phase Noise

As shown, the noise outside of the PLL bandwidth is determined by VCO phase noise, namely:

$$S_0(s) = S_{VCO}\left(\frac{1}{1+k_{PD}G_{LPF}k_{VCO}/Ns}\right)^2 \underset{s\to\infty}{\to} S_{VCO} \qquad (15.29)$$

Besides flicker noise and thermal noise, part of S_{VCO} comes from the spurious noise components principally introduced by variation at the VCO control line, power supply, package, and substrate coupling. In a feedback-based model, Leeson developed a concise expression of single-sideband VCO noise power spectral density [Leeson 1966]. Accuracy is improved by various extensions to Leeson's model [Lee 2000]. A widely accepted formula for the ratio of sideband power in a 1-Hz bandwidth at offset frequency Δf to total signal power is:

$$S_{VCO}(\Delta f) = 10\log\left\{\frac{FkT}{2P_s}\left[1+\left(\frac{f_0}{2\Delta f\cdot Q_L}\right)^2\right]\cdot\left[1+\frac{f_c}{|\Delta f|}\right]\right\} \qquad (15.30)$$

where f_0 is the oscillator output frequency and f_c is the boundary frequency between $1/\Delta f^2$ and $1/\Delta f^3$ regions, namely, the so-called flicker frequency. It should be pointed out that f_c is an empirical parameter dependent on processing. Q_L is the loaded Q of the resonant circuit, ranging from 5 to 20 for an on-chip resonator and 40 to 80 for an off-chip tank. P_s is the average signal power at output of the oscillator active device, and F is the oscillator effective noise figure, which is also an empirical fitting parameter. The term $\left(\frac{f_0}{2\Delta f\cdot Q_L}\right)^2$ shows that the voltage frequency response rolls off as $1/f$, which contributes only to the close-in phase noise. The term $2FkT$ represents the thermal white noise floor. Recall that the thermal noise equivalent power is $\overline{P} = 4kT\cdot BW$. For a unit bandwidth $BW = 1\,\text{Hz}$, the thermal noise power $\overline{P} = 1.66\times10^{-20}$[W], namely, $kT = -174\,\text{dBm/Hz}$. Assuming $P_s = 0\,\text{dBm}$ (this delivers about $224\,\text{mV}_{\text{rms}}$ into a 50 ohm load), all the values in [dBm/Hz] are thus equivalent to the values in [dBc/Hz]). For $F = 3\,\text{dB}$, we obtain the thermal noise floor at $-174\,\text{dBc/Hz}$.

15.2.6.3 Optimal Loop Setting

In the preceding sections, we discussed various sources of phase noise in a PLL. For PLL testing with minimal phase noise, we need to optimally choose PLL loop parameters such as the reference frequency, the loop division ratio, and the loop bandwidth. Based on the phase noise analysis, we summarize the PLL setting and testing rules as follows:

Division ratio N. The division ratio of the feedback divider, N, has a big impact on loop noise performance. First, the in-band phase noise is magnified by $20\log N$ dB. Second it worsens the PLL's capability of rejecting in-band VCO noise because $\left(\frac{1}{1+G/N}\right) = 1-\frac{1}{N}\left(\frac{G}{1+G/N}\right)$. Therefore, we should choose N as small as possible to minimize the in-band phase noise. However, a small N is not suitable to synthesize high frequency, because the output frequency of the synthesizer is given by $f_{out} = N\frac{f_{ref}}{R}$. Note that the minimal synthesizer step size is f_{ref} for $R = 1$. Therefore, choosing a large reference frequency to synthesize a high output frequency is also not acceptable when a small step size is required. This dilemma can be solved by using a fractional-N scheme with a fractional number of the step size. The major drawback of a fractional-N synthesizer is the fractional spurs that have to be canceled either by a high order loop filter or a sigma-delta noise shaper.

Reference frequency. The reference frequency influences the loop division ratio, the synthesizer step size, and the loop noise performance. Choosing a large reference frequency reduces the overall division ratio N for synthesizing the same LO frequency. On the other hand, a larger reference frequency results in a lower Q of the reference resonator and, therefore, higher reference noise. A small reference frequency with a high Q reference source is desirable only when very close-in phase noise (normally <1 kHz offset) or synthesizer step size is an issue.

Loop bandwidth. Loop bandwidth influences the loop settling time, stability, and loop noise performance. The wider the loop bandwidth is, the shorter the loop settling time is. As discussed before, the PLL has a low-pass filtering effect on the in-band phase noise from the reference, phase detector, loop filter, and frequency dividers. Therefore, narrow loop bandwidth benefits the in-band noise filtering. On the other hand, the PLL demonstrates a high-pass filtering effect on the out-band phase noise from the VCO. Therefore, wide loop bandwidth benefits the out-band noise filtering. To minimize the total PLL phase noise, the optimal loop bandwidth should be chosen around the cross point of the in-band phase noise spectral density curve and the out-band (VCO) phase noise spectral density curve. Moreover, loop filter bandwidth should be properly chosen so that the fractional spur can be canceled efficiently.

15.2.7 DAC Nonlinearity Testing

A DAC is one of the critical components in a mixed-signal system. The DAC is a nonlinear system with nonlinearity, distortion, and quantization noise resulting from finite phase and amplitude resolutions. The effect of finite amplitude word length generates random quantization noise. According to Equation (15.8), at the Nyquist rate, the integrated quantization noise-to-carrier ratio resulting from finite word length, D, is given by $3/2 \times 2^{2D} = 6.02 \times D + 1.76$ in units of dB.

On the other hand, because of the finite phase word length, a DAC shows up with the IMD effect that clock frequency f_s, output frequency f_o, and their harmonics tend to mix with each other and alias back to the Nyquist band. Generally speaking, those discrete aliased images cause the spurs at the following frequencies:

$$f_{image} = m \cdot f_S \pm n \cdot f_O \tag{15.31}$$

where m and n are integers. However, this is not always the case because the frequencies of the discrete spurs and their amplitude are also dependent on the ratio of the generated frequency, f_o, to the sampling clock frequency, f_S. For example, if f_S is an integer multiple of f_o, there will not be any spurs at the DAC's output caused by these image frequencies. Those spurs will be attenuated by the deglitch filter transfer function and the DAC's sample-hold function, $sinc(\pi \cdot f_o/f_s)$. In other words, the DAC's zero-order sample-and-hold imposes a *sinc* attenuation envelope to the fundamental, images, and harmonics of the DAC output as:

$$\text{Attenuation} = 20\log\left[\sin\left(\frac{\pi f_o}{f_S}\right)\Big/\frac{\pi f_o}{f_S}\right] \tag{15.32}$$

While considering the quantization noise resulting from both the finite phase bits and the finite amplitude bits, the worst-case spur magnitude at the DAC output can be estimated and given by [Dai 2006b]:

$$\text{Spur Magnitude[dBc]} = 10\log\frac{2}{3}\cdot\left(2^{-2P}+2^{-2D}\right) \tag{15.33}$$

where P is the number of phase bits and D is the number of DAC input bits. It should be noted that all of the preceding analysis is based on the perfect mathematical model of a DAC. But the real DAC is even more complicated and the signal quality of its output also depends on the quality of the DAC and filter designs as well as the phase noise of the clock frequency. Besides those, real DACs still suffer nonlinearities because of process mismatches, imperfect bit-weight scaling circuits, nonideal switching characteristics, etc.

Because spurs are such an important issue for a DAC, DAC tests are also concerned with output characteristics such as the SINAD and the ***spurious-free-dynamic-range*** (SFDR), defined as the ratio between the fundamental signal and the highest spurs.

15.3 ANALOG AND MIXED-SIGNAL TEST ARCHITECTURES

We now turn our attention to test architectures that can be implemented to test analog circuits or the analog portion of mixed-signal circuits. ***Digital signal processing*** (DSP) techniques have been used since the 1980s for fast and accurate testing of analog and mixed-signal circuits [Burns 2000]. In addition, the IEEE 1194.4 standard for a mixed-signal test bus [IEEE 1149.4-1999] can be implemented to improve controllability and observability of analog circuitry as well as to support mixed-signal BIST structures [IEEE1149.4 1999]. Therefore, the focus of the remainder of the chapter will be representative BIST approaches that can be implemented in mixed-signal SOCs to test analog circuitry. However, these test architectures can also be implemented on printed circuit boards for board or system-level testing.

The ***oscillation BIST*** (OBIST) approach [Arabi 1997] represents one of the rare cases where a test technique was originally developed to test analog circuitry and was later migrated to test digital circuitry. The basic idea of the OBIST approach is to reconfigure the analog circuit under test to form an oscillating circuit and to evaluate the frequency of oscillation in order to determine the faulty/fault-free status. The basic OBIST architecture is illustrated in Figure 15.11 where the circuits

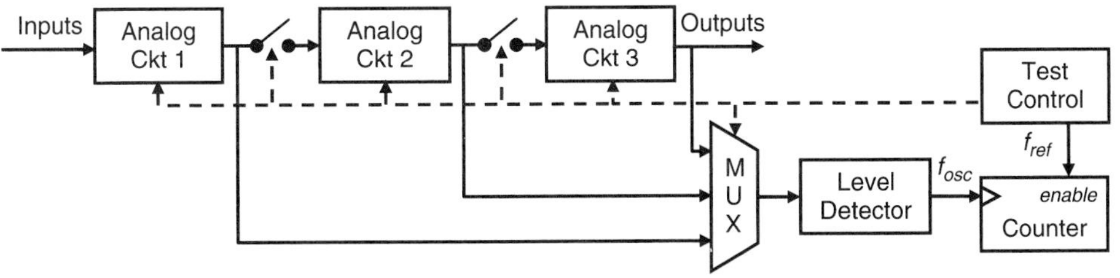

■ **FIGURE 15.11**

Oscillation BIST architecture.

to be tested are partitioned using programmable switches. Each analog circuit contains multiplexers that are used to create a feedback path in the circuit such that it will oscillate in the test mode. The output of each circuit under test is selected in turn by the multiplexer where the oscillations are converted by a level crossing detector to a square wave clock signal, which is, in turn, used to clock an M-bit counter. The test controller produces the enable to the counter. The resulting value in the counter is a function of the frequency of oscillation (f_{osc}) of the circuit under test with respect to the reference frequency (f_{ref}). The frequency of oscillation is a function of the structure and component values of each circuit under test. Once the acceptable range of counter values has been established for each circuit under test, faults are detected when the resultant counter value falls outside the good circuit range for that circuit under test.

The OBIST technique has been shown to give high fault coverage for structural (defect-oriented) testing [Arabi 1997]. Note that although the overall approach is inherently mixed-signal because of the digital test controller and counter, the approach does not require the implementation of DACs and ADCs in the system function; instead the ADC function is performed by the level crossing detector. In addition, there is no ***test pattern generator*** (TPG), because the circuit under test autonomously produces an output as a result of oscillations. However, there is additional circuitry, not shown in Figure 15.11, required in each circuit under test to reconfigure the circuit into an oscillatory condition. This circuitry must be applied on a case-by-case basis and could impact the performance of the circuit in its system mode of operation.

A typical mixed-signal BIST architecture is shown in Figure 15.12 where the DAC and ADC are assumed to be an existing and integral part of the mixed-signal system application functionality. The majority of the BIST circuitry is then added to the digital portion of the mixed-signal circuitry. The digital BIST circuitry includes TPG and ***output response analyzer*** (ORA) functions as well as a test controller. The analog loopbacks are analog multiplexers (denoted MUX1 and MUX2 in the figure) and are the only components associated with the BIST approach to be inserted in the analog domain. As a result, this minimizes the impact of the BIST circuitry on the operation and performance of the analog circuitry. The purpose of the analog loopback is to facilitate a return path for the test signals from the TPG, through

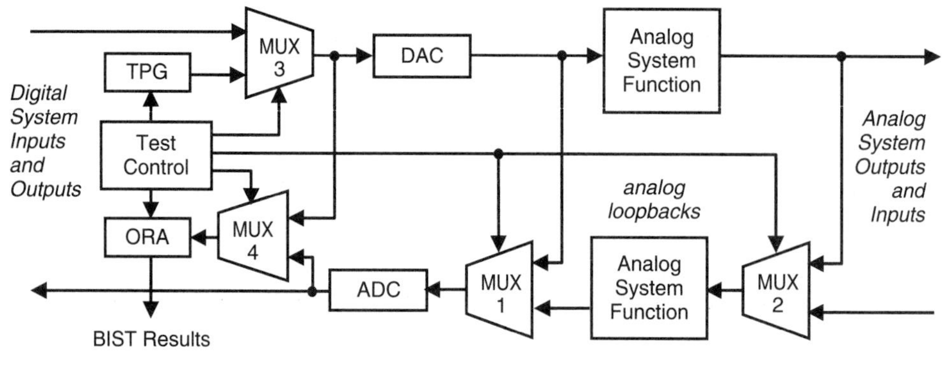

■ **FIGURE 15.12**

Typical mixed-signal BIST architecture.

the analog circuitry under test, and back to the ORA. An additional multiplexer (denoted MUX3 in the figure) is required for the insertion of the digital test patterns into, and isolation of unknown system data from, the input data stream to the DAC. Because the target circuitry under test is the analog system circuits, including the DACs and ADCs, the digital TPG and its associated multiplexer are incorporated immediately before the digital inputs of the DAC. Similarly, the digital ORA is incorporated at the output of the ADC. This basic architecture has been referred to as the ADC/DAC loopback BIST [Burns 2000], or simply ADC/DAC BIST, and has been implemented in a number of applications including those discussed in the subsequent sections.

Faults can be effectively isolated to a given section of analog circuitry within the diagnostic resolution of the analog loopback multiplexers. For example, with analog loopback MUX1 in Figure 15.12 activated, any faults detected are isolated to that path from the TPG to the ORA through the DAC and ADC. If the first BIST sequence indicates a good circuit, then analog loopback MUX1 can be deactivated while the analog loopback MUX2 is activated and the BIST sequence is reexecuted. Faults detected during this second BIST sequence would be isolated to the analog circuitry in the right-hand portion of Figure 15.12. Therefore, by adding more analog loopback multiplexers at strategic locations, the diagnostic resolution can be improved at the expense of area overhead and possible performance penalties in the analog portion of the mixed-signal system. However, it is also important to include the ability to drive the ORA with the outputs of the TPG, via MUX4, to verify that the digital portion of the BIST circuitry is fault-free before testing the analog portion of the mixed-signal system. A digital ORA facilitates reading BIST results directly through a system processor interface without the need for an ADC to retrieve the test results from the analog portion of the mixed-signal system.

To make the BIST circuitry usable in a system for offline testing and system diagnostics, the BIST circuitry must be capable of proper initialization of the analog circuitry under test, isolation of system data inputs, and reproducible results from

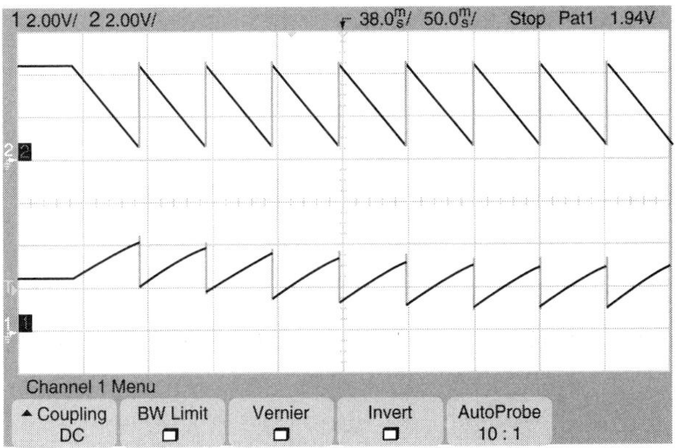

■ **FIGURE 15.13**

Example analog test waveform and output response.

one execution of the BIST sequence to the next. This functionality is typically performed by the test controller where control of the start and length of the BIST sequence is implemented by enabling output response compaction in the ORA. For example, the ability to specify the number of test waveform cycles used for initialization as well as the number of test waveform cycles used for the BIST sequence facilitates testing transient or steady-state responses in the analog circuit under test. This is illustrated in Figure 15.13 where a saw-tooth test waveform is applied to an inverting high-pass filter. As can be seen from the oscilloscope picture of the actual TPG waveform at the output of the DAC (top waveform) and the output response at the input to the ADC (bottom waveform), approximately seven cycles of the input test waveform are required before steady-state conditions are achieved. With an initialization count of zero cycles and a BIST sequence count of seven cycles, the entire transient response can be tested. Conversely, with an initialization count of seven, the BIST sequence only tests the steady-state response of the circuit.

An advantage of mixed-signal BIST approaches that use the architecture of Figure 15.12 is the ability to implement a parameterized VHDL or Verilog model of the BIST circuitry. This in turn provides quick and easy incorporation and subsequent synthesis of the BIST circuitry with the digital portion of any mixed-signal system and helps to minimize any adverse performance effects on the analog domain. The BIST circuitry can be easily customized for any particular system application through the specification of the model parameters. In addition to the standard cell based digital portion of a mixed-signal ***application specific integrated circuit*** (ASIC), the BIST circuitry can also apply to FPGA-based mixed-signal systems and mixed-signal SoC implementations that incorporate embedded FPGA cores. In these programmable logic cases, the BIST circuitry can be configured into the FPGA core only during offline testing to eliminate any area or performance penalties during normal system operation because the intended system function can be reconfigured in the FPGA core once BIST of the analog circuitry is complete.

15.4 DEFECT-ORIENTED MIXED-SIGNAL BIST APPROACHES

Early defect-oriented BIST approaches for the analog portion of mixed-signal systems attempted to make use of digital TPG and ORA functions that were typically used for testing digital circuitry. In the first mixed-signal BIST approach, the TPG consisted of a ***linear feedback shift register*** (LFSR) and the ORA consisted of an accumulator [Agrawal 1987]. Similarly, the second mixed-signal BIST approach used an LFSR-based TPG but also used an LFSR-based ORA for signature analysis [Ohletz 1991]. However, in traditional digital circuit signature analysis, the good circuit signature is based on the assumption that an exact output response sequence is obtained for every fault-free execution of the BIST sequence. In a mixed-signal system, the quantization (sampling) noise in the DAC and ADC as well as processing (*e.g.*, tolerances) and environmental (*e.g.*, temperature and voltage) variations in the analog circuitry can prevent an exact output response sequence from one execution of the BIST sequence to the next. As a result, a set of good circuit BIST signatures is difficult to obtain for the fault-free circuit with traditional signature analysis. The accumulator-based ORA [Agrawal 1987], on the other hand, sums the magnitude of the analog circuit output response and facilitates determination of a range of good circuit BIST signatures to account for acceptable changes in the output response caused by component, voltage, and temperature variations as well as quantization noise in the DAC and ADC.

LFSR-based TPGs generate pseudo-random digital patterns that look similar to white noise when passed through a DAC, considered by many to be a universal waveform for testing analog circuits [Pan 1995]. However, ramp input signals have been used in analog testing and have been found to provide good fault detection results and, in some cases, better results than sinusoidal test signals [Chatterjee 1996]. In addition, it has been observed that the detection of faults with respect to the input test signal can vary with the type of analog circuit under test [Balivada 1996]. Therefore, a variety of test waveforms is needed to provide a high probability of fault detection in a wide range of analog application circuits. As a result, some later mixed-signal BIST approaches implemented TPGs that provide a variety of test waveforms including pseudo-random, ramp, saw-tooth, triangular, step, pulse, and DC waveforms, as well as frequency sweeps using a square wave.

An example of this type of TPG is illustrated in Figure 15.14a, which includes a binary up/down counter that also functions as an LFSR [Stroud 2003]. The counter can be used to produce ramp, saw-tooth, and triangular waveforms. The bit reversal multiplexer reverses the order of the bits from the counter to the DAC (MSB becomes LSB and vice versa) and has the effect of producing test patterns with high frequency components that look like noise riding on lower frequency waveforms. When combined with a shift register and count value holding register, the counter can produce square waveforms that sweep through a frequency range. During the frequency sweep waveform generation, the counter starts counting from different initial values loaded from the count value holding register each time the counter completes a count cycle and generates a carry-out pulse. Simultaneously, the count value holding register is reloaded from the counter when the carry-out is shifted

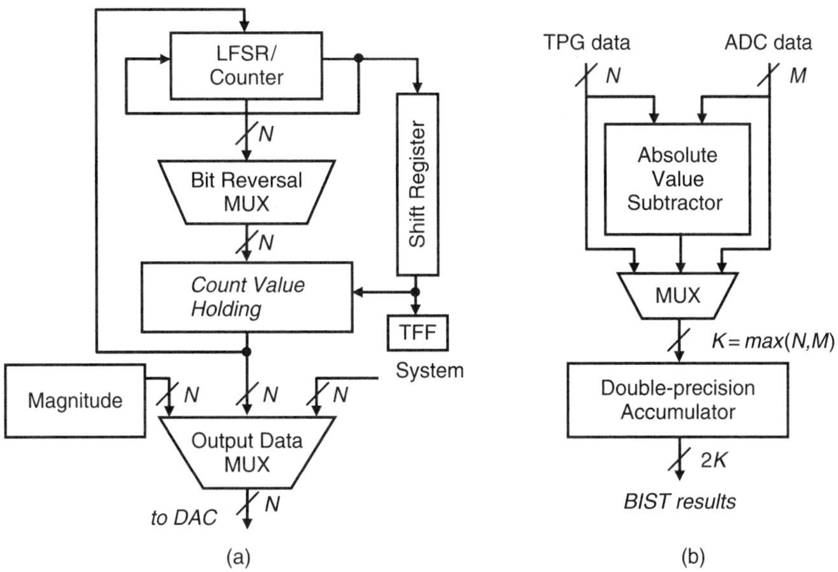

■ **FIGURE 15.14**

TPG and ORA block diagrams: (a) TPG and (b) ORA.

through the N-bit shift register. As a result, the contents of the count value holding register differs from the count value by a increment of N. For example, for a 5-bit shift register, an 8-bit counter will count 0–255, 5–255, 10–255, … , 245–255, and 250–255. A larger number of bits in the shift register will result in a faster frequency sweep. The output data multiplexer sets the magnitude for the square wave whenever the output of the toggle flip-flop (denoted TFF in the figure) is a logic one; otherwise the magnitude is zero. Enabling bit reversal during a frequency sweep will load nonsequential values into the count value holding register and make the square wave frequencies appear pseudo-random in nature.

Improvements in ORA implementations were also made in later mixed-signal BIST approaches to include summing the absolute value of the difference between the input test waveform and the output response of circuit under test [Stroud 1997]. An example of this type of ORA is shown in Figure 15.14b. This facilitates detection of faults that result in noise riding on an otherwise good output response signal, phase shifts, and overshoot or ringing. This is illustrated in Figure 15.15 where the output response to a step function has a delay that can be measured via the resultant sum of the absolute value of the difference between input and output waveforms. As a result, a fault causing excessive delay in the output response will be detected. Similarly, faults that result in undershoot or overshoot/ringing in the output response will also be detected as illustrated in Figure 15.15.

The fault detection capability of these BIST approaches is illustrated in Figure 15.16 taken from an actual BIST implementation with a LPF circuit under test [Stroud 2003]. The input test waveform is a saw-tooth produced by a binary count-up TPG function and is shown as the upper waveform in the oscilloscope

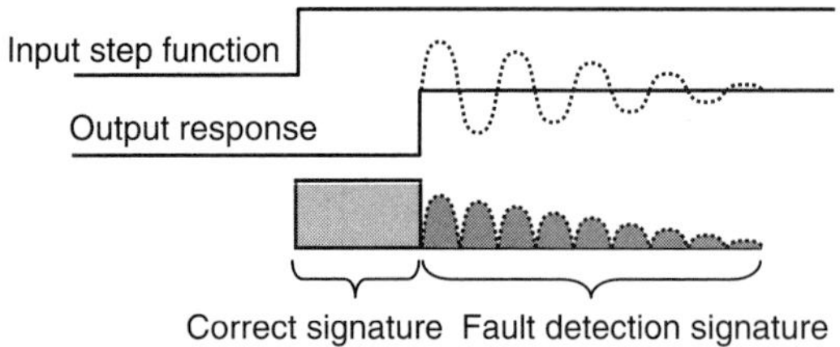

■ **FIGURE 15.15**

Absolute value summing.

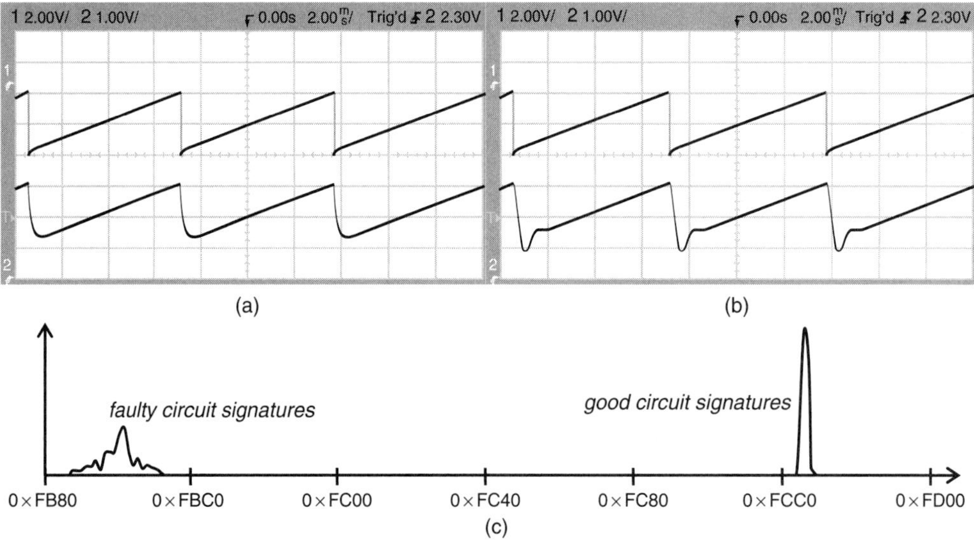

■ **FIGURE 15.16**

BIST fault detection in a low-pass filter: (a) faulty circuit response, (b) good circuit response, and (c) signature distributions.

pictures for a faulty (see Figure 15.16a) and a fault-free (see Figure 15.16b) LPF with the output response for each circuit shown as the lower waveform. The fault inserted in the LPF was an open feedback capacitor on the op-amp. The ORA in this case is a 16-bit accumulator, which provides a range of possible signature values from 0x0000 to 0xFFFF in hexadecimal. For 1000 executions of the BIST sequence, the good circuit signatures ranged from 0xFCCB to 0xFCCE while the faulty circuit signatures ranged from 0xFB92 to 0xFBA5 as shown in the signature distributions in Figure 15.16c. This example illustrates the fact that a range of good circuit signatures is needed to allow for acceptable signal variations in the analog circuit and shows that this particular fault is always detected by the saw-tooth test waveform because all faulty circuit signatures fall outside the good circuit signature range. If the good and faulty circuit signature distributions were to overlap, then the fault would be potentially detected with the probability of detection proportional to the percentage of faulty circuit signatures that lie outside the good circuit signature range [Stroud 2002].

15.5 FFT-BASED MIXED-SIGNAL BIST

Fourier analysis can be used to determine an analog signal's frequency content. Similarly, ***discrete Fourier transform*** (DFT) or ***fast Fourier transform*** (FFT) techniques can be used to determine the frequency response of discretely sampled analog signals. This section describes an FFT technique for testing analog components in a mixed-signal system.

15.5.1 FFT

An ADC can be used to sample an analog signal and produce a quantized version of the signal value at a discrete instant of time. The N-point DFT operation can be used to determine the frequency, or spectral, content of an N-element series of discretely (in-time) sampled signal values. The DFT operation is based on the following equation:

$$X[k] = \sum_{n=0}^{N-1} x[n] W_N^{kn} \tag{15.34}$$

where the output $X[k]$ is the complex value of the DFT at digital spectral frequency k, $x[n]$ is the complex value of the input signal at discrete time n, N is the total number of input signal points, and $W_N = e^{-j2\pi/N}$ is a complex function of N. Figure 15.17 shows an example of a quantized, discretely sampled sinusoidal time domain signal and the magnitude of its spectrum appears at its frequency illustrated in Figure 15.18 as calculated by the DFT. This relationship is key to spectral-based FFT analysis and signal generation.

To calculate the complex spectral frequency values of $X[k]$ using Equation (15.34) requires N complex multiplications and various other complex operations. Using Equation (15.34) to calculate the values for a series of M spectral frequencies will require $O(M \cdot N)$ complex operations (multiplications, additions, and subtractions).

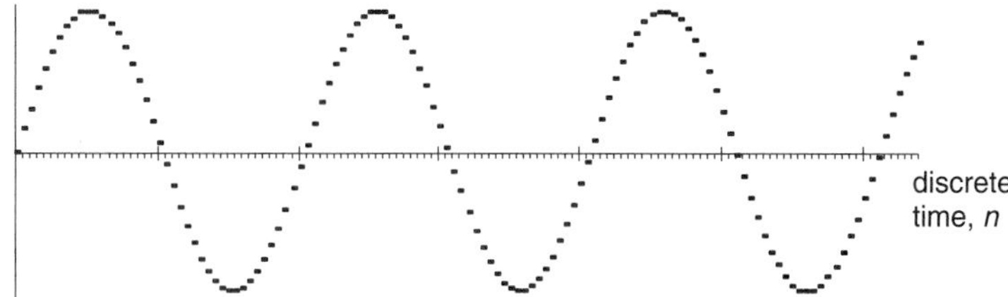

■ **FIGURE 15.17**

Example of discretely sampled input signal points.

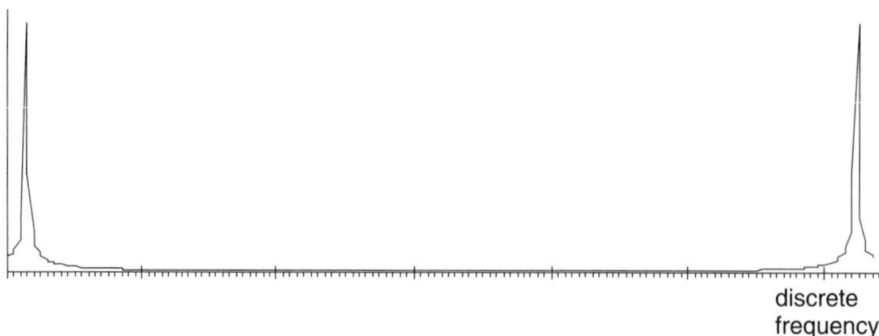

■ **FIGURE 15.18**

FFT magnitude plot for the signal in Figure 15.17.

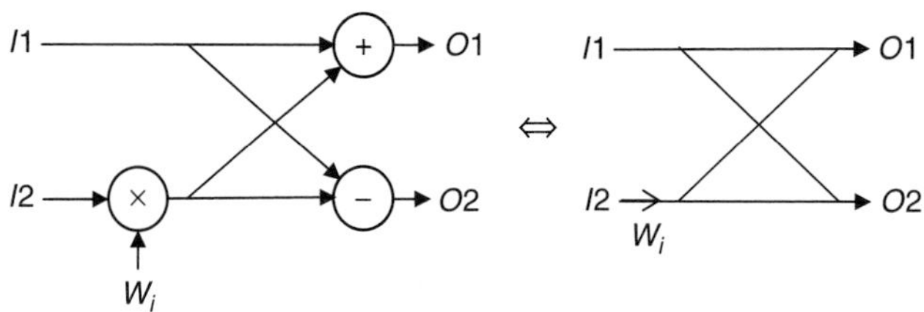

■ **FIGURE 15.19**

FFT butterfly processor.

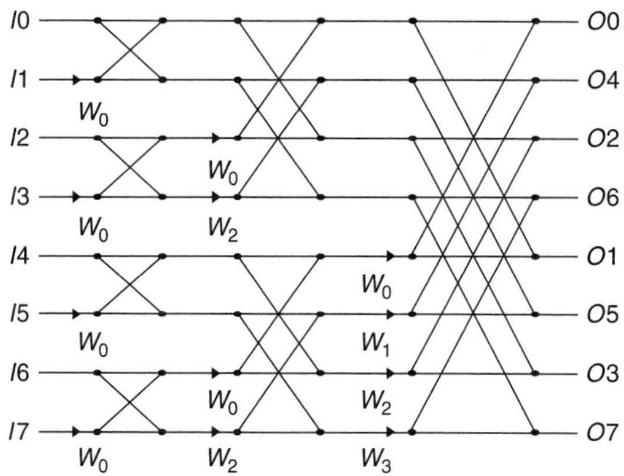

■ **FIGURE 15.20**

Example $N = 8$ point FFT butterfly network.

Often the value of M and N are the same, and the required complex operations reduces to $O(N^2)$. Cooley and Tukey developed the N-point FFT to eliminate duplicate complex operations and reduce the complexity to $O(N\log_2 N)$ [Cooley 1965]. Figure 15.19 illustrates two equivalent symbolic representations of the butterfly processor, which is the basic processing component of the FFT algorithm. The butterfly has three complex inputs and two complex outputs. The twiddle factor W_i is a function of W_N from Equation (15.34) as well as its location in the butterfly processing network, namely, $W_i = W_N^i$. Figure 15.20 shows an example how an 8-point FFT is performed using a butterfly processing network.

15.5.2 Inverse FFT

The ***inverse fast Fourier Transform*** (IFFT) operation is important to the FFT-based TPG and is based on the ***inverse discrete Fourier transform*** (IDFT) expressed as:

$$x[n] = \frac{1}{N}\sum_{k=0}^{N-1} X[k]W_N^{-kn} \tag{15.35}$$

where the output $x[n]$ is the complex value of the IFFT at discrete time n, $X[k]$ is the complex spectrum at digital frequency k, and N is the total number of spectral frequency values for the IDFT equation. By applying the IDFT operation to the spectrum shown in Figure 15.18, the original signal can be recovered as in Figure 15.17. Similar to the FFT, the IFFT also reduces the time complexity of the IDFT to $O(N\log_2 N)$.

15.5.3 FFT-Based BIST Architecture

Figure 15.21 shows the architecture for the FFT-based mixed-signal BIST approach where the digital TPG generates an m-bit digital test pattern. This test pattern is converted into an analog test signal by the DAC to stimulate the DUT. On the ORA side, the analog output of DUT is sent to the ADC and converted to an n-bit digital signal that is collected and analyzed by the digital ORA circuitry. The core component of the system is the digital TPG and ORA circuitry.

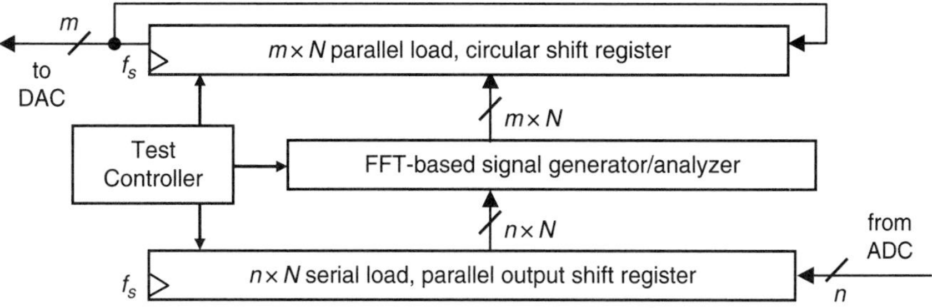

FIGURE 15.21

Example digital TPG/ORA architecture for FFT-based BIST.

The test controller sets the modes of operation and test to be performed, and handles test sequencing. It also controls the FFT processing block working in either TPG mode or ORA mode. The explicit connections for the controller are not shown in the figure.

In the TPG mode, the FFT-based signal generator/analyzer block performs the IFFT operation to create a coherent, multitone digital test signal, which is then loaded into the parallel load circular shift register with N words of m-bit resolution. Once the register is loaded with a coherent, multitone signal, the test control circuitry puts the register in shift mode, such that every clock cycle an m-bit digital test signal is shifted out to the DAC to generate the analog test stimuli and, at the same time, circularly shifted in the other side of the register.

In the ORA mode, the serial load parallel output shift register (with N words of n-bit resolution) locks in N samples of the digital test response signal from the ADC. These N samples are then used to perform the spectral analysis with the FFT operation by the FFT-based signal generator/analyzer block. Analysis of the spectral content of the sampled test response signal provides a wealth of information on the health of the analog circuit under test.

15.5.4 FFT-Based Output Response Analysis

For analog component ORA the FFT-based signal generator/analysis module is set to FFT mode to determine the frequency content of an analog signal. Typically, the input to the ORA comes from an ADC that is sampling the analog signal at frequency f_s, and the time delay between sampled points (sampling period) is $T_s = 1/f_s$. In this case, the imaginary part of each of the N complex inputs of the FFT is set to zero, so the N discrete inputs to the FFT are all real. The frequency ***bandwidth*** (BW) covered by the FFT is a function of the sampling frequency, f_s, of the ADC, and Nyquist defines the base BW of the FFT as $\text{BW} = f_s/2$. For a discrete N-point FFT the frequency spectrum is divided into N discrete frequencies or spectral bins. The FFT calculates a complex value for each spectral bin. The value in each bin represents how much signal is present in the frequency range for that bin. Each spectral bin k (where $k \in \{0, 1, 2, \ldots N/2 - 1\}$) has a frequency range $k \cdot f_s/N - f_s/N/2 \leq f_k \cdot \leq k \cdot f_s/N + f_s/N/2$. For real valued inputs, the values in the upper $N/2$ out of the N spectral bins are the complex conjugates of the values in the lower $N/2$ spectral bins. So usually, only the complex values in the lower $N/2$ out of N spectral bins are calculated. With this information the frequency resolution or the width of each spectral bin, f_s/N, can be determined. The frequency resolution can be improved by increasing the number of points in the FFT calculations. For example, if $f_s = 1$ GHz and $N = 256$, then the base BW is $f_s/2 = 500$ MHz and the frequency resolution is $f_s/N \cong 3.91$ MHz. To improve frequency resolution N can be increased to 512 and the frequency resolution becomes $f_s/N \cong 1.95$ MHz.

Spectral-based techniques allow use of relative signal values, such that accuracy of the signal source and measurement devices is not as critical as it is with absolute measurement techniques. These test techniques can be used to measure parameters like SNR, gain-tracking, frequency response, distortion, and group delay, and, as a result, apply to most processing type of analog components (such as filters, amps, etc.). Therefore, many spectral tests can be performed using the FFT-based BIST approach as follows.

FFT-based BIST can be used to measure a circuit's SNR. Hardware can be saved by eliminating the need for the square root function required to determine the actual magnitude; however, the resultant values will have a wider variation.

By controlling the magnitude of the input signal, FFT-based BIST can be used to determine the gain, $G = S_o/S_i$, where S_o is the output signal magnitude, and S_i is the input signal magnitude. The input signal, S_i, is generated at some frequency of interest, f. The magnitude of S_o is measured as the input signal magnitude, S_i, is increased. The spectral components of the output signal are used for gain comparison.

By using a single tone input signal whose frequency, f, is swept across a range of frequencies or a multitone input signal with frequencies $f_1, f_2, \ldots, f_n$, the circuit's frequency response can be measured. Using a multitone signal can save test time by eliminating the multiple runs required for a frequency sweep.

IMD tests are particularly well suited for the FFT-based technique. Using a two-tone input signal with frequencies f_1 and f_2, an analog component's IMD products, as shown in Figure 15.4 can be determined using spectral components of the output signal S_o. Group delay can also be determined using a two-tone test. By processing the real and imaginary components in two separate spectral bins at different points in time, the phase change, $\Delta\phi$, can be calculated. The group delay can then be calculated using $\Delta\phi/\Delta f$.

15.5.5 FFT-Based Test Pattern Generation

The architecture presented in this section makes use of the FFT not only for signal analysis but also for on-chip multitone signal generation [Emmert 2003, 2005, 2006]. For digital systems, the on-chip implementation of an FFT circuit can be costly relative to circuit area or data processing time; however, many mixed-signal circuits (especially for DSP applications) already have an FFT function available. Even if the FFT is not available, there are many robust approximation techniques that minimize the area overhead required for on-chip implementation. The rest of the section describes FFT-based multitone signal generation for on-chip BIST applications.

The basis for the FFT-based signal generation is the IFFT operation, which can be realized with the same functional block for FFT operation. Comparing Equation (15.34) and Equation (15.35), there are two differences between FFT and IFFT. The first difference is the scaling factor $1/N$ in front of the summation, and the second difference is a minus sign in the exponent of the kernel function. In fixed-point arithmetic, the scaling factor can be fulfilled with a $\log_2 N$ right shift of the bits and no extra hardware is required. For the second difference, it should be noted that all FFT (or IFFT) kernel points are equally spaced around the complex unit circle. It does not matter if we go clockwise or counter clockwise around the complex circle; we end up with the same points. So the extra minus sign for signal generation can be ignored. Therefore, the exact same FFT block can be used to perform the IFFT operation for signal generation as well.

To generate a multitone signal with j-tones, a series of j complex values with magnitude of 1 is created. It should be noted that this value of 1 is arbitrary, and to change the relative contribution of a tone to the overall signal, its relative magnitude in the series can be varied. Additionally, the phases of different tones can be controlled by real and imaginary parts of each of the j complex values.

Next, an N-element complex series is created with every element initialized to 0. Depending on the actual frequency bins of the j tones, the corresponding elements of the series are then assigned with the other series given in the previous section and used as input to the FFT for signal generation. Figure 15.22 shows examples of the real and imaginary inputs as well as the resulting digitized signal for a $j =$ two-tone setting with frequencies at bin 1 and bin 4. Because a digitized j-tone signal repeats itself every period, it can be loaded into a fast circular shift register to create continuous high-speed digitized test stimuli. Once this is done, the FFT is free to process the digitized output of the analog circuit under test in ORA mode.

Figure 15.23 shows an example coherent signal generated using frequencies corresponding to spectral bins 4 and 12. The generated two-tone digital test

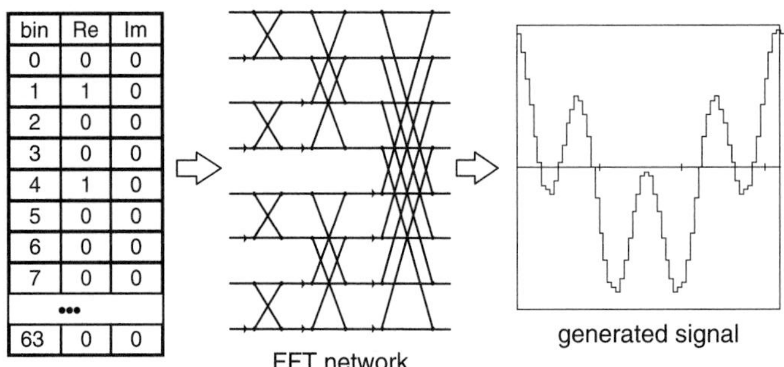

■ **FIGURE 15.22**

Example $N = 64$ point, coherent test signal generated by the IFFT mode.

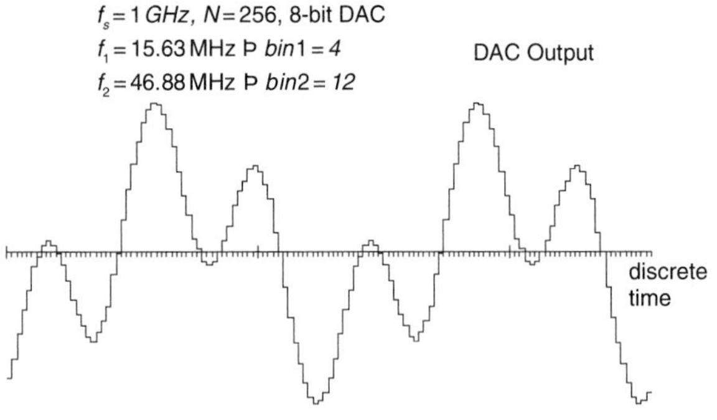

■ **FIGURE 15.23**

Example $N = 256$ point, coherent test signal generated in IFFT mode.

signal is suitable for loading into the high-speed TPG register. As the signal wraps around, the coherent ends match up. If the DAC clock frequency is set to 1 GHz, a continuous signal with frequencies 15.63 MHz and 46.88 MHz is produced.

15.6 DIRECT DIGITAL SYNTHESIS BIST

Direct digital synthesis (DDS) is an important frequency synthesis technique that provides low-cost waveform generation with ultrafine resolution. As shown in Figure 15.24, a conventional DDS includes a digital accumulator that generates the phase word based on the input frequency word. The synthesizer step size is defined as $f_{clk}/2^n$ where n is the number of bits in the accumulator. Fine resolution can thus be achieved using a large accumulator size. The DDS utilizes a ***lookup table*** (LUT) to convert the phase word to a sinusoidal amplitude word, whose length is normally limited by the finite number of input bits of the DAC. Deglitch filters are added after the DAC to remove the spurious components generated in the data conversion process. Although a pure sinusoidal waveform is desired at the DDS output, spurious tones can occur mainly because of the following two nonlinear processes. First, to reduce the ***read-only memory*** (ROM) size of the LUT, the phase word needs to be truncated before being used as the ROM address. This truncation process introduces quantization noise, which can be modeled as a linear additive noise to the phase of the sinusoidal wave. Second, the ROM word length is normally limited by the finite number of bits of the available DAC. In other words, the sinusoidal waveform can be expressed only by words with finite length, which intrinsically contains quantization error additive to the output amplitude. In the case of DDS, the quantization errors are caused by finite phase resolution, e_p, and finite amplitude resolution, e_A, where e_A is the same as e in Section 15.2.4. For e_p, oversampling is employed in DDS, allowing noise-shaping techniques to be used to shift the phase quantization error to a higher frequency band where the noise can

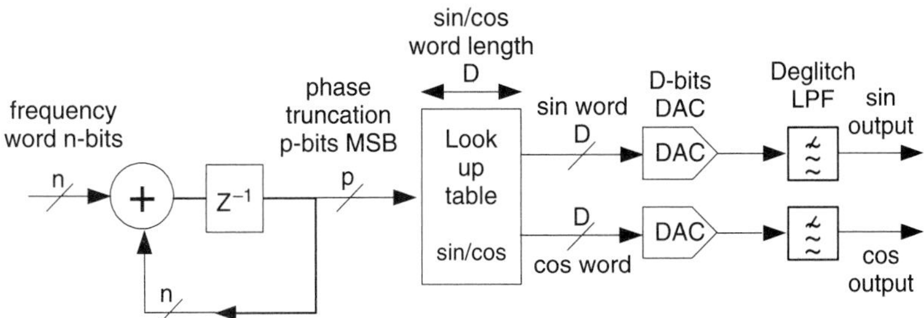

■ **FIGURE 15.24**

Direct digital synthesizer (DDS) for test signal generation.

be removed by the deglitch filter after the DAC [Dai 2004]. Assuming e_p is small relative to the phase, the DDS output can be determined as:

$$\begin{aligned} A_{out} &= A\sin\left(\frac{2\pi Wi}{2^n} + e_p(i)\right) + e_A(i) \\ &\approx A\sin\left(\frac{2\pi Wi}{2^n}\right) + Ae_p(i)\cos\left(\frac{2\pi Wi}{2^n}\right) + Ae_A(i) \end{aligned} \tag{15.36}$$

Analog functional testing requires fine frequency resolution and fast frequency switching time to perform tests such as frequency response and linearity measurements. The resolution and switching speed requirements of an analog BIST system surpass the performance capabilities of conventional analog PLLs. The conventional PLL-based frequency synthesizer has difficulty meeting these requirements because of internal loop delay, low resolution, and limited tuning range of the VCO. In contrast, DDS generates a digitized waveform of a given frequency by accumulating phase changes at a higher clock frequency. Because there is no feedback in a DDS structure, it is capable of extremely fast frequency switching or hopping at the speed of the clock frequency. DDS provides many other advantages including fine frequency-tuning resolution, continuous-phase switching, and various modulations. Thus, it provides a low-cost digital approach to frequency, phase, and amplitude modulations, eliminating the costly analog modulators associated with many analog measurements. The modulated waveform generation is a unique feature of the DDS-based BIST approach.

On the other hand, the DDS has two major deficiencies that are related to the inadequacy of the semiconductor technology. The first deficiency is that the output spectrum of the DDS is normally not as clean as the PLL output. The DDS quantization noise floor is limited by the finite number of amplitude bits (DAC input bits) and the finite number of phase bits. A 12-bit DAC provides a theoretical SNR of 72 dB, which is less than that of a typical PLL synthesizer. The DDS also suffers from a high level of spurious output, derived from the discrete phase accumulation and phase truncation processes as well as the DAC nonlinearity. The second deficiency is that the DDS output frequency is limited by the maximum operation frequency of the DAC and the digital logic. Although DACs with GHz sampling frequencies have been reported, they normally consume a large amount of power with poor resolution. Therefore, the DDS can be used to generate the fine-tuned frequency followed by a RF mixer and PLL used to up-convert the DDS output to the RF frequency. With careful design, the best of both DDS and PLL can be achieved [Dai 2006b].

15.6.1 DDS-Based BIST Architecture

The complete DDS-based BIST architecture is illustrated in Figure 15.25 [Dai 2006a]. The TPG provides precise frequency tone sweeps by controlling the frequency word. It can also generate quadrature phase sinusoidal waveforms simultaneously by shifting the two MSBs of the phase word. As we will show, the DDS can be used to generate the two-tone test stimuli required for IP3 measurements.

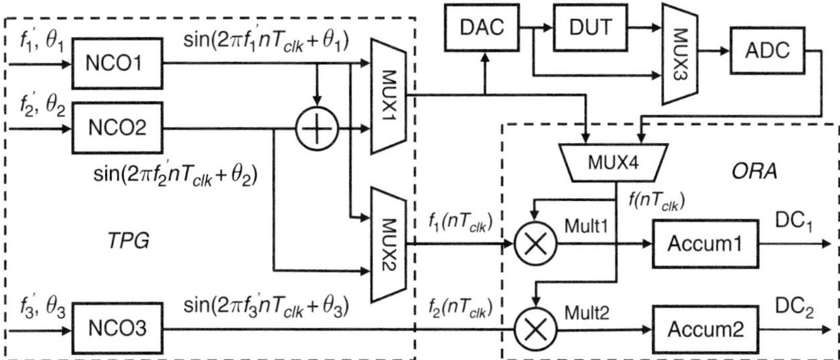

■ **FIGURE 15.25**

DDS-based BIST architecture.

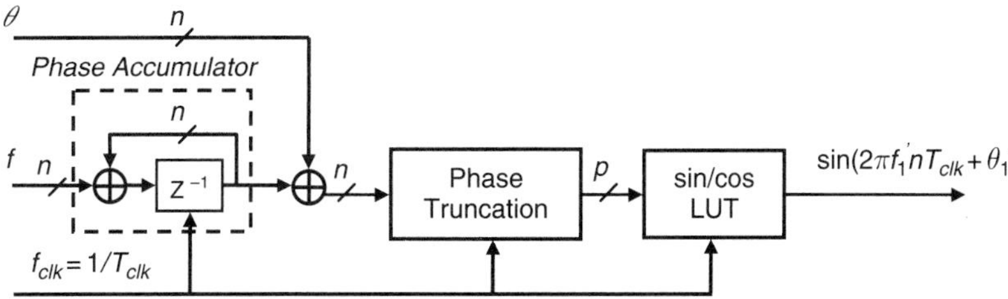

■ **FIGURE 15.26**

Numerically controlled oscillator for TPG.

The area penalty associated with the DDS approach is minimized by the delta-sigma noise shaping scheme [Dai 2006b].

The DDS-based TPG consists of three ***numerically controlled oscillators*** (NCOs) and utilizes the existing DAC from the mixed-signal system to complete the DDS. Figure 15.26 gives a more detailed view of the NCO implementation used in the TPG. The phase accumulator is used to generate the phase word based on the frequency word f and the initial phase word θ. The NCO then utilizes a LUT to convert the truncated phase word sequence to a digital sine wave sequence where the output sine wave frequency is determined as:

$$f' = \frac{f \cdot f_{clk}}{2^n} \tag{15.37}$$

where n is the word width of the phase accumulator. The phase truncation noise introduced in the NCO can be reduced using a sigma-delta modulator. The BIST system shown in Figure 15.25 requires three NCOs to generate three test tones.

To save area, the sine LUT of three NCOs can be shared by time-multiplexing the LUT input and output. Thus, three NCOs contain only three phase accumulators and one sine LUT. It should be noted that the sine LUT consumes the majority of the NCO area.

The ORA incorporates two multiplier/accumulators, each consisting of a D-bit multiplier (where D is the number of bits from the ADC) and a $2D+M$-bit accumulator (where the number of samples to be accumulated is less than 2^M). A 2's complement transformation is performed on negative numbers entering the accumulators, Accum1 and Accum2, such that subtraction is accomplished by the adders in the accumulators. In addition, the DDS input to the two multipliers is converted to a signed magnitude number to remove any DC offset from the DDS output that could affect the accuracy of the measurement. The multipliers in the ORA serve as down-converters to selectively pick the frequency components and down-convert them into DC signals. The DC levels are then compacted by the accumulators for evaluation during various tests and measurements as described in the following subsections.

15.6.2 Frequency Response Test and Measurement

One of the major problems associated with integrated analog filters is the cutoff frequency variation caused by temperature, supply voltage, and process variations. If the cutoff frequency can be monitored on the fly during transmission idle periods, its variation can be compensated using built-in tunable circuitry in LPF designs. In addition to production testing, the frequency response monitoring can also be used to adjust the gain and bandwidth of the amplifier for multiband and multistandard applications. With wireless standards operating in very different frequency bands, market-leading wireless solutions have to offer multimode interoperability with transparent worldwide usage. Thus, the base-band gain stage needs to be tunable for different wireless standards. A DDS-based BIST approach can be used to calibrate the frequency response of the base-band gain stage and LPF in this connection.

Frequency response (both gain and phase response) is the key measure for integrated LPFs and amplifiers. The cutoff frequency of the filters and amplifiers can be found by measuring the passband and stopband amplitude response, whereas the group delay can be determined from the phase response. To test the base-band LPF in a transceiver RFIC, the DDS integrated in the base-band ASIC generates a single frequency tone that loops back from transmitter to receiver through multiplexer controls. The DDS generates frequency tones with fine resolution and can scan the pass and stop bands of the LPF with fine step size to measure the cutoff frequency, pass band, and stop-band ripples of the filter. However, there is normally a phase difference between the external path through the DUT and the internal path from the test generator to the test analyzer, so phase correction needs to be done before the frequency magnitude measurement.

To measure the frequency response, the DDS generates the test tone of $x(t) = A\cos\omega t$ that is applied to the input of an amplifier with transfer function $y(t) = [\alpha_0 + \alpha_1 x(t) + \alpha_2 x^2(t) + \alpha_3 x^3(t) + \cdots]\ |\exp(j\Delta\phi)|$, where $\Delta\phi$ denotes the phase delay through the amplifier and the coefficients α_j are time invariant. Hence, the amplifier output is given by:

$$y(t) = \cos(\Delta\phi)(\alpha_0 + \alpha_1 A\cos\omega t + \alpha_2 A^2\cos^2\omega t + \alpha_3 A^3\cos^3\omega t + \cdots)$$

$$\approx \cos(\Delta\phi)\left\{\alpha_0 + \frac{\alpha_2 A^2}{2} + \left(\alpha_1 A + \frac{3\alpha_3 A^3}{4}\right)\cos\omega t + \frac{\alpha_2 A^2}{2}\cos 2\omega t + \frac{\alpha_3 A^3}{4}\cos 3\omega t\right\} \quad (15.38)$$

Note that if the input signal A is large, the nth harmonic grows approximately in proportion to A^n. Under a small-signal assumption, for example, the input signal A is small, the system is linear, the harmonics are negligible, and the small-signal gain is α_1. For a large signal, nonlinearity becomes evident and the large-signal gain is $\alpha_1 + 3/4(\alpha_3 A^3)$, which varies when the input level changes. If $\alpha_3 < 0$, the output is a "compressive" or "saturating" function of the input signal, namely, the gain is compressed when the input magnitude A increases. For a small input, the linear transfer function of the DUT is:

$$y(t) \approx \cos(\Delta\phi)(\alpha_0 + \alpha_1 A\cos\omega t) \quad (15.39)$$

In the ORA, the amplifier output is mixed (multiplied) with the test frequency. Assume the test tone to be mixed with the DUT output response being of the form A$\cos(\omega t)$. Accumulating the mixer output, we can obtain a DC term given as follows:

$$DC_1 \approx \frac{1}{2}\alpha_1 A^2\cos\Delta\phi\cdot n \quad (15.40)$$

where n is the number of accumulation clock cycles, and $\Delta\varphi$ is the phase difference between the external and internal path. The amplifier output is also mixed (multiplied) with a test tone of $A\sin(\omega t)$. This mixing process produces another DC term:

$$DC_2 \approx \frac{1}{2}\alpha_1 A^2\sin\Delta\phi\cdot n \quad (15.41)$$

Thus, the phase difference $\Delta\phi$ can be determined by:

$$\Delta\phi = \mathrm{tg}^{-1}\frac{DC_2}{DC_1} \quad (15.42)$$

Once the phase difference is measured, the test tone generated by DDS for the frequency response can be phase-adjusted such that the signals at the mixer inputs can be perfectly in-phase. In this connection, DDS should generate test tones in the form of $x(t) = A\cos(\omega t)$ for the DUT and $A\cos(\omega t\text{-}\Delta\phi)$ for the mixer input in the ORA, respectively. Additional phase can be easily added to the phase word in the DDS architecture as shown in Figure 15.26. The amplifier may not have a constant group delay, namely, the delay through the DUT is normally frequency dependent,

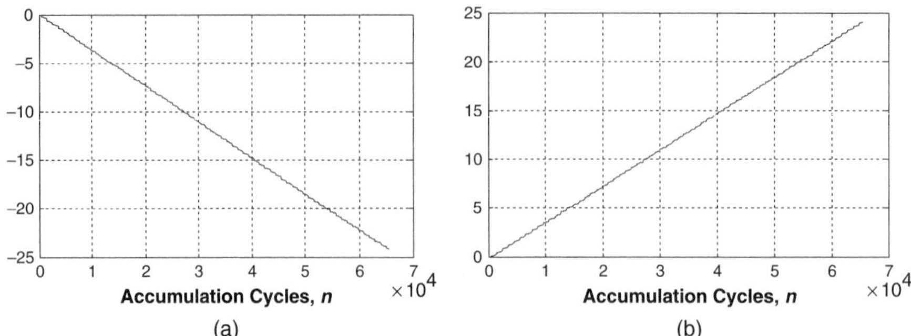

FIGURE 15.27

Accumulated DC_1 and DC_2 components for $\Delta\phi = 135°$: (a) DC_1 output mixed with $A\cos(\omega t)$ and (b) DC_2 output mixed with $A\sin(\omega t)$.

and therefore the phase correction should be performed at each frequency step when DDS generates the test tones that scan the interested band.

Figure 15.27a shows the ORA accumulated DC_1 component of the DUT output mixed with $A\cos(\omega t)$, and Figure 15.27b shows the ORA accumulated DC_2 component of the DUT output mixed with $A\sin(\omega t)$. In both cases, the phase difference $\Delta\phi = 135°$. Notice the slope of DC_1 is negative because of the $\cos\Delta\phi$ term in Equation (15.40), whereas the slope of DC_2 is positive because of the $\sin\Delta\phi$ term in Equation (15.41). Based on the sign of DC_1 and DC_2, we can thus determine the quadrant of the phase difference. Once the phase difference is determined, the actual phase corrected frequency response can be found from either of the DC_1 or DC_2 measurements as:

$$DC = \frac{DC_1}{\cos\Delta\phi} = \frac{DC_2}{\sin\Delta\phi} \tag{15.43}$$

For better accuracy, the DC term with the larger slope is used for calculations. For certain phase differences such as angles close to a multiple of 90°, one of the DC terms will approach zero. In those cases, the relative phase difference $\Delta\phi$ can be adjusted by the desired phase delay θ at the input to the DDS as shown in Figure 15.26.

15.6.3 Linearity Test and Measurement

For linearity measurements with this BIST approach, the following technique is used for the ORA. As Figure 15.4 shows, the closest intermodulation terms to the fundamental are the IM3 terms with frequencies at $2\omega_1 - \omega_2$ and $2\omega_2 - \omega_1$. First, mixing (multiplying) the amplifier output with fundamental tone $A_2\cos\omega_2 t$, produces a DC term:

$$DC_1 = \frac{1}{2}A_2{}^2\left[\alpha_1 + \frac{3}{4}\alpha_3(2A_1{}^2 + A_2{}^2)\right] \approx \frac{1}{2}A_2{}^2\alpha_1 \tag{15.44}$$

where the second term $\frac{3}{4}\alpha_3(2A_1{}^2 + A_2{}^2)$ is normally much smaller than the linear gain, α_1, if the input level is small, such that the amplifier is not desensitized.

Second, mixing (multiplying) the amplifier output with the IM3 tone $A_1\cos(2\omega_2 - \omega_1)t$, produces another DC term:

$$DC_2 = \frac{3}{8}A_1{}^2A_2{}^2\alpha_3 \tag{15.45}$$

Expressing these two DC terms in dB, the difference, ΔP, between the fundamental and the IM3, and thus the IIP3, can be measured using Equation (15.3). Although we can represent dB units using floating-point format, it is not necessary to find the actual IP3 value using real hardware in the ORA for a BIST implementation. Instead, ranges of acceptable values can be precalculated and compared to DC_1 and DC_2 values for a pass-fail BIST indication. For characterization of the circuit, the accumulated values can be read and averaged off-chip to perform the actual IP3 calculation. The complete TPG and ORA for the linearity BIST architecture are illustrated in Figure 15.25. It should be noted that the BIST circuitry for frequency response measurement is a subset of that required for linearity measurement.

15.6.4 SNR and Noise Figure Measurement

For SNR and noise figure measurements, all three DDSs illustrated in Figure 15.25 are used, but these measurements use a subset of the circuitry of the linearity measurement circuitry, as the adder is not needed to superimpose the two test tones. During this measurement, NCO1 produces a constant test waveform at frequency f_1. The resultant sine wave, $\sin(f_1)$, is applied to the DUT for the entire test. The other two DDSs are used to produce a sine wave and cosine wave at frequency f, similar to the frequency response measurement described previously. The output response of the DUT $y(t)$ is multiplied by both $\sin(f)$ and $\cos(f)$ in the multipliers Mult1 and Mult2, respectively. Then the outputs of the multipliers are accumulated in Accum1 and Accum2 to extract DC_1 and DC_2, respectively. The phase and gain of $y(t)$ with respect to $\sin(f)$ is then calculated by the same method as in the frequency response measurement. Again, phase and gain are measured at the same time for frequency f, as DC_1 and DC_2 are extracted simultaneously. This gives the amplitude of the noise produced in the system at frequency f. To obtain the complete signal-to-noise ratio, a series of phase and gain measurements is made by changing the value of f during each measurement such that we sweep through the various frequencies (excluding frequency f_1), averaging the measured noise. A phase and gain measurement is made at frequency f_1 in order to obtain the gain of the signal at frequency f_1 with the final SNR obtained by dividing the gain of the signal by the average of the noise.

15.7 CONCLUDING REMARKS

Although analog ICs and the analog portion of mixed-signal ICs consist of fewer components compared to their digital counterparts, testing analog and mixed-signal devices is more complex regardless of whether the testing approach is specification oriented or defect oriented. This chapter has presented some of the important distinctions between these two approaches and some of the fundamental tests that are applied to analog circuits. In addition, this chapter has presented some representative BIST techniques that have been developed for both functional and structural testing that can be incorporated in mixed-signal IC and SOC implementations to test analog modules. Because the incorporation of BIST circuitry incurs area and associated cost penalties, there are many applications, such as low-cost consumer electronics, in which it may not be worthwhile to consider BIST. On the other hand, high-reliability applications, such as space and medical electronics, are excellent candidates for the incorporation of BIST techniques.

The implementation and operation of the BIST approaches presented in this chapter provide insight into many of the issues that must also be considered for more traditional, non-BIST approaches to analog testing. DSP and FFT-based approaches provide a wealth of test and measurement capabilities and form the core of spectrum analyzers, which, in turn, are a key component of external testing. For SOC implementations, incorporation of an FFT can be costly in terms of area, particularly for high-resolution frequency discrimination. However, when incorporated in an SOC for DSP system applications, not only can the FFT be used for output response analysis, but it can also be extended to test pattern generation during functional testing. The DDS-based approach with multiplier/accumulator ORA provides a considerable reduction area overhead at the expense of test time, as only one frequency component of the analog output response can be analyzed at a time. Otherwise, the FFT and DDS approaches provide similar functional test and measurement capabilities. One of the major advantages of functional testing over structural testing is the ability to calibrate and compensate analog circuit performance in addition to fault detection.

15.8 EXERCISES

15.1 **(Analog Functional Testing)** An amplifier operates at 2 GHz with a gain of 10 dB. A two-tone test with equal power is applied at the input; one is at 2 GHz and another is at 2.01 GHz. At the output, four tones are observed at 1.99, 2.0, 2.01, and 2.02 GHz. The power levels of the tones are −70, −20, −20, and −70 dBm. Determine the IIP3 and 1-dB compression point for this amplifier using the two-tone test.

15.2 **(Analog Functional Testing)** A sampled system samples a 4 MHz video signal with an 8-bit video ADC with sampling clock at 32 MHz. Determine the oversampling ratio and the system SNR.

15.3 **(Defect-Oriented BIST Output Response Analysis)** Assuming an N-bit ADC and an M-bit accumulator for the ORA, what is the maximum number of BIST clock cycles that can be applied without the possibility of accumulator rollover?

15.4 **(FFT BIST Signal Analysis)** Given an ADC with sampling frequency, $f_{sampling} = 250\,\text{MHz}$ and a 256-point FFT, what is the frequency resolution for FFT based signal analysis?

15.5 **(FFT BIST Signal Analysis)** Given an ADC with sampling frequency, $f_{sampling} = 1\,\text{GHz}$, a 256-point FFT, and output peak signal frequency at 125 MHz, what spectral bin should the output test signal fall into?

15.6 **(FFT BIST Signal Generation)** Given a high-speed circular shift register and DAC combination running at $f_{CLK} = 100\,\text{MHz}$ and a 256-point IFFT block, determine which spectral bin numbers to set to nonzero values to create a two-tone analog test signal with frequencies of approximately 5 MHz and 13 MHz.

15.7 **(DDS BIST Phase Delay)** Show that the accumulator values given in Figure 15.27 result in a phase delay measurement of $\Delta\phi = 135°$.

15.8 **(DDS BIST Gain Measurement)** Determine the corrected gain for the frequency response from either DC_1 or DC_2in Figure 15.27.

Acknowledgments

The authors wish to acknowledge Professor John (Marty) Emmert of Wright State University, Department of Electrical Engineering, for contributing the section on FFT-Based Mixed-Signal BIST, as well as Professor Robert Dean and Jie Qin of Auburn University, Department of Electrical and Computer Engineering, for their assistance during preparation of this chapter.

References

R15.0 Books

[Agrawal 1987] V. Agrawal, *BIST at Your Fingertips Handbook*, AT&T, Murray Hill, NJ, 1987.

[Burns 2000] M. Burns and G. Roberts, *Introduction to Mixed-Signal IC Test and Measurement*, Oxford University Press, New York, 2000.

[Larson 1996] L. Larson, editor, *RF and Microwave Circuit Design for Wireless Communications*, Artech House, Boston, 1996.

[Mahoney 1987] M. Mahoney, *DSP-Based Testing of Analog and Mixed-Signal Circuits*, IEEE Computer Society Press, New York, 1999.

[Roberts 1995] G. Roberts and A. Lu, *Analog Signal Generation for Built-In Self-Test of Mixed-Signal Integrated Circuits*, Springer, Boston, 1995.

[Rogers 2006] J. Rogers, C. Plett, and F. Dai, *Integrated Circuit Design for High-Speed Frequency Synthesis*, Artech House, Boston, 2006.

[SIA 2005] SIA, *The International Technology Roadmap for Semiconductors: 2005 Edition*, Semiconductor Industry Association, San Jose, CA (http://public.itrs.net), 2005.

[SIA 2006] SIA, *The International Technology Roadmap for Semiconductors: 2006 Update*, Semiconductor Industry Association, San Jose, CA (http://public.itrs.net), 2006.

[Stroud 2002] C. E. Stroud, *A Designer's Guide to Built-In Self-Test*, Springer, Boston, 2002.

[Vinnakota 1998] B. Vinnakota, editor, *Analog and Mixed-Signal Test*, Prentice Hall PTR, Upper Saddle River, NJ, 1998.

[Wang 2006a] L.-T. Wang, C.-W. Wu, and X. Wen, *VLSI Test Principles and Architectures: Design for Testability*, Morgan Kaufmann, San Francisco, 2006.

R15.1 Introduction

[Kaminska 1997] B. Kaminska, K. Arabi, I. Bell, P. Goteti, J. Heurtas, B. Kim, A. Rueda, and M. Soma, Analog and mixed-signal benchmark circuits: First release, in *Proc. IEEE Int. Test Conf.*, pp. 183–190, November 1997.

R15.2 Analog Functional Testing

[Egan 1990] W. Egan, Modeling phase noise in frequency dividers, *IEEE Trans. Ultrasonics Ferroelectrics and Frequency Control*, 37(4), pp. 307–315, April 1990.

[Kroupa 1982] V. F. Kroupa, Noise properties of PLL systems, *IEEE Trans. Communications*, 30(10), pp. 2244–2252, October 1982.

[Kroupa 2001] V. Kroupa, Jitter and phase noise in frequency dividers, *IEEE Trans. Instrumentation and Measurement*, 50(5), p.1241, May 2001.

[Lee 2000] T. Lee and A. Hajimiri, Oscillator phase noise: A tutorial, *IEEE J. of Solid-State Circuits*, 35(3), pp. 326–336, March 2000.

[Leeson 1966] D. Leeson, A simple model of feedback oscillator noise spectrum, in *Proc. IEEE*, 54(2), pp. 329–330, February 1966.

[Watanabe 2000] Y. Watanabe, T. Okabayashi, S. Goka, and H. Sekimoto, Phase noise measurements in dual-mode SC-cut crystal oscillators, *IEEE Trans. Ultrasonics, Ferroelectrics and Frequency Control*, 47(2), pp. 374–378, February 2000.

R15.3 Analog and Mixed-Signal Test Architectures

[Arabi 1997] K. Arabi and B. Kaminska, Oscillation built-in self-test scheme for functional and structural testing of analog and mixed-signal integrated circuits, in *Proc. IEEE Int. Test Conf.*, pp. 786–795, October 1997.

[IEEE 1149.4-1999] IEEE Std. 1149.4-1999, *IEEE Standard for a Mixed-Signal Test Bus*, IEEE Press, New York, 1999.

R15.4 Defect-Oriented Mixed-Signal BIST Approaches

[Balivada 1996] A. Balivada, J. Chen, and J. Abraham, Analog testing with time response parameters, *IEEE Design & Test of Computers*, 13(2), pp. 18–25, Summer 1996.

[Chatterjee 1996] A. Chatterjee, B. Kim, and N. Nagi, DC built-in self-test for linear analog circuits, *IEEE Design & Test of Computers*, 13(2), pp. 26–33, Summer 1996.

[Ohletz 1991] M. Ohletz, Hybrid built-in self-test for mixed analog/digital integrated circuits, in *Proc. European Test Conf.*, pp. 307–316, March 1991.

[Pan 1995] C. Pan and K.-T. Cheng, Pseudo-random testing and signature analysis for mixed-signal systems, in *Proc. IEEE Int. Conf. on Computer-Aided Design*, pp. 102–107, November 1995.

[Stroud 1997] C. Stroud, P. Karunaratna, and E. Bradley, Digital components for built-in self-test of analog circuits, in *Proc. IEEE Int. ASIC Conf.*, pp. 47–51, September 1997.

[Stroud 2003] C. Stroud, J. Morton, A. Islam, and H. Alassaly, A mixed-signal built-in self-test approach for analog circuits, in *Proc. IEEE Southwest Symp. on Mixed-Signal Design*, pp. 196–201, February 2003.

R15.5 FFT-Based Mixed-Signal BIST

[Cooley 1965] J. Cooley and J. Tukey, An algorithm for the machine computation of complex Fourier series, *Mathematics of Computation*, 19(90), pp. 297–301, June 1965.

[Emmert 2003] J. Emmert and J. Cheatham, A monolithic spectral BIST technique for control or test of analog or mixed-signal circuits, in *Proc. IEEE Int. Symp. on Defect and Fault Tolerance in VLSI Systems*, pp. 303–310, November 2003.

[Emmert 2005] J. Emmert and J. Cheatham, Integrated spectral BIST technique for intelligent radio frequency front end systems, in *Proc. Government Microcircuit Applications and Critical Technology Conf.*, pp. 391–394, April 2005.

[Emmert 2006] J. Emmert, J. Cheatham, H. Axtell, R. Kertis, V. Sokolov, and G. Rash, Analysis and test of a spectral, mixed-signal BIST technique for systems-on-a-chip applications, in *Proc. IEEE North Atlantic Test Workshop*, pp. 130–139, May 2006.

R15.6 Direct Digital Synthesis BIST

[Dai 2004] F. Dai, C. Stroud, D. Yang, and S. Qi, Automatic linearity (IP3) test with built-in pattern generator and analyzer, in *Proc. IEEE Int. Test Conf.*, pp. 271–280, October 2004.

[Dai 2006a] F. Dai, C. Stroud, and D. Yang, Automatic linearity and frequency response tests with built-in pattern generator and analyzer, *IEEE Trans. VLSI Systems*, 14(6), pp. 561–572, June 2006.

[Dai 2006b] F. Dai, W. Ni, Y. Shi, and R. Jaeger, A direct digital frequency synthesizer with single-stage $\Delta\Sigma$ interpolator and current-steering DAC, *IEEE J. of Solid-State Circuits*, 41(4), pp. 839–850, April 2006.

CHAPTER 16

RF Testing

Soumendu Bhattacharya
Georgia Institute of Technology, Atlanta, Georgia

Abhijit Chatterjee
Georgia Institute of Technology, Atlanta, Georgia

ABOUT THIS CHAPTER

In the last few years, ***radiofrequency*** (RF) testing has been gaining importance in industry as more wireless devices are getting infused into the consumer market. However, RF testing is not well understood among engineers because of the various new considerations present in RF testing compared to digital, analog, or mixed-signal testing. Testing RF devices often requires special skills and knowledge of a different set of instruments to successfully perform measurements. Little has been done to bridge this gap so far and as a result, test needs for RF devices are mostly met by in-house training of employees or hiring external vendors to develop and perform the tests.

This chapter explores RF testing in general and points to various aspects that need to be considered to make consistent measurements—be it on the bench or in production. We first discuss RF test methods for specifications of circuit components as well as complex system-level specifications such as I-Q imbalance, modulation error ratio, and ***bit error rate*** (BER). For advanced readers, we further describe innovative test methods proposed by researchers in this domain. We describe common RF instrumentation used in bench and production testing—such as spectrum analyzer, network analyzer, and noise figure meter—to familiarize the reader with their functional details.

Noise is an important factor in RF measurements, and thus we give special emphasis to noise figure measurement. In addition, we explore issues related to making accurate measurements in order to give a clear idea about their implications in a test environment. Finally, we discuss the future directions of RF testing to encourage the reader to conduct a self-study of these advanced topics.

16.1 INTRODUCTION

Radio, derived from the word *radiate,* refers to a device that can transmit or receive electromagnetic signals without any electrical conducting media. With carefully designed systems, such signals can be propagated over long distances. Guglielmo Marconi and J. C. Bose independently invented radio in 1898. Early applications of radio technology were limited to maritime communications, radar, and the like, and the only consumer application was in the form of AM radio. Since then, radio technology has come a long way to become an inseparable part of our present-day lives. In the early days, communication was limited to broadcast mode only, where a base station transmitted the signal and nearby units tuned in to the correct frequency to receive the signal. Such devices were bulky and consumed a lot of power, making them unsuitable for portable applications. With advances in semiconductor manufacturing, it is now possible to integrate an array of transmitter-receiver pairs into a small monolithic device so that the user can send and receive data (*full-duplex communication*) over various frequency bands simultaneously. This has given rise to applications such as pagers, mobile phones, laptop computers, and other devices that are highly portable with a small form factor and that require less power to operate while providing a reliable mode of communication to the end user.

Radiofrequency devices operate at very high frequencies, often in the range of *megahertz* (MHz) to a few *gigahertz* (GHz). Classical definitions limit RF frequencies from 300 MHz to 3 GHz; however the frequencies go well beyond 5 GHz for most modern applications. RF technology has been around for many decades, but only recently has it gained widespread popularity in the consumer market. As wireless products continue to flood into the market at a greater pace, the need to provide better functionality with a lower cost becomes important as manufacturers seek to maintain a profitable market share. With the progress in manufacturing technology following **Moore's law**, production cost of silicon devices has reduced significantly. On the other hand, test costs for RF devices are relatively large compared to their analog counterparts and, unless accounted for, this can considerably offset the overall profit margin for semiconductor manufacturers for the future generation of RF devices. Unlike low-speed analog circuits, RF products pose a different set of challenges for high-volume testing and characterization. In this light, RF testing has drawn considerable attention from design and test communities. To better understand the challenges associated with RF testing, we first discuss the basics of RF technology.

16.1.1 RF Basics

Electrical signals are electromagnetic waves that propagate through a transmission medium and are altered in amplitude, frequency, and phase by the elements of the transmission network. Using Maxwell's four basic equations [Pozar 2005], all such behaviors of a network can be explained. However, it is not always straightforward to use these equations, and solving them may be cumbersome in most cases. Hence, easier analysis methods, such as lumped models, are used to analyze a given circuit.

The *scattering parameter*, popularly known as the *S-parameter*, is one such technique to measure, characterize, and explain the operation of a linear RF device. S-parameters measure the transmitive/reflective properties of a device, such as the input and output impedance, the forward/reverse gain, and are measured using a network analyzer [Agilent NA-2000]. Under linear operating conditions, *s*-parameters represent the reflection and transmission coefficients of a network [Agilent SP-1997]. Hence, for a given input signal, the S-parameters tell us how much of the signal energy is reflected back to the source and by what amount the remaining signal is amplified/attenuated by the network.

Other parameters, such as the *Z-parameter*, *h-parameter*, *Y-parameter* [Pozar 2005], can also be used to model a RF network. Any parameter set can be transformed to any other parameter set using simple matrix transformations. However, computing the S-parameter set exhibits certain advantages over the others and is usually preferred. To determine all other parameters, the input and output ports of the network need to be either opened or shorted. At very high frequencies, this may be difficult to achieve because of the inherent parasitics present in the circuit. To compute S-parameters, the network is connected between a source and a load with a 50 Ω impedance and that makes it easy to perform the measurement even at very high frequencies.

Figure 16.1 illustrates how S-parameters of a two-port network are measured. The parameters a and b, defined next, are used to define S-parameters in terms of the terminal current (I), voltage (V) and the characteristic impedance (Z) (where * represents the complex conjugate) [Kurukowa 1965]:

$$a = \frac{V + ZI}{2\sqrt{|real(Z)|}}, \quad b = \frac{V - Z^{*}I}{2\sqrt{|real(Z)|}} \tag{16.1}$$

Using the a and b values (normalized with respect to characteristic impedance, Z_0) for each port, S-parameters for a network can be computed, where a_1 and a_2 are the voltages generated by the waves incident at ports 1 and 2, respectively, and b_1, b_2 are

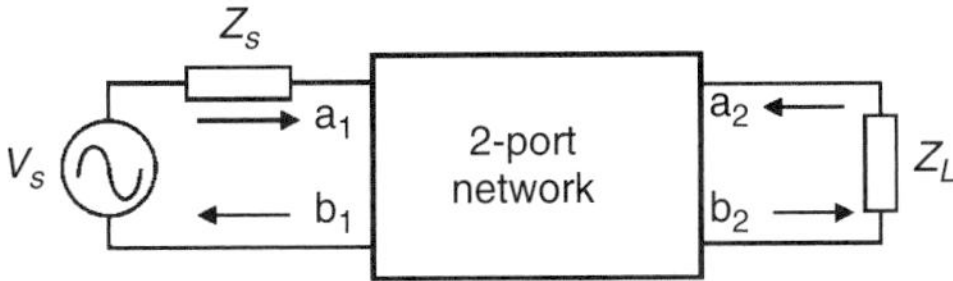

■ **FIGURE 16.1**

Two port network.

the measured voltages generated by the waves reflected from the network at ports 1 and 2, respectively. The *scattering matrix*, of a two-port network is defined as:

$$S = \begin{bmatrix} S_{11} & S_{12} \\ S_{21} & S_{22} \end{bmatrix} \tag{16.2}$$

where

$$S_{11} = \left.\frac{b_1}{a_1}\right|_{a_2=0}, \quad S_{21} = \left.\frac{b_2}{a_1}\right|_{a_2=0}, \quad S_{12} = \left.\frac{b_1}{a_2}\right|_{a_1=0}, \quad \text{and} \quad S_{22} = \left.\frac{b_2}{a_2}\right|_{a_1=0}$$

The preceding discussion holds for linear circuit operation. This is generally true for RF, as the signal levels are usually small and circuits operate in the linear region. However, as the signals are inherently small, there is a need to maximize power transfer between circuit stages and minimize the loss in the form of reflections between the stages. This is done by following the *maximum power (transfer) theorem* [Lee 2003] [Pozar 2005], which states that to obtain *maximum* power delivery from a source with *constant* source impedance, the load impedance must be equal to the complex conjugate of the source impedance. This condition is known as the *matched* condition, where source and load impedances are matched to provide *maximum power transfer* from the source to the load. To achieve impedance matching for a device, first the complex reflection coefficient is computed as follows:

$$\rho = \frac{Z_L - Z_0}{Z_L + Z_0} \tag{16.3}$$

where Z_L is the load impedance and Z_0 is the characteristic impedance of the transmission line connecting the source and the load. Furthermore, to achieve matched conditions, Z_L is adjusted such that ρ is equal to 0. The impedance value that is required to achieve this matching condition can be computed using a *Smith chart*, a handy tool that has been used since 1937 for designing RF circuits. A Smith chart is shown in Figure 16.2. The chart is plotted in the complex plane, representing the complex reflection coefficient computed from Equation (16.3), expressed in polar coordinates. The center of the chart corresponds to the case when the source and the load impedances are matched, and hence the reflection coefficient is zero. The perimeter of the chart corresponds to complete reflection ($|\rho| = 1$), and the angles printed around the perimeter indicate the phase of the reflection coefficient from 0° to 180°, or half a wavelength. The full circles represent the resistive component of the impedance, and the crossing partial circles represent the reactive component. Detailed descriptions of the Smith chart can be found in [Pozar 2005].

16.1.2 RF Applications

Since its emergence in the early 1900s, RF technology has found numerous applications in modern society. Currently, mobile phones are the fastest growing consumer segment with more than 100 million units sold worldwide every year. In addition, RF technology is used for radio communications in various forms, such as citizen

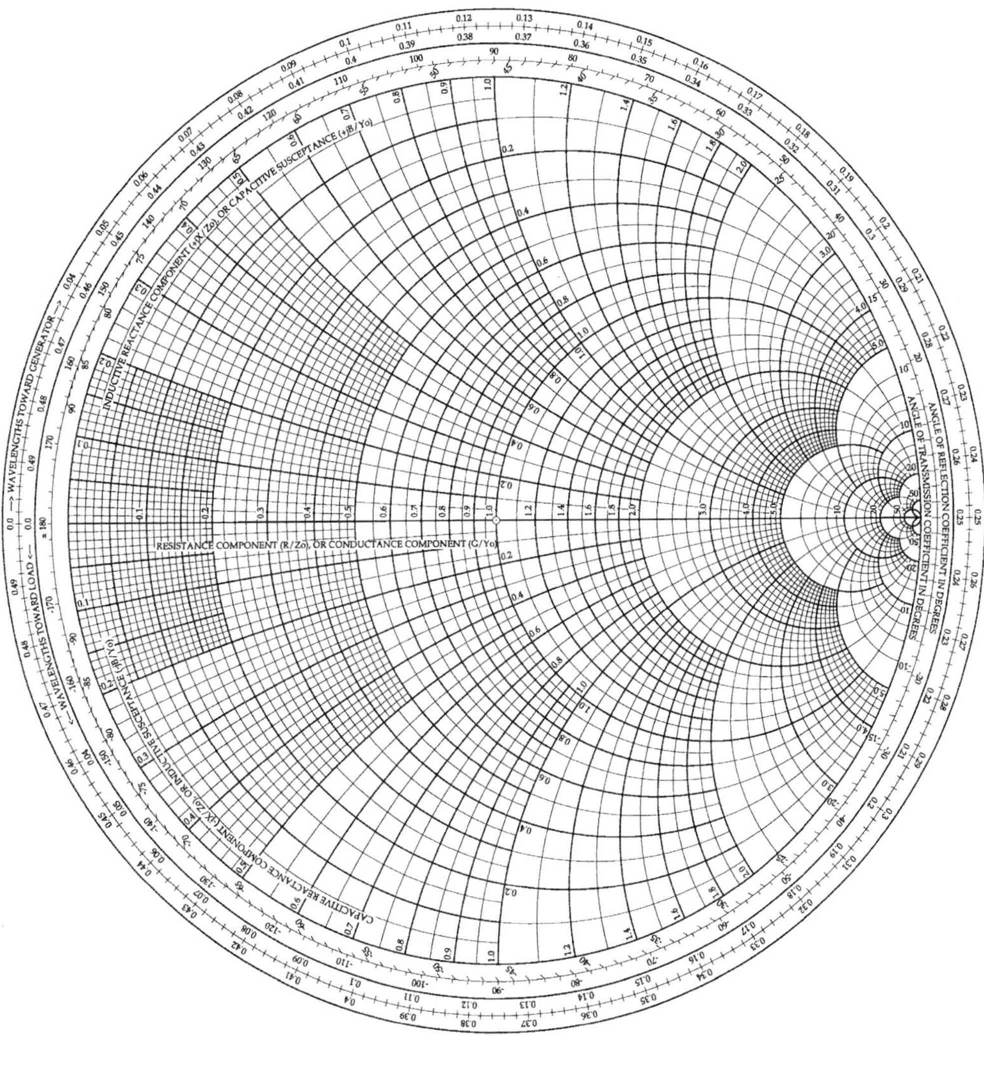

■ **FIGURE 16.2**

Smith chart for impedance match calculations.

FM radio, aviation controls, and private mobile communications (taxi, bus, truck, railways). The ***global positioning system*** (GPS) also uses frequencies in the RF band (1225-MHz, 1375-MHz, and 1685-MHz bands) to communicate between the satellite and the GPS units. Wireless local area networks (WLANs) use unlicensed bands at 2.4 GHz, 5.3 GHz, and 5.8 GHz. Apart from consumer applications, satellite communications and deep space communication are also based on RF technology.

Currently, various applications are being pushed to higher frequencies to obtain efficient propagation, larger bandwidth, immunity to certain types of noise, and reduced antenna size. For example, 2.4-GHz and 5.8-GHz cordless phones have become commonplace in home and office environments. In addition, ultrawideband technology (3.1 to 10.6 GHz) is ready to enter the market in the near future as wireless USB standard [Nekoogar 2005].

16.2 KEY SPECIFICATIONS FOR RF SYSTEMS

A typical RF system consists of various modules, such as amplifiers, frequency converters/mixers, oscillators/frequency synthesizers, and DC bias control units [Razavi 1997]. For reliable operation of the complete system, each module must function reliably as per its performance metrics. Usually, for RF devices, specifications related to gain (conversion gain), nonlinearity, and noise performance are of utmost importance [Kasten 1998]. In addition, oscillator specifications such as phase noise and stability ensure reliable system functionality. Common RF specifications for different device types are listed in Table 16.1. These specifications are discussed in more detail in subsequent sections.

TABLE 16.1 ■ Key Specifications for RF devices

Device Type	Specifications
Amplifiers/low-noise amplifiers (LNA)	Input matching or return loss (S_{11}), voltage standing wave ration (VSWR), gain (S_{21}), output matching (S_{22}), reverse gain (S_{12}), linearity (second-order intercept point [IP2]), third-order intercept point (IP3), noise figure (NF), in-band ripple, stability (gain and phase margins)
Power amplifier (PA)	Apart from the above specifications, output power level, power added efficiency (PAE), harmonics and total harmonic distortion (THD), signal-to-noise ratio (SNR)
Mixer	Conversion gain, IP3, NF, LO leakage, DC offset, image rejection ratio
Frequency synthesizer/oscillator	Frequency resolution, phase noise in dBc/Hz, integrated phase noise in ***root-mean-square*** (RMS) degree, in-band noise, out-band noise, spurs in dBc, reference frequency, tuning range, settling time, loop bandwidth, power consumption, reference feed-through, stability
Voltage controlled oscillator (VCO)	Frequency tuning range, gain in MHz/V, phase noise in dBc/Hz, spurs in dBc, power consumption
Transceiver system	Frequency bands, NF, IP3, dynamic range, sensitivity, maximum receivable power, maximum output power, power consumption, maximum DC offset, SNR, ***adjacent channel power*** (ACP), ***error vector magnitude*** (EVM), ***bit error rate*** (BER), I-Q imbalance (magnitude error, phase error)

16.2.1 Test Instrumentation

At this point, it is worth introducing the reader to the common RF instrumentation used in bench measurements (*i.e.*, characterization) and production testing of RF devices and systems. The common instruments used in RF measurements are the spectrum analyzer [Agilent SA-2005], network analyzer, and noise figure meter [Agilent NF-2006]. Apart from DC tests, it is possible to perform most of RF specification tests using these three instruments.

16.2.1.1 Spectrum Analyzer

The spectrum analyzer is the most commonly used instrument for RF measurements. A *spectrum analyzer* is primarily used to display the spectral content (the power at different frequencies) of a signal over a range of frequencies. Using the spectral data, one can easily visualize both the desired and unwanted frequency tones present in the input signal. For RF testing, the engineer is often interested in measuring spurious tones, especially those originating from third-order intermodulation or the harmonics in the signal. In such cases, usually a spectrum analyzer is used to perform measurements. Apart from displaying frequency domain data of the input signal, modern spectrum analyzers can perform complex measurements automatically, such as *ACP*, *IP3*, and *phase noise*, and with certain modifications in hardware and firmware, they can also accurately measure NF.

Figure 16.3 shows the structure of a spectrum analyzer. Spectrum analyzers use firmware-based control (the embedded software processing unit that interfaces with the user and the display) to interact with the user. The firmware in spectrum analyzers operates at lower frequencies compared to the input signal, as signal processing at such high frequencies becomes extremely difficult. Therefore, the

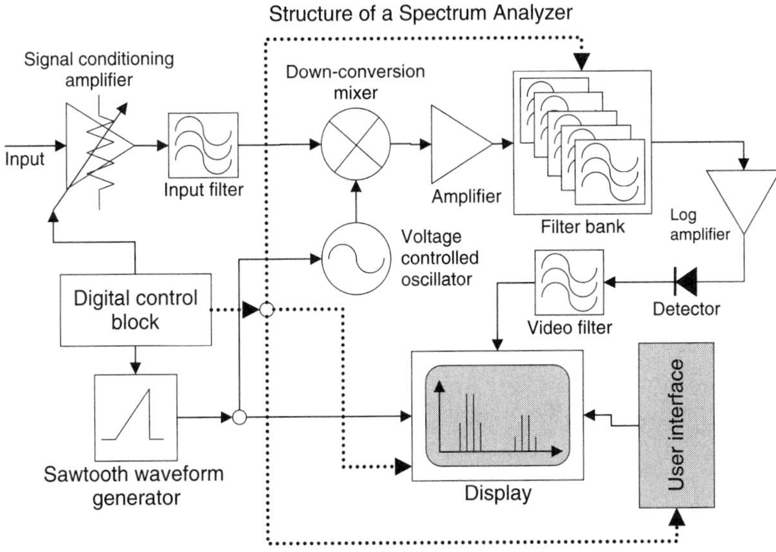

■ FIGURE 16.3

Block diagram of a spectrum analyzer.

input signal is first amplified/attenuated to a preset level and down-converted using a receive chain. Usually, the user specifies a range of frequencies within which the desired signal is expected to lie. Accordingly, the oscillator frequency for down-conversion within the spectrum analyzer is chosen such that the input signal after down-conversion lies at the baseband frequency. For example, if the spectrum analyzer firmware can work up to 20 MHz and the input signal is at 900 MHz, the down-conversion frequency within the spectrum analyzer will be at 920 MHz. Furthermore, based on the range of frequency specified by the user, the oscillator frequency is swept in steps for this range using a *sawtooth waveform generator* and a ***voltage controlled oscillator*** (VCO) (see Figure 16.3). Most modern spectrum analyzers employ two to four down-conversion steps to get the signal down to baseband frequencies.

Next, this signal is passed through a *resolution filter* that controls the resolution of the display. If the resolution filter for the spectrum analyzer is 100 KHz, then the frequency tones 900 MHz and 900.01 MHz in the input signal cannot be distinguished on the display, as they will both pass through the resolution filter. This filter is selected by the digital control block automatically based on the frequency range specified by the user; however, users can choose the filter manually. The filtered signal is passed through an envelope detector. This generates a DC signal proportional to the tone along with some high-frequency components. The high-frequency components are removed by filtering through a video filter. The name *video filter* stems from the fact that it is used to provide a precise DC level to the digital block for display. This is done for all the frequency points within the range and is repeated over and over.

It is interesting to note that if the resolution filter has a wider bandwidth compared to the video filter, some unwanted signal close to the signal tone may pass through the detector and may not be fully eliminated by the video filter. This might lead to incorrect readings on the display of the spectrum analyzer. To prevent this outcome, usually the resolution filter and video filter bandwidths are kept the same. In cases where the *resolution bandwidth* is set too large, a large signal can easily mask any smaller tones in its vicinity. Therefore, it is extremely important to understand the test setup and its implications on the measurement.

16.2.1.2 Network Analyzer

A network analyzer is another versatile instrument that is widely used to characterize two-port and four-port networks. It can be used to measure the S-parameters and all associated specifications (reflection coefficient, VSWR, etc.). In addition, *time-domain reflectometry* (TDR) measurements can also be performed using this instrument.

As previously discussed, S-parameters constitute a significantly important portion of RF test specifications. Usually, network analyzers are used to measure S-parameters. To perform such measurements, four important blocks are present within a network analyzer: (1) a source for stimulus, (2) signal-separation/blocking devices, (3) a down-conversion module, and (4) a ***digital signal processor*** (DSP)

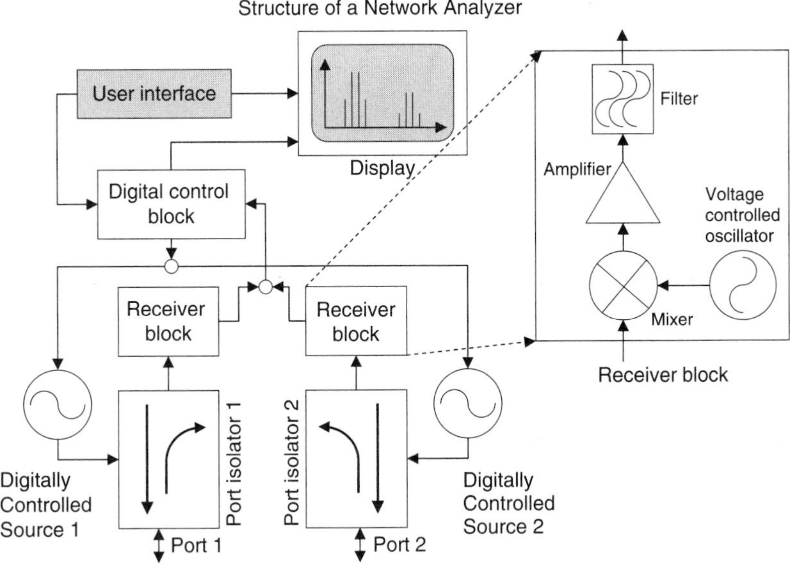

■ **FIGURE 16.4**

Block diagram of a network analyzer.

based firmware unit to manage the user interface and display. Figure 16.4 illustrates a block diagram of a network analyzer.

Usually, a *network analyzer* has two or four ports, through which it can source signal, as well as isolate the reflected signal from the ***device under test*** (DUT). The sourcing is usually done from digitally controlled *frequency synthesizers*. To obtain a reference for S-parameter computation, a part of the stimulus is branched out through a power splitter and fed back to the DSP for reference (not shown in figure). The reflected signal from the DUT into the port is isolated through a *directional coupler* and can be measured separately from the original signal transmitted to the DUT [Pozar 2005]. The directional coupler usually has a low loss with high reverse isolation. With the transmitted and reflected signals from each port, all the S-parameters and associated specifications can be computed using a network analyzer.

16.2.1.3 Noise Figure Meter

A ***noise figure*** (NF) meter is used to accurately measure NF of various types of devices. Usually, a calibrated broadband noise source, also known as the *noise head*, is used to generate noise to perform such measurements.

An NF meter, also known as a *noise figure analyzer*, is used to characterize the noise characteristics of a device. To better understand the functionality of an NF meter, we first discuss the concept of noise temperature. Thermal noise generated by a device increases with increase in physical temperature. At a temperature T and measurement bandwidth of B, the noise power generated by the device is given by $N = kTB$, where k is Boltzmann's constant ($1.3806503 \times 10^{-23}$ kg m²/s²K). This means that any device that generates more noise than kTB at room temperature (298.13 K) is equivalent to being at a higher physical temperature at which it would generate an equal amount of noise power. This is referred to as the noise temperature of the device. Note that if the device is at a higher physical temperature than room temperature, it must be compensated for before determining the actual noise temperature.

Any device output noise has two sources: (1) the noise contribution of the input signal that propagates through the device, and (2) the noise generated by the device itself. NF is a measure of the noise contributed by the DUT only. The equation to measure NF is given by:

$$F = \frac{SNR_{in}}{SNR_{out}} = \frac{S_i/N_i}{S_o/N_o} = \frac{S_i/N_i}{GS_i/(N_{DUT}+GN_i)} = \frac{(N_{DUT}+GN_i)}{GN_i} \tag{16.4}$$

where

$$\text{NF} = 10\log_{10} F$$

where the noise contribution of the device is N_{DUT}, G is the gain of the DUT, and N_i is the input noise power. Note that here the gain is unknown to the measurement system. The input noise is usually thermal noise from the source that equals to $N_i = kTB$, where T is the equivalent noise temperature, and B is bandwidth of the system. Therefore, Equation (16.4) changes to:

$$NF = \frac{N_o}{N_i} = \frac{(N_{DUT}+kTBG)}{kTB} \tag{16.5}$$

Assuming the noise contribution of the device does not change with varying input signal power, the noise contribution of the DUT can be measured by making two measurements with different input noise magnitudes. The output noise power from the DUT scales linearly with changing input noise power (temperature). As shown in the plot on the right-hand side of Figure 16.5, the slope of the line is equal to kBG. The noise contribution of the device is the intercept of the line with output noise power (y-axis) (*i.e.*, the amount of output noise power when there is no input noise).

An NF meter has a calibrated noise source that can provide an accurate noise input to the DUT. The structure of an NF meter is illustrated in Figure 16.5. By obtaining two measurements with known input noise powers, the intercept (*i.e.*, the noise power contribution of the DUT) can be obtained. Usually, the NF meter uses a reverse biased diode to create a steady noise power at the device input (N_{i1}). The output power from this is used as one of the measurement points (N_{o1}). To get the other point, the diode is simply turned off, and the input noise power is at 298.13 K (N_{i2}). From the output power (N_{o2}), the N_{DUT} can be computed as:

$$N_{DUT} = \frac{(N_{o1}N_{i2} - N_{o2}N_{i1})}{(N_{i2} - N_{i1})} \tag{16.6}$$

From the N_{DUT} and the N_0 values, the NF can be further calculated using Equation (16.4).

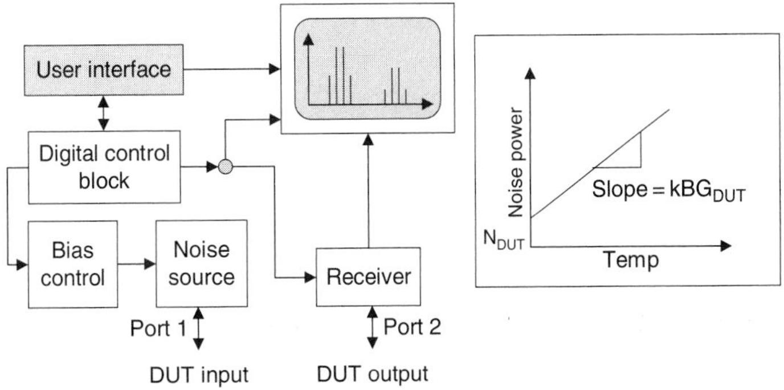

■ **FIGURE 16.5**

Block diagram of an NF meter and calculation of N_{DUT} using NF meter via interpolation method.

16.2.1.4 Phase Meter

Although spectrum analyzers can measure *phase noise,* accurate phase noise measurements can only be made through a phase meter. The measurements from a phase meter are typically better than a spectrum analyzer by 10 to 20 dBc/Hz. Usually, the noise floor of the phase meter is lower than an NF meter. However, the use of phase meters is limited in production testing because measurement times are considerably larger than a spectrum analyzer.

16.2.2 Test Flow in Industry

Next, we discuss the overall test procedure followed in the semiconductor industry. Semiconductor manufacturing is a three-step process: (1) the circuit is designed and fabricated, (2) a prototype is developed and *characterization test/qualification* is implemented, and finally (3) ramping up to high-volume production is performed. This process is shown in Figure 16.6.

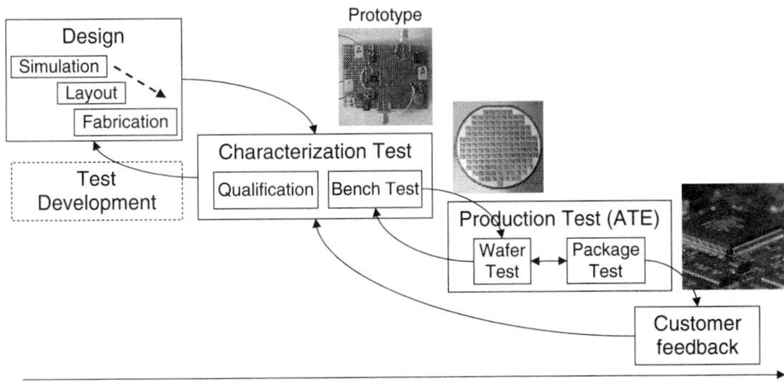

■ **FIGURE 16.6**

Typical semiconductor manufacturing flow.

16.2.2.1 Design and Fabrication

The first step in manufacturing a device is to define a set of specifications for the device, either obtained from a customer directly or developed based on market need. Using ***computer-aided design*** (CAD) software, designers lay out a schematic of the circuit and simulate the design. Once the designers have obtained satisfactory performance from the design, the layout of the schematic is created. The layout is again simulated to incorporate parasitic effects and tweaked to meet the specification needs. Finally, the layout is sent for fabrication. In parallel to the design cycle, the test development process for the device is also initiated. In this phase, the tests that will be performed on the device during high-volume production are developed and coded in the ***automatic test equipment*** (ATE) language. The ATE is commonly referred to as a tester. The test program is usually executed first on a dummy tester and a dummy circuit to detect and debug any serious flaws in the code. To interface the device with the tester, a *load-board* is designed and fabricated to include relevant circuitry and a test *socket*. By the time the device is characterized and fabrication volume is ramped up, the test program and the load-board is ready for use on the tester.

16.2.2.2 Characterization Test

As the first set of devices arrive, they are tested for their specifications using a *bench test setup*, and this process is known as *characterization test*. A typical bench test setup usually consists of accurate (high performance) measurement instruments, such as spectrum analyzers, network analyzers, and high-speed sampling scopes. The devices are tested for nominal specifications, various corners, and a wide range of temperatures. In addition, the load-board and test programs are also tested to diagnose any serious issues. Also, repeatability of the measurement system (*i.e.*, performing the same measurement many times while keeping the measurement setup unchanged) and device-to-device correlation (*i.e.*, the variance of the measurements made on many devices using the same test setup) are computed. All of these steps help in determining the DUT performance and understanding the reliability of the overall test setup.

16.2.2.3 Production Test

After successfully characterizing the devices, fabrication volume is ramped up to production level, where many thousands of units are tested every day. The devices are tested on the ATE using the load-board and test programs created during the development phase. The test instruments used in the tester are, however, not as accurate as the bench setup. This is because these are usually low-cost instruments with lower resolution, making a tradeoff between cost of testing and desired accuracy. Also, the DUT is tested for a subset of specifications during production because of limited tester capability and time constraints. Finally, with satisfactory production test data, the devices are sent to the retailer/end user.

The manufacturing process is a closely monitored system, where any anomalies are taken care of by going back to the previous step. This way, if a device does not perform as expected during characterization, the design is revisited to determine the possible cause of failure and is fixed. However, this is not the case for production testing. During production, the process cannot be interrupted if a problem is detected during testing, and such problems must be dealt with using innovative methods. On the other hand, although characterization is very accurate, it may be extremely inefficient to test a large number of devices in a small amount of time. As a result, characterization and production tests inherently pose different sets of challenges to the manufacturing process, as discussed next.

16.2.3 Characterization Test and Production Test

The scope of characterization (or *qualification*) and production tests differs significantly in the overall manufacturing process, and each poses a different set of challenges [Schaub 2004]. During characterization, the silicon is tested to check the overall performance. Characterization deals with verifying the actual performance of the design; however, production testing focuses on the overall functionality of the silicon to eliminate outliers and defective parts getting shipped to customers. While characterization tests are more complex, production testing tries to achieve 100% yield with a simpler set of tests as the device is ramped up to production volume. Hence, the implications of each of the test processes on the overall manufacturing procedure are different. In this section, we will discuss the differences and advantages of each of the methods.

Often, there are extra steps in the development process, where the device is characterized on ATE as well. This process is similar to the bench test, where the performance of the device at different conditions is measured. This is usually performed on test chips and devices manufactured using advanced fabrication methods.

16.2.3.1 Accuracy

Usually, characterization testing is extremely accurate as the most precise set of instruments is used to perform such tests. This can be attributed to the fact that during characterization, larger averaging is possible, and tests can be performed with the highest resolution settings, such as minimum ***resolution bandwidth*** (RBW) setting in a spectrum analyzer. Production testing is not as accurate as the instruments are combined in a single housing within the tester in the form of line cards interfaced through GPIB, PCI-X, or some other standard. This inherently increases the overall noise generated in the system and affects the measurement accuracy. However, the results from characterization and production must follow the same trend from the obtained test measurements.

16.2.3.2 Time Required for Testing

The goal of production testing is to perform a set of predefined tests in a small amount of time while maintaining an acceptable level of accuracy [Ferrario 2002]. To do this, the tests are usually applied sequentially to the DUT. As different tests have different configurations, these are all implemented together on the load-board, and using a set of control signals generated by the tester, various circuit configurations are achieved via control relays. This reduces the overall test time such that a set of tests for a typical RF amplifier (15 to 20 tests—DC current, gain, IP3, NF, harmonics, etc.) is usually performed in 300 to 500 ms. On the other hand, characterization uses a different set of device boards to measure different specifications, and the process is not automated, thereby increasing the overall test time required for characterizing.

16.2.3.3 Cost of Testing

Production testing is geared toward reducing the overall cost of testing. To achieve this goal, emphasis is placed on an elaborate design of the load-board. This involves a one-time cost, but innovative designs can significantly reduce test time during production, thereby reducing the test cost per device. Also, testing many devices in parallel reduces overall test time. This is not an important concern during characterization, as the test cost is not of major concern.

In summary, one should keep in mind that the goals of production tests and characterization tests are different. Various methods are applied to make each of the test processes work successfully on a bench or a production floor. Although the device is characterized for all or most of its specifications, only a few of the tests are performed during production testing based on test cost and test time constraints. However, the principle of testing each specification remains the same in both cases. Next, we discuss the methods of testing various specifications for RF circuits and systems.

16.2.4 Circuit-Level Specifications

In this section, we discuss a few of the important specifications for stand-alone RF devices that are tested during high-volume production [Wang 2006]. It is important to discuss a new set of units that are commonly used in RF measurements. The unit used to specify gain is ***decibels*** (dB). Numbers are converted to decibels using the following formula:

$$N_{dB} = 20\,log_{10}(N) \tag{16.7}$$

Therefore, a gain of 8 dB essentially means a gain of approximately 2.5. Similarly, 20 dB is 10, and 40 dB is 100. A similar notation, called *dBm*, is used to specify power in logarithmic units, using the following formula:

$$P_{dBm} = 10\,log_{10}(P_{Watts} \times 1000) \tag{16.8}$$

where P_{Watts} is power in Watts. Essentially, dBm measures power with 1mW as the reference such that 1 mW is 0 dBm and 1 W is 30 dBm. These two units are frequently used in RF design and test, and it is important for the reader to be familiar with these units to better understand the subject. The prudent reader might question the need for introducing a new set of units for RF measurements. In RF, the range of power measured is large—from nano-Watts to a few Watts. It can often become cumbersome to manage such a large range of numbers, and calculations become prone to mistakes. Moreover, engineers like to have rounded/integer numbers. Therefore, to make the calculations simple and contain the numbers in a smaller range, the logarithmic transformation of power is used.

16.2.4.1 Gain

Gain represents the linear gain of a device. This is usually specified for amplifiers (power amplifiers, low noise amplifiers, etc.) and mixers. Gain is calculated by providing input signals of known power within the linear range of operation of the device and measuring the output. Usually, a spectrum analyzer or a network analyzer is used to measure gain.

To measure the gain of a device, a single tone input is used. With the DUT powered up and the single tone input applied to the DUT input, the output of the DUT is measured using a spectrum analyzer. Gain is measured by taking the ratio of the input and output power. The amplitude of the input is adjusted so that the output of the DUT does not go into the saturation region. For a device, if the expected gain is A, and the saturation point (*i.e.*, –1 dB compression point, P_{-1dB}, discussed later) is A_{-1dB} dBm, then the input A_{in} must be less than A_{-1dB} – the *expected gain*. The gain of a DUT is given by *measured gain (G)* $= A_{out} - A_{in}$, where the input (A_{in}) and output (A_{out}) signals are given in dBm. Network analyzers can directly measure the gain (S_{21}) of a device by applying an input and monitoring the output as a part of S-parameter measurements.

Gain measurements for mixers work differently because mixers translate the input frequencies to a different range of frequencies. The gain of a mixer is measured as the ratio of amplitudes taken at the input and output frequencies. Because of the inherent frequency translation involved in the process, gain for mixers is often referred as *conversion gain*, also expressed in dB.

16.2.4.2 Harmonics and Third-Order Intercept Point (IP3)

All RF devices are inherently nonlinear. Therefore, for a moderately large signal, the device may show serious nonlinear effects, such as gain compression, desensitization, harmonics, and intermodulation components [Razavi 1997] [Burns 2000] [Cho 2005]. To explain these in greater detail, we first present a typical input-output transfer curve for a device, as shown in Figure 16.7. This can be obtained by sweeping the input power to the DUT and measuring the output power. Because of the DUT behavior, the transfer curve starts to compress at high input power levels, thereby introducing nonlinearity in its output response. To understand the implications of nonlinearity, we fit a polynomial function to the transfer function.

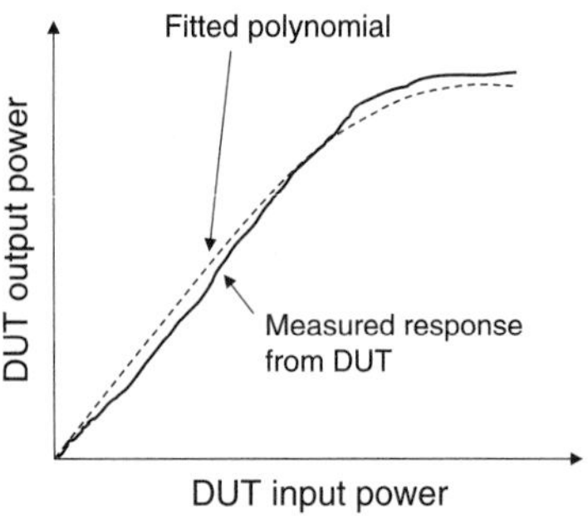

■ **FIGURE 16.7**

Nonlinear response of a DUT.

In the figure, the solid curve represents actual measurements made on a DUT, and the dotted curve is obtained by fitting a polynomial to its input-output response. A generic form of the polynomial is given by:

$$y(t) = a_0 + a_1 x(t) + a_2 x^2(t) + a_3 x^3(t) + \cdots \textit{ higher-order terms} \tag{16.9}$$

where $x(t)$ is the input signal; $y(t)$ is the output; and a_0, a_1, a_2, a_3, etc. represent the various coefficients of the fitted polynomial. If a single tone input (*i.e.*, $x(t) = Asin(2\pi ft)$) is applied to the device, the output not only contains a single tone at frequency f but also other tones at $2f$, $3f$, etc. are created. These tones are known as harmonics. In case of a multitone input with N tones as shown in Equation (16.10), many other tones apart from the harmonics are created close to the input tone frequencies ($f_i \pm f_j$, $2f_i \pm f_j$, ..., where $i, j \vee + \mathrm{I} \subseteq [1, N]$). These tones are known as intermodulation tones. The term *intermodulation* stems from the fact that these tones are created by interactions from two or more fundamental tones. Harmonics are generated for each tone, and they can be eliminated from the output via proper filtering. On the other hand, intermodulation tones are difficult to remove because of their close vicinity to the input tones. Hence, it is important to measure the amount of distortion introduced by intermodulation during production testing:

$$x(t) = \sum_{i=1}^{N} A_i \sin(2\pi f_i t) \tag{16.10}$$

Usually, the second order coefficient (a_2) for RF devices is very small; the most significant component of the nonlinear terms is the third-order coefficient (a_3). For a two-tone input, $x(t) = Acos(\omega_1 t) + Acos(\omega_2 t)$, ignoring the term for a_2, the expression becomes:

$$\begin{aligned} y(t) &= a_0 + k_1 \cos(\omega_1 t) + k_1 \cos(\omega_2 t) \qquad (\leftarrow \text{fundamental terms}) \\ &+ k_2 \cos(2\omega_1 t - \omega_2 t) + k_2 \cos(2\omega_2 - \omega_1 t) \qquad (\leftarrow \text{third-order intermodulation terms}) \\ &+ \text{high-frequency intermodulation terms and harmonics} \end{aligned}$$

where

$$k_1 = a_1A + \frac{9}{4}a_3\,A^3 \approx a_1A \ (A^3\text{is ignored as } A << 1) \quad \text{and} \quad k_2 = \frac{3}{4}a_3A^3 \qquad (16.11)$$

For a more detailed derivation, the reader is referred to Chapter 15.

If we fit a similar model to the response of any DUT, the terms a_1 and a_3 are usually close in magnitude. From this, one can easily deduce that the term relating to the nonlinear behavior of the DUT (k_2) has three times the rate of increase of the linear term (k_1) *(we are talking about logarithmic scale and hence the origin of the term "third-order")*. If the input power is increased continuously in this manner, at one point, the power contributions from the linear and the nonlinear terms would become the same. This is known as the ***third-order intercept point*** (IP3) of the DUT. Although this is not observable in practice as the device saturates long before that, the intercept point can be computed by equating the two terms, k_1 and k_2, with their corresponding slopes and finding their intersection point. However, finding IP3 by this method is computationally intensive, and an easier way exists to compute this during testing. As Figure 16.8 shows a two-tone input is applied to the DUT with input power P_{in}, which is well below the compression region. As the slopes of the fundamental and intermodulation tones bear a ratio of 1:3, the IP3 can be computed as follows (see Figure 16.8):

$$IP3 = \frac{\Delta P}{2} + P_{in} \qquad (16.12)$$

IP3 indicates the input power level for which the linear and nonlinear terms would become equal, and hence the unit for IP3 is dBm (the same as for power). A related specification, known as the *1-dB compression point,* is worth mentioning at this point. It is usually expressed as P_{-1dB} (see Figure 16.8). This indicates the input level where the output of the DUT deviates from its linear behavior by 1 dB and has the same units as that of IP3. The magnitude of P_{-1dB} is much less than IP3 and, the relation $IP3 \approx P_{-1dB} + 10$ (where P_{-1dB} and *IP3* are both expressed in dBm) is frequently used as a rule of thumb.

To quantify the harmonics for a DUT, the power level of a fundamental frequency and its harmonics is presented. A new unit is used for this purpose, known as *dBc,* which indicates the difference in dB value from a reference—in this case, the fundamental frequency. Therefore, −40 dBc means that the tone is 40 dB down compared to the fundamental. Harmonics have frequencies that are multiples of the fundamental tone frequency and are indicated by the multiple numbers. For example, the harmonic at twice the frequency of the fundamental is called the second-order harmonic, etc. Usually, we are interested in harmonics up to the fifth order.

Both IP3 and harmonics are measured using spectrum analyzers. Modern spectrum analyzers have built-in firmware to directly measure IP3 for a two-tone input. However, for high-frequency devices, the harmonics, specifically the higher-order ones, are far apart from the fundamental frequencies. Such measurements require spectrum analyzers to operate over a wide range of frequencies, thus making them expensive to use during production testing.

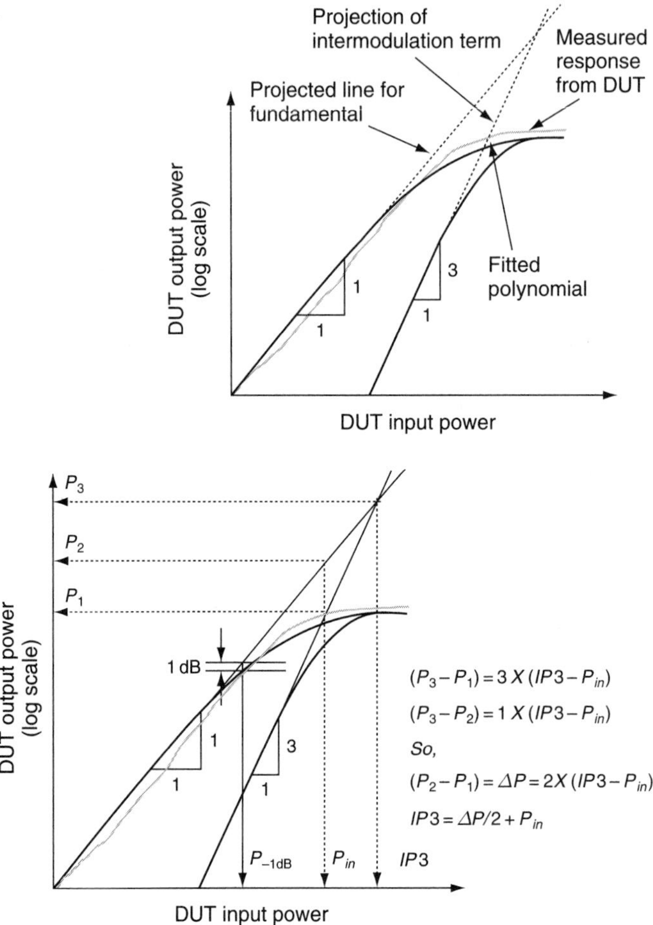

■ **FIGURE 16.8**

IP3 measurement details.

To measure IP3, two tones are applied to the DUT with their frequency values close to each other relative to the overall bandwidth of the DUT. For example, if an amplifier operating at 2.4 GHz has a bandwidth of 20 MHz, the tones applied for IP3 measurement would be 2.3999 MHz and 2.4001 MHz, separated by only 200 kHz. In this case, the intermodulation tones will be located at 2.3997 MHz and 2.4003 MHz. The response of the DUT looks similar to Figure 16.9 where ΔP is used to compute IP3 from Equation (16.12). Harmonics are measured by applying a single tone to the DUT within its frequency range of operation and are specified with respect to the applied frequency. For most practical purposes, second and third harmonics are sufficient; however, some applications require the device to characterize up to the fifth harmonic for various frequencies within the range of operation.

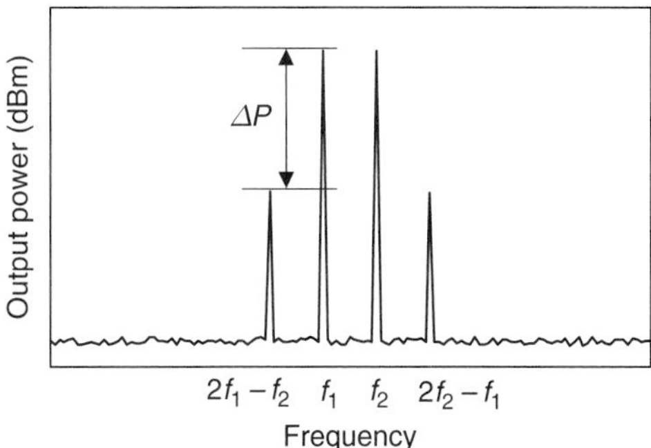

■ **FIGURE 16.9**

IP3 computation from output spectrum.

16.2.4.3 1-dB Compression Point (P_{-1dB})

As previously mentioned, the device output gain reduces with increasing input signal level because of the nonlinear nature of the DUT. The ***1-dB compression point*** indicates the point where the gain drops below the normal gain by 1 dB. Referring to Equation (16.11), the linear response of the device is $k_1 = a_1 A$, whereas nonlinear response equals $k_1 + k_2 = a_1 A + \frac{3}{4} a_3 A^3$. At the 1-dB compression point, if the input is A_{-1dB}, the following equation holds (assuming the frequency of the fundamental and the intermodulation terms are close):

$$20\log_{10}(k_1) - 20\log_{10}(k_1 + k_2) = 1dB$$

$$\text{or, } 20\log\left(\frac{k_1}{k_1 + k_2}\right) = 1dB \text{ leading to } A = \sqrt{0.145\left|\frac{a_1}{a_3}\right|} \tag{16.13}$$

Typically, the 1-dB compression point occurs at −20 to −25 dBm for front-end amplifiers.

16.2.4.4 Total Harmonic Distortion (THD)

THD is a measure commonly used for RF devices. To measure THD, a single tone is applied to the DUT with an input power within the linear range of operation, and the response is captured using a spectrum analyzer (see Figure 16.10). THD indicates the ratio of total output harmonic power and the output power at the fundamental frequency:

$$THD = \frac{(P_{2nd} + P_{3rd} + \cdots + P_{5th})}{P_{fundamental}} \tag{16.14}$$

Because harmonics only up to the fifth order are usually considered, the relative contribution of the harmonics above the fifth order is not significant. However, for

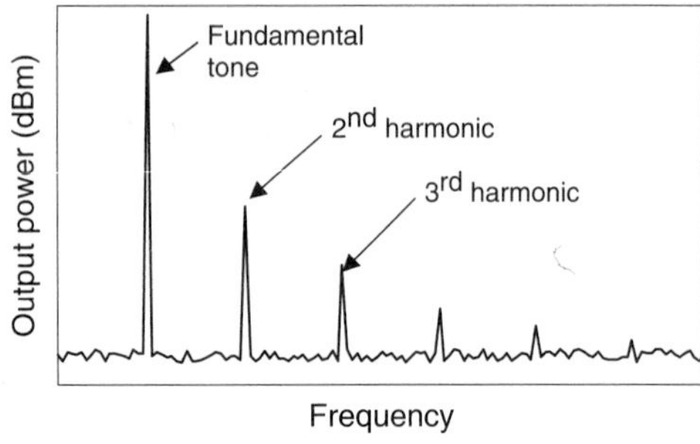

■ **FIGURE 16.10**

Fundamental signal and its harmonics.

a device operating at 2 GHz, the spectrum analyzer needs to be able to measure up to 10 GHz, which is a wideband measurement. Such a wideband measurement is prone to external noise, and as a result such measurements are usually performed via discrete measurements at different frequency bands and require significant test time. As THD is a ratio, it is dimensionless. However, it is often represented in dB to avoid dealing with a very small number.

16.2.4.5 Gain Flatness

Until the late 1990s, gain flatness was not an important specification that was measured during production testing. Recent growth in applications demanding wideband RF amplifiers have made this measurement a necessity during production testing. Gain flatness indicates how much the gain of the DUT varies within its band of operation, as illustrated in Figure 16.11. The specification is represented in dB, which indicates the difference between the maximum and minimum gain values.

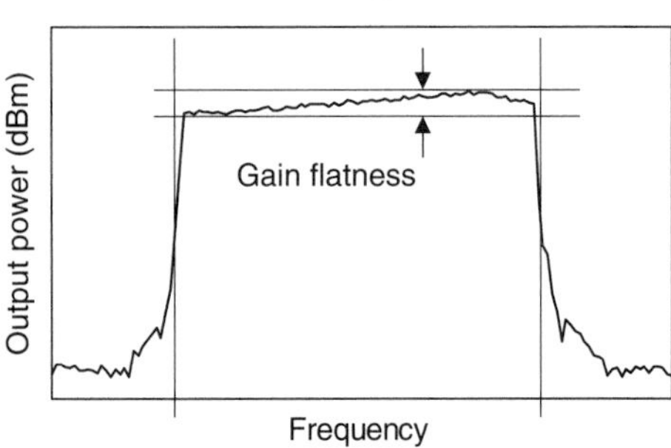

■ **FIGURE 16.11**

Gain flatness of the DUT within the bandwidth.

16.2.4.6 Noise Figure

Any electronic system inherently generates noise. Similarly, an RF circuit also introduces noise to the input signal. Although this cannot be eliminated completely, the noise contribution of the circuit is minimized through careful design techniques. As explained before, NF is a measure (ratio) of the amount of noise added by the device, and it is measured by comparing the input and output ***signal-to-noise ratio*** (SNR) as follows:

$$NF = \frac{SNR_{in}}{SNR_{out}} = \frac{Noise_{out}}{Gain \times Noise_{in}} \tag{16.15}$$

There are mainly three different methods to measure NF, and the choice of the measurement method depends on the range of NF measured and the desired accuracy [Maxim 2003]. The methods include (1) an NF meter, (2) gain method using a spectrum analyzer, and (3) the Y-factor method [Agilent Y-2004].

The first and the most common method uses an NF meter to perform NF measurements. It is very accurate and is capable of measuring very low noise levels. Using an external ***local oscillator*** (LO) source, it is possible to measure the NF of mixers as well. However, for mixers and other devices having large NF values (NF > 10 dB), this method is not accurate. In such cases, the gain method is useful. The gain method uses the following formula:

$$NF = N_{out} - (N_t + G) = N_{out} - (10\log(kTB) + G) \tag{16.16}$$

where N_{out} is the measured output noise, N_t is the thermal noise at the input of the DUT, and G is the system gain in dB. Once again, kTB represents the thermal noise at the input at temperature T. The principle of NF measurement remains the same as before. To measure NF, we merely subtract the noise contribution of the input times the gain of the DUT from the total measured output noise power. However, this method has two limitations. First, it assumes that the input is clean and thermal noise is the sole contributor at the input node. Second, as evident from the equation, we must know the gain of the DUT a priori to compute the NF.

The output noise power can be measured using a spectrum analyzer; however, the minimum noise density measurable in a spectrum analyzer is around −150 dBm/Hz. This means for a system with a bandwidth of 25 MHz and a gain of 15 dB (which is typical for most LNAs), the minimum NF that can be measured is:

$$NF = (-150 + 10\ \log(25 \times 10^6)) - (-174 + 10\ \log(25 \times 10^6) + 15) = 9dB \tag{16.17}$$

Most modern LNA circuits have NF less than 2 dB, which clearly shows that this method is not suitable for measuring small NF values, unless the DUT gain is large enough.

The third method is the Y-factor method. First, we introduce the concept of ***excess noise ratio*** (ENR) and Y-factor. Both ENR and Y-factor denote the characteristics of a noise source, defined as:

$$Y = \frac{P_{On}}{P_{Off}} \quad \text{and} \quad ENR = \frac{(N_{On} - N_{Off})}{N_0} \tag{16.18}$$

P_{On} and P_{Off} are the output power of the DUT when a noise source at the input is turned on and turned off, respectively, and N_{on} and N_{off} are the measured output noise power of the source when it is turned on and off. N_0 is the output noise power of the noise source at room temperature. Some people like to refer to the noise power as noise temperatures for ease of analysis. In this case, ENR is defined as follows:

$$ENR = \frac{(T_{On} - T_{Off})}{T_0} \tag{16.19}$$

We now describe how ENR and Y-factor can be used to efficiently compute the NF of the DUT. One can observe that:

$$P_{off} = N_{DUT} \times KTB \times G$$

and

$$P_{on} = (N_{DUT} + ENR) \times KTB \times G \tag{16.20}$$

Hence, we can write:

$$Y = \frac{P_{On}}{P_{Off}} = \frac{(N_{DUT} + ENR)}{N_{DUT}} \tag{16.21}$$

Rearranging Equation (16.21), we derive the NF of the device as follows:

$$N_{DUT} = \frac{ENR}{Y - 1} \tag{16.22}$$

ENR and Y-factor are usually given in dB units, such that they must be converted to ratio values before using Equation (16.22). Hence, another form of NF using ENR and Y-factor is given as:

$$NF = 10\log_{10}\left(\frac{ENR}{Y-1}\right) = ENR_{dB} - 10\log_{10}\left(10^{\left(\frac{Y_{dB}}{10}\right)} - 1\right) \tag{16.23}$$

The Y-factor can be used to measure a wide range of noise figures and is commonly used to characterize a large variety of devices.

A related topic of interest is the method to compute the NF of cascaded devices. This is extremely important for RF receiver front-end characterization. For such systems, as illustrated in Figure 16.12, the total output noise is given by:

$$N_{out} = N_{in}G_1G_2 + N_1G_2 + N_2 \tag{16.24}$$

where the gain and NF values for the ith stage are given G_i and N_i, respectively. Also, the output signal is given by:

$$S_{out} = S_{in} \times G_1 \times G_2 \tag{16.25}$$

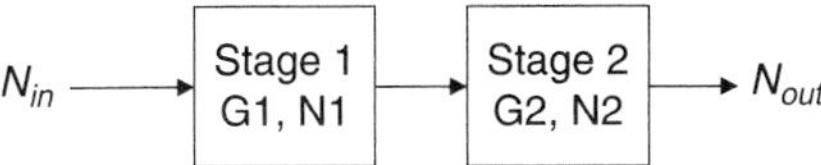

■ **FIGURE 16.12**

Computing noise figure of cascaded stages.

Equations (16.24) and (16.25) can be used to compute NF for a cascaded system:

$$F_{system} = \frac{S_{in}}{N_{in}} \Big/ \frac{S_{out}}{N_{out}} = \frac{S_{in}N_{out}}{S_{out}N_{in}} = 1 + \frac{N_1}{G_1 N_{in}} + \frac{N_2}{G_1 G_2 N_{in}} \tag{16.26}$$

This equation can be reduced to:

$$F_{system} = F_1 + \frac{F_2 - 1}{G_1} \tag{16.27}$$

where $F_1 = 1 + \frac{N_1}{G_1 N_{in}}$ and $F = 1 + \frac{N_2}{G_2 N_{in}}$. A more general equation can be derived for larger systems as follows:

$$F_{system} = 1 + \sum_{i=1}^{n} \frac{(F_i - 1)}{N_{in} \prod_{j=i+1}^{n} G_j} \tag{16.28}$$

16.2.4.7 Sensitivity and Dynamic Range

Sensitivity of the DUT is defined as the minimum signal level that the DUT can detect and operate on with an acceptable level of SNR. This depends on various factors, such as bandwidth, noise figure, and noise power at the input port of the DUT, and can be derived from Equation (16.15):

$$\begin{aligned} NF &= \frac{S_{in}/N_{in}}{SNR_{out}} \Rightarrow S_{in} = SNR_{out} \times NF \times N_{in} \\ S_{in}|_{\min} &= SNR_{out} \times NF \times kTR_s \times B \end{aligned} \tag{16.29}$$

or

$$S_{in}|_{\min} = SNR_{out} + NF + P_{Rs} + 10\log_{10}(B)[\text{with units in dB}] \tag{16.30}$$

where $S_{in}|_{\min}$ is the sensitivity of the DUT, SNR_{out} is the desired SNR of the DUT, R_s is the input source impedance with noise power P_{Rs}, and B is the DUT bandwidth.

A related term, *dynamic range,* represents the ratio of sensitivity and the maximum signal level that the DUT can handle. RF devices typically have high dynamic range, often in excess of 80 dB. Hence, the test system also needs to be capable of performing measurements over such large ranges.

16.2.4.8 Local Oscillator Leakage

In mixers, a ***local oscillator*** (LO) signal is provided that either up-converts or down-converts the input signal. Consider the case of up-conversion, where the input signal is at an intermediate frequency, f_{IF}, and the applied LO signal is at a frequency of f_{LO}. The up-conversion mixer generates the RF frequencies that are located in the output signal at $f_{LO} - f_{IF}$ and $f_{LO} + f_{IF}$, also known as mixer sidebands. However, poor isolation between the mixer ports may cause part of the LO signal energy to propagate (leak) to the output port. Hence, an additional tone at f_{LO} appears in the output signal (circled in Figure 16.13). For RF systems, the difference between the RF and LO frequencies is often small, and hence this unwanted LO signal cannot be eliminated by simple filtering. For this reason, mixers are specified for their LO leakage, which is usually 20 to 30 dB below the mixer sidebands. This is tested using a spectrum analyzer during characterization but is rarely measured during production testing.

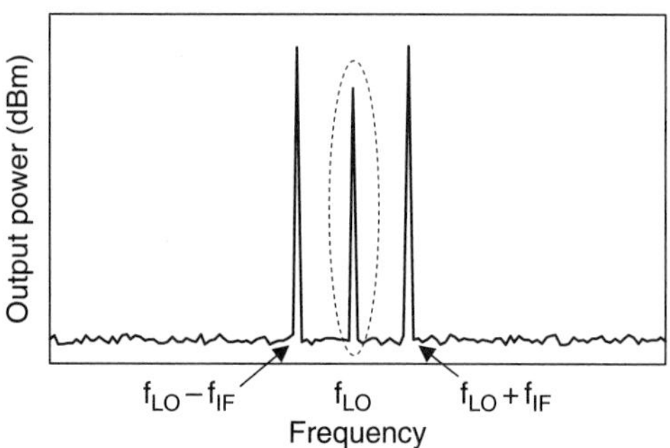

■ FIGURE 16.13

Leaked LO signal within the DUT signal bandwidth.

16.2.4.9 Phase Noise

Mixers need a clean LO signal to reliably perform the frequency translation operation. Usually, LO signals are single-tone, except for few advanced communication protocols where the LO is modulated. Because the LO acts as a reference for frequency translation, it is expected to be stable in terms of both frequency and amplitude. However, all such signals generated from oscillator circuits are nonideal; they exhibit variations in both amplitude and phase/frequency:

$$\begin{aligned} y_{ideal}(t) &= A.\sin(\omega t + \theta)(\text{ideal signal}) \\ y_{real}(t) &= A(t).\sin(\omega t + \theta(t)) \text{ (signal with phase noise)} \\ &= A(1 + \alpha(t)).\sin[\omega t + m.(1 + \sin(\theta_m t) + \theta_{random}] \end{aligned} \quad (16.31)$$

In Equation (16.31), an ideal signal and a signal with phase noise are included; $\alpha(t)$ denotes the amplitude modulation part, whereas m and θ_m denote the phase modulation part. θ_{random} represents the random phase noise portion of the signal.

The origin of such modulations can be attributed mostly to the power supply noise, flicker noise or 1/*f* noise, and thermal noise inherent to the system. The phase modulation term and the random term, together known as phase noise, cause the oscillator frequency to shift with time. This term is often described statistically because of the inherent randomness involved in its origin. For RF systems involving oscillators, it is extremely important to measure phase noise to obtain the desired performance from the system. Chapter 15 elaborates on this topic in greater detail.

Phase noise is measured using a spectrum analyzer. Most modern spectrum analyzers have built-in options to measure phase noise. Phase noise is represented in dBc/Hz, which indicates the noise level below the fundamental frequency of the oscillator within a 1 Hz bandwidth at a specific frequency offset. For example, –90 dBc/Hz @ 10 kHz indicates that the average noise power for the fundamental tone at an offset of 10 kHz is 90 dB below the carrier power. If the carrier power is 10 dBm, the mean noise power measured at 10 kHz offset from the carrier will be −80 dBm.

16.2.4.10 Adjacent Channel Power Ratio

Another important specification related to the nonlinearity and the out-of-band noise of the devices is ACPR. Typically, for cellular communication, modulated data are transmitted in specific frequency bands, also known as channels. It is thus imperative that the signal in one channel does not leak to the adjacent frequency channels and that the out-of-band phase noise of the device will not degrade the signal in the adjacent channel. However, when a modulated signal is fed to an amplifier, some intermodulation frequencies fall into the adjacent bands because of the nonlinearity of the device. The ratio of the signal energy falling into the adjacent bands to the total energy within the desired band is called ACPR of the DUT [RS 1EF40-1998]:

$$ACPR = \frac{E_{adj}}{E_{total}} \tag{16.32}$$

where E_{adj} is the energy in the adjacent band, and E_{total} is the total energy of the modulated signal. As the computed value is a ratio of two terms, it is usually represented in dB. Modern spectrum analyzers are capable of directly measuring ACPR using the built-in firmware. Depending on the communication protocol and the modulation technique, ACPR can be automatically computed for various standards. For production testing, ACPR is usually measured with a multitone input given to the DUT while measuring the amount of power in the sidebands. The spectrum of a DUT output response that is used to measure ACPR is shown in Figure 16.14. For some communication standards, ACPR is measured using a pseudo-random bit stream from the baseband processor while measuring the total leaked power in adjacent bands.

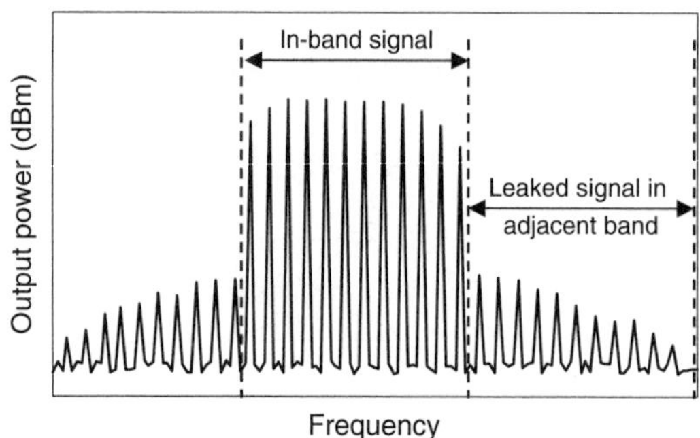

■ **FIGURE 16.14**

ACPR measurement from DUT output spectrum.

16.2.5 System-Level Specifications

A typical RF transceiver is tested for many other specifications apart from the ones discussed earlier. Usually, these specifications are targeted at system-level performance metrics such as the clarity of the received modulated signal, ***quality of service*** (QoS), etc. A few of these specifications are discussed next.

16.2.5.1 I-Q Mismatch

Modern RF communication systems employ complex modulation schemes where data bits are combined at different phases—in-phase (I) and quadrature (Q) are added to increase the effective data rate and efficiency of the modulation process. However, in all practical systems, such modulations introduce imbalances in their phase relations, and this often leads to errors during demodulation. This phenomenon is known as I-Q mismatch. In a system, I and Q are usually represented as follows:

$$I(t) = A_I \cos(\omega t)$$

and

$$Q(t) = A_Q \sin(\omega t) \tag{16.33}$$

Typically, A_I and A_Q are kept same for both channels. However, because of mismatch between the two channels, the modulations often have amplitude and phase imbalances that appear as:

$$I(t) = \alpha_I A \cos(\omega t) + \beta_I$$

and

$$Q(t) = A \sin(\omega t + \theta_Q) + \beta_Q \tag{16.34}$$

where the errors in amplitude and phase have been assumed to be lumped in one of the channels. The amplitude error is represented as α_I, and the phase error is represented as θ_Q. The DC offsets in I and Q channels are β_I and β_Q, respectively. Usually, the offsets can be easily removed by long-term averaging for each channel. However, the mismatches in amplitude and phase are hard to remove and must be characterized carefully for the DUT to operate reliably. I-Q mismatch contributes significantly to the EVM and BER (discussed later) of a DUT.

To measure the mismatch values, we need an ideal source that can phase lock to the signal that we need to measure. For such communication-based measurements where modulated data need to be transmitted, a spectrum analyzer is not sufficient. For this reason, both amplitude and phase imbalance values are not measured and specified during production testing because of the long test times involved and the expensive test hardware required to perform such measurements. Therefore, this is primarily a bench measurement. In most cases, a communication analyzer [Agilent Rx-2002] with the required modulation/demodulation capability is used for this measurement. These can provide a trigger to start the modulation and demodulate the data from the DUT to detect the frames. Usually, many frames of modulated data are looped through the device to estimate the specification values.

16.2.5.2 Error Vector Magnitude

EVM is one of the most important specifications that are measured to assess the quality of transmission of a complete system. EVM is measured for phase-modulated systems such as *quadrature phase shift keying* (QPSK) and *16-quadrature amplitude modulation* (16-QAM). It presents the difference between actual and ideal (modulated) signals [Agilent EVMb-2000].

To measure EVM, all the symbols obtained at the demodulator output for the received waveform are compared against the ideal symbol locations [Agilent EVMa-2005]. The root-mean-square (RMS) sum of the error vectors (which includes the phase error values) is then used to compute EVM over a set of N symbols. As Figure 16.15 shows, the symbol decision from the demodulator is given by $\tilde{y}$, whereas the ideal symbol is given by $\tilde{n}$. Therefore, the error vector, $\tilde{a}$, is given by $(\tilde{y} - \tilde{n})$. EVM for N symbols can be calculated as shown in Equation (16.35):

$$EVM = \sqrt{\frac{1}{N}\sum_{i=1}^{N}(\tilde{y}_i - \tilde{n}_i)^2 \bigg/ \frac{1}{N}\sum_{i=1}^{N}\tilde{n}i} \tag{16.35}$$

Typical EVM plots for QPSK modulation are shown in Figure 16.16. By looking at constellation plots, one can determine the origin of the errors in the device caused by I-Q modulator amplitude/phase imbalance, or phase noise.

Measuring EVM requires test equipment that can demodulate the signal and handle time synchronization for data extraction. Moreover, a large number of data bits are transmitted to measure EVM, which incurs a large test time. Hence, EVM testing is generally not performed during production. However, it is a mandatory

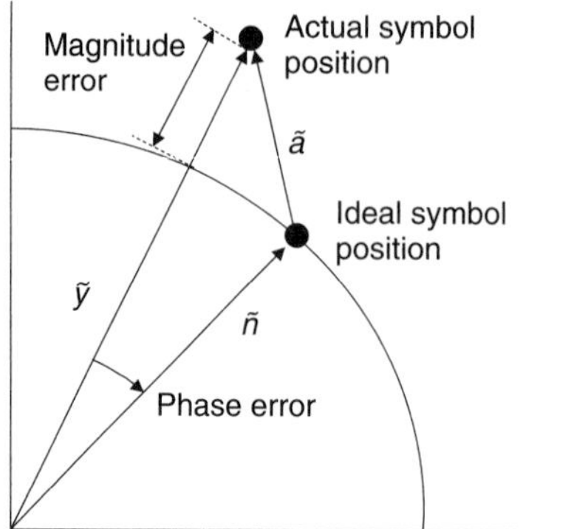

■ **FIGURE 16.15**

Computation of EVM from constellation points.

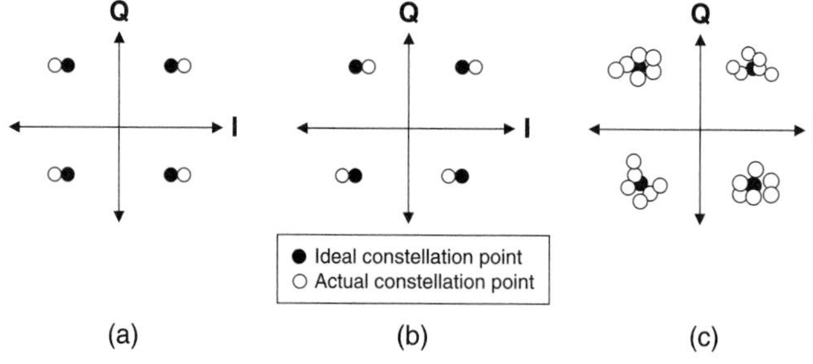

■ **FIGURE 16.16**

Smeared constellation for QPSK for various I-Q mismatch effects: (a) amplitude imbalance ($A_I > A_Q$), (b) phase imbalance ($\theta_Q \neq 0$), and (c) noise (phase noise + random noise) [refer to Equations (16.33) and (16.34)].

test that is performed during the characterization phase. Modern spectrum analyzers and ***vector signal analyzers*** (VSA) with demodulation capabilities can measure EVM. Other techniques for EVM measurements have also been proposed to reduce the overall test time [Halder 2005].

16.2.5.3 Modulation Error Ratio

The SNR of a digitally modulated signal is called the modulation error ratio (MER). SNR is generally used in the context of analog signals, whereas MER is defined for digital modulated signals only. It is usually expressed in dB. For N symbols, MER is defined as:

$$MER = \frac{\sum_{j=1}^{N} (I_{rj}^2 + Q_{rj}^2)}{\sum_{j=1}^{N} \left[(I_{tj} - I_{rj})^2 + (Q_{tj} - Q_{rj})^2 \right]} \tag{16.36}$$

where I_{rj} and I_{tj} are the received and transmitted in-phase components, and Q_{rj} and Q_{tj} are the quadrature counterparts, respectively.

16.2.5.4 Bit Error Rate

The BER test indicates how many bit errors occurred when a certain number of bits (usually a pseudo-random sequence) is passed through a network system. BER indicates the fidelity of the data transmission through an entire network, or a subsystem of the network. In addition, BER measurements also serve as *quality of service* (QoS) agreements between network end users and service providers [RS 7BM03-2002].

A bit error occurs when the receiver circuitry incorrectly decodes a bit. There are many possible causes for BER, such as low signal power, noise within the system, phase noise of oscillators and associated jitter, crosstalk, and EMI from radiated emissions, to name a few. A BER tester, also known as a BERT, is used to test BER. The BERT contains a pseudo-random pattern generator and an error counter that counts the number of bits in error by comparing with the reference bits. Hence, BER is defined as:

$$BER = \frac{N_{err}}{N} \tag{16.37}$$

where N_{err} is the number of error bits, and N is the total number of bits. Obviously, the questions are how long should we transmit bits, and at what value of bit errors will the test be stopped? The rule of thumb is that at least 100 error bits are needed to deem the test process reliable. Therefore, if the BER of a system is expected to be 10^{-6}, which means that 1 error in every 1 million bits is erroneous, then we need to transmit at least 100 million bits to get a reliable measure of BER for that system. Therefore, the lower the BER of the system, the longer it takes to measure BER. During production testing it might take hours, even days, to characterize and measure BER of the system reliably. However, to mitigate this problem, another technique, known as the Q factor method, is used to quickly measure BER. With this technique, instead of comparing bits, the histograms of the amplitudes of the two logic levels are plotted, and a Gaussian curve is fitted to those histograms. The mean and standard deviation of these two histograms can be found by moving the threshold for error detection. From these mean and standard deviations, the Q-factor can be determined as follows:

$$Q = \frac{\mu_1 - \mu_2}{\sigma_1 + \sigma_2} \tag{16.38}$$

where μ_1, μ_2 are the mean and σ_1, σ_2 are the standard deviation values of the two logic levels. From the Q value, the BER can be directly obtained as follows:

$$BER = \left[\frac{1}{Q\sqrt{2\pi}}\right] e^{\left(\frac{-Q^2}{2}\right)} \tag{16.39}$$

This method is valid under the assumption that the error distributions are Gaussian. This might not be the case if periodic patterns are present. Various other methods for efficiently measuring BER have been proposed, such as those based on introducing a limiting filter in loopback test mode or rotating constellation points for efficient test [Bhattacharya 2005].

16.2.6 Structure of RF Systems

RF designers usually employ the well-known super-heterodyne architecture for implementation because of its subtle advantages over other architectures. However, certain applications, such as pagers, use the less-popular homodyne architecture to balance the various requirements of the system. A typical RF system can take a signal from a baseband unit, up-convert to a higher frequency suitable for transmission, further amplify the signal and transmit via an antenna, or do the reverse (*i.e.*, receive the signal from antenna and down-convert to baseband frequencies). In addition to transmit and receive functions, RF systems usually consist of oscillators/frequency synthesizers to synchronize data transmission between the transceiver pair.

Figure 16.17 shows block diagrams of the super-heterodyne and homodyne transceiver architectures [Razavi 1997]. Although various architectures are available to build a system, RF system designers typically use one of these two architectures for realizing a transceiver. The choice of architecture is mainly driven by dynamic range, NF, and linearity of the complete system. Before going into the details of the architectures, it would be useful to discuss the problems of image rejection in RF receivers. Assume, for example, a receiver working at 900 MHz and a down-converted frequency, also known as ***intermediate frequency*** (IF), of 70 MHz; then the oscillator would be working at 970 MHz. The oscillator may also work at 830 MHz, but this frequency is usually not used as it would require a higher tuning range from the oscillator (we will come back to this later). If the input of this receiver has a signal at 1040 MHz (970 MHz + 70 MHz), the receiver would also down-convert it to 70 MHz. This would cause interference to the original signal; this unwanted signal is known as *image* signal. Often, the image is removed by filtering the input signal before down conversion (the first filter after the receiver antenna in Figure 16.17a); however, a very low IF would require a very high-Q filter at RF frequency, which is often difficult to implement. Therefore, the IF frequency is selected to be as high as possible to ease the front-end design of the receiver. It is common in communication systems to use the same oscillator for both the receiver and the transmitter.

In receiver design, the oscillator is often chosen to be higher than the RF frequency. Let us assume that an oscillator needs to tune over a range of 20 MHz, with an RF of 400 MHz—for a fixed IF of 70 MHz. In this case, two possible oscillator frequencies can be used: 330 MHz and 470 MHz. The metric we are trying to reduce is the ratio of tuning range of the oscillator. The higher side oscillator needs to vary from 460 to 480 MHz with a tuning ratio of 1.04, whereas the lower side oscillator would need to tune from 320 to 340 MHz with a ratio of 1.06. This ratio directly

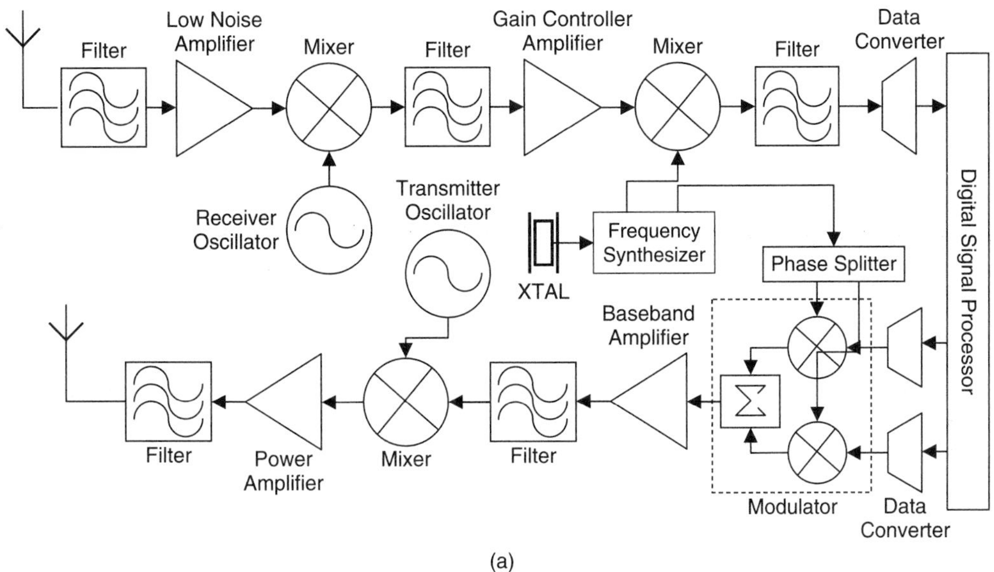

(a)

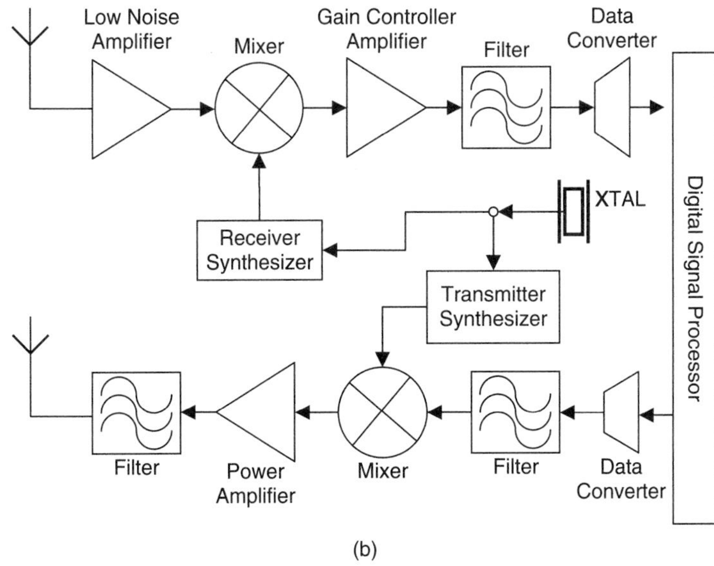

(b)

■ **FIGURE 16.17**

Various RF transceiver architectures: (a) super-heterodyne and (b) homodyne or direct conversion.

translates to the chip area and, thus, it is better to choose an architecture with a higher side oscillator [Lee 2003].

Before comparing the architectures, it is also worthwhile to discuss the various test access ports available for these architectures. Direct-conversion architectures usually have limited access to internal nodes; the only nodes accessible are the RF port connecting to the antenna/duplexer and the baseband port. In most cases, the LO is also internal to the DUT, implemented via a frequency synthesizer. The tuning of LO can be done only via the digital interface, but access to the RF signal within the DUT is usually unavailable. Super-heterodyne, on the other hand, allows implementation of external filters, and hence the RF signals at the output ports of the mixer, LNA and PA, are generally accessible. This allows testing of individual specifications of these blocks and, at the same time, characterization of the complete system for end-to-end specifications.

The architectures illustrated in Figure 16.17 pose different sets of advantages and disadvantages. The super-heterodyne architecture employs an image-rejection architecture using two LO signals in quadrature to cancel the image, and this modified version is known as the *weaver architecture*. Direct conversion does not face the problem of image-rejection as the signal and LO are at the same frequency. It thus reduces the filtering requirements on the receiver. Also, direct-conversion architecture does not require accurate I-Q matching as compared to super-heterodyne.

However, direct conversion has certain limitations compared to super-heterodyne. Direct conversion is prone to producing large DC offsets for small nonlinearities present in the device. Even-order harmonics tend to show up as DC offsets at the output of the receiver, which cannot be removed by simple AC coupling. This forces the designers to use differential signaling, thereby doubling the power requirements of the system. Also, such architectures need isolation from the LO input to the mixer input, as the LO leakage appears as a DC signal at the output of the receiver. Finally, $1/f$ noise is of greater concern in direct conversion, as this increases the settling time of the receiver and may well cause loss of packets for standard modulation schemes.

With the limitations posed by the direct-conversion architecture, super-heterodyne has gained widespread popularity in modern RF devices and will continue to remain popular until a new innovative, low-power architecture is proposed.

16.3 TEST HARDWARE: TESTER AND DIB/PIB

In this section, we discuss ATE and the ***device interface boards*** (DIB) that are used to interface the DUT to the tester. Figure 16.18 shows a generalized block diagram of a tester. Usually, testers perform a series of tests under the control of a test program developed for production test. Any tester usually has three main blocks: *source unit*, *capture unit*, and *clock and synchronization unit*. Next, we discuss each of these submodules of the tester.

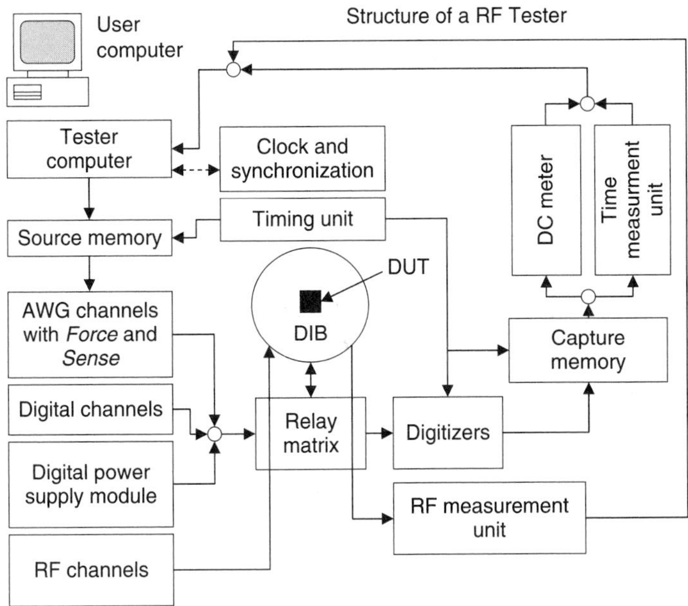

■ **FIGURE 16.18**

A typical tester with RF options.

Power supply (PS) modules. Testers have accurate power supply modules that can source/sink large voltages/currents. Typically, these are digitally controlled with 8 to 12 bit resolution. The voltage sourcing capability of testers can go up to 60 to 80 V, while current capabilities can be 5 to 10 A per channel.

Digital channel (DCH). A tester usually has a large number of digital channels (as many as 256 to 512 channels). The digital channels provide digital signal input to the DUT through a relay matrix (discussed later). The tester computer and the clock unit usually govern the speed of operation. Usually, the speed varies from 1 to 500 MHz, based on the purpose of the tester.

Analog channel (ACH). Each tester has mixed-signal capabilities provided through analog channels. The analog channels provide sinusoids up to a frequency of 100 to 200 MHz as limited by the tester digitizer unit. The sinusoidal signals are usually generated by using a DAC and corresponding filtering stages. For this reason, the number of analog channels is usually limited compared to digital channels because each channel needs to have DAC/ADC and source/capture memory associated with it.

Arbitrary waveform generator (AWG) channel. Apart from digital inputs, testers have an AWG unit that can source any arbitrary waveform. The stimulus to be applied to the DUT is programmed into the source memory by the tester program and AWG sources it to the DUT via the relay matrix.

RF channel. RF testers have a RF signal generator unit apart from the digital and AWG channels. However, the RF channels are not connected through the relay matrix to maintain impedance matching (see Figure 16.18). It is important to maintain a low-loss connection to the DUT from the RF source. Also, test sockets need to be carefully chosen to reduce parasitic impedance and maintain a low reflection coefficient. The range of a typical RF signal generator varies from 1 to 8 GHz that can usually source up to 30 dBm.

Capture unit. Testers have a capture unit that consists of a digitizer and a RF measurement unit. The digitizer captures all the digital and analog outputs and stores them in the capture memory. Depending on the type of measurement to be performed on the DUT, the DC meter or the time measurement unit carries out computations on the captured waveforms. The RF measurements are usually performed in a separate dedicated unit. Although not shown in the figure, there is a digitizer block within the RF measurement unit that converts the results to a digital format.

Clock and synchronization. The clock and synchronization unit controls the signal flow through all the modules in the tester. The clock is derived from the tester CPU or an onboard oscillator/PLL capable of generating a wide range of frequencies.

DIB and Probe Interface Board (PIB). The tester and the DUT are interfaced through a DIB or a ***probe interface board*** (PIB), depending on the type of test. For wafer test, a PIB is used because the tester interfaces to the device via probes touching the pads on the wafer. However, such contact is pressure limited, and probes often form a Schottky contact with the wafer and become limited in signal capabilities (*i.e.*, frequency and current). Hence, only supply current tests and continuity/leakage related tests are performed using PIBs. To test all the devices on the wafer, a stepper machine moves the wafer chuck to align the next die with the PIB. As the probes are thin and fragile, they are usually fixed with a horizontal chuck.

DIBs are often more sturdy and versatile compared to PIBs. Hence, a larger number of tests can be performed using a DIB during production testing, when the part is already packaged. Typically the DIB is a multilayer ***printed circuit board*** (PCB), through which all tester inputs and outputs are connected to the DUT. To reduce noise coupling between the different tester channels, the analog and digital ground planes are kept separate and connected directly beneath the DUT. Furthermore, signal planes are isolated from each other by inserting ground planes between them.

Various active and passive components, such as resistors, capacitors, buffers, and relays, are soldered onto the DIB and PIB to aid in testing and switching between various test modes. Hence, the measured performance of the DUT by the tester is a combined effect of the DUT and the DIB components. Hence, the DIB components must be of small tolerance to produce reliable data and need to be tested for their functionality before production testing. As a result, various diagnostic tests are run on a DIB before using it in production tests.

A socket is used to interface with the DUT, with various types of sockets available for different classes of devices and package types. Whereas close-lid sockets are commonly used for analog/digital/mixed-signal devices, RF parts require carefully designed and characterized sockets to maintain proper impedance matching from the RF port of the tester to the DUT pin. RF sockets are typically ***zero-insertion force*** (ZIF) sockets that press the DUT on the DIB to make reliable contact.

Most of the tester resources are available to the DIB via the pogo pins. The tester-to-DIB interface is made through pogo pins that are connected to the relay matrix on the tester side. However, RF connections are made directly to the DIB through SMA/SMB connectors. Most of the RF traces are routed directly on the top layer. The RF traces on the PCB are of the microstrip type, with the width and trace length carefully selected to achieve minimum power loss. As mentioned before, maintaining a 50-ohm matching is critical in any RF measurement. This is also true for the DIB, and manufacturers use various techniques, such as ***time-domain reflectometry*** (TDR), to check the trace impedances on the DIB.

Test program. A test program is a piece of software developed to interact with the tester using the tester controller CPU based on a high-level language such as C or Pascal. However, each tester has its own set of functions/subroutines that aid in the overall test development process. These functions include programming the different components of the tester, test application, and test response data capture and analysis. Another important function of a test program is to set the global clock and synchronize the test process. This is usually done via global clock pin assignments through the tester code.

16.4 REPEATABILITY AND ACCURACY

Any test process must be verified for its accuracy, precision (repeatability), measurement error, and resolution of the test setup. For RF tests, all these factors are extremely important to obtain reliable measurement data. We next describe each of these factors and discuss its implication on the test process. Refer to Figure 16.19 for the following discussion on the various factors that contribute toward making reliable measurements.

Measurement error. Measurement errors can typically be classified in two types, systematic and random. Systematic errors appear consistently for consecutive measurements. The test process can be easily calibrated to reduce the effect of such errors. Random errors, on the other hand, are more difficult to remove or reduce. Although there can be innumerable sources of random error, the largest contributors are usually thermal noise and noise injected within the system from the power supply. Although it is theoretically impossible to eliminate random errors in measurements, filtering is commonly used to mitigate the effects. In addition, DSP techniques such as smoothing/averaging can be used to minimize random errors. In certain cases, measurement noise is introduced as a result of the human factor involved in the measurement system. A common example is the use of improper resolution bandwidth settings by the user in a spectrum analyzer, which can easily result in 0.5 to 1 dBm error in the measurement.

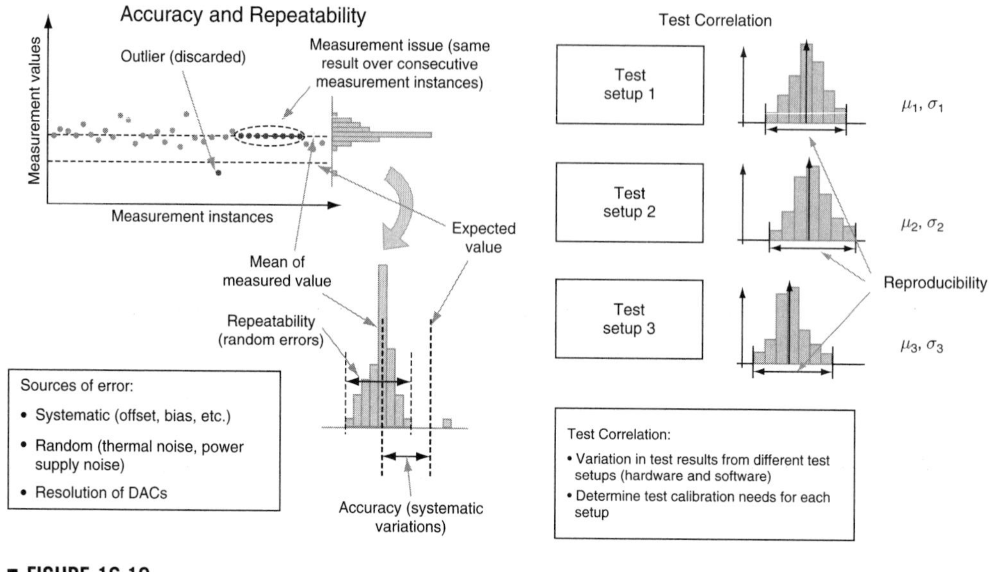

■ **FIGURE 16.19**

Measurement accuracy and various related terms.

Resolution. Another well-known source of error in a measurement system is the resolution of the data converter system. Data converter systems usually consist of ADCs and DACs. As in all other testers, RF testers also digitize the analog signals with ADCs before the tester can process the measurements. In the process of digitization, quantization error is introduced into the measurement system. For an N-bit digitizer with a full-scale voltage of V volts, the quantization noise introduced is $V/(2^N - 1)$.

Accuracy. Accuracy of a test system is defined as the difference between the average of a set of measurements performed under the same test conditions and the ideal outcome, provided the ideal outcome of a test system is known a priori. A test system can consist of a single instrument, a set of instruments, or the tester as a whole. Accuracy is usually affected by systematic variations. Hence, calibration methods can be applied to increase the overall accuracy of the measurement.

Repeatability. If the same measurement is repeated many times on a single test setup, then the variation in the measurements is known as the repeatability of the test system. Variations in the measurement values are caused by random noise inherent to the system. Mathematically, repeatability is represented as the standard deviation of the measurement values obtained. In some cases, repeatability is also defined as the variation in measured values over time or different measurement conditions (*i.e.*, temperature and humidity). This indicates the degree of stability of the measurement system.

Correlation. The variation in the measured data obtained from different test setups is called the degree of correlation. The test setups may consist of different pieces of the same hardware, software, or both. Different test setups will exhibit different mean and variance in the measured data. Obtaining a high correlation of measured values from various test setups in a production test environment is extremely important. Another term that is closely related to correlation is *reproducibility*, which is an indicator of the variation in a particular measurement from different test setups. While correlation is defined over a number of measurements, reproducibility is defined for one measurement set. This is different from *repeatability*, which is the variation in measurement data obtained from a single test. In an RF testing environment, accuracy and repeatability-related issues are extremely important for performing reliable measurements. Correlating the measurements across various test setups (*i.e.*, instruments and testers), different versions of test programs may be necessary to ensure a reliable test and high reproducibility of the test measurement data in a production test environment.

Gauge R&R. During production testing, another important factor that is measured is ***gauge repeatability and reproducibility*** (GR&R). A study of GR&R provides information on the overall test system performance (including ATE and DIB) by analyzing measurement errors from various sources. Typically, the sources of error are from part-to-part variations, operator variations (various operators performing the same test on different parts), and gauge or equipment variations. In certain types of analysis, interaction between parts and operators can provide additional information about the gauging process. In most GR&R studies, multiple operators collect data from the same devices in random order by performing the same set of measurements. Most studies use two or three ATE/DIB combinations and five to ten parts (assuming the test data collection process is automated). The formula shown in Equation (16.40) is used to compute GR&R.

$$GR\&R = \left(\frac{5.15\sigma_c}{c(USL - LSL)} \right) \tag{16.40}$$

where

$$\sigma_c = \sqrt{\frac{1}{N}\sum_{i=1}^{N}\sigma_i^{\,2}}$$

In the above equation, σ_i indicates the standard deviation for each of the parts or devices, and N indicates the total number of parts tested. *USL* and *LSL* are usually the specification limits associated with the individual tests and the value of c is determined based on the number of parts used in the study. A GR&R value of less than 0.1 indicates very high repeatability and reproducibility from the system, while a value of more than 0.3 is usually unacceptable. Values of

GR&R ranging from 0.1 to 0.3 usually indicate that some repair/calibration is needed on the system.

Calibration. Calibration is a method by which test instruments are correlated against standard measurements and are adjusted accordingly to meet the desired values. All the variations, such as accuracy and repeatability, cannot be eliminated completely; however, their effects can be minimized using calibration methods. All test instrumentation (*i.e.*, tester and the bench instruments) are calibrated at regular intervals. The calibration standards are provided either by a standards society, such as the National Institute of Standards and Technology (NIST) or the *original equipment manufacturer* (OEM) itself. Most testers have standards built into them that are stabilized for wide variations in environmental conditions. Although calibration can be for both hardware and software, software calibration is more popular because of its ease of implementation. This essentially adjusts the different weights for measurements and applies them during runtime. Similar calibrations are also performed for DIB boards during production testing.

It is worthwhile to mention that during manufacturing, as the ATE tests thousands of parts continuously, the tester performance is likely to degrade over time. As a result, the tester might erroneously pass defective parts, leading to huge revenue losses. To avoid this outcome, testers are also calibrated frequently during production testing, using *golden devices* (*i.e.*, devices that are known to be good) interfaced through customized calibration boards to quickly diagnose possible faults in the tester.

16.5 INDUSTRY PRACTICES FOR HIGH-VOLUME MANUFACTURING

The test process is well defined in industry and starts at an early stage of product development. After the device specifications are defined, a test plan is usually formed if there are a large number of specifications. Production test, also known as high-volume manufacturing test, has two parts: (1) choice of tester and development of tester programs for that specific tester, and (2) design and fabrication of the DIB/PIB. The choice of tester needs to be carefully made because the tester needs to be capable of performing all the tests for that device during high-volume manufacturing.

As the device is designed and fabricated, first a schematic of the DIB is created. This schematic contains the various test configurations on the same board, and the different tests are performed via switching the circuit configuration through relays. Hence, a successful design is the key to performing tests reliably during production. This schematic goes through thorough design reviews to ensure proper functionality of the DIB board. As this procedure is completed, the test program is developed and the board is fabricated.

The first silicon usually goes to a group of engineers who perform characterization tests on the device. Characterization tests use accurate bench test equipments to test the part. In addition, this process is time consuming as little automation is involved during characterization. Therefore, some complex tests, such as I_{out} *versus* V_{in} curve, variation of bandgap output voltage over temperature, or BER over a range of SNR values, are only performed during characterization, as conducting these tests during production would be impossible because of time constraints and the hardware resources available.

After the device passes the characterization tests, the next step is verifying the test program and the DIB on a tester. This usually involves debugging the test program and finding any possible mistakes within the DIB. Another important aspect is test scheduling, which controls the flow of the tests to obtain optimal performance in a minimum amount of time. Moreover, the idea of scheduling is extended to test instruments, even testers, where the test program is arranged to best utilize the instruments with the least idle time.

Next, the test is correlated between testers and DIBs. To do this, the same set of parts is tested on one tester using different DIB boards manufactured for that device and different testers. This gives an idea of the amount of variation that will be observed during high-volume manufacturing. In addition, tests are performed to measure bin-to-bin correlations. Binning is a process of separating devices with different performances. One such example is binning the IDD leakage failure devices together—as these devices are often expected to fail nonlinearity tests. These correlation values give important feedback to the designers and the fabrication engineers about the quality of the manufacturing process. Once the test program runs properly on the tester (*i.e.*, the tests give proper measurement data for devices) and test time is acceptable, the device is released for high-volume testing.

16.5.1 Test Cost Analysis

Test cost is an important factor that drives the overall cost of the device and may reduce the profit margin significantly. Various factors contribute to the test cost. The major contributor to the cost of testing is the ATE or the tester cost. As devices continue to grow more complex, the test capabilities need to be constantly improved. Also, the speed of the tester needs to increase because constant device scaling since the mid-1980s has pushed the device speeds significantly higher. Manufacturers are constantly looking for low-cost testers that can reliably test a complex, high-speed device during high-volume production testing.

Apart from the cost of testers, large test time is a major factor for increased test costs. Typically, test time for wireless devices ranges from a few seconds to a few minutes. During production, when millions of devices are tested, even such apparently small test times can create a bottleneck. A simple example will make the problem more obvious. Suppose a device test time required during production is 60 seconds. Therefore, the number of devices that can be tested is only 1440 each day ($=24 \times 3600/60$). Therefore, testing is normally carried out on many testers in parallel. If the manufacturer devotes 10 testers solely to this example device, then to release a million devices to the customer will require 70 days. This clearly shows that a small reduction in test time can increase the throughput significantly. Therefore, there is a constant need in the test community to reduce production test time.

Aside from the two major factors just discussed, various other factors contribute to the overall test cost. Additional costs come from engineering errors or other human factors. For example, an improperly designed DIB or a bug in the test program can significantly increase the time required to release a product. This can cause the manufacturer to lose significant market share for that product. Such factors can be fatal for small businesses, and the success of the manufacturer relies heavily on the test process.

16.5.2 Key Trends

Having discussed the various aspects that drive the test cost, we now point out a few methods that manufacturers adopt to reduce the cost of testing. Of the various techniques, the major concern for semiconductor manufacturers is to reduce the cost of testers. This can be achieved by reducing the amount of resources available on the tester used during production testing. However, this may lead to inefficient testing procedures. Another way that seems more feasible is to use low-cost testers that run slower. Such testers are often inexpensive (~$300 K) and can still perform most of the tests for the device.

The major tester manufacturers for RF and wireless devices are Teradyne, Advantest, Agilent, and Roos, among others. Certain manufacturers are also trying to build in-house testers that can support a majority of the devices they manufacture. For smaller industries, maintaining a tester is usually expensive and, as a result, they rely on external test vendors to test their devices.

Apart from using a low-cost tester, manufacturers try to test as many devices in parallel as possible, known as multisite testing. This can be done in a relatively easy manner for low-frequency devices. However, for RF devices, multisite testing can be difficult, as many issues need to be addressed. As discussed earlier, impedance matching is a big concern for RF measurement; a significant amount of signal may be lost if improper impedance matching exists between the tester output port and the DUT input pin. Therefore, the DIB must be carefully designed to avoid such problems. In case of multisite testing, apart from maintaining impedance matching, it becomes important to consider crosstalk between different sites. The current trend in industry is to use a quad-site DIB for RF devices. However, there is also effort in the industry to test 8 or even 16 sites in parallel. This puts stringent requirements on the tester, such as signal sourcing capabilities for all sites, and may be difficult to obtain in the current generation of testers. Innovative methods are used, such as using a single channel, amplifying it, and then sourcing it to multiple sites using power splitters. This requires careful characterization of the DIB before porting it to the tester. Another issue that becomes extremely important for RF testing is providing a clean power supply to the DUT. Most present-day RF devices rely on a clean supply voltage to obtain a steady bias point. Small deviations in power supply voltage may often lead to a nonreproducible test setup and unreliable test data.

Advances in SOC technology have enabled integration of RF and baseband in a single chip. In most SOCs, the RF section is usually controlled by internal DSP signals. Hence, the DSP needs to be programmed during testing to sequentially perform the tests during production. This has increased the problems related to testing because the tester may have limited test resources that can not be allocated for programming the DSP. To address these problems, onboard controllers such as ***field programmable gate arrays*** (FPGA) or ***complex programmable logic devices*** (CPLD) are used to program the DSP within the DUT.

In some cases, SOCs provide limited access to internal RF nodes. Hence, characterization or production testing of embedded modules becomes difficult [Lau 2002]. To ensure proper testing and characterizing of such devices, innovative methods in the form of ***built-in self-test*** (BIST) for ***design for testability*** (DFT)—[Dabrowski 2003] [Ferrario 2003] [Sylla 2003] [Yin 2005]—are often used. This involves adding extra circuit components within the DUT to aid in testing. BIST is well known and has been extensively used for digital devices; however, BIST is becoming more popular for mixed-signal and RF devices [Veillette 1995]. RF BIST is still in its infancy as there is a huge gap between the test and design community. Considerable research is currently under way to bridge this gap and make RF BIST a reality [Voorakaranam 2002] [Bhattacharya 2004] [Cherubal 2004] [Valdes-Garcia 2005].

16.6 CONCLUDING REMARKS

As RF technology advances, it is becoming more important to resort to innovative techniques to keep testing costs down. The scope of test engineering is not limited to the knowledge of test software and tester resources, and it has become imperative for engineers to be well versed with the system design and modeling aspects so that they can make educated decisions to create an efficient test plan. As test development proceeds, it is necessary to verify the software even before the devices are fabricated. This requires modeling of the DUT in order to verify the software beforehand.

In the nanometer era that we are entering, the process variation is greater than ever before. Currently, designers attempt to reduce the effect of this variation for each specification; however, the specification spreads are much larger than that of their earlier counterparts. Hence, the test setup and measurement system needs to be robust enough to work over this wider range of specifications.

With wireless devices becoming ubiquitous, the amount of interference that a present-day RF SOC experiences is extremely large compared to earlier devices. Therefore, the devices must be characterized for harsh conditions where overall SNR is low and most of the dynamic range is occupied by the elevated noise floor. Such effects cannot be guaranteed by simulation only and require the manufacturer to apply a complex and elaborate field test to the DUT. Of course, this increases the test costs significantly. To mitigate these elevated test costs, test, design, and system engineers need to work in unison to develop a strategy that makes the devices testable. This may lead to the development of new protocols with advanced coding schemes at higher RF frequencies. One example is the development of 5.8-GHz cordless telephones that replaced the existing 2.4-GHz units. The technology needs to evolve around the needs of the semiconductor industry, which can be from a consumer perspective, relate to manufacturing issues, or involve test requirements. As we move toward *third-generation* (3G) wireless standards and beyond, new challenges will develop, and we will need innovative test methods for those devices that will test, diagnose, and correct themselves, leading to a new generation of truly adaptive, self-correcting devices.

16.7 EXERCISES

16.1 **(Test Instrumentation)** In a test setup, the minimum spacing between the tones of a multitone waveform is 100 KHz, and the largest amplitude difference is 10 dB. If the available RBW settings in the spectrum analyzer are 30 KHz, 100 KHz, and 300 KHz with Q of 20, 10, and 5 respectively, which RBW setting would you use to accurately measure the waveform and why?

16.2 **(Test Instrumentation)** In a noise figure measurement, three noise measurements are made to determine the NF of a DUT. If the output noise values measured for input noise of −70 dB, −60 dB, and −50 dB are −66.78 dB, −56.32 dB, and −46.91 dB, respectively, what is the NF of the DUT? (*Hint:* Use least-square fit to estimate NF.) What is the maximum accuracy that can be obtained from this instrument?

16.3 **(Circuit-Level Specifications)** From Equation (16.9), derive an expression for IP3 in terms of a_1 and a_3 (assume a_0 and $a_2 = 0$).

16.4 **(Circuit-Level Specifications)** For an instrument, the ENR is given as 1.32, and the measured values of P_{On} and P_{Off} are 13.51 dB and 11.3 dB, respectively. What is the NF of the DUT?

16.5 **(System-Level Specifications)** Measurements made on a receiver system with QPSK modulation show gain mismatch of 0.5 dB and phase mismatch of $\pi/15$ radians. What would be the average EVM value measured on a large number of devices?

16.6 **(System-Level Specifications)** A digital bit stream was measured using a logic analyzer, and the logic levels were captured using a high-speed sampling oscilloscope. The average output high level, VOH, was measured as 1.5 V with a spread of 22%. VOL was 0.32 V with a spread of 31%. What is the range of BER values possible for this device?

16.7 **(Repeatability and Accuracy)** In a measurement setup, an engineer obtained the following data made on a DUT. What is the accuracy of the measurement system? What are the possible anomalies present in the data? What can the engineer do to clean the data?

Instance	Measurement	Instance	Measurement
1	10.65064916	21	10.74065448
2	10.31679807	22	10.50663072
3	10.12127354	23	11.91165271
4	10.81673331	24	11.50850651
5	4.08778264	25	10.55634287
6	10.36293711	26	10.17476899
7	10.5445455	27	10.38719247
8	10.63695599	28	11.54113781
9	10.96387893	29	17.72616125
10	11.02340429	30	10.39783598
11	11.29191079	31	10.46899635
12	11.29191079	32	11.62494034
13	11.29191079	33	11.62740999
14	11.29191079	34	11.70090761
15	10.15647102	35	11.00929383
16	10.13851337	36	11.10099202
17	10.19516033	37	10.74165817
18	11.40144598	38	10.27407421
19	11.52970227	39	10.04422281
20	11.34563046	40	10.93843566

16.8 **(Repeatability and Accuracy)** Explain the differences between accuracy and reproducibility.

16.9 **(Test Cost Analysis)** In a production line, the test time for a DUT is 60 seconds. The device needs to reach the customer in 60 days with a 200-K volume. What level of parallelization is needed in testing (multisite) if the production testers can be run 16 hours a day?

Acknowledgments

The authors wish to acknowledge Professor F. Foster Dai of Auburn University; Professor Sule Ozev of Duke University; Dr. Pramod Variyam and Tushar Jog of WiQuest; Dr. Qing Zhao, Dr. Ajay Kumar, and Dr. Dong Hoon Han of Texas Instruments; and Rajarajan Senguttuvan of the Georgia Institute of Technology for their careful comments and reviews in preparation of this chapter. We would like to extend our special thanks to Professor Charles E. Stroud of Auburn University for proofreading and help in finalizing the chapter.

References

R16.0 Books

[Burns 2000] M. Burns and G. Roberts, *An Introduction to Mixed-Signal IC Test and Measurement*, Oxford University Press, New York, 2000.

[Lee 2003] T. Lee, *The Design of CMOS Radio Frequency Integrated Circuits*, Cambridge University Press, London, United Kingdom, 2003.

[Nekoogar 2005] F. Nekoogar, *Ultra-Wideband Communications: Fundamentals and Applications*, Pearson Education, Upper Saddle River, NJ, 2005.

[Pozar 2005] D. Pozar, *Microwave Engineering*, John Wiley & Sons, Hoboken, NJ, 2005.

[Razavi 1997] B. Razavi, *RF Microelectronics*, Prentice Hall, Upper Saddle River, NJ, 1997.

[Schaub 2004] K. Schaub and J. Kelly, *Production Testing of RF and System-on-a-Chip Devices for Wireless Communications*, Artech House, Boston, 2004.

[Wang 2006] L.-T. Wang, C.-W. Wu, and X. Wen, editors, *VLSI Test Principles and Architectures: Design for Testability*, Morgan Kaufmann, San Francisco, 2006.

R16.1 Introduction

[Agilent NA-2000] Agilent Technologies, www.educatorscorner.com/media/SLDPRE_BTB_2000Network.ppt, 2000.

[Agilent SP-1997] Agilent Technologies, S-parameter techniques, AN95-1, App. Note, 1997.

[Kurukowa 1965] K. Kurukowa, Power waves and the scattering matrix, *IEEE Trans. Microwave Theory and Techniques*, 13(2), pp. 194–202, March 1965.

R16.2 Key Specifications for RF Systems

[Agilent EVMa-2005] Agilent Technologies, Eight hints for making and interpreting EVM measurements, 5989-3144EN, App. Note, 2005.

[Agilent EVMb-2000] Agilent Technologies, Using error vector magnitude measurements to analyze and troubleshoot vector-modulated signals, 5965-2898E PN 89400–14, App. Note, 2000.

[Agilent NF-2006] Agilent Technologies, Fundamentals of RF and microwave noise figure measurement, AN 57-1 5952–8255E, App. Note, 2006.

[Agilent Rx-2002] Agilent Technologies, Testing and troubleshooting digital RF communications receiver designs, AN 1314 5968-3579E, App. Note, 2002.

[Agilent SA-2005] Agilent Technologies, Spectrum analyzer basics, AN 150 5952–0292, App. Note, 2005.

[Agilent Y-2004] Agilent Technologies, Noise figure measurement accuracy: The Y-factor method, 5952-3706 E, App. Note, 2004.

[Bhattacharya 2005] S. Bhattacharya, R. Senguttuvan, and A. Chatterjee, Production test technique for measuring BER of ultra-wideband (UWB) devices, *IEEE Trans. Microwave Theory and Techniques*, 53(11), pp. 3474–3481, November 2005.

[Cho 2005] C. Cho, W. Eisenstadt, B. Stengel, and E. Ferrer, IIP3 estimation from the gain compression curve, *IEEE Trans. Microwave Theory and Techniques*, 53(4), Part 1, pp. 1197–1202, April 2005.

[Ferrario 2002] J. Ferrario, R. Wolf, and S. Moss, Architecting millisecond test solutions for wireless phone RFICs, in *Proc. IEEE Int. Test Conf.*, pp. 1151–1158, October 2002.

[Halder 2005] A. Halder and A. Chatterjee, Low-cost alternate EVM test for wireless receiver systems, in *Proc. IEEE VLSI Test Symp.*, pp. 255–260, April 2005.

[Kasten 1998] J. S. Kasten and B. Kaminska, An introduction to RF testing: Device, method and system, in *Proc. IEEE VLSI Test Symp.*, pp. 462–468, April 1998.

[Maxim 2003] Maxim Integrated Products, Three methods of noise figure measurement, AN2875, App. Note, 2003.

[RS 1EF40-1998] Rhode & Schwarz GmbH & Co, Measurement of adjacent channel power on wideband CDMA signals, 1EF40-0E, App. Note, 1998.

[RS 7BM03-2002] Rhode & Schwarz GmbH & Co, Bit error ratio BER in dVb as a function of S/N, App. Note, 2002.

R16.5 Industry Practices for High-Volume Manufacturing

[Bhattacharya 2004] S. Bhattacharya and A. Chatterjee, Use of embedded sensors for built-in-test of RF circuits, in *Proc. IEEE Int. Test Conf.*, pp. 801–809, October 2004.

[Cherubal 2004] S. Cherubal, R. Voorakaranam, A. Chatterjee, J. McLaughlin, J. L. Smith, and D. M. Majernik, Concurrent RF test using optimized modulated RF stimuli, in *Proc. IEEE Int. Conf. on VLSI Design*, pp. 1017–1022, January 2004.

[Dabrowski 2003] J. Dabrowski, Loopback BIST for RF front-ends in digital transceivers, in *Proc. Int. Symp. on System-on-Chip*, pp. 143–146, November 2003.

[Ferrario 2003] J. Ferrario, R. Wolf, S. Moss, and M. Slamani, A low-cost test solution for wireless phone RFICs, *IEEE Commun. Mag.*, 41(9), pp. 82–88, September 2003.

[Lau 2002] W. Lau, Measurement challenges for on-wafer RF-SOC test, in *Proc. Int. Electronics Manufacturing Technology Symp.*, pp. 353–359, July 2002.

[Sylla 2003] I. Sylla, Building an RF source for low cost testers using an ADPLL controlled by Texas Instruments digital signal processor (DSP) TMS32OC5402, in *Proc. IEEE Int. Test Conf.*, pp. 659–664, October 2003.

[Valdes-Garcia 2005] A. Valdes-Garcia, R. Venkatasubramanian, R. Srinivasan, J. Silva-Martinez, and E. Sanchez-Sinencio, A CMOS RF RMS detector for built-in testing of wireless transceivers, in *Proc. IEEE VLSI Test Symp.*, pp. 249–254, April 2005.

[Veillette 1995] B. Veillette and G. Roberts, A built-in self-test strategy for wireless communication systems, in *Proc. IEEE Int. Test Conf.*, pp. 930–939, October 1995.

[Voorakaranam 2002] R. Voorakaranam, S. Cherubal, and A. Chatterjee, A signature test framework for rapid production testing of RF circuits, in *Proc. IEEE Design, Automation and Test in Europe Conf.*, pp. 186–191, March 2002.

[Yin 2005] Q. Yin, W. Eisenstadt, R. Fox, and T. Zhang, A translinear RMS detector for embedded test of RF ICs, *IEEE Trans. Instrumentation and Measurement*, 54(5), pp. 1708–1714, October 2005.

CHAPTER 17

Testing Aspects of Nanotechnology Trends

Mehdi B. Tahoori
Northeastern University, Boston, Massachusetts

Niraj K. Jha
Princeton University, Princeton, New Jersey

R. Iris Bahar
Brown University, Providence, Rhode Island

ABOUT THIS CHAPTER

As ***complementary metal oxide semiconductor*** (CMOS) devices are scaled down into the nanometer regime, new challenges at both the device and system levels are arising. Many of these problems were discussed in the previous chapters. Though the semiconductor industry and system designers will continue to find innovative solutions to many of these problems in the short term, it is essential to investigate CMOS alternatives. New devices and structures are being researched with vigor within the device community. They include resonant tunneling diodes, quantum-dot cellular automata, silicon nanowires, single electron transistors, and carbon nanotubes. Each of these devices promises to overcome the fundamental physical limitations of lithography-based *silicon* (Si) ***very-large-scale integration*** (VLSI) technology.

Although it is premature to predict which device will emerge as a possible candidate to either replace or augment CMOS technology, it is clear that most of these nanoscale devices will exhibit high defect rates. Consequently, one key question system designers will have to address is how to build reliable circuits, architectures, and systems from unreliable devices. New testing and defect tolerance methodologies at all levels of the design hierarchy will be needed to target these devices.

The focus of this chapter is to provide a brief overview of test technology trends for circuits and architectures composed of some of these emerging nanoscale devices. In particular, defect characterization, fault modeling, and test generation of circuits based on resonant tunneling diodes and quantum cellular automata are discussed. This provides a perspective on circuit-level test generation for nanoscale devices.

To complement this discussion and provide a perspective on system-level testing, testing of architectures based on carbon nanotubes and silicon nanowires is also presented. Finally, techniques to tolerate manufacturing imperfections and variations in logic circuits implemented using carbon nanotube field effect transistors are presented.

17.1 INTRODUCTION

Continued improvement in chip manufacturing technology has resulted in an explosive increase in speed and complexity of circuits. The results are multigigahertz clock rates and billion-gate chips. As CMOS devices are scaled down into the nanometer regime, new challenges at both the device and system level are arising. Some of the challenges at the device level are manufacturing variability, subthreshold leakage, power dissipation, increased circuit noise sensitivity, and cost/performance improvement. At the system level, some of the challenges are effective utilization of more than a billion gates, system integration issues, power, and performance. Historically, the semiconductor industry has overcome many hurdles through innovate solutions, the current being the transition to ***silicon-on-insulator*** (SOI) technology [Cellar 2003]. The ever-persistent need to develop new materials (*e.g.*, high-K and low-K dielectrics (SETs) [Luryi 2004]), shrink device geometries (*e.g.*, dual-gate or Fin field effect transistor (FinFET) devices [Wong 2002]), manage power, decrease supply voltages, and reduce manufacturing costs poses ever-increasing challenges with each shrink in the CMOS technology node.

Although temporary solutions to these challenges will continue to be found, alternative devices need to be explored for possible replacement of or integration within CMOS. Some of the emerging candidates include ***carbon nanotubes*** (CNTs) [Iijima 1991] [Fuhrer 2000] [Rueckes 2000] [Bachtold 2001], ***silicon nanowires*** (NWs) [Kamins 2000] [Huang 2001], ***resonant tunneling diodes*** (RTDs) [Chen 1996], **single electron transistors** (SETs) [Likharev 1999], and ***quantum-dot cellular automata*** (QCA) [Tougaw 1994]. The goal is to try to introduce some of these devices at the 22-nm node or below [Wong 2006]. These devices promise to overcome many of the fundamental physical limitations of lithography-based silicon VLSI technology. Some of these include thermodynamic limits (power), gate oxide thickness, and changes in dopant concentrations that affect device behavior (*i.e.*, variability). Owing to their small size, it is projected that these nanoscale devices will be able to achieve densities of 10^{12} devices per cm^2 and operate at terahertz frequencies [Butts 2002].

Although it is premature to predict how far and how fast CMOS will downscale and which of the afore-mentioned nanoscale devices will eventually enter production, it is certain that nanoscale devices will exhibit high manufacturing defects. Furthermore, it is also clear that the supply voltage, V_{DD}, will be aggressively scaled down to reduce power consumption. The ***International Technology Roadmap for Semiconductors*** (ITRS) published by the Semiconductor Industry Association (SIA) in 2004 predicts V_{DD} will be at 0.5 V for low-power CMOS in

2018 [SIA 2004]. However, V_{DD} at 0.3 V has also been predicted in [Iwai 2004]. This reduction in noise margins will further reduce the circuit reliability and expose computation to high transient error rates (*e.g.*, **soft errors**). In addition, the economics of manufacturing devices at the nanoscale might dictate the use of regular layout or self-assembled structures for a large proportion of the design. This would be a stark paradigm shift from CMOS where the designer has significantly more control over the placement and configuration of individual components on the silicon substrate.

Today's integrated circuits are designed using a top-down approach where lithography imposes a pattern. Unnecessary bulk material is then etched away to generate the desired structure. An alternative bottom-up approach, which avoids the sophisticated and expensive lithographic process, utilizes **self-assembly**, in which nanoscale devices can be self-assembled on a molecule-by-molecule basis. Examples of such devices include QCA, silicon nanowires, and carbon nanotubes. Self-assembly processes promise to lower manufacturing costs considerably, but at the expense of reduced control of the exact placement of these devices. Without fine-grained control, these devices will certainly exhibit higher defect rates.

In addition to lacking control of precise placement of nanoscale devices, designers will also need to consider the effect of using a fabrication process that yields devices that are only a few atoms in diameter. For instance, the contact area between silicon nanowires is a few tens of atoms. With such small cross-sectional and contact areas, fragility of these devices will be orders of magnitude higher than devices currently being fabricated using conventional lithographic techniques. This will result in higher susceptibility to static and transient faults.

Integrated circuits, in general, require thorough testing to identify defective components. If a circuit is found to be defective, it can simply be discarded. However, to improve yield, **defect tolerance** techniques can be applied. For example, spare rows and columns are added in memories in case other rows or columns are found to be defective. Defect tolerance schemes often require high-resolution diagnosis to precisely locate defective resources. Only then can repair be attempted. To improve overall system reliability, periodic testing may also be employed during system operation to identify defects that appear later because of breakdown from device wearout. Finally, defect tolerance can also be used to detect temporary failures (*i.e.*, failures caused by transient or intermittent faults).

Figure 17.1a shows how the various test and diagnosis techniques are used. Application-dependent test and diagnosis techniques are useful for defect tolerance and also for detection, location, and repair of permanent faults during the normal operation of a fault-tolerant reconfigurable system. Application-independent test and diagnosis techniques are used after manufacturing, mainly for identifying defective parts and also for defect tolerance. Test and diagnosis during system operation are complex tasks; however, they help detect permanent and transient faults and hence improve the overall system reliability (see Figure 17.1b).

Given that nanoscale circuits will have higher rates of faults and defects, several works have stressed the need for aggressive defect and fault tolerance for such circuits [Collier 1999] [Butts 2002] [Goldstein 2002] [DeHon 2003a] [Mishra 2003]

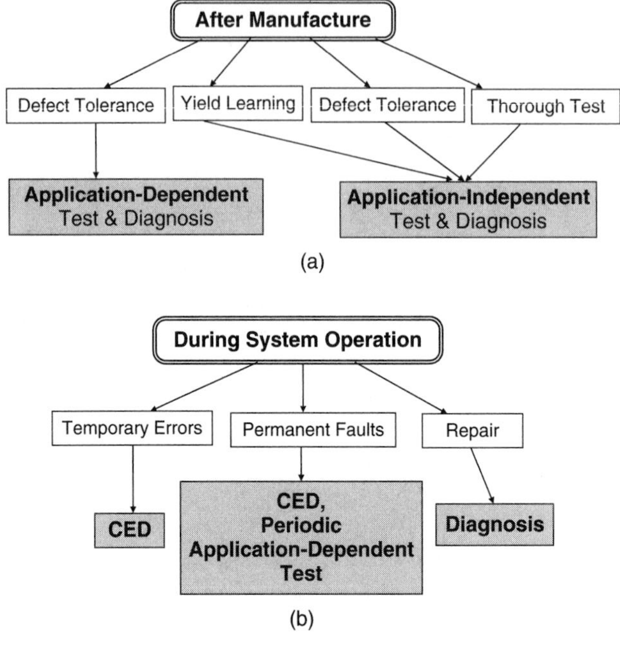

■ **FIGURE 17.1**

Test and diagnosis for defect and fault tolerance.

[Mitra 2004]. Different fault models, test generation, and fault tolerance schemes will need to be developed for these nanoscale devices. In this chapter, we review some issues and trends arising from nanoscale computing and the related test and defect tolerance challenges. In particular, we focus on three promising emerging devices: RTDs and QCA, as well as carbon nanotubes and silicon nanowires. We discuss test methodologies for each of the devices. Also, we discuss defect tolerance by presenting techniques to tolerate imperfections and variations in carbon nanotube transistors. We conclude this chapter with a discussion of future challenges and trends in nanoscale computing.

17.2 RESONANT TUNNELING DIODES AND QUANTUM-DOT CELLULAR AUTOMATA

Although CMOS technology is not predicted to reach fundamental scaling limits for another decade, alternative emerging technologies are being researched in hopes of launching a new era in nanoelectronics. In this section, we concentrate on two promising nanotechnologies: ***resonant tunneling diodes*** (RTDs) and ***quantum-dot cellular automata*** (QCA). A combination of RTDs and ***heterostructure field-effect transistors*** (HFETs) can implement threshold gates, and QCA can implement majority gates, which are a special type of threshold gate. Threshold and majority network design was an active area of research in the 1950s and 1960s. Because of the emergence of these nanotechnologies, interest in this area has been revived. Although nanotechnologies are expected to bring us very high logic density, low power consumption, and high performance, as mentioned earlier, they are also expected to suffer from high defect rates. This makes efficient test generation for such circuits a necessity.

In this section, we first discuss testing of threshold networks with application to RTDs. Next, we describe testing of majority networks with application to QCA. An example is presented to illustrate testing of stuck-at faults. Furthermore, testing of bridging faults is required for QCA majority gates and interconnects. Finally, a test generation methodology is presented to target QCA defects in majority gates and interconnects.

17.2.1 Testing Threshold Networks with Application to RTDs

A **threshold function** is a multi-input function in which each digital input, $x_i, i \in \{1,2,\ldots,n\}$, is assigned a weight w_i such that the output function assumes the value 1 if and only if the weighted sum of the inputs equals or exceeds the value of the function's threshold, T [Muroga 1971]—that is:

$$f(x_1,x_2,\ldots,x_n) = \begin{cases} 1 & if \quad \sum_{i=1}^{n} w_i x_i \geq T \\ 0 & if \quad \sum_{i=1}^{n} w_i x_i < T \end{cases} \tag{17.1}$$

A threshold gate is a multiterminal device that implements a threshold function. We will use the weight-threshold vector $< w_1, w_2, \ldots, w_n;\ T >$ to denote the weights and threshold of a threshold gate.

A threshold function can be realized by a ***monostable-bistable transition element*** (MOBILE), which is composed of RTDs and HFETs, such as the one shown in Figure 17.2a [Chen 1996]. Figure 17.2b shows a MOBILE's equivalent threshold gate representation. The modulation current, ΔI, applied at the output node determines what digital state the device transitions to [Pacha 1999]. The modulation current is obtained from **Kirchoff's current law** and is given as:

$$\Delta I = \sum_{i=1}^{N_p} w_i I(V_{gs}) - \sum_{i=1}^{N_n} w_i I(V_{gs}) \tag{17.2}$$

where N_p and N_n are the number of positive and negative weighted inputs, respectively, and $I(V_{gs})$ is the peak current of a minimum-sized RTD. The net RTD current for the load and driver is $I_T = TI\ (V_{gs})$. Consequently, the output is logic high if $\Delta I - I_T$ is positive and logic low otherwise.

Research in Boolean testing has flourished since the 1960s [Jha 2003]. On the other hand, there has been virtually no research in testing arbitrary threshold networks. The bulk of research in threshold logic was done in the 1950s and 1960s and

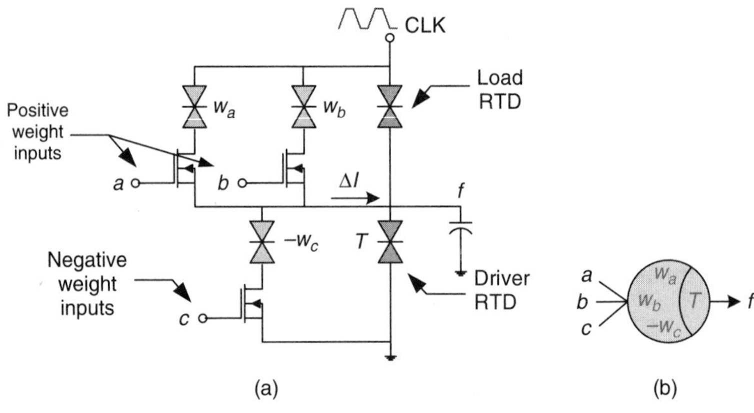

■ **FIGURE 17.2**

(a) A threshold gate implemented using a MOBILE and (b) its schematic representation.

focused primarily on the synthesis of threshold networks [Muroga 1971]. A practical methodology for synthesis of multilevel threshold networks has been presented [Zhang 2004a]. A survey of technologies capable of implementing threshold logic can be found in [Beiu 2003].

The first step in test generation is to decide which fault model to test for [Gupta 2004]. To obtain a fault model, it is important to evaluate the impact of cuts (opens) and shorts on MOBILEs. Figure 17.3a shows the cuts and shorts in a MOBILE that can be modeled as ***single stuck-at faults*** (SSFs) at the logic level. A cut (sites 1, 2, and 3) on an HFET or on a line connecting the RTD and HFET will render it permanently nonconducting and is modeled as a *stuck-at-0* (SA0) fault. Similarly, a short across an RTD (site 4) or the driver RTD (site 8) is also modeled as an SA0 fault because in the former, the input weight will become zero, whereas in the latter, there will be a direct connection between the output and ground. A cut at site 6 represents either SA0 or *stuck-at-1* (SA1) fault depending on the threshold of the gate. If the threshold is less than zero, then the cut is modeled as an SA1 fault. Otherwise, it is modeled as an SA0 fault. On the other hand, faults at sites 5 and 7 are modeled as SA1 faults. A short across the HFET will make it conduct permanently, whereas a direct connection between the output and bias voltage will exist in the presence of a short across the load RTD, making the fault appear as an SA1 fault when the MOBILE is active. These fault models have been verified through HSPICE simulations. HSPICE models for RTD-HFET gates are available from [Prost 2000].

The next step in test generation is redundancy identification and removal. In Boolean testing, irredundant networks are intricately related to circuit testability [Jha 2003]. The same is true for irredundant threshold networks. If no test vector exists for detecting fault s in a threshold network G, then s is redundant. In such a case, the corresponding node or edge in the network can be removed without affecting the functionality of G.

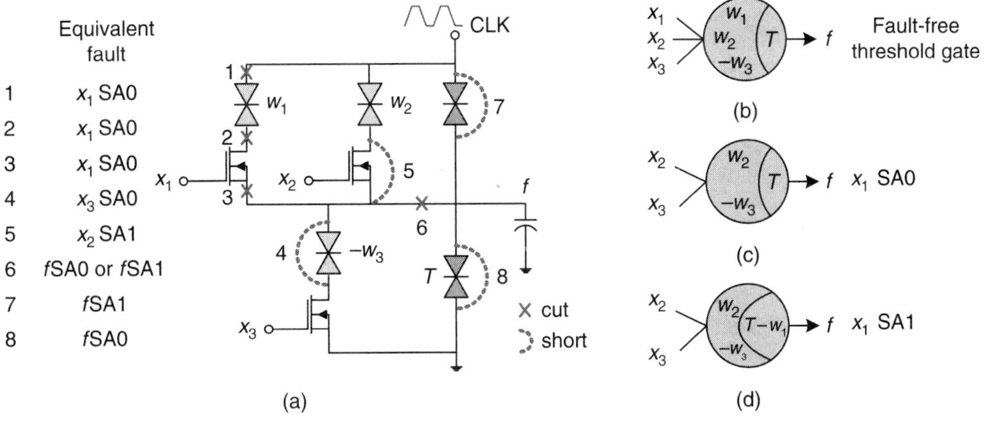

■ **FIGURE 17.3**

(a) Fault modeling of a threshold gate with (b) no faults, (c) an SA0 fault, and (d) an SA1 fault.

The rules for removing a redundant fault in a threshold network are as follows:

1. If an SA0 fault on an edge is redundant, the edge in the network can be removed, as shown in Figure 17.3c.
2. If an SA1 fault on an edge is redundant, the edge in the network can be removed, as shown in Figure 17.3d. In addition, the threshold of the nodes in the edge's fanout must be lowered by the weight of the removed edge.

Furthermore, all nodes and edges in the subnetwork that do not fan out and are in the transitive fanin of the removed edge can be removed from the network in both cases.

The next step is the actual test generation step. To find a test vector for a fault at input x_i of a threshold gate, it is necessary that the term $w_i x_i$ in Equation (17.1) be the dictating factor in determining the output value of the function. The following theorem gives the conditions the test vector has to satisfy:

Theorem 17.1

To find test vectors for x_i SA0 and x_i SA1 in a threshold gate implementing the threshold function $f(x_1, x_2, \ldots, x_n)$, we must find an assignment on the remaining input variables such that one of the following inequalities is satisfied:

$$T - w_i \leq \sum_{j=1, j \neq i}^{n} w_j x_j < T \quad (17.3)$$

or

$$T \leq \sum_{j=1, j \neq i}^{n} w_j x_j < T - w_i \quad (17.4)$$

If an assignment exists, then $< x_1, x_2, \ldots, x_n = 1, \ldots, x_n >$ and $< x_1, x_2, \ldots, x_n = 0, \ldots, x_n >$ are test vectors for x_i SA0 and x_i SA1 faults, respectively. If no assignment exists, then both faults are untestable and, therefore, redundant.

The following theorem shows how to reduce the test generation burden by obtaining the test vector of one fault directly from that of another:

Theorem 17.2

In a threshold gate implementing the threshold function $f(x_1, x_2, \ldots, x_n)$, if there exist two (or more) inputs x_j and x_k such that $w_j = w_k$, then test vectors to detect x_k SA0 and x_k SA1 can be obtained simply by interchanging the bit positions of x_j and x_k in the SA0 and SA1 test vectors for x_j, respectively, assuming they exist.

The preceding theorems can best be illustrated by the following example. Consider the threshold gate that realizes the threshold function, $f(x_1, x_2, x_3) = x_1x_2 + x_1x_3$, with weight-threshold vector $< 2, 1, 1; 3 >$. To test for x_1 SA0, the inequalities to be satisfied are $1 \leq (w_2x_2 + w_3x_3) < 3$ or $3 \geq (w_2x_2 + w_3x_3) < 1$. This leads to three test vectors, namely 101, 110, and 111. The test vectors for x_1 SA1 can be easily obtained by replacing $x_1 = 1$ with $x_1 = 0$ in the original test vectors. Thus, vectors 001, 010, and 011 detect x_1 SA1. Finally, given that vector 110 is a test for x_2 SA0 and because $w_2 = w_3$, a test vector that detects x_3 SA0 is obtained by interchanging the bit positions of x_2 and x_3 in 110 to get 101.

The *D*-algorithm for threshold networks depends on the concepts of ***primitive D-cube of a fault*** (PDCF), propagation *D*-cubes, and singular covers [Jha 2003] [Wang 2006]. A PDCF for a stuck-at fault in a threshold gate can be obtained simply by obtaining a test vector for the fault that results in a D or D' at its output, using the theorems given previously.

Propagation *D*-cubes are used to sensitize a path from the fault site to one (or more) primary outputs. Knowing the threshold function that is implemented by a threshold gate, we can use algebraic substitution to determine the propagation *D*-cubes by using the *D*-notation. For example, to determine the propagation *D*-cubes from x_1 in $f(x_1, x_2, x_3) = x_1x_2 + x_1x_3$, substituting D for x_1 in f we get $Dx_2 + Dx_3$. For the fault to propagate, it is required that only the cubes containing D (or D') get "activated" in f. In this case, because both cubes contain D, activating either or both cubes will result in a propagation *D*-cube. Thus, the propagation *D*-cubes for x_1 are $\{D10D, D01D, D11D\}$. Of course, $\{D'10D'; D'01D'; D'11D'\}$ are also propagation *D*-cubes.

Singular covers are used in test generation to justify the assignments made to the output of a threshold gate. They are easily obtained from the threshold function of the threshold gate. Consider the threshold network in Figure 17.4. Suppose we want to derive a test vector for x_1 SA1. The PDCF for this fault in gate G_1 is $0000D'$. Using the propagation *D*-cube of gate G_2 as shown, the fault effect can be propagated to circuit output f_1. This requires 1 to be justified on line c_2 through the application of the relevant singular cube to gate G_3 as shown. Thus, a test vector for the above fault is $(0, 0, 0, 0, \phi, 1, 0, 0, \phi, \phi)$.

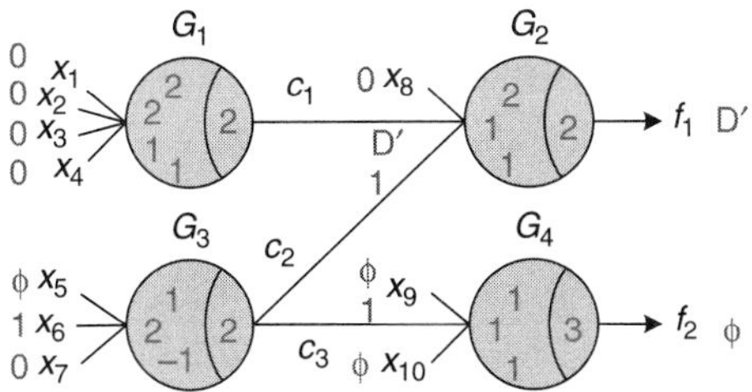

■ **FIGURE 17.4**

Testing for x_1 SA1.

To reduce the run-time of test generation, it is necessary to reduce the number of faults in the fault list. This *fault collapsing* can be done by exploiting fault dominance relationships:

Theorem 17.3

The following fault dominance relationships hold in a threshold gate that implements the threshold function $f(x_1, x_2, \ldots, x_n)$:

1. An output f SA0 (SA1) fault dominates an x_i SA0 (SA1) fault if Equation (17.3) is satisfied.
2. An output f SA1 (SA0) fault dominates an x_i SA0 (SA1) fault if Equation (17.4) is satisfied.

To demonstrate this theorem, consider the threshold function $f(x_1, x_2, x_3) = x_1x_2 + x_1x_3$ again. Applying the theorem, we see that f SA0 (SA1) dominates x_1 SA0 (SA1), x_2 SA0 (SA1), and x_3 SA0 (SA1). Hence, f SA0 and f SA1 can be discarded from the fault list. Exploiting this theorem leads to the following theorem on test generation for irredundant combinational threshold networks, which is similar to the one used for Boolean testing:

Theorem 17.4

In an irredundant combinational threshold network G, any test set V that detects all SSFs on the primary inputs and fanout branches detects all SSFs in G.

17.2.2 Testing Majority Networks with Application to QCA

We next present a test generation framework for majority networks for application to the testing of QCA circuits [Gupta 2006]. QCA is a nanotechnology that has attracted significant recent attention and shows immense promise as a viable future technology [Amlani 1998, 1999] [Lieberman 2002] [Tougaw 1994]. In QCA, logic states, rather than being encoded as voltage levels as in conventional CMOS technology, are represented by the configuration of an electron pair confined within a quantum-dot cell. QCA promises small feature size and ultralow power consumption. It is believed that a QCA cell of a few nanometers can be fabricated through molecular implementation by a self-assembly process [Hennessy 2001]. If this does hold true, then it is anticipated that QCA can achieve densities of 10^{12} devices/cm^2 and operate at THz frequencies [Tahoori 2004a].

Since its initial proposal, QCA has attracted significant attention. Consequently, various researchers have addressed different ***computer-aided design*** (CAD) problems for QCA. In [Zhang 2004a], the authors have developed a majority logic synthesis tool for QCA called **MALS**. Majority logic synthesis has also been addressed in [Zhang 2004b]. A tool called **QCADesigner** for manual layout of QCA circuits has been presented in [Walus 2004]. This tool also offers various simulation engines, each offering a tradeoff between speed and accuracy, to simulate the layout for functional correctness. The authors of [Tahoori 2004a] and [Tahoori 2004b] characterized in detail the types of defects that are most likely to occur in the manufacturing of QCA circuits.

We first introduce some basic concepts. A QCA cell, shown in Figure 17.5, contains four quantum dots positioned at the corner of a square and two electrons that can move to any quantum dot within the cell through electron tunneling [Tougaw 1994]. Because of Coulombic interactions, only two configurations of the electron pair exist that are stable. Assigning a polarization P of -1 and $+1$ to distinguish between these two configurations leads to a binary logic system.

The fundamental logic gate in QCA is the majority gate. The output of a three-input majority gate M is logic 1 if two or more of its inputs are logic 1. That is:

$$M(A,B,C) = AB + AC + BC \tag{17.5}$$

From this point onward, a three-input majority gate will be simply referred to as a majority gate.

Computation in a QCA majority gate, as shown in Figure 17.6a, is performed by driving the device cell to its lowest energy state. This is achieved when the device cell assumes the polarization of the majority of the three input cells. The reason why the device cell always assumes a majority polarization is because it is in this polarization state that the Coulombic repulsion between electrons in the input cells

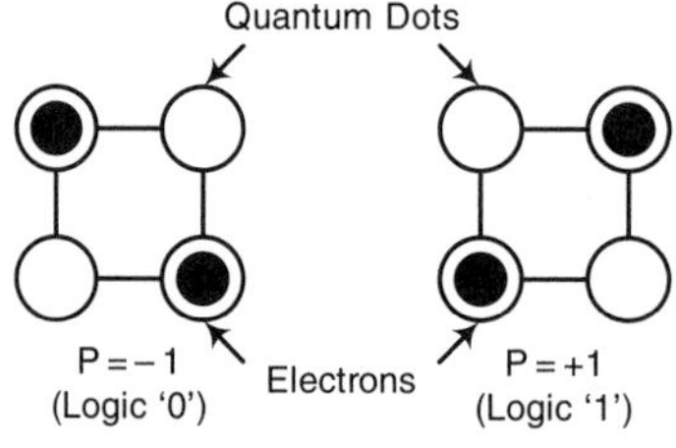

FIGURE 17.5

A QCA cell. The logic states are encoded in the electron pair configuration.

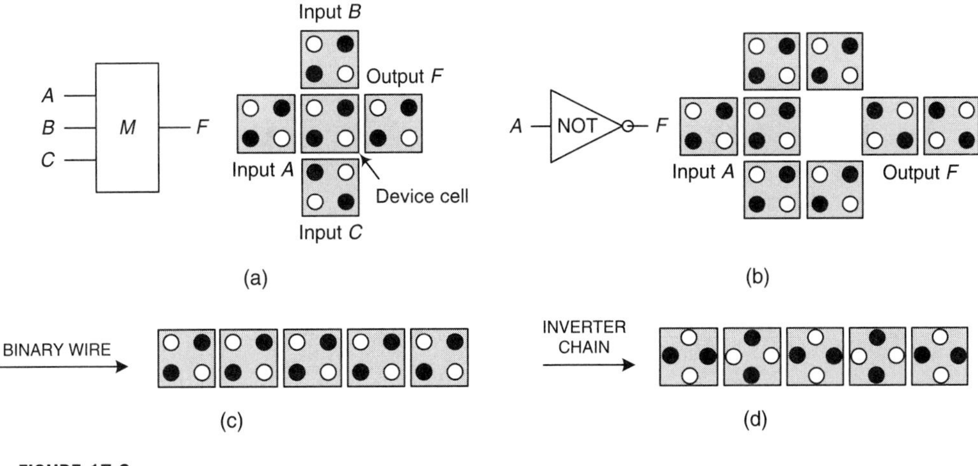

■ **FIGURE 17.6**

A QCA (a) majority gate, (b) an inverter, (c) binary wire, and (d) inverter chain.

is minimized [Tougaw 1994]. The polarization of the device cell is then transferred to the output cell.

The schematic diagrams of an inverter and interconnects are also shown in Figure 17.6. In the QCA binary wire shown in Figure 17.6c, information propagates from left to right. An inverter chain, shown in Figure 17.6d, can be constructed if the QCA cells are rotated by 45°. Furthermore, it is possible to implement two-input AND and OR gates by permanently fixing the polarization of one of the input cells of a majority gate to -1 (logic 0) and $+1$ (logic 1), respectively. Finally, the majority gate and inverter constitute a functionally complete set (*i.e.*, they can implement any arbitrary Boolean function).

The types of defects that are likely to occur in the manufacturing of QCA devices have been investigated in [Momenzadeh 2004], [Tahoori 2004a], and [Tahoori 2004b] and are illustrated in Figure 17.7. They can be categorized as follows:

1. In a **cell displacement defect**, the defective cell is displaced from its original position. For example, in Figure 17.7b the cell with input *B* is displaced to the north by Δnm from its original position (see Figure 17.7a).
2. In a **cell misalignment defect**, the direction of the defective cell is not properly aligned. For example, in Figure 17.7c the cell with input *B* is misaligned to the east by Δnm from its original position.
3. In a **cell omission defect**, the defective cell is missing as compared to the defect-free case. For example, in Figure 17.7d the cell with input *B* is not present.

In QCA interconnects, cell displacement and omission defects on binary wires and inverter chains were also considered in [Momenzadeh 2004]. These defects could be

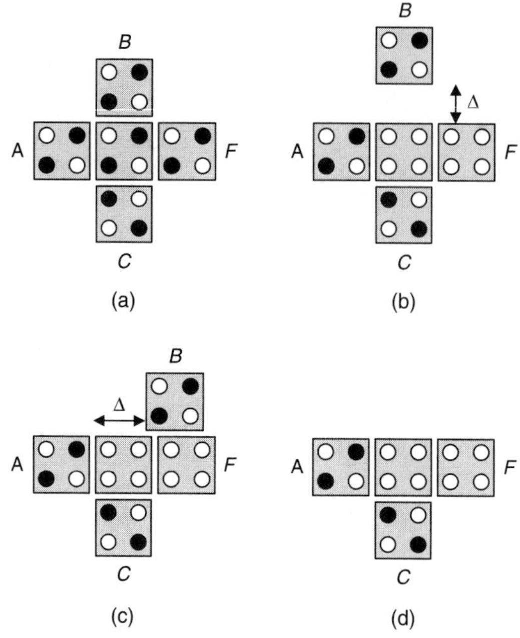

■ **FIGURE 17.7**

(a) Defect-free majority gate, (b) displacement defect, (c) misalignment defect, and (d) omission defect.

better modeled using a dominant fault model in which the output of the dominated wire is determined by the logic value on the dominant wire. Many scenarios could occur in the presence of a bridging fault, as illustrated in Figure 17.8. In the first scenario shown in Figure 17.8b, the second cell is displaced to the north from its original position by Δnm. In this case, the dominated wire O_2 will have a logic value equal to that of the dominant wire O_1. However, if the fourth cell is displaced, as shown in Figure 17.8c, then O_2 will have a logic value equal to the complement of the logic value on O_1 (*i.e.*, $O_2 = O_1'$). Finally, if multiple cells are displaced, as shown in Figure 17.8d, then O_2 will also equal O_1'.

As a motivational example, consider the majority circuit shown in Figure 17.9a, which contains three majority gates ($M_1 - M_3$), seven primary inputs ($A - G$), and one primary output (O). Because there are 10 lines in this circuit, there are 20 SSFs. However, we only need to target SA0/SA1 faults at the primary inputs in this circuit to guarantee detection of all 20 SSFs. This is because majority gates are also threshold gates with weight-threshold vector $< 1, 1, 1; 2 >$. Hence, Theorem 17.4 is also applicable to majority networks. Because there are 7 primary inputs, there are 14 SSFs that need to be targeted during test generation. Figure 17.9b shows a minimal *single* SSF test set for the circuit. For each test vector in Figure 17.9b, all the SSFs detected by it are also shown.

Given the test set in Figure 17.9b, Figure 17.9c shows the input vectors that the individual majority gates of the circuit receive in the fault-free case. For example,

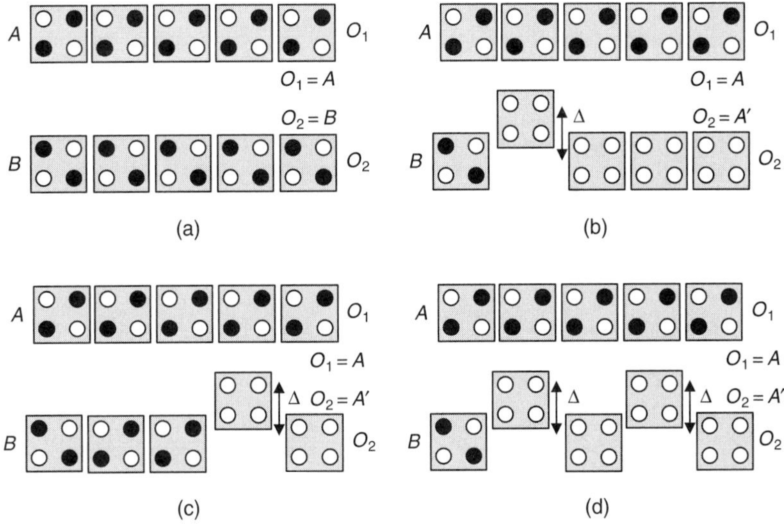

FIGURE 17.8

(a) Fault-free binary wire and (b)–(d) binary wire with a bridging fault.

note that M_2 receives input vectors 001, 011, 100, and 101. Now, consider fault I SA0, which is detected by three test vectors, namely 1100110, 0110110, and 0001011. Given the application of vectors 1100110 and 0110110, M_3 will receive input vector 110 in the fault-free case. If vector 0001011 is applied, M_3 will receive input vector 011. In all these cases, the expected fault-free logic value at output O is logic 1. However, in the presence of I SA0, output O becomes logic 0, thus detecting the fault.

Figure 17.9c shows that M_1receives input vectors 000, 001, 011, 100, and 110. Even though these vectors form a *complete* SSF test set for a majority gate, they are *not* a complete test set for detecting all simulated defects that were presented in [Momenzadeh 2004], [Tahoori 2004a], and [Tahoori 2004b]. In fact, five defects (three misalignments and two displacements on the QCA cells in M_1) cannot be detected by these input vectors. If we add vector 0100110 to our original test set, then M_1 will also receive input vector 010. Note that the effect of the last four of the preceding five vectors is also propagated from the output of $M_1$to output O. This is now a complete test set for all simulated defects in M_1. Gates M_2 and M_3 already receive input vectors, which form a complete defect test set and hence require no additional test vectors (the effect of these vectors is also propagated to output O).

Figure 17.9d shows a possible QCA layout for the circuit in Figure 17.9a (different shades in QCA cells indicate different phases of a four-phase clock that is typically used for QCA circuits [Hennessy 2001]). Consider the QCA cell displaced from the binary wire of input D so that there exists a bridging fault between inputs C and D. In this case, C dominates D. Furthermore, it is unknown whether the bridging fault will result in $D = C$ or $D = C'$. To detect this defect, we need vectors to test for two of four possible conditions. Condition D SA0 with $C = 0$ or D SA1 with $C = 1$

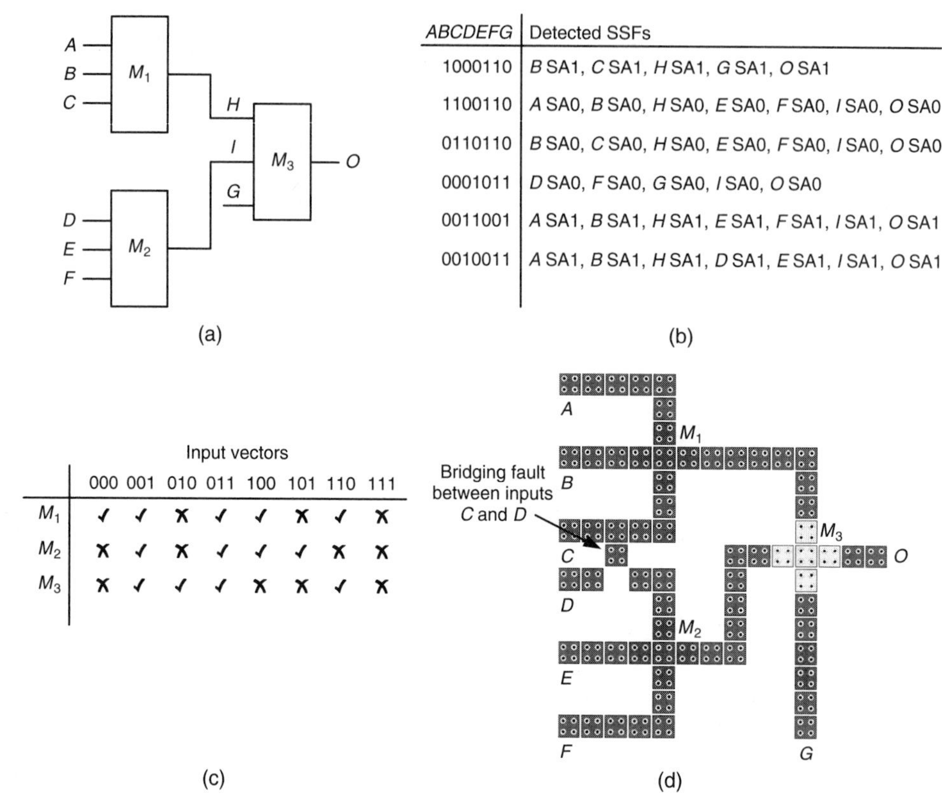

ABCDEFG	Detected SSFs
1000110	*B* SA1, *C* SA1, *H* SA1, *G* SA1, *O* SA1
1100110	*A* SA0, *B* SA0, *H* SA0, *E* SA0, *F* SA0, *I* SA0, *O* SA0
0110110	*B* SA0, *C* SA0, *H* SA0, *E* SA0, *F* SA0, *I* SA0, *O* SA0
0001011	*D* SA0, *F* SA0, *G* SA0, *I* SA0, *O* SA0
0011001	*A* SA1, *B* SA1, *H* SA1, *E* SA1, *F* SA1, *I* SA1, *O* SA1
0010011	*A* SA1, *B* SA1, *H* SA1, *D* SA1, *E* SA1, *I* SA1, *O* SA1

	000	001	010	011	100	101	110	111
M_1	✓	✓	X	✓	✓	X	✓	X
M_2	X	✓	X	✓	✓	✓	X	X
M_3	X	✓	✓	✓	X	X	✓	X

■ **FIGURE 17.9**

(a) Example majority circuit, (b) minimal SSF test set, (c) input vectors received by each majority gate, and (d) QCA layout of the circuit with a bridging fault between inputs *C* and *D*.

must be tested, and condition *D* SA1 with $C = 0$ or *D* SA0 with $C = 1$ must be tested. Note that we will explain how we obtained these conditions in the next section. The first and second conditions can be tested by vectors 0001011 and 0010011 that are already present in our test set. However, we currently have no vector that can test either of the latter two conditions. Therefore, additional test generation is required, and vectors 0000011 and 0011011 are derived as tests for the third and fourth conditions, respectively. Adding either of these two vectors is sufficient as we only need to satisfy either the third or fourth condition for fault detection. The bridging fault between *C* and *D*, when *C* is dominant and *D* is dominated, can now be completely tested for all simulated defects.

The preceding example illustrates that the SSF test set of a circuit cannot guarantee the detection of all simulated defects in the QCA majority gates. Therefore, test generation will be required to cover the defects not covered by the SSF test set. In addition, test generation will also be needed to cover bridging faults on QCA interconnects. For a majority gate, there are nine minimal SSF test sets that contain four test vectors each. The minimal test sets were applied to a QCA majority gate,

SSF Test Set	100% Defect Coverage?	SSF Test Set	100% Defect Coverage?	SSF Test Set	100% Defect Coverage?
{001, 010, 011, 101}	✓	{001, 100, 101, 110}	✓	{010, 011, 100, 101}	✗ 3 Uncovered Defects
{001, 011, 100, 101}	✓	{010, 100, 101, 110}	✓	{001, 010, 011, 110}	✓
{001, 010, 101, 110}	✗ 1 Uncovered Defect	{001, 011, 100, 110}	✗ 5 Uncovered Defects	{010, 011, 100, 110}	✓

■ **FIGURE 17.10**

Minimal SSF test sets for a majority gate. Some of these test sets, however, are not 100% defect test sets for a QCA majority gate.

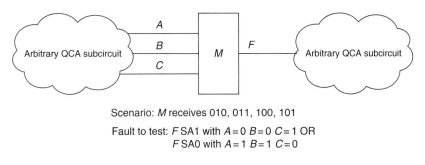

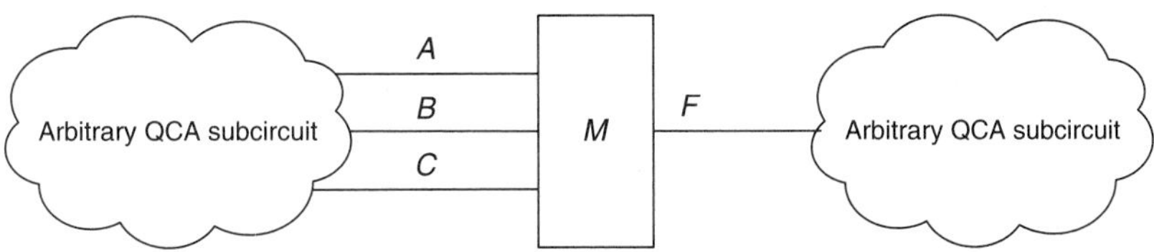

■ **FIGURE 17.11**

Testing of defects in a QCA majority gate.

and all defects described in [Momenzadeh 2004], [Tahoori 2004a], and [Tahoori 2004b] were simulated using QCADesigner [Walus 2004]. Figure 17.10 shows the results of this experiment. Of the nine minimal test sets, three test sets were unable to detect all the simulated defects.

Consider the majority gate *M* in Figure 17.11 that is embedded in a larger network. Suppose that after SSF test generation for the entire circuit, it is determined that *M* receives input vectors 010, 011, 100, and 101. According to Figure 17.10, this is not a complete defect test set as three defects remain uncovered. We can make it complete by ensuring that *M* also receives either 001 or 110 as its input vector.

It is very important to test for defects on QCA interconnects because it is predicted that interconnects will consume the bulk of chip area in future nanotechnologies. As Figures 17.8b to 17.8d show, a displacement of the QCA cells on a wire will result in a bridging fault between the two wires. Using QCADesigner [Walus 2004], it can be verified that such defects can be modeled using the dominant bridging fault model [Tahoori 2004a]. However, depending on the displacement distance, the lower wire will assume either the upper wire's logic value or its complement.

Figure 17.12 shows the possible scenarios that can result if a bridging fault is present between two wires. The scenario that will occur depends on the displacement distance Δ of the defective cell or the number of cells that get displaced.

Scenario 1: Δ is such that $B = A$

Fault-free AB	Faulty AB	Equivalent Condition
0 0	0 0	-
0 1	0 0	B SA0 with $A=0$
1 0	1 1	B SA1 with $A=1$
1 1	1 1	-

Note: Assumption is that B is dominated by A.

(a)

Scenario 2: Δ is such that $B = A'$

Fault-free AB	Faulty AB	Equivalent Condition
0 0	0 1	B SA1 with $A=0$
0 1	0 1	-
1 0	1 0	-
1 1	1 0	B SA0 with $A=1$

Note: Assumption is that B is dominated by A.

(b)

FIGURE 17.12

Modeling of bridging faults in a QCA binary wire.

It also shows the conditions that must be satisfied in order to detect the particular scenario. In the first scenario in Figure 17.12a, the lower wire's logic value is equal to that of the upper wire, whereas in Figure 17.12b, the lower wire's logic value is equal to the complement of that of the upper wire. If the first scenario occurs, then a vector is required to test for one of the two conditions shown in Figure 17.12a. Similarly, if the second scenario occurs, a vector is required to test for either of the two conditions shown in Figure 17.12b. In reality, it will not be known which scenario occurred in the presence of the defect because Δ will not be known beforehand. Consequently, we need to have vectors that test for both scenarios to obtain a test that can completely detect this fault.

As an example, consider once again the majority gate in Figure 17.11. Assume there is a bridging fault between inputs A and B and it is not known whether the defective cell is on wire A or wire B. In addition, Δ is unknown. To test for this fault, four conditions need to be satisfied (the first two result from A dominating B, and the last two result from B dominating A). The conditions are as follows:

$$
\begin{aligned}
&B \text{ SA0 with } A=0 \text{ or } B \text{ SA1 with } A=1\\
&B \text{ SA1 with } A=0 \text{ or } B \text{ SA0 with } A=1\\
&A \text{ SA0 with } B=0 \text{ or } A \text{ SA1 with } B=1\\
&A \text{ SA1 with } B=0 \text{ or } A \text{ SA0 with } B=1
\end{aligned}
$$

If the test set contains vectors that can satisfy the preceding conditions, then this bridging fault can be detected. Otherwise, the fault is not completely testable.

Given a QCA circuit with n lines, there are at most $n(n-1)$ possible bridging faults involving two wires. This is based on the assumption that layout information is not available. Consequently, $2n(n-1)$ conditions will need to be satisfied in order to obtain a test set for all bridging faults. However, it should be noted that at least $n(n-1)$ (*i.e.*, 50%) of these conditions will already be satisfied, given any complete SSF test set. This is because, given a pair of wires, when testing for a SA0/SA1 fault on one wire, the other line must have a value of 0 or 1. If the QCA layout of the

TABLE 17.1 ■ Singular Cover of a Majority Gate

A	B	C	F
—	1	1	1
1	—	1	1
1	1	—	1
—	0	0	0
0	—	0	0
0	0	—	0

circuit is available, then the designer can find out a priori which pairs of adjacent wires to test for bridging faults.

Because a majority network is also a threshold network, the *D*-algorithm steps outlined earlier for threshold networks are also applicable here. For example, to obtain the propagation *D*-cubes of x_1 SA1 in $M = x_1x_2 + x_1x_3 + x_2x_3$, substituting *D* for x_1 in *M* we get $Dx_2 + Dx_3 + x_2x_3$. For the fault to propagate, it is required that only the cubes containing *D* (or *D′*) get "activated" in *M*. Thus, the propagation *D*-cubes for x_1 SA1 are *D*10*D* and *D*01*D*. Of course, *D′*10*D′* and *D′*01*D′* are also propagation *D*-cubes. The propagation *D*-cubes for x_2 and x_3 are {0*D*1*D*, 1*D*0*D*, 0*D′*1*D′*, 1*D′*0*D′*} and {01*DD*, 10*DD*, 01*D′D′*, 10*D′D′*}, respectively. These can be stored in a table and used for fault propagation. The singular cover of a majority gate is shown in Table 17.1. Thus, it is possible to employ *D*-algorithm to perform test generation targeting defects in QCA majority gates as well.

In summary, we have discussed testing of threshold and majority networks and showed how RTD- and QCA-based circuits can be tested using these approaches. *D*-algorithm can be extended to test both types of networks. Other test generation approaches, such as ***path oriented decision making*** (PODEM) and satisfiability-based algorithms, are also possible. Although there are some similarities with testing of Boolean networks, traditional test generation approaches need to be augmented to take into account the different logic primitives (threshold and majority gates) as well as the inadequacy of the SSF model in the case of QCA.

17.3 CROSSBAR ARRAY ARCHITECTURES

Today's approach to designing integrated circuits uses a top-down methodology. That is, layers are added on top of a silicon wafer, requiring hundreds of steps before the final circuit is complete. Although this process has allowed the manufacture of reliable circuits and architectures, future scaling will make the production of reliable mask sets extremely expensive. In the near future, a major shift from top-down lithography-based fabrication may be needed in order to cost-effectively fabricate devices at true nanoscale dimension.

As an alternative, bottom-up approaches rely on self-assembly for defining feature size and may offer opportunities to drastically reduce the number of steps required to produce a circuit. However, the biggest impact in going from top-down designs to bottom-up designs is the inability to arbitrarily determine placement of devices or wires. Without fine control of the design, devices made from self-assembly techniques tend to be restricted to simple structures, such as two-terminal devices. Because these devices are usually nonrestoring, one design challenge is providing signal restoration between nanoscale logic stages. Furthermore, this self-assembly approach also lends itself to defect rates orders of magnitude higher than traditional top-down approaches, so that fabricating defect-free circuits will be virtually impossible. Therefore, some means of defect or fault tolerance (whether at the circuit, logic, or architecture level) must be incorporated into the design if reliable computation is to be achieved. Testing also will take on a new role, not so much to sort out defective parts but rather to identify faulty devices/cells within the circuit and avoid them during operation mode.

Bottom-up assembly techniques require fabrication regularity. In addition, taking a hybrid approach that uses self-assembled structures as an add-on to a CMOS subsystem may create a design framework where fault-tolerance techniques can be more effectively applied. These photo-lithographically manufactured components may be built from regular structures as well, in order to lend themselves more easily to reconfigurable architectures. That is, the desired circuit may be designed by configuring around faulty structures; because all structures are identical, one faulty element can be easily swapped out and replaced with an operational one, thereby creating a reliable system out of an unreliable substrate. On the testing side, the challenge now becomes how to quickly identify faulty structures so that they can be avoided when configuring the desired circuit.

Of the molecular-scale devices being developed using these self-assembly techniques, the nonvolatile programmable switch has gained much attention. With these bottom-up techniques, it is possible to build features (*e.g.*, wires and programmable switches) without relying on lithography. Recent work shows how to build nanoscale ***programmable logic arrays*** (PLAs) using the bottom-up synthesis techniques being developed by physical chemists [Goldstein 2001] [Luo 2002] [DeHon 2004]. The molecular switches can provide connection and disconnection states at the crosspoints of vertical and horizontal wires in a crossbar, thereby providing a path to continue the advance of field-programmable technology beyond the end of the traditional lithographic roadmap [SIA 2005]. Such a switch can be fabricated using two layers of parallel nanowires, with the two layers perpendicular to each other, forming a 2-D array. At every crosspoint, the wires are connected together via a two-terminal nanodevice formed by the layering [Huang 2001]. These crossbar arrays are similar to PLAs and can be used as building blocks for implementing logic. The programmable feature of such crossbars can serve the purpose of making the circuits fault tolerant [Huang 2004] [DeHon 2005a] [Tahoori 2006]. The array can then be interconnected, using CMOS circuitry, as part of a hybrid nanoscale/CMOS design architecture. Recent developments suggest both plausible fabrication techniques and viable architectures for building crossbars using nanowires or nanotubes and molecular-scale switches. We describe some of these hybrid architectures in Section 17.3.1.

Comparing CMOS-scale and nanoscale crossbar-based circuits shows fundamental differences in terms of defects occurring during fabrication. The transistor is considered to be the basic element in CMOS-scale circuits, whereas the crossbar is the main component in reconfigurable nanoscale circuits. The type of faults extracted from these two components is significantly different as well. A crossbar is composed of wires and switches; its type of defects is limited only to these two components. In CMOS technology, faults are seen in both transistors and interconnects, and they manifest themselves as stuck-at, open, short, and bridging faults. These faults in CMOS technology are commonly targeted during manufacturing testing.

The two more probable defects in crossbars are (1) defects in programmable crosspoints, and (2) defects in wires. The defective nanowires can be easily detected with the procedure suggested in [Tahoori 2005] and ***field programmable gate array*** (FPGA) literature [Stroud 1998]. The time required to test the wires of each array is linear in the code space size of the stochastic address decoder. The wire fault model includes broken wire, stuck-at 0, and stuck-at 1.

Defects in programmable switches are caused by the structure of the junctions, which is a sandwich of bistable molecules between two layers of wires. In each crosspoint, there are only a few molecules [Chen 2003]. The programmability of a crosspoint comes from the bistable attribute of the molecules located in the crosspoint area. For instance, if there are too few molecules at the crosspoint, then the junction may never be able to be programmed *closed*, or the *closed* state may have higher resistance than the designed threshold chosen for correct operation and timing of the crossbar.

In general, the model can be abstracted into a simple crosspoint defect model. Crosspoints will be in one of these two states:

- **Stuck-closed.** Crosspoint cannot be programmed to an open state; its associated vertical and horizontal wires are always connected. Crosspoints that cannot be programmed into a suitable *open* state will result in the entire horizontal and vertical nanowires being unusable.
- **Stuck-open.** Crosspoint cannot be programmed to a closed state; its associated vertical and horizontal wires are always disconnected.

Figure 17.13 shows two logic functions (AND and OR) implemented on a simple crossbar and its equivalent diode-resistor logic. No defects are considered in these crossbars. Figure 17.14, however, shows two implementations of the function $f = ab + b'c$ on a defective crossbar-based *programmable logic array* (PLA). The PLA, shown in the figure, is a combination of an AND plane and an OR plane and can be implemented based on diode logic by using a crossbar of nanowires and configurable molecular switches on the crosspoints of the wires. The PLA may also be implemented using switching properties of FETs created with carbon nanotubes.

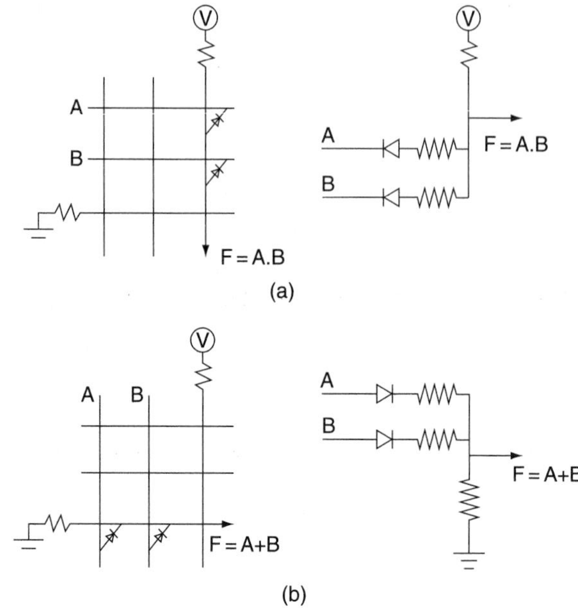

■ **FIGURE 17.13**

Simple logic functions using diode-resistor logic: (a) AND gate implemented on a crossbar and its diode-resistor equivalent circuitry and (b) OR gate implemented on a crossbar and its diode-resistor equivalent circuitry.

As seen in the figure, even in the presence of a number of defects in the crossbar, there are several choices for fault-free implementations of function f. This is because there is a high amount of inherent redundancy in a crossbar and a large number of resources (switches and wires) available for implementing function f on the crossbar.

17.3.1 Hybrid Nanoscale/CMOS Structures

In this section, we provide two examples of hybrid nanoscale/CMOS circuits and architectures being recently proposed. These hybrid designs combine nanoscale devices and nanowires with larger CMOS components. The main advantage of these approaches is that the CMOS subsystem can serve as a reliable medium for connecting nanoscale circuit blocks, providing long interconnects and I/O functions.

17.3.1.1 The nanoPLA

In [DeHon 2004] and [DeHon 2005a], the authors proposed a programmable interconnect architecture built from hybrid components. The main building block, called

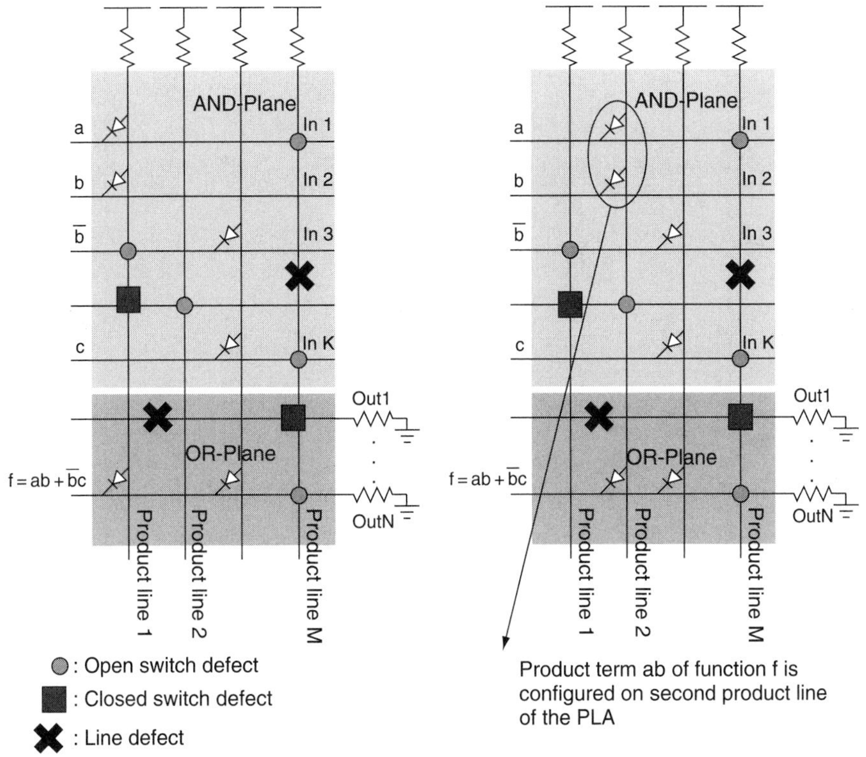

■ **FIGURE 17.14**

Two different implementations of $f = ab + b'c$ on a defective crossbar. Complements of input signals are also provided because it is diode-resistor logic.

the ***nano programmable logic array*** **(nanoPLA)**, is built from a crossed set of N-type and P-type nanowires. An electrically switchable diode is formed at each crosspoint. The diodes then provide a programmable wired-OR plane that can be used to configure or program arbitrary logic into the PLA. The nanoPLA is programmed using lithographic-scale wires along with stochastically coded nanowire addressing [DeHon 2003b].

The nanoPLA block is shown in Figure 17.15. The block is composed of two stages of programmable crosspoints. The first stage defines the logical product terms (*pterms*) by creating a wired-OR of appropriate inputs. The outputs of this wire-OR plane are restored through field-effect controlled nanowires that invert the outputs (thus creating the logical NOR of the selected input signals). These restored signals are then sent to the inputs of the next stage of programmable crosspoints. Each nanowire in this plane computes the wired-OR of one or more restored *pterms*. The outputs of the stage are then restored in the same manner as

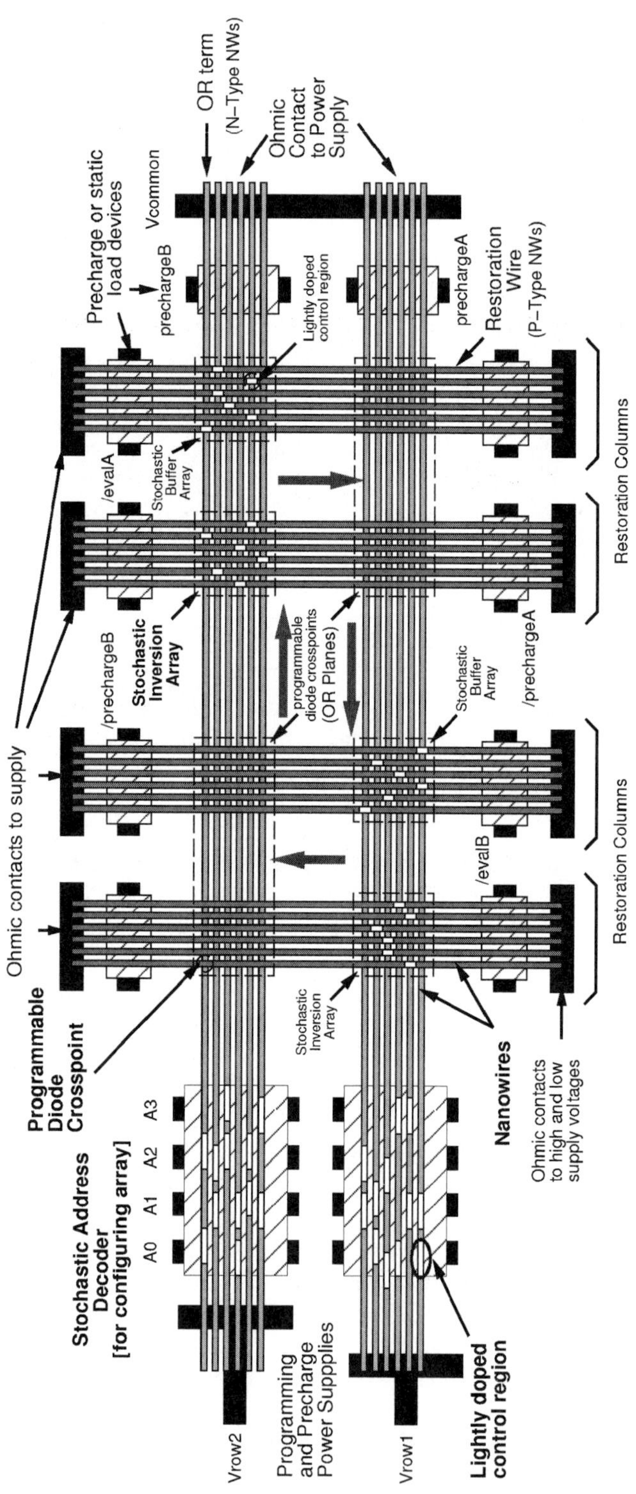

■ **FIGURE 17.15**

A simple nanoPLA block (taken from [DeHon 2005b]).

the first stage. The two stages together provide NOR-NOR logic (equivalent to a conventional PLA) [DeHon 2005a].

The nanoPLA blocks are interconnected by overlapping the restored output nanowires from each block with the wired-OR input region of adjacent nanoPLA blocks. This organization allows each nanoPLA block to receive inputs from a number of different nanoPLA blocks. With multiple input sources and outputs routed in multiple directions, the nanoPLA block can also serve as a switching block by configuring the overlap appropriately. Their experiments mapping benchmark circuits onto the proposed architecture have suggested that device density could be one to two orders of magnitude better than what is projected for the 22nm roadmap node [DeHon 2005a].

Testing of the nanoPLA needs to be done through a process of probing and discovery. A testing process is required to identify a working set of address lines to access crossbars in a nanoscale device. Once a crossbar becomes accessible, defective nanowires and crosspoints can then be identified. This information can then be stored in a *defect map* and used during reconfiguration.

Restoration columns (as shown in Figure 17.15) are used to identify useful addresses. The gate side supply (top lithographic wire contacts in the figure) is driven high, and by sensing the voltage change on the opposite supply line (bottom set of lithographic wire contacts in the figure), the presence of a restored address can be deduced. Broken nanowires, or those with high resistance, will not be able to pull up quickly enough the contact of the bottom supply. This probing of addresses is repeated until enough live wires are discovered. The live addresses can then be used to program a single junction in a diode-programmable OR plane. For additional discussion on this architecture and how it is tested, see [DeHon 2005b].

Note that the programmability of the nanoPLA allows defective devices to be avoided as a means of fault tolerance. The authors in [DeHon 2005c] have shown that when 20% of devices (*i.e.*, crossbar diodes) were defective, only a 10% overhead in devices was needed to correctly configure the array around the defects. The lithographic circuitry and wiring that the nanoPLA is built on top of provide a reliable means of probing for defects and configuring the logic.

17.3.1.2 Molecular CMOS (CMOL)

The **molecular CMOS** (CMOL) circuits proposed in [Likharev 2005] and [Ma 2005] are designed using the same crossbar array structure as the nanoPLA design consisting of two levels of nanowires. The main difference with CMOL is how the CMOS/nanodevices are interfaced. Pins are distributed over the circuit in a square array, on top of the CMOS stack, to connect to either lower or upper nanowire levels. The nano crossbar is turned by some angle less than 90° relative to the CMOS pin array.

By activating two pairs of perpendicular CMOS lines, two pins together with the two nanowires they contact are connected to the CMOS lines (see Figure 17.16). Each nanodevice may be uniquely accessed using this approach. That is, each device may be switched ON/OFF by applying some voltage to the selected nanowires such that the total voltage applied to the device exceeds the switching threshold of the selected nanodevices. By angling the nanoarray, the nanowires do not need to be precisely aligned with each other and the underlying CMOS layer in order to be able to uniquely access a nanodevice.

The most straightforward application of CMOL would be for memories (embedded or stand-alone). The authors project that a CMOL-based memory chip about 2×2 cm in size will be able to store about 1Tb (terabits) of data [Ma 2005]. To improve the reliability of the memory array, the authors proposed adding spare lines and ***error correcting code*** (ECC), a standard procedure in memory array design, to improve yield.

The CMOL circuits have also been proposed for building FPGA-like architectures for implementing random logic [Strukov 2005]. A CMOS cell, composed of an inverter and two pass transistors, is connected to the nanowire crossbar via two pins, as shown in Figure 17.16. This essentially creates a ***configurable logic block*** (CLB) structure, similar to that found in an FPGA. The CMOS cell is then programmed by disabling the inverter and selectively switching devices ON in the crossbar array. After configuration, the pass transistors act as pulldown resistors while the nanodevices programmed to be in the ON state serve as pullup resistors. In this way, wired-NOR gates may be formed within a CMOS cell. Note that the inverter provides signal restoration. Any arbitrary Boolean function (represented as a product-of-sums) may be implemented as a connection of two or more CMOS cells. Further, the idea is to have many nanodevices per CMOS cell. This allows gates with high fanin or high fanout to be formed, with extra devices available as "spares" for reconfiguring around faulty devices.

The testing of this FPGA-like CMOL architecture is connected to its reconfigurable programming. However, before the circuit can be programmed on the FPGA

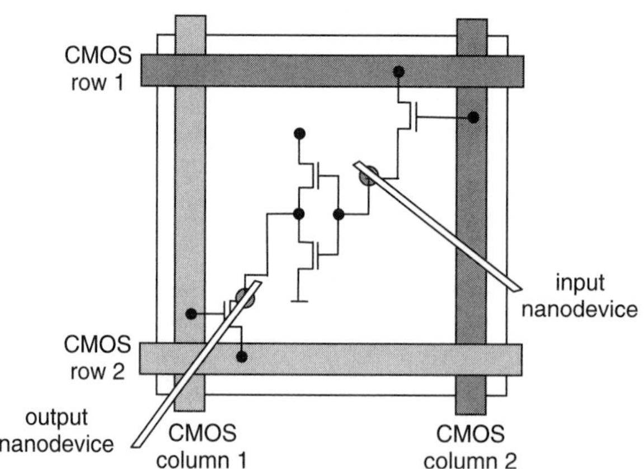

FIGURE 17.16

The CMOS logic cell consisting of two pass transistors and an inverter (taken from [Strukov 2005]). The function is determined by how the overlaying nanowires are programmed. Note that typically, there are many nanowires (and thus many nanodevices) available per CMOS cell.

fabric, there is an implicit assumption made by the authors that each nanodevice in a cell can be tested to determine if it is faulty or not. The authors consider only "stuck-on-open" (stuck-open) faults occurring from the absence of nanodevices at certain nanowire crosspoints. The CMOL FPGA configuration is then carried out at two stages. The first stage maps the desired circuit onto the FPGA cells, assuming a defect-free CMOL fabric. During the second stage, defective components may be reconfigured around, as necessary, until a defect-free mapping is found. The algorithm sequentially attempts to move each gate from a cell with bad input or output connections to a new cell, while keeping the gates in their immediate fanin/fanout in fixed positions. The main goal is to reassign cells for gates such that the interconnect length is minimized. A more detailed description of the reconfiguration algorithm can be found in [Strukov 2005] where their Monte Carlo simulations of a 32-bit Kogge-Stone adder demonstrated that this simple configuration procedure may allow them to achieve 99% circuit yield with as many as 22% defective nanodevices. Similarly, simulations on a 64-bit fully connected crossbar switch have shown a defect tolerance of about 25% [Strukov 2005].

17.3.2 Built-In Self-Test

In the previous section, we reviewed two examples using hybrid nanoscale/CMOS structures. Although testing is required for these designs, the authors focused more on the architecture itself and its configuration. In this section, we turn more toward the testing aspects of these architectures, and in particular, focus on ***built-in self-test*** (BIST) approaches [Wang 2005] [Tehranipoor 2007].

In a BIST scheme, specific components of the architecture are configured to generate the test vectors and observe the outputs in order to test and diagnose the defective components. Here, we consider an island-style architecture for nanodevices containing nanoblocks and switchblocks, as shown in Figure 17.17. Each nanoblock (*e.g.*, crossbar or PLA) is configured as either a ***test pattern generator*** (TPG) or ***output response analyzer*** (ORA). Because of the reconfigurability of nanodevices, no extra BIST hardware is required to be permanently fabricated on-chip. Moreover, dedicated on-chip nanoscale test hardware can be highly susceptible to defects itself.

Test configuration generation is first performed externally and then delivered to the nanodevice during each *test session*. The test session is referred to as one particular test architecture with configured nanoblocks and switchblocks. Note that the nanoblock and switchblock are constructed similar to the crossbar architecture, but the former is used to implement the logic function and the latter for routing. The architecture used in the BIST procedure is called the ***test architecture*** (TA). Each TA includes *test groups* and each test group contains one TPG, one ORA, and one switchblock associated with the TPG and ORA. Note that this BIST scheme is similar to that used for traditional FPGAs discussed in Chapter 12. The TPG tests itself and sends a pattern through a switchblock to the ORA to test it, and then the response is generated and read back by the programming device or tester.

TAs are generated based on the detection of faults in each nanoblock and switchblock. Therefore, several TAs are generated for each test group, and several test configurations are generated to detect faults under consideration. During BIST, all

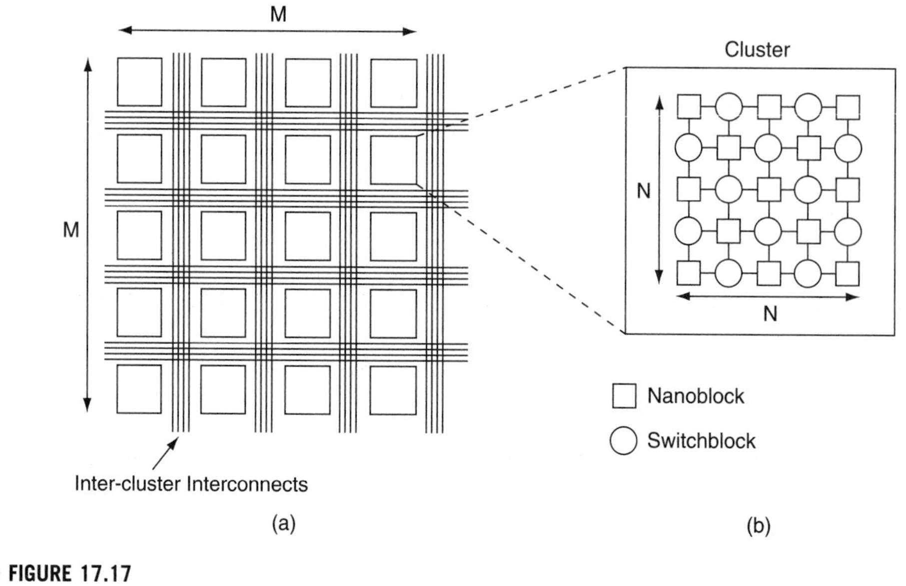

■ **FIGURE 17.17**

(a) Island-style architecture and (b) cluster structure in the nanoarchitecture.

TAs are configured similarly (*i.e.*, the same faults are targeted within test groups in each test session).

The programming device can configure the nanoblocks and switchblocks as required by the test architectures and configurations. Test results should be read back from the ORA blocks using a tester (on/off chip). Interconnect resources provided in the nanodevice architecture should be used for transmitting these results to the tester. Because a fine-grained architecture with small clusters is considered, each cluster is composed of a small number of nanoblocks. Therefore, the number of test groups in a cluster will be small and interconnect resources of the nanodevice are assumed to be sufficient to implement the read-back mechanism. Hence, when a test session is done, the output of ORAs is read for evaluation and analysis by the tester.

The BIST procedure can be performed using an on-chip tester (microprocessor or dedicated BIST hardware) implemented in reliable CMOS-scale circuitry on the substrate of the nanodevice. This will reduce the test time since external devices are generally slower than on-chip testers. The on-chip tester can execute the BIST procedure and collect the test results. It may also eliminate the need to store the defect map on-chip because it can find the faulty blocks before configuration in each cluster. Figure 17.18 shows two different TAs for testing nanoblocks and switchblocks [Wang 2005]. Figure 17.19 shows a test configuration for detecting SA1 faults on all vertical and horizontal lines [Tehranipoor 2007]. By applying 0 and then 1 to the inputs of the crossbar, stuck-open (broken line) faults can be detected. After identifying the location of all faults, they are stored in a defect map and will be used during configuration.

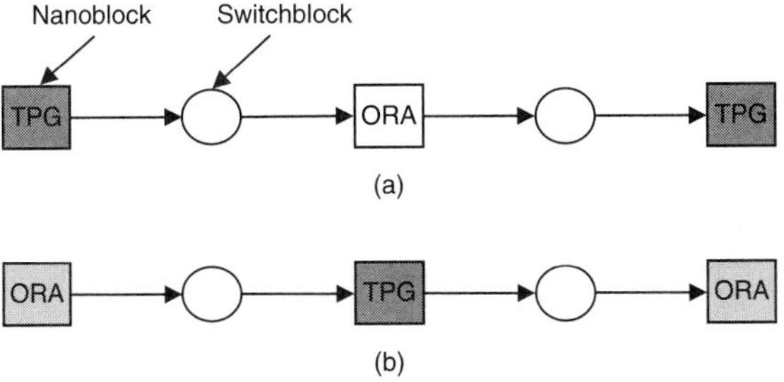

■ **FIGURE 17.18**

Two different test architectures (TAs): (a) nanoblock is used as a TPG in one TA and (b) ORA in another.

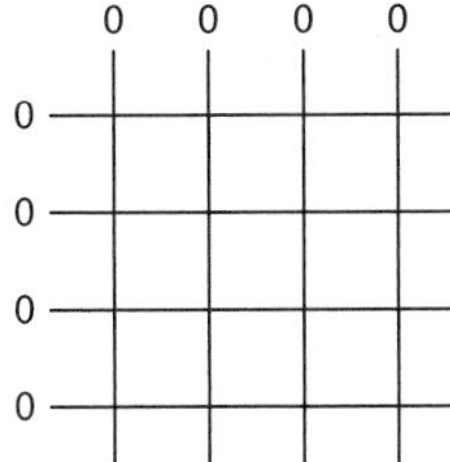

■ **FIGURE 17.19**

A test configuration for detecting SA1 faults on all vertical and horizontal lines.

17.3.3 Simultaneous Configuration and Test

In the examples described in Section 17.3.1, the defect tolerance of the circuits designed from these fabrics relies on some test procedure, completed as a preprocessing step, which identifies the exact location of the defects. This is a nontrivial process since locating all defects in a reconfigurable architecture with a high defect density is a challenging and time-consuming task. Implementing an on-chip BIST circuit or using an on-chip microprocessor will significantly speed up this process, but it will still be a time-consuming process to find the location of all defects in a chip with an extremely large number of blocks. Furthermore, even if these techniques can be applied, storing this information onto a defect map will require large (defect-tolerant) memories. Recall that the configuration process required for nanoPLA or CMOL architectures requires that this defect map be accessed repeatedly during reconfiguration so defective cells can be avoided during the mapping. We cannot use the on-chip nanoscale resources as memory to store the defect map because they are unreliable. On the other hand, using on-chip CMOS-scale memory for the defect map will result in a considerable area overhead. Therefore, it will be impractical to store this large defect map on-chip. It will also be impractical to ship the defect map of each chip along with the chip to the customer. Lastly, the **aging faults** need to be considered as well (*i.e.*, the defect map must be updated regularly). Another important issue for defect-map-based approaches is that different chips have different defect maps, which result in different performance for chips. This also requires per-chip placement and routing, which is prohibitively expensive [Rad 2006].

An alternative to having separate test and configuration stages is to combine them as part of the same step. This approach would also eliminate the need for storing a defect map. The combined test and configuration method that avoids the time-consuming process of locating all defects is called ***simultaneous configuration and test*** (SCT) [Rad 2006]. SCT assumes that the crossbar array architecture offers rich interconnect resources and is able to provide efficient access to its logic blocks through its input/output interfaces. The method is conceptually similar to those proposed for FPGAs, except that the TPG and ORA are components of the BIST circuit to provide test patterns and analyze the responses, respectively. A key conceptual difference of this method compared with other BIST approaches is that the goal of testing is *not* to confirm the correct functionality of a ***block under test*** (BUT) for *every* possible function. Instead, the goal is to make sure that each function (f_i) of an application configured into a block is working correctly. Hence, the test patterns should be applied for testing that function only.

Instead of testing all resources of a reconfigurable architecture to locate all defects in the device, each block of the architecture is tested using the SCT method for a specific function (f_i), after f_i is configured into a block of the fabric. The applied test here just checks the correct functionality of the configured f_i, rather than diagnosing all defects of the block. Hence, there might be defects in molecular switches or wires of the block, but as long as those defects do not cause any malfunction, function f_i is identified as fault-free. In other words, creating function f_i on a block b_j requires just a subset of all wires and switches of that block. Hence, if the defective components of the block are not used for configuring f_i into that block, then the function can operate without a fault. Therefore, the defects of the block are tolerated.

Using the SCT procedure, the application is divided into m-input functions; each function (f_i) should be configured into a block of the fabric and the input and output lines from the BIST circuit to f_i must also be configured. Finally, the same function f_i should be configured into the ***look-up table*** (LUT) of the BIST circuit (see Figure 17.20). Next, the BIST circuit can simply apply an exhaustive set of 2^m test patterns to the function and test its functionality. If the implemented function passes the test, then it can be reliably used in the circuit. The process of selecting a function, mapping it to a block of the fabric, creating connections between that function and the BIST circuit, and testing the function will be repeated for all functions of the application. If a function fails the test, then it must be mapped onto another block and the test process should be repeated.

Note that methods and tools for configuring nanoscale reconfigurable architectures are similar to those used for FPGAs; however, some modifications may be required because of architectural differences. The BIST circuit shown in Figure 17.20 [Rad 2006] is composed of an m-bit counter, an m-input LUT, and a comparator, resulting in low BIST area overhead. The BIST circuit is assumed to be implemented in reliable CMOS-scale circuitry.

The low area overhead of BIST circuits provides an opportunity for parallel implementation of these circuits on-chip so at any time more than one function can be implemented and tested simultaneously, as shown in Figure 17.21.

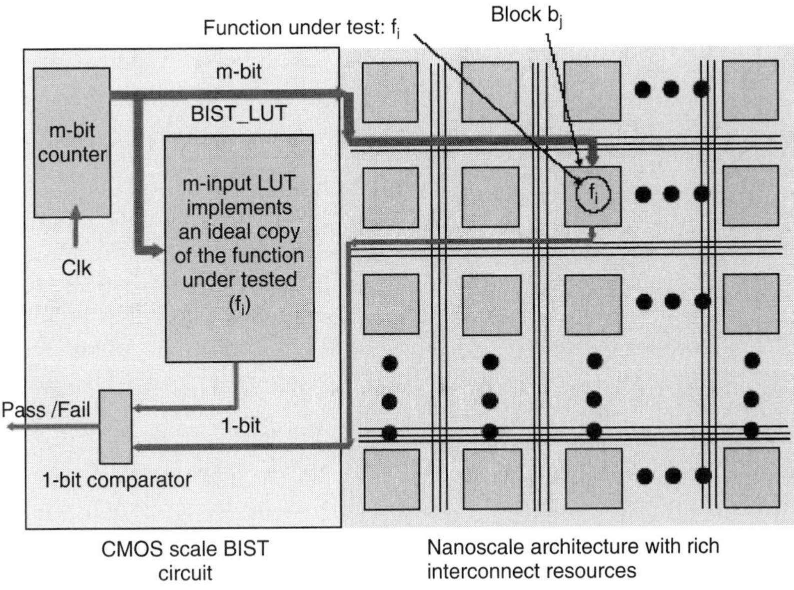

■ **FIGURE 17.20**

CMOS BIST circuit used to test nanodevices.

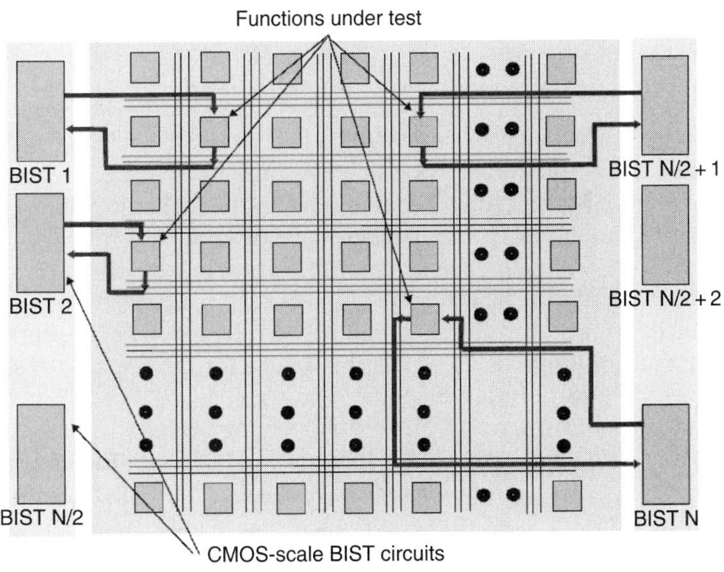

■ **FIGURE 17.21**

Parallel use of multiple BIST circuits for testing nanodevices.

When multiple BIST circuits are implemented, the test time is significantly reduced. In this case, more than one function and more than one block of the device should be selected at any time. Appropriate methods based on meeting placement and routing constraints can be devised for such selections.

17.4 CARBON NANOTUBE (CNT) FIELD EFFECT TRANSISTORS

Carbon nanotubes (CNTs) have been the subject of much research in recent years because of their unique electrical properties. In particular, their fine pitch and ballistic transport conduction mechanism enables fabrication of ***carbon nanotube field effect transistors*** (CNFETs) with excellent CV/I device performance and high transition speeds [Wong 2002, 2003, 2006]. Consequently, CNFETs are regarded as a promising extension to CMOS to facilitate IC scaling beyond the limitations currently projected by the *International Technology Roadmap for Semiconductors* (ITRS) [SIA 2006].

For CNFETs to be used for mainstream VLSI circuits, the impact of imperfections must be understood and, ideally, controlled. Two major sources of imperfections dominate CNFET circuit design: (1) **misaligned carbon nanotubes** that can result in incorrect logic implementations, and (2) **metallic carbon nanotubes** that can result in incorrect logic implementations or variations [Patil 2007].

In this section, we address the topic of reliable and robust carbon nanotube circuits in the presence of these imperfections. First, we present a robust CNFET logic design technique in the presence of a large number of misaligned CNTs. Next, we discuss modeling and analysis of CNFET circuits in the presence of metallic nanotubes.

17.4.1 Imperfection-Immune Circuits for Misaligned CNTs

Many imperfections inhibit the proper functionality of CNFET gates. For example, the carbon nanotubes may terminate too early before reaching the contacts, or the nanotubes themselves may have a point break, electrically severing the tube into two. Both of these defects only vary the effective number of semiconducting CNTs; thus, by simply synthesizing an appropriately higher density of CNTs, the yield loss caused by these defects can be reduced.

On the other hand, one of the most problematic imperfections commonly found in CNFET gates is misaligned CNTs. In this case, CNTs may not run straight from transistor source to drain under the gate. The transistor gate is used as the mask for source/drain doping, so any CNT segments not under a gate will be heavily doped and thus highly conductive (see Figure 17.22). Consequently, a misaligned CNT may bend, which either could prevent the gate channel from forming (causing a short between the nodes) or could establish a path without passing under the correct gate (resulting in possibly incorrect logic) or any gate at all (causing a short between the nodes). As an example, Figure 17.22 illustrates a NAND gate with ideal (aligned) CNTs, whereas Figure 17.23a shows the same NAND layout but with a critically misaligned CNT. This misaligned CNT shorts the output to power and causes the NAND to implement faulty logic, namely the output is always 1.

Thus, misalignment can significantly increase the defect rate of CNFETs and exponentially decrease the yield of a VLSI chip.

A layout design can be assessed as ***misaligned-CNT immune*** (MCI) or ***misaligned-CNT vulnerable*** (MCV). An MCI design is guaranteed to implement the correct logical function regardless of misaligned CNTs, whereas an MCV design may implement incorrect logic in the presence of misaligned CNTs. In [Patil 2007], a method for verifying an MCI design is demonstrated using concepts from graph theory. Fundamentally, the problem is to examine all possible paths (as if there were misaligned CNTs) from power or ground to the output. Each path has a corresponding logical (Boolean) expression consisting of the inputs or its complements, which represent the conduction condition for the path. For an MCI design, the Boolean OR of all path expressions must be equivalent to the intended conduction expression of the network; otherwise, the design is MCV. However, a sufficient condition to prove a design is MCV is to look for paths with conduction expressions that are true when the intended network conduction expression is false. The design is proven to be MCV when the first such path is found, because in this case the Boolean OR of all path expressions cannot possibly be equivalent to the intended conduction expression.

Consider the layout in Figure 17.22. The pullup network has CNTs, which pass under gate A only and under gate B only. This corresponds to the logical terms *A* and *B*, respectively. The desired logic function of the NAND pullup network is *A OR B*, and this particular cell instance implements the logic correctly. However, this layout is MCV, as shown in Figure 17.23a. There is a path that does not pass under any gate because of a misaligned CNT. This causes a short from *Vdd* to *Out*, which represents a logical 1 because the path is always conducting. Because 1 is

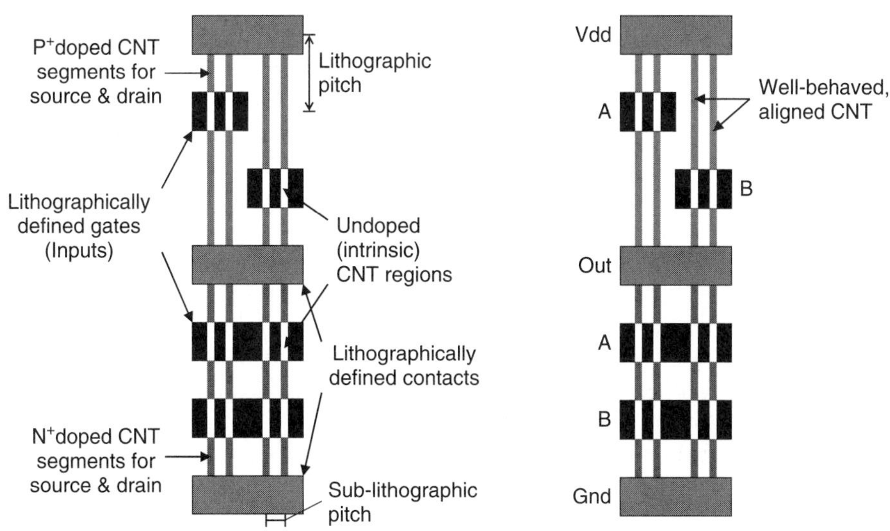

FIGURE 17.22

CNFET NAND cell example showing lithographically defined features as well as sublithographic CNTs with doped and intrinsic segments.

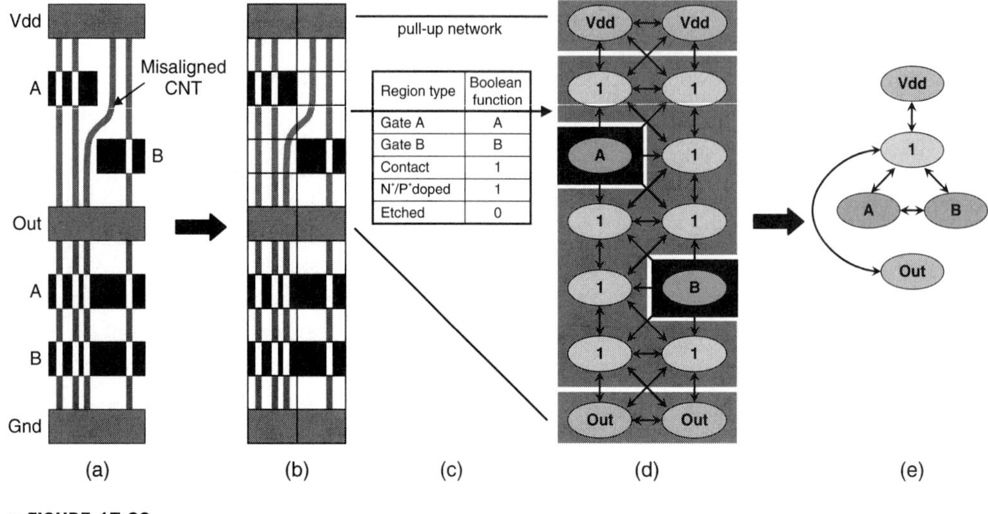

■ **FIGURE 17.23**

(a) A misaligned-vulnerable NAND cell with a critically misaligned CNT. (b) The overlay grid applied to the layout and (c) the conduction condition table for each node are used to create (d) the full path graph, and (e) its equivalent, reduced graph for the pullup network (pulldown network graph omitted).

always true even when the intended pullup network expression is false, this design is MCV and this path is said to induce the vulnerability. A similar analysis can be done with the pulldown network to ensure correct functionality.

Using graph theory, all possible paths can be formalized and algorithmically tested. Figure 17.23d shows the *path graph,* representing all possible paths from *Vdd* to *Out*, of the NAND pullup network. The graph is generated by dividing the layout into a fine grid and determining the Boolean function representing the conduction condition at each square (see Figures 17.23b and 17.23c). The graph edges connect to all the neighbors in eight directions, as a CNT can pass through a square into any of the eight adjacent squares. Also shown is the reduced, equivalent graph in Figure 17.23e. This graph can then be used to derive all possible paths between the supply node (*Vdd*) and the output (*Out*). The Boolean expression corresponding to each path is then the Boolean AND of each of the Boolean functions for the path nodes. For example, from the reduced graph, there is a potential path from node *Vdd* to node *Out*, which only passes through a 1 node (heavily doped source/drain region that always conducts). This corresponds to the same misalignment path shown in Figure 17.23a and indicates that the layout design is MCV. This formal derivation of paths and logical expressions using graph theory allows for an automated MCI design checker for future VLSI.

Not all misaligned CNTs lead simply to shorts from power to output. Consider the layout of a complex logic cell in Figure 17.24. The intended conduction expression of the network is (A AND B) OR (C AND D). However, a misaligned CNT has created a new path gated by gates A and D, and the actual logic function implemented is (A AND B) OR (C AND D) OR (A AND D). Thus, this layout consists of an extra (potential) conduction path, and, hence is MCV. Other forms of incorrect logic

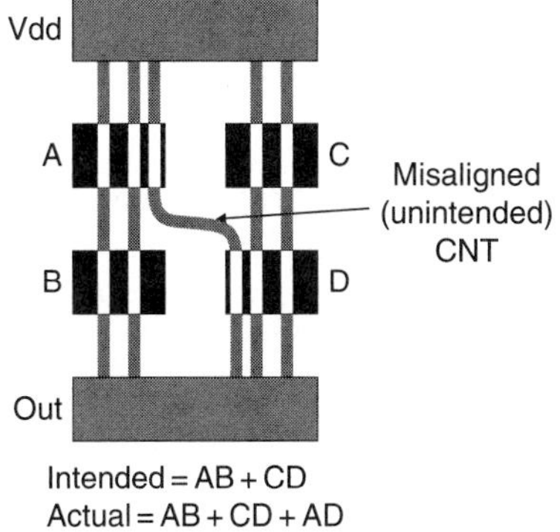

FIGURE 17.24

An example of an MCV pullup network layout design resulting in incorrect logic implementation.

implementation because of misaligned CNTs are also possible in various other logic circuit layouts.

A robust design method for eliminating gate defects because of misaligned CNTs has been developed by [Patil 2007]. By following the design rules, circuit layouts are guaranteed to be MCI. The insight gained from the graph theory analysis presented earlier is that an MCI design must yield a graph, which does not have any additional paths that contribute extra *minterms* (extra 1's that are not in the desired logic table) for the network. Thus, by designing for a graph, which inherently removes such extra paths (and only such extra paths), an MCI layout design can be achieved.

The design method described in [Patil 2007] uses lithographically etched regions to achieve a robust MCI design. Figure 17.25a shows an MCI NAND cell. A region

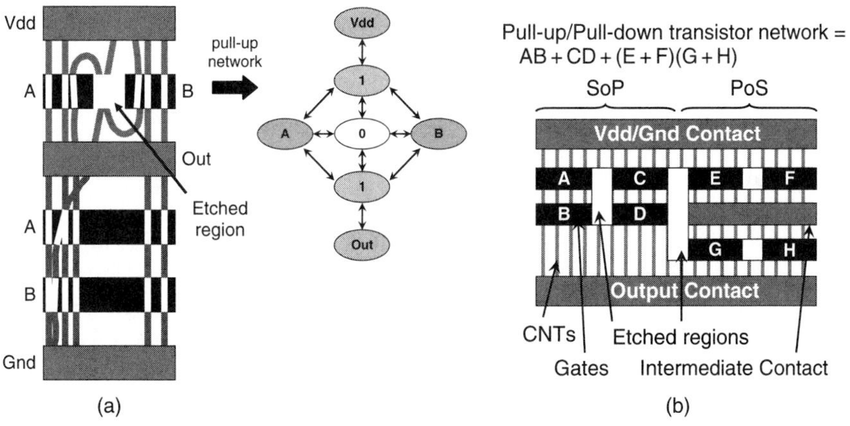

FIGURE 17.25

(a) A misaligned-tolerant (MCI) NAND cell and the corresponding reduced path graph for its pullup network (pulldown network graph omitted). (b) An MCI layout design illustrating how to use "etched regions" for a general logic network in Sums of Products (SoP) or Product of Sums (PoS) form.

in the pullup network is etched by lithographically patterning an etch mask. This causes any and all misaligned nanotubes in this region to be etched and thus nonconducting through the associated node (see Figure 17.23c). Figure 17.25a also shows the corresponding reduced path graph for the pullup network, which no longer has the logical 1 path from *Vdd* to *Out* seen in Figures 17.23a and 17.23e. Figure 17.25b generalizes the MCI design methodology for any logic network.

By developing a robust MCI design and verification methodology, logical defects that result from misaligned CNTs can be completely eliminated in layout design and rigorously checked. Although an MCV design may fail in the presence of misaligned nanotubes, an MCI design is guaranteed to implement the correct logic functionality. Such an MCI design significantly increases the yield and reliability of CNFET VLSI circuits.

17.4.2 Robust Circuits for Metallic CNTs

Variations in CNFETs arise mainly from variations in the physical properties of the carbon nanotubes, which constitute the semiconducting transistor channel. For example, the CNTs may have varying source/drain doping levels or diameters, leading to varying drive current. However, the largest source of variation comes from the varying numbers of semiconducting CNTs in the device caused by the random fraction of undesired metallic nanotubes from growth. In theory, roughly a third of the CNTs will be metallic if the growth process does not have preferential selectivity [Saito 1998]; however, [Li 2004] has reported a preferential growth technique that yields as low as 10% metallic nanotubes.

Metallic nanotubes, unlike **semiconducting nanotubes**, are highly conductive, regardless of the applied gate voltage. Consequently, they behave like resistors in parallel with the transistors, rather than semiconducting channels controlled by the gate. Metallic nanotubes will short the source and drain of the transistor, so they must be removed for proper gate functionality; for example, via plasma etching and electrical breakdown techniques presented in [Collins 2001] and [Zhang 2006]. Assuming metallic nanotubes can be completely removed via a perfected removal process, then for a given number of starting nanotubes the resulting number of semiconducting nanotubes forming the channel will vary depending on the number of metallic nanotubes.

In a study by [Deng 2007], this metallic nanotube-related variation was found to be the dominant form of performance variation compared to CNT doping or diameter variation. The study used the Stanford University CNFET model [Stanford 2007] presented in [Deng 2006], which includes practical nonidealities such as parasitics and screening effects. Using this model allowed them to compare the performance of a CMOS inverter *versus* a CNFET inverter in the presence of CNFET variation. An example CNFET inverter layout and the simulation results of the study are shown in Figures 17.26a and 17.26b, respectively. In the ideal case without variations, the CNFET inverter (eight nanotubes) exhibits 2.6X energy per cycle advantage and 5.1X delay advantage over 32 nm Si CMOS. However, with metallic tube variation, the CNFET inverter advantages reduce to 2.3X and 3.7X, respectively, compared to 2.5X and 4.6X from doping and diameter variations only [Deng 2007].

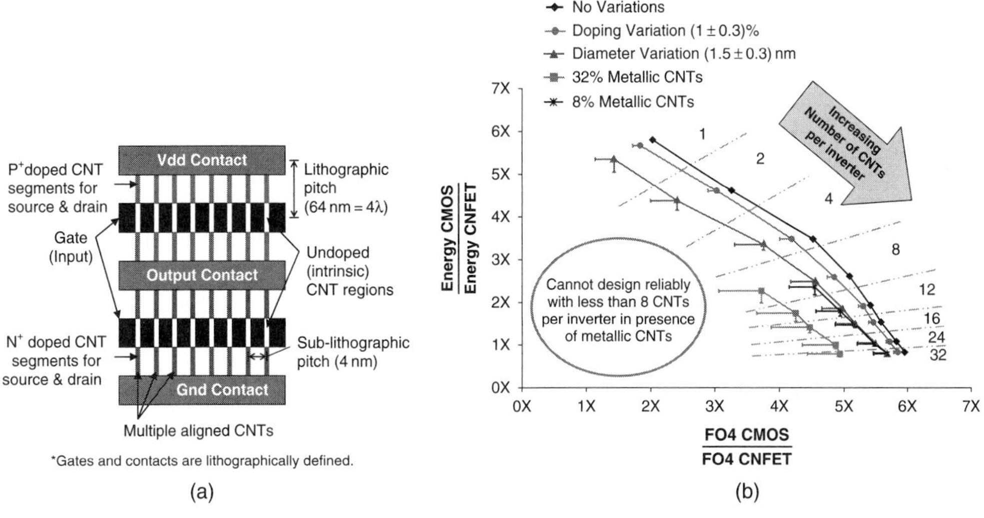

■ **FIGURE 17.26**

Simulating CNFET variation: (a) the layout of a CNFET inverter with multiple CNTs and (b) energy per cycle and FO4 delay improvement of CNFET inverter at 3σ points (bars indicate 6σ variation) compared to the 32-nm CMOS FO4 inverter.

Figure 17.26b illustrates an important conclusion. Although CNT doping and diameter variations can result in performance degradation, the variation resulting from 32% metallic nanotubes causes far more degradation in both delay and energy. In addition, if preferential semiconducting CNT growth techniques can be improved to yield only 8% metallic nanotubes, then the performance degradation from metallic nanotube variation becomes comparable to other sources of variation, such as CNT diameter variation. Consequently, this study hints that CNT growth and the metallic nanotube removal processes deserve great attention for future CNFET VLSI. In addition, if it is assumed that one third of the nanotubes are metallic, the authors in [Deng 2007] noted that devices cannot be reliably designed if too few CNTs are used per device. The study recommends designing for 8 or more semiconducting CNTs (*i.e.*, 12 or more total CNTs) per transistor as a design guideline to reduce the probability of no semiconducting nanotubes in a transistor to about one in a million. Following these design guidelines, a 2.4X energy per cycle advantage and 4.5X delay advantage can be achieved from CNFET inverters, which is considerably closer to the ideal case.

Lastly, practical metallic CNT removal techniques bring about an additional issue for CNFET circuit design. An ideal metallic CNT removal technique etches out all metallic CNTs, while leaving all semiconducting CNTs electrically intact to provide drive current. However, in practice, the selectivity of the removal process is not perfect, and even a good removal technique will remove most, but not all, metallic CNTs and leave most, but not all, semiconducting CNTs. The few remaining metallic

CNTs will likely contribute significant leakage current, as they are very conductive, making leakage power a dominant issue in CNFET circuits. In this case, the leakage power will be dominated by the metallic CNT removal rate (the percentage of metallic CNTs successfully removed). Imperfect removal processes will also affect other circuit performance metrics such as the on-off current ratio. The exact requirements on selectivity and other removal process metrics for reliable CNFET circuit design is currently an area of ongoing research. A complete understanding of the imperfections and variations in CNFETs will aid in creating guidelines for designs. Ideally, these design guidelines will help limit the detrimental effects of imperfections and variations on performance to an acceptable level and thus aid in propelling CNFETs forward as a promising extension to CMOS scaling into the nanometer era.

17.5 CONCLUDING REMARKS

Emerging nanoscale devices will enable an extremely high density of components to be integrated onto a single substrate. In this chapter, we reviewed some of the most promising devices, namely *resonant tunneling diodes* (RTDs), *quantum-dot cellular automata* (QCA), carbon nanotubes/silicon nanowires, and *carbon nanotube field effect transistors* (CNFETs). We discussed some test challenges and presented test generation techniques for these devices. In particular, we presented defect characterization, fault modeling, and test generation of circuits based on RTDs and QCA. We discussed *built-in self-test* (BIST) of carbon-nanotubes-based crossbar array architectures. We also presented imperfections and variations tolerance in logic circuits implemented by CNFETs.

Regardless of which devices make their way into the mainstream of nanoscale computing, there is a general consensus that testing will be a key issue, as these devices are expected to have high defect rates. Consequently, some sort of defect and fault tolerance schemes will have to be built into nanoscale circuits, systems, and architectures. Ultimately, the testability and reliability of these devices will need to become an additional constraint during system design.

Acknowledgments

The authors wish to thank Professor Pallav Gupta of Villanova University for contributing a portion of the Resonant Tunneling Diodes and Quantum-Dot Cellular Automata section, Professor Mohammad H. Tehranipoor of University of Connecticut for contributing a portion of the Crossbar Array Architectures section, Albert Lin and Nishant Patil of Stanford University for contributing the Carbon Nanotube Field Effect Transistors section, Professor Krishnendu Chakrabarty of Duke University for contributing to the overall structure of the chapter, and Professor Subhasish Mitra and Professor H.-S. Philip Wong of Stanford University for reviewing the Carbon Nanotube Field Effect Transistors section.

References

R17.0 Books

[Jha 2003] N. K. Jha and S. Gupta, *Testing of Digital Systems*, Cambridge University Press, Cambridge, United Kingdom, 2003.

[Iwai 2004] H. Iwai, The future of CMOS downscaling, pp. 23–33, in *Future Trends in Microelectronics: The Nano, the Giga, and the Ultra*, S. Luryi, J. Xu, and A. Zaslavsky, editors, John Wiley, New York, 2004.

[Likharev 2005] K. Likharev and D. Strukov, *Introducing Molecular Electronics*, Springer, Berlin, Germany, 2005.

[Luryi 2004] S. Luryi, J. M. Xu, and A. Z. eds., *Future Trends in Microelectronics: The Nano, the Giga, and the Ultra*, John Wiley, New York, 2004.

[Muroga 1971] S. Muroga, *Threshold Logic and Its Applications*, John Wiley, New York, 1971.

[Saito 1998] R. Saito, G. Dresselhaus, and M. Dresselhaus, *Physical Properties of Carbon Nanotubes*, Imperial College Press, London, 1998.

[Wang 2006] L.-T. Wang, C.-W. Wu, and X. Wen, editors, *VLSI Test Principles and Architectures: Design for Testability*, Morgan Kaufmann, San Francisco, 2006.

R17.1 Introduction

[Bachtold 2001] A. Bachtold, P. Harley, T. Nakanishi, and C. Dekker, Logic circuits with carbon nanotube transistors, *Science*, 294(5545), pp. 1317–1320, November 9, 2001.

[Butts 2002] M. Butts, A. DeHon, and S. Goldstein, Molecular electronics: Devices, systems and tools for gigagate, gigachips, in *Proc. Int. Conf. on Computer-Aided Design*, pp. 433–440, November 2002.

[Celler 2003] G. Celler and S. Cristoloveanu, Frontiers of silicon-on-insulator, *J. Applied Physics*, 93, pp. 4955–4978, May 2003.

[Chen 1996] K. Chen, K. Maezawa, and M. Yamamoto, Inp-based high-performance monostable-bistable transition logic elements (MOBILE's) using integrated multiple-input resonant-tunneling devices, *IEEE Electron Device Letters*, 17(3), pp. 127–129, March 1996.

[Collier 1999] C. Collier, E. Wong, M. Belohradsky, F. Raymo, J. Stoddart, P. Kuekes, R. Williams, and J. Heath, Electronically configurable molecular-based logic gates, *Science*, 285(5426), pp. 391–394, July 16, 1999.

[Dehon 2003a] A. DeHon, Array-based architecture for FET-based nanoscale electronics, *IEEE Trans. Nanotechnology*, 2(1), pp. 23–32, March 2003.

[Fuhrer 2000] M. Fuhrer, J. Nygard, L. Shih, M. Forero, Y. Yoon, M. Mazzoni, H. Choi, J. Ihm, S. Louie, A. Zettl, and P. McEuen, Crossed nanotube junctions, *Science*, 288(5465), pp. 494–497, April 21, 2000.

[Goldstein 2002] S. Goldstein and D. Rosewater, Digital logic using molecular electronics, in *Proc. IEEE Int. Solid-State Circuits Conf.*, pp. 204–205, February 2002.

[Huang 2001] Y. Huang, X. Duan, Y. Cui, L. Lauhon, K. Kim, and C. Lieber, Logic gates and computation from assembled nanowire building blocks, *Science*, 294(5545), pp. 1313–1317, November 9, 2001.

[Iijima 1991] S. Iijima, Helical microtubules of graphitic carbon, *Nature*, 354, pp. 56–58, November 7, 1991.

[Kamins 2000] T. Kamins, R. Williams, Y. Chen, Y.-L. Chang, and Y. Chang, Chemical vapor deposition of Si nanowires nucleated by TiSi2 islands on Si, *Applied Physics Letters*, 76(5), pp. 562–564, January 31, 2000.

[Likharev 1999] K.Likharev, Single electron devices and their applications, *Proceedings of the IEEE*, 87(4), pp. 606–632, April 1999.

[Mishra 2003] M. Mishra and S. Goldstein, Defect tolerance at the end of the roadmap, in *Proc. IEEE Int. Test Conf.*, pp. 1201–1210, September 2003.

[Mitra 2004] S. Mitra, W. Huang, N. Saxena, S. Yu, and E. McCluskey, Reconfigurable architecture for autonomous self-repair, *IEEE Design & Test of Computers*, 21(3), pp. 228–240, May/June 2004.

[Rueckes 2000] T. Rueckes, K. Kim, E. Joselevich, G. Tseng, C. Cheung, C. Lieber, Carbon nanotube-based nonvolatile random access memory for molecular computing, *Science*, 289(5476), pp. 94–97, July 7, 2000.

[SIA 2004] SIA, *The International Technology Roadmap for Semiconductors: 2004 Update*, Semiconductor Industry Association, San Jose, CA (http://public.itrs.net), 2004.

[Tougaw 1994] P. Tougaw and C. Lent, Logical devices implemented using quantum cellular automata, *Applied Physics*, (3), pp. 1818–1825, February 1994.

[Wong 2002] H.-S. P. Wong, Beyond the conventional transistor, *IBM J. Research and Development*, 46(2–3), pp. 133–168, March/May 2002.

[Wong 2006] H.-S. P. Wong, Device and technology challenges for nanoscale CMOS, in *Proc. Int. Symp. on Quality Electronic Design*, pp. 515–518, March 2006.

R17.2 Resonant Tunneling Diodes and Quantum-Dot Cellular Automata

[Amlani 1998] I. Amlani, A. Orlov, G. Snider, and C. Lent, Demonstration of a six-dot quantum cellular automata system, *Appl. Phys. Lett.*, (17), pp. 2179–2181, April 1998.

[Amlani 1999] I. Amlani, A. Orlov, G. Toth, C. Lent, G. Bernstein, and G. Snider, Digital logic gate using quantum-dot cellular automata, *Science*, 284(5412), pp. 289–291, April 1999.

[Beiu 2003] V. Beiu, J. Quintana, and M. Avedillo, VLSI implementations of threshold logic: A comprehensive survey, *IEEE Trans. Neural Networks*, 14(5), pp. 1217–1243, September 2003.

[Chen 1996] K. Chen, K. Maezawa, and M. Yamamoto, Inp-based high-performance monostable-bistable transition logic elements (MOBILE's) using integrated multiple-input resonant-tunneling devices, *IEEE Electron Device Letters*, 17(3), pp. 127–129, March 1996.

[Gupta 2004] P. Gupta, R. Zhang, and N. Jha, An automatic test pattern generation framework for combinational threshold networks, in *Proc. Int. Conf. on Computer Design*, pp. 540–543, October 2004.

[Gupta 2006] P. Gupta, N. Jha, and L. Lingappan, Test generation for combinational quantum cellular automata (QCA) circuits, in *Proc. Design, Automation and Test in Europe Conf.*, pp. 311–316, March 2006.

[Hennessy 2001] K. Hennessy and C. Lent, Clocking of molecular quantum-dot cellular automata, *J. Vac. Sci. Technol.*, 19(5), pp. 1752–1755, September/October 2001.

[Lieberman 2002] M. Lieberman, S. Chellamma, B. Varughese, Y. Wang, C. Lent, G. Bernstein, G. Snider, and F. Peiris, Quantum-dot cellular automata at a molecular scale, *Ann. New York Acad. Sci.*, 960, pp. 225–239, April 2002.

[Momenzadeh 2004] M. Momenzadeh, M. Tahoori, J. Huang, and F. Lombardi, Quantum cellular automata: New defects and faults for new devices, in *Proc. Int. Parallel and Distributed Processing Symp.*, pp. 207–214, April 2004.

[Pacha 1999] C. Pacha, P. Glosekotter, K. Goser, W. Prost, U. Auer, and F. Tegude, Resonant tunneling device logic circuits, Technical report, University of Dortmund and Gerhard-Mercator University of Duisburg, July 1999.

[Prost 2000] W. Prost, U. Auer, F. Tegude, C. Pacha, K. Goser, G. Janssen, and T. van der Roer, Manufacturability and robust design of nanoelectronic logic circuits based on resonant tunneling diodes, *Int. J. Circ. Theory Appl.*, 28(6), pp. 537–552, December 2000.

[Tahoori 2004a] M. Tahoori, M. Momenzadeh, J. Huang, and F. Lombardi, Defects and faults in quantum cellular automata at nanoscale, in *Proc. VLSI Test Symp.*, pp. 291–296, April 2004.

[Tahoori 2004b] M. Tahoori, J. Huang, M. Momenzadeh, and F. Lombardi, Testing of quantum cellular automata, *IEEE Trans. Nanotechnology*, 3(4), pp. 432–442, December 2004.

[Tougaw 1994] P. Tougaw and C. Lent, Logical devices implemented using quantum cellular automata, *J. Applied Physics*, 75(3), pp. 1811–1817, February 1994.

[Walus 2004] K. Walus, T. Dysart, G. Jullien, and R. Budiman, QCADesigner: A rapid design and simulation tool for quantum-dot cellular automata, *IEEE Trans. Nanotechnology*, 3(1), pp. 26–31, March 2004.

[Zhang 2004a] R. Zhang, P. Gupta, L. Zhong, and N. Jha, Synthesis and optimization of threshold logic networks with application to nanotechnologies, in *Proc. Design, Automation and Test in Europe Conf.*, pp. 904–909, February 2004.

[Zhang 2004b] R. Zhang, K. Walus, W. Wang, and G. Jullien, A method of majority logic reduction for quantum cellular automata, *IEEE Trans. Nanotechnology*, 3(4), pp. 443–450, December 2004.

R17.3 Crossbar Array Architectures

[Chen 2003] Y. Chen. G.-Y. Jung, D. Ohlberg, X. Li, D. Stewart, J. Jeppesen, K. Nielsen, J. Stoddart, and R. Williams, Nanoscale molecular-switch crossbar circuits, *Nanotechnology*, 14(14), pp. 462–468, March 2003.

[DeHon 2003b] A. DeHon, P. Lincoln, and J. Savage, Stochastic assembly of sublithographic nanoscale interfaces, *IEEE Trans. Nanotechnology*, 2(3), pp. 165–174, September 2003.

[DeHon 2004] A. DeHon and M. Wilson, Nanowire-based sublithographic programmable logic arrays, in *Proc. Int. Symp. Field-Programmable Gate Arrays*, pp. 123–132, February 2004.

[DeHon 2005a] A. DeHon, Design of programmable interconnect for sublithographic programmable logic arrays, in *Proc. Int. Symp. Field-Programmable Gate Arrays*, pp. 127–137, February 2005.

[DeHon 2005b] A. DeHon, Nanowire-based programmable architectures, *ACM J. on Emerging Technologies in Computer Systems*, 1(2), pp. 109–162, July 2005.

[DeHon 2005c] A. DeHon and H. Naeimi, Seven strategies for tolerating highly defective fabrication, *IEEE Design & Test of Computers*, 22(4), pp. 306–315, July 2005.

[Goldstein 2001] S. Goldstein and M. Budiu, Nanofabrics: Spatial computing using molecular electronics, in *Proc. Int. Symp. on Computer Architecture*, pp. 17, 189, June 2001.

[Huang 2001] Y. Huang, X. Duan, Q. Wei, and C. Lieber, Directed assembly of one-dimensional nanostructures into functional networks, *Science*, 291(5504), pp. 630–633, January 26, 2001.

[Huang 2004] J. Huang, M. Tahoori, and F. Lombardi, On the defect tolerance of nano-scale two dimensional crossbars, in *Proc. Int. Symp. on Defect and Fault Tolerance in VLSI Systems*, pp. 96–104, October 2004.

[Luo 2002] Y. Luo, P. Collier, J. Jeppesen, K. Nielsen, E. Delonno, G. Ho, J. Perkins, H. Tseng, T. Yamamoto, J. Stoddart, and J. Heath, Two-dimensional molecular electronics circuits, *Chemphyschem.*, 17(3), pp. 519–525, June 2002.

[Ma 2005] X. Ma, D. Strukov, J. Lee, and K. Likharev, Afterlife for silicon: CMOL circuit architectures, in *Proc. IEEE Conf. on Nanotechnology*, pp. 175–178, July 2005.

[Rad 2006] R. Rad and M. Tehranipoor, SCT: An approach for testing and configuring nanoscale devices, in *Proc. VLSI Test Symp.*, pp 370–377, May 2006.

[SIA 2005] SIA, *The International Technology Roadmap for Semiconductors: 2005 Edition*, Semiconductor Industry Association, San Jose, CA (http://public.itrs.net), 2005.

[Stroud 1998] C. Stroud, S. Wijesuriya, C. Hamilton, and M. Abramovici, Built-in self-test for FPGA interconnect, in *Proc. Int. Test Conf.*, pp. 404–413, October 1998.

[Strukov 2005] D. Strukov and K. Likharev, CMOL FPGA: A reconfigurable architecture for hybrid digital circuits with two-terminal nanodevices, *Nanotechnology*, 16(6), pp. 888–900, June 2005.

[Tahoori 2005] M. Tahoori, Defects, yield, and design in sublithographic nano-electronics, in *Proc. Int. Symp. on Defect and Fault Tolerance in VLSI Systems*, pp. 3–11, October 2005.

[Tahoori 2006] M. Tahoori, Application-independent defect tolerance of reconfigurable nanoarchitectures, *ACM J. on Emerging Technologies in Computing Systems*, 2(3), pp. 197–218, July 2006.

[Tehranipoor 2007] M. Tehranipoor and R. Rad, Built-in self-test and recovery procedures for molecular electronics-based nanofabrics, *IEEE Trans. Computer Aided Design*, 26(5), pp. 943–958, May 2007.

[Wang 2005] Z. Wang and K. Chakrabarty, Using built-in self-test and adaptive recovery for defect tolerance in molecular electronics-based nanofabrics, in *Proc. Int. Test Conf.*, pp. 477–486, November 2005.

R17.4 Carbon Nanotube (CNT) Field Effect Transistors

[Collins 2001] P. G. Collins, M. S. Arnold, and P. Avouris, Engineering carbon nanotubes and nanotube circuits using electrical breakdown, *Science*, 292(5517), pp. 706–709, April 27, 2001.

[Deng 2006] J. Deng and H.-S. P. Wong, A circuit-compatible SPICE model for enhancement mode carbon nanotube field effect transistors, in *Proc. Int. Conf. Simulation of Semiconductor Processes and Devices*, pp. 166–169, September 2006.

[Deng 2007] J. Deng, N. Patil, K. Ryu, A. Badmaev, C. Zhou, S. Mitra, and H.-S. P. Wong, Carbon nanotube transistor circuits: Circuit-level performance benchmarking and design options for living with imperfections, in *Proc. Int. Solid-State Circuits Conf.*, pp. 17, 70–71, February 2007.

[Li 2004] Y. Li, D. Mann, M. Rolandi, W. Kim, A. Ural, S. Hung, A. Javey, J. Cao, D. Wang, E. Yenilmez, Q. Wang, J. F. Gibbons, Y. Nishi, and H. Dai, Preferential growth of semiconducting single-walled carbon nanotubes by a plasma enhanced CVD method, *Nano Letters*, 4, p. 317, January 22, 2004.

[Patil 2007] N. Patil, J. Deng, H.-S. P. Wong, and S. Mitra, Automated design of misaligned-carbon-nanotube-immune circuits, to be published in *Proc. ACM/IEEE Design Automation Conf.*, June 2007.

[SIA 2006] SIA, *The International Technology Roadmap for Semiconductors: 2006 Update*, Semiconductor Industry Association, San Jose, CA (http://public.itrs.net), 2006.

[Stanford 2007] Stanford University, CNFET Model online at the Stanford Nanotechnology Group website, Stanford University, Stanford, CA (http://www.stanford.edu/group/nanoelectronics/model_downloads.htm), 2007.

[Wong 2002] H.-S. P. Wong, Beyond the conventional transistor, *IBM J. Research and Development*, 46(2–3), pp. 133–168, March/May 2002.

[Wong 2003] H.-S. P. Wong, J. Appenzeller, V. Derycke, R. Martel, S. Wind, and P. Avouris, Carbon nanotube field effect transistors: Fabrication, device physics, and circuit implications, in *Proc. Int. Solid-State Circuits Conf.*, pp. 370–371, February 2003.

[Wong 2006] H.-S. P. Wong, Device and technology challenges for nanoscale CMOS, in *Proc. Int. Symp. on Quality Electronic Design*, pp. 515–518, March 2006.

[Zhang 2006] G. Zhang, P. Qi, X. Wang, Y. Lu, X. Li, R. Tu, S. Bangsaruntip, D. Mann, L. Zhang, and H. Dai, Selective etching of metallic carbon nanotubes by gas-phase reaction, *Science*, 314(5801), pp. 974–979, November 10, 2006.

Index